Biological Oceanography of the Baltic Sea

Pauline Snoeijs-Leijonmalm
Hendrik Schubert · Teresa Radziejewska
Editors

Biological Oceanography of the Baltic Sea

Springer

Editors
Pauline Snoeijs-Leijonmalm
Department of Ecology, Environment
and Plant Sciences
Stockholm University
Stockholm
Sweden

Hendrik Schubert
Department of Biological Sciences
University of Rostock
Rostock
Germany

Teresa Radziejewska
Faculty of Geosciences
University of Szczecin
Szczecin
Poland

ISBN 978-94-024-1317-5 ISBN 978-94-007-0668-2 (eBook)
DOI 10.1007/978-94-007-0668-2

Softcover reprint of the hardcover 1st edition 2017

Printed on acid-free paper

This Springer imprint is published by Springer Nature. The registered company is Springer Science+Business Media B.V.
The registered company address is: Van Godewijckstraat 30, 3311 GX Dordrecht, The Netherlands

To all past, present and future Baltic Marine Biologists

Preface

The Baltic Sea is a unique ecosystem characterised by pronounced environmental gradients, particularly with respect to salinity and climate. The biological and evolutionary features of *Mare Balticum* as a large meeting place for marine and freshwater organisms have fascinated scientists for centuries. The ecological interest in the Baltic Sea has been additionally fuelled by anthropogenic pressures on the ecosystem structure and functioning exerted by economic activities in the drainage area and by global-scale processes. The collective knowledge on Baltic Sea biology and ecology is available in a wealth of scientific journal articles and books, and the volume of this literature has been increasing almost exponentially over the last 50 years. Consequently, the Baltic Sea is renowned as one of the most intensively studied ecosystems in the world. It comes then as a paradox that a comprehensive international book on the basic biology and ecology of the Baltic Sea has not been available to generations of students interested in the Baltic Sea. The need for a modern science-based textbook has been discussed for more than 15 years, primarily within the Baltic Marine Biologists (BMB), a non-governmental organisation credited with arranging the biannual Baltic Marine Biologists symposia since 1968 (from 2001 under the umbrella of the biannual Baltic Sea Science Congresses). The idea born in the BMB has now materialised in the form of this textbook.

The sheer broadness of the subject matter makes it essentially impossible for only one or a few persons to write such a book. Therefore, the approach chosen was to pool the knowledge of 92 authors, with all of them active scientists and experts in their respective fields, and nearly all of them teaching at the universities around the Baltic Sea. Rather than writing a series of review papers, the overall objective was to create an integrated book that would introduce a student at the MSc and PhD levels to the basic knowledge, facts and processes important for understanding life in the Baltic Sea. This knowledge should, on the one hand, give a student (and an interested reader) an overview of the Baltic Sea ecosystem's structure and function and, on the other, become a starting point from which to begin exploring individual aspects of the Baltic Sea ecosystem structure and function in more detail.

Together we have made a concerted effort to produce a book we wished existed when we were students ourselves. Although the book's contents are based on our own perception of what is important to know about the biology and ecology of the Baltic Sea, the process of putting this book together was not only a matter of writing down what we teach our students. As scientists we are trained to not only generate new knowledge but also critically discuss scientific results. Although the knowledge presented in this book is mostly based on published peer-reviewed scientific literature, in the course of the editorial work many issues emerged that needed to be discussed, between the authors and between the authors and the editors. This is because we work in different parts of the Baltic Sea, as well as in nine different countries, with even more languages and with somewhat different scientific traditions. Given this background, it came as no surprise that many of the authors not only contributed with their respective parts of a book chapter but also participated in an "intercalibration exercise" with respect to the terminology used and scientific opinions expressed. Although it was not always possible to fully agree, the contents of the book have greatly benefitted from these fruitful discussions as well as from an extensive peer-review procedure applied to all the chapters.

The overall focus of the book is on knowledge about the conditions for life in brackish water and the functioning of the Baltic Sea ecosystem. It highlights biological variation along the unique environmental gradients of the brackish Baltic Sea Area (the Baltic Sea, Belt Sea and Kattegat), especially those of salinity and climate. Throughout the book, we follow an ecosystem approach. Rather than focus on separate topics regarding *e.g.* plankton, fish or birds, or environmental issues such as eutrophication or fisheries, the chapters focus on processes and subsystems. Ecological compartments and environmental issues are treated jointly with the processes and subsystems when relevant. Some overlap between chapters has been allowed when appropriate, e.g. with respect to an issue viewed from different angles or in a different context.

Part I of the book presents the challenges for life processes and ecosystem dynamics that result from the Baltic Sea's highly variable recent geological history and geographical isolation. Part II explains interactions between organisms and their environment, including biogeochemical cycles, patterns of biodiversity, genetic diversity and evolution, biological invasions and physiological adaptations. In Part III, the subsystems of the Baltic Sea ecosystem–the pelagic zone, the sea ice, the deep soft seabeds, the phytobenthic zone, the sandy coasts, estuaries and coastal lagoons–are treated in detail with respect to the structure and function of communities and habitats and consequences of natural and anthropogenic constraints, such as climate change, discharges of nutrients and hazardous substances. Finally, Part IV discusses monitoring and ecosystem-based management to deal with contemporary and emerging threats to the ecosystem's health.

In addition to the main text, the book also contains numerous figures, photographs and boxes with additional information about e.g. ecological principles, methodology or certain species. Each chapter ends with five review questions and five discussion questions for use in discussion seminars. The review questions serve as a control of how the subject matter of the chapter has been understood and assimilated by the reader. The discussion questions are designed to trigger critical thinking about the chapter contents in a wider perspective.

Throughout the book, the usage of geographical names is based primarily on the comprehensive version of the "Times Atlas of the World" (Times Books, London, 9th Edition 1994), in which the names are most often spelled in original languages for water bodies within countries but in English for international waters. This is the reason why we write about e.g. the river Wisła rather than "Vistula" and the Storebælt rather than the "Great Belt", but we do use "Bothnian Bay" and not the Swedish or Finnish name for this water area because it is shared by Sweden and Finland. The taxonomy is based mainly on the following databases: World Register of Marine Species (WoRMS, http://www.marinespecies.org), the database of terrestrial, marine and freshwater alga "AlgaeBase" (http://www.algaebase.org), the plant species database "The Plant List" of the Kew Botanical Garden (http://www.theplantlist.org) and the bacterial database "Bacterial Diversity" of the Leibnitz Institute DSMZ-German Collection of Microorganisms and Cell Cultures (https://www.dsmz.de).

We dedicate this book to all past, present and future Baltic Marine Biologists to honour all those people on whose work the contents of this book are based and to provide a strong encouragement to new generations of scientists who will discover things about the Baltic Sea we do not yet know today.

Stockholm, Sweden — Pauline Snoeijs-Leijonmalm
Rostock, Germany — Hendrik Schubert
Szczecin, Poland — Teresa Radziejewska

Editors

Pauline Snoeijs-Leijonmalm is Professor of Marine Ecology at Stockholm University, Sweden. Her research interests are biodiversity and food-web interactions. Since 1982 she has worked on the coastal and pelagic systems of the Baltic Sea and also since 2002 on sea-ice ecology in the Central Arctic Ocean. For more than 25 years she has taught courses on Baltic Sea ecology at the Universities of Uppsala and Stockholm.

Hendrik Schubert is Professor of Aquatic Ecology at the University of Rostock, Germany. His research interest is brackish water ecology with a focus on the ecophysiology of autotrophs. Since 1998 he has also been working on the biogeography of aquatic macrophytes, exploiting their potential for bioindication of the ecological status of aquatic ecosystems. He has taught aquatic ecology at the Universities of Greifswald and Rostock for over 20 years.

Teresa Radziejewska is Associate Professor at the Faculty of Geosciences, University of Szczecin, Poland. Her research interests are centred on the ecology and diversity of benthic communities, with a particular focus on the meiobenthos, in different habitats (from coastal waters to deep-sea areas) as well as on bentho-pelagic couplings, environmental monitoring and conservation issues.

Acknowledgements

We are grateful to all those who helped us to achieve this book, the publication of which has required the assistance and comments of many contributors. First of all, we wish to acknowledge the assistance of Martine van Bezooijen from Springer who enthusiastically supported the idea of compiling this book and who in the early writing phase met many of the authors at the 8th Baltic Sea Science Congress in Sankt-Petersburg in 2011. Martine sadly died in an accident shortly after that. For excellent help in finally realising the book we are grateful to Judith Terpos and Alexandrine Cheronet from Springer. The editors warmly thank their spouses, Jorma Leijonmalm, Almuth Schubert and Stefan Matalewski for all their support provided in many ways.

Our special thanks go to Kai Myrberg, Ilppo Vuorinen, Kjell Larsson and the students of the 2015 course "The Baltic Sea Ecosystem" (10 ECTS), taught by Erik Bonsdorff and Christoffer Boström at Åbo Akademi University, for excellent comments on various aspects of the book's manuscript, and to Ezra Alexander, Centre for Academic English at Stockholm University, for carefully checking the grammar.

Our thanks also go to the following colleagues for their constructive criticism when reviewing the chapters during the various stages of editing: Eyvind Aas, Jerzy Bolałek, Erik Bonsdorff, Angel Borja, Christoffer Boström, Deborah Bronk, David Delaney, Christiane Fenske, Valentina Galtsova, Rolf Gradinger, Helena Höglander, Marianne Holmer, Herman Hummel, Jeroen Ingels, Jonne Kotta, Christopher Krembs, Erkki Leppäkoski, Anton McLachlan, Klaus Meiners, Juha Merilä, Felix Müller, Raisa Nikula, Susanne Qvarfordt, David Raffaelli, Karsten Reise, Heye Rumohr, Doris Schiedek, Michael Tedengren, Brygida Wawrzyniak-Wydrowska, Rutger de Wit, Zbigniew Witek.

We acknowledge the following organisations for the use of data and figures in several chapters of the book: the Baltic Marine Environment Protection Commission (HELCOM), the International Council for the Exploration of the Sea (ICES), the Swedish Meteorological and Hydrological Institute (SMHI), the National Aeronautics and Space Administration (NASA), and the Baltic Nest Institute (BNI), which has developed and maintains the Baltic Environmental Database (BED) as one of the core facilities funded by the Swedish Agency for Marine and Water Management. The full list of organisations that have contributed to BED is found at http://nest.su.se/bed/

The following 86 photographers have generously made available 556 photographs for this book: Elinor Andrén (8 photographs), Piotr Bałazy (11), Ulf Bergström (5), Wolfgang Bettighofer (1), Claes Björkholm (1), Irmgard Blindow (2), Erik Bonsdorff (2), Michael Borgiel (1), Christoffer Boström (3), Jan-Erik Bruun (1), Martynas Bučas (1), Gertrud Cronberg (2), Sven Dahlke (1), Piotr Domaradzki (1), Finnish Coast Guard (1), Jens Gercken (2), Antje Gerloff (1), Stephan Gollasch (7), Andrey Gusev (1), Camilla Gustafsson (2), Regina Hansen (41), Regina Hansen and Rainer Bahlo (3), Klaus Hausmann (1), Marianne Hielm Pedersén (1), Annelies Hommersom (1), Helena Höglander (1), Urszula Janas (1), Hans Kautsky (73), Lena Kautsky (1), Lech Kotwicki (1), Susanne Kratzer (1), Anette Küster (1), Maria Laamanen (2), Matthias Labrenz (1), Kjell Larsson (7), Ilkka Lastumäki (2), Sirpa Lehtinen (2), Maiju Lehtiniemi (1), Teemu Lehtiniemi (1), Tore Lindholm (1), Rafal Michon (1), Ekaterina Mironova (2), Gerd Niedzwiedz (2), Lies Van Nieuwerburgh (1), Henn Ojaveer (1), Sergej

Olenin (15), Tiina Palme (1), Svante Pekkari (2), Teresa Radziejewska (5), Janne-Markus Rintala (8), Joanna Rokicka-Praxmajer (4), Sara Rydberg (1), Bernd Sadkowiak (1), Soili Saesmaa (1), Heide Sandberg (4), Mariusz Sapota (1), Ralf Schaible (1), Ralph Schill (2), Kristian Schoning (2), Dirk Schories (21), Hendrik Schubert (22), Andrius Šiaulys (1), Ursula Siebert (1), Pauline Snoeijs-Leijonmalm (180), Bo Tallmark (13), Irena Telesh (4), David N. Thomas (6), Anna Ulanova (1), Jacek Urbański (1), Risto Väinölä (5), Jari Uusikivi (2), Brygida Wawrzyniak-Wydrowska (4), Benjamin Weigel (1), Stanislaw Węsławski (3), Gary Wife (1), Maria Włodarska-Kowalczuk and Piotr Bałazy (12), Piotr Wysocki (11), Ingrid Wänstrand (1), Anastasija Zaiko (2), as well as the following organisations: the Baltic Marine Biologists (BMB) (1), the European Space Agency (ESA) (1), the German Oceanographic Museum (1), the International Council for the Exploration of the Sea (ICES) (1), the Leibnitz Institute for Baltic Sea Research (IOW), Warnemünde, Germany (1), Marine Monitoring, Lysekil, Sweden (4), and Uppsala University (6).

Chapter 2: The authors are grateful to Thomas Andrén for valuable comments on an early draft of the manuscript. PS acknowledges the support from the Swedish Research Councils Formas and VR, the Knut and Alice Wallenberg Foundation, Carl Tryggers Stiftelse för Vetenskaplig Forskning and the Swedish Environmental Protection Agency. EA acknowledges the support from the Stockholm County Council. EA dedicates Sect. 2.5 "Geological and climatic background" to the memory of Professor Urve Miller for her never-ending interest in quaternary geology.

Chapter 3: Some of the authors are engaged in the newly formed "Baltic Earth–Earth System Science for the Baltic Sea Region" network–the extension of the former BALTEX programme–and consider their contributions as part of the network activities. KK acknowledges the support from the Polish National Science Centre (Grant UMO-2014/13/D/ST10/01885).

Chapter 4: The author is grateful to Pieter Provoost (OBIS Data Manager, UNESCO), who kindly provided the data used in Table 4.2, and acknowledges the support from the Swedish Research Councils Formas and VR, the Knut and Alice Wallenberg Foundation, Carl Tryggers Stiftelse för Vetenskaplig Forskning and the Swedish Environmental Protection Agency.

Chapter 5: SO and AZ acknowledge the support from Klaipėda University and the project "Development of innovative tools for understanding marine biodiversity and assessing good environmental status" (DEVOTES) of EU FP7 (Grant 308392). ML and AZ acknowledge the support from the BONUS project BIO-C3 of EU FP7, the Academy of Finland and the Research Council of Lithuania. The authors dedicate this chapter to Erkki Leppäkoski, one of the founders of bioinvasion biology in the Baltic Sea.

Chapter 6: RV acknowledges the support from the Academy of Finland. KJ acknowledges the support from the Swedish Research Council Formas, the BONUS projects BaltGene and Bambi of EU FP7, and the scientific input from research colleagues in these projects.

Chapter 7: HS, IT and SS prepared this chapter within the framework of the Ulrich Schiewer Laboratory for Experimental Aquatic Ecology (the USELab Project). HS acknowledges the support from the German Research Council (Grant Schu 983/5-1). IT acknowledges the support from the Russian Foundation for Basic Research (Grant 15-29-02706). SS acknowledges the support from the Russian Foundation for Basic Research (Grant 13-04-00703). MN acknowledges the support from Academy of Finland (Grant 258078).

Chapter 8: The authors acknowledge the Swedish Research Council FORMAS, the Swedish research environment ECOCHANGE, the Swedish Research Council, the Finnish Environment Institute SYKE and the German Science Foundation (DFG).

Chapter 9: The authors are grateful to Klaus Meiners and Rolf Gradinger for valuable comments on an early draft of the manuscript. The authors acknowledge the Walter and André Nottbeck Foundation for supporting much of the research that underpins this chapter. DT acknowledges the support from the Academy of Finland FiDiPro scheme. LT acknowledges the support from the Maj and Tor Nessling Foundation (Project 2010357) and the Estonian

Science Foundation (Project MJD62). MG acknowledges the support from the Academy of Finland graduate school scheme, Kone Foundation, Walter and Lisi Wahls Stiftelse för Naturvetenskaplig Forskning, Oskar Öflunds Stiftelse and Jenny and Antti Wihuri Foundation.

Chapter 10: The authors are indebted to numerous colleagues for sharing their experience, knowledge and references that made it possible to cover a wide range of aspects on the deep soft sediments of the Baltic Sea in this chapter. HELCOM and Alf Norkko generously allowed us to use data and figures. The chapter is in part written within the framework of the BONUS Programme (Research Projects HYPER, COCOA, BIO-C3) of EU FP7 and the Nordic Centre of Excellence (NorMER). EB acknowledges the support from the Åbo Akademi University Endowment, UJ acknowledges the support from Gdańsk University, TR acknowledges the support from the Fund for Statutory Activities of the Palaeoceanology Unit, University of Szczecin. The authors dedicate this chapter to the memory of one of the Great Men of Baltic Sea biology, the late Heye Rumohr.

Chapter 11: The authors are grateful to Christoffer Boström for valuable comments on an early draft of the manuscript. HK acknowledges the support from the Swedish Environmental Protection Agency. PS acknowledges the support from the Swedish Research Council Formas, the Swedish Research Council, the Knut and Alice Wallenberg Foundation, Carl Tryggers Stiftelse för Vetenskaplig Forskning and the Swedish Environmental Protection Agency.

Chapter 12: The authors are grateful to Dariusz Wysocki for providing valuable information regarding seabirds on the southern Baltic Sea coast. TR acknowledges the support from the Fund for Statutory Activities of the Palaeoceanology Unit, University of Szczecin. JK acknowledges the institutional research funding IUT02-20 of the Estonian Research Council and the BONUS project BIO-C3 of EU FP7 and the Estonian Research Council.

Chapter 13: This chapter was prepared within the framework of the Ulrich Schiewer Laboratory for Experimental Aquatic Ecology (the USELab Project). HS acknowledges the support from the German Research Council (Grant Schu 983/5-1). IT acknowledges the support from the Russian Foundation for Basic Research (Grant 15-29-02706).

Chapter 14: MLZ and AD acknowledge the support from the Federal Agency for Nature Conservation (BfN) and the Federal Maritime and Hydrographic Agency (BS) for the projects that partly provided the German data presented in Chapter 14.

Chapter 15: The authors are grateful to Gerald Moore for valuable comments on an early draft of the manuscript. SK acknowledges the support from the Swedish National Space Board, the Swedish MISTRA programme, the Knut and Alice Wallenberg Foundation, the European Space Agency and EU FP5, FP6 and FP7. PK and SS acknowledge the support from the EU-project "Satellite Monitoring of the Baltic Sea Environment" (Grant POIG 01.01.02-22-011/09).

Chapter 16: CB acknowledges funding from the Swedish Radiation Safety Authority (SSM), the Foundation for Baltic and East European Studies, the Swedish Research Council Formas, and the EC FP7 projects STAR and COMET.

Chapter 17: The authors are grateful to Janus Larsen (AU-BIOS, Denmark) for assistance with Fig. 17.1 and Laura Meski (HELCOM Secretariat) for assistance with Figs. 17.2, 17.3 and 17.6.

Chapter 18: The authors thank the Baltic Marine Biologists (BMB) and the International Council for the Exploration of the Sea (ICES) for providing the photographs of their meetings in 1968 and 1904, respectively, in Box 18.1. JH acknowledges the support from the Polish Ministry of Science and Higher Education.

Contents

Chapter authors and box authors

Jesper H. Andersen NIVA Denmark Water Research, Copenhagen, Denmark

Agneta Andersson Department of Ecology and Environmental Sciences, Umeå University, Umeå, Sweden

Åsa Andersson World Wildlife Fund, Ulriksdals Slott, Solna, Sweden

Eugeniusz Andrulewicz National Marine Fisheries Research Institute, Gdynia, Poland

Elinor Andrén School of Natural Sciences, Technology and Environmental Studies, Södertörn University, Huddinge, Sweden

Thomas Andrén School of Natural Sciences, Technology and Environmental Studies, Södertörn University, Huddinge, Sweden

Birgitta Bergman Department of Ecology, Environment and Plant Sciences, Stockholm University/Science for Life Laboratory, Solna, Sweden

Lena Bergström Department of Aquatic Resources, Institute of Coastal Research, Swedish University of Agricultural Sciences, Öregrund, Sweden

Anders Bignert Swedish Museum of Natural History, Stockholm, Sweden

Jaanika Blomster Department of Environmental Sciences, University of Helsinki, Turku, Finland

Erik Bonsdorff Faculty of Science and Engineering, Environmental and Marine Biology, Åbo Akademi University, Turku, Finland

Christoffer Boström Faculty of Science and Engineering, Environmental and Marine Biology, Åbo Akademi University, Turku, Finland

Clare Bradshaw Department of Ecology, Environment and Plant Sciences, Stockholm University, Stockholm, Sweden

Katja Broeg Baltic Sea Centre, Stockholm University, Stockholm, Sweden

Alexander Darr Leibniz Institute for Baltic Sea Research, Rostock-Warnemünde, Germany

Olaf Dellwig Leibniz Institute for Baltic Sea Research, Rostock, Germany

Anna Edlund J. Craig Venter Institute, California, USA

Martin Ekman Department of Ecology, Environment and Plant Sciences, Stockholm University/Science for Life Laboratory, Solna, Sweden

Eeva Eronen-Rasimus Marine Research Centre Finnish Environment Institute, Helsinki, Finland

Stephan Gollasch Gollasch Consulting (GoConsult), Hamburg, Germany

Mats A. Granskog Norwegian Polar Institute, Fram Centre, Tromsø, Norway

Regina Hansen Leibniz Institute for Baltic Sea Research, Warnemünde, Germany

Marianne Hielm-Pedersén Department of Ecology, Environment and Plant Sciences, Stockholm University, Stockholm, Sweden

Jan Horbowy National Marine Fisheries Research Institute, Gdynia, Poland

Karolina Ininbergs Department of Ecology, Environment and Plant Sciences, Stockholm University/Science for Life Laboratory, Stockholm, Sweden

Urszula Janas Institute of Oceanography, University of Gdańsk, Gdynia, Poland

Kerstin Johannesson Department of Marine Sciences-Tjärnö, University of Gothenburg, Strömstad, Sweden

Veijo Jormalainen Department of Biology, University of Turku, Turku, Finland

Klaus Jürgens Leibniz Institute for Baltic Sea Research, Warnemünde, Rostock, Germany

Hermanni Kaartokallio Marine Research Centre Finnish Environment Institute, Helsinki, Finland

Hans Kautsky Department of Ecology, Environment and Plant Sciences, Stockholm University, Stockholm, Sweden

Samuli Korpinen Marine Research Centre, Finnish Environment Institute, Helsinki, Finland

Jonne Kotta Estonian Marine Institute, University of Tartu, Tallinn, Estonia

Lech Kotwicki Institute of Oceanology, Polish Academy of Sciences, Sopot, Poland

Piotr Kowalczuk Institute of Oceanology, Polish Academy of Sciences, Sopot, Poland

Susanne Kratzer Department of Ecology, Environment and Plant Sciences, Stockholm University, Stockholm, Sweden

Karol Kuliński Institute of Oceanology, Polish Academy of Sciences, Sopot, Poland

Harri Kuosa Marine Research Centre Finnish Environment Institute, Helsinki, Finland

Maria Laamanen Ministry of the Environment, Government, Finland

Matthias Labrenz Leibniz Institute for Baltic Sea Research, Warnemünde, Rostock, Germany

Kjell Larsson Linnaeus University, Kalmar Maritime Academy, Kalmar, Sweden

Sirpa Lehtinen Marine Research Centre, Finnish Environment Institute, Helsinki, Finland

Maiju Lehtiniemi Marine Research Centre, Finnish Environment Institute, Helsinki, Finland

Kari K. Lehtonen Marine Research Centre, Finnish Environment Institute, Helsinki, Finland

Tomasz Linkowski National Marine Fisheries Research Institute, Gdynia, Poland

Markus Majaneva Department of Natural History, Norwegian University of Science and Technology, Trondheim, Norway

Georg Martin Estonian Marine Institute, University of Tartu, Tallinn, Estonia

Johanna Mattila Department of Aquatic Resources, Institute of Coastal Research, Swedish University of Agricultural Sciences, Öregrund, Sweden

Pauli Merriman World Wildlife Fund, Ulriksdals Slott, Solna, Sweden

Kai Myrberg Finnish Environment Institute, Marine Research Centre, Helsinki, Finland

Mikko Nikinmaa Department of Biology, Laboratory of Animal Physiology, University of Turku, Turku, Finland

Raisa Nikula Finnish Museum of Natural History, University of Helsinki, Helsinki, Finland

Sergej Olenin Marine Science and Technology Center, Klaipeda University, Klaipeda, Lithuania

Jens Olsson Department of Aquatic Resources, Institute of Coastal Research, Swedish University of Agricultural Sciences, Öregrund, Sweden

Anders Omstedt Department of Marine Sciences, University of Gothenburg, Göteborg, Sweden

Jonna Piiparinen Marine Research Centre Finnish Environment Institute, Helsinki, Finland

Joanna Piwowarczyk Institute of Oceanology, Polish Academy of Sciences, Sopot, Poland

Falk Pollehne Leibniz Institute for Baltic Sea Research, Rostock, Germany

Teresa Radziejewska Faculty of Geosciences, University of Szczecin, Szczecin, Poland

Gregor Rehder Leibniz Institute for Baltic Sea Research, Rostock, Germany

Janne-Markus Rintala Department of Environmental Sciences, University of Helsinki, Turku, Finland

Sara Rydberg Department of Ecology, Environment and Plant Sciences, Stockholm University, Stockholm, Sweden

Sławomir Sagan Institute of Oceanology, Polish Academy of Sciences, Sopot, Poland

Sigrid Sagert Forschungszentrum Jülich, Rostock, Germany

Mariusz Sapota Institute of Oceanography, University of Gdańsk, Gdynia, Poland

Oleg Savchuk Baltic Nest Institute, Stockholm University, Stockholm, Sweden

Doris Schiedek DONG Energy Wind Power, Gentofte, Denmark

Bernd Schneider Leibniz Institute for Baltic Sea Research, Rostock, Germany

Dirk Schories Instituto de Biologia Marina, Campus Isla Teja, Universidad Austral de Chile, Valdivia, Chile

Hendrik Schubert Department of Biological Sciences, University of Rostock, Rostock, Germany

Uwe Selig Forschungszentrum Jülich, Rostock, Germany

Outi Setälä Finnish Environment Institute, Marine Research Centre, Helsinki, Finland

Ursula Siebert Institute for Terrestrial and Aquatic Wildlife Research, University of Veterinary Medicine Hannover, Foundation, Büsum, Germany

Kaarina Sivonen Department of Applied Chemistry and Microbiology, University of Helsinki, Helsinki, Finland

Sergei Skarlato Institute of Cytology of the Russian Academy of Sciences, Sankt-Petersburg, Russia

Krzysztof Skóra University of Gdańsk, Hel, Poland

Pauline Snoeijs-Leijonmalm Department of Ecology, Environment and Plant Sciences, Stockholm University, Stockholm, Sweden

Nardine Stybel Leibniz Institute for Baltic Sea Research, Rostock-Warnemünde, Germany

Timo Tamminen Marine Research Centre, Finnish Environment Institute, Helsinki, Finland

Letizia Tedesco Marine Research Centre Finnish Environment Institute, Helsinki, Finland

Irena Telesh Zoological Institute of the Russian Academy of Sciences, Sankt-Petersburg, Russia

David N. Thomas School of Ocean Sciences, Bangor University, Anglesey, UK

Ottilia Thoreson World Wildlife Fund, Ulriksdals Slott, Solna, Sweden

Jouni Vainio Finnish Meteorological Institute, Helsinki, Finland

Markku Viitasalo Marine Research Centre, Finnish Environment Institute, Helsinki, Finland

Ilppo Vuorinen University of Turku, Archipelago Research Institute, Turku, Finland

Risto Väinölä Finnish Museum of Natural History, University of Helsinki, Helsinki, Finland

Jan Warzocha Department of Oceanography and Marine Ecology, Sea Fisheries Institute, Gdynia, Poland

Jan Marcin Węsławski Institute of Oceanology, Polish Academy of Sciences, Sopot, Poland

Anastasija Zaiko Marine Science and Technology Centre, Klaipeda University, Klaipeda, Lithuania

Michael L. Zettler Leibniz Institute for Baltic Sea Research, Rostock-Warnemünde, Germany

Ulla Li Zweifel Swedish Institute for the Marine Environment, Göteborg, Sweden

Abbreviations

ADP	Adenosine Diphosphate
AIS	Automatic Identification System
AOP	Apparent Optical Properties
ATP	Adenosine Triphosphate
BDE	Brominated Diphenyl Ethers
BFR	Brominated Flame Retardants
BMAA	β-Methylamino-L-Alanine
BMB	Baltic Marine Biologists
BMG	Baltic Marine Geologists
BMP	Baltic Monitoring Programme (HELCOM)
BNI	Baltic Nest Institute
BOLD	Barcode of Life Database
BP	Before Present
BPL	Biopollution Level Index
BQI	Benthic Quality Index
BRP	Biological Reference Points
BSAP	Baltic Sea Action Plan (HELCOM)
BSASD	Baltic Sea Alien Species Database
BSII	Baltic Sea Impact Index (HELCOM)
BSPI	Baltic Sea Pressure Index (HELCOM)
BSRAC	Baltic Sea Advisory Council (EU)
BSSC	Baltic Sea Science Congress (BMB, CBO, BMG)
BTA	Biological Traits Analysis
BWQD	Bathing Water Quality Directive (EU)
CART	Country-Allocated Reduction Targets
CAS	Chemical Abstracts Service (American Chemical Society)
CBD	Convention on Biological Diversity
CBO	Conferences of Baltic Oceanographers
CDOM	Coloured Dissolved Organic Matter
CEC	Contaminants of Emerging Concern
CFD	Computational Fluid Dynamics
CFP	Common Fisheries Policy (EU)
CHASE	Hazardous Substances Status Assessment Tool (HELCOM)
Chl *a*	Chlorophyll *a*
CI	Condition Index
COI	Cytochrome *c* Oxidase I
CWA	Chemical Warfare Agents
DDE	Dichlorodiphenyldichloroethylene
DDT	Dichlorodiphenyltrichloroethane
DIC	Dissolved Inorganic Carbon (C_T)
DIN	Dissolved Inorganic Nitrogen
DIP	Dissolved Inorganic Phosphorus
DOC	Dissolved Organic Carbon
DOM	Dissolved Organic Matter
DON	Dissolved organic nitrogen
DPSIR	Driving Force-Pressure-State Impact-Response Indicator
DSi	Dissolved Silica
dw	dry weight
EAC	Environmental Assessment Criteria
EBM	Ecosystem-Based Management
ECHA	European Chemicals Agency (EU)
EDA	Effect-Directed Analysis
EDC	Endocrine Disrupting Chemicals
EEZ	Exclusive Economic Zone
EQR	Ecological Quality Ratio
ERA	European Research Area (EU)
ESA	European Space Agency
ESG	Ecosystem Goods
ESS	Ecosystem Services
EU	European Union
FWE	Food Web Efficiency
GES	Good Environmental Status
GIS	Geographic Information Systems
HBCDD	Hexabromocyclododecane
HD	Habitats Directive (EU)
HEAT	HELCOM Eutrophication Assessment Tool
HELCOM	Helsinki Commission, Baltic Marine Environment Protection Commission
HIF	Hypoxia-Inducible Factor
HIS	Health Status Index (HELCOM)
HNF	Heterotrophic Nanoflagellates
HOC	Hydrophobic Organic Compounds
HPLC	High-Performance Liquid Chromatography
HSP	Heat Shock Proteins
IAS	Invasive Alien Species
IBAS	Integrated Biomarker Assessment Score
IBAT	Integrated Biomarker Assessment Tool
IBR	Integrated Biomarker Response
IBSFC	International Baltic Sea Fishery Commission (1974–2005)

ICES	International Council for the Exploration of the Sea
ICZM	Integrated Coastal Zone Management
IMO	International Maritime Organization (UN)
IOC	Intergovernmental Oceanographic Commission (UNESCO)
IOP	Inherent Optical Properties
IPCC	Intergovernmental Panel on Climate Change (WMO and UNEP)
ISUM	Integrated Sea Use Management
LC	Liquid Chromatography
LME	Large Marine Ecosystems
LOEC	Lowest Observed Effect Concentration
LOQ	Limit of Quantification
LPO	Lipid Peroxidation
lw	lipid weight
MA	Millennium Ecosystem Assessment
MAAs	Mycosporine-like Amino Acids
MAI	Maximum Allowable Inputs
MARPOL	International Convention for the Prevention of Pollution from Ships (IMO)
MBI	Major Baltic Inflow
MDS	Multi-Dimensional Scaling
MN	Micronuclei
MS	Mass Spectrometry
MSFD	Marine Strategy Framework Directive (EU)
MSP	Marine Spatial Planning
MSY	Maximum Sustainable Yield
NAO	North Atlantic Oscillation
NASA	National Aeronautics and Space Administration (United States Federal Government)
NBS	National Bureau of Standards (United States Department of Commerce)
NGO	Non-Governmental Organisation
NGS	Next-Generation Sequencing
NIS	Non-Indigenous Species
NOEC	No Observed Effect concentration
NPQ	Non-Photochemical Quenching
OBIS	Ocean Biogeographic Information System (UNESCO)
OM	Organic Matter
OSPAR	Oslo-Paris Convention for the Protection of the Marine Environment of the North-East Atlantic
OTU	Operational Taxonomic Unit
PAH	Polycyclic Aromatic Hydrocarbons
PAR	Photosynthetically Active (or "Available") Radiation
PCB	Polychlorinated Biphenyls
PCI	Penis Classification Index
PEC	Predicted Effect Concentration
PFAA	Perfluorinated alkyl acids
PFOS	Perfluorooctane Sulfonate
PNEC	Predicted No-Effect Concentration
POC	Particulate Organic Carbon
POM	Particulate Organic Matter
POP	Persistent Organic Pollutants
PSI	Photosystem I
PSII	Photosystem II
PSS	Practical Salinity Scale
PSU	Practical Salinity Units
QTL	Quantitative Trait Loci
REACH	Registration, Evaluation, Authorization and Restriction of Chemical Substances (EU)
ROS	Reactive Oxygen Species
RPD	Redox Potential Discontinuity
SBL	Safe Biological Limits
SCUBA	Self-Contained Underwater Breathing Apparatus
SMHI	Swedish Meteorological and Hydrological Institute
SNP	Single Nucleotide Polymorphism
SOD	Superoxide Dismutase
SPIM	Suspended Particulate Inorganic Matter
SPM	Suspended Particulate Matter
SPOM	Suspended Particulate Organic Matter
SSB	Spawning Stock Biomass
STECF	Scientific, Technical and Economic Committee for Fisheries (EU)
syn.	synonym
TA	Total alkalinity (A_T)
TAC	Total Allowable Catch (EU)
TBT	Tributyltin
TEEB	The Economics of Ecosystems and Biodiversity (UNEP)
TIE	Toxicity Identification and Evaluation (United States Environmental Protection Agency)
TOSC	Total Oxygen-Scavenging Capacity
UN	United Nations
UNCHE	United Nations Conference on the Human Environment
UNCLOS	United Nations Convention of the Law of the Sea
UNCED	United Nations Conference on Environment and Development
UNEP	United Nations Environment Programme
UNESCO	United Nations Educational, Scientific and Cultural Organization
VHC	Volatile Halocarbon
VSF	Volume Scattering Function
WFD	Water Framework Directive (EU)
WHO	World Health Organization (UN)
WMO	World Meteorological Organization (UN)
ww	wet weight (fresh weight)
WWF	World Wildlife Fund
WWTP	Wastewater Treatment Plants

List of genera and species with author names

(including some frequently used synonyms, for synonymy see text)

Abra alba (Wood 1802)
Abramis brama (Linnaeus 1758)
Abramis Cuvier 1816
Acanthocorbis Hara & Takahashi 1984
Acanthocorbis cf. *apoda* (Leadbeater) Hara & Takahashi 1984
Acanthocorbis cf. *asymmetrica* (Thomsen) Hara & Takahashi 1984
Acartia Dana 1846
Acartia bifilosa (Giesbrecht 1881)
Acartia longiremis (Liljeborg 1853)
Acartia tonsa Dana 1849
Achnanthes brevipes var. *intermedia* (Kützing) Cleve 1895
Achnanthes taeniata Grunow 1880
Acinetobacter Brisou & Prévot 1954
Acipenser baerii Brandt 1869
Acipenser gueldenstaedtii Brandt & Ratzeburg 1833
Acipenser oxyrinchus Mitchill 1815
Acipenser ruthenus Linnaeus 1758
Acipenser sturio Linnaeus 1758
Acrosiphonia centralis (Lyngbye) Kjellman 1893
Actinocyclus Ehrenberg 1837
Actinocyclus octonarius Ehrenberg 1837
Aegagropila linnaei Kützing 1843
Aeolidia papillosa (Linnaeus 1761)
Aglaothamnion halliae (Collins) Aponte et al. 1997
Ahnfeltia plicata (Hudson) Fries 1836
Akashiwo sanguinea (Hirasaka) Hansen & Moestrup *in* Daugbjerg et al. 2000
Alca torda Linnaeus 1758
Alexandrium minutum Halim 1960
Alexandrium ostenfeldii (Paulsen) Balech & Tangen 1985
Alisma wahlenbergii (Holmberg) Juzepczuk 1933
Alosa alosa (Linnaeus 1758)
Alosa fallax (Lacepède 1803)
Alosa pseudoharengus (Wilson 1811)
Ameira divagans Nicholls 1939
Ammodytes tobianus Linnaeus 1758
Amoeba Ehrenberg 1830
Ampelisca Krøyer 1842
Amphibalanus improvisus (Darwin 1854)
Amphiura Forbes 1843
Amphora Ehrenberg *ex* Kützing 1844
Amphora pediculus (Kützing) Grunow *ex* A. Schmidt 1875
Anabaena Bory *ex* Bornet & Flahault 1886
Anas platyrhynchos Linnaeus 1758
Ancistrumina limnica Raabe 1947
Ancylus fluviatilis Müller 1774
Anguilla anguilla (Linnaeus 1758)
Anguillicoloides crassus (Kuwahara et al. 1974)
Anisus leucostoma (Millet 1813)
Anodonta Lamarck 1799
Antithamnion boreale (Gobi) Kjellman 1883
Antomicron Cobb 1920
Apedinella radians (Lohmann) Campbell 1973
Aphanizomenon flos-aquae Ralfs *ex* Bornet & Flahault 1886
Aphanizomenon Morren *ex* Bornet & Flahault 1888
Aphanothece clathrata West & G.S. West 1906
Aphanothece paralleliformis Cronberg 2003
Arctica islandica (Linnaeus 1767)
Ardea cinerea Linnaeus 1758
Arenicola marina (Linnaeus 1758)
Aristichthys nobilis (Richardson 1845)
Artemia Leach 1819
Ascolaimus Ditlevsen 1919
Ascophyllum Stackhouse 1809
Ascophyllum nodosum (Linnaeus) Le Jolis 1863
Askenasia cf. *stellaris* Blochmann 1895
Aspidogaster von Baer 1826
Asplanchna Gosse 1850
Astarte Sowerby 1816
Astarte borealis (Schumacher 1817)
Asterias Linnaeus 1758
Asterias rubens Linnaeus 1758
Aulacoseira islandica (Otto Müller) Simonsen 1979
Aurelia aurita (Linnaeus 1758)
Aythya fuligula (Linnaeus 1758)
Aythya marila (Linnaeus 1761)
Bacillus Cohn 1872
Bacteroides Castellani & Chalmers 1919 *emend.* Shah & Collins 1989
Balanion Wulff 1919
Balanion comatum Wulff 1922
Balanus crenatus Bruguière 1789
Balanus improvisus Darwin 1854
Baltidrilus costatus (Claparède 1863)

Bangia atropurpurea (Mertens *ex* Roth) C.A. Agardh 1824
Bathyporeia pilosa Lindström 1855
Batillipes mirus Richters 1909
Battersia Reinke *ex* Batters 1890
Battersia arctica (Harvey) Draisma et al. 2010
Beggiatoa Trevisan 1842
Berkeleya Greville 1827
Berkeleya rutilans (Trentepohl *ex* Roth) Grunow 1880
Beroe Browne 1756
Beroe gracilis Künne 1939
Biecheleria baltica Moestrup et al. 2009
Bithynia tentaculata (Linnaeus 1758)
Bolinopsis Agassiz 1860
Bonnemaisonia hamifera Hariot 1891
Bosmina Baird 1845
Bosmina longispina subsp. *maritima* (P.E. Müller 1867)
Branta canadensis (Linnaeus 1758)
Brebissonia lanceolata (C.A. Agardh) Mahoney & Reimer 1984
Brissopsis Agassiz 1840
Brongniartella byssoides (Goodenough & Woodward) Schmitz 1893
Buccinum undatum Linnaeus 1758
Bucephala clangula (Linnaeus 1758)
Bucephalus polymorphus von Baer 1827
Bursaria O.F. Müller 1773
Bylgides sarsi (Kinberg *in* Malmgren 1866)
Calanus Leach 1816
Calanus finmarchicus (Gunnerus 1770)
Calidris alpina (Linnaeus 1758)
Calliacantha natans (Grøntved) Leadbeater 1978
Calliacantha simplex Manton & Oates 1979
Callitriche Linnaeus 1753
Callitriche hermaphroditica Linnaeus 1755
Calocaris macandreae Bell 1853
Caloplaca Th. Fries 1860
Calothrix C.A. Agardh *ex* Bornet & Flahault 1886
Calothrix scopulorum C.A. Agardh *ex* Bornet & Flahault 1886
Campylodiscus clypeus (Ehrenberg) Ehrenberg *ex* Kützing 1844
Candidatus Nitrosopumilus maritimus Konneke et al. 2005
Candona neglecta G.O. Sars 1887
Canthocamptus Westwood 1836
Capitella capitata (Fabricius 1780)
Carassius carassius (Linnaeus 1758)
Carassius gibelio (Bloch 1782)
Carcinus Leach 1814
Carcinus maenas (Linnaeus 1758)
Catostomus catostomus (Forster 1773)
Caulerpa taxifolia (Vahl) C.A. Agardh 1817
Centropages hamatus (Liljeborg 1853)
Centropages typicus Krøyer 1849
Cepphus grylle (Linnaeus 1758)
Ceramium Roth 1797
Ceramium diaphanum (Lightfoot) Roth 1806
Ceramium rubrum C.A. Agardh 1811
Ceramium tenuicorne (Kützing) Wærn 1952
Ceramium virgatum Roth 1797
Cerastoderma glaucum (Poiret 1789)
Cerataulina H. Peragallo *ex* Schütt 1896
Cerataulina pelagica (Cleve) Hendey 1937
Ceratium Schrank 1793
Ceratium fusus (Ehrenberg) Dujardin 1841
Ceratium lineatum (Ehrenberg) Cleve 1899
Ceratium tripos (O.F. Müller) Nitzsch 1817
Ceratophyllum demersum Linnaeus 1753
Cercopagis pengoi (Ostroumov 1891)
Chaetoceros Ehrenberg 1844
Chaetoceros castracanei Karsten 1905
Chaetoceros holsaticus Schütt 1895
Chaetoceros mitra (Bailey) Cleve 1896
Chaetoceros neogracilis Van Landingham 1968
Chaetoceros wighamii Brightwell 1856
Chaetogammarus ischnus (Stebbing 1899)
Chaetogammarus warpachowskyi (G.O. Sars 1897)
Chaetomorpha linum (O.F. Müller) Kützing 1845
Chalinula limbata (Montagu 1814)
Chamaepinnularia margaritiana (Witkowski) Witkowski 1996
Chara Linnaeus 1753
Chara aspera Willdenow 1809
Chara baltica Bruzelius 1824
Chara buckellii Allen 1951
Chara canescens Loiseleur 1810
Chara connivens Salzmann *ex* Braun 1835
Chara corallina Klein *ex* Willdenow 1805
Chara horrida Wahlstedt 1862
Chara tomentosa Linnaeus 1753
Chara vulgaris Linnaeus 1753
Charadrius alexandrinus Linnaeus 1758
Charadrius hiaticula Linnaeus 1758
Chelicorophium curvispinum (G.O. Sars 1895)
Chlamydomonas Ehrenberg 1833
Chlamydomonas raudensis Ettl 1976
Chondrus crispus Stackhouse 1797
Chorda filum (Linnaeus) Stackhouse 1797
Chroococcus Nägeli 1849
Chrysochromulina Lackey 1939
Chrysochromulina birgeri Hällfors & Niemi 1974
Chydorus Leach 1843
Ciona Fleming 1822
Ciona intestinalis (Linnaeus 1767)
Cladophora Kützing 1843
Cladophora aegagropila (Linnaeus) Trevisan 1845
Cladophora glomerata (Linnaeus) Kützing 1843
Cladophora rupestris (Linnaeus) Kützing 1843
Clangula hyemalis (Linnaeus 1758)

Clithrocytheridea sorbyana (Jones 1857) Schweyer 1949
Clupea harengus Linnaeus 1758
Clupea harengus subsp. *membras* Wulf 1765
Cocconeis Ehrenberg 1836
Cocconeis pediculus Ehrenberg 1838
Coccotylus truncatus (Pallas) Wynne & Heine 1992
Codium fragile (Suringar) Hariot 1889
Codosiga Clark 1866
Coelosphaerium minutissimum Lemmermann 1900
Coleofasciculus chthonoplastes (Thuret *ex* Gomont) Siegesmund et al. 2008
Colpomenia peregrina Sauvageau 1927
Colwellia Deming et al. 1988
Conchophthirus acuminatus (Claparède & Lachmann 1858)
Corallina officinalis Linnaeus 1758
Corbula gibba (Olivi 1792)
Cordylophora caspia (Pallas 1771)
Coregonus Linnaeus 1758
Coregonus lavaretus (Linnaeus 1758)
Coregonus maraena (Bloch 1779)
Cornigerius maeoticus (Pengo 1879)
Corophium volutator (Pallas 1766)
Coscinodiscus granii Gough 1905
Coscinodiscus wailesii Gran & Angst 1931
Cosmoeca Thomsen 1984
Cottus gobio Linnaeus 1758
Crangon crangon (Linnaeus 1758)
Crassostrea gigas (Thunberg 1793)
Crepidula fornicata (Linnaeus 1758)
Cribroelphidium excavatum (Terquem 1875)
Cryothecomonas Thomsen et al. 1991
Cryothecomonas aestivalis Drebes et al. 1996
Cryothecomonas armigera Thomsen et al. 1991
Cryothecomonas longipes Schnepf & Kühn 2000
Ctenolabrus rupestris (Linnaeus 1758)
Ctenophora pulchella (Ralfs *ex* Kützing) Williams & Round 1986
Cyanea capillata (Linnaeus 1758)
Cyanea lamarckii Péron & Lesueur 1810
Cycas micronesica K.D. Hill 1994
Cyclidium O.F. Müller 1773
Cyclops Müller O.F. 1785
Cyclopterus lumpus Linnaeus 1758
Cygnus olor (Gmelin 1789)
Cymbomonas tetramitiformis Schiller 1913
Cyprinus carpio Linnaeus 1758
Cytheropteron montrosiense Brady Crosskey & Robertson 1874
Cytherura gibba (Mueller 1785) G.O. Sars 1866
Cytophaga Winogradsky 1929 *emend.* Nakagawa & Yamasato 1996
Dactyliosolen fragilissimus (Bergon) Hasle *in* Hasle & Syvertsen 1996
Danio rerio (Hamilton 1822)
Daphnia O.F. Müller 1785
Daptonema Cobb 1920
Dasya baillouviana (Gmelin) Montagne *in* Barker-Webb & Berthelot 1841
Dasysiphonia japonica (Yendo) Kim 2012
Debaryomyces (Zopf) Lodder & Kreger-van Rij 1984
Debaryomyces hansenii (Zopf) Lodder & Kreger-van Rij 1984
Delesseria sanguinea (Hudson) Lamouroux 1813
Dendrodoa grossularia (Van Beneden 1846)
Deshayesorchestia deshayesii (Audouin 1826)
Desulfovibrio Kluyver & van Niel 1936 *emend.* Loubinoux et al. 2002
Diaphanoeca grandis Ellis 1930
Diaphanoeca sphaerica Thomsen 1982
Diastylis rathkei (Krøyer 1841)
Diatoma bottnica Snoeijs in Snoeijs & Potatova 1998
Diatoma constricta (Grunow) Williams 1985
Diatoma vulgaris Bory 1824
Dichromadora Kreis 1929
Dictyocha speculum Ehrenberg 1839
Dictyosiphon chordaria Areschoug 1847
Dictyosiphon foeniculaceus (Hudson) Greville 1830
Dictyosphaerium Nägeli 1849
Didinium Stein 1859
Didinium gargantua Meunier 1910
Dilsea carnosa (Schmidel) Kuntze 1898
Dinobryon Ehrenberg 1834
Dinophysis Ehrenberg 1839
Dinophysis acuminata Claparède & Lachmann 1859
Dinophysis acuta Ehrenberg 1839
Dinophysis norvegica Claparède & Lachmann 1859
Dinophysis rotundata Claparède & Lachmann 1859
Ditylum brightwellii (West) Grunow in Van Heurck 1885
Dolichospermum Thwaites 1850
Dreissena polymorpha (Pallas 1771)
Drepanocladus Müller Halle 1898
Dumontia contorta (Gmelin) Ruprecht 1850
Ebria tripartita (Schumann) Lemmermann 1899
Echiniscoides sigismundi (Schultze 1865)
Echinococcus multilocularis Leuckart 1863
Ecrobia ventrosa (Montagu 1803)
Ectocarpus Lyngbye 1819
Ectocarpus siliculosus (Dillwyn) Lyngbye 1819
Einhornia crustulenta (Pallas 1766)
Elachista fucicola (Velley) Areschoug 1842
Electra crustulenta (Pallas 1766)
Ellerbeckia arenaria (Moore *ex* Ralfs) Crawford 1988
Elodea canadensis Michaux 1803
Elphidium excavatum (Terquem 1875)
Embletonia pallida Alder & Hancock 1854
Emiliania huxleyi (Lohmann) Hay & Mohler *in* Hay et al. 1967
Encyonema Kützing 1834

Encyonema silesiacum (Bleisch) Mann *in* Round et al. 1990
Enoplolaimus de Man 1893
Ensis Schumacher 1817
Ensis directus (Conrad 1843)
Enterococcus (*ex* Thiercelin & Jouhaud 1903) Schleifer & Kilpper-Bälz 1984
Enteromorpha ahlneriana Bliding 1944
Enteromorpha flexuosa (Wulfen) J. Agardh 1883
Enteromorpha intestinalis (Linnaeus) Nees 1820
Ephydatia fluviatilis (Linnaeus 1759)
Ephydra Fallén 1810
Erignathus barbatus (Erxleben 1777)
Eriocheir sinensis H. Milne Edwards 1853
Escherichia Castellani & Chalmers 1919
Escherichia coli (Migula 1895) Castellani & Chalmers 1919
Esox lucius Linnaeus 1758
Eubosmina coregoni subsp. *maritima* (P.E. Müller 1867)
Eucheuma denticulatum (Burman) Collins & Hervey 1917
Eudesme virescens (Carmichael *ex* Berkeley) J. Agardh 1882
Eunotia Ehrenberg 1837
Euplotes O.F. Müller 1786
Eurotium Link 1809
Eurotium rubrum König et al. 1901
Eurytemora Giesbrecht 1881
Eurytemora affinis (Poppe 1880)
Evadne Lovén 1836
Evadne anonyx G.O. Sars 1897
Evadne nordmanni Lovén 1836
Fallacia tenera (Hustedt) Mann *in* Round et al. 1990
Fejervarya cancrivora (Gravenhorst 1829)
Fissidens fontanus Steudel 1824
Flavobacterium Bergey et al. 1923 *emend.* Bernardet et al. 1996 *emend.* Dong et al. 2013 *emend.* Kang et al. 2013 *emend.* Kuo et al. 2013
Flavobacterium gelidilacus Van Trappen et al. 2003 *emend.* Joung et al. 2013
Fontinalis Hedwig 1801
Fontinalis antipyretica Hedwig 1801
Fontinalis dalecarlica Bruch & Schimper in B.S.G. 1846
Fragillariopsis cylindrus (Grunow) Helmcke & Krieger 1954
Fritillaria borealis Lohmann 1896
Fucellia tergina (Zetterstedt 1845)
Fucus Linnaeus 1753
Fucus cottonii Wynne & Magne 1991
Fucus evanescens C.A. Agardh 1820
Fucus radicans Bergström & Kautsky *in* Bergström et al. 2005
Fucus serratus Linnaeus 1753
Fucus spiralis Linnaeus 1753
Fucus vesiculosus Linnaeus 1753
Fulica atra Linnaeus 1758
Furcellaria lumbricalis (Hudson) Lamouroux 1813
Gadus morhua Linnaeus 1758
Gammarus Fabricius 1775
Gammarus duebeni Liljeborg 1852
Gammarus locusta (Linnaeus 1758)
Gammarus oceanicus Segerstråle 1947
Gammarus salinus Spooner 1947
Gammarus tigrinus Sexton 1939
Gammarus zaddachi Sexton 1912
Gasterosteus aculeatus Linnaeus 1758
Gavia arctica (Linnaeus 1758)
Gavia stellata (Pontoppidan 1763)
Gelochelidon nilotica (Gmelin 1789)
Gloeocapsa crepidinum (Thuret) Thuret 1876
Gloeocapsopsis crepidinum (Thuret) Geitler *ex* Komárek 1993
Gmelinoides fasciatus (Stebbing 1899)
Gobius niger Linnaeus 1758
Gomphonema Ehrenberg 1932
Gomphonema olivaceum (Hornemann) Brébisson 1838
Goniomonas Stein 1878
Gracilaria Greville 1830
Gracilaria vermiculophylla (Ohmi) Papenfuss 1967
Graphiola Poitou 1824
Graphiola phoenicis (Mougeot) Poiteau 1824
Gymnocephalus cernuus (Linnaeus 1758)
Gymnodinium Stein 1878
Gymnodinium catenatum Graham 1943
Gymnodinium corollarium Sundström et al. *in* Sundström et al. 2009
Gyraulus acronicus (Férussac 1807)
Gyrodinium Kofoid & Swezy 1921
Gyrodinium fusiforme Kofoid & Swezy 1921
Haematococcus pluvialis Flotow 1844
Haematopus ostralegus Linnaeus 1758
Haliaeetus albicilla (Linnaeus 1758)
Halichoerus grypus (Fabricius 1791)
Haliclona limbata (Montagu 1814)
Haliclona oculata (Linnaeus 1759)
Halicryptus spinulosus von Siebold 1849
Halidrys siliquosa (Linnaeus) Lyngbye 1819
Halosiphon tomentosus (Lyngbye) Jaasund 1957
Hediste Malmgren 1867
Hediste diversicolor (O.F. Müller 1776)
Helicostomella Jörgensen 1924
Hemimysis anomala G.O. Sars 1907
Heterocapsa arctica subsp. *frigida* Rintala & Hällfors 2010
Heterocapsa triquetra (Ehrenberg) Stein 1883
Heterocyprideis sorbyana (Jones 1857) Elofson 1941
Heterocypris salina (Brady 1868)
Heteroleibleinia cf. *kützingii* (Schmidle) Compère 1986
Heteromastus Eisig 1887
Heterosigma akashiwo (Hada) Hada *ex* Hara & Chihara 1987
Hildenbrandia rivularis (Liebmann) J. Agardh 1851
Hildenbrandia rubra (Sommerfelt) Meneghini 1841
Hinia reticulata (Linnaeus 1758)

Holophrya Ehrenberg 1831
Huso huso (Linnaeus 1758)
Hyalella azteca (Saussure 1858)
Hydrobia ulvae (Pennant 1777)
Hydrobia ventrosa (Montagu 1803)
Hydrocharis morsus-ranae Linnaeus 1753
Hydrodyction reticulatum (Linnaeus) Bory 1824
Hyperoplus lanceolatus (Le Sauvage 1824)
Hyphomonas (ex Pongratz 1957) Moore et al. 1984 *emend.* Weiner et al. 2000 *emend.* Li et al. 2014
Hypophthalmichthys molitrix (Valenciennes 1844)
Idotea Fabricius 1798
Idotea balthica (Pallas 1772)
Idotea chelipes (Pallas 1766)
Idotea granulosa Rathke 1843
Isoetes echinospora Durieu 1861
Isoetes lacustris Linnaeus 1753
Kappaphycus alvarezii (Doty) Doty *ex* Silva *in* Silva et al. 1996
Karenia mikimotoi (Miyake & Kominami *ex* Oda) Hansen & Moestrup *in* Daugbjerg et al. 2000
Katablepharis cf. *remigera* (Vørs) Clay & Kugrens 2000
Keratella Bory 1822
Keratella cochlearis subsp. *baltica* (Sokolova 1927)
Keratella quadrata (Müller 1786)
Keratella quadrata subsp. *platei* Jägerskiöld 1894
Kurtiella bidentata (Montagu 1803)
Labyrinthula zosterae Porter & Muehlstein *in* Muehlstein & Short 1991
Lacrymaria Ehrenberg 1830
Lacrymaria rostrata Kahl 1935
Laminaria saccharina (Linnaeus) Lamouroux 1813
Lampetra fluviatilis (Linnaeus 1758)
Lamprothamnium papulosum (Wallroth) Groves 1916
Lamprothamnium succinctum (Braun) Wood 1962
Larus argentatus Pontoppidan 1763
Larus canus Linnaeus 1758
Larus fuscus Linnaeus 1758
Larus marinus Linnaeus 1758
Larus ridibundus Linnaeus 1766
Lemna Linnaeus 1753
Lemna gibba Linnaeus 1753
Lemna minor Linnaeus 1753
Leptodora kindti (Focke 1844)
Leptolaimus elegans (Stekhoven & De Coninck 1933) Gerlach 1958
Leptolaimus papilliger de Man 1876
Leptolegnia baltica Höhnk & Vallin 1953
Leucocryptos marina (Braarud) Butcher 1967
Licmophora debilis (Kützing) Grunow *ex* Van Heurck 1881
Licmophora gracilis var. *angelica* (Kützing) H. Peragallo & M. Peragallo 1901
Limanda limanda (Linnaeus 1758)
Limnocalanus macrurus G.O. Sars 1863
Limnomysis benedeni Czerniavsky 1882
Limosella aquatica Linnaeus 1753
Liparis liparis (Linnaeus 1766)
Littorella uniflora Bergius 1768
Littorina Férussac 1822
Littorina littorea (Linnaeus 1758)
Littorina saxatilis (Olivi 1792)
Lohmaniella Leegaard 1915
Loktanella Van Trappen et al. 2004 *emend.* Moon et al. 2010 *emend.* Lee 2012 *emend.* Tsubouchi et al. 2013
Lumpenus lampretaeformis (Walbaum 1792)
Lutra lutra (Linnaeus 1758)
Lyngbya aestuarii Liebman *ex* Gomont 1892
Macoma Leach 1819
Macoma balthica (Linnaeus 1758)
Macoma balthica subsp. *rubra* (da Costa 1778)
Macoma calcarea (Gmelin 1791)
Maeotias marginata (Modeer 1791)
Manayunkia aestuarina (Bourne 1883)
Marenzelleria Mesnil 1896
Marenzelleria arctia (Chamberlin 1920)
Marenzelleria neglecta Sikorski & Bick 2004
Marenzelleria viridis (Verrill 1873)
Martyana atomus (Hustedt) Snoeijs 1991
Martyana schulzii (Brockmann) Snoeijs 1991
Mastogloia Thwaites *ex* W. Smith 1856
Mastogloia baltica Grunow in van Heurck 1880
Mastogloia smithii Thwaites *ex* W. Smith 1856
Mastogloia smithii var. *amphicephala* Grunow in Van Heurck 1880
Meganyctiphanes norvegica (M. Sars 1857)
Melanitta fusca (Linnaeus 1758)
Melanitta nigra (Linnaeus 1758)
Melosira arctica Dickie 1852
Melosira moniliformis (O.F. Müller) C.A. Agardh 1824
Mergus serrator Linnaeus 1758
Merlangius merlangus (Linnaeus 1758)
Mertensia ovum (Fabricius 1780)
Mesocyclops G.O. Sars 1914
Mesodinium rubrum (Lohmann 1908)
Metacystis Cohn 1866
Metridium dianthus (Ellis 1768)
Micrococcus Cohn 1872 *emend.* Stackebrandt et al. 1995 *emend.* Wieser et al. 2002
Microcystis Beck 1837
Microcystis aeruginosa (Kützing) Kützing 1846
Microlaimus de Man 1880
Mnemiopsis Agassiz 1860
Mnemiopsis leidyi A. Agassiz 1865
Monodinium Fabre-Domergue 1888
Monoporeia affinis (Lindström 1855)
Monoraphidium contortum (Thuret) Komárková-Legnerová *in* Fott 1969
Monosiga Kent 1878
Monostroma Thuret 1854
Mustela vison (Schreber 1761)

Mya Linnaeus 1758
Mya arenaria Linnaeus 1758
Myoxocephalus quadricornis (Linnaeus 1758)
Myoxocephalus scorpius (Linnaeus 1758)
Myrionecta rubra Lohmann 1908
Myriophyllum spicatum Linnaeus 1753
Mysis Latreille 1802
Mysis mixta Liljeborg 1853
Mysis relicta Lovén 1862
Mysis salemaai Audzijonyte & Väinölä 2005
Mysis segerstralei Audzijonyte & Väinölä 2005
Mytilopsis leucophaeata (Conrad 1831)
Mytilus Linnaeus 1758
Mytilus edulis Linnaeus 1758
Mytilus trossulus Gould 1850
Nais elinguis Müller 1774
Najas marina Linnaeus 1753
Nannochloropsis limnetica Krienitz et al. 2000
Nassarius reticulatus (Linnaeus 1758)
Navicula Bory 1822
Navicula pelagica Cleve 1896
Navicula perminuta Grunow *in* van Heurck 1880
Navicula ramosissima (C.A. Agardh) Cleve 1895
Navicula vanhoeffenii Gran 1897
Nemalion multifidum (Lyngbye) Chauvin 1842
Neogobius melanostomus (Pallas 1814)
Neomysis integer (Leach 1814)
Neosiphonia harveyi (Bailey) Kim et al. 2001
Neovison vison (Schreber 1777)
Neptunea antiqua (Linnaeus 1758)
Nerophis ophidion (Linnaeus 1758)
Nitella C.A. Agardh 1824
Nitellopsis obtusa (Desvaux) Groves 1919
Nitokra spinipes Boeck 1865
Nitrobacter Winogradsky 1892
Nitrosomonas Winogradsky 1892
Nitrosopumilus maritimus Konneke et al. 2006
Nitzschia Hassall 1845
Nitzschia frigida Grunow 1880
Nitzschia inconspicua Grunow 1862
Nitzschia paleacea (Grunow) Grunow *in* Van Heurck 1881
Nitzschia sigmoidea (Nitzsch) W. Smith 1853
Nodularia Mertens *ex* Bornet & Flahault 1886
Nodularia baltica Komárek et al.1993
Nodularia litorea Thuret *ex* Komárek et al. 1993
Nodularia spumigena Mertens *ex* Bornet & Flahault 1886
Nostoc Vaucher *ex* Bornet & Flahault 1886
Nuphar lutea (Linnaeus) Smith 1809
Nyctereutes procyonoides Gray 1834
Nymphon grossipes (Fabricius 1780)
Obesogammarus crassus (G.O. Sars 1894)
Odobenus rosmarus (Linnaeus 1758)
Odonthalia dentata (Linnaeus) Lyngbye 1819
Oikopleura dioica Fol 1872
Oithona similis Claus 1866
Oncorhynchus gorbuscha (Walbaum 1792)
Oncorhynchus keta (Walbaum 1792)
Oncorhynchus mykiss (Walbaum 1792)
Ondatra zibethicus (Linnaeus 1766)
Oocystis heteromucosa Hegewald in Hegewald et al. 1980
Ophiura albida Forbes 1839
Ophrydium versatile (Müller 1786) Bory 1824
Ophryoglena Ehrenberg 1831
Orchestia Leach 1814
Orconectes limosus (Rafinesque 1817)
Orconectes virilis (Hagen 1870)
Osmerus eperlanus (Linnaeus 1758)
Osmundea truncata (Kützing) Nam & Maggs *in* Nam et al. 1994
Ostreococcus tauri Courties & Chrétiennot-Dinet 1995
Pachysphaera Ostenfeld 1899
Pacifastacus leniusculus (Dana 1852)
Pagophilus groenlandicus (Erxleben 1777)
Palaemon Weber 1795
Palaemon adspersus Rathke 1837
Palaemon elegans Rathke 1837
Palaemonetes Heller 1869
Palmaria palmata (Linnaeus) Weber & D. Mohr 1805
Paludestrina jenkinsi (Smith 1889)
Paracalanus parvus (Claus 1863)
Paracyprideis fennica (Hirschmann 1909)
Paramacrobiotus richtersi Murray 1911
Paramecium O.F. Müller 1773
Paramysis lacustris (Czerniavsky 1882)
Paraphysomonas De Saedeleer 1930
Parlibellus Cox 1988
Patella Linnaeus 1758
Pauliella Round & Basson 1997
Pauliella taeniata (Grunow) Round & Basson 1997
Penilia avirostris Dana 1849
Pentapharsodinium tyrrhenicum (Balech) Montressor et al. 1993
Perca fluviatilis Linnaeus 1758
Perccottus glenii Dybowski 1877
Peridiniella Kofoid & Michener 1911
Peridiniella catenata (Levander) Balech 1977
Peringia ulvae (Pennant 1777)
Petromyzon marinus Linnaeus 1758
Phalacrocorax carbo (Linnaeus 1758)
Phalacrocorax carbo subsp. *sinensis* Staunton 1796
Phalacroma rotundatum (Claparéde & Lachmann) Kofoid & Michener 1911
Phoca groenlandica Erxleben 1777
Phoca hispida Schreber 1775
Phoca vitulina Linnaeus 1758
Phocoena phocoena (Linnaeus 1758)
Phormidium Kützing *ex* Gomont 1892
Photobacterium Beijerinck 1889
Phoxinus phoxinus (Linnaeus 1758)
Phragmites australis (Cavanilles) Trinius *ex* Steudel 1841

Phyllodistomum Braun 1899
Phyllophora crispa (Hudson) Dixon 1964
Phyllophora pseudoceranoides (Gmelin) Newroth & Taylor 1971
Physa fontinalis (Linnaeus 1758)
Pisidium Pfeiffer 1821
Planktonema lauterbornii Schmidle 1903
Platichthys flesus (Linnaeus 1758)
Pleopis Dana 1852
Pleopis polyphemoides (Leuckart 1859)
Pleurobrachia pileus (O.F. Müller 1776)
Pleuronectes platessa Linnaeus 1758
Pleurosira inusitata (Hohn & Hellerman) Desianti et al. 2015
Pleurosira laevis fo. *polymorpha* Compère 1982
Plumaria plumosa (Hudson) Kuntze 1891
Podiceps auritus (Linnaeus 1758)
Podon Liljeborg 1853
Podon intermedius Liljeborg 1853
Podon leuckartii (G.O. Sars 1862)
Polarella glacialis Montresor et al. 1999
Polyarthra Ehrenberg 1834
Polydora Bosc 1802
Polyides rotundus (Hudson) Gaillon 1828
Polysiphonia Greville 1823
Polysiphonia brodiei (Dillwyn) Sprengel 1827
Polysiphonia fibrillosa (Dillwyn) Sprengel 1827
Polysiphonia fucoides (Hudson) Greville 1824
Polytoma papillatum Pascher 1927
Pomatoschistus microps (Krøyer 1838)
Pomatoschistus minutus (Pallas 1770)
Pontogammarus robustoides (G.O. Sars 1894)
Pontoporeia Krøyer 1842
Pontoporeia femorata Krøyer 1842
Porphyra umbilicalis Kützing 1843
Portlandia arctica (Gray 1824)
Posidonia König 1805
Potamogeton crispus Linnaeus 1753
Potamogeton filiformis Persoon 1805
Potamogeton pectinatus Linnaeus 1753
Potamogeton perfoliatus Linnaeus 1753
Potamopyrgus antipodarum (Gray 1843)
Prasiola (C.A. Agardh) Meneghini 1838
Praunus flexuosus (Müller 1776)
Priapulus caudatus Lamarck 1816
Proboscia Sundström 1986
Proboscia alata (Brightwell) Sundström 1986
Prorocentrum Ehrenberg 1834
Prorocentrum cordatum (Ostenfeld) J. D. Dodge 1975
Prorocentrum micans Ehrenberg 1834
Prorocentrum minimum (Pavillard) J. Schiller 1933
Protaspis Bryant 1933
Protoperidinium Bergh 1881
Protoperidinium divergens (Ehrenberg) Balech 1974
Protoperidinium pentagonum (Gran) Balech 1974
Prymnesium parvum Carter 1937
Prymnesium polylepis (Manton & Parke) Edvardsen et al. *in* Edvardsen et al. 2011
Pseudocalanus Boeck 1872
Pseudocalanus acuspes (Giesbrecht 1881)
Pseudochattonella verruculosa (Hara & Chihara) Tanabe-Hosoi et al. 2007
Pseudofallacia tenera (Hustedt) Liu et al. in Liu et al. 2012
Pseudolithoderma rosenvingei (Wærn) Lund 1959
Pseudomonas Migula 1894 *emend.* Yang et al. 2013
Pseudonitzschia H. Peragallo 1900
Pseudosolenia calcar-avis (Schultze) Sundström 1986
Pseudostaurosira brevistriata (Grunow) Williams & Round 1988
Psychromonas Mountfort et al. 1998 *emend.* Nogi et al. 2002
Pteropus mariannus Desmarest 1822
Pterothamnion plumula (Ellis) Nägeli *in* Nägeli & Cramer 1855
Pungitius pungitius (Linnaeus 1758)
Pusa hispida (Schreber 1775)
Pygospio elegans Claparède 1863
Pylaiella littoralis (Linnaeus) Kjellman 1872
Pyramimonas Schmarda 1849
Pyramimonas gelidicola McFadden et al. 1982
Quadricilia rotundata (Skuja) Vørs 1992
Radix balthica (Linnaeus 1758)
Radix labiata (Rossmässler 1835)
Radix ovata (Draparnaud 1805)
Radix peregra (Müller 1774)
Rangia cuneata (Sowerby 1832)
Ranunculus fluitans Lamarck 1778
Ranunculus palustris Garsault 1764
Rhithropanopeus harrisii (Gould 1841)
Rhizosolenia Brightwell 1858
Rhodochorton purpureum (Lightfoot) Rosenvinge 1900
Rhodomela confervoides (Hudson) Silva 1952
Rhodotorula Harrison 1928
Rhodotorula mucilaginosa (Jörgensen) Harrison 1928
Rhoicospenia abbreviata (C.A. Agardh) Lange-Bertalot 1980
Rivularia atra Roth *ex* Bornet & Flahault 1886
Ruppia cirrhosa (Petagna) Grande 1918
Ruppia maritima Linnaeus 1753
Ruppia spiralis Dumortier 1879
Rutilus rutilus (Linnaeus 1758)
Sabatieria Rouville 1903
Sabatieria pulchra (Schneider 1906)
Saccharina latissima (Linnaeus) Lane et al. 2006
Saduria entomon (Linnaeus 1758)
Salmo salar Linnaeus 1758
Salvelinus alpinus (Linnaeus 1758)
Sander lucioperca (Linnaeus 1758)
Sargassum muticum (Yendo) Fensholt 1955
Sarscypridopsis aculeata (Costa 1847)

Savillea micropora (Norris) Leadbeater 1975
Scagelothamnion pusillum (Ruprecht) Athanasiadis 1996
Sclerotium Tode 1790
Schizonema C.A. Agardh 1824
Scoloplos armiger (Müller 1776)
Scophthalmus maximus (Linnaeus 1758)
Scophthalmus rhombus (Linnaeus 1758)
Scrippsiella Balech *ex* Loeblich 1965
Scrippsiella hangoei (Schiller) Larsen 1995
Scrobicularia plana (da Costa 1778)
Scytosiphon lomentaria (Lyngbye) Link 1833
Serratia Bizio 1823
Shewanella MacDonell & Colwell 1986
Shewanella baltica Ziemke et al. 1998
Shewanella frigidimarina Bowman et al. 1997
Skeletonema Greville 1865
Skeletonema grevillei Sarno et al. 2005
Skeletonema marinoi Sarno et al. 2005
Snowella Elenkin 1938
Solea solea (Linnaeus 1758)
Somateria mollissima (Linnaeus 1758)
Spartina anglica Hubb 1978
Spermothamnion repens (Dillwyn) Rosenvinge 1924
Sphacelaria arctica Harvey 1858
Sphaerium Scopoli 1777
Sphaerolaimus Bastian 1865
Spinachia spinachia (Linnaeus 1758)
Spirodela polyrhiza Hegelmaier 1868
Spirogyra Link 1820
Spirulina Turpin *ex* Gomont 1892
Spongomorpha aeruginosa (Linnaeus) van den Hoek 1963
Sprattus sprattus (Linnaeus 1758)
Spumella Cienkowsky 1870
Stephanodiscus neoastraea Håkansson & Hickel 1986
Stephanoeca Ellis 1930
Sterkiella histriomuscorum (Foissner et al. 1991) Foissner et al. 1991
Stictyosiphon tortilis (Gobi) Reinke 1889
Strobilidium Schewiakoff 1892
Strombidium Claparède & Lachmann 1859
Stuckenia filiformis (Persoon) Börner 1912
Stuckenia pectinata (Linnaeus) Börner 1902
Subularia aquatica Linnaeus 1753
Sulfurimonas Inagaki et al. 2003 *emend.* Takai et al. 2006 *emend.* Labrenz et al. 2013
Sulfurimonas gotlandica Labrenz et al. 2013
Surirella brebissonii Krammer & Lange-Bertalot 1987
Synchaeta Ehrenberg 1832
Synchaeta baltica Ehrenberg 1834
Synchaeta cf. *littoralis* Rousselet 1902
Synechococcus Nägeli 1849
Synechocystis Sauvageau 1892
Syngnathus typhle Linnaeus 1758
Tabularia tabulata (C.A. Agardh) Snoeijs 1992
Talitrus saltator (Montagu 1808)
Teleaulax acuta (Butcher) Hill 1991
Telonema Griessmann 1913
Temora Baird 1850
Temora longicornis (O.F. Müller 1785)
Tenellia adspersa (Nordmann 1845)
Terebellides stroemii M. Sars 1835
Thalassionema nitzschioides (Grunow) Mereschkowsky 1902
Thalassiosira Cleve 1873
Thalassiosira anguste-lineata (A. Schmidt) Fryxell & Hasle 1977
Thalassiosira baltica (Grunow) Ostenfeld 1901
Thalassiosira hyperborea (Grunow) Hasle 1989
Thalassiosira hyperborea var. *lacunosa* (Berg) Hasle 1989
Thalassiosira levanderi van Goor 1924
Thalassiosira oestrupii (Ostenfeld) Proschkina-Lavrenko *ex* Hasle 1960
Thalassiosira punctigera (Castracane) Hasle 1983
Theodoxus fluviatilis (Linnaeus 1758)
Thiobacillus Beijerinck 1904
Thiomargarita namibiensis Schulz et al. 1999
Thymallus thymallus (Linnaeus 1758)
Thysanoessa inermis (Krøyer 1846)
Tintinnidium Kent 1881
Tintinnopsis Stein 1867
Tolypella glomerata (Desvaux) Leonhardi 1863
Tolypella nidifica (O.F. Müller) Leonhardi 1864
Trachelomonas Ehrenberg *emend.* Deflandre 1926
Travisia forbesii Johnston 1840
Tryblionella scalaris (Ehrenberg) Siver & Hamilton 2005
Tubifex costatus (Claparède 1863)
Tubificoides benedii (Udekem 1855)
Ulothrix Kützing 1833
Ulothrix zonata (Weber & Mohr) Kützing 1843
Ulva Linnaeus 1753
Ulva flexuosa Wulfen 1803
Ulva intestinalis Linnaeus 1753
Ulva lactuca Linnaeus 1753
Ulva linza Linnaeus 1753
Ulvopsis grevillei (Thuret) Gayral 1964
Uria aalge (Pontoppidan 1763)
Urospora penicilliformis (Roth) Areschoug 1866
Ursus maritimus Phipps 1774
Urticina felina (Linnaeus 1761)
Valvata piscinalis (O.F. Müller 1774)
Verrucaria maura Wahlenberg 1803
Verrucomicrobium Schlesner 1988
Vertebrata lanosa (Linnaeus) Christensen 1967
Vibrio Pacini 1854
Vorticella Linnaeus 1767
Vorticella anabaena Stiller 1940

Xestoleberis aurantia (Baird 1838)
Yoldia arctica (Gray 1824)
Zannichellia palustris Linnaeus 1753
Zannichellia palustris subsp. *pedicellata* (Rosén & Wahlenberg) Hook 1892
Zoarces viviparus (Linnaeus 1758)
Zostera Linnaeus 1753
Zostera marina Linnaeus 1753
Zostera noltei Hornemann 1832

Part I

The Baltic Sea environment

1 Brackish water as an environment

Hendrik Schubert, Dirk Schories, Bernd Schneider, and Uwe Selig

Abstract

1. Water is the most abundant compound on the surface of the Earth and the chemical basis for life on Earth.
2. The strong polarity of the water molecule assigns special physical and chemical properties to water as the direct environment in which aquatic organisms live, propagate and interact.
3. The surface tension of water creates the pleuston habitat and the viscosity of water requires adaptations, but it is also utilised by organisms for their life functions.
4. Water remains liquid over a broad range of temperatures, and the density anomaly of water makes ice float, which allows life to exist below the ice even when the water surface freezes.
5. In the brackish water of the Baltic Sea the ionic composition and the marine carbonate system deviate from marine water, which requires physiological adaptations of the organisms living in the Baltic Sea.
6. Natural brackish waters are classified according to ecologically relevant salinity ranges.

Keywords
Brackish water • Density anomaly • Ionic composition • Marine carbonate system • Water properties

H. Schubert (✉)
Department of Biological Sciences, University of Rostock, Albert-Einstein-Straße 3, 18051 Rostock, Germany
e-mail: hendrik.schubert@uni-rostock.de

D. Schories
Instituto de Biologia Marina, Campus Isla Teja, Universidad Austral de Chile, Valdivia, Chile
e-mail: dirk.schories@gmx.de

B. Schneider
Leibniz Institute for Baltic Sea Research, Seestraße 15, 18119 Rostock, Germany
e-mail: bernd.schneider@io-warnemuende.de

U. Selig
Forschungszentrum Jülich, Schweriner Straße 44, 18069 Rostock, Germany
e-mail: u.selig@fz-juelich.de

P. Snoeijs-Leijonmalm et al. (eds.), *Biological Oceanography of the Baltic Sea*,
DOI 10.1007/978-94-007-0668-2_1

1.1 Life depends on water

1.1.1 The origin of life

The Greek philosopher Anaximander (~611–547 BC) started a line of reasoning about the origin of life by suggesting that life on Earth emerged "spontaneously" from elements in the universe as opposed to divine creation (Lloyd 1974). Today it is widely accepted that life originated in the ocean, and that water is the "cradle of life". In the past, a number of theories about the exact habitat in which life emerged have been launched, *e.g.* warm shallow pools or lagoons of the ancient ocean.

Nowadays the range of possible locations for the origin of life has widened to also include deep-sea hydrothermal vents (Gold 1999) and even outer space (Seckbach 2012). Because of their unique combination of physical and chemical factors, hydrothermal vents provide plausible clues for hypotheses on the origin of ancestral forms of life. In a classical experiment, Stanley L. Miller demonstrated that synthesis of amino acids, the building blocks of proteins, can under certain conditions occur in an atmosphere of molecular hydrogen (H_2), methane (CH_4), ammonia (NH_3) and water vapour (H_2O) (Miller 1953).

In the 1980s it was suggested that life started in deep-sea vent-like environments because of the abundance of CH_4 and NH_3 present in hydrothermal vent regions, a condition that was not provided by the Earth's primitive atmosphere (Wächtershäuser 1988). Other theories about the origin of life have been put forward as well, but regardless of the differences between them, it is a fact that all known forms of life depend on the availability of water in its liquid form.

1.1.2 The role of water

Living cells require water as (1) a solvent for organic and inorganic compounds, (2) a medium for the transport of compounds, including absorption of nutrients into the cell and release of waste products from the cell, (3) a reagent in metabolic processes, and (4) a support for cellular structures, especially those of proteins.

In the absence of water any metabolic activity stops. To persist during periods without water, or to withstand other severe environmental fluctuations, many smaller organisms are able to form resting stages. Some of them can, in their resting stage, lose up to 99 % of body water and remain viable even after several decades of persistence in a state called anhydrobiosis (Rebecchi et al. 2007). The ability to form resting stages decreases when the complexity of organisms increases, and the dependence on water becomes stronger. No animal larger than 5 mm is known to tolerate complete desiccation which is defined as <0.1 g H_2O per g dry weight.

In fluctuating aquatic environments, prolonged dormancy of phyto- or zooplankton resting stages in sediments is common. The most complex organisms that are able to switch between dormancy and active life as adults are probably the tardigrades, commonly known as "water bears" or "moss piglets" (Fig. 1.1). Their metabolic activity almost stops when environmental conditions become unfavourable, but is resumed when conditions become favourable again. The Baltic Sea supports only a few tardigrade species (*e.g. Batillipes mirus* and *Echiniscoides sigismundi*) which mainly live on sandy bottoms, but they can also be found associated with the blue mussel *Mytilus trossulus* or the bladderwrack *Fucus vesiculosus* (Rahm 1928).

1.2 Physical properties of water

1.2.1 The structure of the water molecule

The essential roles of water are strongly related to its physical properties. Water consists of two hydrogen atoms

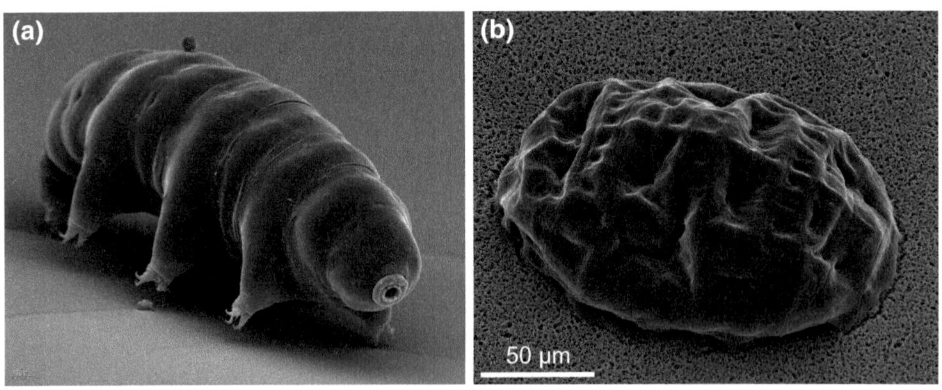

Fig. 1.1 Scanning electron microscopy images of the tardigrade *Paramacrobiotus richtersi* which is able to survive space flight conditions in dehydrated form with a survival rate of 75–95 %, similar to that in the ground control (Rebecchi et al. 2007). (**a**) A hydrated specimen. (**b**) A dehydrated specimen. Photo: © Ralph Schill

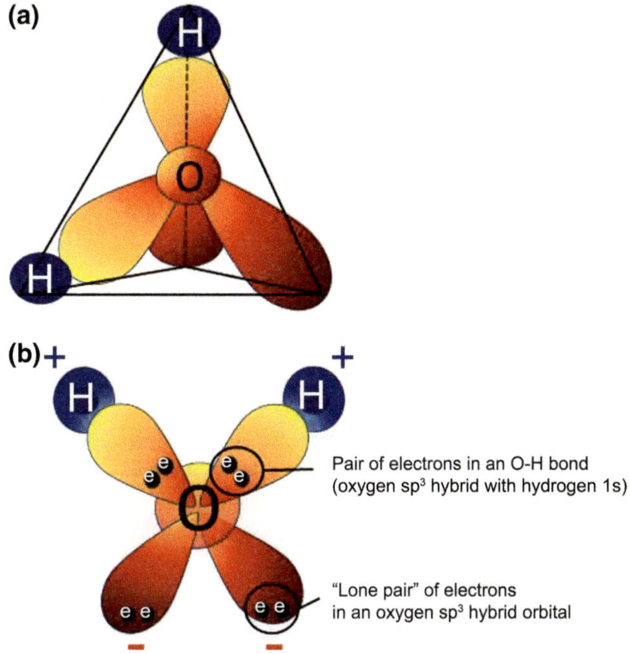

Fig. 1.2 Characteristics of the liquid water molecule. The oxygen s- and p-orbitals are hybridising to a set of 4 sp^3-hybrid orbitals giving the molecule a tetrahedral structure. Two of the hybrid orbitals bind to hydrogen, leaving two "lone pairs" of electrons in the remaining two sp^3-hybrid orbitals. (**a**) Structure of the liquid water molecule. (**b**) Charge location and dipole moment of the liquid water molecule. Figure: © Hendrik Schubert

and one oxygen atom. This simple combination results in a number of properties (Table 1.1). In particular, water has two unique features: it remains liquid over a broad temperature range (0–100 °C) and its density decreases when it freezes. The reason for this is the structure of the water molecule, or, more precisely, its dipole moment of 1.84 Debye.

In most molecules, positive and negative charges do not completely overlap, although their total charge is zero. These molecules are called "polar" because they have a permanent dipole moment. The water molecule is an example of such a polar compound. In its liquid form, the hydrogen atoms are arranged at an H-O-H angle of 104.45°, and the distance between the oxygen atom and each of the hydrogen atoms is 0.9584 Å (Fig. 1.2a).

Chemical orbital theory helps to explain this structure. The outer-shell oxygen electrons of a water molecule are located in four sp^3-hybrid orbitals (Fig. 1.2b), formed by hybridisation of the s and the three p-orbitals of oxygen. Two of the six outer (s and p) electrons will be left unpaired in their orbital because of this hybridisation. At these sites, hydrogen atoms can bind, pairing their single s-electron with the respective unpaired sp^3-electrons of oxygen. Binding to such sp^3-hybrid orbitals means that the two hydrogen atoms are not in line with the oxygen atom (Fig. 1.2).

The resulting angle between H-O-H is between 104° and 110°, depending on pressure and temperature. On the opposite corners of the tetrahedral structure, there exist "lone pairs" of electrons, *i.e.* sp^3-hybrid orbitals with two electrons, but uncovered by a hydrogen atom. This asymmetry creates a dipole moment, defined as the magnitude of charge (Q) multiplied by the distance (r) between the charges:

$$\mu = Q \cdot r \tag{1.1}$$

In this formula, Q is given in coulombs (C) and r as distance in metres (m) between the electron and proton. Instead of C·m, the unit "Debye" is used. This unit is named in honour of the 1963 Nobel Prize winner Peter Debye (1884–1966) who developed a theory for the dipole moment (Debye 1954).

1.2.2 The three states of aggregation

Water is a polar molecule; it binds to itself and to other polar substances with hydrogen-bridge bonds. The strength of these bonds depends on pressure and temperature. At surface atmospheric pressure of 1,013 mbar (1 atm), water occurs in one of three phases – ice, liquid or gas – depending on the temperature.

Below 0 °C, water has a crystalline structure (Fig. 1.3a), in which the thermal energy is so low that the dipole moment-mediated hydrogen-bridge bonds are almost as strong as chemical bonds. The H-O-H angle in ice is ~109.4°, which is wider than in the liquid and gaseous forms (104.45°). This affects the density (Fig. 1.4), so that 10 litres of water correspond to almost 11 litres of ice.

Between 0 °C and 100 °C, water is in a liquid state (Fig. 1.3b), in which the dipole moment-mediated hydrogen-bridge bonds are still effective, but are not as strong as in ice. The crystalline structure of ice is now

Table 1.1 Properties of water. Data from Schwoerbel (1977) and UNESCO (1981)

Property	Unit	Value
Melting point at 1 atm	°C	0.0
Melting energy	$kJ \cdot mol^{-1}$	6.0104
Boiling point at 1 atm	°C	100.0
Boiling energy	$kJ \cdot mol^{-1}$	40.66
Specific heat capacity at 15 °C	$J \cdot kg^{-1} \cdot K^{-1}$	4,186.0
Thermal conductivity at 25 °C	$W \cdot m^{-1} \cdot K^{-1}$	0.569
Density at 25 °C	$kg \cdot m^{-3}$	997.075
Density-maximum temperature	°C	3.98
Dynamic viscosity at 25 °C	$kg \cdot m^{-1} \cdot s^{-1}$ (Pas)	$0.89 \cdot 10^{-3}$
Kinematic viscosity at 25 °C	$m^2 \cdot s^{-1}$ (Stokes)	$0.89 \cdot 10^{-6}$
Surface tension at 25 °C	$N \cdot m^{-1}$	$71.97 \cdot 10^{-3}$

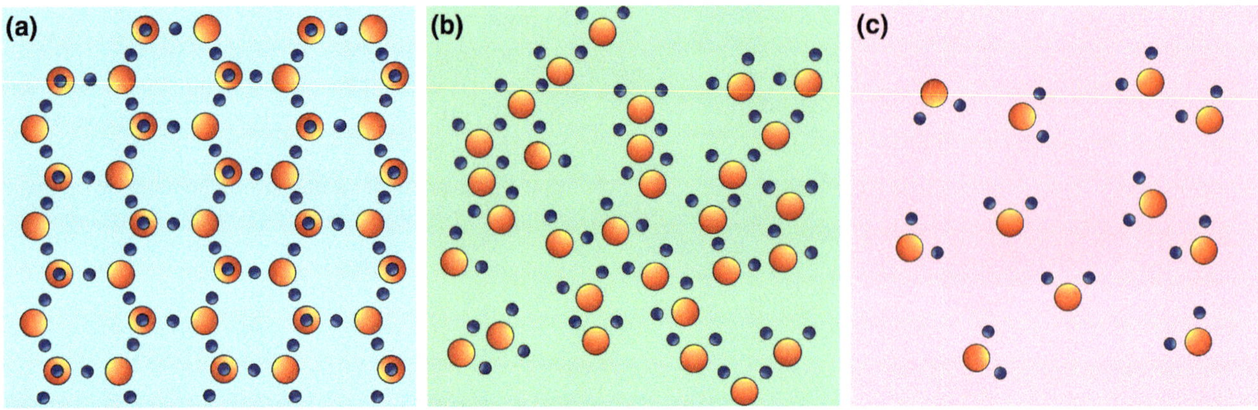

Fig. 1.3 The three different structures of water at 1,013 mbar (1 atm). (**a**) Ice has a three-dimensional tetrahedral structure with an O-H-O angle of 109.4° in which each water molecule is connected to the next upper molecule layer of the crystal (with the blue hydrogen atom drawn on top of the oxygen atom) or the next lower molecule layer of the crystal (no hydrogen atom drawn on top of the oxygen atom). (**b**) In liquid water, the molecules are in a semi-ordered state, forming clusters of different lengths with hydrogen bonds between the molecules. This structure of liquid water allows for "free molecules" between the clusters, and consequently liquid water has a higher density than ice. (**c**) Water vapour has a random molecular structure. Figure: © Hendrik Schubert

broken down into clusters, each consisting of a number of molecules still connected by electrostatic forces between molecules as well as by chemical bonds within the molecules. Between the clusters, "free" molecules occur that are not bound to any cluster. Whereas the crystalline structures are more or less fixed in orientation and distance, the free molecules are able to fit into empty spaces within and between the clusters. Therefore, the density of water increases abruptly by ~8.5 % when ice melts.

Above 100 °C, water is in a gaseous state (Fig. 1.3c) in which the molecules are fully separated from each other because the high thermal energy overcomes the attraction by the dipole moment-mediated hydrogen-bridge bonds.

1.2.3 The density anomaly of water

Usually, liquids contract upon freezing and expand upon melting, but water is an exception to this rule and ice floats on water. Over the short temperature range from 0 to 4 °C, water actually contracts and becomes denser to reach its maximum density at 4 °C (Fig. 1.4). This effect is known as the "density anomaly of water". Above 0 °C, the crystalline structure of ice becomes transformed into individual clusters, and between 0 and 4 °C the increase of thermal energy is not yet able to overcome the effect of free molecules filling the spaces between and within them. When the temperature of the water rises above 4 °C, the increase of the thermal energy dominates and pure water expands like any other matter.

The density anomaly of water prevents natural water bodies from freezing solid and allows life to exist even when the water surface freezes. Aquatic organisms can survive in deeper parts of lakes and rivers in wintertime because the water around them is above the freezing point. The density difference between water masses of different temperatures increases the further they deviate from the density maximum at 4 °C (Fig. 1.5). Stratification between layers of different water temperatures in summer is therefore far more persistent than that in colder seasons (*cf.* Sect. 2.4.3).

In saline water the situation is more complex than in freshwater because the maximum density of water depends not only on the temperature but also on the salinity. Both the freezing point and the maximum density temperature of water decrease with higher salinity, but the maximum density decreases faster than the freezing point (*cf.* Fig. 2.17b).

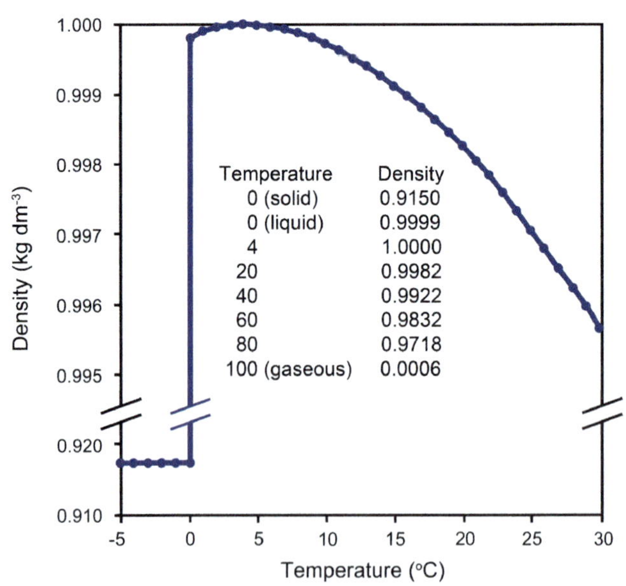

Fig. 1.4 The density anomaly of water, shown by the relationship between temperature and the density of pure water at 1,013 mbar (1 atm). Figure modified from Schwoerbel (1977)

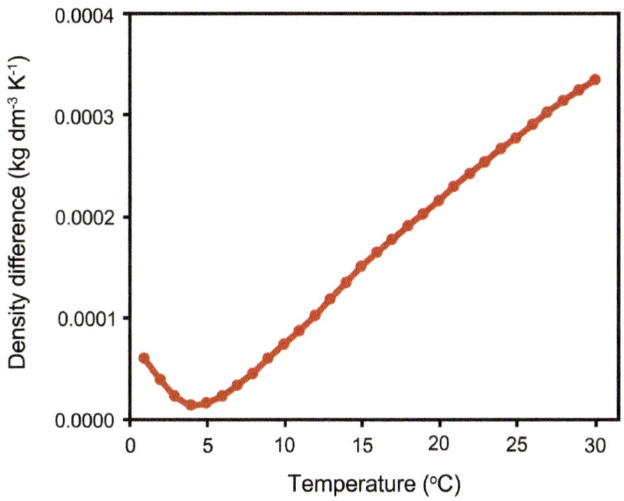

Fig. 1.5 The density difference of water per 1 K temperature decrease. With increasing temperature, the density difference between water layers with different temperatures increases. As a consequence, the stability of the thermocline increases during the summer months. Figure: © Hendrik Schubert

Above salinity 24.7, the maximum density temperature falls below the freezing point. Consequently, the density anomaly does not occur in fully marine waters, but it does in the brackish Baltic Sea where surface-water salinity is always well below 24.7 (*cf.* Sect. 2.4.2). Pressure affects the density in deep water also, but in the shallow Baltic Sea this effect is marginal.

1.2.4 Specific heat capacity

Another extraordinary property of liquid water is its high specific heat capacity, *i.e.* the capacity to absorb and store heat. Specific heat capacity is defined as the amount of energy required to change a substance's temperature by a given temperature unit per kg substance. To heat up one litre of water at 15 °C requires 4.2 $kJ{\cdot}K^{-1}$ as a result of the extensive hydrogen-bridge bonding between the water molecules (Table 1.1). Only ammonia gas requires more energy to change its temperature under the same conditions (5.2 $kJ{\cdot}K^{-1}$).

In contrast, the specific heat capacity of terrestrial materials, such as sand, stone and soil, is below 1 $kJ{\cdot}K^{-1}$, which implies that terrestrial materials absorb and release heat much faster than water. This is the reason why the temperature amplitude of terrestrial systems is much larger than that of aquatic systems at the same latitude. Large-scale monsoon phenomena or, on a smaller scale, daily changes of wind direction in coastal regions such as seaward wind in the morning and landward breeze in the afternoon, are direct consequences of the difference in specific heat capacity between land and sea. On a global scale, this difference contributes to moderating the Earth's climate by buffering diurnal as well as seasonal temperature fluctuations. The thermal energy stored by seawater influences the temperature regime of coastal regions by dampening the annual amplitude. Thus, coastal regions generally have cooler summers and milder winters than areas further inland.

Due to the high specific heat capacity of water, a considerable amount of energy must be released from a water body before freezing can begin. For the formation of sea ice, 6 $kJ{\cdot}mol^{-1}$ (=334 $kJ{\cdot}kg^{-1}$) of melting energy must be released before the water molecules can be ordered into the crystalline structure of ice (Box 1.1, Table 1.1), which can take a long time. However, in the Baltic Sea region, cold air in winter is mostly of continental or polar origin and therefore quite dry. Consequently, evaporation takes place, requiring 2,500 $kJ{\cdot}kg^{-1}$ of energy (enthalpy of vaporisation at 1 atm at 0 °C). This cools the water and explains why an ice layer may form quite fast in the Baltic Sea.

1.2.5 Surface tension

A further consequence of the dipole moment of the water molecule is surface tension which, at 25 °C, is as high as 72 $mN{\cdot}m^{-1}$ (Fig. 1.6, Table 1.1). A comparable surface tension is observed for glycerol (63 $mN{\cdot}m^{-1}$), while mercury has the highest surface tension known so far (487 $mN{\cdot}m^{-1}$).°

Polar and charged substances, which themselves are attracted by dipole molecules such as those of water are easily incorporated (dissolved) into liquid water. These substances are called "hydrophilic", while inert substances such as lipids that cannot overcome the forces between the water molecules are called "hydrophobic". Similarly, whether or not a surface is repelling water or becoming moist depends on an interplay between cohesive forces (attractive forces between molecules of the same substance), between water molecules and adhesive forces (attractive forces between molecules of the different substances), and between water and surface charges at the other substance/water interface.

A consequence of the high surface tension of water is the formation of a layer at the surface of water bodies where hydrophobic materials from the atmosphere are prevented from sinking even when their density is higher than that of water. This layer forms a specific habitat called the "pleustal", where organisms feed on other organisms and hydrophobic organic particles that accumulate at the water surface. The organisms living in this habitat, collectively called the "pleuston", are adapted to the environmental conditions in the air/water interface habitat created by

Box 1.1: The formation of sea ice

Water reaches its maximum density at 4 °C. When water freezes into ice (Box Fig. 1.1), it becomes ~9 % less dense than liquid water, and therefore ice floats on the water surface. In contrast to icebergs, which are large pieces of freshwater ice that have broken off from land ice masses, sea ice is formed from saltwater. During the freezing process, all salt is expelled from the crystalline fraction. Consequently, the solid phase of sea ice consists of freshwater, while channels in the ice are filled with brine (*cf.* Sect. 9.2.4). The formation of sea ice is a multi-step process starting with tiny discs growing into hexagonal crystals which are broken and re-agglutinated to a smooth, thin form of ice ("grease ice"). The grease ice finally grows into a thin continuous ice sheet ("nilas"). The formation of nilas is the starting point for downward ice growth, resulting in the typical plain-surface sea ice of the Arctic regions, which also dominates in the Baltic Sea.

Box Fig. 1.1 Sea ice near the German island of Hiddensee in the Arkona Sea (southern Baltic Sea). Photo: © Irmgard Blindow

surface tension. The term "neuston" refers to the microscopic component of the pleuston. Further subdivisions depend on whether the organisms live above the water surface (epipleuston, epineuston) or immediately underneath it (hypopleuston, hyponeuston).

The occurrence of pleuston depends on calm conditions, since waves disrupt the water surface tension. An example of adaptation to the marine pleustal is provided *e.g.* by the genus *Halobates* (sea skaters) belonging to the insect family Gerridae (water striders). These insects move on the water surface by using the surface tension in combination with water-repellent properties of their feet. Several *Halobates* species occur in coastal habitats of the Baltic Sea and some species even occur in the open ocean, a habitat that in general contains very few insect species (Andersen and Cheng 2004).

1.2.6 Viscosity and the Reynolds number

Viscosity comprises the "internal resistance" of a liquid against flow or deformation. It is defined as the tangential force needed to translocate one kg over one metre per second, and is expressed in Pascal seconds (1 Pa·s = 1 $kg{\cdot}m^{-1}{\cdot}s^{-1}$). The viscosity of water is relatively low and strongly temperature-dependent. At 20 °C, the viscosity is 1 mPa·s, but at 0 °C it is almost twice as much (1.8 mPa·s). The higher viscosity at 0 °C is the result of the increased length of the water molecule clusters at lower temperature.

Viscosity is an important feature for any particle that changes its position in the water medium. The resistance of the water also determines the energetic effort an organism needs to make to move in water. On the other hand, viscosity counteracts sinking of plankton organisms. To minimise the risk of sinking, plankton organisms have evolved a rich array of morphological adaptations, such as cell shape, colony formation and spines, so that they can utilise the water viscosity even better. As viscosity is temperature-dependent, the sinking risk increases with higher temperature, and therefore morphological adaptations against sinking are more important in warm than in cold waters.

When water moves over a surface, cohesive forces between the water molecules and adhesive forces between the water and surface molecules interact to form a boundary layer (the "Prandtl layer"). The adhesive forces keep the layers of water molecules in their position relative to the surface. With increasing distance from the surface, the adhesive forces lose their strength and cohesion becomes dominant. When water velocity increases, the boundary layer becomes more compressed until cohesion is

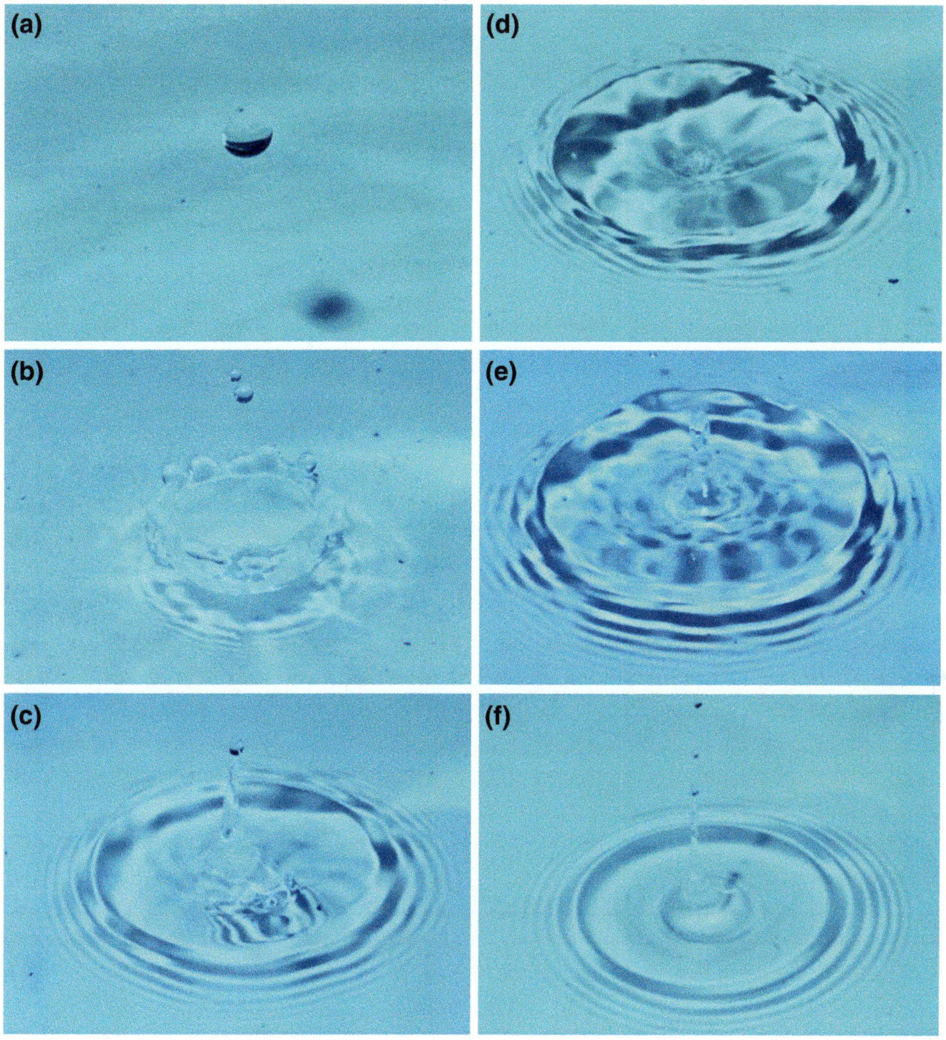

Fig. 1.6 Breaking the surface tension of water. (**a**) When a water drop falls on a motionless water surface it forms a sphere because of the water's high surface tension. (**b**) As the drop hits a larger volume of water, it deflects the water surface. **c** A moment later, the drop breaks the surface tension and creates a hole. (**d**) Water from the sides of the hole quickly refills it. (**e, f**) The refilling is so fast that the water shoots upwards. Photo: © Dirk Schories

no longer able to translocate enough energy to keep the whole system moving in parallel. This is the point at which a laminar flow becomes turbulent. A laminar flow is interrupted by circulation cells in which part of the water is moving backwards against the general direction of the current.

The velocity at which this switch from laminar to turbulent flow happens depends on the relationship between viscosity and inertial forces (drag), and is described by the Reynolds number (Re) which has no unit:

$$\mathrm{Re} = \frac{\rho VL}{\mu} = \frac{VL}{\nu} \tag{1.2}$$

In this formula V is the mean velocity of the object relative to the fluid, and ρ is the density of the fluid. For practical purposes, the viscosity, also called the dynamic viscosity (μ), is often replaced by kinematic viscosity (ν, with the unit Stokes), which is the ratio of dynamic viscosity and mass density ($kg \cdot m^{-1} \cdot s^{-1}$ / $kg \cdot m^{-3}$ = $m^2 \cdot s^{-1}$). The characteristic length (L) is not always easy to define, *e.g.* for tubes or rivers this represents the diameter rather than the length, whereas for a whale or a ship it is the length. In any case, the Re is robust, covering many orders of magnitude, and a rough estimate is sufficient to determine whether laminar or turbulent conditions prevail. The nature of the flow in the vicinity of an organism is determined by its body Re, which is strongly linked to body size (Fig. 1.7). A Re <1 typically denotes laminar conditions while Re >10,000 indicates fully turbulent conditions.

1.2.7 Living in a laminar world

Small organisms, with their low Reynolds number (Fig. 1.7), are surrounded by a highly persistent layer of

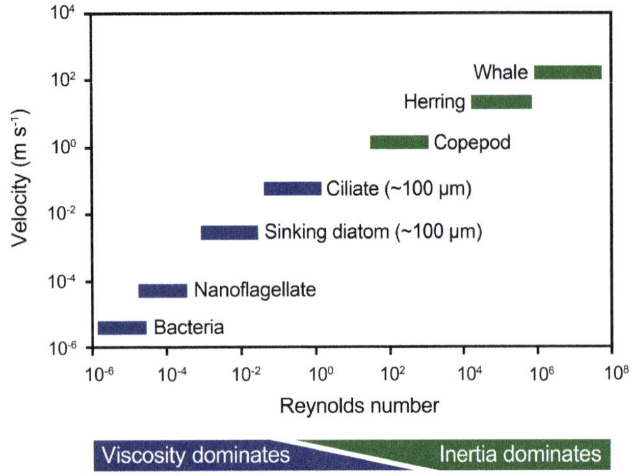

Fig. 1.7 Ranges of the Reynolds number for different groups of organisms. Viscosity dominates at low Reynolds numbers, while inertia dominates at high Reynolds numbers. Figure modified from Schwoerbel (1977)

water. Nutrient uptake by these organisms depletes the nutrient concentration in the persistent layer, resulting in the formation of a concentration gradient along which replenishment of nutrients occurs via diffusion. Diffusion is a slow process, and the rate of diffusion strongly depends on the steepness of the concentration gradient.

Increasing movement of organisms relative to the surrounding water reduces the thickness of the boundary layer and, consequently, raises the nutrient concentration. Therefore, mobility has been proposed to effectively improve nutrient availability. A couple of experiments and many model simulations have been performed to investigate to what extent movement would improve nutrient availability. In general, the results show that body size and relative speed are the main determinants because nutrient supply increases the larger the organisms are and the faster they move.

For sinking diatoms of ~10 µm in diameter, nutrient availability was increased by 10 % compared to non-sinking diatoms, but when a diatom was only ~5 µm in diameter, this positive effect was reduced to less than 1 % (Karp-Boss et al. 1996). For actively moving flagellates of ~5 µm with a speed of ~100 µm s^{-1}, nutrient availability was 30 % higher than for non-moving cells. However, the results obtained from different modelling approaches are not uniform. For example, Langlois et al. (2009) came to the conclusion that the effect of movement on reduction of the boundary layer is negligible with regard to nutrient supply.

In any case, moving in water is a complicated task for small organisms (Purcell 1977). To get an idea about "life at low Reynolds numbers", imagine that you swim in a liquid with a viscosity eight orders of magnitude higher than that of water. The viscosity of peanut butter or ketchup is only three orders of magnitude higher than that of water, so you should think of swimming in a pitch (resin) with a viscosity of $2.3{\cdot}10^8$.

Large organisms, in their turbulent world, may employ the reactive principle of pushing a volume of water behind and using inertia to move forward. A more sophisticated concept is adopted by fish that create vortices of water and use them as pillars to push or drag their body forward. In the case of small organisms in a laminar world, all these concepts are doomed to fail since for them inertia is far too small compared to viscosity (Fig. 1.7). Their type of movement is possible in a screw-like motion by means of a rotating flagellum, often supported by a second flagellum to hamper the resulting rotation momentum (Purcell 1977).

1.3 Chemical properties of water

1.3.1 Salinity

Since its strong dipole moment makes water an excellent solvent for all electrically charged and polar compounds, salts (which consist of cations and anions) are easily dissolved in water without increasing the water volume. Consequently, the density of water increases with increasing concentrations of salts.

The amount of salts present in water is collectively termed "salinity". The salinity of seawater is a result of the transport of salts from terrestrial systems to the oceans and of their accumulation and distribution in the oceans. The salts originate from erosion of the Earth's crust, which releases compounds such as sodium, magnesium, potassium and calcium. Other compounds, including chlorine, bromine, sulphur and fluorine, as well as carbon dioxide, are assumed to originate from the primordial atmosphere and by degassing in the early geological history of the Earth. The salinity of the ocean is rather stable (34–35), but locally it may be higher because of evaporation or lower due to dilution with freshwater from riverine (fluvial) runoff or precipitation.

The definition of salinity has varied over time, as have the ways to measure and express it (Box 1.2). In its very traditional sense, salinity meant grams of dissolved salts per kilogram of water or parts per thousand (‰). This was determined by weighing the solid remains of a sample after evaporation of the water. However, this technique was not fully reliable because some salts retained water in their crystalline structures and others (*e.g.* $MgCO_3$) were transformed by heat and released as gases. Moreover, organic particles, bacteria and phytoplankton organisms could not be separated from the salts by drying.

To overcome these problems, a commission of the International Council for the Exploration of the Sea (ICES) agreed in 1889 on a definition of salinity as "the total amount of solid materials in grams contained in one kilogram of seawater when all the carbonate has been converted to oxide, the bromine and iodine replaced by chlorine, and all organic matter completely oxidised" (Lewis and Perkin 1978). This measurement of salinity yielded reproducible results, but was in practice almost impossible to carry out in the field. Therefore, chlorinity, the total amount of chlorine in seawater which can be measured accurately by a simple analytical method, was determined in practice and salinity became redefined as "the mass of silver required to completely precipitate the halogens in 0.3285234 kg of a seawater sample, assuming a constant proportion of chloride to total salinity" (*cf.* Wallace 1974).

Later on, the definitions of salinity based on chemical analyses were replaced by a definition based on the electrical conductivity of seawater, which depends on salinity and temperature and is more practical to implement (Talley et al. 2011). In practice, the salinity values derived from chlorinity/salinity measurements are identical to those based on conductivity within the temperature range of −2 to 35 °C and pressures equivalent to water depths ranging from 0 to 10,000 m for a given ionic composition.

1.3.2 Ionic composition

The practical salinity scale used to measure salinity (Box 1.2) reflects a conductivity ratio and gives no information about the ionic composition in the water. However, the ionic composition does matter for *e.g.* the nutrient uptake and ionic regulation of organisms (*cf.* Sect. 7.3). In marine water, nine ions account for >99.9 % of the total salinity (Table 1.2). Nitrogen, phosphorus and iron, *i.e.* nutrients indispensable for autotrophic growth, are not major components of seawater. For example, iron, an essential component of enzymes, is readily available in soil, but not in seawater which contains only ~0.0034 ppm of iron.

While marine water has a more or less stable ionic composition, brackish water systems may deviate considerably for several reasons (Kremling 1995): (1) impacts of freshwater runoff with different chemical composition, (2) lower-salinity water has a lower total amount of ions and therefore biological processes can have a larger impact on their relative concentrations, (3) brackish waters are more often subject to stagnation, whereby compounds accumulate in deep water where ions can be reduced under anoxia, and (4) incomplete mixing of water masses of different origin.

During a five-year survey, Kremling (1972, 1995) discovered positive anomalies for the ratios of calcium and hydrogen carbonate (bicarbonate) to chlorinity, when compared to marine water, as a result of high concentrations of these components in riverine runoff to the Baltic Sea. The ratios of sodium, potassium and bromide to chlorinity in the Baltic Sea showed no significant deviations from marine conditions, while those of magnesium, borate and fluoride were slightly higher than in the ocean. The average sulphate to chlorinity ratio did not indicate an anomaly, but relatively high standard deviations suggest large local variation, which may reflect a high contribution of riverine runoff to the sulphate pool of the Baltic Sea. The formation of hydrogen sulphide in the deep basins of the Baltic Sea could theoretically cause a decrease in the ratio as most of the sulphide

Table 1.2 Summary of the ionic composition of the marine and Baltic Sea water for the major ions. Since the salinity of the Baltic Sea differs along the large-scale Baltic Sea gradient (*cf.* Figs. 2.15 and 4.2), the ion composition is given as rX, which represents the ratio of the mass fraction of the respective salt constituent to chlorinity. The calcium and bromide contents of the Baltic Sea water were calculated with the formulae Ca^{2+} = a × Cl‰ + b and Br^- = a × Cl‰ + b, respectively (Kremling 1972); n.d. = not determined. Data from Kremling (1972) and Wallace (1974)

Component	Average concentration in marine seawater (g kg^{-1})	Relative concentration in marine seawater (%)	rX for marine seawater	rX for Baltic Sea water
Chloride (Cl^-)	19.345	55.03	0.998	0.998
Sodium (Na^+)	10.752	30.59	0.556	0.555
Sulphate (SO_4^{2-})	2.701	7.68	0.140	0.141
Magnesium (Mg^{2+})	1.295	3.68	0.0669	0.0670–0.0674
Calcium (Ca^{2+})	0.416	1.18	0.0216	a: 0.0203–0.0228 b: 0.0199–0.0228
Potassium (K^+)	0.390	1.11	0.0206	0.0204–0.206
Bicarbonate (HCO_3^-)	0.145	0.41	0.0075	n.d.
Bromide (Br^-)	0.066	0.19	0.00347	a: 0.00341–0.00335 b: 0.00010–0.00009
Borate [$B(OH)_3$]	0.027	0.08	$6.7 \cdot 10^5$	$6.83 \cdot 10^5$ – $8.05 \cdot 10^5$

Box 1.2: Ways to measure and express salinity

Before 1978: chlorinity

A method for measuring salinity based on the "law of constant proportions" formulated by Marcet (1819) was generally accepted before 1978. This law postulates that regardless of how the salinity may vary from place to place, the ratios between the amounts of the major ions in the waters of the open ocean are nearly constant. Forch et al. (1902) therefore concluded that salinity can be calculated by determining the amount of only one ion, that of chloride. The amount of chloride in a sample represents its chlorinity. It is determined by the titration of seawater with $AgNO_3$, which precipitates the chlorides in the form of AgCl. The amount of chlorides is then determined after drying and weighing the precipitate, and expressed as ‰ of the original sample weight. To ensure the uniformity and accuracy of the measurements, the results have to be compared with a reference, the so-called "Copenhagen Water", an artificial seawater manufactured to serve as a world standard. To convert chlorinity to salinity (the amount of all salts), the following equation, based on the results published by Knudsen (1901), was proposed by Forch et al. (1902):

Salinity (‰) = 1.805 · chlorinity (‰) + 0.03

This equation was commonly used in oceanography until 1966, when, to take into account the ionic anomaly of the Baltic Sea water, it was redefined by a "Joint Panel on Oceanographic Tables and Standards" (Wooster et al. 1969) to

Salinity (‰) = 1.80655 · chlorinity (‰)

The practical salinity scale (PSS)

Chlorinity can only be measured in the laboratory and, *e.g.* to analyse depth profiles, requires many working hours. To obtain detailed information about the oceanic current system – a major task in oceanography – a robust and fast field method became necessary. The subsequent wide use of CTD probes, instruments that simultaneously measure conductivity (C), temperature (T) and depth (D) for salinity determination (*cf.* Box 3.1), caused calibration problems. It is possible to convert electric conductivity data to salinity, but conductivity depends on both the temperature and the salinity of the water. To overcome this problem, Lewis and Perkin (1978) proposed a new system of salinity determination called the "practical salinity scale" (PSS). This scale is based on the ratio between the conductivity of a sample and that of a standard KCl solution at 15 °C, and was adopted by the Intergovernmental Oceanographic Commission (IOC 2010) of UNESCO in 1978 (PSS-78, UNESCO 1981). Since then, this method has been recommended for use in oceanography (Fofonoff and Lewis 1979). However, this scale is based on the law of constant proportions, and it is therefore of limited value in a thermodynamic sense.

Absolute salinity

"Absolute salinity" is defined as an SI unit of concentration (g salt kg^{-1}). Spatial variations of the composition of seawater mean that absolute salinity is not simply proportional to "practical salinity" (based on conductivity). The Thermodynamic Equation of Seawater – 2010 (TEOS-10; http://www.teos-10.org) correctly expresses the thermodynamic properties of seawater as functions of absolute salinity. TEOS-10 is based on a Gibbs function formulation from which all thermodynamic properties of seawater (density, enthalpy, entropy, sound speed, etc.) can be derived in a thermodynamically consistent manner. During the 25th Assembly of the IOC in June 2009, TEOS-10 was adopted as the official description of seawater and ice properties in marine science. While absolute salinity is the salinity variable that is needed in order to calculate density and other seawater properties, the salinity that should be archived in national databases continues to be the measured salinity variable, practical salinity (PSS-78). To avoid confusion while the use of practical salinity in scientific research publications is phased out, published values of salinity should be specifically identified as being either "practical salinity" with the symbol "SP" or "absolute salinity" with the symbol "SA". In this book measurements of practical salinity are generally used.

Calibration and units

Conductometers for salinity measurements are calibrated to the so-called "normal standard seawater" (P-series) prepared from the North Atlantic water. Each standard sample is calibrated with a standard KCl solution and labelled with its conductivity ratio. In practice, these salinity measurements are identical to those derived from chlorinity/salinity measurements. Sometimes the conductometrically determined salinity is expressed in "PSS units": practical salinity units (PSU). However, as it represents a ratio of values, it has in fact no measurement unit. In most contemporary oceanographic literature (as in this book), salinity values are therefore unitless, but it is still possible to find salinity expressed as ‰ or PSU.

sulphur originates from dissolved sulphate, but no such effect has been observed.

Anomalies in the chemical composition of the Baltic Sea water may also be related to oxygen deficiency. Under anoxia, concentrations of redox-dependent elements such as iron, manganese and cobalt can increase by up to 100 times, whereas other elements (*e.g.* arsenic and tin) do not change their total concentration, but their chemical form changes (HELCOM 1990).

1.3.3 Dissolved organic matter

The concentration of dissolved organic matter (DOM) in the Baltic Sea is high. DOM is supplied to aquatic ecosystems from external (allochthonous) and internal (autochthonous) sources. Differences in the origin of DOM, in combination with dissimilar transformations during transport, may lead to significantly different behaviour in autochthonous and allochthonous DOM (Benner 2002).

Most of the low-molecular weight DOM can be metabolised by microorganisms, and its concentration is therefore usually low. In contrast, high-molecular weight DOM, particularly when containing phenolic groups, such as humic substances, may be accumulated and occur at high concentrations. Phenolic groups in humic matter absorb light in the short (blue) wavelength region and colour the water yellowish. This material is therefore termed coloured dissolved organic matter (CDOM), also known as "chromophoric DOM", "gilvin" or "yellow substance" (*cf.* Sect. 15.2.6). It originates primarily from incompletely transformed lignin supplied to the seawater from especially forested land and has been reported to affect the ionic composition of seawater (van den Berg 1995).

CDOM is quite resistant to microbial degradation and stays in the water for a long time. Following the mixing of riverine runoff with seawater, most of the initial decrease of CDOM is likely due to precipitation, *e.g.* with calcium. Coastal water bodies that receive substantial riverine runoff have high CDOM concentrations, which affect the light transmission-related properties of seawater (*e.g.* light attenuation). For example, in his global classification of water bodies based on light attenuation, Jerlov (1951) distinguished between coastal and open waters because of the difference in their short wavelength absorbance characteristics related to CDOM.

In the Baltic Sea, CDOM is strongly related to salinity, both on a large scale along the Baltic Sea salinity gradient (*cf.* Fig. 2.22) and in coastal areas with substantial CDOM-containing terrestrial runoff (Blümel et al. 2002, *cf.* Sect. 15.2.6). CDOM is important for two reasons: (1) it affects light transmission in the water column by strongly attenuating the short wavelength (blue) range of the photosynthetically active radiation (PAR) and can therefore limit the depth distribution of phytoplankton and macrophytes (Schubert et al. 2001), and (2) it is capable of forming stable complexes with transition metals such as copper, iron and nickel, and also with nutrients, *e.g.* phosphates. It is possible that most of the iron found in the Baltic Sea water is bound in CDOM complexes, for which reason the biological availability of iron may be greatly reduced (van den Berg 1995).

1.3.4 Gases in water

The establishment of an equilibrium between the partial pressures of a gas in water and in the atmosphere is described by Henry's Law, which is given as $p = k_H \cdot c$, where p is the partial pressure of the solute in the gas above the solution, c is the concentration of the solute and k_H is a constant with the dimensions of $p \cdot c^{-1}$. The value of k_H depends on the solute, the solvent and the temperature.

Once equilibrium has been established, the partial pressure of a gas in water is balanced by its partial pressure in the atmosphere above the water. This is why nitrogen (N_2 ~78 % in air) and oxygen (O_2 ~21 % in air) are the dominant gases both in the air and in the water. Carbon dioxide (CO_2 ~0.038 % in air) is a low-concentration gas in both air and water. However, processes consuming or releasing gases (*e.g.* respiration and photosynthesis), as well as temperature changes which also affect gas solubility, may shift the partial pressure of a gas in the liquid phase.

The solubility of all gases decreases with increasing temperature. When the temperature increases from 0 to 25 °C, the gases N_2, O_2 and CO_2 become ~40–50 % less soluble. For example, the solubility of O_2 at salinity 35 is reduced from 8.05 to 4.73 mL L^{-1} when the temperature increases from 0 to 25 °C (*cf.* Table 2.5). Consequently, temperature increase leads to deviations in the atmospheric equilibrium and results in gas exchange across the air/water interface. The atmosphere, a large and well-mixed system, can be considered as a constant, with gas concentrations not affected on a short time scale. Conversely, the water column is less mixed and more viscous, which restricts internal gas transport to slow diffusive processes and may lead to large fluctuations in the concentrations of dissolved gases.

1.3.5 Non-equilibria of gases

The behaviour of dissolved gases in brackish water is not basically different from that in marine water or freshwater. The gas composition in water is similar to that in the atmosphere and, on a long-term scale, an equilibrium between water and air is usually maintained irrespective of short-term (daily or seasonal) fluctuations. However, there are two special situations in which an equilibrium can hardly exist: (1) in the deeper parts of strongly stratified water bodies, and (2) in nutrient-rich water bodies where biological processes release or consume gases at a faster rate than that of gas exchange at the air/water interface. Both situations occur in the Baltic Sea.

The first situation occurs because the deep-water layers of the Baltic Sea are cut off from the water surface by salinity stratification (*cf.* Sect. 2.4.3), and gases can only be supplied to deeper water through diffusion. However, diffusion is too slow to keep gases in equilibrium with the atmosphere throughout the whole water column. As a result, the gas composition of the deep-water layer may substantially differ from that of the surface-water layer and the atmosphere. This is best shown by the presence, in occasionally very high concentrations, of hydrogen sulphide (H_2S) and methane (CH_4) in the deep water. These gases are almost absent from the atmosphere.

The partial pressure disequilibria in the Baltic Sea are caused by biological processes and temperature changes, while such disequilibria in the ocean can also result from underwater volcanism or other geological phenomena. In the Baltic Sea, H_2S and CH_4 are released into the water as a result of anaerobic bacterial decomposition of organic matter (*cf.* Sect. 3.6). Both gases are relatively energy-rich compounds, and a number of organisms can use them as a substrate and oxidise them in the presence of O_2. However, in the deep parts of the Baltic Sea, the O_2 in the water column has already been depleted and H_2S and CH_4 are released in even higher concentrations. Therefore, the gas supply from the sediment to the water column in the Baltic Sea is dominated by sulphate reduction and methanogenesis.

The second situation is that biological processes in nutrient-rich water bodies release or consume gases at a rate faster than that of gas exchange at the air/water interface. This can be illustrated by a lack of an equilibrium for O_2 and CO_2, *i.e.* two major gases involved in biological transformation cycles. Oxygen is released during photosynthesis and taken up during respiration, while the opposite happens with CO_2. Photosynthesis and respiration can be so intense that these processes cause substantial deviations from normal gas equilibrium conditions. Because of the light dependence of photosynthesis, the deviations exhibit a pronounced day/night rhythm the amplitude of which is directly related to the net primary production (biomass produced per unit of time). The diurnal O_2 variations as well as O_2 depth profiles are frequently regarded as proxies of primary productivity (the rate at which photosynthesis occurs), which is enhanced by the eutrophication of the Baltic Sea (*cf.* Sects. 8.2.9, 11.15.3, 13.3.6).

However, the influence of temperature on gas equilibria needs to be taken into consideration. The partial pressure of a gas increases with increasing temperature, resulting in an increased relative saturation, *i.e.* the ratio (expressed in %) between the actual partial pressure and the partial pressure under equilibrium conditions. At equilibrium, the relative saturation is 100 %. When temperature increases during daytime, the relative saturation of gases increases as well. When it exceeds 100 %, the gas will be released to the atmosphere. Because an identical effect is produced by photosynthesis, physical exchange processes must therefore be accounted for when analysing O_2 amplitudes to estimate primary productivity (Wesslander et al. 2010).

1.3.6 Carbon dioxide reacts with water

Unlike the major atmospheric constituents, N_2 and O_2, the minor atmospheric constituent CO_2 undergoes chemical reactions with water when dissolved in seawater (Millero 1995). When CO_2 reacts with water, initially carbonic acid is formed:

$$CO_2 + H_2O \Leftrightarrow H_2CO_3 \tag{1.3}$$

H_2CO_3 is a relatively strong acid and in a first step one hydrogen ion (proton) dissociates and is transferred to a water molecule:

$$H_2CO_3 \Leftrightarrow HCO_3^- + H^+ \tag{1.4}$$

Because the concentrations of H_2CO_3 are much smaller than those of CO_2, and because H_2CO_3 has no significant biogeochemical importance, it is not explicitly considered in the marine carbonate system. Instead the sum of CO_2 and H_2CO_3 is used:

$$[CO_2] + [H_2CO_3] = [CO_2^*] \tag{1.5}$$

Combining the reactions in Equations 1.4 and 1.5 yields the chemical equilibrium for the first dissociation step:

$$\frac{[HCO_3^-] \cdot [H^+]}{[CO_2^*]} = k_1 \tag{1.6}$$

In a second dissociation step, hydrogen carbonate ions may transfer another hydrogen ion and carbonate ions are formed:

$$HCO_3^- \Leftrightarrow CO_3^{2-} + H^+ \tag{1.7}$$

and this equilibrium condition reads:

$$\frac{[CO_3^{2-}] \cdot [H^+]}{[HCO_3^-]} = k_2 \tag{1.8}$$

The first and second dissociation constants (k_1 and k_2) are functions of temperature, salinity and pressure.

1.3.7 Carbon dioxide equilibria

Another important property of dissolved CO_2 is its equilibrium partial pressure (pCO_2). The pCO_2 is related to the concentration of CO_2^* (H_2CO_3 + CO_2) by the solubility constant (k_o), which is a function of temperature, salinity and pressure (water depth):

$$[CO_2^*] = k_0 \cdot pCO_2 \qquad (1.9)$$

If surface water pCO_2 is identical with atmospheric CO_2 partial pressure, the system sea/atmosphere is at equilibrium (Fig. 1.8a). However, in most cases this equilibrium does not exist because, in particular, biologically induced CO_2^* changes are much faster than the re-equilibration by the slow CO_2 gas exchange. As a consequence, biological consumption or production of CO_2 leaves a signal in the pCO_2 that is preserved for a considerable time.

Finally, the equilibria that control the solubility of calcium carbonate play an important role in the marine carbonate system. At equilibrium with $CaCO_3$, *e.g.* in a saturated solution, the Ca^{2+} and CO_3^{2-} ions occur at concentrations that are prescribed by the solubility product (k_{sp}), which again is a function of temperature, salinity and pressure (water depth):

$$k_{sp} = \left([Ca^{2+}] \cdot [CO_3^{2-}]\right)_{at\,equilibrium} \qquad (1.10)$$

If the actual concentration product is larger than k_{sp}, the water is oversaturated with regard to $CaCO_3$ and *vice versa*. The degree of saturation (Ω) can be expressed by:

$$\Omega = \frac{[Ca^{2+}] \cdot [CO_3^{2-}]}{k_{sp}} \qquad (1.11)$$

The formation of $CaCO_3$-containing hard shells by calcifying organisms is favoured if Ω >1. A non-biogenic precipitation of $CaCO_3$ does not occur in seawater because this is impeded by the high concentrations of magnesium. At Ω <1, the seawater is undersaturated and $CaCO_3$ will undergo dissolution which can, however, be delayed or even prevented if the $CaCO_3$-containing shell is protected by an organic coating.

1.3.8 Variables of the marine carbonate system

Four variables of the marine carbonate system, also known as the marine CO_2 system, can be determined analytically. These are the hydrogen ion concentration (expressed as pH), the equilibrium partial pressure of carbon dioxide (pCO_2), the total dissolved inorganic carbon concentration (C_T or DIC) and the total alkalinity (A_T or TA). If any two of these four parameters are known, all parameters of the carbonate system (pCO_2, $[CO_2]$, $[HCO_3^-]$, $[CO_3^{2-}]$, pH, C_T and A_T) can be calculated for a given temperature, salinity and pressure (water depth). Two components of the system, CO_2 and HCO_3^-, are of special interest because they are the carbon forms that are taken up by photosynthetic cells to be transformed to organic carbon (Falkowski and Raven 2007).

1.3.9 pH and the marine carbonate system

Among the variables included in the equilibrium conditions in Equations 1.6 and 1.8, the hydrogen ion concentration is the only one that can be determined directly. This is

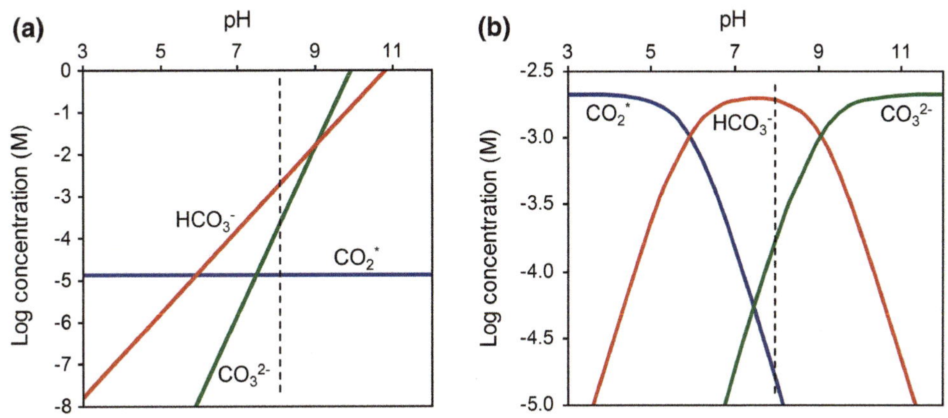

Fig. 1.8 Change occurs in the concentrations of the dissolved inorganic carbon species in the marine carbonate system as a function of pH during a hypothetical titration with an acid or base to decrease or increase the pH, respectively, under two conditions. (**a**) At instantaneous equilibration with atmospheric CO_2 (400 µatm) by gas exchange; (**b**) without gas exchange at a constant total carbon concentration of C_T = 2,100 µmol kg^{-1}. The calculations were performed for salinity 35 and temperature 20 °C. Dashed lines indicate the average ocean surface-water pH. Figure: © Hendrik Schubert

achieved by measuring the pH. Different pH scales can be defined, and in the past the NBS scale was the standard scale defined by the National Bureau of Standards (NBS).

Using this scale, buffers prepared in pure water are used for the potentiometric standardization and the pH refers to the hydrogen ion activity, which represents the thermodynamically effective concentration. In the last two decades an alternative scale, the "total hydrogen ion concentration scale" (or "total pH scale"), has become the standard scale for characterising the marine carbonate system. On the total scale the pH refers to the sum of the "free" hydrogen ions, which actually exist as protonated water molecules, and hydrogen sulphate (HSO_4^-) ions, which act as a strong acid:

$$pH = -\log([H^+] + [HSO_4^-]) \quad (1.12)$$

Buffers dissolved in artificial seawater provide the basis for standardization of the total scale. On average, pH values on the total scale are ~0.15 units lower than those on the NBS scale.

1.3.10 pCO_2 and the marine carbonate system

The partial pressure of carbon dioxide (pCO_2) is related to the concentration of CO_2^* by the solubility constant k_o in Equation 1.9. As the partial pressure is defined for the gas phase only, the pCO_2 of a seawater sample refers to the partial pressure of CO_2 in a gas phase, which is in equilibrium with that seawater sample at a given temperature and salinity. The principle of pCO_2 measurement is based on the equilibration of a small volume of carrier gas phase with a large seawater sample and subsequent determination of the CO_2 volume mixing ratio in the carrier gas with an infrared gas analyser (Körtzinger et al. 1996; Frankignoulle and Vieira Borges 2001).

1.3.11 Total carbon and the marine carbonate system

The concentration of total dissolved inorganic carbon in seawater, abbreviated as C_T or total DIC, and also referred to as "dissolved total CO_2", is defined by:

$$C_T = [CO_2^*] + [HCO_3^-] + [CO_3^{2-}] \quad (1.13)$$

The contributions of CO_2^*, HCO_3^- and CO_3^{2-} to C_T can be described as a function of pH. The relative distributions of these dissolved inorganic carbon species change if a seawater sample is titrated with an acid or a base in order to generate a lower or higher pH, respectively. The results of such a hypothetical titration are shown in Fig. 1.8a for a case when the seawater sample is in contact with the atmosphere and continuously at equilibrium with the atmospheric CO_2 during the titration.

At a pH of ~8, which is typical of ocean surface water, the HCO_3^- ion is the dominating carbon species. However, a considerable C_T fraction also exists as CO_3^{2-} ions, and these become more important with higher pH. CO_2^* stays at a constant level because there is an equilibrium with the atmospheric CO_2, while C_T increases with increasing pH. This is a consequence of the CO_2 input from the atmosphere and the shifting of the equilibria in Equations 1.4 and 1.7 towards HCO_3^- and CO_3^{2-}, respectively, with higher pH. Another hypothetical situation is that gas exchange would be inhibited during the hypothetical titration. In this case C_T would remain constant. The relative distributions of the carbon species would again show that the HCO_3^- concentration constitutes the major component around pH 8 (Fig. 1.8b). At higher pH, the CO_3^{2-} ion concentration increases at the expense of HCO_3^-, whereas CO_2^* and pCO_2 decrease and approach zero. The distributions shown in Fig. 1.8 are slightly modified by changes in temperature and salinity. However, it must be taken into account that the distributions shown in Fig. 1.8 are hypothetical cases because the pH of seawater is not changed by the addition of an acid or base, but by the consumption or production of CO_2 by photosynthesis or respiration, respectively

During photosynthesis, either CO_2^* or HCO_3^- are taken up (Falkowski and Raven 2007). The uptake of HCO_3^- must be accompanied by the uptake of H^+ or the release of OH^- in order to maintain the charge balance in an algal cell. This means that the net uptake is that of CO_2^* (H_2CO_3 or CO_2), which results in a change of C_T. The range of the pH change generated through photosynthesis and respiration is generally much smaller than that given in Fig. 1.8 and more realistic relationships between pH, pCO_2 and C_T are shown in Fig. 1.9. This figure summarises the changes in pH and pCO_2 during the consumption of ~60 µmol C_T L^{-1}, which approximates the average photosynthetic carbon uptake during the phytoplankton spring bloom in the Baltic Sea proper. During this spring bloom the pH increases by 0.35 units, whereas the pCO_2 is reduced by almost 300 µatm, which causes a considerable CO_2 undersaturation of the surface water. In shallow coastal regions with high productivity of floating algal mats, such biologically induced changes may be larger.

Different methods have been used to determine C_T. Currently, the most accurate method is based on the coulometric determination of CO_2 released from a sample by acidification with phosphoric acid (Johnson et al. 1993).

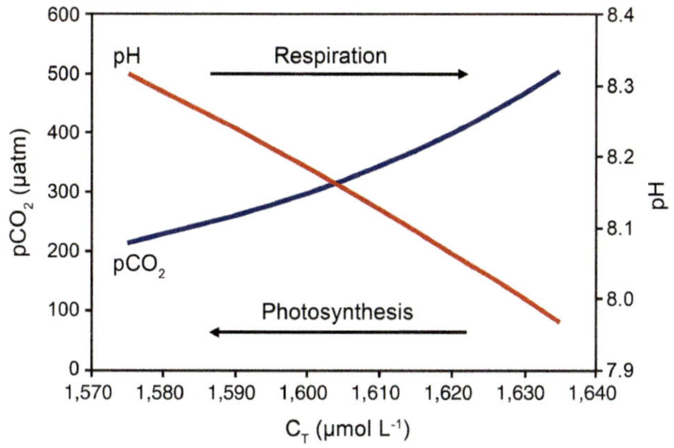

Fig. 1.9 The estimated effect of photosynthesis on the CO_2 partial pressure and the pH during a phytoplankton spring bloom in the Baltic Sea proper. The change in the total dissolved inorganic carbon concentration (C_T) corresponds to a carbon uptake of ~60 µmol C_T L^{-1}. The calculations were performed for salinity 7 and temperature 10 °C. Figure: © Bernd Schneider

1.3.12 Alkalinity and the marine carbonate system

The total alkalinity (A_T) is an important variable because it acts as a buffer for hydrogen ions derived from the uptake of CO_2 or from the input of any acidic substances. Total alkalinity is defined by the excess of proton acceptors over proton donors:

$$A_T = [HCO_3^-] + 2[CO_3^{2-}] + [B(OH)_4^-] + [OH^-] - ([H^+] + [HSO_4^-]) \quad (1.14)$$

This definition of A_T refers to oxic waters and includes the major proton acceptors. Under anoxic conditions, monohydrogen sulphide (HS^-) and ammonia (NH_3) constitute further significant contributions to A_T. Proton donors are the protonated water molecules, which are symbolised by $[H^+]$ and hydrogen sulphate ions.

For CO_2 dissolved in pure water, the alkalinity is zero because the production of HCO_3^- and CO_3^{2-} ions must be accompanied by the generation of an equivalent amount of H^+. The situation is different for seawater, which receives large amounts of A_T via river water input. The A_T in river water originates from weathering processes in the drainage area during which atmospheric CO_2 and CO_2 generated by the mineralisation of soil organic matter reacts with limestone and silicates to form HCO_3^- and CO_3^{2-}. This leads to a shift in the pH from the acidic values for CO_2 in pure water to basic values of ~8 in seawater. Another consequence is the increase in total dissolved inorganic carbon in river water and in seawater because CO_2 is absorbed during the weathering process. A potentiometric titration is used to determine A_T.

1.4 The classification of brackish waters

1.4.1 What is brackish water?

The Baltic Sea is brackish, but how is brackish water defined? Physically, this is very simple: at the cross-section of maximum density and the freezing point at salinity 24.7 (*cf.* Fig. 2.17b). However, this is not a biologically relevant limit for the distribution of species.

Some scientists (*e.g.* Redeke 1922) used the term brackish only for a mixture of seawater and freshwater. Others also incorporated hypersaline waters (salinity >35) into "brackish water", based on the occurrence of the same organisms in both hyposaline and hypersaline water bodies, but not in the sea (Bayly 1967). Again, others consider the high temporal variability of salinity (*e.g.* in tidal estuaries) as the main characteristic of brackish water.

A solution for the many controversies in trying to define brackish water was in the 20th century sought in so-called "classification systems" based on biological observations at different salinities. The most commonly used concept for brackish-water typology is a classification system based on salinity ranges (Fig. 1.10).

1.4.2 From Redeke to Välikangas

In certain areas, *e.g.* in estuaries and lagoons, salinity may fluctuate extensively and the biological communities are adapted to the salinity range rather than to the average salinity (*cf.* Box 13.3). Limits for individual salinity classes were proposed by Redeke already in 1922, based on distinct changes in biological community composition observed in Dutch estuaries (Fig. 1.10).

Some years later, Välikangas outlined a classification system based on plankton studies in the eastern part of the Baltic Sea. In his first version, Välikangas (1926) proposed salinity limits almost identical to those of Redeke (1922). For this reason, the classification of Välikangas was mainly considered as a corroboration of the general applicability of Redeke's system. However, in his second version, Välikangas (1933) used the same data as he did in 1926 to develop a system with class limits that were different from Redeke's original classification (Fig. 1.10).

1.4.3 The Venice System

In 1958, the so-called "Venice System" was internationally adopted during a special meeting of the International Association of Limnology in Venice, Italy (Anonymous

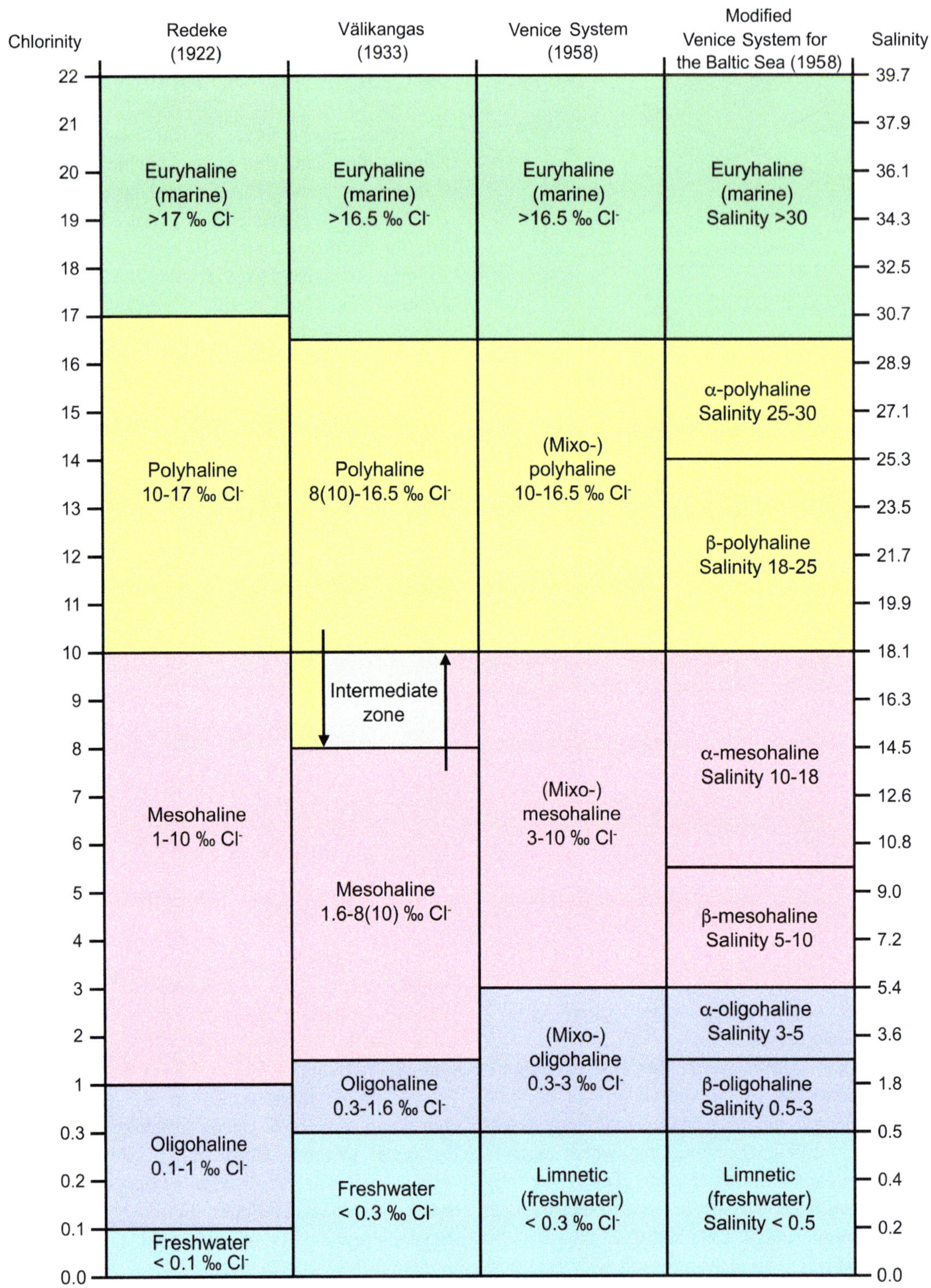

Fig. 1.10 Comparison of several classification systems for brackish water based on salinity ranges. These systems were published before 1978, so salinity is still expressed as chlorinity (*cf.* Box 1.2). For comparison with the chlorinity scale (to the left), the salinity scale is shown as well (to the right), and the limits of the modified Venice System for the Baltic Sea are given as salinity. The chlorinity and salinity scales are proportional, except for the chlorinity interval 0–1 ‰ (salinity 0–1.8), which has an enlarged scale. Figure based on data in Redeke (1922), Välikangas (1933), Anonymous (1958) and den Hartog (1964)

1958, 1959). Immediately after this meeting, the Venice System was heavily debated. Its drawbacks were summarised by den Hartog (1964), who pointed especially to the facts that (1) the Venice System lacks, in contrast to the earlier classification systems proposed by Redeke (1922) and Välikangas (1926, 1933), a biological basis, and (2) the Venice System shifted biologically meaningful limits only to achieve a "compromise" instead of being based on scientific research results.

Compared to the classification systems of Redeke (1922) and Välikangas (1933), the lower part of the mesohaline zone (salinity ~2–5) was shifted to the oligohaline zone in the Venice System (Fig. 1.10). This change was made to account for ecological conditions encountered in southern Europe, southern Africa and some other areas (Anonymous 1958, 1959). Den Hartog (1964) concluded that the changes made the system inapplicable for northern and western Europe. He proposed that a multifactorial, hierarchical system for the classification of brackish-water biota should be developed, based on statistical treatment of already existing data – a tedious task that has not yet been performed.

Despite its drawbacks, the Venice System has anyway become an instrument that provides a common ground in biological oceanography. To be biologically meaningful, the system often needs local adjustments. For example, to account for the biological observations in the Baltic Sea Area, the three brackish classes of the Venice System were divided into subclasses already in 1958 (Anonymous 1958).

Most of the Baltic Sea proper falls within the β-mesohaline subclass with salinity 5–10, whereas large parts of the Gulfs of Riga, Finland and Bothnia fall within the α-oligohaline subclass with salinity 3–5 (Fig. 1.10). The salinity interval between 5 and 8 is also defined as the "horohalinicum" (Kinne 1971) because in this interval organisms seem to experience the strongest physiological stress (*cf.* Sect. 7.3.2).

1.4.4 Classification by habitat type

In 1964, Cees den Hartog proposed a classification system with focus on habitats instead of salinity ranges, based on biological data collected worldwide (den Hartog 1964). Nine categories were distinguished in this system and the Baltic Sea belongs to Category 1:

1. Brackish seas: large water bodies with small annual salinity fluctuations
2. River mouths not subjected to lunar tides: river mouths characterised by a continuous transition between freshwater and seawater, with only seasonal salinity fluctuations
3. River mouths subjected to lunar tides (estuaries *sensu stricto*): river mouths characterised by a continuous transition between freshwater and seawater, with salinity fluctuations mainly governed by the tidal rhythm
4. Shock habitats: areas characterised by a sudden transition between freshwater and seawater, *e.g.* small streams trickling directly into the sea
5. Epilittoral (supralittoral) pools: small collections of water within the reach of spring tides and storm waves, characterised by a temporary to almost permanent isolation from the sea and highly irregular salinity fluctuations as a result of evaporation, precipitation and episodic seawater influence
6. Semi-isolated brackish waters (lagoons, ponds, étangs): water areas characterised by a very restricted contact with seawater and with seasonal salinity fluctuations
7. Relict brackish waters: water bodies characterised by a highly aberrant ionic composition, but still maintaining marine elements in their biological communities
8. The littoral border environment: the intertidal (eulittoral) and epilittoral (supralittoral) zones of the coast
9. Coastal ground water: collections of water characterised by the absence of light

Review questions

1. What is the structure of the water molecule and how are water molecules organised in the solid, liquid and gaseous state?
2. How does an ice cover form at low and high salinities?
3. What does the Reynolds number express and what are the consequences of living at low Reynolds numbers for nutrient uptake?
4. Explain the total inorganic carbon composition of seawater and its pH and salinity dependence.
5. How can brackish water bodies be classified?

Discussion questions

1. As observed from the density anomaly of water, the structure of ice is not densely packed. Do other structures of ice exist? Under what conditions can they be found?
2. Discuss consequences of the high specific heat capacity of water for seasonality in coastal regions, including specific phenomena occurring on smaller and larger scales.

3. Ionic composition, the different forms of dissolved inorganic carbon, alkalinity and biological activity affect the pH of seawater in a rather complex way. Discuss how the effects of increased atmospheric partial pressure of CO_2 on the pH of the water can be assessed in a brackish-water system.
4. If the salts in the seawater originate from terrestrial erosion, the salinity of the oceans should increase with time. Discuss processes that are involved in balancing out such an increase. Has the salinity of the oceans changed in the past?
5. Given that 21 % of the atmosphere consists of oxygen, it seems at first sight unlikely that anoxia would occur in aquatic systems. What are the conditions required for the development of anoxia?

References

Andersen NM, Cheng L (2004) The marine insect *Halobates* (Heteroptera: Gerridae): biology, adaptations, distribution and phylogeny. Oceanography and Marine Biology: An Annual Review 42:119–180

Anonymous (1958) The venice system for the classification of marine waters according to salinity. Limnology and Oceanography 3: 346–347

Anonymous (1959) Final resolution of the symposium on the classification of brackish waters. Archo Oceanography Limnology 11(Supplement):243–245

Bayly IAE (1967) The general biological classification of aquatic environments with special reference to those of Australia. In: Weatherley AH (ed) Australian inland waters and their fauna. Australian National University Press, Canberra, pp 78–104

Benner R (2002) Chemical composition and reactivity. In: Hansell D, Carlson CA (eds) Biogeochemistry of dissolved organic matter. Academic Press, London, pp 59–90

Blümel C, Domin A, Krause JC, Schubert M, Schiewer U, Schubert H (2002) Der historische Makrophytenbewuchs der inneren Gewässer der deutschen Ostseeküste. Rostocker Meeresbiologische Beiträge 9:5–111 [in German]

Debye (1954) The collected papers of Peter J. W. Debye. Interscience, New York-London 700 pp

den Hartog C (1964) Typologie des Brackwassers. Helgoländer Meereswissenschaftliche Beiträge 10:377–390 [in German]

Falkowski PG, Raven JA (2007) Aquatic photosynthesis, 2nd edn. Princeton University Press, Princeton NJ 512 pp

Fofonoff NP, Lewis EL (1979) A practical salinity scale. Journal of the Oceanographical Society of Japan 35:63–64

Forch C, Knudsen M, Sörensen SPL (1902) Berichte über die Konstantenbestimmungen zur Aufstellung der Hydrographischen Tabellen. Kongelige Danske Videnskabernes Selskabs Skrifter Naturvidenskabelige og Mathematiske Afhandlinger 7(1):1–151 [in German]

Frankignoulle M, Vieira Borges A (2001) Direct and indirect pCO_2 measurements in a wide range of pCO_2 and salinity values (the Scheldt Estuary). Aquatic Geochemistry 7:267–273

Gold T (1999) The deep hot biosphere: the myth of fossil fuels. Springer, New York, NY 243 pp

HELCOM (1990) Second periodic assessment of the state of the marine environment of the Baltic Sea Area, 1984–1988. Baltic Sea Environment Proceedings 35B:1–432

IOC SCOR IAPSO (2010) The international thermodynamic equation of seawater—2010: calculation and use of thermodynamic properties. Intergovernmental Oceanographic Commission, UNESCO Manuals and Guides 56:1–196

Jerlov NG (1951) Optical studies of ocean water. Reports of the Swedish Deep-Sea Expedition 1947–1948, 3:73-97

Johnson KM, Wills KD, Buttler DB, Johnson WK, Wong CS (1993) Coulometric total carbon dioxide analysis for marine studies: maximizing the performance of an automated gas extraction system and coulometric detector. Marine Chemistry 44:167–187

Karp-Boss L, Boss E, Jumars PA (1996) Nutrient fluxes to planktonic osmotrophs in the presence of fluid motion. Oceanography and Marine Biology: An Annual Review 34:71–107

Kinne O (1971) Salinity: animals—invertebrates. In: Kinne O (ed) Marine ecology, vol 1., Environmental factors, Part 2 John Wiley and Sons, New York NY, pp 821–996

Knudsen M (1901) Hydrographical tables. GEC Gad, Copenhagen 63 pp

Körtzinger A, Thomas H, Schneider B, Gronau N, Mintrop L, Duinker JC (1996) At sea intercomparison of two newly designed underway pCO_2 system—encouraging results. Marine Chemistry 52:133–145

Kremling K (1972) Untersuchungen über die chemische Zusammensetzung des Meerwassers aus der Ostsee III, Frühjahr 1969—Herbst 1970. Kieler Meeresforschungen 28:99–118 [in German]

Kremling K (1995) Ionenanomalien. In: Rheinheimer G (ed) Meereskunde der Ostsee, 2nd edn. Springer, Berlin, pp 88–91 [in German]

Langlois VJ, Andersen A, Bohr T, Visser AW, Kiørboe T (2009) Significance of swimming and feeding currents for nutrient uptake in osmotrophic and interception-feeding flagellates. Aquatic Microbial Ecology 54:35–44

Lewis EL, Perkin RG (1978) Salinity: its definition and calculation. Journal of Geophysical Research 83:466–478

Lloyd GER (1974) Early greek science: thales to aristotle. WW Norton and Company, New York NY 156 pp

Marcet A (1819) Über das specifische Gewicht, die Temperatur und die Salze des Meerwassers in verschiedenen Theilen des Weltmeers und in eingeschlossenen Meeren. Annals of Physics 10:113–158 [in German]

Miller SL (1953) A production of amino acids under possible primitive earth conditions. Science 117:528–529

Millero FJ (1995) Thermodynamics of the carbon dioxide system in the oceans. Geochimica et Cosmochimica Acta 59:661–677

Purcell EM (1977) Life at low Reynolds numbers. American Journal of Physics 45:3–11

Rahm G (1928) Tardigrada. In: Grimpe O (ed) Die Tierwelt der Nord- und Ostsee Leipzig, 11(1b):1–25 [in German]

Rebecchi L, Altiero T, Guidetti R (2007) Anhydrobiosis: the extreme limit of desiccation tolerance. Invertebrate Survival Journal 4:65–81

Redeke HC (1922) Zur Biologie der niederländischen Brackwassertypen. Bijdragen tot de Dierkunde 22:329–335 [in German]

Schubert H, Sagert S, Forster RM (2001) Evaluation of the different levels of variability in the underwater light field of a shallow estuary. Helgoland Marine Research 55:12–22

Schwoerbel J (1977) Einführung in die Limnologie, 3rd edn. Gustav Fischer, Jena, 191 [in German]

Seckbach J (ed) (2012) Genesis—in the beginning: precursors of life, chemical models and early biological evolution. Springer, Berlin 934 pp

Talley LD, Pickard GL, Emery WJ, Swift JH (2011) Descriptive physical oceanography: an introduction, 6th edn. Academic Press, London 560 pp

UNESCO (1981) Tenth report of the joint panel on oceanographic tables and standards. UNESCO Technical Papers in Marine Science 36:1–25

Välikangas I (1926) Planktologische Untersuchungen im Hafengebiet von Helsingfors. Acta Zoologica Fennica 1:1–298 [in German]

Välikangas I (1933) Über die Biologie der Ostsee als Brackwassergebiet. Verhandlungen der Internationalen Vereinigung für theoretische und angewandte Limnologie 6:62–112 [in German]

van den Berg CMG (1995) Evidence for organic complexation of iron in seawater. Marine Chemistry 50:139–157

Wächtershäuser G (1988) Pyrite formation, the first energy source for life: a hypothesis. Systematic and Applied Microbiology 10:207–210

Wallace WJ (1974) The development of the chlorinity/salinity concept. Elsevier, Amsterdam 227 pp

Wesslander K, Omstedt A, Schneider B (2010) Inter-annual and seasonal variations of the air-sea CO_2 balance in the southern Baltic Sea and the Kattegat. Continental Shelf Research 30:1511–1521

Wooster WS, Lee AJ, Dietrich G (1969) Redefinition of salinity. Limnology and Oceanography 14:437–438

Why is the Baltic Sea so special to live in?

2

Pauline Snoeijs-Leijonmalm and Elinor Andrén

Abstract

1. Geographical position, geological development, hydrographical features, climate and physical drivers together create the Baltic Sea environment.
2. Baltic Sea water is brackish and characterised by pronounced salinity gradients, both in horizontal and vertical directions, because of the large volume of freshwater runoff from over 100 rivers, which mixes with the saline water from the Kattegat that enters the Baltic Sea via narrow shallow straits.
3. Being a semi-enclosed continental sea with a large drainage area compared to its water volume, the Baltic Sea ecosystem is heavily impacted by the surrounding landmasses.
4. The water residence time in the Baltic Sea is long (30–40 years), and therefore discharged nutrients and toxic compounds circulate within the sea for a long time, which contributes to its vulnerability to eutrophication and chemical contamination by hazardous substances.
5. The Baltic Sea Area is geologically young and the Baltic Sea ecosystem is extremely young in an evolutionary perspective. Only few macroscopic species are fully adapted to its low-salinity environment.
6. Chief factors that affect species distributions in the Baltic Sea along local, regional and ecosystem-wide gradients are salinity, climate, ice cover, currents, permanent salinity stratification, hypoxia, and benthic substrate types (rock, sand, mud).
7. Environmental drivers vary either in time or space or both and contribute to the north-south "large-scale Baltic Sea gradient", along which many species experience physiological stress, lose the ability to reproduce sexually and reach the ecological limit of their occurrence.
8. In an ecosystem-wide perspective, the large-scale Baltic Sea gradient is the principal ecological characteristic of the Baltic Sea.

Keywords

Ecological characteristics • Environmental gradients • Geography • Geology • Human impacts • Hydrography

P. Snoeijs-Leijonmalm (✉)
Department of Ecology, Environment and Plant Sciences, Stockholm University, 10691 Stockholm, Sweden
e-mail: pauline.snoeijs-leijonmalm@su.se

E. Andrén
School of Natural Sciences, Technology and Environmental Studies, Södertörn University, 14189 Huddinge, Sweden
e-mail: elinor.andren@sh.se

P. Snoeijs-Leijonmalm et al. (eds.), *Biological Oceanography of the Baltic Sea*,
DOI 10.1007/978-94-007-0668-2_2

2.1 The Baltic Sea in perspective

2.1.1 Baltic Sea, East Sea and West Sea

The brackish Baltic Sea is an arm of the Atlantic Ocean, extending from northern Germany and Poland in the south almost to the Arctic Circle in the north (Fig. 2.1). It is a continental mediterranean (inland) sea, *i.e.* a semi-enclosed sea that has limited water exchange with the ocean, in which the water circulation is dominated by salinity and temperature differences rather than by winds. Although the salinity of the Baltic Sea is far below 30, and thus not within the marine range (*cf.* Fig. 1.10), it is classified as one of the world's 66 "large marine ecosystems" (http://www.lme.noaa.gov). The marine influence on the Baltic Sea ecosystem is large, but it is strongly influenced by freshwater as well.

The Baltic Sea separates the Scandinavian Peninsula from the rest of continental Europe and is surrounded by nine countries. The Latin name "Mare Balticum" (Baltic Sea) was already used in the 11th century and may refer to the Danish straits (known as "belts") or because the sea stretches through the land as a belt. It is also possible that "baltas" refers to the white colour of ice and snow.

In four riparian countries of the Baltic Sea, its name has a component sounding similar to "Baltic": in Latvia (Baltijas jūra), Lithuania (Baltijos jūra), Poland (Morze Bałtyckie) and Russia (Baltiyskoye Morye). In four other countries, the Baltic Sea is known as the "East Sea": in Denmark (Østersøen), Germany (Ostsee), Sweden (Östersjön), and even in Finland (Itämeri). The latter name does not reflect the sea's geographic position in relation to Finland, but is the result of the long common history of Sweden and Finland as one country (until ~200 years ago). In Estonia the name is, geographically correctly, the "West Sea" (Läänemeri).

2.1.2 The large-scale Baltic Sea gradient is unique

The Baltic Sea exhibits gradients of critical environmental drivers, created by the sea's semi-enclosed location and the strong influences of the surrounding landmasses and climate. The large-scale Baltic Sea gradient from temperate marine to subarctic limnic is unique in the world with respect to the broad ranges of the environmental drivers in combination with the large geographical size of the area.

Surface-water salinity, the strength of the halocline (a jump layer in salinity) and surface-water temperature decrease northwards in the Baltic Sea, while the influence of a winter ice cover increases. The depth of the halocline increases from the Kattegat to the Baltic Sea proper. Nutrient dynamics, superimposed by human-made nutrient emissions, generate low productivity in the north, high productivity in the south and eutrophication in the east.

The large-scale Baltic Sea gradient strongly affects the open-water and coastal systems. However, the coastal areas are also influenced by local gradients in salinity, temperature, depth of the photic zone and ice cover, as well as by

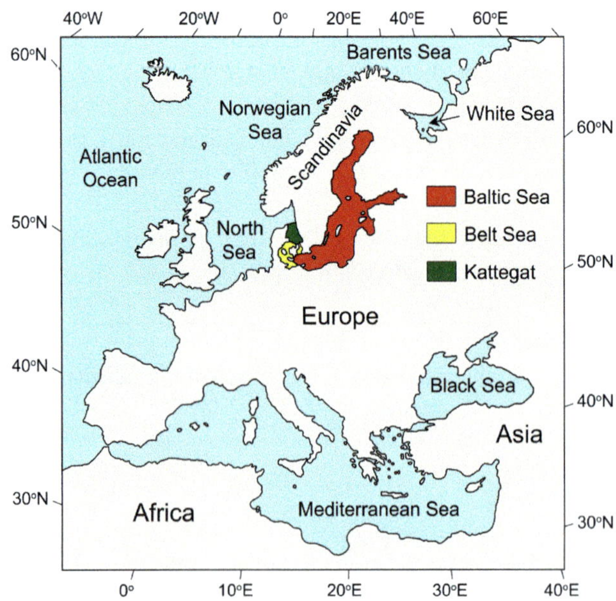

Fig. 2.1 Map of Europe, showing the semi-enclosed position of the Baltic Sea and its transition zone to the North Sea (Belt Sea and Kattegat) in northern Europe. Figure: © Pauline Snoeijs-Leijonmalm

other factors such as bottom type and water movement. Taken together, all these environmental drivers create a mosaic of habitats, supporting different types of organisms, populations and communities.

2.1.3 Not an estuary

While the Baltic Sea and estuaries on oceanic coasts share the common feature of being brackish transitional waters between freshwater and marine systems, the Baltic Sea is not an estuary (Elliott and McLusky 2002; McLusky and Elliott 2007). An estuary is a partly enclosed coastal body of water with one or a few rivers or streams flowing into it, and with a free connection to the open sea. An estuary usually involves the outflow from one river system only. The Baltic Sea receives outflows from over 200 rivers, is much larger and deeper than an average estuary and has narrow and shallow connections with the ocean. In the latter respect, the Baltic Sea would be more similar to a gigantic threshold fjord with a series of subbasins.

An estuary is subject to large daily and seasonal fluctuations in salinity due to factors such as freshwater runoff, tides and winds (*cf.* Sect. 13.1.2). The Baltic Sea, on the contrary, has a long, stable salinity gradient. Many organisms living in the Baltic Sea belong to species of estuarine origin. However, the Baltic representatives differ from their estuarine conspecifics by adaptations to Baltic Sea conditions, *e.g.* to a permanently submerged life in water with constant low salinity.

2.1.4 Comparisons with other water bodies

The Baltic Sea is one of the world's largest brackish water bodies, together with the Black Sea and the Caspian Sea. The Black Sea has a 25 % larger surface area than the Baltic Sea, while that of the Caspian Sea is about the same (Table 2.1). As they are deeper, the respective water volumes of the Black Sea and the Caspian Sea are 26 and 4 times larger than that of the Baltic Sea. However, since only ~30 % of the Black Sea is located on the continental shelf and the Caspian Sea is in fact a lake (being fully enclosed by land), it is safe to say that the Baltic Sea is the world's largest continental brackish-water sea.

The coastlines of the continents feature numerous semi-enclosed inland waters and some of them show certain similarities with the Baltic Sea (Table 2.1). Although located in a totally different climate zone and having a salinity higher than that of the ocean due to evaporation, the Red Sea and the Persian Gulf suffer, like the Baltic Sea, from large-scale hypoxia (>0 mL O_2 L^{-1} and <2 mL O_2 L^{-1}) and anoxic conditions (≤0 mL O_2 L^{-1}) because of limited water exchange with the ocean. However, the most hypoxic water body is the strongly stratified Black Sea; ~90 % of its water and ~75 % of its bottoms are anoxic because of its limited and shallow connection with the intercontinental Mediterranean Sea (Murray et al. 1989). In some areas of the Black Sea, hypoxia is also closely related to anthropogenic nutrient inputs. For example, between 1990 and 2000 the oxygen conditions on the northwestern shelf part of the Black Sea improved substantially when nutrient loads from the Danube river decreased because of the end of intensive farming through economic decline as a result of the dissolution of the Soviet Union (Rabalais et al. 2010). The tidal North Sea, the White Sea and the Hudson Bay have open connections with the ocean and in these water bodies hypoxia is local and mainly of anthropogenic origin.

The Chesapeake Bay, the Gulf of Ob and the Caspian Sea are examples of water bodies with salinity levels close to that of the Baltic Sea (Table 2.1). When organisms from such areas are introduced to the Baltic Sea, they may establish persistent populations and impact the ecosystem. As a brackish-water lake with a long evolutionary heritage, the Caspian Sea is an important source area for non-indigenous species introductions to the Baltic Sea (*cf.* Sect. 5.1).

2.1.5 Humans and the Baltic Sea

Human colonisation of the Baltic Sea Area started a few centuries after the end of the last glaciation (~15,000 years ago) and has continued without any notable interruption until today (Jöns 2011). The different developmental stages of the Baltic Sea since the ice age have provided food and means of transportation to the people living around them: palaeolithic reindeer hunters during the Baltic Ice Lake and Yoldia Sea stages, mesolithic and early neolithic hunters-gatherers and fishermen during the Ancylus Lake and Littorina Sea stages, and seamen and traders during the post-Littorina stage (Jöns 2011).

In the course of the 20th century, the Baltic Sea ecosystem degraded as a result of imprudent anthropogenic activities such as eutrophication, chemical pollution and overfishing. Some environmental conditions are improving (*e.g.* eutrophication, some forms of contamination), while new threats emerge (*e.g.* climate change, new forms of contamination). Today, the people and governments of the Baltic Sea countries are well aware of the ecosystem services provided by the Baltic Sea and the importance of wise ecosystem management for sustainable use of our common resource (*cf.* Sect. 18.5).

Table 2.1 Comparison between the physical properties of the Baltic Sea and those of some other semi-enclosed seas, large estuaries and lakes

Water body	Location centre	Climate type	Average winter ice cover (% of surface area)	Surface area (10^3 km^2)	Water volume (10^3 km^3)	Average depth (m)	Freshwater budget	Average surface-water salinity	Classification	Connection with the ocean	Lunar tides	Hypoxia
Chesapeake Bay	38°N, 76°E	Temperate	<5 %	12	0.17	6.5	+	15	Estuary	Open	Strong	In summer
Gulf of Ob	73°N, 74°E	Arctic	100 %	41	0.49	12	+	5	Estuary	Open	Strong	Local
North Sea	56°N, 03°E	Temperate	0 %	700	67	95	+	35	Semi-enclosed sea	Open	Strong	Local
Hudson Bay	60°N, 85°E	Subarctic-Arctic	100 %	1,233	123	100	+	30	Semi-enclosed sea	Limited	Strong	Local
White Sea	66°N, 37°E	Subarctic-Arctic	100 %	90	5	60	+	30	Semi-enclosed sea	Limited	Strong	Local
Mediterranean Sea	37°N, 17°E	Mediterranean	0 %	2,500	3,750	1,500	−	38	Semi-enclosed sea	Very limited	Weak	Local
Persian Gulf	27°N, 52°E	Desert	0 %	239	9	36	−	40	Semi-enclosed sea	Very Limited	Weak	Large-scale
Red Sea	22°N, 38°E	Desert	0 %	438	215	490	−	40	Semi-enclosed sea	Very limited	Nearly absent	Large-scale
Baltic Sea	60°N, 20°E	Subarctic-temperate	~50 %	369	21	57	+	5–8	Semi-enclosed sea	Very limited	Nearly absent	Large-scale
Black Sea	43°N, 35°E	Mediterranean-continental	~20 %	461	552	1,197	+	20	Semi-enclosed sea	Very limited	Absent	Extreme
Caspian Sea	43°N, 50°E	Continental	~20 %	371	78	211	0	12	Brackish lake	None	Absent	Local
Lake Superior	48°N, 88°E	Continental	~60 %	83	12	149	0	<0.1	Freshwater lake	None	Absent	Local

2.1.6 A vulnerable ecosystem

Owing to its natural geographical, hydrographical, geological and climatic features, the Baltic Sea may be considered an ecosystem with low ecological resilience, *i.e.* a system susceptible to change (*cf.* Sect. 4.8 and 17.2). For example, many organisms in the Baltic Sea live at the limit of their salinity distributions in species-poor communities with low functional diversity (*cf.* Sect. 4.5). Key species that dominate their functional group, and thus have crucial roles in the Baltic Sea ecosystem, are, in the pelagic zone, the piscivorous fish the Atlantic cod *Gadus morhua* (*cf.* Fig. 4.12) and the two planktivorous clupeids: the Atlantic herring *Clupea harengus* (*cf.* Fig. 4.12) and the European sprat *Sprattus sprattus* (*cf.* Fig. 4.12). In the benthic zone, key species are the habitat-forming macrophytes: the bladderwrack *Fucus vesiculosus* (*cf.* Fig. 4.27b) on rocky coasts and common eelgrass *Zostera marina* (*cf.* Fig. 12.13) on sandy coasts, in addition to the filter-feeding animals: the blue mussel *Mytilus trossulus* (*cf.* Fig. 4.29) on rocky coasts and the sand gaper *Mya arenaria* (*cf.* Box Fig. 5.1) on sandy coasts. Even if these species run a low risk of being lost from the entire Baltic Sea, there may be no other species capable of fulfilling their role in areas where they would become extinct. Threats to these species include *e.g.* overexploitation of the fish stocks by humans, habitat destruction in coastal areas and chemical pollution.

2.2 Geography

2.2.1 Geography and the distribution of organisms

The geographical position of the Baltic Sea creates environmental gradients that affect the distribution of species and impinge on evolutionary processes. The shallow sills at the entrance of the Baltic Sea, and between its subbasins (Figs. 2.2 and 2.3), influence species distributions by modifying environmental drivers such as salinity and oxygen supply. Insolation, sediment composition and activities on the land within the drainage area vary along the ~2,000 km long large-scale Baltic Sea gradient from the Skagerrak to the northern Bothnian Bay and eastern Gulf of Finland.

The sandy coasts that dominate in the south of the Baltic Sea host different biological communities than those found on the rocky coasts that dominate in the north. The densely populated southern drainage area, which is dominated by agricultural land-use, influences the sea biota in a different way compared to the sparsely populated northern part, which is dominated by boreal forests.

Typical of the ~8,000 km long Baltic Sea coastline are archipelagos with numerous skerries stretching out from the coastline into the open sea, as well as estuaries and lagoons which extend landward from the coastline. Estuaries, and often also lagoons and archipelagos, receive freshwater runoff from land, which creates local, and often temporally variable, gradients in *e.g.* salinity, nutrient availability, water movement and sedimentation all around the Baltic Sea coasts. These local environmental gradients also strongly influence species distributions.

2.2.2 The boundaries of the Baltic Sea

The "Baltic Sea Area" was defined by HELCOM (1993) as "the Baltic Sea proper with the Gulf of Bothnia, the Gulf of Finland and the entrance of the Baltic Sea bounded by the parallel of the Skaw in the Skagerrak at 57°44.8′ N". This is the boundary between the Kattegat and the Skagerrak, also known as the "Skagerrak-Kattegat front" (Fig. 2.2). Thus, the Baltic Sea Area includes the whole sea area that is significantly influenced by the brackish-water outflow of the Baltic Sea. The relatively deep Skagerrak (average depth 210 m) north of the boundary is part of the North Sea.

The shallow Kattegat and Belt Sea (average depths 23 m and 14 m, respectively) comprise the transition zone between the North Sea and the Baltic Sea (HELCOM 1996). The Kattegat and Belt Sea are part of the Baltic Sea Area (as defined by HELCOM), but not of the Baltic Sea *sensu stricto* (Fig. 2.1, Table 2.2). The transition zone is heavily influenced by marine water inflow from the Skagerrak as well as by brackish-water outflow from the Baltic Sea. Geographically, the whole Belt Sea, or the Belt Sea south of the Lillebælt sill and the Storebælt sill, are often considered part of the Baltic Sea. In the latter case, the southern part of the Belt Sea is referred to as the "Western Baltic Sea". In a biological sense, as already recognised by *e.g.* Remane (1934), the southern Belt Sea typically belongs to the transition zone to the North Sea because many marine organisms can still live there.

The Baltic Sea (as used in this book) is the area east of the Belt Sea, being located between the outflow of the Szczecin Lagoon at the Polish-German border (latitude 53°55′ N) and the estuary of the Torne älv at the Finnish-Swedish border (65° 48′ N) (Fig. 2.1). From west to east, it stretches from the lower tip of the Danish island of Falster (longitude 11°59′ E) to the inner Neva Bay in Russia (30°59′ E). The western boundaries of the Baltic Sea consist of the shallowest sill between Denmark and Sweden in the Öresund (the Drogden sill, 8 m water depth) and the shallowest sill between Denmark and Germany (the Darß sill, 18 m, Lemke et al. 1994) (Figs. 2.2 and 2.3). The Drogden and Darß sills are also the natural biological

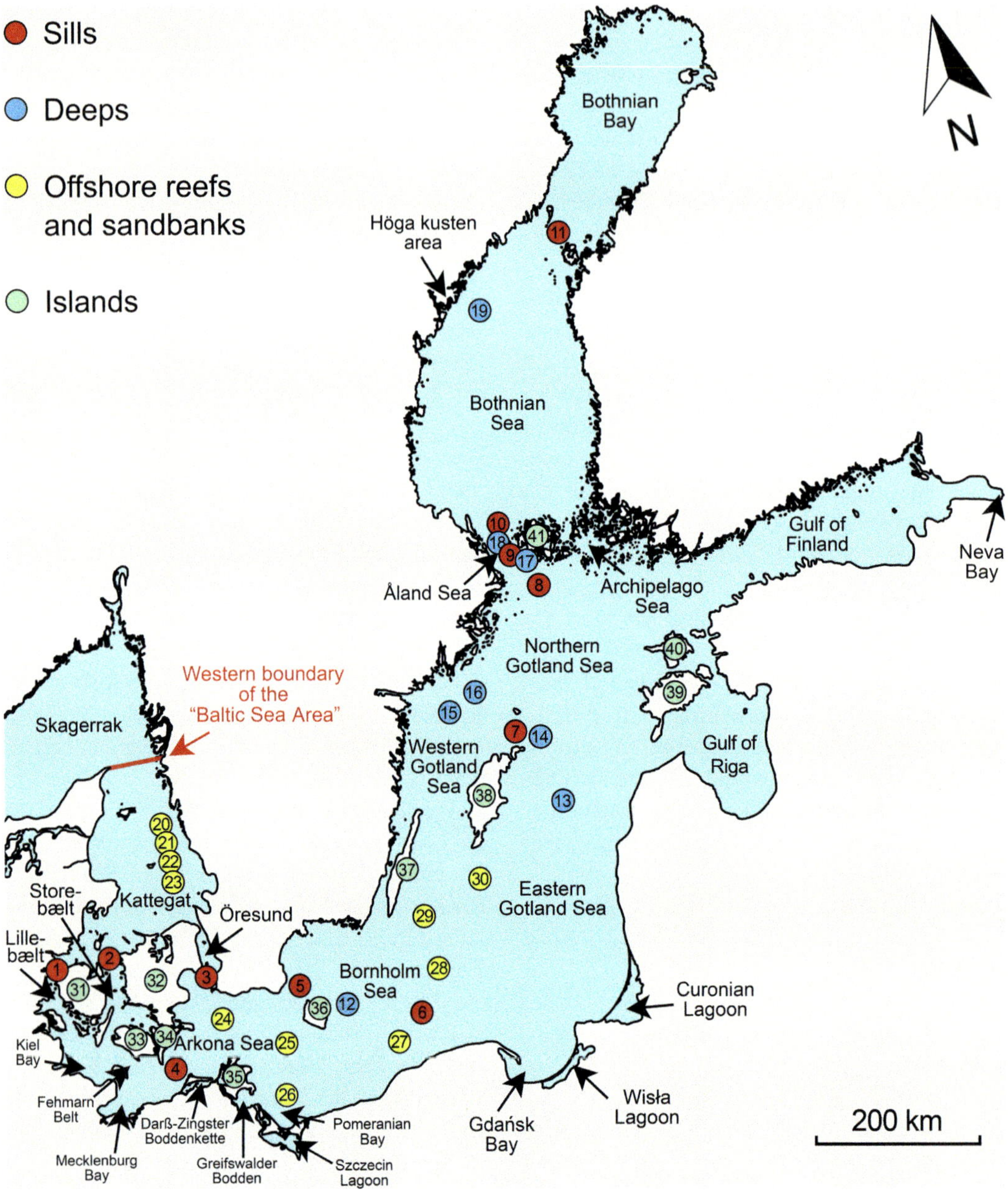

Fig. 2.2 Map of the Baltic Sea Area, showing its subregions. The Arkona Sea, Bornholm Sea, Gdańsk Bay and Gotland Sea are together called "the Baltic Sea proper". The Arkona Sea, Bornholm Sea and the southern part of the Eastern Gotland Sea are together called "the southern Baltic Sea proper". The Western Gotland Sea and the northern part of the Eastern Gotland Sea are together called "the central Baltic Sea proper". The Northern Gotland Sea is called "the northern Baltic Sea proper". The Bothnian Sea and the Bothnian Bay are together called "the Gulf of Bothnia". Straits, channels and sills are indicated in red: 1 = Lillebælt (maximum depth 81 m), 2 = Storebælt (maximum depth 60 m), 3 = Drogden sill (8 m), 4 = Darß sill (18 m), 5 = Bornholmsgattet (maximum depth 45 m), 6 = Słupsk channel (maximum depth 56 m), 7 = Fårö sill (115 m), 8 = Southern Åland sill (70 m), 9 = Middle Åland sill (70 m), 10 = Södra Kvarken sill (100 m), 11 = Norra Kvarken sill (25 m). The major deeps are indicated in blue: 12 = Bornholm deep (105 m), 13 = Gotland deep (249 m), 14 = Fårö deep (205 m), 15 = Norrköping deep (205 m), 16 = Landsort deep (459 m), 17 = Lågskär deep (220 m), 18 = Åland deep (301 m), 19 = Ulvö deep (293 m). The major offshore stone reefs and sand banks are indicated in yellow: 20 = Fladen, 21 = Lilla Middelgrund, 22 = Morups bank, 23 = Stora Middelgrund, 24 = Kriegers flak, 25 = Adlergrund, 26 = Odra bank, 27 = Słupsk bank, 28 = Södra Midsjö bank, 29 = Norra Midsjö bank, 30 = Hoburgs bank. The major islands are indicated in green: 31 = Fyn (2984 km^2), 32 = Sjælland (7031 km^2), 33 = Lolland (1243 km^2), 34 = Falster (514 km^2), 35 = Rügen (926 km^2), 36 = Bornholm (588 km^2), 37 = Öland (1347 km^2), 38 = Gotland (3184 km^2), 39 = Saaremaa (2922 km^2), 40 = Hiiumaa (1023 km^2), 41 = Åland (1552 km^2). Figure: © Pauline Snoeijs-Leijonmalm

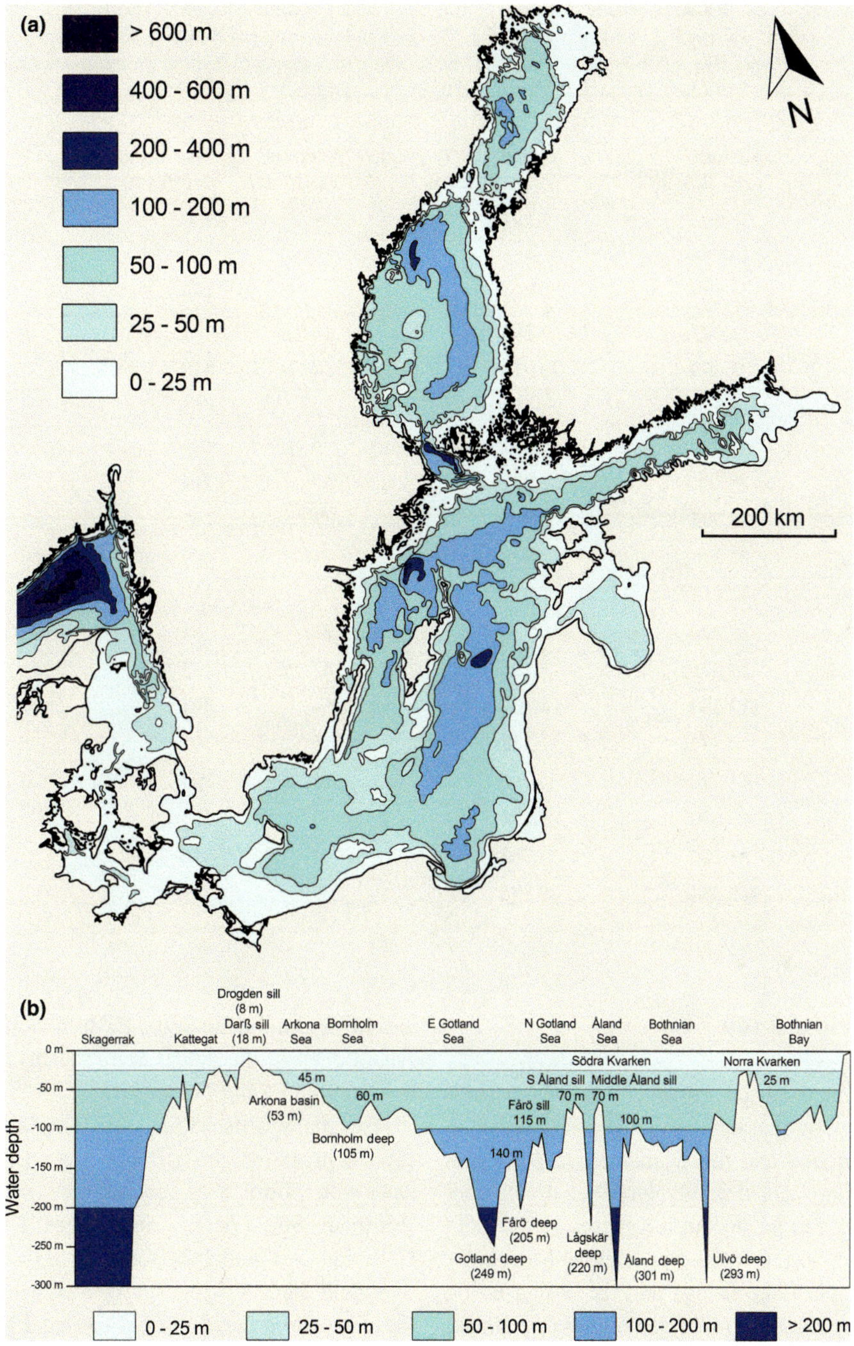

Fig. 2.3 Water depth in the Baltic Sea Area. (**a**) Geographical map showing water depth. (**b**) Schematic cross-section from the Skagerrak to the Bothnian Bay via the Eastern Gotland Sea, showing the maximum depths of the major deeps and the minimum depths of the major sills (*cf.* Fig. 2.2). Figure (**a**) modified from Bernes (2005), (**b**) modified from Sjöberg (1992)

boundaries, *e.g.* for species distributions, between the the Belt Sea (with strong marine influence) and the low-salinity brackish water of the Baltic Sea (*cf.* Sect. 4.2.2).

Altogether, the Baltic Sea covers a water area of ~369,000 km^2 with an average volume of ~21,000 km^3 (Table 2.2). The largest part of the Baltic Sea, containing 64 % of its total water volume, is called the "Baltic Sea proper" (Fig. 2.2), which is sometimes also referred to as the "Baltic Proper". The Gulfs of Riga, Finland and Bothnia are connected to the northern and northeastern Baltic Sea proper.

Table 2.2 Dimensions of the Baltic Sea Area (Baltic Sea and the transition zone) and the Skagerrak (North Sea). *The Belt Sea includes the Danish straits (Storebælt 7,765 km^2, Lillebælt 2,942 km^2, Öresund 2,278 km^2) and the southern Belt Sea, also known as the "Western Baltic Sea" (Kiel Bay 3,382 km^2 and Mecklenburg Bay 4,636 km^2). **The Baltic Sea *sensu stricto* does not include the transition zone. Surface area data from HELCOM (2012), water depth data from Leppäranta and Myrberg (2009), and surface-water salinity data (shown as long-term averages 1970–2008) from Andersen et al. (2015)

Subregion	Surface area (km^2)	Water volume (km^3)	Average depth (m)	Maximum depth (m)	Surface-water salinity
Skagerrak (North Sea)	33,400	7,281	218	700	34–35
Transition zone	**43,105**	**801**	**19**	**109**	**9.6–30.2**
Kattegat	22,102	508	23	130	12.2–30.2
Belt Sea*	21,003	293	14	81	9.6–22.9
Baltic Sea proper	**208,253**	**13,313**	**64**	**459**	**5.0–11.3**
Arkona Sea	16,502	380	23	53	7.6–11.3
Bornholm Sea	41,970	1,931	46	105	4.3–8.1
Gdańsk Bay	5,806	331	57	114	
Eastern Gotland Sea	74,985	5,774	77	249	
Western Gotland Sea	27,876	1,979	71	459	5.0–7.5
Northern Gotland Sea	41,114	2,919	71	150	
Gulf of Riga	**18,796**	**432**	**23**	**51**	**4.1–6.2**
Gulf of Finland	**29,901**	**1,106**	**37**	**123**	**1.2–5.6**
Gulf of Bothnia	**112,384**	**6,106**	**54**	**301**	**1.8–6.6**
Åland Sea	4,433	332	75	301	
Archipelago Sea	11,077	210	19	104	3.8–6.6
Bothnian Sea	63,650	4,201	66	293	
Bothnian Bay	33,224	1,362	41	146	1.8–3.9
Baltic Sea**	**369,334**	**20,958**	**57**	**459**	**1.8–11.3**

2.2.3 A semi-enclosed sea

The Baltic Sea is located on the Eurasian continental shelf and is almost completely enclosed by landmasses made up by the European mainland and the Scandinavian peninsula (Fig. 2.1). It is artificially linked to the North Sea by the Kiel Canal (via the Belt Sea), to the northern Kattegat by the Göta Canal through Sweden, to the White Sea (Beloye Morye) by the White Sea Canal via Lake Onega and the Volga-Baltic Waterway, and even to the Ponto-Caspian region (the Black, Azov and Caspian Seas) via a ramified network of inland waterways including human-made canals. Water exchange through these waterways is negligible, but the building of the canals has provided possibilities for non-indigenous species to reach the Baltic Sea (*cf.* Sect. 5.3.2).

2.2.4 A shallow sea

The average depth of the Baltic Sea is only ~57 m (Fig. 2.3, Table 2.2). The deepest place is the Landsort deep with a recorded depth of 459 m (Figs. 2.2 and 2.3). Other deep sites are the Gotland deep in the Eastern Gotland Sea (249 m), the Åland deep in the Åland Sea (301 m) and the Ulvö deep in the Bothnian Sea (293 m).

The major sills within the Baltic Sea are the Słupsk sill (56 m) in the southern Baltic Sea proper and those associated with Södra Kvarken between the Åland Sea and the Bothnian Sea (70 m) and Norra Kvarken between the Bothnian Sea and the Bothnian Bay (25 m). There is no threshold between the Baltic Sea proper and the Gulf of Finland, while the Gulf of Riga is enclosed by large islands and shallow waters. The Gulf of Riga (average depth 23 m) is shallower than the Gulf of Finland (37 m), which is shallower than the Bothnian Sea (66 m) and the Bothnian Bay (41 m).

The largest islands in the Baltic Sea are, from north to south: Åland (Finland), Hiiumaa and Saaremaa (Estonia), Gotland and Öland (Sweden), Bornholm, Lolland, Falster, Fyn and Sjælland (Denmark) and Rügen (Germany).

In the Baltic Sea and the shallow Kattegat, there are also a number of offshore stone reefs and sand banks. These shallow areas, with water depths of 5–20 m and surrounded by deeper waters, are not directly affected by terrestrial runoff, and if undisturbed by trawling they host a pristine macroalgal vegetation with associated fauna. The offshore stone reefs and sand banks are important foraging areas for waterbirds and act as refuges for endangered coastal species. The largest banks inside the Baltic Sea are Kriegers flak, Adlergrund, Odra bank, Słupsk bank, Södra Midsjö bank, Norra Midsjö bank and Hoburgs bank (Fig. 2.2). In the Kattegat, the largest banks are Fladen, Lilla Middelgrund and Stora Middelgrund.

2.2.5 A large drainage area

The drainage area (syn. drainage basin or catchment area) of the Baltic Sea is located between latitudes 49 °N and 69 °N and between longitudes 8 °E and 37 °E (Fig. 2.4). Its climate varies from mild in the south to subarctic in the north and from oceanic in the west to continental in the east. The size of the Baltic Sea drainage area is 1.7 million km^2, which is more than four times larger than the water surface area. Compared with most other seas, the Baltic Sea water surface area is large in relation to its volume. Water renewal time (turnover time) in the Baltic Sea is slow, around 30 years in the southern part and 40 years in the northern part, because of the shallow sills in the southwest. Altogether, this means that the Baltic Sea is heavily influenced by runoff from the surrounding landmasses and by anthropogenic activities on land.

The Baltic Sea is strongly affected by activities in the nine countries that are directly bordering it: Sweden, Finland, Russia, Estonia, Latvia, Lithuania, Poland, Germany and Denmark (Fig. 2.5a). In six of these countries, most of the land surfaces belong to the drainage area of the Baltic Sea, but those of Denmark, Germany and the Russia only partially. The nine Baltic Sea countries, together with the European Union, co-operate in monitoring the environmental status of their common sea and jointly devise measures for its adequate management. The international governmental body coordinating these activities is the Baltic Marine Environment Protection Commission (Helsinki Commission, HELCOM, *cf.* Sect. 17.8.4). Five more countries, Norway, Belarus, Ukraine, Slovakia and the Czech Republic, are partially located within the drainage area, and runoff from these countries affects the Baltic Sea after passing through one of the riparian countries.

2.2.6 A large human population in the drainage area

Altogether, ~85 million people live in the Baltic drainage area, almost 18 % of them within 10 km of the coast (HELCOM 2010a). The population is unevenly distributed, with roughly 10 % in the northern part, 15 % in the eastern part and 75 % in the southern part (Fig. 2.5b). The large human population around the Baltic Sea is served by some of the busiest shipping routes in the world. Around 2,000 large vessels are normally at sea at any one time, including cargo ships (>50 % of the large vessels in 2009), oil tankers (~18 %) and passenger ferries (~11 %) (HELCOM 2010b).

According to the Oslo/Paris Convention for the Protection of the Marine Environment of the North-East Atlantic (OSPAR 1998, *cf.* Box 14.1), the eutrophication of marine waters refers to the enrichment of water by nutrients, which causes an accelerated growth of algae and plants that produce an undesirable disturbance in the balance of organisms present in the water and to the quality of the water concerned. Hence, eutrophication includes the undesirable effects resulting from anthropogenic nutrient inputs. In the southern part of the drainage area, ~65 % of the land area is used for agriculture (Fig. 2.5c), which is the major cause of eutrophication of the Baltic Sea. In the northern part, agriculture uses less than 5 % of the land area and over 80 % is covered by boreal forest.

The southern part of the drainage area of the Baltic Sea also contains the largest number of "hotspots" (HELCOM 2010a), *i.e.* sites known for large discharges of hazardous substances (toxic, persistent and bioaccumulating compounds), which cause chemical pollution of the Baltic Sea (Fig. 2.5d). Even though a large number of these hotspots have disappeared during the last decades, the Baltic Sea water still contains high amounts of contaminants, one of the main reasons being the long water renewal time.

2.2.7 Patterns in sedimentation

The bedrock in the Baltic basin is superimposed by gradually younger sediments (Fig. 2.6, Table 2.3). On top of the bedrock, tills (unsorted sediments deposited during the ice ages) fill up its troughs and other depressions. Glaciofluvial deposits of sand, gravel and stones (eskers and river deltas) are occasionally found as remnants of glaciofluvial systems. On top of the tills and coarser glaciofluvial deposits, glacial clays drape the bottom topography, hence glacial and post-glacial deposits can be well over 100 m thick. The youngest sediment consists of post-glacial mud, a fine-grained, organic-rich material with a high capacity to bind various types of chemical pollutants. The mean annual sediment accumulation rate in the open Baltic Sea is ~0.5–2 mm, but there is considerable local variation.

The seabed can be subdivided into two sedimentation types: non-depositional, where erosion or transportation of sediment occurs, and depositional, where sediments accumulate. In-between the depositional and non-depositional bottoms there are areas where sediments are

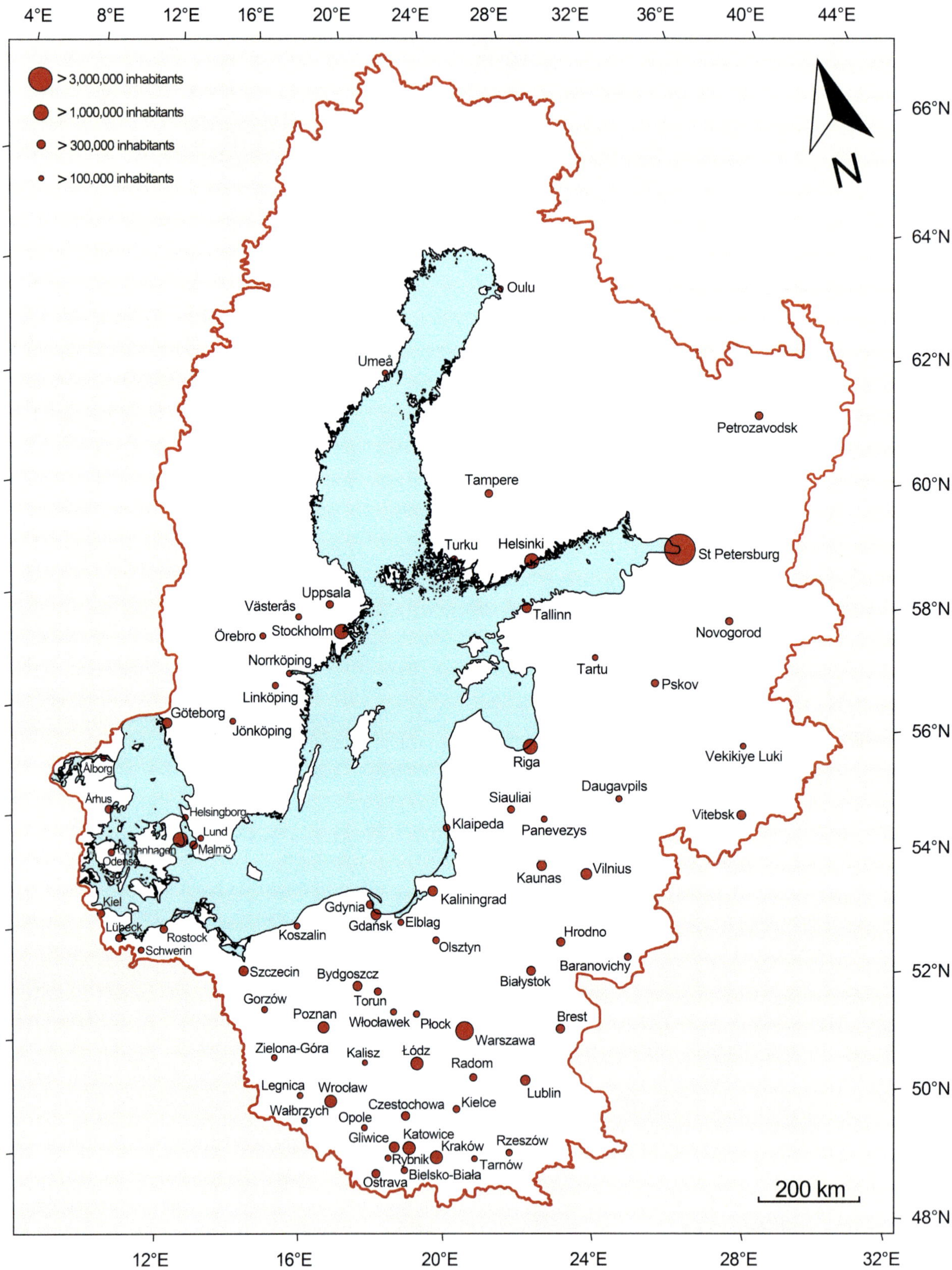

Fig. 2.4 Map of the 1.7 million km^2 drainage area of the Baltic Sea Area (indicated by the red line), showing the positions of the major cities (>100,000 inhabitants). Drainage area outline according to HELCOM (2010a). Figure: © Pauline Snoeijs-Leijonmalm

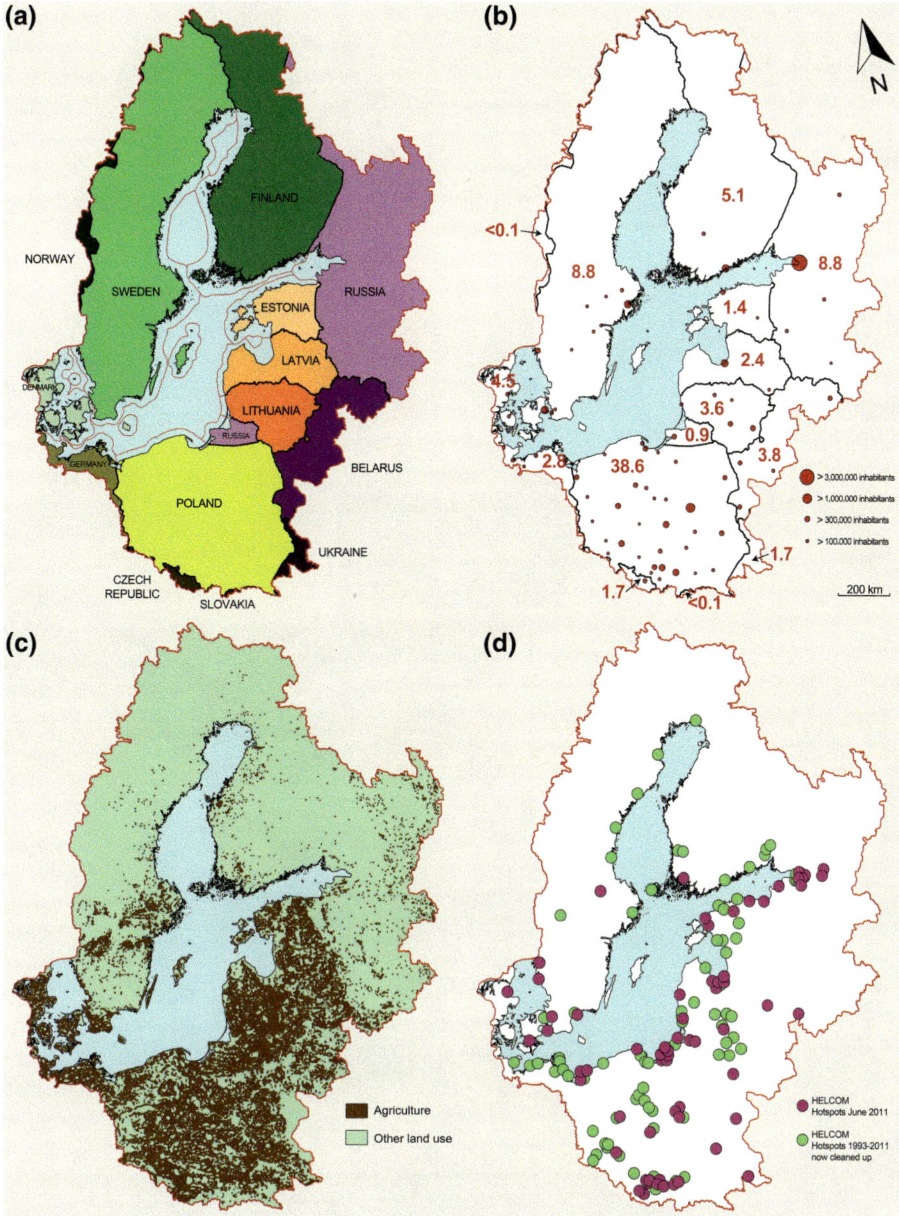

Fig. 2.5 Maps of the drainage area of the Baltic Sea Area, showing (**a**) The drainage areas of the nine countries with coastlines on the Baltic Sea with their territorial waters (thin red lines) and the five additional countries with runoff to the Baltic Sea. (**b**) The size of the population living in each of the 14 countries within the drainage area in millions people, and the locations of the cities with more than 100,000 inhabitants (for city names, see Fig. 2.4). (**c**) Agricultural land use. (**d**) HELCOM hotspots in June 2011 and previous HELCOM hotspots that were cleaned up by June 2011. Figure (**a**) © Pauline Snoeijs-Leijonmalm, (**b**) based on population data in Hannerz and Destouni (2006), (**c**) modified from Bernes (2005), (**d**) modified from HELCOM (2010a), updated with data from the HELCOM web site (http://www.helcom.fi) until June 2011

more dynamic, occasionally accumulating, but they can be resuspended and transported easily. Resuspension of sediments occurs naturally as a result of wave action, currents and mass-movements affecting steep slopes, and as a result of land uplift or land subduction. However, it is also induced by anthropogenic activities such as trawling and dredging.

The boundaries between deposition and non-deposition on the seabed are not constant, especially in the northern part of the Baltic Sea where on-going land uplift shifts the baseline of erosion and new bottom areas emerge above the wave base. About half the Baltic Sea bottoms are of the non-depositional type. Erosion results in bottoms consisting of exposed bedrock, hard coarse deposits such as gravel and

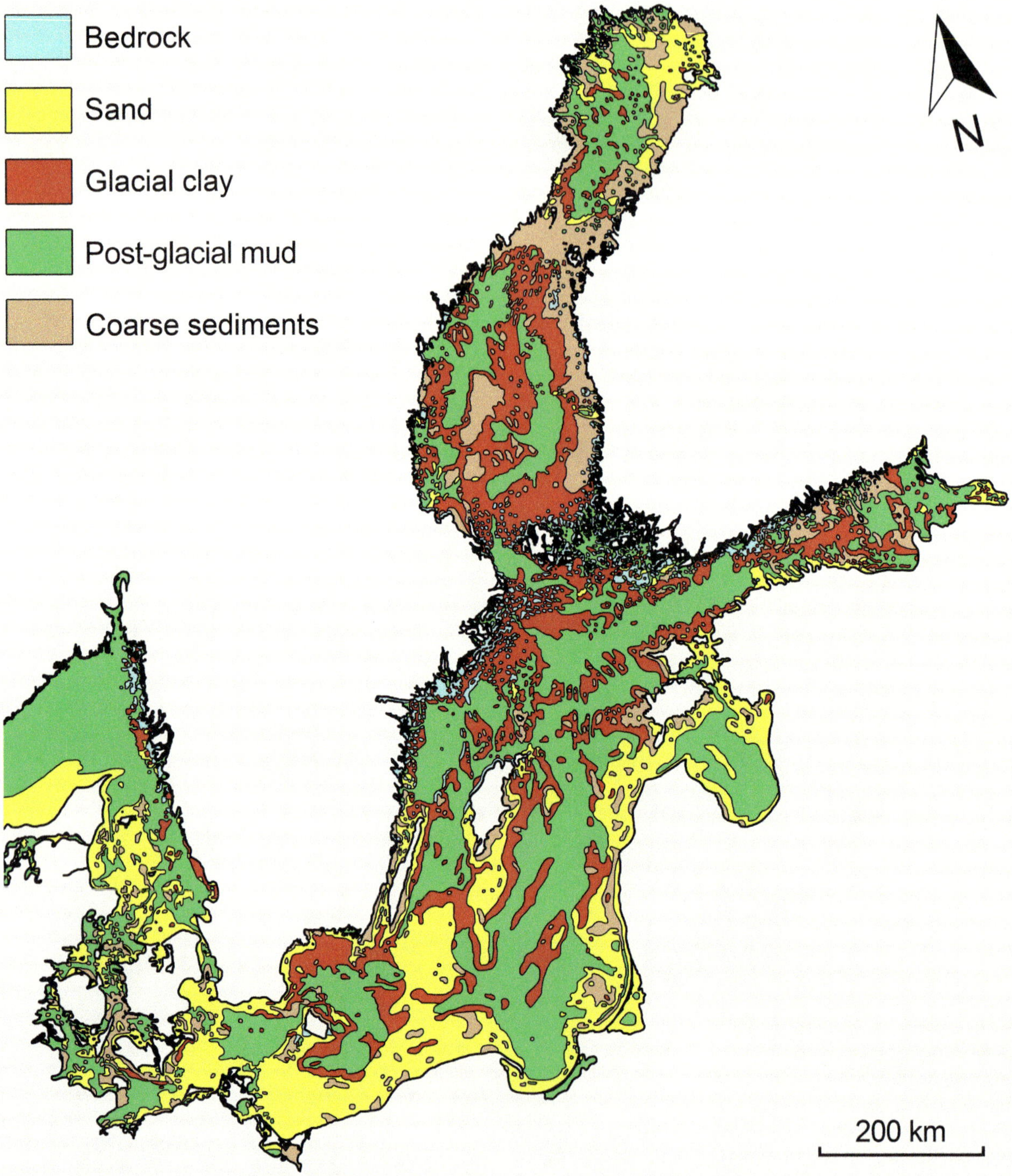

Fig. 2.6 Map of the Baltic Sea Area, showing its bottom sediment composition. Figure modified from Al-Hamdani and Reker (2007)

stones or resistant glacial clays. This implies that sea bottom does not necessarily consist of recent sediment but can have deposits with an origin of thousands of years.

Sediment transport in the Baltic Sea takes place in both vertical and lateral directions and roughly follows the water depth contour, resulting in an accumulation of sediment beneath the halocline (compare Figs. 2.3a and 2.6). Post-glacial mud accumulates in the deeper areas, which function as the so-called "depocentres", but can also accumulate in sheltered coastal embayments. Glacial clay is

Table 2.3 Simplified grain size table after Wentworth (1922). For sand and silt each scale value is half the size of the scale value above. The logarithmic phi scale (Krumbein 1934) is a modification that allows grain size data to be expressed in units of equal value for the purpose of statistical analyses and graphical plotting.

Substrate type	Wentworth size class	Lower grain size limit (mm)	Phi (Φ)
Gravel	Boulder	256	−8.0
	Cobble	64	−6.0
	Pebble	4	−2.0
	Granule	2.00	−1.0
Sand	Very coarse sand	1.00	0.0
	Coarse sand	0.50 (1/2)	1.0
	Medium sand	0.25 (1/4)	2.0
	Fine sand	0.125 (1/8)	3.0
	Very fine sand	0.0625 (1/16)	4.0
Silt	Coarse silt	0.031 (1/32)	5.0
	Medium silt	0.0156 (1/64)	6.0
	Fine silt	0.0078 (1/128)	7.0
	Very fine silt	0.0039 (1/256)	8.0
Mud	Clay	0.00006	14.0

found at the fringes of these deeper areas where erosion or transportation of sediments prevail. Sand and coarse sediments occur in shallow areas, mainly along the coasts. Sand dominates in the Kattegat, the southern and eastern Baltic Sea proper and the northeastern corner of the Bothnian Bay (Fig. 2.7). Coarse sediment is a generic name for gravel of different size (Table 2.3), which was deposited by ice sheets as till or is of glaciofluvial origin. These coarse deposits dominate along the rocky coasts of Sweden and Finland.

2.2.8 Rocky shores and sandy beaches

The Baltic Sea coasts consist of many different coast types formed along varying time scales and by various factors, such as tectonic movements, glacial erosion, glacial deposition and on-going wind- and coastal processes. According to a simplified system, the entire coast of the Baltic Sea Area can be divided into five coast types (Fig. 2.8). While the southern and southeastern Baltic Sea coast is dominated by open low sandy beaches or lagoon and bodden coasts (Fig. 2.7), the northwestern part is dominated by rocky archipelago coasts (Fig. 2.9). Fjord and klint coasts are rare.

The grain size of rocks and sediments is an important environmental driver, *e.g.* it determines the type of macrophyte vegetation and its accompanying fauna (*cf.* Sect. 11.1). The following grain size limits are commonly used: clay/mud has a grain size <3.9 µm, silt 3.9–63 µm, sand 63–2,000 µm, gravel >2 mm. Gravel is subdivided into different sizes of stones: granules 2–4 mm, pebbles 0.4–6.4 cm, cobbles 6.4–25.6 cm and boulders >25.6 cm (Table 2.3).

2.2.9 Fjords and fjärds

High fjord-like coasts are in the Baltic Sea Area found only in one small area in the western Bothnian Sea (the Höga Kusten area). The fjord-like inlets here mainly consist of water-filled valleys in exposed bedrock with no or minor coverage of glacial deposits. This is different from the "true" fjords of the type found in Norway, Greenland, Alaska and Chile, which are long, narrow inlets with steep walls formed by glacial erosion. Fjords typically have deep basins in the inner part due to glacial erosion, and since the glacier deposited gravel and sand where it met the sea, a threshold partly separates the fjord from the sea. There are no fjords of this type in the Baltic Sea Area; the closest one is the ~25 km long and 1–3 km wide Gullmarsfjorden in Bohuslän on the Swedish Skagerrak coast, with a threshold of ~40 m and a maximum depth of ~120 m in the inner part. The estuary of the river Ångermanälven in the Höga Kusten area is a fjord-like estuary. With its length of ~35 km, width of 1–3 km, threshold of ~20 m and maximum depth of ~100 m, it has a structure similar to a true fjord, but with less steep walls.

However, along the rocky archipelago coasts of Sweden and Finland there are many "fjärds", which are drowned shallow glacial valleys that form large inlets of the Baltic Sea. For example, Bråviken on the Swedish coast of the Baltic Sea proper near Norrköping is over 50 km long. Fjärds are broader and shallower than fjords and lack steep walls. Eroded local materials were deposited into the fjärds, and after the last glaciation they were filled with seawater during the eustatic sea level rise. Contrary to fjords, fjärds may contain mud flats,

Fig. 2.7 Beaches of the southern and eastern Baltic Sea. (**a**, **b**) Views from the air. (**c**) Sandy beach with dispersed boulders. (**d**) Stony beach below a chalk cliff. (**e**) Sandy beach with empty shells and collected driftwood. (**f**) Construction for protection against beach erosion. Photo: (**a**, **b**, **d–f**) © Hendrik Schubert, (**c**) © Hans Kautsky

salt marshes and flood plains. The Swedish word "fjärd" is also used for *e.g.* broad open water areas between archipelago islands away from the mainland because the word simply means "fare way" (like the Norwegian word "fjord").

A third type of inlets of glacial origin is found in the Belt Sea area, and to complicate the matter, they are also called "fjords" in Danish and "Förden" in German. However, the glacial mechanics were different from those

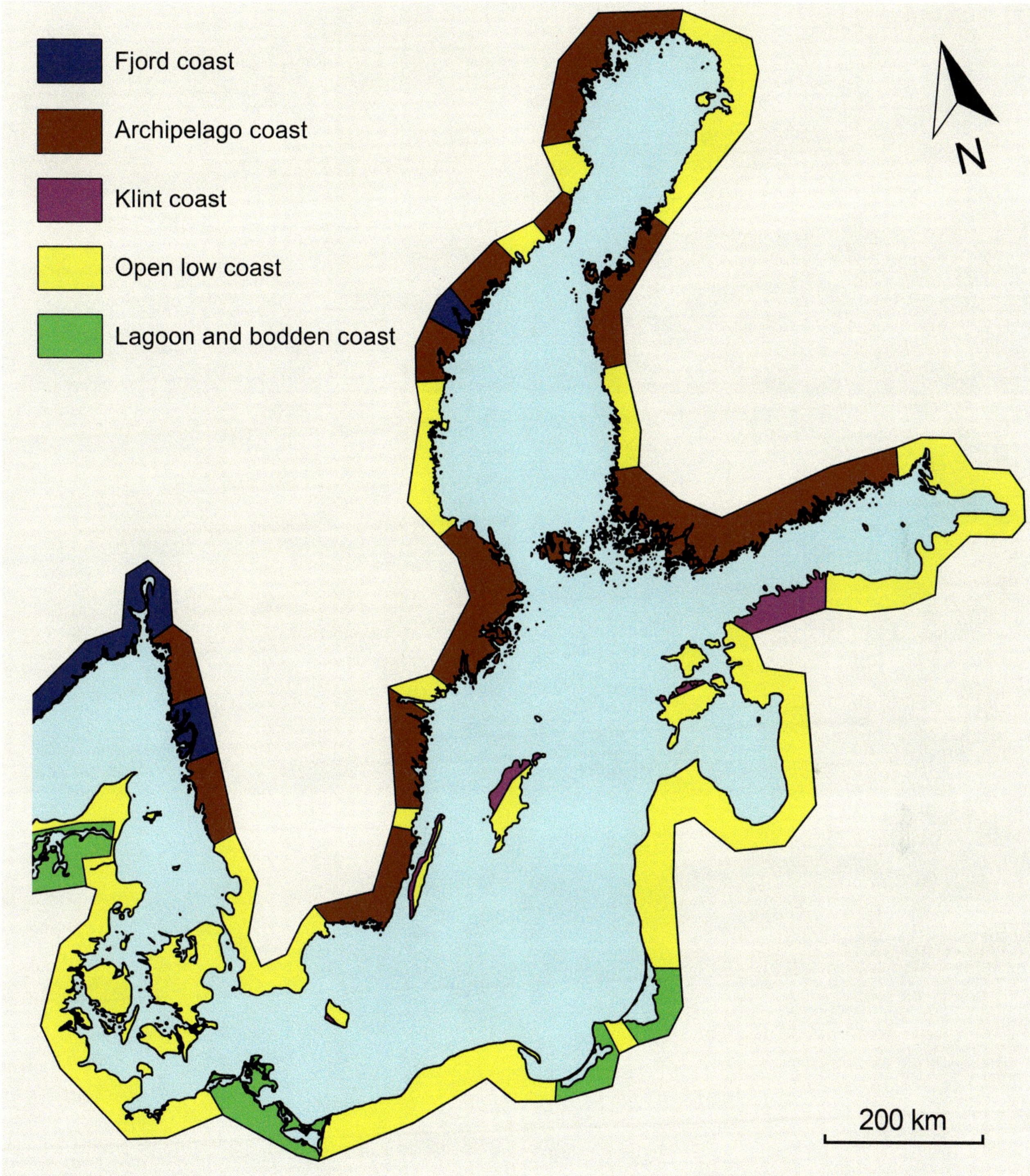

Fig. 2.8 Map of the Baltic Sea Area, showing the large-scale pattern of where different coast types dominate. Note that this only refers to the coastline (not inland geology) and that when zooming in to certain areas there is often a mosaic of different coast types. Figure based on coastal data in Boedeker and Knapp (1995), Lampe (1995) and Soesoo and Miidel (2007). Figure: © Pauline Snoeijs-Leijonmalm

of Norwegian fjords and Swedish and Finnish fjärds. In the Belt Sea area, the movement of the Weichselian ice sheet's edge carved out depressions in the land, which were later filled with seawater during the eustatic sea level rise.

2.2.10 Archipelago coasts and klint coasts

Archipelago coasts, rocky coasts with numerous islands, skerries (small islands) and fjärds prevail in the northwestern part of the Baltic Sea Area and the eastern Kattegat

Fig. 2.9 Rocky coasts and archipelagos of the Baltic Sea. (**a**) Bornholm, southern Baltic Sea proper. (**b**) Ekenäs archipelago, Gulf of Finland. (**c**, **d**) Kråkelund, Baltic Sea proper. (**e**) Inner Stockholm archipelago, northern Baltic Sea proper. (**f**) Sheltered bay at Oskarshamn, Baltic Sea proper. Photo: © Pauline Snoeijs-Leijonmalm

(Figs. 2.8 and 2.9). The largest archipelago areas are the Stockholm archipelago (>30,000 islands and skerries) and the Archipelago Sea (>50,000 islands and skerries) between Åland and the Finnish mainland. Archipelago coasts can be formed by different processes, *e.g.* large-scale tectonic movements in the bedrock and glacial erosion creating fissure valley landscapes where the fissures are water-filled, creating abundant islands.

Larger areas with klint coasts, with up to 50 m high cliffs of stratified sedimentary limestone bedrock, occur in the southern Gulf of Finland, on the Estonian coast and on the islands of Gotland and Bornholm (Fig. 2.8). Klint coasts are created when waves undercut a cliff, which results in rock-fall, and the talus masses (accumulations of broken rock debris) are subsequently washed away by waves. The klint on the coasts of Öland and northern Estonia belongs to the "Baltic klint ridge", while the klint on the coasts of Gotland and Saaremaa belongs to the "Silurian Klint" ridge. These ridges continue on the seafloor across the Baltic Sea proper.

2.2.11 Low open coasts and bodden coasts

Low and open coasts, consisting mainly of sand, prevail in the southeastern Baltic Sea (Figs. 2.7 and 2.8). The shore profile is usually low because the continent is old and mature compared to the Scandinavian mountain range, but chalk cliffs occur at some places (Fig. 2.7d). The sand, derived from the weathering of bedrock, was carried to the Baltic basin by fluvial transport from the continent for several hundred thousand years, with only relatively short breaks of glacial coverage. Sandy beaches are easily eroded, by both waves and winds, which creates coasts featuring sand dunes and spits. However, in the southern Baltic Sea there is also a considerable redeposition of sediments as a result of a transgressive water level.

Areas with low coasts interrupted by shallow coastal lagoons (boddens) and with bottoms covered by muddy sediments, often forming chains, occur in the southern Baltic Sea, *e.g.* the Darß-Zingster Boddenkette in northern Germany. Boddens have only narrow connections with the sea as they are semi-enclosed by peninsulas and are strongly affected by freshwater runoff. Most of the peninsulas were originally formed as till ridges at the ice sheet margins, and have later been modified by wind erosion and coastal processes in areas subjected to land subsidence. Some of the peninsulas are pure sand deposits.

2.3 Hydrography

2.3.1 Hydrography and the distribution of organisms

Species distributions in the Baltic Sea Area strongly depend on water circulation. The Baltic Sea basically lacks an intertidal zone because there are no regular daily water level fluctuations between high and low water level. However, irregular water level changes make the upper littoral of the Baltic Sea a highly dynamic zone with periods of total desiccation. This has forced the perennial Atlantic intertidal species that can penetrate into the brackish Baltic Sea to adapt to a permanently submerged life. Freshwater transports eroded materials and nutrients from the drainage area to the sea. This fertilises the sea and enhances the growth of primary producers.

The weak water exchange with the ocean promotes oxygen deficiency in the deeper areas of the Baltic Sea. This generates oxygen stress for organisms and affects animal life as well as the functional diversity of bacteria. The net water budget and long water residence time in the Baltic Sea create a nearly stable salinity gradient, along which species are distributed according to their salinity optima and tolerances.

For many species in the Baltic Sea, a major prerequisite for life involves large inflows of saline water from the Kattegat (Box 2.1). When an inflow occurs, salinity stress for marine species living in the Baltic Sea decreases and oxygen stress for animals living in the deeper areas is alleviated, at least temporarily. The surface-water currents in the Baltic Sea affect the transport-related processes, such as the dispersal of organisms.

2.3.2 Currents

The average net surface-water circulation in the main subregions of the Baltic Sea is counter-clockwise (Fig. 2.10). These currents are induced by complex interactions between the Coriolis force, wind stress at the sea surface, sea level tilt, thermohaline gradient of water density, the (minimal) tidal forces, bottom topography and friction. The Baltic Sea surface-water currents are, on average, rather weak ($\sim 5\ \text{cm s}^{-1}$), but during storms, wind-driven currents can reach $50\ \text{cm s}^{-1}$ and up to $100\ \text{cm s}^{-1}$ in straits (Leppäranta and Myrberg 2009). The counter-clockwise surface-water circulation transports salt, heat, nutrients, contaminants, sediments, plankton and propagules of species (including non-native ones) and thus affects many aspects of the Baltic Sea ecosystem. For example, the circulation continuously shapes the large sandy beaches in the southeastern Baltic Sea.

Zooplankton and fish species that live in deeper water can be transported from the Kattegat into the Baltic Sea with saltwater inflows. Their dispersal is thus mainly controlled by the baroclinic flow field and bottom topography (Box 2.1). Hydrodynamic drift modelling has shown that the potential dispersion of *e.g.* comb jellies follows the deep-water currents from the Bornholm Sea towards the north and the east of the Baltic Sea and is limited by topographic features and low advection velocities (Lehtiniemi et al. 2012). However, if such species are new invaders in the area (*cf.* Sect. 5.1), and the conditions for growth and reproduction are favourable in the Baltic Sea, they will be able to form stable populations despite the fact that most individuals are hampered by the hydrodynamics of the deep water.

Box 2.1: Major Baltic inflows (MBIs)

Kai Myrberg

Barotropic inflows

Since the mid-1970s, the frequency and intensity of barotropic MBIs (based on sea level differences between the Baltic Sea and the North Sea) have decreased (*cf.* Fig. 2.13a). They were completely missing between 1983 and 1993 (Lass and Matthäus 1996; Matthäus et al. 2008), which was the longest break of MBIs ever since the measurements started in 1890s. After another 10 years without a large MBI between 1993 and 2014, a strong event occurred in December 2014 (Mohrholz et al. 2015; Gräwe et al. 2015). Several theories have been proposed to explain the decrease of the frequency of MBIs, which are *e.g.* coupled to changing riverine runoff or meteorological patterns (*e.g.* Leppäranta and Myrberg 2009).

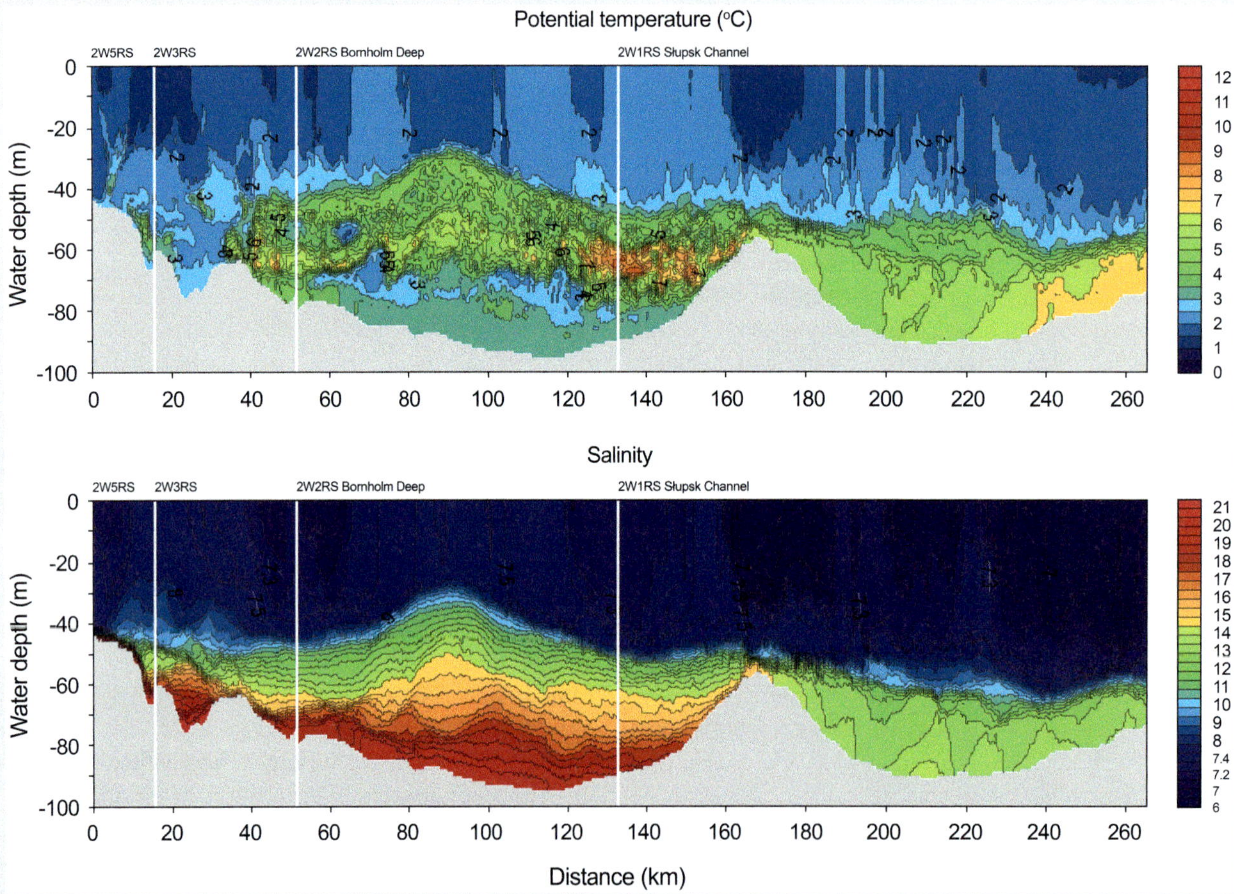

Box Fig. 2.1 Potential temperature and salinity on 16–18 February 2003, along the axis Arkona basin—Bornholmsgattet—Bornholm deep—Słupsk channel—Gdańsk deep. Figure reprinted from Piechura and Beszczyńska-Möller (2004) with permission from Oceanologia (Institute of Oceanology PAN, Sopot, Poland)

Theories explaining the frequency of barotropic inflows: riverine runoff

From long-term observations it becomes evident that there is a good fit between the minimum deep-water salinity and the maximal riverine runoff if there is a time interval of six years between salinity and runoff measurements (Launiainen and Vihma 1990). From the dynamic viewpoint, there are two main mechanisms driven by the riverine runoff variability and counteracting the inflows (Matthäus and Schinke 1999). Firstly, the low-salinity outflowing water mixes in the surface layer with the more saline lower layer which penetrates into the Baltic Sea in the near-bottom layer of the sill areas. Increased water supply from rivers reduces the salinity of the outflowing water and strongly dilutes the inflowing waters. Secondly, an increase in the freshwater supply to the Baltic Sea produces larger outflow, reduces or impedes the inflow of saline water and causes unfavourable conditions for MBIs.

Theories explaining the frequency of barotropic inflows: meteorology

On a larger scale, strong wintertime westerly winds (associated with a high NAO-index) transport intensively moist air masses from the North Atlantic to Europe, resulting in increased precipitation in the Baltic Sea region, with lower evaporation and increased riverine runoff. Above-normal Baltic Sea levels occur frequently for long periods, which hampers saltwater inflows (Zorita and Laine 2000). Increases of stagnation periods can be due to the high salinity in the bottom waters. High bottom salinities in the 1950s and 1960s may have been caused by the MBI in 1951 (Meier et al. 2006), filling the Baltic deeps with highly saline water and making their replacement by later inflows difficult. There are indications of changes in the Baltic Sea local wind climatology. An anomalous west wind component at the Kap Arkona station was found between August and October during 1951–1990 for seasons without a MBI as compared with the corresponding years with MBIs. In the years without MBIs, the period with easterly winds is shortened. Such changes of local wind patterns may cause variations in long-term salinity patterns, which cannot be explained by accumulated freshwater inflow or by low-frequency variability of the zonal wind (Lass and Matthäus 1996). Lehmann and Post (2015) examined atmospheric circulation conditions necessary to force large volume changes (LVCs, total volume changes of the Baltic Sea of at least 100 km^3). MBIs can be considered as subset of LVCs transporting additionally a large amount of salt into the Baltic Sea. An LVC is a necessary condition for a MBI, but an LVC alone is not sufficient. Lehmann and Post (2015) confirmed earlier conclusions about the importance of the pre-inflow period when prevailing easterly winds increase the Baltic Sea brackish water outflow, lower the mean sea level and hinder the inflow of Kattegat water through the Danish straits (Storebælt, Lillebælt and Öresund).

Baroclinic inflows

Warm baroclinic inflows into the Baltic Sea (based on water-density differences between the Baltic Sea and the North Sea) also occur. Such inflows regularly take place in late summer and autumn in the southern Baltic Sea Area, as shown by the mean long-term annual temperature cycles in the deep water of the Arkona, Bornholm and Gdańsk basins (Matthäus 2006). Inflows with exceptionally warm waters recently occurred, in 2002 and 2003 (Box Fig. 2.1). Two types of such inflows have been observed with specific dynamic mechanisms (Matthäus 2006). The first type is caused by heavy westerly gales which pass over the Darß and the Drogden sills. The second type is a long-lasting baroclinic inflow, which only passes over the Darß sill, caused by calm weather conditions over central Europe. In such a situation the inflows are driven by baroclinic pressure gradients, especially caused by horizontal salinity differences (Feistel et al. 2006). Warm baroclinic inflows can transport large volumes of exceptionally warm water into the deeper layers of the Gotland Sea. However, these inflows in fact import oxygen-deficient waters, although they seem to be important for ventilation of intermediate layers in the Eastern Gotland basin deep water through entrainment (Feistel et al. 2006). On the other hand, warm water inflows (baroclinic or barotropic) do transport less oxygen to the Baltic Sea than cold-water inflows, and higher temperatures increase the rate of oxygen consumption in the deep water and facilitate formation of hydrogen sulphide (Matthäus 2006).

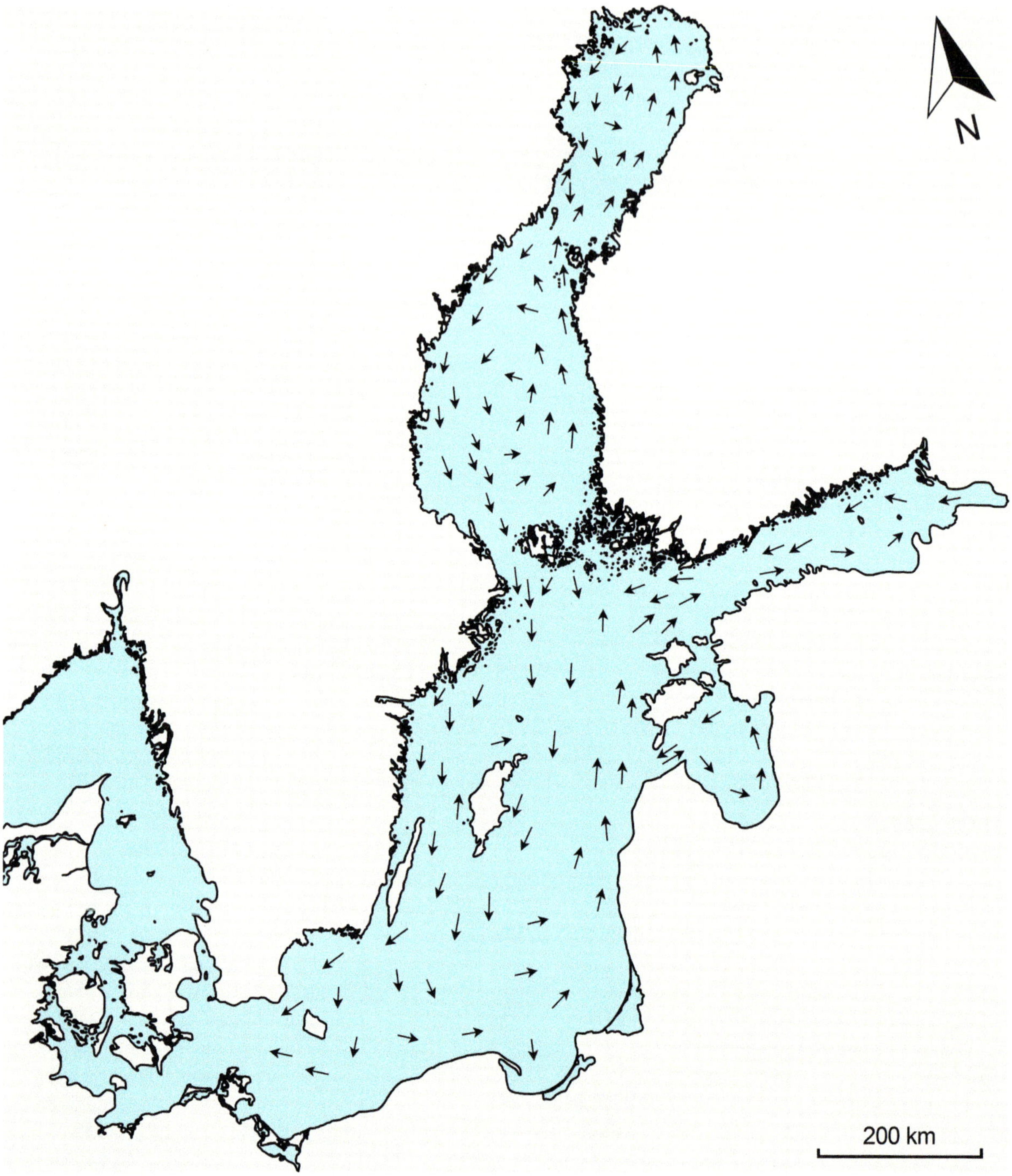

Fig. 2.10 Map of the Baltic Sea Area, showing the average (net) directions of the surface-water circulation based on measurements obtained from drifters deployed and followed by lightships. Figure modified from Leppäranta and Myrberg (2009)

2.3.3 A microtidal sea

The Baltic Sea is "microtidal", which is defined as a tidal amplitude <2 m (Hayes 1979). In fact, the tidal amplitude in the Baltic Sea is much smaller than that. In the Belt Sea, the tidal amplitude in sea level is ~10 cm, over most of the Baltic Sea it is 2–5 cm and only from the eastern Gulf of Finland amplitudes of >10 cm have been reported (Leppäranta and Myrberg 2009). These tidal changes are so small that the difference between high and low tide is basically undetectable anywhere in the Baltic Sea because they are masked by the much larger water level fluctuations caused by air-pressure changes and winds (Novotny et al. 2006).

Tides involve the rise and fall of the water level caused by the combined effects of the gravitational forces exerted by the moon and the sun and the rotation of the Earth. The subbasins of the Baltic Sea themselves are too small and shallow to have their own significant tides. Instead, the few cm of tidal amplitude in the southern Baltic Sea originate from the tidal waves coming in from the North Sea via the Skagerrak and the transition zone.

The North Sea tidal amplitude depends on counter-clockwise surface-water circulation (due to the Coriolis force) and distance from central amphidromic points (sites with zero amplitude in the open sea). Since the North Sea gets narrower southwards to the Strait of Dover, the tidal amplitude increases southward along the British coast to macrotidal (>4 m of tidal amplitude). When the water masses move northward again along the shallow sandy Belgian, Dutch and Danish coasts, the tidal waves are reduced by friction. Finally, when the circulating water reaches the Skagerrak, the tidal amplitude is only a few dm, and has decreased to <10 cm when it enters the Baltic Sea over the the Darß and Drogden sills.

2.3.4 Sea level changes

Sea level changes in the Baltic Sea are predominantly controlled by meteorological forcing. In any location, the water level is directly influenced by the local air pressure. High air pressure produces low water levels, and *vice versa*. Strong winds also affect the water level by pushing the water up against the coast or by pressing the water away from the coast depending on the wind direction. These weather-dependent sea level changes are largest when the combination of wind and air pressure results in storm surges, especially in shallow sea areas and near the coast, and can be up to 1–2 m (*cf.* Fig. 11.22b). Before the construction of the Neva Bay dam outside Sankt-Petersburg, extremes of up to 4 m in the easternmost Gulf of Finland were experienced.

As a consequence of the weather dependence, the irregular water level changes in the Baltic Sea can be fast and drastic, *e.g.* in some areas it can increase or decrease by more than 1 m over one day. For example, in winter there can suddenly be a 1-m thick air layer between the sea ice and the water in the Bothnian Bay (*cf.* Fig. 11.22c).

A low water level in one part of the Baltic Sea raises the level in another part, a phenomenon known as "seiche". For example, in the northern Baltic Sea, autumn and winter water levels tend to be higher because of an increased frequency of conditions with low air pressure and strong westerly winds. In spring and summer, high air pressure and gentle winds dominate and water levels tend to be lower. However, the pattern is different in other parts of the Baltic Sea and seasonal patterns may be obscured by *e.g.* stochastic extreme weather conditions or large freshwater inflows.

The water level of the Baltic Sea is also subject to two slow long-term trends: (a) the land uplift in the north of the area (*cf.* Fig. 2.26b) which decreases the water level by up to ~1 cm per year and (b) the global sea level increase due to the melting of the Earth's glaciers and expansion of the seawater volume by global warming. As a result of the latter trend, the average water level of the Baltic Sea has increased by 11 cm in the 80 years between 1890 and 1970 and by another 11 cm in the 45 years between 1970 and 2015 as measured by the Swedish Meteorological and Hydrological Institute (SMHI, http://www.smhi.se).

2.3.5 Coastal zonation

Oceanic coasts are typically subdivided into the supralittoral zone (splash zone), the eulittoral (intertidal) zone and the sublittoral (subtidal) zone that extends from below low tide to the edge of the continental shelf (Levinton 2010). Since tides are negligible in the Baltic Sea, and the whole sea is located on the continental shelf, its shores have only an epilittoral zone and a sublittoral zone. The epilittoral zone is not covered by water but receives saline sea spray, while the sublittoral zone is the submerged part of the shore.

Subdivisions of the sublittoral zone have been made according to regional conditions in different parts of the Baltic Sea Area, but none of these are applicable everywhere. The most widely used subdivision is that of a ~0.5–1 m wide "hydrolittoral" zone in-between the epilittoral zone and the permanently submerged sublittoral zone The hydrolittoral zone is the part of the littoral that is most affected by the irregular sea level changes of the Baltic Sea and is defined as the zone that extends from the annual minimum water level up to the mean summertime water level (Du Rietz 1930). The hydrolittoral zone is subject to longer periods of desiccation (*cf.* Fig. 11.22a) and is inhabited by ephemeral algae and their accompanying fauna (Wærn 1952).

2.3.6 More than 200 rivers discharge into the Baltic Sea

Freshwater enters the Baltic Sea from over 200 rivers discharging along its coastline (Fig. 2.11). In the north, the runoff is usually largest in May from snow melt, and smallest in January and February when the air temperature is below 0 °C.

Twenty-eight major rivers together cover 80 % of the drainage area (Table 2.4). Twelve of these rivers are classified as "eutrophic". In addition to nutrients from natural sources on land, they transport excess nitrogen and phosphorus from agricultural land to the sea. These 12 rivers mainly discharge into the Baltic Sea proper and the Gulf of

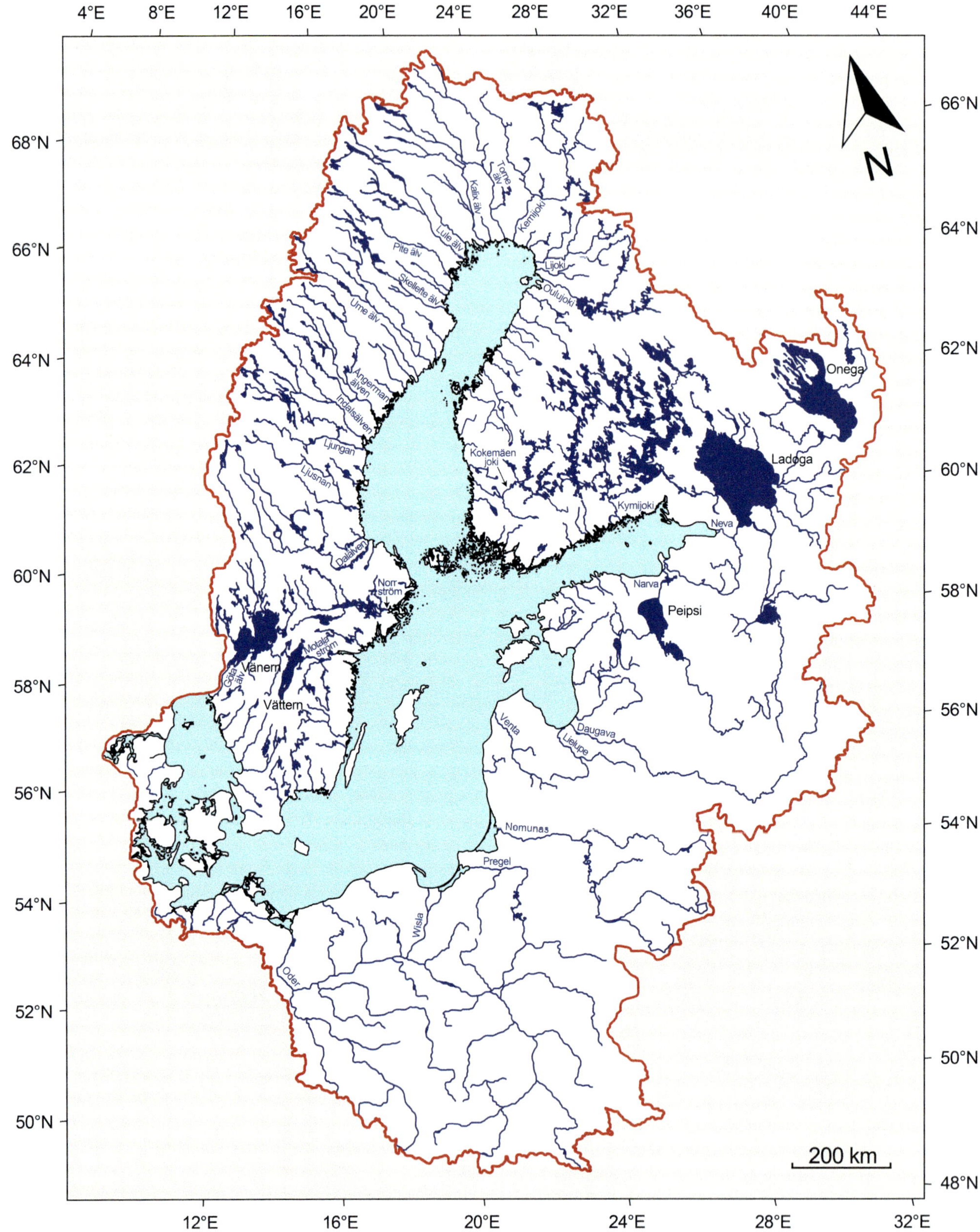

Fig. 2.11 Map of the Baltic Sea drainage area, showing the major lakes and rivers discharging into the Baltic Sea. Drainage area outline according to HELCOM (2010a). Figure: © Pauline Snoeijs-Leijonmalm

Table 2.4 List of the 28 major rivers discharging into the Baltic Sea, together covering 80 % of the drainage area, the Baltic Sea subregion to which they discharge, their drainage areas, land use in the drainage area and nutrients discharged. For the locations of these rivers, see Fig. 2.11. Monthly observations of river nutrient data (Si, N and P) and hydrological data based on measurements accessed from the major databases around the Baltic Sea with the decision support system Baltic Nest (http://nest.su.se, Wulff et al. 2013). These measurements represent river mouth data. The nutrient data were discharge-weighted, and averaged using monthly data from 1980 to 2000. Data from Humborg et al. (2007)

River	Subregion	Drainage area (km^2)	Forest (%)	Agriculture (%)	Bare, water, wetlands (%)	Discharge (km^3 year^{-1})	Nitrogen (tonnes year^{-1})	Phosphorus (tonnes year^{-1})	Silicate (tonnes year^{-1})
Eutrophic agricultural									
Neva	Gulf of Finland	285,835	68	15	17	80	56,996	3,707	17,927
Wisła	Baltic Sea proper	192,899	33	64	3	34	123,021	6,126	111,851
Odra	Baltic Sea proper	117,589	33	64	3	17	73,754	6,305	58,257
Nemunas	Baltic Sea proper	92,104	37	61	2	16	56,261	1,360	31,423
Daugava	Gulf of Riga	85,853	46	51	3	21	39,112	1,301	35,321
Narva	Gulf of Finland	56,665	40	52	8	14	9,367	507	19,549
Göta älv	Kattegat	48,214	67	13	20	18	15,098	328	8,586
Norrström	Baltic Sea proper	22,534	64	22	14	5	3,761	183	3,363
Lielupe	Gulf of Riga	17,835	28	70	2	3	18,436	327	6,696
Motala ström	Baltic Sea proper	15,544	60	20	20	3	2,535	105	2,414
Pregel	Baltic Sea proper	14,749	22	75	3	3	5,301	140	6,648
Venta	Baltic Sea proper	11,597	37	62	1	3	7,816	122	5,617
Regulated boreal									
Kemijoki	Bothnian Bay	50,918	84	3	13	19	6,989	401	55,486
Kymijoki	Gulf of Finland	36,522	78	3	19	10	5,683	215	10,255
Ångermanälven	Bothnian Sea	31,421	85	3	12	17	4,727	209	23,316
Dalälven	Bothnian Sea	28,873	86	5	9	12	5,319	221	23,432
Ume älv	Bothnian Sea	26,737	83	1	16	15	3,518	183	20,663
Kokenmäenjoki	Bothnian Sea	26,667	79	10	11	8	9,695	474	15,167
Indalsälven	Bothnian Sea	25,458	79	4	17	15	4,581	138	14,614
Lule älv	Bothnian Bay	24,934	62	1	37	17	3,506	155	19,780
Oulujoki	Bothnian Bay	22,825	86	2	12	9	3,205	184	10,902
Ljusnan	Bothnian Sea	19,751	89	3	8	8	3,155	116	19,149
Iijoki	Bothnian Bay	14,264	92	2	6	6	2,240	137	15,673
Ljungan	Bothnian Sea	13,042	89	2	9	4	1,627	80	8,605
Skellefte älv	Bothnian Bay	11,577	76	1	23	5	1,418	50	6,467
Pite älv	Bothnian Bay	11,209	74	0	26	6	1,528	89	10,640
Unperturbed boreal									
Torne älv	Bothnian Bay	39,613	73	1	26	14	5,085	375	32,630
Kalix älv	Bothnian Bay	17,674	72	1	27	10	3,612	246	22,286

Finland. The mouths of the five largest rivers, the Neva, Wisła (Vistula), Odra (Oder), Nemunas and Daugava, are major point sources of nutrient emissions to the Baltic Sea (*cf.* Sect. 18.8). The other 16 major rivers are boreal rivers transporting less nitrogen and phosphorus, but relatively more humic substances, from forested areas. Most of these northern rivers discharge into the Gulf of Bothnia.

Recent human interferences in the drainage area resulted in, *inter alia*, reduced inputs of dissolved silica (DSi) to the northern Baltic Sea. Only two of the large boreal rivers, the Torne älv and the Kalix älv, are unperturbed, while the other 14 have been used for hydroelectric power generation since 1920–1970 (Table 2.4). Damming reduces the DSi input to the sea due to longer residence times for water in the river systems. DSi inputs to the sea are also reduced by eutrophication of rivers through biogenic silica production (by diatoms) and subsequent sedimentation along the river system. Overall, the river-borne DSi loads entered into the Baltic Sea were estimated to have dropped by 30–40 % during the last century (Humborg et al. 2007). This may ultimately decrease the biogenic silica production, notably that of pelagic diatoms, in offshore parts of the Baltic Sea proper.

2.3.7 Limited water exchange with the ocean

The narrow and shallow thresholds at the entrance of the Baltic Sea hamper water exchange with the North Sea and this, together with the large freshwater discharge into the Baltic Sea, is the main reason why the Baltic Sea is brackish. Water exchange between the Baltic Sea and the North Sea is governed by sea level differences, wind-driven currents and water density, the latter depending on salinity and temperature (Lass and Matthäus 2008).

The large riverine runoff to the Baltic Sea causes an increase in the water level from the entrance of the Baltic Sea to the Gulf of Bothnia by ~25 cm. This permanent sea level tilt is the result of the water density decrease with lower salinity towards the north. The less dense water in the northern Baltic Sea occupies a larger water volume than the denser water in the south. The outflow from the Baltic Sea increases with easterly winds and decreases with westerly winds and is also dependent on air pressure. In years with heavy precipitation relative to evaporation in the Baltic Sea drainage area, the water outflow increases as well.

Like the Baltic Sea, the North Sea is a continental sea, but unlike the Baltic Sea it is not a semi-enclosed sea since it has a wide and deep connection to the Atlantic Ocean. Therefore, the salinity of the North Sea is close to that of the ocean, although its surface-water salinity (especially in the northeastern North Sea) is influenced by the brackish-water outflow from the Baltic Sea, which represents ~60 % of all freshwater entering the North Sea (Leppäranta and Myrberg 2009).

If not interrupted by wind-driven currents, a continuous inflow of saline water from the Skagerrak forms the deep water of the Baltic Sea, while the less saline Baltic Sea water outflow occurs at the surface. This is because the more saline water has a higher density and is thus heavier (*cf.* Sect. 1.3.1). The inflow of the saline deep water is mainly governed by the bottom topography with sills and deeper channels (Fig. 2.3). Approximately 70–80 % of the saline inflow enters the Arkona Sea via the Storebælt and Lillebælt and the southern Belt Sea over the 18 m deep Darß sill, and 20–30 % via the Öresund over the 8 m deep Drogden sill (Mattsson 1996; Jakobsen and Trébuchet 2000). From the Arkona Sea, the saline water moves via Bornholmsgattet to the Bornholm deep and via the Słupsk channel to the Gotland deep in the Eastern Gotland Sea. A shallow connection with the Baltic Sea proper prevents the deep water from entering the Gulf of Riga. From the Gotland deep, the water flow continues to the Fårö deep and further on to the northern Baltic Sea proper. The Åland Sea sills (Södra Kvarken) and the shallowness of the Archipelago Sea prevent the deep water from entering the Gulf of Bothnia from the Baltic Sea proper. Thus, it flows southward into the Western Gotland Sea with the Landsort deep and the Norrköping deep and eastward into the Gulf of Finland.

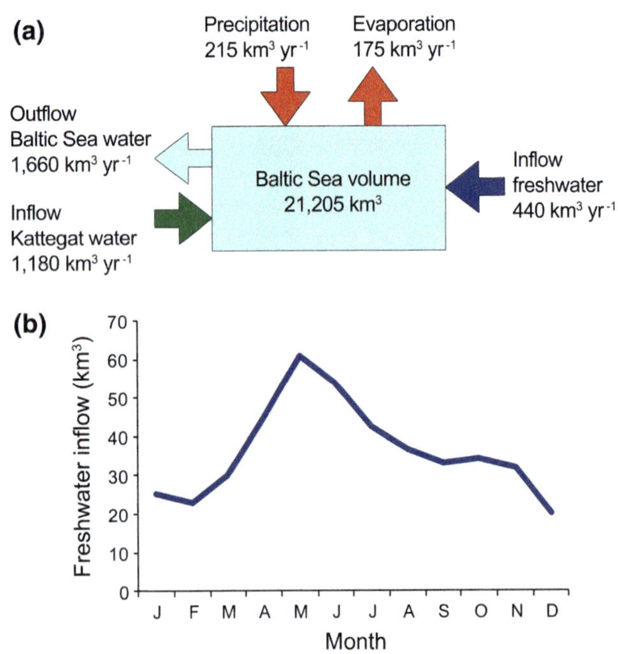

Fig. 2.12 Water budget of the Baltic Sea. (**a**) The average annual Baltic Sea water budget, showing freshwater runoff from land to the right, water exchange at the entrance to the Baltic Sea to the left and precipitation and evaporation at the top. According to this budget, the sum of water input and output is zero, leaving a water volume in the Baltic Sea (including the southern Belt Sea) of 21,205 km^3. The net outflow to the Kattegat is 480 km^3 year^{-1}. (**b**) The average monthly riverine runoff (inflow of freshwater) to the Baltic Sea for the time period 1901–1990. Figure (**a**) based on water budget data in Leppäranta and Myrberg (2009), (**b**) modified from Cyberski and Wróblewski (2000)

2.3.8 The Baltic Sea water budget

The annual average outflow from the Baltic Sea is estimated at 1,660 km^3 (~52,600 m^3 s^{-1}). The estimated annual inflow is very similar, 1,620 km^3, which is the sum of 1,180 km^3 inflow from the Kattegat and 440 km^3 freshwater inflow in the form of riverine runoff from the drainage area (Fig. 2.12a). This results in an annual net outflow from the Baltic Sea of 480 km^3 (~15,000 m^3 s^{-1}) and constitutes ~60 % of the total freshwater supply to the North Sea. The riverine runoff from the drainage area to the Baltic Sea is largest in May-June and lowest in December-February (Fig. 2.12b). In long-term reconstructions for the time period 1500–1995 no significant long-term changes in the total riverine runoff to the Baltic Sea were detected, although decadal and regional variability was large and the runoff is sensitive to temperature decreasing by 3 % (450 m^3 s^{-1}) per degree Celsius increase (Hansson et al. 2011a).

The outflow from the Baltic Sea equals the sum of the freshwater discharges into the Baltic Sea and the difference between precipitation and evaporation over the Baltic Sea. Precipitation is slightly higher than evaporation (Omstedt et al. 1997). Precipitation is 500–600 mm per year, with the lowest monthly values of 25–50 mm in December-May and the highest monthly values in July–September (50–75 mm). Evaporation is 450–500 mm per year with minimum values in the spring (10–20 mm during May–June) when the surface-water temperature is low, and maximum values in late autumn (70–80 mm during October–December) when the turbulent air-sea exchange is extensive (Leppäranta and Myrberg 2009).

A simple calculation of the water renewal time in the Baltic Sea, based on the total volume of 20,958 km^3 and 480 km^3 of freshwater runoff, yields 43.7 years. However, this is not a full estimate because water entering the Baltic Sea from the Kattegat may flow out again within a short time. Thus, water masses closer to the entrance of the Baltic Sea tend to stay for a relatively shorter time in the sea compared to water masses farther away from the entrance. More accurate renewal times for the surface water in different parts of the Baltic Sea were estimated at 26–30 years for the Bornholm Sea, 28–34 years for the Gotland Sea, 34–38 years for the Bothnian Sea and 38–42 years for the Bothnian Bay (Meier 2007).

2.3.9 Major inflows from the Kattegat are rare

The normal water exchange with the Kattegat as described above is not strong enough to renew hypoxic and anoxic water masses in the deeper parts of the Baltic Sea, which is a prerequisite for animal life in and on deep soft seabeds (*cf.* Sect. 10.11). Renewal of deep waters only occurs during very strong and intensive inflow events of Kattegat water, which are, however, rare (Fig. 2.13a).

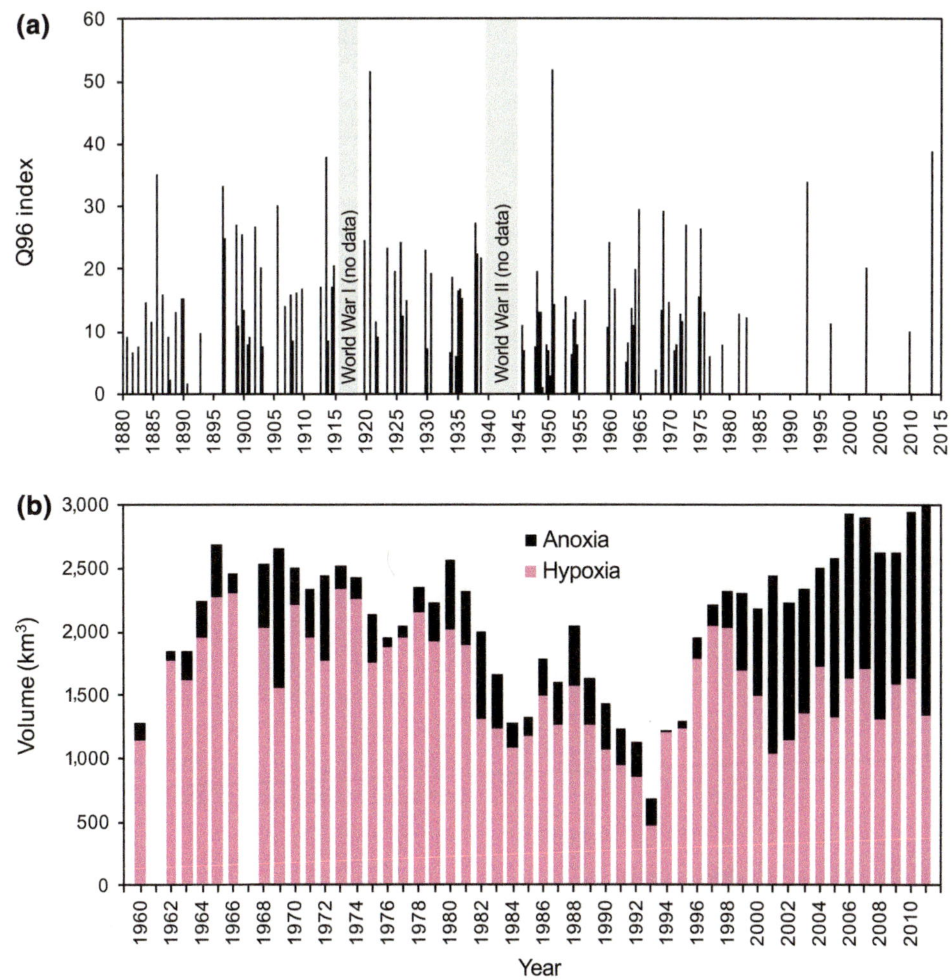

Fig. 2.13 Major inflows of saline water from the North Sea into the Baltic Sea and oxygen conditions in the Baltic Sea. (**a**) Inflow intensity from 1880 to 2014, expressed as "Q96", which is an intensity index based on the total amount of salt transported through the Belt Sea and the Öresund into the Baltic Sea during an inflow event (Fisher and Matthäus 1996). (**b**) Oxygen conditions, shown as the total volumes of hypoxic water (>0 mL O_2 L^{-1} and <2 mL O_2 L^{-1}) and anoxic water (≤ 0 mL O_2 L^{-1}) in the Baltic Sea 1960–2011. Figure (**a**) modified from Matthäus (2006) and Mohrholz et al. (2015), (**b**) modified from Hansson et al. (2011b)

Such major Baltic inflows (MBIs, Box 2.1) are infrequent because the required weather conditions are rare. They occur only when, at first, strong easterly winds dominate over the entrance to the Baltic Sea for several weeks, which then are followed by persistent, strong westerly winds (Lass and Matthäus 2008; Leppäranta and Myrberg 2009). The easterly winds increase the outflow of surface water and push the water level in the Baltic Sea down to a minimum level. The subsequent westerly winds force the Kattegat water to pile up in the Belt Sea and the Öresund, push the Baltic Sea water eastwards and press the Kattegat water over the sills into the Baltic Sea. During an MBI, the water level of the Baltic Sea can rise by one metre. Finally, the water masses with high salinity (and density) sink into the deeper areas of the Baltic Sea proper.

A strong MBI took place in December 2014. Together with the 1,913 MBI, this was the third largest one recorded since 1880. It was estimated that the total inflow of highly saline oxygen-rich water during the 2014 MBI was ~198 km^3 (~4 Gt salt), of which 138 km^3 (2.60 Gt salt) entered through the Storebælt and Lillebælt and 60 km^3 (1.38 Gt salt) through the Öresund (Mohrholz et al. 2015). While the MBI events in 1993 and 2003 interrupted the anoxic bottom conditions in the Baltic Sea only temporarily, the large 2014 MBI may have induced a longer-lasting improvement of oxygen levels. During 2015, this large new water inflow was slowly spreading northward and this has the potential to turn (most of) the hypoxic and anoxic deep water of the Baltic Sea proper into oxic conditions, with substantial consequences for marine life and biogeochemical cycles. However, it is not certain that this will actually happen.

Table 2.5 The solubility of oxygen in water (in mL L^{-1}) at different temperatures for salinities relevant for the Baltic Sea Area at a pressure of 1 atm. The data were calculated with the Oceanographic Calculator of the International Council for the Exploration of the Sea (http://ocean.ices.dk/Tools/Calculator.aspx)

Salinity	0 °C	5 °C	15 °C	25 °C
0	10.22	8.93	7.05	5.77
3	10.01	8.75	6.92	5.67
5	9.88	8.64	6.84	5.61
7	9.74	8.53	6.75	5.54
10	9.54	8.36	6.63	5.45
15	9.22	8.09	6.43	5.30
25	8.61	7.57	6.05	5.00
35	8.05	7.09	5.69	4.73

2.3.10 Hypoxia

Surface waters are always rich in O_2 because they are in contact with the atmosphere and mixed by winds. In addition, oxygen is produced from water during photosynthesis by primary producers in the upper part of the water column. In deep water, aerobic bacteria consume oxygen when they respire during the decomposition of organic material that sinks to the bottom from the upper water column. If there is a surplus of organic material in the system, the oxygen in the deep water is depleted and anaerobic bacteria take over (*cf.* Sect. 3.6).

Hypoxia (oxygen stress) occurs when there is a mismatch between oxygen supply and the demand for it. Hypoxia thresholds vary widely across marine benthic organisms (Vaquer-Sunyer and Duarte 2008). Hypoxia is usually defined as an oxygen concentration <2 mL O_2 L^{-1}, which seems most relevant as a threshold for organisms that have evolved in normoxic waters without severe hypoxia as an evolutionary stress (Diaz and Rosenberg 1995). This threshold of 2 mL O_2 L^{-1} equals 2.9 mg O_2 L^{-1} or 91 μM O_2. Other commonly used thresholds are 2 mg O_2 L^{-1} (1.4 mL O_2 L^{-1} or 63 μM O_2) or 30 % O_2 saturation (Rabalais et al. 2010).

Since the solubility of oxygen in water decreases with increasing salinity and temperature (Benson and Krause 1984, Table 2.5), the % oxygen saturation increases with increasing salinity and temperature. For example, a concentration of 1.4 mL O_2 L^{-1} generates 14 % O_2 saturation at salinity 5 and 0 °C, but 30 % O_2 saturation at salinity 35 and 25 °C. At a concentration of 2 mL O_2 L^{-1} the same salinity-temperature combinations generate 20 % and 42 % oxygen saturation, respectively. Because of the different thresholds applied and the different ways to express oxygen concentrations, it may be difficult to compare different studies on hypoxia.

2.3.11 Hypoxia and anoxia in the Baltic Sea

Hypoxia in the deeper basins of the Baltic Sea has been occurring since the Littorina Sea stage some 8,000 years ago (Zillén et al. 2008). In the geological development of the Baltic Sea, hypoxia has increased during warmer periods and decreased during colder periods. The average size of the seafloor in the Baltic Sea proper that during the last 40 years was affected by hypoxia and anoxia is ~49,000 km^2 (23 %), including practically all deep bottoms (Conley et al. 2009; Hansson et al. 2011b). While anoxia is typical of stratified semi-enclosed seas (*e.g.* ~90 % of the Black Sea bottoms are anoxic), hypoxia and anoxia in the Baltic Sea have - during the last century - increased drastically as a result of anthropogenic activities in the drainage area.

The major cause of the increased oxygen deficiency in the deep basins of the Baltic Sea proper is excess nutrient loading, which induces algal and cyanobacterial blooms and subsequently increases the sedimentation of organic material (eutrophication). An additional cause is the reduced incidence of MBIs of saline and oxygen-rich Kattegat water (Fig. 2.13a), so that bottom waters are not renewed. After large inflows of water from the Kattegat, the extent of hypoxic and anoxic bottoms decrease (Figs. 2.13b and 2.14). The relative importance of physical forcing (MBIs) *versus* eutrophication for hypoxia is still debated. A recent estimate is that hypoxia has increased 10-fold during the last 115 years, which is primarily linked to nutrient inputs from land (Box 2.2), although increased respiration due to higher temperatures during the last two decades has probably also contributed to worsening oxygen conditions (Carstensen et al. 2014).

In the Gulf of Bothnia, bottom oxygen conditions are much better than in the Baltic Sea proper because the saline below-halocline water of the Baltic Sea proper cannot enter the Bothnian Sea over the Södra Kvarken sill. Thus, the Gulf of Bothnia has a much weaker halocline (*cf.* Sect. 2.4.3) and lower primary production in the photic zone as a result of lower nutrient concentrations, compared with the Baltic Sea proper. In the more eutrophic Gulf of Finland, hypoxia occurs because there is no sill between the Baltic Sea proper and the Gulf of Finland and the bottom oxygen conditions vary (Hansson et al. 2011b).

Anoxic bottoms are also widespread in the shallow coastal areas of the Baltic Sea (Conley et al. 2011). The lack of oxygen leads to the death of organisms that live in and on the bottom and weakens the function of the coastal zone as a nursery habitat for fish. Increases in the number and size of both small- and large-scale hypoxic areas in the Baltic Sea may be attributed to elevated nutrient levels resulting from activities in the drainage area: the excessive use of fertilisers, the presence of large animal farms with intensive livestock production, the burning of fossil fuels, discharges of effluents from *e.g.* municipal wastewater treatment plants, industrial point sources, fish farms and the disappearance of wetlands that act as nutrient traps.

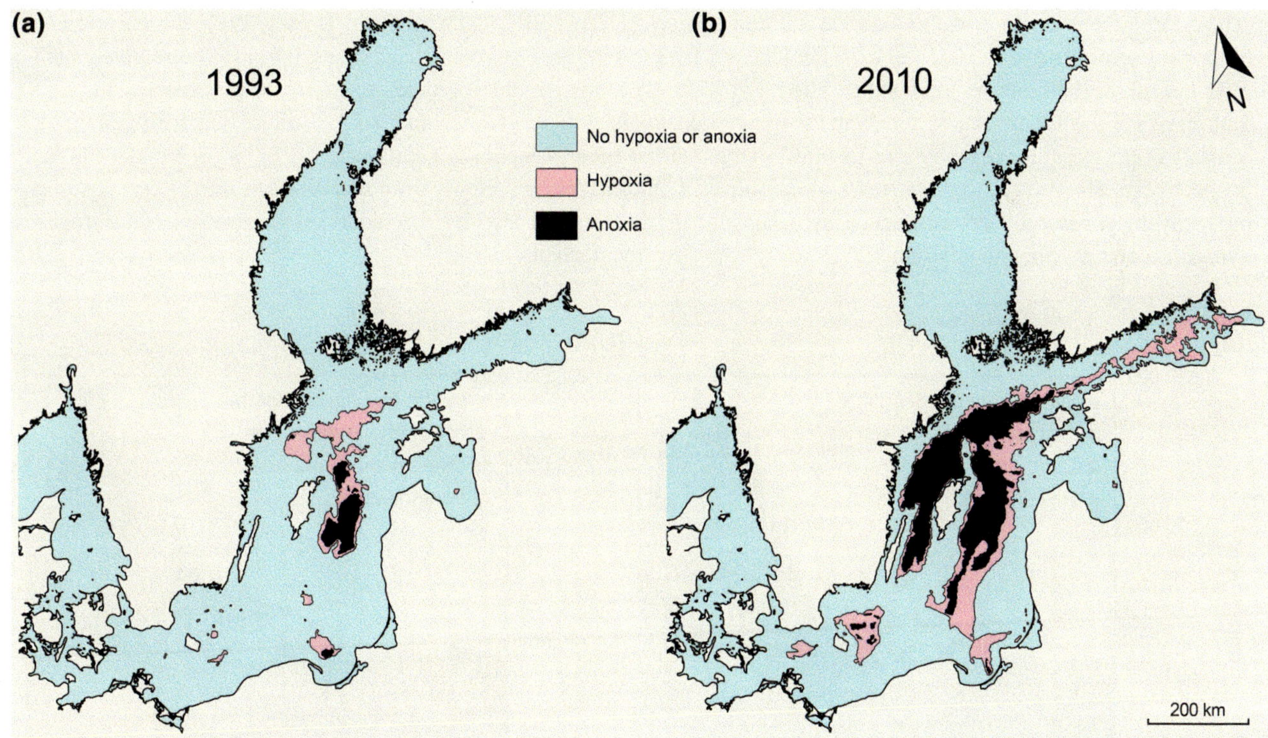

Fig. 2.14 Comparison of the extent of hypoxic conditions (>0 mL O_2 L^{-1} and <2 mL O_2 L^{-1}) and anoxic conditions (≤0 mL O_2 L^{-1}) in the deep waters of the Baltic Sea with different inflows of saline water from the North Sea into the Baltic Sea (*cf.* Fig. 2.13a). (**a**) Oxygen conditions in the deep waters of the Baltic Sea in autumn 1993 after the large inflow in January 1993. (**b**) Oxygen conditions in the deep waters of the Baltic Sea in autumn 2010 after a long period of stagnation. Figure modified from Hansson et al. (2011b)

Box 2.2: External nutrient inputs to the Baltic Sea

Oleg Savchuk

Misbalance in nutrient cycles

Eutrophication can be considered as the result of a misbalance in biogeochemical cycling in which more nutrients enter the system than leave it. Knowledge of the past and present nutrient inputs to the Baltic Sea is important, both for understanding the development of eutrophication (*cf.* Sect. 17.4), and for designing remedial measures in ecosystem management (*cf.* Sect. 18.5).

History of eutrophication in the Baltic Sea

The human eutrophication impact on the Baltic Sea became explicit after the 1950s (Box Fig. 2.2, Zillén et al. 2008). Current estimates of the loads of terrestrial origin and atmospheric deposition in 1970–2006 are based on good data with sufficiently high coverage and resolution (Savchuk et al. 2012a). However, because of the lack of reliable information on how historical nutrient inputs were generated by natural and human-made mechanisms within the entire drainage area and atmospheric deposition area, the temporal dynamics for 1850–1970 were reconstructed by a linear interpolation between loads calculated for as few as four points in time: 1850, 1900, 1950 and 1970 (Savchuk et al. 2012b). The water exchange and nutrient imports from the North Sea (Skagerrak) were estimated from reconstructed sea level variations and measurements of nutrient concentrations at the entrance of the Baltic Sea (Gustafsson et al. 2012), assuming a 15 % linear decrease of the concentrations from modern time back to 1900 (Savchuk et al. 2008 and references therein).

Four major sources of nutrient inputs

The relative contributions of the different nutrient sources follow well-known patterns. During 1977–2006, the most important sources of both N and P were the loads of terrestrial origin by rivers discharging into the Baltic Sea, which supplied about half of the total inputs (Box Fig. 2.2). The significance of other sources differs between the two nutrients: point sources, atmosphere, and the Skagerrak supply ~5 %, ~23 % and ~13 % of the total nitrogen input, respectively, while the contributions of these sources to the total phosphorus input are ~14 %, ~7 % and ~34 %, respectively. The pronounced decreases of the nitrogen and phosphorus inputs from the 1980s to the 2000s is caused not only by a naturally driven decline in freshwater discharges, but also by the reduction of atmospheric nitrogen emissions and by phosphorus removal in wastewater treatment plants.

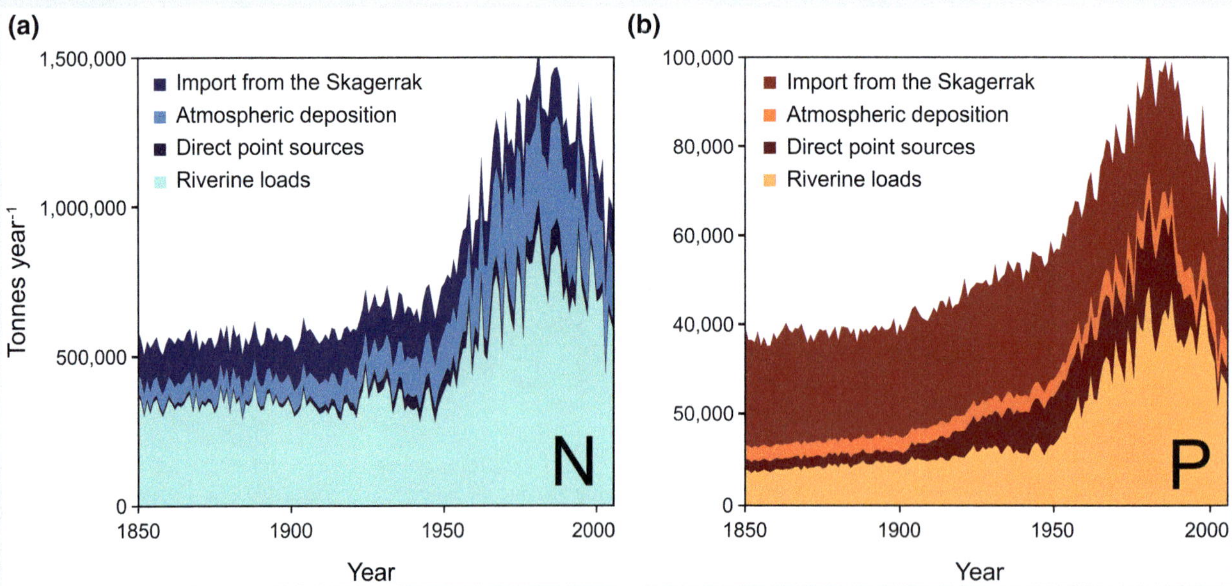

Box Fig. 2.2 External inputs of nutrients to the entire Baltic Sea. (**a**) Total nitrogen. (**b**) Total phosphorus. Figure based on a reconstruction in Savchuk et al. (2012b). Figure: © Oleg Savchuk

2.4 Environmental gradients

2.4.1 Environmental gradients and the distribution of organisms

The geographical position and hydrographical features of the Baltic Sea together generate strong environmental gradients, which affect the distributions of organisms. Physical factors such as salinity, temperature, light and pH directly affect the performance and survival of organisms.

Differences in salinity produce the strongest environmental gradients in the Baltic Sea. Marine species from the North Sea and Atlantic Ocean enter the Baltic Sea via the transition zone and meet their lower salinity limit somewhere along the large-scale Baltic Sea gradient. Freshwater species enter the Baltic Sea all along its coasts and meet their higher salinity limit at some point along the inshore-offshore salinity gradient. Most of the Baltic Sea water column has a permanent halocline, which prevents water mixing and creates oxygen stress for the organisms living in the deeper areas, especially in the Baltic Sea proper. Areas with weak or no haloclines are the Gulf of Riga, the eastern Gulf of Finland and the Gulf of Bothnia.

The species distributions in the Baltic Sea also strongly depend on the climatic gradient, which stretches over 12 degrees of latitude from a temperate climate in the southern Arkona Sea to a subarctic (boreal) climate in the northern Bothnian Bay. Cold-adapted stenotherm species are abundant in the north and species requiring higher temperature populate the south. The long periods with ice cover in the north promote the occurrence of sympagic (ice-dependent) organisms, while the shading and scouring by ice hampers the growth of perennial macrophytes in the littoral zone.

Nutrient dynamics, which determine productivity, differ between the subbasins of the Baltic Sea. The Bothnian Bay is phosphorus-limited, similar to most temperate limnic environments, while the Baltic Sea proper is nitrogen-limited as are most seas. When considering both nitrogen and phosphorus, the Gulf of Riga is the most nutrient-rich area while the Gulf of Bothnia the most nutrient-poor area. Many environmental gradients (*e.g.* climate and nutrients) co-vary in some way with salinity and together they form the "large-scale Baltic Sea gradient".

2.4.2 Large-scale and local Baltic Sea salinity gradients

The brackish water of the Baltic Sea is a mixture of marine North Sea water and freshwater from rivers and precipitation (Fig. 2.12). This creates the ~2,000 km large-scale salinity gradient between the Bothnian Bay and the Skagerrak, as well as similar gradients from the inner Gulfs of Finland and Riga to the Skagerrak. The Baltic Sea surface-water salinity is 0 in the Neva Bay (easternmost Gulf of Finland), ~2 in the northern Bothnian Bay and the eastern Gulfs of Finland and Riga and ~10 in the western Arkona Sea (Fig. 2.15a, c). In the major part of the Baltic Sea, the year-round stable surface-water salinity varies between 5 and 8 along the gradient (Fig. 2.15a). Coastal sites in the Baltic Sea may experience considerable local salinity fluctuations which are governed by the intensity of land-runoff, and in such sites species are more adapted to salinity fluctuation than to a specific stable salinity.

Normally, the estimated daily and decadal variations of the surface-water salinity in the open Baltic Sea are below 0.1 and 0.5, respectively (Omstedt and Axell 2003; Meier and Kauker 2003). In the transition zone, the gradient is less stable and the surface-water salinity varies between 10 and 23 in the Belt Sea and between 12 and 30 in the Kattegat (Table 2.2). The transport of water masses in and out of the Baltic Sea is to a large extent controlled by winds and the surface-water salinity of both the Belt Sea and the Kattegat can vary strongly from day to day.

The Baltic Sea bottom waters display a horizontal salinity gradient as well, but with higher salinity than that of the surface waters (Fig. 2.15b, c). The high-salinity bottom water originates from the water flowing in from the Kattegat. Vertical salinity increases from the water surface to the seafloor are found in all subbasins of the Baltic Sea, but the salinity difference between surface and bottom waters decreases from south to north, depending on distance from the transition zone and bottom topography. For example, the saline deep water in the Baltic Sea proper is blocked by the Åland sills and the shallow Archipelago Sea (Figs. 2.2 and 2.3), and cannot enter the Bothnian Sea (Fig. 2.15c). The bottom-water salinity is ~15 in the southern Baltic Sea proper, ~12 in the Gotland deep (but higher after major inflow events) and only ~6.5 in the Bothnian Sea.

2.4.3 Variation of pycnoclines in space and time

The density of water is a function of salinity and temperature (*cf.* Sect. 1.2). High-salinity water and colder water are heavier than low-salinity and warmer water. In the Baltic Sea, the surface-water salinity is lower than that of the bottom water (Fig. 2.15), and in summer the surface water is warmer than the bottom water (Fig. 2.16). When water masses with different densities are not fully mixed, the denser water moves down and the less dense water moves up, and stratification ensues.

Between the water masses with different densities there is a pycnocline, a relatively thin layer in which the water

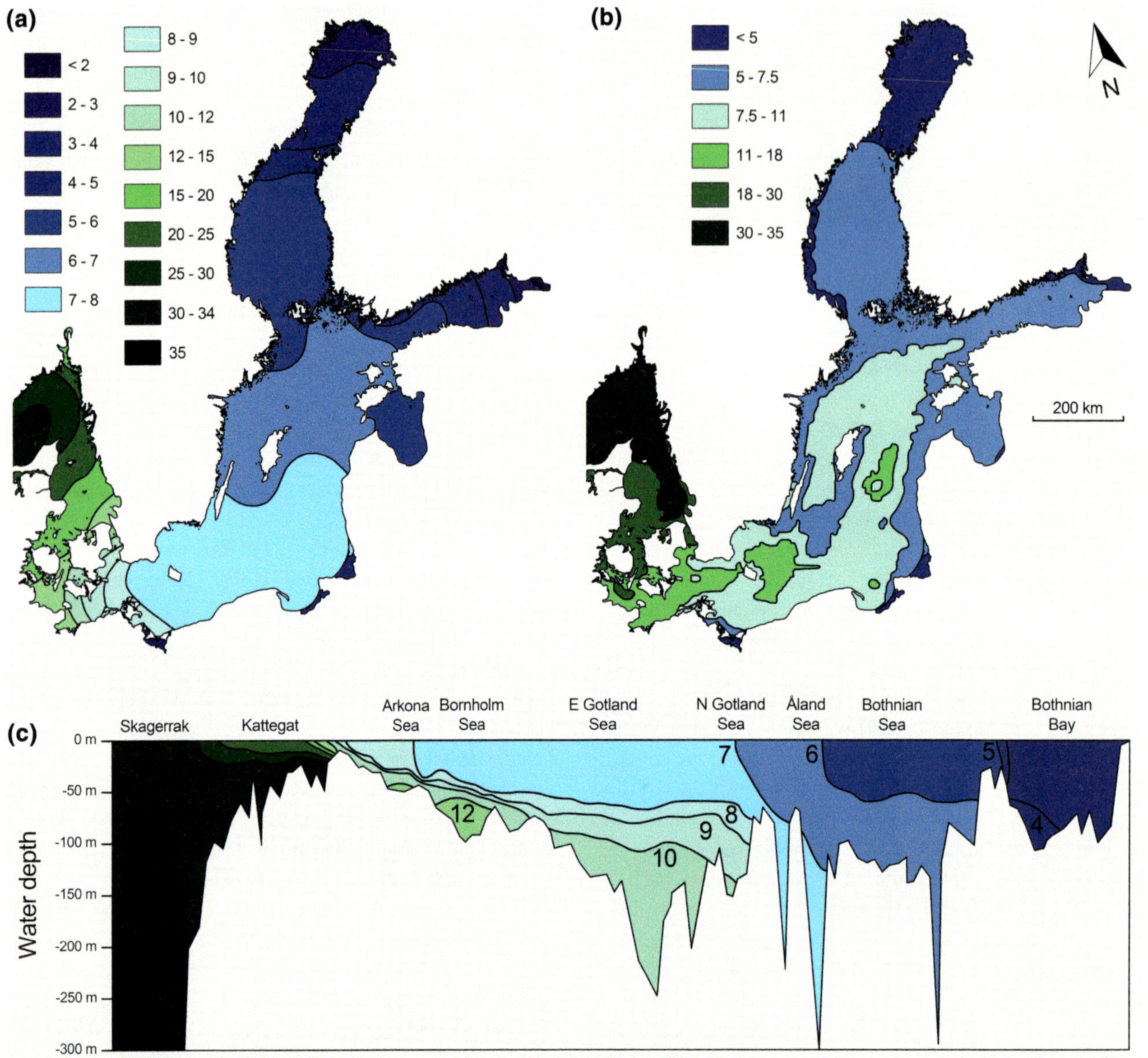

Fig. 2.15 The salinity gradient of the Baltic Sea Area. (**a**) Average surface-water salinity. (**b**) Average bottom-water salinity. (**c**) Schematic cross-section from the Skagerrak to the Bothnian Bay via the Eastern Gotland Sea, showing the vertical distribution of the salinity gradient. The legend for (**c**) is given in (**a**). Figure (**a**) modified from Bernes (2005), (**b**) modified from Al-Hamdani and Reker (2007), (**c**) modified from Sjöberg (1992)

density increases rapidly with depth. Pycnoclines constitute a barrier preventing homogeneous distribution of salt and temperature, but also of other compounds, *e.g.* nutrients and oxygen. Such a strong separation between different water masses can thus have a large impact on the distribution of organisms in the open Baltic Sea. In the coastal waters the water mass is generally well-mixed.

All basins of the Baltic Sea have a thermocline in the warm season (Fig. 2.16c). It is coincident with the pycnocline separating the warmer surface layer (upper 15–30 m) from the colder water below. In the southern Baltic Sea, the seasonal thermocline starts building up in the beginning of May and in the Bothnian Bay this starts in June. Along the thermocline, the temperature drops by as much as 10 °C over a distance of a few metres. The surface layer is well-mixed and the vertical temperature change does not exceed 0.1 °C m^{-1}. In autumn, the surface water cools down, starts sinking and finally the thermocline disappears. During summer, wind-induced upwelling can locally produce a drastic drop in coastal surface-water temperatures (by up to 10 °C). Cooler and saltier deep water then flows up to replace wind-displaced surface water, bringing nutrient-rich

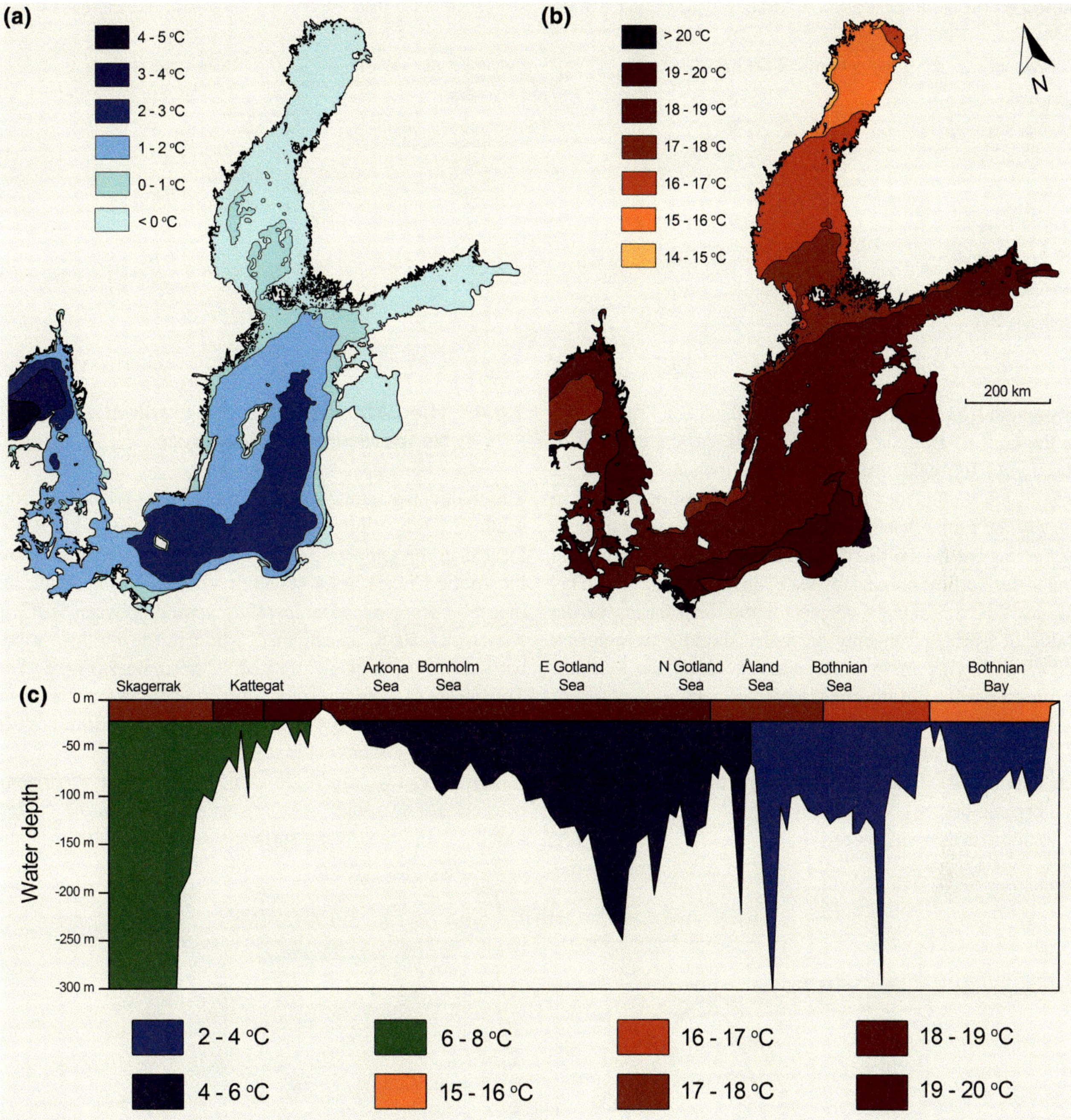

Fig. 2.16 The temperature gradient of the Baltic Sea Area. (**a**) Average surface-water temperature in February 1990–2005. (**b**) Average surface-water temperature in August 1990–2005. (**c**) Schematic cross-section from the Skagerrak to the Bothnian Bay via the Eastern Gotland Sea, showing the approximate vertical distribution of the temperature gradient in summer of the mixed surface-water layer with the thermocline above a 15–30 m water depth and the much lower water temperatures below 15–30 m. Figure (**a**, **b**) modified from Siegel et al. (2008), (**c**) © Figure: Pauline Snoeijs-Leijonmalm

water to the photic zone (Myrberg and Andrejev 2003; Lehmann and Myrberg 2008), and this affects the growth and species composition of the biota in the sea.

The large volume of freshwater entering the Baltic Sea forms a low-salinity layer on top of the saltier bottom water throughout the sea, and not only in estuarine areas as is usually the case in fully marine seas. A permanent halocline, which is a pycnocline between water masses of different salinities, separates the upper and lower layers in the Baltic Sea. This halocline is found at different depths in different parts of the Baltic Sea (Fig. 2.15c). A halocline occurs all the way to the Skagerrak, as long as the Baltic outflow water forms a layer with lower salinity on top of more saline deeper water. The halocline increases in depth and decreases

Table 2.6 The depth of the halocline and salinity ranges in the different subregions of the Baltic Sea Area. Data from Leppäranta and Myrberg (2009) and Andersen et al. (2015)

Subregion	Depth of the halocline (m)	Surface-water salinity	Deep-water salinity
Transition zone	15–20	10–30	30–34
Arkona Sea	20–30	8–11	10–15
Bornholm Sea	50–60	4–8	13–17
Gotland Sea	60–80	5–8	9–13
Gulf of Riga	20–30	4–6	6–7
Gulf of Finland	60–80	1–6	3–9
Bothnian Sea	60–80	4–7	6–7
Bothnian Bay	50–60	2–4	4–5

in strength from the Kattegat to the Gotland Sea (Table 2.6). In the Gulf of Bothnia, the halocline is extremely weak or absent. As the halocline is rather shallow in the Kattegat (15–20 m), it coincides with the summer thermocline, but in most of the Baltic Sea the halocline is located deeper than 50 m, *i.e.* far below the thermocline. Like a lid, the halocline limits the vertical mixing of water in the Baltic Sea. This implies that the oxygen content of the deep basins of the Baltic Sea proper can only be replenished by oxygen-rich saltwater flowing in from the Kattegat along the seafloor, and not by diffusion of atmospheric oxygen.

In the deep basins of the Baltic Sea, at water depths >100 m, a second pycnocline, a redoxcline, may occur (*cf.* Sect. 3.6.4). Below the redoxcline the water is anoxic and organic matter is oxidised by sulphate, which is reduced to toxic hydrogen sulphide.

2.4.4 The Baltic Sea climatic gradient: temperature and ice cover

The Baltic Sea stretches over more than 1,400 km from south to north. While the southern part of the Baltic Sea is located in the temperate zone, the northern part (just below the Arctic Circle) has a subarctic climate. In the south, the monthly average air temperature varies between 0 °C in winter and 20 °C in summer, while in the north it varies between −9 °C in winter and 18 °C in summer (Fig. 2.17a). This yields an average temperature gradient in the Baltic Sea surface waters of 4–5 °C between south and north, both in summer and in winter (Fig. 2.16a, b).

However, the largest differences in surface-water temperature are seasonal. February is, on average, the coldest month of the year with surface-water temperatures in the

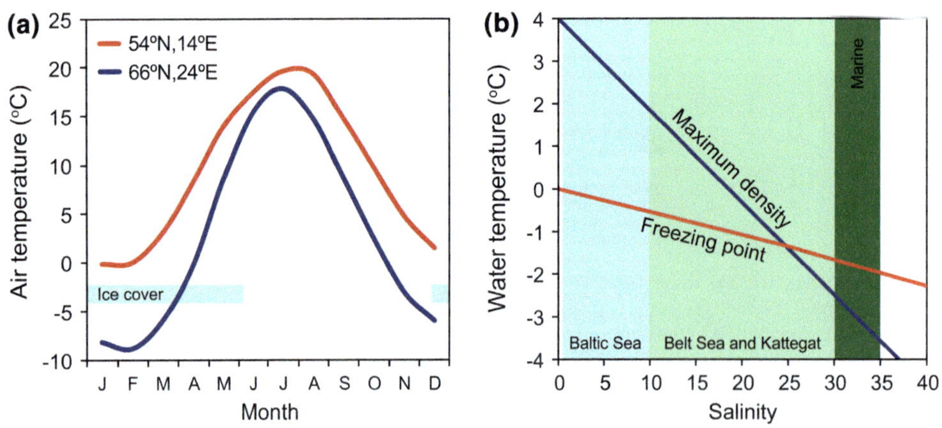

Fig. 2.17 Relationships between temperature and ice with respect to the Baltic Sea Area. (**a**) Annual variation of the monthly air temperature in the northernmost (66 °N, 24 °E) and southernmost (54 °N, 14 °E) Baltic Sea (averages for 1979–2011, n = 33 years). The horizontal light-blue line indicates the approximate time of ice coverage of the northernmost Baltic Sea. (**b**) The relationship between the temperature of maximum density and the freezing point for water of different salinity. At the intercept of the maximum density temperature and the freezing point (at salinity 24.7 and temperature −1.3 °C), the freezing properties of water change fundamentally. Water of salinity <24.7 is defined as brackish water from a purely physical point of view while in ecology water of salinity <30 is considered brackish (*cf.* Fig. 1.10). Figure (**a**) based on data extracted from the National Oceanic and Atmospheric Administration database (http://www.noaa.gov; 20th Century Reanalysis V2c dataset), accessed on 27 November 2015. Figure: © Pauline Snoeijs-Leijonmalm

range of <0 to 3 °C. August is, on average, the warmest month of the year with surface-water temperatures in the range of 15 to 20 °C. Bottom-water temperature is rather stable throughout the year, ranging from 4–6 °C in the Baltic Sea proper to 2–4 °C in the Gulf of Bothnia (Fig. 2.16c).

Because of the density anomaly of its water (*cf.* Sect. 1.2.3), the Baltic Sea behaves more like a freshwater lake than an ocean with respect to the formation of a winter ice cover (Fig. 2.18). Salt decreases the freezing point, and marine water of salinity 35 freezes around −1.9 °C, but its maximum density is reached below that, at −3.25 °C (Fig. 2.17b). This means that vertical convection does not cease in an ocean like it does in a freshwater lake, and seawater does not freeze so easily, even if the air is cold. When a freshwater lake cools down below +4 °C (the maximum density of pure water), its vertical convection ceases and a thin upper water layer with lower density is formed, which freezes when the temperature decreases to 0 °C. Most of the Baltic Sea has the surface-water salinity between 5 and 8 (Fig. 2.15), and this water freezes around −0.4 °C, while its maximum density is reached around +2.5 °C (Fig. 2.17b). Therefore, the water remains at the surface when its temperature is between −0.4 and +2.5 °C, which speeds up further cooling and facilitates the formation of ice.

The surface water of the Bothnian Bay and the easternmost Gulf of Finland freezes every year (Fig. 2.19) and in the northern Bothnian Bay the ice cover persists for about half a year (Fig. 2.17a). Roads for car- and snow-scooter traffic are built on the ice in coastal areas, and previously this even made it possible to cross the Bothnian Bay between Sweden and Finland. The maximum extent of the ice cover in the Baltic Sea is normally reached in February-March.

More to the south, ice conditions vary extensively from one year to another. About once every decade in the period 1956–2005, the ice cover was so large that only a small area in the southern Baltic Sea remained ice-free. During this time period, the maximum extent of the ice cover varied between ~50,000 and 293,000 km^2, *i.e.* 13–100 % of the whole Baltic Sea, while the yearly average was ~218,000 km^2. During World War II several winters in a row were severe and the Baltic Sea froze over completely.

The melting season starts in March-April in the south, but in the northern Bothnian Bay the last sea ice usually does not disappear before June (Fig. 2.17a).

2.4.5 The Baltic Sea insolation gradient

For primary production, the number of daylight hours is important because this determines for how many hours per day photosynthesis is possible. The light-harvesting step of photosynthesis depends on light energy (until the saturation level, *cf.* Box Fig. 11.9), while carbon fixation is a temperature-dependent enzymatic process. However, primary production rates of phytoplankton are the outcome of the balance between light availability (including its penetration depth in the water column), temperature and the amount of available nutrients. Thus, longer days alone do not warrant higher primary production.

Because of the Baltic Sea's long latitudinal gradient, between 53°55′ N and 65°48′ N, day length varies significantly between the south and the north throughout the year. In the south, mid-winter day length is ~7.5 hours while mid-summer day length is ~17 hours (Fig. 2.20a). In the northernmost Baltic Sea, the seasonal difference is much more extreme, with ~3 hours of daylight around mid-winter and 24 hours of daylight around mid-summer. Between the equinoxes on 21 March and 23 September (with 12 hours of light and 12 hours of darkness everywhere), the north of the Baltic Sea receives more daylight hours than the south, while the south receives more daylight hours during the cold half of the year. Despite the longer days in the north, the total daily amount of light energy in summer does not vary much between the north and the south of the Baltic Sea, but in mid-winter it is ~90 % less in the north than in the south (Fig. 2.20b, c).

2.4.6 The photic zone

The photic zone is the upper part of the water column where the solar radiation is sufficient for photosynthesis to take place. The lower limit of the photic zone is generally defined as the depth to which 1 % of the sunlight penetrates (Kirk 2011). In clear ocean waters, the photic zone can be up to 200 m deep, but in the Baltic Sea it is only about one-tenth of that (Fig. 2.21).

Light penetration in natural waters is attenuated by a combination of coloured dissolved organic matter (CDOM), phytoplankton pigments and scattering by particles (*cf.* Sect. 15.2.3). The Baltic Sea is comparatively rich in CDOM, which mainly consists of humic substances, such as tannins, released from decaying plant detritus. The supply of CDOM from land is high, especially in the northern part of the Baltic Sea where boreal forests and bogs cover most of the land. Along the salinity gradient of the Baltic Sea, CDOM is inversely related to salinity (Fig. 2.22). However, the surface waters of the Kattegat and the Skagerrak are still significantly affected by CDOM in the outflow from the Baltic Sea (Stedmon et al. 2010).

In the Baltic Sea, the photic zone is around 20–25 m deep in the central parts of the subbasins and thins down to <5 m in coastal areas (Fig. 2.21). The photic zone is usually shallower near the coasts, partly due to the CDOM and sediment particles in the land runoff, but mainly because

Fig. 2.18 Winter conditions in the Baltic Sea. (**a**) The beginning of ice cover formation. (**b**) Snow-covered ice with open water in-between the ice sheets. (**c**) Snow-covered solid ice sheet; in winter, roads for car and snow-scooter traffic are made on the ice, especially in the Bothnian Bay. (**d**) A snow-covered beach in the southern Baltic Sea; the sea is covered with ice near the shore. (**e**) On board R/V Argos for seawater sampling in early March. (**f**) Diving to study perennial macroalgae in winter. Photo: (**a–c**, **e**) © Pauline Snoeijs-Leijonmalm, (**d**) © Hendrik Schubert, (**f**) © Lies Van Nieuwerburgh

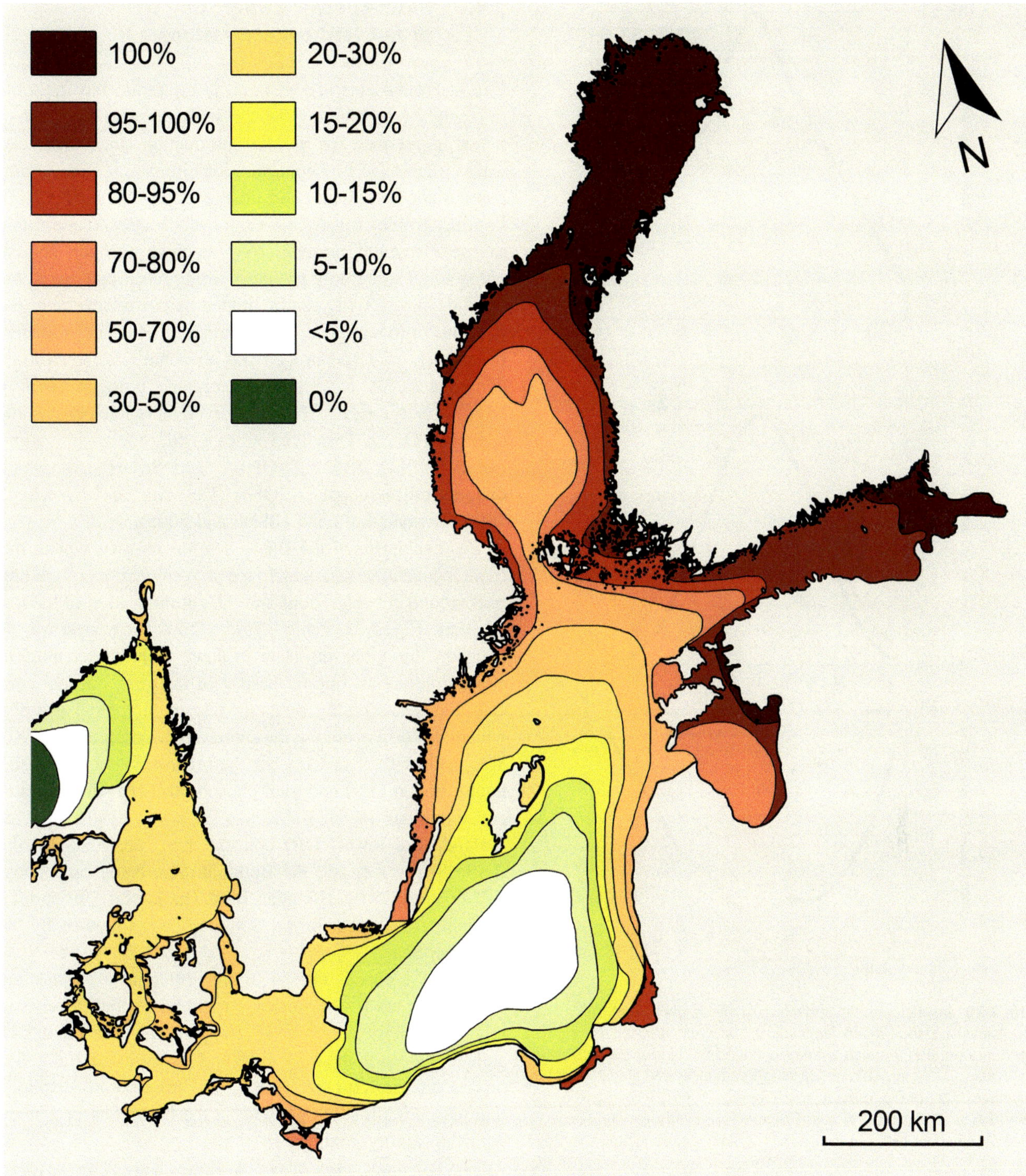

Fig. 2.19 The ice cover gradient of the Baltic Sea Area, shown as the average maximum winter ice coverage (in % of years) during 50 years (1956–2005). Figure modified from Schmelzer et al. (2008)

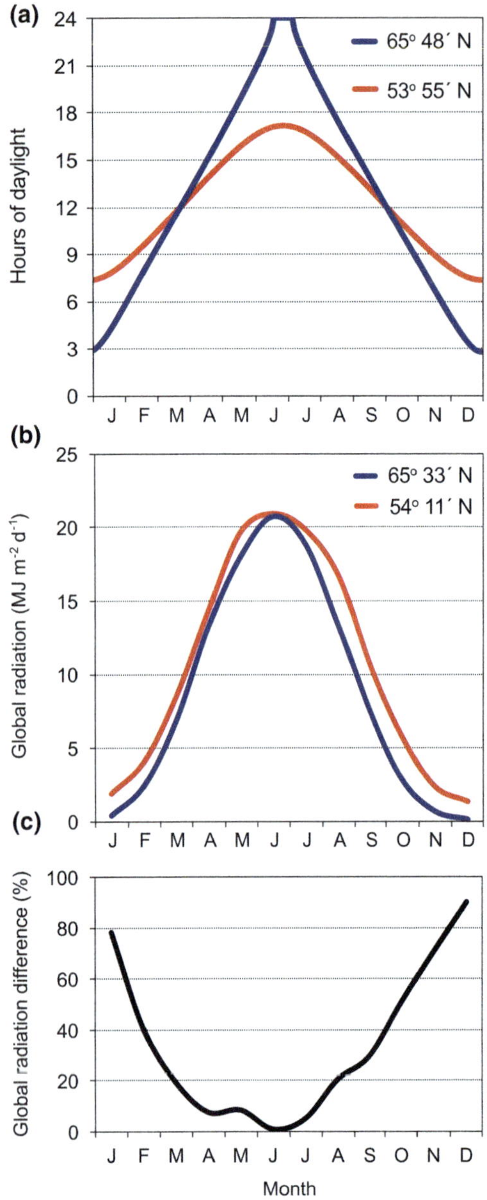

Fig. 2.20 Insolation in the northernmost and southernmost Baltic Sea. (**a**) Annual variation in day length during one year at 65°48′ N near the outflow from the Torne älv estuary and at 53°55′ N near the outflow of the Szczecin Lagoon. (**b**) Annual variation in measured global radiation (averages for 1965–2010, n = 46 years) at 65°33′ N near Luleå (Sweden) and at 54°11′ N near Kołobrzeg (Poland). (**c**) Difference in global radiation between Luleå and Kołobrzeg based on (**b**). Figure (**a**) based on data calculated with the Solar Calculator of the National Oceanic and Atmospheric Administration (http://www.noaa.gov), (**b**) based on data extracted from the Global Energy Balance Archive database (http://www.geba.ethz.ch), accessed on 25 November 2015. Figure: © Pauline Snoeijs-Leijonmalm

phytoplankton biomass is usually higher near the coasts than in the open waters as a result of nutrient emissions from land. The photic zone in shallow coastal areas also decreases when winds and waves whirl up particles from the sea bottom.

2.4.7 Basin-specific patterns of nutrient concentrations

The macronutrients nitrogen and phosphorus, and for some microalgae (*e.g.* diatoms and chrysophytes) also silica, are of major importance for biomass production since these elements often have low concentrations in natural waters compared to the other major constituents of organisms (C, O, H). For diazotrophic (nitrogen-fixing) cyanobacteria that are able to use elemental nitrogen (N_2) as their nitrogen source, phosphorus is the only growth-limiting macronutrient.

Thus, productivity can be limited by one or several of the nutrients nitrogen, phosphorus and silica. Primary producers most easily take up nitrogen and phosphorus as small ionic compounds, *i.e.* as dissolved inorganic nitrogen (DIN = NO_3^- + NO_2^- + NH_4^+) and dissolved inorganic phosphorus (DIP = PO_4^{3-}). Dissolved silica (DSi) occurs in natural waters as $SiO(OH)^{3-}$, $Si(OH)_2^{2-}$ and $Si(OH)_4$, of which diatoms utilise mainly $Si(OH)_4$ (Del Amo and Brzezinski 2000; Thamatrakoln and Hildebrand 2008).

Each subbasin of the Baltic Sea has its own typical nutrient concentrations, which are a combination of natural background concentrations and the eutrophication of the last 60 years (Fig. 2.23, Table 2.7, Box 2.2). The easiest way to compare the subbasins is to evaluate their winter nutrient concentrations as, during the rest of the year, nutrient concentrations are highly dynamic due to biological activity, which also differs among the subbasins (Granéli et al. 1990). The shallow Gulf of Riga has the highest winter DIN, DIP, total nitrogen (TN) and total phosphorus (TP) surface-water concentrations in the Baltic Sea, followed by the Gulf of Finland. The lowest DIN concentrations are found in the Baltic Sea proper and the Bothnian Sea, being only about half of those in the Bothnian Bay. The lowest DIP and TP concentrations occur in the Bothnian Bay, followed by the Bothnian Sea.

The DSi concentrations of the different subbasins are roughly related to the freshwater discharge they receive in relation to their water volume (compare Tables 2.2 and 2.4), and are highest in the three gulfs and lowest in the Baltic Sea proper and the Kattegat. Thus, DSi concentrations display a large-scale north-south gradient in the Baltic Sea in concert with the salinity gradient.

The DIN concentrations in the Kattegat are about twice as high as those of the Baltic Sea proper, while TN concentrations are similar (Fig. 2.23a, c). This is related to the higher CDOM levels in the Baltic Sea proper compared to the Kattegat (Fig. 2.22). CDOM contains dissolved organic nitrogen (DON), which photochemically can be made available to both heterotrophic and autotrophic plankton (Vähätalo and Järvinen 2007). Other processes that influence nitrogen-cycling, and which vary among the Baltic Sea subbasins, are nitrification and denitrification (*cf.* Sect. 3.6).

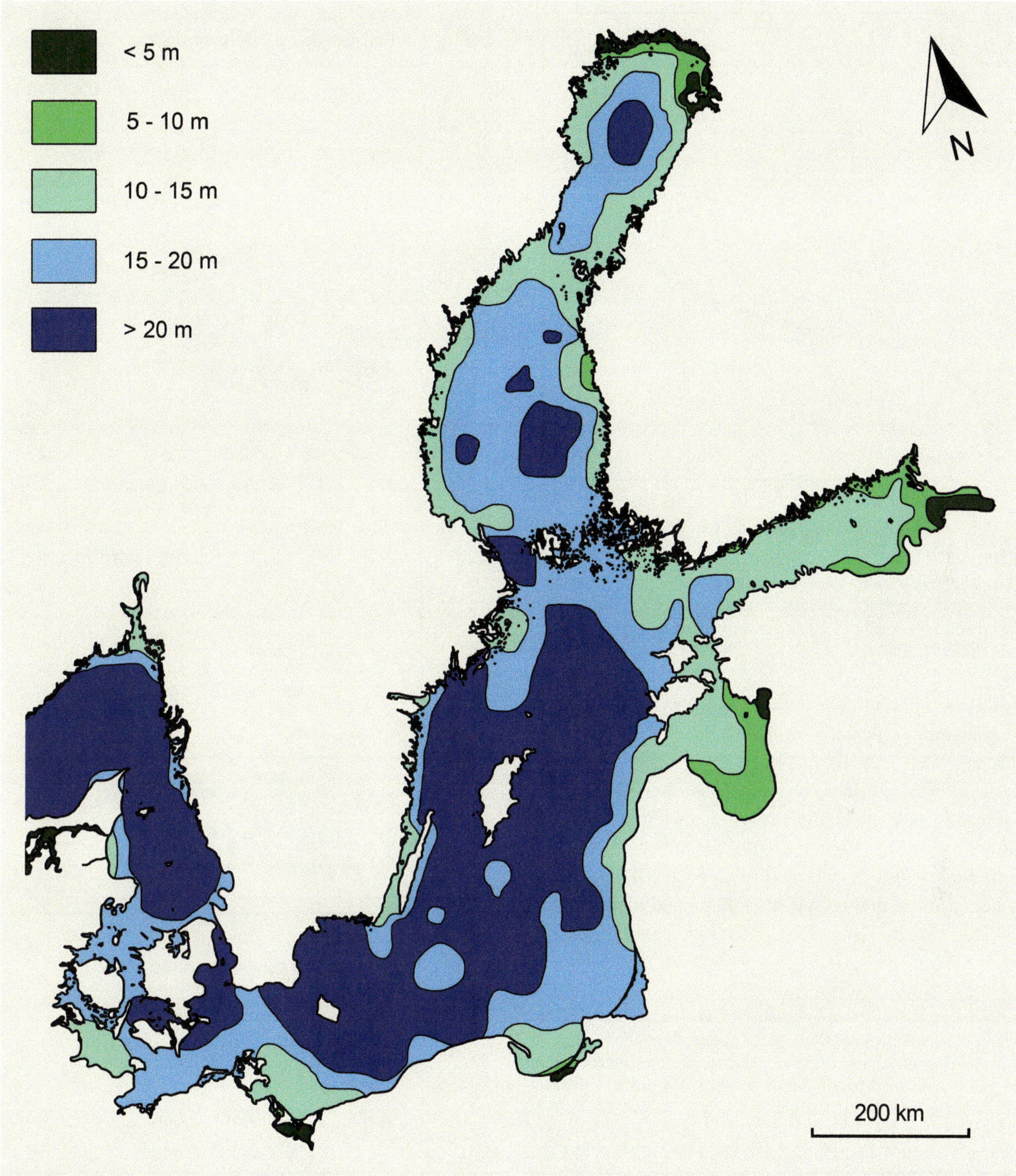

Fig. 2.21 Distribution of the photic zone depth (derived from Secchi depth data) in the Baltic Sea Area. Figure modified from Cameron and Askew (2011)

2.4.8 Basin-specific patterns of nutrient stoichiometry

Redfield (1934, 1958) discovered that the composition of marine particulate matter is relatively uniform and matching the metabolic demands of "average" phytoplankton. In due course, these findings were generalised to the universally accepted rule that the C:N:P molar ratio in phytoplankton is 106:16:1 (Redfield et al. 1963), known as the "Redfield ratio" for optimal phytoplankton growth (*cf.* Sect. 3.2.3). Later on this ratio was complemented with silicate to C:N:Si:P = 106:16:15:1 (Brzezinski 1985), known as the "Redfield-Brzezinski ratio" for optimal diatom growth. These ratios are widely used as reference levels for the assessment of nutrient depletion and nutrient repletion in primary producers and for the assessment of nutrient availability in aquatic

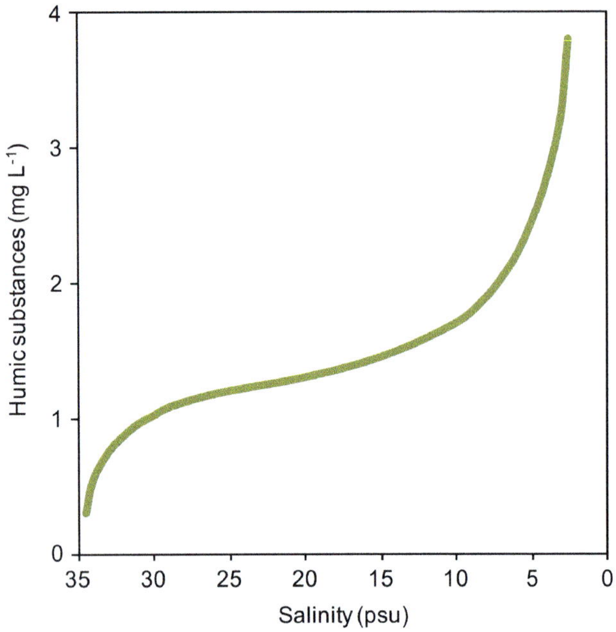

Fig. 2.22 The relationship between salinity and humic substances (CDOM) along the large-scale Baltic Sea gradient from salinity ~34 in the Skagerrak to ~2 in the Bothnian Bay. Figure modified from salinity Fonselius (1995)

environments, although the ratios may vary quite significantly in nature, especially in lakes (Hecky et al. 1993; Deutsch and Weber 2012).

Besides typical nutrient concentrations, the subbasins of the Baltic Sea also have their own characteristic nutrient stoichiometry, and – together with the absolute nutrient concentrations – this results in the chlorophyll *a* concentration being lower in the Gulf of Bothnia and higher in the Baltic Sea proper (Fig. 2.24). Eutrophication due to

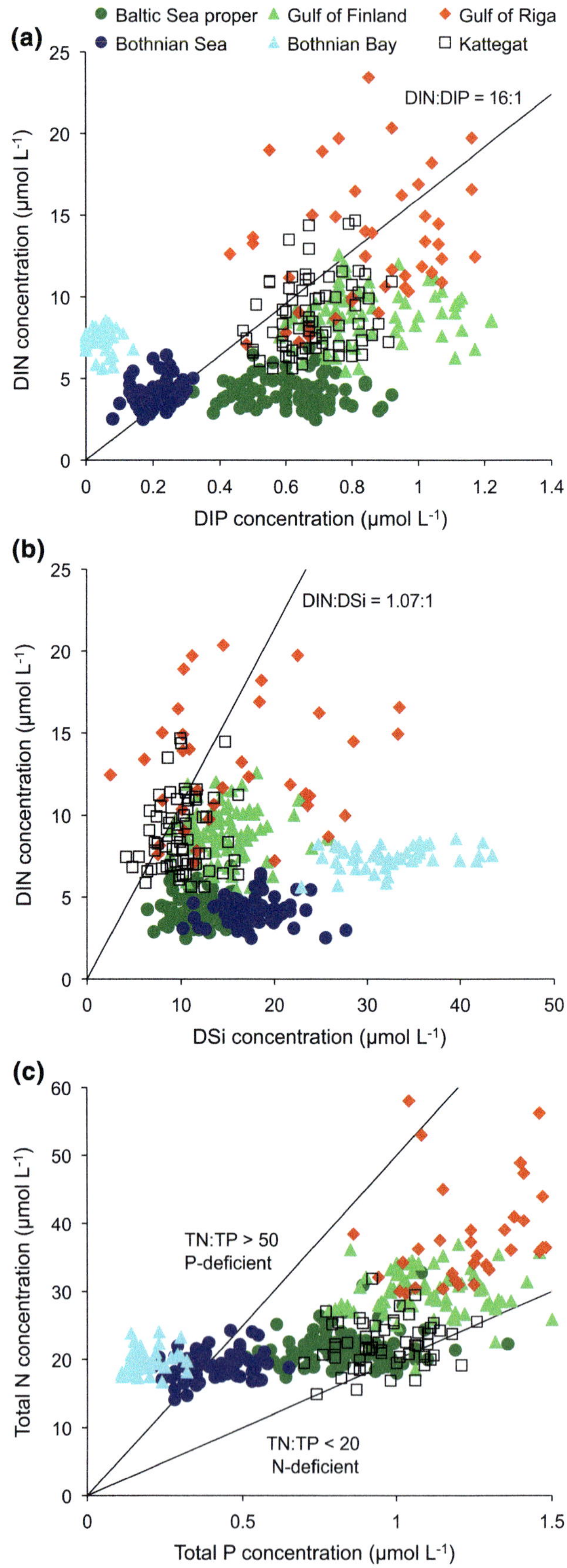

Fig. 2.23 The relationship between winter surface-water nutrient concentrations in the different subbasins of the Baltic Sea Area in 1980–2012. Note that these concentrations change throughout the year. Each score represents the highest reported winter concentration (January, February or March) for one year. For each subbasin, representative monitoring stations are included (Kattegat: GF4, GF8, Fladen; Baltic Sea proper: BY1, BY15, BY20; Gulf of Riga: G1, 119, 123; Gulf of Finland: F1, LL5, LL7; Bothnian Sea: F18, F26, EB1; Bothnian Sea: F3, B03). (**a**) Dissolved inorganic nitrogen (DIN) and dissolved inorganic phosphorus (DIP). The line represents the Redfield ratio (Redfield et al. 1963). (**b**) Dissolved inorganic nitrogen (DIN) and dissolved inorganic silica (DSi). The line represents the Redfield-Brzezinski ratio for optimal diatom growth (Brzezinski 1985). (**c**) Total inorganic and organic nitrogen, including particulate forms (TN) and tota; inorganic and organic phosphorus, including particulate forms (TP). The lower line represents TN:TP = 20, the upper line represents TN:TP = 50 indicating N- and P-deficiency limits, respectively, according to Guildford and Hecky (2000). Figure based on measurements accessed from the major databases around the Baltic Sea with the decision support system Baltic Nest (http://nest.su.se; Wulff et al. 2013). Figure: © Pauline Snoeijs-Leijonmalm

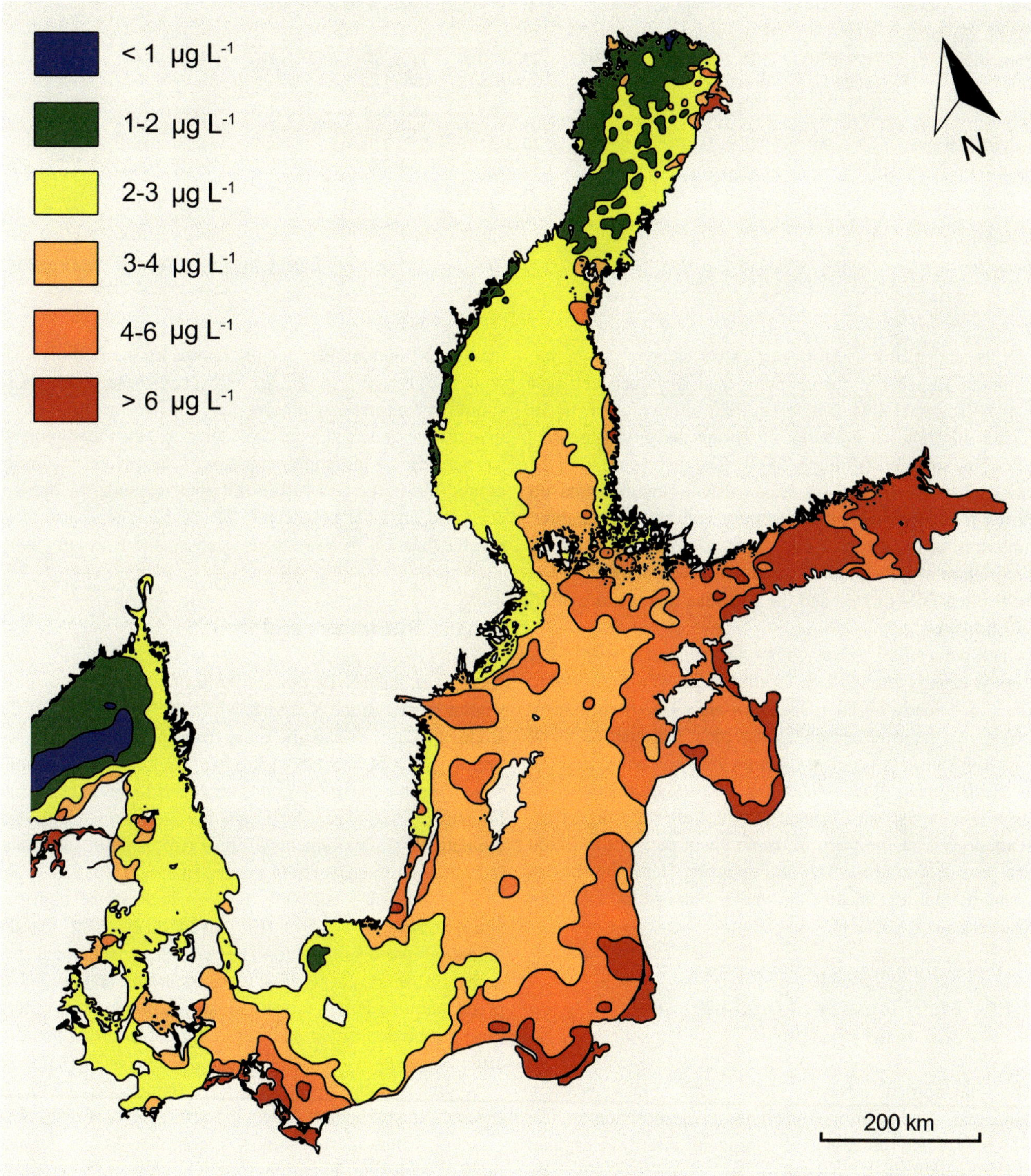

Fig. 2.24 Average chlorophyll *a* concentrations in the Baltic Sea Area in July–August 2004. Figure modified from chlorophyll *a* concentrations from the SeaWiFS satellite (http://oceancolor.gsfc.nasa.gov)

enrichment with nitrogen and phosphorus, which is superimposed on the natural nutrient dynamics, is most pronounced in the Gulf of Finland and the Gulf of Riga. Low winter surface-water DIN:DIP ratios of 7 and 11 indicate nitrogen limitation in the Baltic Sea proper and the Gulf of Finland, respectively (Table 2.7, Fig. 2.23a). In contrast, the average winter DIN:DIP ratio of 203 in the Bothnian Bay indicates a strong phosphorus limitation and in this subbasin

Table 2.7 Average winter nutrient concentrations and stoichiometry in the surface waters in six subregions of the Baltic Sea Area based on the data shown in Fig. 2.23

Subregion	DIN ± SD (μmol L^{-1})	DIP ± SD (μmol L^{-1})	DSi ± SD (μmol L^{-1})	TN ± SD (μmol L^{-1})	TP ± SD (μmol L^{-1})	DIN:DIP	DIN:DSi	TN:TP
Kattegat	8.9 ± 2.3	0.69 ± 0.11	10 ± 3	22 ± 3	0.99 ± 0.21	13 ± 3	0.94 ± 0.32	23 ± 5
Baltic Sea proper	4.2 ± 0.8	0.61 ± 0.13	12 ± 2	22 ± 3	0.83 ± 0.16	7 ± 2	0.36 ± 0.10	27 ± 6
Gulf of Riga	14.7 ± 7.5	0.86 ± 0.21	17 ± 9	39 ± 11	1.25 ± 0.26	18 ± 8	1.08 ± 0.80	31 ± 8
Gulf of Finland	8.9 ± 1.6	0.85 ± 0.21	15 ± 4	30 ± 3	1.13 ± 0.21	11 ± 3	0.62 ± 0.17	27 ± 5
Bothnian Sea	4.1 ± 0.8	0.21 ± 0.05	18 ± 3	19 ± 2	0.41 ± 0.10	20 ± 6	0.24 ± 0.06	50 ± 12
Bothnian Bay	7.4 ± 0.7	0.05 ± 0.03	32 ± 5	20 ± 2	0.19 ± 0.06	203 ± 111	0.23 ± 0.04	116 ± 30

DIN is underutilised. However, estuarine systems in the Bothnian Bay are more complex than the open sea and display seasonal switches in nutrient limitation with in general phosphorus limitation in spring and nitrogen limitation in summer (Conley 2000). The average DIN:DIP ratios of the Gulf of Riga, Kattegat and Bothnian Sea lie closest to the Redfield ratio. The most variable of these three subbasins is the Gulf of Riga with a more or less equal distribution of DIN and DIP limitation, while the Kattegat is more often DIN-limited and the Bothnian Sea is more often DIP-limited.

The winter DSi values show that, compared with DIN, there is usually enough DSi for diatom growth in the Baltic Sea, since nearly all DIN:DSi ratios are <1.07 (Fig. 2.23b). DSi shows the same seasonal variation as the other nutrients with low values in summer and high values in winter, except in the Bothnian Bay, where low primary production and large DSi inputs via freshwater cannot alter the high concentrations over the year. In the southern part of the Baltic Sea proper, the situation is the opposite. Here, the diatom spring bloom can consume the whole winter period DSi pool (Wulff et al. 1990).

2.4.9 Stoichiometry of total nitrogen and total phosporus

The total nutrient reservoir in the water column is often expressed as total nitrogen (TN) and total phosphorus (TP), which includes the dissolved inorganic, dissolved organic and the particulate forms of N and P, respectively. The particulate forms include the N and P bound in plankton organisms. A large meta-analysis comparing the TN and TP ratios in lakes and oceans with respect to nutrient limitation found that N-deficient growth is apparent at TN:TP <20 and P-deficient growth at TN:TP >50, while at intermediate TN:TP ratios either N or P can become deficient (Guildford and Hecky 2000).

The winter TN and TP concentrations in the Baltic Sea subbasins fall mostly within the intermediate range, except for the whole Bothnian Bay and partly also for the Bothnian Sea with TN:TP >50 (Fig. 2.23c). However, these observations should be interpreted with some caution. There are two aspects in particular that need to be considered: (1) the measurements were made by different laboratories in several countries around the Baltic Sea (using different methods for nutrient analyses), and (2) in general, TP measurements are more reliable than TN measurements (Hansen and Koroleff 1999).

2.4.10 Phosphate and iron

Altogether, the TN:TP ratio of the Baltic Sea proper is most similar to that of the Kattegat, while the TN:TP ratio of the Bothnian Bay deviates the most from the rest of the Baltic Sea and resembles that of a lake (Fig. 2.23c). In the temperate zone, near-neutral freshwater lakes tend to be phosphorus-limited and coastal seas tend to be nitrogen-limited, although the nutritional requirements of the phytoplankton are similar in both environments (Hecky and Kilham 1988).

This difference can partly be explained by the high sulphate content of sea salt (Blomqvist et al. 2004). In the oxidative hydrolysis of iron (Fe) and the concomitant precipitation of PO_4^{3-}, at least two Fe atoms are needed to precipitate one PO_4^{3-} ion. In anoxic marine bottom waters, Fe:P <2 predominates and some PO_4^{3-} is left in solution after oxygenation due to a shortage of dissolved Fe for PO_4^{3-} co-precipitation by iron oxyhydroxide [FeO(OH)]. In contrast, anoxic bottom waters in most freshwater lakes have Fe:P >2, which allows an almost complete PO_4^{3-} removal upon oxygenation. Thus, the general bottom-water chemistry in the habitat gradient from limnic to marine shows higher phosphorus availability in marine waters, primarily because of an enhanced iron sequestration by sulphides.

The regional variation in phosphate concentrations between the Baltic Sea subbasins can be further explained by eutrophication-driven microbial processes in sediments. In the nutrient-loaded Gulf of Finland and Baltic Sea proper, the sediments appear to have reached a state in which sulphate reduction is the dominant mineralisation pathway (Lehtoranta et al. 2008). This implies that the capacity of the sediments to retain phosphorus is limited and high amounts of bioavailable PO_4^{3-} occur in the water column (*cf.* Sect. 3.6.7). The bottom sediments of the Gulf of Bothnia are still in a state in which iron reduction and coupled cycling of iron and phosphorus prevail in the surface sediments.

2.4.11 Patterns of primary production

Even if it partly depends on the light climate in the water, the concentration of the photosynthetic light-harvesting pigment chlorophyll *a* (Chl *a*) is often used as a proxy for the total biomass of primary producers. The phytoplankton Chl *a* concentration can be estimated from the colour of the sea by satellites, and can thus be monitored over large geographical areas at relatively short time intervals (*cf.* Sect. 15.1). As a result of changes in primary production, the Chl *a* concentration varies between seasons and years (*cf.* Sect. 8.2), and it can even vary significantly on a weekly basis.

The Baltic Sea displays an obvious spatial pattern of Chl *a* concentration, which is mainly shaped by natural nutrient conditions and anthropogenic nutrient inputs. In the Gulf of Bothnia, the Chl *a* concentration is generally low, while it is high in the Gulf of Finland, the Gulf of Riga, the Curonian Lagoon, the Gdańsk Bay and the northeastern coast of Germany (Fig. 2.24). Usually, the Chl *a* concentration is higher in coastal waters than in the open sea because of nutrient emissions from point sources (*e.g.* river discharges, cities) and diffuse sources (*e.g.* land erosion, agriculture). Another factor that influences the spatial distribution of the Chl *a* concentration is the water circulation pattern in the Baltic Sea (Fig. 2.10). For example, along the western coast of the Bothnian Sea the concentration is normally lower than along its eastern coast. This is partly caused by the transport of nutrient-rich water northwards along the Finnish coast by average currents, while nutrient-poor water is transported southwards along the Swedish coast (Fig. 2.24).

Given the climatic gradient of the Baltic Sea, the growing season is much shorter in the Bothnian Bay than in the southern Baltic Sea. This is, next to low nutrient concentrations (Fig. 2.23), a major factor that limits the annual phytoplankton primary production in the Bothnian Bay, which is as low as 17–28 g C m^{-2} $year^{-1}$ (Fig. 2.25). The values for the Baltic Sea proper (67–163 g C m^{-2} $year^{-1}$) overlap with the average net primary production in the ocean, which is estimated at 140 g C m^{-2} $year^{-1}$ (Field et al. 1998). In comparison with estuarine-coastal systems worldwide, the average net phytoplankton primary production in the Baltic Sea region (112 g C m^{-2} $year^{-1}$) is lower than the average in estuarine-coastal systems (252 g C m^{-2} $year^{-1}$, Cloern et al. 2014). Using the classification of Nixon (1995), the Bothnian Bay and the Bothnian Sea are

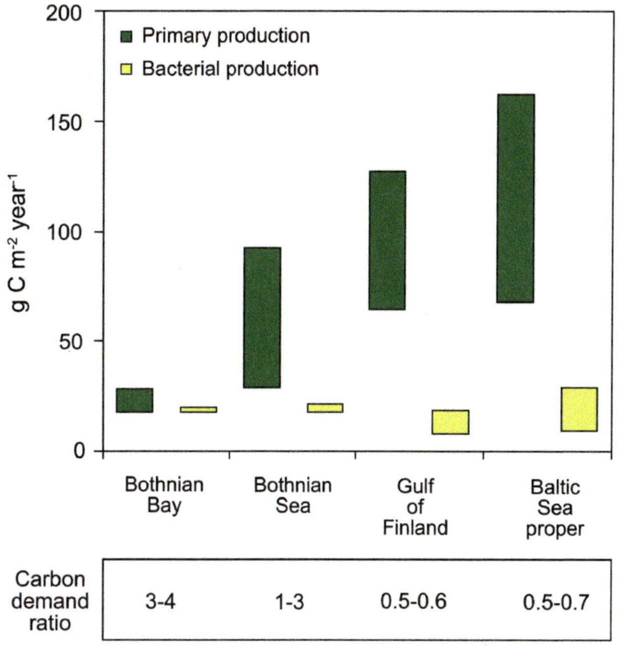

Fig. 2.25 Ranges of Primary production, bacterial production and carbon demand ratio (bacterial carbon demand:primary production, calculated with a bacterial growth efficiency of 25 %) in the pelagic zones of the four major basins of the Baltic Sea. Figure based on data calculated from Table 7.2 in Hagström et al. (2001). Figure: © Pauline Snoeijs-Leijonmalm

oligotrophic (<100 g C m^{-2} $year^{-1}$) while the rest of the Baltic Sea is mesotrophic (100–300 g C m^{-2} $year^{-1}$), although eutrophic (300–500 g C m^{-2} $year^{-1}$) and hypertrophic (>500 g C m^{-2} $year^{-1}$) conditions may occur locally in coastal areas with a high nutrient load.

2.4.12 Symptoms of eutrophication

Increases in primary production strongly influence ecosystem functioning (Nixon and Buckley 2002). For example, the eutrophication of the Baltic Sea that started in the 1950s (Box 2.2) seems to be the main reason for an average 15-fold (median 4-fold) increase of benthic animal biomass above the halocline (Cederwall and Elmgren 1980) and an eight-fold increase of fish biomass in the Baltic Sea with peak values in the 1970s–1980s (Thurow 1997), in addition to decreased benthic animal biomass (Cederwall and Elmgren 1980), increased supply of organic matter and a decline in oxygen concentrations below the halocline (Carstensen et al. 2014). Simultaneously, the frequency and intensity of potentially toxic cyanobacterial blooms (*cf.* Box 16.4) in summer have increased since the 1960s (Finni et al. 2001).

In the case of the Baltic Sea, the nutrient inputs exceed the natural processing capacity of the ecosystem with an accumulation of nutrients, while it is naturally susceptible to nutrient enrichment due to a combination of long retention times and stratification restricting ventilation of deep waters (Diaz and Rosenberg 2008; Andersen et al. 2015).

2.4.13 Heterotrophy dominates in the Gulf of Bothnia

Organic matter of terrestrial origin may serve to compensate for the low primary production in the northern Baltic Sea, *i.e.* in the Bothnian Bay and the coastal Bothnian Sea. The carbon budgets for these areas show a higher bacterial carbon demand than can be supported by primary production (Kuparinen et al. 1994; Zweifel et al. 1995). In the Bothnian Bay, the bacterial carbon demand is four times that of the available carbon produced by autotrophs on an annual basis (Fig. 2.25). In fact, the Bothnian Bay resembles a subarctic lake more than a marine environment and should be considered as net heterotrophic (Hagström et al. 2001). In the Bothnian Sea, the carbon demand ratio is lower than in the Bothnian Bay, and in the Baltic Sea proper the primary production supports the bacterial carbon demand.

2.5 Geological and climatic background

2.5.1 Geology, climate and the distribution of organisms

The geological development of the Baltic basin, along with profound climate changes during the Holocene, has left its traces in the biota of the Baltic Sea in the form of evolutionary adaptations. This is especially apparent for the "glacial relicts", which are descendants of species that survived in the area after glaciation events (*cf.* Sect. 4.6.2). Since ~20,000 years before present (BP), the Baltic basin has experienced total glacial coverage and several alternating freshwater and brackish-water stages thereafter, including a brackish sea with higher salinity than the present Baltic Sea.

Using a variety of methods (Box 2.3), geologists have been able to reconstruct the environment during the different geological stages of the Baltic basin. Throughout the geological development, organisms invaded and became extinct or adapted to the environmental conditions characteristic of the different stages.

During the last 3,000 years, the environmental gradients in the Baltic basin have been rather stable, although some geological processes are still ongoing (notably the isostatic rebound) and some natural climatic fluctuations have been rather prominent (*e.g.* the "Medieval Climate Anomaly" and the "Little Ice Age"). Today's large-scale changes in the Baltic Sea environment caused by anthropogenic activities (*cf.* Sect. 17.2) occur incredibly fast compared to most natural geological, climatic and evolutionary changes.

2.5.2 Eustatic sea level rise and isostatic rebound

The late glacial and post-glacial development that shaped the present Baltic Sea was governed by interactions between the eustatic sea level rise and isostatic rebound. Eustasy refers to changes in the amount of water in the oceans. With increasing temperature, the water volume of the oceans increases by the melting of ice and snow as well as by thermal expansion, and *vice versa*. Isostasy refers to the gravitational equilibrium between the lithosphere and the viscous mantle of the Earth on which the tectonic plates are "floating". When a heavy glacier (ice sheet) presses down the lithosphere by its weight, it takes a long time for the lithosphere to rebound to isostasy (the land rising) when the glacier melts, and even after it has melted.

As a consequence of the processes of eustatic sea level rise and isostatic rebound, the Baltic basin went through different stages during a geologically and evolutionarily short period of time (Björck 2008). When the climate became warmer ~15,000 years BP, the eustatic sea level rise and isostatic rebound followed the gradual melting of the Weichselian ice sheet that covered northern Europe during the latest ice age. Melting of the Earth's ice sheets caused a 120-m sea level rise (Fig. 2.26a). The isostatic rebound after the heavy load of the ice sheet on Scandinavia is still ongoing, by up to ~9 mm per year in the northern part of the Baltic Sea (Fig. 2.26b). The forces in the lithosphere that cause the land uplift in the northern part of the Baltic basin result at the same time in a land subduction of ~1 mm per year in the southern part of the basin.

Since the sea level rise and land uplift opened and closed the connection with the ocean several times, the salinity of the water in the Baltic basin fluctuated between that of freshwater and brackish water during the last 12,000 years (Fig. 2.26c). The shoreline displacement, caused by an interaction between the isostatic rebound and eustatic sea level rise, in southern Sweden over the last 14,000 years BP has been estimated at ~70 m, while in the north the highest shoreline is situated 285 m above the sea level (Fig. 2.26d).

2.5.3 The Eemian interglacial

During the Eemian interglacial (~130,000–115,000 years BP), the Baltic basin developed in a manner comparable to that during the present interglacial Holocene (~11,600–0 years BP). However, the Saalian ice sheet (preceding the Eemian interglacial) was thicker and heavier than the Weichselian ice sheet (succeeding the Eemian interglacial), and the subsequent isostasy was larger. This resulted in a predecessor of the Baltic Sea, the Eemian Sea, which had a short-lived connection to the Barents Sea via the White Sea

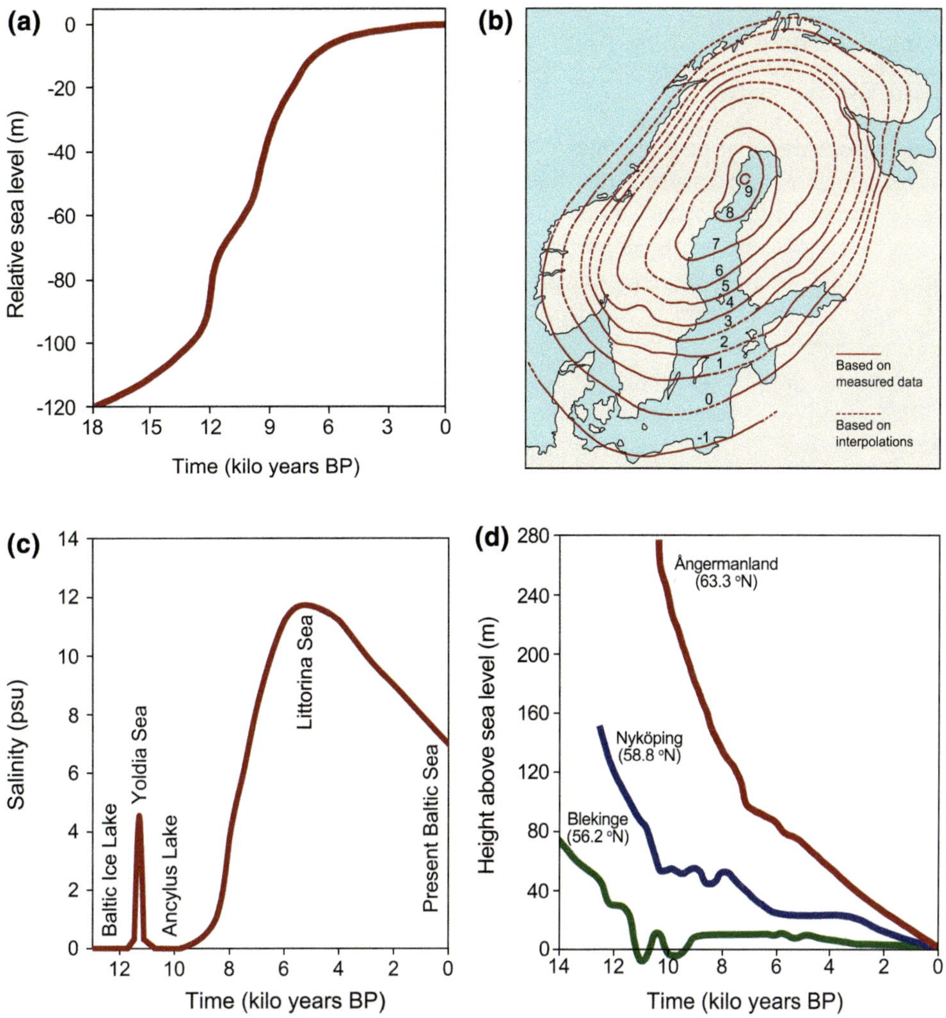

Fig. 2.26 Late-glacial and post-glacial changes in environmental conditions. (**a**) The eustatic sea level rise measured at Barbados. (**b**) The apparent post-glacial uplift of Fennoscandia, here shown for the period 1892–1991 in mm per year, based on sea level records, lake-level records and repeated high-precision levellings. (**c**) Tentative reconstructed surface-water palaeosalinity in the open Baltic Sea proper during the last 13,000 years, based on diatom frustules (data: Elinor Andrén) and different published proxy data, *e.g.* diatom, mollusc, silicoflagellate and cyanobacteria distributions, geochemical measurements (Gustafsson and Westman 2002) and strontium isotopes (Widerlund and Andersson 2006). (**d**) Shoreline displacement over time at three places in Sweden at different latitudes. Figure (**a**) modified from Fairbanks (1989), (**b**) modified from Ekman (1996), (**c**) Figure: © Elinor Andrén, (**d**) modified from Fredén (1994)

in addition to the westerly connection with the Atlantic Ocean. At that time, the Baltic basin experienced more marine conditions than today (Andrén et al. 2011). Since then, the Baltic basin has not been connected to the Arctic Ocean, but there have been several connections with the Atlantic Ocean in the west.

2.5.4 The Baltic Ice Lake: a dammed melt-water lake

The Weichselian ice sheet had its maximum areal extension around 20,000 years BP and covered at that time the whole Baltic basin and surrounding lands, from the northern half of today's Great Britain and Ireland up to Svalbard, Novaya Zemlya and Franz Josef Land (Svendsen et al. 2004). When the warming of the global climate resulted in deglaciation, a melt-water lake, the Baltic Ice Lake, was formed in the Baltic basin in front of the receding ice. The Baltic Ice Lake (Fig. 2.27a) lasted from ~16,000 until ~11,700 years BP (Andrén et al. 2011).

The Baltic Ice Lake was a dammed freshwater lake in which glacial clays and silts were deposited. Seasonal cyclicity in the melt-water discharge, with high discharge in spring-summer and no melt water in winter, resulted in the deposition of laminated clays in the areas proximal to the ice sheet. These clays normally have thick, silty light-coloured summer layers and thin, clayey dark-coloured winter layers.

Box 2.3: Geological methods

Elinor Andrén and Thomas Andrén

Different types of data witness of past changes

The present understanding of the post-glacial development of the Baltic basin is based on a variety of multiproxy data derived from a wide range of sedimentological, geochemical and biological techniques. The deglaciation of the Weichselian ice sheet was initially investigated from terrestrial traces of the ice sheet's movement and its melting such as glacial erratics, glacial striation, moraine ridges, eskers and the lithological composition of tills. A time scale following the Weichselian deglaciation in Sweden, known as the "Swedish Time Scale", was constructed by measuring, counting and correlating the clay varves deposited in front of the receding ice sheet. By studying raised beach ridges and varved glacial clays, which contained fossil molluscs (Box Fig. 2.3), an account of the early Baltic Sea stages with shifting salinity was revealed in the late 19th century. As technology developed, it became possible to investigate bedrock and seabed sediments from research vessels by using hydroacoustic and coring equipment (Box Fig. 2.4), and this is the major approach used today.

How to find coring sites

The use of hydroacoustic profiling (*e.g.* with seismics, echo sounder or side scan sonar) enables us to follow the areal extension of lithological units below the seafloor and to identify the most suitable coring locations. Sediment cores serve as archives of past events and are used to study changes in the palaeo-environment and climate. Essential for all stratigraphical research are correct age determination and models. To achieve this, a number of different dating methods are available, ranging from radiometric (*e.g.* radiocarbon, uranium-series, ^{137}Cs, ^{210}Pb, optically stimulated luminescence) and annually laminated records (*e.g.* varved glacial clays) to tephrochronology and palaeo-magnetism.

Sediment properties

Multi-sensor core logging is a non-destructive geophysical method used on sediment cores, enabling continuous measurements of γ-ray attenuation, p-wave velocity, and magnetic susceptibility. Sediment properties such as density, porosity and water content can then be calculated, and this provides information on the origin of the sediments and if sedimentation is continuous. Such data can also be used for making correlations between different sediment cores.

Box Fig. 2.3 Huge amounts of marine shells were deposited in the nutrient-rich and turbulent environment at the narrow threshold area between the Yoldia Sea and the Kattegat ~11,000 years ago. The deposits, now situated outside the town of Uddevalla in western Sweden, were visited by Carl von Linné in 1746 (Linnaeus 1747). He wrote: "Skalbärgen räknas med rätta ibland et af de största Bohus-Läns under, ty de ligga uppå Landet, nästan hela quarten på somliga ställen, ifrån hafwet. Desse Skalbärgen bestå af Snäcke- och Mussle-skal, som här äro samlade i den myckenhet, at man kan undra det så många lif lefwat i werlden". This roughly translates as: "The shell hills are rightly considered one of the largest wonders of Bohuslän as they are located on land far from the sea. These shell hills consist of sea snails and bivalves, which are gathered in such high abundance that one can wonder if there ever have been so many lives living in the world". These fossil deposits are today protected in a nature reserve and a small museum informs the public of the site's extraordinary past development. Photo: © Elinor Andrén

Biological and geochemical proxies

Several biological proxies, fossil remains of organisms preserved in the sediments, are used to assess changing environmental conditions over time, *e.g.* salinity, water depth, primary production and climate. A wide range of fossils can be found preserved in sediments, *e.g.* diatoms, molluscs, ostracods, foraminifers, silicoflagellates, dinoflagellate cysts, chrysophyte cysts, plant remains, fish skeletons and otoliths. The most widely used fossils for palaeo-environmental studies in sediments from the offshore Baltic Sea are diatom frustules. Geochemical measurements, such as the contents of organic carbon, biogenic silica and lipid and pigment biomarkers, are used to reveal *e.g.* changes in vegetation dynamics, primary production, salinity and water temperature.

Radioactive and stable isotopes

Isotopes are atoms of the same element, *i.e.* with the same number of protons in their nucleus, but with a different number of neutrons. Isotopes are radioactive when they have an unstable nucleus with excess energy, and these isotopes are subject to radioactive decay with time. Other isotopes have a stable nucleus and they do not change over time. A number of radioactive and stable isotopes are essential in geological research. Since the 1950s, radiocarbon dating has been used to determine the age of organic materials of up to ~50,000 calendar years old. Radiocarbon (^{14}C) forms in the upper atmosphere through the interaction between nitrogen and neutrons from cosmic rays. Plants fix all isotopes of atmospheric carbon in photosynthesis and the level of ^{14}C in living matter is in equilibrium with the ^{14}C levels in the atmosphere. However, after death of an organism the radioactive decay of ^{14}C into ^{14}N starts with a half-life of ~5,730 years and the ^{12}C:^{14}C ratio increases. By measuring the ^{12}C:^{14}C ratio in a material with organic origin (*e.g.* fossils, sediment and wood) it is possible to estimate the age of the material. Radiocarbon years need to be calibrated since the concentration of radiocarbon in the atmosphere has varied over time and is expressed as calendar years before present (BP), which corresponds to years before 1 January 1950. This date was agreed upon as a standard since nuclear weapons tests have changed the proportion of carbon isotopes in the atmosphere during the last ~60 years.

Isotopic fractionation

The deviating atomic weights of stable isotopes can cause isotopic fractionation. Processes in nature (*e.g.* varying temperature) can affect the relative abundance of isotopes of the same element and the ratio between heavy and light isotopes can be used to trace the process in nature. The stable oxygen ($\delta^{18}O$ = the ^{18}O:^{16}O ratio) and carbon ($\delta^{13}C$ = the ^{13}C:^{12}C ratio) isotope tracers in foraminifers are among the most important proxies in global palaeo-oceanography and powerful tools for palaeo-climatic reconstructions. Stable oxygen isotopes enable *e.g.* the reconstruction of past ocean temperatures, global ice volume, ocean circulation, river discharge, and surface-water salinity. Stable carbon ($\delta^{13}C$) and nitrogen ($\delta^{15}N$ = the ^{15}N:^{14}N ratio) isotopes are used to trace palaeo-primary production and carbon and nitrogen sources (*e.g.* to assess the proportion of material with terrestrial origin). Several other stable isotopes such as silica, strontium and sulphur are also used and show potential to improve palaeo-environmental interpretations and reconstructions.

Box Fig. 2.4 A gravity corer is used for sampling the most recent soft sediments. The number of weights on the corer can be adjusted to fit the softness of the sediment. When longer sediment sequences are required, a piston corer is used to secure coring of undisturbed sediments. (**a**) The corer is lifted up from the water. (**b**) The lower end of the corer is closed with a lid. (**c**) The sediment core is retrieved from the corer. Photo: © Elinor Andrén

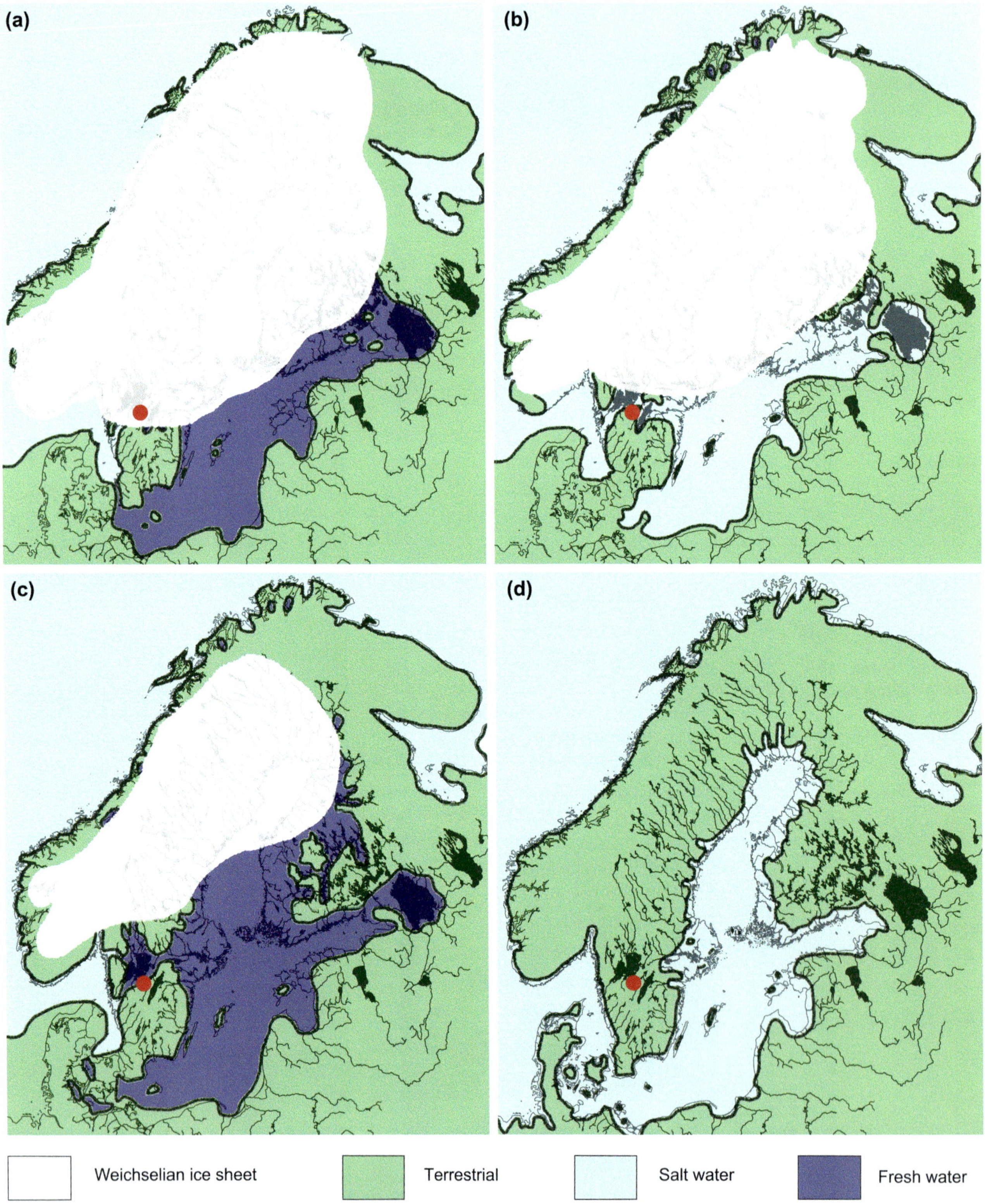

Fig. 2.27 Palaeo-geographic maps of the Baltic basin with the configuration of the present Baltic Sea and the major lakes and rivers in the present drainage area shown as background layer. (**a**) The Baltic Ice Lake just prior to its maximum extension and final drainage at ~11,700 years BP. (**b**) The Yoldia Sea at the end of the brackish phase at ~11,100 years BP. (**c**) The Ancylus Lake during its maximum transgression at ~10,500 years BP. (**d**) The Littorina Sea during the most saline phase at ~6,500 years BP. The red dot indicates the location of Mount Billingen. Figure modified from Andrén et al. (2011)

One clay varve (a sequence of one summer and one winter layer) reflects the deposition during one year. These varved glacial clays have been used to construct a deglaciation chronology named the "Swedish Time Scale" (De Geer 1912). In areas farther away from the ice margin, the seasonality was less pronounced and homogeneous clays were deposited.

The water transparency in the Baltic Ice Lake was probably low due to the heavy load of sediments from the melting ice sheet. Light conditions in the water, and possibly also a lack of nutrients, were unfavourable for the growth of photosynthetic organisms. Consequently, the Baltic Ice Lake was a barren water body with extremely low biological production (Winterhalter 1992). This is recorded in the sediments deposited during this time as a low organic carbon content and the absence of fossils.

During the early stage of the Baltic Ice Lake, its water level was the same as that of the ocean, which was ~100 m lower than at present due to the enormous amount of water still bound by the global ice sheets (Fig. 2.26a). The Baltic Ice Lake had an early outlet to the ocean in the Öresund area (Fig. 2.27a), where the easily eroded Quaternary deposits were superimposed on chalk. As soon as the erosion of the threshold in Öresund reached the bedrock ~14,000 years BP, the erosion ceased and the water level in the Baltic Ice Lake started to rise (Björck 2008).

2.5.5 Several drainages of the Baltic Ice Lake

The Baltic Ice Lake was dammed by the Scandinavian ice sheet at the northern point of Mount Billingen in south-central Sweden (Fig. 2.27a). When the ice sheet receded northwards, a connection between the lake, now at an altitude of ~10 m above the sea level, and the North Sea was created in south-central Sweden. This first drainage of the Baltic Ice Lake, possibly subglacial, occurred ~13,000 years BP (Björck 1995). There is no evidence of a marine inflow into the Baltic basin at that time (Andrén et al. 2011). The drainage ceased when a climatic cold event, the Younger Dryas, ~12,800 years BP, caused a re-advance of the ice sheet and closed the drainage path at Mount Billingen, so that the Baltic Ice Lake became dammed once again.

Because of the isostatic rebound, the threshold area was rising faster than the eustatic sea level, and the Baltic Ice Lake was drained by a waterfall in the Öresund area (Björck 2008). At the very end of the Younger Dryas cold event at ~11,700 years BP, a second, much more dramatic, drainage of the Baltic Ice Lake took place at the northern point of Mount Billingen, and lowered the water level of the Baltic Ice Lake by 25 m down to the oceanic water level within 1–2 years (Andrén et al. 2011).

2.5.6 The Yoldia Sea: at the level of the world oceans

The next stage in the development of the Baltic basin is named the Yoldia Sea (~11,700–10,700 years BP) (Fig. 2.27b) after the Arctic nutclam *Yoldia arctica* (syn. *Portlandia arctica*, Box Fig. 2.5), which is a common fossil in the Yoldia-stage sediments. After the sudden drainage of the Baltic Ice Lake, a passage between the Baltic basin (including Lake Vänern in western Sweden) and the Skagerrak was created in south-central Sweden. The passage included the area from today's Göta älv river valley to Uddevalla and the Otteid/Steinselva straits at the border between Sweden and Norway. Even though there was an open connection between the Baltic basin and the ocean, it was initially only the cold melt water from the still receding ice sheet that passed through it. It took 300 years before the melt-water outflow had reduced enough for marine water to enter the Baltic basin.

The water exchange with the ocean resulted in a three-phase Yoldia Sea, consisting of a short (~150–350 years long) brackish-water phase, which coincided with the cold Preboreal climate oscillation, in-between two freshwater phases (Andrén et al. 2011). It has been suggested that the incoming marine water caused flocculation and subsequent rapid sedimentation of clay particles, which had previously prevented the sunlight from penetrating the water body (Winterhalter 1992). Thus, conditions for phytoplankton growth improved with the increasing depth of the photic zone, and the nutrient-rich marine water entering the Baltic basin served as a fertiliser.

High abundances of fossil diatom silica frustules deposited during this time provide evidence of an increased primary production in the pelagic zone during the short brackish-water phase. Brackish-water diatom species, such as *Thalassiosira baltica* (Box Fig. 2.7), indicate slightly brackish conditions in the open basin (Andrén et al. 2000). In the narrow threshold straits in south-central Sweden, fossils of ostracods, such as *Cytheropteron montrosiense* and *Paracyprideis fennica*, and the foraminifer *Cribroelphidium excavatum* (syn. *Elphidium excavatum*) (Box Fig. 2.6), indicate a weak marine influence (Schoning and Wastegård 1999). The inflow of marine water can also be traced from sulphide bands or stains and weakly developed varved clays in the sediment, which suggests the development of a weak halocline in the Baltic basin.

2.5.7 The end of the Yoldia Sea

The isostatic rebound continued when the melting ice sheet retreated northwards. The threshold straits in south-central

Box 2.4: Shells used in palaeo-ecological studies of the Baltic basin

The molluscs that gave the geological stages of the Baltic basin their names

The Yoldia Sea is named after the Arctic nutclam *Portlandia arctica* (syn. *Yoldia arctica*), a marine infaunal detritivore, which lives mainly in the sublittoral zone on silty sediments in the Arctic Ocean (Holte and Guliksen 1998). Adults of this species, which does not occur alive in the present Baltic Sea, are usually 15–18 cm in length. The shell of *Portlandia arctica* (Box Fig. 2.5a, b) is a common fossil in the Yoldia-stage sediments and indicates cold marine conditions. The Ancylus Lake is named after the freshwater limpet *Ancylus fluviatilis*, a pulmonate gastropod, which is widespread throughout Europe. It typically occurs on stone surfaces in running waters with high dissolved oxygen concentrations. It is ~4–8 mm in size and seems to prefer diatoms as food (Calow 1973). The shell of *Ancylus fluviatilis* (Box Fig. 2.5c) is a common fossil in the Ancylus-stage sediments and indicates freshwater conditions. The Littorina Sea is named after the common periwinkle *Littorina littorea*, which occurs on marine intertidal rocky shores and is native to the northeastern Atlantic Ocean. It occurs in the Kattegat and the Belt Sea, but in this microtidal area it is mainly found in the sublittoral zone (Lauckner 1984). The food preference of *Littorina littorea* are foliose green ephemeral algae such as *Ulva* spp., and adults are usually 2–4 cm in size (Watson and Norton 1985). The snail itself is considered a delicacy in *e.g.* Scotland, Ireland and Belgium. The shell of *Littorina littorea* (Box Fig. 2.5d, e) is a common fossil in the Littorina-stage sediments and indicates marine influence. The geological stage of the present Baltic Sea is called the *Mya* Sea after the sand gaper *Mya arenaria* (Box Fig. 2.5f), which is a very common cryptogenic species in the Baltic Sea (*cf.* Box 5.2).

Small fossils in Yoldia-stage sediments

The Arctic ostracod (seed shrimp) *Cytheropteron montrosiense* is a small crustacean (~0.5 mm in body size) with a bivalve-like calcareous perforate shell (carapace) that protects its body. In Yoldia-stage sediments, *Cytheropteron montrosiense* carapaces (Box Fig. 2.5a) indicate cold marine influence (Schoning 2001). The foraminifer *Cribroelphidium excavatum* is a common infaunal herbivore and detritivore of marine coasts and occurs in the Belt Sea (Schönfeld and Numberger 2007). This ~0.5 mm-sized amoeboid protozoan has a hard external skeleton (test) made

Box Fig. 2.5 Shells of the four mollusc species after which the geological stages of the Baltic basin have been named. (**a**) *Portlandia arctica* from Yoldia Sea deposits, 23 × 15 mm. (**b**) *Portlandia arctica* from Yoldia Sea deposits, 22 × 14 mm. (**c**) *Ancylus fluviatilis* from freshwater in central Sweden, 7 × 5 mm (recent material). (**d**) *Littorina littorea* from Littorina Sea deposits, 13 × 11 mm. (**e**) *Littorina littorea* from Littorina Sea deposits, 12 × 10 mm. (**f**) *Mya arenaria* from the Baltic Sea proper, 33 × 22 mm (recent material). Images photographed from the shell collection at the Museum of Natural History, Stockholm. Photo: © Pauline Snoeijs-Leijonmalm of calcium carbonate which is preserved in sediments. In brackish-phase Yoldia-stage sediments, *Cribroelphidium excavatum* tests (Box Fig. 2.6b) indicate marine influence (Schoning 2001).

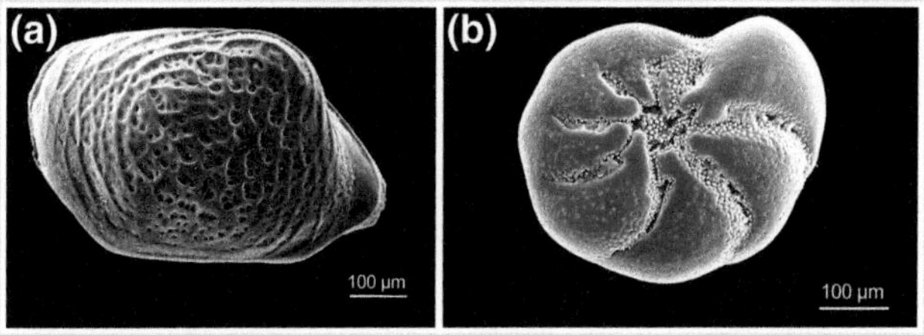

Box Fig. 2.6 Scanning electron micrographs of small shells from Yoldia-stage sediments. **a** Carapace of the ostracod *Cytheropteron montrosiense*. **b** Test of the foraminifer *Cribroelphidium excavatum* (syn. *Elphidium excavatum*). Photo: © Kristian Schoning

Diatom silica frustules

The centric diatom *Thalassiosira baltica* is a common phytoplankton species in the Baltic Sea with a diameter of ~55 µm. Diatoms have a silicified cell wall (frustule), which is preserved in sediments. In Yoldia-stage sediments, *Thalassiosira baltica* frustules (Box Fig. 2.7a, b) indicate slightly brackish surface water conditions. Benthic pennate diatom species belonging to the genus *Mastogloia* occur along the coasts of the Baltic Sea, mainly in places less exposed to wave action. These species are ~20–40 µm long and they often live in a gelatinous matrix together with other diatom species. High abundances of *Mastogloia* frustules (Box Fig. 2.7c–e) are typical of the Initial Littorina Sea (originally named the "Mastogloia Sea") and indicate slightly brackish conditions in coastal environments. Typical of *Mastogloia* are the distinct marginal chambers of their silica frustules. The epipelic diatom *Campylodiscus clypeus* occurs on soft bottoms in nutrient-rich brackish-marine lagoonal areas of the Baltic Sea. Its frustule (Box Fig. 2.7f) indicates such conditions in Littorina-stage sediments as suitable environments arose when lagoons were formed in uplift areas. The species is extant today, but may have been more common in subfossil times (Poulícková and Jahn 2007). The centric diatom *Pseudosolenia calcar-avis* is a marine phytoplankton species. Its frustule (Box Fig. 2.7g) is an indicator of the warmer and more marine conditions in Littorina-stage sediments during the Holocene Thermal Maximum (~8,000–4,000 years BP) and the Medieval Climate Anomaly (~1,000 years BP).

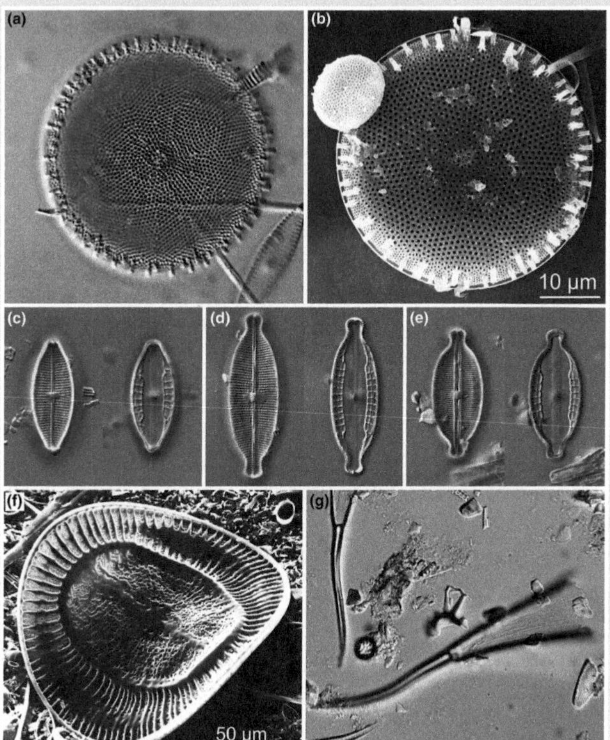

Box Fig. 2.7 Diatom silica frustules that indicate different environmental conditions in the geological stages of the Baltic basin. (**a**) Light micrograph of a *Thalassiosira baltica* frustule. (**b**) Scanning electron micrograph of a *Thalassiosira baltica* frustule, with in the upper left a frustule of the smaller species *Thalassiosira levanderi*. (**c**) Light micrographs of a *Mastogloia smithii* frustule with a different focus. The valve chambers are visible in the micrograph to the right. (**d**) Light micrographs of a *Mastogloia smithii* var. *amphicephala* frustule with a different focus. The valve chambers are visible in the micrograph to the right. (**e**) Light micrographs of a *Mastogloia baltica* frustule with a different focus. The valve chambers are visible in the micrograph to the right. (**f**) Scanning electron micrograph of *Campylodiscus clypeus*. (**g**) Remnants of *Pseudosolenia calcar-avis* frustules in sediment from the Gotland Sea. Photo: (**a–f**) © Pauline Snoeijs-Leijonmalm, (**g**) © Elinor Andrén

Sweden became more and more shallow and, due to the high outflow of melt water, no marine water could enter the Baltic basin anymore at the end of the Yoldia Sea stage. However, the very short (only a few hundred years long) brackish influence had a strong impact on the immigration of marine organisms, so-called "glacial relicts", to the Baltic basin (De Geer 1932), *e.g.* the harp seal *Pagophilus groenlandicus* (syn. *Phoca groenlandica*), the ringed seal *Pusa hispida* (syn. *Phoca hispida*, *cf.* Box 4.13), the whiting *Merlangius merlangus*, the Arctic char *Salvelinus alpinus* and the benthic isopod *Saduria entomon* (*cf.* Fig. 4.25d), some of which still occur in the Baltic Sea. Furthermore, the land uplift and the still low sea level of the ocean resulted in a large land bridge between Scandinavia and the rest of the European continent, which facilitated the immigration of terrestrial plants and animals, including humans, following the retreat of the ice sheet.

2.5.8 The Ancylus Lake: dammed once again

Eventually, the threshold areas west of Lake Vänern rose, and a transgression (rising water level) occurred in the southern part of the Baltic basin, south of the line between Stockholm and Helsinki, causing a flood. North of this line, however, the isostasy was so large that a regression (decreasing water level) occurred. These conditions created a freshwater lake in the Baltic basin (including the basin of Lake Vänern), which is named the Ancylus Lake (Fig. 2.27c) after the freshwater limpet *Ancylus fluviatilis* (Box Fig. 2.5). The Ancylus Lake lasted from ~10,700 until ~9,800 years BP.

Sediments deposited during this stage consist of homogeneous clays, which suggests a fairly oligotrophic environment with low organic carbon accumulation and a diatom flora characterised by species typical of large clear-water lakes, such as *Aulacoseira islandica* and *Stephanodiscus neoastraea* in the pelagic zone and *Ellerbeckia arenaria* in the littoral zone. The Ancylus clays sometimes contain sulphide bands or stains, but this does not indicate the presence of a halocline. These sulphide traces are rather a secondary effect due to hydrogen sulphide diffusion from the overlaying younger organic sediments (Sohlenius et al. 2001). In sediments from the open basin, the transition to the Ancylus Lake is visible as a small increase in the organic carbon content, an effect of the damming when old reworked carbon from the transgressed coasts was discharged and redeposited in the deeper areas (Andrén et al. 2000).

The Ancylus transgression, estimated to have raised the water level by 10 m above that of the ocean (Björck et al. 2008), is today visible as raised beach ridges along the coasts of Estonia, Latvia, the Swedish mainland and the island of Gotland. It is also documented by the simultaneous flooding of pine forest along the coasts of the southern Baltic basin. A new outlet to the ocean, called the Dana river, was created in the Storebælt area. This was a complex river system with various channels and lakes (Björck et al. 2008). After a few hundred years, the Ancylus Lake was at the level of the ocean as a result of the worldwide eustatic sea level rise caused by climatic warming and the subsequent melting of ice sheets. The end of the Ancylus Lake stage is defined by records of the first weak inflows of marine water into the Baltic basin.

2.5.9 The Initial Littorina Sea: between fresh and brackish

The next transitional phase with slightly brackish water conditions was originally named the "Mastogloia Sea" after the diatom genus *Mastogloia* (Box Fig. 2.7), but today it is known as the "Initial Littorina Sea" or the "Early Littorina Sea". This stage lasted from ~9,800 until ~7,500 years BP. The transformation from the Ancylus Lake into a brackish sea did not proceed simultaneously in different parts of the Baltic basin. It started close to the narrow inlet in the south and gradually spread northwards during a period of ~2,000 years. This implies that while the southern part of the Baltic basin was already experiencing slightly brackish Initial Littorina Sea conditions, the central part still had an Ancylus Lake setting (Andrén 1999).

The inflowing marine water contributed to the development of a halocline and brought nutrients into the system, which resulted in pronounced cyanobacterial blooms as early as ~8,000 years ago (Bianchi et al. 2000). Distinctive of the Initial Littorina Sea stage is the combination of high cyanobacterial production and low diatom abundance in the open basin. However, in the coastal zones a diverse brackish diatom flora, the so-called "*Mastogloia* flora", was established during this transitional phase (Miettinen 2002).

2.5.10 Hypoxia in the Initial Littorina Sea

The salinity stratification, together with the decreased oxygen saturation of the warmer water and increased primary production, initiated periods of deep-water hypoxia in the open Baltic basin. This is evident from findings of extended areas of laminated gyttja-clay deposits in the sediment record (Zillén et al. 2008). Increased upward transport of phosphorus, released from the anoxic bottoms to the photic zone, has been suggested as an explanation of the enhanced primary production observed at the Ancylus/Littorina transition (Sohlenius et al. 2001).

Diazotrophic cyanobacterial blooms have been proposed to have played a role as a trigger of eutrophication during this period. The stable nitrogen isotope is indicative of the origin of nitrogen (Box 2.3) and $\delta^{15}N$ measurements indicate that the bloom-forming cyanobacteria of the Initial Littorina Sea were actually fixing nitrogen (Borgendahl and Westman 2007). The sediments deposited during the Initial Littorina Sea show a sudden increase in the total organic carbon content from ~1 % to 4–8 %. This is reflected in a characteristic ecosystem-wide change from the homogeneous Ancylus clay to the homogeneous or laminated Littorina gyttja-clay (Winterhalter 1992).

2.5.11 The Littorina Sea: the most saline stage

The Littorina Sea (Fig. 2.27d), named after the common periwinkle *Littorina littorea* (Box Fig. 2.5), is the stage during the Holocene development of the Baltic basin with the largest marine influence. It is time-transgressive, *i.e.* it occurred at different times in different parts of the Baltic basin, and is estimated to have lasted from ~7,500 until ~3,000 years BP in the central part of the basin.

The current knowledge about the Baltic palaeo-salinity is mainly based on fossilised organisms such as diatoms and molluscs. Early attempts to reconstruct Baltic palaeo-salinities by using the minimum and maximum salinity tolerances of the mollusc fauna found in raised beach ridges date back to the late 19th century (*e.g.* Lindström 1886). The classical view of the palaeo-salinity in the Baltic basin during the Littorina Sea stage is that brackish conditions were established ~7,500 years BP and that the salinity increased until it reached a maximum level during ~6,000–4,000 years BP, after which it gradually decreased down to the present level (Fig. 2.26c).

Attempts to quantify palaeo-salinity by using species distribution optima (*e.g.* molluscs, diatoms), species morphology (spine length of dinoflagellate cysts) or measurements of pore water and stable isotopes (oxygen, carbon, strontium, Box 2.3) have produced varying results. At present, there is no consensus as to the exact magnitude of the past salinity changes. However, we do know that the horizontal salinity gradient in the Baltic basin was not as pronounced during the Littorina Sea as it is today because the sills separating the subbasins were situated deeper (Gustafsson and Westman 2002). The halocline was probably found at a similar water depth as it is today.

2.5.12 The Holocene Thermal Maximum and the Littorina Sea

The Holocene Thermal Maximum in northwestern Europe during ~8,000–4,000 years BP was characterised by high temperatures and low humidity (Seppä et al. 2009). The warm climate led to a high global eustatic sea level, which increased the depth of the sills between the ocean and the Baltic basin and enhanced the inflow of marine water (Gustafsson and Westman 2002). The high salinity strengthened the halocline, which together with declining oxygen saturation in a warmer sea created hypoxic bottoms that released phosphorus. This in turn increased the abundance of diazotrophic cyanobacterial blooms (Sohlenius et al. 2001).

Models of the fluctuating salinity in the Littorina Sea until present show that freshwater supply has been the main driver, and that sill-depth changes in the transition zone only partly contributed to the variations in salinity (Gustafsson and Westman 2002). The low humidity during the thermal maximum resulted in higher evaporation and decreased freshwater discharge, resulting in higher salinity in the Baltic basin.

2.5.13 Shoreline displacement in the Littorina Sea

In the coastal zone, the shoreline displacement (Fig. 2.26d), and consequently the configuration of the Baltic basin, was affected by both eustatic and isostatic components. The Littorina stage is characterised by a fluctuating water level, but the number and magnitude of transgressions on the Baltic coasts varied from one location to another due to a higher isostatic rebound in the northern part of the basin where the ice sheet had been thicker and the duration of glaciation longer (Pirazzoli 1991).

As a result of shoreline oscillations, the sediments deposited in the coastal zone were resuspended and thereby they contributed to nutrient cycling. This made the coastal zone of the Littorina Sea a highly dynamic environment, which was naturally rich in nutrients. In near-shore areas, the diatom flora was very diverse and consisted of the so-called "*Clypeus* flora", named after the diatom *Campylodiscus clypeus* (Box Fig. 2.7), also known as "the lagoonal flora". This diatom flora is indicative of shallow nutrient-rich bays, which were gradually lifted up due to the isostatic rebound, and is common along the Finnish and Swedish coasts (Miettinen 2002).

Slowly, the Littorina Sea turned into the Baltic Sea of today, also known as the "Mya stage", named after the bivalve *Mya arenaria* (Box Fig. 2.5), whose appearance is considered as an early example of human-mediated species introductions to the North Sea-Baltic Sea Area (Strasser 1999).

2.5.14 Laminated sediments, palaeo-production and hypoxia

Sediment cores that cover the whole time-span from the Littorina Sea until the present Baltic Sea are characterised by homogeneous gyttja-clays interlayered by laminated gyttja-clays. The homogeneous layers are indicative of oxygenated bottoms with burrowing animals, which perturb the seasonal deposits. Laminated sediments are formed when seasonal sedimentation is not disturbed by animal life since animals are absent due to oxygen stress. Therefore, laminated sediments are used to trace past bottom-water hypoxia (Zillén et al. 2008).

The laminated sediments of the Baltic Sea have a high organic carbon content, which could be interpreted as a sign of a high primary production. However, it has been debated whether a high organic carbon content is exclusively coupled to changes in primary production or whether it may also partly be due to a better preservation of carbon under anoxic conditions (Sohlenius et al. 2001). Enhanced preservation of carbon could explain why variations in the geographical distribution of laminae are correlated with water depth (Zillén et al. 2008). Thus, the formation of laminae could partly be triggered by increased stratification due to increased salinity in a warmer climate, and not exclusively by increased primary production.

Apart from the laminated sediments that are deposited today, the sediment record of the open Baltic Sea shows the presence of laminated sediments deposited during two previous periods. The first period occurred ~8,000–4,000 years BP and correlates with the Holocene Thermal Maximum (Zillén et al. 2008). The pelagic microfossils in these sediment layers show a high abundance of resting stages of the diatom genus *Chaetoceros* and the silicoflagellate *Dictyocha speculum*, both of which are indicators of high nutrient concentrations. Simultaneously, diatom taxa indicative of warmer and more marine conditions occur as well, *e.g.* *Pseudosolenia calcar-avis* (Box Fig. 2.7), *Thalassionema nitzschioides*, *Thalassiosira oestrupii* and *Chaetoceros mitra* (Andrén et al. 2000).

The second period with laminated sediments is dated to ~2,000–700 years BP and correlates with the Roman Warm Period around 2,000 years BP and the Medieval Climate Anomaly around 1,000 years BP (Zillén et al. 2008; Seppä et al. 2009). The microfossil record from the Medieval Climate Anomaly contains resting stages of the diatom genus *Chaetoceros* and the ebridian *Ebria tripartita*, both indicating high nutrient concentrations (Andrén et al. 2000). During this warm event, the pelagic diatoms were dominated by *Pseudosolenia calcar-avis*. This species gradually disappeared with the succeeding cold event known as the "Little Ice Age" (~500–100 years BP), and the diatoms shifted to the dominance of *Thalassiosira hyperborea* var. *lacunosa* and varieties of *Actinocyclus octonarius*.

2.6 A changing ecosystem

2.6.1 Fast changes and regime shifts

Over the past 1,000 years the Baltic Sea has changed dramatically from the hypoxic Medieval Climate Anomaly to the oxic Little Ice Age to the present day, again with hypoxia. These changes were primarily climate-driven, without any major interferences by humans. Today the large human population living in the Baltic Sea drainage area exerts strong pressures on the sea and its well-being increasingly depends on how it is managed (*cf.* Sect. 18.5).

The Baltic Sea ecosystem is not the same as it was 100 years ago, *i.e.* with the same species composition, food webs and productivity, and this could be expected of a dynamic ecosystem. However, the changes caused by anthropogenic pressures are large and take place very fast. Some of the large-scale changes that are listed as major threats to the diversity of the Baltic Sea ecosystem are multifaceted. While there is a risk with each non-indigenous species invading the Baltic Sea, some have enriched its functional diversity. While eutrophication has many negative effects on the ecosystem, it does increase fish productivity. While climate change compromises the existence of the cold-water species in the Baltic Sea, it increases the length of the growing season and increases productivity. The Baltic Sea is one of the large marine ecosystems with the highest recorded temperature increases during the past century (Belkin 2009). This temperature increase is consistent with the anthropogenic climate change signal (Bhend and von Storch 2009; Rutgersson et al. 2014). The global warming of the Earth's surface since the 1980s has been highest at latitudes above 50 °N (Fig. 2.28) where the Baltic Sea is situated between latitude 53°55′ N and 65°48′ N.

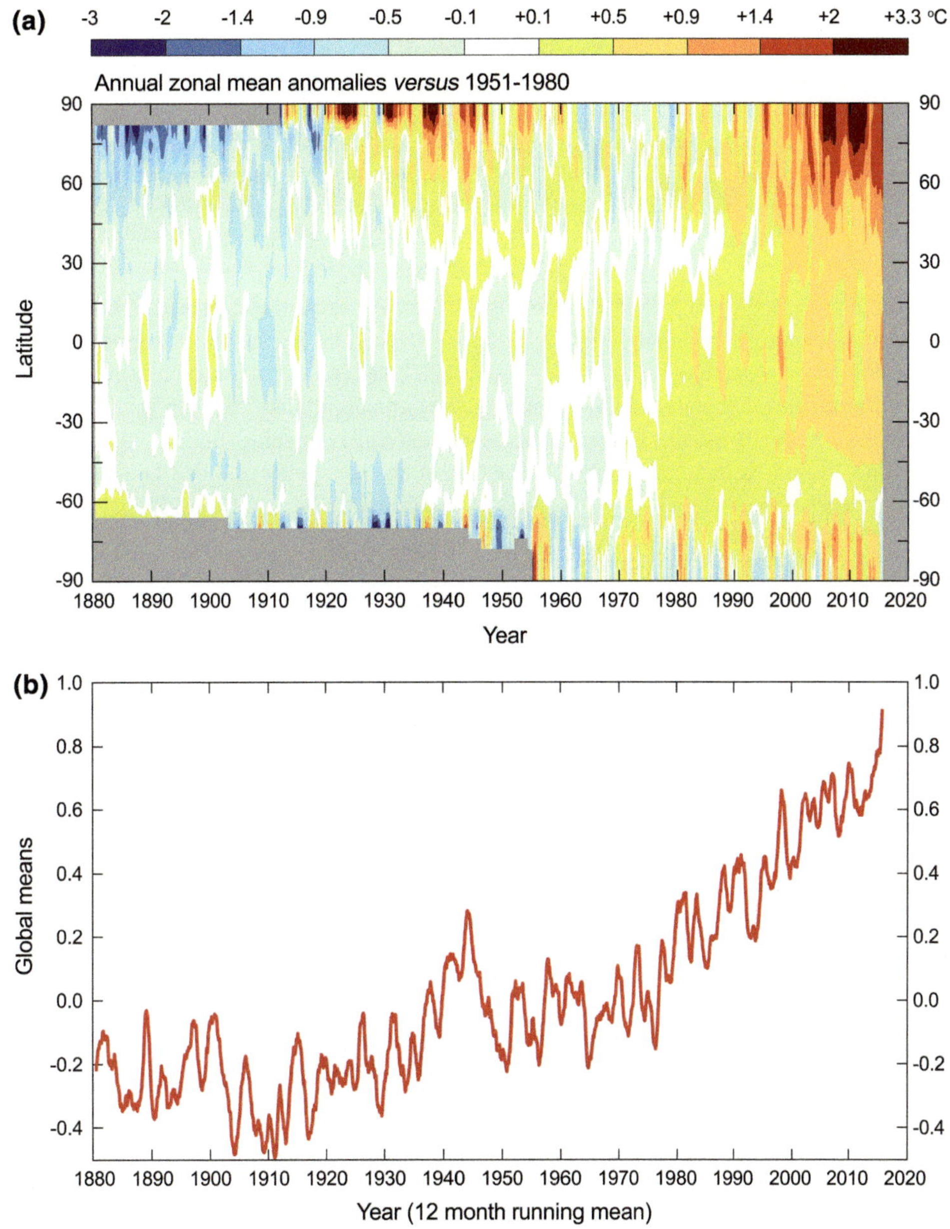

Fig. 2.28 Summary of global warming based on the Earth surface temperature Land-Ocean Temperature index (ERSST v4), whereby the temperature for each year is compared with the mean temperature for the years 1951–1980. (**a**) Annual zonal mean anomalies from 1951 to 1980. (**b**) Global mean anomalies from 1951 to 1980. Figures printed with permission from the National Aeronautics and Space Administration (NASA), USA (http://data.giss.nasa.gov/gistemp/time_series.html)

One of the typical features of the Baltic Sea ecosystem is that at least the northern part is ice-covered in winter. With climate change the ice cover extension, duration and thickness will decrease, which is expected to lead to significant changes in the ecosystem. The ice cover affects physical processes, *e.g.* the water masses are less affected by winds and the ice cover changes air-sea heat fluxes. An example of ice cover impacts on biological processes is the timing of the onset of the pelagic spring bloom and its species composition (*cf.* Sect. 9.6.4).

Many environmental and biological factors interact and it is not possible to understand ecosystem change by monitoring one or a few factors only. An integrated, simultaneous analysis of many factors shows that they change in concert, and that some change more than others (Box 2.5). Powerful instruments for understanding ecosystem functioning and

Box 2.5: Regime shifts

Threshold-like ecological shifts

Abrupt and rapid shifts in food web and community structure, so-called "regime shifts", are increasingly being reported for large marine ecosystems around the world (Kraberg et al. 2011). Such a threshold-like ecological shift occurred in the pelagic system of the Baltic Sea proper at the end of the 1980s and changed the composition of the zooplankton and fish communities (Box Fig. 2.8). Almost synchronous shifts were recorded in the other subbasins of the Baltic Sea (Diekmann and Möllmann 2010), as well as in the North Sea and some other seas in the northern hemisphere (Möllmann and Diekmann 2012). These changes could partly be coupled to the North Atlantic oscillation (NOA) and other climatic phenomena, which modified local temperature regimes. Human-induced trophic cascades, triggered by the removal of predators such as seals and cod (Österblom et al. 2007; Casini et al. 2009), as well as eutrophication and introductions of non-indigenous species, are also coupled to regime shifts as they create possible "tipping points" for food web functioning. Multiple drivers potentially interact in a way that one driver (*e.g.* overfishing) undermines resilience and another one (*e.g.* climate change) provides the final impulse for an abrupt change (Möllmann and Diekmann 2012).

Wasp-waist trophic structure

Like many other aquatic ecosystems, the Baltic Sea exhibits a characteristic so-called "wasp-waist" trophic structure (Cury et al. 2000; Bakun 2006), in which one or a few species of small planktivorous fish entirely dominate their trophic level where bottom-up and top-down processes meet. In the Baltic Sea, the cod collapsed during the 1980s in concert with the establishment of the ecosystem's dominance by the cod's wasp-waist prey, the sprat.

The future of the Baltic Sea ecosystem is in our hands

The future of the Baltic Sea ecosystem is largely down to how it will be managed even if climate change sets some limits to possible outcomes. Two different future cod fishing and eutrophication scenarios were investigated by combining three regional biogeochemical models with an Ecopath model including the Ecosim food web procedure by Niiranen et al. (2013). The results of this modelling study showed that by the end of the 21st century, the combination of intensive cod fishing and high nutrient loads projected a strongly eutrophicated sprat-dominated sea, whereas low cod fishing in combination with low nutrient loads would result in a cod-dominated system with eutrophication levels close to present.

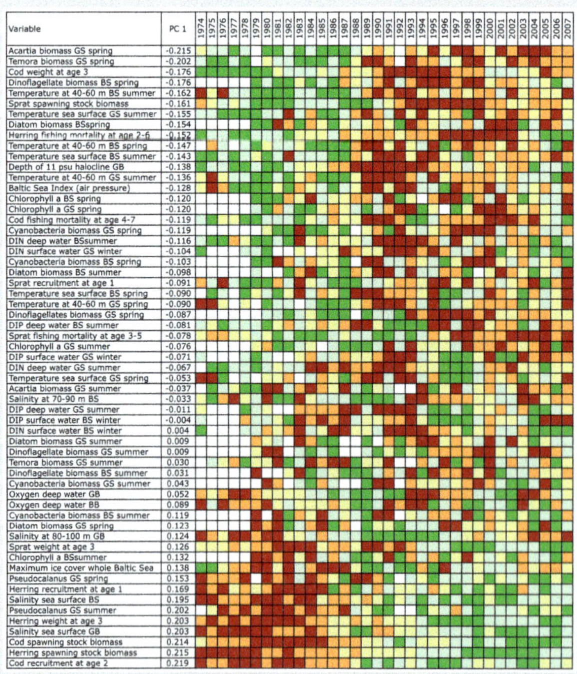

Box Fig. 2.8 Regime shifts can only be detected in data sets that include relevant measurements and cover a substantial period of time including data from before, during and after a potential regime shift. This figure shows a "traffic-light plot" of the temporal development of environmental and biological variables in the Baltic Sea proper from 1974 to 2007, and documents a sudden "regime shift" in the end of the 1980s. GS = Gotland Sea, BS = Bornholm Sea. Dark green = very low values, light green = low values, yellow = intermediate values, orange = high values, red = very high values. The variables are sorted according to their scores on the first axis (PC1) of a principal components analysis. Figure modified from Diekmann and Möllmann (2010)

predicting future changes in the Baltic Sea ecosystem are being developed, *e.g.* by the coupling of oceanographic, climatic and biological models (Box 2.6).

Several recent studies have suggested that the on-going changes in the Baltic Sea ecosystem form a directional trend with a sudden "regime shift" in the end of the 1980s (Box 2.5, *cf.* Sect. 17.2.4), rather than being the result of normal inter-annual fluctuations (Österblom et al. 2007; Möllmann et al. 2009). Regime shifts are abrupt, substantial and persistent changes in the state of natural systems. The 1980s regime shift was recorded not only in the Baltic Sea and the North Sea (Alheit et al. 2005), but on a planetary scale, and can be explained by changing climatic factors through interactions between major volcanic eruptions and anthropogenic climate change as the main forcing factors (Beaugrand et al. 2015; Reid et al. 2016). Regionally, other environmental drivers may coincide and interact with global warming. In the case of the Baltic Sea, the 1980s regime shift was also strongly related to an overfishing-induced trophic cascade from a cod-dominated to a sprat-dominated food web in the pelagic zone (Möllmann and Diekmann 2012).

2.6.2 How to turn negative trends

A scenario with intensive cod fishing and high nutrient loads projects a strongly eutrophicated sprat-dominated sea by the end of the 21st century, whereas a scenario with low cod fishing in combination with low nutrient loads is expected to result in a cod-dominated system with eutrophication levels close to present (Box 2.5). Although ecosystem-based management is the agreed principle today, in practice the various environmental problems are still handled separately, since we still lack both basic ecological knowledge and appropriate governance structures for managing them together, in a true ecosystem approach (Elmgren et al. 2015). Proper ecosystem-based management is mandatory for maintaining a well-functioning ecosystem that can provide goods and services to human society in a sustainable way (*cf.* Sect. 18.5). Modelling results have shown that regional management is likely to play a major role in determining the future of the Baltic Sea ecosystem (Box 2.5, Niiranen et al. 2013).

Some improvements of the Baltic Sea environment have been achieved as a result of changes in the management of the sea's resources through actions such as financial investments, legislation and international agreements. As a result of the reduction of atmospheric nitrogen emissions (*e.g.* from traffic) and the reduction of phosphorus emissions by building a large number of modern wastewater treatment plants (especially in the eastern part of the drainage area), nutrient inputs to the Baltic Sea have declined since 1980 (Box Fig. 2.2). Some species that previously were endangered by contaminants and/or hunting, such as the white-tailed eagle *Haliaeetus albicilla* (*cf.* Box Fig. 4.19), the common guillemot *Uria aalge* (*cf.* Fig. 16.2), the grey seal *Halichoerus grypus* (*cf.* Box Fig. 4.21b) and the ringed seal *Pusa hispida* (*cf.* Box Fig. 4.21c) have recovered thanks to legislation forbidding the use of certain chemicals (*cf.* Sect. 16.1) and protecting them as red-listed species (*cf.* Sect. 18.6.2). Fisheries regulations are the main reason why the eastern Baltic cod stock has shown some signs of recovery after more than two decades of low biomass and productivity (Cardinale and Svedäng 2011). This recovery was mainly driven by a sudden reduction in fishing mortality despite the absence of any exceptionally large year classes.

These examples show that proper management can have positive effects on ecosystem health on a relatively short time scale. Necessary actions include decreasing nutrient inputs, chemical pollution bans, habitat protection and control of fisheries, so that key functions of the ecosystem can be operational and the Baltic Sea can be a fully functional ecosystem delivering ecosystem goods and services to society. Since the Baltic Sea is surrounded by countries that live in peace and are rather wealthy in comparison with many other areas in the world, the Baltic Sea countries have the potential to invest in adequate ecosystem-based management, *i.e.* to set clear operational goals focused on long-term ecological sustainability (Thrush and Dayton 2010). Also the political willingness to improve the Baltic Sea environment seems to be present, *e.g.* through well-organised international meetings at different levels. However, in practice the necessary governance processes are unfortunately remarkably slow while a variety of pressures on the sea, from ship traffic to climate change, are increasing.

2.6.3 Will we overcome the "tragedy of the commons"?

The "tragedy of the commons" (Hardin 1968) is a theory that emerged in the 1960s from the growing concern about the rapid human population growth on Earth. The subtitle of Hardin's paper in the journal "Science" is "The population problem has no technical solution; it requires a fundamental extension in morality". According to this theory each human individual acts rationally and independently, conforming to his/her own interests, in contrast to the general long-term interest of society, by depleting common resources. Pollution of the environment is one of the examples that Hardin raised to illustrate his theory.

Today we know even better than 50 years ago that environmental resources, like clean water and clean air, are not endless. However, while public awareness has grown, the morality of the average human individual may not have changed much, as the theory claims. The "tragedy of the commons" is also applicable to countries sharing a common resource, *e.g.* when negotiating about fishing quota and nutrient emissions. The Baltic Sea is such a common resource that has been overexploited and polluted for a long time, mainly by its nine riparian countries.

Box 2.6: Modelling coastal seas

Anders Omstedt

Computational fluid dynamics

The use of computational fluid dynamics (CFD) to analyse and predict environmental changes has increased considerably in recent decades. Mathematical models are now standard tools in research, as well as in a wide range of practical applications. Intensifying concern about human influence on climatic and environmental conditions has increased the need for multidisciplinary modelling efforts, including numerical modelling of oceans, lakes, land surfaces, ice, rivers, and the atmosphere. Scientists have traditionally developed specialised models limited to application within their own disciplines. Today, increasing efforts are being made to develop Earth System models that include major processes that one needs to consider when dealing with climate change and other environmental changes. In general, the models rely on conservation laws, including many processes that are not known in detail.

Parameterisations

These rather unknown processes then need to be parameterised in different ways. For example, turbulence, which is always present in coastal seas, is poorly understood and needs to be parameterised in the models. This is also true for a number of chemical and biological processes. For example, there is no standard parameterisation yet available for biological processes such as plankton growth and ecosystem change, and instead most parameterisations are based on some available observations. However, one can regard the new and updated models as a systematic collection of the present available knowledge. The models can therefore help us to identify gaps in understanding and where new research programmes need to be developed. CFD cannot, however, profess adequately without reference to experimental and field validation. This is also a good reason why models, field experiments and monitoring programmes need to be strongly linked to each other. Coastal sea models not only include codes for solving the conservation laws, they also include initial data (*e.g.* salinity and temperature) and forcing data (*e.g.* weather conditions, riverine runoff, atmosphere and land emissions from nutrients and carbon components).

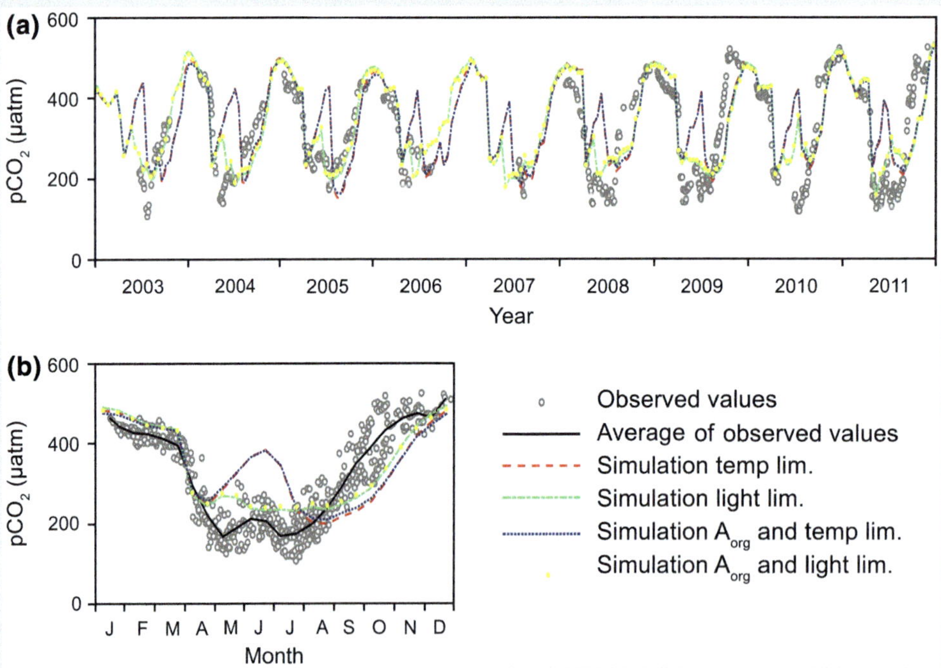

Box Fig. 2.9 Variation of pCO_2 in the surface water of the Gotland Sea (Baltic Sea proper). (**a**) Observed and simulated pCO_2 during nine years (2003–2011). (**b**) Observed and simulated seasonal average pCO_2 in 2003–2011. The red lines indicate the results from a model simulation with the original parameterisation for growth of cyanobacteria ("temperature limitation"). The green lines indicate the results from a simulation with the new parameterisation for growth of cyanobacteria ("light limitation"). The blue lines and yellow dots indicate the results from simulations including organic alkalinity (A_{org}) with temperature and light limitation, respectively. Figure modified from Gustafsson et al. (2015)

Different types of Baltic coastal sea models

The available coastal sea models vary from simple box models to coupled three-dimensional atmosphere-land-ocean models. They are often developed for different applications and use different kinds of forcing fields. One class of models is process-based models (*e.g.* Omstedt 2015) and another class is three-dimensional models (*e.g.* Meier et al. 2006). Both types of models have been used in many Baltic Sea applications. The strength of process-based models is here exemplified by a study that analyses CO_2 dynamics in the Baltic Sea (Gustafsson et al. 2015). In this study, the modelling focused on how air-water CO_2 fluxes respond to parameterisations of organic alkalinity, gas transfer, phytoplankton growth, and changes in river loads. The forcing data in the study included the most complete compilation of Baltic river loads for dissolved inorganic and organic carbon (DIC and DOC), as well as total alkalinity. One result demonstrated how air-water CO_2 fluxes depend on the river load of carbon. If the river load of total alkalinity decreases, the CO_2 buffer capacity is reduced, which has the effect that the partial CO_2 pressure (pCO_2) in the water increases and changes the air-water CO_2 fluxes. By analysing different aspects in the modelling, the calculations were getting closer to the observations. For example, a new parameterisation of cyanobacteria (removing strict temperature dependence and instead letting the growth of cyanobacteria be more strongly controlled by light intensity) significantly improved the seasonal development of pCO_2, although the values were overestimated in summer and underestimated in autumn (Box Fig. 2.9).

Model comparisons

The accuracy of Baltic Sea models to produce biogeochemical parameters was compared by Meier et al. (2012). This is one of the first studies that develop the concept of adding the results from different models together in a so-called "ensemble calculation". In general, the results show that the models were not too far from the observations, and that the best representation of the data was the ensemble mean (Box Fig. 2.10). Thus the community of scientists working with models now enters a new era, as not a single model should be used to give management advice. Instead, a number of different models needs to be used. This is in line with the modelling developments within global climate change, illustrating that management actions in the future need to be closely linked to assessment activities.

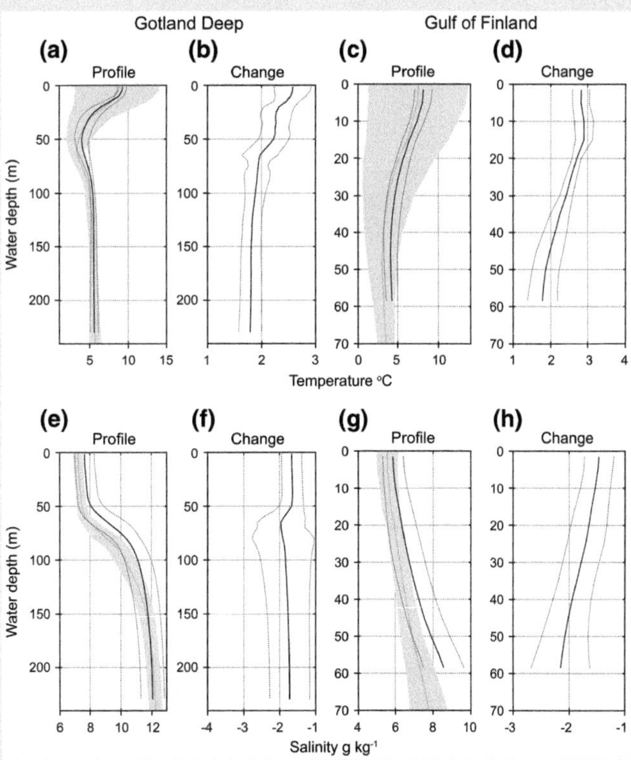

Box Fig. 2.10 Vertical profiles and predicted changes in temperature and salinity in the Gotland deep (left panels) and the Gulf of Finland (right panels). (**a**, **c**, **e**, **g**) Profiles for the control period 1978–2007, showing the average (green line) ± 1 standard deviation (grey shaded area) of observations, and the ensemble average of the results of different models (black line) ± 1 standard deviation (dotted line). (**b**, **d**, **f**, **h**) Ensemble average changes between 1978–2007 and 2069–2098, showing the predicted increase of temperature and the predicted decrease in salinity for the Baltic Sea by the end of the century. Figure modified from Meier et al. (2012)

Review questions

1. What factors influence the salinity of the Baltic Sea?
2. How does geography affect the sensitivity of the Baltic Sea ecosystem to external impacts?
3. What were the causes of the changes between the different stages of the Baltic basin during its late- and post-glacial environmental history?
4. How and why has the geographical extent of hypoxic sea bottoms varied through times?
5. Summarise the conditions that make the Baltic Sea of today such a special environment for different types of species.

Discussion questions

1. What do the Baltic Sea, the Black Sea and the Caspian Sea have in common, and how is this related to their geography and geological developments?
2 What do the Baltic Sea and the Mediterranean Sea have in common with respect to water exchange with the ocean? What are the differences?
3. In what way is knowledge about the geological development of the Baltic basin relevant when discussing contemporary environmental problems?
4. What do you think the Baltic Sea would be like today if no anthropogenic impact had occurred for the last 10,000 years?
5. Will we overcome the "tragedy of the commons"? How?

References

Al-Hamdani Z, Reker J (eds) (2007) Towards marine landscapes in the Baltic Sea. BALANCE Interim Report 10:1–116 [http://balance-eu.org/publications/index.html]

Alheit J, Möllmann J, Dutz J, Kornilovs G, Loewe P, Mohrholz V, Wasmund N (2005) Synchronous ecological regime shifts in the central Baltic and the North Sea in the late 1980s. ICES Journal of Marine Science 62:1205–1215

Andersen JH, Carstensen J, Conley DJ, Dromph K, Fleming V, et al. (2015) Long-term temporal and spatial trends in eutrophication status of the Baltic Sea. Biological Reviews [http://onlinelibrary.wiley.com/doi/10.1111/brv.12221/epdf]

Andrén E (1999) Holocene environmental changes recorded by diatom stratigraphy in the southern Baltic Sea. Meddelanden från Stockholms universitets institution för geologi och geokemi 302:1–22 [PhD Thesis]

Andrén E, Andrén T, Kunzendorf H (2000) Holocene history of the Baltic Sea as a background for assessing records of human impact in the sediments of the Gotland basin. Holocene 10:687–702

Andrén T, Björck S, Andrén E, Conley DJ, Zillén L et al (2011) The development of the Baltic Sea basin during the last 130,000 years. In: Harff J, Björck S, Hoth P (eds) The Baltic Sea basin. Springer, Berlin, pp 75–97

Bakun A (2006) Wasp-waist populations and marine ecosystem dynamics: navigating the "predator pit" topographies. Progress in Oceanography 68:271–288

Beaugrand G, Conversi A, Chiba S, Edwards M, Fonda-Umani S et al (2015) Synchronous marine pelagic regime shifts in the Northern Hemisphere. Philosophical Transactions of the Royal Society B 370:20130272

Belkin IM (2009) Rapid warming of large marine ecosystems. Progress in Oceanography 81:207–213

Benson BB, Krause D Jr (1984) The concentration and isotopic fractionation of oxygen dissolved in freshwater and seawater in equilibrium with the atmosphere. Limnology and Oceanography 29:620–632

Bernes C (2005) Change beneath the surface—an in-depth look at Sweden's marine environment. Monitor 19:1–192. Swedish Environmental Protection Agency, Stockholm

Bhend J, von Storch H (2009) Is greenhouse gas forcing a plausible explanation for the observed warming in the Baltic Sea catchment area? Boreal Environment Research 14:81–88

Bianchi TS, Engelhaupt E, Westman P, Andrén T, Rolff C et al (2000) Cyanobacterial blooms in the Baltic Sea: natural or human-induced? Limnology and Oceanography 45:716–726

Björck S (1995) A review of the history of the Baltic Sea. 13.0–8.0 ka BP. Quaternary International 27:19–40

Björck S (2008) The late Quaternary development of the Baltic Sea. BACC Author Team. Assessment of climate change for the Baltic Sea basin. Regional Climate Studies, Springer, Berlin, pp 398–407

Björck S, Andrén T, Jensen JB (2008) An attempt to resolve the partly conflicting data and ideas on the Ancylus-Littorina transition. Polish Geological Institute Special Papers 23:21–26

Blomqvist S, Gunnars A, Elmgren R (2004) Why the limiting nutrient differs between temperate coastal seas and freshwater lakes: a matter of salt. Limnology and Oceanography 49:2236–2241

Boedeker D, Knapp HD (1995) Ökologie der Salzwiesen, Dünen und Schären. In: Rheinheimer G (ed) Meereskunde der Ostsee, 2nd edn Springer, Berlin, pp 222–229 [in German]

Borgendahl J, Westman P (2007) Cyanobacteria as a trigger for increased primary productivity during sapropel formation in the Baltic Sea—a study of the Ancylus/Littorina transition. Journal of Paleolimnology 38:1–12

Brzezinski MA (1985) The Si-C-N ratio of marine diatoms—interspecific variability and the effect of some environmental variables. Journal of Phycology 21:347–357

Calow P (1973) The food of *Ancylus fluviatilis* (Müll.), a littoral stone-dwelling herbivore. Oecologia 13:113–133

Cameron A, Askew N (eds) (2011) EUSeaMap—preparatory action for development and assessment of a European broad-scale seabed habitat map: final report. Technical appendix No. 1: Light data and thresholds [http://jncc.gov.uk/euseamap]

Cardinale M, Svedäng H (2011) The beauty of simplicity in science: Baltic cod stock improves rapidly in a 'cod hostile' ecosystem state. Marine Ecology Progress Series 425:297–301

Carstensen J, Andersen JH, Gustafsson BG, Conley DJ (2014) Deoxygenation of the Baltic Sea during the last century. Proceedings of the National Academy of Sciences of the USA 111: 5628–5633

Casini M, Hjelm J, Molinero JC, Lövgren J, Cardinale M et al (2009) Trophic cascades promote threshold-like shifts in pelagic marine ecosystems. Proceedings of the National Academy of Sciences of the USA 106:197–202

Cederwall H, Elmgren R (1980) Biomass increase of benthic macrofauna demonstrates eutrophication of the Baltic Sea. Ophelia Supplement 1:287–304

Cloern JE, Foster SQ, Kleckner AE (2014) Phytoplankton primary production in the world's estuarine-coastal ecsystems. Biogeosciences 11:2477–2501
Conley DJ (2000) Biogeochemical nutrient cycles and nutrient management strategies. Hydrobiologia 410:87–96
Conley DJ, Björck S, Bonsdorff E, Carstensen J, Destouni G et al (2009) Hypoxia-related processes in the Baltic Sea. Environmental Science and Technology 43:3412–3420
Conley DJ, Carstensen J, Aigars J, Axe P, Bonsdorff E et al (2011) Hypoxia is increasing in the coastal zone of the Baltic Sea. Environmental Science and Technology 45:6777–6783
Cury P, Bakun A, Crawford RJM, Jarre A, Quinones RA, Shannon LJ et al (2000) Small pelagics in upwelling systems: patterns of interaction and structural changes in "wasp-waist" ecosystems. ICES Journal of Marine Science 57:603–618
Cyberski J, Wróblewski A (2000) Riverine water inflows and the Baltic Sea water volume 1901–1990. Hydrology and Earth System Sciences Discussions, European Geosciences Union 4:1–11
De Geer G (1912) A geochronology of the last 12,000 years. XIth International Geological Congress, Stockholm 1:241–253
De Geer G (1932) Stockholmstraktens kvartärgeologi: beskrivning till kvartärgeologisk karta över Stockholmstrakten. Sveriges geologiska undersökning, Serie Ba, 89 pp [in Swedish]
Del Amo Y, Brzezinski MA (2000) The chemical form of dissolved Si taken up by marine diatoms. Journal of Phycology 35: 1162–1170
Deutsch C, Weber T (2012) Nutrient ratios as a tracer and driver of ocean biogeochemistry. Annual Review of Marine Science 4: 113–141
Diaz RJ, Rosenberg R (1995) Marine benthic hypoxia: a review of its ecological effects and the behavioural responses of benthic macrofauna. Oceanography and Marine Biology 33:245–303
Diaz RJ, Rosenberg R (2008) Spreading dead zones and consequences for marine ecosystems. Science 321:926–929
Diekmann R, Möllmann C (eds) (2010) Integrated ecosystem assessments of seven Baltic Sea areas covering the last three decades. ICES Cooperative Research Report 302:1–90
Du Rietz GE (1930) Algbälten och vattenståndsväxlingar vid svenska östersjökusten. Botaniska Notiser 1930:421–432 [in Swedish]
Ekman M (1996) A consistent map of the postglacial uplift of Fennoscandia. Terra Nova 8:158–165
Elliott M, McLusky DS (2002) The need for definitions in understanding estuaries. Estuarine, Coastal and Shelf Science 55:815–827
Elmgren R, Blenckner T, Andersson A (2015) Baltic Sea management: successes and failures. AMBIO 44 (Supplement):S335–S344
Fairbanks RG (1989) A 17,000-year glacio-eustatic sea level record: influence of glacial melting rates on the younger Dryas event and deep-ocean circulation. Nature 342:637–642
Feistel R, Nausch G, Hagen E (2006) Unusual Baltic inflow activity in 2002–2003 and varying deep-water properties. Oceanologia 48:21–35
Field CB, Behrenfeld MJ, Randerson JT, Falkowski P (1998) Primary production of the biosphere: integrating terrestrial and oceanic components. Science 281:237–240
Finni T, Kononen K, Olsonen R, Wallström K (2001) The history of cyanobacterial blooms in the Baltic Sea. AMBIO 30:168–172
Fisher H, Matthäus W (1996) The importance of the Drogden sill in the sound for major Baltic inflows. Journal of Marine Systems 9:137-157
Fonselius S (1995) Västerhavets och Östersjöns Oceanografi. Swedish Meteorological and Hydrological Institute (SMHI), Norrköping, 200 pp [in Swedish]
Fredén C (ed) (1994) Berg och jord. Sveriges Nationalatlas. SNA Förlag, Stockholm, 208 pp [in Swedish]
Granéli E, Wallström K, Larsson U, Granéli W, Elmgren R (1990) Nutrient limitation of primary production in the Baltic Sea Area. AMBIO 19:142–151
Gräwe U, Naumann M, Mohrholz V, Burchard H (2015) Anatomizing one of the largest saltwater inflows into the Baltic Sea in December 2014. Journal of Geophysical Research 120:7676–7697
Guildford SJ, Hecky RE (2000) Total nitrogen, total phosphorus, and nutrient limitation in lakes and oceans: is there a common relationship? Limnology and Oceanography 45:1213–1223
Gustafsson BG, Schenk F, Blenckner T, Eilola K, Meier HEM et al (2012) Reconstructing the development of Baltic Sea eutrophication 1850–2006. AMBIO 41:534–548
Gustafsson BG, Westman P (2002) On the causes for salinity variations in the Baltic Sea for the last 8500 years. Paleoceanography 17: 1–14
Gustafsson E, Omstedt A, Gustafsson BG (2015) The air-water CO_2 exchange of a coastal sea—a sensitivity study on factors that influence the absorption and outgassing of CO_2 in the Baltic Sea. Journal of Geophysics Research, Oceans 120:5342–5357
Hagström Å, Azam F, Kuparinen J, Zweifel UL (2001) Pelagic plankton growth and resource limitations in the Baltic Sea. In: Wulff FV, Rahm L, Larsson P (eds) A systems analysis of the Baltic Sea. Springer, Berlin, Ecological Studies 148:177–210
Hannerz F, Destouni G (2006) Spatial characterization of the Baltic Sea drainage basin and its unmonitored catchments. AMBIO 35:214–219
Hansen HP, Koroleff F (1999) Determination of nutrients. In: Grasshoff K, Kremling K, Erhardt M (eds) Methods of seawater analysis, 3rd edn Wiley-VCH Verlag, Weinheim, pp 159–228
Hansson D, Eriksson C, Omstedt A, Chen D (2011a) Reconstruction of river runoff to the Baltic Sea, AD 1500–1995. International Journal of Climatology 31:696–703
Hansson M, Andersson L, Axe P (2011b) Areal extent and volume of anoxia and hypoxia in the Baltic Sea, 1960–2011. Report Oceanography 42:1–63. Swedish Meteorological and Hydrological Institute, Norrköping
Hardin G (1968) The tragedy of the commons. Science 162:1243–1248
Hayes MO (1979) Barrier island morphology as a function of tidal and wave regime. In: Leatherman SP (ed) Barrier islands. Academic Press, New York NY, pp 1–28
Hecky RE, Campbell P, Hendzel LL (1993) The stoichiometry of carbon, nitrogen and phosphorus in particulate matter of lakes and oceans. Limnology and Oceanography 38:709–724
Hecky RE, Kilham P (1988) Nutrient limitation of phytoplankton in freshwater and marine environments: a review of recent evidence on the effects of enrichment. Limnology and Oceanography 33:796–822
HELCOM (1993) Convention of the protection of the marine environment of the Baltic Sea Area, 1974 (Helsinki Convention), with amendments to its Annexes adopted by the Helsinki Commission in 1983, 1987, 1989, 1990, 1992 and 1993. Printed by HELCOM, Helsinki, 27 pp
HELCOM (1996) Water balance of the Baltic Sea—a regional cooperation project of the Baltic Sea states international summary report. Baltic Sea Environment Proceedings 16:1–174
HELCOM (2010a) Hazardous substances in the Baltic Sea—an integrated thematic assessment of hazardous substances in the Baltic Sea. Baltic Sea Environment Proceedings 120B:1–116
HELCOM (2010b) Maritime activities in the Baltic Sea—an integrated thematic assessment on maritime activities and response to pollution at sea in the Baltic Sea region. Baltic Sea Environment Proceedings 123:1–64
HELCOM (2012) Checklist of Baltic Sea macro-species. Baltic Sea Environment Proceedings 130:1–203

Holte B, Guliksen B (1998) Common macrofaunal dominant species in the sediments of some north Norwegian and Svalbard glacial fjords. Polar Biology 19:375–382

Humborg C, Smedberg E, Rodriguez-Medina M, Mörth CM (2007) Changes in dissolved silicate loads to the Baltic Sea—the effects of lakes and reservoirs. Journal of Marine Systems 73: 223–235

Jakobsen F, Trébuchet C (2000) Observations of the transport through the Belt Sea and an investigation of the momentum balance. Continental Shelf Research 20:293–311

Jöns (2011) Settlement development in the shadow of coastal changes—case studies from the Baltic rim. In: Harff J, Björck S, Hoth P (eds) The Baltic Sea basin. Springer, Berlin, pp 301–336

Kirk JTO (2011) Light and photosynthesis in aquatic ecosystems, 3rd edn Cambridge University Press, Cambridge 662 pp

Kraberg AC, Wasmund N, Vanaverbeke J, Schiedel D, Wiltshire KH, Mieszkowska (2011) Regime shifts in the marine environment: the scientific basis and political context. Marine Pollution Bulletin 62:7–20

Krumbein WC (1934) Size frequency distributions of sediments. Journal of Sedimentary Petrology 4:65–77

Kuparinen J, Leonardsson K, Mattila J, Wikner J (1994) Food web structure, carbon flow and trends in the Gulf of Bothnia, Baltic Sea. Vatten 50:201–219

Lampe R (1995) Küstentypen. In: Rheinheimer G (ed) Meereskunde der Ostsee, 2nd edn Springer, Berlin, pp 17–25 [in German]

Lass HU, Matthäus W (1996) On temporal wind variations forcing salt water inflows into the Baltic Sea. Tellus A 48:663–671

Lass HU, Matthäus W (2008) General oceanography of the Baltic Sea. In: Feistel R, Nausch G, Wasmund (eds) State and evolution of the Baltic Sea, 1952–2005—a detailed 50-year survey of meteorology and climate, physics, chemistry, biology, and marine environment. John Wiley and Sons, Hoboken, NJ, pp 5–43

Lauckner G (1984) Brackish-water submergence of the common periwinkle, *Littorina littorea*, and its digenean parasites in the Baltic Sea and in the Kattegat. Helgoländer Meeresuntersuchungen 37:177–184

Launiainen J, Vihma T (1990) Derivation of turbulent surface fluxes—an iterative flux-profile method allowing arbitrary observing heights. Environmental Software 5:113–124

Lehmann A, Myrberg K (2008) Upwelling in the Baltic Sea. Journal of Marine Systems 74:S3–S12

Lehmann A, Post P (2015) Variability of atmospheric circulation patterns associated with large volume changes of the Baltic Sea. Advances in Science and Research 12:219–225

Lehtiniemi M, Lehmann A, Javidpour J, Myrberg K (2012) Spreading and physico-biological reproduction limitations of the invasive American comb jelly *Mnemiopsis leidyi* in the Baltic Sea. Biological Invasions 14:341–354

Lehtoranta J, Ekholm P, Pitkänen H (2008) Eutrophication-driven sediment microbial processes can explain the regional variation in phosphorus concentrations between Baltic Sea subbasins. Journal of Marine Systems 74:495–504

Lemke W, Kuijpers A, Hoffmann G, Milkert D, Atzler R (1994) The Darß sill, hydrographic threshold in the southwestern Baltic: Late Quaternary geology and recent sediment dynamics. Continental Shelf Research 14:847–870

Leppäranta M, Myrberg K (2009) Physical oceanography of the Baltic Sea. Springer, Berlin 378 pp

Levinton JS (2010) Marine biology: function, biodiversity, ecology, 3rd edn Oxford University Press, Oxford 588 pp

Lindström G (1886) Om postglaciala sänkningar af Gotland. Geologiska Föreningens i Stockholm Förhandlingar 102 VIII:251–281 [in Swedish]

Linnaeus, C. (1747) Wästgöta-resa, på riksens högloflige ständers befallning förrättad år 1746. Med anmärkningar uti oeconomien, naturkunnogheten, antiquiteter, invånarnes seder och lefnads-sätt. Lars Salvius, Stockholm [in Swedish]

Matthäus W (2006) The history of investigation of salt water inflows into the Baltic Sea: from the early beginning to recent results. Baltic Sea Research Institute Warnemünde. Germany. Marine Science Reports 65:1–74

Matthäus W, Nehring D, Feistel R, Nausch G, Mohrholz V, Lass HU (2008) The inflow of highly saline water into the Baltic Sea. In: Feistel R, Nausch G, Wasmund (eds) State and evolution of the Baltic Sea, 1952–2005—a detailed 50-year survey of meteorology and climate, physics, chemistry, biology, and marine environment. John Wiley and Sons, Hoboken, NJ, pp 265–309

Matthäus W, Schinke H (1999) The influence of river runoff on deep water conditions of the Baltic Sea. Hydrobiologia 393:1–10

Mattsson J (1996) Some comments on the barotropic flow through the Danish straits and the division of the flow between the Belt Sea and the Öresund. Tellus A 48:456–464

McLusky DS, Elliott M (2007) Transitional waters: a new approach, semantics or just muddying the waters? Estuarine, Coastal and Shelf Science 71:359–363

Meier HEM (2007) Modeling the pathways and ages of inflowing salt- and freshwater in the Baltic Sea. Estuarine, Coastal and Shelf Science 74:610–627

Meier HEM, Feistel R, Piechura J, Arneborg L, Burchard H et al (2006) Ventilation of the Baltic Sea deep water: a brief review of present knowledge from observations to models. Oceanologia 48:133–164

Meier HEM, Kauker F (2003) Modeling decadal variability of the Baltic Sea: 2. Role of freshwater inflow and large-scale atmospheric circulation for salinity. Journal of Geophysical Research 108 (C11):3368

Meier HEM, Müller-Karulis B, Andersson HC, Dieterich C, Eilola K et al (2012) Impact of climate change on ecological quality indicators and biogeochemical fluxes in the Baltic Sea: a multi-model ensemble study. AMBIO 41:558–573

Miettinen A (2002) Relative sea level changes in the eastern part of the Gulf of Finland during the last 8,000 years. Annales Academiae Scientarum Fennicae Geologica-Geographica 162:1–102

Mohrholz V, Naumann M, Nausch G, Krüger S, Gräwe U (2015) Fresh oxygen for the Baltic Sea—an exceptional saline inflow after a decade of stagnation. Journal of Marine Systems 148:152–166

Möllmann C, Diekmann R (2012) Marine ecosystem regime shifts induced by climate and overfishing: a review for the northern hemisphere. Advances in Ecological Research 47:303–347

Möllmann C, Diekmann R, Müller-Karulis B, Kornilovs G, Plikshs M, Axe P (2009) Reorganization of a large marine ecosystem due to atmospheric and anthropogenic pressure: a discontinuous regime shift in the central Baltic Sea. Global Change Biology 15:1377–1393

Murray JW, Jannasch HW, Honjo S, Anderson RF, Reeburgh WS et al (1989) Unexpected changes in the oxic/anoxic interface in the Black Sea. Nature 338:411–413

Myrberg K, Andrejev O (2003) Main upwelling regions in the Baltic Sea—a statistical analysis based on three-dimensional modelling. Boreal Environment Research 8:97–112

Niiranen S, Yletyinen J, Tomczak MT, Blenckner T, Hjerne O et al (2013) Combined effects of global climate change and regional ecosystem drivers on an exploited marine food web. Global Change Biology 19:3327–3342

Nixon SW (1995) Coastal marine eutrophication: a definition, social causes, and future concerns. Ophelia 41:199–219

Nixon SW, Buckley BA (2002) "A strikingly rich zone"—nutrient enrichment and secondary production in coastal marine ecosystems. Estuaries 25:782–796

Novotny K, Liebsch G, Lehmann A, Dietrich R (2006) Variability of sea surface heights in the Baltic Sea: an intercomparison of observations and model simulations. Marine Geodesy 29:113–134

Omstedt A (2015) Guide to process-based modelling of lakes and coastal seas, 2nd edn Springer-Praxis books in Geophysical Sciences, Springer, Berlin 273 pp

Omstedt A, Axell LB (2003) Modeling the variations of salinity and temperature in the large gulfs of the Baltic Sea. Continental Shelf Research 23:265–294

Omstedt A, Meuller L, Nyber L (1997) Interannual, seasonal and regional variations of precipitation and evaporation over the Baltic Sea. AMBIO 26:484–492

OSPAR (1998) OSPAR Agreement 1998–18, Annex 1. OSPAR strategy to combat eutrophication. [http://www.bmu.de/files/pdfs/allgemein/application/pdf/ospar_strategy3_eut.pdf]

Österblom H, Hansson S, Larsson U, Hjerne O, Wulff F et al (2007) Human-induced trophic cascades and ecological regime shifts in the Baltic Sea. Ecosystems 10:877–889

Piechura J, Beszczyńska-Möller A (2004) Inflow waters in the deep regions of the southern Baltic Sea—transport and transformations (corrected version). Oceanologia 46:113–141

Pirazzoli PA (1991) World Atlas of Holocene sea level changes. Elsevier, 291 pp

Poulícková A, Jahn R (2007) *Campylodiscus clypeus* (Ehrenberg) Ehrenberg ex Kützing: typification, morphology and distribution. Diatom Research 22:135–146

Rabalais NN, Díaz RJ, Levin LA, Turber RE, Gilbert D et al (2010) Dynamics and distribution of natural human-caused hypoxia. Biogeosciences 7:585–619

Redfield AC (1934) On the proportions of organic derivatives in seawater and their relation to the composition of plankton. In: James Johnstone Memorial Volume. University of Liverpool, pp 176–192

Redfield AC (1958) The biological control of chemical factors in the environment. American Scientist 46(205–221):230A

Redfield AC, Ketchum BH, Richards FA (1963) The influence of organisms on the composition of seawater. In: Hill MN (ed) The sea. Wiley-Liss, New York NY, pp 26–77

Reid PC, Hari RA, Beaugrand G, Livingstone DM, Marty C et al (2016) Global impacts of the 1980s regime shift. Global Change Biology 22:682–703

Remane A (1934) Die Brackwasserfauna. Verhandlungen der Deutschen Zoologischen Gesellschaft 36:34–74 [in German]

Rutgersson A, Jaagus Schenk JF, Stendel M (2014) Observed changes and variability of atmospheric parameters in the Baltic Sea region during the last 200 years. Climate Research 61:177–190

Savchuk OP, Eilola K, Gustafsson BG, Rodríguez Medina M, Ruoho-Airola T (2012a) Long-term reconstruction of nutrient loads to the Baltic Sea, 1850–2006. Baltic Nest Institute Technical Report Series 6:1–9

Savchuk OP, Gustafsson BG, Rodríguez Medina M, Sokolov AV, Wulff F (2012b) External nutrient loads to the Baltic Sea, 1970–2006. Baltic Nest Institute Technical Report Series 5:1–19

Savchuk OP, Wulff F, Hille S, Humborg C, Pollehne F (2008) The Baltic Sea a century ago—a reconstruction from model simulations, verified by observations. Journal of Marine Systems 74:485–494

Schmelzer N, Seinä A, Lundquist JA, Sztobryn M (2008) Ice. In: Feistel R, Nausch G, Wasmund N (eds) State and evolution of the Baltic Sea 1952–2005—a detailed 50-year survey of meteorology and climate, physics, chemistry, biology, and marine environment. John Wiley and Sons, Hoboken, NJ, pp 199–240

Schönfeld J, Numberger L (2007) Seasonal dynamics and decadal changes of benthic foraminiferal assemblages in the western Baltic Sea (NW Europe). Journal of Micropalaeontology 26:47–60

Schoning K (2001) The brackish Baltic Sea Yoldia Stage—palaeoenvironmental implications from marine benthic fauna and stable oxygen isotopes. Boreas 30:290–298

Schoning K, Wastegård S (1999) Ostracod assemblages in late Quaternary varved glaciomarine clay of the Baltic Sea Yoldia stage in eastern middle Sweden. Marine Micropaleontology 37:313–325

Seppä H, Bjune AE, Telford RJ, Birks HJB, Veski S (2009) Last nine-thousand years of temperature variability in Northern Europe. Climate of the Past 5:523–535

Siegel H, Gerth M, Tschersich G (2008) Satellite-derived sea surface temperature for the period 1990–2005. In: Feistel R, Nausch G, Wasmund N (eds) State and evolution of the Baltic Sea 1952–2005: a detailed 50-year survey of meteorology and climate, physics, chemistry, biology, and marine environment. John Wiley and Sons, Hoboken, NJ, pp 241–265

Sjöberg B (ed) (1992) Hav och kust. Sveriges Nationalatlas. SNA Förlag, Stockholm, 127 pp [in Swedish]

Soesoo A, Miidel A (2007) North Estonian Klint. MTÜ GEOGuide Baltoscandia. Tallinn, 2007, 29 pp

Sohlenius G, Emeis KC, Andrén E, Andrén T, Kohly A (2001) Development of anoxia during the fresh-brackish water transition in the Baltic Sea. Marine Geology 177:221–242

Stedmon CA, Osburn CL, Kragh T (2010) Tracing water mass mixing in the Baltic—North Sea transition zone using the optical properties of coloured dissolved organic matter. Estuarine, Coastal and Shelf Science 87:156–162

Strasser M (1999) *Mya arenaria*—an ancient invader of the North Sea coast. Helgoländer Meeresuntersuchungen 52:309–324

Svendsen JI, Alexandersson H, Astakhov VI, Demidov I, Dowdeswell JA et al (2004) Late Quaternary ice sheet history of northern Eurasia. Quaternary Science Reviews 23:1229–1271

Thamatrakoln K, Hildebrand M (2008) Silicon uptake in diatoms revisited: a model for saturable and nonsaturable uptake kinetics and the role of silicon transporters. Plant Physiology 146:1397–1407

Thrush SF, Dayton PK (2010) What can ecology contribute to ecosystem-based management? Annual Review in Marine Science 2:419–441

Thurow F (1997) Estimation of the total fish biomass in the Baltic Sea during the 20th century. ICES Journal of Marine Science 54:444–461

Vähätalo AV, Järvinen M (2007) Photochemically produced bioavailable nitrogen from biologically recalcitrant dissolved organic matter stimulates production of a nitrogen-limited microbial food web in the Baltic Sea. Limnology and Oceanography 52:132–143

Vaquer-Sunyer R, Duarte CM (2008) Thresholds of hypoxia for marine biodiversity. Proceedings of the National Academy of Sciences of the USA 105:15452–15457

Wærn M (1952) Rocky-shore algae in the Öregrund archipelago. Acta Phytogeographica Suecica 30:1–298

Watson DC, Norton TA (1985) Dietary preferences of the common periwinkle, *Littorina littorea* (L.). Journal of Experimental Marine Biology and Ecology 88:121–193

Wentworth CK (1922) A scale of grade and class terms for clastic sediments. The Journal of Geology 30:377–392

Widerlund A, Andersson PS (2006) Strontium isotopic composition of modern and Holocene mollusc shells as a palaeosalinity indicator for the Baltic Sea. Chemical Geology 232:54–66

Winterhalter B (1992) Late-Quaternary stratigraphy of Baltic Sea basins—a review. Bulletin of the Geological Society of Finland 64:189–194

Wulff F, Sokolov A, Savchuk (2013) Nest—a decision support system for management of the Baltic Sea, a user manual. Baltic Nest Institute, Stockholm University Baltic Sea Centre, Technical Report 10:1–70

Wulff F, Stigebrandt A, Rahm L (1990) Nutrient dynamics of the Baltic Sea. AMBIO 19:126–133

Zillén L, Conley DJ, Andrén T, Andrén E, Björck S (2008) Past occurrences of hypoxia in the Baltic Sea and the role of climate variability, environmental change and human impact. Earth-Science Review 91:77–92

Zorita E, Laine A (2000) Dependence of salinity and oxygen concentrations in the Baltic Sea on large-scale atmospheric circulation. Climate Research 14:25–41

Zweifel UL, Wilmer J, Hagström A, Lundberg E, Norrman B (1995) Dynamics of in a coastal ecosystem. Limnology and Oceanography 40:299–305

Part II

Ecological processes in the Baltic Sea

Biogeochemical cycles

3

Bernd Schneider, Olaf Dellwig, Karol Kuliński, Anders Omstedt, Falk Pollehne, Gregor Rehder, and Oleg Savchuk

Abstract

1. The internal cycles of carbon, nitrogen and phosphorus in the Baltic Sea are, like in other seas, mainly controlled by biological production and degradation of organic matter (OM).
2. Biological activity also modulates the acid/base balance (pH), which is mainly a function of alkalinity and the total CO_2 concentration.
3. Particulate organic matter (POM) produced in the photic zone sinks into deeper water layers and is deposited on the sediment surface, where it is mineralised. Mineralisation is a form of microbial oxidation and thus leads to oxygen depletion. Due to its semi-enclosed position and its bottom topography, large-scale oxygen depletion of deep bottoms is common in the Baltic Sea.
4. Under anoxic conditions, the burial of phosphorus bound to ferric oxide is inhibited and the availability of phosphate for incorporation in new OM production increases.
5. In stagnant waters, the oxic/anoxic interface may migrate from the sediment into the water column, forming a pelagic redoxcline. Such a redoxcline occurs in large areas of the Baltic Sea.
6. At oxygen concentrations close to zero, nitrate acts as an oxidant and is reduced to elemental nitrogen (denitrification). After the exhaustion of both oxygen and nitrate, OM is oxidised by sulphate, which is reduced to toxic hydrogen sulphide.
7. The final step in the mineralisation process is the microbial formation of methane in deeper sediment layers, which reflects the internal oxidation/reduction of OM.
8. A significant fraction of the organic carbon, nitrogen and phosphorus escapes mineralisation and is permanently buried in the sediment. On a long-term basis, this loss, together with export to the North Sea and internal sinks, is mainly balanced by riverine inputs and atmospheric deposition to the Baltic Sea.

Keywords

Biogeochemistry • Carbon cycle • Human impacts • Nitrogen cycle • Organic matter • Phosphorus cycle

B. Schneider (✉) · O. Dellwig · F. Pollehne · G. Rehder
Leibniz Institute for Baltic Sea Research, Seestraße 15, 18119 Rostock, Germany
e-mail: bernd.schneider@io-warnemuende.de

O. Dellwig
e-mail: olaf.dellwig@io-warnemuende.de

F. Pollehne
e-mail: falk.pollehne@io-warnemuende.de

G. Rehder
e-mail: gregor.rehder@io-warnemuende.de

K. Kuliński
Institute of Oceanology, Polish Academy of Sciences, ul. Powstańców Warszawy 55, 81712 Sopot, Poland
e-mail: kroll@iopan.gda.pl

A. Omstedt
Department of Marine Sciences, University of Gothenburg, P.O. Box 460, 40530 Göteborg, Sweden
e-mail: anders.omstedt@marine.gu.se

O. Savchuk
Baltic Nest Institute, Stockholm University, 10691 Stockholm, Sweden
e-mail: oleg.savchuk@su.se

P. Snoeijs-Leijonmalm et al. (eds.), *Biological Oceanography of the Baltic Sea*,
DOI 10.1007/978-94-007-0668-2_3

3.1 Biogeochemical cycling

3.1.1 The cycling of matter in the sea

The biogeochemical cycling of matter in the sea refers to both internal mutual transformations between different chemical and physical forms and to inputs from land and subsequent burial in the sediment. The term "cycling" is commonly used for the transformations and flows of matter even though there are no closed loops between different forms of any element, because mutual transformations interfere with internal losses and external inputs.

Biological production and degradation of organic matter (OM) are the main drivers of internal biogeochemical cycling. This includes the cycling of carbon, nitrogen and phosphorus, which are essential elements for the development of living OM. The biogeochemistry of these elements in the Baltic Sea is closely linked to processes in its drainage area, which is connected with the Baltic Sea mainly by inflowing river water and atmospheric deposition. However, the Baltic Sea is also embedded in global biogeochemical cycles, through the atmosphere and the exchange with ocean water. The cycling of matter in the sea occurs at time scales ranging from seconds for fast biochemical reactions to millions of years for geological processes. Here we mainly consider cycling over seasonal to decadal time scales.

Although many technical developments have opened new horizons for biogeochemical studies, *e.g.* autonomous underwater vehicles, moored profilers, equipment on "ships of opportunity", most of the work is still performed by oceanographers working on board research vessels (Box 3.1).

3.1.2 The Gibbs free energy controls chemistry, supported by biology

Both biogeochemical transformations and the input/output of matter control the composition of seawater. In a closed system in which there is no exchange of matter at the boundaries, the concentrations of any seawater constituent are entirely prescribed by chemical reactions. Chemical reactions are driven by a decrease in the Gibbs free energy and the reactions proceed until the energy has reached a minimum. This state is called "chemical equilibrium", which is connected with a characteristic relationship between the concentrations of the reacting substances and those of the reaction products.

A prominent and important example of chemical equilibrium is the dissociation reaction of an acid. For example, carbonic acid (H_2CO_3) is formed when carbon dioxide (CO_2) is dissolved in seawater. H_2CO_3 transfers hydrogen ions (protons) to water molecules, producing hydronium and hydrogen carbonate ions:

$$H_2CO_3 + H_2O \Rightarrow H_3O^+ + HCO_3^- \tag{3.1}$$

At equilibrium, the concentrations of the involved species are related to each other by an equilibrium constant, which is a function of temperature, pressure and salinity and in this case is called the "acid dissociation constant" (k_a) (note that in many cases H_3O^+ is depicted in a simplified manner by H^+):

$$k_a = \frac{[HCO_3^-] \cdot [H_3O^+]}{[H_2CO_3]} \tag{3.2}$$

Since this reaction takes place in an aqueous solution, changes in the H_2O concentration due to proton transfer are extremely small and can be ignored.

Whereas such acid/base equilibrium reactions proceed almost spontaneously, this is not true for most of the relevant biogeochemical processes. An example is the oxidation of OM which, according to the change in Gibbs free energy, should occur in the presence of oxygen. However, in an abiotic environment almost nothing happens at ambient temperatures because the activation energy that controls the reaction rate constant (and thus the reaction velocity) is extremely high. Bacteria possess chemical tools, such as enzymes, by which they can circumvent the activation energy barrier, allowing the oxidation (mineralisation) of OM to take place. Such microbe-induced reactions play a major role in many biogeochemical transformations.

3.1.3 The sea is not a closed vessel

It is evident that neither the Baltic Sea, nor any other sea, is a closed vessel in which the chemical composition is controlled by chemical reactions only. Instead, incessant inputs and outputs of various substances are superimposed on the internal transformations. At longer time scales, *e.g.* years, many of these systems have the tendency to reach a "steady state". This means that external inputs by river water or atmospheric deposition balance the outputs, which includes the removal by internal transformations and by transport into the sediment or to adjacent sea areas. The result is that the concentrations of these substances remain constant.

Inputs (sources) and outputs (sinks) are specified by the variable "transport" (P), which has the dimension mass per unit time, *e.g.* tonnes year^{-1}, or the variable "flux" (F), which relates P to the area of the considered system, *e.g.* tonnes year^{-1} m^{-2}. Concentrations (c) are given as the mass of a seawater constituent per volume or mass of seawater, *e.g.* g L^{-1} or g kg^{-1}. If we multiply c by the total volume or mass, we obtain the "inventory" (I) of a system in mass units.

In many cases it is useful to use molar concentrations instead of mass unit concentrations because these are directly related to the stoichiometry of the chemical reactions. Another useful variable is the "residence time" (τ), which may be interpreted as the mean time during which a molecule or particle stays in the system before it is removed by any output mechanism. In a steady state system, τ is obtained by dividing the inventory of a seawater constituent by the input or the identical output given in transport units:

$$\tau = \frac{I}{P_{in}} = \frac{c \cdot V}{P_{in}} \tag{3.3}$$

or, divided by the area:

$$\tau = \frac{c \cdot z}{F_{in}} \tag{3.4}$$

where z is the mean depth of the system.

3.1.4 A self-regulating system

Since it is unlikely that independent inputs and outputs balance each other, steady state requires feedback between inputs and outputs and thus constitutes a self-regulating system. The feedback results from a relationship between the output and the concentration of the corresponding substance. This can be simply illustrated for a hypothetical substance that is introduced into the Baltic Sea with a constant input flux and can only be removed by burial in the sediment. We consider a box containing well-mixed seawater in which the initial (t = 0) concentration of the substance is zero. With the onset of the constant input flux (F_{in}), the concentration begins to increase.

However, at the same time the removal process starts through uptake by particles, subsequent sedimentation and final burial. Particle interactions may consist of incorporation or adsorption and are assumed to increase proportionally to the concentration of the dissolved substance. This implies that the output flux continuously increases in parallel with the increasing concentrations of the substance because of the permanent input. This process is stopped when, after a characteristic time, the output flux equals the input flux, at which point a steady state is established. The temporal development of the concentration follows an exponential function (Fig. 3.1):

$$c = c_{ss} \cdot [1 - \exp(-t/t_r)] \tag{3.5}$$

where c_{SS} is the concentration at steady state and t_r is the time constant for steady state development. In this simple case, in which a linear relationship between the output flux and the

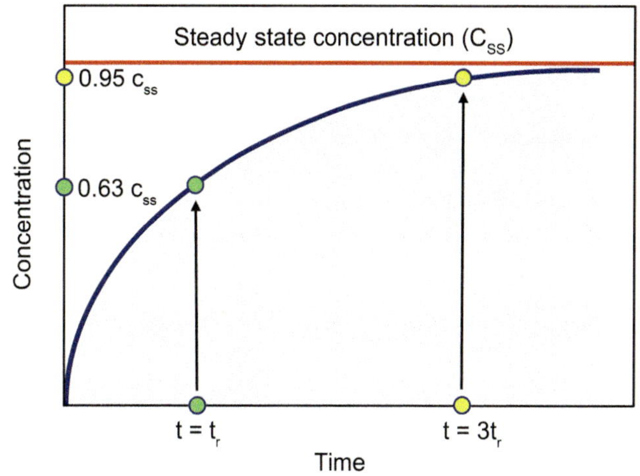

Fig. 3.1 Temporal development towards a steady state concentration (c_{SS}) according to Equation 3.5. The arrows show the concentrations reached when the time elapsed is equal to the response time (t_r, green) and to three-fold the response time (3 t_r, yellow), respectively. Figure: © Bernd Schneider

concentration is assumed, the time constant t_r is identical to the residence time τ. The time constant t_r is also called the "response time" of the system because it also controls the transition to a new steady state following a change in F_{in}.

Due to the mathematical idealisation of the processes involved, the establishment of the perfect steady state concentration requires an infinite time. However, the response time is a useful measure for estimating the time that is necessary to generate a steady state. If the elapsed time after the start or the change of an input flux is equal to the response time ($t = t_r$), ~63 % of the final steady state concentration is reached ($c = 0.63\ c_{SS}$). Steady state conditions are practically fulfilled (95 %) after the three-fold response time has elapsed (Fig. 3.1).

3.1.5 A "steady state" exists only at longer time scales

The assumption of a steady state for the seawater constituents in the Baltic Sea is the basis for many mass balance calculations, which facilitate the determination of biogeochemical fluxes and/or transformation rates that are not directly accessible by experimental methods. However, the steady state hypothesis must refer to an appropriate time scale. In the Baltic Sea, and in many other marine systems, biogenic production of particulate organic matter (POM) is the main driver of the removal of biogeochemically relevant seawater constituents and their burial in the sediment. Since the production of POM shows a pronounced seasonality (*cf.* Sect. 8.2), the minimum time scale for the assumption of a steady state for the surface water is one year.

Box 3.1: Oceanographic field work

Sampling and on-board measurements

Biogeochemical measurements are usually made from research vessels (Box Fig. 3.1a, b), either by direct measurements in the water column or from analysing seawater samples taken at different water depths. Water samples are usually taken with Niskin bottles attached to an oceanographic instrument that measures conductivity, temperature and depth (CTD), which is lowered into the sea from a research vessel (Box Fig. 3.1c–e). Niskin bottles can be opened at both ends. The open bottle is lowered into the sea on a CTD until it reaches a certain, predestined depth, at which the

Box Fig. 3.1 Oceanographic field work at research vessels. (**a**) The former Swedish research vessel R/V Argos (built 1974), (**b**) The Finnish research vessel R/V Aranda (built 1989). (**c**) Different types of bottles for water sampling. (**d**) A CTD with Niskin bottles attached in a circular rosette sampler is lifted out of the sea with a crane and brought on deck. (**e**) Water samples for gas analyses are taken from the Niskin bottles immediately after the rosette has come up from the sea. (**f**) On-board equipment for filtering phytoplankton from water samples for chlorophyll *a* analyses. The samples need to be kept in the dark so that light does not destroy the pigments. Photo: © Pauline Snoeijs-Leijonmalm

Box Fig. 3.2 During oceanographic field work on board research vessels, equipment needs to be firmly attached because of the ship's movements. (**a**) Oceanographer operating a CTD. (**b**) Oceanographer taking a water sample. (**c**) On-board equipment for filtering phytoplankton from water samples for metagenomic analyses. The phytoplankton from more than 100 L of seawater is sequentially filtered on large (diameter 29 cm) filters with pore sizes of 3 μm, 0.8 μm and 0.1 μm. (**d**) Equipment for fluorometric measurement of photosynthetic capacity. (**e**) Titration equipment is used for measuring *e.g.* the O_2 concentration in seawater. (**f**) Gas chromatography and mass spectrometry is used for measuring *e.g.* the concentrations of volatile halocarbons in seawater. Photo: © Pauline Snoeijs-Leijonmalm bottle is closed automatically and a water sample has been taken. However, there are also many different designs of Niskin-type bottles which can be operated by hand (Box Fig. 3.2b). A CTD can also be combined with other sensors, *e.g.* for measuring fluorescence (Box Fig. 3.2a). Some chemical analyses of seawater samples can be performed later in a land-based laboratory, while others require immediate on-board analysis. Phytoplankton samples are preferably filtered on board and the filters are frozen, *e.g.* for later chlorophyll *a* analysis (Box Fig. 3.1f) or metagenomics (Box Fig. 3.2c). Measurements of photosynthesis and certain gases need to be carried out immediately (Box Fig. 3.2d–f).

CO_2 measurements on board "ships of opportunity"

Some types of oceanographic measurements can be made by an automated measurement system installed on board the so-called "ships of opportunity", *e.g.* cargo ships and passenger ferries that frequently follow the same route. This is a relatively cheap way to collect a large amount of data, both in space and time. Measurements of the partial pressure of CO_2 in surface waters are made by an automated measurement system on board the cargo ship "Finnmaid" (Finnlines Shipping Company), which commutes regularly at time intervals of two to three days between Lübeck at the Mecklenburg Bay (Belt Sea) and Helsinki in the Gulf of Finland (Box Fig. 3.3). Seawater is continuously pumped into an equilibrator, while at the same time air is circulating in a closed loop through the water column in the equilibrator and equilibrates with the CO_2 in the seawater. The CO_2 partial pressure of the equilibrated air is detected by an infrared gas analyser. Due to the high temporal resolution and the spatial coverage of the measurements, the data can be used to identify and quantify plankton bloom events in different regions of the Baltic Sea proper. Furthermore, the data facilitate the calculation of the gas exchange between the Baltic Sea water and the atmosphere.

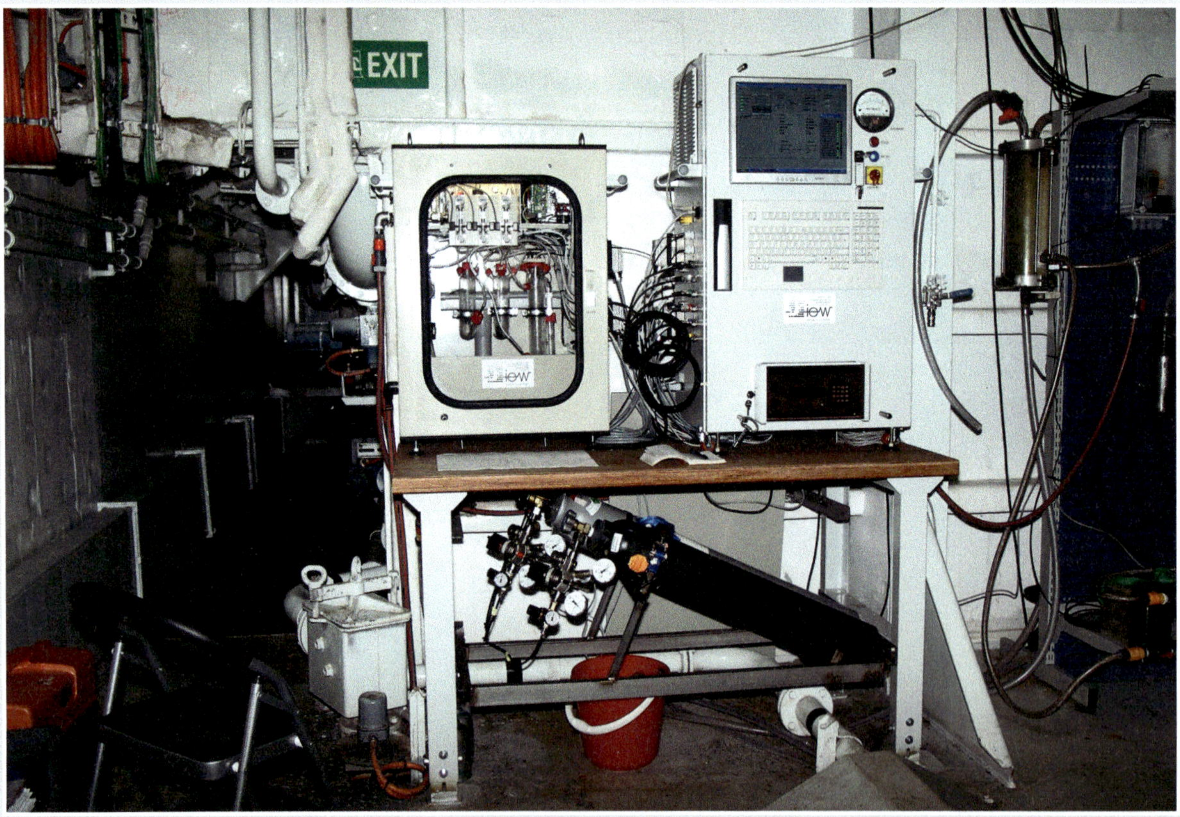

Box Fig. 3.3 Instrumentation for CO_2 measurements on board the cargo ship "Finnmaid", showing the equilibrator to the left and the infra-red CO_2 analyser to the right. Photo: © Bernd Sadkowiak

The situation is different for the deep water in some of the Baltic Sea's major basins, such as the Gotland Sea or Bornholm Sea. Due to periods of stagnation, which can last for many years, the products of OM mineralisation may accumulate continuously until a water renewal event occurs. In such cases, the prerequisites for a steady state are only fulfilled if concentrations and input/output fluxes are averaged over many decades.

3.2 Principles of organic matter production

3.2.1 Biogeochemical cycles start with photosynthesis

Primary production is based on photosynthesis, *i.e.* a complex sequence of biochemical reactions during which dissolved CO_2 is transformed into OM. Water plays a vital role in this process, as well as dissolved inorganic nitrogen and phosphorus compounds. The latter two compounds are called "inorganic nutrients" and they are essential. The net photosynthetic process and the subsequent biochemical reactions, which are necessary to generate the building blocks of a living cell, can be expressed based on a rough approximation of the bulk composition of OM (Redfield et al. 1963):

$$106CO_2 + 16NO_3^- + HPO_4^{2-} + 122H_2O + 18H^+ \Rightarrow (CH_2O)_{106}(NH_3)_{16}H_3PO_4 + 138O_2 \quad (3.6)$$

This reaction requires the input of energy, which is delivered by solar radiation. Chlorophyll *a* and other pigments absorb light in the visible region of the spectrum (400–700 nm), which is called photosynthetically active radiation (PAR). The light energy is transferred into chemical energy via several enzymatically catalysed electron-transfer reactions that lead to the formation of adenosine triphosphate (ATP).

After undergoing several complex biochemical reactions, CO_2 is transformed into carbohydrates such as sugars and polysaccharides, which constitute major components of living OM. Hence, the carbon in CO_2 is reduced from an oxidation number of +4 to zero. During this process the O^{2-} ions of the water molecules have acted as electron donors and are thus oxidised to elemental oxygen.

3.2.2 Living cells need nitrogen and phosphorus

Another major class of organic compounds produced in living cells, in addition to sugars and polysaccharides, comprises amino acids and proteins. Nitrogen is the key element in these compounds and is available for primary producers (phytoplankton) in the surface water mainly as nitrate (NO_3^-), but also as ammonium (NH_4^+). Since the nitrogen in nitrate ions has an oxidation number of +5, a reduction step must take place in order to obtain an oxidation number of −3, which is characteristic of amino groups in OM. Again, the O^{2-} ions of water molecules act as electron donors and additional oxygen is released.

Ammonium has the oxidation number −3. Hence, its use is energetically favoured because there is no need for any reduction/oxidation reactions before it can serve as a nitrogen source for OM production. However, surface-water ammonium concentrations are generally low and the use of ammonium as a nutrient plays a role only during high zooplankton abundance, *i.e.* at the late stage of a plankton bloom. Zooplankton organisms graze on phytoplankton and excrete ammonium, which fuels the so-called "regenerated primary production".

In the absence of dissolved inorganic nitrogen (DIN: nitrate + nitrite + ammonium), the nitrogen supply for primary production is ensured by nitrogen fixation. During this process elemental nitrogen dissolved in seawater is used for the formation of nitrogen-containing OM. Some organisms are equipped with the enzyme systems needed to accomplish this energy-demanding process. The main diazotrophic (nitrogen-fixing) organisms in the Baltic Sea are cyanobacteria, which become abundant after the complete exhaustion of DIN and contribute significantly to the nitrogen budget of the Baltic Sea (Larsson et al. 2001; Wasmund et al. 2001).

Finally, organic phosphorus compounds play an important role in the structure and function of living cells. Phosphorus is contained mainly in ATP and the phospholipids that form cell membranes. In seawater, phosphorus is present as phosphate ions, which at the typical seawater pH of ~ 8 prevail mainly as monohydrogen phosphate, HPO_4^{2-}, although that is commonly depicted as PO_4^{3-} or dissolved inorganic phosphorus (DIP). The incorporation of HPO_4^{2-} into OM does not require any reduction/oxidation reaction.

Although nitrogen and phosphorus are the central elements for living cells, many other elements are involved in biochemical metabolism and may act as limiting micronutrients. A prominent example is iron, which *e.g.* limits plankton growth in large areas of the Southern Ocean (Olson et al. 2000).

3.2.3 The elemental composition of organic matter may vary

Another important aspect of primary production is the stoichiometry of POM production because it is related to the limitation of productivity by the lack of an essential nutrient. Pioneering studies on the elemental ratios in plankton were performed by Alfred C. Redfield (Fig. 3.2) who found

Fig. 3.2 Plaque commemorating Alfred C. Redfield (1890–1983) placed at the entrance to the main building of the Woods Hole Oceanographic Institute, Woods Hole, Massachusetts (USA). Photo: © Teresa Radziejewska

average molar C:N:P ratios of 106:16:1 (Redfield et al. 1963). These ratios are known as "Redfield ratios", and they are still used in many budget calculations and biogeochemical models (see Equation 3.6).

The term "Redfield ratio" is also used to describe the relationship between dissolved nitrate and phosphate in the water, which in the ocean is mostly close to the average N:P ratio of 16:1 in POM. However, this is not the case in the Baltic Sea because nitrate and phosphate are not recycled in a closed loop of POM production and decomposition. Rather, their concentrations (*cf.* Fig. 2.23) are strongly influenced by external inputs and internal sinks and sources.

Here we use the Redfield ratio mainly with respect to the composition of POM. However, Redfield himself already recognised that these ratios are not universal constants and that substantial deviations are common. This can happen *e.g.* when the molar inorganic N:P ratio of the nutrient pool significantly deviates from the Redfield ratio. In this case, nitrogen-deficient cells or phosphorus-deficient cells are formed. POM compositions that differ from the Redfield ratios are also found in the Baltic Sea (Schneider et al. 2003). For example, during the pelagic spring bloom slightly nitrogen-deficient POM develops with molar C:N ratios of up to 9 (Redfield ratio C:N = 6.6) and N:P ratios as low as ~12 (Redfield ratio N:P = 16). However, these observations do not necessarily reflect a shift in the elemental ratios during uptake. POM consists not only of living cells but also of detritus that, during bacterial decomposition, may have been subjected to the preferential mineralisation of nitrogen and/or phosphorus over carbon.

A more drastic change in the elemental composition of POM occurs with the onset of production that is fuelled by nitrogen fixation coinciding with low PO_4^{3-} concentrations. In this case, the C:P and N:P ratios of POM easily exceed the Redfield ratios by a factor of four or even more, while the C:N ratio remains around 6.6. However, the nature of the POM responsible for these extreme elemental ratios is still unclear.

3.2.4 Oxygen is involved in POM production and mineralisation

The relationship between POM production and O_2 release is of high biogeochemical importance because it controls also oxygen consumption during the mineralisation of POM. According to the simplified description of the photosynthesis reaction in Equation 3.6, only carbohydrates are produced and the ratio between O_2 release and organic carbon production is 138:106 = 1.30, or 1.00 if the reduction of NO_3^- is disregarded.

However, since living cells consist of a mixture of different types of organic compounds (proteins, carbohydrates, lipids, nucleic acids), alternative compositions of POM, containing additional hydrogen-carbon bonds and thus being in a more reduced state, have been suggested (Anderson 1995). Consequently, the ratio between O_2 release and the formation of organic carbon increases and may reach values of approximately 150:106, which therefore also characterises O_2 consumption during the mineralisation of POM.

3.2.5 Silicate and calcium are needed for the synthesis of cell covers

Many plankton species have hard cell covers that consist of either silicate or calcium carbonate. Diatoms, *e.g. Coscinodiscus granii* (Fig. 3.3a), produce a variety of siliceous structures from silicates dissolved in seawater. Diatoms are the dominant component of plankton during the early spring bloom in the Baltic Sea. The silicate demand of diatom cells roughly corresponds to that of nitrogen (Si:N ≈ 1:1.07, *cf.* Sect. 2.4.8).

Calcifying organisms in the plankton, *e.g.* the haptophyte *Emiliania huxleyi* (Fig. 3.3b), use calcium and carbonate ions for the formation of their calcium carbonate ($CaCO_3$) shells, in the form of either aragonite or calcite. The main difference between these two crystalline modifications is their solubility in seawater, which for aragonite is almost twice that for calcite. In the Baltic Sea, the abundance of calcifying plankton is significant only in the transition zone to the North Sea.

The reason for the virtual absence of calcifying organisms in the pelagic zone of the Baltic Sea is not entirely clear, but may be related to the low salinity of these waters, which

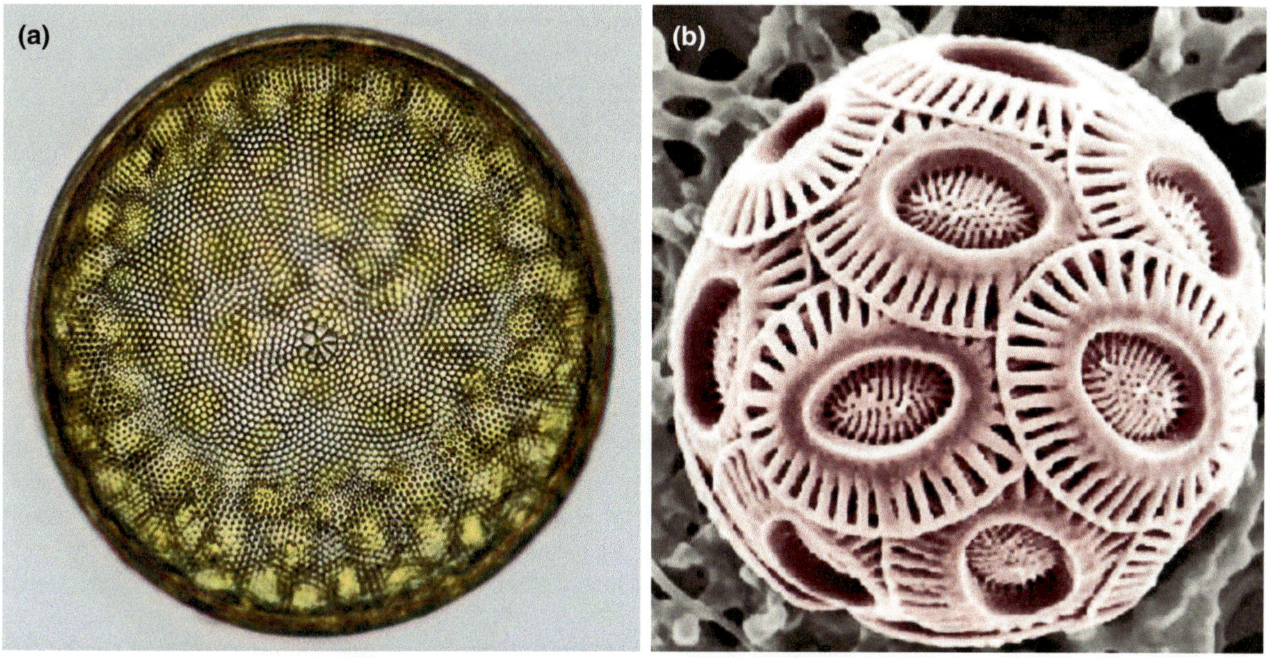

Fig. 3.3 Plankton algae from the Baltic Sea Area. (**a**) Light micrograph of the diatom *Coscinodiscus granii*, (**b**) Scanning-electron micrograph of the calcifying haptophyte *Emiliania huxleyi*. Photo: (**a**) © Pauline Snoeijs-Leijonmalm, (**b**) © Regina Hansen

implies low calcium concentrations. Together with the typical low carbonate ion concentrations during winter and at the beginning of the spring bloom, this may inhibit the formation of $CaCO_3$ (Tyrrell et al. 2008).

3.2.6 DOM contributes to the cycling of carbon and nutrients

Also of significance, dissolved organic matter (DOM) participates in the biogeochemical cycles of C, N and P. In the surface water of the Baltic Sea proper, the concentrations of dissolved organic carbon (DOC) are in the range of 300–350 µmol L^{-1} (Nausch et al. 2008). About 70 % of the DOC is of terrestrial origin and consists mainly of long-lived, refractory humic substances (*cf.* Sect. 15.2.6). The remaining ~30 % is made up of exudates, mainly low-molecular-weight organic compounds available for bacterial metabolism, that are released during the production and decomposition of POM.

Both terrestrial and marine DOM contains nitrogen and phosphorus. In the Baltic Sea proper, the concentrations of dissolved organic nitrogen (DON) and dissolved organic phosphorus (DOP) are ~15–20 µmol L^{-1} and ~0.3–0.5 µmol L^{-1}, respectively (Nausch et al. 2008). Laboratory experiments and field studies have shown that some of the compounds contained in DOM are used as nutrients during primary production (*e.g.* Asmala et al. 2013). This may also occur under natural conditions when the corresponding inorganic nutrient pools are depleted.

3.2.7 Nutrient uptake rates are limited

Nutrients needed for the autotrophic production of OM are taken up by algal cells from the ambient water at a rate dependent on nutrient concentrations:

$$V_{upt} = \frac{V_{max} \cdot N}{k_c + N} \tag{3.7}$$

In Equation 3.7, V_{upt} and V_{max} are the respective actual and maximum nutrient uptake rates (in µmol L^{-1} d^{-1}) at the concentration N (*e.g.* NH_4^+, NO_2^-, NO_3^-, HPO_4^{2-}, SiO_4^{4-}). The constant k_c is the half-saturation constant and represents the concentration at which $V_{upt} = V_{max}/2$. For a given V_{max}, the half-saturation constant determines the steepness of the uptake curve (Fig. 3.4) and is thus a measure of the uptake rate at low concentrations: low uptake at high k_c and high uptake at low k_c.

V_{max} is reached asymptotically at high nutrient concentrations and is the natural limit for the rate of nutrient consumption, determined by species-specific physiological limits on the rate of cell division or limits in the availability of light energy required for inorganic nutrient assimilation. Similar curves characterise enzyme kinetics (the Michaelis-Menten equation: Michaelis and Menten 1913; Johnson and Goody 2011), growth of microorganisms (the Monod equation: Monod 1942, 1949) and surface adsorption (the Langmuir equation: Langmuir 1916, 1917). All of these processes are involved in nutrient uptake and give rise to the empirical relationship described by Equation 3.7.

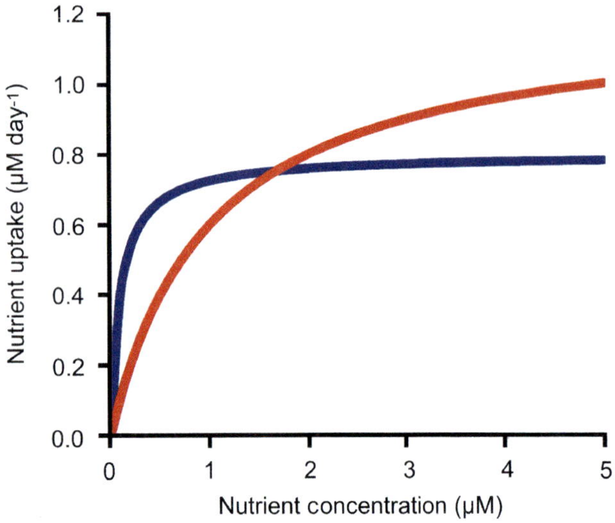

Fig. 3.4 Examples of nutrient-uptake curves, illustrating interspecific competition between two primary producers under initial nutrient-replete conditions. The species marked in red ($k_c = 1.0$, $V_{max} = 1.2$) has an advantage at high nutrient concentrations (above ~1.8 μM), whereas the species marked in blue ($k_c = 0.1$, $V_{max} = 0.8$) takes over when nutrient concentrations decrease. k_c = the half-saturation constant, V_{max} = the maximum nutrient uptake rate, see Equation 3.7. Figure: © Oleg Savchuk

Since nitrogen occurs in OM in reduced forms of amines and amide groups, ammonium is preferentially taken up relative to oxidised nitrogen compounds (nitrate and nitrite). This preferential uptake is regulated by the availability of nitrate and nitrite reductases, as the synthesis of these nitrogen-reducing enzymes is suppressed by ammonium at concentrations in the range of 0.5–1.0 μM (Mulholland and Lomas 2008).

In addition to inorganic compounds, algae may utilise the nitrogen and phosphorus contained in dissolved organic compounds, such as urea, amino acids, nucleotides (ATP) and phospholipids, by employing a variety of mechanisms, ranging from the active uptake of low-molecular-weight substances to enzymatic mineralisation at the algal cell surface prior to uptake.

3.2.8 Measurement of POM production

Different production terms are used to characterise the efficiency of primary production and, accordingly, there are different methods for its quantification. Production rates were traditionally determined by measurements of oxygen release in enclosed seawater samples. This method has been largely replaced by the ^{14}C method, which yields production rates at the time scale of hours. Seawater samples are incubated with ^{14}C-labelled hydrogen carbonate, and the uptake of ^{14}C by the plankton is determined by measuring the radioactivity in POM. The production rates obtained in this way represent instantaneous production and may be used to study the related controlling factors, such as nutrient concentrations and light conditions. Important to note is that such rates do not represent the net production over longer time scales since the interplay between POM production and respiration is not accounted for.

Studies on the impact of primary production on biogeochemical cycles and budgets are mostly based on net effects occurring at time scales of a day or longer. Different methods, mainly based on simple budget calculations, have been introduced to estimate net production (Wasmund et al. 2005a). Traditionally, the loss of nutrients in the surface water during the productive period in combination with the corresponding Redfield ratio is considered to be a measure of production. For example, nitrate depletion, together with Redfield's or any other appropriate C:N ratio, provides a measure of the net production during the early spring bloom. However, this method fails if production is fuelled by nitrogen fixation. In this case the net production can be estimated from phosphate consumption.

Because of highly variable C:P ratios, which may deviate from the Redfield C:P ratio (106:1) by a factor of four or higher (*cf.* Sect. 3.2.3), this method is associated with a large uncertainty. Alternatively, the generation of oxygen during OM production has been used to estimate net production. However, this approach is based on a mass balance in which O_2 gas exchange with the atmosphere is a major budget term, and this also causes substantial uncertainty.

This problem is partly circumvented when the loss of CO_2 during OM production, rather than the gain in O_2, is considered (Schneider et al. 2006). The time for equilibration with the atmosphere is about five to ten times longer for CO_2 than for O_2. Consequently, gas exchange plays only a minor role in the mass balance of CO_2 and relatively reliable POM production estimates can be obtained.

3.3 Nutrient control of primary production in the Baltic Sea

3.3.1 Growth limitation during the spring bloom

The concept of nutrient limitation was recognised by agricultural scientists already in the first half of the 19th century and had implications for the use of fertilisers (Hignett 1985). In what is now known as "Liebig's Law of the Minimum", organisms are understood to require different nutrients at a characteristic relationship and cease to grow if one of the essential nutrients is no longer available. This nutrient is then the limiting factor in production. While oceans are predominately nitrogen-limited and freshwater bodies are predominately phosphorus-limited, the brackish Baltic Sea has gradients of nutrient limitation at both regional and local scales.

To assess nutrient limitation for production in the Baltic Sea, it is convenient to consider the concentrations of DIN and DIP that have accumulated during the dark cold winter (Fig. 3.5a, b) as these constitute the starting conditions for the pelagic spring bloom. The ratio between winter DIN and DIP concentrations in the surface waters indicates which of these nutrients limits phytoplankton growth in spring. Ratios that exceed the phytoplankton N:P demand, according to the Redfield ratio (16:1), indicate that the system is phosphorus-limited because phosphate will be exhausted while nitrate is still available. This is the case in the Bothnian Bay and in many coastal areas that are directly influenced by the input of nutrients from land (Fig. 3.5c). By contrast, the winter DIN:DIP ratios in the Baltic Sea proper and in large parts of the Gulf of Finland are below 16 and thus indicate nitrogen limitation during the spring bloom.

Since the spring bloom is dominated by diatoms, silicate concentrations must also be taken into account when assessing limiting factors for phytoplankton growth. The silicate demand of diatoms in relation to nitrate uptake corresponds to a Si:N ratio of 0.8–1.2 (Brzezinski 1985). Since the winter DSi:DIN ratio in the surface waters is higher in the open Baltic Sea (*cf.* Fig. 2.23b), silicate is not exhausted and silicate limitation of the spring bloom should not occur. Nonetheless, there are several indications that the silicate uptake is inhibited if DSi concentrations drop below a certain threshold value (*e.g.* Azam and Chrisholm 1976).

3.3.2 Production continues after nitrate depletion

The pelagic spring bloom in the Kattegat may start already in February and propagates northwards in the Baltic Sea and from coastal to offshore waters. Nutrient uptake by plankton results in pronounced seasonal dynamics of nitrate and phosphate in the photic zone. In the Eastern Gotland Sea, the spring phytoplankton bloom occurs in March-April and results in the consumption of the nutrients accumulated during winter, approximately in Redfield ratio proportions (*e.g.* Ptacnik et al. 2010; Walve and Larsson 2010). At the same time, CO_2 concentrations in the photic zone decrease (*cf.* Sect. 3.4.2).

There is almost no nitrate left by mid-April and phytoplankton growth becomes nitrogen-limited, while ~0.2–0.3 μM of phosphate is still available (Fig. 3.6). During the following weeks, from mid-April to June, this remaining phosphate, which amounts to almost 50 % of the winter phosphate pool, and the surface water total CO_2 are depleted (Schneider et al. 2009). This clearly indicates net OM production which, however, requires nitrogen, the presence of which can be explained by several hypotheses. For example, nitrogen demand could be satisfied by the utilisation of dissolved organic nitrogen (DON) compounds (Bronk 2002). In addition, it cannot be ruled out that nitrate from deeper water layers is transported up to the photic zone by the active vertical migration of phytoplankton (Höglander et al. 2004; Laanemets et al. 2004). Finally, nitrogen fixation may take place already in the cold spring water by some unknown species, such as members of the picocyanobacteria or heterotrophic bacteria. These hypotheses have yet to be substantiated by direct observations.

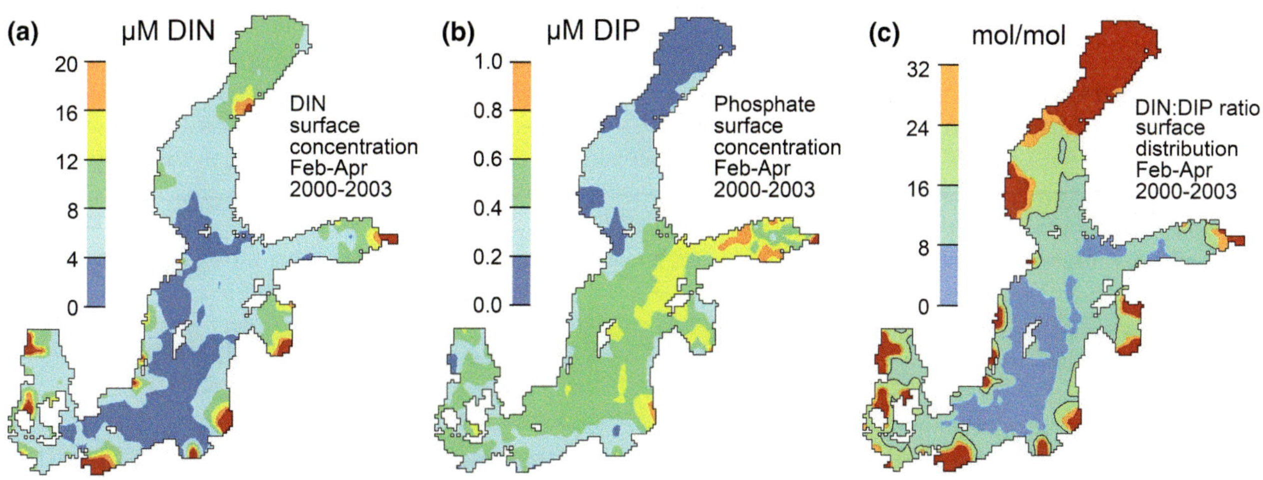

Fig. 3.5 Distribution of late-winter inorganic nutrient concentrations in the surface water of the Baltic Sea Area. (**a**) Dissolved inorganic nitrogen (DIN), (**b**) Dissolved inorganic phosphorus (DIP), (**c**) the resultant DIN:DIP ratio. Red colour indicates >20 μM DIN in (**a**), >1.0 μM DIP in (**b**), and DIN:DIP >32 on a molar basis in (**c**). The images of the distributions were drawn from average three-dimensional fields reconstructed with the Data Assimilation System (DAS) from data contained in the Baltic Environmental Database (BED) for February-April 2000–2003. Figure: © Oleg Savchuk

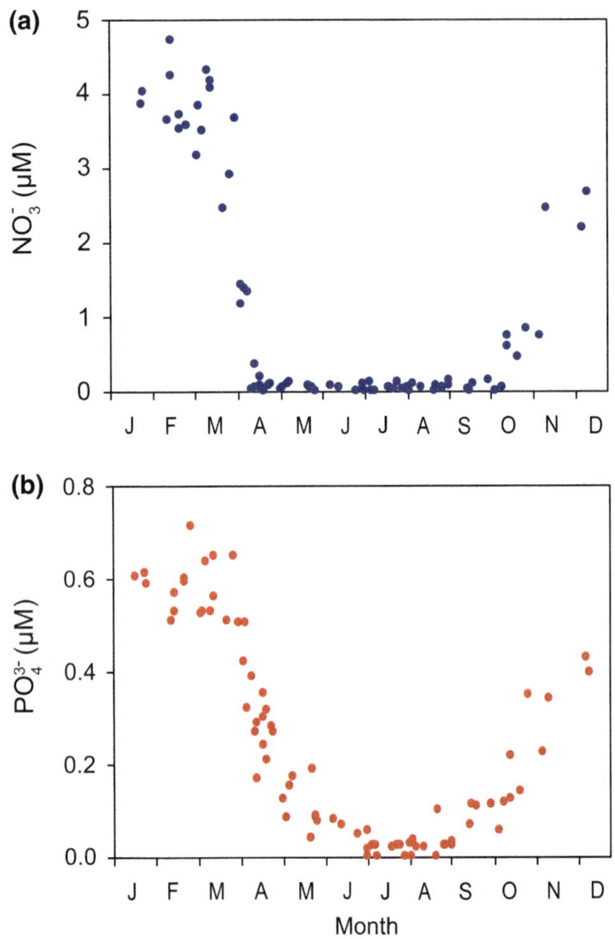

Fig. 3.6 Seasonality of inorganic nutrient concentrations in the surface water of the Eastern Gotland Sea 2004–2009. (**a**) Nitrate concentrations. (**b**) Phosphate concentrations. Figure based on nutrient data from the Finnish Alg@line monitoring by the Marine Research Centre of the Finnish Environment Institute (http://www.syke.fi). Figure: © Bernd Schneider

3.3.3 Cyanobacteria circumvent nitrogen limitation

From late June to early August, at water temperatures above 14–16 °C, diazotrophic filamentous cyanobacteria fix elemental nitrogen in a large part of the Baltic Sea. During calm weather and at high water temperatures this results in the formation of spectacular cyanobacterial blooms (*cf.* Fig. 8.5). The fixed nitrogen also becomes available to other members of the phytoplankton community via release from the cyanobacterial cells and by heterotrophic regeneration carried out by bacteria and zooplankton. The intense biomass production during cyanobacterial blooms is also reflected by a peak in POM sedimentation in the middle of summer (*cf.* Sect. 3.5.2). Cyanobacterial blooms are mainly confined to the Baltic Sea proper and the Gulf of Finland, where phosphate is still available after the nitrogen-limited spring bloom. However, also in these regions phosphate concentrations are low at the start of the summer bloom in late June-July (Fig. 3.6b).

There are several hypotheses that can explain the source of the phosphorus supply enabling the production of high cyanobacterial biomass (POM). Among these are the use of organic phosphorus compounds and phosphate fluxes to the surface water by upwelling (Nausch et al. 2009). Yet, the extremely high C:P and N:P ratios of the POM, which may exceed 400 and 60, respectively, indicate that phosphorus demand for POM production is much lower than could be expected from the Redfield ratio.

The lack of measurements performed at adequate spatial and temporal scales still hamper reliable estimates of annually integrated nitrogen fixation rates. Nevertheless, nitrogen fixation undoubtedly contributes significantly to the nitrogen budget of the Baltic Sea and thus to eutrophication. Depending on the methodological approach and the year considered, values from 170 kilotonnes year^{-1} to 792 kilotonnes year^{-1} have been reported (Larsson et al. 2001; Wasmund et al. 2005b; Rolff et al. 2007), which is comparable to both atmospheric nitrogen deposition and nitrogen input from land (*cf.* Sect. 3.8.4).

3.3.4 The autumn bloom concludes the productive season

Cooling of the surface water and increasing winds result in the gradual erosion of the seasonal thermocline in September-November. Consequently, vertical mixing transports nutrients from deeper layers to the surface waters and gradually restores winter nutrient concentrations. This stimulates the last productive period, the phytoplankton autumn bloom. In this case, production is not limited by nutrient concentrations, which increase continuously as the mixed layer deepens. Rather, shorter days and the increased mixing reduce the exposure of plankton cells to solar radiation, and limit primary production at this time of year.

3.4 Interactions between the CO_2 system and biological processes

3.4.1 Atmospheric CO_2 and alkalinity control mean seawater C_T

At a given atmospheric partial pressure of CO_2 (pCO_2), alkalinity controls the mean total of dissolved inorganic carbon concentration (C_T) in the water (*cf.* Sect. 1.3.11). Unlike in ocean waters where the alkalinity/salinity ratio is almost constant, the relationship between total alkalinity (A_T) and salinity (S) in the Baltic Sea is much more complex. Plotting

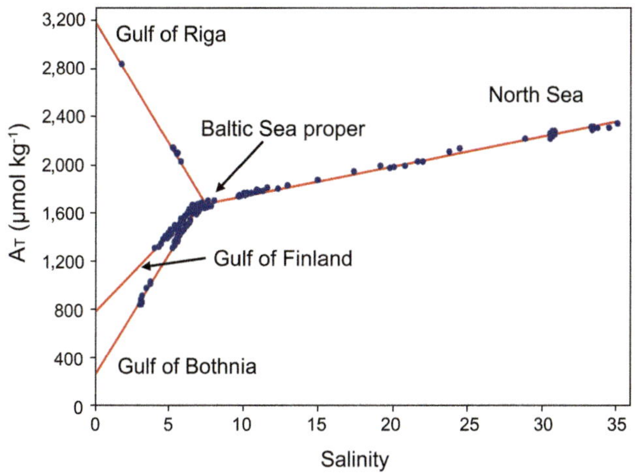

Fig. 3.7 The relationship between salinity and alkalinity in different regions of the Baltic Sea and in a gradient from the Baltic Sea proper to the North Sea. Figure based on salinity and alkalinity data in Schneider (2011). Figure: © Bernd Schneider

the surface-water alkalinity from different regions as a function of salinity is a convenient method to characterise the alkalinity distribution in the Baltic Sea (Fig. 3.7).

Extrapolating A_T on a regional basis to salinity = 0 yields the mean A_T of the rivers discharging into the corresponding sea area (Fig. 3.7). The low river-water A_T of Scandinavian rivers discharging into the Gulf of Bothnia reflects the prevailing igneous rocks, which are resistant to weathering, in this part of the Baltic Sea drainage area. In contrast, the limestone-rich part of the drainage area in continental Europe, which is more affected by weathering, provides high A_T values in rivers such as the Daugava that flows into the Gulf of Riga (*cf.* Fig. 2.11). The A_T in the Baltic Sea proper is a mixture of the different A_T contributions including that from the northern Atlantic Ocean, which can be estimated by extrapolating the A_T for the transition zone between the Baltic Sea and the North Sea to S = 35 (Fig. 3.7).

The large-scale distribution of C_T along a transect from the Kattegat through the Western Gotland Sea to the Bothnian Bay shows large horizontal and vertical gradients (Fig. 3.8). Since C_T and A_T are related (*cf.* Sect. 1.3.12), the surface-water C_T continuously decreases towards the low A_T water in the north. The higher C_T of deeper water layers is caused by two factors: (1) higher salinities originating from the inflow of Kattegat water imply a higher A_T and thus a higher C_T, and (2) deeper water layers may be strongly enriched in C_T because of the mineralisation of POM derived from either sinking particles from the surface or lateral particle transport.

3.4.2 Biology induces daily and seasonal cycles of pCO_2 and pH

CO_2 consumption during primary production superimposes a distinct seasonal C_T cycle on the A_T-controlled background C_T. This is clearly reflected in the seasonal variability of the surface-water partial pressure of CO_2 (pCO_2) in the Baltic Sea proper (Fig. 3.9). With the start of spring bloom production, the surface-water pCO_2 drops below the atmospheric pCO_2 and reaches a minimum by mid-May.

In the following weeks, the pCO_2 increases slightly because the effect of rising temperatures dominates over that

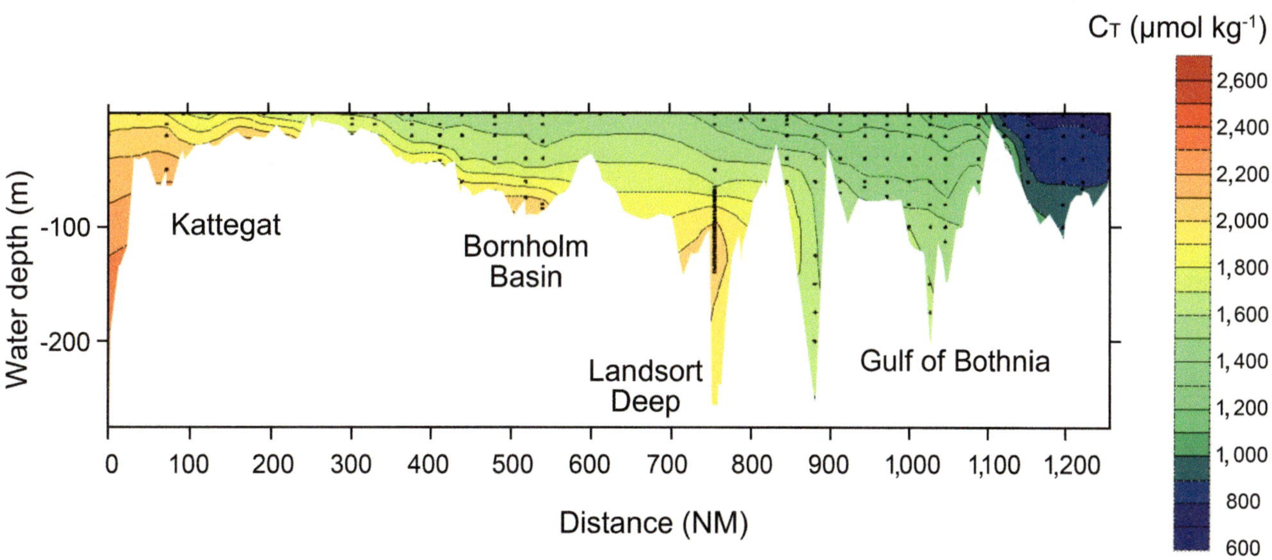

Fig. 3.8 Large-scale distribution of the total dissolved inorganic carbon (C_T) concentration along a transect from the Kattegat to the Bothnian Bay. Black dots indicate sampling positions and depths. The distance is given in nautical miles (NM). Figure modified from Bełdowski et al. (2010)

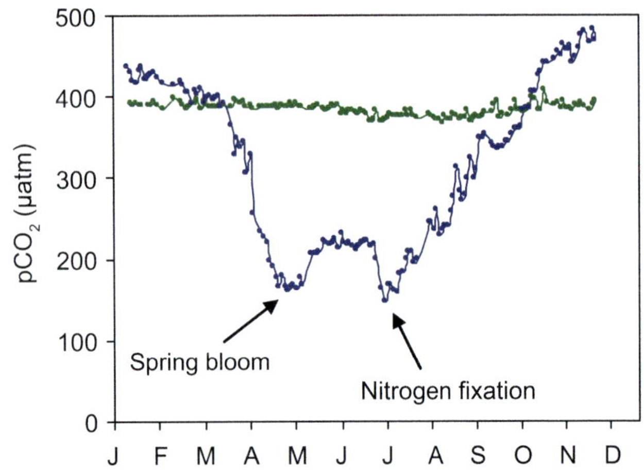

Fig. 3.9 Typical seasonal development of atmospheric pCO_2 (green) and surface water pCO_2 (blue) in the Baltic Sea proper. The data were obtained from an automated pCO_2 measurement system deployed on a cargo ship. Figure modified from Schneider (2011)

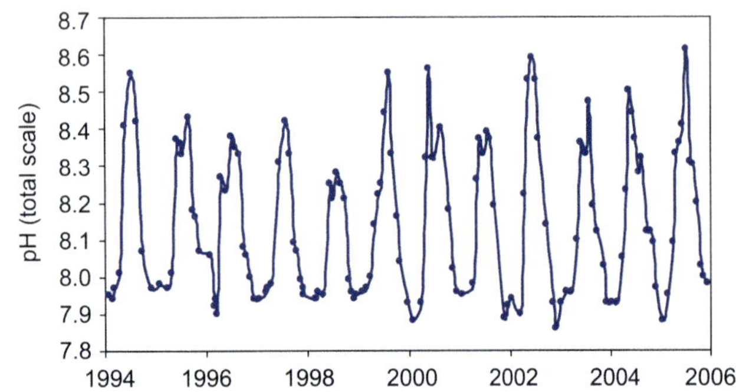

Fig. 3.10 Time-series data of surface-water pH (total scale, *cf.* Sect. 1.3.9) in the Eastern Gotland Sea 1994–2006. Figure based on pH data from the Swedish National Monitoring Programme of the Swedish Meteorological and Hydrographical Institute (SMHI). Figure: © Bernd Schneider

of OM production, which is less efficient during this period. However, this trend stops by mid-June, when the pCO_2 decreases again, and a second minimum develops by mid-July. This reflects a second intense production phase fuelled by nitrogen fixation of cyanobacterial blooms. In the following months, the shallow summer stratification gradually erodes and deeper CO_2-enriched water layers are transported to the surface. The pCO_2 in the surface water therefore increases and finally reaches maximum values well above the atmospheric pCO_2.

Since the pCO_2 and the hydrogen ion concentration are closely linked (*cf.* Sect. 1.3.9), the surface-water pH exhibits a distinct seasonality that is mainly controlled by biological production and the input of CO_2-enriched water by vertical mixing. Maximum pH values of 8.5–8.6 are characteristic of the Baltic Sea proper during the productive periods in spring and summer (Fig. 3.10). The pH decreases with the onset of the deep mixing in autumn and winter, reaching a minimum of around 7.9. There is also considerable inter-annual variability, with the peak-to-peak amplitude varying between 0.34 and 0.73 pH units (Fig. 3.10). Furthermore, the day/night cycle of OM production and respiration is also reflected in the pH, and during the spring bloom the difference between day and night may exceed 0.1 pH units.

3.4.3 Increasing atmospheric CO_2 acidifies the sea

The long-term pH of seawater will drop because of the dissolution of additional CO_2 due to the rising atmospheric CO_2 levels. This process is called "ocean acidification" and the possible ecological consequences are currently the subject of numerous research initiatives. The increase in atmospheric CO_2 from the pre-industrial level of 280 ppm to the current values of ~400 ppm has the potential to decrease the pH of the Baltic Sea water by ~0.15 units. A further decrease by almost 0.30 units can be expected if a doubling of the present atmospheric CO_2 will occur, which is a conceivable scenario for the next 100 years according to the estimates of the Intergovernmental Panel on Climate Change (IPCC, Stocker et al. 2013).

However, these estimates are based on the assumption of no further changes in any of the other variables affecting the acid/base system. This refers especially to total alkalinity (A_T), which strongly interacts with the pH (Fig. 3.11). An increase in A_T, *e.g.* by increased weathering in the drainage

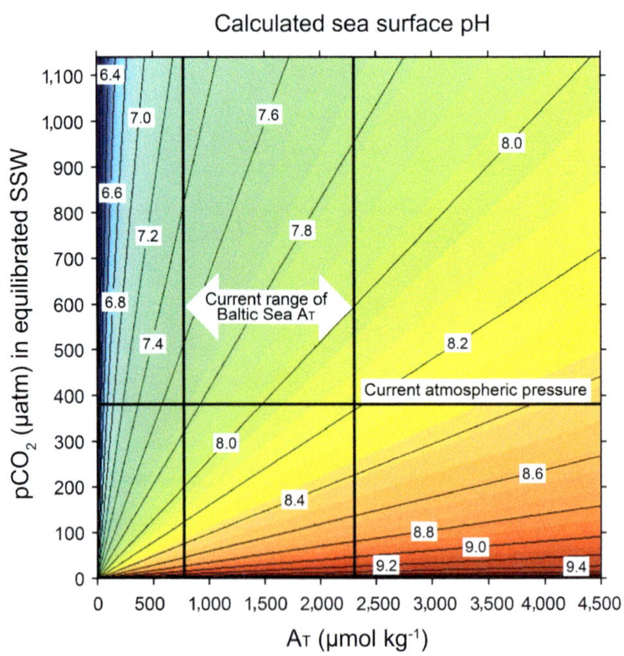

Fig. 3.11 Surface-water pH as a function of the partial pressure of CO_2 (pCO_2) and total alkalinity (A_T). SSW = surface seawater. Figure modified from Omstedt et al. (2010)

area due to the higher atmospheric CO_2, and by intensified CO_2 generation due to increasing mineralisation of soil OM, may counteract ocean acidification. Conversely, a decrease in A_T, *e.g.* by acidic precipitation, will enforce a pH decrease.

Ocean acidification influences calcium carbonate saturation since it reduces the concentrations of carbonate ions. Currently, the Baltic Sea surface water is undersaturated in both aragonite and calcite during autumn and winter (Tyrrell et al. 2008). However, because of the low pCO_2 and the high pH during OM production, the sea is clearly oversaturated during spring and summer. The consequence of increasing dissolved CO_2 is a longer and more intense period of undersaturation, which may affect the survival of calcifying organisms.

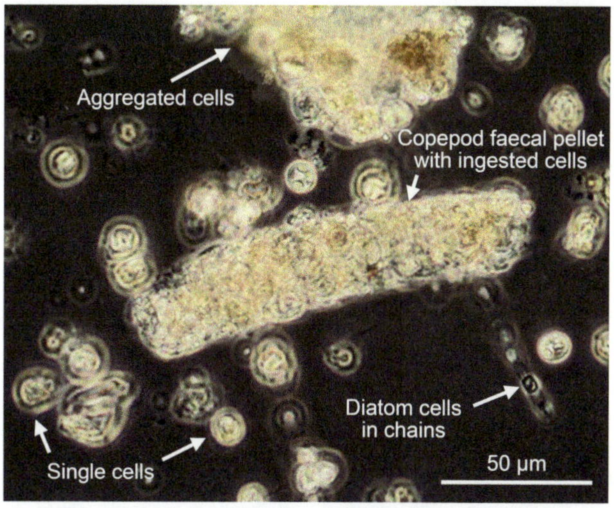

Fig. 3.12 Different carriers of sinking diatoms (mainly *Thalassiosira levanderi*). The material was collected from a sediment trap at a 180 m water depth in the Gotland Sea (Baltic Sea proper) after a spring bloom. Photo: © Regina Hansen

3.5 Sedimentation of particulate organic matter

3.5.1 The transport of particles depends on their physical properties

The sinking of particles provides the major connection between the surface water and the deep water in a stratified sea like the Baltic Sea. The loss of organic particles from the surface-water layer is the initial step in an important chain of consecutive biological and biogeochemical processes, both in the water column and on the seafloor. The vertical flux of particles (1) feeds pelagic and benthic organisms, (2) forms the organic part of the sediments and (3) creates gradients of dissolved gases (oxygen, carbon dioxide, nitrous oxide and methane) and other dissolved constituents (hydrogen sulphide, phosphate, ammonium, nitrate and DOC) through biological degradation processes.

Particles in the sea are subject to gravity and accordingly they sink. Thus, their means of transport differs from that of dissolved matter, which moves with the surrounding water mass and is dispersed by diffusion, turbulence and currents in the sea. This is the main reason why the products of surface-layer primary production, such as organic particles and oxygen, end up in different locations and why vertical gradients develop between the different gaseous, dissolved and particulate constituents within the water column and between water and sediment.

Sinking phytoplankton organisms occur in different states of aggregation (Fig. 3.12). For example, cells belonging to the same diatom species may sink as single cells, chains of cells, larger aggregates or enclosed in the faecal pellets of zooplankton. These differences create patterns in microbial decomposition in the water column depending on particle type. Given the large difference in sinking rate between small single algal cells in the 3–10 µm size range ($\sim$0.5 m d^{-1}) and aggregates or faecal pellets >50 µm (up to 100 m d^{-1}), material of the same origin will reach different water layers in a different state of conservation. Thus, in addition to all other determinants, the mode of sinking of primary producers is of great importance for the environmental structure of an aquatic ecosystem.

3.5.2 Particle flux mirrors the seasonal production pattern

Biological production in the surface layer is both qualitatively and quantitatively reflected in the seasonality of the particle flux (Walger et al. 1987). Particle flux can be studied by means of sediment traps (Buesseler et al. 2007). Advanced funnel-shaped instruments collect sinking particles in a battery of collectors, which are sequentially activated at pre-set time intervals (Fig. 3.13) and thus enable high-resolution flux measurements. The particle flux in the Gotland Sea at a $\sim$150 m water depth shows two maxima that coincide with the spring bloom and the summer cyanobacterial bloom, respectively (Fig. 3.14a).

The large particle export from the mixed layer in July and August is based on the growth of diazotrophic filamentous cyanobacteria, mainly of the genera *Aphanizomenon* and *Nodularia* (*cf.* Fig. 8.2) which take up phosphate or exploit stored phosphate when nitrogen nutrients are already depleted. The $\delta^{15}N$ stable isotope signature (*cf.* Box 2.3) in the sedimenting particles provides valuable information on the source of the nitrogen in these particles because nitrogen in nitrate contains a larger proportion of ^{15}N than elemental nitrogen that is utilised by cyanobacteria (Capone et al.

Fig. 3.13 Deployment of a sediment trap in the Baltic Sea. The funnel-shaped trap is covered with a grid to avoid turbulent exchange with surrounding water. A set of bottles at the bottom collect the trapped particles at pre-set times. Photo: © Leibnitz Institute for Baltic Sea Research (IOW), Warnemünde, Germany

2008; Leipe et al. 2008). This is reflected in the seasonal development of $\delta^{15}N$ in the sediment trap material, in which there is a shift from higher ^{15}N values in spring to lower ^{15}N values in summer (Fig. 3.14b).

The calculated values of the annual carbon flux in the Baltic Sea proper are in the range of 0.4 mol C m^{-2} $year^{-1}$ ($\sim$5 g C m^{-2} $year^{-1}$), of which more than half may be supplied by diazotrophic cyanobacteria in the summer period (Struck et al. 2004). Diatoms dominate the export of organic particles in spring and autumn and cyanobacteria in summer, whereas dinoflagellates also occur at background levels over the whole growth period. An interesting exception is the high flux of the diatom *Nitzschia paleacea*, which in some years is present in late summer at the peak of cyanobacterial development (Fig. 3.15). This diatom species is an epiphyte on the large aggregates of cyanobacteria. In some years, *Nitzschia paleacea* significantly contributes to the transfer of silica, which is incorporated in its frustules, from the surface layer to the sediment in late summer.

3.5.3 Sinking mineral particles contribute to sediment composition

While POM is important as a driver of biogeochemical processes, the mineral fraction of the sinking particles supplies the structural source material of the sea sediments (Broecker et al. 1982). Minerals such as clays and feldspars are formed in the geological cycle and are in the sea derived from weathering processes on land. Other minerals are biogenically formed in the water column. The silica frustules of diatoms are the most prominent example of this type of mineral particles, but calcite (calcium carbonate), barite (barium sulphate) and manganese oxides are also common in some regions of the Baltic Sea Area. The flux of minerals is closely coupled to the pelagic production cycle as even imported detrital minerals, such as river-derived clay, tend to sink together with biogenic particles.

3.6 Mineralisation and associated processes

3.6.1 Mineralisation involves different oxidants

The use of OM as substrate for bacterial growth or as food for zooplankton reverses its formation during photosynthesis and thus generates CO_2, ammonium and phosphate. This microbial oxidation process is called OM mineralisation. In

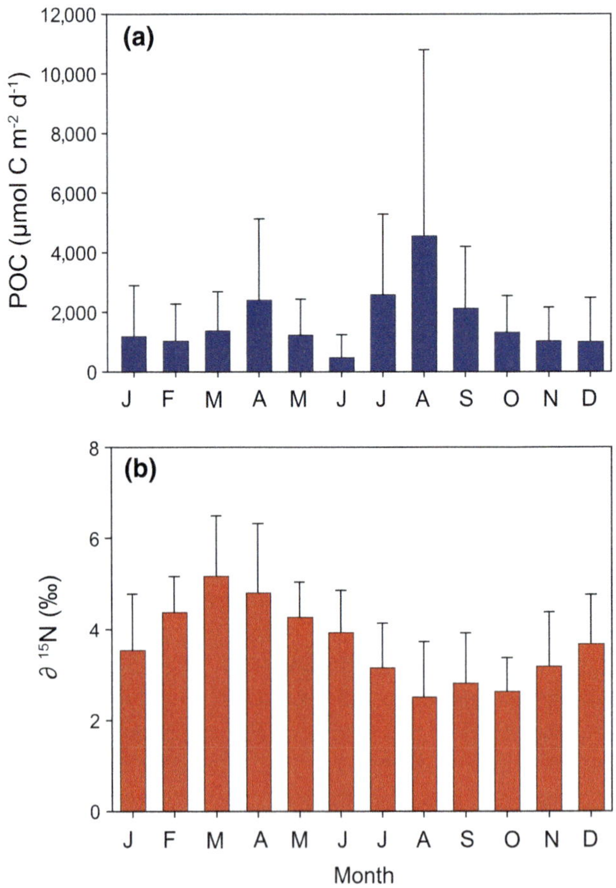

Fig. 3.14 Seasonal patterns of sedimentation recorded in sediment traps at a $\sim$150 m water depth in the central Gotland Sea. (**a**) Particle flux expressed as the daily vertical particulate organic carbon (POC) flux. (**b**) Isotopic signature of nitrogen ($\delta^{15}N$). Data are monthly averages for the period 1995–2006. Error bars indicate the standard deviation. Figure modified from Leipe et al. (2008)

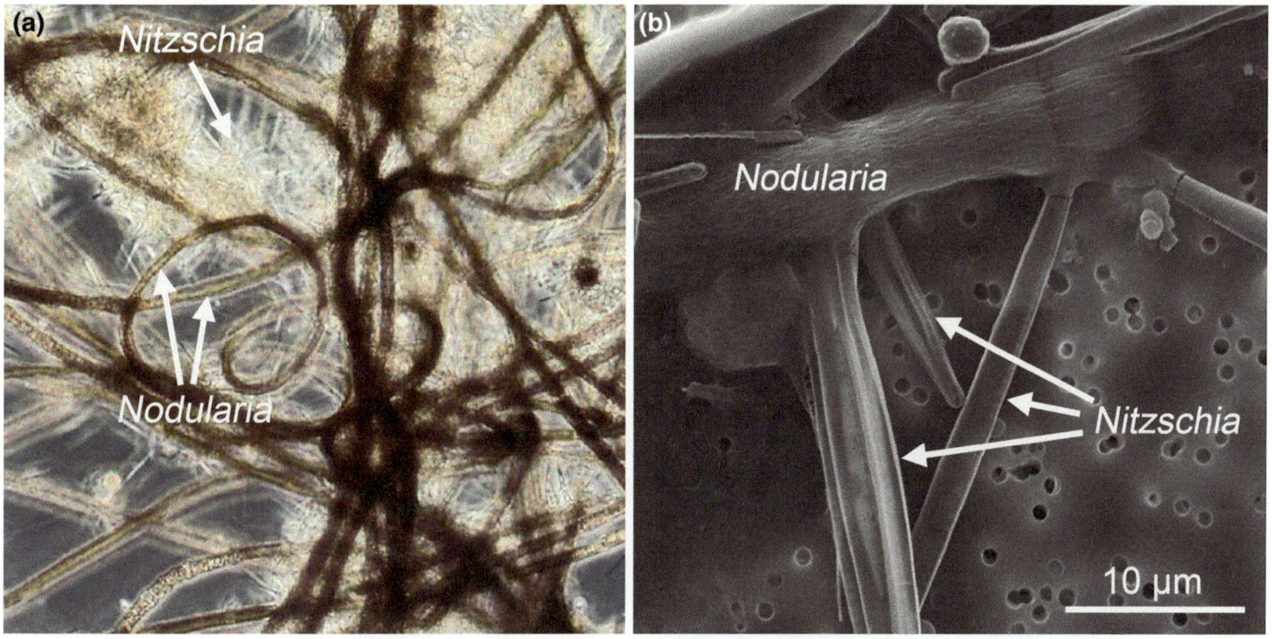

Fig. 3.15 The cyanobacterium *Nodularia spumigena* with its diatom epiphyte *Nitzschia paleacea* in a sample from the Baltic Sea. (**a**) Light microscopy. (**b**) Scanning electron microscopy (SEM). Photo: © Regina Hansen

the photic zone, OM mineralisation competes with primary production, and thus it prevails during night time. The mineralisation that occurs below the photic zone, is of far-reaching importance for biogeochemical cycling on longer time scales. This refers especially to the water layers below the permanent halocline (*cf.* Sect. 2.4.3) and to the surface sediments.

The use of different oxidants (electron acceptors), such as oxygen, nitrate, manganese oxides, iron oxides and sulphate for the microbially induced mineralisation of OM follows a sequence that is dictated by the change in Gibbs free energy (Table 3.1). The succession of the different oxidation processes refers to mineralisation both in the water column and in the sediment. However, these oxidation processes are most clearly displayed in the vertical distribution of the involved reactants in the pore water of sediments (Hensen et al. 2003). This is because the progress in the oxidation process is related to the sediment depth and because there is little vertical mixing of the pore water.

An idealised vertical distribution pattern of the chemical species involved in the mineralisation process is presented in Fig. 3.16 for sediment that is in contact with an oxic water column and where OM, manganese and iron oxides are deposited at the oxic surface. In the upper few cm of the sediment, oxygen is rapidly consumed by OM oxidation, while nitrate transiently increases due to the oxidation of ammonium (nitrification) before its removal by denitrification associated with OM oxidation.

Almost at the same depth at which denitrification occurs, the increase of Mn^{2+} and, somewhat deeper, that of Fe^{2+} indicate that manganese and iron oxides are taking over the oxidation of OM. Having reached a maximum, Fe^{2+}

Table 3.1 Gibbs free energy (Δ_rG^o) for a sequence of mineralisation reactions (r) at standard conditions (temperature = 25 °C, pressure = 1 atm, pH = 7), while the concentrations of other dissolved reactants involved are 1 mol L^{-1} (Emerson and Hedges 2008). The values refer to 1 mole carbon bound in carbohydrates. Interactions of the mineralisation products with the acid/base system are not taken into account. Note that the effective gain in ΔG of the reactions depend on the stoichiometry (*e.g.* influence of pH) and on the real concentrations. This is the reason why mineralisation starts with the use of oxygen despite the slightly lower Δ_rG^o of the denitrification reaction.

Process	Reaction	Δ_rG^o [kJ (mol $CH_2O)^{-1}$]
Oxic mineralisation	$CH_2O + O_2 \Rightarrow CO_2 + H_2O$	−506.6
Denitrification	$5CH_2O + 4NO_3^- + 4H^+ \Rightarrow 5CO_2 + 2N_2 + 7H_2O$	−508.4
Mn(IV) reduction	$CH_2O + 2MnO_2 + 4H^+ \Rightarrow CO_2 + 2Mn^{2+} + 3H_2O$	−502.7
Fe(III) reduction	$CH_2O + 4FeOOH + 8H^+ \Rightarrow CO_2 + 4Fe^{2+} + 7H_2O$	−281.4
Sulphate reduction	$2CH_2O + SO_4^{2-} \Rightarrow 2CO_2 + S^{2-} + 2H_2O$	−147.1
Methanogenesis	$2CH_2O \Rightarrow CO_2 + CH_4$	−93.6

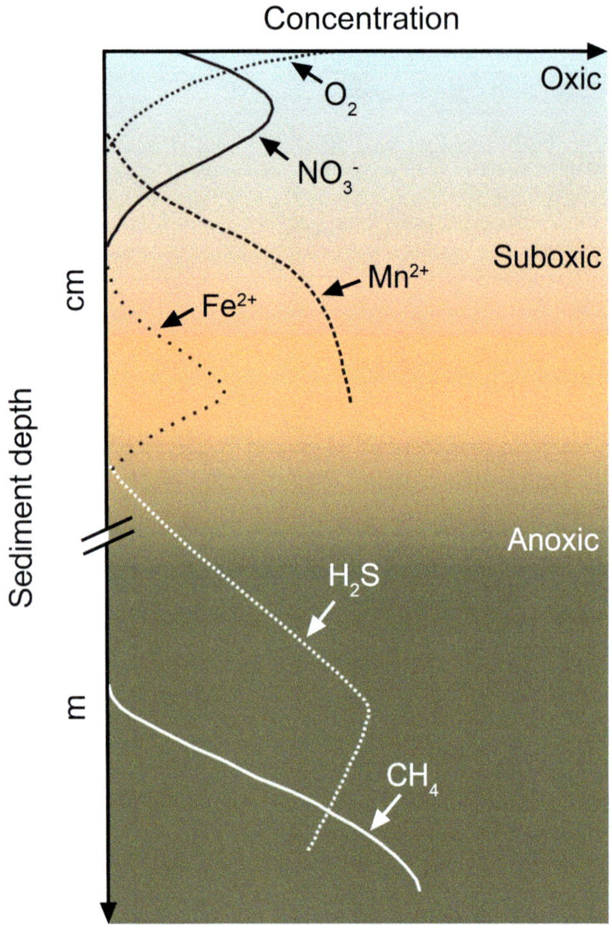

Fig. 3.16 Succession in the use of different oxidants for the mineralisation of organic matter in sediments, as indicated by the pore-water profiles of the concentrations of the oxidants (O_2, NO_3^-) and the reduction products (Mn^{2+}, Fe^{2+}, H_2S, CH_4). Figure: © Gregor Rehder

concentrations decrease again and gradually approach zero. This can be explained by the use of sulphate as an oxidant (sulphate reduction), which generates H_2S and precipitates iron sulphide. Subsequently, there is a continuous increase in H_2S until sulphate is entirely reduced. The exhaustion of sulphate, which only occurs in sediments, is the prerequisite for methanogenesis, which constitutes an internal oxidation/reduction (disproportionation) of OM molecules that generates methane and CO_2.

3.6.2 Oxygen is used to mineralise organic carbon

Both autochthonous OM that is produced within the marine system and allochthonous OM brought into the system from outside are subject to oxidation (mineralisation) by the microbial community. OM mineralisation reverses the process of production and takes place continuously, both in the water column and in the sediments. Under oxic conditions, the mineralisation of OM occurs according to the equation:

$$(CH_2O)_{106}(NH_3)_{16}H_3PO_4 + 106O_2 \Rightarrow 106CO_2 + 16NH_3 + H_3PO_4 + 106H_2O \qquad (3.8)$$

The mineralisation products on the right side of Equation 3.8 interact with the marine acid/base system and are transformed mainly to HCO_3^-, NH_4^+ and HPO_4^{2-}.

While OM mineralisation occurs through a complex chain of chemical catalytic reactions that generate several intermediate products, the kinetics of mineralisation are in most cases described as a first-order reaction in which the decay rate depends only on the OM concentration (OM) and the rate constant k_{OM}:

$$\frac{dOM}{dt} = -k_{OM} \cdot OM \qquad (3.9)$$

Integration of Equation 3.9 yields an exponential function that describes OM decay:

$$OM(t) = OM(0) \cdot e^{-k_{OM} \cdot t} \qquad (3.10)$$

As the rate of mineralisation in Equation 3.9 depends on the OM concentration only, other factors that potentially affect the decay rates, such as temperature and the O_2 concentration, are implicitly included in the rate constant k_{OM}. Since the mineralisation process is a part of bacterial metabolism, bacterial abundance and activity may also influence the OM decay rate. However, the high diversity of the microbial community and the high bacterial growth rates result in rapid development of specific microbial communities that are adapted to the level of OM availability.

Hence, limitation of the mineralisation rates by bacterial abundance is unlikely. A precise theoretical description of the kinetics of the mineralisation process is complicated by the fact that OM consists of many different compounds with a wide spectrum of chemical properties. Some more labile compounds are readily oxidised whereas others are more refractory and resist oxidation for a long time. Thus, each of the OM compounds has its own degradation rate constant, and the k_{OM} in Equations 3.9 and 3.10 represents a mean value for a given OM composition. However, this composition, and thereby also the k_{OM}, changes over time because of the different degradation rates of the individual OM constituents.

3.6.3 Regenerated ammonium consumes oxygen

Both ammonium and phosphate regenerated during mineralisation can be rapidly consumed by autotrophs in the photic zone in the surface waters. In deeper, dark oxygenated waters ammonium does not accumulate but is

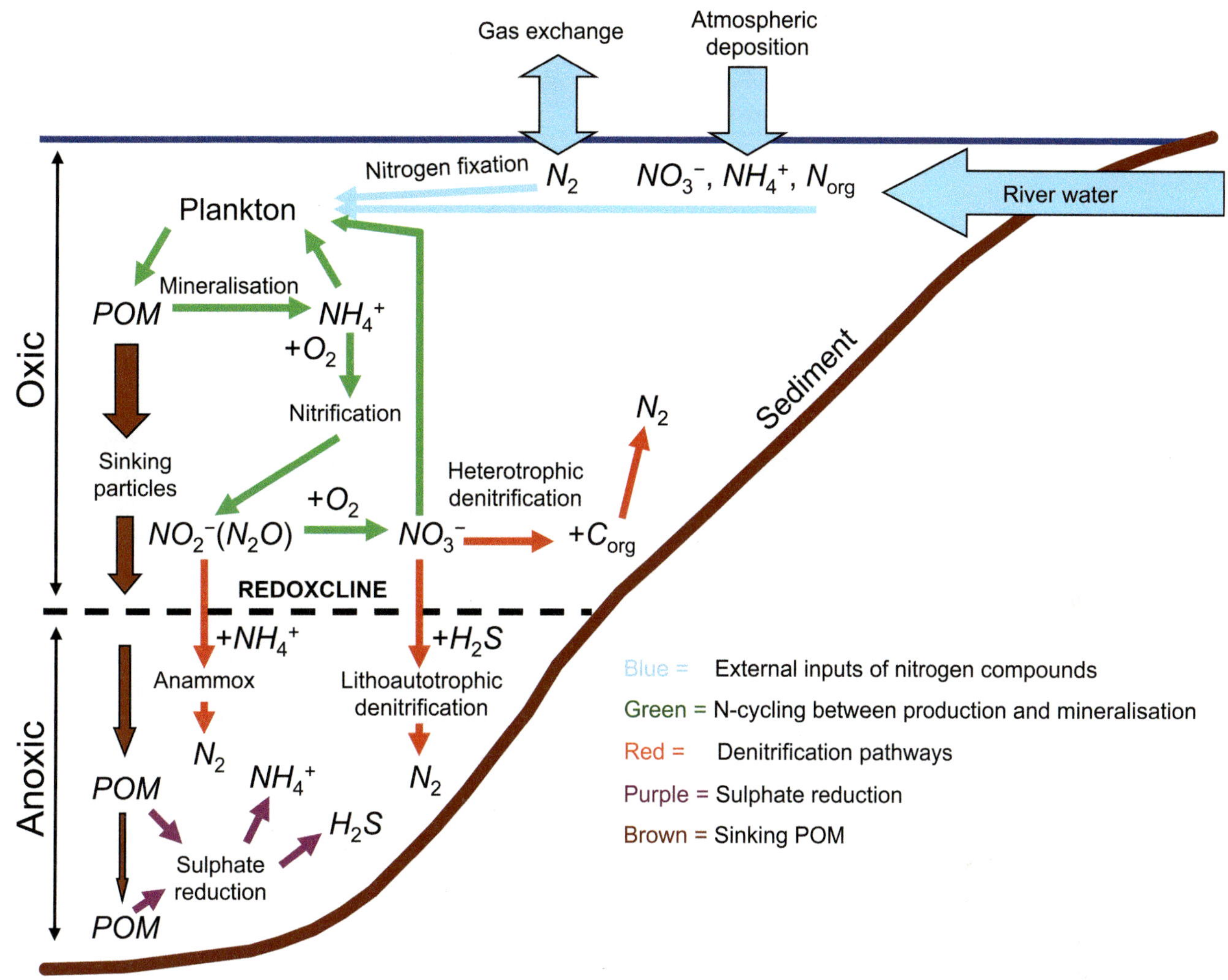

Fig. 3.17 Nitrogen transformations and fluxes in a sea basin with a pelagic redoxcline. Figure: © Bernd Schneider

further oxidised, first to nitrite (NO_2^-) and then to nitrate (NO_3^-) in a two-step process called nitrification (Fig. 3.17):

$$2NH_4^+ + 3O_2 \Rightarrow 2NO_2^- + 4H^+ + 2H_2O \tag{3.11}$$

$$2NO_2^- + O_2 \Rightarrow 2NO_3^- \tag{3.12}$$

Since both steps of the nitrification process are energy-yielding reactions, the entire process is readily used by autotrophic microorganisms as an energy source, *e.g.* by the ammonium-oxidising bacteria of the genus *Nitrosomonas* (Equation 3.11) and the nitrite-oxidising bacteria of the genus *Nitrobacter* (Equation 3.12). These processes can also be described as first-order reactions. Together with the mineralisation of nitrogen contained in OM, they can be expressed by a system of equations describing nitrogen regeneration under oxic conditions:

(1) Production of ammonium by mineralisation (k_{No} = mineralisation rate constant, N_{org} = organic nitrogen):

$$\frac{dNH_4^+}{dt} = -\frac{dN_{org}}{dt} = k_{No} \cdot N_{org} \tag{3.13}$$

(2) Production of nitrite by the oxidation of ammonium (k_{N1} = ammonium oxidation rate constant):

$$\frac{dNO_2^-}{dt} = k_{N1} \cdot NH_4{}^+ \tag{3.14}$$

(3) Production of nitrate by the oxidation of nitrite (k_{N2} = nitrite oxidation rate constant):

$$\frac{dNO_3^-}{dt} = k_{N2} \cdot NO_2^- \tag{3.15}$$

Nitrification comprises not only nitrate regeneration but also oxygen consumption. According to Equations 3.11 and 3.12, the oxidation of 16 mol of NH_3 requires 32 mol of O_2. Since 16 mol of NH_3 are released from 106 mol of OM, which require 106 mol of O_2 for their mineralisation, the relationship between total O_2 consumption and OM during mineralisation under oxic conditions is given by the ratio of 138:106. Taking into account that OM contains not only carbohydrates but also structural groups that require more O_2 for oxidation (*cf.* Sect. 3.2.4), this ratio may be as high as 150:106.

3.6.4 Nutrient loss by denitrification

If the rate of oxygen consumption is persistently higher than the rate of oxygen supply, the oxygen concentration may decrease to levels at which it can no longer satisfy the demand for electron acceptors in OM oxidation. In this situation, the role of an electron acceptor is increasingly taken on by the reduction of oxidised nitrogen compounds in a chain of biologically mediated transformations, namely from nitrate (NO_3^-) to nitrite (NO_2^-) to nitric oxide (NO) to nitrous oxide (N_2O) and finally to dinitrogen (N_2) (Fig. 3.17). Taking into account only the organic carbon part of OM that is considered to consist mainly of carbohydrate units (CH_2O), the denitrification is chemically described by:

$$5CH_2O + 4NO_3^- + 4H^+ \Rightarrow 5CO_2 + 2N_2 + 7H_2O \quad (3.16)$$

This use of nitrogen compounds for OM oxidation is known as heterotrophic denitrification. It can be performed by a multitude of microbial species, almost all of which are facultative anaerobes. However, denitrification starts even if oxygen is still available. An exact O_2 threshold value cannot be given, but observations indicate an onset of denitrification at O_2 concentrations of $\sim$0.1–0.3 mL L^{-1} (Capone et al. 2008). Denitrification is also accomplished by chemolithoautotrophic bacteria. The energy source for CO_2 assimilation by these organisms derives from the oxidation of hydrogen sulphide by nitrate, which yields elemental nitrogen (Fig. 3.17):

$$5HS^- + 8NO_3^- + 3H^+ \Rightarrow 4N_2 + 5SO_4^{2-} + 4H_2O \quad (3.17)$$

This process is the major denitrification pathway at the pelagic redoxcline, where anoxic waters mix with nitrate-containing oxic waters. Under these conditions, also anaerobic ammonium oxidation (anammox) takes place. During anammox, chemolithoautotrophic bacteria oxidise ammonium using nitrite (Equation 3.18), which is an intermediate of the nitrification process that becomes available by vertical mixing at the redoxcline (Fig. 3.17).

$$NH_4^+ + NO_2^- \Rightarrow 2H_2O + N_2 \quad (3.18)$$

From the ecosystem point of view, the most important aspect of denitrification is that it effectively removes nitrogen from the nutrient pool and thus counteracts both nitrogen fixation and external loading from the land and atmosphere.

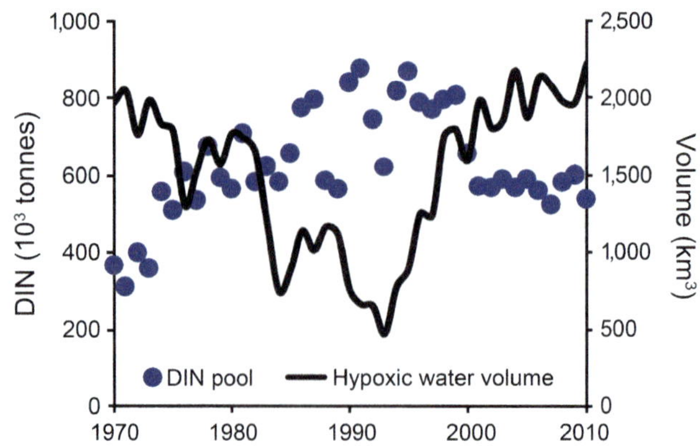

Fig. 3.18 Long-term dynamics of the DIN pool and the hypoxic (<2 mL O_2 L^{-1}) water volume for the years 1970–2010, shown as annual averages across the whole Baltic Sea proper. The time series was computed with the Data Assimilation System (DAS) from three-dimensional oxygen, ammonium, nitrite, and nitrate fields reconstructed from measurements provided by all the major databases around the Baltic Sea. Figure: © Oleg Savchuk

Denitrification pathways require hypoxic (<2 mL O_2 L^{-1}) waters or the presence of a redoxcline. In most areas of the Baltic Sea, hypoxic conditions are found in the sediments whereas pelagic redoxclines are found only in the deep basins. An implicit indication for a large-scale relationship between denitrification and oxygen conditions is given by the correlation between the total DIN pool of the Baltic Sea proper and hypoxic water volumes (Fig. 3.18, *cf.* Sect. 2.3.11).

3.6.5 Organic matter mineralisation produces hydrogen sulphide

After the complete consumption of oxygen, nitrate and Mn(IV) and Fe(III) oxides, the sulphur contained in sulphate ions acts as an electron acceptor for the microbial oxidation of OM. Considering carbohydrates as a proxy for OM, the following net reaction takes place:

$$2CH_2O + SO_4^{2-} \Rightarrow 2CO_2 + S^{2-} + 2H_2O \quad (3.19)$$

Hence, two CO_2 molecules are produced during the reduction of sulphate to sulphide. Because of its strong basic character, S^{2-} takes up hydrogen ions. Thus, at a typical pH of around 7 in anoxic waters, HS^-ions and H_2S are formed with shares of $\sim$70 % and $\sim$30 %, respectively. The sum of sulphide species is commonly referred to as "hydrogen sulphide" or "H_2S". The nitrogen contained in OM is not oxidised by sulphate and is released as ammonia which, at a pH below 9, is almost entirely transformed into ammonium ions. Likewise, organic phosphorus compounds are not subjected to oxidation and phosphorus is released as phosphate.

A prerequisite for sulphate reduction and the formation of hydrogen sulphide is the accumulation of OM and the absence of oxygen and nitrate. In general, these conditions are found in surface sediments where OM is deposited and the input of O_2 by mixing with the overlying water is limited. However, in the Baltic Sea proper anoxic conditions also regularly develop in the water column below the halocline of the deep basins. This is caused by periods of stagnation, which may last for many years, during which lateral water exchange and thus the input of oxygen are inhibited.

At the same time, the continuous input and mineralisation of OM leads to oxygen depletion and finally to the formation of hydrogen sulphide. The process is interrupted when O_2-rich water is laterally transported into the deep basins. These water renewal events are driven by special meteorological conditions in the Belt Sea and Kattegat area and occur irregularly. The frequency of the renewal events through major Baltic inflows (MBIs) has decreased significantly during the last decades (Schinke and Matthäus 1998, *cf.* Fig. 2.13a and Box 2.1).

Hydrogen sulphide is primarily formed at the sediment surface, where OM is accumulated. It is partly precipitated as iron(II) sulphide, but the major fraction of the hydrogen sulphide is transported upwards by vertical mixing and a pelagic redoxcline is formed. In the Gotland Sea, a redoxcline may be found at water depths of ~100 m after a prolonged stagnation period. A further upward propagation of the hydrogen sulphide interface (redoxcline) does not occur because of the O_2 input across the halocline and the increasing lateral water exchange above ~100 m.

In addition to its formation through long-term anoxic conditions in the deep basins of the Baltic Sea, hydrogen sulphide may be formed in coastal regions during the summer. The resulting seasonal anoxia is caused by enhanced production and a strong thermocline that prevents mixing of the water column and thereby the input of O_2 from the upper water column.

3.6.6 Mineralisation affects the CO_2 system and pH

The major oxidation product released during the mineralisation of OM, CO_2, accumulates in the deeper water layers of the Baltic Sea where a permanent halocline inhibits mixing with the surface layer. This prevents the escape of the excess CO_2 into the atmosphere by gas exchange. This is illustrated in Fig. 3.19a, which shows the vertical profile of C_T recorded after a prolonged stagnation period. The C_T is almost constant down to 50 m because the measurements were made in winter when convective mixing leads to a homogeneous surface layer.

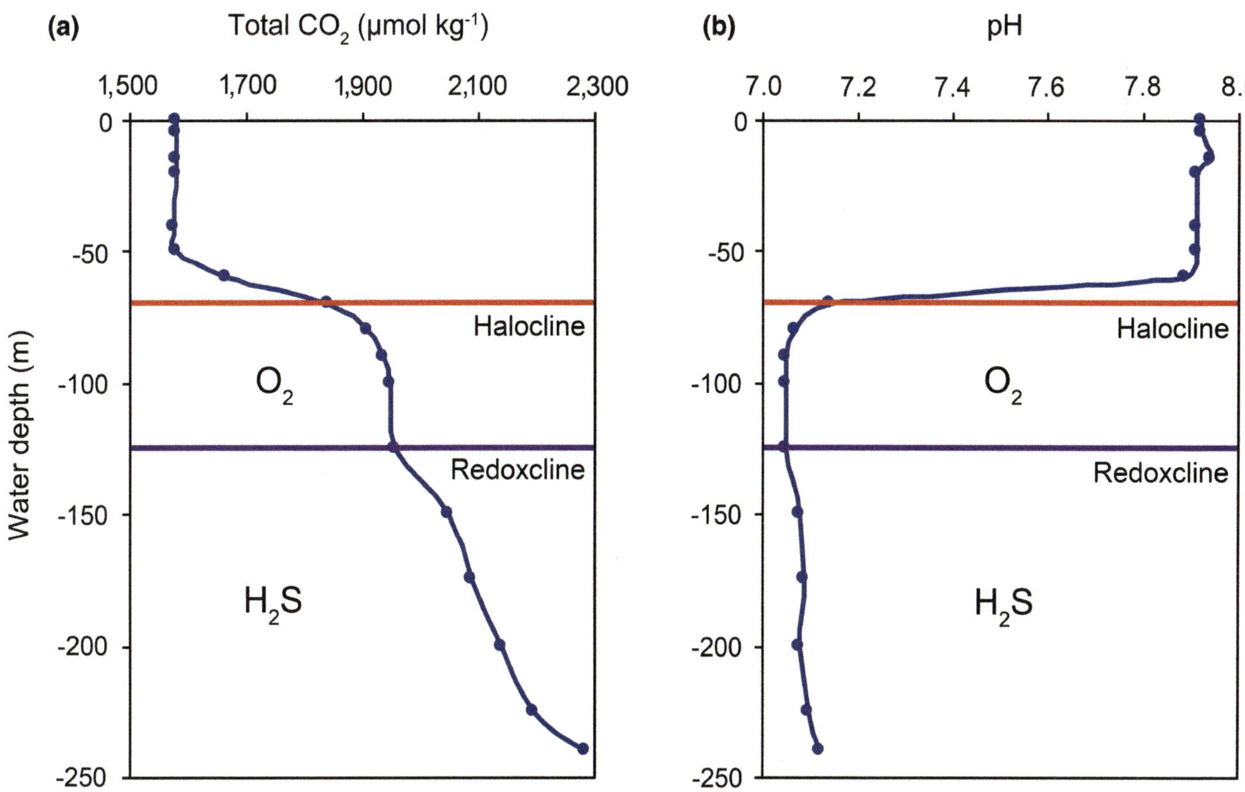

Fig. 3.19 Vertical profiles measured in the Eastern Gotland Sea in winter after a period of ~10 years without a major water renewal event. (**a**) Total CO_2 concentration (C_T). (**b**) pH (total scale, *cf.* Sect. 1.3.9). Figure modified from Hammer (2012)

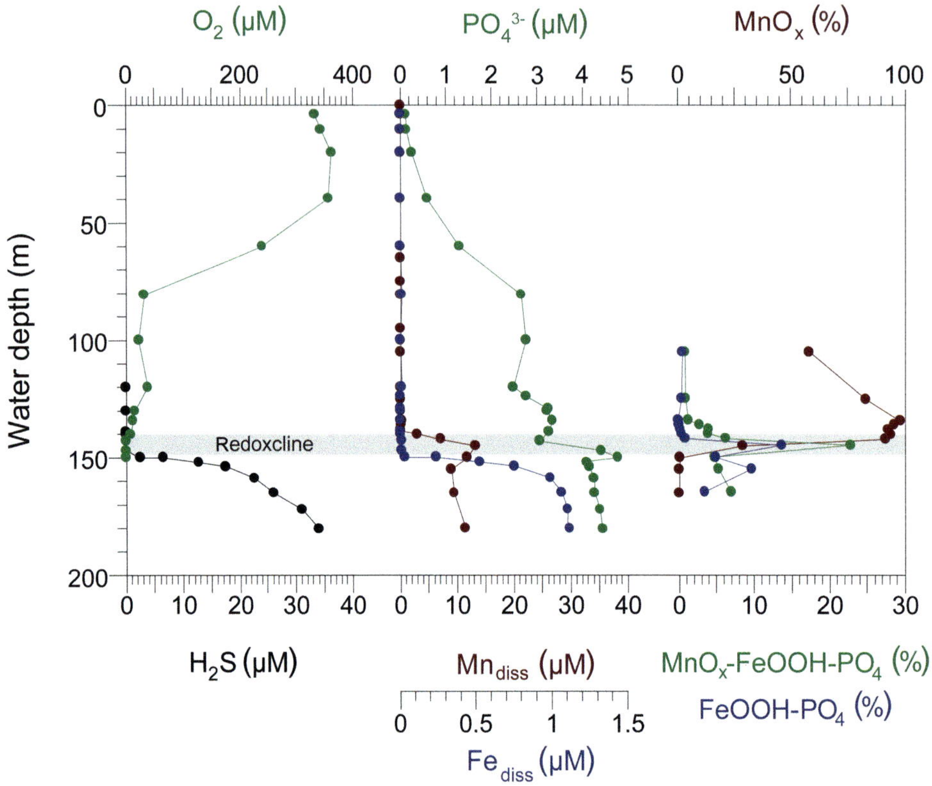

Fig. 3.20 Vertical profiles in the Gotland Sea of dissolved O_2 and H_2S concentrations (to the left), dissolved PO_4^{3-}, Mn and Fe concentrations (in the middle) and the relative abundances of authigenic Mn-, Fe- and P-containing particles (to the right). The grey bar indicates the approximate position of the redoxcline, *i.e.* the suboxic zone where O_2 and H_2S concentrations fall below the detection limits of commonly used methods. Figure modified from Dellwig et al. (2010)

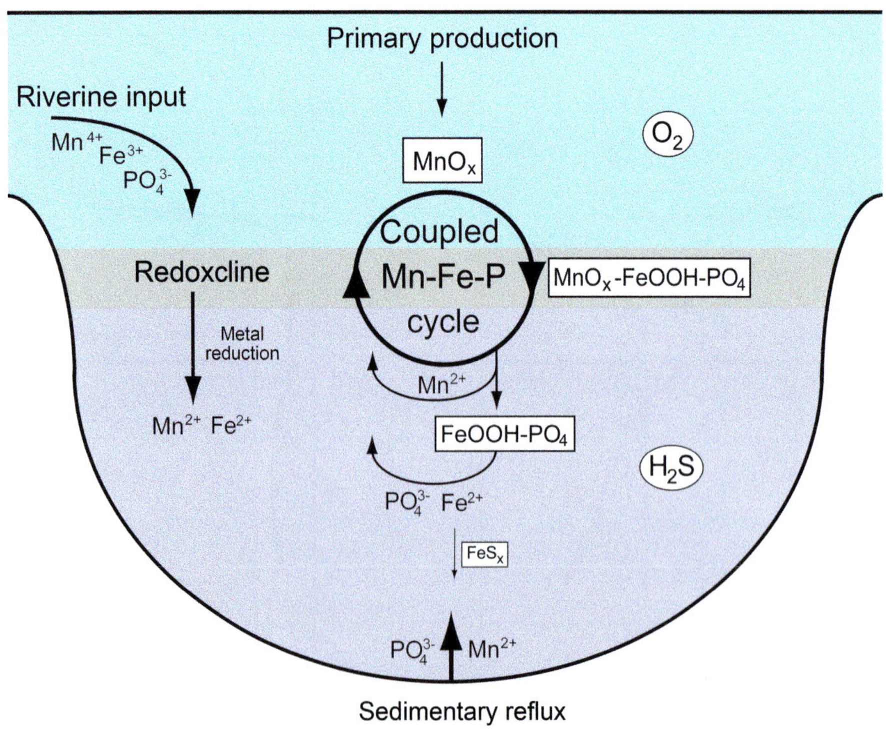

Fig. 3.21 A conceptual model of the “Mn-Fe-P shuttle” at the pelagic redoxcline of anoxic basins. Figure modified from Dellwig et al. (2010)

However, there is a steep C_T gradient across the halocline, resulting in a C_T increase by more than 300 μmol kg^{-1}. This has a strong impact on the pH, which decreases from 7.92 in the surface water to 7.05 below the halocline (Fig. 3.19b). In addition to the high C_T, nitrification also slightly contributes to the pH decrease because the formation of nitrate implies the release of hydrogen ions (Equation 3.11), which reduces the alkalinity and thus the pH.

Denitrification takes place close to the redoxcline and the previously formed nitrate participates in oxidising the OM to CO_2. During this process, hydrogen ions are consumed (Equation 3.16) and the associated increase in A_T counteracts the pH decrease by increasing CO_2. However, the effect is small and is not visible in the vertical pH distribution.

In contrast, CO_2 production by sulphate reduction is accompanied by a strong increase in alkalinity. Sulphide ions are strong proton acceptors and the formation of each S^{2-} ion adds two alkalinity units to the system. Alkalinity is further increased by the release of ammonia, which also acts as a proton acceptor. Therefore, the strong pH decrease with depth is stopped below the redoxcline despite the strong increase in C_T (Fig. 3.19a). Instead, the pH below the redoxcline may slightly increase and stabilises at ~7.1 (Fig. 3.19b).

3.6.7 Phosphorus release by mineralisation depends on redox conditions

The death of marine organisms results in the sinking of nutrient-derived phosphorus bound to organic particles and aggregates (particulate organic phosphorus) down to the seafloor. Depending on the water depth and the residence time of the particulate organic phosphorus in the water column, during which the sinking OM decomposes, only a certain proportion of this particulate organic phosphorus finally reaches the sediments. After the deposition on the seafloor, intense mineralisation takes place within the "fluffy layer", *i.e.* the thin transition layer between the water column and the sediments.

The remaining particulate organic phosphorus becomes part of the sediment and constitutes the major fraction of the phosphorus therein (Mort et al. 2010). Mineralisation continues in the sediments and releases phosphate (PO_4^{3-}) into the pore water, as indicated by increasing PO_4^{3-} concentrations with sediment depth (*e.g.* Jilbert et al. 2011). Hence, a diffusive PO_4^{3-} flux out of the sediment takes place. However, this PO_4^{3-} does not necessarily enter the water column because it may be bound to ferric oxide minerals at the oxic sediment surface. Since the early work of Einsele (1938), numerous studies have focused on the strong affinity of PO_4^{3-} to ferric oxides in various ecosystems under changing redox conditions (*e.g.* Gunnars and Blomqvist 1997). Binding to ferric oxides strongly depends on the redox conditions because the reduction of Fe^{3+} (ferric oxides) to Fe^{2+} under anoxic conditions by Fe-reducing bacteria or H_2S leads again to dissolution of PO_4^{3-}.

As a result of the variability in redox conditions, PO_4^{3-} fluxes differ significantly in the Baltic Sea. The highest fluxes occur temporarily in seasonally anoxic bottom waters, where PO_4^{3-}-rich iron oxides first are formed under oxic conditions and then they dissolve during the establishment of anoxic conditions. Intermediate fluxes characterise the anoxic/sulphidic basins, but are, however, comparatively stable throughout the year. After MBIs (*cf.* Fig. 2.13a), which may cause the complete oxygenation of the sulphide bottom waters and thus elevated precipitation of particulate Fe-P mineral phases, the re-establishment of reducing conditions initially leads to an extreme PO_4^{3-} release. The fluxes are lowest at permanently oxic sites (Mort et al. 2010), where iron oxides are able to bind a significant fraction of pore water PO_4^{3-}. This effect may be reinforced by bioturbation and bioventilation by the zoobenthos in soft bottoms (*cf.* Sect. 10.10). These animals increase the penetration depth of oxygen into the sediment and thus increase the thickness of the iron-oxide layer that is capable of binding PO_4^{3-}.

3.6.8 Phosphate is trapped below the pelagic redoxcline

Although constant pore water fluxes are responsible for distinct PO_4^{3-} enrichments of the bottom waters of anoxic basins, upwardly diffusing PO_4^{3-} has problems passing the pelagic redoxcline. As a result of the shift in redox conditions, the vertical distributions of PO_4^{3-} and dissolved Fe^{2+} and Mn^{2+} reveal certain similarities. The concentrations of all three are elevated in deeper anoxic waters, but decrease drastically at the redoxcline (Fig. 3.20).

Yet, while dissolved Fe^{2+} and Mn^{2+} concentrations virtually approach zero in oxygen-containing waters, PO_4^{3-} shows some pronounced anomalies at the redoxcline. Distinct PO_4^{3-} concentration minima and maxima occur, which is a typical feature of any anoxic basin (Shaffer 1986; Yakushev et al. 2007). The decreasing concentrations of dissolved PO_4^{3-}, Fe^{2+} and Mn^{2+} are paralleled by enrichments in the particulate fraction, indicating a transfer from the dissolved to the particulate phase at the redoxcline (Fig. 3.20).

This coupling of Mn, Fe and phosphorus is described in a conceptual model (Fig. 3.21). The initial reaction is the bacterial oxidation of sediment-derived and upwardly diffusing Mn^{2+}, leading to enrichments of particulate

Mn-oxides (MnO_x) just above the redoxcline. Once these MnO_x particles have formed they sink in the water column, reaching a zone where dissolved Fe^{2+} is present. At this stage, the sinking MnO_x particles efficiently oxidise Fe^{2+} to Fe^{3+} and are reduced to soluble Mn^{2+}.

The resulting Fe oxyhydroxides adsorb PO_4^{3-} and give rise to an increasing abundance of particles consisting of mixed Fe-Mn oxides (MnO_x-FeOOH-PO_4^{3-}) and almost pure iron hydroxo-phosphate (FeOOH-PO_4^{3-}). After sinking to greater depths, these particles dissolve because of the reduction of FeOOH and MnO_x by increasing H_2S concentrations. Except for the possible formation of iron sulphides (FeS_x), which appears to be relevant only for deeper anoxic basins like the Black Sea, all the elements are able to re-enter the cycle. This "Mn-Fe-P shuttle" thus traps PO_4^{3-} below the redoxcline and enhances its storage in deep anoxic basins (Dellwig et al. 2010).

3.6.9 Finally, mineralisation generates methane

Once all electron acceptors (*i.e.* oxygen, nitrate, manganese, iron and sulphate) are utilised, OM can be further degraded by methanogenesis, which is usually confined to deeper sediment layers (Fig. 3.16). The formation of methane is mediated by methanogenic archaea. From the biogeochemical point of view, methanogenesis is a disproportionation reaction. Organic carbon, with an average oxidation number of 0, is partly oxidised to carbon dioxide (oxidation number +4) and partly reduced to methane (oxidation number −4). Consequently, only about half of the organic carbon in the methanogenic zone can be transformed to methane.

Methanogenesis is a strictly anaerobic process in which even minor amounts of oxygen are not tolerated. Methane has properties that are very distinct from those of almost all the other constituents generated during the mineralisation of OM. It is a non-polar, poorly soluble gas that is about ten times less soluble than CO_2. Thus, methane concentrations can easily exceed saturation within the methanogenic zone of organic-rich sediments and may even form a free gas phase, which has the potential to migrate upwards in the sediments as bubbles.

3.6.10 Only small amounts of methane escape into the atmosphere

Both aerobic and anaerobic methanotrophic (*i.e.* methane-oxidising) microbiological pathways strongly limit the amount of methane that escapes from the seafloor, or from the water column, into the atmosphere. Methane oxidation by a consortium of sulphate-reducing bacteria and methane-oxidising archaea (Orphan et al. 2001) is now known to mediate the anaerobic oxidation of methane, and is believed to be responsible for the oxidation of more than 90 % of all methane produced in the marine environment (Jørgensen and Kasten 2006). This effective reaction leads to the coexistence of methane and sulphate in only a very narrow horizon of the sediments, known as the sulphate-methane transition layer (Fig. 3.16).

However, a net methane flux into the water column might still result from incomplete oxidation or enhanced flux because of the advective transport of reduced pore water fluids or free (methane) gas along geological weakness zones. Methane in the water column can be further oxidised anaerobically (in anoxic parts of the water column) or aerobically. The highest turnover rates are observed in the Baltic Sea proper due to aerobic oxidation in the transition layer between the upper oxic and lower anoxic water body (Jakobs et al. 2013). However, turnover times for the oxidation of methane are still on the order of several months to several years. Thus, transport into the atmosphere is possible when water transport brings methane-enriched water to the surface layer within a time frame that is shorter or comparable to that of methane oxidation in the water column.

3.6.11 Methane distribution in the Baltic Sea

In the Baltic Sea, the production of methane at a rate allowing the formation of free gas is mainly limited to the upper young (Littorina) OM-rich mud deposits overlying older glacial and post-glacial strata with very low OM content. The recent deposition and content of OM is controlled by primary production and current-driven lateral transport. OM degradation and the fraction thereof that is mineralised before the onset of methanogenesis depends on the availability of oxygen, sulphate and other electron acceptors, which varies along the salt and redox gradients inherent to the Baltic Sea. A survey of the methane distribution in the water column of the Baltic Sea in all major basins showed that strongly enhanced concentrations are generally encountered below the permanent halocline (if occurring) and that they are closely coupled to oxygen depletion (Fig. 3.22).

Little ventilation with the atmosphere because of stratification, strong vertical redox gradients and regional bottom-wateranoxia obviously impose strong controls on the methane inventory of the open waters of the Baltic Sea. Ultimately, other hydrographical processes, such as upwelling or seasonal variation of the mixed layer depth, govern the fate of methane released from the sediments into the lower water column and its potential as a source for atmospheric methane (Gülzow et al. 2013). In general, the more limnic basins in the northern part of the Baltic Sea (the Bothnian Sea and the Bothnian Bay) are characterised by

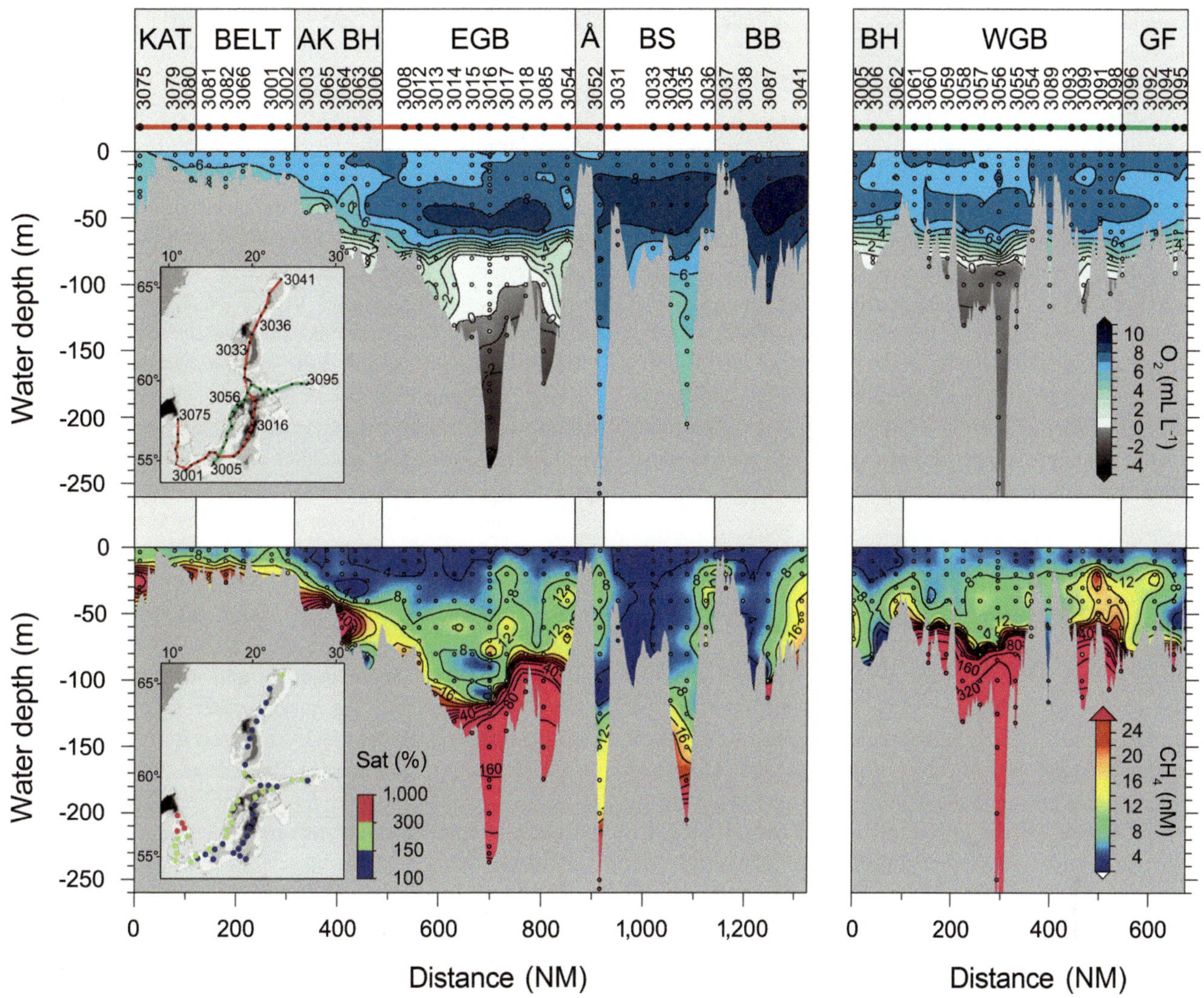

Fig. 3.22 The distributions of oxygen (upper panel) and methane (lower panel) along two transects surveyed in 2008 from the Kattegat to the Bothnian Bay (the red transect in the upper insert) and from the Bornholm Sea to the Gulf of Finland (the green transect in the upper insert). The insert in the lower panel shows the % saturation of methane in the surface water (0–5 m of water depth). The distance is given in nautical miles (NM). KAT = Kattegat, BELT = Belt Sea, AK = Arkona Sea, BH = Bornholm Sea, WGB = Western Gotland basin, EGB = Eastern Gotland basin, Å = Åland Sea, BS = Bothnian Sea, BB = Bothnian Bay, GF = Gulf of Finland. Figure modified from Schmale et al. (2010)

lower water column methane concentrations and surface water saturation values close to the atmospheric equilibrium, between 106 % and 116 % (Fig. 3.22).

3.6.12 Only a small fraction of the sinking POM is buried in the sediments

The burial of POM is mainly related to particle export from the water column to the bottom sediments and constitutes a final sink for carbon, nitrogen and phosphorus. The settling material in the Baltic Sea is a mixture of marine and terrestrial particulate matter that varies in composition depending on the physical and biogeochemical characteristics of the different basins (Leipe et al. 2008). Marine OM is thought to be much more labile than OM originating from land. POM often forms aggregates with mineral particles and is subjected to mineralisation and decomposition already during vertical transport in the water column. After its deposition on the seafloor, POM accumulates in the sediment. The accumulation rate of any constituent in the sediment is defined as the product of the total mass accumulation rate and the concentration of the constituent considered in the bulk sediments:

$$A = MAR \cdot C \qquad (3.20)$$

where A is the accumulation rate of the constituent (in g constituent m^{-2} $year^{-1}$), MAR is the total mass accumulation rate (in g total mass m^{-2} $year^{-1}$) and C is the concentration of the constituent in the bulk accumulated material (in g constituent per g bulk accumulated material).

The mineralisation and decomposition of POM continue in the surface sediments. Soluble products of these processes are released into the interstitial water and diffuse back into the water column as the so-called "return flux". The transformation of OM in surface sediments (early diagenesis) is driven by different, mainly microbial processes that depend on the redox conditions (Fig. 3.16). Since labile POM is mineralised preferentially, early diagenesis stabilises OM biochemically by making it more refractory. The fraction of OM accumulated in the bottom sediments that escapes mineralisation is buried in subsurface sediments. This results in the elimination of carbon, nitrogen, phosphorus and other elements contained in POM from short- and medium-term cycling (Rullkötter 2006).

3.6.13 POM accumulation in sediments is not homogeneous

Sedimentation of POM produced in the surface water is a major source of carbon, nitrogen and phosphorus in the sediment. Thus, the accumulation and burial of these elements should be directly related to the biological production that occurs in the photic zone. However, there are large differences in the accumulation and burial rates that cannot be explained by regional differences in surface-water productivity. Rather, POM accumulation patterns are related to the sediment type and the depth of the basin. Low-level accumulations are characteristic of the sandy sediments typical of shallow areas (Leipe et al. 2011). This is due to the strong near-bottom currents, which cause the resuspension, lateral transport and finally the redeposition of POM, mostly in the deep basins which are less affected by strong bottom currents and act as depositional areas in the Baltic Sea.

It is estimated that 2–3 times more POM enters the sediments of the Eastern Gotland Sea laterally from surrounding areas than vertically from the water column (Leipe et al. 2008). Intense POM accumulation is also observed in some areas located close to river mouths where terrigenous material, often rich in POM, is deposited along a gradient from these sites to the open sea.

Together with differences in surface-water POM production, the physically controlled efficiency of the sedimentation process leads to large inhomogeneities in the spatial distribution of POM in the sediments. Low accumulation rates of ~ 1.7 mol C m^{-2} $year^{-1}$ are found in the Arkona Sea sediments, whereas much higher values, potentially exceeding 5 mol C m^{-2} $year^{-1}$, are typical of sediments in the Gdańsk deep and the northeastern Gulf of Finland (Leipe et al. 2011).

3.6.14 Sedimentation removes nitrogen and phosphorus from the nutrient cycle

In the Baltic Sea sediments, nitrogen occurs mainly in organic forms. Thus, the nitrogen distribution in sediments usually follows the distribution of POM. Molar ratios of organic carbon to nitrogen do not significantly differ between sediments from different depositional areas and vary between 8.4 and 9.5 (Leipe et al. 2011). These values correspond approximately to the C:N ratios of the surface-water POM pool and indicate the absence of significant C-N fractionation during mineralisation.

High concentrations of phosphorus, similar to the distribution of carbon and nitrogen, are found in the fine-grained sediments typical of deep depositional areas. In general, the organic fraction of phosphorus dominates over the inorganic one. Unlike for nitrogen, the C:P molar ratios of POM show large regional variations, with values ranging from close to the Redfield ratio (106) up to more than 200. In the Gotland Sea, high C:P ratios have been found in sediments younger than 25–35 years (Hille 2006). This coincided with the more frequent and intense anoxic conditions of the Gotland Sea bottom water. It was speculated that phosphate is preferentially released during mineralisation under anoxic conditions (Jilbert et al. 2011). However, it is equally possible that the high C:P ratios reflect the decelerated mineralisation of organic carbon under anoxic conditions.

In addition to organic phosphorus, inorganic phosphorus associated with iron, calcium and aluminium minerals is present in the sediments (Mort et al. 2010). Among these, iron hydroxo-phosphates are the most important species and constitute an important phosphorus sink in some areas of the Baltic Sea. These Fe-P minerals are formed in oxic surface sediments and may accumulate in regions with permanently oxic bottom waters. The Gulf of Bothnia is a prominent example of the accumulation of Fe-P minerals in the surface sediments that contributes to the low phosphate concentrations in the water column (*cf.* Sect. 2.4.10).

3.7 Interactions between biogeochemistry and hydrography

3.7.1 Biogeochemical cycling differs temporally and regionally

Biogeochemical processes, and thus also the distributions of oxygen, hydrogen sulphide, nitrate and phosphate, exhibit temporal variability over different time scales. The seasonal and decadal dynamics of these compounds are to a great extent controlled by hydrographical conditions, *e.g.* mixing, stratification, lateral transport and vertical transport.

Since there are large regional differences in hydrographical conditions within the Baltic Sea Area, the temporal changes of the depth distributions of O_2, H_2S, nitrate and phosphate also vary regionally. For example, the Kattegat, the Baltic Sea proper and the Bothnian Bay are three very different sea areas. Each of these areas is characterised by its typical hydrographical conditions, which are reflected in distinct distribution patterns of biogeochemically reactive seawater constituents.

3.7.2 The Kattegat: ruled by water exchange

The hydrographical conditions in the Kattegat are strongly influenced by the exchange of water between the North Sea and the Baltic Sea (*cf.* Sect. 2.3.7). The dynamics through the Öresund and the Belt Sea are highly transient, oscillating from 0 to 100,000 $m^3\ s^{-1}$ in both directions. The strait flows are mainly driven by differences in sea level across the entrance area. The outflow water from the Kattegat meets North Sea water at the Skagerrak-Kattegat front.

The surface water of the Kattegat is dynamic, with fronts and eddies where mixing takes place, and this forms the water that, during favourable weather conditions, enters the Baltic Sea as dense bottom currents. The Kattegat is strongly salinity-stratified, with a low-salinity surface layer that coincides with the summer thermocline at a 10–15 m water depth. The surface water becomes increasingly saline northward (*cf.* Fig. 2.15).

3.7.3 Seasonal cycles of O_2 and nutrients in the Kattegat

The distribution pattern of O_2 in the Kattegat shows a distinct seasonality, which is most pronounced in the surface layer down to a ~15 m water depth (Fig. 3.23a). The

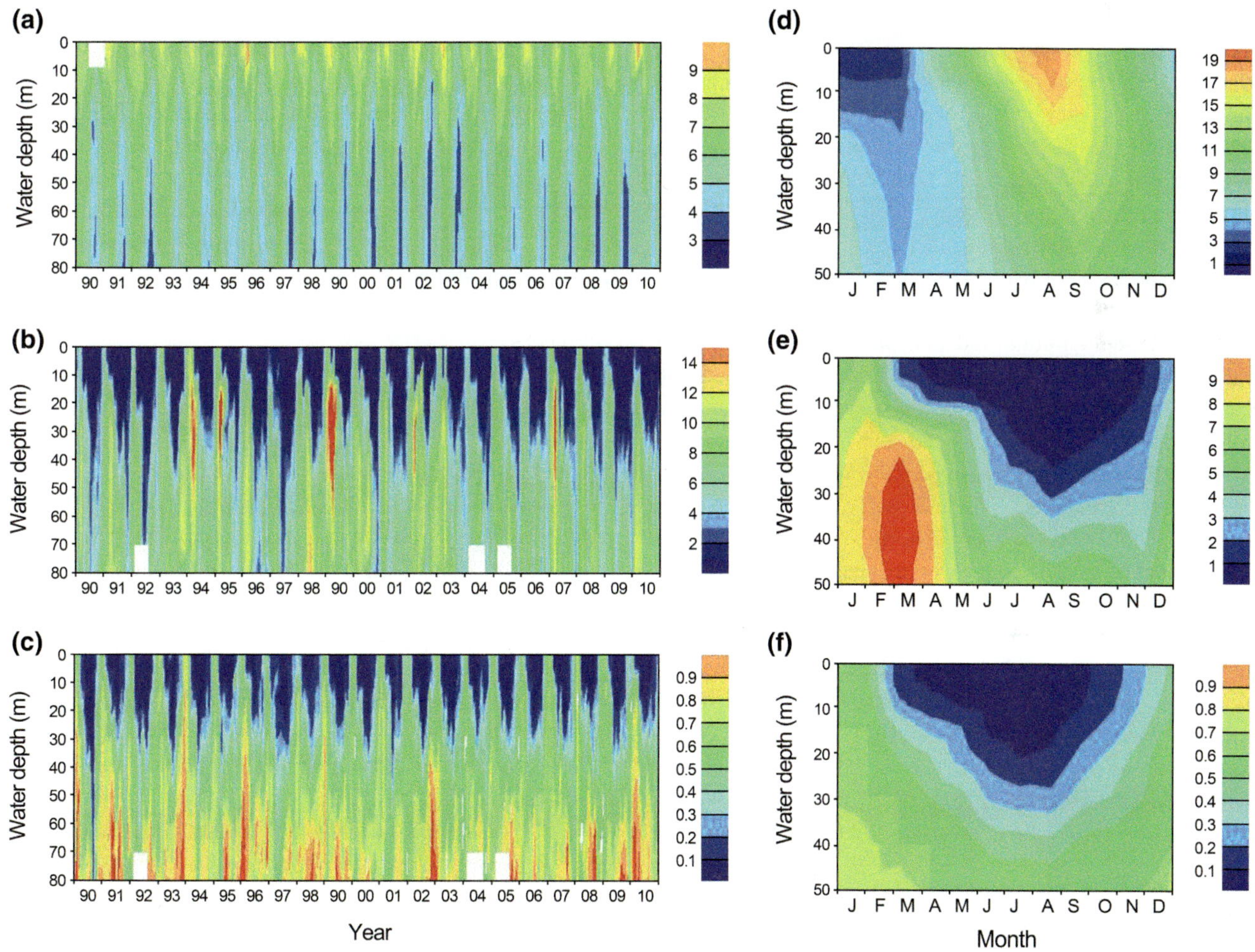

Fig. 3.23 Depth distributions of oxygen (mL L^{-1}), temperature (°C) and inorganic nutrients (µmol L^{-1}) in the central Kattegat 1990–2010. (**a**) Oxygen concentrations over time. (**b**) Nitrate concentrations over time. (**c**) Phosphate concentrations over time. (**d**) Average seasonal water temperature. (**e**) Average seasonal nitrate concentrations. (**f**) Average seasonal phosphate concentrations. White areas denote missing data. Figure based on measurements accessed from the major databases around the Baltic Sea with the decision support system Baltic Nest (http://nest.su.se; Wulff et al. 2013). Figure: © Oleg Savchuk

surface-water O_2 concentrations are mainly controlled by temperature. The lowest concentrations occur in summer because of the lower solubility of O_2 at higher temperatures (*cf.* Table 2.5), and accordingly O_2 concentrations are higher during winter.

A seasonality of the O_2 concentrations is also characteristic of the deeper water layers in the Kattegat because of the continuous entrainment of surface water from the North Sea. However, below the depth of ~15 m O_2 concentrations clearly decrease, despite the generally lower temperature below the surface layer. These lower O_2 levels in deeper water are indicative of O_2 consumption by mineralisation of OM, resulting in oxygen concentrations as low as ~3 mL L^{-1} in summer. However, this is still above the limit below which O_2 conditions are considered to be hypoxic.

The concentrations of nitrate and phosphate in the Kattegat also show clear seasonal patterns (Fig. 3.23b). Biological production results in the almost complete exhaustion of nitrate and phosphate in the surface water from spring to autumn. Nearly every year, the nitrate depletion extends much deeper than the photic zone (0 to ~30 m in the Kattegat). This deep nitrate depletion, which also occurs in the Baltic Sea proper, may be explained by the weak thermal stratification in early spring, which facilitates vertical mixing at a rate that allows algal cells both to take up nitrate from deeper water layers and to spend enough time in the photic zone for photosynthesis.

With the exception of a few years with extreme deep nutrient depletion, the deep-water layers in the Kattegat are enriched in both nitrate and phosphate (Fig. 3.23c). This partly reflects the mineralisation of OM. However, in contrast to the O_2 concentrations, the seasonality of nutrient concentrations is less pronounced and interannual comparisons of nitrate and phosphate peaks with O_2 minima show only weak correlations. This indicates that the deep-water nutrient concentrations are mainly controlled by the entrainment of nutrient-rich water originating from the North Sea.

Also the winter nutrient concentrations in the Kattegat surface water are strongly influenced by mixing with North Sea water and typical concentrations are ~6–7 µmol nitrate L^{-1} and ~0.5 µmol phosphate L^{-1}. Hence, in the Kattegat the average molar N:P ratio of the nutrient pool at the start of the spring bloom (~13) is much closer to the Redfield N:P ratio than is the case in other parts of the Baltic Sea Area.

3.7.4 The Baltic Sea proper: ruled by stagnation and inflow events

After passing the sills at the Baltic Sea entrance area, the inflowing Kattegat water forms bottom currents that may penetrate into the deeper parts of Baltic Sea proper. This results in a strong stratification in the Baltic Sea proper despite a decrease in density of the bottom water through mixing with the surrounding ambient Baltic Sea water. The water column of the Baltic Sea proper is permanently stratified with a halocline at the depth of ~60 m (*cf.* Table 2.6). The thermocline depth is typically 20–30 m and reduces the density in the upper water column in summer. The permanent halocline effectively isolates the deeper O_2-poor water layers from the O_2-rich surface water (Fig. 3.24a).

Frequently occurring weak inflows of more saline denser water may interfere with the permanent halocline and ventilate the water at depths of ~100 m. However, ventilation of the deeper parts takes place only through major inflows of saline water from the Kattegat into the Baltic Sea, which have a large impact on O_2 conditions and nutrient distributions (Savchuk 2010, compare Fig. 3.24 with Fig. 2.13a). Less favourable meteorological conditions may cause long periods of stagnation. The water budget also plays a role: when riverine runoff is large the freshwater outflow is strong and may hamper MBIs (Schinke and Matthäus 1998).

The dynamics of the major inflows into the Baltic Sea's deep areas indicate that saline water from the Bornholm Sea flows over into the Słupsk channel mostly in the form of short pulses, and that the deep flow pattern in the Słupsk channel strongly depends on winds (Piechura and Beszczynska-Möller 2004).

3.7.5 Seasonal cycles of O_2, H_2S and nutrients in the Baltic Sea proper

In contrast to the Kattegat area, the temperature-controlled O_2 seasonality in the Baltic Sea proper is confined to the surface layer of ~60–70 m above the permanent halocline (Fig. 3.24a). Below the halocline, which hampers water exchange with oxygenated surface water, the O_2 concentration rapidly decreases because oxygen is consumed during OM mineralisation. In the time period 1970–2006, O_2 concentrations were generally hypoxic (<2 mL L^{-1}) below the lower boundary of the halocline (~80 m). The O_2 depletion increases with depth and in many years the continuous mineralisation of OM finally results in the occurrence of H_2S at depths of only 100–120 m. These conditions are enhanced by long-lasting periods of stagnation with virtually no lateral input of oxygenated water.

The nutrient concentrations in the surface water show a distinct seasonal signal (Fig. 3.24). Winter concentrations are in the range of 3–5 µmol nitrate L^{-1} and 0.5–0.7 µmol phosphate L^{-1} and approach zero after the main productive period in late summer. The average molar N:P ratio of ~6–7 indicates that the phytoplankton spring bloom

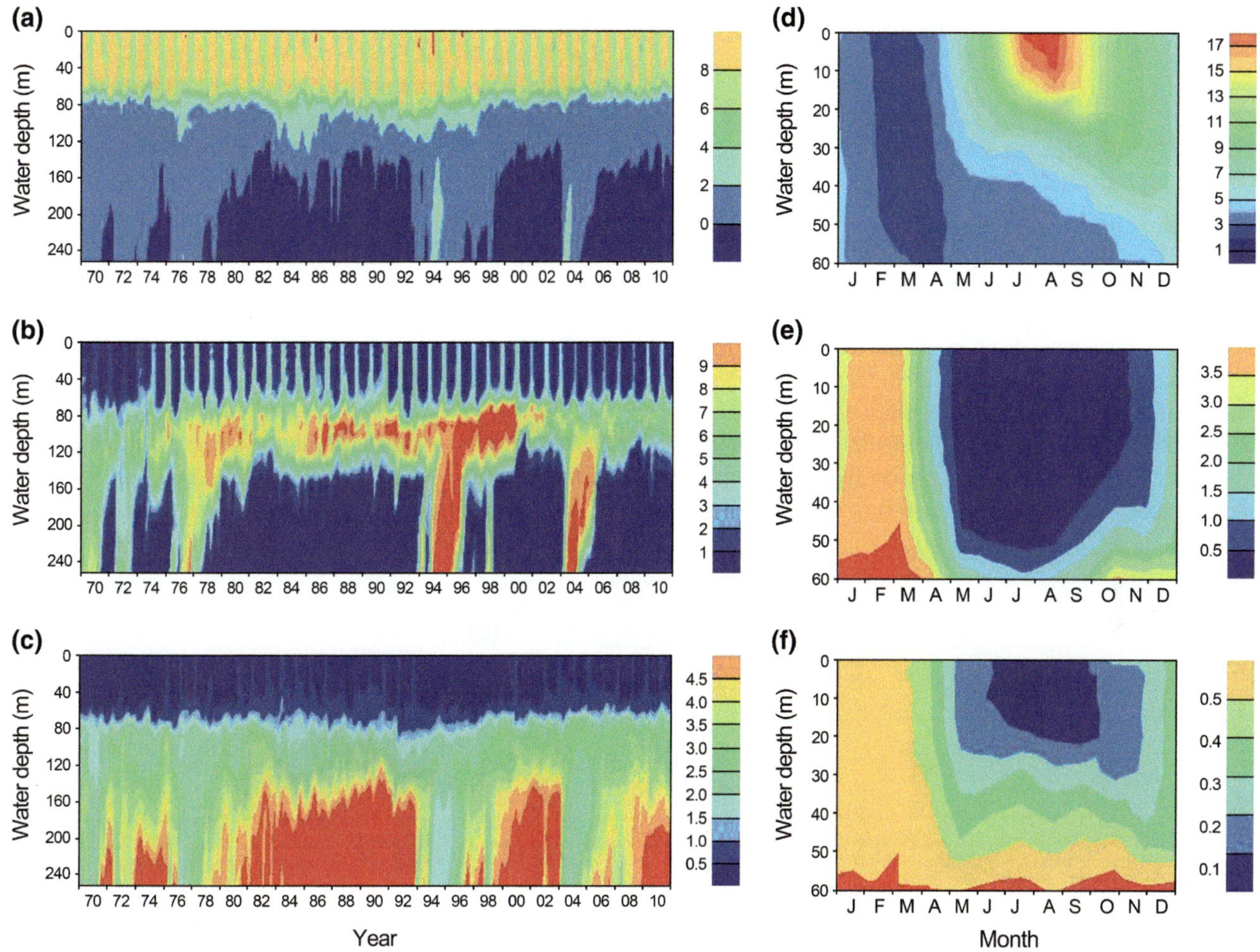

Fig. 3.24 Depth distributions of oxygen (mL L^{-1}), temperature (°C) and inorganic nutrients (µmol L^{-1}) in the eastern Gotland deep 1970–2010. (**a**) Oxygen concentrations over time with H_2S concentrations presented as negative oxygen equivalents (dark blue). (**b**) Nitrate concentrations over time. (**c**) Phosphate concentrations over time. (**d**) Average seasonal water temperature. (**e**) Average seasonal nitrate concentrations. (**f**) Average seasonal phosphate concentrations. Figure based on measurements accessed from the major databases around the Baltic Sea with the decision support system Baltic Nest (http://nest.su.se; Wulff et al. 2013). Figure: © Oleg Savchuk

is nitrogen-limited. Nitrate depletion extends down to the halocline because of a weak thermocline with frequent deeper mixing during the initial phase of the spring bloom. Consequently, almost the entire nitrate pool above the halocline is made available for primary production. Phosphate depletion is confined to a shallower layer because the excess phosphate available after nitrate depletion is consumed at a later stage of the spring bloom when a fully developed thermal stratification prevents vertical mixing.

Below the halocline, large amounts of ammonium are released by the mineralisation of OM. Under oxic conditions this ammonium is oxidised to nitrate by nitrifying microbes. Accordingly, high nitrate concentrations occur just below the halocline and in deeper water layers after water renewal events (Fig. 3.24b). However, due to denitrification either at the sediment surface or by mixing with anoxic water, nitrate concentrations are much lower than estimated from oxygen consumption and approach zero at the redoxcline. Hence, there is no nitrate below the redoxcline, where ammonium accumulates as the primary mineralisation product.

The fate of phosphate generated by mineralisation is also influenced by the redox conditions. The distribution pattern of phosphate is almost congruent with that of H_2S (Fig. 3.24a, c) because phosphate is readily released by the mineralisation of OM and by the reduction and dissolution of Fe(III)-hydroxo-phosphates under anoxic conditions. Hence, phosphate accumulates during periods of stagnation when anoxic conditions progressively develop. This process is interrupted and partly reversed when the deep water becomes oxygenated by a water renewal event, as was the case in 1993 and 2003. Iron sulphides that were previously formed at the sediment surface are oxidised to Fe(III)-hydroxo-oxides, which bind a large part of the deep-water phosphate pool. However, this process constitutes only a temporary phosphate sink because the Fe(III)-oxides will be reduced and then dissolved over the course of the next stagnation period.

3.7.6 The Bothnian Bay: ruled by low salinity and deep mixing

Three large gulfs, the Gulf of Bothnia, the Gulf of Finland and the Gulf of Riga are connected to the Baltic Sea proper (*cf.* Fig. 2.2). These gulfs are strongly influenced by large amounts of freshwater from numerous rivers (*cf.* Fig. 2.11) and by water exchange with the Baltic Sea proper. The deep waters of the gulfs are formed from the inflowing more saline (denser) surface water from the Baltic Sea proper, which mixes with freshwater from the rivers. The inflowing surface water is oxygen-rich and regularly ventilates the deep waters of the gulfs. The northern part of the Gulf of Bothnia, the Bothnian Bay, is connected to the Bothnian Sea through the Norra Kvarken strait. During inflows to the Bothnian Bay, surface water from the Bothnian Sea enters and ventilates the deep water of the Bothnian Bay. The surface- and deep-water salinities of the Bothnian Bay are ~3 and ~4, respectively (*cf.* Fig. 2.15c), and the halocline is very weak. Sea ice forms every year and the ice season typically lasts for 5–6 months (*cf.* Fig. 2.17a).

3.7.7 Seasonal cycles of O_2 and nutrients in the Bothnian Bay

The temperature-controlled surface-water O_2 concentrations in the Bothnian Bay show an irregular seasonal depth amplitude (Fig. 3.25a). This may be artifactual because the seasonal resolution of the O_2 and nutrient concentration data

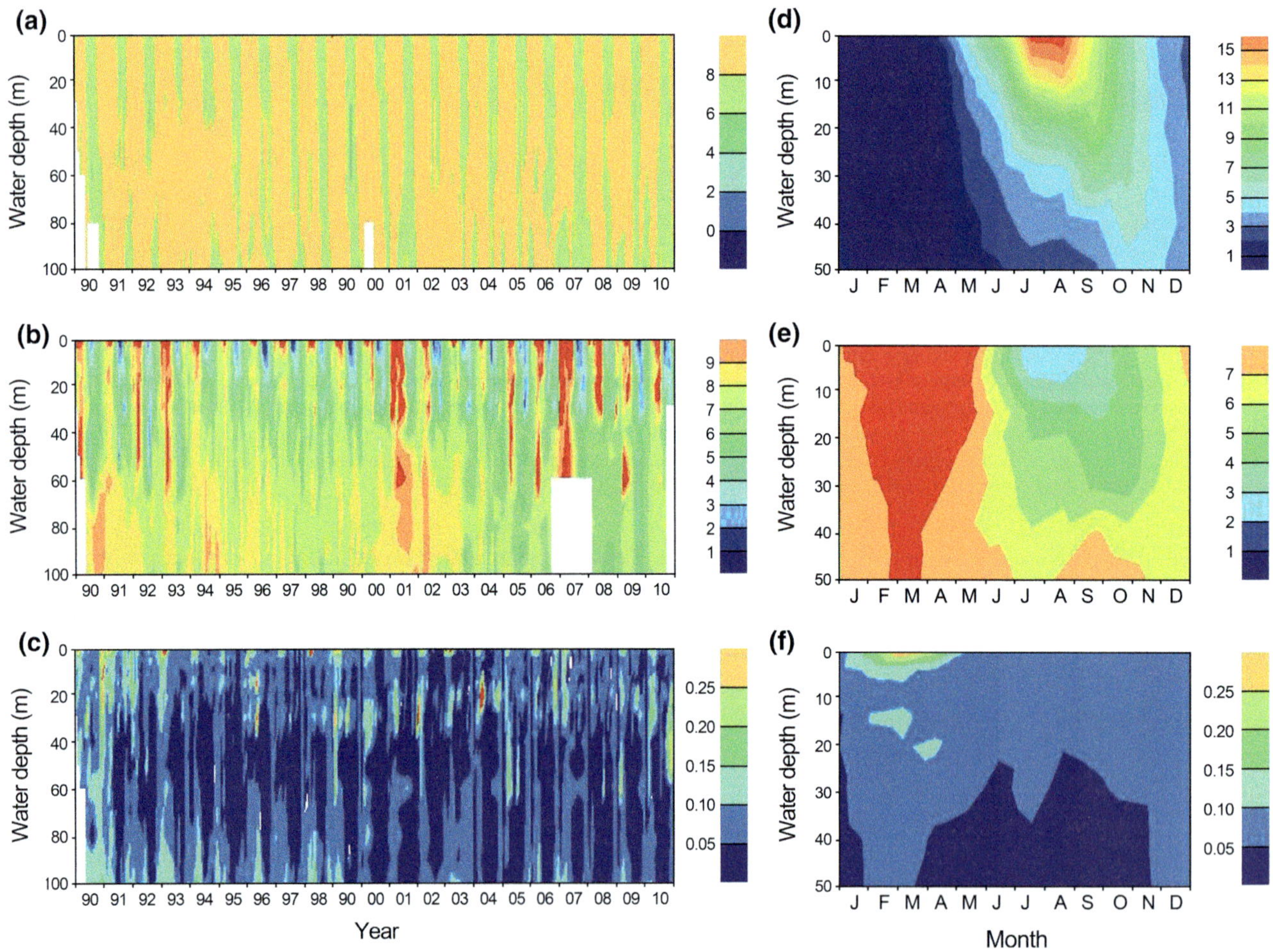

Fig. 3.25 Depth distributions of oxygen (mL L^{-1}), temperature (°C) and inorganic nutrients (µmol L^{-1}) in the Bothnian Bay 1990–2010. (**a**) Oxygen concentrations over time. (**b**) Nitrate concentrations over time. (**c**) Phosphate concentrations over time. (**d**) Average seasonal water temperature. (**e**) Average seasonal nitrate concentrations. (**f**) Average seasonal phosphate concentrations. Elevated phosphate concentrations (above ~0.1 µM) are influences from the coastal zone and from inflows from the Bothnian Sea. White areas denote missing data. Figure based on measurements accessed from the major databases around the Baltic Sea with the decision support system Baltic Nest (http://nest.su.se; Wulff et al. 2013). Figure: © Oleg Savchuk

is lower for the Bothnian Bay than for other areas of the Baltic Sea. This is particularly the case in winter, when the ice cover hampers measurements from the ordinary ships, which can only partly be replaced by those made from helicopters and ice-breakers. The O_2 depletion in the deeper water layers is only moderate, with minimum values of ~7 mL L^{-1}. This is a consequence of both low OM input from the low-productive surface layer and frequent water renewal. The latter may occur either by the entrainment of surface water from the Bothnian Sea or by sporadic complete mixing of the water column.

Winter surface nitrate concentrations in the Bothnian Bay range between 7 and 9 µmol L^{-1} (Fig. 3.25b). During biological production in spring and summer, nitrate concentrations decrease without becoming entirely exhausted because production is limited by low phosphate concentrations. Deeper water layers are generally significantly enriched in nitrate because of mineralisation, and the data suggest a tendency towards higher nitrate concentrations during the last three decades. However, it is unclear whether this is caused by enhanced production in the surface water and the subsequent sedimentation of OM or by a decrease in deep-water renewal.

The phosphate concentrations in the surface water of the Bothnian Bay are extremely low (Fig. 3.25c). With the exception of coastal areas and the vicinity of the Norra Kvarken sill (*cf.* Fig. 2.2), the maximum winter concentration over the vast offshore regions does not exceed 0.1 µmol L^{-1}, and phosphate is completely exhausted during the subsequent months by biological production. According to the Redfield N:P ratio, this would require a corresponding nitrate uptake of 1.6 µmol L^{-1}. However, nitrate loss during the spring-summer production amounts to ~4 µmol L^{-1}, corresponding to a phosphorus demand of ~0.25 µmol L^{-1}. Thus far, it is unclear whether there is an alternative phosphorus source, such as dissolved organic phosphorus, or whether highly phosphorus-depleted OM is produced.

A similar mismatch characterises the release of phosphate and nitrate by mineralisation in deeper water layers. The high N:P ratio indeed suggests that the OM produced is strongly depleted in phosphorus. However, it is also likely that phosphate has been partly removed from the water column by the formation of Fe(III)-hydroxo-phosphate in the oxic bottom water (*cf.* Sect. 2.4.10).

3.8 Carbon and nutrient budgets

3.8.1 Mass balance budgets as tools

Mass balance budgets are simplified tools for analysing and synthesising what is known about biogeochemical systems. They constitute balances comprising all inputs (sources) and outputs (sinks) that may occur by water transport, exchange with the atmosphere, sedimentation and internal processes. In many cases, mass balance budgets are based on the assumption of a steady state and they are used to examine the consistency of inputs and outputs. However, they can also provide a basis for estimating the sinks or sources that result from an imbalance in the budget and which otherwise cannot be quantified. For many substances, the transport by river water is the main pathway into the Baltic Sea, but the exchange with the North Sea may also contribute significantly to the budgets. Therefore, the budgets of carbon and nutrients are closely related to the Baltic Sea water budget (*cf.* Fig. 2.12).

3.8.2 The carbon budget

The carbon budget encompasses both inorganic and organic carbon. Their individual budgets are interrelated because transformations between these forms occur in seawater. The inorganic fraction (IC) is equivalent to the dissolved total CO_2 ($CO_2 + H_2CO_3 + HCO_3^- + CO_3^{2-}$) because the abundance of particulate inorganic carbon (carbonates) is negligible in the Baltic Sea (*cf.* Fig. 1.8). The organic fraction (OC) refers to both dissolved organic carbon (DOC) and particulate organic carbon (POC) compounds. The major sinks and sources of IC and OC for the Baltic Sea are river inflow, water exchange with the North Sea, sediments and the atmosphere (Fig. 3.26). The transport rates used in this budget were obtained from concentration measurements in

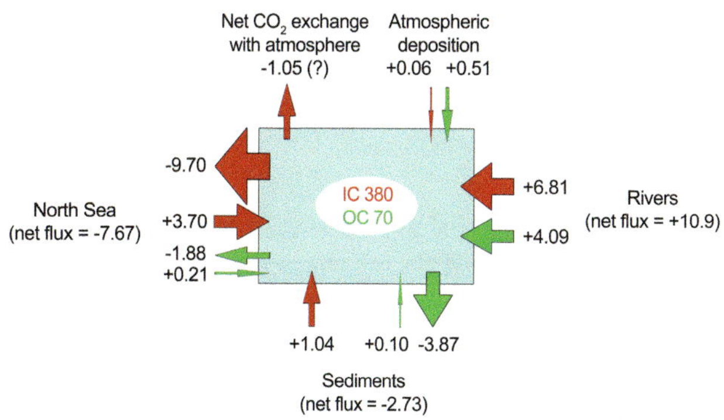

Fig. 3.26 The Baltic Sea carbon budget. Transport rates are given in million tonnes $year^{-1}$. Red arrows represent transport rates for inorganic carbon (IC) and green arrows represent transport rates for organic carbon (OC). The estimate for the net CO_2 exchange with the atmosphere, indicated with (?), is relatively uncertain (see text). Minor sinks (fisheries, −0.06 million tonnes OC $year^{-1}$) and point sources (+0.04 million tonnes OC $year^{-1}$) are not shown. The numbers in the centre of the box show estimates of the inventories of IC and OC in the Baltic Sea water (in million tonnes). Figure based on carbon data from the beginning of the 21st century in Kuliński and Pempkowiak (2011). Figure: © Bernd Schneider

the corresponding compartments, which were combined with water flow data. The budget terms, given as million tonnes per year, are average estimates that refer to the last few decades (Kuliński and Pempkowiak 2011).

According to this budget, large amounts of terrestrial OC ($\sim$4.09 million tonnes year^{-1}) are transported to the Baltic Sea by rivers (Fig. 3.26). However, only $\sim$40 % of the riverine contribution is removed by net OC exchange with the North Sea. Hence, assuming a steady state, there must be an internal OC sink, which could be the burial of terrestrial POC in the Baltic Sea sediments.

However, most of the terrestrial OC input consists of DOC, which is not subjected to sedimentation. Thus, a substantial fraction of the DOC must have been removed by mineralisation and act as an internal source of IC. It is not possible to quantitatively unravel these processes because data for terrestrial POC in river water and in sediments are not available. The transport of IC to the Baltic Sea via river water ($\sim$6.81 million tonnes year^{-1}) only exceeds the net IC export to the North Sea by $\sim$0.81 million tonnes year^{-1}. This supports the assumption that the IC loss due to OM production and final burial in the sediments ($\sim$3.87 million tonnes year^{-1}) of the Baltic Sea is partly compensated for by the mineralisation of the riverine OC input.

Also CO_2 gas exchange with the atmosphere contributes to the IC budget of the Baltic Sea. Based on the assumption of a steady state, the mass balance budget (Fig. 3.26) indicates that the Baltic Sea, on average, acts as a slight source ($\sim$1 million tonnes year^{-1}) of atmospheric CO_2. However, the estimated rate is questionable since it amounts to <10 % of the major input/output rates and is within the range of uncertainty. CO_2 gas exchange based on measurements of the differences in CO_2 partial pressure at the sea surface has been estimated for different subregions of the Baltic Sea. These data clearly indicate that the Baltic Sea proper is a distinct sink for atmospheric CO_2 (Schneider et al. 2014).

3.8.3 Different residence times for organic and inorganic carbon

The average residence times of IC and OC can be calculated by dividing the inventories of these two carbon forms in the Baltic Sea by the input rates (*cf.* Sect. 3.1.3, Fig. 3.26). The IC inputs via the Kattegat and river discharge must be complemented by the mineralisation of OC as an internal source. Since it is not known how much of the internal OC loss is explained by the burial of terrestrial POC, we ignore this fraction and consider the limiting case that the difference between the riverine OC input and the export to the North Sea (2.42 million tonnes year^{-1}) has occurred by mineralisation and thus constitutes an IC source. Adding up all source terms, including the mineralisation term, then yields a total input rate of 13.97 million tonnes IC year^{-1} and a residence time of 27 years. This is slightly less than the mean residence time of salt and water in the Baltic Sea (*cf.* Sect. 2.3.8). The situation is different for OC, in which the total input of 4.91 million tonnes year^{-1} yields a residence time of only 16 years. This is a consequence of large internal OC losses, probably caused mainly by OM mineralisation.

3.8.4 Nutrient budgets are affected by natural and anthropogenic variations

The compilation of nutrient budgets for the Baltic Sea extends back to the 1960s (*e.g.* Savchuk 2005). Budgets published over the years have differed in many important characteristics, from time intervals to processes accounted for. Nutrient inputs into the Baltic Sea are subject to both natural variations and anthropogenic impacts.

Variation in the freshwater balance occurs when a few consecutive years are wetter or drier than the adjacent years. This results in higher or lower waterborne inputs and thus in variations of riverine loads and atmospheric depositions. Superimposed on these natural variations are anthropogenic changes in land use, wastewater treatment and nitrogen emission to the atmosphere (Fig. 3.27a). Redox alterations of biogeochemical cycles caused by the major saltwater inflows result in significant changes in the nutrient pools residing in the sea (Savchuk 2010) and in transport flows across its boundary.

The nitrogen budget and phosphorus budget for the Baltic Sea presented here (Fig. 3.28) are based on a few simple principles and are meant to provide a general overview of the quantities involved. The long-term interval of 1985–2005, following the eutrophication phase (*cf.* Box 2.2) and apparently lacking long-term trends in nutrient pools, was chosen here in order to level out both natural fluctuations and anthropogenic impacts occurring at shorter time scales.

These budgets have been compiled only for external inputs into the Baltic Sea and outputs from the Baltic Sea. They do not take into account internal biogeochemical processes and exchange between the different basins of the Baltic Sea. Assuming steady state, the difference between integral input and output is defined as the internal sink term. Data on nutrient loads from point sources situated directly on the coastline are mainly available as inputs of total nitrogen and total phosphorus and are here assumed to be equally split between inorganic and organic fractions. Nutrient exchange at the lateral boundary was estimated by multiplying the annual water flows between the Belt Sea and Arkona Sea by the respective nutrient concentrations. These were computed as the long-term (1985–2005) ecosystem-wide annual averages from thousands of

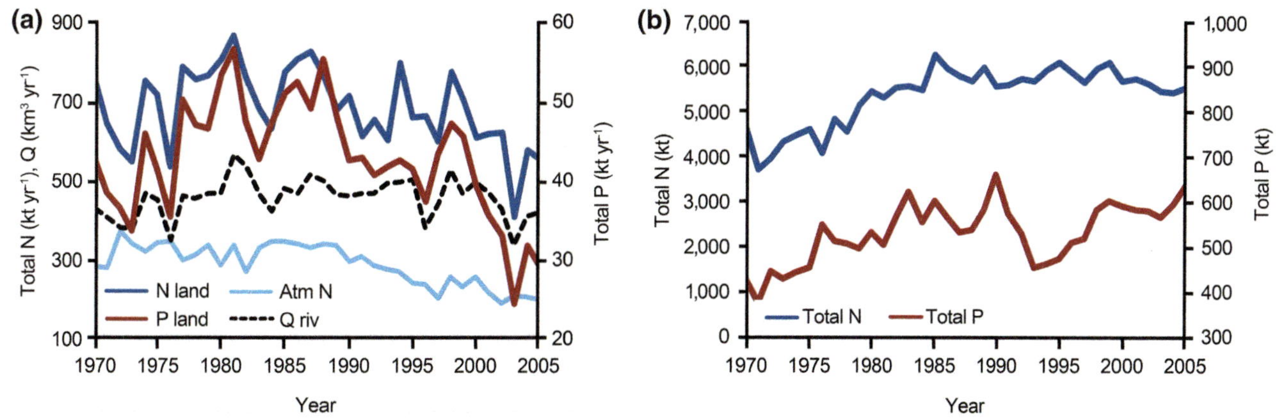

Fig. 3.27 Data necessary for calculating the N and P budgets of the Baltic Sea. (**a**) River water discharge (Q riv) and nutrient inputs into the Baltic Sea for the time period 1970–2005 in kilotonnes year^{-1}. N land = nitrogen from riverine inputs and direct point sources, P land = phosphorus from riverine inputs and direct point sources, Atm N = atmospheric DIN input. (**b**) Nutrient pools in the Baltic Sea for the time period 1970–2005 in kilotonnes. Note the different scales on the y-axes. Figure based on measurements accessed from the major databases around the Baltic Sea with the decision support system Baltic Nest (http://nest.su.se, Wulff et al. 2013). Figure: © Oleg Savchuk

measurements collected in the Baltic Environmental Database (BED) maintained at the Baltic Nest Institute (BNI).

3.8.5 The sediments are the major sink for phosphorus

The most prominent characteristic of the phosphorus budget (Fig. 3.28b) is that net exchange through the Danish straits removes only ~23 % of the external phosphorus inputs from the Baltic Sea compared to ~40 % of the external carbon inputs (Fig. 3.26). Except for variations determined by the reversible hypoxia-related phosphorus exchange between the water column and bottom sediments, there are no recent long-term trends in the phosphorus pool, which means that eventually ~30–40 kilotonnes of phosphorus should be buried in the sediments, mainly in the form of organic phosphorus compounds, but also as iron-III-hydroxo-phosphate in oxic surface sediment layers. Note also that this annual sink is relatively small compared to the total phosphorus pool of ~564 kilotonnes residing in the water column of the Baltic Sea. This sink contributes to a rather long phosphorus residence time of ~12 years and implies a high phosphorus buffer capacity of the Baltic Sea.

3.8.6 Denitrification is a major budget term

The contribution of organic nitrogen to the transports between the Baltic Sea and the North Sea is several times larger than that of inorganic nitrogen (Fig. 3.28a) despite the higher proportion of inorganic nitrogen inputs from external

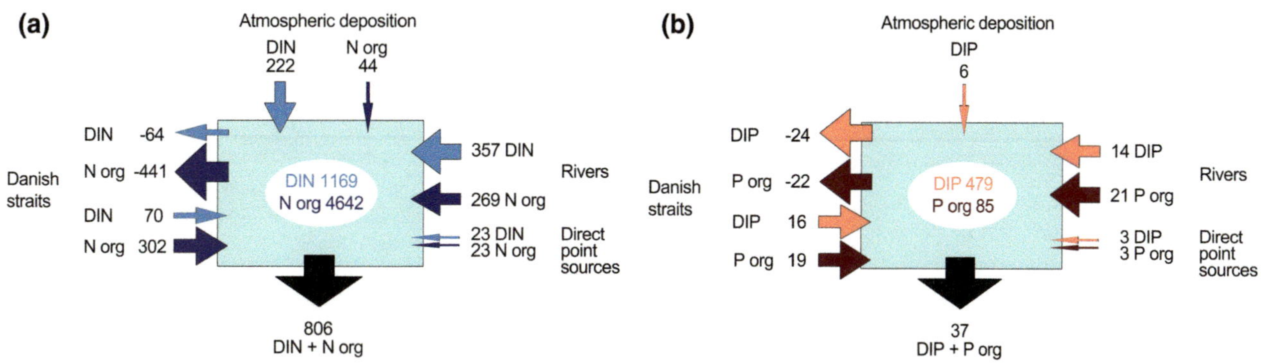

Fig. 3.28 Annual N and P budgets (in kilotonnes year^{-1}) and pools (in kilotonnes) of the Baltic Sea. (**a**) Inorganic nitrogen (DIN) and organic nitrogen (N org), based on average data for the time period 1985–2005. (**b**) Inorganic phosphorus (DIP) and organic phosphorus (P org), based on average data for the time period 1985–2005. In contrast to the carbon budget (*cf.* Fig. 3.26), the downward arrows at the bottom indicate a sink, defined as the difference between inputs and outputs rather than a burial rate. The sink term is estimated independently from observations. "Danish straits" include the Öresund. The loads from land and atmospheric depositions were reconstructed at the Baltic Nest Institute (BNI) using data from a variety of sources, including publications and technical reports, as well as data from the European Monitoring and Evaluation Programme (EMEP) and from the Baltic Marine Environment Protection Commission (HELCOM). Figure: © Oleg Savchuk

sources. The reason for this difference is the high proportion of CDOM in the organic nitrogen fraction in the Baltic Sea. Another important feature of the budget is the nearly equal export and import of DIN to and from the Belt Sea, which means that the entire net nitrogen transport from the Baltic Sea proper to the Belt Sea consists of organic nitrogen, again with a significant contribution of CDOM.

Uncertainty arises with interpretation of the integral sink term of 806 kilotonnes (kt) of N determined as the difference between inputs and outputs. In reality, this term sums up not only true sinks such as burial in sediments and denitrification but also nitrogen sources in the system that are provided by nitrogen fixation. To illustrate the quantities involved, we can try to untangle these processes by making use of independent estimates for the rates of the processes involved. Assuming that denitrification, which occurs both in the sediments and in the water column, removes 600–900 kilotonnes N year^{-1} from the biotic cycle, while nitrogen fixation contributes with 300–400 kilotonnes N year^{-1}, the resulting misbalance of 300–500 kilotonnes N year^{-1} would require the additional annual burial of 300–500 kilotonnes N to match the estimated sink of 806 kilotonnes N year^{-1}. With an estimated average carbon burial rate of 2,730 kilotonnes C year^{-1} (Fig. 3.26) and a mean molar C:N ratio for the buried material of 9 (corresponding to a C:N mass ratio of 7.7), the nitrogen burial rate should be ~350 kilotonnes N year^{-1}, which is close to the lower boundary of the estimated range of the burial rate of 300–500 kilotonnes N. The lower estimates of nitrogen burial are also supported by a relatively low empirical phosphorus burial rate of 37 kilotonnes P year^{-1}, as determined from the P budget (Fig. 3.28b).

A corollary of the lower nitrogen burial flux are the higher denitrification fluxes, which are close to the upper boundary of its range. Considering the uncertainties involved in such estimates, including significant year-to-year variations in the denitrification and nitrogen fixation rates, this reasonable matching can be regarded as an additional confirmation of the consistency of the numbers involved. Compared with phosphorus, the nitrogen cycle is characterised by a higher removal efficiency, which is caused by denitrification. As a result, the mean residence time of nitrogen is shorter than that of phosphorus and amounts to ~6 years.

Review questions

1. The input of which nutrient must be reduced in order to combat oxygen depletion in deep water?
2. How do sinking particles affect the vertical distribution and the air-sea exchange of dissolved gases?
3. What is the most important process controlling sedimentary phosphate reflux into the open water column?
4. What happens to ammonium in anoxic water after a saltwater inflow?
5. What is the effect of calcifying organisms on pH and pCO_2?

Discussion questions

1. How would you describe an ecosystem in "steady state"? Discuss theory *versus* reality.
2. How does climate change affect productivity and oxygen depletion in the Baltic Sea?
3. Is there a chance for storage of anthropogenic CO_2 in the Baltic Sea? How?
4. Which aspects of the water exchange between the Baltic Sea and the North Sea are of major biogeochemical importance? Why?
5. Are the sediments sinks or sources for carbon, nitrogen and phosphorus? Why?

References

Anderson LA (1995) On the hydrogen and oxygen content of marine phytoplankton. Deep Sea Research 42:1675–1680

Asmala E, Autio R, Kaartokallio H, Pitkänen L, Stedmon CA et al (2013) Bioavailability of riverine dissolved organic matter in three Baltic Sea estuaries and the effect of catchment land use. Biogeosciences 10:6969–6986

Azam F, Chrisholm SW (1976) Silicic acid uptake and incorporation by natural marine phytoplankton populations. Limnology and Oceanography 21:427–435

Bełdowski J, Löffler A, Schneider B, Joensuu L (2010) Distribution and biogeochemical control of total CO_2 and total alkalinity in the Baltic Sea. Journal of Marine Systems 81:252–259

Broecker WS, Peng TH, Beng Z (1982) Tracers in the sea. Columbia University, Lamont-Doherty Geological Observatory 690 pp

Bronk DA (2002) Dynamics of DON. In: Hansell DA, Carlson CA (eds) Biogeochemistry of marine dissolved organic matter. Academic Press—Elsevier, St. Louis, pp 153–249

Brzezinski MA (1985) The Si:C:N ratio of marine diatoms: interspecific variability and the effect of some environmental variables. Journal of Phycology 21:347–357

Buesseler KO, Antia AN, Chen M, Fowler SW, Gardner WD et al (2007) An assessment of the use of sediment traps for estimating upper ocean particle fluxes. Journal of Marine Research 65:345–416

Capone DG, Bronk DA, Mulholland MR, Carpenter EJ (2008) Nitrogen in the marine environment, 2nd edn. Academic Press—Elsevier, Burlington, 1729 pp

Dellwig O, Leipe T, März C, Glockzin M, Pollehne F et al (2010) A new particulate Mn-Fe-P-shuttle at the redoxcline of anoxic basins. Geochimica et Cosmochimica Acta 74:7100–7115

Einsele W (1938) Über chemische und kolloidchemische Vorgänge in Eisen-Phosphat-Systemen unter limnochemischen und limnogeologischen Gesichtspunkten. Archiv für Hydrobiologie 33:361–387 [in German]

Emerson SR, Hedges JI (2008) Chemical reactions in marine sediments. In: Emerson SR, Hedges JI (eds) Chemical oceanography and the marine carbon cycle. Cambridge University Press, Cambridge, pp 404–444
Gülzow W, Rehder G, Schneider B, von Deimling J, Seifert T, Tóth Z (2013) One year of continuous measurements constraining methane emissions from the Baltic Sea to the atmosphere using a ship of opportunity. Biogeosciences 10:81–99
Gunnars A, Blomqvist S (1997) Phosphate exchange across the sediment-water interface when shifting from anoxic to oxic conditions – an experimental comparison of freshwater and brackish-marine systems. Biogeochemistry 37:203–226
Hammer K (2012) Erprobung und erste Anwendung einer spektrophotometrischen pH-Messvorrichtung. University of Rostock. [Diploma Thesis, in German]
Hensen C, Zabel M, Pfeifer K, Schwenk T, Kasten S et al (2003) Control of sulfate pore-water profiles by sedimentary events and the significance of anaerobic oxidation of methane for the burial of sulfur in marine sediments. Geochimica et Cosmochimica Acta 67:2631–2647
Hignett TP (ed) (1985) Fertilizer manual. Springer, Developments in Plant and Soil Science, Volume 15:1–363
Hille S (2006) New aspects of sediment accumulation and reflux of nutrients in the Eastern Gotland basin (Baltic Sea) and its impact on nutrient cycling. University of Rostock, 119 pp [PhD Thesis]
Höglander H, Larsson U, Hajdu S (2004) Vertical distribution and settling of spring phytoplankton in the offshore NW Baltic Sea proper. Marine Ecology Progress Series 283:15–27
Jakobs G, Rehder G, Jost G, Kießlich K, Labrenz M, Schmale O (2013) Comparative studies of pelagic microbial methane oxidation within the redox zones of the Gotland deep and Landsort deep (central Baltic Sea). Biogeosciences 10:7863–7875
Jilbert T, Slomp CP, Gustafsson BG, Boer W (2011) Beyond the Fe-P-redox connection: preferential regeneration of phosphorus from organic matter as a key control on Baltic Sea nutrient cycles. Biogeosciences 8:1699–1720
Johnson KA, Goody RS (2011) The original Michaelis constant: translation of the 1913 Michaelis-Menten paper. Biochemsitry 50:8264–8269
Jørgensen BB, Kasten S (2006) Sulfur cycling and methane oxidation. In: Schulz HD, Zabel M (eds) Marine geochemistry, 2nd edn. Springer, Berlin, pp 271–309
Kuliński K, Pempkowiak J (2011) The carbon budget of the Baltic Sea. Biogeosciences 8:3219–3230
Laanemets J, Kononen K, Pavelsen J, Poultanen EL (2004) Vertical location of seasonal nutriclines in the western Gulf of Finland. Journal of Marine Systems 52:1–13
Langmuir I (1916) The constitution and fundamental properties of solids and liquids, Part I: solids. Journal of the American Chemical Society 38:2221–2295
Langmuir I (1917) The constitution and fundamental properties of solids and liquids, Part II: liquids. Journal of the American Chemical Society 39:1848–1906
Larsson U, Hajdu S, Walve J, Elmgren R (2001) Baltic Sea nitrogen fixation estimated from the summer increase in upper mixed layer total nitrogen. Limnology and Oceanography 46: 811–820
Leipe T, Harff J, Meyer M, Hille S, Pollehne F et al (2008) Sedimentary records of environmental changes and anthropogenic impacts during the past decades. In: Feistel R, Nausch G, Wasmund N (eds) State and evolution of the Baltic Sea 1952–2005: a detailed 50-year survey of meteorology and climate, physics, chemistry, biology, and marine environment. John Wiley and Sons, Hoboken, NJ, pp 295–439
Leipe T, Tauber F, Vallius H, Virtasalo J, Uścinowicz S et al (2011) Particulate organic carbon (POC) in surface sediments of the Baltic Sea. Geo-Marine Letters 31:175–188
Michaelis L, Menten ML (1913) Die Kinetik der Invertinwirkung. Biochemische Zeitschrift 49:333–3869 [in German]
Monod J (1942) Recherches sur la croissance des cultures bactériennes. Hermann, Paris, 210 pp [Doctoral Thesis, in French]
Monod J (1949) The growth of bacterial cultures. Annual Review of Microbiology 3:371–394
Mort HP, Slomp CP, Gustafsson BG, Andersen TJ (2010) Phosphorus recycling and burial in Baltic Sea sediments with contrasting redox conditions. Geochimica et Cosmochimica Acta 74:1350–1362
Mulholland M, Lomas MW (2008) Nitrogen uptake and assimilation. In: Capone DG, Bronk DA, Mulholland M, Carpenter EJ (eds) Nitrogen in the marine environment, 2nd edn. Academic Press – Elsevier, Burlington, pp 303–384
Nausch G, Nehring D, Nagel K (2008) Nutrient concentrations, trends and their relation to eutrophication. In: Feistel R, Nausch G, Wasmund N (eds) State and evolution of the Baltic Sea 1952–2005: a detailed 50-year survey of meteorology and climate, physics, chemistry, biology, and marine environment. John Wiley and Sons, Hoboken, NJ, pp 337–366
Nausch M, Nausch G, Lass HU, Mohrholz V, Nagel K et al (2009) Phosphorus input by upwelling in the Eastern Gotland basin (Baltic Sea) in summer and its effects on filamentous cyanobacteria. Estuarine, Coastal and Shelf Science 83:434–442
Olson RJ, Sosik HM, Chekalyuk AM, Shalapyonok A (2000) Effects of iron enrichment on phytoplankton in the Southern Ocean during late summer: active fluorescence and flow cytometric analysis. Deep-Sea Research 47:3181–3200
Omstedt A, Edman M, Anderson LG, Laudon H (2010) Factors influencing the acid-base (pH) balance in the Baltic Sea: a sensitivity analysis. Tellus 62B:280–295
Orphan VJ, House CH, Hinrichs KU, McKeegan KD, DeLong EF (2001) Methane-consuming archaea revealed by directly coupled isotopic and phylogenetic analysis. Science 293(5529):484–487
Piechura J, Beszczyńska-Möller A (2004) Inflow waters in the deep regions of the southern Baltic Sea – transport and transformations. Oceanologia 46:113–141
Ptacnik P, Andersen T, Tamminen T (2010) Performance of the Redfield ratio and a family of nutrient limitation indicators as thresholds for phytoplankton N *versus* P limitation. Ecosystems 13:1201–1214
Redfield AC, Ketchum BH, Richards FA (1963) The influence of organisms on the composition of sea-water. In: Hill MN (ed) The sea: ideas and observations on progress in the study of the seas, vol 2., The composition of seawater: comparative and descriptive oceanographyInterscience Publishers, New York, NY, pp 26–77
Rolff C, Almesjö L, Elmgren R (2007) Nitrogen fixation and abundance of the diazotrophic cyanobacterium *Aphanizomenon* sp. in the Baltic Proper. Marine Ecology Progress Series 332:107–118
Rullkötter J (2006) Organic matter: the driving force for early diagenesis. In: Schulz HD, Zabel M (eds) Marine geochemistry, 2nd edn. Springer, Berlin, pp 125–168
Savchuk OP (2005) Resolving the Baltic Sea into seven subbasins: N and P budgets for 1991–1999. Journal of Marine Systems 56:1–15
Savchuk OP (2010) Large-scale dynamics of hypoxia in the Baltic Sea. In: Yakushev EV (ed) Chemical structure of pelagic redox interfaces: observation and modelling. Springer-Verlag, Berlin, pp 137–160
Schinke H, Matthäus W (1998) On the causes of major Baltic inflows – an analysis of long time series. Continental Shelf Research 18:67–97

Schmale O, Schneider von Deimling J, Gülzow W, Nausch G, Waniek J, Rehder G (2010) Distribution of methane in the water column of the Baltic Sea. Geophysical Research Letters 37:L12604

Schneider B (2011) The CO_2 system of the Baltic Sea: biogeochemical control and impact of anthropogenic CO_2. In: Schernewski G, Hofstede J, Neumann T (eds) Global change and Baltic coastal zones. Springer, Dordrecht, Coastal Research Library 1:33–50

Schneider B, Gülzow W, Sadkowiak B, Rehder G (2014) Detecting sinks and sources of CO_2 and CH_4 by ferrybox-based measurements in the Baltic Sea: three case studies. Journal of Marine Systems 140:13–25

Schneider B, Kaitala S, Maunula P (2006) Identification and quantification of plankton bloom events in the Baltic Sea by continuous pCO_2 and chlorophyll *a* measurements on a cargo ship. Journal of Marine Systems 59:238–248

Schneider B, Nausch G, Nagel K, Wasmund N (2003) The surface water CO_2 budget for the Baltic Proper: a new way to determine nitrogen fixation. Journal of Marine Systems 42:53–64

Schneider B, Kaitala S, Raateoja M, Sadkowiak B (2009) A nitrogen fixation estimate for the Baltic Sea based on continuous pCO_2 measurements on a cargo ship and total nitrogen data. Continental Shelf Research 29:1535–1540

Shaffer G (1986) Phosphate pumps and shuttles in the Black Sea. Nature 321:515–517

Stocker TF, Qin D, Plattner GK, Tignor M, Allen SK, et al. (eds) (2013) The physical science basis – contribution of Working Group I to the Fifth Assessment Report of the Intergovernmental Panel on Climate Change. Cambridge University Press, Cambridge, 1535 pp

Struck U, Pollehne F, Bauerfeind E, von Bodungen B (2004) Sources of nitrogen for the vertical particle flux in the Gotland Sea (Baltic Proper) – results from sediment trap studies. Journal of Marine Systems 45:91–101

Tyrrell T, Schneider B, Charalampopoulou A, Riebesell U (2008) Coccolithophores and calcite saturation state in the Baltic and Black Seas. Biogeosciences 5:1–10

Walger E, Zeitzschel B, Rumohr J (eds) (1987) Seawater – sediment interactions: an interdisciplinary study in Kiel Bight. Springer, Berlin 338 pp

Walve J, Larsson U (2010) Seasonal changes in Baltic Sea seston stoichiometry: the influence of diazotrophic cyanobacteria. Marine Ecology Progress Series 407:13–25

Wasmund N, Nausch G, Schneider B (2005a) Primary production rates calculated by different concepts – an opportunity to study the complex production system in the Baltic Proper. Journal of Sea Research 54:244–255

Wasmund N, Nausch G, Schneider B, Nagel K, Voss M (2005b) Comparison of nitrogen fixation rates determined with different methods: a study in the Baltic Proper. Marine Ecology Progress Series 297:23–31

Wasmund N, Voss M, Lochte K (2001) Evidence of nitrogen fixation by non-heterocystous cyanobacteria in the Baltic Sea and recalculation of a budget of nitrogen fixation. Marine Ecology Progress Series 214:1–14

Wulff F, Sokolov A, Savchuk (2013) Nest – a decision support system for management of the Baltic Sea, a user manual. Baltic Nest Institute, Stockholm University Baltic Sea Centre, Technical Report 10:1–70

Yakushev EV, Pollehne F, Jost G, Kuznetsov I, Schneider B, Umlauf L (2007) Analysis of the water column oxic/anoxic interface in the Black and Baltic Seas with a numerical model. Marine Chemistry 107:388–410

Patterns of biodiversity

4

Pauline Snoeijs-Leijonmalm

Abstract

1. More than 4,400 known species live in the brackish Baltic Sea. Of these, 4 % are cyanobacteria, 51 % unicellular eukaryotes (protists), 8 % macrophytes, 32 % invertebrates and 5 % vertebrates.
2. In the Baltic Sea Area (Baltic Sea and the transition zone), the species richness of these five groups is >6,600, 50 % higher than in the Baltic Sea alone, while the water volume increases by only 4 %.
3. The higher richness in the transition zone is caused by North Sea species that still occur in the Kattegat and Belt Sea but cannot survive in the low salinity of the Baltic Sea. Unicellular organisms may be especially diverse in the transition zone as they move with the water masses of different salinities from the Skagerrak and the Baltic Sea that mix here.
4. The true number of species is much higher than the diversity reported from both the Baltic Sea and the transition zone since most archaean and bacterial species, as well as many protists, fungi and small invertebrates, are still unknown.
5. The dominant species in the Baltic Sea proper are mainly hardy, estuarine species, accompanied by a number of glacial relicts, freshwater species and ~130 (non-indigenous) brackish-water species. In the three large gulfs of the Baltic Sea (the Gulfs of Bothnia, Finland and Riga), and near large freshwater discharges along the entire coasts, freshwater species dominate below a salinity of ~4.
6. The species richness of cyanobacteria, heterotrophic bacteria and benthic diatoms is not impeded in the Baltic Sea. These groups are highly diverse in both marine and freshwater and enter the Baltic Sea from both habitats.
7. Macroscopic organisms show a species minimum at salinity 5–7. There are very few "true" brackish-water species in the Baltic Sea, and the loss of marine species, *e.g.* macroalgae, polychaetes, crustaceans and molluscs, along the large-scale Baltic Sea gradient is poorly compensated for by species entering the Baltic Sea from freshwater such as charophytes, vascular plants, oligochaetes and insect larvae.
8. Despite a pool of >1,500 macroscopic species, the evenness of the communities in the Baltic Sea proper is low, as they are typically dominated by mass occurrences of a few macroscopic species that build simple food webs in a highly productive system.
9. With few species in each functional group (*e.g.* habitat-forming macrophytes, filter-feeding animals, pelagic fish), there is a high risk that the loss or drastic reduction of a single key species may alter functions that are important for the maintenance of the ecosystem, such as provision of habitats, balanced food webs and resilience.

Keywords

Biodiversity gradients • Evenness • Functional diversity • Human impacts • Low-diversity ecosystem • Resilience

P. Snoeijs-Leijonmalm (✉)
Department of Ecology, Environment and Plant Sciences,
Stockholm University, 10691 Stockholm, Sweden
e-mail: pauline.snoeijs-leijonmalm@su.se

P. Snoeijs-Leijonmalm et al. (eds.), *Biological Oceanography of the Baltic Sea*,
DOI 10.1007/978-94-007-0668-2_4

4.1 Concepts of biodiversity

4.1.1 A measure of biological variation

The term biodiversity refers to the degree of variation within and between living organisms. This includes much more than merely counting the number of taxonomic units (richness). Richness does not account for the proportional abundances (evenness) or the functional aspects (traits) of the taxa in a community, habitat or ecosystem, and therefore reflects the actual – rather than the effective – amount of biological variability. In most cases, the evenness and trait components of diversity are of much higher importance for the relationship between diversity and ecosystem functioning than the richness component. Richness may exaggerate the role of rare taxa that contribute little to function unless they are "keystone species", *i.e.* species that, relative to their abundances, have a disproportionately large effect on community structure; these are usually predators (Paine 1995).

Patterns of biodiversity in space and time are central aspects of ecology and conservation biology (Naeem et al. 1994; Naeem 2002; Hooper et al. 2005), and in ecosystem management biodiversity is increasingly used as a measure of ecosystem health (*cf.* Sect. 17.7). To be able to compare the biodiversity of different sites, or to record biodiversity changes over time, it is crucial that diversity data are collected and interpreted in appropriate and consistent ways (Hillebrand and Matthiessen 2009). This requires the combination of taxonomic and ecological skills.

Biodiversity patterns occur within species, communities, habitats, regions, ecosystems, biomes and the entire Earth. The within-species diversity is studied as genetic diversity (*cf.* Sect. 6.1) and refers to the total number of characteristics in the genetic composition of a species. Genetic diversity enables populations to adapt to changing environments. When the genetic variation within a population is large, it is more likely that some individuals will possess alleles that are suited for a particular environment. These individuals have a larger chance to survive in that environment and to produce offspring bearing those alleles. Genetic diversity and species diversity are interdependent, *i.e.* diversity within species is necessary to maintain diversity among species, and *vice versa* (Lankau and Strauss 2007). For example, a decrease in species diversity may cause changes in the environment that lead to genetic adaptation of the remaining species, while a decrease in genetic diversity may lead to loss of species.

4.1.2 Diversity of what?

The collection of biodiversity data requires defining which organisms are included in a survey. Diversity is often considered at the "community level", with "community" defined as a group of organisms that live together in space and time and interact, as opposed to an "assemblage" of organisms without biotic interactions. However, organism body size (Table 4.1) and methodological issues put limits on the analysis of entire communities, *e.g.* macroscopic species are usually easy to identify while microscopic species are not and may require more detailed examination such as molecular identification.

Thus, community analyses rarely include all trophic levels from primary producers and microbial decomposers to top predators, *i.e.* all interactions within a biological community. In practice, as long as the boundaries of the community are clearly defined, a community can refer to all the components in a specific habitat's food web (*e.g.* benthic community) or only to a part of the food web (*e.g.* zoobenthic community). A community can also refer to all organisms belonging to a specific taxonomic group (*e.g.* ciliate community) or to all organisms of a specific body size (*e.g.* microbial community) (Box 4.1). Even when the limit between groups of organisms is rather artificial, such as that between meiofauna and macrofauna (the mesh size of a sieve, which *e.g.* can separate juveniles from adults of the same species), it is common to limit ecological studies, including the analysis of diversity, to "meiobenthic communities" and "macrozoobenthic communities".

When entire communities are compared, it may turn out that the diversity of different groups of organisms responds to environmental change in different ways related to the body size and ecological functions of the organisms. Even within physiologically uniform diatom communities, taxa of different size may respond differently to environmental change (Snoeijs et al. 2002). One way to avoid having to interpret such confusing patterns is to analyse the diversity of ecological traits in a community rather than the diversity of taxonomic units (*cf.* Sect. 4.1.6).

4.1.3 Recording and comparing diversity: a matter of scale

The collection of biodiversity data should define the geographical scale at which a survey is carried out. With reference to the Theory of Island Biogeography (MacArthur

Table 4.1 Relationships between taxonomic groups and organism body size in the sea. In practical work the communities are often subdivided into different size fractions by filters, sieves and nets. Blue = a major proportion of the organisms of this functional group are within this size fraction. Grey = a minor proportion of the organisms of this functional group are within this size fraction. For plankton the limits of the size fractions are generally accepted, but for the benthos size fractions different standards are still used in different countries and/or laboratories. a = The size classes of microbenthos and meiobenthos overlap, traditionally "microbenthos" is used for autotrophic protists (microalgae) and "meiobenthos" for heterotrophic protists (protozoa), although the limit between these two groups of protists is diffuse.

Plankton size group	**Body size**	**Viruses Archaea**	**Heterotrophic bacteria**	**Cyanobacteria**	**Protists (unicellular algae and protozoa)**	**Fungi**	**Invertebrates (metazoan plankton)**	**Vertebrates**
Femtoplankton	0.02–0.2 μm							
Picoplankton	0.2–2 μm							
Nanoplankton	2–20 μm							
Microplankton	20–200 μm							
Mesoplankton	0.2–2 mm							
Macroplankton	2–20 mm							
Megaplankton	>20 mm							
Benthic size group	**Body size**	**Viruses Archaea**	**Heterotrophic bacteria**	**Cyanobacteria**	**Protists (unicellular algae and protozoa)**	**Fungi**	**Invertebrates**	**Macroalgae Vascular plants Vertebrates**
Femtobenthos	0.02–0.2 μm							
Picobenthos	0.2–2 μm							
Nanobenthos	2–20 μm							
Microbenthos[a]	20–200 μm							
Meiobenthos[a]	0.044(0.063) – 0.5(1) mm							
Macrobenthos	>0.5(1) mm							

and Wilson 1967), it has long been recognised that habitat size, habitat variability and distance between similar habitats need to be taken into account when interpreting biodiversity patterns. This has practical consequences when comparing the biodiversity of different areas, *e.g.* the different subregions of the Baltic Sea Area.

The biodiversity of different habitats can be compared if the way the taxa were recorded, the sample scale and the biodiversity index are the same. Whittaker (1960) introduced the terms α-, β- and γ-diversity for local within-habitat, between-habitat and regional diversity, respectively. Both α-diversity and γ-diversity represent inventories of species diversity, but at different spatial scales. On the other hand, β-diversity has developed into a measure for the extent or rate of change in taxonomic composition. β-diversity in its simplest form, according to Whittaker's (1960) original definition, is $\beta = \gamma/\alpha$, where γ-diversity is the total species diversity of a region, and α-diversity is the average species diversity per habitat. β-diversity as a measure of species change (turnover) can be expressed as $\beta = (\gamma - \alpha)/\alpha = \gamma/\alpha - 1$ or $\beta = (\gamma - \alpha)/\gamma = 1 - \alpha/\gamma$. If two samples are compared, and presence-absence data are used, these two equations resemble the Sørensen similarity coefficient and the Jaccard similarity coefficient, respectively (Tuomisto 2010a, b).

Since Whittaker (1960) did not clearly define what should be considered "local", "habitat" and "regional", later scientists have had a hard time agreeing on the quantitative interpretation of β-diversity. Again, this is a matter of scale. When the area sampled increases, the observed species diversity also increases. This is due to the facts that more individuals are included in the sample and large areas are environmentally more heterogeneous than small areas. In practice, γ-diversity is often used for the total diversity observed in a dataset consisting of several samples, α-diversity is the average diversity for all samples and β-diversity expresses the differences in diversity between the samples (Tuomisto 2010a–c). When samples are taken along a spatial or temporal scale, β-diversity is a measure of the gains and losses (turnover) of species from place to place or from time to time, respectively.

If comparisons are made, the sample size (sampling effort) must also be the same since the number of species increases with sample size (Whittaker 1972), although there

Box 4.1: Microbes are everywhere

Anna Edlund

What are microbes?

A microbe is any organism that can only be seen with a microscope, *i.e.* microbes are smaller than ~100 μm in cell size (Kirchman 2008). However, a few species belonging to different microbial groups have exceptionally large cells that can be detected even with the naked eye. Examples are the largest known bacterium, the gram-negative coccoid bacterium *Thiomargarita namibiensis* (phylum Proteobacteria) with a cell size up to 750 μm in diameter, and the diatom *Tryblionella scalaris*, which in the Baltic Sea can be over 650 μm long. Furthermore, some microbes can be visible with the naked eye as a group of cells (colonies) or as more complex multicellular organisms such as filamentous cyanobacteria. Some scientists also regard viruses as microbes, but since viruses are only "alive" and able to multiply inside the cells of other organisms, other scientists argue that they are non-living.

Microbe abundance

Microbes (including viruses) are the most abundant organisms on Earth. In one litre of surface seawater it may be possible to find 10^9–10^{10} viruses, 10^8–10^9 bacteria, 10^6–10^7 cyanobacteria and 10^4–10^5 protists (unicellular eukaryotes). Their abundances are negatively related to their cell size (Box Fig. 4.1). Relative to bacteria, archaea are particularly abundant in deeper water layers.

Genetic diversity

Microbes comprise the highest genetic diversity in all ecosystems on Earth. Four billion years of evolution in the global oceans have shaped a tremendous richness of microbial species within all three domains of life (Archaea, Bacteria and Eukarya). Especially the archaea and bacteria have developed an outstanding capacity to adapt to almost every environment, ranging from anaerobic deep-sea sediments to aerobic surface waters and biofilms covering rocks and organism surfaces. In many extreme environments, where no other organisms can survive, archaea and bacteria are still abundant. Variation in physiological adaptations in the sea is lower among the marine microbial eukaryotes than that among bacteria and archaea, but still higher than in most terrestrial environments. For example, microalgal protists belong to genetically and physiologically very different groups such as the Chlorophyta, Chrysophyta, Cryptophyta, Dinophyta, Euglenophyta and Haptophyta, while all terrestrial plants are closely related to the Chlorophyta. Microalgal protists are mainly autotrophs and occur in the photic zone. However, all microalgal phyla also contain heterotrophic and mixotrophic species (*cf.* Table 4.4). Much of our knowledge regarding microbial diversity and ecological activities has been obtained during the past 50 years. Only recently have hypotheses depicting the evolutionary relationships among the major microbial clades attained some degree of consensus.

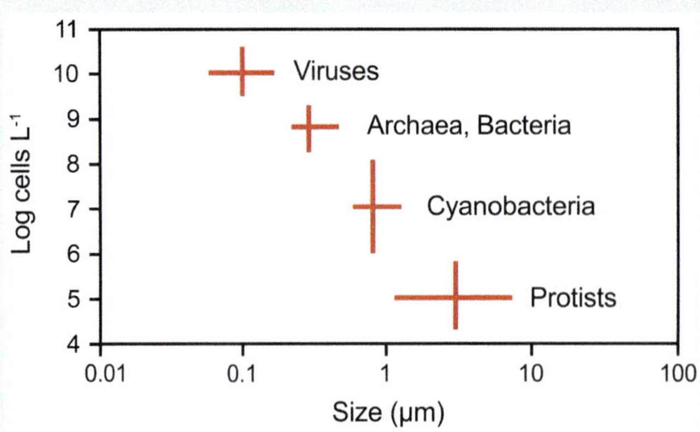

Box Fig. 4.1 The approximate relationship between the abundance and cell size of the major microbial groups in surface waters of aquatic habitats. Figure modified from Kirchman (2008)

Microbes and nutrient cycling

Much of the organic material that accumulates in the marine environment is transformed back (mineralised) to its inorganic starting material, most importantly CO_2, NH_4^+ and PO_4^{3-}, by microbes. However, there are certain restrictions for mineralisation processes to occur, such as the availability of electrons for cellular respiration. From a thermodynamic perspective, the reduction of molecular oxygen (O_2) is energetically the most favourable process and occurs first. When all O_2 is consumed, other compounds are used for cell respiration, and the sediment or water column becomes anoxic. A small sediment particle or marine snow particles ($\sim$1–2 mm in size) may maintain an anoxic centre, even when the particle is surrounded by air or oxygenated water (Fenchel and Finlay 2008). Therefore, anoxic marine environments are not necessarily isolated from their oxic surroundings. Some of the most anaerobic active habitats in the marine environment occur as islands in a micro-aerobic matrix where they are only temporarily anaerobic. Thus, it is important to consider the boundaries between aerobic and anaerobic habitats. Heterotrophic organisms inhabiting these zones catalyse the restoration of chemical equilibrium through the oxidation of reduced carbon produced by photosynthetic organisms. When O_2 is depleted, nitrate (NO_3^-) will serve as an electron acceptor, followed by manganese dioxide (MnO_2), iron oxide (FeOOH), sulphate (SO_4^{2-}) and carbon dioxide (CO_2) (*cf.* Table 3.1).

Ecological functions

New genetic tools exist to identify microbial organisms and to determine their biogeographical distributions and ecological functions. Studies of microbial diversity provide insight on how their community composition contributes to ecosystem functioning. Major functional groups comprise primary producers, photoheterotrophs, heterotrophs, grazers, viruses, diazotrophs, nitrifiers and denitrifiers (Box Fig. 4.2). The data obtained can be used in predictive models that describe how microbial communities will respond to natural or anthropogenically-mediated changes in environmental conditions. Microbial communities comprise a wide range of metabolic processes and play key roles in the production and mineralisation of organic matter. The ecological roles of microbes in the Baltic Sea are still far from being well understood and the least is known about viruses and archaea. However, our knowledge has greatly increased during the last decade, and is increasing exponentially, with the help from the rapid developments in molecular research technologies, including genomics, transcriptomics and proteomics (*cf.* Box 4.3).

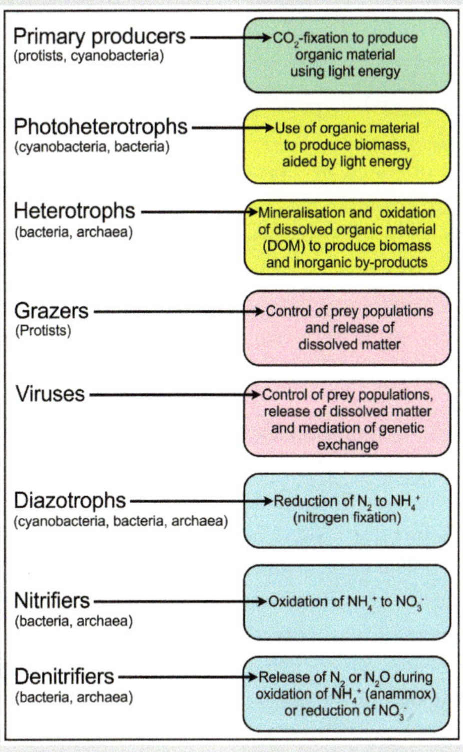

Box Fig. 4.2 Functional groups of microbes in aquatic environments. Figure based on data in Kirchman (2008)

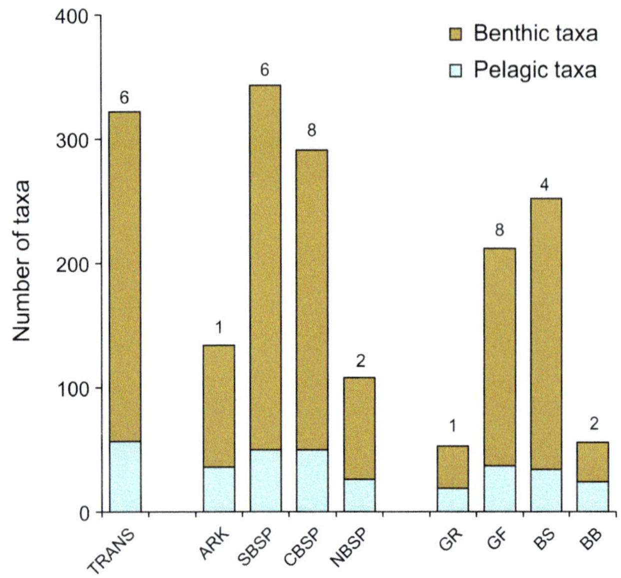

Fig. 4.1 The number of recorded diatom taxa reflects the number of sampling sites in the subregions of the Baltic Sea Area. The number of sampling sites is indicated above the bars. In the Bothnian Sea (BS) the number of taxa is relatively high because more within-site samples were studied at one of the sites. The taxa are subdivided into taxa with their main distribution in the pelagic zone (pelagic taxa) and taxa with their main distribution in the benthic zone (benthic taxa). TRANS = the transition zone (Kattegat and Belt Sea), ARK = Arkona Sea, SBSP = Southern Baltic Sea proper, CBSP = Central Baltic Sea proper, NBSP = Northern Baltic Sea proper, GR = Gulf of Riga, GF = Gulf of Finland, BS = Bothnian Sea, BB = Bothnian Bay. Data calculated from the checklist of Snoeijs et al. (1993–1998). Figure: © Pauline Snoeijs-Leijonmalm

are ways to address this (*e.g.* rarefaction analysis). Typically, sample size includes all individuals that are associated with a certain surface area (in the benthic zone) or found in a certain volume of water (in the pelagic zone). Sample size can also be a defined number of individuals or a defined biomass. Finally, diversity between different areas is difficult to compare if the sampling effort is not equal, *e.g.* some checklists of Baltic Sea organisms clearly show in which part of the Baltic Sea sampling has been most intensive and where it should be intensified (Fig. 4.1).

Checklists are often compiled to assess the total richness of a geographic area. They are also a useful aid in the identification of species, especially when they are illustrated. These checklists are usually based on literature surveys and/or a given author's own observations. The compilation of species checklists is a dynamic process and will never provide a full inventory. Checklists have a number of drawbacks that need to be taken into account when using them for making comparisons between geographical areas. There may be bias related to scale (size of the areas, number of samples per area, etc.) and quality of species identification if different people have collected the data and taxon recognition is in many cases subjective.

4.1.4 Taxonomic resolution

Taxonomic richness is the number of different taxa, usually at the species level, represented in a sample. Instead of species, some studies use the term "operational taxonomic unit" (OTU), which is used mainly in molecular biology for *e.g.* DNA sequences. Taxonomic expertise is often a bottleneck when collecting biodiversity data. In the case of turbellarians (and other soft-bodied meiofauna such as *e.g.* gastrotrichs or naked rotifers), taxon identification is impeded not only by the lack of taxonomic expertise, but also by preservation – they cannot be identified to the species level after preservation with formalin or ethanol. Taxonomists with extensive knowledge of the morphological identification of specific groups of organisms are becoming extinct worldwide. This problem is partly solved by species identification with the help of DNA barcoding (Box 4.2), massive parallel sequencing (Box 4.3) or biochemical analyses, *e.g.* of algal pigments (Box 4.4).

Standardisation of nomenclature is a necessity for scientists to be able to communicate. It is especially important when taxonomic data collected by different people are pooled. Within monitoring programmes, such as that carried out by HELCOM (*cf.* Sect. 17.8.4) in the Baltic Sea Area, intercalibration (synchronisation) of species identification is an on-going process. Intercalibration is achieved through meetings of the people analysing the samples and the publication of species sheets, checklists and databases. On a global scale, the taxonomy of marine species is standardised in the "World Register of Marine Species" (WoRMS; http://www.marinespecies.org) with the aim to provide an authoritative and comprehensive list of names of marine organisms, including information on synonymy. This is a valuable resource for the intercalibration of species identification as well as for the use of a universal nomenclature because species names are constantly being changed as a result of taxonomic revisions.

Molecular techniques, from microsatellites and fingerprinting techniques to (meta-) genomics, transcriptomics and proteomics involving massive parallel sequencing, provide excellent possibilities to identify species with DNA sequences as OTUs, and even to assess the taxonomic and functional diversity of entire communities (Box 4.3). However, with the exception of archaean, bacterial and fungal communities, most biodiversity surveys today are still based on morphological identifications of the organisms collected. Biodiversity datasets may therefore contain a "false" number of taxonomic units because some species are recorded at the species level while other species are lumped into higher taxonomic units. Thus, compilations of datasets collected by researchers with different levels of expertise may produce "false" biodiversity patterns, *e.g.* some checklists of Baltic Sea organisms clearly show in which part of the Baltic Sea certain taxonomists with exceptionally high expertise have been active (Nielsen et al. 1995; Hällfors 2004).

4.1.5 Biodiversity indices

Basic biodiversity data can be recorded as presence-absence data, *i.e.* a taxon is present or it is not ("qualitative data"), or by relative or absolute abundance related to a defined surface area, volume, biomass, etc. ("quantitative data"). When a quantitative measure for each taxon in a sample is available, it is possible to not only assess the taxonomic richness, but also obtain a measure of taxon evenness in the sample.

Besides richness and evenness, other biodiversity indices – often combining richness and evenness in different ways – are used to express the biodiversity of a sample, *e.g.* the Shannon index, the Simpson index and the Gini-Simpson index (Box 4.5). All these indices summarise different things, and their values are therefore not directly comparable, but they complement each other. These indices are of theoretical value, but are often not suitable for practical use in environmental monitoring where other aspects, *e.g.* the total biomass of a community, may be more important (*cf.* Sect. 14.5.2).

4.1.6 Classification by functional traits

Diversity can be also expressed in terms of the different life functions of organisms in a community or an area, in which case it is termed the "functional diversity". Taxonomical identification is not necessary for assessing functional diversity, but thorough knowledge of the ecology of the organisms is required. Species belonging to the same functional group can be lumped together to assess functional diversity within a community, habitat or ecosystem. For example, microbes can be lumped together according to the gene expression of metabolic functions (Boxes 4.1 and 4.3), micro- and macrophytes according to growth form (Box 4.6, *cf.* 11.5.2) and meio- and macrozoobenthos according to feeding type (Box 4.7, *cf.* Sect. 4.7.2).

4.2 What types of species live in the Baltic Sea?

4.2.1 What is a Baltic Sea species?

To be able to assess the total richness, *i.e.* the species pool (γ diversity) of a geographic area, it is necessary to define the limits of the area and decide which species are to be included. A species that is found in the Baltic Sea is not necessarily a "Baltic Sea species". Only species that form sustainable populations in the Baltic Sea are considered "true" Baltic Sea species. However, this is not always easy to determine, especially not for microbes belonging to the "rare biosphere", *i.e.* small species that occur in very low numbers (Pedrós-Alio 2012). Species distributions in the Baltic Sea follow many gradients, the most prominent ones being salinity and climate, but they are also influenced by currents, human-induced introductions of non-indigenous species and stochastic events, which may bring species from elsewhere to the Baltic Sea. The latter category includes *e.g.* rare visits by dolphins, large whales and other species that cannot survive in the Baltic Sea for a long time due to the prevailing environmental conditions (low salinity, ice cover) and/or lack of food. These occasional visitors usually find their way out of the Baltic Sea rather fast and are not considered Baltic Sea species. Non-indigenous species are considered Baltic Sea species if they are able to reproduce and form stable populations in the Baltic Sea, and even if their inability to reproduce is masked by continuous multiple introductions (*cf.* Sect. 5.2.2).

Marine organisms enter the Baltic Sea with currents from the Kattegat, and freshwater organisms enter the Baltic Sea with freshwater inflows from rivers, streams and lake systems all around the Baltic Sea coast (*cf.* Fig. 2.11). Continuous multiple introductions may mask their true environmental preferences. They are part of the Baltic Sea plankton communities, they photosynthesise, respire and participate in food web interactions, and are therefore considered Baltic Sea species. However, their condition may not be optimal due to the low (for marine species) or high (for freshwater species) salinity in the Baltic Sea, and they are not able to survive for a long time. These types of marine species are more abundant near the entrance to the Baltic Sea (Arkona Sea), and these types of freshwater species are more abundant in coastal areas than in the open Baltic Sea.

4.2.2 Baltic Sea species are adapted to stable low salinity

The expression "large-scale Baltic Sea gradient" implies that this is a continuum and that the species distributions in the Baltic Sea would change gradually along the gradient. However, in reality the gradient involves a series of subbasins with different salinity regimes shaped by barriers for water exchange. Thus, the distributions of many species along the gradient change stepwise in concert with the stepwise changes in the salinity conditions.

The Baltic Sea Area starts at the Skagerrak-Kattegat front where the surface-water salinity becomes significantly influenced by outflows from the Baltic Sea and falls within the definition of brackish water (salinity below $\sim$30, *cf.* Fig. 1.10). Salinity fluctuations in the Kattegat and the Belt

Box 4.2: Species identification by DNA barcoding

What is DNA barcoding?

DNA barcoding is a taxonomic method that uses short DNA sequences for species identification. The sequences are chosen on the basis of taxonomic distinction between species, not from a phylogenetic or functional gene perspective. The method is useful when different species cannot be distinguished from each other with traditional methods (*e.g.* morphology), which is the case in many small species, and even in macrofauna organisms such as *Marenzelleria* spp. (*cf.* Box 5.3) and chironomid larvae (*cf.* Fig. 4.31). Barcoding then enables the study of a species' geographical range, the detection of overlooked species and even the discovery of new species if unknown barcodes are detected. DNA barcoding is also a useful method in *e.g.* fisheries management as it enables the identification of fish species, and (at the subspecies level) the stock it was harvested from, by DNA barcoding of fish fillets or other body parts available at food markets. Thus, DNA barcoding can serve as an effective modern tool in marine biodiversity assessment and conservation.

Methodology

A prerequisite for the positive identification of a species by DNA barcoding is that its barcode is known. The most extensive barcode database is the Barcode of Life Database (BOLD, http://www.barcodinglife.org), which contains barcodes for >550,000 species (accessed 2 April 2016). For many organisms the target for DNA barcoding is a 648 base pair region of the mitochondrial cytochrome *c* oxidase I (COI) gene. This DNA region has been the standard for vertebrates since 2003 (Hebert et al. 2003) while a variety of alternative regions have been proposed for invertebrates and protists (Pawlowski et al. 2012). The barcode region of the COI gene can also be a useful diagnostic character for the species-level identification of invertebrates and protists, *e.g.* dinoflagellates (Stern et al. 2010) and copepods (Blanco-Bercial et al. 2014), in both individual specimens and bulk samples. However, it is difficult to find a single set of molecular markers that works for all protist lineages as the genetic divergence observed between and within major protistan groups greatly exceeds that found in plants, fungi, and animals. Scientists are still seeking universal criteria for barcode-based species identification in protists (Pawlowski et al. 2012). A universal DNA barcode for protists coupled with group-specific barcodes will enable an explosion of taxonomic research that will catalyse diverse applications, such as faster and cheaper alternative to cell counts and higher taxonomic resolution in environmental monitoring of phytoplankton communities.

Barcoding fish

The international research collaboration FISH-BOL (the Fish Barcode of Life Campaign) aims at constructing a standardised reference DNA sequence library for all the Earth's fish (Ward et al. 2009). Barcodes for >17,000 fish species are already available in the BOLD database. Several specimens with divergent barcode sequences have been confirmed by integrative taxonomic analysis as new species. Concerns in relation to the use of fish barcoding for species discrimination include hybridisation, recent radiations, regional differentiation in barcode sequences and nuclear copies of the barcode region. However, research results indicate that these issues are of low importance for the great majority of specimens (Ward et al. 2009).

Barcoding invertebrates

There are many invertebrate groups of organisms with a body size <1 mm, the diversity of which can only be guessed while only a fraction of the species pool has been barcoded. For example, barcodes for only 493 nematode species are published in the BOLD database, while the number of described species is quoted as up to 40,000 (including many parasites), but the real species richness can be >1,000,000 (Blaxter et al. 2005). A relatively greater number of barcodes is available for copepods, with 431 barcoded species in the BOLD database *versus* a total estimated number of ~2,500 copepod species. The exceptional morphological conservation of the copepods, with numerous sibling species groups, makes the morphological identification of species challenging, even for expert taxonomists, and DNA barcoding is a suitable alternative (Blanco-Bercial et al. 2014).

Barcoding and massive parallel sequencing

DNA barcoding has become a well-funded, global enterprise since its proposition as a technique for species identification, delimitation and discovery. Genomic studies in conjugation with DNA barcoding can be very effective in the assessment of global biodiversity. However, to produce a taxonomically comprehensive database of barcode sequences, continued integrative morphological-molecular taxonomic analysis is needed. On the other hand, DNA barcoding may become irrelevant due to the rapid development of massive parallel sequencing (*cf.* Box 4.3) because of the speed with which it generates large volumes of genomic data (Taylor and Harris 2012). The advancement of high-throughput sequencing can be useful in analysing bulk environmental samples (Creer et al. 2010) at lower analysis costs (Stein et al. 2014).

Sea are large (Fig. 4.2). The Drogden sill and the Darß sill are the shallowest natural barriers for water exchange between the transition zone and the Baltic Sea (Fig. 4.3), and this is where the first major stepwise change along the Baltic Sea gradient takes place: that from strongly fluctuating brackish conditions to relatively stable low-salinity conditions.

In the Belt Sea, the average surface-water salinity varies from 11 to 18 depending on the distance from the Darß sill, but fluctuates between ~8 and ~27 depending on the direction of the currents (Fig. 4.2). Thus, the Belt Sea is the main area where North Sea and Baltic Sea species mix (Zettler et al. 2007), especially those species that are passively transported with the currents. The salinity fluctuations in the Kattegat and Belt Sea are erratic, but with an average seasonal cycle of high salinity in winter when the outflow from the Baltic Sea is low (Fig. 4.4a).

The Baltic Sea *sensu stricto*, east of the Darß sill, has a much more constant surface-water salinity, <10 (Figs. 4.2 and 4.4a, b). This is a salinity at which only the most euryhaline North Sea species can survive, having additionally been forced to adapt to stable low rather than to fluctuating salinity. The deeper waters of the Baltic Sea (below the halocline at 60–80 m depth, *cf.* Table 2.6) have slightly higher salinity, *e.g.* up to ~15 below ~60 m in the Bornholm Sea and up to ~12 below ~100 m in the Eastern Gotland Sea (*cf.* Fig. 2.15), which affects the living conditions of the deep-water archaea, bacteria and fauna.

The surface-water salinity in the Baltic Sea proper is stable around 7 all the way down to the halocline. In the Gulf of Finland (without a threshold to the Baltic Sea proper), the surface-water salinity decreases gradually to ~5 while in the Gulf of Riga (connected to the Baltic Sea proper by a shallow area) it decreases a little more steeply down to ~5. However, at the two thresholds of the Gulf of Bothnia, the Södra Kvarken sill and the Norra Kvarken sill, salinity decreases abruptly to ~5 and ~3, respectively (Fig. 4.2). The sudden change between the Baltic Sea proper and the Bothnian Sea is characterised by a shift in dominance from marine to freshwater species, while the Norra Kvarken sill forms the distributional limit for the former. In the innermost parts of all three gulfs, close to river mouths, salinity goes down to 0.

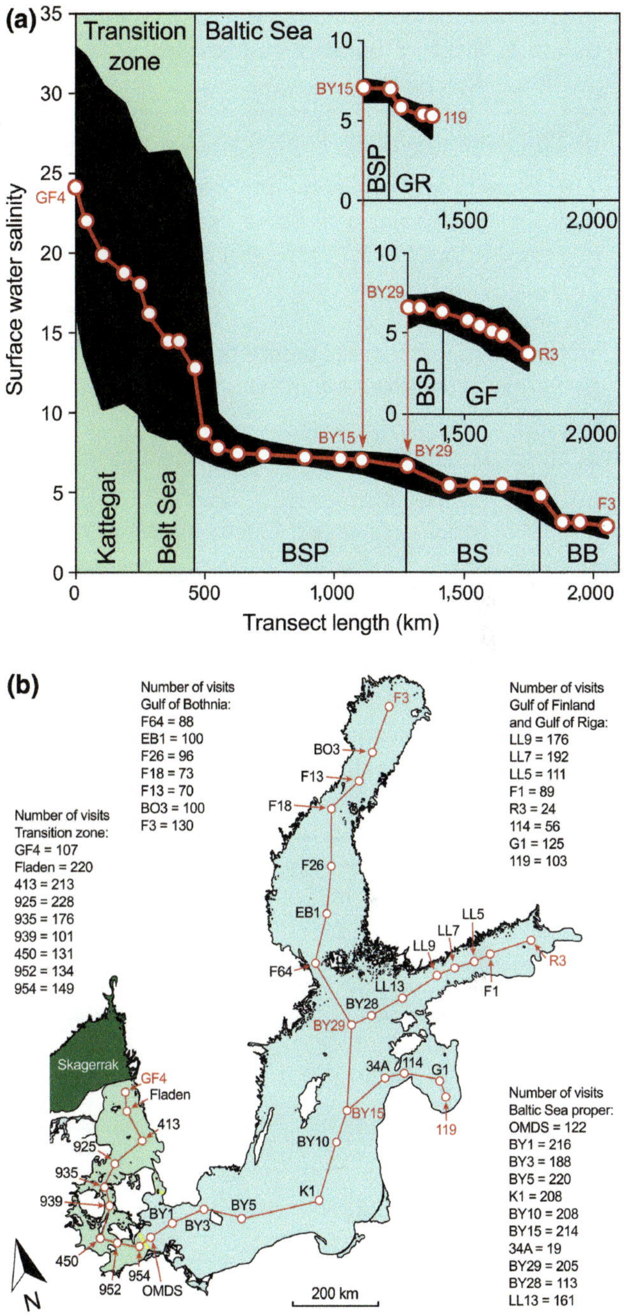

Fig. 4.2 Summary of surface-water salinity variation along the large-scale Baltic Sea gradient over 20 years (1996–2015), based on 4,866 monthly measurements, each representing a unique visit with a research vessel to one of 35 monitoring stations. (**a**) The average (red circles) and range (black areas) of the year-round surface-water salinity in the Baltic Sea Area along transects from the northern Kattegat to the northern Bothnian Bay (BB), the southern Gulf of Riga (GR) and the eastern Gulf of Finland (GF). BSP = Baltic Sea proper, BS = Bothnian Sea. (**b**) The geographic positions of the 35 monitoring stations and the number of times each station was visited. Figure based on measurements accessed from the major databases around the Baltic Sea with the decision support system Baltic Nest (http://nest.su.se, Wulff et al. 2013). Figure: © Pauline Snoeijs-Leijonmalm

4.2.3 Many Baltic Sea species are salinity-stressed

Very few species in the Baltic Sea are fully adapted to the salinity of the water they live in. Along the large-scale Baltic Sea gradient, this is clearly manifested in the fact that many marine species, algae as well as animals, and invertebrates as

Box 4.3: Massive parallel sequencing as a tool for identifying and monitoring Baltic Sea microbes and their functions

Birgitta Bergman, Martin Ekman, and Karolina Ininbergs

The diversity and role of aquatic microbes

For a long time, bacteria in oceans were thought to be merely "passive riders" if present at all. They were challenging to study due to their extremely small cell size and difficulties in growing them in the laboratory. Still less than 1 % of the bacteria on Earth are cultured. Thanks to the recent introduction of massive parallel sequencing technologies, also referred to as "next-generation sequencing" (NGS), it is today estimated that microorganisms (viruses, archaea, bacteria, cyanobacteria and protists) make up ~90 % of the biomass in the oceans and that they are critically linked to global productivity and health. Thus, aquatic microbiology has evolved into a vibrant research field.

Why massive parallel sequencing of microbes?

Through massive sequencing of DNA (metagenomics) or RNA (metatranscriptomics) isolated from microorganisms living in mixed, often complex microbial communities, information about community diversity can reach a magnitude never experienced with traditional methodologies. As a multi-faceted and organism-oriented approach, it will substantially widen our knowledge bank on microbes and open new research avenues. It will provide information not only on the genetic identity of the organisms but also on their functional capacity (*i.e.* presence and expression of functional genes) in time and space. For example, the detection of *nif* genes (DNA) and *nif* transcripts (RNA) verifies the presence of nitrogen-fixers and suggests ongoing nitrogen fixation, respectively. In addition, the collection of metadata (temperature, pH, salinity, nutrient concentrations, *etc.*) may be a crucial part in the sampling procedure as it allows for the interpretation of the data in an ecological and environmental context. As billions of nucleotides and millions of peptides from a massive number of different types of microbes will typically be obtained, the many steps in the workflow (Box Fig. 4.3) require substantial computational resources and bioinformatic tools to analyse the generated datasets, an area under constant development and improvement. Hence, final steps in the workflow may focus on the biodiversity of the organisms in a given environment or on the exploration of ongoing cellular pathways and processes. Adding microbial network and statistical analyses will further reveal how members of the microbial community interact with each other and respond to prevailing ecological parameters.

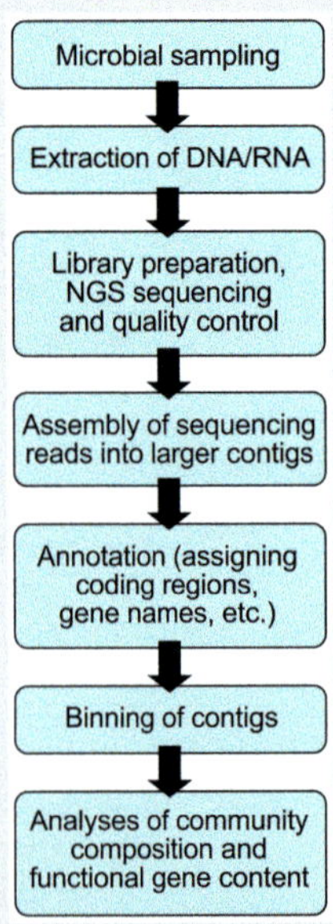

Box Fig. 4.3 A typical workflow using massive parallel sequencing to examine diversity and function of microbial communities. The procedure may differ depending on research questions, *e.g.* not all analyses require the assembly and annotation of data. Contigs = longer stretches of overlapping sequencing reads

Diversity and roles of microbes in the Baltic Sea

The Baltic Sea "microbiome" was recently captured using massive parallel sequencing generating metagenomic and metatranscriptomic databases (Dupont et al. 2014; Ininbergs et al. 2015). Salinity and macronutrients were the overall drivers for shaping the microbial community. This was also reflected in core metabolic processes, some of which were distinctly different between microbes at low and high salinities, which may explain the dramatic divide between marine and limnic populations (Box Fig. 4.4). Unlike higher organisms (macroalgae, macrofauna), the microbes adapt readily to the brackish-water gradients offered, although "signature" microbes were found at both low and high salinity regimes, respectively. However, SAR11, the world's most common bacterium, was present throughout the entire Baltic Sea. Besides bloom-forming colonial cyanobacteria, unicellular *Synechococcus* spp. were common and widespread. Other bacteria included proteorhodopsin-containing photoheterotrophs that use sunlight to drive proton-pumping rhodopsins. An extensive variety of genetic signatures of both beneficial and disease-causing bacteria, as well as a rich repertoire of viruses, were identified. This included microbes involved in basic nutrient cycling and with the potentials to produce vitamins, chelate metals and degrade polluting substances. The Baltic Sea water column showed different biodiversity patterns with water depth, *e.g.* typical communities, including many archaea, occurred close to the anoxic bottom layers of the Landsort deep (Box Fig. 4.4).

Monitoring microbes for a better environment

Today the challenge is to fully understand how the microbiome interacts with the overall health of the Baltic Sea, which stresses the necessity to include microbes in monitoring programmes. As long as a fundamental knowledge bank is "hidden" in the microbial world, management will suffer. Genetic probes can be designed to target important groups (*e.g.* vitamin producers, pathogens) and processes (*e.g.* toxin production) and be used as indicators to assess ecosystem health. Monitoring can be implemented via stationary or free-drifting remote-controlled or automated sampling buoys equipped with *in situ* analysers of environmental parameters, as well as proteins and nucleic acids, and generate an unprecedented level of new information to support future management procedures.

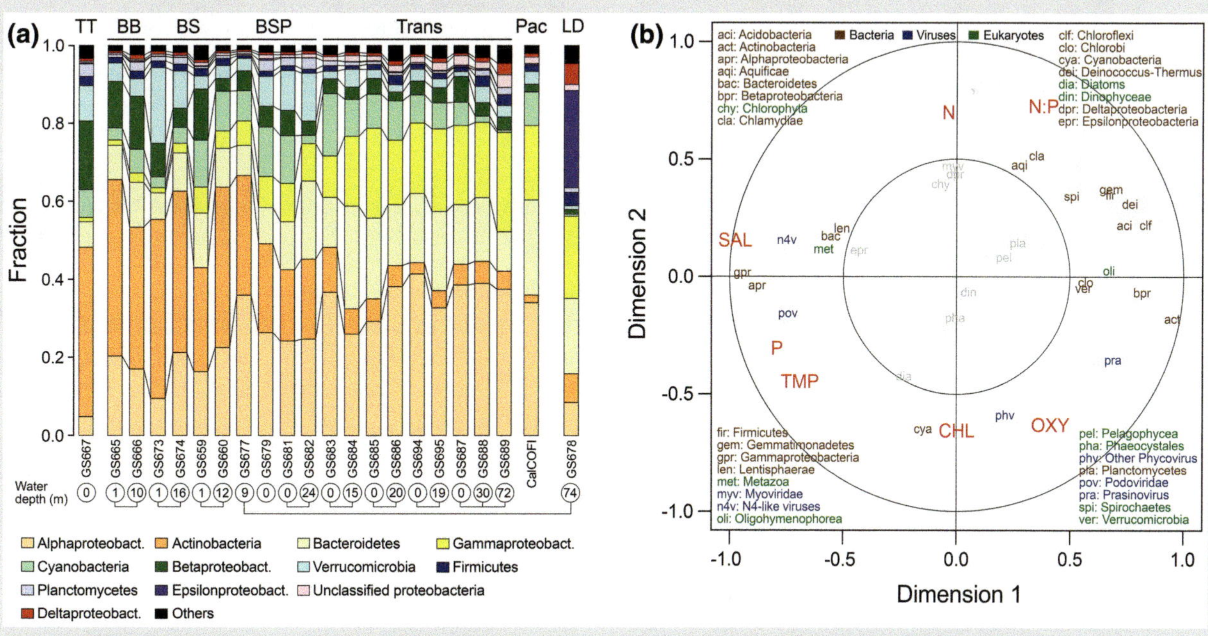

Box Fig. 4.4 Microbial community composition derived from massive parallel sequencing. (**a**) Comparison of Baltic Sea bacterial communities with those in freshwater and marine habitats, based on 31 phylogenetic maker genes. Sampling depth is indicated within circles below sample names, with interconnecting circles indicating samples from the same location. TT = Torneträsk (freshwater), BB = Bothnian Bay, BS = Bothnian Sea, BSP = Baltic Sea proper, Trans = Transition zone (Kattegat and Belt Sea), CalCOFI = a merged set of metagenomes from the coastal eastern Pacific Ocean, LD = the Landsort deep, a site in the Baltic Sea proper with oxygen deficiency in deep water. (**b**) Biplot of a multivariate statistical analysis (RCCA) relating community composition (colour-coded according to taxonomical group: bacteria, viruses or eukaryotes) to environmental data (red). The biplot shows correlations between the phylogenetic and environmental drivers with the corresponding first and second canonical axes. The first two canonical axes explain 65 % of the variance in the environmental drivers and 49 % of the variance in the organismal variables. The outer circle (1) and inner circle (0.5) display the amount of the variance explained by the linear combinations of the variables, respectively; the variance explained for scores within the inner circle (grey) is less than 25 %. Figure modified from Dupont et al. (2014)

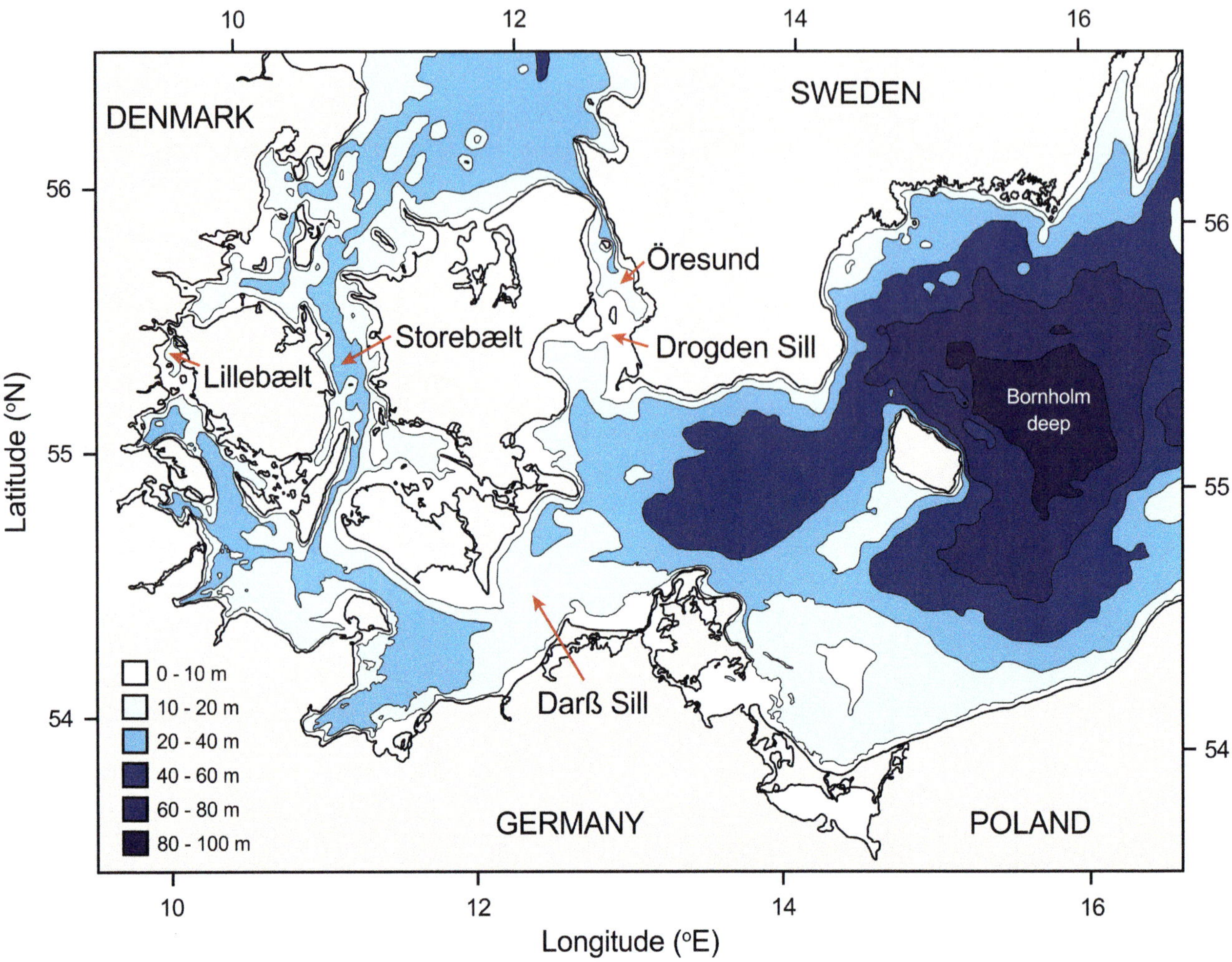

Fig. 4.3 Map showing the situation of the Danish straits (Lillebælt, Storebælt and Öresund) and the two major thresholds for water exchange between the Baltic Sea and the transition zone: the 8-m deep Drogden sill and the 18-m deep Darß sill. Figure: © Pauline Snoeijs-Leijonmalm

well as fish, decrease in body size (Fig. 4.5). Species that do not live under optimal conditions are also more sensitive to diseases and get more easily infested by parasites. Close to their limit of distribution, the species often lose their ability to reproduce sexually, *e.g. Zostera marina* (Reusch et al. 1999). For many animals, this immediately marks the end of their occurrence, while some plants and algae may form gigantic clones (consisting of one individual) at the edge of their existence (Reusch et al. 1999; Bergström et al. 2005; Tatarenkov et al. 2005). Also some freshwater species decrease in body size in the Baltic Sea, *e.g.* the snails *Bithynia tentaculata* and *Theodoxus fluviatilis* (Figs. 4.5 and 4.6).

4.2.4 Many Baltic Sea species are cold-adapted

Typical Baltic Sea species are not only adapted to stable low salinity but also to low temperature and a short growing season. The warm season is about one month shorter in the Baltic Sea proper than in the transition zone, which is mainly due to a later warming of the surface waters in spring (Fig. 4.4c). In the Gulf of Riga and the Gulf of Finland, winter temperatures are lower than in the northern Baltic Sea proper while summer temperatures are similar (Fig. 4.4d). In the Gulf of Bothnia, both winter and summer temperatures are lower than in the northern Baltic Sea proper, and the warm season is shorter due to lower temperature both in spring and in autumn, especially in the Bothnian Bay where an ice cover can occur between November and June.

Examples of cold-adapted species in the Baltic Sea are the bacteria, protists and metazoans that colonise the Baltic Sea ice, forming typical sympagic (ice-associated) communities during the 4–6 month-long ice cover in the north (*cf.* Sect. 9.1). In spring, these communities seed the pelagic spring bloom with cold-water species. A number of typical Baltic Sea species of diatoms, macroalgae and invertebrates are cold-adapted glacial relicts (*cf.* Sect. 4.6.2).

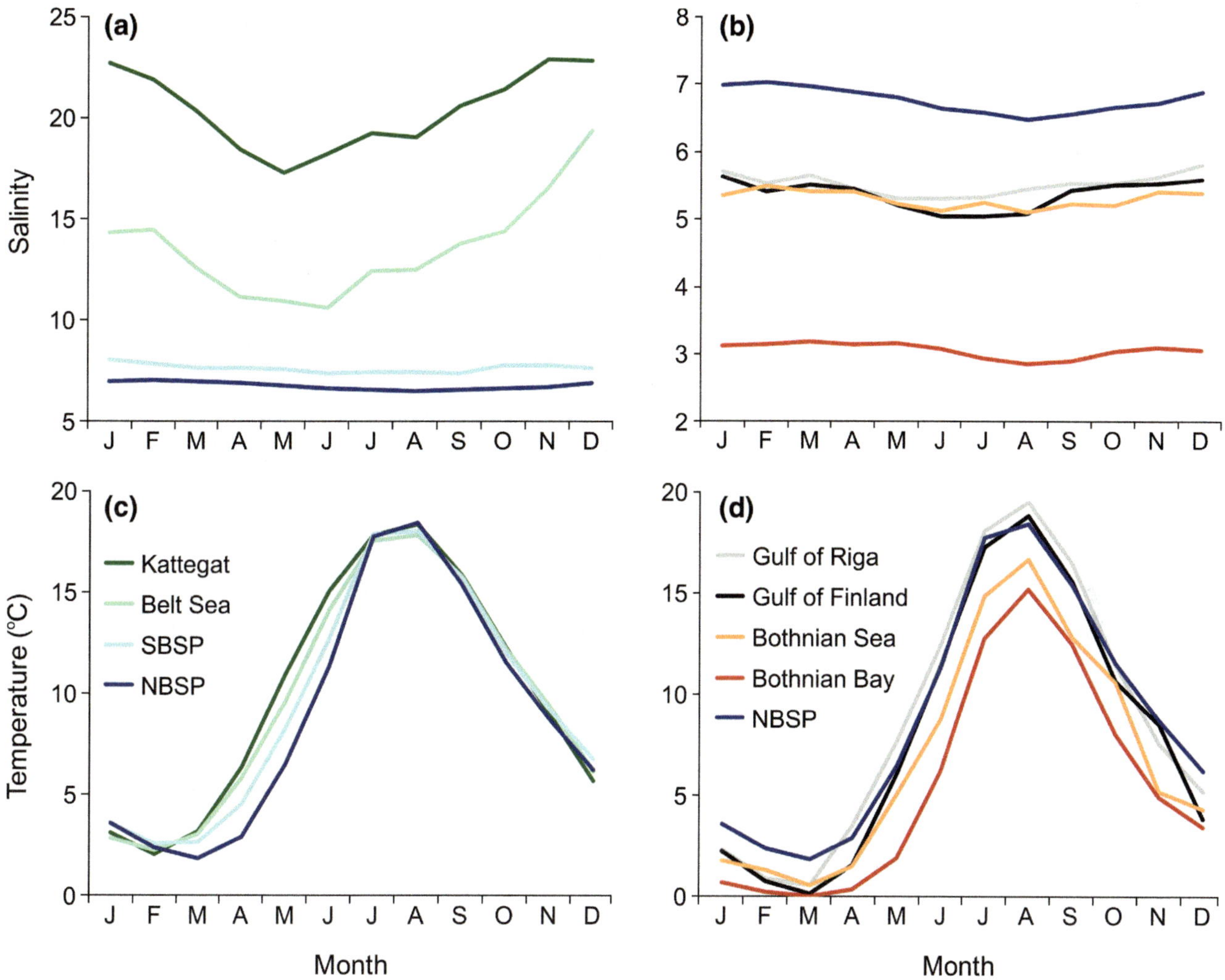

Fig. 4.4 Seasonal variation in surface-water salinity and temperature in different parts of the Baltic Sea Area. The data represent averages per subregion for the open-sea monitoring stations shown in Fig. 4.2b, except for Stations R3 and 34A due to the low number of measurements at these stations. (**a**, **c**) Comparisons between the transition zone (Kattegat and Belt Sea), the southern Baltic Sea proper (SBSP) and the northern Baltic Sea proper (NBSP). (**b**, **d**) Comparisons between the northern Baltic Sea proper (NBSP) and the Gulfs of Riga, Finland and Bothnia. The legend for (**a** and **c**) is given in (**c**), the legend for (**b** and **d**) is given in (**d**). Figure: © Pauline Snoeijs-Leijonmalm

4.2.5 Species adapted to low oxygen levels

Even if oxygen depletion has increased in the deep waters of the Baltic Sea proper, mainly as a result of the large-scale eutrophication process, it is a natural feature of the ecosystem. With decreasing oxygen availability in their surroundings, microbial community composition shifts from aerobic bacteria to sulphur- and methane-metabolising bacteria and archaea (*cf.* Sect. 3.6). Most of the typical zoobenthos and fish species living in the deep waters can tolerate relatively low oxygen levels (*e.g.* the Baltic clam *Macoma balthica* and the isopod *Saduria entomon*), but only a few species are extremely well adapted to these conditions (*e.g.* the polychaete *Bylgides sarsi* and some nematodes), and none of the animals can survive without oxygen (*cf.* Sect. 10.11).

4.2.6 Benthic or pelagic?

Most species in aquatic ecosystems have their main distribution either in the pelagic zone or in the benthic zone. For example, all macrophytes are basically restricted to the benthic zone. However, there is a high degree of benthic-pelagic coupling, which includes the movement of organisms between the two zones. This must be taken into account when recording species richness. For example, the Atlantic herring *Clupea harengus* is a pelagic fish that

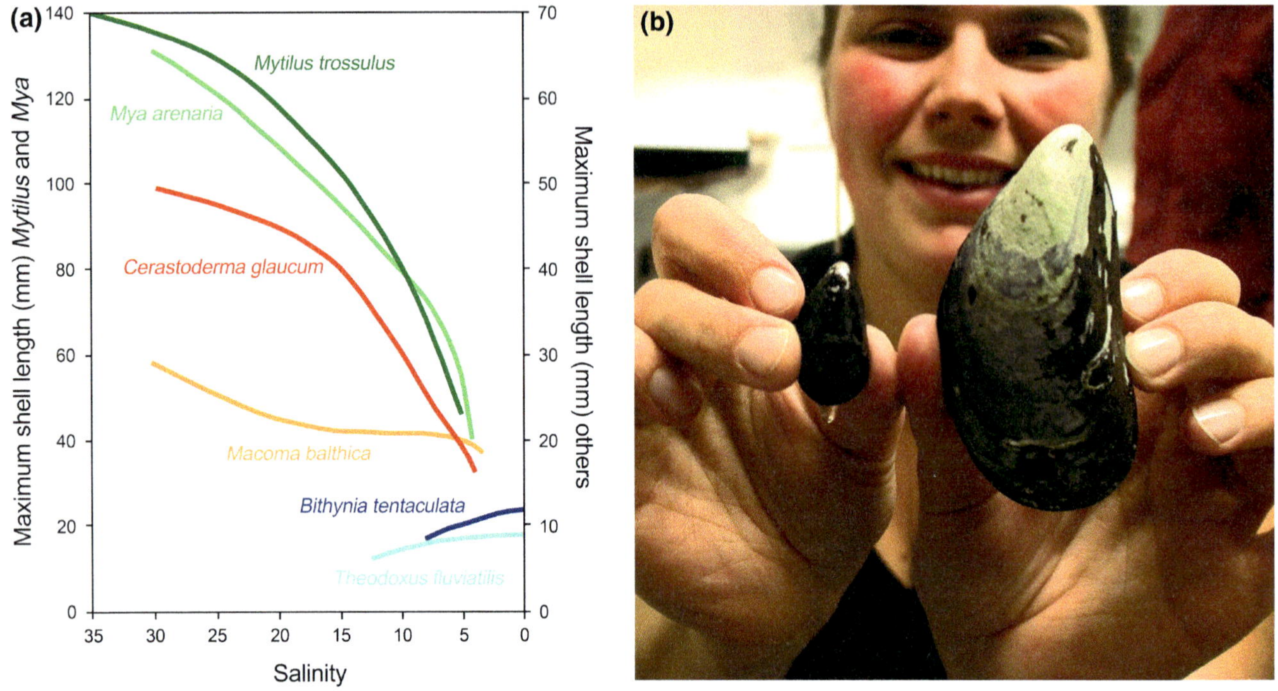

Fig. 4.5 Marine species and freshwater species decrease in body size with decreasing and increasing salinity, respectively. (**a**) The relationship between maximum length and salinity of some molluscs as presented by Remane. Note that *Mytilus trossulus* and *Macoma balthica* may include hybrids (*cf.* Box 6.5). (**b**) A comparison between *Mytilus trossulus* from the Baltic Sea (smaller) and from the transition zone (larger). Figure (**a**) modified from Remane (1934 and 1958, exactly the same figure occurs in both publications). Photo: (**b**) © Hans Kautsky

migrates to the coast for spawning. Moreover, many pelagic microorganisms form cysts that overwinter in the sediments, many benthic invertebrates have pelagic juvenile stages, mysids (opossum shrimps) make daily vertical migrations from bottom water into higher parts of the water column and even macrophytes may become detached and drift while still alive. Benthic bacteria, protists and fungi are easily resuspended by waves and are often encountered in coastal plankton communities. For example, >60 % of ciliate species richness in the coastal zooplankton may consist of benthic species, although pelagic species prevail numerically (Mironova et al. 2014). Similarly, benthic diatoms are common in coastal phytoplankton communities. There are even other ways for benthic protists to be transported into the pelagic zone, *e.g.* they can "hitch-hike" with copepods to the open Baltic Sea (Fig. 4.7).

4.3 How many species live in the Baltic Sea?

4.3.1 Growing biodiversity databases

During the last 20 years, many on-line databases on marine diversity, covering sea areas from local to global scales, have been established. The species lists in these databases are not – and will never be – complete, and distributional data are today still strongly biased by the number of studies and samples included for specific geographical areas. However, some databases that focus on distributional data have become rather comprehensive and may be used to detect biodiversity patterns, although the obtained patterns must be interpreted with care.

4.3.2 Baltic Sea species richness in a global database

The largest global database that provides distributional data for marine species is the UNESCO database "Ocean Biogeographic Information System" (OBIS; http://www.iobis.org), which strives to document the ocean's diversity, distribution and abundance of life, and allows users to search marine species datasets from all of the world's oceans. On 29 September 2015 the OBIS database included 2014 taxa at the species level for the Baltic Sea, 6,981 for the North Sea, 3,196 for the Barents Sea and 1,479 for the Black Sea (Table 4.2). Thus, according to the records in the OBIS database, species richness is lower in the brackish Baltic and Black Seas than in the marine North and Barents Seas, and this holds true for nearly all phyla.

The general trend of a lower number of taxa in brackish water than in marine water may be true, although there are

Fig. 4.6 Four freshwater snails that are abundant in the Baltic Sea with different salinity ranges. (**a**) The most euryhaline species *Theodoxus fluviatilis* occurs all the way from the Kattegat to the Bothnian Bay. (**b**) *Bithynia tentaculata* occurs from the Arkona Sea to the Bothnian Bay. (**c**) *Physa fontinalis* occurs from the northern Baltic Sea proper to the Bothnian Bay. (**d**) *Anisus leucostoma* occurs from the Bothnian Sea to the Bothnian Bay. Photo: © Pauline Snoeijs-Leijonmalm

issues that may bias the OBIS database in the case of the Baltic Sea. For example, the OBIS database focuses on "marine species datasets" while many freshwater species live in the Baltic Sea as well. Furthermore, it is not clear if the OBIS database includes the transition zone (Belt Sea and Kattegat) in its definition of the "Baltic Sea".

4.3.3 More species in regional checklists

The OBIS database contains more taxa of Cyanobacteria, Amoebozoa, Choanozoa, Ciliophora (ciliates), Euglenozoa and Rotifera for the Baltic Sea than for the other seas (Table 4.2). On the other hand, the Rhodophyta (8 and 227

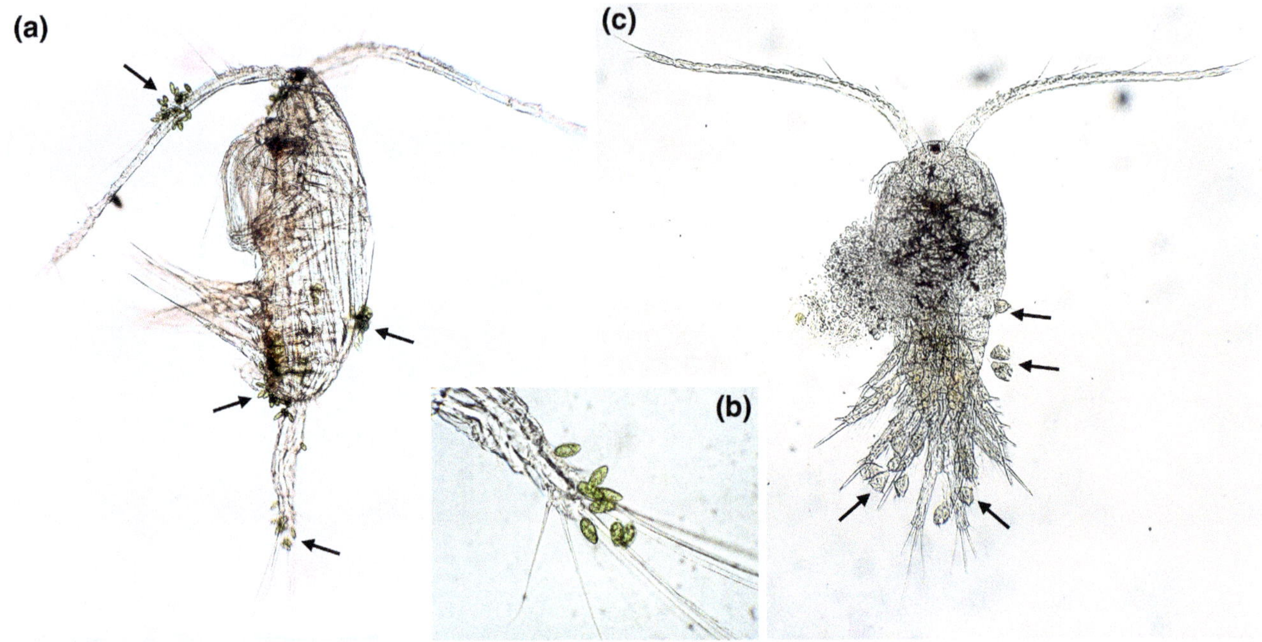

Fig. 4.7 Benthic protists (arrows) can "hitch-hike" with copepods to the pelagic zone. (**a**) Green microalgae growing on a copepod. (**b**) Close-up of the green microalgae in (**a**). (**c**) Ciliates growing on a copepod. Photo: © Pauline Snoeijs-Leijonmalm

taxa, respectively) and Foraminifera (3 and 157 taxa, respectively) seem to be heavily underrepresented in the Baltic Sea compared to the North Sea. The question is: are these trends real or is this because the diversity of some phyla has been studied more or less extensively in the Baltic Sea compared to other seas? This question can partly be answered by studying the diversity of different groups of organisms along the large-scale Baltic Sea gradient in regional checklists. They will reveal that the recorded number of red algal species in the Baltic Sea is not 8 but 69, and that that of foraminifers is not 3 but 15, still lower than in the North Sea but higher than in the OBIS database.

The total richness of Baltic Sea taxa in the OBIS database is low. Compilations of data from the major checklists for the Baltic Sea Area (Baltic Sea, Belt Sea and Kattegat) show the total richness of cyanobacteria, protists, macrophytes and animals in the Baltic Sea to be at least 4,419, more than twice that in the OBIS database, and in the Baltic Sea Area the total species richness is at least 6,647, including many North Sea species (Table 4.3). This does not include parasites, which are estimated at >380 taxa in vertebrates alone (Ojaveer et al. 2010). Including those of invertebrates, the total number of parasites in the Baltic Sea will be much higher. Thus, it can be concluded that there is a need to update the OBIS database for the Baltic Sea, and probably for other seas as well.

4.4 Diversity changes along the large-scale Baltic Sea gradient

4.4.1 Bacterial richness

The species richness of heterotrophic bacteria, as well as archaea, in the Baltic Sea is unknown and can only be estimated. A rough estimate of bacterial species is between 10^3 and 10^6 (Ojaveer et al. 2010). The wide range of this estimate, three orders of magnitude, displays its uncertainty. This is not only a result of the limited number of studies carried out in the Baltic Sea, but also of methodological problems in culturing cells and identifying them down to the species level by molecular techniques.

The bacteria in the pelagic zone of the Baltic Sea have been studied in more detail than those living in the benthic zone. A bacterial inventory produced by 454 pyrosequencing of partial 16S rRNA genes from 60 offshore sampling stations revealed that the surface water OTU richness does not display any clear minimum or maximum along the Baltic Sea salinity gradient (Fig. 4.8). The sum of OTUs that match known freshwater and marine sequences is relatively constant along the salinity gradient ($\sim$90 in 800 randomly picked reads), as is the total number of observed OTUs ($\sim$150 in 800 randomly picked reads).

Table 4.2 The number of taxa (species level and below) reported from the Baltic Sea in the global OBIS database (accessed 29 September 2015) compared to that in three other large marine ecosystems, the North Sea, the Barents Sea and the Black Sea. OBIS = the Ocean Biogeographic Information System of the Intergovernmental Oceanographic Commission of UNESCO. http://www.iobis.org. The Protoctista include the protists together with their multicellular descendants, *e.g.* multicellular algae (Margulis 1990). * = including chrysophytes (including diatoms) and brown algae, ** = including dinoflagellates

	Latitude in centre	**60 °N**	**56 °N**	**75 °N**	**43 °N**
	Surface area	369,000 km^2	700,000 km^2	1,400,000 km^2	461,000 km^2
	Average water depth	57 m	95 m	230 m	1,197 m
	Average salinity	6	35	35	20
Biota group	**Phylum**	**Baltic Sea**	**North Sea**	**Barents Sea**	**Black Sea**
Prokaryotes	Cyanobacteria	57	16	1	31
Protoctista (Algae)	Chlorophyta	60	120	24	51
Protoctista (Algae)	Cryptophyta	18	19	7	8
Protoctista (Algae)	Haptophyta	8	21	9	47
Protoctista (Algae)	Ochrophyta *	246	530	247	209
Protoctista (Algae)	Rhodophyta	8	227	5	50
Protoctista (Protozoa)	Amoebozoa	24	6	1	0
Protoctista (Protozoa)	Choanozoa	60	11	5	2
Protoctista (Protozoa)	Ciliophora	193	88	35	10
Protoctista (Protozoa)	Euglenozoa	17	15	7	7
Protoctista (Protozoa)	Foraminifera	3	157	226	35
Protoctista (Protozoa)	Myzozoa **	148	259	172	263
Protoctista (Protozoa)	Radiozoa	0	29	7	0
Protoctista (Protozoa)	Protoctista incertae sedis	17	17	4	2
Fungi	Ascomycota	5	38	1	1
Animals	Annelida	196	777	446	103
Animals	Arthropoda	271	1,470	874	257
Animals	Bryozoa	36	189	180	1
Animals	Chordata	203	516	218	108
Animals	Cnidaria	61	335	153	12
Animals	Echinodermata	19	105	126	10
Animals	Gastrotricha	5	26	4	0
Animals	Mollusca	130	730	385	98
Animals	Nematoda	86	846	4	134
Animals	Nemertea	11	29	0	3
Animals	Platyhelminthes	33	185	1	1
Animals	Porifera	7	89	22	19
Animals	Rotifera	40	21	1	2
	28 other phyla	52	110	31	15
Total number of phyla		**46**	**53**	**38**	**31**
Total number of species		**2,014**	**6,981**	**3,196**	**1,479**

The total number of cyanobacterial taxa recorded in phytoplankton samples in the Baltic Sea Area is 184, of which 179 are present in the Baltic Sea (Tables 4.3 and 4.4). This includes not only pelagic species (*cf.* Fig. 8.2), but also resuspended benthic species that may occur in coastal plankton communities. Along the large-scale Baltic Sea gradient, the richness distribution is irregular, most probably because of the low sampling effort in some of the areas

Box 4.4: Pigment diversity

Algal pigments

Three major categories of pigments are involved in the process of photosynthesis in algae and cyanobacteria: chlorophylls, phycobilins and carotenoids. The chlorophylls and phycobilins are mainly involved in light harvesting, while the carotenoids modulate the light energy and protect the photosynthetic apparatus and other cellular structures against oxidative damage and have additional functions in *e.g.* cell signalling (Esteban et al. 2015). All vascular plants are closely related to the green algae, and these evolutionarily relatively young groups have a similar basic pigment composition of chlorophylls and carotenoids. However, during the long evolutionary time in the ocean, the photosynthetic systems of cyanobacteria and algae have differentiated, which is reflected in the broad variation of pigment composition between different taxonomic groups (Box Fig. 4.5). Animals also utilise carotenoids as antioxidants, *e.g.* zooplankton transfer β-carotene from their algal food into astaxanthin and canthaxanthin and store them in body tissues, with astaxanthin often esterified with fatty acids (Snoeijs and Häubner 2014).

Pigment signatures reflect phytoplankton community composition

Given that phytoplankton pigments vary in chemotaxonomic specificity (Box Fig. 4.5), their relative abundances reveal the phytoplankton community structure at the level of the major algal groups (Box Fig. 4.6). Examples of

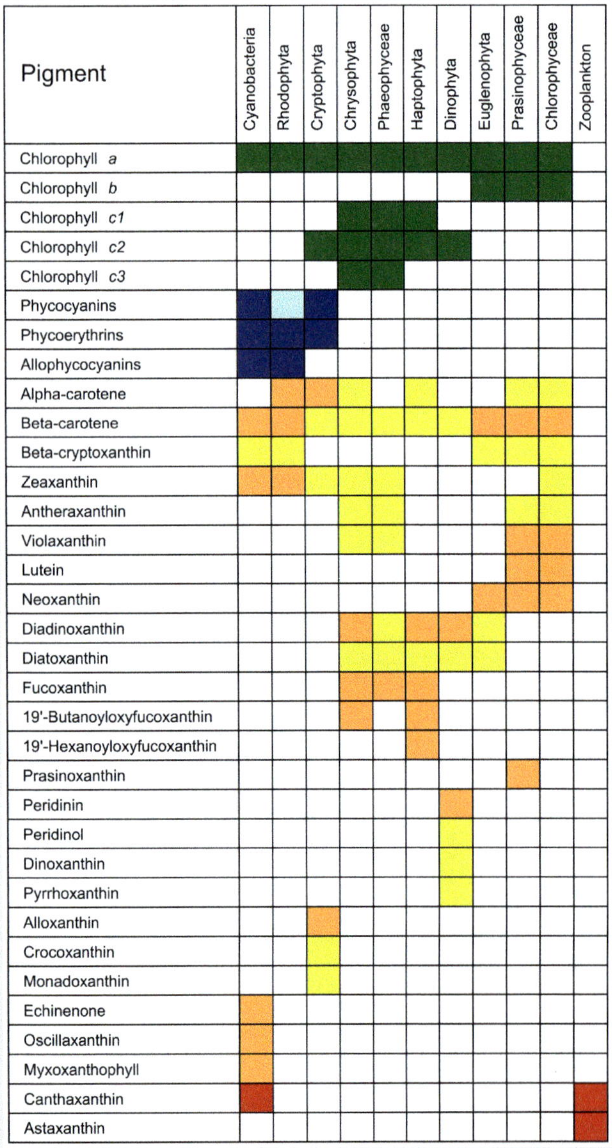

Box Fig. 4.5 Distributions of major and taxonomically significant pigments in different algal groups and zooplankton. This list of pigments is not complete and other pigments may occur in the algal groups as well, *e.g.* dinoflagellates can contain pigments of their symbionts, the green alga *Haematococcus pluvialis* stores astaxanthin in its cysts and some freshwater filamentous cyanobacteria contain echinenone while most marine cyanobacteria do not. Green = chlorophyll, dark-blue = abundant phycobilins, light-blue = low-abundant phycobilins, orange = abundant carotenoid, yellow = low-abundant carotenoid, red = abundant zooplankton carotenoid. Figure based on data in van den Hoek et al. (1995) and Jeffrey and Vesk (1997). Figure: © Pauline Snoeijs-Leijonmalm carotenoid biomarkers for single algal groups are 19′-hexanoyloxyfucoxanthin for haptophytes, prasinoxanthin for prasinophytes, peridinin for dinoflagellates, alloxanthin for cryptophytes and echinenone for filamentous cyanobacteria. With the help of multivariate statistics, a fingerprint can be made of the phytoplankton community composition and be compared with other communities. It is also possible to estimate the relative abundances of the different phytoplankton groups by using a model (*e.g.* CHEMTAX; Mackey et al. 1996). Pigment analyses produce objective high-quality data in terms of repeatability and sample throughput. Compared to cell counts, the method is sensitive as it also includes organisms difficult to identify microscopically such as pico- and nano-sized algae or species easily damaged by sample fixation (Wänstrand and Snoeijs 2005). However, a drawback is that heterotrophic organisms (without pigments) in the phytoplankton community are not recorded.

Pigment signatures reflect physiological status
Besides species composition, the quantitative pigment composition also reflects the physiological status, including responses to environmental signals, of a phytoplankton community or macroalga. For example, the ratio of chlorophylls to chlorophyll degradation products (*e.g.* chlorophyllide, pheophorbide) can indicate the degree of senescence of a phytoplankton bloom. The ratio of carotenoids to chlorophylls as well as the ratios between the carotenoids of the violaxanthin-antheraxanthin-zeaxanthin xanthophyll cycle (*e.g.* in green algae) and those of the diatoxanthin-diadinoxanthin xanthophyll cycle (*e.g.* in diatoms) can indicate the status of protection against reactive oxygen species (ROS; Ursi et al. 2003; Choo et al. 2005; Andersson et al. 2006).

The analysis of pigments
Algal concentrations of chlorophyll *a* and phycobilins are usually measured by simple spectrophotometry or fluorometry. The full range of chlorophyll and carotenoid concentrations in a sample can be measured with high-performance liquid chromatography (HPLC). Phycobilins cannot be measured from the same sample because these pigments are water-soluble while chlorophylls and carotenoids are fat-soluble and therefore extraction procedures differ.

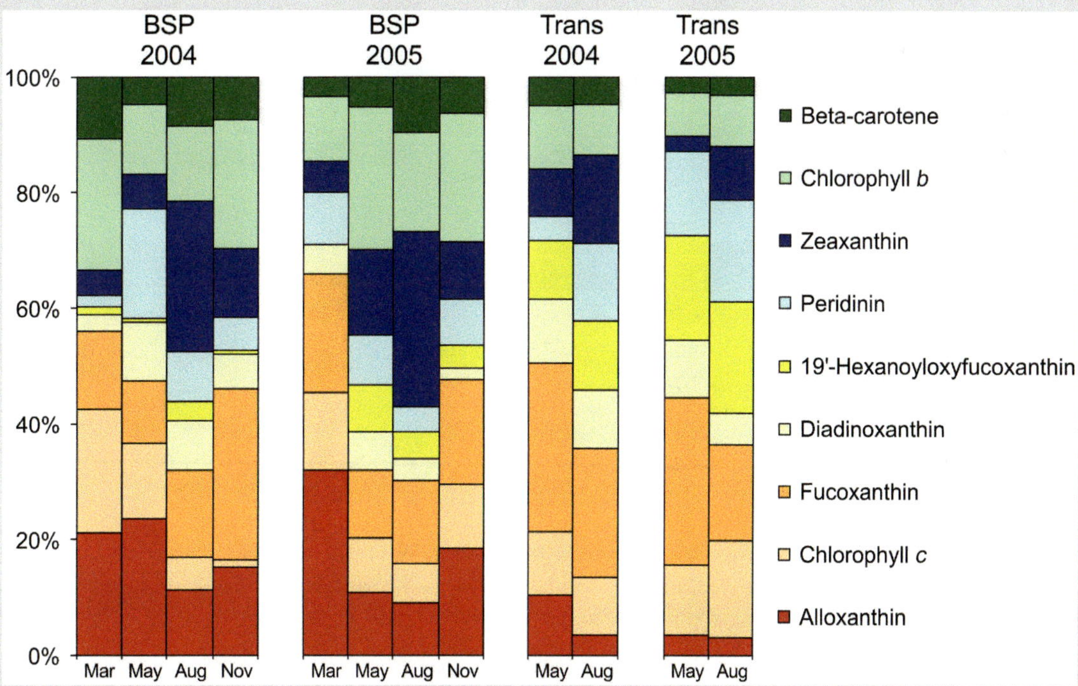

Box Fig. 4.6 Average pigment composition of the phytoplankton in the Baltic Sea proper and the transition zone, excluding chlorophyll *a* (which occurs in all algal groups), expressed as the % of the total pigment weight. The figure is based on pigment data from 94 sampling stations in the Baltic Sea proper and the transition zone visited in 2004 and 2005. During the warm season (May–August), the proportion of cyanobacteria (zeaxanthin), green algae (chlorophyll *b*) and cryptophytes (alloxanthin) was higher in the Baltic Sea than in the transition zone while the proportion of haptophytes (19′-hexanoyloxyfucoxanthin), dinoflagellates (peridinin) and chrysophytes, including diatoms (chlorophyll *c* and fucoxanthin) was higher in the transition zone. Figure modified from Snoeijs and Häubner (2014)

Box 4.5: Biodiversity calculation tools

A biodiversity sample is a collection of individuals selected in a well-defined way, *e.g.* from a certain substrate area or a certain water volume, with individuals subdivided into distinct taxonomic units, *e.g.* species, genera or DNA sequences (OTUs), and recorded using specific criteria, *e.g.* biovolume, biomass or number of individuals per sample. Comparisons can only be made if the samples are collected in exactly the same way.

Richness (S) is the number of different taxa in a sample. Species richness does not take into account the abundances of the taxa or their relative abundance distributions.

Evenness is a measure of the distribution of individuals between the different taxa in a sample. Maximum evenness occurs when all taxa in a sample are equally abundant.

The Shannon index (H′), also known as the "Shannon-Wiener index", the "Shannon-Weaver index" or "Shannon entropy", is a measure of biodiversity, which takes into account both richness and evenness and is calculated as follows:

$$H' = -\sum_{i=1}^{s} (p_1 * \ln p_i)$$

where p_i is the proportion of the sample belonging to the ith taxon. A higher value of H′ denotes higher biodiversity. The Shannon index varies between 0 (only one taxon in the sample) and $H'_{max} = \ln S$ (S taxa in the sample with equal proportions). The logarithm base in the Shannon index can be chosen at will (usually e, 2 or 10), but comparison of samples requires the same base.

Pielou's evenness index (J′) is a measure reflecting the distribution of individuals between the different taxa in a sample and can be calculated as follows:

$$J' = \frac{H'}{H'_{max}}$$

A higher value of J′ denotes higher evenness. The evenness index varies between 0 (only one taxon in the sample) and 1 (each taxon in the sample is represented by the same number of individuals).

The Simpson index (λ) is a measure of biodiversity, which indicates the degree of concentration of taxa (heterogeneity or "dominance"), calculated as follows:

$$\lambda = \sum_{i=1}^{s} p_i^2$$

The Simpson index equals the probability that two entities taken at random from a sample (with replacement) represent the same taxon. A higher value of λ denotes thus higher dominance in the sample. The Simpson index varies between 1/S (the taxa in the sample have equal proportions) and 1 (only one taxon in the sample).

The Gini-Simpson index (E = 1 − λ) equals the probability that two entities taken at random from a sample (with replacement) represent different taxa. A higher value of $1 - \lambda$ denotes higher homogeneity and it varies between 0 (only one taxon in the sample) and approaching 1 (∞ taxa in the sample with equal proportions).

Table 4.3 The number of taxa (species level and below) of cyanobacteria, protists, macrophytes and animals reported from the Baltic Sea and the Baltic Sea Area (Baltic Sea, Belt Sea and Kattegat). For the ciliates and the metazooplankton the transition zone includes only the Mecklenburg Bay (southern Belt Sea). [a] = including heterotrophic flagellates, [b] = estimate based on the literature cited in Ojaveer et al. (2010)

	Baltic Sea	Baltic Sea Area	% Increase	Reference
Water surface area	369,334 km^2	412,439 km^2	12	Table 2.1
Water volume	20,958 km^3	21,764 km^3	4	Table 2.1
Cyanobacteria	**179**	**184**	**3**	Hällfors (2004)
Protists (unicellular eukaryotes)	**2,234**	**2,946**	**32**	
Phytoplankton [a]	929	1,208	30	Hällfors (2004)
Microphytobenthos (diatoms)	817	899	10	Hällfors (2004), Snoeijs et al. (1993–1998)
Microzooplankton/benthos (ciliates)	473	743	57	Mironova et al. (2014)
Microzoobenthos (foraminifers)	15	96	540	Frenzel et al. (2005)
Macrophytes	**339**	**531**	**57**	HELCOM (2012a)
Invertebrates	**1,457**	**2,685**	**86**	
Metazooplankton	160	217	36	Telesh et al. (2009)
Meiobenthos [b]	400	570	43	Ojaveer et al. (2010)
Macrozoobenthos	897	1,898	112	HELCOM (2012a)
Vertebrates	**210**	**301**	**30**	
Fish and lampreys	152	240	58	HELCOM (2012a)
Breeding birds	53	56	6	HELCOM (2012a)
Mammals	5	5	0	HELCOM (2012a)
Total number of reported taxa	**4,419**	**6,647**	**51**	

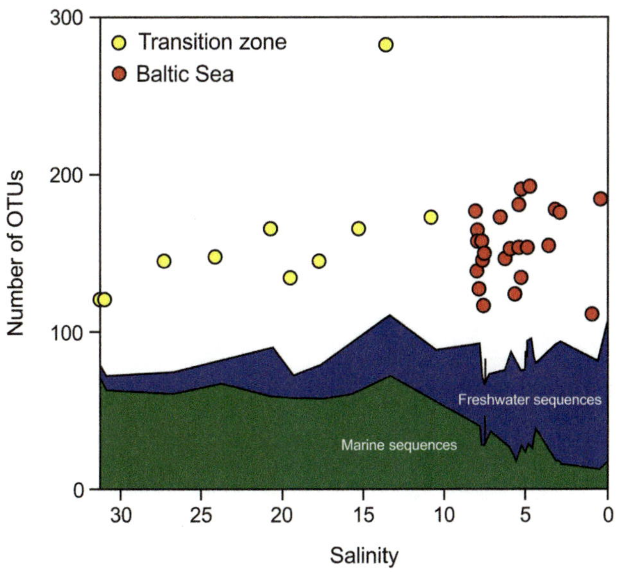

Fig. 4.8 The number of surface water OTUs (operational taxonomic units), each as a sequence of 800 randomly picked reads, along the Baltic Sea salinity gradient (yellow and red spheres). Green = the number of Baltic Sea OTUs that are ≥ 97 % identical to known marine sequences outside the Baltic Sea. Blue = the number of Baltic Sea OTUs that are ≥ 97 % identical to known freshwater sequences outside the Baltic Sea. Figure modified from Herlemann et al. (2011)

(Fig. 4.9a). In four subregions of the Baltic Sea, cyanobacterial richness is higher than in the transition zone and in the other four subregions it is lower.

4.4.2 Fungal richness is still largely hidden

Fungi and fungus-like organisms straddle the realms of microbiology and macrobiology as they include unicellular yeasts and molds as well as multicellular organisms. Similarly to archaea and heterotrophic bacteria, fungal species richness in the Baltic Sea is still largely hidden, and diversity patterns in nature can only be unveiled by molecular techniques. Globally, ~3,000 fungi are known from aquatic habitats, 465 of which were reported from marine waters (Shearer et al. 2007). However, the true number is expected to be much higher than 3,000, and should include many hitherto undescribed species with undescribed functions in the ecosystem. For example, a number of fungal taxa were recently identified from suboxic and anoxic waters in the Gotland deep (Stock et al. 2009).

Fungi have the potential to play crucial roles in material cycling and interactions with other organisms, primarily as

Box 4.6: Diatom traits

Attached or motile, colonial or solitary

The species-rich microphytobenthos growing directly on rocky, sandy and soft bottoms, as well as the micro-epiphyton in the macrophytobenthos, are often dominated by benthic diatoms (Snoeijs 1995; Ulanova et al. 2009). In fact, benthic diatoms grow on any substrate exposed to seawater, including animals, drifting materials and garbage on the seafloor. These diatoms are either motile or firmly attached to the substrate with mucilage structures in different ways, and the cells live solitarily or form colonies (Box Fig. 4.8). Diatom cells are not able to move actively in water, but raphe-bearing pennate species can glide over a surface, driven by the excretion of adhesive mucilage and an actin-myosine motility system (Poulsen et al. 1999), while planktonic diatoms, consisting of pelagic species as well as resuspended benthic species, are passively transported with the movement of the water.

Diatom life forms in trait analysis

The life forms of diatoms can be used in trait analysis. A large data set collected along the Baltic Sea gradient from the southern Baltic Sea proper to the northern Bothnian Bay allowed for the distinction of four trait groups (Box Fig. 4.7):

(1) Cell dimensions: length, surface, volume, surface:volume ratio and a shape index (the aspect ratio of the largest cell dimension divided by the second-largest cell dimension).
(2) Coloniality: colonial or solitary
(3) Vegetation height: High = large colonies visible to the naked eye (up to ~50 cm high) with cells in mucilage tubes or long chains, or in bushes on long, branched mucilage stalks; Medium = medium-sized colonies of ~10–50 cells and/or cells elevated from the substratum on mucilage stalks; Low = solitary cells or small colonies (<10 cells), which are motile, adnate or attached with pads.
(4) Cell mobility: Attached = non-motile species attached to a substrate; Motile = species that may move relatively fast over a substratum (*e.g. Navicula* spp.); Creeping = adnate species that are basically non-motile, but may move very slowly on their substratum (*e.g. Cocconeis* spp.); Floating = non-motile species that are basically pelagic, but may occur in the benthic community.

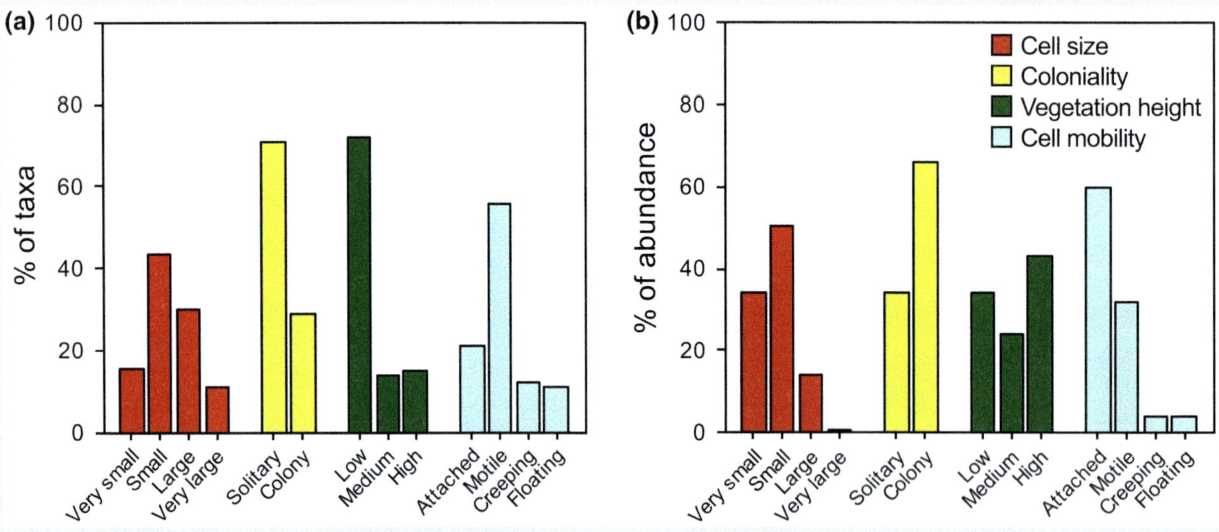

Box Fig. 4.7 The distribution of diatom traits in epilithic spring samples collected along the Baltic Sea gradient from the southern Baltic Sea proper to the northern Bothnian Bay, based on counts of 119,000 diatom valves belonging to 355 taxa, sampled from 119 stations (1,000 valves per station). (**a**) % of diatom taxa. (**b**) % of diatom abundance. The legend for both graphs is given in (**b**). The figure shows that the largest diatom species in the epilithic communities have very low abundances, that there are more solitary than colonial species but colonial species are more abundant, and that there are more motile than attached species but that the attached species are more abundant. Figure based on data in Svensson et al. (2014). Figure: © Pauline Snoeijs-Leijonmalm

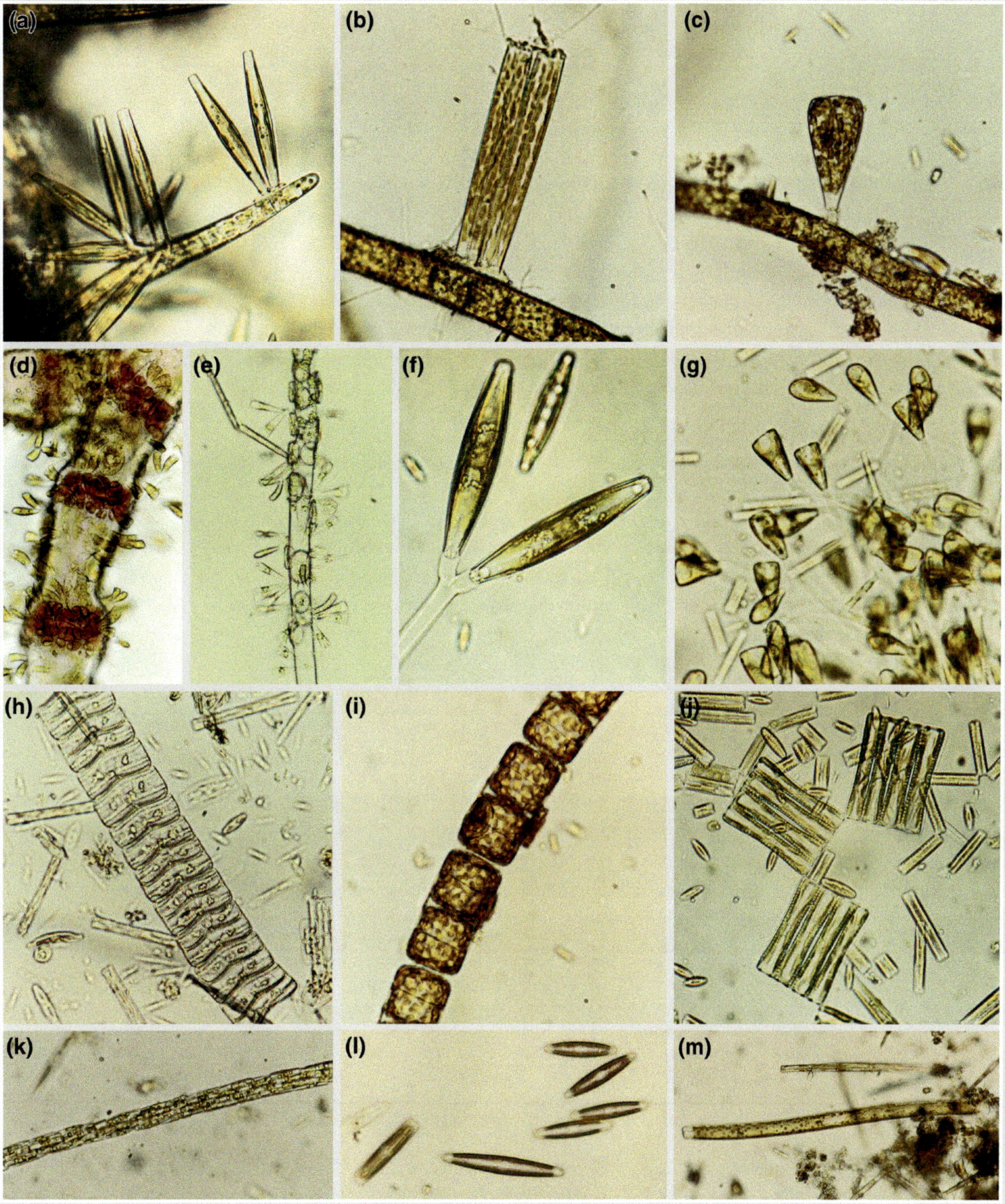

Box Fig. 4.8 Examples of different life forms of benthic diatoms. (**a**) *Ctenophora pulchella* is attached with a pad. (**b**) *Tabularia tabulata* is attached with a pad. (**c**) *Licmophora gracilis* var. *angelica* is attached with a pad. (**d**) *Rhoicosphenia abbreviata* is attached with a short stalk. (**e**) *Licmophora debilis* is attached with a short stalk and *Cocconeis pediculus* is adnate (**f**) *Brebissonia lanceolata* is attached with a long stalk. (**g**) *Gomphonema olivaceum* is attached with a long stalk. (**h**) *Achnanthes brevipes* var. *intermedia* forms in straight chains. (**i**) *Melosira moniliformis* forms in straight chains. (**j**) *Diatoma bottnica* forms in zigzag-shaped chains. (**k**) *Berkeleya rutilans* lives in mucilage tubes. (**l**) *Navicula* sp. is solitary and motile. (**m**) *Nitzschia sigmoidea* is solitary and motile. Photo: © Pauline Snoeijs-Leijonmalm

Box 4.7: Functional diversity in the meiobenthos

Teresa Radziejewska

Towards a functional approach to meiofauna diversity

Like in other marine and brackish systems, meiobenthic organisms in the Baltic Sea represent a vast, albeit incompletely known, component of biodiversity. In addition to the high taxonomic richness of the benthic meiofauna, they are also differentiated based on their adaptations to their (micro-) habitats and functions in the ecosystem. One of those functions is the involvement in carbon/energy flow, implicitly assumed to be expressed via adaptations to a specific mode of feeding (a feeding guild). The best-known illustration of this type of functional diversity can be found among free-living nematodes. The scarcity of direct evidence on the type of food and the mode of feeding led researchers to infer those functions from the nematode buccal cavity structure, on the assumption that the size and shape of the mouth as well as the size and structure of buccal armature (if any) is directly related to the type of food and the feeding mode. This assumption gave rise to the so-called trophic type classification of nematodes, first described by Wieser (1953) and hence commonly referred to as Wieser's classification (Giere 2009).

Wieser's classification of nematode feeding types

Wieser (1953) distinguished between 4 trophic (feeding) types of nematodes:

1A: Selective deposit feeders: a tiny mouth opening without buccal armature (Box Fig. 4.9a)
1B: Non-selective deposit feeders: a larger mouth opening without buccal armature (Box Fig. 4.9b)
2A: Epistrate feeders (feeding on protists and bacteria): the buccal cavity equipped with hard cuticular ridges for scraping sand grain surfaces or with pointed hard tips for piercing algal cells, *e.g.* those of diatoms (Box Fig. 4.9c)
2B: Omnivores/predators: a large buccal cavity featuring large pointed teeth and lancets as the buccal armature (Box Fig. 4.9d)

A gradient in nematode feeding types with distance away from the shore

Although Wieser's original classification was modified by subsequent workers, its main tenets have been retained and are still in use. A good illustration of a shift in the trophic structure of the nematode community with alteration of the sedimentary environment has been provided by Szulwiński et al. (2001). In their study of the meiobenthos off the southern Baltic Sea coast, they demonstrated how the proportion of different nematode trophic types changed with distance away from the shore (and hence with a change in sediment type and the type of food resources available), from deposit and epistrate feeders co-dominating the detritus-enriched sandy bottom close to the shore to deposit feeders dominant on the mud-covered seafloor offshore.

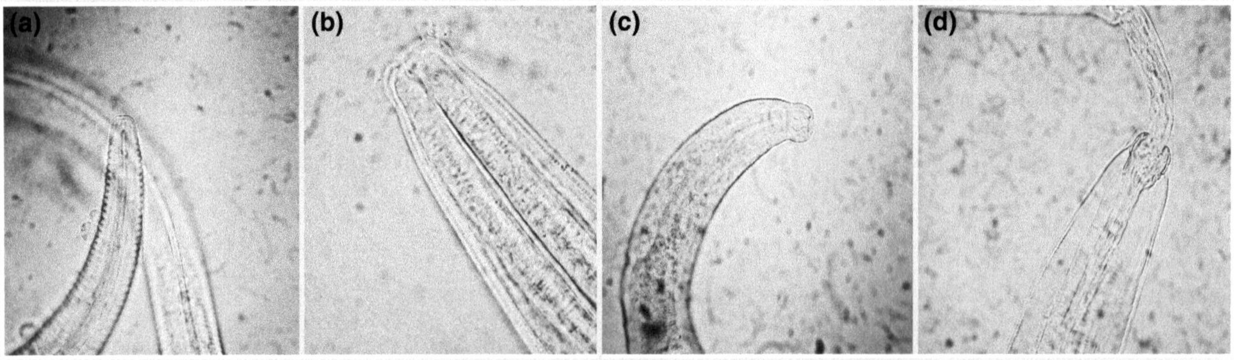

Box Fig. 4.9 Trophic types among the Baltic Sea nematodes. (**a**) *Antomicron* sp., a selective deposit feeder. (**b**) *Sabatieria* sp., a non-selective deposit feeder. (**c**) *Microlaimus* sp., an epistrate feeder. (**d**) *Sphaerolaimus* sp., a predator (here ingesting a prey). Photo: © Joanna Rokicka-Praxmajer

Table 4.4 The number of taxa (species level or below), the % of benthic taxa and the % of heterotrophs, reported from phytoplankton samples taken in the Baltic Sea Area (Baltic Sea, Belt Sea and Kattegat) for different taxonomic groups. Note that this table follows the classification used in Hällfors (2004) with Chrysophya synonymous with Heterokontophyta. Photosynthetic heterokonts are also known as Ochrophyta (*cf.* Table 4.2), a group of which the classification is still being worked out. Data from Hällfors (2004) and Snoeijs et al. (1993–1998). [a] = Heterotrophic flagellates (zooflagellates)

Phylum	Group(s)	Number of taxa	Benthic taxa (%)	Heterotrophs (%)
Cyanobacteria	Chroococcales	92	8	–
Cyanobacteria	Oscillatoriales	56	41	–
Cyanobacteria	Nostocales	36	19	–
Chlorophyta	Charophyceae, Chlorophyceae, Pedinophyceae, Prasinophyceae	361	3	1
Chrysophyta	Chrysophyceae (golden-brown algae)	84	1	31
Chrysophyta	Eupodiscales (centric diatoms)	233	18	–
Chrysophyta	Bacillariales (pennate diatoms)	885	97	–
Chrysophyta	Dictyochophyceae, Raphidophyceae, Synurophyceae, Tribophyceae	55	5	–
Cryptophyta	Cryptophyceae (recoiling algae)	26	–	15
Dinophyta	Dinophyceae (dinoflagellates)	225	1	41
Euglenophyta	Euglenophyceae (euglenoids)	44	–	20
Haptophyta	Prymnesiophyceae (haptophytes)	67	–	6
Zoomastigophora[a]	Choanoflagellidea (choanoflagellates)	54	–	100
Zoomastigophora[a]	Kinetoplastidea	62	–	100
Zoomastigophora[a]	Bicosoecidea, Ebriidea	11	–	100
Total		**2,291**		

decomposers of organic matters, but also as saprophytes, parasites and food sources. Decomposing fungi are especially abundant in areas where organic materials accumulate, *e.g.* in estuaries and lagoons. Parasitic fungi live primarily in marine algae and animals. For example, the oomycete *Leptolegnia baltica*, which produces a branched mycelium with extramatrical filamentous sporangia, can cause massive mortality in the copepod *Eurytemora affinis* in the Bothnian Bay (Höhnk and Vallin 1953).

4.4.3 Protists in the phytoplankton and microbenthos

Phytoplankton is a heterogeneous group of organisms, both taxonomically and functionally, varying from diazotrophic (nitrogen-fixing) cyanobacteria to different microalgal phyla and heterotrophic flagellates (Table 4.4). The organisms in the latter group are by definition not "phytoplankton" because they do not photosynthesise, although they are routinely included in phytoplankton counts. Heterotrophic species are found together with auto- and mixotrophic species within all classical microalgal phyla, Chlorophyta, Chrysophyta, Cryptophyta, Dinophyta, Euglenophyta and Haptophyta (syn. Prymnesiophyta) (Table 4.4). On the other hand, protozoans such as ciliates can retain chloroplasts derived from ingested algal cells and be mixotrophic (Stoecker and Michaels 1991). These are some of the arguments for combining microalgae and protozoa in the paraphyletic group of "protists" (unicellular eukaryotes), a grouping that is maintained for practical reasons (Schlegel and Hülsmann 2007), or in the kingdom Protoctista that denotes a diverse collection of eukaryotic organisms that are neither true plants, animals or fungi (Margulis 1990). The Protoctista include the protists together with their multicellular descendants, *e.g.* the multicellular algae.

The microphytobenthos of the Baltic Sea is usually dominated by benthic diatoms (*cf.* Box Fig. 4.8). These protists live attached to different substrates with mucilage pads or stalks, and are thus *e.g.* epiphytic (on plants and algae), epizoic (on animals), epilithic (on stones) or epipsammic (on sediment grains), but many species are also free-living (epipelic) in sediments, among the vegetation and even within cyanobacterial colonies (Round et al. 1990; Snoeijs and Wakuru-Murasi 2004). Diatoms are not able to move in water, but many species can glide over a surface with the help of mucilage (Box 4.6). Also, a large number of diatom taxa typically form macroscopic colonies with cells arranged in long chains, in bushes on branched mucilage stalks or inside mucilage matrices. Resuspended benthic

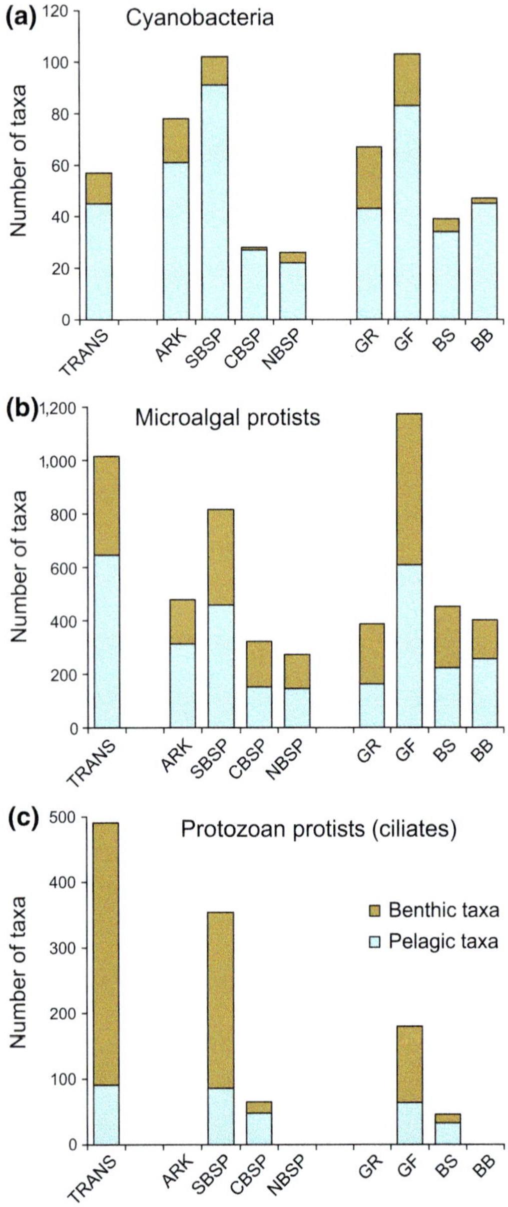

Fig. 4.9 Distributions of cyanobacterial richness and protistan richness (species level and varieties) along the large-scale Baltic Sea gradient as reported in comprehensive checklists with presence-absence data for the subregions of the Baltic Sea Area. The taxa are subdivided into taxa with their main distribution in the pelagic zone (pelagic taxa) and taxa with their main distribution in the benthic zone (benthic taxa). (**a**) Cyanobacterial taxa in phytoplankton samples according to the checklist of Hällfors (2004) (**b**) Microalgal protist taxa, including heterotrophic flagellates according to the checklists of Snoeijs et al. (1993–1998) and Hällfors (2004). (**c**) Protozoan protist taxa (ciliates) with data for only five of the subregions according to the checklist of Mironova et al. (2014). TRANS = the transition zone (Kattegat and Belt Sea, for ciliates only the southern Belt Sea), ARK = Arkona Sea, SBSP = Southern Baltic Sea proper, CBSP = Central Baltic Sea proper, NBSP = Northern Baltic Sea proper, GR = Gulf of Riga, GF = Gulf of Finland, BS = Bothnian Sea, BB = Bothnian Bay. The legend for all graphs is given in (**c**). Figure: © Pauline Snoeijs-Leijonmalm

diatom cells are often recorded in the coastal plankton, usually at low abundances in places with little wave action but at high abundances after stormy winds. However, due to their extraordinarily high species richness, benthic diatoms may greatly enhance the species richness of the coastal plankton.

The microzooplankton and microzoobenthos in the Baltic Sea are dominated by protists belonging to the Ciliophora (ciliates), which occur as free-living cells (*e.g. Strombidium* spp. and *Tintinnopsis* spp.) or attached to a substrate (*e.g. Vorticella* spp.). Another protozoan group that can be abundant in certain areas of the Baltic Sea Area are the Foraminifera, amoeboid protists that use streaming granular ectoplasm for catching food and may possess an external shell (called a "test") made of calcium carbonate or agglutinated sediment particles. However, while freshwater and brackish-water species do exist, most foraminifers are marine (Gupta 1999), and the number of species living in the Baltic Sea is low.

4.4.4 Protistan richness

The most comprehensive survey on microalgal protistan richness in the Baltic Sea Area is the HELCOM phytoplankton checklist for the Baltic Sea Area (Hällfors 2004). This checklist includes 1,830 nano- and micro-sized taxa, 626 (34 %) of which are resuspended benthic diatoms. When these data are combined with 277 additional taxa from the diatom checklist of Snoeijs et al. (1993–1998), the microalgal protists (including heterotrophic flagellates) in the Baltic Sea Area comprise at least 1,208 phytoplankton taxa and 899 benthic diatom taxa, a total of 2,107 (Table 4.3). Of these, 1,746 have been recorded in the Baltic Sea.

Except for the Gulf of Finland, microalgal protistan richness seems to be lower in all subregions of the Baltic Sea than in the transition zone (Fig. 4.9b). The HELCOM checklist is strongly biased toward the Gulf of Finland. A comparison of three areas with the same salinity revealed that 1,192 microalgal protist taxa were recorded in the Gulf of Finland, but only 446 and 544 in the Gulf of Riga and the Gulf of Bothnia, respectively. This difference depends on the inclusion of more benthic taxa and freshwater taxa in the Gulf of Finland. The high number of benthic taxa indicates that relatively more coastal plankton samples were included in the Gulf of Finland. The high number of freshwater taxa, *e.g.* 16 *Eunotia* spp. and 15 *Gomphonema* spp. are absent from all other subregions, and this indicates a higher proportion of coastal samples receiving freshwater discharge in the Gulf of Finland.

Within the ciliates, 743 taxa have been recorded in the Baltic Sea Area and 473 in the Baltic Sea (Table 4.3). Of the 743 taxa, 78 % are benthic species. Ciliate richness seems to decrease along the large-scale Baltic Sea gradient from the transition zone to the gulfs (Fig. 4.9c). Studies on extant foraminifers in the Baltic Sea (*e.g.* Brodniewicz 1965; Frenzel et al. 2005; Schönfeld and Numberger 2007) have been restricted to the western part of the Baltic Sea Area (Kiel Bight) and the southern part of the Baltic Sea proper. In the southern Belt Sea (transition zone), 96 species were recorded, but the nearby Pomeranian Bay (southern Baltic Sea proper) yielded as few as 15, which is in agreement with the general distribution pattern of foraminifers whereby the highest species richness occurs in marine waters (Gupta 1999).

Altogether, we can conclude that the number of protist taxa recorded in the Baltic Sea today is at least 2,234 and 2,946 in the Baltic Sea Area (Table 4.3). However, this inventory of protists is far from complete. For example, the number of benthic diatom taxa could easily be doubled by intensive sampling and scanning electron microscopy studies in combination with taxonomic studies on synonymy and description of new species. Based on the available checklist data, the number of protists increases by 32 % with an increasing water surface area of 12 % and water volume of 4 % when the transition zone is added to the Baltic Sea (Table 4.3). This would indicate a positive effect on overall protistan richness at salinity >10. However, this should be interpreted with care because (1) the protist presence-absence data for some of the subregions of the Baltic Sea are still incomplete (2) Skagerrak and Baltic Sea water masses with protists from both areas meet here.

4.4.5 Macrophyte richness

The distributions of the 531 species of macrophytobenthos in the Baltic Sea Area, *i.e.* multicellular algae and vascular plants, are known rather well. Of these, 339 species have been reported from the Baltic Sea while 192 are marine species occurring only in the transition zone (Table 4.3). In the different subregions of the Baltic Sea species richness is on average only one-third of that found in the transition zone (Fig. 4.10a). Low macrophyte richness prevails especially in the Gulf of Riga and the Bothnian Bay with 73 and 117 recorded species, respectively. Based on the available checklist data, macrophyte richness increases by 57 % when the transition zone is added to the Baltic Sea (Table 4.3). This indicates a positive effect on overall macrophyte richness at salinity >10.

4.4.6 Zooplankton richness

The metazooplankton (meso- and macrozooplankton combined) of the Baltic Sea Area consists mainly of rotifers, cladocerans, copepods, cnidaria and ctenophora in the mesozooplankton size group of 0.2 to 2 mm (Table 4.1, *cf.* Figs. 8.18–8.20). Their juvenile stages (*cf.* Fig. 8.17a) fall within the microzooplankton size range and only a few species in the Baltic Sea can become larger than 2 cm. Larger crustacean plankton, such as krill, is absent from the Baltic Sea. Also juveniles of some benthic animals, *e.g.* crustaceans, molluscs and polychaetes, may occur in the metazooplankton (*cf.* Fig. 8.17d).

Despite the fact that the Baltic Sea metazooplankton has been studied for over 100 years, there is still no consistent data available on the distribution of species along the whole large-scale Baltic Sea gradient. This is probably caused mainly by the fact that good taxonomic skills are necessary for identification at the species level, especially for copepods, while it is impossible to identify most juveniles at the species level. Consequently, regional studies usually only list the dominant taxa, and taxonomic resolution is often at the genus level. Altogether, 217 metazooplankton species have been reported from five subregions in the Baltic Sea Area, 160 of which were found in the Baltic Sea (Table 4.3). In the different subregions of the Baltic Sea, species richness is about half of that in the transition zone (Fig. 4.10b). However, integrated presence-absence data are still missing for half of the Baltic Sea subregions.

4.4.7 Meiobenthos richness

The meiobenthos is defined as benthic protists (mainly ciliates, *cf.* Sect. 4.4.4) and invertebrates that pass through a sieve with a 0.5 mm (or 1 mm) mesh size and are retained on sieves with 0.044 mm (or 0.063 mm) mesh size (Table 4.1, Box 4.8). The meiofauna has a regulating role in biogeochemical processes, *e.g.* it is closely linked with bacteria. In sediments with abundant and diverse meiofauna, denitrification is double that in sediments with low meiofauna abundance, suggesting that meiofauna bioturbation has a stimulating effect on nitrifying and denitrifying microbes (Bonaglia et al. 2014). Three species of ostracods, *Candona neglecta*, *Paracyprideis fennica* and *Clithrocytheridea sorbyana* (syn. *Heterocyprideis sorbyana*), are common in the Baltic Sea proper and often contribute more to the total meiobenthic biomass and the uptake of phytodetritus than any other taxon (Modig et al. 2000). Highly abundant ostracods in the phytobenthic communities of the Bothnian Sea are *e.g. Sarscypridopsis aculeata*, *Heterocypris salina*, *Cytherura gibba* and *Xestoleberis aurantia* (Snoeijs and Mo 1987). A checklist of recent and fossil ostracods for the Baltic Sea and Belt Sea includes 131 taxa (Frenzel et al. 2010).

In the Baltic Sea Area, the largest meiobenthos richness is found within the Turbellaria with 134 species and the

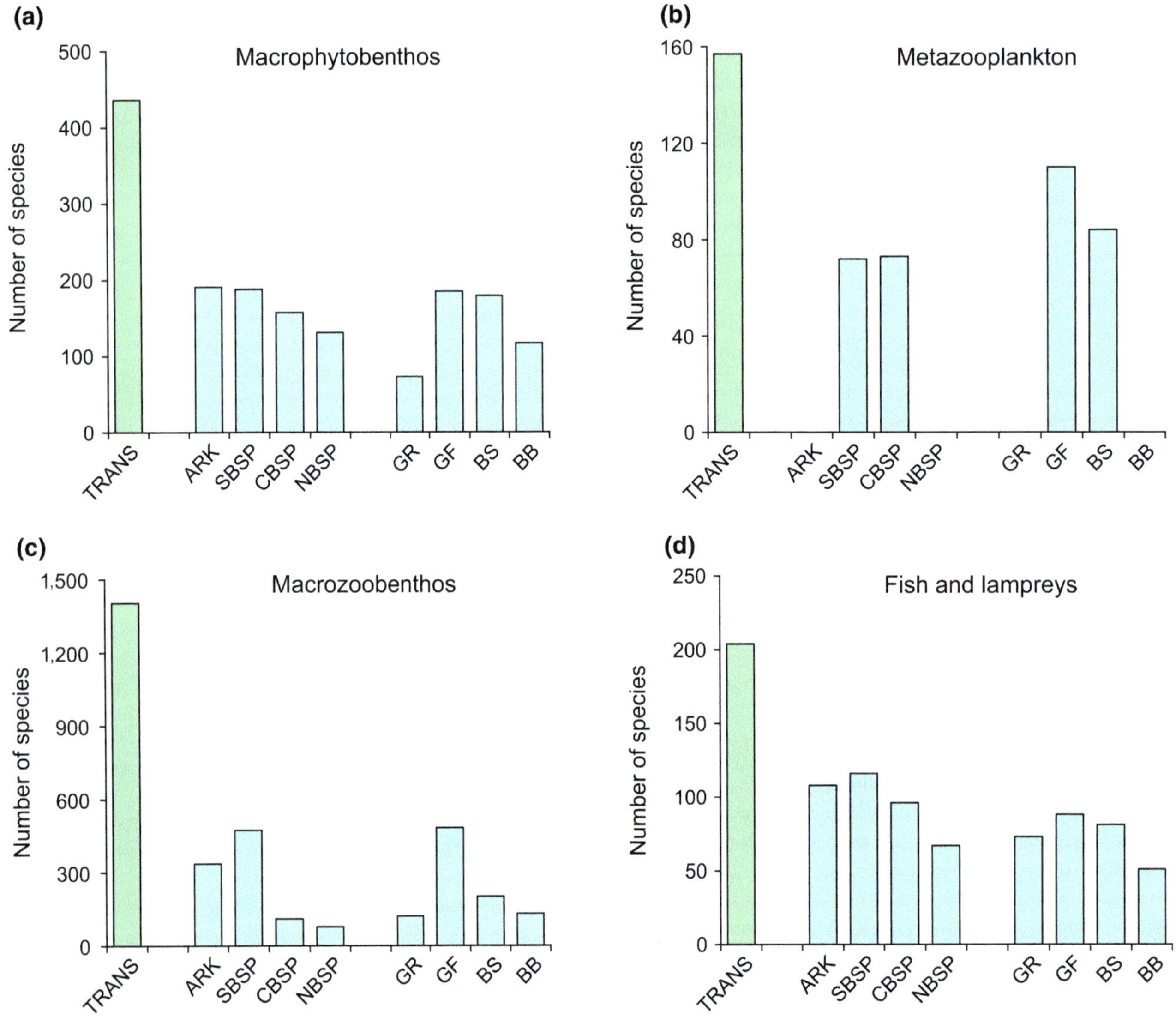

Fig. 4.10 Species richness distributions of macroscopic organisms along the large-scale Baltic Sea gradient as reported in comprehensive checklists with presence-absence data for the subregions of the Baltic Sea Area. (**a**) Macrophytobenthos species according to the checklist of HELCOM (2012a), which for multicellular algae is an extension of the earlier checklist of the Baltic Marine Biologists (Nielsen et al. 1995). This checklist includes all multicellular algae, including microscopic epi- and endophytes. (**b**) Metazooplankton species according to the checklist of Telesh et al. (2009). (**c**) Macrozoobenthos species according to the checklist of HELCOM (2012a). (**d**) Fish and lamprey species according to the checklist of HELCOM (2012a). TRANS = the transition zone (Kattegat and Belt Sea), ARK = Arkona Sea, SBSP = Southern Baltic Sea proper, CBSP = Central Baltic Sea proper, NBSP = Northern Baltic Sea proper, GR = Gulf of Riga, GF = Gulf of Finland, BS = Bothnian Sea, BB = Bothnian Bay. Green bars indicate the transition zone, blue bars indicate the subregions of the Baltic Sea. Figure: © Pauline Snoeijs-Leijonmalm

Nematoda with at least 200 species (Ojaveer et al. 2010). Further groups are Cnidaria (3), Gastrotricha (34), Halacaroidea (14), Harpacticoida (82), Kinorhyncha (2), Oligochaeta (20), Ostracoda (40), Rotifera (36) and Tardigrada (3). According to the 2010 census (Ojaveer et al. 2010), the meiobenthos richness in the Baltic Sea Area is estimated at ~570 species, and more than two-thirds of those occur in the Baltic Sea. As opposed to the macrozoobenthos, no study along the large-scale Baltic Sea gradient has ever been conducted for the meiobenthos.

4.4.8 Macrozoobenthos richness

The macrozoobenthos consists of benthic invertebrates larger than 0.5 mm (or 1 mm) in body size (Table 4.1). The distribution of marine macrozoobenthos species in the Baltic Sea is well-known while that of its freshwater species, occurring in the near-shore shallow areas at salinities below ~6 (*e.g.* in the entire Bothnian Bay), has been investigated much less intensively, even with respect to the most species-rich freshwater group, the insect larvae (Fig. 4.11).

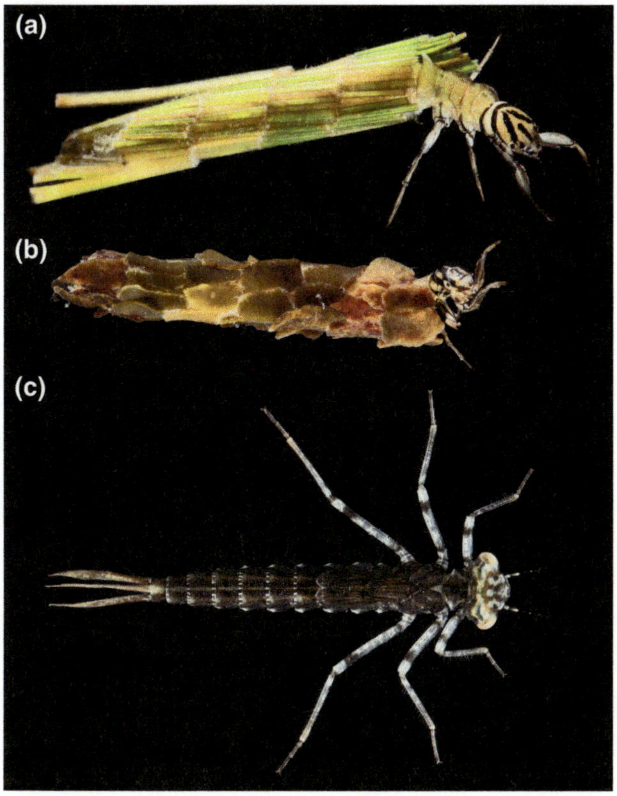

Fig. 4.11 Examples of insect larvae that are common in the northern Baltic Sea. (**a**, **b**) Trichoptera (caddisfly) larvae build protective cases of different materials such as plants, sand or debris. (**c**) Zygoptera (damselfly) larva. Photo: © Nicklas Wijkman/Azote

Wherever they occur, they contribute to nutrient recycling and serve as food for *e.g.* fish. Insect larvae are not restricted to areas with freshwater discharge such as river mouths; *e.g.* in shallow water in the southern Bothnian Sea at salinity 5, the yearly average richness of freshwater taxa is ~80 % of the total species richness, even with insect larvae only identified at the family or order level (Snoeijs 1989).

Of the 1,898 macrozoobenthos species reported from the Baltic Sea Area, 897 were recorded from the Baltic Sea while 1,001 are marine species restricted to the transition zone (Table 4.3). In the different subregions of the Baltic Sea, species richness is on average only one-sixth of that found in the transition zone (Fig. 4.10c). Based on the available checklist data, macrozoobenthos richness increases by 112 % when the transition zone is added to the Baltic Sea (Table 4.3). This is evidence of a strong positive effect of salinity >10 on the overall macrozoobenthic richness.

Especially low macrozoobenthos species richness is found in the northern Baltic Sea proper with 79 recorded species (Fig. 4.10c), and this is a well-investigated area. The highest richness inside the Baltic Sea is found in the Arkona Sea and the southern Baltic Sea proper where a number of marine species still penetrate from the transition zone. Also, in the Gulf of Finland species richness is relatively high, which is explained by the powerful salinity gradient and the shallowness of the water, which provides a diversified environment with an additional influx of freshwater species from the Neva estuary (HELCOM 2012a; Zettler et al. 2014). However, the numbers of freshwater species in the Gulf of Riga and the Gulf of Bothnia are most probably higher than reported, as fewer detailed taxonomic studies (notably of insect larvae) have been carried out there compared to the Gulf of Finland.

4.4.9 Vertebrate richness

Of the 240 fish and lamprey species reported from the Baltic Sea Area, 152 are recorded from the Baltic Sea while 88 are marine species restricted to the transition zone (Table 4.3). In the different subregions of the Baltic Sea, species richness is on average only 42 % of that found in the transition zone (Fig. 4.10d). The highest richness inside the Baltic Sea is found in the Arkona Sea and the southern Baltic Sea proper to where a number of marine species still penetrate from the transition zone. Based on the available checklist data, fish richness increases by 58 % when the transition zone is added to the Baltic Sea (Table 4.3). This underpins a positive effect of salinity >10 on overall fish richness. While marine species dominate the pelagic zone (Fig. 4.12) and the deep-water seafloor (Fig. 4.13a, b), freshwater fish species dominate in the coastal areas of the Baltic Sea (Box 4.9).

Of the 56 breeding waterbirds listed for the Baltic Sea Area (HELCOM 2012a), 51 are reported from the transition zone and 53 from the Baltic Sea (Table 4.3). The bird richness of the Baltic Sea does not seem to change along the large-scale Baltic Sea gradient. Besides *e.g.* gulls (Fig. 4.14), some typical waterbirds living in the Baltic Sea are the common eider *Somateria mollissima* (Box 4.10), the great cormorant *Phalacrocorax carbo sinensis* (Box 4.11) and the white-tailed eagle *Haliaeetus albicilla* (Box 4.12).

The mammalian richness of the Baltic Sea does not change along the large-scale Baltic Sea gradient (HELCOM 2012a). The same five species have been reported from both the transition zone and the Baltic Sea, although there are clear abundance patterns for the three seal species within the Baltic Sea (Box 4.13). The five Baltic Sea mammals listed for the Baltic Sea Area by HELCOM are the harbour seal *Phoca vitulina*, the grey seal *Halichoerus grypus*, the ringed seal *Pusa hispida*, the harbour porpoise *Phocoena phocoena* (Fig. 4.15) and the European otter *Lutra lutra* (Fig. 4.16). The European otter is basically a freshwater species, and when it lives on the Baltic Sea coast it needs access to freshwater to wash the salt from its fur (Kruuk 1995). Other mammals may influence the Baltic Sea ecosystem as well, *e.g.* the introduced American mink *Neovison vison* (syn. *Mustela vison*) and the racoon dog *Nyctereutes procyonoides* (a canid indigenous to

Box 4.8: Meiobenthos

Teresa Radziejewska

What is meiobenthos?

The meiobenthos or benthic meiofauna (*cf.* Sect. 12.6) is a grouping of benthic organisms distinguished originally by a body size of 44(63)–500(1,000) µm (the different limits reflect different definitions). The meiobenthos consists of a variety of taxa, including both protists (unicellular eukaryotes) and metazoans (multicellular eukaryotes of the kingdom Animalia). The body size of the meiofauna overlaps, in the lower end, with that of the microbenthos (20–200 µm, *cf.* Table 4.1). At present, the meiobenthos is regarded not only as a size category but also as an ecological category distinguished on account of life cycle traits (reproduction and development), evolution, specific adaptations to the sedimentary environment, its role in partitioning material and energetic resources of the sediment, and its contribution to the energy flow in benthic communities (Giere 2009).

A poorly studied benthos compartment

The meiobenthos is a relatively poorly studied compartment of the benthos worldwide, and the Baltic Sea is no exception (Ojaveer et al. 2010). Therefore, any study of meiobenthic communities in the Baltic Sea, whether with a taxonomic, molecular or ecological focus, is a fertile ground bound to bring about new and interesting data (*e.g.* Lokko 2014). Although some progress has been made through molecular techniques (*e.g.* Leasi and Norenburg 2014, *cf.* Box 4.2), identification of most meiobenthic taxa requires specialised taxonomic knowledge and dedicated study. Therefore, most of the literature dealing with the meiobenthos, including in the Baltic Sea, considers its members at the level of the so-called "higher taxa", *i.e.* taxa at the supra-species level: genus (seldom), family, order, or even phylum.

The Baltic Sea meiobenthos

A conservative estimate of the taxonomic richness of the meiobenthos in the Baltic Sea Area points to ~1,300 species reported and/or described so far, including ~600 benthic ciliates and ~100 foraminifers (*cf.* Sect. 4.4.4), and ~600 metazoans (Ojaveer et al. 2010). The meiobenthos in the Baltic Sea is usually abundant, particularly at the seafloor below the halocline where the abundance and biomass of the macrozoobenthos are greatly reduced due to hypoxia and anoxia. The sediment of the anoxic bottoms frequently features meiobenthic nematodes, the only benthic organisms capable of overcoming severe oxygen deficiency stress. The low salinity prevents a number of major meiobenthic taxa from occurring in the Baltic Sea and restricts their occurrence to the transition zone (Kattegat and Belt Sea).

Meiofaunal protists

Among the meiofaunal protists that are prominent and widespread in shallow coastal areas of the Baltic Sea Area are ciliates (*cf.* Fig. 4.7c). Also relatively common, albeit extremely poorly investigated, are the foraminifers, particularly those representing agglutinating taxa (*e.g.* members of the family Saccamminidae, Box Fig. 4.10a) and representatives of the soft-walled taxon Allogromiida (Box Fig. 4.10b) (Radziejewska, unpublished data).

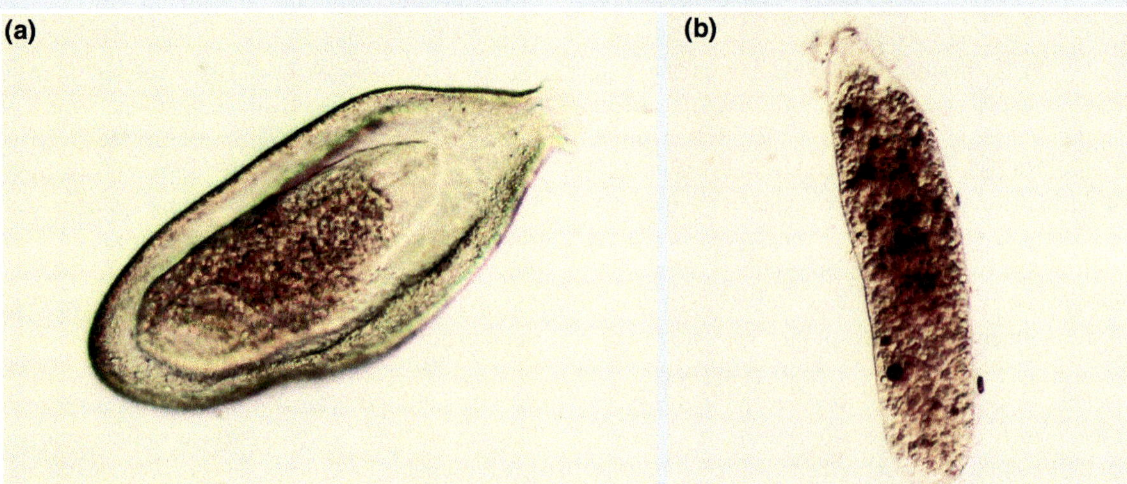

Box Fig. 4.10 Examples of meiobenthic foraminifers (protists) in the Baltic Sea. (**a**) A saccamminid individual of 0.7 mm in length. (**b**) An allogromiid individual of 0.6 mm in length. Photo: © Teresa Radziejewska

Meiofaunal metazoa

Metazoan invertebrates are represented in the meiobenthos of the Baltic Sea Area by members of the Cnidaria, Turbellaria, Nematoda (Box Fig. 4.11a), Gastrotricha (Box Fig. 4.11b), Rotifera, Kinorhyncha, Tardigrada, Ostracoda (Box Fig. 4.11c), Copepoda Harpacticoida (Box Fig. 4.11d), and Halacaroidea. Members of these taxa, for the most part, exhibit meiobenthic traits (primarily the body size) throughout their life span and are therefore termed the "permanent meiobenthos". In addition, the meiofaunal compartment of the benthos also frequently contains small, young forms of macrozoobenthic species (oligochaetes, polychaetes, bivalves, gastropods). These are termed the "temporary meiobenthos" and may even dominate the meiofauna in the warm season (summer - early autumn), particularly in shallow coastal areas of the Baltic Sea (Radziejewska, unpublished data).

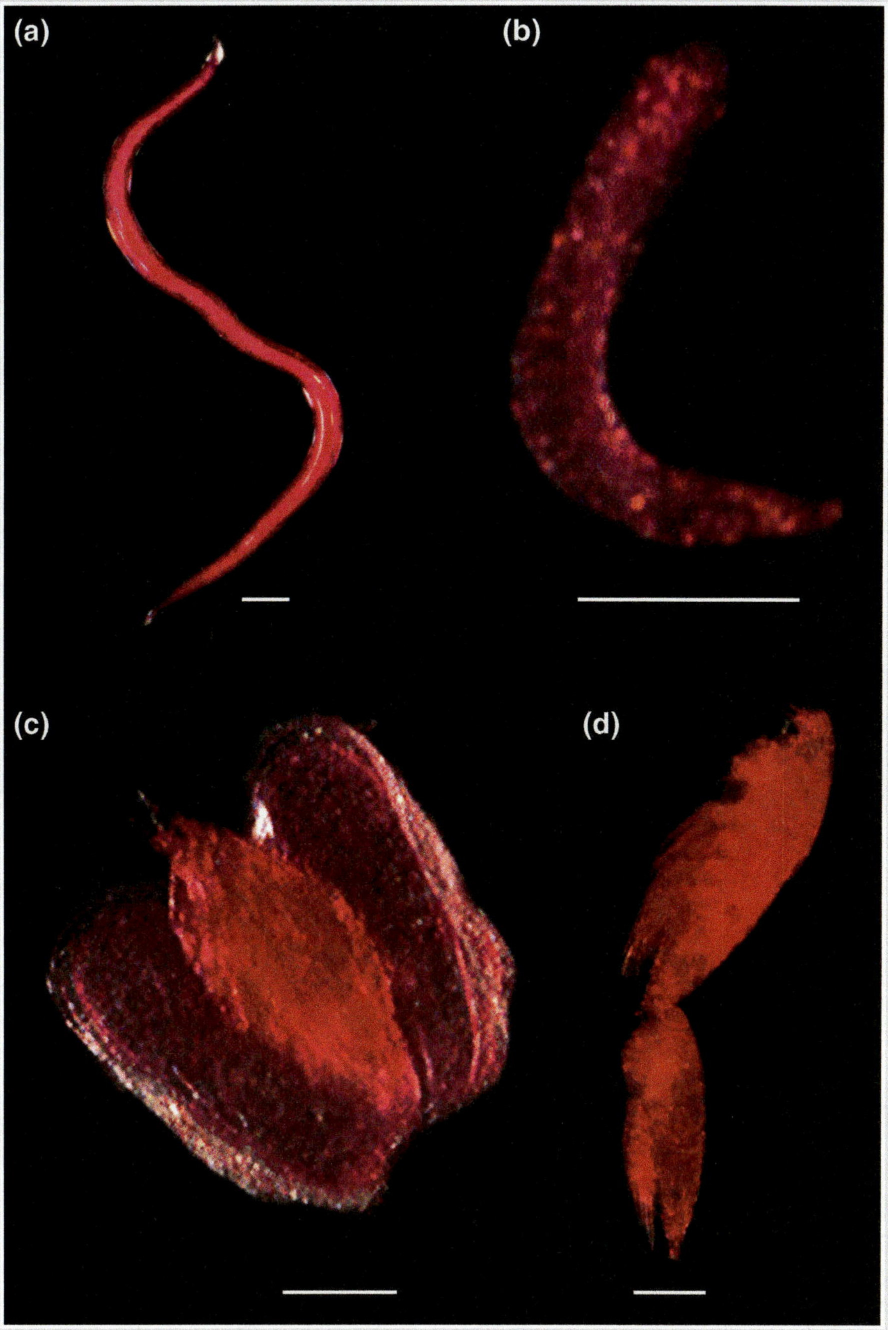

Box Fig. 4.11 Examples of meiobenthic metazoans in the Baltic Sea. The specimens are photographed under incident light and the red hue is imparted by Rose Bengal (a pigment commonly used to stain preserved meiofauna). (**a**) A nematode. (**b**) A gastrotrich. (**c**) An ostracod. (**d**) Two harpacticoid copepods. Scale bar = 100 μm in all images. Photo: © Brygida Wawrzyniak-Wydrowska

Fig. 4.12 The three dominant fish species in the pelagic zone of the Baltic Sea, the Atlantic cod *Gadus morhua* and its two prey species: the Atlantic herring *Clupea harengus* and the European sprat *Sprattus sprattus*. An easy way to separate the two clupeids is that an imaginary line between the central dorsal and ventral fins of the sprat is at a 90° angle to the body axis from nose to tail while this is not the case in the herring. Photo: © Pauline Snoeijs-Leijonmalm

Fig. 4.13 Some deep-water benthic fish that are widely distributed in the Baltic Sea. (**a**) The European plaice *Pleuronectes platessa* regularly reproduces south of the northern Baltic Sea proper. (**b**) The European flounder *Platichthys flesus* regularly reproduces throughout the Baltic Sea except for the Bothnian Bay. (**c**) The fourhorn sculpin *Myoxocephalus quadricornis*, a glacial relict, regularly reproduces in the Baltic Sea north of the Bornholm Sea. Photo: (**a**) © Tobias Dahlin/Azote, (**b**) © Tony Holm/Azote, (**c**) © Bent Christensen/Azote

Two additional seal species, 14 additional whale species, as well as six dolphin species are reported in the OBIS database from the North Sea. Thus, if the large-scale Baltic Sea gradient is extended all the way to the North Sea, there is a large loss of species toward lower salinity. The North Sea/Atlantic mammals sometimes pay short visits to the Baltic Sea, but they cannot survive in the Baltic Sea for a long time due to lack of food and/or inappropriate environmental conditions.

4.5 What underlies the patterns in species richness?

4.5.1 Different patterns of species richness

Salinity is the principal environmental driver that determines the presence or absence of both marine and freshwater species at a certain place along the large-scale Baltic Sea gradient, but only when this place is not affected by local

Fig. 4.14 Some water birds of the family Laridae (gulls) common in the Baltic Sea Area. (**a**) The lesser black-backed gull *Larus fuscus*. (**b**) The European herring gull *Larus argentatus*. (**c**) The great black-backed gull *Larus marinus*. (**d**) The common gull *Larus canus*. Photo: © Bo Tallmark

freshwater runoff from land so that freshwater species become more locally abundant. In coastal areas, land-runoff is the major "disturbing factor" of the basic pattern of species distributions along the large-scale Baltic Sea gradient. Thus, it is not possible to relate the presence-absence data from checklists for coastal species directly to average salinity for the subregions of the Baltic Sea because freshwater species reported from *e.g.* the transition zone were probably recorded near freshwater discharges. Although such discrepancies may blur possible patterns inferred from the available checklists, checklist data can still be used to discover general large-scale geographical trends for the subregions of the Baltic Sea.

Two basic patterns of species richness distribution along the Baltic Sea gradient from the transition zone via the Baltic Sea proper to the gulfs can be distinguished. The first pattern is that there is no pronounced decrease in species richness with lower salinity along the large-scale Baltic Sea gradient. This is the case for cyanobacteria, heterotrophic bacteria, microphytobenthos (benthic diatoms) and phytoplankton (Fig. 4.17). For cyanobacteria, there may even be an increase in species richness towards lower salinity, but this trend is not certain; the cyanobacteria in the transition zone may be underinvestigated.

The second pattern is that decreases in species richness occur in concert with lower salinity along the large-scale Baltic Sea gradient. This is the case for macrophytobenthos, macrozoobenthos and fish (Fig. 4.17). This pattern was already discovered a long time ago and represents the standard model for the loss of macroscopic species along the large-scale Baltic Sea gradient. Especially the macrozoobenthos shows an abrupt decrease in species richness over a small geographical distance at the Darß sill, followed by a gradual decrease towards the northern Baltic Sea proper and a levelling out of the decrease in the gulfs when freshwater species join. An extreme form of this second pattern is displayed by foraminifers, which are almost a strictly marine group of organisms. The drop in species richness is very steep (Table 4.3), and since there are no or only very few potential freshwater species to

Box 4.9: Coastal fish communities

Lena Bergström and Jens Olsson

Species composition

Coastal fish in the Baltic Sea Area spend the main part of their life cycle in shallow near-shore areas at a depth of less than 20 m. The Baltic Sea salinity gradient has a strong effect on local species composition, which is a mix of marine and freshwater species. Marine species, such as the Atlantic cod *Gadus morhua* and labrids such as the goldsinny wrasse *Ctenolabrus rupestris* (Box Fig. 4.12) characterise the coastal areas of the transition zone (Kattegat and Belt Sea), whereas freshwater species predominate in the coastal areas of the Baltic Sea. Typical freshwater species include the European perch *Perca fluviatilis* (Box Fig. 4.13), northern pike *Esox lucius* (Box Fig. 4.14), roach *Rutilus rutilus* (Box Fig. 4.15), whitefish *Coregonus maraena*, Eurasian ruffe *Gymnocephalus cernuus*, breams *Abramis* spp. and three-spined stickleback *Gasterosteus aculeatus* (HELCOM 2012b). Due to the occurrence of these freshwater species in the coastal area, coastal species richness is typically higher than that on shallow offshore banks where mainly species of marine origin prevail. Whereas most freshwater species live in the coastal area year-round, most marine species migrate between coastal and offshore areas. Examples of marine species that regularly occur in coastal habitats are the Atlantic herring *Clupea harengus*, which uses coastal areas for spawning, as well as the Atlantic cod and the European flounder *Platichthys flesus*, which come to the coast to forage. Herring is common in all subregions of the Baltic Sea, but cod and flounder are mainly found in the Baltic Sea proper.

Population structure and recruitment

The coastal fish communities of the Baltic Sea are primarily composed of populations with limited migration distances. Hence, many species have a pronounced genetic substructure (Laikre et al. 2005). Their population dynamics are

Box Fig. 4.12 The goldsinny wrasse *Ctenolabrus rupestris* in the transition zone. Photo: (**a**) © Ulf Bergström, (**b**) © Dirk Schories

influenced by variation in local environmental drivers, in addition to large-scale changes in the environment. Many species show high inter-annual variability in recruitment success, so that strong year-classes have a large influence on population structure for several years thereafter. For example, freshwater species such as perch, pike and roach, spawn in shallow coastal bays and in freshwater tributaries, and their recruitment success is dependent on favourable temperature and feeding conditions during the critical period when fish larvae hatch, typically in early spring. Hence, local topography and other physiographic factors affecting habitat quality are indirectly of high importance for coastal fish productivity in a given coastal area (Sundblad et al. 2013). Most freshwater species are favoured during periods with milder temperatures, whereas *e.g.* whitefish and many marine species predominate during time periods with colder temperatures (Olsson et al. 2012).

Ecosystem role of coastal fish

The coastal fish communities contribute to coastal zone functioning as well as to fisheries. Commercial fishing on coastal fish is widespread in the Baltic Sea and an important source of livelihood. However, recreational and household fisheries are of manifold higher magnitude in many countries. Coastal fish also provide an important food source for top predators. Predation by other fish species, waterbirds and seals may be of high significance for structuring coastal fish populations locally, and may also result in competitive interactions with the fisheries (Boström et al. 2012; Östman et al. 2013). Most coastal fish species feed mainly on invertebrate fauna and smaller-sized fish, although the share of species with a planktivorous or omnivorous diet increases in concert with changes in species composition with decreasing salinity along the Baltic Sea salinity gradient. By their feeding behaviour, the coastal fish contribute to regulating the local abundances of their prey species by top-down control. Examples of the opposite situation are observed in the case of overfishing. When populations of important piscivorous species are depleted, a release from predation pressure on the prey species may follow and in turn cause increased abundances of meso-predatory fish and invertebrate prey (Eriksson et al. 2011). Such cases manifest food web disruptions with consequences for food web and ecosystem functioning, and are also good examples of how the different parts of the coastal food web are connected to each other via biotic interactions (Östman et al. 2013).

Threats to coastal fish communities

Other anthropogenic pressures on coastal fish besides fishing mortality, are eutrophication and habitat degradation. Eutrophication may impair the quality of the recruitment and feeding areas or alter the competitive balance between species so that some species benefit over others. For example, cyprinids (*e.g.* roach) and pike-perch *Sander lucioperca* may be favoured by eutrophic conditions, whereas spawning habitats for flounder and perch are disturbed (Bergström et al. 2013; Snickars et al. 2015). Since the availability and quality of nursery and spawning habitats are of key

Box Fig. 4.13 The European perch *Perca fluviatilis* is a typical freshwater coastal resident species in the Baltic Sea. (**a**) A perch caught feeding on herring, which is a marine species that spawns in the coastal area. (**b**) Egg strands of perch in a shallow coastal habitat. Photo: © Ulf Bergström importance for the coastal fish productivity, the populations are also affected by habitat degradation (Sundblad and Bergström 2014). The essential habitats for coastal fish often coincide with areas favoured for coastal developments, such as tourism, housing and industries, leading to a continuous degradation and risks for the long-term viability of coastal fish populations. Another threat to the native coastal fish communities is competition with non-indigenous fish species. For example, the Ponto-Caspian round goby *Neogobius melanostomus* has already completely changed coastal fish communities in the southern Baltic Sea and is spreading northwards (*cf.* Box 5.6).

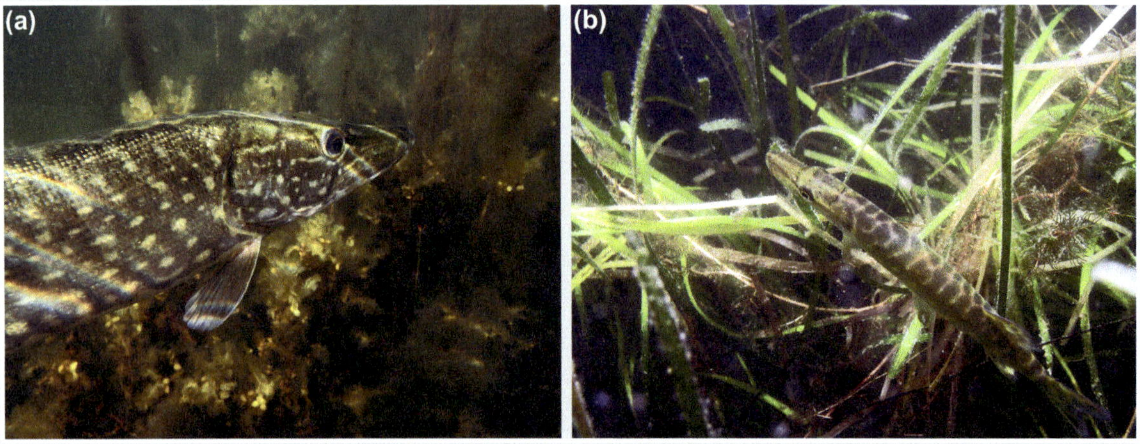

Box Fig. 4.14 The northern pike *Esox lucius* is a typical freshwater coastal resident species in the Baltic Sea. (**a**) A pike foraging among a *Fucus vesiculosus*-dominated vegetation. (**b**) A young pike hiding in a *Zostera marina*-dominated vegetation. Photo: © Ulf Bergström

Box Fig. 4.15 The roach *Rutilus rutilus* is a typical freshwater coastal resident species in the Baltic Sea. Photo: © Pauline Snoeijs-Leijonmalm

Fig. 4.15 The harbour porpoise *Phocoena phocoena*. (**a**) A harbour porpoise swimming away from the photographer. (**b**) Portrait of a harbour porpoise. The growth rates of the harbour porpoise subpopulations in Kattegat, Belt Sea and Baltic Sea proper are negative, but the exact rate of decline is not known. The Baltic Sea proper subpopulation probably contains <250 individuals and is critically endangered (Härkönen et al. 2013). Photo: © Sven-Erik Arndt/Azote

meet up in the Baltic Sea, the group perishes in the Baltic Sea proper and few species occur in the gulfs. However, the foraminifers of the Baltic Sea have not been extensively investigated (Box 4.8) and, while the trend may be true, the numbers may change when research in this field progresses.

The species richness of ciliates and the metazooplankton seems to comply with the second distributional pattern (Fig. 4.17). However, the available checklists still do not cover all subregions of the Baltic Sea. In coastal areas with freshwater discharge in the southern Belt Sea, the number of ciliate species in plankton samples was reported to peak at salinity 5–8 (Telesh et al. 2011a, b), and although questioned (Ptacnik et al. 2011), this may be an effect of the intensive mixing of water masses with different salinity, carrying different ciliate communities, in the Belt Sea.

4.5.2 Marine and freshwater species meet in the Baltic Sea

The unique characteristic of the Baltic Sea as a large meeting place of marine and freshwater species is shown by the typical distributions of taxonomic groups with either decreasing or increasing species richness along the large-scale Baltic Sea gradient (Fig. 4.18). Like in an estuary, the species richness of the Baltic Sea depends on the availability of species that are able to penetrate into brackish water from marine and freshwater habitats. Among the inhabitants of the latter, only the freshwater group Hirudinea (leeches) shows a pronounced species maximum in the Baltic Sea (Fig. 4.18c) compared to other freshwater species. This, however, may be a result of research intensity, whereby leeches received more attention than other groups of small freshwater invertebrates (Jueg and Zettler 2015).

From the transition zone, "marine" taxonomic groups enter the Baltic Sea proper, with the most species-rich groups being the dinoflagellates, centric diatoms, haptophytes, red, brown and green macroalgae, polychaetes, crustaceans, molluscs, bryozoans, cnidarians and perch-like fish (Fig. 4.18). From the gulfs, "freshwater" taxonomic groups enter the Baltic Sea proper, with the most species-rich groups being the green microalgae, cyanobacteria, chrysophytes, vascular plants, charophytes (stoneworts), insects, oligochaetes, and cyprideid (carp-like), acipenserid (sturgeon-like) and salmonid fish. However, species from nearly all groups make it all the way from the transition zone to the gulfs or *vice versa*. Only four "marine groups": ascidians (Fig. 4.19), echinoderms (Fig. 4.19), sharks and rays, do not reach the gulfs, while the aquatic mosses are the only "freshwater group" that does not reach the transition zone.

The obvious conclusion is that the richness of taxonomic groups that have diversified more in the marine environment decreases along the large-scale Baltic Sea gradient while that of taxonomic groups that have diversified more in freshwater environments increases along the gradient. For example, the polychaetes have mainly developed in the sea, and very few species occur in freshwater. Insects, on the other hand, are a typical terrestrial/freshwater group that

Fig. 4.16 The European otter *Lutra lutra* eats mainly fish. This species was widespread in the Baltic Sea before the 1950s when it declined due to chemical contamination by organochlorines, and became classified as an endangered species. Minimum population levels were reached in the 1960s–1990s, after which it started to recover. The populations have expanded from freshwater habitats to the coasts and the Baltic Sea populations may still be largely supported by freshwater populations. Photo: © Sven-Erik Arndt/Azote

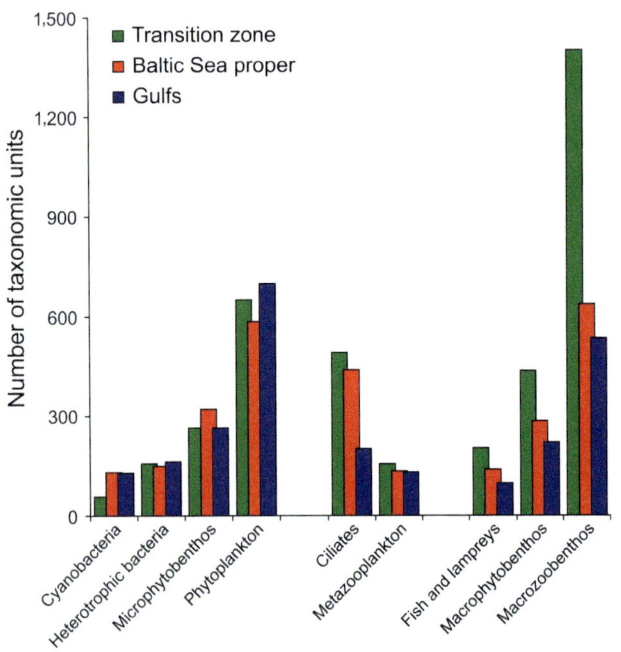

Fig. 4.17 Two different patterns of the distribution of species richness along the large-scale Baltic Sea gradient from the transition zone via the Baltic Sea proper to the gulfs (Gulf of Riga, Gulf of Finland and Gulf of Bothnia). The data were extracted from Fig. 4.6 (heterotrophic bacteria) and the checklists referred to in Table 4.3. Pattern 1 (to the left): Species richness does not decrease along the gradient for cyanobacteria, heterotrophic bacteria, microphytobenthos and phytoplankton. Pattern 2 (to the right): Species richness decreases along the gradient for fish and lampreys, macrophytes and macrozoobenthos. Note that this figure is based on the checklists that are available today, *e.g.* those of ciliates and metazooplankton (in the middle) still do not cover all subregions of the Baltic Sea Area. Figure: © Pauline Snoeijs-Leijonmalm

developed on land; they originated from crustaceans ~480 million years ago at about the same time that terrestrial plants appeared (Misof et al. 2014), and marine insect species are few. In accordance with their evolutionary background, polychaetes and insects enter the Baltic Sea from opposite ends (Fig. 4.18c).

4.5.3 Some limitations of freshwater organisms

The macrozoobenthos is the ecological compartment that loses most of its species richness in the Baltic Sea, followed by the macrophytobenthos (Fig. 4.10). The reason for this is that the loss of marine species is not fully compensated for by the increase of freshwater species. For example, the respiratory system of insects evolved on land, and they cannot live in deep water because their air-filled respiratory system would collapse under high pressure. Thus insects are restricted to shallow water and cannot take advantage of vacant niches not occupied by deep-water marine invertebrates due to the Baltic Sea's low salinity. Similarly, vascular plants cannot take advantage of vacant niches on rock surfaces that are not filled by marine macroalgae. While macroalgae absorb nutrients from the water and can attach to hard substrates, vascular plants need to be rooted in sandy or soft substrates to be able to take up nutrients. As a group, the phytoplankton is more balanced than the macrophytobenthos, with freshwater green microalgae, chrysophytes and cyanobacteria matching the loss of marine dinoflagellates, centric diatoms and haptophytes.

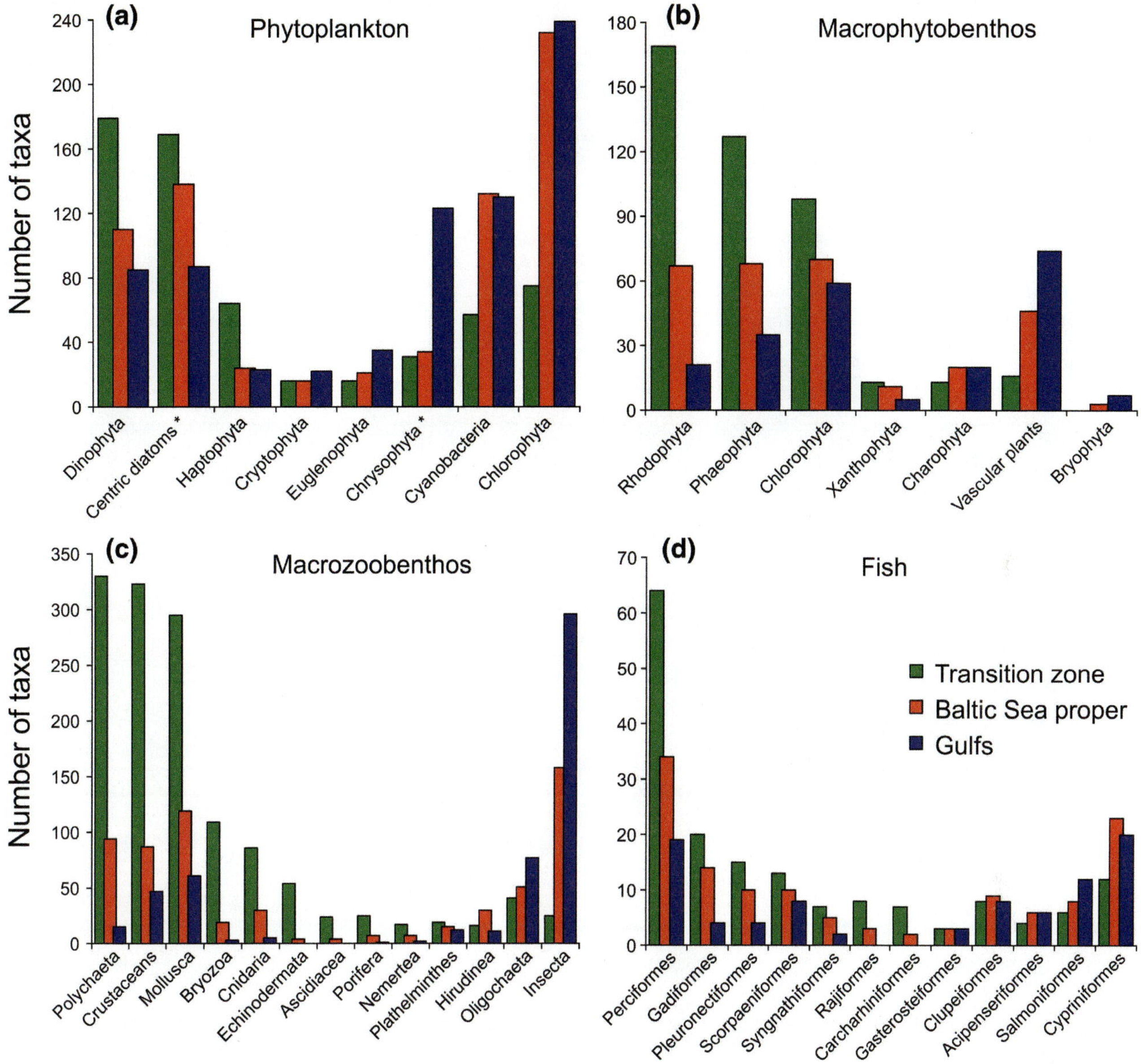

Fig. 4.18 Differences in richness within four groups of organisms along the large-scale Baltic Sea gradient extracted from the checklists referred to in Table 4.3. (**a**) Based on the distributions of all phytoplankton taxa in the dataset (excluding resuspended benthic diatoms). * = All Chrysophyta except for the diatoms, pelagic diatoms are shown separately. (**b**) Based on the distributions of all macrophytobenthos species in the dataset. (**c**) Based on the distributions of 97 % of the macrozoobenthos species in the checklist (15 rare groups were excluded). The legend for all graphs is given in (**d**). (**d**) Based on the distributions of 78 % of the fish species in 12 orders (23 rare fish and lamprey orders were excluded). Green = transition zone (Belt Sea and Kattegat) with surface-water salinity usually 9–25 (in deep water up to 35), red = Baltic Sea proper with surface-water salinity ~7 (in deep water up to 15), blue = gulfs with surface-water salinity 3–5.5 (in deep water up to 7). Figure: © Pauline Snoeijs-Leijonmalm

4.5.4 Benthic diatoms: a successful meeting in the Baltic Sea

Unlike the insects and vascular plants, the species-rich diatoms have successfully evolved in both marine and freshwaters, although they are believed to be ancestrally marine (Sims et al. 2006). Although several diatom genera, especially raphid diatoms (*e.g. Amphora*, *Navicula* and *Nitzschia*), are widely distributed in both habitats (Mann 1999), >90 % of the genera are either marine or freshwater, and only a few species "leak" into the opposite habitat (Round and Sims 1981; Round et al. 1990; Potapova 2011).

Fig. 4.19 Examples of marine animal species that occur in the transition zone but not in the Baltic Sea, except for perhaps single specimens in the easternmost Arkona Sea. (**a**) The lion's mane jellyfish *Cyanea capillata* (phylum Cnidaria). (**b**) The nudibranch sea slug *Aeolidia papillosa* (phylum Mollusca). (**c**) The brittle star *Ophiura albida* (phylum Echinodermata). (**d**) The lugworm *Arenicola marina* (phylum Annelida). (**e**) The sea anemone *Metridium dianthus* (phylum Cnidaria) together with the sea squirt *Dendrodoa grossularia* (Class Ascidiacea). (**f**) The sea squirt *Ciona intestinalis* (Class Ascidiacea), and within the centre of the *Ciona* colony, the sponge *Haliclona oculata* (Phylum Porifera). (**g**) The two sea anemone species *Urticina felina* and *Metridium dianthus* (Phylum Cnidaria). Photo: © Dirk Schories

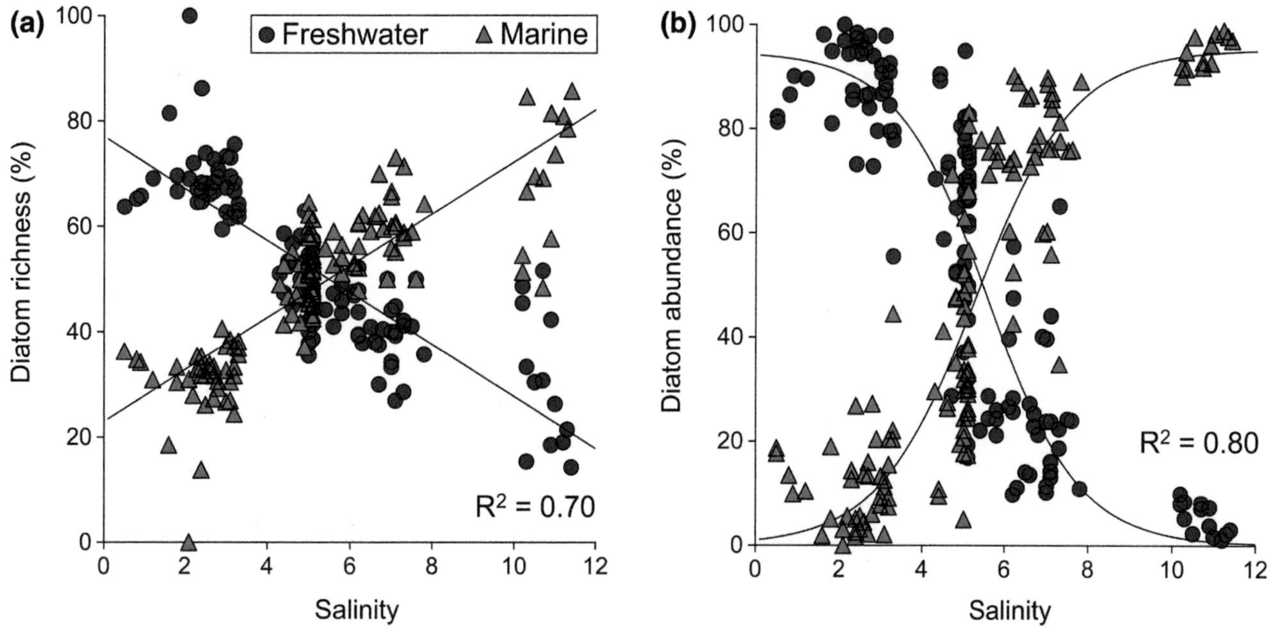

Fig. 4.20 The proportions of epilithic diatoms with marine and freshwater affinities in microphytobenthic communities at 132 sampling stations along the Baltic Sea salinity gradient from the northern Bothnian Bay (salinity 0–2) to the northern Öresund (salinity 10–12). (**a**) The % of marine and freshwater taxa based on diatom richness in the samples. (**b**) The % of marine and freshwater taxa based on diatom abundance in the samples. The data represent counts of 1,000 diatom valves per sampling station. Figure based on data in Ulanova et al. (2009). Figure: © Pauline Snoeijs-Leijonmalm

This makes it quite easy to assign the diatoms of the Baltic Sea to either marine or freshwater taxa, even if the species display different degrees of euryhalinity (Snoeijs et al. 1993–1998; Snoeijs 1995). The mainly pelagic centric diatoms in the Baltic Sea Area contain more marine taxa (Fig. 4.18a), while the pennate diatom taxa are in general equally distributed in the transition zone, the Baltic Sea proper and the gulfs.

Thus, the pennate diatoms that dominate the microphytobenthic communities in the Baltic Sea are an excellent example of a taxonomic group entering the Baltic Sea from both the marine and freshwater side with high species richness. This can explain their relatively stable occurrence along the large-scale Baltic Sea gradient and corroborates previous observations that benthic diatom richness is stable along local salinity gradients, *e.g.* for the Schlei Estuary in the Belt Sea (Hustedt 1925; Simonsen 1962; Wendker 1990), as well as for the large-scale Baltic Sea gradient (Snoeijs 1995; Ulanova et al. 2009).

An interesting question is then how far the physiological modifications of marine and freshwater diatom species would stretch along the salinity gradient of the Baltic Sea. The results of a comprehensive study including 132 sampling sites along the large-scale Baltic Sea gradient show that the exact "meeting point", *i.e.* the mixture of 50 % marine diatoms and 50 % freshwater diatoms, is at salinity 5.5 (Fig. 4.20). The meeting point is at exactly the same salinity for the % of species richness as for the % of individuals (abundance). However, in the entire salinity range between 4 and 8, the diatom communities consist of a mixture of marine and freshwater species building typical brackish-water diatom communities. This salinity range equals that of the photic zone in practically the entire Baltic Sea proper, the Gulf of Riga, the major part of the Gulf of Finland and the Bothnian Sea. Outside this range the microphytobenthic communities can be classified as either freshwater (Bothnian Bay) or marine (transition zone) communities.

Heterotrophic bacteria, another group with a stable taxonomic richness along the large-scale Baltic Sea gradient, show a smooth gradual change from communities dominated by freshwater taxa to those dominated by marine taxa between salinity 0 and 13 (Fig. 4.8). The brackish waters of the Baltic Sea seem to be occupied by a diverse combination of freshwater clades and marine clades of bacteria that appears to have adapted to the brackish conditions of the Baltic Sea and forms an autochthonous brackish microbiome (Herlemann et al. 2011). The reason for this may be that the bacteria are physiologically more diverse than the diatoms, which are clearly subdivided into two physiological groups in relation to salinity.

Box 4.10: The common eider *Somateria mollissima*

Kjell Larsson

Distribution

The common eider *Somateria mollissima* has a circumpolar breeding distribution (Waltho and Coulson 2015). Different migratory and sedentary populations are found in Arctic and temperate regions in northern Europe, northern Asia and North America. In Europe, common eiders breed in the Baltic Sea, Wadden Sea and along the coasts of Norway, Britain, Iceland, Svalbard and northern Russia. In the Baltic Sea, the core breeding areas are situated in the Swedish and Finnish archipelagos of the northern Baltic Sea proper. The wintering areas are situated in the Danish and German parts of the Baltic Sea and in the Wadden Sea.

Breeding behaviour

In late winter and spring, the females feed intensively to build up a storage of fat, proteins and other nutrients necessary for egg production and incubation. Egg laying starts in April or in the beginning of May, earlier in the southern parts of the Baltic Sea and later in the northern parts. Common eiders nest on islands, either solitarily or in groups, and often close to other colonial coastal birds such as gulls and terns. Feeding ceases during egg laying and during the incubation period, which lasts for ~25 days, and the females rarely leave the nest (Box Fig. 4.16). From pre-laying to hatching, a female may lose up to 40 % of her body weight. Males leave the incubating females and the breeding sites after the start of incubation and form feeding flocks with other males and non-breeders. After hatching, the females lead their young to shallow waters where they can feed themselves. Several broad-rearing females may form crèches, that is, groups of females which together guard their young.

Feeding

Common eiders feed mainly on molluscs, especially the blue mussel *Mytilus trossulus*, which they swallow whole. Since it is only the soft parts that are of nutritional value, adult common eiders must consume very large quantities of blue mussels each day, up to two kilograms per individual, as well as get rid of large quantities of crushed shells, to maintain their energy balance (*cf.* Sect. 11.13.9). The newly hatched young feed mainly on small invertebrates.

Population development

The Baltic common eider population increased in numbers from the 1950s and reached a peak at the beginning of the 1990s. Thereafter, the population declined rapidly. From 1992–1993 to 2007–2009 the estimated number of wintering common eiders in the Baltic Sea decreased by 50 % from ~1,000,000 to ~500,000 birds (Skov et al. 2011). The number of nesting birds in the core breeding areas in Sweden and Finland has also continued to decrease after 2009. The male-skewed sex ratio has also become more pronounced in recent decades. There are several causes for the fluctuations in numbers. The increase up to the 1990s has been hypothesised to be connected to the general eutrophication of the Baltic Sea. More phytoplankton led to higher densities of plankton filter-feeding blue mussels, that is, more food for common eiders in both the wintering and breeding areas. The recent rapid decline can most likely be explained by a combination of factors including reversed eutrophication processes in the wintering areas, other ecosystem changes affecting the quality and quantity of food, and increased predation pressures on adult nesting females due to the return of the white-tailed eagle *Haliaeetus albicilla* (Kilpi et al. 2015).

Box Fig. 4.16 The common eider *Somateria mollissima*. (**a**) An incubating female. Female common eiders do not feed during incubation but they take occasional recesses to drink to prevent dehydration. During the egg laying period some females may approach and lay eggs in nests of other females. The frequency of intra-specific nest parasitism differs between years and sites. (**b**) A male. The males leave the incubating females after the start of incubation. In recent decades the sex ratio has become more skewed. There are considerably more males than females in the population. Photo: © Kjell Larsson

Box 4.11: The great cormorant *Phalacrocorax carbo sinensis*

Kjell Larsson

Population development

Surveys performed in the year 2012 showed that the continental subspecies of the great cormorant *Phalacrocorax carbo sinensis* (Box Fig. 4.17) bred in almost all European countries (Bregnballe et al. 2014). The population size was estimated at ~215,000 breeding pairs within the 28 EU member states. A large part of the European population, ~168,000 pairs, was found to breed in the Baltic Sea region (Box Fig. 4.18). The other subspecies in Europe, *Phalacrocorax carbo carbo*, breed along the North Atlantic coast in Norway, the UK, France, Ireland and Iceland. The population size of the latter was estimated at ~42,000 breeding pairs. Cormorants have not always been as common in the Baltic Sea as they are today. The continental subspecies *Phalacrocorax carbo sinensis* was rare in Europe up until the beginning of the 1970s. In the early 1960s the total population in the Netherlands, Germany, Denmark, Sweden and Poland consisted of only ~4,000 breeding pairs. The population started to increase in the 1970s and during the 1980s the species started to expand its breeding range towards the northern and eastern parts of the Baltic Sea. Since 2006, the total number of breeding pairs in the Baltic Sea has stabilised around 160,000 (Bregnballe et al. 2014; Herrmann et al. 2014). The population growth at the end of the 20th century has been possible due to increased protection of breeding colonies, a decrease in the use of pesticides and increased abundance of small prey fish because of eutrophication. Although the majority of the Baltic breeding cormorants of the subspecies *Phalacrocorax carbo sinensis* migrate to southern Europe in winter, an increasing proportion has started to overwinter in the region (Herrmann et al. 2015). A part of the cormorant population of the subspecies *Phalacrocorax carbo carbo* that breed in Norway also winters in Danish waters.

Feeding

The great cormorant is an opportunistic fish predator feeding on a large number of different fish species. They prefer sheltered coasts and highly eutrophic estuaries and lagoons. They are rarely found feeding in sea areas deeper than 10 m. The great cormorant is a highly gregarious species. The cormorants often feed in flocks, and they often breed in colonies consisting of 1,000 pairs or more.

Management actions

Because of conflicts with fisheries, several countries have set up management actions to control or reduce the number of great cormorants. Such control actions have included the shooting of adult birds, scaring birds away from colonies before egg-laying and the destruction of eggs and nests (Herrmann et al. 2014). When disturbed, cormorants often move to other nearby areas.

Box Fig. 4.17 The great cormorant *Phalacrocorax carbo sinensis* is breeding in colonies in the Baltic Sea region. Photo: © Kjell Larsson

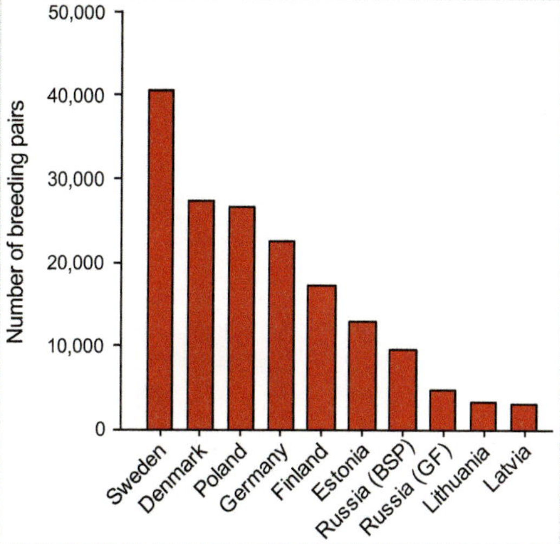

Box Fig. 4.18 Estimated number of breeding pairs of the great cormorant *Phalacrocorax carbo sinensis* in the Baltic Sea region in 2012. Figure based on data in Bregnballe et al. (2014)

Box 4.12: The white-tailed eagle *Haliaeetus albicilla*

Kjell Larsson

Breeding pairs

The white-tailed eagle, or the white-tailed sea eagle, *Haliaeetus albicilla* (Box Fig. 4.19), breeds along the coast as well as at inland lakes in all countries surrounding the Baltic Sea. Since the beginning of the 1990s, the population has grown rapidly in numbers and expanded its breeding range. Surveys have shown that the total Baltic Sea population increased from ~670–680 breeding pairs in 1991 to 2,070–2,200 breeding pairs in 2007, and the population has since then continued to grow (Herrmann et al. 2011).

Declines

In the beginning of the 20th century the species was very rare or even extinct in several countries because of persecution. The population started to recover when protection measures were put in place, but from the 1950s to the 1980s the population suffered heavily from the harmful effects of chemical pollutants in the environment. Substances such as DDT, PCB and mercury (*cf.* Table 16.1) had strong negative effects on the birds' survival and reproductive success.

Recovery

After the ban of DDT, PCB and other pesticides the mean reproductive success returned to almost normal levels in the mid-1990s (Helander et al. 2015). The white-tailed eagle feeds on waterbirds, fish and mammals. The mammalian prey often consists of carcasses. Individuals may also specialise on certain types of prey. The return of the white-tailed eagle along the Baltic Sea coast has had effects on waterbird populations. For example, the decline of the Baltic Sea common eider population in the core breeding areas in Finland and Sweden can to a significant extent be ascribed to increased predation on adult nesting female eiders by white-tailed eagles (Kurvinen et al. 2016).

Box Fig. 4.19 The white-tailed eagle *Haliaeetus albicilla*. Photo: © Lars-Eric Sellberg/Azote

4.5.5 Euryhalinity is at the species level

No obligate marine or freshwater species occur in the brackish Baltic Sea; all its species are euryhaline. Some Baltic Sea species exhibit a high degree of euryhalinity. For example, 37 fish, 41 macrozoobenthos, 45 macrophyte and 105 microalgal protistan species occur all the way from the transition zone to the Bothnian Bay and are reported from at least five of the seven other subregions of the Baltic Sea. These most extreme euryhaline species in the Baltic Sea Area are not taxonomically distinct, *i.e.* they belong to many different taxonomic groups. For example, the 37 euryhaline fish species belong to 14 different orders. Thus, euryhalinity seems to occur mainly at the species level and less at higher taxonomic levels.

4.5.6 Remane and the Baltic Sea gradient

One of the most famous conceptual models in brackish-water biology is the "Remane diagram" (Fig. 4.21). The diagram illustrates how macroscopic invertebrate species richness changes from freshwater via brackish water to marine water. Its most profound feature is that minimum overall species richness occurs at salinity 5–7, which is the critical salinity for physiological stress in many marine and freshwater animals (*cf.* Sect. 7.3.2).

The Remane diagram is based on species-distribution data from the North Sea and the Baltic Sea, but is in general often used as a model for the distribution of species in brackish water of different salinity, including tidal estuarine gradients. This had led to many modifications of the model, as well as debates about its general validity (Attrill and Rundle 2002; Whitfield et al. 2012). For example, in tidal estuaries on Atlantic coasts there are no brackish-water adapted invertebrate species; the species present in the mid-estuary are simply marine or freshwater taxa at the extreme edges of their range (Barnes 1989; Attrill and Rundle 2002).

Dahl (1956) and Deaton and Greenberg (1986) proposed explanations for the low number of macroscopic species in estuarine areas with strong salinity fluctuations observed globally. Dahl (1956) posits that "adaptation to a change of salinity, at least when affected by means of active osmotic and ionic regulation, make demands upon the general metabolism of the body, and the more abrupt the change, the greater is the output of energy required to meet it as well as the directly harmful effect, especially in the neighbourhood of the tolerance limits. It is not surprising therefore that the fauna and flora inhabiting unstable estuarine waters is as a rule a good deal poorer with respect to the number of species than corresponding habitats in more stable brackish waters". Deaton and Greenberg (1986), also concerned with palaeo-ecological data, found that the two ecological explanations for the occurrence of the species minimum – a species-area effect and the stability-time hypothesis – are inconsistent with published data on species distributions in brackish waters. They conclude that the low macrofaunal species diversity in brackish water may be explained, in part, by two factors: (1) few animals evolve the physiological

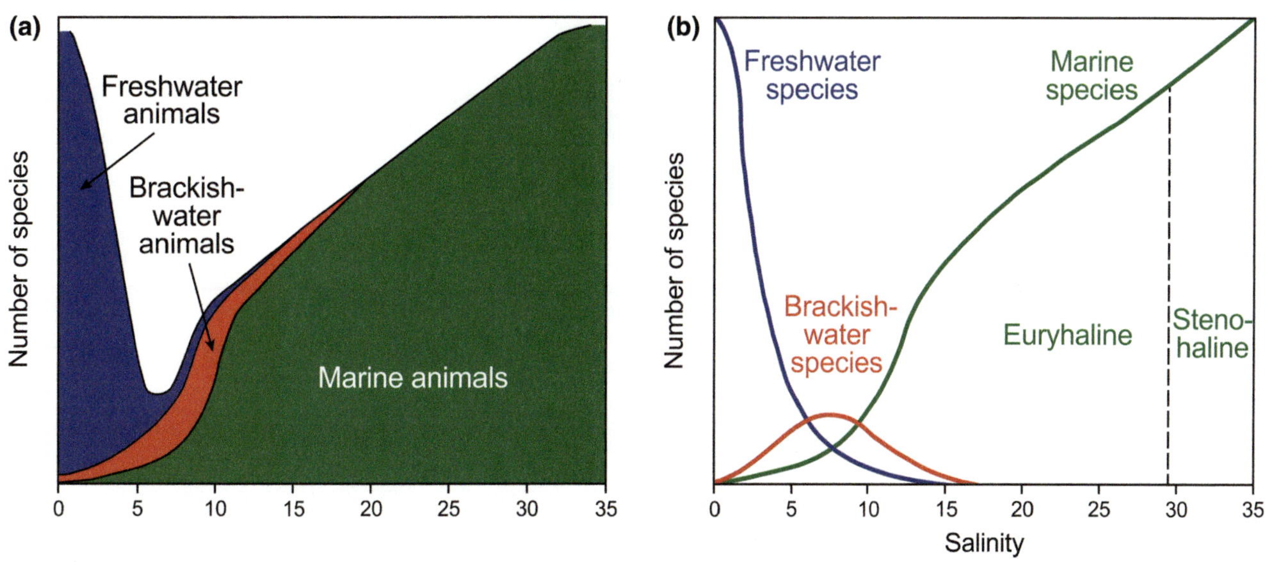

Fig. 4.21 The "Remane diagram". (**a**) The original Remane diagram based on the occurrence of macroscopic animals at different salinities in the North Sea and Baltic Sea Area, with minimum species richness of macroscopic animals at salinity 5–7. (**b**) A variation of the Remane diagram showing a model of the occurrence of species groups according to their salinity tolerances. Figure (**a**) modified from Remane (1934 and 1958, exactly the same figure occurs in both publications), (**b**) modified from Barnes RSK (1974)

Box 4.13: The three seal species of the Baltic Sea

Species distributions

Three seal species, with different geographical distributions, reproduce in the Baltic Sea Area (Box Fig. 4.20). The harbour seal *Phoca vitulina* (Box Fig. 4.21a) has a wide distribution along the temperate and Arctic marine coasts of the northern hemisphere. In the Baltic Sea Area, it occurs only in the Kattegat, Belt Sea and the southeastern part of the Baltic Sea with its northernmost population in the Kalmarsund (the sound between the island of Öland and the Swedish mainland). The grey seal *Halichoerus grypus* (Box Fig. 4.21b) is a North Atlantic species, which occurs in the whole Baltic Sea but does not form functional populations in the Kattegat. It has not been breeding in the Kattegat since the 1930s except for a few observations made in recent years (Härkönen et al. 2015). The ringed seal *Pusa hispida* (Box Fig. 4.21c) is the most common seal in the Arctic Ocean, and in the Baltic Sea it is an Arctic relict. Due to its dependence on sea ice for reproduction, it does not occur in the Atlantic Ocean or in the transition zone and the southern Baltic Sea. However, subspecies of the ringed seal occur in some large and cold freshwater lakes: Lake Saimaa in Finland and Lake Ladoga in northern Russia. These subspecies have evolved since the last ice age ~11,000 years ago. Adult ringed seals are slightly smaller (up to 1.75 m long and up to 120 kg in weight) than harbour seals (1.85 m/130 kg) while the grey seal is the largest species (females 2.1 m/180 kg, males 3 m/300 kg). Ringed seals can become almost 50 years old while harbour and grey seal males only get maximally ~25 years old and females ~35 years.

Population development

The grey seals in the Baltic Sea decreased from 88,000–100,000 individuals in 1900 to only ~4,000 in the late 1970s, initially due to excessive hunting and from the 1950s mainly due to organochlorine-caused sterility (Harding and Härkönen 1999). In the same period, the ringed seal population decreased from 190,000–220,000 to ~5,000 individuals. The populations have been slowly recovering after a ban of the use of organochlorines (DDTs, PCBs, (*cf.* Table 16.1) was introduced in the 1970s. These reductions in the seal populations have modified the trophic structure of the Baltic Sea food webs (*cf.* Sect. 17.2.4). Active species conservation measures to restitute the grey seal population were undertaken in several Baltic Sea countries, *e.g.* since the early 1980s at Forsmark (Sweden) by the Swedish Museum of Natural History and since the early 1990s at the seal research centre of the University of Gdańsk Marine Research Station at Hel (Poland). Seal surveys were carried out, and grey seals reared from captive parents fed

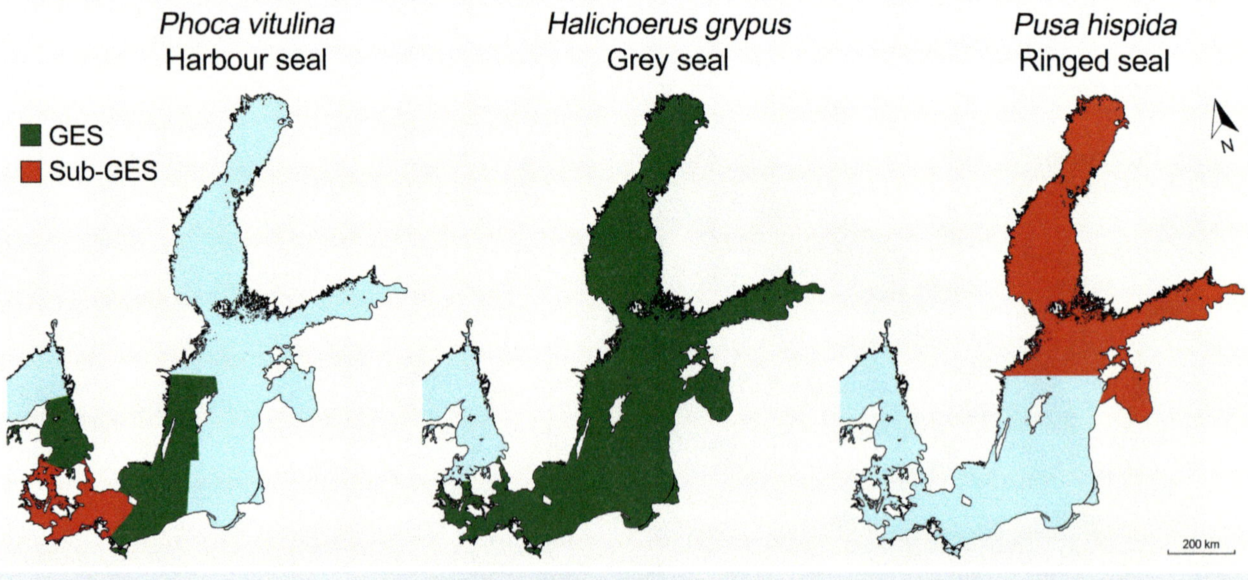

Box Fig. 4.20 Areas of reproduction of the three seal species in the Baltic Sea Area. GES = "good environmental status" for seals (GES, *cf.* Sect. 17.8.1). Figure modified from Härkönen et al. (2015)

with non-polluted fish were released into the Baltic Sea. In 2012, the population sizes were estimated at ~10,000 harbour seals and ~28,000 grey seals in the Baltic Sea Area (Härkönen et al. 2013). The ringed seal population in the Bothnian Bay has been increasing at a rate of 4.5 % per year since 1988 and consists of ~3,000 individuals, which is less than half the intrinsic capacity. However, the 5-year mean of the growth rate shows very high variability, probably reflecting the uncertainty of population censuses (Härkönen et al. 2013). In the southern breeding areas, the Gulf of Riga, the Gulf of Finland and the Archipelago Sea, no increasing trends have been observed (Karlsson et al. 2007).

Current status of the seal populations

Today the grey seal population growth has reached the level of "good environmental status" (GES, *cf.* Sect. 17.8.1), as has the harbour seal population in some areas (Box Fig. 4.20). The HELCOM core indicator used for this estimation has three components for each species: distribution on haul-out sites (sites on land where seals are to be found when they temporarily leave the water), breeding sites and foraging areas. GES is achieved when the distribution of seals is close to pristine conditions (*e.g.* such as those prevailing 100 years ago), or where appropriate when all currently available haul-out sites are occupied, and when no decrease in area of occupation occurs (Härkönen et al. 2015). The ringed seal population growth rate is considerably below GES levels and is considered alarming in some areas. As a winter ice cover is critical for the breeding success of the ringed seal, climate change is a serious threat to this species in the Baltic Sea. Climatological modelling predicts sea ice to diminish and the ice season to become shorter in the future. This will likely result in the extirpation (local extinction) of the ringed seal population in the Gulf of Riga and will severely reduce the population growth rate in the Gulf of Finland and the Bothnian Bay (Sundqvist et al. 2012).

Conflicts between seals and fishermen

Seals can destroy fishery gears and eat the fish caught in them. This causes a conflict between seals and fishermen even if today's seal populations constitute only a fraction of the historical natural population sizes. Although the seals in the Baltic sea are protected by European regulations in the Habitats Directive (*cf.* Table 18.3), if no other means can deter the seals from attacking fishing gears, the responsible management authorities can give limited hunting permission in some parts of the Baltic Sea (*cf.* Sect. 18.6.1). On the other hand, a few hundred seals are killed each year as by-catches of fisheries in the Baltic Sea.

Box Fig. 4.21 Portraits of the three seal species that occur in the Baltic Sea. (**a**) The harbour seal *Phoca vitulina*. (**b**) The grey seal *Halichoerus grypus*. (**c**) The ringed seal *Pusa hispida*. Photo: (**a**, **b**) © Sven-Erik Arndt/Azote, (**c**) © Charlotta Moraeus/Azote mechanisms required for life in the variable habitat, and (2) these species, which are very eurytopic, have low rates of speciation.

The Remane diagram (Fig. 4.21) has often been interpreted in terms of biodiversity in an ecological sense, but it is essentially more about evolution. Remane simply pooled the number of zoobenthos species within five defined salinity ranges, subdivided them into freshwater, brackish and marine species and smoothed the curves to achieve a conceptual model. Thus, the diagram shows the γ diversity for different "regions", which are represented by salinity ranges, but it does not show that one m^2 of the seafloor, or one litre of water in the pelagic zone, has higher or lower biodiversity at salinity 5 than at salinity 1 or 15. Thus, information about α- and β-diversity, which are ecologically relevant for community and ecosystem functioning, is absent.

Furthermore, the Remane diagram has been misinterpreted as a model that would accommodate all groups of organisms. Along the large-scale Baltic Sea gradient, it can be fitted to macrofauna, macrophytes and fish richness distributions along the Baltic Sea, but not to *e.g.* bacteria, benthic diatoms, phytoplankton, waterbirds and mammals. Remane (1958) himself already realised that some microbial groups did not seem to have minimum species richness in brackish water, which he based on publications on diatom and ciliate distributions along salinity gradients (Hustedt 1925; Kahl 1928).

4.5.7 Is evolution filling vacant niches?

An evolutionary feature of the Remane diagram is that "true" brackish-water species would occur in the Baltic Sea. For over 100 years, biologists have recognised that many species in the Baltic Sea have deviant morphologies compared to their conspecifics in areas outside the Baltic Sea. Especially, the Baltic Sea specimens are often smaller in body size than those in marine habitats as a result of salinity stress. In the first half of the 20th century, many scientists tried to explain this from an evolutionary perspective by following the development from physiological and/or morphological adaptations via ecotypes to endemic species. For example, up to ~40 endemic species of Baltic Sea macroalgae were recognised at the time Remane made his diagram (Lakowitz 1929; Russell 1988). Later, the existence of endemic species in the Baltic Sea became disputed (*e.g.* Hoffmann 1950), and all previously described endemics lost their specific status and were united with their marine or freshwater counterparts.

Today, molecular evidence supports that speciation is going on in the Baltic Sea. The pronounced large-scale Baltic Sea gradient exerts a strong selective pressure upon the organisms in a remote and effectively isolated corner of the ocean, and adaptive evolutionary changes are taking place. This makes the Baltic Sea an ideal place for evolutionary studies. The recently described species *Fucus radicans* is considered an endemic brackish-water species of the Baltic Sea (*cf.* Sect. 6.3.5), and *Diatoma bottnica* (Fig. 4.22, Snoeijs and Potapova 1998) is an example of a possible endemic diatom species. Thus, evolution seems to fill the vacant niches in the Baltic Sea, but very slowly.

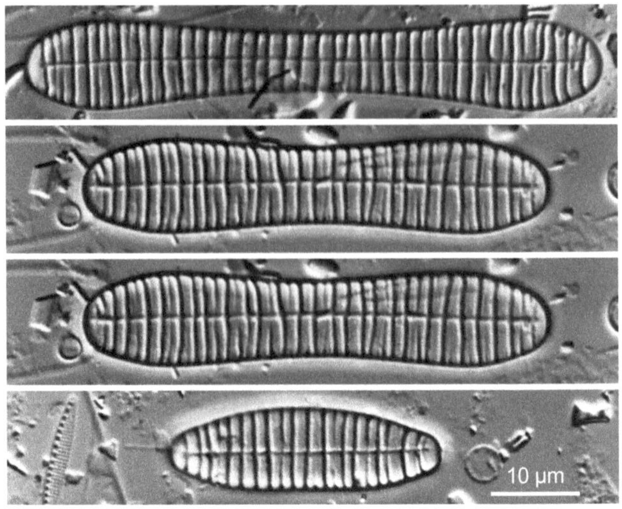

Fig. 4.22 Four valves of the benthic araphid diatom *Diatoma bottnica* that forms substrate-associated zig-zag shaped colonies in the Gulf of Bothnia (*cf.* Box Fig. 4.8j). This may be an endemic diatom species in the Baltic Sea related to *Diatoma constricta* and *Diatoma vulgaris* with which it co-occurs. Photo: © Pauline Snoeijs-Leijonmalm

To illustrate the evolutionary time scale needed for an ecosystem to become filled with brackish-water organisms, three large brackish-water ecosystems with different evolutionary developments – the Caspian Sea (which in fact is a lake), the Black Sea and the Baltic Sea – can be compared (*cf.* Table 2.1, Fig. 4.23). The organisms in the Caspian Sea, which became land-locked ~5.5 million years ago, have been adapting to brackish water for millions of years, and therefore the Caspian Sea contains most brackish-water species. The Baltic Sea has been brackish for only ~8,000 years, including the Littorina Sea stage (*cf.* Fig. 2.26c), and the Baltic Sea contains the fewest brackish-water species. A number of brackish-water species living in the Black Sea and the Caspian Sea are remnants of the Pliocene low-salinity Pontian Sea-Lake (Finenko 2008). These species are referred to as "Ponto-Caspian relicts". In the late Pontian stage the Earth's crust began to rise in the northern Caucasus, gradually isolating the Caspian Sea from the Pontian basin. From that period onwards the Caspian Sea and the Black Sea developed separately, although temporary links between them were formed from time to time. More and more endemic brackish-water species evolved in the Caspian Sea (Karpinsky et al. 2005) while the Black Sea became influenced by marine Mediterranean Sea species. Thus, the Black

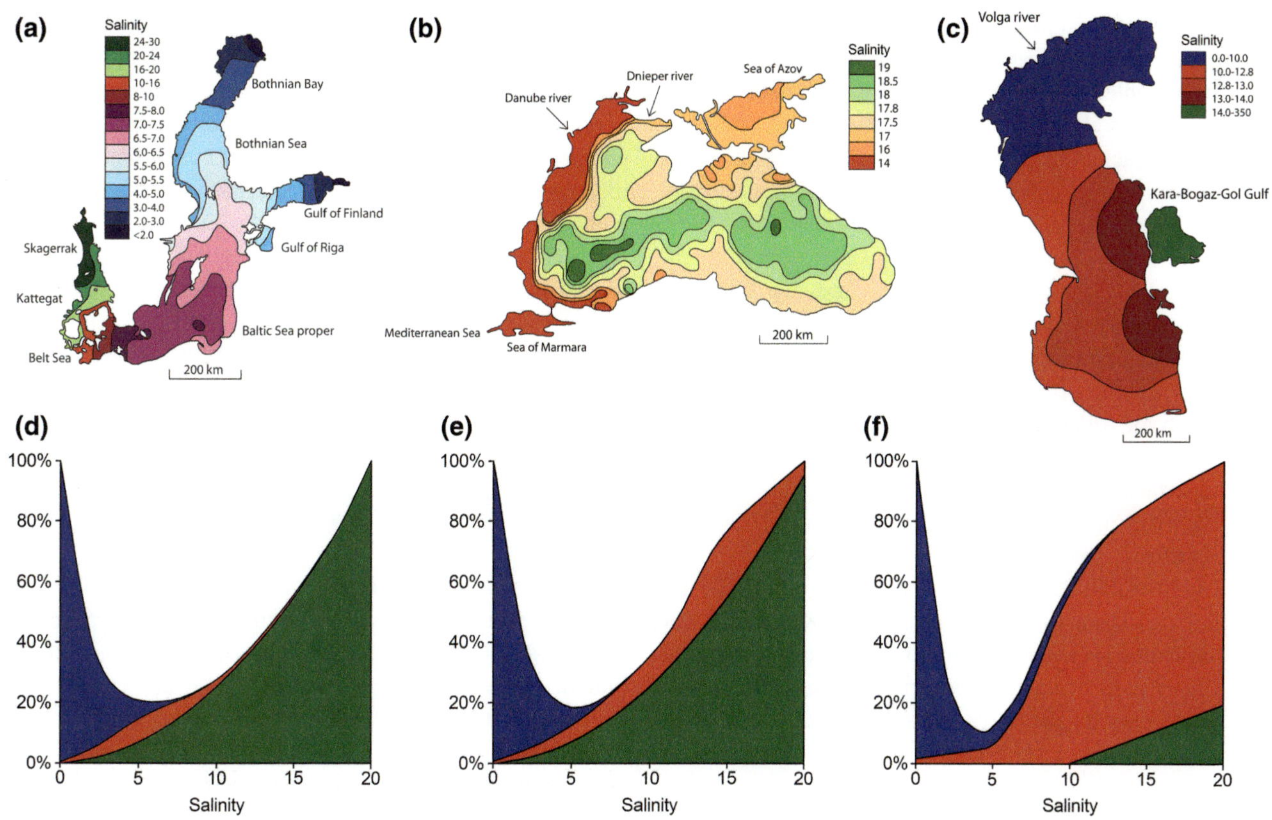

Fig. 4.23 Theoretical Remane-type diagrams with the relative species richness of marine, brackish-water and freshwater animals plotted against salinity for (**a**, **d**) the Baltic Sea, (**b**, **e**) the Black Sea, (**c**, **f**) the Caspian Sea. Figure modified from Snoeijs and Weckström (2010)

Sea only has a limited number of brackish-water species compared to the Caspian Sea (Fig. 4.23).

4.5.8 Faster than evolution: non-indigenous species

Primary successional processes are still on-going in the Baltic Sea following the latest ice age, and numerous ecological niches remain available for immigration (Bonsdorff 2006). Human assistance in moving species between ecosystems is speeding up this immigration, and the many vacant niches for macroscopic species in the Baltic Sea provide grounds for invasions of non-indigenous species, and ~130 are established here today (*cf.* Sect. 5.1). These invasions are often called "biological contamination", but for the species-poor Baltic Sea the non-indigenous macroscopic species may also be positive enrichments to the functional biodiversity of the ecosystem.

Of the 51 non-indigenous zoobenthos species in the Baltic Sea, ~33 % originate from marine waters and ~67 % originate from inland waters (48 % Ponto-Caspian species and 19 % freshwater species) (Zettler et al. 2014). Brackish-water species are rare among the invaders from freshwater or estuaries outside the Baltic Sea, but most of the newcomers from the warm-water Ponto-Caspian region are "true" brackish-water species. The Ponto-Caspian fauna is especially rich in the estuaries and lagoons of the southern Baltic Sea. Only a few non-indigenous species occur in the Bothnian Bay. In the northern Baltic Sea, their distribution is likely to be limited by low temperatures.

The proportion of non-indigenous zoobenthos species in the Baltic Sea is negatively correlated with salinity and total zoobenthic diversity (Zettler et al. 2014). This is in agreement with the hypothesis that areas with low native species richness are more susceptible to invaders. The spread of non-indigenous species poses a major threat to both the structure and functioning of natural biodiversity worldwide as the invaders may outcompete native species. An example of such an invasive species in the Baltic Sea is the Ponto-Caspian round goby *Neogobius melanostomus* (*cf.* Box 5.6).

Several non-indigenous species have provided success stories regarding their respective invasions as they have managed to take over and become established in vacant niches in the Baltic Sea. These species have enriched the functional diversity of the Baltic Sea ecosystem (Olenin and Leppäkoski 1999; Paavola et al. 2005). Some examples are the following:

(1) The New Zealand mud snail *Potamopyrgus antipodarum* (*cf.* Box Fig. 5.11), a surface deposit feeder on extremely soft bottoms where the native mud snails do not occur.

(2) The zebra mussel *Dreissena polymorpha* (*cf.* Box 5.4), a filter feeder in oligohaline and freshwater parts of coastal inlets where the native blue mussel *Mytilus trossulus* is absent.
(3) The barnacle *Amphibalanus improvisus* (*cf.* Fig. 5.12), a suspension filter feeder in the uppermost hydrolittoral zone, the only barnacle that is widely distributed in the entire Baltic Sea. The other barnacle species in the Baltic Sea is a spill-over species from the transition zone, *Balanus crenatus*, and occurs only up to the Bornholm Sea (HELCOM 2012a).
(4) The spionid polychaetes *Marenzelleria* spp. (*cf.* Box 5.3), which are capable of deeper bioturbation of the sediment than the native species and constitute an important new food source for benthic fish.
(5) The hydrozoan *Cordylophora caspia*, a sessile raptorial suspension feeder.
(6) The decapods *Rhithropanopeus harrisii* (*cf.* Fig. 5.15) and *Eriocheir sinensis* (*cf.* Box 5.8), which are epibenthic invertebrate predators and scavengers in the diluted parts of the inlets where native marine decapod crustaceans do not occur.

4.6 A sea of dominants

4.6.1 Brackish water of salinity 5–7

A large part of the Baltic Sea proper, the Gulf of Riga and parts of the Gulfs of Finland and Bothnia have brackish water with salinity 5–7 (Fig. 4.2). This salinity interval was once thought to represent sharp changes in the ionic composition of seawater diluted with freshwater and to constitute a physico-chemical barrier between marine and freshwater faunas in estuaries (Khlebovich 1968). Based on such premises, the segment of the salinity gradient between 5 and 8 was defined as the "horohalinicum" (Kinne 1971). When the ionic composition of diluted seawater was reinvestigated, it turned out that the "horohalinicum" does not exist because there is no physico-chemical barrier (Deaton and Greenberg 1986).

Nevertheless, the richness of macroscopic species is lowest at salinity 5–7 in both estuaries and the Baltic Sea, although the reduction is not explained by a physico-chemical barrier. Actually, the richness of macroscopic species in the Baltic Sea proper, excluding the Arkona Sea, with >1500 species of macrophytes, metazooplankton, zoobenthos and fish, is low compared to fully marine areas, but not extremely low. It is the evenness component of diversity that is extremely low at salinity 5–7, *e.g.* typical of the Baltic Sea proper are monotonous algal belts with only one structurally important species in each belt (*cf.* Sects. 11.7–11.10). Thus, the major feature of the Baltic Sea biodiversity pattern is not in the first place a low species richness but low evenness through the dominance of good competitors under the stable brackish conditions at which most species are osmotically stressed and therefore weak competitors. Typical of the Baltic Sea proper are mass occurrences of only a few macroscopic species, building simple food webs in a highly productive system, while the communities are more diverse in the Gulf of Bothnia, at least with respect to evenness (*e.g.* Fig. 4.24), but productivity is low (*cf.* Figs. 11.37 and 11.38).

4.6.2 Origins of the dominant species in the Baltic Sea proper

The majority of the dominant species in the Baltic Sea proper are post-glacial marine immigrants that are able to form large populations at the edge of their salinity distribution. However, some euryhaline freshwater species such as the green alga *Cladophora glomerata* and the snail *Theodoxus fluviatilis* are dominant in the upper sublittoral zone, and some cold-adapted stenotherm glacial relicts are abundant in the deep soft-bottom system.

Glacial relicts are species that immigrated to the Baltic Sea in early post-glacial times and remained after the Baltic Ice Lake and Yoldia Sea stages (Segerstråle 1962). Some glacial relicts, *e.g.* the crustaceans *Limnocalanus macrurus*, *Monoporeia affinis* (*cf.* Fig. 4.30), *Mysis relicta* (*cf.* Box 6.3) and *Saduria entomon* (Fig. 4.25d), and the fourhorn sculpin *Myoxocephalus quadricornis* (Fig. 4.13c), are also found in cold, deep lakes in Fenno-Scandia and Russia (Audzijonyte and Väinölä 2005). Other glacial relicts, *e.g.* the pennate diatom *Pauliella taeniata* (syn. *Achnanthes taeniata*, Round and Basson 1997), the brown macroalga *Battersia arctica* (*cf.* Fig. 11.28) and the ringed seal *Pusa hispida* (*cf.* Box Fig. 4.21c) occur both in the Baltic Sea and the Arctic Ocean.

4.6.3 Dominant species in the pelagic zone

Only three fish species dominate the pelagic biomass of the Baltic Sea proper: two clupeids – the European sprat *Sprattus sprattus* and the Atlantic herring *Clupea harengus* – and their predator the Atlantic cod *Gadus morhua* (Fig. 4.12). Another pelagic predatory fish is the Atlantic salmon *Salmo salar* (Fig. 4.26), but its total biomass in the Baltic Sea is much lower than that of the cod. Widespread benthic flatfish in the Baltic Sea are the European plaice *Pleuronectes platessa* (Fig. 4.13a), European flounder

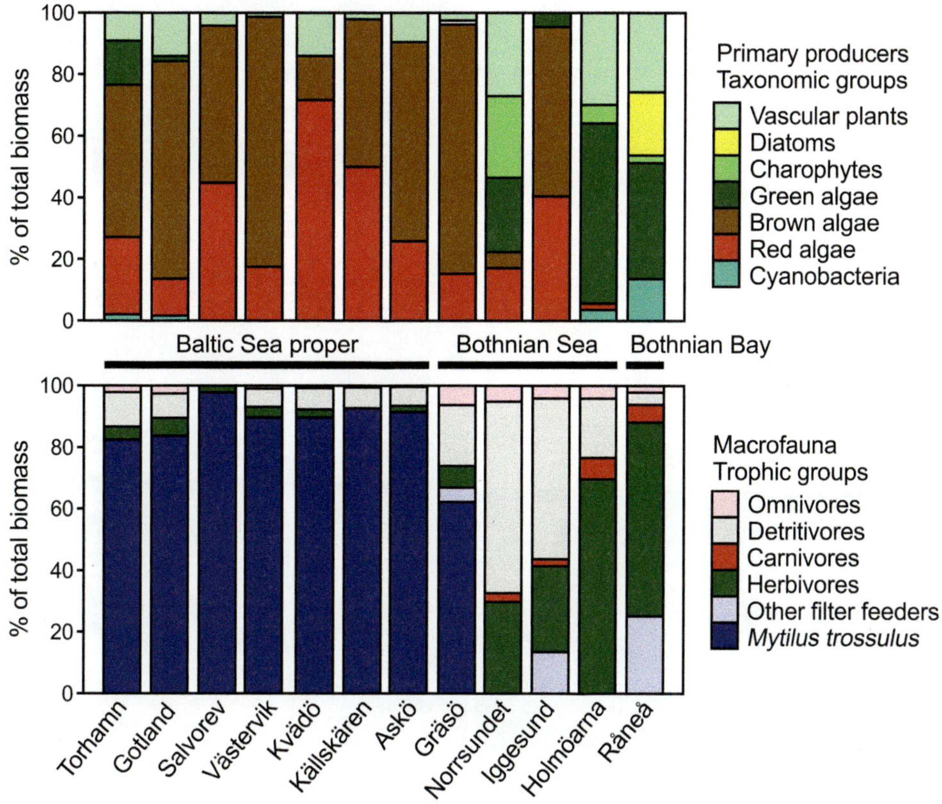

Fig. 4.24 Changes in the community composition in the phytobenthic zone along the Swedish coast from Torhamn in the southern Baltic Sea proper to Råneå in the northern Bothnian Bay, showing the dominance of *Fucus vesiculosus* (constituting most of the brown algal biomass), *Furcellaria lumbricalis* (constituting most of the red algal biomass) and *Mytilus trossulus* in the Baltic Sea proper. The category "Other filter feeders" refers to all filter feeders other than *Mytilus trossulus*. Figure modified from Kautsky (1995)

Platichthys flesus (Fig. 4.13b) and turbot *Scophthalmus maximus* while the brill *Scophthalmus rhombus* and the dab *Limanda limanda* occur in the southern Baltic Sea only, and more obligate marine species such as the common sole *Solea solea* are restricted to the transition zone and are only occasionally found the Baltic Sea. The zooplankton is dominated by only five species, the copepods *Acartia longiremis* (*cf.* Fig. 8.19), *Pseudocalanus acuspes* (*cf.* Fig. 8.20a), *Temora longicornis* (*cf.* Fig. 8.20c) and the cladocerans *Bosmina longispina maritima* (*cf.* Fig. 8.18d) and *Evadne nordmanni* (*cf.* Fig. 8.18c) (Table 4.5), which constitute the main food for the clupeid fish.

Conspicuous primary producers in the pelagic zone of the Baltic Sea during summer-autumn are the blooms of the diazotrophic filamentous cyanobacteria *Aphanizomenon*, *Dolichospermum* (syn. *Anabaena*) and *Nodularia* (*cf.* Fig. 8.2). Analyses of the 16S rRNA gene have shown that the three morphologically distinct pelagic *Nodularia* strains in the Baltic Sea, which have been described as different species (*Nodularia baltica*, *Nodularia litorea* and *Nodularia spumigena*), are in fact the same species. This confirms the view that there is only one pelagic *Nodularia* species, *Nodularia spumigena* (Laamanen et al. 2001). Likewise, the only *Aphanizomenon* species is *Aphanizomenon flos-aquae* (Laamanen et al. 2002). The genus *Dolichospermum* seems to have a much higher genetic diversity than *Aphanizomenon* in the pelagic zone of the Baltic Sea (Halinen et al. 2008). It has been proposed that salinity is a limiting factor for the genetic diversity of the *Aphanizomenon flos-aquae* population in the Baltic Sea, since this species is more diverse in freshwater environments (Laamanen et al. 2002). However, the genetic diversity of the Baltic Sea *Dolichospermum* is comparable to that of the freshwater *Dolichospermum* strains, which are widely distributed in the 16S rRNA gene tree, and thus salinity does not seem to have any obvious limiting effect. Still, despite the high genetic and morphological diversity within *Dolichospermum*, only one genetically valid species, referred to as *Dolichospermum* sp., occurs in the plankton of the Baltic Sea (Halinen et al. 2008)

In winter, the glacial-relict diatom *Pauliella taeniata* (*cf.* Fig. 8.4a) is dominant in ice-covered regions and can make up >90 % of the diatom valves in the deep-water upper sediment of the Baltic Sea proper and the Gulf of Finland (Hällfors and

Fig. 4.25 Some crustaceans that are abundant in the Baltic Sea. (**a**) The native Baltic Sea shrimp *Palaemon adspersus*. (**b**) The non-indigenous shrimp *Palaemon elegans*. (**c**) The mysid *Neomysis integer*. (**d**) The Arctic relict isopod *Saduria entomon*. (**e**) The brown shrimp *Crangon crangon*. Photo: (**a**, **d**) © Pauline Snoeijs-Leijonmalm, (**b**) © Stefan Beskow/Azote, (**c**) © Nicklas Wijkman/Azote, (e) © Gunnar Aneer/Azote

Niemi 1975). Other bloom-forming phytoplankton species are *e.g.* the diatoms *Coscinodiscus granii* (*cf.* Fig. 3.3a) and *Skeletonema marinoi* (*cf.* Fig. 8.4a) in spring and the dinoflagellates *Dinophysis norvegica* (*cf.* Fig. 8.7b) and *Prorocentrum cordatum* (syn. *Prorocentrum minimum*) (*cf.* Fig. 8.7b) in late summer-autumn (*cf.* Sect. 8.2).

4.6.4 Dominant species in the benthic zone

On hard bottoms, the benthic zone of the Baltic Sea proper consists of rather monotonous belts of a few macroalgae and one mussel. The benthic biomass is completely dominated by the brown alga *Fucus vesiculosus* (Fig. 4.27b), the red

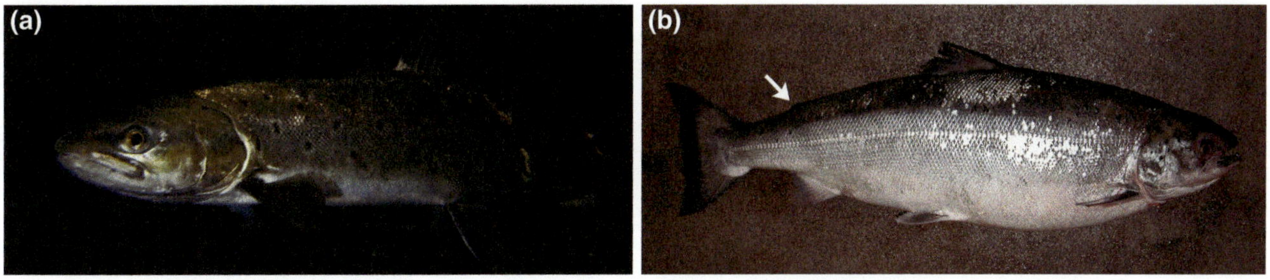

Fig. 4.26 The Atlantic salmon *Salmo salar* in the Baltic Sea. (**a**) A young salmon. (**b**) A salmon caught in the Baltic Sea – this specimen has escaped from aquaculture because its adipose fin has been removed (indicated by the arrow), which is routinely done for cultured fish to be able to distinguish them from wild ones. Photo: (**a**) © Piotr Bałazy, (**b**) © Pauline Snoeijs-Leijonmalm

alga *Furcellaria lumbricalis* (Fig. 4.28), and one filter-feeding animal, the blue mussel *Mytilus trossulus* (Fig. 4.29). These species are important habitat builders, and their occurrence is a prerequisite for a large number of smaller organisms that live on and between them. Additionally, the green alga *Cladophora glomerata* (Fig. 4.27c) is a habitat-forming species on rocky substrates in the upper metre, as is the common eelgrass *Zostera marina* (*cf.* Fig. 11.2b) on soft and sandy bottoms in the photic zone.

In deep soft seabeds the polychaete *Bylgides sarsi* (*cf.* Fig. 10.2a), the bivalve *Macoma balthica* (*cf.* Box 13.5), the deposit-feeding amphipods *Monoporeia affinis* and *Pontoporeia femorata* (Fig. 4.30), and their predator, the up to 9 cm large isopod *Saduria entomon* (Fig. 4.25d) dominate (Laine et al. 1997, Laine 2003). In the eastern Baltic Sea *Saduria entomon* reaches densities of up to 180 individuals m^{-2} (Haahtela 1990). In shallow bays of the Bothnian Sea, the macrozoobenthos is dominated by midges of the subfamily Orthocladiinae (Fig. 4.31), which can reach densities of >25,000 individuals m^{-2}. Other abundant taxa in these habitats are oligochaetes with maximum densities of >12,000 individuals m^{-2}, gammarids with >2,500 individuals m^{-2} and the small freshwater snail *Theodoxus fluviatilis* (Fig. 4.6a) with >2,000 individuals m^{-2}.

The zoobenthos of the Baltic Sea is numerically dominated by its meiobenthic component, with free-living

Table 4.5 The dominant zooplankton taxa along the large-scale Baltic Sea gradient. Adapted from Postel (1995). TRANS = the transition zone (Kattegat and Belt Sea, for ciliates, only the southern Belt Sea), ARK = Arkona Sea, SBSP = Southern Baltic Sea proper, CBSP = Central Baltic Sea proper, GF = Gulf of Finland, BS = Bothnian Sea, BB = Bothnian Bay

Taxon	Taxonomic group	TRANS	ARK	SBSP	CBSP	GF	BS	BB
Acartia tonsa	Copepoda							
Calanus finmarchicus	Copepoda							
Centropages typicus	Copepoda							
Oithona similis	Copepoda							
Paracalanus parvus	Copepoda							
Carnivorous cladocerans	Cladocera							
Oikopleura dioica	Appendicularia							
Centropages hamatus	Copepoda							
Pseudocalanus spp.	Copepoda							
Acartia spp.	Copepoda							
Temora longicornis	Copepoda							
Bosmina longispina maritima	Cladocera							
Evadne nordmanni	Cladocera							
Eurytemora affinis	Copepoda							
Acartia bifilosa	Copepoda							
Synchaeta spp.	Rotatoria							
Limnocalanus macrurus	Copepoda							
Fritillaria borealis	Appendicularia							
Pleurobrachia pileus	Ctenophora							
Keratella spp.	Rotatoria							
Daphnia spp.	Cladocera							

Fig. 4.27 Three abundant macroalgae in the Baltic Sea. (**a**) The marine brown alga *Fucus serratus*, which occurs only in the southern Baltic Sea proper. (**b**) The marine brown alga *Fucus vesiculosus*, which occurs in the whole Baltic Sea at salinity >4. (**c**) The freshwater green alga *Cladophora glomerata*, which occurs in the entire Baltic Sea. Photo: © Pauline Snoeijs-Leijonmalm

nematodes as the most abundant constituent. Meiofauna has a high tolerance capacity and, as a result, distributional ranges of many marine and freshwater meiofauna in the Baltic Sea characteristically extend more widely into critical brackish zones compared to macrozoobenthos (Ólafsson and Elmgren 1997; Giere 2009). The domination among the

Fig. 4.28 The marine red alga *Furcellaria lumbricalis* covered with herring eggs and one specimen of the blue mussel *Mytilus trossulus*. Photo: © Sergej Olenin

nematode taxa is strongly dependent on the sediment type. For example, the shallow (1.5 m water depth) sandy habitats of the Gdańsk Bay off the Hel Peninsula (southern Baltic Sea proper) are dominated by the genera *Ascolaimus*, *Enoplolaimus*, *Daptonema* and *Dichromadora* (Urban-Malinga et al. 2006) while *Leptolaimus elegans* and *Leptolaimus papilliger* dominate among nematodes in the organic-enriched shallow sandy habitats of the Gulf of Riga (Pallo et al. 1998) and *Sabatieria pulchra* dominates in the shallow sandy-mud habitats of the inshore Pomeranian Bay (Rokicka-Praxmajer and Radziejewska 2002). The latter species was also found to be dominant in a shallow ~27 m deep area enriched by a sedimented cyanobacteria bloom off the Swedish coast in the Baltic Sea proper (Nascimento et al. 2008) as well as in the muds of the Gotland deep and the Gdańsk deep devoid of macrozoobenthos (Kotwicki et al. 2016).

4.6.5 Some species-rich communities in the Baltic Sea

The microorganisms of the Baltic Sea are not species-poor, with some exceptions consisting of organisms that are confined to marine conditions as a group (*e.g.* foraminifers). Bacteria and protists in the pelagic and benthic zones quickly adapt to changes in ambient environmental conditions through changes in species composition at a scale ranging from days to weeks. At this level, there is no difference compared to fully marine ecosystems, such as the North Sea.

Fig. 4.29 The marine blue mussel *Mytilus trossulus*, which occurs throughout the Baltic Sea at salinity >4. (**a**) A monoculture on a sea traffic buoy lifted from the water in the southern Baltic Sea proper. (**b**) A small colony from the Gulf of Finland. **c** A filter-feeding blue mussel in the Belt Sea. Photo: (**a**, **b**) © Pauline Snoeijs-Leijonmalm, (**c**) © Dirk Schories

The highest diversity of animals (invertebrates and fish) in the Baltic Sea is found in the coastal zone in habitats with high biomass of the macrophytes *Fucus vesiculosus* and *Zostera marina* (*cf.* Sects. 11.8 and 11.11). The three-dimensional structures formed by these large macrophytes are colonised by other macrophytes, protists and bacteria. The animals that find shelter and food in these habitats represent many different species occurring at high abundances.

4.7 Functional diversity

4.7.1 *r*- and *K*-selected species

As a simplified description of life-history characteristics in ecology, *r*- and *K*-strategies are sometimes assigned to different species, which then are called *r*- and *K*-selected species, respectively (Box 4.14). These two strategies are the extremes of the life-history continuum with the *r*-strategy representing small-sized, fast-growing, short-lived generalists and the *K*-strategy representing large-sized, slow-growing, long-lived specialists. There are many examples demonstrating that the *r*/*K* ratio increases along the Baltic Sea gradient from the transition zone to the inner gulfs. For example, larger perennial algae are replaced by smaller filamentous algae (*cf.* Sect. 11.5.3), and deep-burrowing animals such as large polychaetes and echinoderms are replaced by smaller polychaetes and nematodes (*cf.* Sect. 10.6.2, Box 4.8). Another example is that specialists are replaced by generalists, *e.g.* the fauna associated with *Zostera marina* (*cf.* Sect. 11.11.3) and parasites (Zander and Reimer 2002).

4.7.2 Feeding types

The concept of functional groups was originally proposed to fuse different species that have the same function in the food web in order to obtain a better understanding of food web functioning (*cf.* Sect. 4.1.6). These functions are primary producers, herbivores, primary carnivores and top carnivores. Further categories are omnivores, which consume both primary producers and animals, scavengers, which consume dead algae, plants and animals, and detritivores, which consume decomposing organisms as well as faeces. Animals can also be subdivided according to how and where they eat, *e.g.* suspension feeders (including filter feeders), deposit feeders (including grazers) and predators. This latter subdivision gives a clue about where a species occurs (how they eat), whereas the first approach concerns the species in their role in the ecosystem (what they eat).

4.7.3 Biological traits analysis (BTA)

Functional diversity tackles the complexity of communities and ecosystems by constructing simple taxon-independent models of structure and dynamics based on the ecological functions of the constituting species. The functions

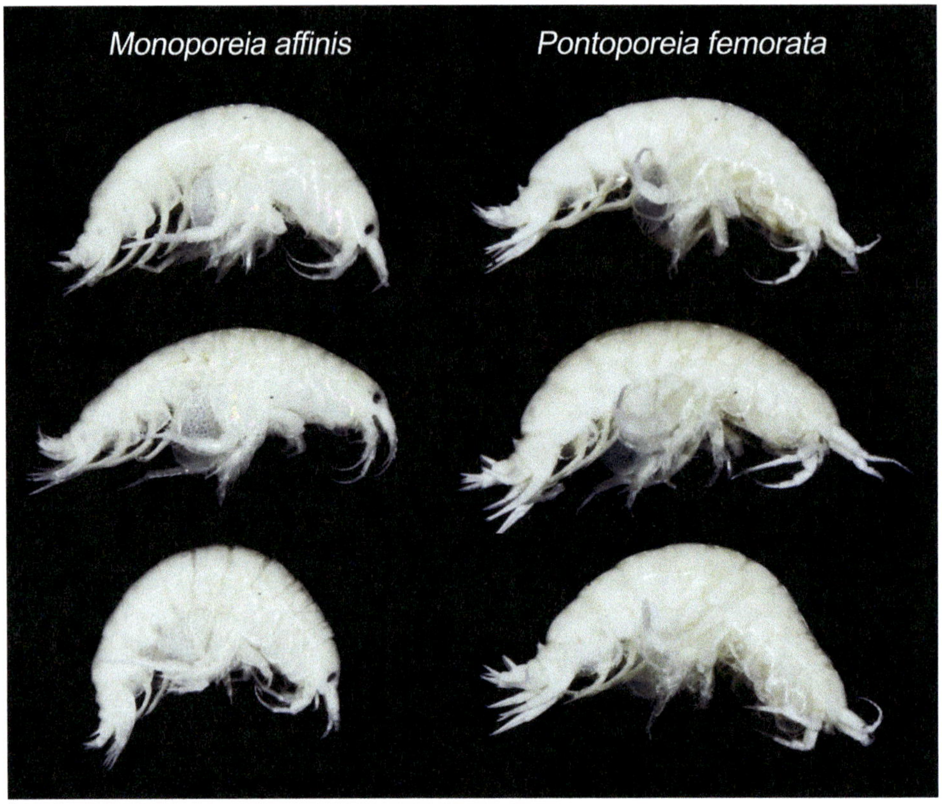

Fig. 4.30 Two abundant benthic amphipods in the deep-bottom habitats of the Baltic Sea, the Arctic relict *Monoporeia affinis* that lives in the Baltic Sea (from the Arkona Sea to the Bothnian Bay), the Arctic Ocean and cold Fennoscandian lakes, and the marine species *Pontoporeia femorata* that lives both in the Baltic Sea and the transition zone. *Monoporeia affinis* lives closer to the sediment surface and *Pontoporeia femorata* deeper down in the sediments (Hill and Elmgren 1987). The species can be distinguished by *e.g.* the colour of their eyes: black in *Monoporeia affinis* and light-red in *Pontoporeia femorata*. In the photographed ethanol-conserved material the red eye-colour of *Pontoporeia* has bleached. Photo: © Pauline Snoeijs-Leijonmalm

Fig. 4.31 The larvae of nonbiting midges (Chironomidae) are highly abundant in shallow bays of the Baltic Sea at salinity <6. Figure: © Robert Kautsky/Azote

performed by the species are important for the regulation of ecosystem processes, and these functions are determined by the biological traits of the species. Thus, by defining species according to their traits, *i.e.* body size, form, life-history, behavioural characteristics, etc., communities with different species can be compared through the common currency of their traits in a "biological traits analysis" (BTA). BTA uses multivariate ordination to describe patterns of biological trait composition over communities and ecosystems, *i.e.* the types of traits present and the relative frequency with which they occur (Bremner et al. 2006a, b).

The selection of biological traits to be used in BTA is important (*e.g.* Boxes 4.6 and 4.7). A wide variety of traits are available for describing ecological functioning, but they may not all be equally useful. Different traits can describe different aspects of ecological functioning, and some are intimately linked to particular functions, whereas others serve only as indirect indicators (Lavorel and Garnier 2002). Development of BTA must therefore also include an assessment of which traits provide the most useful description of ecological functioning so that the trait selection is optimised (Bremner et al. 2006a, b).

4.7.4 Traits analysis of microorganisms

For unicellular organisms, cell size is considered to be the "master trait". Cell size determines to a large extent how the organisms respond to changes in the environment by affecting several crucial ecological processes such as light harvesting, nutrient uptake, growth rate, and predator avoidance. Thus, in a wide perspective, cell size impacts carbon sequestration and energy transfer to higher trophic levels.

Along the Baltic Sea salinity gradient, diatom cell size increases with increasing salinity, apparently mediated by nutrient stoichiometry (Svensson et al. 2014). This is in accordance with fundamental differences in size distributions of marine and freshwater diatoms, with marine diatoms being significantly larger than freshwater species (Litchman et al. 2009). Cell sizes of phytoplankton and microphytobenthos determine the matter and energy flows to higher trophic levels and, hence, the structure and functioning of aquatic food webs (Smetacek 1999). For example, when a specific grazer selects food of a particular small size, an increase in food size can alter the food web.

4.7.5 Traits analysis of Baltic invertebrates

The decrease in species richness of the macrozoobenthos along the large-scale Baltic Sea gradient leads to an overall reduction in the number of functional groups (*cf.* Fig. 10.8). However, functional richness remains comparatively high even at the lowest level of taxonomic richness; a comprehensive study (Törnroos et al. 2015) found a decrease from 151 taxa and 50 trait categories in the transition zone to 6 taxa and the still high number of 33 trait categories in the Bothnian Bay. Primarily, the reduction in species richness altered trait categories related to feeding, life form and movement, and thus potentially could have an effect on various ecosystem processes. This highlights the importance of recognising ecosystem multifunctionality, especially at low taxonomic richness. A system can be species-poor but still rich in functions (Törnroos and Bonsdorff 2012).

Fig. 4.32 Large predators such as crabs and starfish are absent from the Baltic Sea, except for perhaps single specimens in the easternmost Arkona Sea. (**a**) The crab *Carcinus maenas* feeding on a clam in the transition zone. (**b**) The common starfish *Asterias rubens* feeding on the blue mussel *Mytilus trossulus* in the transition zone. (**c**) *Asterias rubens* feeding on the bivalve *Mya arenaria* in the transition zone. Photo: © Dirk Schories

4.7.6 Consequences of missing functions

The high abundances of a few marine filter-feeding bivalve species in the Baltic Sea (*Mytilus trossulus*, *Mya arenaria*) is explained by their extreme tolerance of low salinity, but also by the absence of predators (except for waterbirds such as the common eider). The limit of distribution regarding larger bivalve-consuming predators such as the starfish *Asterias rubens* and the crab *Carcinus maenas* (Fig. 4.32), both common species in the transition zone, is in the easternmost Arkona Sea where they live on their salinity margin. A predator of this significance is not found among the

Box 4.14: *r*- and *K*-strategies as extremes of the life-history continuum

Veijo Jormalainen

What are *r*- and *K*-strategists?

The grouping of species into *r*- and *K*-strategists represents a simplified description of their life-history characteristics. Their life-history traits are thought to have evolved either to maximise the population growth rate (*r*-strategists) or to allow the population to subsist at the carrying capacity of the environment (*K*-strategists). The designation of the strategies follows from the parameterisation of the logistic population growth model, with *r* defining the intrinsic population growth rate and *K* setting the carrying capacity of the environment (Box Fig. 4.22). Traits maximizing *r*, such as a high number of small-sized offspring, small body-size, early age of maturity, semelparity and short lifespan, characterise *r*-strategists. Accordingly, traits supporting existence under keen density-dependent competition, such as parental care, a small number of large-sized offspring, iteroparity and long lifespan, characterise *K*-strategists.

Extremes on the continuum of potential life-histories

Evolution of the strategies was attributed to *r*- and *K*-selection, forms of natural selection driving evolution at the two alternative directions. This selective dichotomy was put forward first by MacArthur and Wilson (1967) in their classic book on the Theory of Island Biogeography, particularly with respect to dispersal ability and differences in environmental stability and predictability. In quickly changing and unpredictable environments, traits supporting high population growth rate and dispersal ability are beneficial, and thus *r*-selection is assumed to predominate. On the other hand, in stable, predictable environments populations are limited close to their carrying capacity by intra- and interspecific competition, and *K*-selection should favour traits supporting high competitive ability. Pianka (1970) further explicitly attributed the *r*/*K* reasoning to the evolution of life-histories by predicting how life-history traits would evolve in response to the selection for high population growth rates, on the one hand, and in response to high population densities on the other (Reznick et al. 2002). Different kinds and often opposing selective pressures need to be compromised; the same traits do not confer fitness benefits in low- and high-density environments. However, Pianka (1970) has already pointed out that "certainly, no organism is completely "*r*-selected" or completely "*K*-selected", but all must reach some compromise between the two extremes". *r*- and *K*-strategies represent the extremes on the continuum of potential life-histories, setting the realms of possibilities for all kinds of strategies to evolve.

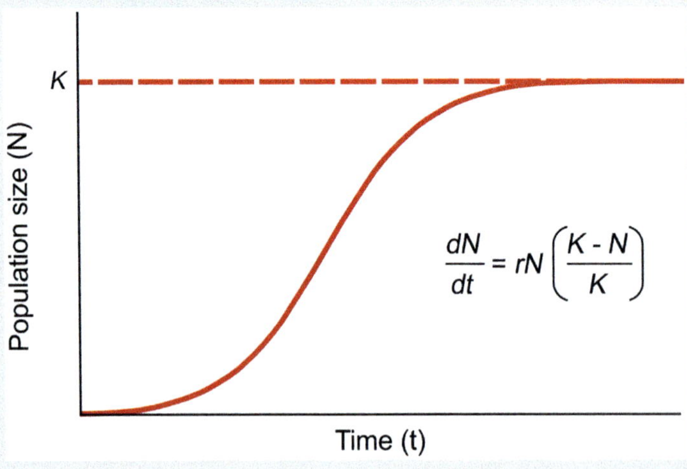

Box Fig. 4.22 The logistic equation and the corresponding population growth curve. Population size increases with the rate determined simultaneously by the intrinsic capacity to increase, *r*, and the carrying capacity of the environment, *K*. While the value of *K* sets the upper asymptote of the population size, the value of *r* determines how quickly the asymptote is approached.

Towards a more general life-history theory

This *r*/*K* selection theory was the first predictive model of life-history evolution. It has been influential in providing a heuristic context between the environment and life-history evolution as well as in stimulating life-history research. However, evidence showing that the theory was too simplistic had accumulated by the early 1980s. For example, the mortality regime among different life-history stages turned out to be important for the evolution of life-history traits such as the timing of maturity, reproductive effort and the number of offspring (Stearns 1977). Stearns further criticised the theory noting that while *r* can be defined as a function of life-history traits, the same is not true for the *K* that represents a composite parameter of a population, its resources and their interaction; therefore, there is no common currency for *r* and *K*. The *r*/*K* selection theory became replaced by a more general life-history theory (Stearns 1992) that incorporated demographic mortality patterns as causes for life-history trait evolution (Reznick et al. 2002). However, the themes in the heart of the *r*/*K* selection theory – environmental stability, resource availability and density dependent population regulation – have also remained important in the current life-history theory (Reznick et al. 2002).

Discrepancies of *r*/*K* dichotomy in the macrophytobenthos

The early proponents of *r*/*K* theory were zoologists, but the theory has been widely used to characterise both plant and animal life-histories. For example, in the macrophyte communities of the Baltic Sea, the annual or transient, opportunistic species, typically filamentous algae, can be described as *r*-strategists with high growth rates and efficient dispersal ability. On the other hand, perennial, habitat-forming species such as macroalgae and seagrasses resemble *K*-strategists with a long life-span and an ability to maintain dense populations under low nutrient availability. While the *r*/*K* dichotomy casts light on the differences in competitive strategies of these groups, it fails to explain many life-history traits. For example, strict semelparity is seldom found in filamentous algae. Their competitive ability outperforms that of perennial ones in nutrient-rich environments and perennial algae often produce huge amounts of small-sized recruits.

Other life-history theories

There are other approaches to describe the life-histories of plants. The growth-differentiation – balance hypothesis (Herms and Mattson 1992) is founded on the premise of competition, with herbivory being the major biotic factor affecting plant fitness and, thus, shaping plant life-histories. Owing to the fundamental physiological trade-off between growth (cell division and enlargement) and differentiation (cell maturation and specialisation), a continuum of strategies from growth-dominated to differentiation-dominated plant species evolves, depending on the relative importance of competition and herbivory. The traits predicted to be important for growth- and differentiation-dominated plants resemble those of *r*- and *K*-strategists, respectively (Herms and Mattson 1992). These strategies are related to another plant life-history classification, the three strategies suggested by Grime (1977). According to this theory, plant biomass is primarily limited by two factors: stress, defined as all conditions limiting production, and disturbance, including all abiotic and biotic factors destructing plant biomass. Variation in stress and disturbance set the stage for the evolution of three viable strategies: (1) competitive strategy, with high competitive ability and high growth-rate in low-stress and low-disturbance environments, (2) ruderal strategy, with fast growth and good colonisation ability in highly disturbed but potentially productive environments, (3) stress-tolerant strategy, with a slow growth rate, low palatability and long lifespan in stressful environments. The first two strategies correspond with the growth-dominated strategy and the third one with the differentiation-dominated strategy by Herms and Mattson (1992). Grime's three strategies also relate to the *r*/*K* continuum: the ruderal and stress-tolerant strategies correspond to the *r* and *K* extremes, respectively, while the competitive strategy remains somewhere in between. These newer approaches to plant life-histories are founded on the fundamental importance of growth rate and competitive ability, but differ from the *r*/*K* dichotomy by acknowledging explicitly the selective environment of plants (herbivory, stress) as well as the crucial role of environmental disturbance.

Box 4.15: The M74 syndrome

A reproductive disorder in the Baltic Sea salmon

For over 40 years the Atlantic salmon *Salmo salar* stocks in the Baltic Sea have suffered from a reproductive disorder, which is mainly manifested by high yolk-sac fry mortality (Box Fig. 4.23). This phenomenon was discovered in 1974 and named the "M74 syndrome", in which the "M" stands for the Swedish word "miljöbetingad" (environmentally-induced). When the M74 fry mortality peaked in the 1990s, on average >60 % of the wild salmon fry in Swedish hatcheries died (Box Fig. 4.24). After this, the incidence of M74 has fluctuated between 5 and 30 %, in 2012 it was negligible, but in 2016 it has increased again.

A similar problem in the Laurentian Great Lakes

In the Laurentian Great Lakes area of North America, there are similar reproductive disturbances in salmonid fish. These are the Early Mortality Syndrome and the Cajuga syndrome, which were found to be related to thiamine (vitamin B_1) deficiency. This could be explained by the introduction of the alewife *Alosa pseudoharengus*, a non-indigenous clupeid fish with high thiaminase activity that became a major prey for the salmonid fish in the Great Lakes (Honeyfield et al. 2005). However, in the Baltic Sea no new prey species for the salmon had been introduced when the M74 syndrome emerged, and its diet of mainly Atlantic herring *Clupea harengus* and European sprat *Sprattus sprattus* had remained the same (Hansson et al. 2001).

Thiamine deficiency and oxidative stress

Over the years, the research on M74 in the Baltic Sea has generated many hypotheses regarding what the cause of the syndrome might be, from population genetics to chemical pollution (Bengtsson et al. 1999). However, when it was discovered that the symptoms of the salmon fry can be alleviated by the administration of thiamine (Bylund and Lerche 1995), a thiamine-related explanation of the M74 syndrome seemed plausible. The M74-affected yolk-sac fry dies of thiamine (vitamin B_1) deficiency because the amount of thiamine that the female transfers to her eggs are too low for the fry to live through the yolk-sac stage (Amcoff et al. 1998). M74-affected brood fish, eggs and fry all show evidence of oxidative stress (Vouri and Nikinmaa 2007), including lowered levels of antioxidants (Pettersson and Lignell 1999) and increased levels of oxidised fatty acids (Pickova et al. 1998). M74-impacted salmon eggs can be identified visually by their pale colour, which is caused by low concentrations of the red

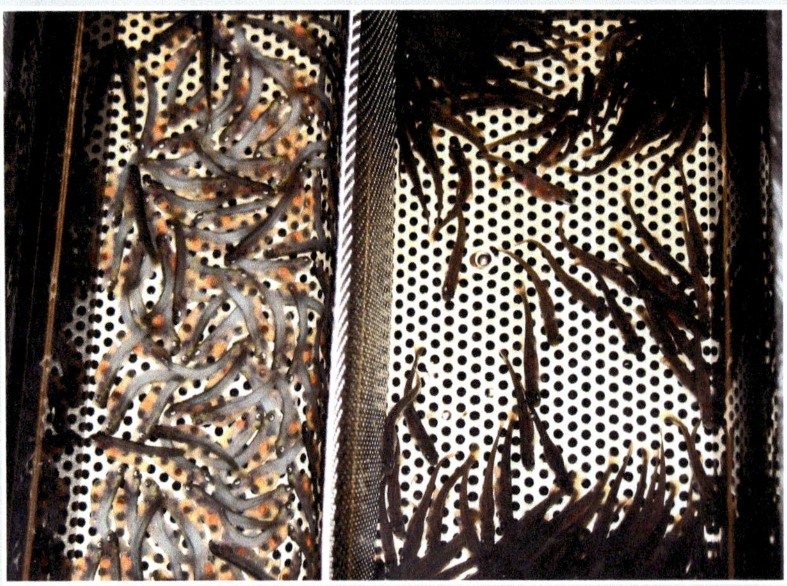

Box Fig. 4.23 Egg-yolk fry in the salmon hatchery in Älvkarleby (Sweden), showing healthy fry to the right and fry affected by the reproductive disorder M74 to the left. Symptoms shown by M74-affected fry are pale skin, retained yolk sack, slow growth, lethargy, lack of coordination and disturbed swimming patterns. Photo: © Ingrid Wänstrand carotenoid astaxanthin, a strong antioxidant (*cf.* Sect. 4.7.6). Salmon egg batches with an astaxanthin content of <0.15 μg egg^{-1} all resulted in M74 fry. Batches with >0.22 μg egg^{-1} resulted in healthy fry and batches with intermediate levels of astaxanthin (0.15–0.22 μg egg^{-1}) resulted in healthy or partially M74-affected fry batches (Pettersson and Lignell 1999).

Thiamine in the Baltic Sea food web

The thiamine and carotenoids in the salmon diet are produced by phytoplankton, and thiamine also by bacteria, at the base of the food web. The observed shifts in phytoplankton production and community composition in the Baltic Sea, as a result of large-scale changes in the environment such as eutrophication and climate warming, may result in decreased pelagic production of these compounds, especially when dinoflagellates and nanoflagellates dominate the phytoplankton communities (Sylvander 2013). However, the cyanobacterial blooms that also increase with eutrophication, as well as diatoms that are still abundant, are good producers of thiamine and the cause of M74 may also partly be found higher up in the food web. In a detailed study comparing the main prey species of the salmon that originate from the northern Gulf of Bothnia rivers – which are the herring and the sprat in the Baltic Sea proper and the Bothnian Sea – it was found that the thiamine concentration of both prey species is lowest in the youngest age groups (Keinänen et al. 2012). Because the average fat content and energy density are higher in sprat than in herring, with the highest being found in the youngest sprat, the supply of thiamine per unit of energy is least in a diet consisting of young sprat. Also, the greater the supply of fat from sprat in the southern Baltic Sea proper the preceding year, the lower the concentration of thiamine in salmon eggs. This study suggests that the thiamine deficiency in the salmon eggs results from an unbalanced diet that is abundant in fatty prey fish, such as young sprat, from which the supply of thiamine is insufficient in proportion to the supply of energy and unsaturated fatty acids, as salmon must undergo a long pre-spawning fasting period (Keinänen et al. 2012). This explanation is in agreement with the dramatic increase of the Baltic Sea proper's sprat stock when the M74 syndrome emerged (ICES 2012). Thus, the M74 syndrome with may be at least partly due to a higher proportional consumption of young sprat by the salmon.

Difficult to provide evidence for the cause of the M74 syndrome

Although it seems a plausible explanation, is not possible to prove that a larger consumption of young sprat is the (only) reason for the occurrence of M74. In the brackish Baltic Sea, the Atlantic salmon cannot obtain its optimal food, *i.e.* a high proportion of crustaceans (*cf.* Sect. 4.7.6) and may therefore suffer from low levels of several micronutrients such as vitamins and antioxidants, not only thiamine. This would explain its high general level of oxidative stress (Vuori and Nikinmaa 2007). Since the Atlantic salmon is already living on the edge of survival in the Baltic Sea, different types of additional stress may cause a tip-over. When living in nature the salmon is affected by many factors at the same time and it is therefore difficult to find hard proof of a single environmental or nutritional factor (natural or anthropogenic) that might cause the M74 syndrome.

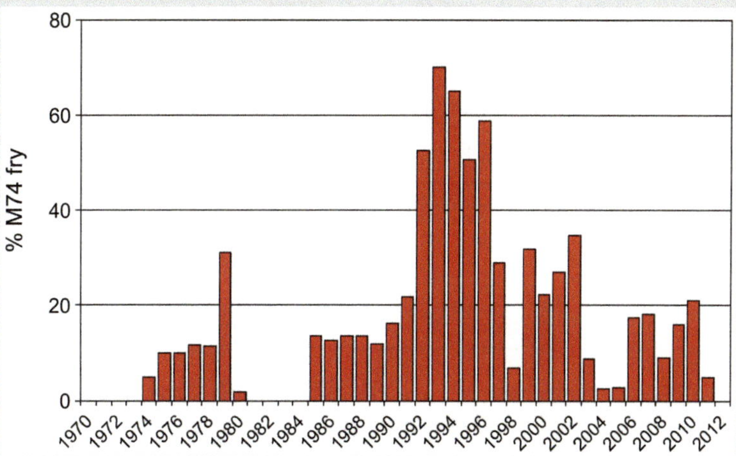

Box Fig. 4.24 The average proportion (%) of M74-affected batches from seven wild salmon stocks from the Baltic Sea in hatcheries. Figure based on data in Fiskhälsan (2007), updated until 2012

Box 4.16: The intermediate disturbance hypothesis

Ilppo Vuorinen

What is the intermediate disturbance hypothesis?
The intermediate disturbance hypothesis is a model, with its roots in community succession, that describes the relationship between disturbance and species diversity. The hypothesis states that species diversity is low at the lower end of a gradient of increasing disturbance (*e.g.* environmental stress, grazing pressure), highest at intermediate disturbance levels and low again at high disturbance levels (Box Fig. 4.25). Its main explanation is that more species can coexist at intermediate disturbance levels because disturbance reduces species' densities, thereby weakening competition and preventing the competitive exclusion that would otherwise occur.

A classic example of the development of a scientific idea
In a review called "The disturbing history of intermediate disturbance", the development of the intermediate hypothesis was summarised by Wilkinson (1999). By the 1940s, the prevalent idea held was that disturbance was in some way important in controlling species richness in succession. During the 1960s classic studies showed that disturbance through predation could maximise the coexistence of species. This led to a full statement of the intermediate disturbance hypothesis, complete with a hump-backed graphical model (Box Fig. 4.25), first by Grime (1973a, b), then Horn (1975) and finally Connell (1978). While Grime was the first to provide a model for the relationship, Horn was the first to explicitly state the hypothesis and Connell is generally cited in text books and journals as the founder of the hypothesis, although the credit should go to Grime. The idea behind the intermediate disturbance hypothesis is rather simple, and it could be argued that nature is more complicated than this. Furthermore, in practice the predicted hump-back model can be difficult to find in the field (Fox 2013). However, it is still generally accepted that competition-colonisation mechanisms can produce stable coexistence, and peaks in diversity at intermediate disturbance levels (Sheil and Burslem 2013).

Box Fig. 4.25 A generalised version of the hump-backed graphical model of the intermediate disturbance hypothesis. Figure modified from Wilkinson (1999)

non-indigenous species that have invaded the Baltic Sea either. Thus, the large filtration capacity of the Baltic Sea may be an advantage for the ecosystem, which is partly created by predator absence.

Another example of a missing function can explain the pale colour and low fat content of the muscle of the Atlantic salmon *Salmo salar* in the Baltic Sea (Nie et al. 2011). In fully marine areas, an adult salmon's food consists of up to 80 % of larger pelagic crustaceans (krill) such as *Meganyctiphanes norvegica* and *Thysanoessa inermis* (Jacobsen and Hansen 1996), but these are absent from the Baltic Sea due to the low salinity. The nectobenthic mysids *Mysis mixta* and *Mysis relicta* are small (1.5–2 cm) and have low abundances in the main feeding area of the salmon, the Baltic Sea proper, because of the lack of oxygenated deep water (Salemaa et al. 1990). Thus, because appropriate crustaceans are not available in the Baltic Sea proper, the salmon is forced to feed on monotonous food, *i.e.* the planktivorous clupeid fish herring and sprat. The carotenoid astaxanthin is the red pigment in crustaceans, salmon muscle and fish gonads. However, crustaceans mainly contain the trans-isomer of astaxanthin esterified to fatty acids while the astaxanthin of the clupeid gonads is the cis-isomer in its free form (Nie et al. 2011). Since the latter form of astaxanthin leaves the guts of the salmon with the faeces, *i.e.* without being taken up in the blood, the salmon obtains its astaxanthin mainly from copepods and cladocerans in the clupeids' stomachs. This explains the extremely low astaxanthin concentrations in the Baltic Sea salmon, which may have consequences for its general condition because astaxanthin is a strong antioxidant. For example, the occurrence of the M74 syndrome in the Baltic Sea salmon (Box 4.15) is correlated with the astaxanthin content in the eggs.

4.8 Diversity change and ecosystem resilience

4.8.1 Diversity, stability and resilience

The relationship between diversity and stability has fascinated ecologists for a long time. Before the 1970s, it was generally accepted that more diverse communities would enhance stability, *i.e.* the community would basically stay the same with only smaller fluctuations around a stable mean (McCann 2000). A fundamental problem in this context is that "stability" can have many different definitions (Ives and Carpenter 2007). The two classic theoretical ones are (1) the capacity of an ecosystem to withstand a disturbance without loss (then called "community resistance") and (2) the capacity to recover from a disturbance after incurring losses (then called "community resilience"). A more modern view is combining the two as resilience being the capacity of a community or ecosystem to respond to a perturbation or disturbance by resisting damage and recovering quickly, or in other words: resilience is the capacity of a system to continually change and adapt yet remain within critical thresholds (Folke et al. 2004, 2010).

The intermediate disturbance hypothesis predicts that more species can co-exist in an environment that is subject to intermediate disturbance levels (Box 4.16). An unstable estuarine environment characterised by salinity fluctuations (*e.g.* the surface waters of the transition zone, *cf.* Figs. 4.2 and 4.4) could be considered an intermediately disturbed environment where theoretically a large number of species could coexist. However, with respect to macroscopic species the opposite is the case: species richness is lower than in both marine and freshwaters. In the Baltic Sea, with a much more stable salinity than in the transition zone, macroscopic species diversity is considerably lower than in the transition zone, which would be in accordance with the intermediate disturbance hypothesis when compared to the transition zone as well as to freshwater.

4.8.2 Resilience of the Baltic Sea ecosystem

A much debated question is "How vulnerable is the Baltic Sea ecosystem to human-induced additional stress?" It has been suggested that the system should be less sensitive to additional stress because the dominant species are hardy, estuarine species that are the last to survive under polluted conditions in more diverse marine ecosystems (Jernelöv and Rosenberg 1976). On the other hand, the glacial relicts are sensitive to disturbance (Elmgren and Hill 1997), and the osmotic stress increases the susceptibility of Baltic Sea species to other perturbations (Tedengren and Kautsky 1986).

Despite its simpler food webs, the productivity of the Baltic Sea proper equals that of the more diverse North Sea. This seems to counter the assumption that diversity enhances productivity and stability in communities through a more efficient use of resources (Worm et al. 2006; Cardinale et al. 2006; Gross and Cardinale 2007). The Baltic Sea's case demonstrates that a high diversity is not a prerequisite of ecosystem functioning. On the other hand, as shown by *e.g.* the phytoplankton of the Baltic Sea, the efficiency of resource use, and thus carbon fixation, is directly linked to the diversity of phytoplankton communities (Ptacnik et al. 2008; Olli et al. 2014). This would suggest that the bacterial communities and protistan communities, which are not species-poor in the Baltic Sea proper, are able to uphold the productivity of the system.

However, with fewer species in each functional group of macroscopic organisms in the Baltic Sea, there is a higher risk that the loss or drastic reduction of a single key species, *e.g.* by a disease, may alter functions that are important for the maintenance of the ecosystem, such as the provision of habitats and resilience. The magnitude of these effects is ultimately determined by the identity of species that would become extinct and the geographical area of the extinction.

4.8.3 Consequences of human-induced diversity change

Human-induced diversity change, reduction in particular, is quoted as a major threat to ecosystems worldwide. The Baltic Sea provides ample proof that changes in dominance patterns can have serious ecosystem-wide consequences. For example, the reduction of the cod's role in the pelagic system through overfishing (*cf.* Sect. 17.6.4) and damage to the bladderwrack belts through eutrophication (*cf.* Sect. 11.15.3) may result in drastic changes at the ecosystem level (Zweifel and Laamanen 2009).

Even a diversity change in the phytoplankton of the Baltic Sea may have profound effects on ecosystem functioning. Two major changes that have been observed are a change in the dominance of the spring bloom from diatoms to dinoflagellates and increases in the magnitude of the summer bloom of cyanobacteria (*cf.* Sect. 8.2). Such changes in community composition may have negative effects on the quality of the phytoplankton as food for higher trophic levels with regard to *e.g.* vitamins and fatty acids (Ahlgren et al. 2005; Sylvander 2013). The spring-bloom change from diatoms to dinoflagellates also has a large effect on remineralisation. Diatoms sink to the seafloor fast, whereas dinoflagellates lyse before reaching the sediment (or produce resistant cysts that are not degraded in the sediment). Thus, the shift from diatoms to dinoflagellates increases the nutrient pools in the water column. Together with eutrophication this enhances the cyanobacterial blooms, which may release toxins into the water (*cf.* Box 16.4) and consume oxygen during the bacterial degradation of their large biomasses, causing oxygen depletion especially in deep water.

Review questions

1. Why is diversity something more than just the number of species?
2. How do insects add to the biodiversity and ecosystem functioning of the Baltic Sea?
3. What are the five most important key species in the Baltic Sea and why?
4. Give three examples of non-indigenous species that have enriched the biodiversity of the Baltic Sea in a positive way. What do they have in common?
5. What do you know about the glacial relict species living in the Baltic Sea? What species are they? Why are they here? Where do they live?

Discussion questions

1. Does it make any sense to learn how to identify species if we have barcodes, genomics and transcriptomics to identify them and to know what they are doing?
2. Why have protists and multicellular eukaryotes such different patterns of species richness in the Baltic Sea Area?
3. What would happen to the Baltic Sea ecosystem if the cod disappeared completely?
4. Are there any rare species living in the Baltic Sea that need to be protected in order to keep the ecosystem functioning? If so, why?
5. Would it be a good idea to artificially introduce more species to the Baltic Sea to raise biodiversity? If you were to decide which species to introduce, what type of species would you choose?

References

Ahlgren G, van Nieuwerburgh L, Wänstrand L, Peredsén M, Boberg M, Snoeijs P (2005) Imbalance of fatty acids in the base of the Baltic Sea food web – a mesocosm study. Canadian Journal of Fisheries and Aquatic Sciences 62:2240–2253

Amcoff P, Börjesson H, Eriksson R, Norrgren L, McDonald G, et al. (1998) Effects of thiamine treatments on survival of M74-affected feral Baltic salmon. In: McDonald G, Fitzsimons JD, Honeyfield DC (eds) Early life stage mortality syndrome in fishes of the Great Lakes and Baltic Sea. American Fisheries Society Symposium 21:31–40

Andersson M, Schubert H, Pedersén M, Snoeijs P (2006) Different patterns of carotenoid composition and photosynthesis acclimation in two tropical red algae. Marine Biology 149:653–665

Attrill MJ, Rundle SD (2002) Ecotone or ecocline: ecological boundaries in estuaries. Estuarine and Coastal Shelf Science 55:929–936

Audzijonyte A, Väinölä R (2005) Diversity and distributions of circumpolar fresh- and brackish-water *Mysis*. Hydrobiologia 544:89–141

Banks PB, Nordström M, Ahola M, Salo P, Fey K, Korpimäki E (2008) impacts of alien mink predation on island vertebrate communities of the Baltic Sea archipelago: review of a long- term experimental study. Boreal Environment Research 13:3–16

Barnes RSK (1974) Estuarine Biology. Edward Arnold Ltd., London, 76 pp

Barnes RSK (1989) What, if anything, is a brackish-water fauna? Transactions of the Royal Society of Edinburgh: Earth Sciences 80:235–240

Bengtsson BE, Hill C, Bergman Å, Brandt I, Johansson N et al (1999) Reproductive disturbances in Baltic fish: a synopsis of the FiRe Project. Ambio 28:2–8

Bergström L, Tatarenkov A, Johannesson K, Jonsson RB, Kautsky L (2005) Genetic and morphological identification of *Fucus radicans* sp. nov. (Fucales, Phaeophyceae) in the brackish Baltic Sea. Journal of Phycology 41:1025–1038

Bergström U, Sundblad G, Downie AL, Snickars M, Bostrom C, Lindegarth M (2013) Evaluating eutrophication management scenarios in the Baltic Sea using species distribution modelling. Journal of Applied Ecology 50:680–690

Blanco-Bercial L, Cornils A, Copley N, Bucklin A (2014) DNA barcoding of marine copepods: assessment of analytical approaches to species identification. PLoS Currents 6 [http://dx.doi.org/10.1371/currents.tol.cdf8b74881f87e3b01d56b43791626d2]

Blaxter M, Mann J, Chapman T, Thomas F, Whitton C et al (2005) Defining operational taxonomic units using DNA barcode data. Philosophical Transactions of the Royal Society B 360: 1935–1943

Bonaglia S, Nascimento FJA, Bartoli M, Klawonn I, Brüchert V (2014) Meiofauna increases bacterial denitrification in marine sediments. Nature Communications 5:5133

Bonsdorff E (2006) Zoobenthic diversity-gradients in the Baltic Sea: continuous post-glacial succession in a stressed ecosystem. Journal of Experimental Marine Biology and Ecology 333:383–391

Boström MK, Östman Ö, Bergenius MAJ, Lunneryd SG (2012) Cormorant diet in relation to temporal changes in fish communities. ICES Journal of Marine Science 69:175–183

Bregnballe T, Lynch J, Parz-Gollner R, Marion L, Volponi S et al (2014) Breeding numbers of great cormorants *Phalacrocorax carbo* in the Western Palearctic, 2012-2013. IUCN Wetlands International Cormorant Research Group Report. Scientific Report from DCE – Danish Centre for Environment and Energy 99:1–223

Bremner J, Rogers SI, Frid CLJ (2006a) Matching biological traits to environmental conditions in marine benthic ecosystems. Journal of Marine Systems 60:302–316

Bremner J, Rogers SI, Frid CLJ (2006b) Methods for describing ecological functioning of marine benthic assemblages using biological traits analysis (BTA). Ecological Indicators 6: 609–622

Brodniewicz I (1965) Recent and some Holocene Foraminifera of the southern Baltic Sea 10:131–248

Bylund G, Lerche O (1995) Thiamine therapy of M74 affected fry of Atlantic salmon *Salmo salar*. Bulletin of the European Association of Fish Pathology 15:93–97

Cardinale BJ, Srivastava DS, Duffy JE, Wright JP, Downing AL et al (2006) Effects of biodiversity on the functioning of trophic groups and ecosystems. Nature 443:989–992

Choo KS, Nilsson J, Pedersén M, Snoeijs P (2005) Photosynthesis, carbon uptake and antioxidant defence in two coexisting filamentous green algae under different stress conditions. Marine Ecology Progress series 292:127–138

Connell JH (1978) Diversity in tropical rain forests and coral reefs. Science 199:1302–1310

Creer S, Fonseca VG, Porazinska DL, Giblin-Davis RM, Sung W (2010) Ultrasequencing of the meiofaunal biosphere: practice, pitfalls and promises. Molecular Ecology 19:4–20

Dahl E (1956) Ecological salinity boundaries in poikilohaline waters. Oikos 7:1–21

Deaton LE, Greenberg MJ (1986) There is no horohalinicum. Estuaries 9:20–30

Dupont CL, Larsson J, Yooseph S, Ininbergs K, Goll J et al (2014) Functional trade-offs underpin salinity-driven divergence in microbial community composition. PLoS ONE 9(2):e89549

Elmgren R, Hill C (1997) Ecosystem function at low biodiversity – the Baltic example. In: Ormond R, Gage J, Angel M (eds) Marine biodiversity, patterns and processes. Cambridge University Press, Cambridge, pp 319–336

Eriksson BK, Sieben K, Eklöf J, Ljunggren L, Olsson J et al (2011) Effects of altered offshore food webs on coastal ecosystems emphasize the need for cross-ecosystem management. AMBIO 40:786–797

Esteban R, Moran JF, Becerril JM, García-Plazaola JI (2015) Versatility of carotenoids: an integrated view on diversity, evolution, functional roles and environmental interactions. Environmental and Experimental Botany 119:63–75

Fenchel T, Finlay B (2008) Oxygen and the spatial structure of microbial communities. Biological Review 83:553–569

Finenko ZZ (2008) The Black Sea Environment. In: Kostoanoy AG, Kosarev AN (eds) The handbook of environmental chemistry. Springer, Berlin, pp 351–374

Fiskhälsan (2007) Produktion av lax och havsöring baserad på vildfisk från Östersjön och Västerhavet – kontrollprogram för vissa smittsamma sjukdomar samt utfallet av M74. Fiskhälsan FH AB, Älvkarleby 10 pp

Folke C, Carpenter S, Walker B, Scheffer M, Chapin T, Rockström J (2010) Resilience thinking: integrating resilience, adaptability and transformability. Ecology and Society 15(4):20

Folke C, Carpenter S, Walker B, Scheffer M, Elmqvist T et al (2004) Regime shifts, resilience, and biodiversity in ecosystem management. Annual Review of Ecology, Evolution and Systematics 35:557–581

Fox JW (2013) The intermediate disturbance hypothesis should be abandoned. Trends in Ecology and Evolution 28:86–92

Frenzel P, Tech T, Bartholdy J (2005) Checklist and annotated bibliography of recent Foraminifera from the German Baltic Sea coast. Studia Geologica Polonica 124:67–86

Frenzel P, Keyser D, Viehberg F (2010) An illustrated key and (palaeo) ecological primer for postglacial to recent Ostracoda (Crustacea) of the Baltic Sea. Boreas 39:567–575

Giere O (2009) Meiobenthology – the microscopic motile fauna of aquatic sediments, 2nd edn. Springer, Berlin Heidelberg 527 pp

Grime JP (1973a) Competitive exclusion in herbaceous vegetation. Nature 242:344–347

Grime JP (1973b) Control of species density in herbaceous vegetation. Journal of Environmental Management 1:151–167

Grime JP (1977) Evidence for existence of three primary strategies in plants and its relevance to ecological and evolutionary theory. American Naturalist 111:1169–1194

Gross K, Cardinale BJ (2007) Does species richness drive community production or *vice versa*? – reconciling historical and contemporary paradigms in competitive communities. American Naturalist 170:207–220

Gupta BKS (1999) Modern foraminifera. Kluwer Academic Publishers, Dordrecht 384 pp

Haahtela I (1990) What do Baltic studies tell us about the isopod *Saduria entomon*? Annales Zoologici Fennici 27:269–278

Halinen K, Fewer DP, Sihvonen LM, Lyra C, Eronen E, Sivonen K (2008) Genetic diversity in strains of the genus *Anabaena* isolated from planktonic and benthic habitats of the Gulf of Finland (Baltic Sea). FEMS Microbiology Ecology 64:199–208

Hällfors G (2004) Checklist of Baltic Sea phytoplankton species (including some heterotrophic protistan groups). Baltic Sea Environment Proceedings 95:1–208

Hällfors G, Niemi Å (1975) Diatoms in surface sediment from deep basins in the Baltic Proper and the Gulf of Finland. Merentutkimuslaitos Julkaisu 240:71–77

Hansson S, Karlsson L, Ikonen E, Christensen O, Mitans A et al (2001) Stomach analyses of Baltic salmon from 1959–1962 and 1994–1997: possible relations between diet and yolk-sac-fry mortality (M74). Journal of Fish Biology 58:1730–1745

Harding and Härkönen (1999) Development in the Baltic grey seal (*Halichoerus grypus*) and ringed seal (*Phoca hispida*) populations during the 20th century. Ambio 28:619–627

Härkönen T, Galatius A, Bräeger S, Karlsson O, Ahola M (2013) Distribution of Baltic seals. HELCOM Core Indicator Report 2013, 34 pp [http://www.helcom.fi]

Härkönen T, Galatius A, Harding K, Karlsson O, Ahola M, Avellan L (2015) Distribution of Baltic seals. HELCOM Core Indicator Report, December 2015, 18 pp [http://www.helcom.fi]

Hebert PD, Cywinska A, Ball SL, DeWaard JR (2003) Biological identifications through DNA barcodes. Proceedings of the Royal Society B 270:313–321

Helander B, Bignert A, Herrmann C, Stjernberg T (2015) White-tailed eagle productivity. HELCOM Core Indicator Report. [http://www.helcom.fi]

HELCOM (2012a) Check-list of Baltic Sea macro-species. Baltic Sea Environment Proceedings 130:1–203

HELCOM (2012b) Indicator-based assessment of coastal fish community status in the Baltic Sea 2005–2009. Baltic Sea Environment Proceedings 131:1–88

Herlemann DPR, Labrenz M, Jürgens K, Bertilsson BS, Waniek JJ, Andersson AF (2011) Transitions in bacterial communities along the 2000 km salinity gradient of the Baltic Sea. The ISME Journal 5:1571–1579

Herms DA, Mattson WJ (1992) The dilemma of plants: to grow or defend? The Quarterly Review of Biology 67:283–335

Herrmann C, Bregnballe T, Larsson K, Rattiste K (2014) Population development of great commorant. HELCOM Baltic Sea Environment Fact Sheet. [http://www.helcom.fi]

Herrmann C, Krone O, Stjernberg T, Helander B (2011) Population development of Baltic bird species: white-tailed sea eagle (*Haliaeetus albicilla*). HELCOM Baltic Sea Environment Fact Sheet. [http://www.helcom.fi]

Herrmann C, Wendt J, Köppen U, Kralj J, Feige KD (2015) Changes in the migration pattern of the great cormorant *Phalacrocorax carbo sinensis* from the 1930s until today. Vogelwarte 53:139–154

Hill C, Elmgren R (1987) Vertical distribution in the sediment in the co-occurring amphipods *Pontoporeia affinis* and *P. femorata*. Oikos 49:221–229

Hillebrand H, Matthiessen B (2009) Biodiversity in a complex world: consolidation and progress in functional biodiversity research. Ecology Letters 12:1405–1419

Hoffmann C (1950) Über das Vorkommen endemischer Algen in der Ostsee. Kieler Meeresforschung 7:24–34

Höhnk W, Vallin S (1953) Epidemisches Absterben von *Eurytemora* im Bottnischen Meerbusen verursacht durch *Leptolegnia baltica* nov. spec. Veröffentlichungen des Institutes für Meeresforschung in Bremerhaven 2:215–223

Honeyfield DC, Hinterkopf JP, Fitzsimons JD, Tillitt DE, Zajicek JL, Brown SB (2005) Development of thiamine deficiencies and early mortality syndrome in lake trout by feeding experimental and feral fish diets containing thiaminase. Journal of Aquatic Animal Health 17:4–12

Hooper DU, Chapin FS, Ewel JJ, Hector A, Inchausti P et al (2005) Effects of biodiversity on ecosystem functioning: a consensus of current knowledge. Ecological Monographs 75:3–35

Horn HS (1975) Markovian properties of forest succession. In: Cody ML, Diamond JM (eds) Ecology and evolution of communities. Belknap Press, Cambridge, MA, pp 196–211

Hustedt F (1925) Bacillariales aus den Salzgewässern bei Oldesloe in Holstein. Mitteilungen der Geographischen Gesellschaft und des Naturhistorischen Museums in Lübeck 30:84–120

ICES (2012) Report of the Baltic salmon and trout assessment working group (WGBAST), 15–23 March 2012, Uppsala, Sweden. ICES CM 2012/ACOM:08. 347 pp

Ininbergs K, Bergman B, Larsson J, Ekman M (2015) Microbial metagenomics in the Baltic Sea: recent advancements and prospects for environmental monitoring. AMBIO 44(Supplement):S439–S450

Ives AR, Carpenter SR (2007) Stability and diversity of ecosystems. Science 317:58–62

Jacobsen JA, Hansen LP (1996) The food of Atlantic salmon, Salmo salar L., north of the Faroe Islands. Conference and Meeting Document M 10. Anadromous and Catadromous Fish Committee, ICES Annual Science Conference, 1996, Reykjavik, Iceland

Jeffrey SW, Vesk M (1997) Introduction to marine phytoplankton and their pigments. In: Jeffrey SW, Mantoura RFC, Wright SW (eds) Phytoplankton pigments in oceanography. Unesco Publishing, Paris, pp 37–84

Jernelöv A, Rosenberg R (1976) Stress tolerance of ecosystems. Environmental conservation 3:43–46

Jueg U, Zettler ML (2015) Distribution and ecology of leeches (Hirudinea) in brackish waters of the German Baltic. Ecologica Montenegrina 2:42–50

Kahl A (1928) Ciliata der Oldesloer Salzwasserquellen. Archiv für Hydrobiologie 19:50–123 [in German]

Karlsson O, Härkönen T, Bäcklin BM (2007) Sälar på uppgång. Havet 2007:84–89 [in Swedish]

Karpinsky M, Shiganova TA, Karunin DN (2005) The Caspian Sea Environment. In: Kostoanoy AG, Kosarev AN, The handbook of environmental chemistry. Springer, Berlin, pp 159-173

Kautsky H (1995) Quantitative distribution of sublittoral plant and animal communities in the Baltic Sea gradient. In: Eleftheriou A, Ansell AA, Smith CJ (eds) Biology and ecology of shallow coastal waters. Olsen and Olsen, Fredensborg, pp 23–30

Keinänen M, Uddström A, Mikkonen J, Casini M, Pönni J et al (2012) The thiamine deficiency syndrome M74, a reproductive disorder of Atlantic salmon (*Salmo salar*) feeding in the Baltic Sea, is related to the fat and thiamine content of prey fish. ICES Journal of Marine Science 69:516–528

Khlebovich VV (1968) Some peculiar feature of the hydrochemical regime and the fauna of mesohaline waters. Marine Biology 2:47–49

Kilpi M, Lorentsen SH, Petersen IK, Einarsson A (2015) Trends and drivers of change in diving ducks. TemaNord 2015:516, Nordic Council of Ministers, Copenhagen 2015

Kinne O (1971) Salinity: animals – invertebrates. In: Kinne O (ed) Marine ecology, vol 1., Environmental factors, Part 2. John Wiley and Sons, New York NY, pp 821–996

Kirchman DL (2008) Microbial diversity of the oceans, 2nd edn. John Wiley and Sons, Hoboken NJ 593 pp

Kotwicki L, Grzelak K, Bełdowski J (2016) Benthic communities in chemical munitions dumping site areas within the Baltic deeps with special focus on nematodes. Deep-Sea Research Part II: Topical Studies in Oceanography 128:123–130

Kruuk H (1995) Wild otters – predation and populations. Oxford University Press, Oxford, p 290

Kurvinen L, Kilpi M, Nordström M, Öst M (2016) Drivers of decline and changed nest-site preference of the Baltic eider: an island-level analysis from southwestern Finland. Ornis Fennica 93:55–66

Laamanen MJ, Forsström L, Sivonen K (2002) Diversity of *Aphanizomenon flos-aquae* (cyanobacterium) populations along a Baltic Sea salinity gradient. Applied and Environmental Microbiology 68:5296–5303

Laamanen MJ, Gugger MF, Lehtimäki JM, Haukka K, Sivonen K (2001) Diversity of toxic and nontoxic *Nodularia* isolates (Cyanobacteria) and filaments from the Baltic Sea. Applied and Environmental Microbiology 67:4638–4647

Laikre L, Palm S, Ryman N (2005) Genetic population structure of fishes: implications for coastal zone management. AMBIO 34:111–119

Laine AO (2003) Distribution of soft-bottom macrofauna in the deep open Baltic Sea in relation to environmental variability. Estuarine, Coastal and Shelf Science 57:87–97

Laine AO, Sandler H, Andersin AB, Stigzelius J (1997) Long-term changes of macrozoobenthos in the Eastern Gotland basin and the

Gulf of Finland (Baltic Sea) in relation to the hydrographical regime. Journal of Sea Research 38:135–159
Lakowitz CW (1929) Die Algenflora der gesamten Ostsee (ausschl. Diatomeen). R. Friedländer, Danzig, 474 pp [in German]
Lankau RA, Strauss SY (2007) Mutual feedbacks maintain both genetic and species diversity in a plant community. Science 317:1561–1563
Lavorel S, Garnier E (2002) Predicting changes in plant community composition and ecosystem functioning from plant traits: revisiting the Holy Grail. Functional Ecology 16:545–556
Leasi F, Norenburg JL (2014) The necessity of DNA taxonomy to reveal cryptic diversity and spatial distribution of meiofauna, with a focus on Nemertea. PLoS ONE 9(8):e104385
Litchman E, Klausmeier CA, Yoshiyma K (2009) Contrasting size evolution in marine and freshwater diatoms. Proceedings of the National Academy of Sciences of the USA 106:2665–2670
Lokko K (2014) Seasonal and spatial variability of zoopsammon communities in relation to environmental parameters. University of Tartu Press, Dissertationes Biologicae Universitatis Tartuensis 263:1–129 [PhD Thesis]
MacArthur R, Wilson EO (1967) The Theory of Island Biogeography (reprint 2001). Princeton University Press, 224 pp
Mackey MD, Mackey DJ, Higgins HW, Wright SW (1996) CHEMTAX – a program for estimating class abundances from chemical markers: application to HPLC measurements of phytoplankton. Marine Ecology Progress Series 144:265–283
Mann DG (1999) Crossing the Rubicon: the effectiveness of the marine/freshwater interface as a barrier to the migration of diatom germplasm. In: Mayama S, Idei M, Koizumi I (eds) Proceedings of the 14th International Diatom Symposium. Koeltz, Königstein, pp 1–21
Margulis L (ed) (1990) Handbook of Protoctista – the structure, cultivation, habitats, and life histories of the eukaryotic microorganisms and their descendants exclusive of animals, plants, and fungi: a guide to the algae, ciliates, foraminifera, sporozoa, water molds, slime molds, and the other protoctists. Jones and Bartlett Publishers, Boston, MA, 914 pp
McCann KS (2000) The diversity-stability debate. Nature 405:228–233
Mironova E, Telesh I, Skarlato S (2014) Ciliates in plankton of the Baltic Sea. Protistology 8:81–124
Misof B, Liu S, Meusemann K, Peters RS, Donath A et al (2014) Phylogenomics resolves the timing and pattern of insect evolution. Science 346:763–767
Modig H, van den Bund WJ, Olafsson E (2000) Uptake of phytodetritus by three ostracod species from the Baltic Sea: effects of amphipod disturbance and ostracod density. Marine Ecology Progress Series 202:125–134
Naeem S (2002) Ecosystem consequences of biodiversity loss: the evolution of a paradigm. Ecology 83:1537–1552
Naeem S, Thompson LJ, Lawler SP, Lawton JH, Woodfin RM (1994) Declining biodiversity can alter the performance of ecosystems. Nature 368:734–737
Nascimento FJA, Karlson AML, Elmgren R (2008) Settling blooms of filamentous cyanobacteria as food for meiofauna assemblages. Limnology and Oceanography 53:2636–2643
Nie XP, Zie J, Häubner N, Tallmark B, Snoeijs P (2011) Why Baltic herring and sprat are weak conduits for astaxanthin from zooplankton to piscivorous fish. Limnology and Oceanography 56:1155–1167
Nielsen R, Kristiansen A, Mathiesen L, Mathiesen H (1995) Distributional index of the benthic macroalgae of the Baltic Sea Area. Acta Botanica Fennica 155:1–51
Nordström M, Högmander J, Laine J, Nummelin J, Laanetu N, Korpimäki E (2003) Effects of feral mink removal on seabirds, waders and passerines on small islands in the Baltic Sea. Biological Conservation 109:359–368
Ojaveer H, Jaanus J, MacKenzie BR, Martin G, Olenin S et al (2010) Status of Biodiversity in the Baltic Sea. PLoS ONE 5(9):e12467
Ólafsson E, Elmgren R (1997) Seasonal dynamics of sublittoral meiobenthos in relation to phytoplankton sedimentation in the Baltic Sea. Estuarine and Coastal Shelf Science 45:149–164
Olenin S, Leppäkoski E (1999) Non-native animals in the Baltic Sea: alteration of benthic habitats in coastal inlets and lagoons. Hydrobiologia 393:233–243
Olli K, Ptacnik R, Andersen T, Trikk O, Klais R et al (2014) Against the tide: recent diversity increase enhances resource use in a coastal ecosystem. Limnology and Oceanography 59:267–274
Olsson J, Bergström L, Gårdmark A (2012) Abiotic drivers of coastal fish community change during four decades in the Baltic Sea. ICES Journal of Marine Science 69:961–970
Östman Ö, Boström MK, Bergström U, Andersson J, Lunneryd SG (2013) Estimating competition between wildlife and humans – a case of cormorants and coastal fisheries in the Baltic Sea. PloS ONE 8(12):e83763
Paavola M, Olenin S, Leppäkoski E (2005) Are invasive species most successful in habitats of low native species richness across European brackish water seas? Estuarine and Coastal Shelf Science 64:738–750
Paine RT (1995) A conversation on refining the concept of "keystone species". Conservation Biology 9:962–964
Pallo P, Widbom B, Ólafsson E (1998) A quantitative survey of the benthic meiofauna in the Gulf of Riga (eastern Baltic Sea), with special reference to the structure of nematode assemblages. Ophelia 49:117–139
Pawlowski J, Audic S, Adl S, Bass D, Belbahri L et al (2012) CBOL Protist Working Group: barcoding eukaryotic richness beyond the animal, plant, and fungal kingdoms. PLoS Biology 10(11): e1001419
Pedrós-Alio C (2012) The rare bacterial biosphere. Annual Review of Marine Science 4:449–466
Pettersson A, Lignell P (1999) Astaxanthin deficiency in eggs and fry of Baltic salmon (*Salmo salar*) with the M74 syndrome. Ambio 28:43–47
Pianka ER (1970) On *r* and *K* selection. American Naturalist 104: 592–597
Pickova J, Kiessling A, Pettersson A, Dutta PC (1998) Comparison of fatty acid composition and astaxanthin content in healthy and by M74 affected salmon eggs from three Swedish river stocks. Comparative Biochemistry and Physiology 120B:265–271
Postel L (1995) Zooplankton. In: Rheinheimer G (ed) Meereskunde der Ostsee, 2nd edn. Springer, Berlin, pp 150–160 [in German]
Potapova M (2011) Patterns of diatom distribution in relation to salinity. In: Seckback J, Kociolek JP (eds) The diatom world. Springer, Dordrecht, pp 313–332
Poulsen NC, Spector I, Spruck TP, Schultz TF, Wetherbee R (1999) Diatom gliding is the result of an actin-myosin motility system. Cell Motility and the Cytoskeleton 44:23–33
Ptacnik R, Olli K, Lehtinen S, Tamminen T, Andersen T (2011) Does plankton diversity peak at intermediate salinities? Comment on Telesh et al. (2011) Marine Ecology Progress Series 432:291–292
Ptacnik R, Solimini AG, Andersen T, Tamminen T, Brettum P et al (2008) Diversity predicts stability and resource use efficiency in natural phytoplankton communities. Proceedings of the National Academy of Sciences of the USA 105:5134–5138
Remane A (1934) Die Brackwasserfauna. Verhandlungen der Deutschen Zoologischen Gesellschaft 36:34–74 [in German]
Remane A (1958) Ökologie des Brackwassers. In: Remane A, Schlieper C (eds) Die Biologie des Brackwassers. Binnengewässer 22:1–216
Reusch TBH, Boström C, Stam WT, Olsen JL (1999) An ancient eelgrass clone in the Baltic. Marine Ecology Progress Series 183:301–304
Reznick D, Bryant MJ, Bashey F (2002) *r*- and *K*-selection revisited: the role of population regulation in life-history evolution. Ecology 83:1509–1520

Rokicka-Praxmajer J, Radziejewska T (2002) Free-living nematodes of the Pomeranian Bay (Southern Baltic). A preliminary analysis of nematode variability in an area affected by direct river runoff. Acta Scientiorum Polonorum. Series Piscaria 1:85–104

Round FE, Basson PW (1997) A new monoraphid genus (*Pogoneis*) from Bahrain and the transfer of previously described species *A. hungarica* and *A. taeniata* to new genera. Diatom Research 12:71–81

Round FE, Crawford RM, Mann DG (1990) The diatoms: biology and morphology of the genera. Cambridge University Press, Cambridge 747 pp

Round FE, Sims PA (1981) The distribution of diatom genera in marine and freshwater environments and some evolutionary considerations. In: Ross R (ed) Proceedings of the Sixth International Diatom Symposium. Koeltz, Königstein, pp 301–320

Russell G (1988) The seaweed flora of a young semi-enclosed sea, the Baltic – salinity as a possible agent of flora divergence. Helgoländer Meeresuntersuchungen 42:243–250

Salemaa H, Vuorinen I, Välipakka P (1990) The distribution and abundance of *Mysis* populations in the Baltic Sea. Annales Zoologici Fennici 27:253–257

Schlegel M, Hülsmann N (2007) Protists – a textbook example for a paraphyletic taxon. Organisms, Diversity and Evolution 7:166–172

Schönfeld J, Alve E, Geslin E, Jorissen F, Korsun S et al (2012) The FOBIMO (FOraminiferal BIo-MOnitoring) initiative – towards a standardised protocol for soft-bottom benthic foraminiferal monitoring studies. Marine Micropaleontology 94–95:1–13

Schönfeld J, Numberger L (2007) Seasonal dynamics and decadal changes of benthic foraminiferal assemblages in the western Baltic (NW Europe). Journal of Micropaleontology 26:47–60

Segerstråle SG (1962) The immigration and prehistory of the glacial relicts of Eurasia and North America – survey and discussion of modern views. Internationale Revue der gesamten Hydrobiologie 47:1–25

Shearer CA, Descals E, Kohlmeyer B, Kohlmeyer J, Marvanová L et al (2007) Fungal biodiversity in aquatic habitats. Biodiversity and Conservation 16:49–67

Sheil D, Burslem FRP (2013) Defining and defending Connell's intermediate disturbance hypothesis: a response to Fox. Trends in Ecology and Evolution 28:571

Simonsen R (1962) Untersuchungen zur Systematik und Ökologie der Bodendiatomeen der Westlichen Ostsee. Internationale Revue der gesamten Hydrobiologie, Systematische Beihefte 1:1–144

Sims PA, Mann DG, Medlin LK (2006) Evolution of diatoms: insights from fossil, biological and molecular data. Phycologia 45:361–402

Skov H, Heinänen S, Žydelis R, Bellebaum J, Bzoma S et al (2011) Waterbird populations and pressures in the Baltic Sea. TemaNord 2011:550. Nordic Council of Ministers, Copenhagen, p 201

Smetacek V (1999) Diatoms and the ocean carbon cycle. Protist 150:25–32

Snickars M, Weigel B, Bonsdorff E (2015) Impact of eutrophication and climate change on fish and zoobenthos in coastal waters of the Baltic Sea. Marine Biology 162:141–151

Snoeijs P, Mo K (1987) Macrofauna on rocky substrates in the Forsmark Biotest basin, March 1984-March 1985. Swedish Environmental Protection Board, Report 3397:1–55

Snoeijs P (1989) Effects of increasing water temperatures and flow rates on epilithic fauna in a cooling-water discharge basin. Journal of Applied Ecology 26:935–956

Snoeijs P (1995) Effects of salinity on epiphytic diatom communities on *Pylaiella littoralis* (Phaeophyceae) in the Baltic Sea. Ecoscience 2:382–394

Snoeijs P, Busse S, Potapova M (2002) The importance of diatom cell size in community analysis. Journal of Phycology 38:265–272

Snoeijs P, Häubner N (2014) Astaxanthin dynamics in Baltic Sea mesozooplankton communities. Journal of Sea Research 85: 131–143

Snoeijs P, Potapova M (1998) Ecotypes or endemic species? – a hypothesis on the evolution of *Diatoma* taxa (Bacillariophyta) in the northern Baltic Sea. Nova Hedwigia 67:303–348

Snoeijs P, Vilbaste S, Potapova M, Kasperoviciene J, Balashova J (1993–1998) Intercalibration and distribution of diatom species in the Baltic Sea, Volumes 1-5. Opulus Press, Uppsala, 645 pp

Snoeijs P, Wakuru-Murasi L (2004) Symbiosis between diatoms and cyanobacterial colonies. Vie Milieu 54:163–170

Snoeijs P, Weckström K (2010) Diatoms and environmental change in large brackish-water ecosystems. In: Smol JP, Stoermer EF (eds) The diatoms: applications for the environmental and earth sciences, 2nd edn. Cambridge University Press, Cambridge, pp 287–308

Stearns SC (1977) Evolution of life-history traits – critique of theory and a review of data. Annual Review of Ecology, Evolution and Systematics 8:145–171

Stearns SC (1992) The evolution of life histories. Oxford University Press, Oxford, UK 262 pp

Stein EC, Martinez MC, Stiles S, Miller PE, Zakharov EV (2014) Is DNA Barcoding actually cheaper and faster than traditional morphological methods: results from a survey of freshwater bioassessment efforts in the United States? PLoS ONE 9(4):95525

Stern RF, Horak A, Andrew RL, Coffroth MA, Andersen RA et al (2010) Environmental barcoding reveals massive dinoflagellate diversity in marine environments. PLoS ONE 5(11):e13991

Stock A, Jürgens K, Bunge J, Stoeck Y (2009) Protistan diversity in suboxic and anoxic waters of the Gotland deep (Baltic Sea) as revealed by 18S rRNA clone libraries. Aquatic Microbial Ecology 55:267–284

Stoecker DK, Michaels AE (1991) Respiration, photosynthesis and carbon metabolism in planktonic ciliates. Marine Biology 108:441–447

Sundblad G, Bergström U (2014) Shoreline development and degradation of coastal fish reproduction habitats. AMBIO 43:1020–1028

Sundblad G, Bergström U, Sandström A, Eklov P (2013) Nursery habitat availability limits adult stock sizes of predatory coastal fish. ICES Journal of Marine Science 71:672–680

Sundqvist L, Härkönen T, Svensson CJ, Harding KC (2012) Linking climate trends to population dynamics in the Baltic ringed seal – impacts of historical and future winter temperatures. Ambio 41:865–872

Svensson F, Norberg J, Snoeijs P (2014) Diatom cell size, coloniality and motility: trade-offs between temperature, salinity and nutrient supply with climate change. PLoS ONE 9(10):e109993

Sylvander P (2013) Thiamine dynamics in the pelagic food web of the Baltic Sea. Stockholm University, 35 pp [PhD Thesis]

Szulwiński M, Radziejewska T, Drgas A (2001) Trophic structure of free-living nematode assemblages along a southern Baltic transect. Folia Universitatis Agriculturae Stetinensis 28:141–150

Tatarenkov A, Bergström L, Jonsson RB, Serrão EA, Kautsky L et al (2005) Intriguing asexual life in marginal populations of the brown seaweed *Fucus vesiculosus*. Molecular Ecology 14:647–651

Taylor HR, Harris WE (2012) An emergent science on the brink of irrelevance: a review of the past 8 years of DNA barcoding. Molecular Ecology Resources 12:377–388

Tedengren M, Kautsky N (1986) Comparative study of the physiology and its probable effect on size in blue mussels (*Mytilus edulis* L.) from the North Sea and the northern Baltic Proper. Ophelia 25: 147–155

Telesh I, Postel L, Heerkloss R, Mironova E, Skarlato S (2009) Zooplankton of the open Baltic Sea: extended atlas. Leibniz Institute for Baltic Sea Research. Marine Science Reports 76: 1–290

Telesh I, Schubert H, Skarlato S (2011a) Revisiting Remane's concept: evidence for high plankton diversity and a protistan species

maximum in the horohalinicum of the Baltic Sea. Marine Ecology Progress Series 421:1–11
Telesh I, Schubert H, Skarlato S (2011b) Protistan diversity does peak in the horohalinicum of the Baltic Sea: reply to Ptacnik et al. (2011). Marine Ecology Progress Series 432:293–297
Törnroos A, Bonsdorff E (2012) Developing the multitrait concept for functional diversity: lessons from a system rich in functions but poor in species. Ecological Applications 22:2221–2236
Törnroos A, Bonsdorff E, Bremner J, Blomqvist M, Josefsson AB et al (2015) Marine benthic ecological functioning over decreasing taxonomic richness. Journal of Sea Research 98:49–56
Tuomisto H (2010a) A diversity of beta diversities: straightening up a concept gone awry. Part 1. Defining beta diversity as a function of alpha and gamma diversity. Ecography 33:2–22
Tuomisto H (2010b) A diversity of beta diversities: straightening up a concept gone awry. Part 2. Quantifying beta diversity and related phenomena. Ecography 33:23–45
Tuomisto H (2010c) A consistent terminology for quantifying species diversity? Yes, it does exist. Oecologia 4:853–860
Ulanova A, Busse S, Snoeijs P (2009) Coastal diatom-environment relationships in the brackish Baltic Sea. Journal of Phycology 45:54–68
Urban-Malinga B, Hedtkamp SIC, van Beusekom JEE, Wiktor J, Węsławski JM (2006) Comparison of nematode communities in Baltic and North Sea sublittoral, permeable sands – diversity and environmental control. Estuarine, Coastal and Shelf Science 70: 224–238
Ursi S, Pedersén M, Plastino E, Snoeijs P (2003) Intraspecific variation of photosynthesis, respiration and photoprotective carotenoids in *Gracilaria birdiae* (Gracilariales: Rhodophyta). Marine Biology 142:997–1007
van den Hoek C, Mann DG, Jahns HM (1995) Algae: an introduction to phycology. Cambridge University Press, Cambridge 623 pp
Vouri KAM, Nikinmaa M (2007) M74 syndrome in Baltic salmon and the possible role of oxidative stresses in its development: present knowledge and perspective for future studies. Ambio 36:168–172
Waltho J, Coulson J (2015) The common eider. T and A D Poyser, London, 352 pp
Wänstrand I, Snoeijs P (2005) Phytoplankton community dynamics assessed by ships-of-opportunity sampling in the northern Baltic Sea: a comparison of HPLC pigment analysis and cell counts. Estuarine, Coastal and Shelf Science 66:135–146
Ward RD, Hanner R, Hebert PD (2009) The campaign to DNA barcode all fishes, FISH-BOL. Journal of Fish Biology 74:329–356
Wendker S (1990) Untersuchungen zur subfossilen und rezenten Diatomeenflora des Schlei-Ästuars (Ostsee). Bibliotheca Diatomologica 20:1–268
Whitfield AK, Elliott M, Basset A, Blaber SJM, West RJ (2012) Paradigms in estuarine ecology – a review of the Remane diagram with a suggested revised model for estuaries. Estuarine and Coast Shelf Science 97:78–90
Whittaker RH (1960) Vegetation of the Siskiyou Mountains, Oregon and California. Ecological Monographs 30:279–338
Whittaker RH (1972) Evolution and measurement of species diversity. Taxon 21:213–251
Wieser W (1953) Die Beziehung zwischen Mundhöhlengestalt, Ernahrungsweise und vorkommen bei freilebenden marinen Nematoden. Arkiv för Zoologi 2:439–484 [in German]
Wilkinson DM (1999) The disturbing history of intermediate disturbance. Oikos 84:145–147
Worm B, Barbier EB, Beaumont N, Duffy JE, Folke C et al (2006) Impacts of biodiversity loss on ocean ecosystem services. Science 314:787–790
Wulff F, Sokolov A, Savchuk (2013) Nest – a decision support system for management of the Baltic Sea, a user manual. Baltic Nest Institute, Stockholm University Baltic Sea Centre, Technical Report 10:1–70
Zander CD, Reimer LW (2002) Parasitism at the ecosystem level in the Baltic Sea. Parasitology 124:S119–S135
Zettler ML, Karlsson A, Kontula T, Gruszka P, Laine AO et al (2014) Biodiversity gradient in the Baltic Sea: a comprehensive inventory of macrozoobenthos data. Helgoland Marine Research 68:49–57
Zettler ML, Schiedek D, Bobertz B (2007) Benthic biodiversity indices *versus* salinity gradient in the southern Baltic Sea. Marine Pollution Bulletin 55:258–270
Zweifel UL, Laamanen M (eds) (2009) Biodiversity in the Baltic Sea – an integrated thematic assessment on biodiversity and nature conservation in the Baltic Sea. Baltic Sea Environment Proceedings 116B:1–188

Biological invasions

5

Sergej Olenin, Stephan Gollasch, Maiju Lehtiniemi, Mariusz Sapota, and Anastasija Zaiko

Abstract

1. The term "non-indigenous species" (NIS) represents a biogeographical category, which indicates human involvement in the introduction of a certain species to a particular ecosystem and has nothing to do with putting "good" or "bad" tags on these species.
2. A biological invasion is the spread of a NIS or a cryptogenic species (of uncertain or unknown origin) to an area where it did not previously occur.
3. About 130 NIS and cryptogenic species have been introduced to the Baltic Sea Area by anthropogenic activities.
4. Most NIS have arrived to the Baltic Sea during recent decades due to intensification of global trade, human mobility and removal of custom barriers, although the first introductions are thought to have taken place already centuries ago.
5. The NIS in the Baltic Sea mainly originate from the coastal waters of three source areas (the North American east coast, the Ponto-Caspian region and East Asia), which are connected to the Baltic Sea by a number of introduction pathways, such as shipping and human-made canals.
6. In the Baltic Sea, NIS are represented by many taxonomic groups, from unicellular plankton organisms to crustaceans, molluscs, fish, waterbirds and mammals.
7. Many of the NIS in the Baltic Sea have increased functional diversity, bringing new and unusual functions to the species-poor Baltic Sea ecosystem.
8. Some NIS may spread, highly increase in abundance and cause an adverse impact on biological diversity, ecosystem functioning, socio-economic values and/or human health. These NIS are called "invasive alien species".
9. As it cannot be predicted which NIS will become invasive and cause harm in a particular ecosystem, a precautionary approach, preventing the arrival of new NIS in general, is advisable.

Keywords

Biodiversity • Functional diversity • Human-mediated introductions • Invasive alien species • Non-indigenous species

S. Olenin (✉) · A. Zaiko
Marine Science and Technology Center, Klaipeda University, Herkaus Manto 84, 5808 Klaipeda, Lithuania
e-mail: sergej.olenin@jmtc.ku.lt

A. Zaiko
e-mail: anastasija.zaiko@jmtc.ku.lt

S. Gollasch
Gollasch Consulting (GoConsult), Grosse Brunnenstraße 61, 22763 Hamburg, Germany
e-mail: sgollasch@gmx.net

M. Lehtiniemi
Marine Research Centre, Finnish Environment Institute, P.O. Box 140, 00251 Helsinki, Finland
e-mail: maiju.lehtiniemi@ymparisto.fi

M. Sapota
Institute of Oceanography, University of Gdańsk, Al. M. Piłsudskiego 46, 81378 Gdynia, Poland
e-mail: ocems@univ.gda.pl

P. Snoeijs-Leijonmalm et al. (eds.), *Biological Oceanography of the Baltic Sea*,
DOI 10.1007/978-94-007-0668-2_5

5.1 Who is "alien" and who is not?

5.1.1 Non-indigenous species and cryptogenic species

Generally, the term "non-indigenous species" (NIS) is used for a species that through human interference has been moved from its native dispersal range to a new area (Box 5.1). Synonyms used for NIS are "alien", "exotic", "non-native", "allochthonous" and "introduced" species. However, the true native area of a species may be uncertain or unknown, especially for unicellular organisms, and therefore they cannot be classified as either indigenous or non-indigenous. Such species are called "cryptogenic species" (Carlton 1996).

Vagrant species, such as fish or planktonic organisms with a high dispersal capacity, may spread to areas outside their normal reproductive range by natural phenomena such as currents. This can even happen on a continuous basis and such species have often been mistakenly depicted as NIS. Moreover, climate change alters species distributions (Parmesan and Yohe 2003). However, fluctuations in distributional ranges due to gradual changes in temperature or ocean currents do not qualify a species to be a NIS either.

About 130 NIS and cryptogenic species have been introduced to the Baltic Sea Area by anthropogenic activities. The list of Baltic Sea NIS changes continuously. Mostly species are added when new introductions occur, but sometimes species are deleted from the list as well, *e.g.* when it appears that a species has been misidentified. The AquaNIS information system always provides the latest updated list of NIS in the Baltic Sea (AquaNIS 2015).

5.1.2 Biological invasions and invasive species

Biological invasions (bioinvasions), in the broad sense, are the movements of organisms to areas where they did not previously occur. This includes natural slow gradual spread and natural rapid expansion due to unusual geological or climatic events. Such invasions took place well before anthropogenic activities began to play any notable role in changing the biogeography of the marine realm (*cf.* Sect. 4.6.2).

However, the strong increase of biological invasions in recent decades is almost entirely caused by the spread of NIS due to anthropogenic activities. This is why the term "biological invasions" is increasingly used to designate the spread of NIS and cryptogenic species to areas where they did not previously occur, and this is also how we use the term here (Box 5.1). The term "invasive alien species" (IAS) is reserved for non-indigenous and cryptogenic species that have an adverse effect on biological diversity, ecosystem functioning, socio-economic values and/or human health.

Invasion biology is a complex interdisciplinary scientific research area, which involves both fundamental and applied aspects (Fig. 5.1). Biological invasions offer a unique opportunity to study fundamental processes in population,

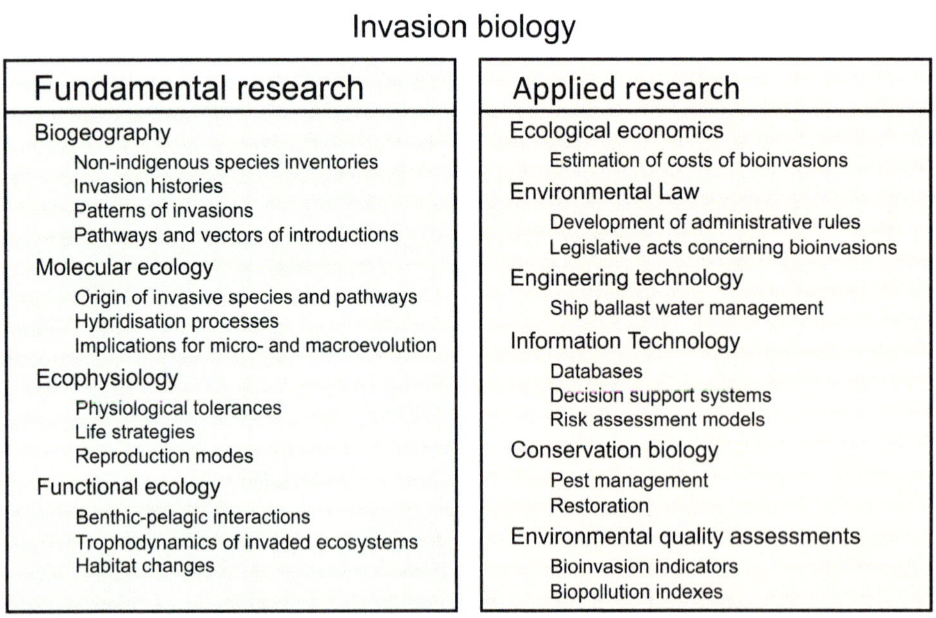

Fig. 5.1 Examples of fundamental and applied reseach directions in the rapidly growing interdisciplinary research field of invasion biology. Figure: © Sergej Olenin

community, ecosystem and evolution across taxonomic groups. Many ecological disciplines perform bioinvasion studies. Population ecologists investigate the dynamics of NIS populations, their arrival, establishment, expansion and/or decrease. Community ecologists study the interactions between NIS and long-time residents, including the effects of the invaders on the diversity of communities and the responses of communities to the invaders. These interactions have implications for invasion resistance and ecosystem resilience. Invasions are put in a societal context in more applied disciplines like conservation biology, restoration ecology and pest management.

Since invasion biology is a fast-developing discipline, its terminology also evolves rapidly and is influenced by a variety of sometimes contrasting academic, cultural and even political views (Elliott 2003; Carlton 2009; Olenin et al. 2010). The key definitions are centred around the opposing categories "native *versus* non-native", "spreading *versus* non-spreading", "impacting *versus* non-impacting" and "harmful *versus* harmless" (Box 5.1).

5.1.3 Continuous net immigration to the Baltic Sea

Like in other areas, the species pool of the Baltic Sea is a product of immigration and extinction. Being a young and environmentally unstable sea in a geological time perspective, the Baltic Sea has been exposed to several waves of species immigrations and extinctions during its respective geological stages (*cf.* Sect. 2.5). The current geological stage of the Baltic Sea is named the "Mya" stage, after the sand gaper *Mya arenaria* (Box 5.2), which occurs abundantly in the Baltic Sea sandy habitats. After it was recognised that *Mya arenaria* is in fact a cryptogenic species, probably introduced to the Baltic Sea by humans a long time ago (Strasser 1999), the name of this geological stage seems even more appropriate; it not only illustrates a dominant species in the Baltic Sea ecosystem but also an era of human-induced changes in biodiversity.

Most species living at present in the Baltic Sea are post-glacial immigrants that have extended their native range from adjacent marine or freshwater regions (*cf.* Sect. 4.6.2). It is commonly accepted to consider these species as native to the Baltic Sea. However, there is a continuous net immigration of species into the Baltic Sea through both natural dispersal and human-mediated introduction of species, and that is why scientists sometimes call the Baltic Sea "a sea of invaders" (Leppäkoski et al. 2002a). The rate of new arrivals has greatly increased in recent decades due to the intensification of global trade, human mobility and removal of former custom barriers.

5.1.4 Are non-indigenous species "good" or "bad"?

Categorising species into "indigenous" and "non-ndigenous" has nothing to do with putting "good" or "bad" tags on them. The term "non-indigenous" represents a biogeographical category, which indicates human involvement in the introduction of a certain species to a particular ecosystem. Most NIS do not cause harm to the biological diversity and ecosystem functioning of the Baltic Sea, nor to socio-economic values or human health associated with it, although for a large number of NIS in the Baltic Sea their impacts are still unknown (Ojaveer and Kotta 2015).

Adding a new NIS increases species richness, yet an invaded ecosystem loses its biogeographical peculiarities (Leppäkoski and Olenin 2001). From the beginning of the 19th century and up to the 1970s, intentional species introductions were in many European countries a popular measure to "improve nature" or compensate for destroyed stocks of native species (Leppäkoski et al. 2002b). Some of these acclimatisation experiments have had commercial success, *e.g.* by increased shellfish production, whereas others have caused devastating effects on local fauna and economic losses (Westman 2002). In recent decades our knowledge on ecosystem functioning has increased, and it is recognised that it is advisable to abstain from intentional introductions because of their often unpredictable consequences.

Since the brackish Baltic Sea is a species-poor system, at least at the level of multicellular organisms (*cf.* Fig. 4.10), its functional diversity is also low. Some of the NIS are known to bring new functions and increase the functional diversity of the Baltic Sea ecosystem, contributing to the circulation of nutrients and fostering the energy flow from the pelagic to the benthic system (Olenin and Leppäkoski 1999; Kotta et al. 2003; Norkko et al. 2012). However, faster turnover of nutrients does not *a priori* mean that an ecosystem becomes "better" and that a NIS that contributes with a new function is "good". The same NIS can outcompete native species, release extra nutrients and/or chemical pollutants from bottom sediments into the water column, or cause economic losses (Leppäkoski 2002; Gren et al. 2009).

There are no "ultimately good or bad" species or functions. When making evaluations, we should always ask the question "good" for what or "bad" for what? In general, we should avoid judgment of NIS in a moralistic context, but rather objectively assess their impacts and role in the ecosystem on a case-by-case basis.

Finally, there are many documented examples worldwide showing that some NIS may become pests and pose serious threats to biodiversity, the economy and even human health. This is why bioinvasions remain high on the environmental conservation agenda and are the subject of

Box 5.1: Key definitions in invasion biology

Non-indigenous species (NIS) is the preferred term used for species, subspecies or lower taxa (such as a variety or form), which are introduced outside of their past or present natural range and outside of their natural dispersal potential (Olenin et al. 2010). This includes any propagule of a NIS, such as a gamete, seed or resting stage, a gravid female, a pair of individuals of different sexes (in species with sexual reproduction) or a vegetative reproductive organ or section of tissue (in species with asexual reproduction), which might survive, reproduce and subsequently form a population. NIS also include hybrids between non-indigenous and native species, fertile polyploid organisms and artificially hybridised species, irrespective of their natural range or dispersal potential. The presence of a NIS in a given region is always due to intentional or unintentional introduction resulting from anthropogenic activities. Natural shifts in distributional ranges (*e.g.* due to climate change or dispersal by ocean currents) do not qualify a species to be a NIS. However, secondary spread of NIS from the area(s) of their first arrival may occur without human involvement due to spread by natural means.

Cryptogenic species are species of uncertain or unknown origin, *i.e.* it cannot be reliably demonstrated if they are introduced or native (Carlton 1996). For example, the true origin of a species may remain obscure because of insufficient taxonomic knowledge or due to a lack of records from the time before they were possibly introduced to a certain area. Unicellular organisms with wide global distributions are especially often considered as cryptogenic.

Biological invasion is the spread of a NIS or a cryptogenic species to an area where it did not previously occur.

Invasive alien species (IAS) is a commonly accepted term to indicate a subset of established NIS and/or cryptogenic species, which have spread, are spreading or have demonstrated their potential to spread elsewhere, and have an adverse impact on biological diversity, ecosystem functioning, socio-economic values and/or human health in the invaded region (Olenin et al. 2010). Sometimes the term "invasive species" is used as a synonym to all NIS. This is not correct, because only a small part of NIS may actually reach high abundances and cause harm. Also, the term IAS should not be used to indicate native species, which can reach high abundances and thereby cause the same type of adverse effects as IAS. Such native species are included in the definitions of "pests" and "outbreak-forming species".

Biological pollution is the adverse effects of IAS on the quality of the environment by impacts at several levels of biological organisation: an individual organism (*e.g.* internal biological pollution by parasites or pathogens), a population (*e.g.* hybridisation of native species with IAS or shifts in size/age structure due to predation by an IAS), a community (*e.g.* structural shifts such as replacement or total elimination of native species by IAS), a habitat (*e.g.* modification of physico-chemical conditions by IAS) or an ecosystem (*e.g.* changes in energy and organic matter flows caused by IAS). Commonly used synonyms of biological pollution are "biopollution", "biological invasion impact" and "bioinvasion impact". Biological pollution may also cause economic losses and impacts on human health.

Pests are harmful native, cryptogenic or non-indigenous species living in places where they are unwanted and have an adverse impact on biological diversity, ecosystem functioning, socio-economic values and/or human health in the invaded regions.

Outbreak-forming species (OFS) are native, cryptogenic or non-indigenous species with pulse-like, short-term (days to a few months) exponential population growth, during which they have an adverse impact on biological diversity, ecosystem functioning, socio-economic values and/or human health in invaded regions.

intensive research. Since it cannot be predicted which NIS will become invasive and cause harm in a particular ecosystem, a precautionary approach, preventing the arrival of new NIS in general, is advisable.

5.1.5 Which non-indigenous species have the largest impacts?

The NIS with the largest identified impacts on the Baltic Sea ecosystem are the polychaete worms *Marenzelleria* spp., the zebra mussel *Dreissena polymorpha*, the cladoceran *Cercopagis pengoi*, the amphipods *Gammarus tigrinus*, *Obesogammarus crassus* and *Pontogammarus robustoides* and the fish *Neogobius melanostomus* (Zaiko et al. 2011; Ojaveer and Kotta 2015). These species were introduced to the Baltic Sea in 1975–1992, except for *Dreissena polymorpha*, which has been present in the Baltic Sea already for some centuries (Leppäkoski et al. 2002a).

Fig. 5.2 The Ponto-Caspian zebra mussel *Dreissena polymorpha* is a notorious ecosystem engineer, capable of forming dense aggregates and mussel beds. It only needs a small patch of stable substrate to form an attached aggregate. The new habitat created by *Dreissena polymorpha* serves as an "island" of high biodiversity on the otherwise rather monotonous soft bottoms in low-salinity areas of the southeastern Baltic Sea and facilitates the establishment of other species. Photo: © Sergej Olenin

5.1.6 Impacts of *Marenzelleria* spp. and *Dreissena polymorpha*

The benthic invertebrates *Marenzelleria* spp. (Box 5.3) and *Dreissena polymorpha* (Box 5.4) alter the physical habitat, nutrient cycling and trophic interactions. *Marenzelleria* spp. can reach densities of >30,000 individuals m^{-2} and a biomass of >400 g wet weight m^{-2} (Zettler 1996). These non-indigenous worms compete with native species, *e.g.* they reduce the abundance of the native polychaete *Hediste diversicolor* and the native amphipod *Monoporeia affinis* (Kotta et al. 2006). Bioturbation by dense *Marenzelleria* spp. populations may lead to an enhanced release of soluble nutrients and hazardous substances from the sediments to the water column (Hedman et al. 2011). On the other hand, they may also aid in the recovery of oxygen conditions on the seafloor (Norkko et al. 2012).

Improved bottom-water oxygen conditions in the coastal areas of the northern Baltic Sea coincide with increased abundances of *Marenzelleria* spp., which bioturbate a relatively thick upper sediment layer. Using a reactive-transport model, Norkko et al. (2012) demonstrated that the long-term bioirrigation activities of high, but natural, abundances of *Marenzelleria* (>3,000 individuals m^{-2}) lead to a substantial increase in the iron-bound phosphorus content of sediments while reducing the concentration of labile organic carbon. In contrast to short-term laboratory experiments, the model simulations, which covered a 10-year period, showed that *Marenzelleria* has the potential to enhance long-term phosphorus retention in muddy sediments. This may facilitate the switch from a seasonally hypoxic system (<2 mL O_2 L^{-1}) back to a normoxic system by reducing the potential for sediment-induced eutrophication in the upper water column.

Dreissena polymorpha beds (Fig. 5.2) occupy ~300 km^2 in the part of the Curonian Lagoon that is directly exposed to the Nemunas river outflow, which is ~20 % of the lagoon's total bottom area (Daunys et al. 2006). In this area, the 14–20 mm body size class of *Dreissena polymorpha* dominates, the density varies from only 40 up to 57,000 individuals m^{-2} and

Box 5.2: The sand gaper *Mya arenaria*

Invasion history

Mya arenaria (Box Fig. 5.1) is a typical cryptogenic species, which cannot reliably be ascribed as introduced or native. Originating in the Pacific Ocean ~12 million years ago (middle Miocene), it was already present on the west and east coasts of the Atlantic Ocean in the Pliocene. However, in the eastern Atlantic Ocean it died out during the Pleistocene glaciations (Strasser 1999). From analyses of ancient kitchen middens and marine shell deposits from that period, it was concluded that *Mya arenaria* was not present in northwestern European coastal waters until ~500 years ago (Hessland 1946; Bernard 1979; Petersen et al. 1992, 2005). Later investigations have shown that the first shell deposits of *Mya arenaria* appeared on the Danish coast around the 13th century. Based on these observations, it is assumed that the species may have been transferred from the Atlantic coast of North America to Europe already before Columbus by the Vikings. *Mya arenaria* may have served as a food item during early ship travels crossing the Atlantic Ocean. Surprisingly, it is not exploited as a human food item in Europe, while it is highly valued as such in the USA and Canada. Nowadays, *Mya arenaria* inhabits the entire Baltic Sea, except for the Bothnian Bay and the eastern Gulf of Finland where salinity is too low for its survival.

Invasive traits

Mya arenaria possesses several traits that enable it to colonise new habitats, including high fecundity, pelagic larval development, rapid growth, and tolerance of a wide range of environmental conditions. An additional key feature of this species is its long life span; individuals older than 27 years have been observed. Despite these features, *Mya arenaria* is unexpectedly characterised by a relatively low level of genetic polymorphism compared to other marine bivalves, both in its native and introduced range (Strasser and Barber 2009).

Biology

Mya arenaria is an infaunal species, *i.e.* it lives buried in sediments beneath the surface of the seafloor. After burying itself, an individual will stay in the same place for the rest of its life. During its first year the burial depth is only 5–10 cm, but >10 years old *Mya arenaria* can live down to ~40 cm deep in the sediment. This mode of life provides an excellent defence against predators and also against freezing during severe winters. *Mya arenaria* has a high filtration capacity; one individual of 6–7 cm shell length can filtrate 1–10 L seawater h^{-1} (Jørgensen and Riisgård 1988; Riisgård and Seerup 2003). In the southern Baltic Sea, population filtration rates of *Mya arenaria* can be >8 m^3 m^{-2} day^{-1} (Forster and Zettler 2004). Water passes in and out of the clam through two siphons that reach to the sediment surface (Box Fig. 5.1b). Populations of *Mya arenaria* often consist of adult specimens only, which suggests that long episodes without recruitment of new individuals are common. Sexual maturity usually sets in when the oval shells reach a length of 2–5 cm (Strasser 1999). Fertilisation is external and larvae are pelagic and planktotrophic for 10–35 days. Mortality is up to 90 % during the first year, and even higher after mild winters when predation pressure on newly settled clams is high. It is, however, lower after cold winters when predation pressure is lower (Beukema 1982).

Box Fig. 5.1 The marine-brackish suspension feeder *Mya arenaria*. (**a**) Empty shells on a beach. (**b**) Water passes in and out of a clam through two tubes (siphons), which are fused to form a brownish-coloured, thick structure that is oval in cross-section. Photo: © Sergej Olenin

Impacts in the Baltic Sea
Being an ancient invader, *Mya arenaria* has already passed through all invasion stages (*cf.* Fig. 5.3), and is so well established in the Baltic Sea ecosystem that it is difficult to identify its impacts. However, when *Mya arenaria* invades a new area it still shows its invasive properties. For example, a salinity increase from ~9 to ~12 in the Ringkøbing Fjord on the Danish west coast in the 1990s caused a shift in the dominating pathway of organic matter production from pelagic turnover to benthic-pelagic coupling through new recruitment and growth of *Mya arenaria* (Petersen et al. 2008).

the biomass varies from 0.5 to 5,000 g shell-free dry weight m^{-2}. Similarly dense *Dreissena polymorpha* beds occur in other places with a large freshwater input, *e.g.* the inner Neva Estuary (Orlova et al. 2004). The presence of this species, which contributes up to 95 % of total community biomass, has caused an essential redistribution of the native benthic communities in the Curonian Lagoon (Zaiko et al. 2011).

5.1.7 Impacts of *Cercopagis pengoi*

The Ponto-Caspian fish-hook water flea *Cercopagis pengoi* (Box 5.5) can fundamentally change community composition and food web structure in the pelagic zone. It adds an extra trophic level to the food web as a zooplankton predator on smaller zooplankton. *Cercopagis pengoi* has caused a decline in the native cladocerans *Bosmina longispina maritima*, *Evadne nordmanni* and *Pleopis polyphemoides*, probably by direct predation (Ojaveer et al. 2004; Kotta et al. 2006). *Cercopagis pengoi* itself is used as a food source by several fish species. In late summer and early autumn, it can constitute a large proportion of the diets of the major planktivorous fish species of the Baltic Sea, *e.g.* the sticklebacks *Gasterosteus aculeatus* and *Pungitius pungitius*, the Atlantic herring *Clupea harengus* and the European sprat *Sprattus sprattus* (Gorokhova et al. 2004; Ojaveer et al. 2004). As *Cercopagis pengoi* tends to attach to fishing gear and clog nets and trawls, it may cause substantial economic losses for fishermen (Leppäkoski and Olenin 2000).

5.1.8 Impacts of non-indigenous amphipods

The North American amphipod *Gammarus tigrinus* and the Ponto-Caspian amphipod *Pontogammarus robustoides* are able to fundamentally change community composition in the phytobenthic zone. Especially in the lagoons of the southeastern Baltic Sea proper and the eastern Gulf of Finland, they outcompete their native relatives *Gammarus duebeni* and *Gammarus zaddachi* (*cf.* Sect. 11.4.3), probably because the introduced amphipods are more versatile feeders (Orlova et al. 2006). In many places, *Gammarus tigrinus* completely dominates the nektobenthos, *e.g.* in some sections of the Wisła Lagoon, where it has replaced not only the native species, but also the previously introduced amphipod *Pontogammarus robustoides* (Grabowski et al. 2006).

5.1.9 Impacts of *Neogobius melanostomus*

In the Gdańsk Bay, a trophic cascade has occurred as a result of the introduction of the Ponto-Caspian round goby *Neogobius melanostomus* (Box 5.6). The great cormorant *Phalacrocorax carbo sinensis*, a top predator, has shifted its diet from the European eel *Anguilla anguilla* and European sprat *Sprattus sprattus* to *Neogobius melanostomus*, which caused population increases in eel and sprat. In turn, sprat feeding has reduced zooplankton biomass and the subsequent reduced zooplankton grazing on the phytoplankton has caused an increase in the phytoplankton biomass (Corkum et al. 2004). The expanding populations of *Neogobius melanostomus* in the coastal areas of the southeastern Baltic Sea may have also reduced the blue mussel *Mytilus trossulus* population, since the non-indigenous fish preys upon these mussels (Karlson et al. 2007).

5.2 The invasion process

5.2.1 Dispersal from the source area

A human-mediated biological invasion process of a NIS includes several consecutive stages (Fig. 5.3). The process starts in a source (donor) area when a species interacts with an introduction pathway (*e.g.* shipping). The source area of a NIS in the Baltic Sea may be the native region of the species, *e.g.* the Caspian Sea for *Cercopagis pengoi*, but it can also be an area to which it has already been introduced. Examples of the latter case are the Pacific oyster *Crassostrea gigas* (Fig. 5.4) and the common slipper shell *Crepidula fornicata* (Fig. 5.5). Both these species were first introduced from their native areas to the North Sea (primary introduction), and from there they have later spread to the Baltic Sea (secondary introduction).

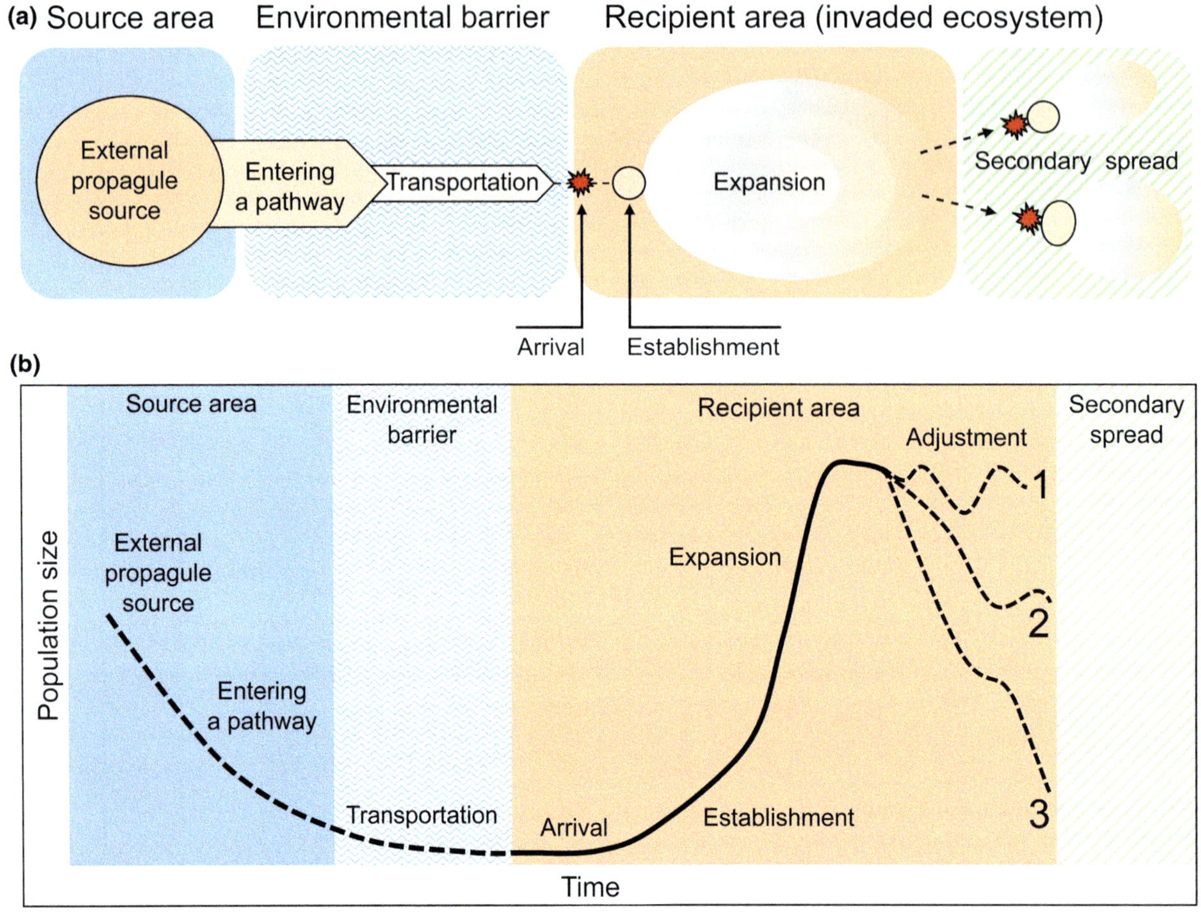

Fig. 5.3 The consecutive stages in a human-mediated primary introduction of a non-indigenous species (NIS), including direct dispersal from its native source area by crossing an environmental barrier with the help of a human-mediated vector, and its subsequent arrival and establishment in the recipient area, possibly followed by a secondary spread to other areas. In principle, a secondary spread follows the same stages as the primary introduction, but pathways may be human-mediated vectors and/or natural processes. (**a**) The dispersal of propagules from the source area to secondary spread. (**b**) The relative development of the population size during the different stages of (**a**). Population size usually declines during transportation, but during the expansion phase it is often (temporarily) higher in the recipient area than in the source area. During the adjustment phase three different scenarios are possible: 1 = the population size remains high, 2 = the population size declines to a lower level after which it remains more-or-less stable, 3 = the population becomes extinct. Figure modified from Olenin et al. (2011)

Fig. 5.4 The bivalve mollusc *Crassostrea gigas* (Pacific oyster) is a native species in estuarine and coastal marine waters in Japan and Southeast Asia. (**a**) Seven individuals, showing the sharp edges of the shells. (**b**) Open shells, showing the oyster's soft body parts, ready to eat. *Crassostrea gigas* has been introduced throughout the world for use in aquaculture because it grows fast and tolerates a wide range of environmental variation. In the Baltic Sea Area, this oyster is only found in the westernmost Arkona Sea, the Belt Sea and the Kattegat. Photo: © Pauline Snoeijs-Leijonmalm

Fig. 5.5 The oval shells of the North American common slipper shell *Crepidula fornicata* are up to 5 cm long and commonly build curved chains of up to 12 animals. The species is usually found attached to shells and stones on soft substrates in the upper littoral zone. *Crepidula fornicata* was first observed in Europe in 1872 on the west coast of Great Britain. In the Baltic Sea Area, it has only been found in the Belt Sea and the Kattegat, to where it probably arrived as a secondary spread from the North Sea. Photo: © Sergej Olenin

The number of species involved in a pathway is always higher than the number of species that manage to survive transportation over an environmental barrier (Fig. 5.3). With respect to primary introductions of NIS to the Baltic Sea, the environmental barriers between the Baltic Sea and the source areas (brackish or limnic systems elsewhere) are either land masses or vast open ocean spaces. Secondary introductions of NIS to the Baltic Sea may occur through human-mediated vectors but also by natural processes such as tidal movements, alongshore drift, flooding events, turbidity currents, and transfer by wind and animals. These natural processes may also greatly facilitate the dispersal of NIS within a recipient ecosystem.

In general, there are comparatively few primary introductions recorded in the Baltic Sea. Some examples include mainly Ponto-Caspian NIS such as *Cercopagis pengoi*, *Dreissena polymorpha* and *Neogobius melanostomus* (Fig. 5.6). Secondary introductions from both adjacent inland waters and the North Sea have historically been, and still are, more common than primary introductions.

5.2.2 Propagule pressure

The potential of a species to establish a stable population in an area where it previously did not occur is called "propagule pressure". Propagule pressure differs from settlement or recruitment because it represents the potential for an introduction, not the realised introduction (Johnston et al. 2009). For species with sexual reproduction, a propagule may be a gamete, a seed or a resting stage, a gravid female or a pair of individuals of different sex. For species with asexual reproduction, this may be a vegetative reproductive organ or a tissue section. NIS propagules also include hybrids between non-indigenous and indigenous species, fertile polyploid organisms and artificially hybridised species (Box 5.1).

The propagule pressure of a NIS can be calculated as the number of its propagules released into a region that they are not able to reach naturally (*i.e.* without transport by humans), multiplied by the number of discrete release events. When the number of propagules or the number of releases increases, the propagule pressure also increases. Thus, species that are constantly being introduced in large quantities are more likely to survive in the recipient area, provided they tolerate the environmental conditions in this area, and can utilise the habitats and energy resources present, whereas species introduced in small numbers with only one or a few release events are more likely not to establish.

However, one introduction on one occasion may be enough for a NIS to establish in the recipient area. There is a well-documented case of such a single introduction event from the Mediterranean Sea, where the green alga *Caulerpa taxifolia* began to spread rapidly after release from the Monaco aquarium (Jousson et al. 1998). For the Baltic Sea, such evident cases are not known with certainty, but several invasions may have started from single-event introductions. For example, the wedge clam *Rangia cuneata* (Box 5.7) was most likely transferred from the Belgian or Dutch waters of the North Sea by Dutch ladder-dredge boats, which were dredging the waterway of the harbour of Kaliningrad in 2008, although transfer by ship ballast water cannot be ruled out completely (Rudinskaya and Gusev 2012).

5.2.3 Arrival, establishment and expansion

The successful invasion of a NIS into a recipient area always begins with one or more incidences of arrival, followed by the establishment of a small group of reproducing individuals, which may proceed into an expansive phase (Fig. 5.3; Reise et al. 2006). During an expansion peak the impacts of a NIS on the recipient ecosystem are the strongest. Generally, the expansion of planktonic species (*e.g. Cercopagis pengoi* and *Prorocentrum cordatum*) is more rapid and covers larger areas than the expansion of nektobenthic crustaceans (*e.g. Hemimysis anomala*) or demersal fish (*e.g. Neogobius melanostomus*) (Fig. 5.6).

Some species enter the expansion phase almost immediately after arrival. An example of this is *Cercopagis pengoi*, which was first found in 1992 in the Gulfs of Finland and Riga. It spread rapidly to the Baltic Sea proper, and by 2002 this cladoceran reached the Gulf of Bothnia, including the Bothnian Bay in the north and by 2004 also the German coast in the south (Fig. 5.6a).

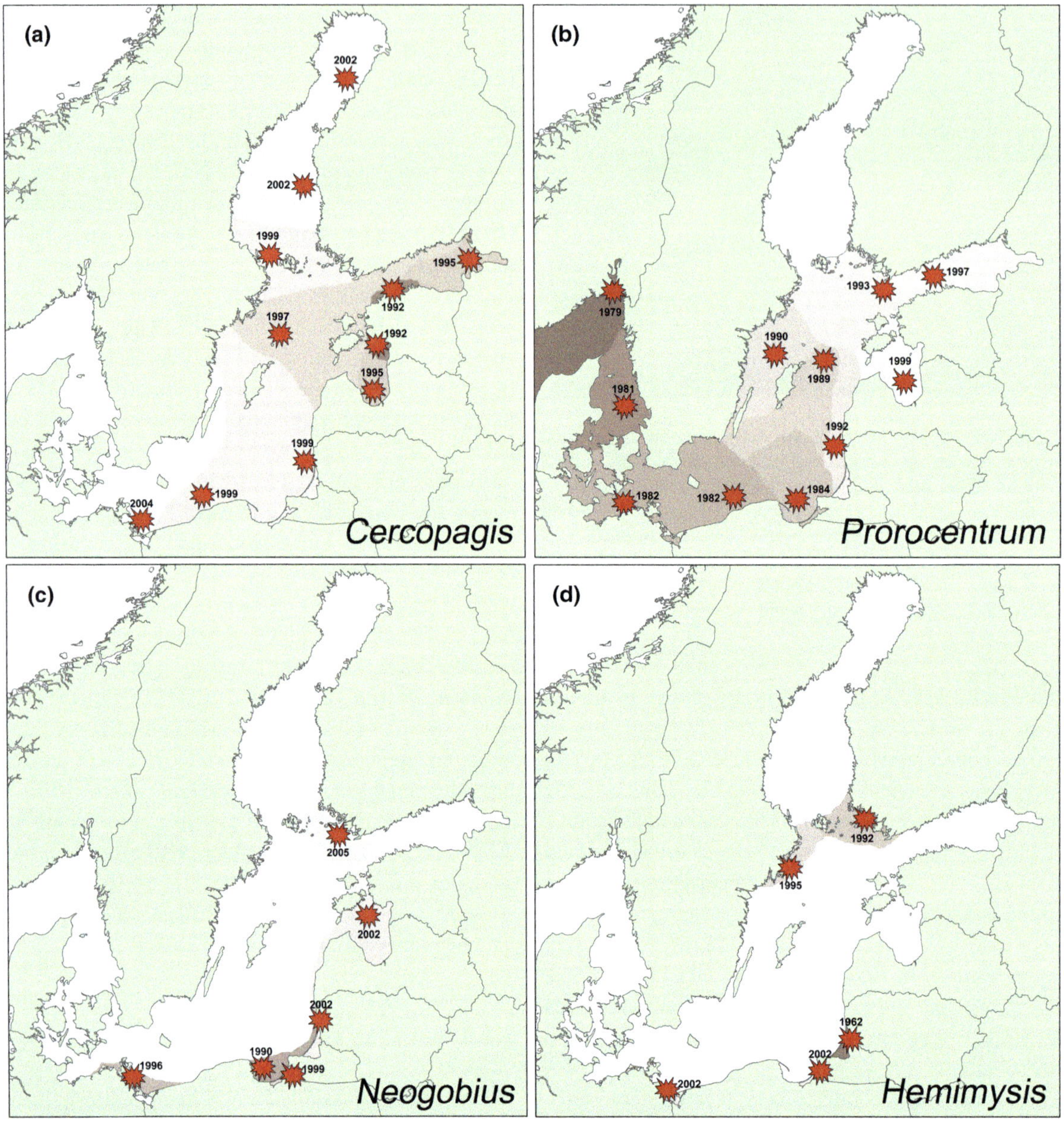

Fig. 5.6 Introduction and secondary spread of non-indigenous species in the Baltic Sea: numbers indicate the year of the first record and shaded sea areas show the potential directions of spread with the darkest colour indicating the primary recipient area. (**a**) Secondary spread of the planktonic cladoceran *Cercopagis pengoi* from the Gulf of Finland and the Gulf of Riga from 1992 to 2004. (**b**) Gradual incursion of the planktonic dinoflagellate *Prorocentrum cordatum* from the North Sea into the inner parts of the Baltic Sea from 1979 to 1999. (**c**) Secondary spread of the round goby *Neogobius melanostomus* from the Gdańsk Bay between 1990 and 2005. (**d**) Intentional introduction of the Ponto-Caspian mysid *Hemimysis anomala* into the Curonian Lagoon and its secondary spread to other coastal regions from 1962 to 2002. Figure: © Anastasija Zaiko

Other NIS may be "sleepers" for decades and then expand when conditions become favourable. For example, the Chinese mitten crab *Eriocheir sinensis* (Box 5.8), which was first found in Germany ~100 years ago, was also recorded in the less saline northern Gulf of Finland in the 1930s. Altogether, 25 individuals were found in an area extending from the Archipelago Sea in the west to Vyborg Bay in the east. From the 1930s until the early 2000s, an average of 1–2 individuals was reported annually from this area. However, in 2002 *Eriocheir sinensis* suddenly expanded and at least 103 individuals were documented during 2002–2004; several individuals were often caught together (Ojaveer et al. 2007). The reason for the apparent increasing occurrence and abundance of *Eriocheir sinensis* may be related to increased surface-water temperature and relatively mild winters, which may reduce the environmental stress experienced by *Eriocheir sinensis* in the low-saline parts of the Baltic Sea.

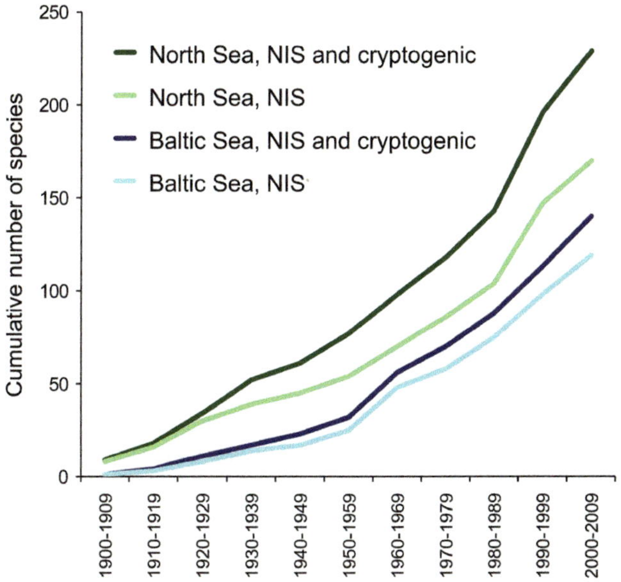

Fig. 5.7 Cumulative numbers of non-indigenous species (NIS) and cryptogenic species recorded in the North Sea and the Baltic Sea between 1900 and 2009. Figure modified from Olenin et al. (2014)

An expansion phase is usually followed by a phase of stasis or decline, which may be termed an adjustment or accommodation phase (Zaiko et al. 2014). There are numerous examples for such "boom and bust" phenomena of NIS in the Baltic Sea, but often the actual causes of the declines cannot be identified. Possible causes include a lower availability of the resources that initially allowed for rapid population expansions or that more predators and/or pathogens become focused on the invading species.

Besides more intensive human-mediated transport, the increase in the numbers of NIS introductions to the Baltic Sea during the past three decades (Fig. 5.7) may also reflect a higher awareness and larger research efforts. Nevertheless, the number of known Baltic Sea NIS is still only about one-sixth of that recorded in the Mediterranean Sea and almost one-third of that recorded on the European Atlantic coast. This difference is not only due to the smaller size of the Baltic Sea, but also to the hostility of its brackish waters for both marine and freshwater species, a comparatively lower trans-oceanic shipping activity and fewer species used in aquaculture.

5.3 Pathways and vectors

5.3.1 How do non-indigenous species cross environmental barriers?

A pathway is the route a NIS takes to invade a non-native ecosystem, and by definition the pathway for a primary introduction of a NIS is always human-mediated. The variety of pathways known worldwide may be classified into nine principal categories (Table 5.1). The main pathways to the Baltic Sea are shipping, human-made canals and fisheries

Table 5.1 Summary of the main pathways and vectors for primary introductions of non-indigenous species known worldwide. The first three pathways are the most important ones in the Baltic Sea Area. Table modified from Minchin et al. (2009)

Pathway	Human-mediated vectors
Shipping	Ballast-tank water and sediments, ship's hull and bilges, wells, tanks and cargo, anchors and anchor chains, lockers, fenders, portable moorings, deck recesses and snagged materials related to vessels, platforms, buoys and other floating structures
Canals	Water flows and transmissions, tidal exchanges and other shifts in water level like lock-flushing in inland canals
Fisheries	Intentional stock movements, population re-establishment, discharges of by-catch, disease agents from processing live, fresh and frozen foods, live bait releases, movements of retrieved fishing equipment, releases of organisms intended as living food supplements, discharges of packaging materials, releases of transported water
Aquaculture	Unintentional or unauthorised releases of NIS (including genetically modified organisms), releases and movement of stock-associated water, movements of nets, cages, lines, pumping equipment, discarded or lost nets, floats, traps, contaminated containers, discharges of packaging materials
Ornamental species and live seafood	Unintentional or unauthorised releases of NIS from aquaria, untreated waste discharges, unauthorised releases of imported living food organisms, releases of organisms associated with rock, gravels and sediments ("living rock"), discharges of packaging materials, releases of transported water, dumping of hobby aquariums
Marine leisure and tourism	Unintentional or unauthorised transport and release of angling catch, live bait movements, water sport equipment (diving, angling gear), live souvenirs, stocking for angling, discharges of packaging materials
Research and education	Unintentional or unauthorised releases of NIS used in experiments or as demonstration materials, releases/escapes of caged organisms used for water-quality monitoring, wastewater and biological waste discharges, field and experimental gear movement (including diving gear)
Habitat restoration and management	Soil stabilisation/reclamation using rock barriers, sediments and plants, use of filter-feeding invertebrates for managing water quality
Biological control	Releases of NIS to reduce diseases or parasites and to control invasive species or pests

Box 5.3: The red-gilled mud worms *Marenzelleria* spp.

Invasion history

Two morphologically very similar spionid polychaetes, *Marenzelleria neglecta* and *Marenzelleria viridis*, have recently been introduced to the Baltic Sea from the east coast of North America (Box Fig. 5.2). A third species, *Marenzelleria arctia*, is of Arctic origin. After the first record of *Marenzelleria* in the southern Baltic Sea in 1985, it was thought that only one species had invaded the Baltic Sea as a secondary spread from the North Sea to where it had been introduced earlier. However, after more detailed studies, involving scanning electron microscopy and genetic analysis (Blank et al. 2008), it was revealed that two independent introduction events of two species had taken place: one to the North Sea and another one to the Baltic Sea, probably by transfer of ballast water (Bastrop et al. 1995). At the same time, *Marenzelleria neglecta* was identified as a new species to science in both its native and invaded range. The specific epithet "*neglecta*" indicates that the species has previously been misidentified and overlooked. Currently, all three *Marenzelleria* spp. are widely spread in the Baltic Sea, from coastal waters, estuaries and shallow bays to oxygen-deficient deep-water zones where most other macrozoobenthos species cannot survive. There are, however, differences in the distributions of the species based on differences in their salinity tolerance ranges and preferred substrate. *Marenzelleria neglecta* and *Marenzelleria viridis* may co-occur in the more saline southwestern part of the Baltic Sea, but *Marenzelleria neglecta* has a wider distribution at lower salinity inside the Baltic Sea. In the northern part of the Baltic Sea the distribution of *Marenzelleria neglecta* partly overlaps with that of the *Marenzelleria arctia* (Blank et al. 2008).

Invasive traits

The invasion success of *Marenzelleria* spp. is favoured by their broad feeding strategy as they are both deposit and suspension feeders. Their dispersal potential is large since they have planktonic larvae and adult specimens are able to swim. *Marenzelleria* spp. also have a broad spectrum of habitat preferences and the ability to cope with low oxygen levels (Schiedek 1993; Fritzsche and von Oertzen 1995).

Biology

These infaunal polychaetes are up to 16 cm long and dwell in burrows in sediments beneath the surface of the seafloor (Box Fig. 5.2b). Typical faecal pellet strings are deposited near the openings of the burrows. Gametogenesis of *Marenzelleria neglecta* occurs in spring. After ~20 weeks the gametes reach maturity and the animals spawn in autumn. During spawning, the larval density near the coast can be as high as 21 million individuals m^{-3}. The development of the pelagic larvae into juvenile benthic worms depends largely on water temperature and takes between 4 and 12 weeks. The larvae occur in the coastal water column mainly from September to November, but they can be found up to March (Bochert 1997; Sikorski and Bick 2004).

Impacts in the Baltic Sea

Marenzelleria spp. have a negative impact on the abundances of the native polychaete *Hediste diversicolor* and the native amphipod *Monoporeia affinis* (Kotta et al. 2001; Kotta and Ólafsson 2003). On the other hand, *Marenzelleria* spp. have positive impacts on *Zostera marina*, a species of high conservation value, as the worms bury the seeds of the vascular plant and thereby reduce seed predation and facilitate seed germination (Delefosse and Kristensen 2012). Since *Marenzelleria* spp. have high burrowing activity they improve oxygen circulation in the sediments, but burrowing may also lead to an enhanced release of nutrients and hazardous substances from the sediments into the water column (Kotta et al. 2001; Hedman et al. 2011; Norkko et al. 2012). *Marenzelleria* spp. have become a food source for demersal fish such as European plaice *Pleuronectes platessa* and European flounder *Platichthys flesus* (Winkler and Debus 1997).

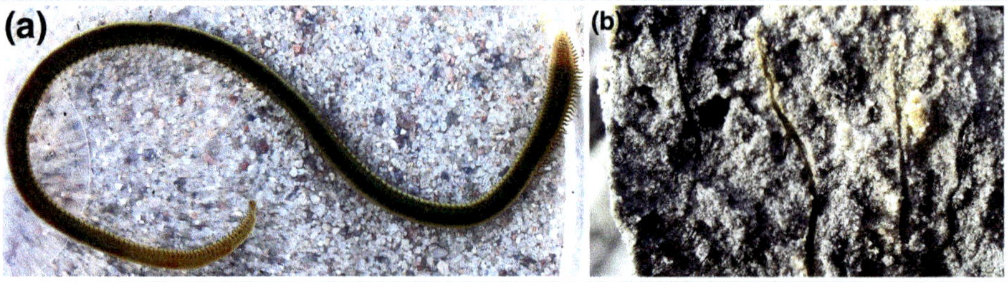

Box Fig. 5.2 All three *Marenzelleria* species in the Baltic Sea have an elongated greenish body with rows of short chaeta along both sides, but without dorsal scales. (**a**) *Marenzelleria neglecta*. (**b**) Mucus-lined burrows of *Marenzelleria* spp. have a maximum diameter of 2 mm and can penetrate down to ~35 cm in the sediment. This activity brings oxygen to the sediments, as shown by the yellowish colour of oxygenated sediments around the middle burrow. Photo: (**a**) © Andrius Šiaulys, (**b**) © Sergej Olenin

Box 5.4: The zebra mussel *Dreissena polymorpha*

Invasion history

The native area of *Dreissena polymorpha* (Box Fig. 5.3) is the Ponto-Caspian region. It is likely that this species already occurred in the Baltic Sea Area during the last interglacial period, became extinct and re-established in the early 1800s (Buynevich et al. 2011). This re-establishment was probably related to the building of canals between rivers, which opened new inland waterways for ship traffic between eastern and central Europe in the beginning of the 19th century (*cf.* Fig. 5.11). *Dreissena polymorpha* may have been attached to timber rafts that reached the Curonian Lagoon via the Dnepr-Nemunas Canal system. The zebra mussel is today one of the most common species in the oligohaline southern and eastern coastal lagoons and inlets of the Baltic Sea (Kotta et al. 1998; Orlova et al. 2004; Daunys et al. 2006).

Invasive traits

The invasion success of *Dreissena polymorpha* is favoured by its high fecundity (up to 1.5 million eggs per female per year), rapid growth, a planktonic stage that is easily dispersed and wide environmental tolerances, *e.g.* a salinity range of 0 to ~4 and water temperature up to 29 °C. *Dreissena polymorpha* usually spawns in May-July and fertilised eggs give rise to veliger larvae of up to 100 µm in body size. Before the larvae form byssus and attach to a substrate, they live planktonically for 2–4 weeks and are then able to move by means of a velum (Orlova 2002).

Biology

Dreissena polymorpha attaches by its byssus to a variety of human-made and natural stable surfaces, such as rocky substrates, macrophytes, animals, garbage (Box Fig. 5.3b) and fishing gear. It can, however, also build mussel beds in soft bottom areas because it only needs a small patch of stable substrate to form an attached aggregate. The triangular-shaped shells of adult *Dreissena polymorpha* are ~3 cm long and have a characteristic prominent banding pattern to which the name "zebra mussel" refers. Its specific epithet "polymorpha" is derived from the many variations in shell colour, pattern and shape depending on substrate, depth and density of aggregation.

Impacts in the Baltic Sea

Dreissena polymorpha is a notorious ecosystem engineer, capable of modifying the physical, morphological, biological and bio-geochemical properties of bottom habitats as it is able to form dense colonies and beds of living individuals and empty shells (Zaiko et al. 2009). Such modified habitats serve as "islands" of high biodiversity on the otherwise rather monotonous soft bottoms in low-salinity areas of the southeastern Baltic Sea and facilitate the establishment of other species, both native and non-indigenous. *Dreissena polymorpha* may overgrow native unionid bivalves and seriously decrease the abundances of the native species in recently invaded areas. Large *Dreissena polymorpha* beds have a high filtration capacity. They can process huge amounts of particulate organic matter and release dissolved inorganic nutrients (Orlova et al. 2004). A positive effect of *Dreissena polymorpha*, owing to its efficient filtering of the water, is that it may help mitigate eutrophication, increase water transparency and ameliorate growth conditions for benthic macrophytes. Through biodeposition *Dreissena polymorpha* increases the density of benthic deposit feeders and the zebra mussel itself can be an important food item for some fish, crayfish and waterbirds. Economic losses caused by *Dreissena polymorpha* are fouling of intake pipes, ship hulls, navigational constructions and cages used in aquaculture, as well as injuries to bathers from the sharp edges of the shells (Minchin et al. 2002). As a powerful filter feeder, *Dreissena polymorpha* is known to bioaccumulate chemical pollutants and toxins.

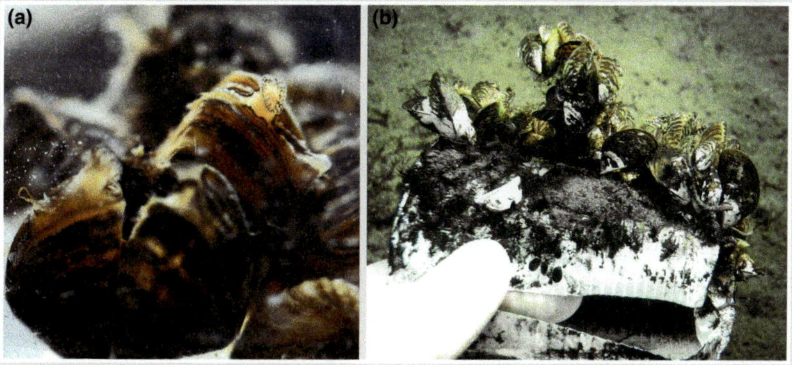

Box Fig. 5.3 The freshwater zebra mussel *Dreissena polymorpha*. (**a**) Suspension-feeding individuals. (**b**) Shells attach to anything, here to a plastic mug that was disposed of by someone who was using the Baltic Sea as a garbage can. Photo: (**a**) © Anastasija Zaiko, (**b**) © Sergej Olenin

Box 5.5: The fish-hook water flea *Cercopagis pengoi*

Invasion history

The first records of the carnivorous cladoceran *Cercopagis pengoi* (Box Fig. 5.4) in the Baltic Sea are from the Gulf of Finland and the Gulf of Riga in 1992 (Kotta et al. 2006). Most probably it arrived to the Baltic Sea with ships from its native area, the Ponto-Caspian region. By 2004, *Cercopagis pengoi* had expanded to the whole Baltic Sea proper, the Pomeranian Bay and the northern parts of the Gulf of Bothnia. The introduction of *Cercopagis pengoi* to the Laurentian Great Lakes in North America, where it was first recorded in 1999, was most likely a secondary spread from the Baltic Sea through shipping (Cristescu et al. 2001).

Invasive traits

Cercopagis pengoi tolerates a wide salinity range (0–17) and temperature (3–38 °C). It is a generalist predator, capable of feeding on a variety of prey species of different sizes, such as small cladocerans, larvae of the bay barnacle *Amphibalanus improvisus* and adults of the copepods *Eurytemora affinis* and *Acartia* spp. (Laxson et al. 2003; Plichlová-Ptácniková and Vanderploeg 2009).

Biology

The most conspicuous body parts of *Cercopagis pengoi* (Box Fig. 5.4a) are the head with one large compound eye, a well-developed second pair of antennae, four pairs of thoracic legs with the first pair larger than the others, an abdomen, a brood pouch in females, and a caudal process (posterior extension of the body) that in the summer has a distinctive loop-like curvature at the end (Grigorovich et al. 2000; Kotta et al. 2006). The body length, without the caudal process, is larger in females ($\sim$1.2–2.0 mm) than in males ($\sim$1.1–1.4 mm). The caudal process is usually 5–7 times longer (up to $\sim$10 mm long) than the main body, but this seems to vary regionally (Grigorovich et al. 2000). *Cercopagis pengoi* is a cyclic parthenogen, which mainly reproduces asexually in summer when the parthenogenetic young develop in a brood pouch that ruptures to release them (Mordukhai-Boltovskoi and Rivier 1971, 1987). In early autumn, parthenogenetic females produce eggs that develop into males and gametogenic females, which copulate. Sexual reproduction intensifies in late autumn when water temperature declines. This results in resting eggs that are released when the brood pouch ruptures and overwinter in the sediment. These eggs hatch in spring-summer when water temperature increases after winter. Resting eggs guarantee survival during unfavourable environmental

Box Fig. 5.4 The carnivorous cladoceran *Cercopagis pengoi*. (**a**) One individual, showing the head with one large compound eye and the caudal process with a distinctive loop-like curvature at the end. (**b**) Aggregates of *Cercopagis pengoi* form cotton-like masses that can clog fishing gear. Photo: (**a**) © Soili Saesmaa, (**b**) © Teemu Lehtiniemi conditions such as low winter temperatures (Katajisto et al. 2013). Resting eggs may also act as an effective means of dispersal for *Cercopagis pengoi* as they can withstand extreme conditions during transport in ballast water tanks.

Impacts in the Baltic Sea

Cercopagis pengoi is one of the most impactive invasive species in the Baltic Sea. The enormous expansion of the distribution area of *Cercopagis pengoi* in the Baltic Sea has caused changes in the pelagic food web and increased competition for food. In some areas of the Baltic Sea, *e.g.* in the Gulf of Riga, the population of the native water flea *Bosmina longispina maritima* has drastically decreased as a result of the *Cercopagis pengoi* invasion, probably by direct predation (Ojaveer et al. 2004). In spring and summer, *Cercopagis pengoi* competes for food with small planktivorous fish, which enhances eutrophication because of heavy predation on phytoplankton-grazing zooplankton. Aggregates of *Cercopagis pengoi* form cotton-like masses (Box Fig. 5.4b). Biofouling of fishing equipment by *Cercopagis pengoi* is a problem and the clogging of nets and trawls by the species causes substantial economic losses for fishermen (Leppäkoski and Olenin 2000).

while aquaculture, so far, is less important compared to other European seas. The remaining five pathways are of lower significance or not (yet) known to be involved in NIS transfer to the Baltic Sea Area.

Each pathway contains several vectors (Table. 5.1 and Fig. 5.8). A vector is the actual transfer mechanism, the direct physical means, by which a NIS is transported from one geographical region to another. Several vectors within a pathway may be involved in the transfer of one species, *e.g.* transfer by the pathway "shipping" may include several vectors such as a ship's ballast water, its hull and its anchoring equipment.

5.3.2 Shipping and canals

Ships from more than 50 countries arrive directly to ports in the Baltic Sea. It has been estimated that ~2,000 large ships (excluding pleasure boats) travel across the Baltic Sea each day (HELCOM 2010). Most of these ships transport cargo and passengers between ports within the Baltic Sea or to and from ports in western Europe. Other shipping routes connect to areas further away, such as the Mediterranean Sea, the Ponto-Caspian region, Northwest Africa, the North American east coast and Asia. Even assuming low numbers of propagules in the total volume of the ship ballast tanks,

Fig. 5.8 Interactions of various introduction vectors in estuarine and coastal areas: 1= Arrival of NIS with shipping. 2 = Range expansion through canal systems. 3 = Transfer of fouling organisms on small craft and to marina sites from sea and overland transport of boats. 4 = Stocking of organisms to provide leisure pursuits or for fisheries management, 5 = Releases from aquaria or from water ponds. 6 = Releases of organisms intended as live food, 7 = Releases by anglers or from their equipment. 8 = Aquaculture escapees. 9 = Discharges of wastes following processing, 10 = Movements associated with fishing gear or discards. Figure based on vector data in Minchin et al. (2006). Figure: © Vitalija Gasiunaite

Box 5.6: The round goby *Neogobius melanostomus*

Invasion history

The native area of *Neogobius melanostomus* (Box Fig. 5.5) is the Ponto-Caspian region. The first Baltic Sea individuals were caught near the tip of the Hel Peninsula (Poland) in 1990, but judging from the age of these specimens, it is likely that they have been inhabiting the Gdańsk Bay since at least 1987. *Neogobius melanostomus* rapidly dispersed over large areas in the Gdańsk Bay, *e.g.* in the surroundings of piers >350 individuals larger than 8 cm per 100 m^{-2} have been observed (Sapota and Skóra 2005). It was most probably brought to the Baltic Sea in ballast water as fertilised eggs or larvae. The population in the Gdańsk Bay was probably the base for the introduction of *Neogobius melanostomus* into other regions of the Baltic Sea. This is assumed because this fish's range of migration is small, comprising a distance of some hundred m. The longest migrations (up to some km) take place in late autumn and early spring when parts of the population move to and from deeper waters. Currently, *Neogobius melanostomus* is present in all countries surrounding the Baltic Sea. Its comparatively rapid secondary spread within the Baltic Sea may have been aided by shipping. For example, in Lithuania it was first found in the harbour of Klaipėda before it spread inside the Curonian Lagoon and further along the open coast. At the Swedish coast, *Neogobius melanostomus* was first reported in Karlskrona in 2008, and by 2014 it spread to Öland, Gotland and the Stockholm archipelago (data: Swedish Agricultural University). Further records from the Archipelago Sea, Gulf of Finland and the coast of southern Denmark (Kornis et al. 2012) confirm that this species is becoming well-established throughout the Baltic Sea. The introduction of *Neogobius melanostomus* to the Laurentian Great Lakes in North America took place at about the same time as that to the Baltic Sea (Corkum et al. 2004; Sapota and Skóra 2005).

Invasive traits

The invasion success of *Neogobius melanostomus* is favoured by its long spawning period. It is a multiple spawner, *i.e.* the females release their eggs in portions throughout the reproductive season (April to September). Adults aggressively defend their spawning sites and they can thus prevent the native fish from occupying prime spawning areas. *Neogobius melanostomus* has a broad diet of benthic animals, ranging from chironomids, amphipods and isopods to bivalves such as *Dreissena polymorpha*, *Macoma balthica* and *Mytilus trossulus* (Corkum et al. 2004; Sapota 2004; Rakauskas et al. 2013).

Biology

Neogobius melanostomus can live in limnic and brackish-water environments. They are typically found near sandy, stony bottoms, marine structures, sunken objects and mussel beds. The eggs are deposited in nests guarded by males, and several females can use the same nest. The nests are built on solid substrate such as stones, rocks, wood, roots of vascular plants or human-made constructions such as piers (Tomczak and Sapota 2006). *Neogobius melanostomus* has a relatively large head and its pelvic fins are fused to form a suctorial disk that is used for anchoring to substrates, especially in running waters. The fish usually stays in one place with only limited repositioning with its pectoral fins

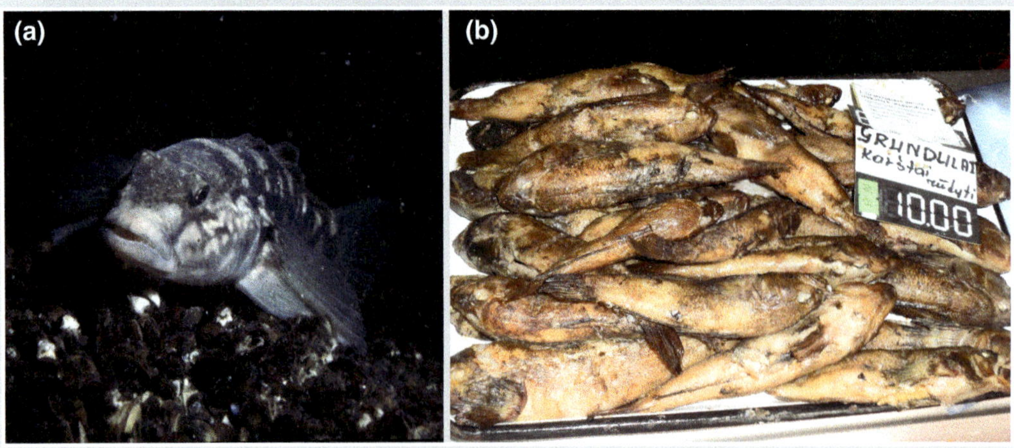

Box Fig. 5.5 The round goby *Neogobius melanostomus*. (**a**) An individual above a mussel bed in the Baltic Sea. (**b**) Smoked *Neogobius melanostomus* at a local fish market in Palanga, Lithuania. Photo: (**a**) © Mariusz Sapota, (**b**) © Henn Ojaveer

(Box Fig. 5.5a). *Neogobius melanostomus* can be distinguished from the black goby *Gobius niger*, which is native to the Baltic Sea, by a distinct black spot on the first dorsal fin of the invader. Sexual dimorphism is marked in *Neogobius melanostomus*, with males having a larger body size with age, enlarged cheeks and darker brownish-grey colour (Kornis et al. 2012). Breeding males are black with white-edged caudal fins. The maximum length of *Neogobius melanostomus* in the Gdańsk Bay is ~24 cm in males and ~18 cm in females (Sokołowska and Fey 2011).

Impacts in the Baltic Sea

In the southern Baltic Sea, *Neogobius melanostomus* seems to have a longer life span (up to 6 years) and a larger body size with age than its Ponto-Caspian and North American conspecifics. This suggests that this invasive species has found favourable conditions and a vacant niche in the Baltic Sea, including food resources, suitable habitats and spawning grounds. Where *Neogobius melanostomus* is numerous it can seriously affect populations of benthic invertebrates and may outcompete native benthivorous fish such as Eurasian ruffe *Gymnocephalus cernuus* and European flounder *Platichthys flesus* (Karlson et al. 2007; Rakauskas et al. 2013). *Neogobius melanostomus* may also feed on eggs and fry of native fish species. *Neogobius melanostomus* has been shown to significantly contribute (by 7–18 %) to the diet of piscivorous fish such as zander *Sander lucioperca* and the European perch *Perca fluviatilis* and waterbirds such as the great cormorant *Phalacrocorax carbo sinensis* and the grey heron *Ardea cinerea* (Rakauskas et al. 2013). In the Ponto-Caspian region *Neogobius melanostomus* is commonly caught for human consumption, and even in the Baltic Sea it has begun to be commercially exploited (Box Fig. 5.5b).

billions of NIS propagules are released into the Baltic Sea ecosystem each year (Fig. 5.9).

The Baltic Sea is connected to other sea regions by human-made canals that interconnect natural inland waterways. For example, the Kiel Canal (Fig. 5.10) connects the mouth of the Elbe river in the North Sea to the Kiel Bay in the southwestern Belt Sea. This allows the spread of NIS from the North Sea to the Baltic Sea, either by natural means or by shipping. A second connection with the North Sea is via the Limfjorden system, a natural waterway that cuts across the northern part of the Jylland peninsula (Denmark) to the Kattegat. Many NIS that were primarily introduced to the North Sea have arrived in the Baltic Sea by secondary spread through these two inland waterways or by travelling around the northern tip of Denmark.

In the eastern part of the Gulf of Finland, NIS can penetrate into the Baltic Sea through a ramified network of inland waterways and human-made canals (Fig. 5.11). The Volga-Baltic Waterway is a series of canals and rivers in Russia, which links the Baltic Sea to the Volga river basin. Ultimately, this connects the Neva Bay in the vicinity of Sankt-Petersburg (Russia) to the large Ponto-Caspian region, which comprises the Black Sea, the Sea of Azov and the Caspian Sea. Some organisms have gradually spread to the Baltic Sea via the Volga river due to the removal of previous environmental barriers and the emergence of new suitable habitats, *e.g.* cold hypolimnions in water reservoirs or the opposite, thermal discharges from power plants. The White Sea Canal connects the White Sea with Lake Onega (Russia), which is further connected to the Baltic Sea via the Volga-Baltic Waterway. This connection is also a potential pathway for NIS spread to the Baltic Sea.

Two additional entrance points for Ponto-Caspian species to the Baltic Sea are situated in the southeastern part of the Baltic Sea proper and were also opened by linking rivers with canals (Fig. 5.11). These canals are the Dnepr-Bug Canal to the Gdańsk Bay and the Dnepr-Nemunas Canal to the Curonian Lagoon. While the former is still being used for shipping, the latter has been closed for navigation since World War II (Karatayev et al. 2008). The role of the Dnepr-Nemunas Canal was especially important for the transfer of NIS in the 19th century when several species, including *Dreissena polymorpha*, penetrated into the coastal lagoons of the southeastern Baltic Sea.

5.3.3 Fisheries and aquaculture

The import of stocking material for fisheries has been an important introduction vector for non-indigenous fish in the Baltic Sea. Particularly in the 1950s–1970s, a number of fish species (including nine salmonid and four sturgeon species) were introduced intentionally, but none of them managed to establish self-reproducing populations (AquaNIS 2015). Living food supplements for commercial fish were intentionally introduced during the 1950s–1970s as well. For example, the Ponto-Caspian mysids (*Hemimysis anomala*, *Limnomysis benedeni* and *Paramysis lacustris*) and amphipods (*Chaetogammarus ischnus*, *Chaetogammarus warpachowskyi*, *Chelicorophium curvispinum*, *Obesogammarus crassus* and *Pontogammarus robustoides*) were transferred from Dnepr water reservoirs into the inland waters of Lithuania and the Curonian Lagoon in 1960 (Arbaciauskas 2002). All these crustaceans have

Fig. 5.9 Examples of the shipping pathway of non-indigenous species. (**a**) Ballast water release in a harbour. (**b**) Ballast water overflow on deck. (**c**) Ballast water release in a dock at night. (**d**) Sediment in a ballast water tank after release of the water. (**e**) Collection of organisms from a ship's hull (bow propeller). Photo: © Stephan Gollasch

successfully acclimatised and most of them have later spread to other coastal areas in the Baltic Sea.

Compared to many other European marine and inland water bodies, aquaculture is less developed in the Baltic Sea and is thus of minor importance as a potential vector for introductions of NIS. Besides native species such as the Atlantic salmon *Salmo salar* and the European eel *Anguilla anguilla*, fish aquaculture in the Baltic Sea Area commonly relies on the North American rainbow trout *Oncorhynchus mykiss*, which is not able to reproduce in the Baltic Sea, but

Fig. 5.10 A ship entering the Kiel Canal from the North Sea, an important pathway for the introduction of non-indigenous species from the North Sea to the Baltic Sea. Photo: © Stephan Gollasch

may occur in the wild through continuous escapes from fish farms. Several cultivation experiments with the Pacific oyster *Crassostrea gigas* (Fig. 5.4) were conducted in the Kiel Bay (southwestern Belt Sea), although without success (Meixner and Gerdener 1976). The first free-living Pacific oysters in the Baltic Sea were found in 2009, and they are thought to have been dispersed from the Kattegat by natural means (Wrange et al. 2010).

5.3.4 Certainty of pathways and vectors

In order to take appropriate management decisions, it is necessary to identify the active vector(s) for a specific NIS. Such knowledge can help to prevent other NIS using the same transfer mechanism in the future. The highest certainty level ("direct evidence") is provided when the transport of a NIS to a particular locality is clearly associated with a specific vector, such as the intentional stock movements of Ponto-Caspian mysids and amphipods to the Curonian Lagoon (Arbaciauskas 2002). The "very likely" level of certainty is applied if a NIS has appeared for the first time in a locality where a single pathway/vector is known to operate and the conclusion is deduced from the analysis of the

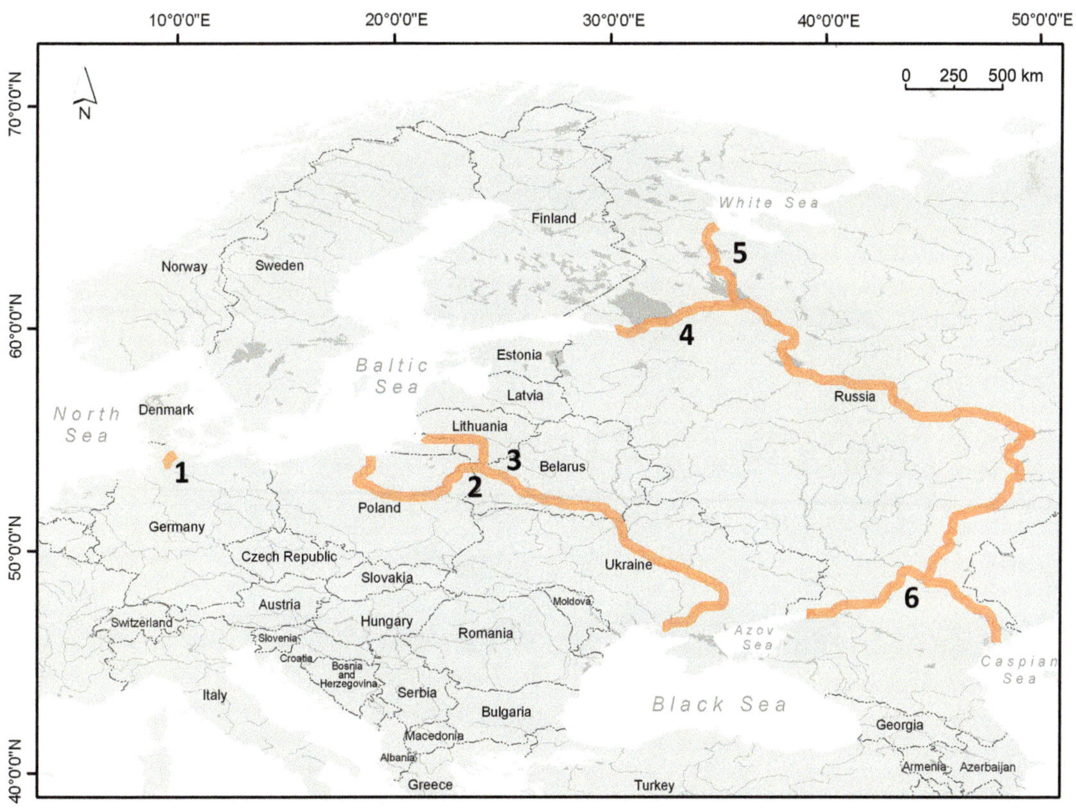

Fig. 5.11 Map showing the positions of the six major human-made canals (1–6) that interconnect natural inland waterways and thereby link the Baltic Sea with other marine and estuarine regions: 1 = the Kiel Canal, 2 = the Dnepr-Bug Canal, 3 = the Dnepr-Nemunas Canal, 4 = the Volga-Baltic Waterway, 5 = the White Sea Canal, 6 = the Volga-Don Canal. Figure: © Anastasija Zaiko

Box 5.7: The North American wedge clam *Rangia cuneata*

Invasion history

One of the most recent newcomers to the eastern Baltic Sea coast, *Rangia cuneata* (Box Fig. 5.6), is considered to be native to the Gulf of Mexico (North America), where it mainly occurs in brackish-water estuaries (Wakida-Kusunoke and MacKenzie 2004; Verween et al. 2006). In the Baltic Sea, *Rangia cuneata* was first recorded in the Russian part of the Wisła Lagoon in 2010 (Rudinskaya and Gusev 2012). Its pathway to the Wisła Lagoon was attributed to ladder-dredge boats, which operated in the sea channel of the harbour of Kaliningrad in 2008, from areas where the species had been introduced earlier (*e.g.* the harbour of Antwerpen, Belgium). There are indications that *Rangia cuneata* has begun to spread to adjacent Polish and Lithuanian coastal waters (Warzocha and Drgas 2013).

Invasive traits

Rangia cuneata is highly tolerant to varying environmental conditions. It has high fecundity and a planktonic larval stage, which provides the clam with a good dispersal ability. Conditions that are unfavourable for many native species in the Baltic Sea, like sudden salinity fluctuations, may trigger an outburst of *Rangia cuneata*.

Impacts in the Baltic Sea

Although there is not enough information on ecosystem-wide impacts of *Rangia cuneata*, the rapid spread of this species within the Wisła Lagoon, and its ability to reach high abundances fast, suggest that it may induce adverse transformations in the local benthic communities. By modifying soft-bottom habitats and restructuring the benthic communities, *Rangia cuneata* might affect food webs of the coastal areas and therefore the resource availability for economically important fish species. However, in its native area *Rangia cuneata* is harvested for human consumption and is considered a valuable economic resource. This aspect should be taken into account when planning management and mitigation measures for *Rangia cuneata* in the Baltic Sea.

Box Fig. 5.6 Shell of *Rangia cuneata* found in the Wisła Lagoon where recorded shell length was up to 40 mm (Rudinskaya and Gusev 2012). Photo: © Andrey Gusev

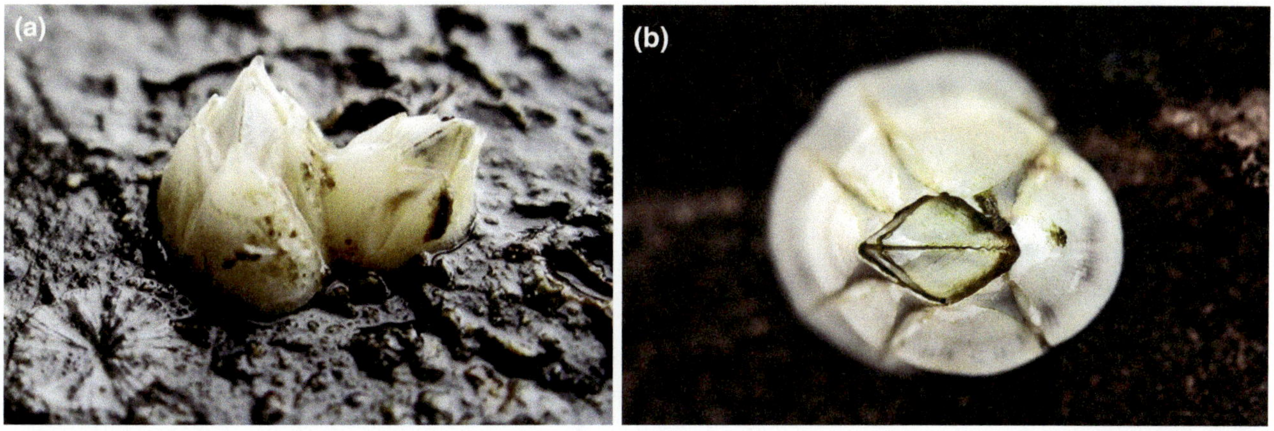

Fig. 5.12 Adults of the bay barnacle *Amphibalanus improvisus* usually grow up to ~10 mm in size. This species may have arrived to the Baltic Sea from North America as adults attached to ship hulls, or may have spread secondarily as larvae by sea currents or in ballast water from the neighbouring North Sea. This species was first observed in the Baltic Sea in the 1880s. Photo: (**a**) © Sergej Olenin, (**b**) © Pauline Snoeijs-Leijonmalm

introduction event. For example, since the zooplankton crustacean *Cercopagis pengoi* was first found in the harbour areas of Tallinn and Pärnu it can be assumed that this NIS was most likely introduced with ballast water (Leppäkoski and Olenin 2000).

In many cases, the introduction of a NIS cannot be convincingly ascribed to a single pathway/vector, because more than one pathway could be involved and/or different life stages of the same species may be transported by different vectors of the same pathway. In such cases, the "possible" level of certainty is applied. For example, the bay barnacle *Amphibalanus improvisus* (Figs. 5.12 and 5.13) may have arrived to the Baltic Sea from North America as adults attached to ship hulls or spread as larvae by sea currents or in ballast water from the neighbouring North Sea. The overlap between pathways and vectors can be even more complicated (Fig. 5.8; Minchin et al. 2006). For example, a species that initially arrived via canals could be further transported within the recipient area via the shipping pathway or fishing gears, or have naturally spread to adjacent waters with currents.

Fig. 5.13 Two non-indigenous species in the Baltic Sea that originate from North America, *Orconectes limosus* and *Amphibalanus improvisus*. The freshwater spiny-cheek crayfish *Orconectes limosus* is up to 12 cm long and was actively introduced to Germany for aquaculture in 1890. It has established in freshwaters in Europe, as well as in the Baltic Sea (Leppäkoski and Olenin 2000; Jaszczołt and Szaniawska 2011). In this photograph, *Orconectes limosus* is covered by the brackish-water bay barnacle *Amphibalanus improvisus* (*cf.* Fig. 5.12). Photo: © Sergej Olenin

The assumed pathway by which a species arrives is often based on known anthropogenic activities in the area. The role of different pathways and vectors may shift due to climate change (*e.g.* warm-water species may more easily survive in northern areas when these areas become warmer), changes in environmental quality (*e.g.* new ecological niches may emerge due to construction of artificial habitats or degradation of native biota), political and social-economic events (removal of custom control, closure of acclimatisation programmes, changes in aquacultural practices), management policy (*e.g.* ballast water management), and the emergence of new trading routes.

5.4 Origin and distribution

5.4.1 Source areas

NIS that have been introduced to the Baltic Sea originate from coastal marine (brackish) and freshwater bodies in many regions of the world. The most important source areas are the Ponto-Caspian region, the North American east coast and East Asia (Fig. 5.14).

The Ponto-Caspian species have evolved in the watersheds and estuarine areas of brackish water bodies: the Black Sea, the Sea of Azov and the Caspian Sea (*cf.* Table. 2.1), and are therefore well adapted to the brackish-to-limnic conditions of the estuarine systems of the Baltic Sea (*cf.* Sect. 13.2). Most of these species have settled in the Gulf of

Box 5.8: The Chinese mitten crab *Eriocheir sinensis*

Invasion history

The native area of *Eriocheir sinensis* (Box Fig. 5.7) is East Asia in the South and East China Seas (Gollasch 2009). In Europe, the crab was first found in a tributary of the Weser river (Germany), approximately a century ago. It was probably introduced by ballast water. *Eriocheir sinensis* is known to actively migrate over long distances (hundreds of km), and about a decade after the first record it was found in the German part of the Baltic Sea where it possibly arrived via active migration through the Kiel Canal. Currently, *Eriocheir sinensis* has spread all over the Baltic Sea, including the inner parts of the Gulf of Bothnia and the Gulf of Finland. It is also found in rivers flowing into the Baltic Sea (Ojaveer et al. 2007).

Biology

Eriocheir sinensis reproduces in marine water and juveniles migrate inland by travelling upstream in rivers, or along the Baltic Sea coast east- and northwards to lower salinity and freshwater habitats. Adult crabs migrate back to the sea for reproduction. Mass developments of crabs occur every 15–30 years. Such population oscillations do occur in the Baltic Sea, but they are more pronounced in North Sea estuaries. As the reproduction is limited to more saline waters (salinity >10), it is believed that specimens of *Eriocheir sinensis* found north and east of the Baltic Sea proper have migrated here from distant places.

Impacts in the Baltic Sea

Eriocheir sinensis is an active predator that feeds on benthic organisms, but also on fish caught in traps and on nets, which damages fishing gear. The crabs can clog industrial constructions such as water intake filters. Since *Eriocheir sinensis* burrows in sediments, it can destabilise sediment structure and increase the recirculation of nutrients. The burrowing activity also increases the erosion of dikes, as well as river and lake embankments. *Eriocheir sinensis* is the second intermediate host for the human lung fluke parasite in Asia, but this parasite has not been recorded in European crabs yet. In Asia, *Eriocheir sinensis* is served in restaurants as a delicacy.

Box Fig. 5.7 The carapax of *Eriocheir sinensis* is up to 5 cm in size and brownish in colour. It has characteristic mitten-like "fur" on its claws. (**a**) Adult *Eriocheir sinensis* in an aquarium. (**b**) An individual caught in the harbour of Klaipėda, Lithuania). Photo: (**a**) © Stephan Gollasch, (**b**) © Anastasija Zaiko

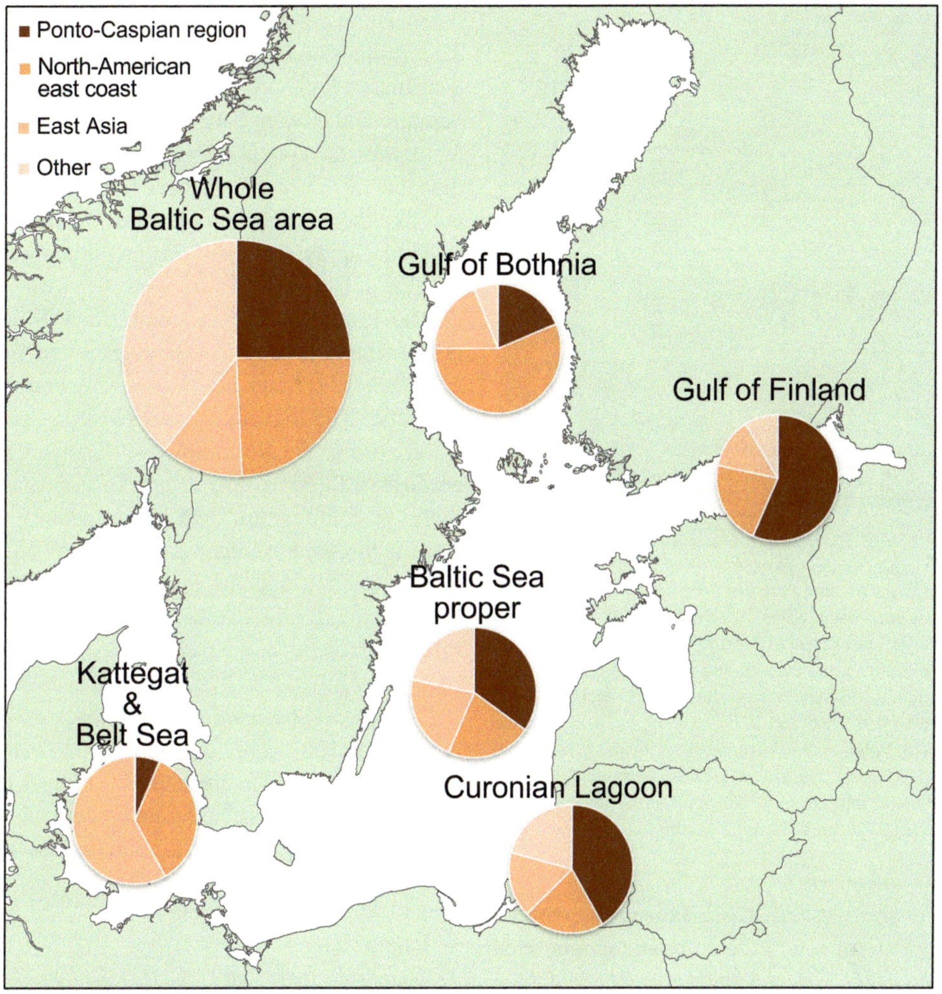

Fig. 5.14 Native areas of the non-indigenous species (NIS) established in the Baltic Sea Area by 2010. The proportion of Ponto-Caspian NIS is larger in the eastern Baltic Sea while NIS originating from the North Amerian East Coast and East Asia dominate in the western part of the Baltic Sea Area. The category "Other" includes species from other parts of the world, as well as cryptogenic species. Figure: © Anastasija Zaiko

Finland, particularly in the Neva Bay off Sankt-Petersburg, and in the large coastal lagoons of the southeastern Baltic Sea proper (the Curonian Lagoon) and southern Baltic Sea proper (the Wisła Lagoon and the Szczecin Lagoon).

The proportion of Ponto-Caspian NIS diminishes westward and also northward in the Baltic Sea (Fig. 5.14). Only three NIS belonging to this group (*Dreissena polymorpha*, *Neogobius melanostomus* and the hydrozoan *Cordylophora caspia*) are found in the Kattegat and the Belt Sea, where they occur in limnic and/or oligohaline coastal conditions. For most of the Ponto-Caspian NIS, the Baltic Sea is the area of primary introduction outside their native range, and some of them have continued to spread from the Baltic Sea to other regions of the world. For example, *Cercopagis pengoi*, *Cordylophora caspia*, *Dreissena polymorpha* and *Neogobius melanostomus* have become established in the Laurentian Great Lakes in North America (Mills et al. 1993; Cristescu et al. 2001).

In contrast to the introductions from the Ponto-Caspian region, the occurrences of most NIS originating from North America, East Asia as well as from other parts of the world (South America, Africa and the Indo-Pacific region) are the results of secondary spread from the North Sea or other western European sea areas as primary recipients. Most of

Fig. 5.15 The white-fingered mud crab *Rhithropanopeus harrisii* has a carapace up to 2 cm wide. The native area of this small brackish-water crab is the east coast of North America. It shows high fecundity, a long planktonic larval period, and a wide tolerance range for several environmental drivers, which has likely facilitated its invasion success. It was first observed in Europe in 1874 and in the Baltic Sea in 1951 (in Poland). *Rhithropanopeus harrisii* has been reported from coastal areas of the Baltic Sea in Germany, Poland, Lithuania, Estonia, Finland and Sweden (Hegele-Drywa and Normant 2009; Fowler et al. 2013). Photo: © Sergej Olenin

these NIS occur in the western, more saline and warmer parts of the Baltic Sea Area, *e.g.* the common slipper shell *Crepidula fornicata* (Fig. 5.5) and the white-fingered mud crab *Rhithropanopeus harrisii* (Fig. 5.15), and practically all of them are results of unintentional introductions.

Only a few NIS of North American origin have been introduced intentionally to the Baltic Sea region, *e.g.* three freshwater crayfish species (*Orconectes limosus* (Fig. 5.13), *Orconectes virilis* and *Pacifastacus leniusculus*), the Canada goose *Branta canadensis* (Fig. 5.16) and the mammals: American mink *Neovison vison* and muskrat *Ondatra zibethicus* (Westman 2002; Jaszczołt and Szaniawska 2011). Ten species have been intentionally imported into the Baltic Sea and adjacent water bodies from Siberian and Russian Far East inland waters, but only two of those (the Baikalian amphipod *Gmelinoides fasciatus* and fish *Perccottus glenii*) have established self-reproducing populations within the Baltic Sea.

5.4.2 The Baltic Sea bioinvasion gradient

The distinct environmental gradients of the Baltic Sea (*cf.* Sect. 2.4) determine the boundaries of spread and colonisation potential for both native and non-native species. The primary factor shaping the large-scale geographical distributions of NIS is salinity (Paavola et al. 2005). Temperature and oxygen concentrations are additional significant factors for the spread of NIS, but their roles are less known than that of salinity. On a local scale, the distributions of NIS are, like those of native organisms, modified by factors such as food supply, competition, predators, and availability of suitable substrates.

The lowest number of established NIS is found in the northernmost part of the Baltic Sea, the Bothnian Bay (19 NIS), where salinity is low and temperature conditions are subarctic. The highest number (37 NIS) occurs in the transition zone to the North Sea (Belt Sea and Kattegat), mainly because of the proximity to the North Sea and intensive ship traffic in combination with higher salinity and milder winters. In this area a larger proportion of NIS originate from North America and the Pacific Ocean (Fig. 5.14). In contrast, in the Baltic coastal lagoons with pronounced local salinity gradients and ice cover in winter, the Ponto-Caspian NIS prevail.

The lowest species richness of macroscopic organisms in brackish waters occurs in salinity of 5–7 (Remane 1934; *cf.* Sect. 4.5.6), which is the salinity in most of the Baltic Sea (*cf.* Fig. 2.15). Thus, the human-mediated species introductions of NIS from brackish source areas to the Baltic Sea

Fig. 5.16 The Canada goose *Branta canadensis* is the only non-indigenous coastal bird species in the Baltic Sea region. Centuries ago it was intentionally introduced to Europe as an ornamental species and for hunting. *Branta canadensis* is a summer visitor, transit migrant and irregular winterer on Baltic Sea coasts, where it has been observed since the 1930s. Strictly speaking, this species is not a typical waterbird as it prefers open, grassy habitats where it feeds on grasses, herbs, and plant roots, but it can also feed on aquatic plants. The Canada goose can hybridise with native species and is considered a sanitary problem at *e.g.* bathing sites. Photo: © Sergej Olenin

flatten the "Remane diagram" (*cf.* Fig. 4.21) by filling in the trough between fully limnic and fully marine waters. For example, Paavola et al. (2005) analysed the distributions of 84 NIS belonging to 15 phyla that are established in different salinity zones of the three large European brackish water bodies: the Baltic Sea, the Black Sea (including the Sea of Azov) and the Caspian Sea. They found that the majority of these 84 NIS (72–83 %) tolerate more than one salinity zone, and nearly half of them occur in at least three salinity zones in all three water bodies. It turned out that most NIS are well adapted to the salinities holding the lowest species richness already in their native area, and that a NIS richness maximum in brackish water bodies occurs in the salinity intervals of the native species richness minimum. This predictable pattern in the salinity range of NIS provides a tool for the initial risk assessment of future invasions in brackish water bodies, especially when mapping potential source and recipient areas.

Since Elton's (1958) seminal work, there has been a general belief that diverse native communities use resources to a larger extent and thus leave fewer opportunities for potential invaders. Seemingly, this concept may hold true for the Baltic Sea since NIS tend to proliferate in areas of this naturally species-poor brackish ecosystem. However, many studies have also shown the opposite: the degree of invasion seems to be more often positively correlated with the species richness of natives, particularly in systems where the biodiversity distribution is largely determined by environmental drivers (Zaiko et al. 2007). Interestingly, along the Baltic Sea bathymetric gradient, from coastal areas to deeper basins, the species richness of both native and non-native species declines rapidly. So far, species-poor and oxygen-deficient subhalocline areas have been invaded only by three spionid polychaete species, all belonging to the genus *Marenzelleria* (Norkko et al. 2012).

5.4.3 Invasion "hotspots"

The invasion success of a NIS, and its further distribution in the Baltic Sea, is determined by the similarity in environmental conditions between the source and recipient areas, the proximity to shipping routes and/or inland waterways, and the level of anthropogenic or/and invasive disturbance. Therefore, many NIS are abundant or even dominant in coastal areas, especially in lagoons and inlets with intensive anthropogenic activities such as shipping routes, harbour areas, marinas and hydrotechnical constructions.

The facilitative effect of environmental modifications that promote new NIS invasions may be asserted through physical or biological mechanisms. An example of a physical mechanism is the provision of hard substrate in an area with natural sandy beaches by *e.g.* harbour constructions or wind farms. In these cases, a sessile NIS that needs a hard substrate for attachment is likely to experience only low competition for space and resources from local organisms. Biological mechanisms include altered habitat conditions caused by an already established NIS so that the invasion of a new NIS is favoured. This may create a feedback system that accelerates the accumulation of NIS and forms an invasion "hotspot" (Simberloff and Von Holle 1999). For example, dense aggregates of the zebra mussel *Dreissena polymorpha* provide an attractive habitat for numerous native and non-native benthic fauna in coastal lagoons with large freshwater influences in the southeastern Baltic Sea (Zaiko et al. 2007).

Also, natural or anthropogenic disturbances may create new ecological niches that favour NIS invasions. For example, the outcompeting of native gammarid populations by the North American *Gammarus tigrinus* in many places along the coasts of the Wisła Lagoon may be partly due to eutrophication and chemical contamination by hazardous substances, which may strengthen the competitive capacity of the invader (Grabowski et al. 2006). Another example of disturbed environments being favourable for species invasions are the discharge areas of cooling water from nuclear power plants into the Baltic Sea (Box 5.9). These discharges create habitats with continuously elevated water temperatures on a scale of a few km^2. Some notorious NIS, such as the New Zealand mud snail *Potamopyrgus antipodarum* and Conrad's false mussel *Mytilopsis leucophaeata*, thrive in these environments exceptionally well and/or were first introduced to these environments and spread from there, *e.g.* the eel parasite *Anguillicoloides crassus*.

A generalised model of an "invader-friendly" habitat where invasion "hotspots" may be found can be defined by the following features:

1. The habitat has favourable physical conditions for maintaining diverse communities in general. In this case native species richness can be considered as an indicator of a habitat's invasibility.
2. The habitat has an increased amount of usable resources, *e.g.* through anthropogenic nutrient inputs of nitrogen and/or phosphorus. Both spatial and temporal variation in the availability of resources facilitates NIS invasion by providing resource pools to new colonists.
3. The habitat is severely disturbed by natural or anthropogenic stressors, *e.g.* a heavy storm or bottom dredging. Every additional disturbance event may promote a new surge of NIS invasions.

Box 5.9: Thermal discharges and non-indigenous species

Pauline Snoeijs-Leijonmalm

Thermal discharges stimulate the spread of non-indigenous species

Nuclear power plants discharge large volumes of cooling water into the coastal environment at several places in the Baltic Sea. This creates locally disturbed habitats that are notorious for high abundances of some non-indigenous species such as the diatom *Pleurosira inusitata* (syn. *Pleurosira laevis fo. polymorpha*), the Conrad's false mussel *Mytilopsis leucophaeata* and the mud snail *Potamopyrgus antipodarum*. The vicinities of large cooling-water outlets are not only hotspots where non-indigenous species can build up high-density populations, but they can also be "first bridgeheads" for non-indigenous species that later on invade other areas of the Baltic Sea. The year-round increased water temperature in the cooling-water discharge areas may promote the adaptation and spread of the newcomers to a new habitat. For example, the polychaete *Marenzelleria* sp. and the amphipod *Gammarus tigrinus* were first observed in heated water at Loviisa (Ilus 2009) before they spread to other Finnish coastal areas in the eastern Gulf of Finland. Similarly, high numbers of the eel parasite *Anguillicoloides crassus* were first discovered in heated water at Oskarshamn (Höglund and Andersson 1993) before it invaded the rest of the Baltic Sea.

Large cooling-water discharges in the Baltic Sea Area

Five large Swedish and Finnish nuclear power plants with 14 reactor units use brackish seawater to dispose of waste heat in the Baltic Sea Area (Box Fig. 5.8). Other coastal nuclear power plants (*e.g.* in Sankt-Petersburg, Russia) use cooling towers. About one-third of the energy produced in a nuclear power plant, by either a boiling-water reactor or a pressurised-water reactor, is transferred to electricity. The other two-thirds of the energy produced is excess heat, and thus the reactors need to be cooled down. At full operation, the 14 reactor units together produce 12,226 MW of

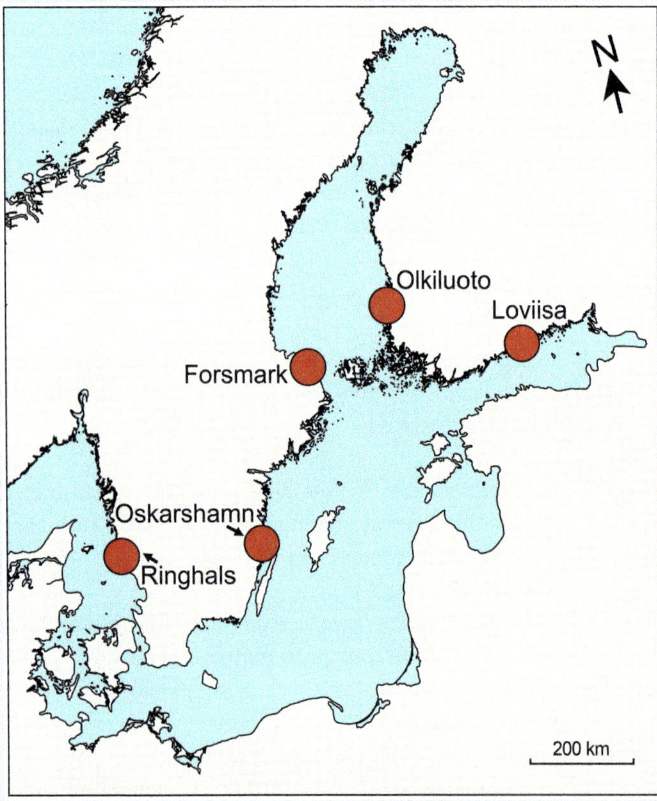

Reactor unit	Start year	MW electricity	MW waste heat
Forsmark 1	1980	1,010	1,918
Forsmark 2	1981	1,010	1,918
Forsmark 3	1985	1,190	2,110
Oskarshamn 1	1972	492	883
Oskarshamn 2	1975	661	1,139
Oskarshamn 3	1985	1,450	2,450
Ringhals 1	1976	865	1,675
Ringhals 2	1975	865	1,787
Ringhals 3	1981	1,047	2,088
Ringhals 4	1983	940	1,835
Olkiluoto 1	1978	860	1,640
Olkiluoto 2	1980	860	1,640
Loviisa 1	1977	488	1,012
Loviisa 2	1977	488	1,012

Box Fig. 5.8 Locations and energy production (electricity and waste heat) of the five nuclear power plants with large cooling-water discharges in the Baltic Sea Area. Figure: © Pauline Snoeijs-Leijonmalm electricity and 23,107 MW of waste heat. For ~1,000 MW (MJ s^{-1}) of electricity production ~45 m^3 s^{-1} of cooling water is needed. The water is heated by ~10 °C when it is returned to the sea, which creates a coastal area of a few km^2 with significantly increased water temperature. For example, at Forsmark the warm water keeps 2–3 km^2 of the Bothnian Sea free of ice in winter. On the coasts of the Baltic Sea many other industries use brackish cooling water as well, *e.g.* paper mills and even a large data centre in Finland (Hamina, Gulf of Finland), but these thermal discharges are more limited than those of the nuclear power plants.

The diatom *Pleurosira inusitata*

The large chain-forming diatom *Pleurosira inusitata* (Box Fig. 5.9) was probably introduced to the Forsmark area (Sweden) during an experimental release of eels that had been raised in aquaria in southern Europe. This diatom was never observed in Forsmark before 1989, but since 1990 it forms up to 0.5-meter high colonies in water heated

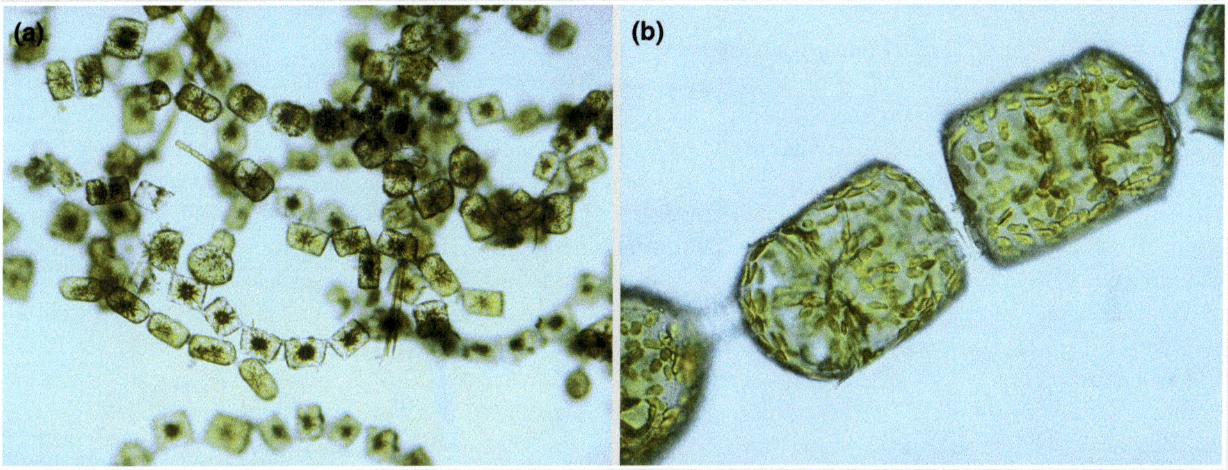

Box Fig. 5.9 *Pleurosira inusitata* (syn. *Pleurosira laevis fo. polymorpha*) is a relatively large centric diatom with many chloroplasts per cell, oval-shaped silica valves averaging 92 × 83 µm and a pervalvar axes averaging 124 µm (measurements from Forsmark). Photo: © Pauline Snoeijs-Leijonmalm

Box Fig. 5.10 The dreissenid bivalve *Mytilopsis leucophaeata*. The individuals in the photograph are 18–20 mm long. Photo: © Pauline Snoeijs-Leijonmalm

by ~10 °C each year in September-November. The colonies are attached to stones and macroalgal vegetation and they hang like fishing nets in the water, especially at sites with slow-flowing water (Snoeijs and Weckström 2010).

Conrad's false mussel *Mytilopsis leucophaeata*
In the cooling water discharge at Loviisa (Finland) a strong recruitment of young dreissenid bivalves of the species *Mytilopsis leucophaeata* (Box Fig. 5.10) was observed in 2003. Already one year later, a dense population with up to 28,000 adult individuals m^{-2} (9.8 kg wet weight m^{-2}) completely covered boulders and stones (Laine et al. 2006). In 2011 the species also arrived at Forsmark and spread fast in the heated water (Florin et al. 2013). *Mytilopsis leucophaeata* has most probably been transported to the Baltic Sea by ballast water. Its body size, shape and habitat are very similar to those of the blue mussel *Mytilus trossulus*, which is native to the Baltic Sea, but in contrast to *Mytilopsis leucophaeata*, *Mytilus trossulus* avoids the heated water in the cooling-water discharges.

The New Zealand mud snail *Potamopyrgus antipodarum*
The native habitat of the deposit-feeding prosobranch snail *Potamopyrgus antipodarum* (syn. *Paludestrina jenkinsi*, Box Fig. 5.11) is freshwater in New Zealand (Ponder 1988). Molecular studies have identified two mitochondrial haplotypes from the North Island of New Zealand that are identical to those found in Europe (Städler et al. 2005). The original introduction to Europe was probably a secondary spread from Australia, and transport may have been in drinking water barrels on board ships. The first European finds of *Potamopyrgus antipodarum* around 1890 were from estuaries and the brackish coasts of the Baltic Sea, from where it spread further to European freshwaters. In the cooling-water discharges at Forsmark, Oskarshamn, Olkiluoto and Loviisa, *Potamopyrgus antipodarum* is a common to dominant species (Snoeijs 1989; Ilus 2009). In most places it lives in soft bottoms, but it colonises rocky shores as well. In sediments at Forsmark it has been observed at densities of almost 30,000 individuals m^{-2} (Sandström 1990) and with densities of 8,000–10,000 individuals m^{-2} on rocky substrates at ~10 °C in winter, as well as at ~28 °C in summer (Snoeijs and Mo 1987). These numbers are from macrofaunal samples (body size >1 mm), but the true densities of the species were much higher since many specimens of *Potamopyrgus antipodarum* are <1 mm in size. The species is viviparous, reproduces year-round and has wide temperature and salinity tolerances, which are traits that explain its invasion success (Snoeijs 1989).

Box Fig. 5.11 The New Zealand mud snail *Potamopyrgus antipodarum* which is usually only up to 4 mm long. (**a**) Dorsal view, note the two tentacles. (**b**) Ventral view with an open operculum and the snail's head. Photo: © Pauline Snoeijs-Leijonmalm

4. The habitat properties are altered due to previous NIS introductions. A successfully established habitat-engineering species should be considered as a powerful facilitative factor for further invasions.

5.5 Diversity and ecology of non-indigenous species

5.5.1 Correct taxonomic identification is crucial

The presence of a NIS often remains unnoticed until it becomes abundant and/or creates trouble because of incomplete taxonomic knowledge at the time of its arrival in the recipient ecosystem. In most cases it is also difficult to check a species' identity afterwards because reference specimens of first introductions have seldom been kept.

For example, there has been a great deal of confusion around the introduced *Marenzelleria* spp. (Box 5.3) in the Baltic Sea Area. These polychaetes can be up to 16 cm long, but still they can only be reliably identified to the species level by genetic analysis (Blank et al. 2008). Three species with different geographical distributions live in the Baltic Sea. *Marenzelleria viridis* occurs from the Skagerrak to the Öresund and has also been observed in the southwestern and southeastern Baltic Sea proper as well as in the Gulf of Riga. *Marenzelleria neglecta* occurs in the whole Baltic Sea proper up to the Åland Sea and *Marenzelleria arctia* occurs in the northern Baltic Sea proper and the Bothnian Sea. Both *Marenzelleria viridis* and *Marenzelleria neglecta* have been introduced from the North American east coast while *Marenzelleria arctia* is an Arctic species.

5.5.2 Taxonomic confusion about *Mnemiopsis leidyi*

Another good example of taxonomic confusion is the case of the American comb jelly *Mnemiopsis leidyi* (Box 5.10) in the Baltic Sea. This species is an actively hunting ctenophore, a hermaphrodite with a translucent body and a length of up to 14 cm, which is native to the American Atlantic coast from Narragansett Bay (USA) in the north to the Valdés Peninsula (Argentina) in the south. *Mnemiopsis leidyi* invaded the Black Sea in the early 1980s (Purcell et al. 2001) and the Caspian Sea in mid 1990s, to which it was likely transported through the Volga-Don Canal in ballast water (Kideys 2002). In 2005, the species appeared in coastal areas of the North Sea (Oliveira 2007; Tendal et al. 2007) and in 2006 on the Swedish west coast (Hansson 2006), with up to 92 individuals per m^3 in the Kiel Bay in the southwestern Belt Sea (Javidpour et al. 2006). In 2007 it was reported that *Mnemiopsis leidyi* had spread to the Bothnian Sea and the Gulf of Finland as well, but molecular evidence proved that the comb jelly observed in these northern areas was in fact the Arctic comb jelly *Mertensia ovum* (Gorokhova et al. 2009). *Mertensia ovum* has a broad Arctic and circumboreal distribution, but had never been reported from the Baltic Sea before.

Thus, the search for a northward expansion of *Mnemiopsis leidyi* in the Baltic Sea had instead yielded increased knowledge of native biodiversity. Later it was shown that while the invader *Mnemiopsis leidyi* would perhaps be able to survive for a short time in vast areas of the northern Baltic Sea, if it would be transported there, its reproduction is prevented by salinity <10 and temperature <12 °C. Thus, due to the combined effect of low salinity and low temperature, it is not probable that *Mnemiopsis leidyi* will establish permanent populations in the central or northern Baltic Sea (Lehtiniemi et al. 2012).

Zooplankton and fish species that live in deep water can be transported from the Kattegat into the Baltic Sea with saltwater inflows. Their dispersal is thus mainly controlled by the baroclinic flow field and bottom topography (*cf.* Box 2.1). Hydrodynamic drift modelling has shown that the potential dispersion of *e.g.* comb jellies follows the deep-water currents from the Bornholm Sea towards the north and the east of the Baltic Sea and is limited by topographic features and low advection velocities (Lehtiniemi et al. 2012). However, if such species are new invaders in the area, and the conditions for growth and reproduction are favourable in the Baltic Sea, they will be able to form stable populations despite the fact that most individuals are hampered by the hydrodynamics of the deep water.

5.5.3 Phytoplankton

At least 13 non-native phytoplankton species, six diatoms, five dinoflagellates and two silicoflagellates have been recorded in the Baltic Sea Area (Olenina et al. 2010; Kownacka et al. 2013). None of these phytoplankton NIS originate from the Ponto-Caspian region and most of them are certain secondary introductions. They account for less than 1 % of the more than 2,000 phytoplankton species that are known from the Baltic Sea (Hällfors 2004). However, since phytoplankton organisms are easily distributed by ships' ballast water, the number of non-indigenous/cryptogenic phytoplankton species in the Baltic Sea is probably underestimated.

Box 5.10: The American comb jelly *Mnemiopsis leidyi*

Invasion history

The native area of the American comb jelly *Mnemiopsis leidyi* (Box Fig. 5.12) is the east coast of North and South America. Outside its native area it was first discovered in the Black Sea in 1982, after which it rapidly invaded the Sea of Azov, the Aegean Sea, and the Marmara Sea. In 1999 it was also found in the Caspian Sea. It reached Northern Europe in 2005 (North Sea) and the Belt Sea in 2006. At present, *Mnemiopsis leidyi* occurs in the transition zone and the southern Baltic Sea with its northernmost established population in the Bornholm Sea. Genetic studies have revealed multiple introductions: the populations in northern Europe originate from the northeastern coast of America while the southern European populations, including the Mediterranean and Ponto-Caspian regions, originate from the Gulf of Mexico.

Invasive traits

Mnemiopsis leidyi has wide salinity and temperature ranges and tolerates low oxygen levels. However, it seems unlikely that *Mnemiopsis leidyi* would establish permanently in the northern or eastern parts of the Baltic Sea as its spread and population growth is limited by low salinity and low temperature (Lehtiniemi et al. 2012).

Impacts in the Baltic Sea

After its introductions to the Black Sea and the Caspian Sea, *Mnemiopsis leidyi* reproduced rapidly and formed very large populations. In these seas it found optimum conditions: plenty of food due to eutrophication, high temperature and overfished populations of potential competitors. Being a highly efficient predator of zooplankton, fish eggs and small larvae, and due to its voracious appetite, *Mnemiopsis leidyi* has caused drastic shifts in the Black Sea and Caspian Sea ecosystems. In the Baltic Sea its impact is less pronounced due to low population density and small body size. While the local fisheries collapsed in the Black Sea and the Caspian Sea during the periods of *Mnemiopsis leidyi* peak abundances, there is no evidence so far of it threatening the Baltic herring, sprat or cod stocks (Jaspers et al. 2011).

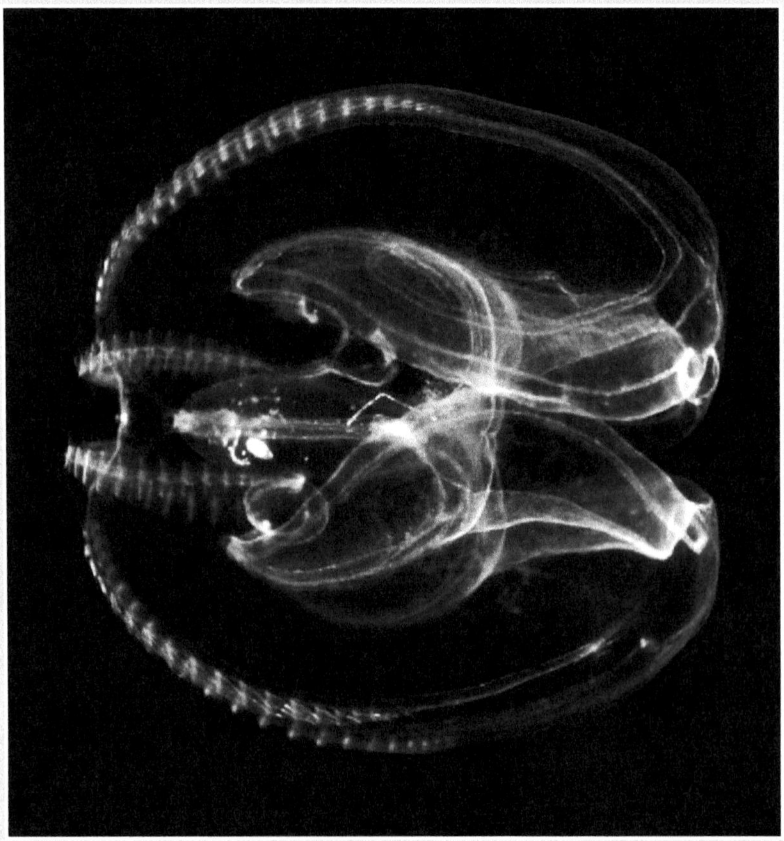

Box Fig. 5.12 *Mnemiopsis leidyi* is a transparent gelatinous ctenophore. It grows up to 14 cm in body size in the Black Sea, but in the colder Baltic Sea Area it does not exceed 8 cm. Photo: © Maiju Lehtiniemi

There is usually a high level of uncertainty when assigning a unicellular plankton organism to NIS (Gómez 2008). In fact, all 13 non-indigenous phytoplankton species recorded in the Baltic Sea are cryptogenic, *i.e.* their native area is uncertain or unknown because they have already spread to many places on Earth. However, for all 13 species it was possible to show that they were new to the Baltic Sea because the phytoplankton community composition in the area has been studied for more than 100 years (Wasmund et al. 2008). If a species can be identified by light microscopy, and is abundant today despite not having been recorded earlier, it has most probably invaded the Baltic Sea. Remnants of some phytoplankton organisms accumulate in the sediments, *e.g.* diatoms and chrysophytes, and in these cases it is even possible to prove the absence of a species from the Baltic Sea for much longer than a time period of 100 years.

A number of notorious IAS-classified phytoplankton species increase in abundance worldwide with negative impacts on biological diversity, ecosystem functioning and socio-economic values (Anderson 2009). The only recognised phytoplankton IAS in the Baltic Sea is the dinoflagellate *Prorocentrum cordatum*. This cryptogenic dinoflagellate spread from the western part of the Baltic Sea up to the Gulf of Finland between 1979 and 1993 (Fig. 5.6b). The summer-autumn blooms of *Prorocentrum cordatum* can have a massive bioinvasion impact on ecosystem functioning; when the species' abundance exceeds 1 million cells L^{-1} it can completely dominate the phytoplankton community (by up to 98 % of the total biomass) and change physical (water transparency) and chemical (nutrient concentrations) properties of seawater (Olenina et al. 2010).

Prorocentrum cordatum also has the potential of forming toxic blooms that can kill crustaceans, fish and other marine organisms, but in the Baltic Sea the species has not been observed to be toxic. Other potentially toxic cryptogenic phytoplankton species in the Baltic Sea are the dinoflagellates *Alexandrium minutum*, *Alexandrium ostenfeldii*, *Gymnodinium catenatum* and *Karenia mikimotoi*, and the silicoflagellates *Heterosigma akashiwo* and *Pseudochattonella verruculosa*. All six species are known to cause "red tides" (large toxic blooms) elsewhere, but this phenomenon has not been recorded in the Baltic Sea so far, although it has been shown that *Alexandrium ostenfeldii* is able to produce paralytic shellfish poisoning toxins on the southwestern coast of Finland (Hakanen et al. 2012).

Another type of damage to the ecosystem can be caused by phytoplankton species that form dense blooms accompanied by copious amounts of mucilage, such as the diatom *Coscinodiscus wailesii*. Because of its comparatively large cell size with a 175–500 μm diameter, *Coscinodiscus wailesii* is inedible to most grazing zooplankton, and when its blooms decay the cells aggregate, sink and may cause anoxia at the seafloor. A direct socio-economic impact of mucilage is the clogging of equipment such as nets and cages used in fisheries and aquaculture.

5.5.4 Macrophytes

Thirteen non-native benthic macrophytes, including 11 algae and two vascular plants, have been recorded in the Baltic Sea Area (AquaNIS 2015). None of them originates from the Ponto-Caspian region and most of them are secondary introductions. There is no macrophyte NIS that has become an IAS in the Baltic Sea like the green alga *Caulerpa taxifolia* in the Mediterranean Sea (Meinesz et al. 2001).

Seven of the Baltic macrophyte NIS, the red algae *Bonnemaisonia hamifera*, *Dasya baillouviana* and *Gracilaria vermiculophylla*, the brown algae *Colpomenia peregrina*, *Fucus evanescens* and *Sargassum muticum*, and the reed *Spartina anglica*, occur inside the Baltic Sea, but only in the more saline areas in the Arkona Sea, the Belt Sea and/or the Öresund (Nyberg 2007). Four other macroalgal NIS, the green alga *Codium fragile* and three red algae (*Aglaothamnion halliae*, *Dasysiphonia japonica* and *Neosiphonia harveyi*) have not entered the Baltic Sea, but occur in the Kattegat (Nyberg 2007). The charophyte *Chara connivens* and the American pondweed *Elodea canadensis* are the only two macrophyte NIS that are restricted to the most limnic parts of the Baltic Sea.

5.5.5 Zooplankton

At least eight zooplankton NIS are established in the Baltic Sea (AquaNIS 2015). Six crustaceans: the cladocerans *Cercopagis pengoi*, *Cornigerius maeoticus*, *Evadne anonyx* and *Penilia avirostris*, and the copepods *Acartia tonsa* and *Ameira divagans*, comprise ~10 % of the total crustacean zooplankton species richness in the Baltic Sea, although this percentage varies somewhat between the different subregions of the Baltic Sea.

Due to a low number of (known) native gelatinous zooplankton species, two non-indigenous gelatinous species (the jellyfish *Maeotias marginata* and the comb jelly *Mnemiopsis leidyi*) represent more than 30 % of the species richness of the jellyfish (Cnidaria) in the northern Baltic Sea and more than 15 % of the comb jellies (Ctenophora) in the southern Baltic Sea.

The principal ecological difference between the two groups of zooplankton NIS (crustaceans and gelatinous) is the way they are utilised as a food source by higher trophic levels. Cladocerans and copepods are often valuable additions to the diet of a range of predators, while gelatinous zooplankton organisms are mainly preyed upon by carnivorous gelatinous top predators that utilise secondary production otherwise consumed by fish (Mills 1995; Boero et al. 2008). Therefore, gelatinous zooplankton organisms are often regarded as "dead ends" in marine food webs (Verity and Smetacek 1996).

The distribution of zooplankton NIS in the Baltic Sea is mainly governed by salinity. *Ameira divagans* and *Penilia avirostris* occur only in the more saline conditions of the Arkona Sea, Belt Sea and Kattegat. *Cornigerius maeoticus* and *Maeotias marginata* have been observed only in the Gulf of Finland, but it is not certain that these two species are absent from the Baltic Sea proper. The other four zooplankton NIS are more widely distributed: *Acartia tonsa* in the entire Baltic Sea, *Cercopagis pengoi* and *Evadne anonyx* in the northern and eastern parts, and *Mnemiopsis leidyi* in the western and southern parts.

Some benthic invertebrate NIS, *e.g.* the mussels *Dreissena polymorpha* and *Mytilopsis leucophaeata* and the barnacle *Amphibalanus improvisus*, have a planktonic larval stage. Also juveniles of the spionid polychaetes *Marenzelleria* spp. may occur in the water column above the sediments. The possession of such free-living life stages is a useful trait for a NIS as it provides an advantage over obligate sessile species when spreading to new areas.

Some zooplankton NIS that have established permanent populations are now part of the pelagic and benthic food webs in the Baltic Sea. In some cases, they have changed the energy flow in the food webs by adding an extra trophic level to the system. For example, *Cercopagis pengoi* and *Evadne anonyx*, predators of smaller zooplankton, and *Acartia tonsa* and *Penilia avirostris*, which graze on phytoplankton, have extended the native food webs (Saiz and Kiørboe 1995; Lehtiniemi and Gorokhova 2008). The zooplankton NIS are also preyed upon by planktivorous fish as well as by carnivorous invertebrates such as mysids and gelatinous zooplankton. *Cercopagis pengoi* is the only NIS in the Baltic Sea that seems to have a strong impact on ecosystem functioning in the pelagic zone. However, the distribution of the other introduced carnivorous cladoceran *Evadne anonyx* is increasing in the Baltic Sea (Põllupüü et al. 2008; Bielecka et al. 2014), and this species may also be a relevant food source for planktivorous fish in late summer when its population peaks.

5.5.6 Benthic and nektobenthic invertebrates

The largest group of NIS recorded in the Baltic Sea (~60 species) are benthic and nektobenthic invertebrates, mainly crustaceans, molluscs and polychaetes. Of these, ~45 species are currently established in the Baltic Sea (*cf.* Fig. 4.18c). The same three taxonomic groups also dominate the native benthic invertebrate fauna of the Baltic Sea, and NIS constitute ~8 % of the crustaceans, ~4 % of the molluscs and ~3 % of the polychaetes.

No shallow hard- or soft-bottom habitat in the Baltic Sea is entirely free from human-mediated benthic invaders anymore. Non-indigenous species can even be abundant or dominant in these habitats, *e.g. Dreissena polymorpha* on hard bottoms and *Marenzelleria* spp. on soft bottoms in the low-salinity lagoons of the southeastern Baltic Sea proper (Leppäkoski et al. 2002a).

Today, the native freshwater amphipods have disappeared from the central freshwater part as well as from the more brackish northern part of the Curonian Lagoon while the Ponto-Caspian amphipods *Obesogammarus crassus* and *Pontogammarus robustoides* proliferate here now, together with the North-American amphipod *Gammarus tigrinus* (Grabowski et al. 2006). High densities of *Pontogammarus robustoides* are associated with a reduced biomass of the green habitat-forming filamentous alga *Cladophora glomerata* (Arbaciauskas and Gumuliauskaite 2005), suggesting a grazing effect.

5.5.7 Fish

About 30 non-indigenous fish species have been introduced to the Baltic Sea and adjacent waters (AquaNIS 2015). Most of them were introduced intentionally between the 1950s and the 1970s. They have added a considerable number of species to the ~120 native marine, freshwater and migratory fish species known from the Baltic Sea. However, the majority of the intentionally introduced fish species have not been able to form self-reproducing populations in the Baltic Sea and their rare encounters in the wild concern specimens that have escaped from fish cultures. Examples of such NIS are the Siberian sturgeon *Acipenser baerii*, the Russian sturgeon *Acipenser gueldenstaedtii*, the sterlet *Acipenser ruthenus*, the spotted silver carp *Aristichthys nobilis*, the longnose sucker *Catostomus catostomus*, the beluga sturgeon *Huso huso*, the silver carp *Hypophthalmichthys molitrix*, the pink salmon *Oncorhynchus gorbuscha*, the chum salmon *Oncorhynchus keta*, and the North American rainbow trout *Oncorhynchus mykiss*.

Three of the intentionally introduced NIS that are able to reproduce in the Baltic Sea are the Chinese sleeper *Perccottus glenii* (introduced in 1916), which occurs in the most diluted low-salinity eastern parts of the Gulf of Finland (Orlova et al. 2006), the Prussian carp *Carassius gibelio* (introduced in the 17th century), which now is common in the Wisła Lagoon (Witkowski and Grabowska 2012) and along the Estonian coast (Vetemaa 2006) and *Cyprinus carpio* (introduced in the 14th century), which is common in the Curonian Lagoon (Virbickas 2000).

The most notable unintentional fish introduction is that of the Ponto-Caspian round goby *Neogobius melanostomus*. After being first recorded in the Gdańsk Bay in 1990 its incursion was reported from several other areas in the Baltic Sea. It is believed that the secondary spread of this species has been facilitated by shipping because in new localities it was first found mainly in or near habours.

In general, the estuarine and inshore waters of the Baltic Sea are more amenable to invasions of non-indigenous fish species than the open sea areas because most of the NIS originate from limnic or brackish-water source areas. Marine non-indigenous fish species are unable to form self-reproducing populations in the Baltic Sea for any longer time due to the, for them, unfavourable low salinity.

5.5.8 Mammals

Three mammal IAS, two native to North America and introduced to Europe in the 1920s–1930s and one native to East Asia, have spread along the Baltic Sea coasts (Nummi 2002). The American mink *Neovison vison*, the racoon dog *Nyctereutes procyonoides* and the muskrat *Ondatra zibethicus* were originally introduced for fur farming and large populations of these three mammals have built up in the Baltic Sea region from escaped and released individuals.

The mink and the racoon dog prey, for example, on eggs in bird nests and on incubating waterbirds in the archipelagos of the Baltic Sea. The mink may also cause losses for fish farms. The muskrat disturbs the structure of the littoral vegetation as it mainly feeds in reed belts and digs for plant roots. This may create floods and mud flats and has a negative impact on macrofauna, fish and bird nests due to habitat destruction. The muskrat also bears a large number of parasites, including the dwarf tapeworm *Echinococcus multilocularis*, which may infect humans (Nummi 2002).

5.5.9 Non-indigenous species associates

An aspect that has only rarely been studied is that one NIS can in fact be more than one. For example, the zebra mussel *Dreissena polymorpha* was shown to carry at least 14 types of parasites and other symbionts within the mantle cavity and/or associated with internal tissues, including ciliates (*Ancistrumina limnica*, *Conchophthirus acuminatus* and *Ophryoglena* sp.), trematodes (Echinostomatidae, *Aspidogaster* sp., *Bucephalus polymorphus* and *Phyllodistomum* sp.), nematodes, oligochaetes, mites, chironomids and leeches (Karatayev et al. 2000). It is complicated to study such associated species because it is difficult to prove where the host became infected: in the source area, on the pathway or in the recipient area.

Transport of the host can be the vector for the introduction of a parasitic NIS that can also infest native species. This has happened *e.g.* with the nematode *Anguillicoloides crassus*, which was probably introduced to Europe with eels imported from Japan (Lefebvre et al. 2012). This parasite feeds on host tissues and reproduces in the swimbladder lumen of eels. In less than three decades, driven by intercontinental eel trade, it has spread over four continents, infecting six of the 20 eel species and subspecies described worldwide, including the European eel *Anguilla anguilla*. In the Baltic Sea, *Anguillicoloides crassus* is distributed from the Kattegat to the Archipelago Sea.

5.6 Environmental quality and invasive species

5.6.1 Biological pollution

NIS can change the biological, chemical and/or physical properties of an aquatic ecosystem and cause a decline in ecological quality. Such changes include, but are not limited to, local elimination of sensitive and/or rare species, alteration of native communities, harmful blooms, modification of the substrate, changes in oxygen and nutrient concentrations, pH, water transparency, and accumulation of hazardous substances. The outcomes of biological invasions that decrease ecological quality are called "biological pollution" or "biopollution" and the species involved are IAS (Box 5.1).

An IAS can affect one or more levels of biological organisation, *e.g.* internal biological pollution by parasites or pathogens, genetic changes (*e.g.* hybridisation) or shifts in the age structure of a prey population at the population level, structural shifts at the community level, modification of physical-chemical conditions at the habitat level and/or alteration of energy and organic material flow at the ecosystem level.

There is a fundamental difference between various forms of pollution. IAS do not respond to remedial efforts in the same way as eutrophication or chemical pollution, which can be diminished if appropriate measures are taken. The risk of biological pollution can be most effectively reduced by a precautionary approach (*e.g.* vector and pathway management) while eradication or control of existing IAS are more challenging. IAS usually expand their distribution and increase their abundance from a local source via processes that are not controllable through management. The spatial extent, rate of spread, and impacts on the environment will depend on the biological traits of a NIS and the environmental conditions within an invaded ecosystem.

5.6.2 Environmental status of the Baltic Sea

The environmental status of marine waters is traditionally evaluated by taking into account the effects of eutrophication, chemical pollution, habitat destruction and overexploitation of fish stocks. However, biological pollution, which may even surpass the impacts of the "traditional" stressors, can also have pronounced effects on the environment, and should be included in environmental assessments. One of the "good environmental status" (GES, *cf.* Sect. 17.8.1) descriptors in the EU Marine Strategy Framework Directive (MSFD, *cf.* Sect. 17.8) specifically addresses the bioinvasion problem: "Non-indigenous species introduced by anthropogenic activities are at levels that do not adversely alter the ecosystem". Thus, the absence or minimal level of biological pollution is one of the goals of achieving a GES of the Baltic Sea.

NIS cause adverse environmental impact and economic losses only after attaining a critical level of abundance and only when occupying a sufficiently large area. To classify the level of bioinvasion impacts, an integrative method called the "biopollution level index" (BPL) was proposed for aquatic ecosystems (Olenin et al. 2007). This index is based on a classification of the abundance and distributional range of NIS and the magnitude of their impacts on native communities, habitats and ecosystem functioning. It includes five BPL classes: 0 = no impact, 1 = weak impact, 2 = moderate impact, 3 = strong impact and 4 = massive impact. An overall bioinvasion impact assessment based on the BPL of the entire Baltic Sea revealed that strong biopollution (BPL 3) often occurs in coastal lagoons, inlets and gulfs, and moderate biopollution (BPL 2) in the open sea areas (Zaiko et al. 2011). However, despite continuously accumulating information, documented ecological impacts are known so far for only one-third of the ~130 NIS in the Baltic Sea. Our understanding of both the direction and magnitude of impacts at the ecosystem level of even the most widespread NIS is still poor (Ojaveer and Kotta 2015).

Bioinvasion impacts may compromise the value of some indicators used for the ecological status assessment of coastal waters. For example, the ability of *Dreissena polymorpha* to modify bottom habitats and to form local patches of elevated biological diversity may bias the results of species richness-based environmental quality assessments by showing a false improvement of ecological status (Zaiko and Daunys 2015). Thus, the assessment may reflect the IAS impact rather than that of anthropogenic pressure.

5.7 Risk assessment and management

5.7.1 Risk assessment of impacts by non-indigenous species

Risk assessment of impacts by NIS includes the prediction of whether a species is capable of spreading from its native or introduced area, as well as the identification of possible impacts it might have in a new area if it were introduced (Gollasch and Leppäkoski 2007; Olenin et al. 2014). Such an assessment contains a high degree of uncertainty due to the lack of information on the probability of a species to be transported and established under certain environmental conditions. Additional uncertainty results from the scarcity of data on effects such species may provoke (David et al. 2013a). Risk analyses are aided by predictive habitat and niche modelling which helps to identify areas susceptible to new introductions. This in turn helps to design and target monitoring efforts and to plan control measures.

However, effective risk assessment requires detailed knowledge on the traits and ecology of the introduced species as well as on their ecological interactions with native species, which are most often poorly known. Moreover, while impacts on the invaded habitat and community structure may be tractable, information about IAS impacts on ecosystem functioning is mostly lacking. Extrapolating information on non-indigenous species impacts from one area to another is often problematic and should be performed with caution.

It is essential to compare the traits of successful and unsuccessful invaders with those of related native species to better understand why some species become pests in some areas or under certain environmental conditions. The traits of NIS vary and their effects may therefore be unpredictable and opposite to impacts of other NIS in the area.

5.7.2 Information support

While biological invasions attract increasing attention from scientists, policy makers and various management authorities, the knowledge base on NIS is continuously expanding. With the implementation of the EU MSFD and similar legislation addressing the problem of biological invasions, the availability of advanced, scientifically validated and up-to-date information support on NIS is essential for aquatic ecosystem assessment and management.

The Baltic Marine Biologists (BMB) initiated the first regional open information source worldwide, the Baltic Sea Alien Species Database (BSASD, online since 1997), which contains detailed information on NIS origin, introduction history, pathways and vectors for the Baltic Sea Area. Now this BSASD is part of a larger, new generation information system (AquaNIS), dealing with aquatic NIS introduced to the marine, brackish and coastal freshwater environments of Europe and adjacent regions. This system is designed to assemble, store and disseminate comprehensive data on NIS and to assist in the evaluation of the progress made towards achieving management goals (AquaNIS 2015).

5.7.3 Early detection and molecular techniques

In order to enhance the opportunities and efficacy of management measures, it is important to detect a NIS at an initial stage of incursion, *i.e.* when a population is still confined to a small area and has low density (Fig. 5.3). Therefore, NIS monitoring and surveys should be prioritised in bridgehead sites and dispersal hubs which are often the first recipient areas for new introductions (Lehtiniemi et al. 2015).

Early detection requires proper species identification, which in many cases depends on explicit taxonomic expertise. Traditional taxonomic approaches are laborious, and often fail to identify cryptic species (two or more species hidden under one species name) or larval stages. Access to the appropriate taxonomic expertise, intercalibration exercises and searchable digital databases with image recognition functions may aid identification and enhance the quality of taxonomic assignment. Increasingly, genetic methods allow tracking of the source population and identifying pathways of the introductions. Population genetics can reveal the relatedness of two populations (*e.g.* native and introduced) and make it possible to roughly estimate the timing of the introduction in order to assess if the introduced species has one or more source regions and its possible pathways.

Rapidly advancing new molecular techniques provide promising tools for species identification from environmental samples. Novel molecular approaches such as metabarcoding have huge potential to provide more accurate and standardised, high-resolution taxonomic data on all organisms present in a sample, including hosts with all their parasites. Metabarcoding allows taxonomical assignment of a specimen based on sequencing of a short standardised DNA fragment (molecular marker or barcode), across entire biological communities (*cf.* Box 4.2). The recent development of high-throughput sequencing offering massive sequencing capacities allows multiple samples to be processed faster and cheaper than can be achieved by traditional morphological approaches (Pochon et al. 2013; Kelly et al. 2014). This new technique is expected to revolutionise NIS surveillance in the near future.

5.7.4 Precaution and mitigation

Thus far, the Baltic Sea has not been exposed to devastating biological pollution to the extent experienced by some other aquatic ecosystems of the world, *e.g.* the Mediterranean Sea, the Black Sea, the Caspian Sea and the Laurentian Great Lakes of North America. Still, this does not mean that large bioinvasion impacts cannot occur in the Baltic Sea in the future. Among the vast spectrum of potential NIS, it is practically impossible to predict which species may become invasive. Therefore, precaution is recommended as species introductions are irreversible and accumulate over time (David et al. 2013a).

No control of IAS without affecting other components of the ecosystem is feasible once an invasion process is underway. Given the severity of the problems that can be caused by IAS, it is mandatory for policy and management to focus on the vectors of introduction to prevent introductions of species in general. Regarding vector management two prime instruments are applied: (1) the Ballast Water Management Convention (BWMC) of the UN International Maritime Organization (IMO), and (2) the Code of Practices of the European Union (EU) and the International Council for Exploration of the Sea (ICES, *cf.* Box 18.1) for planned species introductions in aquaculture.

5.7.5 The Ballast Water Management Convention

The aim of the BWMC is to prevent, minimise and ultimately eliminate the risks associated with species transfers in ballast water (IMO 2004). Ballast water may be managed by either exchanging the water at high sea or by ballast water management systems. Several of the countries around the Baltic Sea have ratified the BWMC, but its entry-into-force requirements have not yet been met. HELCOM (*cf.* Sect. 17.8.4) and OSPAR (*cf.* Box 14.1) countries have voluntary ballast water management measures in place, which are based on ballast water exchange (BWE) (David and Gollasch 2008).

When BWE is applied, coastal ballast water is taken up in a harbour and later exchanged by seawater from a sea area with a water depth of at least 200 m and at least 200 nautical miles away from land. If this is impossible the ballast water is exchanged in a sea area with a water depth of at least 200 m and at least 50 nautical miles away from land. If this also is impossible, the arrival harbour State may, in accordance with IMO Guidelines, designate a ballast water exchange area that may be closer to land and in less deep waters. However, this procedure is of limited efficiency and cannot be applied in shallow seas like the Baltic Sea. This highlights that BWE needs to be phased-out over time and replaced by a more stringent ballast water performance standard.

A ballast water performance standard sets maximum permitted numbers of living organisms in ballast water discharged from ships. This may be achieved by ballast water management systems installed on board. Methods include mechanical separation of objects in the ballast water (*e.g.* filtration) followed by ultraviolet radiation or the use of so-called "active substances" (*e.g.* chemical reagents) (Gollasch et al. 2007; Gollasch and David 2012; David et al. 2013b; David and Gollasch 2015).

5.7.6 Code of Practices in aquaculture

In the EU, the import of living organisms for use in aquaculture is regulated by the EU Council Regulation No. 708/2007 regarding the use of non-indigenous and locally absent species in aquaculture (EU Council Regulation 2007). This instrument applies to both open and closed aquaculture facilities. It contains provisions on which species can be imported and concerns measures intended to combat possible risks of NIS movements. These measures include the requirement to obtain a permit for species movements, risk assessments, quarantine and monitoring. The instrument does not apply to movements of organisms within a EU Member State (except if there is a risk to the environment), pet-shops, garden centres or aquaria where there is no contact with EU waters, and selected species listed in Annex IV to the EU Council Regulation No. 708/2007.

A similar document was developed by ICES as a voluntary instrument (ICES 2005; Gollasch 2007). It indicates that a desk evaluation should be conducted well in advance of the introduction to include the following: (1) any previous known introduction(s) of the species elsewhere, (2) a review of all known diseases, parasites and other pests associated with the species, (3) a review of its physical tolerances and ecological interactions, (4) a determination of whether there are any possible genetic interactions in the new environment, and (5) a determination of the possible consequences of such an introduction and a hazard assessment. The document also prescribes quarantine and monitoring.

Review questions

1. What is the difference between the range expansion of species and biological invasions?
2. What are the main stages of a biological invasion?
3. What are the major pathways of species introductions into the Baltic Sea?
4. Which habitats are most susceptible to biological invasions in the Baltic Sea?
5. What is biological pollution?

Discussion questions

1. What do environmental managers need to know about bioinvasions?
2. Why is it not correct to put "good" or "bad" tags on non-indigenous species?
3. How would you rank prospective areas of bioinvasion research according to their importance for (a) the development of basic science and (b) direct societal applications?
4. What are the differences between biological pollution and other forms of aquatic pollution?
5. What is the most effective management option for (a) *Prorocentrum cordatum*, (b) *Dreissena polymorpha* and (c) *Neogobius melanostomus*, and why?

References

Anderson DM (2009) Approaches to monitoring, control and management of harmful algal blooms (HABs). Ocean and Coastal Management 52:342–347

AquaNIS (2015) Information system on aquatic non-indigenous and cryptogenic species. World Wide Web electronic publication. [http://www.corpi.ku.lt/databases/aquanis; Version 2.36+, viewed 2015-02-16]

Arbaciauskas K (2002) Ponto-Caspian amphipods and mysids in the inland waters of Lithuania: history of introduction, current distribution and relations with native malacostracans. In: Leppäkoski E, Gollasch S, Olenin S (eds) Invasive aquatic species of Europe – distribution, impacts and management. Kluwer Academic Publishers, Dordrecht, pp 104–115

Arbaciauskas K, Gumuliauskaite S (2005) Invasion of the Baltic Sea basin by the Ponto-Caspian amphipod *Pontogammarus robustoides* and its ecological impact. In: Gherardi F (ed) Biological invaders in inland waters: profiles, distribution, and threats. Springer, Berlin, pp 463–477

Bastrop R, Röhner M, Jürss K (1995) Are there two species of the polychaete genus *Marenzelleria* in Europe? Marine Biology 121:509–516

Bernard FR (1979) Identification of the living *Mya* (Bivalvia: Myoida). Venus 38:185–204

Beukema JJ (1982) Annual variation in reproductive success and biomass of the major macrozoobenthic species living in a tidal flat area of the Wadden Sea. Netherlands Journal of Sea Research 16:37–45

Bielecka L, Mudrak-Cegiołka S, Kalarus M (2014) *Evadne anonyx* G. O. Sars, 1897 – the first record of this Ponto-Caspian cladoceran in the Gulf of Gdańsk (Baltic Sea). Oceanologia 56:141–150

Blank M, Laine AO, Jürss K, Bastrop R (2008) Molecular identification key based on PCR/RFLP for three polychaete sibling species of the genus *Marenzelleria*, and the species' current distribution in the Baltic Sea. Helgoland Marine Research 62:129–141

Bochert R (1997) *Marenzelleria viridis* (Polychaeta: Spionidae): a review of its reproduction. Aquatic Ecology 31:163–175

Boero F, Bouillon J, Gravili C, Miglietta MP, Parsons T, Piraino S (2008) Gelatinous plankton: irregularities rule the world (sometimes). Marine Ecology Progress Series 356:299–310

Buynevich IV, Damušyte A, Bitinas A, Olenin S, Mažeika J, Petrošius R (2011) Pontic-Baltic pathways for invasive aquatic species: geoarchaeological implications. GSA Special Papers 473:189–196

Carlton JT (1996) Biological invasions and cryptogenic species. Ecology 77:1653–1655

Carlton JT (2009) Deep invasion ecology and the assembly of communities in historical time. In: Rilov G, Crooks JA (eds) Biological invasions in marine ecosystems – ecological, management, and geographic perspectives. Springer, Berlin, Ecological Studies 204:13–56

Corkum LD, Sapota M, Skóra KE (2004) The round goby, *Neogobius melanostomus*, a fish invader on both sides of the Atlantic Ocean. Biological Invasions 6:173–181

Cristescu MEA, Hebert PDN, Witt JDS, MacIsaac HJ, Grigorovich IA (2001) An invasion history for *Cercopagis pengoi* based on mitochondrial gene sequences. Limnology and Oceanography 46:224–229

Daunys D, Zemlys P, Olenin S, Zaiko A, Ferrarin C (2006) Impact of the zebra mussel *Dreissena polymorpha* invasion on the budget of suspended material in a shallow lagoon ecosystem. Helgoland Marine Research 60:113–120

David M, Gollasch S (2008) EU shipping in the dawn of managing the ballast water issue. Marine Pollution Bulletin 56:1966–1972

David M, Gollasch S (eds) (2015) Global maritime transport and ballast water management – issues and solutions. Springer, Dordrecht, Invading Nature – Springer Series in Invasion Ecology 8:1–306

David M, Gollasch S, Leppäkoski E (2013a) Risk assessment for exemptions from ballast water management – the Baltic Sea case study. Marine Pollution Bulletin 75:205–217

David M, Gollasch S, Pavliha M (2013b) Global ballast water management and the "same location" concept: a clear term or a clear issue? Ecological Applications 23:331–338

Delefosse M, Kristensen E (2012) Burial of *Zostera marina* seeds in sediment inhabited by three polychaetes: laboratory and field studies. Journal of Sea Research 71:41–49

Elliott M (2003) Biological pollutants and biological pollution – an increasing cause for concern. Marine Pollution Bulletin 46:275–280

Elton CS (1958) The ecology of invasions by animals and plants. Methuen, London 181 pp

EU Council Regulation (2007) (No.708/2007 of 11 June 2007) concerning use of alien and locally absent species in aquaculture

Florin AB, Mo K, Svensson F, Schagerström E, Kautsky L, Bergström L (2013) First records of Conrad's false mussel, *Mytilopsis leucophaeata* (Conrad, 1831) in the southern Bothnian Sea, Sweden, near a nuclear power plant. Bioinvasions Records 2:303–309

Forster S, Zettler ML (2004) The capacity of the filter-feeding bivalve *Mya arenaria* L. to affect water transport in sandy beds. Marine Biology 144:1183–1189

Fowler AE, Forsström T, von Numers M, Vesakoski O (2013) The North American mud crab *Rhithropanopeus harrisii* (Gould, 1841) in newly colonized Northern Baltic Sea: distribution and ecology. Aquatic Invasions 8:89–96

Fritzsche D, von Oertzen JA (1995) Metabolic responses to changing environmental conditions in the brackish water – polychaetes *Marenzelleria viridis* and *Hediste diversicolor*. Marine Biology 121:693–699

Gollasch S (2007) International collaboration on marine bioinvasions – the ICES response. Marine Pollution Bulletin 55:353–359

Gollasch S (2009) *Eriocheir sinensis* Milne Edwards, Chinese mitten crab (Varunidae, Crustacea). In: Drake JA (ed) Handbook of alien species in Europe. Springer, Dordrecht, Invading Nature – Springer Series in Invasion Ecology 3:312

Gollasch S, David M (2012) A unique aspect of ballast water management requirements – the same location concept. Marine Pollution Bulletin 64:1774–1775

Gollasch S, David M, Voigt M, Dragsund E, Hewitt C, Fukuyo Y (2007) Critical review of the IMO international convention on the management of ships' ballast water and sediments. Harmful Algae 6:585–600

Gollasch S, Leppäkoski E (2007) Risk assessment and management scenarios for ballast water mediated species introductions into the Baltic Sea. Aquatic Invasions 2:313–340

Gómez F (2008) Phytoplankton invasions: comments on the validity of categorizing the non-indigenous dinoflagellates and diatoms in European Seas. Marine Pollution Bulletin 56:620–628

Gorokhova E, Fagerberg T, Hansson S (2004) Predation by herring (*Clupea harengus*) and sprat (*Sprattus sprattus*) on *Cercopagis pengoi* in a western Baltic Sea bay. ICES Journal of Marine Science 61:959–965

Gorokhova E, Lehtiniemi M, Viitasalo-Frösen S, Haddock SHD (2009) Molecular evidence for the occurrence of ctenophore *Mertensia ovum* in the northern Baltic Sea and implications for the status of the *Mnemiopsis leidyi* invasion. Limnology and Oceanography 54:2025–2033

Grabowski M, Konopacka A, Jażdżewski K, Janowska E (2006) Invasions of alien gammarid species and retreat of natives in the Vistula Lagoon (Baltic Sea, Poland). Helgoland Marine Research 60:90–97

Gren IM, Isacs L, Carlsson M (2009) Costs of alien invasive species in Sweden. AMBIO 38:135–140

Grigorovich IA, MacIsaac HJ, Rivier IK, Aladin NV, Panov VE (2000) Comparative biology of the predatory cladoceran *Cercopagis pengoi* from Lake Ontario, Baltic Sea and Caspian Sea. Archiv für Hydrobiologie 149:23–50

Hakanen P, Suikkanen S, Franzén J, Franzén H, Kankaapää H, Kremp A (2012) Bloom and toxin dynamics of *Alexandrium ostenfeldii* in a shallow embayment at the SW coast of Finland, northern Baltic Sea. Harmful Algae 15:91–99

Hällfors G (2004) Checklist of Baltic Sea phytoplankton species (including some heterotrophic protistan groups). Baltic Sea Environment Proceedings 95:1–208

Hansson HG (2006) Ctenophores of the Baltic and adjacent Seas – the invader *Mnemiopsis* is here! Aquatic Invasions 1:295–298

Hedman JE, Gunnarsson JS, Samuelsson G, Gilbert F (2011) Particle reworking and solute transport by the sediment-living polychaetes *Marenzelleria neglecta* and *Hediste diversicolor*. Journal of Experimental Marine Biology and Ecology 407:294–301

Hegele-Drywa J, Normant M (2009) Feeding ecology of the American crab *Rhithropanopeus harrisii* (Crustacea, Decapoda) in the coastal waters of the Baltic Sea. Oceanologia 51:361–375

HELCOM (2010) Maritime activities in the Baltic Sea – an integrated thematic assessment on maritime activities and response to pollution at sea in the Baltic Sea region. Baltic Sea Environment Proceedings 123:1–64

Hessland I (1946) On the Quaternary Mya period in Europe. Arkiv för Zoologi 37A(8):1–51

Höglund J, Andersson (1993) Prevalence and abundance of *Anguillicola crassus* in the European eel (*Anguilla anguilla*) at a thermal discharge site on the Swedish coast. Journal of Applied Ichthyology 9:115–122

ICES (2005) International Council for the Exploration of the Sea code of practice on the introduction and transfers of marine organisms. Denmark, Copenhagen 14 pp

Ilus E (2009) Environmental effects of thermal and radioactive discharges from nuclear power plants in the boreal brackish-water conditions of the northern Baltic Sea. STUK – Radiation and Nuclear Safety Authority, Helsinki, Finland, 380 pp

IMO (2004) International convention on the control and management of ship's ballast water and sediments. International Maritime Organization, London, UK, 36 pp

Jaspers C, Titelman J, Hansson LJ, Haraldsson M, Ditlefsen CR (2011) The invasive ctenophore *Mnemiopsis leidyi* poses no direct threat to Baltic cod eggs and larvae. Limnology and Oceanography 56:431–439

Jaszczołt J, Szaniawska A (2011) The spiny-cheek crayfish *Orconectes limosus* (Rafinesque, 1817) as an inhabitant of the Baltic Sea – experimental evidences for its invasion of brackish waters. Oceanological and Hydrobiological Studies 40:52–60

Javidpour J, Sommer U, Shiganova T (2006) First record of *Mnemiopsis leidyi* A. Agassiz 1865 in the Baltic Sea. Aquatic Invasions 1:299–302

Johnston EL, Piola RF, Clark GF (2009) The role of propagule pressure in invasion success. In: Rilov G, Crooks JA (eds) Biological invasions in marine ecosystems – ecological, management, and geographic perspectives. Springer, Berlin, Ecological Studies 204:133–151

Jørgensen CB, Riisgård HU (1988) Gill pump characteristics of the soft clam *Mya arenaria*. Marine Biology 99:107–109

Jousson O, Pawlowski J, Zaninetti L, Meinesz A, Boudouresque CF (1998) Molecular evidence for the aquarium origin of the green alga *Caulerpa taxifolia* introduced to the Mediterranean Sea. Marine Ecology Progress Series 172:275–280

Karatayev AY, Burlakova LE, Molloy DP, Volkova LK (2000) Endosymbionts of *Dreissena polymorpha* (Pallas) in Belarus. International Review of Hydrobiology 85:543–559

Karatayev AY, Mastitsky SE, Burlakova LE, Olenin S (2008) Past, current, and future of the central European corridor for aquatic invasions in Belarus. Biological Invasions 10:215–232

Karlson AML, Almqvist G, Skóra KE, Appelberg M (2007) Indications of competition between non-indigenous round goby and native flounder in the Baltic Sea. ICES Journal of Marine Science 64: 479–486

Katajisto T, Karjala L, Lehtiniemi M (2013) Fifteen years after invasion: egg bank of the predatory cladoceran *Cercopagis pengoi* in the Baltic Sea. Marine Ecology Progress Series 482:81–92

Kelly RP, Port JA, Yamahara KM, Martone RG, Lowell N et al (2014) Harnessing DNA to improve environmental management. Science 344:1455–1456

Kideys AE (2002) Fall and rise of the Black Sea ecosystem. Science 297:1482–1483

Kornis MS, Mercado-Silva N, Van der Zanden MJ (2012) Twenty years of invasion: a review of round goby *Neogobius melanostomus* biology, spread and ecological implications. Journal of Fish Biology 80:235–285

Kotta J, Kotta I, Simm M, Lankov A, Lauringson V et al (2006) Ecological consequences of biological invasions: three invertebrate case studies in the northeastern Baltic Sea. Helgoland Marine Research 60:106–112

Kotta J, Ólafsson E (2003) Competition for food between the introduced exotic polychaete *Marenzelleria viridis* (Verrill) and the native amphipod *Monoporeia affinis* Lindström in the Baltic Sea. Journal of Sea Research 50:27–35

Kotta J, Orav H, Kotta I (1998) Distribution and filtration activity of zebra mussel, *Dreissena polymorpha* (Pallas) in the Gulf of Riga and the Gulf of Finland. Proceedings of the Estonian Academy of Sciences, Biology and Ecology 47:32–41

Kotta J, Orav H, Sandberg-Kilpi E (2001) Ecological consequence of the introduction of the polychaete *Marenzelleria viridis* into a shallow water biotope of the northern Baltic Sea. Journal of Sea Research 46:273–280

Kotta J, Orav-Kotta H, Paalme T, Kotta I, Kukk H (2003) Benthos studies in the Estonian coastal sea during 1998-2001. Proceedings of the Estonian Academy of Sciences, Biology and Ecology 52:85–90

Kownacka J, Edler L, Gromisz S, Łotocka M, Olenina I et al (2013) Non-indigenous species *Chaetoceros cf. lorenzianus* Frunow 1863 – a new, predominant components of autumn phytoplankton in the southern Baltic Sea. Estuarine, Coastal and Shelf Science 119:101–111

Laine AO, Mattilla J, Lehikoinen A (2006) First record of the brackish water dreissenid bivalve *Mytilopsis leucophaeata* in the northern Baltic Sea. Aquatic Invasions 1:38–41

Laxson CL, McPhedran KN, Makarewicz JC, Telesh I, MacIsaac HJ (2003) Effects of the non-indigenous cladoceran *Cercopagis pengoi* on the lower food web of Lake Ontario. Freshwater Biology 48:2094–2106

Lefebvre F, Wielgoss S, Nagasawa K, Moravec F (2012) On the origin of *Anguillicoloides crassus*, the invasive nematode of anguillid eels. Aquatic Invasions 7:443–453

Lehtiniemi M, Gorokhova E (2008) Predation of the introduced cladoceran *Cercopagis pengoi* on the calanoid copepod *Eurytemora affinis* in the Gulf of Finland, Baltic Sea. Marine Ecology Progress Series 362:193–200

Lehtiniemi M, Lehmann A, Javidpour J, Myrberg K (2012) Spreading and physico-biological reproduction limitations of the invasive American comb jelly *Mnemiopsis leidyi* in the Baltic Sea. Biological Invasions 14:341–354

Lehtiniemi M, Ojaveer H, David M, Galil B, Gollasch S et al (2015) Dose of truth – monitoring marine non-indigenous species to serve legislative requirements. Marine Policy 54:26–35

Leppäkoski E (2002) Harmful non-native species in the Baltic Sea – an ignored problem. In: Schernewski E, Schiewer U (eds) Baltic coastal ecosystems – structure, function and coastal management. Springer, Berlin, pp 253–275

Leppäkoski E, Gollasch S, Gruszka P, Ojaveer H, Olenin S, Panov V (2002a) The Baltic – a sea of invaders. Canadian Journal of Fisheries and Aquatic Sciences 59:1175–1188

Leppäkoski E, Gollasch S, Olenin S (2002b) Alien species in European waters. In: Leppäkoski E, Gollasch S, Olenin S (eds) Invasive aquatic species of Europe – distribution, impacts and management. Kluwer Academic Publishers, Dordrecht, pp 1–6

Leppäkoski E, Olenin S (2000) Non-native species and rates of spread: lessons from the brackish Baltic Sea. Biological Invasions 2:151–163

Leppäkoski E, Olenin S (2001) The meltdown of biogeographical peculiarities of the Baltic Sea: the interaction of natural and manmade processes. AMBIO 30:202–209

Meinesz A, Belsher T, Thibaut T, Antolic B, Mustapha KB et al (2001) The introduced green alga *Caulerpa taxifolia* continues to spread in the Mediterranean. Biological Invasions 3:201–210

Meixner R, Gerdener C (1976) Untersuchungen über die Möglichkeiten einer Fortpflanzung der eingeführten Pazifischen Auster *Crassotrea gigas* in deutschen Küstengewässern. Archiv für Fischereiwissenschaft 26:155–160 [in German]

Mills CE (1995) Medusae, siphonophores, and ctenophores as planktivorous predators in changing global ecosystems. ICES Journal of marine Science 51:575–581

Mills EL, Leach JH, Carlton JT, Secor CL (1993) Exotic species in the Great Lakes: a history of biotic crises and anthropogenic introductions. Journal of Great Lakes Research 19:1–54

Minchin D, Galil BS, David M, Gollasch S, Olenin S (2006) Overall introduction. In: Gollasch S, Galil BS, Cohen AN (eds) Bridging divides – maritime canals as invasion corridors. Springer, Dordrecht, Monographiae Biologicae 83:1–4

Minchin D, Gollasch S, Cohen AN, Hewitt CL, Olenin S (2009) Characterizing vectors of marine invasion. In: Rilov G, Crooks JA (eds) Biological invasions in marine ecosystems – ecological, management, and geographic perspectives. Springer, Berlin, Ecological Studies 204:109–116

Minchin D, Lucy F, Sullivan M (2002) Zebra mussel: impacts and spread. In: Leppäkoski E, Gollasch S, Olenin S (eds) Invasive aquatic species of Europe – distribution, impacts and management. Kluwer Academic Publishers, Dordrecht, pp 135–146

Mordukhai-Boltovskoi FD, Rivier IK (1971) A brief survey of the ecology and biology of the Caspian Polyphemoidea. Marine Biology 8:160–169

Mordukhai-Boltovskoi FD, Rivier IK (1987) Predatory cladocerans (Podonidae, Polyphemidae, Cercopagidae and Leptodoridae) in the world's fauna. Nauka, Leningrad 1–182 [in Russian]

Norkko J, Reed DC, Timmermann K, Norkko A, Gustafsson BG et al (2012) A welcome can of worms? – hypoxia mitigation by an invasive species. Global Change Biology 18:422–434

Nummi P (2002) Introduced semiaquatic birds and mammals in Europe. In: Leppäkoski E, Gollasch S, Olenin S (eds) Invasive aquatic species of Europe – distribution, impacts and management. Kluwer Academic Publishers, Dordrecht, pp 162–172

Nyberg (2007) Introduced marine macroalgae and habitat modifiers – their ecological role and significant attributes. Göteborg University, 61 pp [PhD Thesis]

Ojaveer H, Gollasch S, Jaanus A, Kotta J, Laine AO et al (2007) Chinese mitten crab *Eriocheir sinensis* in the Baltic Sea – a supply-side invader? Biological Invasions 9:409–418

Ojaveer H, Kotta J (2015) Ecosystem impacts of the widespread non-indigenous species in the Baltic Sea: literature survey evidences major limitations in knowledge. Hydrobiologia 750:171–185

Ojaveer H, Simm M, Lankov A (2004) Population dynamics and ecological impact of the non-indigenous *Cercopagis pengoi* in the Gulf of Riga (Baltic Sea). Hydrobiologia 522:261–269

Olenin S, Alemany F, Cardoso AC, Gollasch S, Goulletquer P et al (2010) Marine Strategy Framework Directive – Task Group 2 Report. Non-indigenous species. EUR 24342 EN. Office for Official Publications of the European Communities, Luxembourg, p 44

Olenin S, Elliott M, Bysveen I, Culverhouse PF, Daunys D et al (2011) Recommendations on methods for the detection and control of biological pollution in marine coastal waters. Marine Pollution Bulletin 62:2598–2604

Olenin S, Leppäkoski E (1999) Non-native animals in the Baltic Sea: alteration of benthic habitats in coastal inlets and lagoons. Hydrobiologia 393:233–243

Olenin S, Minchin D, Daunys D (2007) Assessment of biopollution in aquatic ecosystems. Marine Pollution Bulletin 55:379–394

Olenin S, Narščius A, Minchin D, David M, Galil B et al (2014) Making non-indigenous species information systems practical for management and useful for research: an aquatic perspective. Biological Conservation 173:98–107

Olenina I, Wasmund N, Hajdu S, Jurgensone I, Gromisz S et al (2010) Assessing impacts of invasive phytoplankton: the Baltic Sea case. Marine Pollution Bulletin 60:1691–1700

Oliveira OMP (2007) The presence of the ctenophore *Mnemiopsis leidyi* in the Oslofjorden and considerations on the initial invasion pathways to the North and Baltic Seas. Aquatic Invasions 2:185–189

Orlova M, Golubkov S, Kalinina L, Ignatieva N (2004) *Dreissena polymorpha* (Bivalvia: Dreissenidae) in the Neva Estuary (eastern Gulf of Finland, Baltic Sea): is it a biofilter or source for pollution? Marine Pollution Bulletin 49:196–205

Orlova MI (2002) *Dreissena polymorpha* (Pallas, 1771): evolutionary origin and biological peculiarities as prerequisites of invasion success. In: Leppäkoski E, Gollasch S, Olenin S (eds) Invasive aquatic species of Europe – distribution, impacts and management. Kluwer Academic Publishers, Dordrecht, pp 127–134

Orlova MI, Telesh I, Berezina NA, Antsulevich AE, Maximov AA, Litvinchuk LF (2006) Effects of non-indigenous species on diversity and community functioning in the eastern Gulf of Finland (Baltic Sea). Helgoland Marine Research 60:98–105

Paavola M, Olenin S, Leppäkoski E (2005) Are invasive species most successful in habitats of low native species richness across European brackish water seas? Estuarine, Coastal and Shelf Science 64:738–750

Parmesan C, Yohe G (2003) A globally coherent fingerprint of climate change impacts across natural systems. Nature 421:37–42

Petersen JK, Hansen JW, Laursen MB, Clausen P, Carstensen J, Conley DJ (2008) Regime shift in a coastal marine ecosystem. Ecological Applications 18:497–510

Petersen KS, Rasmussen KL, Heinemeier J, Rud N (1992) Clams before Columbus? Nature 359:679

Petersen KS, Rasmussen KL, Rasmussen P, von Platen-Hallermund F (2005) Main environmental changes since the Weichselian glaciation in the Danish waters between the North Sea and the Baltic Sea as reflected in the molluscan fauna. Quaternary International 133–134:33–46

Plichlová-Ptácniková R, Vanderploeg HA (2009) The invasive cladoceran *Cercopagis pengoi* is a generalist predator capable of feeding on a variety of prey species of different sizes and escape abilities. Fundamental and Applied Limnology 173:267–268

Pochon X, Bott NJ, Smith KF, Wood SA (2013) Evaluating detection limits of Next-Generation Sequencing for the surveillance and monitoring of international marine pests. PLoS ONE 8(9):e73935

Põllupüü M, Simm M, Põllumäe A, Ojaveer H (2008) Successful establishment of the Ponto-Caspian alien cladoceran *Evadne anonyx* G.O. Sars 1897 in low-salinity environment in the Baltic Sea. Journal of Plankton Research 30:777–782

Ponder WF (1988) *Potamopyrgus antipodarum*: a molluscan coloniser of Europe and Australia. Journal of Molluscan Studies 54:271–285

Purcell JE, Shiganova TA, Decker MB, Houde ED (2001) The ctenophore *Mnemiopsis* in native and exotic habitats: U.S. estuaries *versus* the Black Sea basin. Hydrobiologia 451:145–176

Rakauskas V, Putys Z, Dainys J, Lesutiene J, Lozys L, Arbaciauskas K (2013) Increasing population of the invader round goby *Neogobius melanostomus* (Actinopterygii: Perciformes: Gobiidae) and its trophic role in the Curonian Lagoon, SE Baltic Sea. Acta Ichthyologica et Piscatoria 43:95–108

Reise K, Olenin S, Thieltges DW (2006) Are aliens threatening European aquatic coastal ecosystems? Helgoland Marine Research 60:106–112

Remane A (1934) Die Brackwasserfauna. Verhandlungen der Deutschen Zoologischen Gesellschaft 36:34–74 [in German]

Riisgård HU, Seerup DF (2003) Filtration rates in the soft clam *Mya arenaria*: effects of temperature and body size. Sarsia 88: 415–428

Rudinskaya LV, Gusev AA (2012) Invasion of the North American wedge clam *Rangia cuneata* (G.B. Sowerby I, 1831) (Bivalvia: Mactridae) in the Vistula Lagoon of the Baltic Sea. Russian Journal of Biological Invasions 3:220–229

Saiz E, Kiørboe T (1995) Predatory and suspension feeding of the copepod *Acartia tonsa* in turbulent environments. Marine Ecology Progress Series 122:147–158

Sandström (1990) Environmental monitoring at the Forsmark nuclear power plant. Swedish Environmental Protection Board, Report 3860:1–37

Sapota MR (2004) The round goby (*Neogobius melanostomus*) in the Gulf of Gdańsk – a species introduction into the Baltic Sea. Hydrobiologia 514:219–224

Sapota MR, Skóra KE (2005) Spreading of alien (non-indigenous) fish species *Neogobius melanostomus* in the Gulf of Gdańsk (South Baltic). Biological Invasions 7:157–164

Schiedek D (1993) *Marenzelleria viridis* (Verrill) (Polychaeta), a new benthic species within European coastal waters: some metabolic features. Journal of Experimental Marine Biology and Ecology 211:85–101

Sikorski AV, Bick A (2004) Revision of *Marenzelleria* Mesnil, 1896 (Spionidae, Polychaeta). Sarsia 89:253–275

Simberloff D, Von Holle B (1999) Positive interactions of nonindigenous species: invasional meltdown? Biological Invasions 1:21–32

Snoeijs P (1989) Effects of increasing water temperatures and flow rates on epilithic fauna in a cooling-water discharge basin. Journal of Applied Ecology 26:935–956

Snoeijs P, Mo K (1987) Macrofauna on rocky substrates in the Forsmark Biotest basin, March 1984-March 1985. Swedish Environmental Protection Board, Report 3397:1–55

Snoeijs P, Weckström K (2010) Diatoms and environmental change in large brackish-water ecosystems. In: Smol JP, Stoermer EF (eds) The diatoms: applications for the environmental and Earth sciences, 2nd edn. Cambridge University Press, Cambridge, pp 287–308

Sokołowska E, Fey DP (2011) Age and growth of the round goby *Neogobius melanostomus* in the Gulf of Gdánsk several years after invasion. Is the Baltic Sea a new promised land? Journal of Fish Biology 78:1993–2009

Städler T, Frye M, Neiman M, Lively CM (2005) Mitochondrial haplotypes and the New Zealand origin of clonal European *Potamopyrgus*, an invasive aquatic snail. Molecular Ecology 14:2465–2473

Strasser CA, Barber PH (2009) Limited genetic variation and structure in soft-shell clams (*Mya arenaria*) across their native and introduced range. Conservation Genetics 10:803–814

Strasser M (1999) *Mya arenaria* – an ancient invader of the North Sea coast. Helgoländer Meeresuntersuchungen 52:309–324

Tendal OS, Jensen KR, Riisgård HU (2007) Invasive ctenophore *Mnemiopsis leidyi* widely distributed in Danish waters. Aquatic Invasions 2:455–460

Tomczak MT, Sapota MR (2006) The fecundity and gonad development cycle of the round goby (*Neogobius melanostomus* Pallas 1811) from the Gulf of Gdańsk. International Journal of Oceanography and Hydrobiology, Oceanological and Hydrobiological Studies 35:353–367

Verity PG, Smetacek V (1996) Organism life cycles, predation and the structure of marine pelagic ecosystems. Marine Ecology Progress Series 130:277–293

Verween A, Kerckhof F, Vincx M, Degraer S (2006) First European record of the invasive brackish water clam *Rangia cuneata* (G.B. Sowerby I, 1831) (Mollusca: Bivalvia). Aquatic Invasions 1: 198–203

Vetemaa M (2006) Invasion history and population structure of the alien gibel carp *Carassius gibelio* in Estonian marine waters. In: Ojaveer H, Kotta J (eds) Alien invasive species in the northeastern Baltic Sea: population dynamics and ecological impacts. Estonian Marine Institute Report Series 14:30–34

Virbickas J. (2000) Lietuvos žuvys ("Fish of Lithuania"). Vilnius: Trys žvaigždutės, 192 p. [in Lithuanian]

Wakida-Kusunoke AT, MacKenzie CL (2004) *Rangia* and marsh clams, *Rangia cuneata*, *R. flexuosa* and *Polymesoda caroliniana*, in eastern Mexico: distribution, biology and ecology, and historical fisheries. Marine Fisheries Review 66:13–20

Warzocha J, Drgas A (2013) The alien gulf wedge clam (*Rangia cuneata* G.B. Sowerby I, 1831) (Mollusca: Bivalvia; Mactridae) in the Polish part of the Vistula Lagoon (SE Baltic). Folia Malacologica 21:291–297

Wasmund N, Göbel J, von Bodungen B (2008) 100-years-changes in the phytoplankton community of Kiel Bight (Baltic Sea). Journal of Marine Systems 73:300–322

Westman (2002) Alien crayfish in Europe: negative and positive impacts and interactions with native crayfish. In: Leppäkoski E, Gollasch S, Olenin S (eds) Invasive aquatic species of Europe – distribution, impacts and management. Kluwer Academic Publishers, Dordrecht, pp 76–95

Winkler HM, Debus L (1997) Is the polychaete *Marenzelleria viridis* an important food item for fish? In: Andrusaitis A (ed) Proceedings of the 13th Symposium of the Baltic Marine Biologists, Riga, pp 147–151

Witkowski A, Grabowska J (2012) The non-indigenous freshwater fishes of Poland: threats to the native ichthyofauna and consequences for the fishery: a review. Acta Ichthyologica et Piscatoria 42:77–87

Wrange AL, Valero J, Harkestad LS, Strand Ø, Lindegarth S et al (2010) Massive settlements of the Pacific oyster, *Crassostrea gigas*, in Scandinavia. Biological Invasions 12:1145–1152

Zaiko A, Daunys D (2015) Invasive ecosystem engineers and biotic indices: giving a wrong impression of water quality improvement? Ecological Indicators 52:292–299

Zaiko A, Daunys D, Olenin S (2009) Habitat engineering by the invasive zebra mussel *Dreissena polymorpha* (Pallas) in a boreal coastal lagoon: impact on biodiversity. Helgoland Marine Research 63:85–94

Zaiko A, Lehtiniemi M, Narščius A, Olenin S (2011) Assessment of bioinvasion impacts on a regional scale: a comparative approach. Biological Invasions 13:1739–1765

Zaiko A, Minchin D, Olenin S (2014) "The day after tomorrow": anatomy of an "r" strategist aquatic invasion. Aquatic Invasions 9:145–155

Zaiko A, Olenin S, Daunys D, Nalepa T (2007) Vulnerability of benthic habitats to the aquatic invasive species. Biological Invasions 9:703–714

Zettler ML (1996) Successful establishment of the spionid polychaete, *Marenzelleria viridis* (Verrill, 1873), in the Darß-Zingst estuary (southern Baltic) and its influence on the indigenous macrozoobenthos. Archive of Fishery and Marine Research 43:273–284

Genetic diversity and evolution

6

Risto Väinölä and Kerstin Johannesson

Abstract

1. Genetic variability among individuals, populations and species represents the basic level of biodiversity, and is a prerequisite of adaptive evolution.
2. Adaptive evolution is driven by natural selection that acts at the level of individual phenotypes.
3. Genetic variation can also be used as a tool to study the history of species and populations, and to explore their current structure and reproductive strategies.
4. Genetic markers that are presumably neutral to selection are used in measuring connectivity among Baltic populations and their uniqueness compared to those in the neighbouring marine or freshwater habitats. Genetic markers have often revealed the presence of previously unknown cryptic species that are much older than the Baltic Sea.
5. In most taxa studied, some genetic differentiation has arisen post-glacially between the Baltic Sea and North Sea populations, *e.g.* in the Atlantic herring *Clupea harengus* and the Atlantic cod *Gadus morhua*.
6. Despite such differentiation, few of the Baltic organisms are considered as locally evolved endemic taxa. An exception is the partly asexually reproducing brown algal species *Fucus radicans*, which has evolved locally and now coexists with its ancestor *Fucus vesiculosus* in the northern Baltic Sea.
7. The unique blue mussel and Baltic clam populations in the Baltic Sea are closely related to Pacific lineages (*Mytilus trossulus* and *Macoma balthica balthica*) but are distinct from the neighbouring North Sea populations (*Mytilus edulis* and *Macoma balthica rubra*). They have been modified by interbreeding in the transition zone between the Baltic Sea and the North Sea, and now constitute hybrid swarms.
8. A current methodological shift from single-gene approaches to genome-wide studies will help in distinguishing genes and patterns of variation that are affected by selection from those that merely reflect population structure, and in identifying characters that account for the adaptations to the unique Baltic Sea environment.

Keywords

Adaptive evolution • Genetic diversity • Genetic markers • Hybrid swarms • Population structure • Reproductive traits • Speciation

R. Väinölä (✉)
Finnish Museum of Natural History, University of Helsinki,
P.O. Box 17, 00014 Helsinki, Finland
e-mail: risto.vainola@helsinki.fi

K. Johannesson
Department of Marine Sciences-Tjärnö, University of Gothenburg,
45296 Strömstad, Sweden
e-mail: kerstin.johannesson@marine.gu.se

P. Snoeijs-Leijonmalm et al. (eds.), *Biological Oceanography of the Baltic Sea*,
DOI 10.1007/978-94-007-0668-2_6

6.1 Genetic diversity

6.1.1 Biological and adaptive importance

Any biological properties of organisms, whether they are morphological, physiological or behavioural, are encoded in DNA. The functioning of organisms is thus based on the information in their genomes. Consequently, the differences between individuals, populations and species, which make the essence of biodiversity at the lower levels of biological organisation (below that of communities and ecosystems), directly or indirectly have a genetic basis. Particularly, genetic variation forms the basis of adaptation of organisms and populations to different environmental conditions, and provides the raw material for further evolution.

The Baltic Sea organisms inhabit an atypical marginal environment, and most of them have invaded this environment from either a true marine or a freshwater habitat (*cf.* Sect. 4.5.2). By the process of natural selection, most Baltic populations are likely to have undergone genetic changes to adapt to the exceptional conditions, although the exact genomic or genetic basis of such adaptations is as yet rarely known. Nevertheless, for species with a wide physiological tolerance range, an invasion into a new kind of environment could also be successful without major genetic adjustments.

The time frame of the adaptive changes for the organisms now living in the Baltic Sea is the post-glacial period since the start of the current marine phase of the basin (roughly the last 8,000 years, *cf.* Fig. 2.26c), and this is short in the evolutionary perspective. In particular, it is short for the formation of new species that are morphologically distinct and are reproductively isolated from their closest relatives outside the Baltic Sea. Therefore, despite the putative genetic differentiation, few of the Baltic organisms are considered as locally evolved endemic taxa.

6.1.2 Phenotypic traits

The studies of genetic variation within and among populations and species can have various goals. In one main line of research, the interest is in the functional and adaptive significance of the genetically controlled traits themselves and on their evolution. In the other line of research, genetic properties of organisms are used as research tools to study other aspects of their biology, particularly as markers that bear information on the historical relationships and current connections of species and populations, and of the relatedness between individuals.

The adaptive evolution of populations is driven by natural selection, which acts at the level of individual phenotypes.

Fig. 6.1 The isopod *Idotea balthica* exhibits a genetically determined polymorphism in its colouration. The colour morphs co-occur, but are differentially susceptible to predation in sheltered, structurally complex high-diversity habitats *versus* exposed habitats, and are consequently found in different frequencies at different locations, despite gene flow between them (Merilaita 2001). Photo: © Risto Väinölä

However, in relatively few cases have the genetic backgrounds of individual traits of presumably adaptive significance in the Baltic Sea been unveiled. Traits of this type include vision, resistance genes (immunology) and pigmentation (camouflage, mating colouration). For example, conspicuous genetic colour polymorphism occurs in *Idotea balthica* (Fig. 6.1), an abundant isopod associated with phytobenthic communities dominated by the bladderwrack *Fucus vesiculosus* or the common eelgrass *Zostera marina* in the Baltic Sea (*cf.* Sect. 11.13).

In other cases, the variation is not organised as discrete polymorphism but is continuously distributed. Such quantitative traits usually have a partly genetic basis, which by conventional methods could not be directly traced to individual genes. This involves variation in morphological and life-history traits such as size at maturation, and measures of physiological and reproductive performance, such as growth rate and number of offspring.

Assessment of the genetic background of quantitative traits is based on statistical analysis of the resemblance between individuals of known relatedness, or experimental set-ups where environmental influences can be controlled. With this approach one can estimate the heritability of a trait in the particular population and under the particular experimental conditions. In more modern approaches, the heritable component of a trait may now be traced back to certain chromosomal regions, given that information on a genetic map is also available (QTL mapping, referring to quantitative trait loci).

For well-studied genomes, and with the introduction of cost-efficient methods for large-scale DNA sequencing and genotyping (next generation sequencing, NGS), it becomes possible to identify genomic regions containing genetic variation that affect such central properties as individual fitness, and this will ultimately help to identify the genes that themselves are the targets of selection.

6.1.3 Genetic markers of population structure and history

Much of the recent genetic research on marine organisms is not related to the study of visible or measurable phenotypic traits, and is often not even looking for the basis of adaptive differentiation. Instead, the information on the distribution of molecular genetic variation among species, populations and individuals is used to infer the history of these entities and their current relationships.

This approach arises from the availability of tools to study genetic variation at the DNA and protein levels (Box 6.1). In principle, the genomes of organisms contain an almost unlimited number of inter-individual differences and polymorphisms that could be used to characterise biological units at all levels, including species, biogeographically distinct lineages, evolutionary independent populations, or demographically independent populations (which may exchange genes but are still independent in terms of population growth and survival).

Further, genetic markers can be used to identify reproductive units such as clonality and the kinship and genealogical relationships among individual organisms. In these approaches, the functional importance of the variation studied is usually not known, but is assumed to be more or less unimportant and selectively neutral for the most part. Markers are taken to represent the general patterns of genomic variation, which should be shaped by the history and population structure in a similar way for any gene.

6.1.4 Application of genetic markers

The use of marker variation is based on the theoretical and statistical understanding of the processes of molecular evolution and population genetics. The patterns of variation among species, populations and individuals can be theoretically predicted under various historical scenarios, population structures and modes of inheritance. Conversely, from the observed patterns of variation, information on the structure and history of populations can be inferred based on the same theory (Avise 2000; Box 6.2).

An arsenal of genetic markers is available (Box 6.1). Different tools are differently suited to address questions at the various hierarchical levels of biological organisation. The suitability of a marker depends basically on its mutation rate, the mode of inheritance (whether it is clonal, such as for the maternally inherited mitochondria, or sexual and biparental, involving reshuffling of different genes) and the economy of the detection, *e.g.* the possibility to screen large numbers of variable characters simultaneously. Approaches using large numbers of markers distributed across the genome are referred to as population genomics.

The marker traits are basically used at three hierarchical levels of biological organisation:

1. The history of a species, which includes phylogeny (the evolutionary relationships between taxa), speciation, species boundaries (practical species discrimination), hybridisation (the breaking of species boundaries) and phylogeography, which addresses broad-scale population history. The differences that are addressed in these studies usually originated during pre-Baltic Sea times, from 10^4 to 10^7 years ago.
2. Population structure within a species, which includes the contemporary connectedness and substructure of populations and their history. In practice, such studies deal with the time scale of the existence of the Baltic Sea and its recent geological stages (*cf.* Sect. 2.5) during the last 10,000 years.

Box 6.1: Genetic markers and their applicability

Colour polymorphisms: Some animal species show variation in colour patterns that are inherited following simple Mendelian rules. Before the availability of molecular tools, such phenotypic traits could be used as markers of population structure and history, although they are also likely to be affected by natural selection.

Allozymes: Genes encode proteins, and a small proportion of DNA-level differences in a gene cause amino acid differences in the protein. Some 30 % of such differences in protein structure can be separated by their electric charge. Allozymes, the electrophoretically separated variants of metabolic enzymes, were the main marker for studies of population genetic structure in the 1970s–1990s. They were also effective in identifying cryptic species and population histories. In the DNA era, allozymes have been replaced by mtDNA, microsatellites and SNPs.

Mitochondrial DNA (mtDNA): The genome of the animal mitochondrion is generally inherited maternally, as a non-recombining clone. Mitochondrial DNA has been the single most-used genetic marker molecule through the 1990s–2000s. Its advantages are the universal and efficient methodology for its study, the high mutation rate that generates variation, and the small effective population size (*cf.* Box 6.2) that makes it sensitive to loss of variation and the detection of demographic changes. Results from comparative mtDNA gene sequencing can be interpreted as gene genealogies (trees), which are efficient in the reconstruction of phylogenetic and phylogeographic histories on time scales from thousands to millions of years. The mitochondrially encoded COI (cytochrome *c* oxidase I) gene is now promoted as a universal barcode character for animal species identification, but is not useful in plants.

Nuclear gene sequencing: Mitochondrial data alone are susceptible to biases, mtDNA being effectively a single gene with a unique mode of inheritance. For a reliable view of species phylogeny, several independent genes from the standard nuclear genome need to be sequenced.

Microsatellites: Microsatellites are hypervariable co-dominant genotypic markers, in which most individuals may have different genotypes. Microsatellites are used in the study of population structure, but they are also efficient for the exact identification of individuals and of close family relationships (*e.g.* parent-offspring), whereas they are not useful for phylogenies. The differences between alleles are based on varying numbers of tandemly repeated microsatellite motifs in nucleotide sequence [*e.g.* CTACTACTACTA… = (CTA)$_4$...], in which length mutations easily happen. As a drawback, microsatellites are taxon-specific, and primers for their study have to be separately designed for each target species or genus. Microsatellites are mostly considered as neutral markers that are not affected by natural selection, and thus only reflect population structure.

SNPs: Single nucleotide polymorphisms (SNPs) refer to differences based on point mutations (single nucleotide change, *e.g.* A<>G) anywhere in the genome. An individual SNP is less informative than a microsatellite, but there are hundreds of thousands of them throughout the genome, providing an unlimited source of information. They can be studied individually in a targeted way, or by assessing a large proportion of the genome at a time with high-throughput approaches. SNPs are used *e.g.* to probe population structures on a genome-wide basis (population genomics) and in identifying anomalously behaving genome regions (*cf.* Sect. 6.3.4). While the vast majority of SNPs on the genomic scale are themselves effectively neutral, their variation can sometimes reflect that of a physically and closely linked functionally important polymorphism.

Next generation sequencing (NGS): While previous studies were targeting a restricted set of known marker characters, the current NGS technologies can provide extremely large amounts of data even from very small amounts of DNA, and from many specimens in a single sequencing run. Typically, short fragments are sequenced that are randomly distributed throughout the genome. Alternatively, the whole genome of individuals may be resequenced, at low coverage, and compared to a model genome of the species, and data on marker variation are compiled by bioinformatic procedures. The first goal is often to identify SNP variation for high-throughput genotyping approaches. To obtain a reasonable (not too large) amount of data, various genome reduction techniques are used. For example, only the transcriptome (from mRNA) is studied, or a subset of the genome is picked by using specific restriction-cutting sites to define priming sites (RAD sequencing). Genome-wide sequencing can be used *e.g.* for comprehensive population structure analysis, for identification of candidate genes potentially affected by natural selection using outlier detection, and even for phylogenetics. If a well-annotated reference genome of the target species or a closely related species is available, genome-wide sequencing combined with pedigree information can be used to link phenotypic traits to genes of known function by quantitative trait loci (QTL) mapping. A reference genome also makes it more efficient to detect sequence (SNP) variation by whole-genome resequencing. Transcriptome sequencing can also be used to study gene expression, *e.g.* to detect response to environmental change (RNA-seq).

Box 6.2: Dynamics of genetic variation—interpreting information from genetic tools

Population genetics: dynamics of variability

Genetic variation is introduced to a population by mutation, by gene flow (migration) from other populations, or even by hybridisation and introgression from other taxa. Genetic variation will be lost through natural selection that purges ill-adapted gene variants (alleles) and tends to favour well-adapted ones, and by genetic drift—the random loss of variation, which occurs at a rate that is inversely related to the effective population size N_e.

A large population can accumulate and maintain large amounts of genetic variation, whereas small populations will lose approximately a proportion $1/(2N_e)$ of their variation each generation. Therefore, it is thought that episodes of small population size, the so-called population bottlenecks, may reduce the ability of a population to cope with future environmental challenges. In the Baltic Sea, such effects may apply to fish and mammals, or to any populations established by a small number of colonisers, without further contact with the source (*e.g.* in the case of the Baltic harbour seal, *cf.* Sect. 6.2.6).

Population genetics: dynamics of differentiation

Gene flow, the movement of reproductively effective individuals between populations, maintains the common genetic constitution of populations and keeps their allele frequencies uniform. A continued lack of gene exchange, combined with small population size, causes populations to diverge at neutral loci by drift. In practice, observations of population differentiation are used to infer population structure (connectivity or isolation) or population history. Differentiation is impeded by gene flow but also by large population sizes, and large populations only diverge over long time periods. Marine populations are often large, whereas the age of the Baltic Sea is relatively short, and therefore differences usually remain small. In contrast, gene loci that govern adaptive traits may diverge much more rapidly than neutral loci if two populations are under different regimes of selection, *e.g.* in the Baltic Sea and the North Sea. Gene flow is needed to enable the spread of adaptive mutations from one population to another, but the amount of migration (level of connectivity) needed to ensure a strongly favoured gene to spread is very small. On the other hand, the lack of gene flow should enhance maintaining fine-scale adaptive differentiation among localities, such as those between spawning populations of salmon in different rivers.

Molecular evolution: dynamics of divergence

Over long time scales, molecular differences between isolated populations or species should accumulate in DNA at the same rate as neutral mutations occur. Mutation rates are different for different marker genes, and these rates largely determine which molecular markers are suitable to solve particular problems of systematics and population biology. For example, genes of vertebrate mitochondrial DNA may diverge by 1–2 % per million years in the long run. Mitochondrial DNA has long been the most widely used genetic tool for animal population or species history on time scales of 10^4–10^7 years. The more slowly evolving nuclear gene sequences are suitable for longer time scales. Studies on shorter time scales cannot rely on mutation dynamics, but rather on the population genetics of polymorphic nuclear markers such as microsatellites and SNPs (*cf.* Box 6.1).

3. Mating systems and individual relationships within populations, which include the analysis of the transfer of genetic material from one generation to the next, *e.g.* detection of sexual *versus* clonal reproduction, and parentage analyses used in studies of sexual selection.

6.2 Species diversity, evolutionary history and population structure

6.2.1 Species diversity and its geographical origins

In conventional biogeographic thinking, it is assumed that populations of the current Baltic Sea represent direct, post-glacially created extensions either from the adjacent marine populations in the North Sea or from the freshwater populations in the neighbouring lakes and rivers (*cf.* Sect. 4.5.2). Only the species that were later introduced by anthropogenic activities would have added more exotic ingredients to the region's biodiversity (*cf.* Sect. 4.5.8).

These views are based on interpreting distribution patterns of morphologically identified taxa in the light of geological history. Molecular characters now provide tools for a more accurate resolution of systematic diversity. While the traditional concepts are upheld in most cases, in a number of instances the genes have revealed more complex patterns of population relationships. This has led to revised concepts of species diversity and of the initial colonisation history of the Baltic Sea.

Box 6.3: Cryptic species and cryptic invasions

Mysis spp.

The mysid conventionally labelled *Mysis relicta* has been considered a dominant nektobenthic macro-crustacean species in the northern marginal parts of the Baltic Sea. It is a widely distributed circumpolar "glacial relict" taxon, which has its main distribution in relatively large boreal lakes, and is thought to have survived the glaciation in continental ice-marginal lakes. However, molecular characters, both at the protein and DNA levels, recognised two genetically distinct cryptic species (different biological species hidden under one species name) in the Baltic Sea, and a total of four species worldwide. The two Baltic Sea species, *Mysis relicta* and *Mysis salemaai*, which subsequently have also been formally described on the basis of subtle morphological differences, have different salinity-related distributions, but they also coexist across large areas in the Bothnian Bay (Audzijonyte and Väinölä 2005). Both species also live in lakes in Northern Europe (Box Fig. 6.1).

Hediste diversicolor

The ragworm *Hediste diversicolor* (Box Fig. 6.2) is a widespread polychaete of temperate European estuaries, and it is also the principal native large-sized shallow-water polychaete in the Baltic Sea. Molecular data indicate that within the Baltic Sea, *Hediste* comprises two distinct, non-interbreeding cryptic species, which are often found together (in sympatry).

However, on a broader European scale these species have separate distribution areas, which suggests independent invasion events to the Baltic Sea and subsequent mixing. So far, the two cryptic species have not been taxonomically described (Audzijonyte et al. 2008). Also another cryptic invasion of a polychaete complex, the non-indigenous *Marenzelleria* spp., has occurred recently in the Baltic Sea (*cf.* Box 5.3).

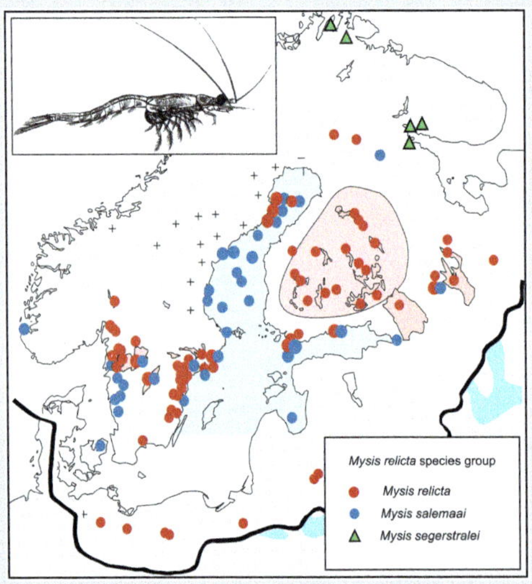

Box Fig. 6.1 Verified distributions of the mysid species *Mysis relicta* and *Mysis salemaai* in the Baltic Sea Area, based on data in Audzijonyte and Väinölä (2005). Records of a third species, *Mysis segerstralei*, are confined to the Barents and White Sea areas. The maximum extent of the latest continental glaciation is shown by the heavy black line. Figure: © Risto Väinölä

Box Fig. 6.2 The ragworm *Hediste diversicolor*. Photo: © Nicklas Wijkman/Azote

Typically, such changes in the concepts of diversity have been necessary in taxa that have broad circumboreal distributions, or are at least widespread in Europe, and which were previously thought to be uniform throughout their ranges. However, those ranges are not, or have not always been, contiguous.

During the Pleistocene climatic and geological history (since 2.6 million years ago) and even earlier, these taxa often became subdivided by more or less stable dispersal barriers, resulting in long-term inter- or transoceanic isolation between populations. Genomic differences accumulated during these isolation periods, and intrinsic reproductive barriers may have evolved. Still, the genetically diverged lineages or taxa remain hard to separate morphologically. When such cryptic, morphologically inseparable taxa then move around and become reshuffled between oceans or continents, and in some instances come to overlap in distribution, the composition of local faunas will be complex and their history is easily misinterpreted. Particularly for invertebrates, there will often be a temporal "blind window" covering the most recent 2–5 million years, during which population and species relationships cannot be resolved from morphology. An important part of the biotic history that took place during the relatively recent evolution and existence of the boreal climate may therefore be untraceable by the traditional morphological approaches.

In several Baltic Sea animals, cryptic taxon boundaries have been recognised using genetic markers, either within the Baltic Sea itself or between Baltic populations and those of the neighbouring waters (Boxes 6.3 and 6.4). Such cryptic subdivisions may be more than 100 times older than the post-glacial development of the Baltic Sea. Molecular differences this old are already strong enough to be easily diagnosed. However, the cryptic taxa that have been revealed represent quite variable biogeographical origins and different time scales of reshuffling. They may involve taxa that evolved in isolation in the Atlantic and Pacific Oceans or in different continental freshwater refugia of the ice-ages. Furthermore, they may represent either early post-glacial arrival or cryptic species recently introduced by man, which then may coexist either with earlier arrivers or other introduced species.

6.2.2 Outcomes of secondary contact

The cryptic taxon boundaries found within or at the border of the Baltic Sea represent secondary contact zones of genetic lineages whose primary divergence took place elsewhere in isolation (allopatry). The genetic outcomes from such secondary contacts in the Baltic Sea, in terms of the potential interbreeding and gene exchange, vary significantly. The outcome can be sympatric coexistence, parapatric contact, often with a hybrid zone, or mixing of lineages and fusion of differentiated genomes.

A common, morphologically uniform taxon has turned out to consist of two or more reproductively isolated sibling species that currently co-occur within the Baltic Sea, *e.g.* in the pelagic mysid *Mysis* and the benthic polychaete worms *Marenzelleria* and *Hediste* (*cf.* Box 5.3, Box 6.3). Such instances highlight a previous underestimation of species diversity. This type of coexistence of sympatric cryptic species occurs in taxa of all possible environmental backgrounds (marine, estuarine, freshwater). The taxa that make up a cryptic species group can represent either recent invaders, original Baltic taxa, or even both as in the case of *Hediste*.

In another pattern of cryptic diversity, the Baltic Sea population of a marine species is genetically deeply diverged from the neighbouring North Sea population. Thus, the North Sea population cannot be the immediate ancestor, but the relationship must be much older. At the same time, the Baltic Sea population is genetically close to a geographically distant relative, *e.g.* one from the northern Pacific or northwestern Atlantic. This is the case in the mollusc genera *Mytilus* and *Macoma*, and the Atlantic sturgeon *Acipenser oxyrinchus* (Box 6.4). The Baltic Sea may also have been very recently invaded by a stock that is different from the older resident North Sea population. This is *e.g.* the case for the prawn *Palaemon elegans*, which recently invaded the Baltic Sea from the Mediterranean Sea or the Black Sea, whereas the North Sea population has not been able to colonise the Baltic Sea and remains in parapatry (Reuschel et al. 2010).

Several freshwater or anadromous fish taxa are now represented in the Baltic Sea and elsewhere in its watershed by two or more lineages of different phylogeographical origins. Those lineages diverged in different parts of continental Europe during several glaciation cycles, and have now come into secondary contact in the brackish Baltic Sea. Examples include the European bullhead *Cottus gobio*, whose eastern and western lineages (also treated as subspecies) mix in Finnish coastal waters (Box 6.5). Similarly, the European perch *Perca fluviatilis* in the Baltic Sea represents a mixture of different continental refugial lineages. The Swedish freshwater grayling *Thymallus thymallus* is of a western lineage that spread from the south and met the eastern lineage in the north, as in the bullhead. Nevertheless, only the eastern grayling lineage is found in the Baltic Sea, where it only thrives at salinities <4. The contact of the western and eastern nine-spined stickleback *Pungitius pungitius* lineages is, in turn, in the Danish straits (Teacher et al. 2011). Also, populations of the Atlantic salmon *Salmo salar* in Baltic Sea rivers are derived from eastern (continental freshwater) and western (anadromous Atlantic) immigrations, although the number and distribution of the refugial origins remain under discussion (Säisä et al. 2005; Bourret et al. 2013).

Box 6.4: Identity and origin of Baltic Sea organisms: cases of rewritten biogeography

Risto Väinölä and Raisa Nikula

Pacific bivalves in the Baltic Sea

The blue mussel *Mytilus* and the Baltic clam *Macoma balthica* are dominant benthic invertebrates in the Baltic Sea, with their typical habitats being hard and soft bottoms, respectively. They are both euryhaline taxa with broad circumboreal distributions, and outside the Baltic Sea they typically dominate intertidal communities, particularly estuarine environments, but also open coasts. Initially, these molluscs arrived from the Pacific Ocean to the Atlantic Ocean along a northern route across the Bering Strait and the Arctic basin some 2–5 million years ago. A subsequent, long-term isolation of lineages between the two oceans allowed molecular differences to accumulate between them, whereas hardly any difference evolved in morphology. Using molecular markers as characters, a taxonomic distinction has been revealed between the Atlantic *Mytilus edulis* and the Pacific *Mytilus trossulus*, as well as between the Atlantic *Macoma balthica rubra* and the Pacific *Macoma balthica balthica*, respectively (Box Fig. 6.3). However, the full zoogeographical picture is more complicated, and occurrences of the Pacific taxa of *Mytilus* and *Macoma* are now found in the Atlantic Ocean along with true Atlantic taxa. This suggests repeated trans-Arctic invasions of Pacific molluscs into the Atlantic basin, even post-glacially. The current Baltic Sea *Mytilus* and *Macoma* populations essentially represent the Pacific taxa *Mytilus trossulus* and *Macoma balthica balthica*, which have strongly diverged from the true Atlantic bivalve taxa that live in the neighbouring North Sea. The Pacific *Macoma balthica* lineage is also found in the White Sea and the Barents Sea (Box Fig. 6.3). *Mytilus trossulus* is in turn present in the Atlantic Ocean off North America along with *Mytilus edulis*, and also occurs in places in Norway, Scotland and northern Russia. Where the Atlantic and Pacific bivalves meet, they tend to hybridise. The Baltic Sea and North Sea populations of *Mytilus* and *Macoma* are connected through clinal hybrid zones across the Öresund (*cf.* Box 6.5), and the interbreeding in these zones has fundamentally affected the Baltic gene pools (Nikula et al. 2007, 2008; Väinölä and Strelkov 2011; Fraïsse et al. 2016).

Early invasion and modern restoration of the North American sturgeon

The Baltic Sea sturgeon represents another case of zoogeographical reshuffling. The European sea sturgeon *Acipenser sturio* is a large anadromous fish that had historically inhabited European coastal rivers, from the Mediterranean Sea to the northern Baltic Sea (as it was thought), but by the end of the 20th century they had been extirpated, except in one French estuary. Unexpectedly, archaeo-genetic data show that the original Baltic Sea sturgeon was actually the North-American sister species *Acipenser oxyrinchus* (the Atlantic sturgeon), and not *Acipenser sturio* which perhaps never occurred in the Baltic Sea. Some *Acipenser sturio* genes are, however, present in the Baltic *Acipenser oxyrinchus* as a result of inter-species gene flow (introgression) (Popovic et al. 2014). Conservation efforts to restore the extirpated Baltic Sea sturgeon population are therefore now based on reintroducing a Canadian stock, rather than the surviving West European one.

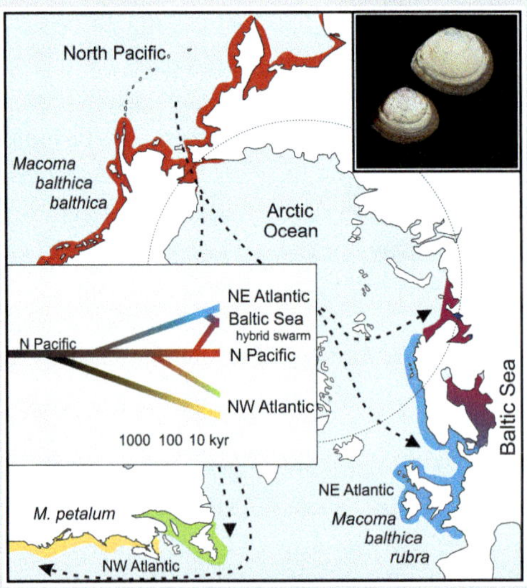

Box Fig. 6.3 Distribution and evolution of cryptic taxa within the *Macoma balthica* complex of bivalve molluscs, as inferred from mitochondrial and nuclear markers. The North-Atlantic *Macoma* diversity represents four separate trans-Arctic invasion waves from the North Pacific, probably since Pliocene times. The latest invasion was post-glacial and brought the Pacific *Macoma balthica balthica* (red) to the Baltic, Barents and White Seas, where it met and mixed with *Macoma balthica rubra* (blue), which had arrived much earlier. Figure based on data in Väinölä (2003) and Nikula et al. (2007, 2008). Photo: © Risto Väinölä, Figure: © Raisa Nikula and Risto Väinölä

If cryptic taxa that have been separated in allopatry come into secondary contact and are still able to interbreed, the mixing of their genetic variants through hybridisation will generate a population with novel recombinant genotypes and unusually high intra-population genetic diversity. In such a population, there is an abundance of material for natural selection to work on (Box 6.5).

6.2.3 Population structure: post-glacial processes

The cryptic species diversity, and the contact zones described above, involve variation that initially arose long before the formation of the current Baltic Sea. Geographical patterns of variation have also been generated during the existence of the basin, both within the Baltic Sea itself and between the Baltic Sea and the North Sea. The patterns in neutral marker genes reflect the demographic structures and histories of populations: population sizes, patterns of their connectedness or isolation, as well as colonisation events.

On the other hand, population differences may arise in non-neutral characters as a result of natural selection, which drives genetic adaptation to the local environmental conditions, and these conditions typically vary strongly in the Baltic Sea Area (*cf.* Sect. 2.4). However, at the same time such environmental differences create physical barriers to dispersal and thereby enhance population differentiation even in neutral genes. The two processes may occur together and amplify each other so that their effects are difficult to disentangle.

6.2.4 Genetic structuring within the Baltic Sea

Even the species that have fairly continuous distributions, such as the primarily freshwater fish: northern pike *Esox lucius*, European perch *Perca fluviatilis* and whitefish *Coregonus* sp. at the northern and western coasts of the Baltic Sea, show some geographical differentiation regarding genetic marker characters within the sea. The main driver of these patterns appears to be the geographic distance: the more distant two populations are from each other, the less exchange of genes there is between them, and the more they will differentiate ("isolation by distance"). The differences reflect limited genetic connectivity and indicate the relative demographic independence of local populations. Such, sometimes rather subtle, geographical differences have been suggested as justification for delineating management units in fisheries regulation (*e.g.* Laikre et al. 2005).

More distinct inter-population differences arise between the stocks of the anadromous Atlantic salmon *Salmo salar* that breed in different rivers. The salmon stocks temporarily mix during their marine feeding phase, but the genetic isolation of the river populations is maintained by strong homing behaviour that returns the fish to breed in their natal river. The genetic differences of breeding populations provide a means with which to assess the proportions of different fish stocks caught during the feeding and migration phases in the sea. This information can be used to help managing the salmon fisheries. A major goal of this management is to protect the remaining naturally breeding stocks; most of the catches are still based on stocks artificially reared in hatcheries. The salmon is by far the most intensively studied Baltic organism as regards genetic structure, and the one whose study has had the most direct societal applications (Nilsson et al. 2001; Koljonen 2001, 2006).

The three-spined stickleback *Gasterosteus aculeatus* is a small fish that is abundant and widespread along the Baltic Sea coasts and inflowing streams. This species was studied more comprehensively using both neutral and presumably non-neutral genetic markers. No geographic structuring was found in the variation of the neutral markers. By contrast, five discrete Baltic subpopulations were identified when analysing markers that are associated with genes involved in freshwater tolerance, suggesting adaptive differentiation. Given that the differential survival of individuals along the salinity gradient will reduce gene flow, evolution may ultimately also result in neutral gene differences (DeFaveri et al. 2013).

6.2.5 Demographic independence and colonisation bottlenecks

The salinity difference between the transition zone (Belt Sea and Kattegat) and the Baltic Sea (*cf.* Fig. 4.2) represents the distributional limit for a large number of marine species (*cf.* Sect. 4.2). For the species whose distribution extends across this zone, the steep change in salinity is thought to present severe physiological challenges and to cause selection pressures, which have led to adaptive genetic changes and the establishment of genetically determined physiological adaptations to different salinity regimes. The transition zone and the Baltic Sea also differ in characteristics other than just salinity, most notably temperature, but also biotic complexity, the Baltic Sea macroscopic biota being much less diverse than that of the North Sea. Such putatively adaptive differences have been recognised in a number of cases, and are now explored in much greater detail using genomic approaches.

More generally, some genetic differentiation between Baltic Sea and Kattegat (or North Sea) populations has been found in most of the species that have been analysed, regardless of the type of marker studied. Most of the studies have dealt with presumably neutral markers such as microsatellites. The steepest genetic change typically takes place in the Danish straits (*e.g.* for the turbot *Scophthalmus maximus* and bivalves) or in some cases, such as for the Atlantic cod *Gadus morhua*, just east of the straits within the southern Baltic Sea.

Box 6.5: Hybrid zones and hybrid swarms

Risto Väinölä and Raisa Nikula

The European bullhead *Cottus gobio*

While genetic data often reveal sympatric cryptic taxa or old lineages that no longer interbreed, in other instances such diverged lineages still can and do hybridise. The eastern and western refugial lineages of freshwater fish such as the European bullhead *Cottus gobio* form hybrid stocks in some coastal areas of the Baltic Sea (Box Fig. 6.4). Such populations have higher genetic variation than either of their parental freshwater lineages, which diverged in different parts of Europe during the glaciations. Hybridisation may bring together new combinations of genes that hypothetically could facilitate adaptations to the new habitat, not typical of the primarily freshwater organisms.

The bivalves *Macoma* and *Mytilus*

The Baltic populations of *Macoma* and *Mytilus* (*cf.* Box 6.4) represent extreme cases of interlineage hybridisation. In the Baltic Sea, the post-glacially invading Pacific lineage *Macoma balthica balthica* has become mixed with the older

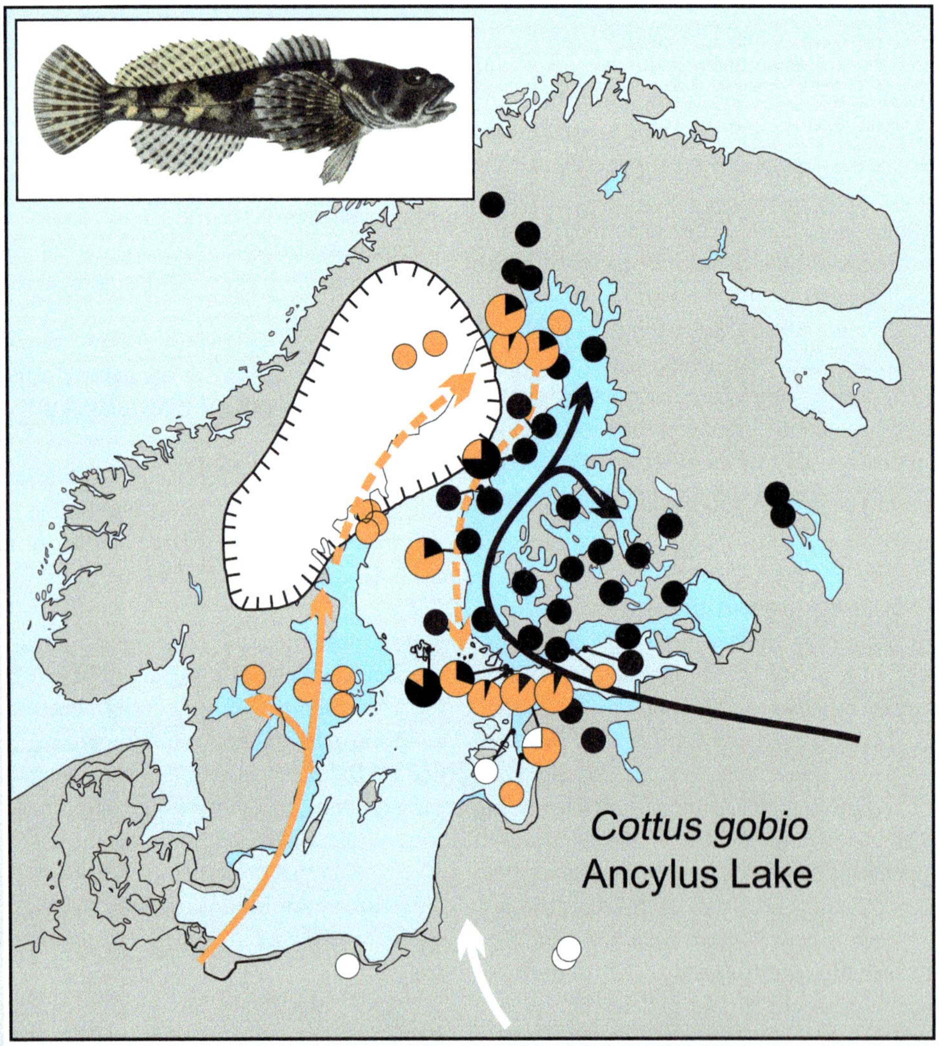

Box Fig. 6.4 Distinct western and eastern refugial lineages of the European bullhead *Cottus gobio* invaded the earlier freshwater Ancylus Lake phase of the Baltic basin (*cf.* Fig. 2.27c) from different directions, and pure stocks then became isolated in Swedish and Finnish lakes, respectively. After the final ice retreat the western and eastern lineages met in the north and were mixed in hybrid populations. Hypothetically, these genetically more diverse fish have then spread further along the Baltic Sea coast, possibly after the saline Littorina Sea phase. The symbols depict the proportions of three distinct mitochondrial lineages in modern samples; nuclear characters and morphological differences follow the same pattern. Figure based on data in Kontula and Väinölä (2001). Figure: © Tytti Kontula

Atlantic taxon, *Macoma balthica rubra* that still lives as a pure population in the North Sea. The Baltic Sea *Macoma* is a thorough mixture of the two genomic origins, with the Pacific genes still in majority. This mixture is observed independently in most of the genes analysed.

The Baltic *Mytilus* represents another kind of mixture, where different parts of the genome are derived from different parental taxa in a mosaic manner. Most nuclear genes are from the Pacific *Mytilus trossulus*, the mitochondrial genome is from the North Sea *Mytilus edulis*, and still other genes have equal contributions of alleles from both.

In both *Macoma* and *Mytilus*, the genome of the Baltic population represents a combination of genes of parental species that have long diverged in isolation from each other. All individuals in these populations inherit a roughly similar proportion of their genome from a given parental species. Although of mixed ancestry, the Baltic populations are judged to be in genetic equilibria at a local and regional level. Such "hybrid swarms" behave genetically as independent local species.

The Baltic bivalve hybrid swarms are in contact with their pure North Sea counterparts through the Kattegat-Belt Sea area, where clinal transition zones have formed (Box Fig. 6.5). In these zones the Baltic and North Sea taxa commonly interbreed, but still retain their genetic differences, probably due to strong selection against the immediate hybrid individuals (Nikula et al. 2008; Väinölä and Strelkov 2011).

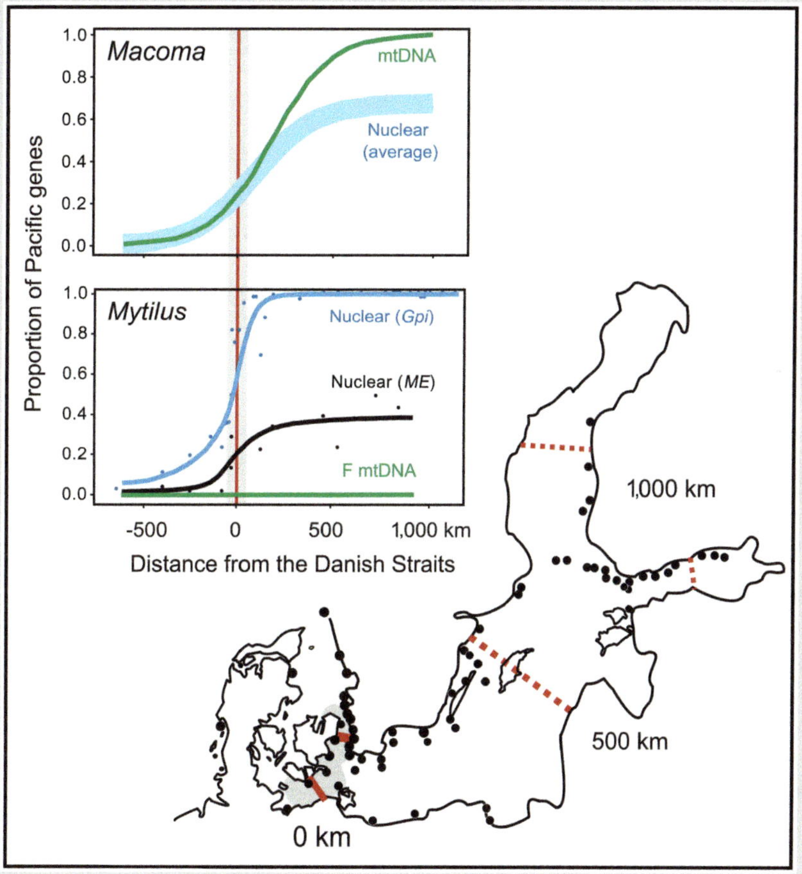

Box Fig. 6.5 Genetic clines across the North Sea – Baltic Sea transects in two marine bivalves in various molecular marker characters. The North Sea and the inner Baltic Sea waters are inhabited by different taxa of each bivalve, representing Atlantic and Pacific ancestries, respectively. In the transition zone, the taxa interbreed in a hybrid zone. Interbreeding has also resulted in introgression (gene leakage) beyond the immediate contact area. The amount of this leakage is different for the mitochondrial and nuclear genomes, and for different nuclear markers. Overall, most nuclear genes of the inner Baltic *Macoma* currently represent a mixture of two origins with an average 30–40 % influence from the North Sea (a hybrid swarm). In *Mytilus*, most genes have predominantly Pacific *Mytilus trossulus* alleles (exemplified by the *Gpi* allozyme locus). Others are mixed (*ME*), while mtDNA is from the Atlantic *Mytilus edulis*. Points on the map represent samples used in any one of the data sets. The zero point of the transect is set at the Darß sill and Drogden sill thresholds. Figure based on data in Nikula et al. (2008), Väinölä and Strelkov (2011) and additional literature sources. Figure: © Raisa Nikula and Risto Väinölä

Such genetic breaks are often not directly related to adaptive differentiation of the marker characters, but rather reflect a long-term demographic independence due to restricted dispersal across the steepest part of the salinity gradient, accompanied by stochastic accumulation of differences between the Baltic Sea and Kattegat stocks. In most species, the levels of intra-population genetic variability (heterozygosity) measured from multiple nuclear markers are lower within the Baltic Sea than in the Kattegat and North Sea, which is another indication of the long-term isolation and independence of the Baltic populations following the initial post-glacial colonisation of the basin (Johannesson and André 2006).

The colonisation process of the Baltic Sea populations, and their subsequent genetic independence from ancestors outside the Baltic Sea, are particularly well characterised by studies of mitochondrial DNA (mtDNA). This is a maternally transmitted, non-recombining molecule whose population-level diversity is much more strongly affected by population bottlenecks than that of nuclear genes. This is because the effective population size of mtDNA molecules is theoretically only one-fourth of that of the biparentally inherited diploid nuclear genes (Box 6.2).

Negligible mtDNA variation within a population, just like low variability at nuclear markers, is a signature of history where the population was founded by a small number of individuals and has not received much gene flow from outside the Baltic Sea since its establishment. For example, both in the widespread European estuarine mysid *Neomysis integer* and in the lagoon cockle *Cerastoderma glaucum*, a single mtDNA variant out of the broad European pool of variants has come to dominate in the entire Baltic Sea population (Remerie et al. 2009; Tarnowska et al. 2010).

6.2.6 Molecular *versus* taxonomical divergence

The relationship between biological (phenotypic) *versus* marker divergence is not always straightforward. Inferences from genetic markers sometimes contradict the conventional views on systematics. For example, the ice-breeding ringed seal of the northern Baltic Sea *Pusa hispida* (syn. *Phoca hispida*) used to be classified as a distinct subspecies, *Phoca hispida botnica*. It is currently separated by a 2,000 km distance from its conspecifics in the Arctic region. Still, there is only weak gene-level differentiation between the Baltic and Arctic ringed seal populations, both of which have historically been quite large (Palo et al. 2001).

Similarly, while the small-sized Baltic form of the Atlantic herring *Clupea harengus* is phenotypically distinct and was once referred to as a separate subspecies *Clupea harengus membras*, the early molecular marker studies did not find a genetic difference from the North Sea herring. For very large populations, such as those of the pelagic herring, the time would have been insufficient for stochastic divergence even if the populations would be independent (Box 6.2). However, more comprehensive genomic studies have later identified adaptive molecular differences in *Clupea harengus* (*cf.* Sect. 6.3.4).

The harbour seal (*Phoca vitulina*) represents a contrasting case of relatively rapid stochastic differentiation in a small population. This species now has only three Baltic colonies in the Kalmarsund on the Swedish east coast, with a few hundred seals in total (*cf.* Box 4.13). The genetic difference between the Baltic Sea *versus* the Kattegat and Skagerrak populations is much larger than that between populations at similar geographic distances elsewhere in the Atlantic Ocean. The difference can be explained by demographic history alone: complete isolation of the Baltic harbour seals since this population was established some 8,000 years ago, and a documented recent bottleneck during which the population size was smaller than 50 individuals (Härkönen et al. 2006).

6.2.7 Variation in reproductive traits: algae and plants

Some Baltic Sea species can reproduce both sexually and asexually (by cloning). The diversity and distribution of marker genotypes among individuals can be used to infer the prevailing reproductive mode in a population. By genotyping using multiple highly variable markers such as microsatellites, it is possible to produce genotype profiles that act as individual fingerprints which differentiate between any two individuals that are not clones. Such genetic information enables detection of clonal propagation in natural populations, and can be used for mapping the geographical distribution of different clonal lines. In practice, a set of five to ten microsatellite loci is sufficient to distinguish any two sexually produced individuals from each other.

Within-species variation in reproductive mode is best documented for macroalgae and marine vascular plants. In *Zostera marina*, *Fucus vesiculosus*, and the red algae *Ceramium tenuicorne* and *Furcellaria lumbricalis* (and also in the polychaete worm *Pygospio elegans*) the prevalence of asexual reproduction is higher in the Baltic Sea than in the Atlantic Ocean. A single clone of *Zostera marina* found in the Åland archipelago (northern Baltic Sea proper) was estimated to be about one thousand years old and to occupy an area of 6,400 m^2, whereas the largest clones elsewhere occupy less than 400 m^2 each (Reusch et al. 1999).

The reason why the plasticity of reproductive modes is important for Baltic populations may be related to the special challenge that the low salinity places on the sensitive early life-history stages (eggs, sperm, larvae), which are associated with sexual reproduction. In fucoid algae, normal fertilisation is achieved only if a single sperm enters the egg. However, the mechanism that prevents multiple sperm from penetrating the egg largely fails at low salinities. High levels of polyspermy in *Fucus* are indeed found at the boundary of its continuous distribution in the Baltic Sea in the northern Bothnian Sea where salinity is only 4–5, which causes an extensive failure of sexual reproduction in these populations (Serrão et al. 1999).

From the distribution of clonal (repeated) and unique genotypes in *Fucus*, it has been inferred that these algae use asexual reproduction as an alternative way of recruiting newly attached individuals in the Baltic Sea. Such clonal recruitment in this genus is not known from outside the Baltic Sea (Tatarenkov et al. 2005). Asexual propagation is most common in the newly identified endemic Baltic species *Fucus radicans* (Bergström et al. 2005; Fig. 6.2), which is known from the Bothnian Sea and from the island of Saaremaa in Estonia (Fig. 6.3). Some clones of *Fucus radicans* are dominant in populations and are found at more than one locality. Many others are rare and occur at single sites. The most predominant clone of *Fucus radicans*, a female, is extremely common over much of the species' distribution in Sweden and Finland (Johannesson et al. 2011a). It extends over a geographic range of 550 km. Individuals of this clone develop receptacles and produce viable eggs, but sexual reproduction is not effective because of a scarcity of males in some local populations, and of the detrimental polyspermy in others where males are abundant. In Estonia, sexual recruitment is, however, common and most individuals (thalli) have unique genotypes (Fig. 6.3).

6.2.8 Reproductive traits and mating strategies: animals

Adjustments of reproductive traits to fit the Baltic Sea brackish water are also known from fish. The European flounder *Platichthys flesus*, which generally spawns pelagic eggs, has adopted a demersal (benthic) spawning habit in the northern Baltic Sea, where its eggs cannot float due to the low density of the water. There is no direct evidence for genetic control of this trait, but marker gene data show that the bottom-spawning populations are distinct from the pelagic ones. This population-genetic difference between breeding populations is likely to be stable in the long term

Fig. 6.2 The Baltic endemic species *Fucus radicans* and its ancestor species *Fucus vesiculosus* growing attached to the same stone. Photo: © Lena Kautsky

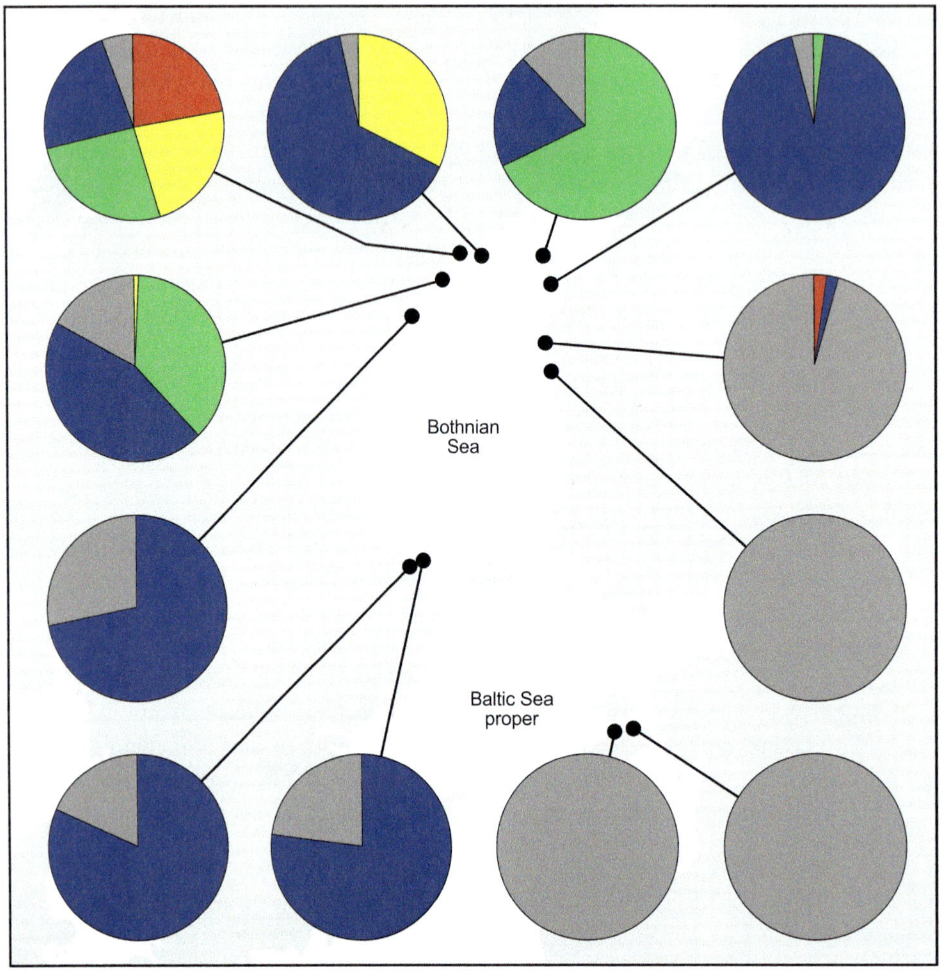

Fig. 6.3 Frequencies of large clones of *Fucus radicans*, two of which are females (blue and yellow) and two of which are males (green and red). Grey indicates the combined proportion of unique genotypes and local, small clones. The clones have been identified by their microsatellite marker genotypes. The sampling sites are representative of the known distribution of *Fucus radicans* in the northern and southern Bothnian Sea and at the Estonian island of Saaremaa in the northeastern Baltic Sea proper. Figure: © Daniel Johansson

(Florin and Höglund 2008). Also, in cod the development of eggs requires that they float neutrally in an oxygen-rich pelagic water layer. In Baltic cod populations, the gravity of the pelagic eggs is adapted to lower salinities than in the North Sea cod populations, which prevents the eggs from sinking to the bottom water where oxygen-poor conditions often prevail (Nissling and Westin 1997).

The highly variable microsatellite markers can also be employed in studies of mating strategies, sexual selection and sexual conflicts, as they enable identification of the father (or mother) when several males have potentially been involved in the production of offspring (multiple mating). Thus, females of the rough periwinkle *Littorina saxatilis*, which is found in the Kattegat-Belt Sea Area, are typically fertilised by more than 20 males each, of which a few account for most of the offspring (Panova et al. 2010). In the sand goby *Pomatoschistus minutus*, where the male protects the offspring in his nest, the so-called "sneaker males" (with extremely large gonads) manage to father a large proportion of the offspring protected by the nest owner (with smaller gonads) (Kvarnemo et al. 2010).

6.3 Adaptive evolution and speciation in a young sea

6.3.1 Adaptation and divergence over the salinity gradient

From the evolutionary point of view, and for understanding the uniqueness of the Baltic Sea ecosystem, the differences that are actually caused by natural selection and reflect genetic adaptation to the Baltic Sea environment may ultimately be of more central interest than those in the selectively neutral characters, which could have arisen and been maintained by demographic forces alone. Although demonstrating the role of selection and the mechanism of adaptation is not always straightforward, the number of examples of post-colonisation genetic adaptations to the low salinity of the Baltic Sea, or to its other physical characteristics, is growing.

Nevertheless, besides genetic adaptation, an alternative strategy to cope with the extremely low salinity and other environmental variations is a wide phenotypic plasticity in the critical physiological traits. Even if the tolerance of individuals to *e.g.* salinity would vary, the differences need not be determined genetically, but they could be induced by cues from the natal environment. An example is provided by the larvae of the vase tunicate *Ciona intestinalis*, whose distribution extends from the North Sea down to the Kiel Bay in the southwestern Belt Sea; they have a broad salinity tolerance, but the larval development is still most successful at a salinity similar to that prevailing in the area where the parents were kept prior to and during reproduction (Renborg et al. 2014).

6.3.2 Common garden and translocation experiments

Common garden experiments and reciprocal translocations are basic approaches when trying to verify a genetic basis and adaptive nature of phenotypic differences between populations from different environments. Individuals from the alternative environmental and phenotypic origins are reared under the conditions of both their native environment and those of the alternative phenotype's environment, either in a controlled laboratory set-up or in nature. The experiments should start with very young individuals (if not embryos) whose performance is preferably monitored through their ontogenetic development. These approaches have been used to demonstrate adaptive differences in Baltic populations of algae, fish and invertebrates.

When small vegetative shoots (adventitious branches) of the brown alga *Fucus vesiculosus* from Baltic and North Sea populations were grown in the laboratory, shoots from the Baltic Sea individuals grew more rapidly at a low (4) compared to high (25) salinity, while the opposite was true for the shoots of North Sea origin. Thus, there is local adaptation of growth rate to low salinity in the Baltic population of this species. Local adaptation of Baltic populations concerning morphological and physiological traits have also been found in other species of macroalgae (*e.g.* in the red algae *Ceramium tenuicorne* and *Delesseria sanguinea*).

Similar to *Fucus vesiculosus*, the two marine amphipod species *Gammarus oceanicus* and *Gammarus locusta* also extend their distributions to the northern Baltic Sea. In reciprocal laboratory translocation experiments, the *Gammarus* populations from the Baltic Sea could no longer survive salinities above 12. Direct interbreeding between Baltic amphipods with those of the ancestral North Sea habitat would thus not be possible anymore, as they cannot survive when brought into a common environment (Kolding 1985). However, gene flow could still take place step-wise through intermediate populations, as long as the geographical change in salinity tolerance is gradual and neighbouring populations retain the ability to interbreed.

In the European flounder *Platichthys flesus*, genetic differentiation along the North Sea-Baltic Sea gradient is very weak in neutral genetic markers. However, in a reciprocal translocation experiment, 5 % of the genes (several thousands were studied) showed differences in gene expression levels in a microarray analysis. A closer look at four particular genes that are associated with salinity regulation showed that Baltic Sea fish have evolved more efficient ways to regulate the activity of these genes than North Sea fish. This is presumably critical in the fluctuating salinity that characterises the southern Belt Sea (Larsen et al. 2008).

6.3.3 Covariation of molecular and environmental differences

For genes or traits with a fairly well-known functional role and mechanism of adaptation, geographical differences that can be correlated with an environmental difference that represents a putative selection pressure have been interpreted as evidence of adaptation.

However, such evidence alone is not considered very strong. The genetically and demographically distinct southern population of Baltic cod (*cf.* Sect. 6.3.7) is also characterised by a different predominant allele of its haemoglobin gene. The strong inter-basin difference in haemoglobin allele frequencies, and the geographical cline between the populations, are a classical example in population genetics, discovered long before the genomic era (Sick 1965). The corresponding haemoglobin protein differs from the common haemoglobin variant of the North Sea cod by two amino acid substitutions. This difference is argued to be adaptive by improving the oxygen-binding properties of haemoglobin in cold waters (Andersen et al. 2009).

A similar steep cline of haemoglobin variants over the Kattegat-Baltic Sea transition is also known from another fish, the viviparous eelpout *Zoarces viviparus* (Christiansen and Frydenberg 1974). The Baltic population of the sand goby *Pomatoschistus minutus* has in turn a genetic variant of a visual pigment in the eye that should enable the fish to see more efficiently in the dim light conditions of the local turbid waters, in comparison with the pigment that is predominant in the fish of the North Sea environment (Larmuseau et al. 2010).

6.3.4 Outliers as evidence of adaptation

The geographical differentiation of neutral gene frequencies is mainly shaped by population history and structure, which affect all genes in a uniform way. The variation patterns of all neutral genes are therefore expected to be similar in the statistical sense. An important approach to finding genes that are affected by other forces, particularly by natural selection, is to try to identify statistical outliers—those that differ from the common variation pattern more than is expected by chance. In the study of geographical variation, this means identifying genes that show either unusually strong or unusually weak differentiation. The same principle is used *e.g.* in gene expression experiments, to identify genes whose expression is unusually strongly affected by an environmental manipulation.

Next generation technologies now allow comparing patterns of differentiation among thousands and thousands of genetic marker loci (Box 6.1). Studies of herring provided one of the early demonstrations of the use and power of comprehensive NGS data in Baltic Sea-Atlantic outlier analysis (Lamichhaney et al. 2012). Earlier research using allozymes and microsatellites had largely failed to find any genetic differences between the Baltic and Atlantic herrings, even though the populations are phenotypically quite distinct (*cf.* Sect. 6.2.6). In accord with this, a genomic analysis of more than 400,000 polymorphic nucleotide sites (single nucleotide polymorphisms, SNPs) showed no differentiation in the vast majority of markers, regardless of whether these markers were from the coding regions of the genome or from the immediately flanking non-coding regions.

Nevertheless, a set of some 4,000 outlier SNPs showed unusually strong differences between the Baltic and Atlantic herrings. Those particular SNPs are inferred to reflect selection and adaptation to the specific environments in genes represented by or closely linked to the respective markers. Such population-level adaptation, manifested as *e.g.* an existence of specific breeding stocks, is also likely to restrict gene flow and keep the populations demographically separate. Despite a demographic independence, the lack of general differentiation over the majority of the genome is understandable in light of the very large effective population sizes of herring. In a large population, stochastic differences among populations will accumulate very slowly, even under complete isolation (Box 6.2).

The next step from such genome scan and outlier detection analyses towards understanding the adaptation process, is to identify the genes that are actually the targets of selection. When an annotated genome sequence or gene map is available, this search will involve evaluating the potential roles of the genes closely linked to the outlier markers and identifying those that could potentially be important for adaptation in the environmental context, on the basis of what is known about their function. This would ideally be followed by direct examination of the variation in the structures, expression and biochemical properties of their products.

6.3.5 Speciation in the Baltic Sea

The accumulation of genetic differences among populations, to the extent that they constitute different species reproductively isolated from each other, is generally thought to take place over evolutionary time scales, which means hundreds of thousands of years at least. For the young post-glacial Baltic Sea, the expectation is that divergence has seldom, if ever, progressed to the level of speciation and that there are few, if any, new endemic species in the Baltic Sea. The criteria by which biologists decide whether two taxa should be considered as distinct species, however, vary. For example, there are different views of whether a speciation event should imply true reproductive incompatibility (inability to interbreed), or just de facto cessation of gene flow associated with a stable genomic difference, or what level of phenotypic divergence should be established.

Nevertheless, there are a few instances regarding the recent origin of new taxa in the Baltic Sea that challenge the traditional wisdom of insufficient time for the evolution of new species here. A prominent case is the endemic brown algal species *Fucus radicans* that has evolved from the widespread bladderwrack *Fucus vesiculosus* quite recently within the Baltic Sea (Fig. 6.2). The ancestral *Fucus vesiculosus* is a North Atlantic species on hard substrates and also a dominant habitat-forming element in the Baltic Sea phytobenthic communities (*cf.* Sect. 11.8). Where present, the new, morphologically distinct *Fucus radicans* often occurs side by side in sympatry with its ancestor. *Fucus radicans* is distributed and dominates in parts of the coasts of the Bothnian Sea and is also found on the island of Saaremaa, Estonia (Fig. 6.3). Genotypic distributions of highly polymorphic microsatellite loci show that the two species remain consistently separate in these areas, with only very few hybrids.

The overall genetic closeness of the two *Fucus* species indicates that they have diverged very recently, within a few thousand or just a few hundred years, and possibly independently in the Bothnian Sea and Estonia (Pereyra et al. 2009, 2013). So far it is not clear what prevents the interbreeding of the two taxa in nature while experimentally crossed hybrid zygotes are readily produced. The two species are distinguished by their morphologies, but they also have physiological and ecological differences. Where they overlap, they may show different degrees of asexual reproduction, and timing of reproduction is also different in some areas (*cf.* 6.2.7).

6.3.6 Unique Baltic Sea diversity from synthesis of old variation

Another avenue to a new systematic entity is inferred from the genomic composition and history of the two dominant Baltic bivalves, the blue mussel *Mytilus trossulus* and the Baltic clam *Macoma balthica* (Box 6.5). Each Baltic bivalve population is distinct from the neighbouring North Sea taxon (*Mytilus edulis* and *Macoma balthica rubra*, respectively), but is more closely related to another species or subspecies distributed in the northern Pacific or in Atlantic North America: *Mytilus trossulus* and *Macoma balthica balthica*, respectively. Nevertheless, there are also differences with respect to those geographically distant ancestors. Those differences result from the extensive hybridisation that has taken place between the Pacific and the adjacent Atlantic (North Sea) taxon and introduced some Atlantic genes to the Baltic Sea (*e.g.* Väinölä 2003; Fraïsse et al. 2016).

In *Macoma* such introgression (genetic leakage) has resulted in a Baltic population where the genome of each individual clam is a mixture of genes from the two ancestral gene pools (60–70 % of the Pacific origin, the remainder being of the Atlantic origin). Such a population that represents an equilibrium mixture of diverged genomes is called a hybrid swarm. As a result of the genomic fusion, the Baltic clam population now possesses much higher levels of internal genetic variation than each of the ancestor taxa. This is in contrast to the pattern found in most other Baltic species, which have lower internal genetic variation than their ancestors (Johannesson and André 2006). Arguably, such a swarm itself can be regarded as a new taxon, since it has a unique and presumably stable overall genomic composition that evolved *in situ* following the post-glacial contact. Compared with evolutionary time scales, such speciation, through a synthesis of the pre-existing variability, is almost instantaneous (Nikula et al. 2008).

The history of the Baltic *Mytilus* is similar to that of *Macoma*, but probably even more complicated (*e.g.* Väinölä and Strelkov 2011; Zbawicka et al. 2014). While a part of the Baltic blue mussel genome represents a mixture of two backgrounds, most parts (genes) seem to come exclusively from the Pacific *Mytilus trossulus* and some parts, such as the matrilineal mitochondrial genome, exclusively from the Atlantic *Mytilus edulis*; still other features seem to have recently evolved within the Baltic Sea. The genome of the Baltic blue mussel is therefore best characterised as a mosaic of different taxonomic backgrounds. As a population, the Baltic blue mussels are still a hybrid swarm, behaving genetically as a single randomly interbreeding population. It remains distinct from the adjacent North Sea *Mytilus edulis*, even though a connection is maintained through a contact zone across the Öresund, where interbreeding continues to take place, just as in the case of *Macoma* (Box 6.5).

6.3.7 The Baltic cod: a model case

In steep environmental gradients, such as the salinity drop between the Kattegat and the Baltic Sea (*cf.* Fig. 4.2), genetic adaptation in a certain character to the specific conditions on either side of the ecotone will have the indirect effect of generally restricting gene flow of any gene between the populations. This is because migrating individuals are at a disadvantage and selected against in the non-native environment. Adaptation thus promotes the demographic independence, which in turn also promotes differentiation in neutral genes. The processes strengthen each other, and the populations eventually behave as effectively independent and ecologically specialised units, even though reproductive incompatibility would not yet have evolved.

The interplay of the processes may be demonstrated by the genetic structure of the Atlantic cod *Gadus morhua*, one of the most comprehensively studied Baltic Sea taxa. This is turning out to be another model organism for marine genomic studies along with the Atlantic salmon *Salmo salar* and the three-spined stickleback *Gasterosteus aculeatus*. In principle, cod could be a dispersive species, but at the entrance to the Baltic Sea, the evolution of complex subdivisions and maintenance of locally adapted distinct stocks are demonstrated by the patterns of variation in several traits at different spatial scales.

Two features, possibly of direct adaptive importance, were already mentioned: the clinally varying haemoglobin that could affect the efficiency of respiration in relation to temperature, and the different buoyancy of eggs in the low-salinity part of the range. Interestingly, the clinal change does not directly correspond to the salinity gradient, but occurs somewhat to the east within the southern Baltic Sea. The main distinction is between the cod stock in the south-eastern Baltic Sea that spawns east of Bornholm *versus* another stock in the Öresund, long known to constitute relatively independent breeding units. Chances of interbreeding are further decreased by separate breeding times, as the eastern Baltic cod spawns during spring or summer and those further west in late winter (Nissling and Westin 1997).

Neutral marker genes also illustrate the effective isolation of these populations. West of Bornholm, some crossbreeding nevertheless takes place. However, this is a hybrid zone, which does not imply extensive gene flow between the populations. Unlike the hybrid zones in bivalve molluscs, which represent secondary contacts of old (sub) species (secondary intergradation, Box 6.5), the cod zone is thought to be a direct consequence of differentiation driven by local adaptation in the population that invaded the post-glacial Baltic Sea from the North Sea, and represents primary intergradation (Nielsen et al. 2003).

More recently, in a genomic analysis of thousands of SNPs, a small subset of the marker loci showed unusually strong differences between the southeastern Baltic Sea and Öresund cod stocks, being correlated with environmental variation relevant for the reproductive success. These outliers are considered signals of recent adaptive evolution. Since much information on the cod genome structure already exists, these markers are known to represent several discrete genomic regions on various chromosomes, where differences between geographical stocks are probably maintained by natural selection. Also, some of the markers are located within or are linked to genes that are involved in osmoregulation and in egg development, which are obvious candidates for targets of divergent natural selection that also promotes isolation (Berg et al. 2015). Some differences also occur among the North Sea-Kattegat-Öresund cod stocks, where the environmental correlations are not so evident, but these are largely concentrated to one particular region of the genome.

6.4 Perspectives of genetic diversity and evolution

6.4.1 Next generation: from genetics to genomics

The examination of a limited number of conventional genetic markers has been useful in resolving many basic patterns of large-scale population genetic structure and biogeographical and evolutionary histories of Baltic Sea species. Observing phenotypic or fitness-related traits in common garden experiments has in turn elucidated the nature of brackish-water adaptation in a limited number of cases.

Nevertheless, the availability of genomic techniques that now enable the examination of thousands of marker genes across the genomes represents a quantum shift in the amount of information that can be retrieved about population structures and histories, and about the targets and mechanisms of selection and adaptation at the molecular level, as well as at other levels of biological organisation. The variations of these methods are numerous and they are evolving fast at the moment, and although they are efficient they are still associated with challenges in both the molecular and bioinformatic analyses. So far, they have been applied to Baltic Sea problems in a few pilot cases.

Besides population genomics, which addresses intra-species population structure and phylogeography, a prime target in Baltic Sea research will be the identification of the actual genes and mechanisms related to brackish-water adaptation. Moreover, when reference genomes of additional Baltic Sea species become available, genome-wide sequencing (*e.g.* RAD and capture sequencing) or whole-genome resequencing of samples of experimental individuals can be used for more specific goals, for example to trace chromosomal rearrangements that are thought to have important roles in population and species evolution.

Analyses of transcriptomes, the expressed parts of the genome, can be performed through resequencing of the messenger RNA (mRNA) that is produced while genes are transcribed (instead of using chips), and used to identify which genes are active during a specific stage of an organism's development in specific tissues or under specific environmental conditions. Sequencing information of methylation patterns will give additional information on epigenetic effects induced by environmental or other external cues.

6.4.2 Conservation and management

Features of adaptive genetic variation are thought to be important for the ability of a population to cope with future environmental changes and challenges. At the same time, even neutral genetic variation is informative of the population structure, such as the existence of demographically isolated and independently evolving population units. Marker variation can be used to estimate population size and the rate of inbreeding, which in turn inform about the population's capacity to hold, or its risk to lose, potentially adaptive variation.

Such information about gene pools and population structure is useful for evaluating conservation concerns about particular species and for managing the commercially exploited stocks. More generally, the genetic resources of Baltic Sea populations are likely to constitute the key to the future of the Baltic Sea ecosystem (Johannesson et al. 2011b). Surveys of population genetics will become increasingly important and constitute a standard tool in conservation and management. With the development of new technologies and lower prices per unit information, such approaches are becoming more efficient.

For example, a basic approach is to relate the amount of genetic differentiation (measured by the gene frequency variance statistic F_{ST}) to predictions from a theoretical model of population structure, and thus translate it into an estimate of the amount of inter-population gene flow per generation. By using an arsenal of several hypervariable loci, or even of thousands of SNP markers over the genome, it is possible not only to discriminate between populations but also to identify the population origins of individual animals.

The practical applications of such approaches are many. The ability to assign an individual fish to the correct stock could give authorities a strong weapon in combatting illegal fishing and trading of fish from protected stocks. The same approach is the basis of mixed stock analysis, where the proportions of various breeding stocks in open-sea catches are estimated when allele frequency data from the baseline (breeding) populations and genotypic data from the mixed catch are available. This has long been applied to fisheries of mixed salmon stocks, but could also be used in the management of the various southwestern Baltic Sea and Kattegat cod stocks and herring stocks that may occur together on the fishing grounds.

Levels of intra-population genetic diversity are thought to reflect the ability of populations to cope with environmental challenges. Measures of intra- and inter-population variability are therefore suggested as indicators in monitoring the favourable conservation status of populations. In general, the concern about sufficient levels of genetic variation may be less relevant for marine than for terrestrial organisms, as marine populations tend to be quite large and therefore not susceptible to rapid loss of variation. Still, there are aspects of specific Baltic Sea populations that make them vulnerable to diversity loss. This is particularly a concern for anadromous fish whose breeding in the native rivers has been prevented by river damming, and whose remaining reproducing populations may be very small. At the same time, stockings of hatchery-reared fish, whose genetic basis can be equally narrow, are used to uphold fishery production. Moreover, the stocking of large numbers of hatchery-reared fish may constitute a risk to the purity and unique properties of the remaining, locally adapted natural breeding stocks.

Another instance where the amount of genetic diversity may be of concern are taxa with high rates of asexual reproduction. The endemic brown alga *Fucus radicans* is largely clonal in some regions of the Baltic Sea (Fig. 6.3), which may render particular populations vulnerable under changing environmental conditions. In such cases, measures to protect the production of sexual offspring in other areas can be critical to generate new genotypes able to cope with future environments. Thus for management purposes, locating sexually recruited populations by genetic inventories would be a necessary first step. Monitoring the genetic variation and following its change over time may become a standard tool in the conservation and management of Baltic populations.

Review questions

1. What are neutral genetic markers, and what can they be used for?
2. Why are there so few endemic species in the Baltic Sea?
3. Which alternative processes could have caused differentiation between Baltic Sea and North Sea populations regarding a certain genetic trait or marker gene?
4. What can happen when two diverged populations or (cryptic) species meet after long-term isolation? Can you give examples from the Baltic Sea?
5. What differences are known between reproductive strategies of Baltic and Atlantic populations of same species?

Discussion questions

1. How would you evaluate and decide whether a Baltic population has diverged enough to be considered a new independent species?
2. How would you choose which genetic marker method to use if the goal is to assess genetic variation in a certain Baltic Sea species? (Choose your species!)
3. How would you decide whether phenotypic differences between Baltic Sea and North Sea populations reflect genetic adaptation (evolution) or just non-genetic physiological plasticity?
4. How can information on genetic variation be useful for the conservation of Baltic Sea species and the Baltic Sea ecosystem?
5. Which features of a (Baltic Sea) species or population would make it vulnerable to genetic threats?

References

Andersen Ø, Wetten OF, De Rosa MC, André C, Carelli Allinovi C et al (2009) Haemoglobin polymorphisms affect the oxygen-binding properties in Atlantic cod populations. Proceedings of the Royal Society B 276:833–841

Audzijonyte A, Väinölä R (2005) Diversity and distributions of circumpolar fresh- and brackish-water *Mysis*. Hydrobiologia 544:89–141

Audzijonyte A, Ovcarenko I, Bastrop R, Väinölä R (2008) Two cryptic species of the *Hediste diversicolor* group (Polychaeta, Nereididae) in the Baltic Sea, with mitochondrial signatures of different population histories. Marine Biology 155:599–612

Avise JC (2000) Phylogeography: the history and formation of species. Cambridge University Press, Cambridge 447 pp

Berg PR, Jentoft S, Star B, Ring KH, Knutsen H et al (2015) Adaptation to low salinity promotes genomic divergence in Atlantic cod (*Gadus morhua* L.). Genome Biology and Evolution 7:1644–1663

Bergström L, Tatarenkov A, Johannesson K, Jonsson RB, Kautsky L (2005) Genetic and morphological identification of *Fucus radicans* sp. nov. (Fucales, Phaeophyceae) in the brackish Baltic Sea. Journal of Phycology 41:1025–1038

Bourret V, Kent MP, Primmer CR, Vasemägi A, Karlsson S et al (2013) SNP-array reveals genome-wide patterns of geographical and potential adaptive divergence across the natural range of Atlantic salmon (*Salmo salar*). Molecular Ecology 22: 532–551

Christiansen FB, Frydenberg O (1974) Geographical patterns of four polymorphisms in *Zoarces viviparus* as evidence of selection. Genetics 77:765–770

DeFaveri J, Jonsson PR, Merilä J (2013) Heterogeneous genomic differentiation in marine three-spined sticklebacks: adaptation along an environmental gradient. Evolution 67:2530–2546

Florin AB, Höglund J (2008) Population structure of flounder (*Platichthys flesus*) in the Baltic Sea: differences among demersal and pelagic spawners. Heredity 101:27–38

Fraïsse C, Belkhir K, Welch JJ, Bierne N (2016) Local interspecies introgression is the main cause of extreme levels of intraspecific differentiation in mussels. Molecular Ecology 25:269–286

Härkönen T, Harding KC, Goodman SJ, Johannesson K (2006) Colonization history of the Baltic harbor seals: integrating archaeological, behavioral, and genetic data. Marine Mammal Science 21:695–716

Johannesson K, André C (2006) Life on the margin: genetic isolation and diversity loss in a peripheral marine ecosystem, the Baltic Sea. Molecular Ecology 15:2013–2029

Johannesson K, Johansson D, Larsson KH, Huenchunir CJ, Perus J et al (2011a) Frequent clonality in fucoids (*Fucus radicans* and *Fucus vesiculosus*; Fucales, Phaeophyceae) in the Baltic Sea. Journal of Phycology 47:990–998

Johannesson K, Smolarz K, Grahn M, André C (2011b) The future of Baltic Sea populations: local extinction or evolutionary rescue? AMBIO 40:179–190

Kolding S (1985) Genetic adaptation to local habitats and speciation processes within the genus *Gammarus* (Amphipoda, Crustacea). Marine Biology 89:249–255

Koljonen ML (2001) Conservation goals and fisheries management units for Atlantic salmon in the Baltic Sea Area. Journal of Fish Biology 59(Supplement A):269–288

Koljonen ML (2006) Annual changes in the proportions of wild and hatchery Atlantic salmon (*Salmo salar*) caught in the Baltic Sea. ICES Journal of Marine Science 63:1274–1285

Kontula T, Väinölä R (2001) Postglacial colonization of Northern Europe by distinct phylogeographic lineages of the bullhead, *Cottus gobio*. Molecular Ecology 10:1983–2002

Kvarnemo C, Svensson O, Manson W (2010) Investment in testes, sperm duct glands and lipid reserves differs between male morphs but not between early and late breeding season in *Pomatoschistus minutus*. Journal of Fish Biology 76:1609–1625

Laikre L, Miller LM, Palme A, Palm S, Kapuscinski AR et al (2005) Spatial genetic structure of northern pike (*Esox lucius*) in the Baltic Sea. Molecular Ecology 14:1955–1964

Lamichhaney S, Barrio AM, Rafati N, Sundström G, Rubin CJ et al (2012) Population-scale sequencing reveals genetic differentiation due to local adaptation in Atlantic herring. Proceedings of the National Academy of Sciences of the USA 109: 19345–19350

Larmuseau MHD, Vancampenhout K, Raeymaekers JAM, Van Houdt JKJ, Volckaert FAM (2010) Differential modes of selection on the rhodopsin gene in coastal Baltic and North Sea populations of the sand goby, *Pomatoschistus minutus*. Molecular Ecology 19:2256–2268

Larsen PF, Nielsen EE, Williams TD, Loeschcke V (2008) Intraspecific variation in expression of candidate genes for osmoregulation, heme biosynthesis and stress resistance suggests local adaptation in European flounder (*Platichthys flesus*). Heredity 101:247–259

Merilaita S (2001) Habitat heterogeneity, predation and gene flow: colour polymorphism in the isopod *Idotea balthica*. Evolutionary Ecology 15:103–116

Nielsen EE, Hansen MM, Ruzzante DE, Meldrup D, Grønkjær P (2003) Evidence of a hybrid-zone in Atlantic cod (*Gadus morhua*) in the Baltic and the Danish Belt Sea revealed by individual admixture analysis. Molecular Ecology 12:1497–1508

Nikula R, Strelkov P, Väinölä R (2007) Diversity and trans-Arctic invasion history of mtDNA lineages in the North Atlantic *Macoma balthica* complex (Bivalvia: Tellinidae). Evolution 61:928–941

Nikula R, Strelkov P, Väinölä R (2008) A broad transition zone between an inner Baltic hybrid swarm and a pure North Sea subspecies of *Macoma balthica* (Mollusca, Bivalvia). Molecular Ecology 17:1505–1522

Nilsson J, Gross R, Asplund T, Dove O, Jansson H et al (2001) Matrilinear phylogeography of Atlantic salmon (*Salmo salar* L.) in Europe and postglacial colonization of the Baltic Sea Area. Molecular Ecology 10:89–102

Nissling A, Westin L (1997) Salinity requirements for successful spawning of Baltic and Belt Sea cod and the potential for cod stock interactions in the Baltic Sea. Marine Ecology Progress Series 152:261–271

Palo JU, Mäkinen HS, Helle E, Stenman O, Väinölä R (2001) Microsatellite variation in ringed seals (*Phoca hispida*): genetic structure and history of the Baltic Sea population. Heredity 86: 609–617

Panova M, Boström J, Hofving T, Areskoug T, Eriksson A et al (2010) Extreme female promiscuity in a non-social invertebrate species. PLoS ONE 5(3):e9640

Pereyra RL, Bergström L, Kautsky L, Johannesson K (2009) Rapid speciation in a newly opened post-glacial marine environment, the Baltic Sea. BMC Evolutionary Biology 9:70

Pereyra RT, Huenchuñir C, Johansson D, Forslund H, Kautsky L et al (2013) Parallel speciation or long-distance dispersal? Lessons from seaweeds (*Fucus*) in the Baltic Sea. Journal of Evolutionary Biology 26:1727–1737

Popovic D, Panagiotopoulou H, Baca M, Stefaniak K, Mackiewicz P et al (2014) The history of sturgeon in the Baltic Sea. Journal of Biogeography 41:1590–1602

Remerie T, Vierstrate A, Weekers PHH, Vanfleteren JR, Vanreusel A (2009) Phylogeography of an estuarine mysid, *Neomysis integer* (Crustacea, Mysida), along the north-east Atlantic coasts. Journal of Biogeography 36:39–54

Renborg E, Johannesson K, Havenhand J (2014) Variable salinity tolerance in ascidian larvae is primarily a plastic response to the parental environment. Evolutionary Ecology 28:561–572

Reusch TBH, Boström C, Stam WT, Olsen JL (1999) An ancient eelgrass clone in the Baltic. Marine Ecology Progress Series 183:301–304

Reuschel S, Cuesta JA, Schubart CD (2010) Marine biogeographic boundaries and human introduction along the European coast revealed by phylogeography of the prawn *Palaemon elegans*. Molecular Phylogenetics and Evolution 55:765–775

Säisä M, Koljonen ML, Gross R Nilsson J, Tähtinen J et al (2005) Population genetic structure and postglacial colonization of Atlantic salmon (*Salmo salar*) in the Baltic Sea Area based on microsatellite DNA variation. Canadian Journal of Fisheries and Aquatic Sciences 62:1887–1904

Serrão EA, Brawley SH, Hedman J, Kautsky L, Samuelson G (1999) Reproductive success of *Fucus vesiculosus* (Phaeophyceae) in the Baltic Sea. Journal of Phycology 35:254–269

Sick K (1965) Haemoglobin polymorphism of cod in Baltic and the Danish Belt Sea. Hereditas 54:1–48

Tarnowska K, Chenuil A, Nikula R, Feral P, Wołowicz M (2010) Complex genetic population structure of the bivalve *Cerastoderma glaucum* in a highly fragmented lagoon habitat. Marine Ecology Progress Series 406:173–184

Tatarenkov A, Bergström L, Jonsson RB, Serrão EA, Kautsky L et al (2005) Intriguing asexual life in marginal populations of the brown seaweed *Fucus vesiculosus*. Molecular Ecology 14:647–651

Teacher AGF, Shikano T, Karjalainen ME, Merilä J (2011) Phylogeography and genetic structuring of European nine-spined sticklebacks (*Pungitius pungitius*)—mitochondrial DNA evidence. PLoS ONE 6(5):e19476

Väinölä R (2003) Repeated trans-Arctic invasions in littoral bivalves: molecular zoogeography of the *Macoma balthica* complex. Marine Biology 143:935–946

Väinölä R, Strelkov P (2011) *Mytilus trossulus* in Northern Europe. Marine Biology 158:817–833

Zbawicka M, Sańko T, Strand J, Wenne R (2014) New SNP markers reveal largely concordant clinal variation across the hybrid zone between *Mytilus* spp. in the Baltic Sea. Aquatic Biology 21:25–36

Physiological adaptations

7

Hendrik Schubert, Irena Telesh, Mikko Nikinmaa, and Sergei Skarlato

Abstract

1. Strategies of aquatic organisms to cope with ambient environmental conditions involve avoidance reactions or more profound behavioural and physiological adjustments, collectively called "adaptations".
2. Modulative (irreversible) and modificative (reversible) adaptations are short-term compensatory changes (acclimations) in an individual in response to environmental change, which are made possible through phenotypic plasticity.
3. Strong triggers for physiological adaptations that are more specific for the Baltic Sea than for most other water bodies are low salinity and low oxygen levels.
4. Mechanisms for adaptation to the salinity of the Baltic Sea, as well as to salinity fluctuations in Baltic coastal regions due to freshwater discharge, involve ion regulation (through ion channels, ion exchange proteins or primary ion pumps) and osmotic adaptation (*e.g.* through intracellular concentrations of osmotically active substances, such as low-molecular carbohydrates, amino acids and nucleic acids).
5. Low oxygen levels are dealt with by avoidance or a more effective energy metabolism.
6. Stress proteins provide cellular and whole-body responses of organisms to a vast range of changes in environmental conditions, *e.g.* water temperature, salinity, acidification, light availability, chemical pollution and hypoxia.
7. The photosynthetic apparatus of autotrophs is designed to cope with variability in irradiance; it becomes more efficient at low irradiance and more protective against excess energy at high irradiance.

Keywords

Behavioural adaptation • Energy metabolism • Ion regulation • Osmotic adaptation • Phenotypic plasticity

H. Schubert (✉)
Department of Biological Sciences, University of Rostock, Albert-Einstein-Straße 3, 18051 Rostock, Germany
e-mail: hendrik.schubert@uni-rostock.de

I. Telesh
Zoological Institute of the Russian Academy of Sciences, Universitetskaya Emb. 1, 199034
Sankt-Petersburg, Russia
e-mail: irena.telesh@zin.ru

M. Nikinmaa
Department of Biology, Laboratory of Animal Physiology, University of Turku, 20014 Turku, Finland
e-mail: miknik@utu.fi

S. Skarlato
Institute of Cytology of the Russian Academy of Sciences, Tikhoretsky Ave. 4, 194064 Sankt-Petersburg, Russia
e-mail: s_skarlato@yahoo.com

P. Snoeijs-Leijonmalm et al. (eds.), *Biological Oceanography of the Baltic Sea*,
DOI 10.1007/978-94-007-0668-2_7

7.1 General principles of adaptation

7.1.1 Homeostasis, tolerance and avoidance

In all natural habitats, organisms are affected by environmental variability and biotic interactions, the intensity of which is constantly changing. The time scales at which such changes act vary widely, from milliseconds in the case of solar radiation (Schubert et al. 2003) to thousands of years in the case of climate variability. All living organisms keep at least some components of their internal environment constant. In a stressed environment, maintaining this constancy ("homeostasis") is the main goal of an organism. Homeostasis is often achieved by simply avoiding a disturbance or by a behavioural change.

Most species can be broadly classified into two groups, conformers and regulators. Conformers may not need to exert themselves to maintain homeostasis as their plasticity ("conformity") allows them to accommodate to environmental change. Regulators need to maintain their internal environment at a more or less constant level, irrespective of the extracellular conditions. However, the subdivision into conformers and regulators is not sharp and absolute because the two categories frequently overlap. A species may be a conformer with respect to one environmental driver but a regulator with respect to another one. For example, fish are temperature-conformers but ion-regulators.

Tolerance is achieved when an organism adapts physiologically and/or morphologically to the variability of environmental (abiotic and biotic) factors and therefore can persist in its habitat. When avoidance is employed, an organism eliminates the impact of the changing factor whenever its physiological tolerance limit is exceeded. Well-known examples of avoidance strategies are migrations of birds, resting stages of protists and dormancy ("diapause") of zooplankton species.

7.1.2 The concept of adaptation

Adaptation is both a central and controversial concept in biology. The term means different things to scientists working in different fields of biology. In genetics and ecology, adaptation usually denotes a process involving heritable changes in the genome, which result in the emergence of functions that enable the organism to live in its ambient environment. Thus, adaptation is used for either a process of selection or a trait resulting from selection.

Evolutionary adaptation is a slow process, usually irreversible, which involves hundreds or thousands of generations (*cf.* Sect. 6.1). The end result of evolutionary adaptation is a habitat-specific genotype (an ecotype or a species), with a tolerance bandwidth adapted to the variability of the environment they live in. How fast a given set of habitat-specific characters evolves depends on the distance over which the respective trait pattern is shifted, the available gene pool, and the strength of the forcing factor. In physiology, the term "adaptation" is often used to describe the responses of an individual to environmental change. In this case, a geneticist or an ecologist would use the term "phenotypic plasticity".

Modulative and modificative adaptations are the short-term compensatory changes of an individual in response to environmental change. Both are the outcome of phenotypic plasticity resulting from evolution. While modulative compensatory changes are irreversible, and can therefore be considered an optimisation of the evolutionarily achieved set of habitat-specific adaptations, modificative compensatory changes are fully reversible and adjust the organism's performance to environmental changes within an individual's life span. More appropriate, at least with respect to modificative compensatory changes, are the terms "acclimation" or "acclimatisation". The term "acclimation" may be reserved for compensatory changes in physiological experiments according to the experiment set-up whereas "acclimatisation" is used to denote a natural process. However, in most cases the distinction between these two terms is not so sharp, and they are largely used as synonyms.

7.2 Environmental variability

7.2.1 Salinity and temperature

In addition to seasonal cycles and climatic gradients (*e.g.* latitudinal), which affect any large ecosystem, the dominant stress factor for all organisms living in the Baltic Sea is the wide salinity range. The Baltic Sea Area, including the Baltic Sea and the transition zone (Belt Sea and Kattegat), features a salinity gradient from 0 to 35 over a distance of >2,000 km (*cf.* Fig. 4.2). While water temperature varies during the year, salinity is relatively stable in most parts of the open Baltic Sea, compared to estuaries (Fig. 7.1). However, in estuaries, as well as in other coastal areas with freshwater discharges into the Baltic Sea, there are pronounced inshore-offshore salinity gradients between the freshwater discharge points and the open Baltic Sea water.

Benthic organisms in the Baltic Sea do not have the opportunity to just wait through a period of suboptimal

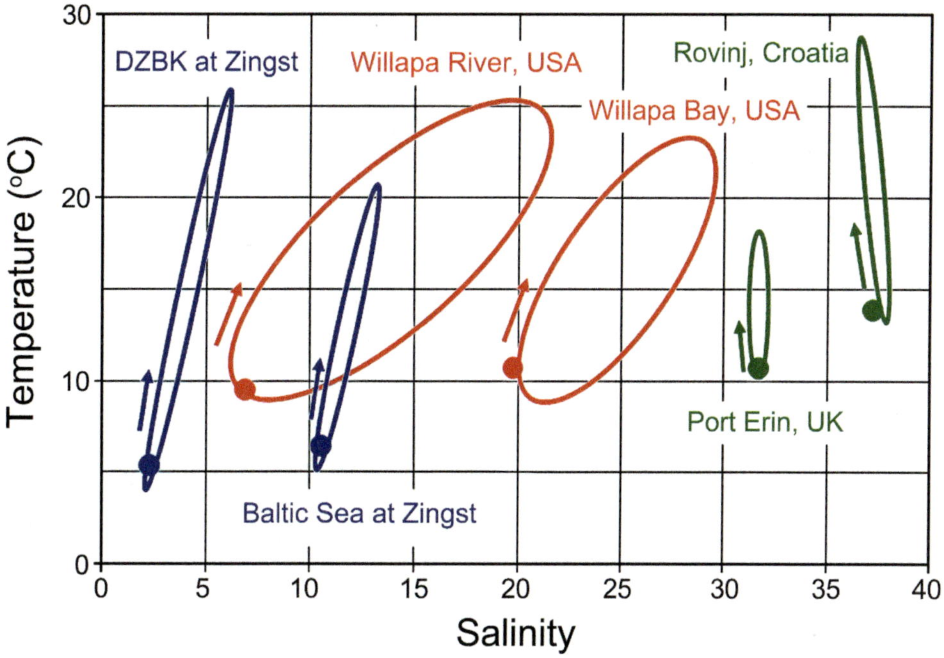

Fig. 7.1 Comparison of seasonal salinity and temperature in a coastal lagoon connected to the Arkona Sea (the Darß-Zingster Boddenkette, DZBK, Germany), an estuarine area in the North Pacific Ocean (the Willapa River and Willapa Bay, USA), and two marine areas (the Irish Sea at Port Erin, Isle of Man, UK, and the Adriatic Sea at Rovinj, Croatia). The values for January are marked with a dot. Arrows indicate the direction of the seasonal cycle. Figure based on data in Hedgpeth (1951) for Willapa River, Willapa Bay, Port Erin and Rovinj and unpublished data for Zingst (H. Schubert, collected in 1995). Figure: © Hendrik Schubert

salinity until conditions improve, as is the case for benthic organisms in the intertidal zone elsewhere. For example, molluscs and barnacles in tidal regions may close their shells and stop filter-feeding until salinity soon reaches optimum conditions again. During periods with low salinity, molluscs in the littoral of tidal seas may also apply avoidance mechanisms whereby the low temperature reduces their metabolic maintenance costs. In the virtually non-tidal Baltic Sea, such avoidance mechanisms are not applicable, and appropriate physiological adaptations of benthic organisms to temperature and salinity variations are of vital importance. The seasonal cycling of solar radiation regulates the heat balance at the water surface (Fig. 7.1), and thus the energetic input to the system, and triggers acclimation in the organisms. Both the energy flow and acclimation reactions have to be taken into account when analysing the biotic effects of the salinity regime.

Salinity adaptations depend on two major components: the ionic composition of the water and the osmotic pressure. No living cell is constantly "iso-ionic" with its environment. All extant organisms must perform ionic regulation, *i.e.* they need to maintain an active control over their intracellular ionic composition, which always differs from that outside the cell. In animals, the actively generated and maintained ion gradients over cell membranes are used for neural conduction, but they are also prerequisites for many cross-membrane transport processes channelling nutrients and other charged compounds into and out of the organism. For example, the intracellular K^+/Na^+ ratio is always kept at a level that is higher than that of the ratio in the marine environment. The reduced intracellular sodium level makes it possible for cells to use the ion gradient both in neural conduction and for substrate transport.

The species that have evolved in marine habitats are often "iso-osmotic". As the resultant intracellular concentration of ions exceeds the physiological demand, a species can stay iso-osmotic over a wide range of decreasing salinities by simply lowering its internal total ionic concentration. However, this so-called "poikilo-osmotic" strategy cannot be maintained down to salinity zero because freshwater is too poor in ions and charged compounds to allow for a sufficient intracellular concentration of osmotically active substances (*e.g.* low-molecular carbohydrates, amino acids and nucleic acids). In addition, the actively maintained ion gradients required for neural conduction cannot occur under these conditions. Thus, at a certain salinity, organisms must be

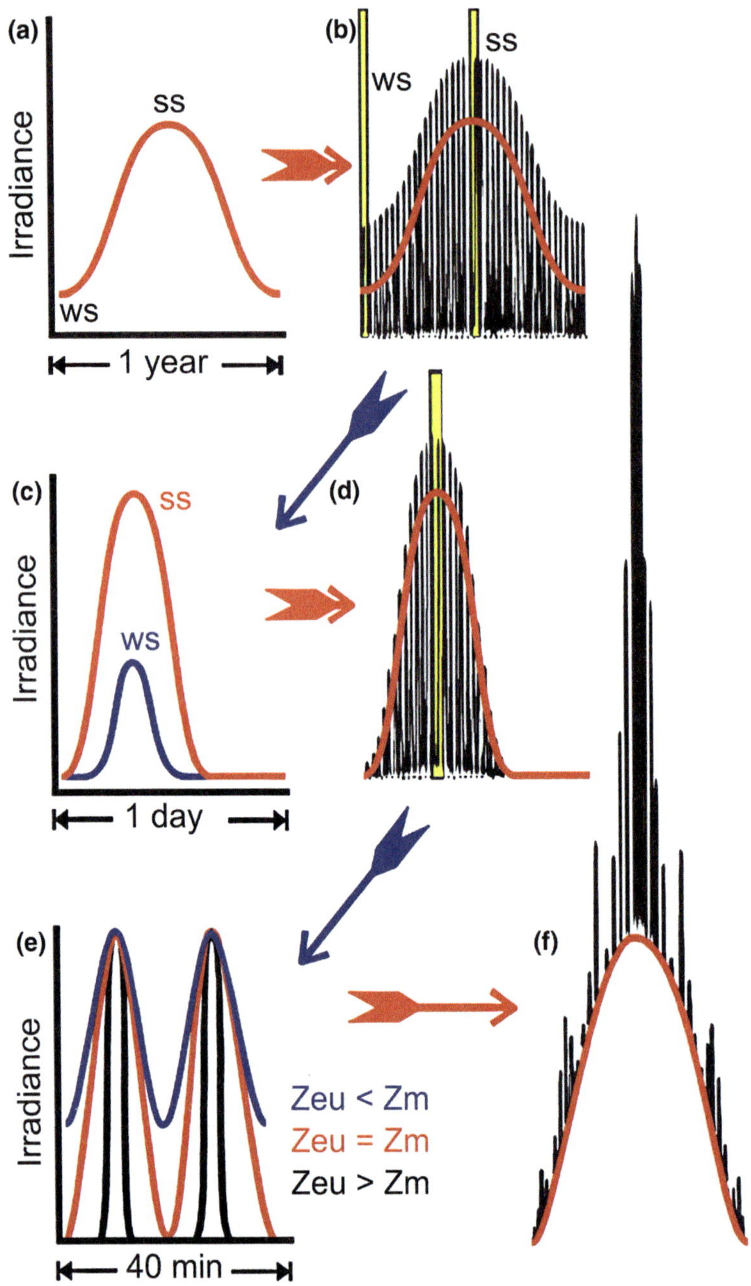

Fig. 7.2 Model of irradiance variability perceived by phytoplankton. (**a**) The seasonal cycle of irradiance, which at 60 °N ranges between 5 and 90 photons m^{-2} day^{-1} between the winter solstice (ws) and the summer solstice (ss), respectively. (**b**) Variability in day length and irradiance amplitude during the 365 day/night cycles of one year. The yellow bars indicate ws and ss. (**c**) Variability in one day/night cycle for a ws day and an ss day shown in blue and red, respectively. (**d**) Variability caused by wind-induced Langmuir circulation; each of the day/night cycles consists, regardless of weather conditions (which, however, are an additional element of variability), of several vertical movements through the upper part of the water column. The yellow bar indicates the set of two cycles shown in more detail in (**e**). (**e**) An individual Langmuir cycle has been measured to take ~20–40 minutes for a ~2 m deep, mixed water layer. Shown here are two cycles of 20 minutes each. Irradiance fluctuates when the photic zone depth (Z_{eu}) is shallower than or equals the mixing depth (Z_m) (**e**, blue and red lines). However, when the turbidity is very high, the $Z_{eu} > Z_m$, irradiance may consist of a series of on/off light cycles (**e**, black line). (**f**) Superimposed on to the Langmuir circulation is wave focusing (the light-focusing effect of the water-surface waves), for which reason irradiance can increase up to five times the mean level measured by conventional radiometers for a couple of milliseconds. The black lines are flashes of light, resulting from wave focusing, on top of the red Langmuir-induced variability for $Z_{eu} = Z_m$. Figure modified from Schubert and Forster (1997)

capable of hypertonic regulation that allows them to maintain an osmotic gradient. This results in a gradient of the water potential (Ψ), which must be balanced either by active water pumping (*e.g.* by contractile vacuoles) or by increasing the inner pressure potential (p). The latter requires either a rigid outer structure such as the cell walls of algae and plants or, in an animal cell, energy supply to make the cell functionally impermeable to sodium ions so that the cell volume can be maintained.

In addition to direct salinity effects, the success of organisms in brackish water may depend on indirect effects such as the influence of salinity on the availability of food items or how water density affects the buoyancy of organisms. This in turn impacts *e.g.* energy use and oxygen availability.

7.2.2 Irradiance and temperature

Both osmotic regulation and ionic regulation in organisms requires energy. The major energy input to ecosystem processes is the photosynthesis-driving solar radiation, in addition to the dissolved and particulate matter from terrestrial runoff. The variability of the underwater irradiance climate is strikingly large (Fig. 7.2). The acclimation capabilities of photosynthetic organisms not only need to meet the challenges posed by seasonal and day/night irradiance cycles, but also challenges posed by mid- and short-term variability in Langmuir circulation (a series of shallow, counter-rotating vortices at the sea surface generated by steadily-blowing winds, Langmuir 1938; Thorpe 2004). Moreover, they also need to deal with the light-focusing effect of waves at the water surface ("wave-focusing", Stramski and Legendre 1992). Photosynthetic organisms have to adapt to irradiance changes involving the quantitative variability in light intensity and the qualitative variability in spectral composition. Both light intensity and spectral composition vary at different frequencies.

As planktonic photoautotrophs are subjected to the full range of irradiance variability, they have to employ light-protection mechanisms and other functions that allow them to enhance their light-use efficiency. This is because, as opposed to *e.g.* nutrients, irradiance cannot be stored internally to dampen external variability. Active avoidance mechanisms in light protection are restricted to comparatively large organisms and calm weather. In contrast to lakes with a rather small wind fetch, the surface layer of the open Baltic Sea and the entire water column of shallow coastal areas are well-mixed most of the time, which prevents small unicellular organisms from actively controlling the depth of their occurrence in the water column.

Directly linked to irradiance are changes in the seawater temperature. Due to the high thermal capacity of water (so-called "thermal inertia"), effects of temperature variation are delayed, compared to the faster changes in irradiance. Because enzyme kinetics is highly temperature-dependent, there is a difference in the degree to which temperature influences photosynthetic light harvesting and the purely enzyme-driven respiratory processes in primary producers. This may lead to a time lag between irradiance input and water temperature, which is manifested by *e.g.* high respiratory rates in the early night hours when water temperature is still high but primary production is limited by low irradiance.

7.2.3 Oxygen

Most organisms require oxygen to serve as the terminal electron acceptor in energy production (respiration) and to drive many redox reactions. The oxygen concentration in the water provides key information about *e.g.* algal and cyanobacterial blooms and ecosystem health in general, and the oxygenation status of the environment (*e.g.* oxygen depletion) can be estimated by measuring dissolved oxygen (DO). Changes in the shape of the DO depth curve (illustrating the vertical distribution of DO in the water column), as well as oxygen deficiency in near-bottom waters, are meaningful indices of eutrophication.

Oxygen availability changes with water depth. It is released during photosynthesis, which is restricted to the upper part of the water column (the photic zone, *cf.* Fig. 2.21). The salinity stratification of the water column hampers deep vertical mixing, and the organic material, accumulated below the halocline as a result of sedimentation, undergoes heterotrophic decomposition which requires oxygen to proceed. Consequently, oxygen is used up, and hypoxic (<2 mL O_2 L^{-1}), or even anoxic (0 mL O_2 L^{-1}), conditions ensue (*cf.* Sect. 3.6). Special adaptations are necessary for organisms if they are to survive such conditions. Hypoxia may occur not only in deep waters but also in shallow eutrophic bays when the amount of light is not adequate for photosynthesis. Under such conditions, hypoxia is usually intermittent, occurring during the night, whereas during the day the conditions may even be hyperoxic (*i.e.* the oxygen partial pressure in the water exceeds that of the air).

7.3 Ionic regulation

7.3.1 The ionic anomaly of brackish water

Generally, the major inorganic ions involved in metabolism can be classified into two groups: (1) metabolisable solutes, especially N- and P-containing compounds, which may exist in the inorganic form in osmotically significant quantities during nutrient sufficiency (luxury accumulation), but more frequently are assimilated into organic molecules, and (2) non-metabolisable ions, principally K^+, Na^+ and Cl^-.

To sustain its interior ionic homeostasis, any cell in a liquid medium needs to employ active and selective ion transport. The ionic composition of the extracellular fluid of truly marine organisms, *e.g.* echinoderms, has been shown to be very similar to that of seawater. In truly marine organisms, almost all ion-regulation mechanisms are concentrated at the cell surface/body fluid border whereby the extracellular fluid buffers the ion-composition variability of the external medium. However, such a mechanism may fail in brackish water where the relative ionic composition exhibits anomalies (Fig. 7.3; Table 7.1).

7.3.2 Effects of the ionic anomaly on biota

The physiological stress at salinities of 5–8 are most likely responsible for the restricted number of macrozoobenthos and macrophyte species in the unique permanent salinity gradient of the brackish Baltic Sea (*cf.* Sect. 4.5). Each species reaches its own physiological salinity limit (McLusky and Elliott 2004; Elliott and Whitfield 2011; Whitfield et al. 2012). Ionic composition and osmotic regulation and variability have been listed among the causes underlying this diversity minimum. For example, measurements of ion concentrations along the salinity gradient from Baltic Sea estuaries to the North Sea have shown that the Ca^{2+}/Cl^- ionic ratio is quite stable within the salinity range of 7–34. However, below salinity ~7 the relative proportion of Ca^{2+} ions increases, which also has consequences for the distribution of species (Khlebovich 1968, 1974).

Ionic composition and osmotic regulation and variability have been extensively studied along the large-scale Baltic Sea gradient. However, the physiological consequences of irregular salinity changes are poorly understood in the Baltic Sea. Without doubt, such variation induces stress because continuous responses of the organisms are required. In shallow

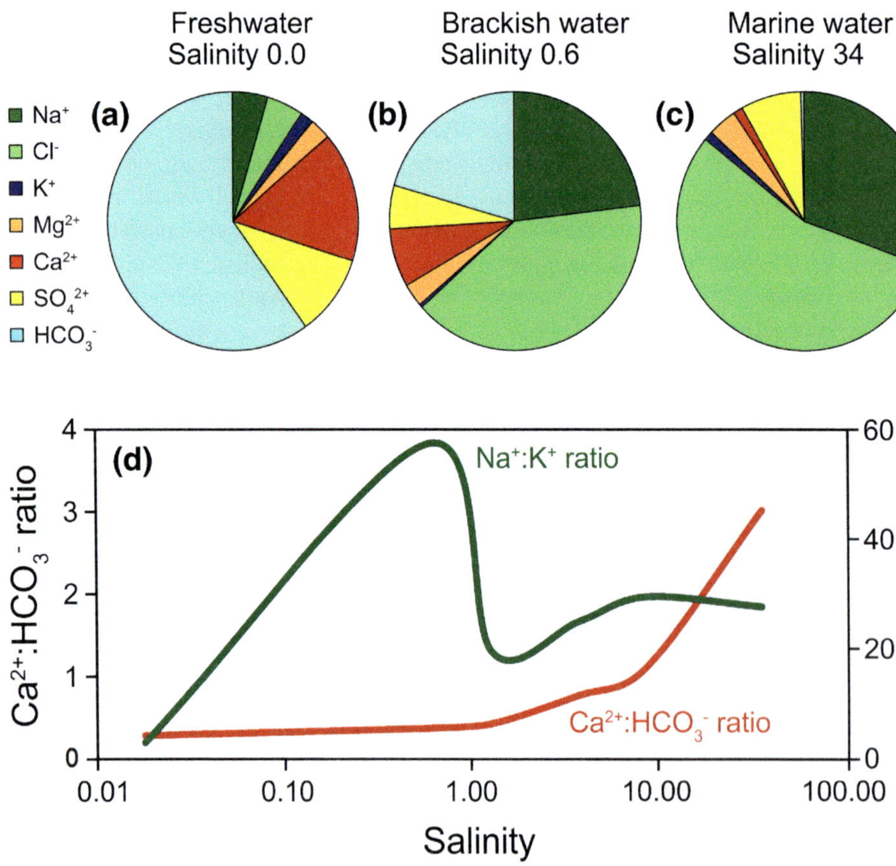

Fig. 7.3 The relative ion composition of water with different salinities. (**a**) Freshwater. (**b**) Baltic Sea water at Recknitz, Germany in the southern Belt Sea. (**c**) Marine water. (**d**) The $Ca^{2+}:HCO_3^-$ and $Na^+:K^+$ ratios in relation to salinity on a logarithmic scale, showing the increase of the $Ca^{2+}:HCO_3^-$ ratio with salinity and the ion anomaly of the $Na^+:K^+$ ratio in brackish water. Figure based on data in Nessim (1980). Figure © Hendrik Schubert

Table 7.1 Salinity (in mg L^{-1}) and selected ionic ratios in freshwater and along a brackish gradient in the Darß-Zingster Boddenkette, a chain of coastal lagoons (Recknitz, Bodstedt, Zingster Strom, Grabow) in northern Germany connected to the Arkona Sea. Data from Nessim (1980)

	Freshwater	Recknitz	Bodstedt	Zingster Strom	Grabow	Marine water
Salinity	0.02	0.6	1.3	3.8	8.6	35
Na^+	2.0	286	464	1,264	2,983	10,700
Cl^-	2.0	500	890	2,100	4,700	19,340
K^+	0.7	5	24	51	102	390
Mg^{2+}	1.1	36	71	127	258	1,290
Ca^{2+}	6.9	91	60	105	139	420
SO_4^{2-}	4.3	67	184	353	698	2,700
HCO_3^-	25	255	146	139	126	140
$Na^+:K^+$	2.72	57.20	19.33	24.78	29.25	27.44
$Ca^{2+}:Mg^{2+}$	6.09	2.53	0.85	0.83	0.54	0.33
$Ca^{2+}:Cl^-$	3.42	0.18	0.07	0.05	0.03	0.02
$Ca^{2+}:HCO_3^-$	0.27	0.36	0.41	0.76	1.10	3.00

coastal waters of the Baltic Sea, salinity may fluctuate erratically within a relatively broad range (Fig. 7.4). The extent to which such salinity changes hamper the adaptation of long-lived macrozoobenthos and macrophyte species, or increase the probability of coexistence (or even prevalence in pelagic biodiversity) of short-lived planktonic organisms – in the sense of the "intermediate disturbance hypothesis" (Grime 1973; Connell 1978) and the "protistan species maximum concept" for the horohalinicum in coastal waters (Telesh et al. 2011, 2013, 2015) – still needs further exploration.

The rates of water and ion fluxes between organisms and their environment largely depend on the body wall permeability for water and ions. These rates are mainly determined by the structure and osmotic properties as well as by electrical charges in the extra- and intracellular compartments, including extracellular matrices or epithelial cell layers (Fig. 7.5). When animals are exposed to a diluted medium, specific loss rates of sodium (in µmol g^{-1} h^{-1}) vary from >1,000 in fully marine crabs and 800–900 in marine intertidal crabs, to 100–200 in estuarine/freshwater crabs and ~5 in fully freshwater crayfish. These marked differences between marine and freshwater species can be generalised to a higher permeability of ions and water in marine organisms compared to freshwater ones.

The ion uptake and efflux can occur in three ways: by ion channels, ion exchange proteins or primary ion pumps (Box 7.1). Where substantial extracellular material (*e.g.* mucus) is present, ion-binding effects complicate the ionic transport processes.

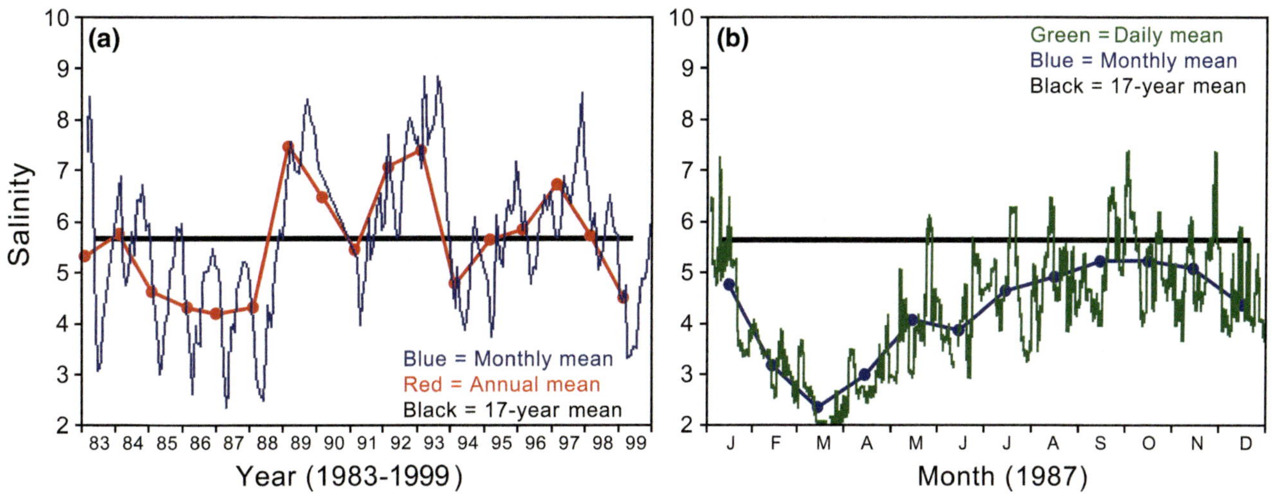

Fig. 7.4 Variability of salinity in the Darß-Zingster Boddenkette (Germany), a coastal lagoon connected to the Arkona Sea, on different time scales. Salinity was recorded every 15 min and was averaged to obtain daily means, which were used to calculate monthly means, annual means and the 17-year mean. (**a**) Monthly means, annual means and the 17-year mean for the time period 1983–1999. (**b**) Daily means and monthly means for the year 1987 and the 17-year mean. Figure modified from Sagert et al. (2008)

Box 7.1: Ion transport pathways

Ion transport

The permeability of a cell membrane to ions (or even to water) can be altered in the short term, within the lifespan of an individual, via acclimation mechanisms. At the level of cell membranes, acclimation involves either addition or removal of ion transporters or regulation of the transport properties of existing proteins. More than 300 types of ion transport pathways occur in living cells. They can be classified according to the ion species that pass through the gates (pores), the location of proteins, the number of gates, the requirement for counter-ions to make transport possible, or the active (with energy costs) or passive (without energy costs) nature of the transport. These transport pathways can be subdivided into ion channels, ion exchange proteins and active transport.

Ion channels

Ion channels (Box Fig. 7.1a) are pore-forming membrane proteins that allow the transport of ions along an electrochemical gradient made up by ion concentrations and membrane potential. They do not use metabolic energy. Thus, transport is by pure diffusion and the size of the channel determines which ion is transported through the channel. Ion channels control the flow of ions down their electrochemical gradient through the membranes that surround all biological cells. In many aquatic invertebrates, salts move relatively freely across the surfaces via such channels.

Ion exchange proteins

Ion exchange proteins carry out either passive (Box Fig. 7.1b) or secondarily active ion transport. In passive ion transport, ions on both sides of the membrane move down their electrochemical potential, *i.e.* transport is pure diffusion. The difference between ion channels and ion exchangers is that the latter require suitable ions for transport on both sides of the membrane whereas ion channels can transport ions without counter-ions. In secondarily active ion transport, the transport of one ion is coupled to the transport of another with a transmembrane gradient maintained actively by primary ion pumps. In this case, both transported ions can be displaced from their passive electrochemical gradient.

Active transport

Active transport is carried out with primary ion pumps (ATPases), which use energy (ATP) directly coupled to the function of the protein. An ATPase is an enzyme that catalyses the decomposition of ATP into ADP and a free phosphate ion, a reaction that releases energy. Some of the primary ion pumps do not need a counter-ion (Box Fig. 7.1c) whereas others do (Box Fig. 7.1d).

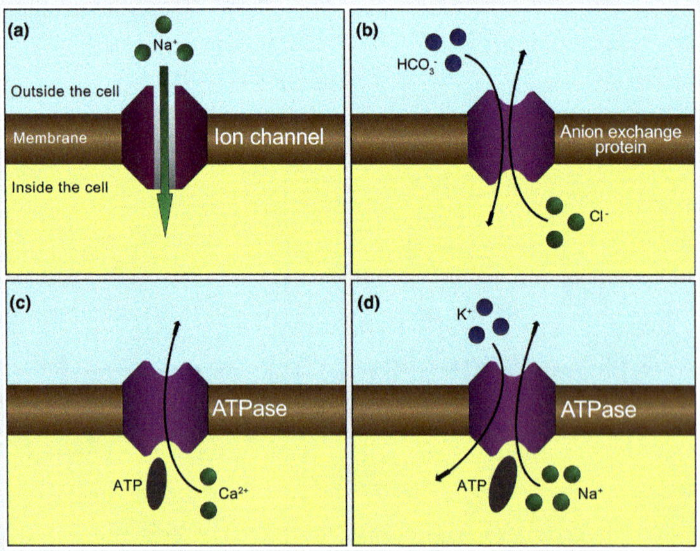

Box Fig. 7.1 Examples of ion transport mechanisms across the cell membrane. (**a**) An ion channel through which Na^+ is transported. (**b**) Ion exchange protein. An example is the anion exchanger by which most algae are able to take up carbon as bicabonate (HCO_3^-) in exchange to chloride in an electroneutral manner. Anion exchangers are also used in the control of pH in most organisms. (**c**) The "Ca^{2+} pump" is a primary ion transport ATPase that serves to remove calcium ions from the cell. (**d**) The "Na^+/K^+ pump" is a primary ion transport ATPase that serves to pump sodium ions out of the cell and potassium into the cell (both ions move against their concentration gradient). This transport regulates the cell volume. Figure: © Irena Telesh

Box 7.2: Maintenance of steady state volume in animal cells

Osmotic pressure

Cells contain impermeable poly-ions and molecules that make the intracellular contents osmotically more active than the extracellular fluid, which generates an osmotic pressure difference between cells and their surroundings (Edwards and Marshall 2013). The presence of a higher number of osmotically active particles within the cell than outside the cell generates an influx of water into the cell for balancing the osmotic pressure difference (Box Fig. 7.2a). The influx of water generates a diffusive imbalance for small permeable ions, whereby they will diffuse into the cell, mostly through membrane protein pores. This again generates osmotic imbalance, and water will enter the cell. Unopposed, this situation would lead to a continuous inflow of water until the cells burst.

The double-Donnan equilibrium

In plant cells, continuous swelling is prevented by a rigid cell wall that counterbalances the osmotic pressure difference. However, animal cells do not possess rigid walls and, consequently, they need another mechanism to maintain the cell volume. This is achieved with the so-called "double-Donnan equilibrium", also known as the "the pump-leak model" (Macknight and Leaf 1977). This system requires that a permeable solute is removed as soon as it enters the cell. Thus, even though the solute is permeable, it is functionally impermeable. In animal cells, the functionally impermeable solute is sodium. As soon as sodium enters the cell via diffusive pathways, it is actively pumped out by the sodium pump (Box Fig. 7.2b). Owing to the functional impermeability of sodium by ATP-requiring extrusion, a steady state volume can be achieved.

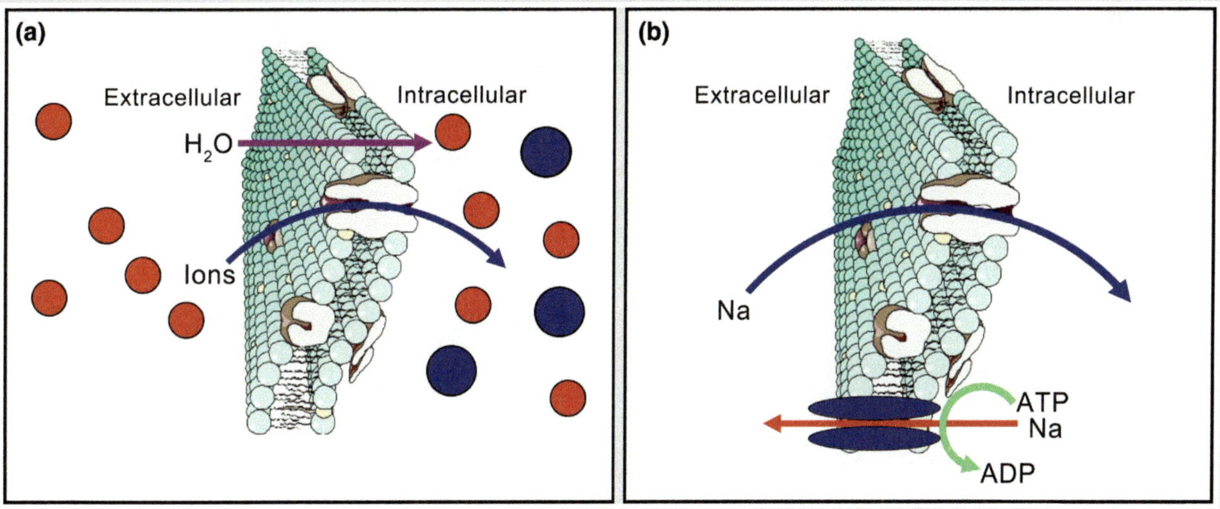

Box Fig. 7.2 Explanation of the double-Donnan equilibrium. (**a**) A cell membrane with an intracellular fluid that is osmotically more active than the extracellular fluid. (**b**) An animal cell membrane with a sodium pump, fuelled by ATP. Images drawn with the help of templates in Smartdraw 7. Figure: © Mikko Nikinmaa

7.4 Osmotic adaptations

7.4.1 Osmotic variability and its effect on biota

Water is a vital constituent of any living cell, tissue, organ or organism (*cf.* Sect. 1.1). The cell membranes are highly permeable to water, which also acts as a solvent for substrates and products of metabolism and a reactant in several basic metabolic processes, including hydrolysis and photosynthesis. Moreover, water is required for regulation and adjustment of cell volume.

Cells of aquatic organisms contain a wide array of intracellular solutes. These have a diversity of functions, but a major role is also to regulate the osmotic pressure within a cell. The minimum intracellular osmotic pressure is ~0.1 MPa, which is consistent with the pressure resulting from the concentration of osmotically active substances required for normal metabolism. In the absence of other forces, immersion of a cell in freshwater would lead to water uptake and cell lysis. However, organisms can prevent this in either of two ways. One is the presence of a rigid cell wall (*e.g.* in plants, prokaryotes and some protists),

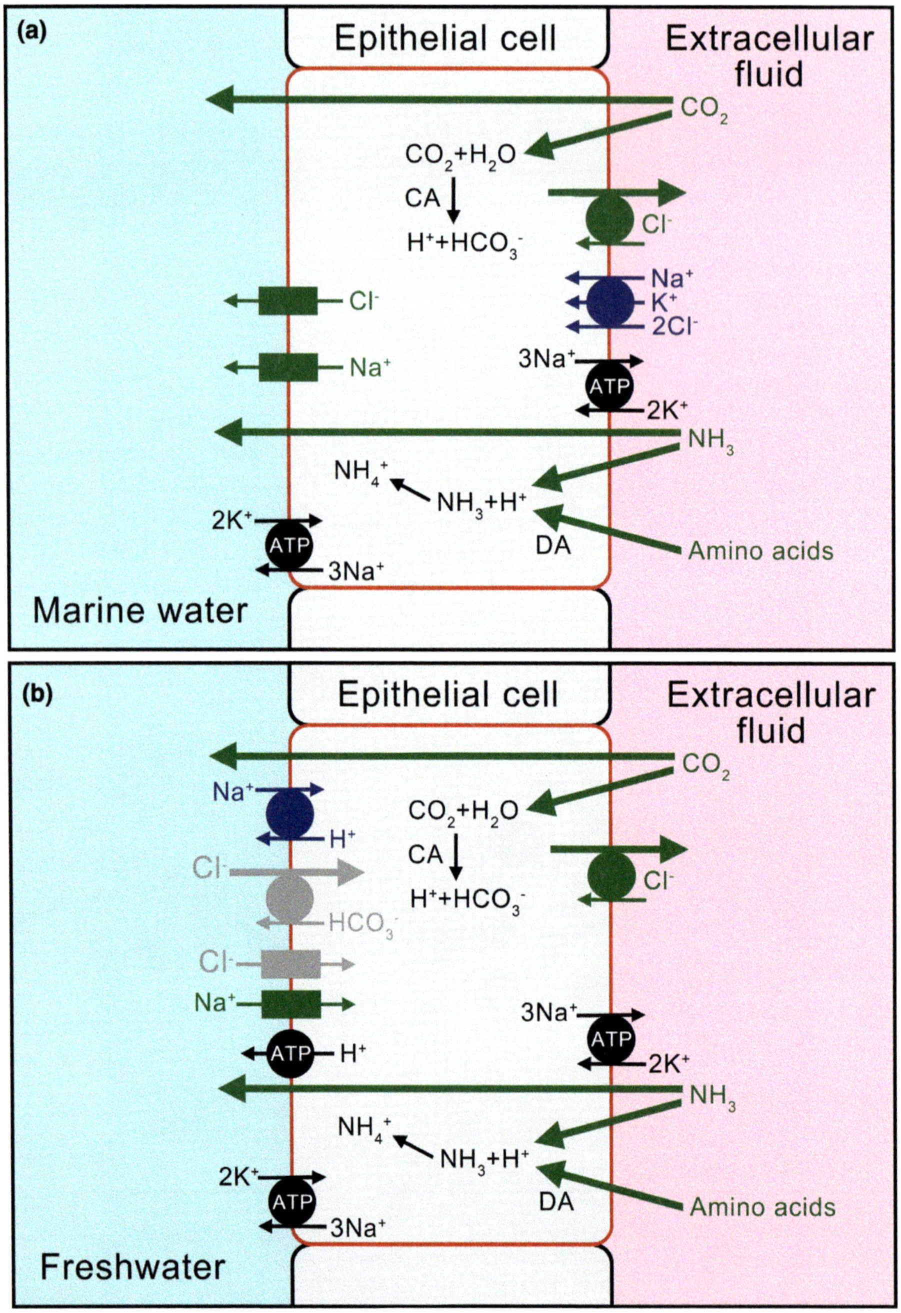

Fig. 7.5 Ion movements between a cell, the extracellular fluid (ECF, blood) and the external medium in (**a**) marine fish and (**b**) freshwater fish. Green colour indicates that the transport is passive, blue that it is secondarily active and black that it is active. The ion transport behaviour of epithelium in brackish-water fish depends on the relationship between the salinity of the water and the animal. If the water salinity is higher than that of the animal, the marine model is followed, and if the salinity of water is lower than that of the animal, the freshwater model is followed. (**a**) In marine water, the animal must excrete salts (especially chloride) actively. The active step is the secondarily active uptake of chloride from the ECF into the epithelial cells via the Na-K-2Cl transporter that utilises the sodium gradient generated by the sodium pump (Na^+/K^+-ATPase). Once in the cell, chloride can be excreted to the environment via passive channels. (**b**) In freshwater, the animal must take up salt from the water (especially sodium and chloride) actively. For sodium, this takes place either via the secondarily active sodium/proton exchange or a passive sodium channel that is functionally coupled to a proton pump. The energetics of the chloride transport pathways, chloride/bicarbonate exchange and chloride channels, are not fully clarified (and therefore shown in grey); both active and secondarily active components have been suggested. Most of the carbon dioxide and ammonia are excreted as such, both in freshwater and seawater, but some carbon dioxide is converted to bicarbonate and some ammonia to ammonium in reactions catalysed by carbonic anhydrase (CA) and deamination (DA) reactions of amino acids. Bicarbonate serves as a counter ion for chloride, and ammonium can be transported via the sodium pump (or via sodium/proton exchange). Figure: © Mikko Nikinmaa

which allows a positive hydrostatic pressure (turgor) to develop, whereby water entry and cell enlargement is prevented. The cell volume in multicellular animals is maintained by the so-called "double-Donnan equilibrium" (Box 7.2). Regardless of the mechanism employed, the importance of maintaining the cell volume is that both condensing and dilution will affect the three-dimensional structure of functional proteins, which may negatively impact their functioning.

Salinity may act as a driver of directional selection, and genetically determined variation in the salinity tolerances (osmo-adaptations) of different species from the same group can be related to the salinity regimes of their habitats. In certain estuarine and marine populations of several species of red algae (*e.g. Bangia atropurpurea* and *Vertebrata lanosa*), the salinity response can be interpreted in terms of intraspecific variation (Reed 1995).

In general, the Baltic Sea organisms are regarded as more euryhaline than organisms in marine waters, although this is not always the case. The lower turgor of estuarine organisms is likely to be an adaptive response, which may be a way to avoid problems associated with high turgor in response to hyposaline stress. Estuarine plants show a lower rate of intracellular solute loss in response to extreme hyposaline stress in Ca^{2+}-deficient media (Reed 1995). For example, when placed in a hypersaline medium (150 % seawater), the red alga *Delesseria sanguinea* from the Baltic Sea showed evidence of damage, as opposed to *Delesseria sanguinea* from the North Sea (Reed 1995). In a survey of the salinity responses of several other algal species, including *Ceramium tenuicorne* and *Rhodomela confervoides*, a hyposaline shift in the halo-tolerance of the Baltic Sea algae compared with their North Atlantic conspecifics was found (Russell 1985).

7.4.2 Osmotic adaptation strategies

In an organism adapted to marine conditions, exposure to hypo-osmotic conditions elicits an immediate response in the form of swelling, resulting from inflow of freshwater (Remmert 1969b). If the organism is unable to release intracellular osmotically active substances and/or actively pump water out of its cells, it may even burst. Exposure to hyper-osmotic conditions, on the contrary, leads to shrinking up to the point when the external and internal water potentials are balanced, which may lead to plasmolysis. Organisms capable of acclimation by releasing osmotically active substances and/or by active pumping, and therefore able to adapt to changing salinity in brackish-water habitats, originate from different ancestral groups, *e.g.* marine, freshwater or terrestrial lineages.

The strategies of adapting to brackish-water conditions differ among organisms (Fig. 7.6). The first, but by no means the "simplest", adaptation mechanism of marine-brackish organisms is to extend their iso-osmotic response compared to primary marine organisms (Fig. 7.6a). This requires complicated adaptations of biochemical pathway kinetics. An example of an organism with this strategy is the snail *Littorina littorea* (Fig. 7.6b).

Other mechanisms, shown by *e.g. Carcinus* sp. and *Gammarus* sp., which originate from the marine intertidal zone, involve hypertonic regulation at low extracellular salinities (Fig. 7.6c). In the epilittoral zone, both hypo- and hypertonic regulation has been found to occur in *e.g. Orchestia* sp. (Fig. 7.6d).

In freshwater, hypotonic regulation is not necessary and this function was therefore lost in most freshwater organisms, which are only able to perform hypertonic regulation, as in *e.g.* many freshwater gastropods (Fig. 7.6e). Some freshwater organisms are capable of a limited degree of both hypo- and hyper-osmotic regulation, *e.g.* freshwater arthropods and vertebrates (Fig. 7.6f).

Species that have returned to brackish and/or marine environments from freshwater or terrestrial habitats have re-adopted hypotonic regulation or re-extended their poikilo-osmotic range (Fig. 7.6g), often by means of osmolytes, which are substances that affect the water potential of a system. Osmolytes can be defined as physiologically compatible substances that increase the osmotic pressure at a low energetic cost, such as urea in sharks (Fig. 7.6g) and in the crab-eating frog *Fejervarya cancrivora* (Fig. 7.6h). Organisms invading brackish-water environments from salt lakes always employ both hypo- and hypertonic regulation, regardless of whether their evolutionary origin is marine, freshwater or terrestrial (Fig. 7.6i).

7.4.3 Physiological mechanisms of osmotic acclimation

In the brackish Baltic Sea water, the salinity range occupied by a species depends on the efficiency of the physiological mechanisms by which it is adapted to changes in ambient salinity. The ability to adapt evolutionarily depends on the generation length of the species. Generally, osmotic regulation is involved in the maintenance of a difference of ionic concentrations inside and outside the cells at appropriate physiological levels. In mobile animals the situation may be somewhat different than in sessile organisms, as they are capable of escaping from conditions of inappropriate salinity. However, numerous species migrate actively between saline water and freshwater (Box 7.3).

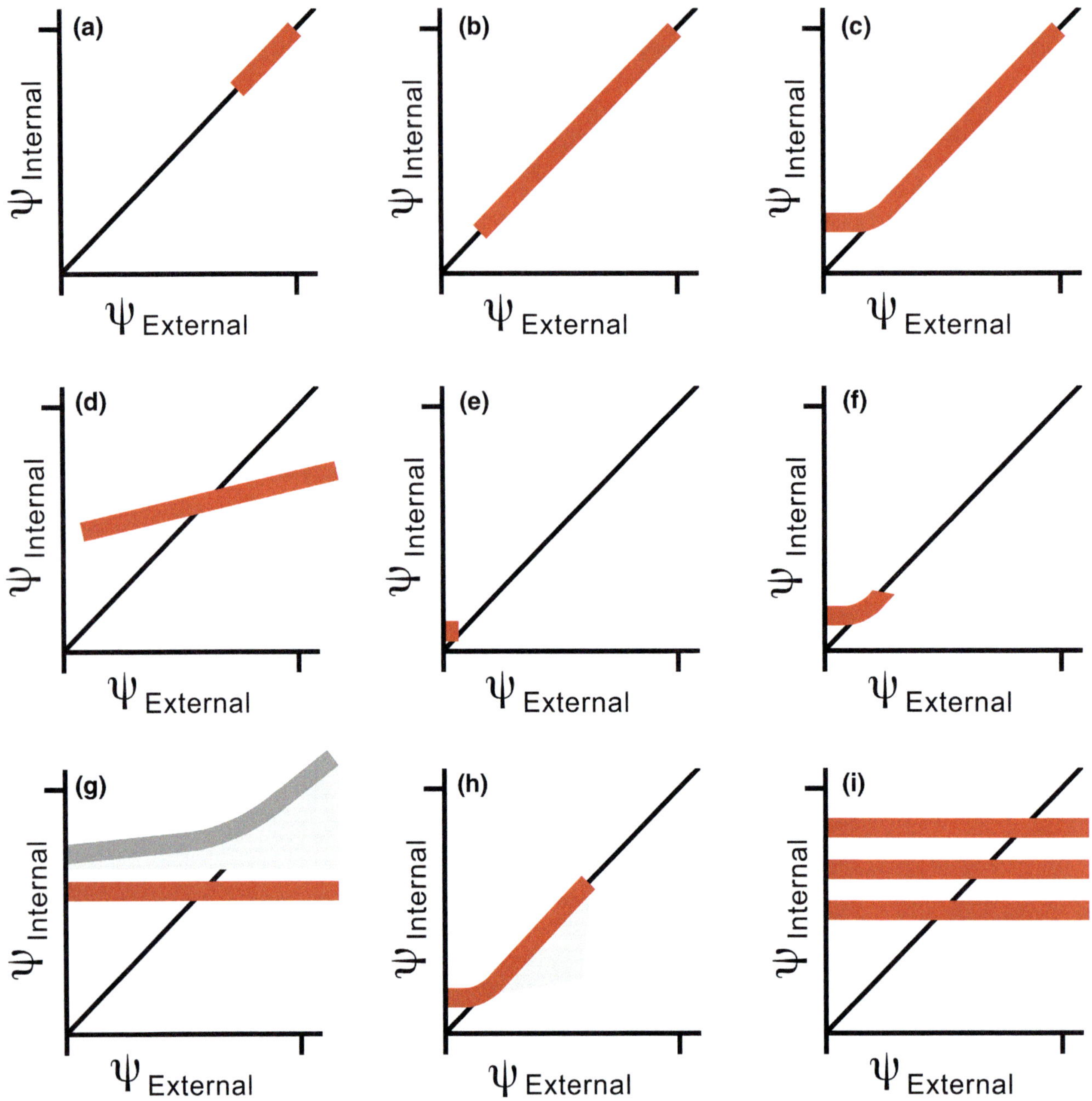

Fig. 7.6 Osmoregulation characteristics of selected species. The external and internal water potentials are drawn inversely (1/Ψ) on the *x*- and *y*-axis, respectively. The equivalents of fully marine conditions (~25 bar at 5 °C) are indicated on both axes. The 45° black line marks iso-osmotic conditions and the red line marks the range of the organism. (**a**) A primary marine organism. (**b**) *Littorina littorea* in the marine intertidal zone and in brackish water. (**c**) *Gammarus duebeni* in the marine intertidal zone and in brackish water. (**d**) *Orchestia* sp. in the epilittoral zone. (**e**) Freshwater gastropods. (**f**) Freshwater vertebrates. (**g**) Marine teleost fish. The light-grey area indicates the range covered by osmolytes (urea) and the dark-grey line that of the Selachimorpha (modern sharks). (**h**) *Fejervarya cancrivora* (crab-eating frog). The light-grey area indicates the range covered by osmolytes (urea). (**i**) Salt-lake organisms, with from top to bottom *Palaemonetes* sp. of marine origin, *Ephydra* sp. of terrestrial origin, *Artemia* sp. of freshwater origin. Figure modified from Remmert (1969a)

Aquatic bacteria, including cyanobacteria (which lack nuclei, mitochondria and chromoplastids), as well as nucleus-bearing protists (algae with chromoplastids, fungi and protozoa with mitochondria), demonstrate high physiological adaptability to changes in salinity. These taxa show extensive adaptive radiation. Protists seem to have retained a considerable evolutionary euryhalinity and are widely distributed, the smallest planktonic representatives of them being particularly diverse under the conditions of the brackish Baltic Sea waters that are stressful for larger sessile organisms (Telesh et al. 2015). This is reflected in high bacterial and maximum protistan species richness in

Box 7.3: Anadromous and catadromous fish

Salmon, lampreys and eels

Salmon and many of its relatives are anadromous fish. They feed in the marine environment and breed in freshwater where the young stay until they smoltify and migrate to sea. In addition to salmonids, lampreys often feed in marine environments and spawn in freshwater, where the young also grow for up to several years. These fish are called anadromous. Eels, in contrast, are catadromous fish: they breed in seawater, and migrate to freshwater to feed. A comprehensive account of the different aspects of anadromous and catadromous life cycles is given in McCormick et al. (2013).

Differences in osmoregulation between marine and fresh waters

In seawater, fish are hypo-osmotic regulators, while in freshwater they are hyper-osmotic regulators. The regulation is very effective and the osmolarity of body fluids changes little when the fish moves from the freshwater to the marine environment and *vice versa*. The development of hypo-osmotic regulation capacity has been studied by following changes accompanying the smoltification of young salmon and their migration from rivers to the marine environment. The development of hyper-osmotic regulation has been studied particularly intensively in eels migrating from the marine environment to the freshwater.

Hormones, drinking rates and urine production

In anadromous and catadromous fish, the development of both hypo- and hyper-osmotic regulation capacities is controlled by hormones. Cortisol and prolactin together enable successful acclimation to freshwater. Freshwater adaptation involves a marked reduction in the permeability of gills and skin to sodium and water, a reduction of the water drinking rate, and ultimately a cessation of drinking and an increase in the urine production rate. Alternatively, successful acclimation to seawater requires action by growth hormone, an insulin-like growth factor, cortisol, and in many cases, thyroid hormone. Important cues triggering hormonal changes include fish body size, photoperiod and temperature. When young salmon (parr) move from freshwater to the sea (smoltification), the direction of passive ion and water fluxes reverse. While the gradients in the freshwater favour water influx and salt efflux, the gradients in the seawater favour water efflux and salt influx. Consequently, animals increase their drinking rate, decrease their urine production, and modify their ion transport systems both in the gills and the intestine. The urine produced is markedly hypotonic compared to body fluids in freshwater, and it changes to virtually isotonic in the seawater. The intestinal uptake of sodium and chloride is facilitated in seawater; intestinal salt uptake is followed by osmotically obliged water enabling the water uptake by the animal to replace the amount lost by diffusion. The salt accumulated is secreted in the gills. Salt secretion occurs via the combined actions of sodium pump and Na-K-2Cl co-transporter. When smoltification occurs, the activity of Na/K ATPase (sodium pump) in the gills increases markedly because of an increase in the number of pump molecules. The number of Na-K-2Cl co-transporters increases as well.

brackish water, especially at critical salinities 5–8 (Telesh et al. 2011).

Animals possess an excretory organ, which is often generically termed "kidney", and it ranges from subcellular and unicellular structures such as contractile vacuoles and flame cells to complex organs with several different constituent tissues. In unicellular organisms, the contractile vacuolar complex is necessarily the primary osmoregulatory system (Box 7.4). In practice, the primary function of these organelles and organs is almost always osmoregulation rather than excretion.

The nitrogenous waste from the body that needs to be eliminated is added almost incidentally to the osmoregulated urine (or not added at all). Despite the large variation in origin, size and complexity of osmoregulatory organs, all of them operate under certain common structural and physiological principles. Nearly all consist of one or many tubular structures and most include an initial collecting area where the primary urine is formed, as well as one or more areas where it is modified by the addition or removal of particular solutes. Many of the organs include a distal area where the urine is more concentrated (in vertebrates, the hyper-osmotic urine can be produced only by mammals) or more diluted (hypo-osmotic) than the body fluids.

In terms of both evolution and contemporary life in brackish waters, osmoregulation is a major problem for which diverse mechanisms have been developed to regulate salt and water content in various groups of animals (Fig. 7.7). However, some marine species have achieved adaptation to low salinities without osmoregulation. They are poikilo-osmotic and allow their body fluids to be

Box 7.4: The contractile vacuolar complex

The contractile vacuolar complex (CVC, Box Figs. 7.3 and 7.4) is a subcellular membrane-bound organelle used to eliminate the excess cytosolic water acquired by osmosis (Hausmann et al. 2003). Typically, the contractile vacuole (cv) fills slowly with a fluid from the narrow collecting channels (diastole), and the fluid is periodically expelled through the contractile vacuole pore (pvc) to the surrounding medium by contractions of the vacuole (systole). Depending on the species and the osmolarity of the environment, the amount of water expelled from the cell and the frequency of contraction may vary considerably. The CVC is found predominantly in freshwater and brackish-water protists that lack a cell wall (*e.g. Amoeba, Paramecium*) and in several types of cells in sponges and fungi. Evolutionarily, the CVC was eliminated in multicellular organisms, but some of its molecular and cellular characteristics are used by multicellular organisms in their own osmoregulatory mechanisms.

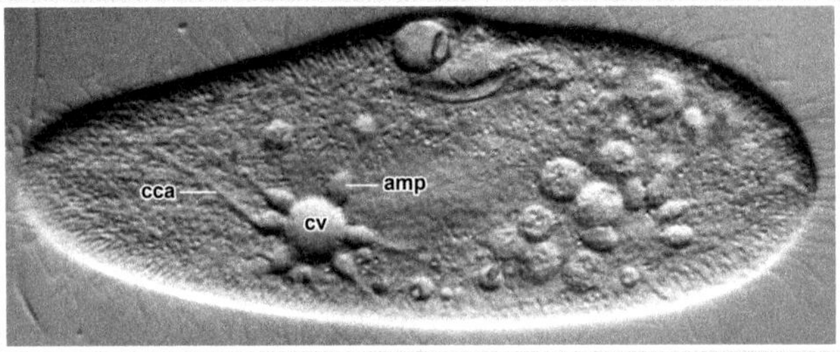

Box Fig. 7.3 Light-microscopic image of a live specimen of *Paramecium* sp. with its contractile vacuolar complex (CVC), showing ampullae (amp), collecting canals (cca) and the contractile vacuole (cv). Photo: © Klaus Hausmann

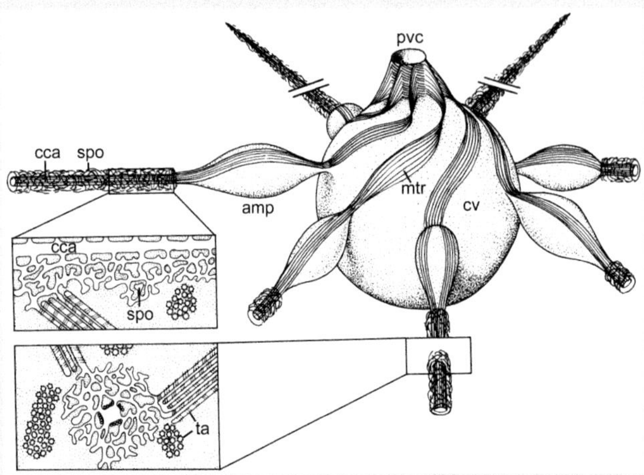

Box Fig. 7.4 Schematic drawing of the contractile vacuolar complex (CVC), showing ampullae (amp), collecting canals (cca), the contractile vacuole (cv) and a pore (pvc). Collecting canals are connected with and surrounded by irregularly arranged spongiomal tubules (spo). Tubular aggregates (ta) are located at a larger distance. The entire CVC is stabilised by several microtubular ribbons (mtr). Figure reprinted from Hausmann et al. (2003) with permission from Schweizerbart'sche Verlagsbuchhandlung

isotonic with the salinity of the external medium. This naturally requires that their cellular constituents are able to function in a wide salinity range. In contrast, the homoioosmotic species, when exposed to minor changes in ambient salinity, tend to retain their initial internal osmotic concentration. The evolution of excretory strategies also involves the availability of water for excretion of nitrogenous wastes.

7.5 Osmotic and ionic adaptations in charophytes

7.5.1 Charophytes have adapted to all salinities

As shown above, ionic and osmotic homoeostasis are interlinked, and non-linear effects are expected because of the ionic anomaly of brackish water (Fig. 7.3). How this affects

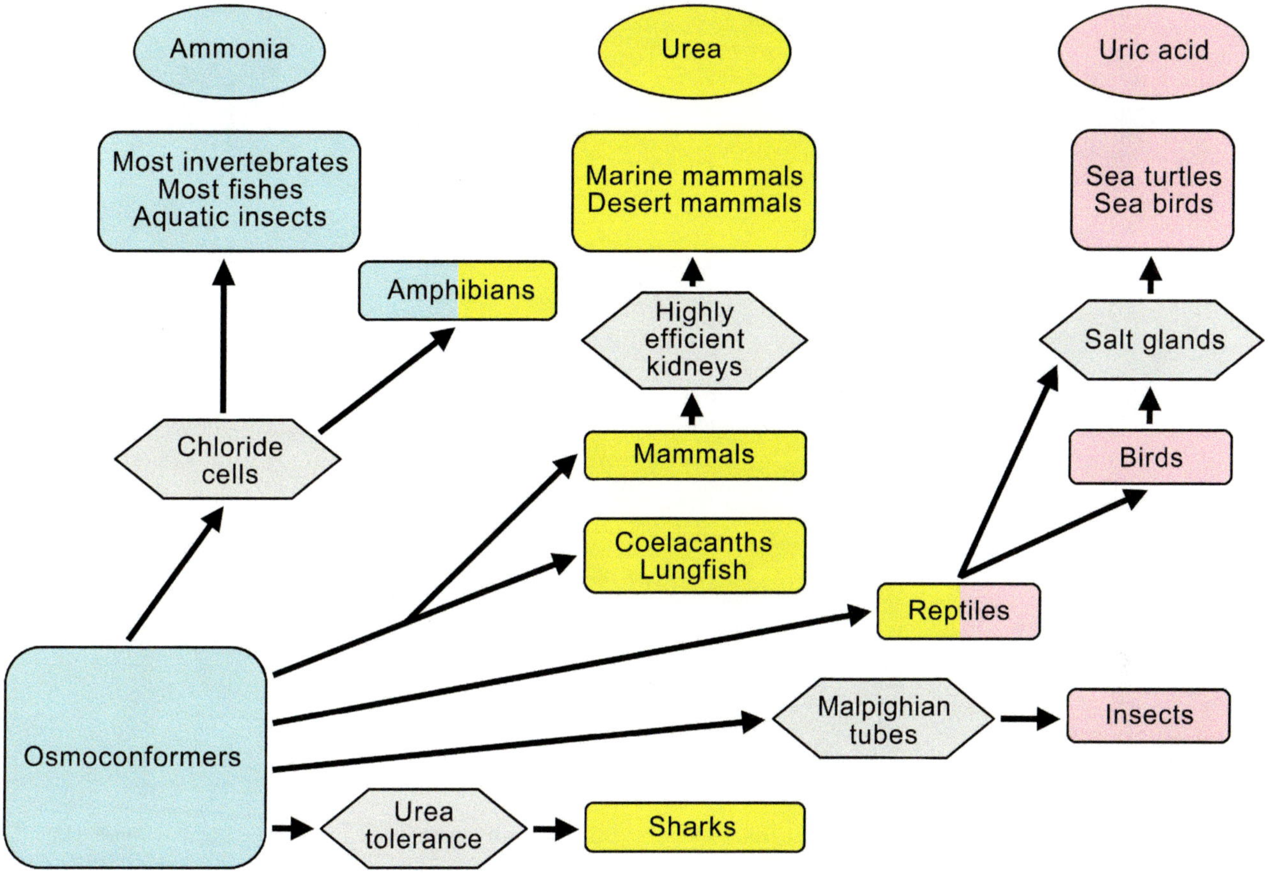

Fig. 7.7 Evolutionary lines of osmoregulation and excretory strategies of nitrogen metabolism waste products in various groups of animals. Note that this figure does not cover all evolutionary pathways or all animal taxa. Figure modified from McShaffrey (2002)

brackish-water organisms has been studied in detail in the charophytes, a group of green algae that have succeeded to adapt to all salinity ranges (Bisson and Kirst 1995).

Charophytes maintain an osmotic potential that is higher than the outer osmotic pressure, which results in a substantial turgor pressure (Winter and Kirst 1990, 1991, 1992). For example, in *Chara vulgaris* the turgor pressure adds up to ~340 mOsmol kg^{-1}, which is equivalent to salinity 13 (Winter and Kirst 1990). When *Chara* species that are unable to acclimate to different osmotic potentials are grown at high salinity, cell elongation rather than the cell division rate is lowered because cell elongation depends on the turgor pressure (Winter and Kirst 1991). With respect to their abilities of osmotic adjustment, four groups of charophytes can be distinguished: freshwater, oligohaline, mesohaline and euryhaline species.

7.5.2 Freshwater and oligohaline charophytes

The first charophyte group consists of all the "purely freshwater species", *e.g. Chara corallina* and *Nitella* spp., which are able to keep their osmotic potential constant mainly by a K^+-regulation system, but which are incapable of adjusting it to ambient salinity changes. In freshwater, this regulation is sufficient to keep the turgor constant (Bisson and Kirst 1995).

The second group contains oligohaline or "halo-tolerant" freshwater species that are able to regulate their turgor via the accumulation of ions such as Na^+ and Cl^- as well as by accumulating osmolytes, especially sucrose. Examples of species in this group are *Chara vulgaris* (Winter et al. 1987; Winter and Kirst 1990) and *Nitellopsis obtusa* (Winter et al. 1999). This mechanism seems to be restricted to low salinities due to the toxic effect of Na^+. The K^+/Na^+ ratio, which usually exceeds 1 in charophytes, decreases with increasing salinity. This results in a "reduced vitality" and competitive disadvantages of oligohaline charophyte species at salinities exceeding 5.

7.5.3 Mesohaline and euryhaline charophytes

The third group of charophytes consists of mesohaline brackish-water species, *e.g. Chara aspera* and *Chara canescens* (Fig. 7.8), which are successful competitors at salinities up to ~15 (Winter and Kirst 1991, 1992). These species exhibit a reduced spectrum of regulation capabilities found in euryhaline species. *Chara canescens* keeps its K^+

Fig. 7.8 The charophyte *Chara canescens* is one of the few species restricted to brackish-water habitats. In the Baltic Sea, only female plants that reproduce by ovoapogamy (apogamy by parthenogenetic formation of the oospores) have been found. In the numerous "Lacken" (small temporary and permanent brackish-water ponds) in the Neusiedler See area (Austria/Hungaria) both ovoapogamic and bisexual lineages occur (Schaible et al. 2011). It is not known why only parthenogens seem to live in the Baltic Sea, but their success may be a result of a higher probability of reproduction. (**a**) Female *Chara canescens* from the Baltic Sea. (**b**) A bisexual *Chara canescens* population from the Neusiedler See area. (**c**) A ripe antheridium from the Neusiedler See area. (**d**) A male individual from the Neusiedler See area. Photo: (**a**), (**b**) © Hendrik Schubert, (**c**) © Anette Küster, (**d**) © Ralf Schaible

concentration constant up to salinity ~4. When salinity increases, the species starts to regulate turgor pressure via K^+ and Na^+ (and Cl^-) accumulation. However, *Chara canescens* seems to be unable to support its turgor regulation by sucrose accumulation, a mechanism observed in *Chara aspera* (Winter and Kirst 1992). At salinities of ~20, the K^+/Na^+ ratio in *Chara canescens* drops below 1, as the pronounced K^+ import typical of euryhaline species seems to be missing. In contrast to *Chara canescens*, which starts turgor regulation only at salinities of ~4, *Chara aspera* regulates turgor pressure in freshwater as well, resulting in a perfect constancy of turgor pressure at salinities between 0.8 and 8 (Winter and Kirst 1991, 1992).

The fourth group of charophytes represents euryhaline species such as *Lamprothamnium papulosum*, *Lamprothamnium succinctum* and *Chara buckellii*, which are able to tolerate a very broad range of salinities. At low salinities (up to ~6), these species keep their K^+ concentration constant and regulate their turgor mainly via the accumulation of Na^+ and Cl^-. At higher salinities (up to ~13), the turgor pressure is regulated by the uptake of both K^+ and Na^+ (and Cl^-). At salinities >13, the turgor regulation is accomplished by the accumulation of mainly K^+ and Cl^- supported by accumulation of sucrose, whereas Na^+ is kept constant (Beilby et al. 1999). The K^+/Na^+ ratio is thus kept at a high level, allowing these charophytes to survive at salinities of up to 70. In the field, these species seem to be poor competitors compared to other macrophytes, and they mainly occur in the salinity range 20–40 (Winter et al. 1996). This kind of salinity regulation also occurs in *Tolypella glomerata* and *Tolypella nidifica*, species which in their natural habitats are restricted to much lower salinity than the other species in this group. This is probably because they fail to develop oogonia at salinities >12, rather than because their growth is reduced (Winter et al. 1996).

7.6 Adaptation to ambient temperature

7.6.1 Temperature ranges in the sea

The overall upper limit of the temperature tolerance range of aquatic invertebrates is ~50 °C (Nguyen et al. 2011), while the lower limit is equal to the freezing point of −1.86 °C for fully marine seawater and about −0.4 °C for the brackish Baltic Sea water (*cf.* Fig. 2.17b). However, most aquatic organisms are seriously affected by temperature change outside their own temperature tolerance limits. With its specific heat capacity of ~3,000 times that of the air, water is a good heat conductor. Consequently, temperature differences in seas are highly buffered, and a considerable heat flux is required in order to modify the water temperature. Thus, even daily water temperature changes are rarely dramatic enough to cause functional changes in aquatic organisms in the sea, although they can be significant for the inhabitants of shallow coastal areas.

7.6.2 Enzymatic adaptations

All organisms use enzymes (proteins) to adapt to changing thermal conditions. The performance of enzymes is affected by temperature in a variety of ways that may be adaptive and extend the thermal range tolerated by the organism. The nature and speed of such modifications vary with the time scale of the temperature change. Enzymatically regulated adaptations can be fast if only adjustments of the existing proteins are required, but they are much slower if *de novo* protein synthesis is needed.

An important mechanism involved in the responses of living organisms to thermal change is the synthesis of stress proteins, which are often referred to as "heat shock proteins" (HSPs, Box 7.5). HSPs increase the thermal tolerance and perform functions essential for cell survival under stressful conditions. These proteins are naturally present in a cell at constitutive levels under normal conditions, but they are expressed at a higher rate when a cell is exposed to a sudden thermal change, as well as to other sudden changes in the environment, *e.g.* salinity or pH (Durante and Colucci 2010; Roberts et al. 2010; Hartl et al. 2011).

7.7 Adaptation to ambient light

7.7.1 Light and aquatic photosynthesis

Water bodies are variable photic environments due to variability in solar elevation and waves (Fig. 7.2) and, additionally, through complex, depth-dependent interactions between light and suspended particles and dissolved matter, which involve the absorption and scattering of light (*cf.* Sect. 15.2). Light is of utmost importance for primary production, as light availability directly affects the growth, survival and coexistence strategies of autotrophs. Changes in the depth distribution of phytoplankton, *e.g.* by vertical movement, may cause dramatic changes in light and nutrient availability over short time scales (seconds to days) and spatial scales (cm to m).

Autotrophs have evolved a broad variety of strategies to acclimate to the complex temporal and spatial variability of irradiance. Probably the best known evolutionary achievements include the construction of light-harvesting antenna systems that increase the energy supply to the photosynthetic reaction centres. There are three lineages of such antenna systems, which differ with respect to their absorbance characteristics: (1) the chlorophyll antenna system, (2) the xanthophyll antenna system, and (3) the phycobilin antenna system (van den Hoek et al. 1995).

Photosynthesis is non-linearly light-dependent *inter alia* because of photoinhibition occurring above a certain irradiance level (Fig. 7.9). With regard to the variability of the underwater light, all aquatic photoautotrophs need fast acclimation mechanisms, except for those living in a few habitats with permanent low irradiance conditions.

7.7.2 Surviving under low irradiance

Under low irradiance, the available photons can be efficiently used, either by increasing the amount of antenna pigments (a λ-neutral mechanism) or by spectral acclimation of the antenna system (Fig. 7.9). Without losing relative absorbance efficiency, organisms employing these strategies are able to acclimate, within a couple of days, to changes in the spectral composition of the underwater light caused by *e.g.* developing phytoplankton blooms (Schubert et al. 1997).

By increasing pigmentation, the absorbance efficiency of the pigments expressed as photons absorbed per unit time will decrease due to the packaging effect. Nevertheless, this kind of acclimation, which can result in rendering the algae almost optically black, is by far the most common strategy. Examples of this strategy in the Baltic Sea can be observed in dark-coloured individuals of red algae such as *Furcellaria lumbricalis* and *Polyides rotundus*. Alternatives to this mechanism are "chromatic acclimations" whereby the absorbance characteristics of the antenna system are adjusted to the spectral composition of the prevailing irradiance. Probably the most sophisticated mechanism of this kind is the so-called "complementary chromatic adaptation", which occurs in some cyanobacteria. In this mechanism, the pigment phycoerythrin, which is absorbed in the green wavelength region and dominates the antenna under green light conditions, is replaced by phycocyanin, an orange region-absorbing light-harvesting pigment.

7.7.3 Dealing with high irradiance

At high irradiance, photoautotrophs are not just energetically "saturated", but they need to be protected from damage by excess light energy (Fig. 7.9). An oversupply of energy to a reaction centre, especially to the reaction centre of

Box 7.5: Stress proteins

Heat shock proteins (HSPs)

At the biochemical level, a basic and evolutionarily most conserved molecular defensive mechanism is the synthesis of stress proteins, often referred to as heat shock proteins (HSPs, Hartl et al. 2011). HSPs perform chaperone function, *i.e.* they assist in refolding proteins that were damaged by stress and stabilise new proteins by ensuring correct protein folding (Box Fig. 7.5). Thus, HSPs provide cellular and whole-body adaptation for all organisms studied so far in a vast range of extreme environmental conditions (Box Fig. 7.6). Their production can be triggered by many natural and human-induced stresses, *e.g.* fluctuations in seawater temperature, salinity, acidification, light availability and pollution levels, hypoxia or hyperoxia, etc. Once induced in response to a particular stress, the HSPs can make the organism more tolerant of other stresses.

HSP families

Stress proteins are represented by a number of families, differing in molecular weight, the nucleotide sequences of the encoding genes, and functions (Hartl et al. 2011). HSPs include both relatively large (*e.g.* HSP60, HSP70, HSP90, HSP100) and small (*e.g.* HSP10, HSP27, ubiquitin) proteins. The structure of most HSP families is conserved even across kingdoms, and their action also seems to be highly conservative. Several HSPs may exist in both constitutive and stress-inducible forms (*e.g.* HSP70).

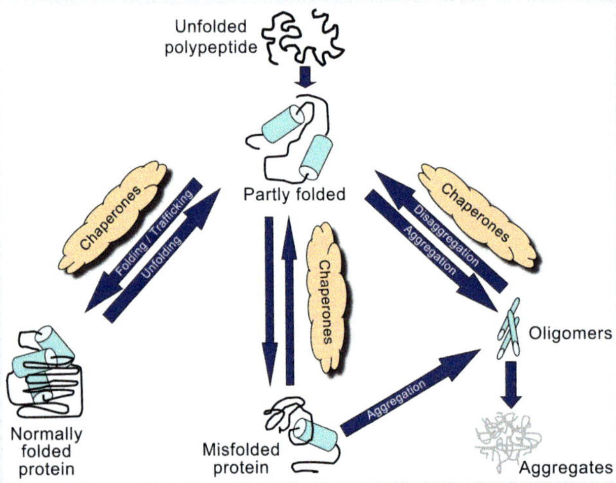

Box Fig. 7.5 Normal protein molecules in living cells are naturally folded into specific configurations, requisite for their proper functioning, but they may unfold in response to various kinds of stress. Such unfolded proteins may then refold wrongly, and may be susceptible to interactions with other cellular components. Stress proteins as molecular chaperones serve to limit these interactions by binding temporarily to the unfolded proteins and thus stabilising their state. Thus, chaperones have important physiological roles through facilitating the synthesis, *de novo* folding, assembly, trafficking, and secretion of specific proteins in various cellular compartments as well as guarding the cellular proteome against misfolding and inappropriate aggregation. Figure modified from Hartl et al. (2011)

Box Fig. 7.6 The epilittoral zone is an especially stressful habitat for both terrestrial and aquatic organisms. (**a**, **b**) *Fucus vesiculosus* in mixed stands with terrestrial plants on the virtually non-tidal coast of the island of Saaremaa (Estonia) in the Baltic Sea. (**c**) *Fucus cottonii* in mixed stands with terrestrial plants on the tidal coast of Ireland (Neiva et al. 2012). While changes on tidal coasts are fairly predictable, even in the epilittoral zone, erratic changes of temperature, water and irradiance on the virtually non-tidal Baltic Sea coast require fast and non-specific stress-protection mechanisms. Photo: © Hendrik Schubert

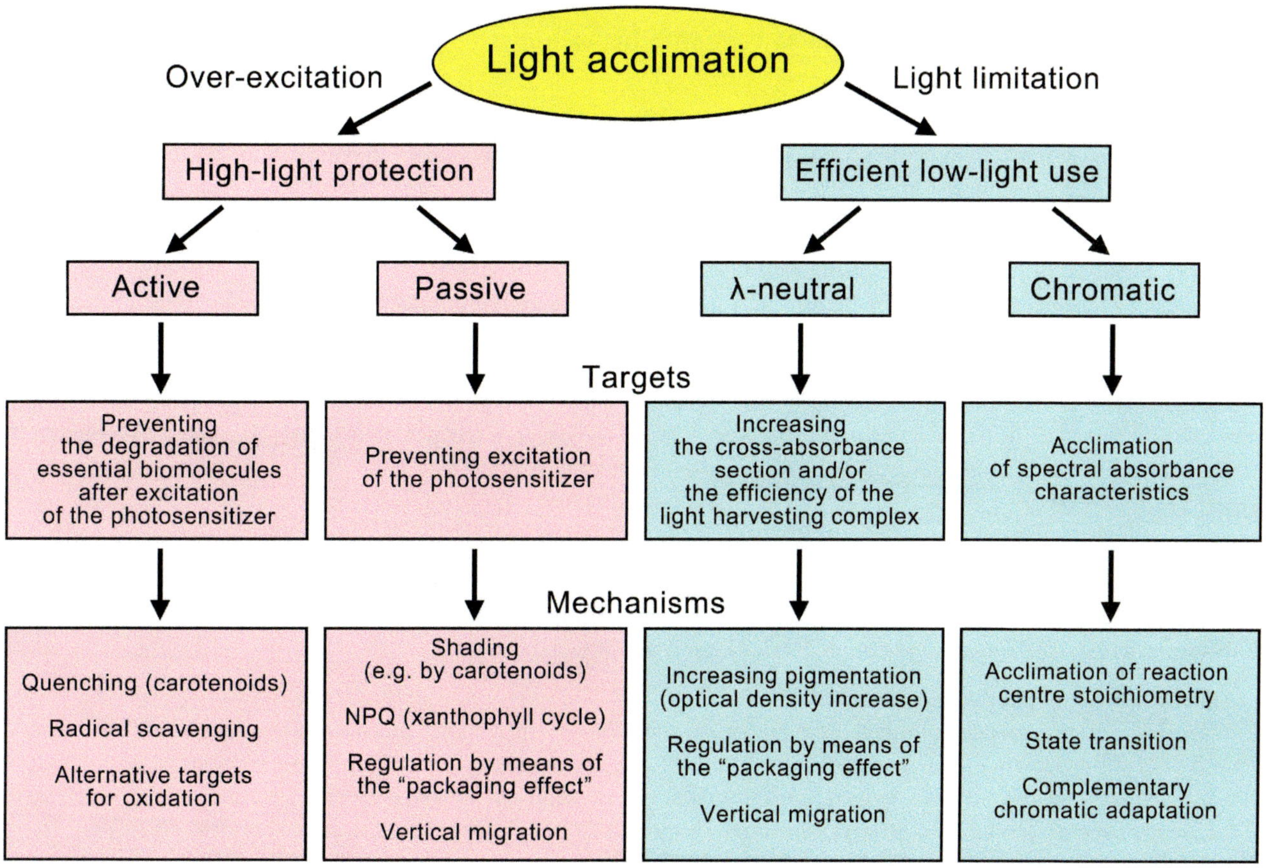

Fig. 7.9 Summary of light acclimation strategies in photoautotrophs, including two directions of acclimation. Due to the wide variation in light conditions in the aquatic environment, photoautotrophs must be able to cope with periods of both overexcitation and light limitation. NPQ = non-photochemical quenching. Figure based on data in Schubert et al. (2004, 2006) and Marquardt et al. (2010)

photosystem II (PSII), increases the probability that chlorophyll transfers excited electrons to oxygen instead of to plastoquinone. PSII is the main target because the rate-limiting step of oxygenic photosynthesis is the regeneration of the plastoquinone molecules at the cytochrome b_6f-complex. Overexcitation of PSII therefore results in acceptor limitation, whereas photosystem I (PSI) lacks electrons (because of donor limitation) and cannot be excited anymore.

Plant cells can employ two main strategies to cope with situations of excessive energy supply. The "active" mechanisms allow excitation of the photosensitiser (chlorophyll), but protect biomolecules from the consequences of acceptor limitation. This can be performed by quenching the potentially harmful triplet excitation states of chlorophyll, *i.e.* before being transferred to oxygen or by quenching the already activated oxygen molecules. Once activated via electron transfer, oxygen soon forms radicals that must be quenched by specific reactions requiring energy input and the biosynthesis of specialised enzymes or alternative targets.

An alternative to this rather sophisticated and energy-demanding strategy of photoprotection is to prevent overexcitation of chlorophyll, which is most easily accomplished by shading pigments. However, this strategy is complicated as well because the irradiance is highly variable on short time scales (Fig. 7.2). Such a "sunscreen" of shading pigments must be turned on and off very fast to be effective, otherwise it would make more sense to just reduce the chlorophyll content in order to solve the problem of the acceptor limitation of PSII. Therefore, it is not surprising that only little evidence for the existence of such a dynamic sunscreen mechanism has been found so far and the question of whether or not xanthophyll cycling can act as such a mechanism is still debated (Masojídek et al. 2004).

However, there are many mechanisms, including the xanthophyll cycle, with which the overall quantum efficiency of photosynthesis under the conditions of PSII acceptor limitation can be reduced. These processes, often collectively termed "non-photochemical quenching" (NPQ), may either reduce the excitation energy transfer to the reaction centre by *e.g.* decoupling the light-harvesting complexes, transferring the already absorbed photon energy into alternative sinks or, alternatively, they may decrease the charge-separation efficiency itself and therefore reduce the extent of the acceptor limitation of PSII.

Box 7.6: Strategies of aquatic animals to cope with hypoxia

Reduction of energy use

When animals experience oxygen limitation, they may respond by reducing their energy use (Hochachka and Somero 2002). This is employed by virtually all hypoxia-tolerant species when they encounter low oxygen levels. For example, oxygen consumption of both the blue mussel *Mytilus trossulus* (Box Fig. 7.7) and the crucian carp *Carassius carassius* is directly dependent on the ambient oxygen tension.

Energy production efficiency

Animals may also respond to oxygen limitation by increasing the efficiency of their energy production. The animals try to keep their aerobic energy production active, as it is much more efficient than anaerobic energy production. Thus, the first response is to increase the water flow past the respiratory epithelium. The increase in the bioventilation rate and amplitude, however, demands an increased use of energy. The positive effect on the oxygen available for energy production vanishes when the increased ventilatory energy use exceeds the potential for aerobic energy production (Dejours 1975; Farrell and Richards 2009).

Respiratory pigments

The oxygen affinity of respiratory pigments is higher in hypoxia-tolerant species than in hypoxia-sensitive species (Weber and Jensen 1988). The amount of respiratory pigments increases under hypoxia in most species. For example, in aquatic vertebrates, erythrocytes are released from storage organs and the production of erythrocytes increases (Nikinmaa 1990; Gallaugher and Farrell 1998; Nilsson and Randall 2010).

Anaerobic energy production

If aerobic energy production cannot be maintained, the animals must resort to anaerobiosis. To extend the time during which anaerobiosis can be maintained, hypoxia-tolerant organisms have much larger supplies of appropriate substrates (*e.g.* glycogen) than hypoxia-sensitive species. In addition, hypoxia-tolerant species use the so-called "alternative energy-producing pathways" that produce more ATP per glucose molecule than traditional glycolysis. The end products include acetate, succinate or malate, depending on the invertebrate species (Grieshaber et al. 1994).

Box Fig. 7.7 The deep regions of the Baltic Sea are typically affected by hypoxia, but hypoxia may occur episodically in shallow-water habitats as well, *e.g.* within dense populations of filter feeders or in phytobenthic communities. (**a**) A *Mytilus trossulus* bed affected by hypoxia. (**b**) A *Zostera marina*-dominated community affected by hypoxia, overgrown by a *Spirogyra* mat. Photo: (**a**) © Hendrik Schubert, (**b**) © Sven Dahlke

Another strategy under the conditions of photosynthesis overexcitation is the so-called "packaging effect" in which chloroplasts are lined up in places along the cell wall receiving the lowest energy input. Yet another strategy is that employed by motile phytoplankton species, which can avoid overexcitation by vertical migration. This mechanism is also employed for "nutrient pumping", allowing for uptake of nutrients in deeper strata of the water column.

7.8 Adaptation to low oxygen levels

7.8.1 Withstanding hypoxia and surviving anoxia

As organisms vary greatly in their oxygen requirements, low dissolved oxygen concentration may result in biodiversity loss (*cf.* Fig. 10.7). However, populations of most animal species living in estuaries and lagoons are able to tolerate short-term exposure to low dissolved oxygen concentrations without noticeable adverse effects. Extended exposure to dissolved oxygen concentrations below 60 % air saturation may cause behavioural modifications, reduced abundance and productivity, negative reproductive effects and mortality. Moreover, there is evidence that hypoxia (<2 mL O_2 L^{-1}) can inhibit immune responses, causing higher mortality than would otherwise occur when organisms are challenged with a pathogen (Burnett and Stickle 2002).

An early behavioural response to hypoxia can be to move toward better-oxygenated water, even when other conditions there might be unfavourable. Under hypoxic conditions an animal may also slow down its swimming and feeding activities, which reduces its need for energy and hence oxygen. However, while reduced activity may render the animal more hypoxia-tolerant for a short period, a lower swimming activity makes the animal more vulnerable to predation, and reduced feeding decreases its growth. If oxygen insufficiency persists, death will ultimately occur in animals using this strategy only.

Many aquatic animals respond to a short period of hypoxia by increasing their efficiency of oxygen transport to cells and mitochondria (Box 7.6). Because of the slow diffusion of oxygen in water relative to that in air, the movement of water across permeable membranes or tissue surfaces for respiratory needs is almost universal among aquatic animals. The pumping process places high energetic demands on the animals and additionally exposes cellular surfaces to osmotic gradients.

Changes in oxygen tension are sensed by haemoproteins. Altogether, 1–2 % of animal genes appear to be directly regulated by oxygen, although many of the molecular responses to hypoxia are still poorly known. In metazoans the major regulation of oxygen-dependent genes occurs via the function of the hypoxia-inducible factor (HIF, Rytkönen et al. 2011). In vertebrates, the most important regulatory hypoxia-inducible factor is HIF-1, which receives signals from the molecular oxygen sensor through redox reactions and/or phosphorylation, and regulates the transcription of a number of hypoxia-inducible genes, including those involved in erythropoiesis, angiogenesis and glycolysis (Wu 2002; Nikinmaa and Rees 2005). Multicellular species have evolved highly complex organs for oxygen uptake (lung), transport (blood), and tissue distribution (cardiovascular system). Ingeniously, the main functional regulator of oxygen homeostasis is the local oxygen partial pressure itself rather than a genetically encoded developmental programme or a central oxygen-measuring regulator (Wenger 2002).

7.8.2 Consequences of hypoxia for biodiversity

In the Baltic Sea, the increasing prevalence of oxygen-depleted bottom water in deep areas has perhaps become the strongest factor influencing the biodiversity of zoobenthic communities (*cf.* Sect. 10.11). The Baltic Sea proper is permanently stratified, consisting of an brackish-water upper layer with a salinity of ~6–8 and lower layers of more saline waters with salinities of ~9–13 (*cf.* Fig. 2.15). A permanent halocline at depths of ~60–80 m (*cf.* Table 2.6) prevents vertical mixing of the water column and the transport of more oxygenated waters to the deeper parts of the basin. The separation between normoxic and moderately hypoxic water masses and hypoxic or anoxic waters creates a temporal and spatial mosaic of stress to benthic animals living at larger depths. Due to low oxygen levels, macrozoobenthic communities at larger depths differ from those living on shallower bottoms (Conley et al. 2009). Hypoxia often eliminates large deep-burrowing, actively bioturbating species because their long generation times prevent the development of viable populations (Solan et al. 2004).

At the ecosystem level, there is a general tendency for suspension feeders to be replaced by deposit feeders, demersal fish by pelagic fish and macrozoobenthos by meiobenthos due to hypoxia in the Baltic Sea. Nanoplankton also tend to dominate in the phytoplankton community in hypoxic environments. Even when species are not entirely lost, they may become functionally extinct due to low abundance. A reduction of bioturbation thus decreases the natural purification capacity and increases the internal nutrient loading of Baltic sediments (Karlson et al. 2007), which increases with the spatial extension of the anoxic/hypoxic zone in the Baltic Sea proper.

Intermittent hypoxia, which may occur in shallow eutrophic bays, will also affect the success of species. HIF-dependent regulation is only known to function in short-term hypoxic events (Rissanen et al. 2006), and this regulation may thus be important in habitats with intermittent hypoxia.

Review questions

1. What are the major variable environmental drivers in the Baltic Sea that require physiological adaptations?
2. Which osmo- and ion-regulation adaptations are typical of the brackish-water conditions in the Baltic Sea?
3. What are the main strategies of osmoregulation in brackish water bodies?
4. What are the main lines of irradiance acclimation?
5. How do organisms deal with hypoxia?

Discussion questions

1. Do phylogenetic relationships mirror the salinity-related distribution of organisms? How would you construct a "tree of life" when combining Fig. 7.7 and recent phylogeny?
2. The obvious lack of macroalgae, except for green algae, in low-salinity and freshwater habitats are hypothesised as being a consequence of the lack of hard substrates. However, in the Baltic Sea there are plenty of low-salinity areas with hard substrates. Which macrophyte species live there? Are there alternative explanations for the absence of red and brown algae? What arguments definitely exclude any kind of osmotic or ionic regulation-based explanation for the absence of red and brown algae?
3. How do organisms adapt to fast changes in environmental drivers? Which mechanism(s) shown in Fig. 7.9 back up the different time scales of irradiance variability shown in Fig. 7.2?
4. The double-Donnan equilibrium contributes to maintaining homeostasis of animal cells lacking rigid cell walls. What problems do the unprotected surfaces (*e.g.* gills) of hypo-osmotic animals, such as marine teleost fish, face? How do they cope with the osmotic problems of their egg and sperm cells? Could this be a reason for anadromous behaviour? Is there a relationship between phylogeny and anadromous/catadromous behaviour? What alternative explanations for reproductive migration exist?
5. Why does a rigid cell wall alleviate osmotic adjustment? What would Fig. 7.5 look like for a plant cell? What consequences can be expected from this for the ability of plants to invade brackish, freshwater and terrestrial environments, compared to animals?

References

Beilby MA, Cherry CA, Shepherd VA (1999) Dual turgor regulation response to hypotonic stress in *Lamprothamnium papulosum*. Plant, Cell and Environment 22:347–359

Bisson MA, Kirst GO (1995) Osmotic acclimation and turgor pressure regulation in algae. Naturwissenschaften 82:461–471

Burnett LE, Stickle WB (2002) Physiological responses to hypoxia. In: Rabalais NN, Turner RE (eds) Coastal hypoxia: consequences for living resources and ecosystems. American Geophysical Union, Washington, DC, Coastal and Estuarine Studies 58: 101–114

Conley DJ, Björck S, Bonsdorff E, Carstensen J, Destouni G et al (2009) Hypoxia-related processes in the Baltic Sea. Environmental Science and Technology 43:3412–3420

Connell JH (1978) Diversity in tropical rain forests and coral reefs. Science 199:1302–1310

Dejours P (1975) Principles of comparative respiratory physiology. Elsevier, Amsterdam, 253 pp

Durante P, Colucci L (eds) (2010) Handbook of molecular chaperones: roles, structures and mechanisms. Nova Biomedical Books, New York NY, 567 pp

Edwards SL, Marshall WS (2013) Principles and patterns of osmoregulation and euryhalinity in fishes. In: McCormick SD, Farrell AP, Brauner CJ (eds) Euryhaline fishes. Elsevier – Academic Press, Amsterdam, Fish Physiology 32:1–44

Elliott M, Whitfield AK (2011) Challenging paradigms in estuarine ecology and management. Estuarine Coastal and Shelf Science 94:306–314

Farrell AP, Richards JG (2009) Defining hypoxia: an integrative synthesis of the responses of fish to hypoxia. In: Richards JG, Farrell AP, Brauner CJ (eds) Hypoxia. Elsevier – Academic Press, Amsterdam, Fish Physiology 27:487–503

Gallaugher P, Farrell AP (1998) Hematocrit and blood oxygen carrying capacity. In: Perry SF, Tufts B (eds) Fish respiration. Academic Press, San Diego, Fish Physiology 17:185–217

Grieshaber MK, Hardewig I, Kreutzer U, Pörtner HO (1994) Physiological and metabolic responses to hypoxia in invertebrates. Reviews of Physiology Biochemistry and Pharmacology 125:43–147

Grime JP (1973) Competitive exclusion in herbaceous vegetation. Nature 242:344–347

Hartl FU, Bracher A, Hayer-Hartl M (2011) Molecular chaperones in protein folding and proteostasis. Nature 475:324–332

Hausmann K, Hülsmann N, Radek R (2003) Protistology, 3rd edn. E. Schweizerbart'sche Verlagsbuchhandlung, Stuttgart, 379 pp

Hedgpeth JW (1951) The classification of estuarine and brackish waters and the hydrographic climate. In: Ladd HS, Gunter G, Lohman KE, Revelle R (eds) National Research Council (Washington, DC) Report of the Committee on a Treatise on Marine Ecology and Paleoecology 11:49–56

Hochachka PW, Somero GN (2002) Biochemical adaptation: mechanism and process in physiological evolution. Oxford University Press, Oxford 466 pp

Karlson K, Bonsdorff E, Rosenberg R (2007) The impact of benthic macrofauna for nutrient fluxes from Baltic Sea sediments. AMBIO 36:161–167

Khlebovich VV (1968) Some peculiar features of the hydrochemical regime and the fauna of mesohaline waters. Marine Biology 2:47–49

Khlebovich VV (1974) The critical salinity of biological processes. Nauka, Leningrad, 236 pp [in Russian]

Langmuir I (1938) Surface motion of water induced by wind. Science 87(2250):119–123

Macknight ADC, Leaf A (1977) Regulation of cellular volume. Physiological Reviews 57:510–573

Marquardt R, Schubert H, Varela DA, Huovinen P, Henriquez L, Buschmann AH (2010) Light acclimation strategies of three commercially important red algal species. Aquaculture 299:140–148

Masojídek J, Kopecký J, Koblízek M, Torzillo G (2004) The xanthophyll cycle in green algae (chlorophyta): its role in the photosynthetic apparatus. Plant Biology 6:342–349

McCormick SD, Farrell AP, Brauner CJ (eds) (2013) Euryhaline fishes. Elsevier – Academic Press, Amsterdam, Fish Physiology 32:1–594

McLusky DS, Elliott M (2004) The estuarine ecosystem: ecology, threats and management, 3rd edn. Oxford University Press, Oxford 214 pp

McShaffrey D (2002) Nitrogen excretion and osmotic regulation. The Sextant [http://www.marietta.edu]

Neiva J, Hansen GI, Pearson GA, Van De Vliet MS, Maggs CA, Serrão EA (2012) *Fucus cottonii* (Fucales, Phaeophyceae) is not a single genetic entity but a convergent salt-marsh morphotype with multiple independent origins. European Journal of Phycology 47:461–468

Nessim RB (1980) Untersuchungen zur Verteilung der Hauptkomponenten des Salzgehaltes im Wasser und Sediment der Darß-Zingst-Boddenkette unter besonderer Berücksichtigung der Ionenanomalie. University of Rostock, Rostock, 149 pp [PhD Thesis, in German]

Nguyen KDT, Morley SA, Lai CH, Clark MS, Tan KS et al (2011) Upper temperature limits of tropical marine ectotherms: global warming implications. PLoS ONE 6(12):e29340

Nikinmaa M (1990) Vertebrate red blood cells. Springer, Berlin, 262 pp

Nikinmaa M, Rees BB (2005) Oxygen-dependent gene expression in fishes. American Journal of Physiology: Regulatory, Integrative and Comparative Physiology 288:R1079–R1090

Nilsson GE, Randall DJ (2010) Adaptation to hypoxia in fishes. In: Nilsson GE (ed) Respiratory physiology of vertebrates: life with and without oxygen. Cambridge University Press, Cambridge, pp 131–173

Reed RH (1995) Solute accumulation and osmotic adjustment. In: Cole KM, Sheath RG (eds) Biology of the red algae. Cambridge University Press, Cambridge, pp 147–170

Remmert H (1969a) Der Wasserhaushalt der Tiere im Spiegel ihrer ökologischen Geschichte. Naturwissenschaften 56:120–124 [in German]

Remmert H (1969b) Über Poikilosmotie und Isoosmotie. Zeitschrift für vergleichende Physiologie 65:424–427 [in German]

Rissanen E, Tranberg HK, Sollid J, Nilsson GE, Nikinmaa M (2006) Temperature regulates hypoxia-inducible factor-1 (HIF-1) in a poikilothermic vertebrate, crucian carp (*Carassius carassius*). Journal of Experimental Biology 209:994–1003

Roberts RJ, Agius C, Saliba C, Bossier P, Sung YY (2010) Heat shock proteins (chaperones) in fish and shellfish and their potential role to fish health: a review. Journal of Fish Diseases 33:789–801

Russell G (1985) Recent evolutionary changes in the algae of the Baltic Sea. British Phycological Journal 20:87–104

Rytkönen KT, Williams TA, Renshaw GM, Primmer CR, Nikinmaa M (2011) Molecular evolution of the metazoan PHD-HIF oxygen-sensing system. Molecular Biology and Evolution 28:1913–1926

Sagert S, Rieling T, Eggert A, Schubert H (2008) Development of a phytoplankton indicator system for the ecological assessment of brackish coastal waters (German Baltic Sea coast). Hydrobiologia 611:91–103

Schaible R, Bergmann I, Schubert H (2011) Genetic structure of sympatric sexually and parthenogenetically reproducing population of *Chara canescens* (Charophyta). ISRN Ecology, 13 pp [doi:10.5402/2011/501838]

Schubert H, Andersson M, Snoeijs P (2006) Relationship between photosynthesis and non-photochemical quenching of chlorophyll fluorescence in two red algae with different carotenoid composition. Marine Biology 149:1003–1013

Schubert H, Forster RM (1997) Sources of variability in the factors used for modelling primary productivity in eutrophic waters. Hydrobiologia 349:75–85

Schubert H, Gerbersdorf S, Titlyanov E, Titlyanova T, Granbom M et al (2004) Circadian rhythm of photosynthesis in *Kappaphycus alvarezii* (Rhodophyta): independence of the cell cycle and possible photosynthetic clock targets. European Journal of Phycology 39:423–430

Schubert H, Matthijs HCP, Forster RM (1997) Occurrence of complementary chromatic adaptation in a eutrophic estuary. The Phycologist 46:28

Schubert H, Schlüter L, Feuerpfeil P (2003) The underwater light climate of a shallow Baltic estuary – ecophysiological consequences. ICES Research Report 257:29–37

Solan M, Cardinale BJ, Downing AL, Engelhardt KAM, Ruesink JL, Srivastava DS (2004) Extinction and ecosystem function in the marine benthos. Science 306:1177–1180

Stramski D, Legendre L (1992) Laboratory simulation of light-focusing by water-surface waves. Marine Biology 114:341–348

Telesh IV, Schubert H, Skarlato SO (2011) Revisiting Remane's concept: evidence for high plankton diversity and a protistan species maximum in the horohalinicum of the Baltic Sea. Marine Ecology Progress Series 421:1–11

Telesh IV, Schubert H, Skarlato SO (2013) Life in the salinity gradient: discovering mechanisms behind a new biodiversity pattern. Estuarine Coastal Shelf Science 135:317–327

Telesh IV, Schubert H, Skarlato SO (2015) Size, seasonality, or salinity: what drives the protistan species maximum in the horohalinicum? Estuarine Coastal and Shelf Science 161:102–111

Thorpe SA (2004) Langmuir circulation. Annual Review of Fluid Mechanics 36:55–79

van den Hoek C, Mann DG, Jahns HM (1995) Algae – an introduction to phycology. Cambridge University Press, Cambridge, 640 pp

Weber RE, Jensen FB (1988) Functional adaptations in hemoglobins from ectothermic vertebrates. Annual Review of Physiology 50:161–179

Wenger RH (2002) Cellular adaptation to hypoxia: O_2-sensing protein hydroxylases, hypoxia-inducible transcription factors, and O_2-regulated gene expression. FASEB J 16:1151–1162

Whitfield AK, Elliott M, Basset A, Blaber SJM, West RJ (2012) Paradigms in estuarine ecology: a review of Remane diagram with a suggested revised model for estuaries. Estuarine Coastal and Shelf Science 97:78–90

Winter U, Kirst GO (1990) Salinity response of a freshwater charophyte, *Chara vulgaris*. Plant, Cell and Environment 13: 123–134

Winter U, Kirst GO (1991) Partial turgor pressure regulation in *Chara canescens* and its implications for a generalized hypothesis of salinity response in charophytes. Botanica Acta 104:37–46

Winter U, Kirst GO (1992) Turgor pressure regulation in *Chara aspera* (Charophyta): the role of sucrose accumulation in fertile and sterile plants. Phycologia 31:240–245

Winter U, Kirst GO, Grabowski V, Heinemann U, Plettner I, Wiese S (1999) Salinity tolerance in *Nitellopsis obtusa*. Australian Journal of Botany 47:337–346

Winter U, Kuhbier K, Kirst GO (1987) Characeen-Gesellschaften im oligohalinen Kuhgrabensee und benachbarten Gewässern. Abhandlungen der Naturwissenschaftlichen Vereine Bremen 40:381–394 [in German]

Winter U, Soulié-Märsche I, Kirst GO (1996) Effects of salinity on turgor pressure and fertility in *Tolypella* (Characeae). Plant, Cell and Environment 19:869–879

Wu RSS (2002) Hypoxia: from molecular responses to ecosystem responses. Marine Pollution Bulletin 45:35–45

Part III

Subsystems of the Baltic Sea ecosystem

The pelagic food web

8

Agneta Andersson, Timo Tamminen, Sirpa Lehtinen, Klaus Jürgens, Matthias Labrenz, and Markku Viitasalo

Abstract

1. Environmental drivers and food web structure in the pelagic zone vary from south to north in the Baltic Sea.
2. While nitrogen is generally the limiting nutrient for primary production in the Baltic Sea, phosphorus is the limiting nutrient in the Bothnian Bay.
3. In the Gulf of Bothnia the food web is to a large extent driven by terrestrial allochthonous material, while autochthonous production dominates in the other parts of the Baltic Sea.
4. Changes in bacterioplankton, protist and zooplankton community composition from south to north are mainly driven by salinity.
5. Bacteria are crucial constituents of the pelagic food web (microbial loop) and in oxygen-poor and anoxic bottom waters where they mediate element transformations.
6. Diatoms and dinoflagellates are the major primary producers in the pelagic zone. Summer blooms of diazotrophic (nitrogen-fixing) filamentous cyanobacteria are typical of the Baltic Sea, especially in the Baltic Sea proper and the Gulf of Finland.
7. The mesozooplankton (mainly copepods and cladocerans) channel energy from primary producers and the microbial food web to fish and finally to the top predators in the pelagic system (waterbirds and mammals).
8. Herring and sprat populations are affected by the foraging intensity of their main predator (cod), and therefore the environmental conditions that affect cod may also influence mesozooplankton due to food web effects "cascading down the food web".
9. Anthropogenic pressures, such as overexploitation of fish stocks, eutrophication, climate change, introduction of non-indigenous species and contamination of top predators by hazardous substances, cause changes in the pelagic food web that may have consequences for the balance and stability of the whole ecosystem.

Keywords

Baltic Sea gradient • Food-web interactions • Human impacts • Microbial loop • Pelagic communities • Productivity • Trophic cascades

A. Andersson (✉)
Department of Ecology and Environmental Sciences,
Umeå University, 90187 Umeå, Sweden
e-mail: agneta.andersson@umu.se

T. Tamminen · S. Lehtinen · M. Viitasalo
Marine Research Centre, Finnish Environment Institute,
P.O. Box 140, 00251 Helsinki, Finland
e-mail: timo.tamminen@ymparisto.fi

S. Lehtinen
e-mail: sirpa.lehtinen@ymparisto.fi

M. Viitasalo
e-mail: markku.viitasalo@ymparisto.fi

K. Jürgens · M. Labrenz
Leibniz Institute for Baltic Sea Research, Seestraße 15,
18119 Warnemünde, Rostock, Germany
e-mail: klaus.juergens@io-warnemuende.de

M. Labrenz
e-mail: matthias.labrenz@io-warnemuende.de

P. Snoeijs-Leijonmalm et al. (eds.), *Biological Oceanography of the Baltic Sea*,
DOI 10.1007/978-94-007-0668-2_8

8.1 Life in the open water

As in most other brackish-water bodies, lakes and seas, the pelagic habitat of the Baltic Sea is inhabited by plankton and fish. Organisms smaller than 100 μm account for ~90 % of the total biomass (Sandberg et al. 2004), while organisms visible to the human eye account for ~10 % of the biomass. Planktonic organisms are classified according to their size into pico- (0.2–2 μm), nano- (2–20 μm), micro- (20–200 μm) and meso- (200–2,000 μm) plankton (*cf.* Table 4.1). In addition, viruses occur in the "dissolved" fraction (<0.2 μm). In general, the smaller an organism is in body size, the higher its abundance (number of individuals) in the seawater (*cf.* Box Fig. 4.1).

Pelagic organisms can be subdivided according to their feeding modes, which determine their role in the food web. They can be autotrophic, heterotrophic or mixotrophic (both autotrophic and heterotrophic). The most common virus types are those that use bacteria as their hosts (bacteriophages), but during phytoplankton blooms, viruses on phytoplankton can be dominant as well.

The organisms in the water column of the open sea are linked to each other by trophic interactions in complicated food webs (Fig. 8.1). The food web structure is controlled by bottom-up (nutrient concentration-dependent) and top-down (predation, including anthropogenic activities) processes.

8.2 Phytoplankton

8.2.1 Phytoplankton: a pragmatic concept

The phytoplankton is composed of unicellular, floating photosynthetic algal protists and cyanobacteria (Box 8.1). They reproduce mainly through asexual cell division, and constitute the main primary producers in the pelagic food web. However, this is only a very general definition and the terms "unicellular", "floating", "photosynthetic" and "asexual reproduction" are all challenged by numerous exceptions.

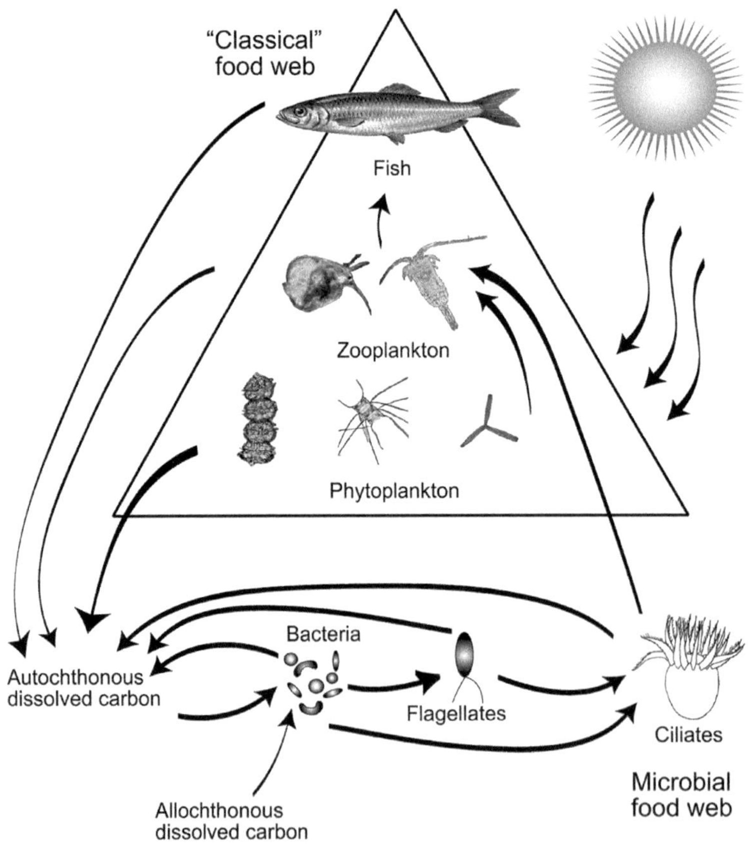

Fig. 8.1 A simplified model of the pelagic food web, showing groups of organisms and the transfer of organic material in the food web. In the "classical pelagic food web", energy is channelled from the phytoplankton (mainly diatoms and dinoflagellates) to mesozooplankton (mainly copepods and cladocerans) and from there to fish. In the microbial food web, energy is channelled from heterotrophic bacteria to heterotrophic flagellates and further to ciliates, then to mesozooplankton and finally to fish. In reality, the pelagic food web is much more complicated, *e.g.* by mixotrophy. Figure: © Kristina Viklund

While being truly unicellular, many phytoplankton species form chains or other types of colonies. While floating within water masses, several species are able to either swim by using flagellae or other motile structures, or to regulate their position in the water column by developing intracellular gas vacuoles. In addition to asexual cell division, several phytoplankton species possess complicated life cycles, including dormant resting stages, cysts that are deposited on the seafloor, and even sexual reproduction (*e.g.* in diatoms).

Most phytoplankton are photosynthetic (autotrophic), but many species are also capable of utilising organic carbon sources, either through uptake of dissolved organic compounds, or through engulfing particulate objects, including other algae or bacteria (phagotrophy), so that these species are in fact mixotrophic. Some organisms that are traditionally counted microscopically in phytoplankton samples (and included in phytoplankton analyses), lack photosynthetic pigments altogether, thus being true heterotrophs (depending solely on organic carbon sources). These different feeding modes may cut across taxonomic groups, *e.g.* among the dinoflagellates there are autotrophic, mixotrophic and heterotrophic species.

Many phytoplankton species are not algae (eukaryotes) but phototrophic prokaryotes, or cyanobacteria, which were previously called "blue-green algae". Globally, the most numerous photosynthetic planktonic organisms are picocyanobacteria that also account for a large share of the primary-producing biomass in the pelagic zone of the Baltic Sea. The most striking pelagic blooms during summer are formed by diazotrophic filamentous cyanobacteria, especially *Aphanizomenon flos-aquae* and *Nodularia spumigena*, which are often accompanied by *Dolichospermum* sp. (Fig. 8.2). Freshwater *Aphanizomenon flos-aquae* produces neurotoxins, but in the Baltic Sea it is non-toxic (Sivonen et al. 1989, 1990; Lehtimäki et al. 1997). Blooms of *Nodularia spumigena* and *Dolichospermum* sp. produce the hepatotoxins nodularin and microcystin, respectively (*cf.* Box 16.4; Sivonen et al. 1989; Repka et al. 2004; Halinen et al. 2007).

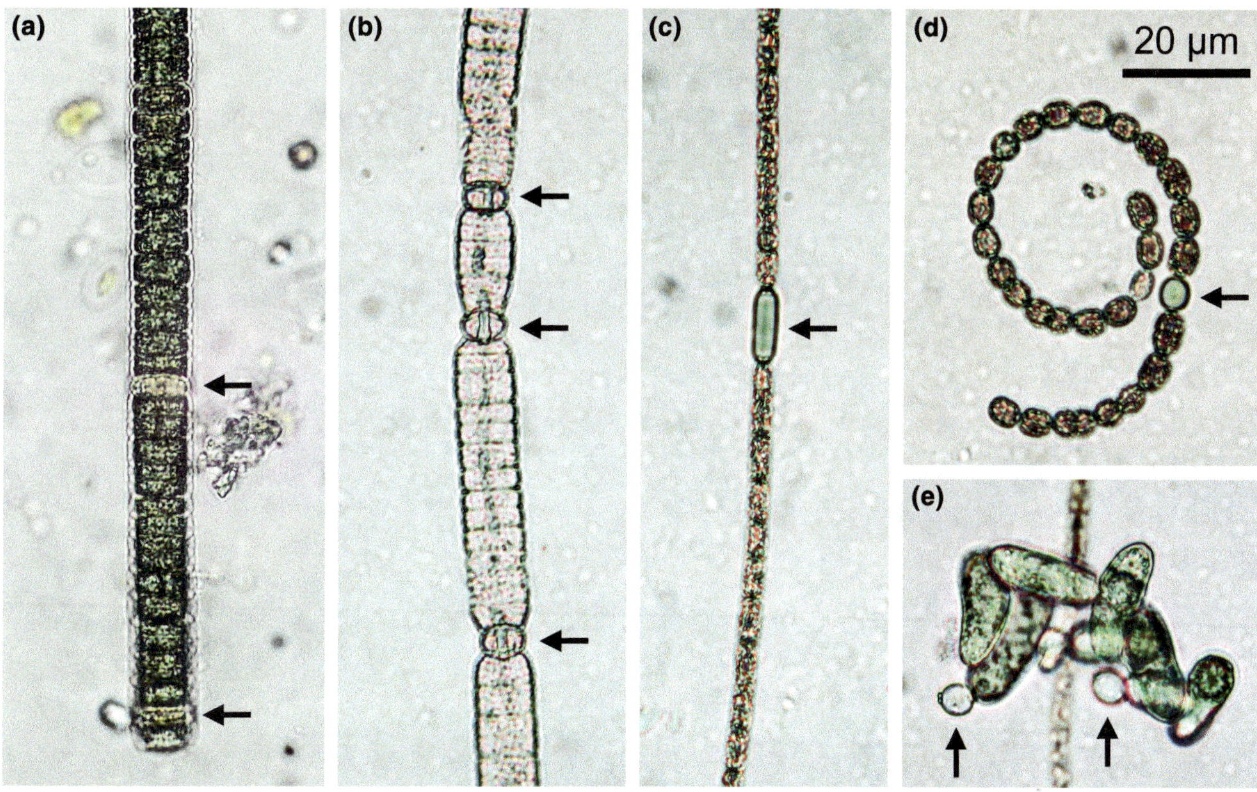

Fig. 8.2 Light micrographs of the three most abundant filamentous cyanobacterial species in the pelagic zone of the Baltic Sea proper in August. Arrows indicate the specialised heterocyst cells that create a microanaerobic environment where nitrogen fixation takes place. Heterocysts are crucial for the ability of these photosynthetic bacteria to fix nitrogen gas (into ammonia, nitrites or nitrates) because they keep the oxygen-labile enzyme nitrogenase away from the photosynthetically produced O_2 (Muro-Pastor and Hess 2012). (**a**) Filament of *Nodularia spumigena*. (**b**) Decaying filament of *Nodularia spumigena* showing the cell wall structure of heterocysts and vegetative cells. (**c**) Filament of *Aphanizomenon flos-aquae*. (**d**) Filament of *Dolichospermum* sp. (syn. *Anabaena* sp.). (**e**) Akinetes (thick-walled dormant cells derived from the enlargement of vegetative cells) of *Dolichospermum* with heterocysts still attached. The filaments of *Nodularia* (*cf.* Fig. 8.4b) and *Dolichospermum* are often curled up, while those of *Aphanizomenon* assemble in bundles with more or less straight filaments (*cf.* Fig. 8.4b). The scale bar for all images is given in (**d**). Photo: © Pauline Snoeijs-Leijonmalm

Box 8.1: Phytoplankton diversity in the Baltic Sea

Regina Hansen

The fascinating world of phytoplankton

The great variety of shapes in phytoplankton has fascinated biologists since Antonie van Leeuwenhoek (1632–1723) discovered the existence of unicellular plankton organisms by using a small hand-held light microscope. Previously, the existence of unicellular organisms was completely unknown and van Leeuwenhoek's discovery was initially met with scepticism by *e.g.* the Royal Society in London. Already in 1674 he described the green filamentous alga *Spirogyra* and various ciliated and flagellated protozoa, and discovered the first diatoms in 1702 (Backer et al. 2014). The larger microplankton (20–200 µm) can be easily identified under a light microscope, but the smaller nanoplankton (2–20 µm), and especially the picoplankton (0.2–2 µm), often cannot be identified to the species or genus level. Despite their smaller cell size, the nano- and picoplankton organisms may be of high ecological relevance through high abundances, fast turn-over rates and participation in the food web, including the microbial loop, as both prey and grazers. Many phytoplankton species are well-known indicators for particular forms of ecosystem functioning. The visual inspection of plankton samples by light microscopy is a traditional, but still highly informative, instrument for ecosystem assessment, and is used in many laboratories around the world.

The major phytoplankton groups

Major constituents of the phytoplankton communities in the Baltic Sea are cyanobacteria (Box Fig. 8.1), diatoms (Box Fig. 8.2) and dinoflagellates (Box Figs. 8.3 and 8.4). Smaller phytoplankton species often belong to other groups such as chrysophytes, cryptophytes, euglenophytes, haptophytes, prasinophytes and raphidophytes (Box Fig. 8.5). Some phytoplankton species are bloom-forming, *e.g.* the diatom *Skeletonema marinoi* in spring, the diazotrophic filamentous cyanobacteria *Nodularia spumigena* and *Aphanizomenon flos-aquae* in summer, as well as the dinoflagellates *Dinophysis norvegica* and the heart-shaped species *Prorocentrum cordatum* (syn. *Prorocentrum minimum*) in late summer–autumn.

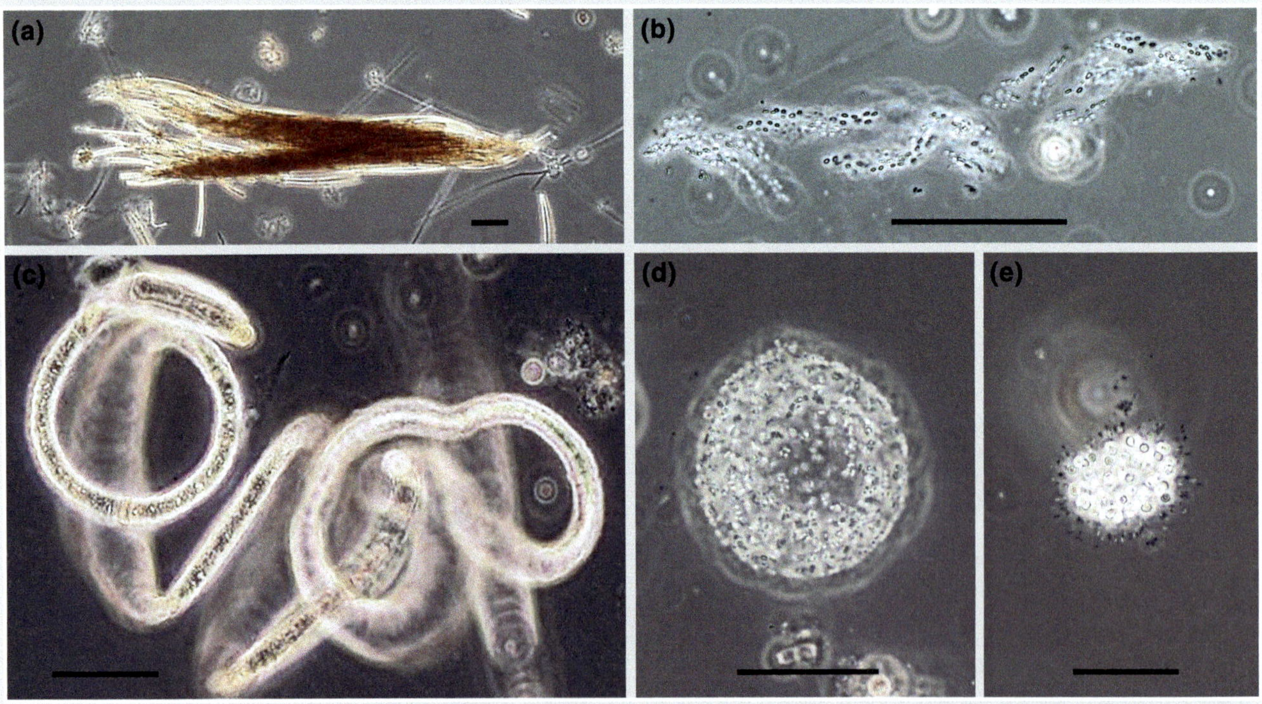

Box Fig. 8.1 Common cyanobacteria in the phytoplankton of the Baltic Sea. (**a**) *Aphanizomenon flos-aquae*. (**b**) *Aphanothece paralleliformis*. (**c**) *Nodularia spumigena*. (**d**) *Coelosphaerium minutissimum*. (**e**) *Snowella* sp. Scale bar = 50 µm. Photo: © Regina Hansen

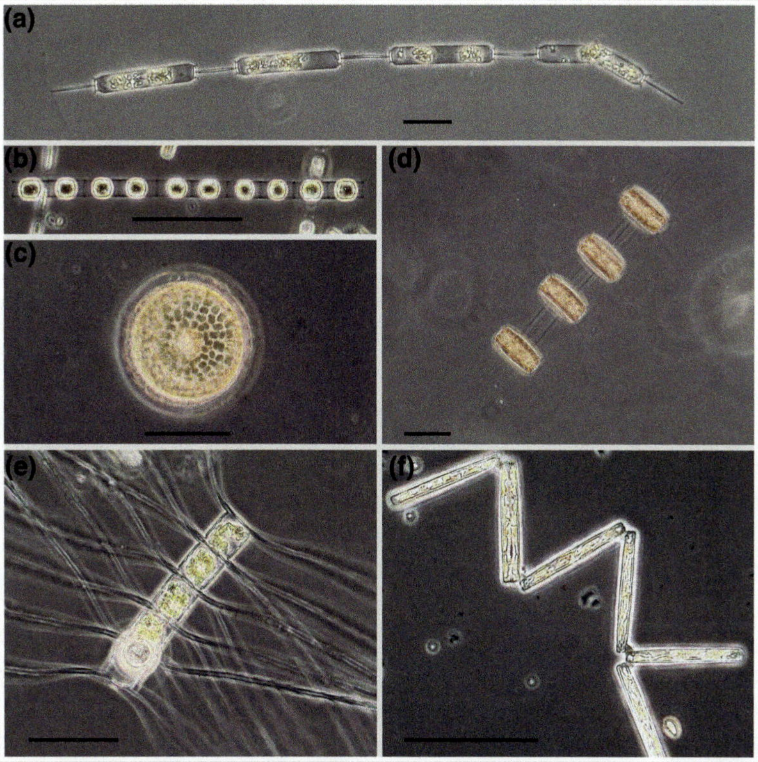

Box Fig. 8.2 Common diatoms in the phytoplankton of the Baltic Sea. (**a**) *Ditylum brightwellii*. (**b**) *Skeletonema marinoi*. (**c**) *Thalassiosira punctigera*. (**d**) *Thalassiosira anguste-lineata*. (**e**) *Chaetoceros castracanei*. (**f**) *Thalassionema nitzschioides*. Scale bar = 50 µm. Photo: © Regina Hansen

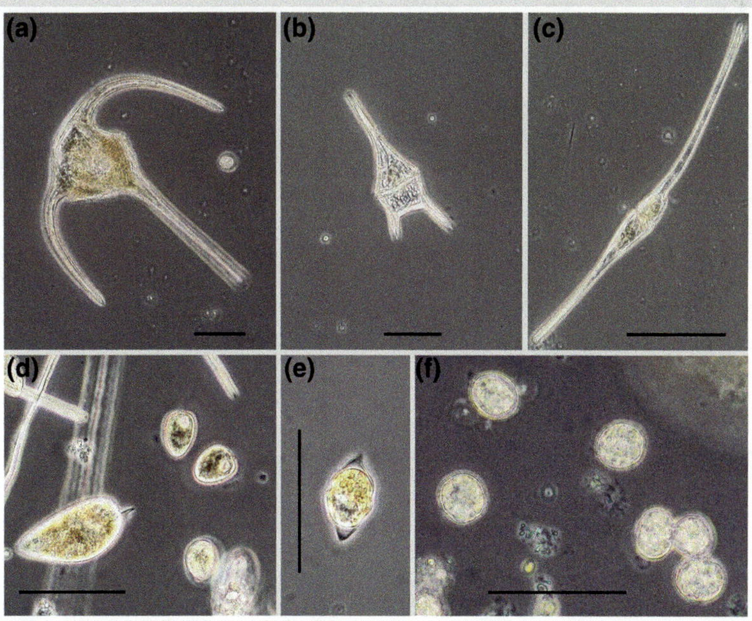

Box Fig. 8.3 Common dinoflagellates in the phytoplankton of the Baltic Sea. (**a**) *Ceratium tripos*. (**b**) *Ceratium lineatum*. (**c**) *Ceratium fusus*. (**d**) *Prorocentrum micans* (a larger individual to the left) and *Prorocentrum cordatum* (three smaller individuals to the right). (**e**) *Heterocapsa triquetra*. (**f**) *Gymnodinium corollarium*. Scale bar = 50 µm. Photo: © Regina Hansen

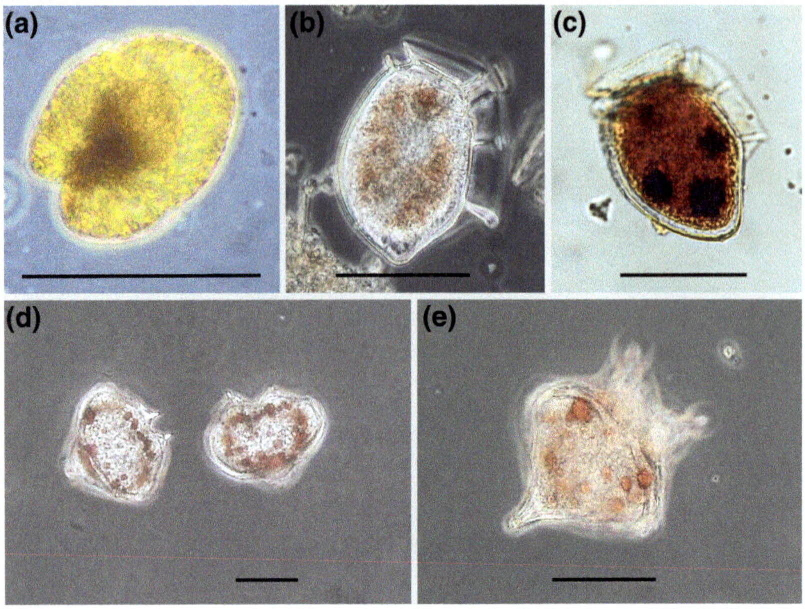

Box Fig. 8.4 Common mixotrophic and heterotrophic dinoflagellates in the phytoplankton of the Baltic Sea. (**a**) *Akashiwo sanguinea*. (**b**) *Dinophysis acuta*. (**c**) *Dinophysis norvegica*. (**d**) *Protoperidinium pentagonum*. (**e**) *Protoperidinium divergens*. Scale bar = 50 µm. Photo: © Regina Hansen

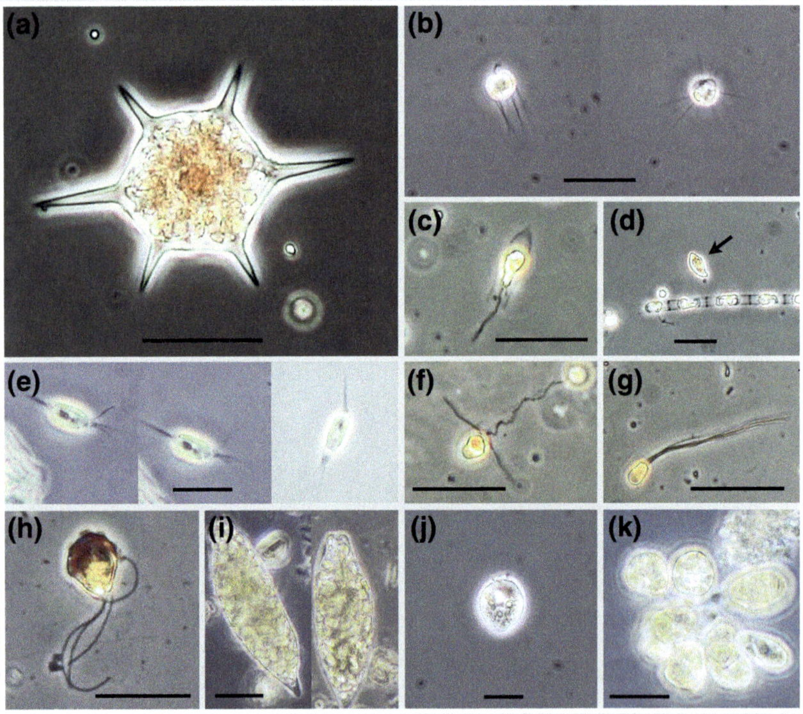

Box Fig. 8.5 Common phytoplankton species in the Baltic Sea representing different taxonomic groups. (**a**) *Dictyocha speculum* (chrysophyte). (**b**) *Apedinella radians* (chrysophyte). (**c**) *Dinobryon* sp. (chrysophyte). (**d**) *Teleaulax acuta* (cryptophyte). (**e**, **f**) *Chrysochromulina* sp. (haptophyte). (**g**) *Pachysphaera* sp. (prasinophyte). (**h**) *Cymbomonas tetramitiformis* (prasinophyte). (**i**) Euglenophyte. (**j**) *Trachelomonas* sp. (euglenophyte). (**k**) *Heterosigma akashiwo* (raphidophyte). Scale bar = 20 µm. Photo: © Regina Hansen

Altogether, there seems to be no rule in the planktonic realm without an exception, and life forms seem to blur into each other across the most basic taxonomic distinctions. "Phytoplankton" is thus best taken as a pragmatic concept, describing the primary producers in the pelagic part of aquatic systems. Nowadays, ~2,000 phytoplankton species are known to occur in the Baltic Sea, and they vary from marine to freshwater species (Hällfors 2004) and range in average cell length from 1 to 335 μm (Olenina et al. 2006). It is more than probable that the current rapid developments in genetics will lead to continuous changes in the atlas of planktonic life, as the traditional phenotypic divisions are being revised and replaced by genotypic ones.

8.2.2 Specific features that influence the Baltic Sea phytoplankton

The Baltic Sea represents an intriguing environment for the study of phytoplankton in several respects. The young geological age of the Baltic Sea suggests that phytoplankton communities are much more dynamic than those found in the vast and stable domains of the world's oceans. The results of evolution and speciation could be more direct and evident here than in most other pelagic habitats, and for the short-lived unicellular phytoplankton, evolutionary adaptations should be far more obvious than for organisms with longer generation times.

Most phytoplankton species in the Baltic Sea have their origin and their closest relatives in either truly marine environments, at salinities four to six times higher than those in the major part of the Baltic Sea (*cf.* Fig. 2.15), or in freshwater environments. The positive water balance of the Baltic Sea (average net outflow to the Kattegat of 480 km^3 $year^{-1}$, *cf.* Fig. 2.12a) indicates that there is a significant, continuous supply of freshwater organisms into the species pool of the Baltic Sea, and the water exchange between the Baltic Sea and the North Sea in the transition zone (Belt Sea and Kattegat) ensures the inflow of marine phytoplankton.

In the temperate zone, the seasonal successional cycle in coastal waters classically begins with a winter-spring diatom bloom that is seasonally replaced by summer communities dominated by dinoflagellates (Smayda and Trainer 2010). However, the algal spring bloom in the Baltic Sea represents an anomaly from the winter-spring bloom patterns worldwide in terms of frequent and recurring dominance of dinoflagellates over diatoms (Klais et al. 2011).

Ecosystem-wide compilations of long-term monitoring data have shown that the Baltic Sea phytoplankton community composition (*cf.* Fig. 3.12) is in a state of on-going change (Olli et al. 2011, 2014). Disentangling the drivers of this change and separating them into specific components, *e.g.* evolutionary changes intermingled with anthropogenic pressures (eutrophication, climate change, hazardous substances, non-indigenous species), will be a major challenge of Baltic Sea ecology research in the coming years.

The Baltic Sea is also an interesting environment for studies of phytoplankton dynamics because it is heavily affected by eutrophication, and phytoplankton is the biological component of aquatic ecosystems that is first to respond to increased nutrient availability by increased biomass production and compositional changes. The eutrophication process is channelled further within the ecosystem through food web interactions and biogeochemical cycles (*cf.* Sect. 3.1). The framework for planktonic primary producers in the Baltic Sea is set by the following specific conditions:

1. The high spatial variability created by the south-north gradient in the annual solar radiation cycle
2. The high spatial variability created by the stable south-north salinity gradient, with surface-water salinity varying between ~10 and 0
3. The high spatial variability due to large differences in nutrient supplies and retention times in the different subbasins
4. The high seasonal variability due to a prominent winter season with a long ice-covered period in the north, and season-specific vertical mixing patterns
5. The on-going human-induced changes in environmental conditions and community composition.

8.2.3 The development of Baltic Sea phytoplankton research

Microscopic cell counting in samples taken with plankton nets was the conventional method used in phytoplankton studies up until the 1920s. The results obtained with this method can hardly be regarded as quantitative data, and only large-sized phytoplankton was counted since the mesh size of the nets was usually ~40–115 μm (Lohmann 1908). Nevertheless, attempts to obtain quantitative plankton data have been made in the Baltic Sea Area since Hensen's (1887) study in the Kiel Bay (southwestern Belt Sea) between 1883 and 1886. Since the 1960s, the Utermöhl technique (Utermöhl 1958) has become the standard for quantitative phytoplankton analyses (Box 8.2).

Box 8.2: Sampling and analysis of plankton communities

Agneta Andersson and Markku Viitasalo

Sampling of pelagic bacteria and protists

Seawater samples for the analysis of bacteria and protists (pico-, nano- and microplankton) are often collected at discrete depths by using a plankton sampler, *e.g.* a "Ruttner" sampler connected to a length-marked line, which is operated from a ship. This Ruttner sampler is very similar to a Niskin bottle used in oceanography (*cf.* Box 3.1), consisting of an open tube with top and bottom lids. After lowering the sampler to the desired water depth a closing mechanism is released, whereby both the upper and lower lids close. The sampler is then lifted out of the water, and while on the deck, seawater samples are collected and *in situ* water temperature is recorded. This plankton sampler is named after Franz Ruttner (1882–1961), the limnologist who originally designed this sampling device. The Ruttner sampler has been further developed for efficiency in offshore samplings on board large ships. For example, a series of Ruttner-type samplers can be connected to a line at different distances. The closing mechanism is released simultaneously for all samplers at different depths. Another advanced technique for offshore samplings is the use of a rosette sampler, which has computerised depth and release control (*cf.* Box 3.1). To obtain integrated water samples, for example from the upper mixed-water layer, samples taken at discrete depths can be pooled. However, since the microorganisms occurring are often vertically stratified, it is difficult to obtain a truly integrated sample when using a pooling method. Instead, a plastic hose sampler can be used to obtain a more representative vertical sample from the upper water column. A long, open plastic hose is lowered vertically into the sea, and when the hose is in the correct position a plug is inserted at the top of the hose. Due to the capillary force the seawater stays in the hose, and on board the ship the hose is unplugged and the seawater is collected in *e.g.* a bucket.

Preservation and counting of bacteria and protists

The preservation of plankton samples is performed directly after sampling, and a variety of fixation methods can be used. For biochemical analyses, *e.g.* of elemental composition (C,N,P), pigments or DNA, the sample can be filtered on to glass fibre filters or membrane filters (Box Fig. 8.6a–c), and be immediately frozen in a freezer or in liquid nitrogen on board (Box Fig. 8.6e). Samples for the analysis of nano- and microplankton community composition can be preserved with Lugol's solution (Box Fig. 8.6d), and later analysed with the Utermöhl technique. Lugol's solution of elemental iodine and potassium iodide in water, and is named after the physician Jean Lugol (1786–1851). The Utermöhl technique was first developed by Hans Utermöhl (1931, 1958) and has become the standard quantitative method for counting phytoplankton worldwide (Paxinos and Mitchell 2000). It uses a chamber into which a preserved phytoplankton water sample is placed and left to settle on to a cover slip, after which the plankton cells are counted using an inverted light microscope. Bacteria and picophytoplankton samples can be preserved with glutaraldehyde and later be analysed with epifluorescence microscopy (using fluorescence to generate an image) or flow cytometry (a laser-based, biophysical technology for cell sorting and counting).

Zooplankton sampling with vertically hauled nets

Crustacean zooplankton, especially calanoid copepods that dominate the mesozooplankton (0.2 to 2 mm) of the world ocean, have traditionally been sampled with large nets, with mesh sizes from 0.1 to 0.5 mm (Box Fig. 8.7a, b). Microzooplankton, *i.e.* ciliates and small rotifers, can be sampled with water samplers (see above). In the Baltic Sea, where the mesozooplankton are slightly smaller (typically 0.5–1.5 mm) than in the ocean, the WP2 net with a mesh size of 0.1 mm is commonly used. The water volume sampled with a plankton net can be calculated from the area of the net opening multiplied by the tow length. For example, the WP2 net has an area 0.25 m^2, and with a 20 m tow 5 m^3 of water is sampled. However, because the net tends to become clogged by phytoplankton and other debris during the haul, the net filters less and less water. Therefore, small propeller-driven current meters are often attached at the net opening, and with the help of these measurements the amount of water actually entering the net can be estimated more accurately. Usually, the nets also include a system that allows the closing of the net at desired depth intervals. In the Baltic Sea, the depth intervals sampled separately are often from the bottom to the halocline, from the halocline to the thermocline, and from the thermocline to the water surface. The Baltic Marine Environment Protection Commission (HELCOM, *cf.* Sect. 17.8.4) has published methodological guidelines that are especially followed when regular monitoring of the Baltic Sea environment is performed (HELCOM 2015).

Box Fig. 8.6 Phytoplankton sample treatment in the Baltic Sea. (**a**) Equipment for filtering phytoplankton for DNA analyses on polycarbonate membrane filters. (**b**) Equipment for filtering phytoplankton for elemental (C,N,P) or pigment analyses on glass fibre filters. (**c**) Close-up of a phytoplankton sample on a GF/F glass fibre filter of 25 mm diameter. When the sample is dominated by diatoms or dinoflagellates the colour is yellowish-brown. (**d**) Water samples preserved with Lugol's solution for identification and counting of phytoplankton. (**e**) Filters can be frozen in liquid nitrogen. This photograph has been taken in a laboratory on land rather than in an on-board one where liquid nitrogen containers must be firmly attached for safety reasons. Photo: © Pauline Snoeijs-Leijonmalm

Other ways to sample zooplankton

In addition to the vertically hauled nets, sampling devices towed horizontally have also been designed. The "Continuous Plankton Recorder" is a device that has been used since 1931 and is being towed by commercial vessels ("ships of opportunity") crossing large sea areas. Here, the plankton enters a metal sampler from a small opening, *e.g.* with a surface area of 1 square inch ($\sim$6.45 cm^2), after which the organisms are trapped between two silks with mesh sizes of 0.27 mm. To avoid clogging the silk, the silks gradually move, with the force of a current propeller attached in the end of the device, hence revealing clean silk as the tow proceeds. On its return to the laboratory, the silk is removed and divided into samples representing 10 nautical miles of tow each. More modern devices have also been developed and are commonly used on board research vessels. The silk movement, or the opening and closing of attached nets, as well as the tow depth, are controlled with electric motors moving the silk inside the device and levers ("fins") attached to the devices. This allows for detailed study of the vertical and horizontal distribution of plankton in the sea. Even more advanced devices exist, *e.g.* video cameras that can observe, count and identify plankton individuals directly in

Box Fig. 8.7 Zooplankton sampling in the Baltic Sea. (**a**) A WP2 zooplankton net is lowered into the sea with a winch from R/V Aranda. (**b**) A newly taken zooplankton sample has just arrived on board the ship. (**c**) Close-up of a zooplankton sample on a 200 μm plankton net. (**d**) Close-up of a zooplankton sample for biochemical analyses on a GF/C glass fibre filter of 25 mm diameter. (**c**, **d**) Both samples are dominated by copepods and cladocerans of 0.5 to 1.0 mm length. The red colour is the carotenoid astaxanthin, a strong antioxidant produced by the crustaceans from β-carotene and zeaxanthin supplied by their phytoplankton food items (*cf.* Sect. 4.7.6). Photo: © Pauline Snoeijs-Leijonmalm

the sea. Such equipment is used in specific research projects, *e.g.* for studying small-scale distribution or the feeding and predation behaviour of plankton *in situ*.

Preservation and analysis of zooplankton samples

After sampling, the zooplankton suspension is poured from the "cod end" of the net on to a small-size net (Box Fig. 8.7c) and from there into a sampling bottle with filtered (<1 μm) seawater. For biochemical analyses *e.g.* the elemental composition (C,N,P), pigments (chlorophylls and carotenoids), vitamins or fatty acids, the sample can be filtered on to a glass fibre filter (Box Fig. 8.7d) and be immediately frozen in a freezer or in liquid nitrogen on board (Box Fig. 8.6e). For analysis of community composition, formaldehyde or Lugol's solution can be added to the sample. Species identification and counting can be performed later under a microscope in a land-based laboratory. Since the samples usually contain several thousands of zooplankton individuals, it is customary to split a sample into smaller units. Several devices can be used for this procedure. For example, a "Folsom splitter" is a narrow cylinder kept sideways, which can be used to split the sample first in half, then in 1:4, 1:8, etc. Another device, a "Kott splitter" is a cylinder kept flat. It is subdivided into sectors and the sample is immediately split in 1:8, then in 1:64, etc. The goal is to divide the original sample so that only ~500 zooplankton individuals need to be identified under the microscope.

Long-term studies are needed to evaluate the current environmental status and any changes that occur; thus comparability of results is essential. With time, methods for sampling, preservation and analysis of phytoplankton (Box 8.2), as well as taxonomic knowledge and education, have been developed and harmonised (intercalibrated) between institutes and countries around the Baltic Sea. In 1991, the Baltic Marine Environment Protection Commission (HELCOM, *cf.* Sect. 17.8.4) established the Phytoplankton Expert Group to work on intercalibration of Baltic Sea phytoplankton issues, and this group has met every year since. In the early 1990s, the joint HELCOM monitoring of phytoplankton around the Baltic Sea was initiated, but some of the time series started already in the late 1970s.

Nowadays, phytoplankton is monitored according to standard guidelines with inverted light microscopes (HELCOM 2015), and different staining techniques, electron microscopy and genetic analyses are used. Techniques involving particle counters and image analysis are under development. In addition to traditional sampling methods, automatic ferry-box sampling is used to take samples along ferry routes (Rantajärvi et al. 1998). Ferry boxes are also used to collect fluorescence data for pigment composition analyses (Seppälä et al. 2007). Algal and cyanobacterial blooms are monitored and the surface-water chlorophyll *a* concentration is estimated by remote sensing (*cf.* Sect. 15.1).

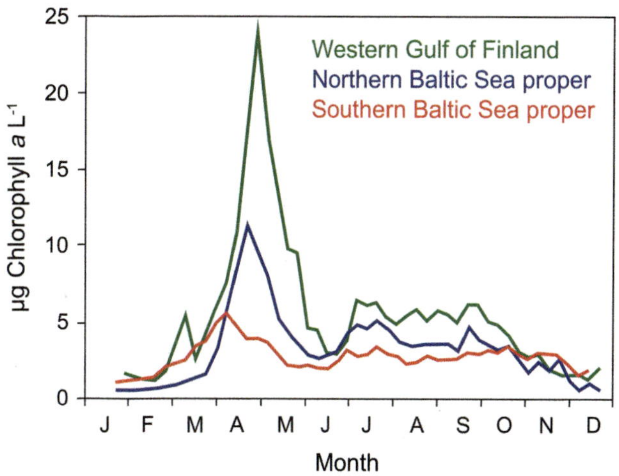

Fig. 8.3 The average seasonal development of phytoplankton biomass for the years 1992–2010, based on measurements of chlorophyll *a* concentration in the western Gulf of Finland, the northern Baltic Sea proper and the southern Baltic Sea proper. Variations between years can be large. Figure modified from Kaitala et al. (2011)

8.2.4 Seasonal succession

The general pattern of seasonal succession in the phytoplankton community shows similar trends across the entire Baltic Sea, even though there are differences in timing and species composition between areas (Fig. 8.3). The annual biomass maximum occurs during the spring bloom, when the amount of light does not restrict growth anymore and there are plenty of dissolved inorganic nutrients in the water. In the southern Baltic Sea proper, the spring bloom starts as early as February-March, and in the northern Baltic Sea proper in March-April. However, further north, the onset of the spring bloom depends on when the ice cover melts, which in the Bothnian Bay may be as late as June.

The spring bloom is dominated by diatoms and dinoflagellates (Fig. 8.4a). Grazers are still sparse, and most of the spring-bloom biomass is lost from the upper water layers through sedimentation (Lignell et al. 1993). The proportion of dinoflagellates in the spring bloom varies between years, but it is usually highest in the Baltic Sea proper and lowest in the Kattegat and the Bothnian Bay. During the last decades, the relative proportion of dinoflagellate biomass in the Baltic Sea has generally increased while that of diatoms has decreased (Klais et al. 2011, 2013). A particular increase in the proportion of dinoflagellates has been observed in the northern Baltic Sea (the Bothnian Bay and the Gulf of Finland). Since there is no evidence of nutrient limitation during the build-up of the spring bloom, the change in the diatom-dinoflagellate relationship has been linked mainly to weather and hydrographical conditions favouring the slow-growing, large, and motile *K*-strategist dinoflagellates (Kremp et al. 2008, *cf.* Box 4.14).

In early summer, when inorganic nutrients have been depleted from the photic water layer and the autotrophic production is mainly based on regenerated nutrients, the phytoplankton biomass is low, and the community consists primarily of small flagellates. In summer, grazers start to play a more important role in regulating phytoplankton biomass, and selective feeding affects the phytoplankton community composition.

After the summer biomass minimum, the late-summer biomass maximum starts to build up. Diazotrophic cyanobacteria (Figs. 8.2 and 8.4b) dominate, except in the Bothnian Bay and the Belt Sea where another diatom peak is typically found in late summer – autumn. Stable weather conditions and high water temperatures stimulate the formation of large diazotrophic cyanobacterial surface blooms (Fig. 8.5). At the beginning of a bloom, *Aphanizomenon flos-aquae* usually starts to increase in abundance first, while *Nodularia spumigena* takes over with increasing water temperature (Laamanen and Kuosa 2005). Observations from the air (Fig. 8.5a) and from space (*cf.* Fig. 15.1) reveal the impressive spatial extent of the cyanobacterial surface accumulations, which may at one time cover thousands of km^2. During recent decades, monitoring by satellites has also disclosed dramatic increases in these cyanobacterial blooms

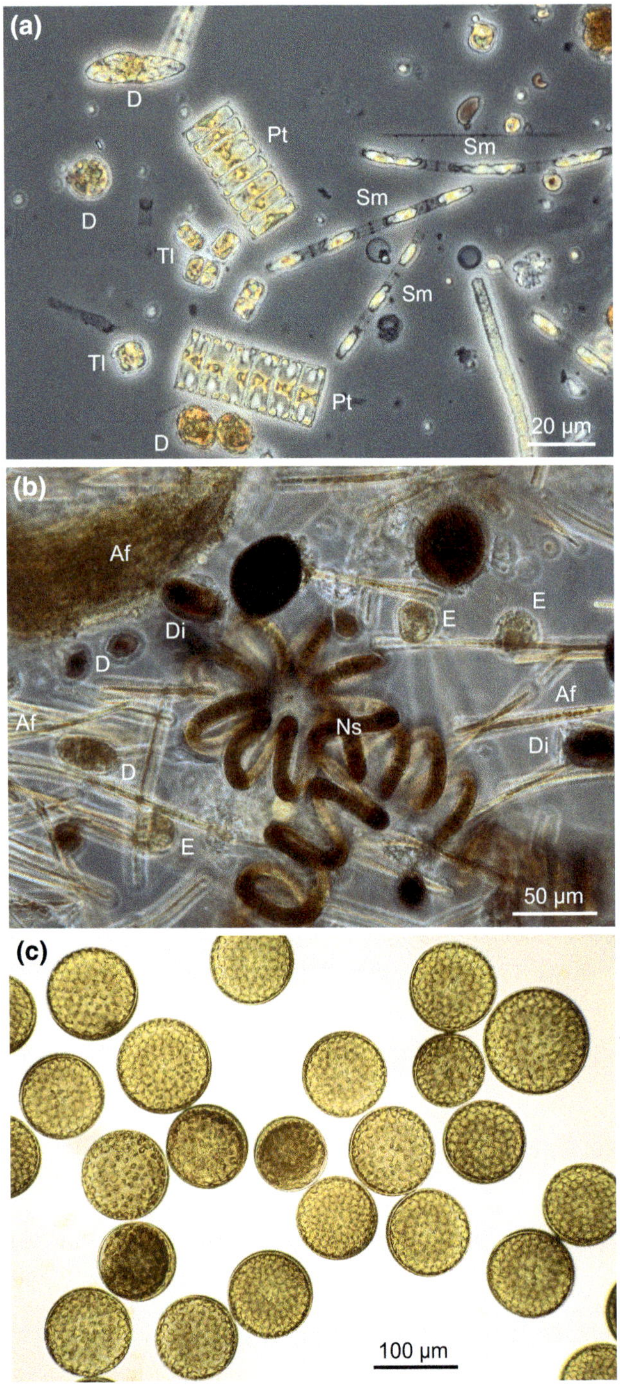

Fig. 8.4 Light micrographs of typical phytoplankton communities in the Baltic Sea. (**a**) Spring community dominated by the diatoms *Pauliella taeniata* (Pt), *Skeletonema marinoi* (Sm) and *Thalassiosira levanderi* (Tl) and dinoflagellates (D). (**b**) Late-summer community dominated by the cyanobacteria *Aphanizomenon flos-aquae* (Af) and *Nodularia spumigena* (Ns), the dinoflagellate *Dinophysis acuminata* (Di) and other dinoflagellates (D), as well as the heterotrophic protist *Ebria tripartita* (E). (**c**) Autumn bloom with total dominance of the large centric diatom *Coscinodiscus granii* in the Baltic Sea proper (Hanöbukten) in November. Photo: (**a**, **b**) © Sirpa Lehtinen, (**c**) © Pauline Snoeijs-Leijonmalm

and their large inter-annual variability in spatial extent (Funkey et al. 2014; Kahru and Elmgren 2014).

When the cyanobacterial blooms decay, diatoms, *e.g. Coscinodiscus granii* (Fig. 8.4c) may reach high abundances. In late autumn, the phytoplankton biomass declines and reaches its annual minimum during winter when both light and temperature limit phytoplankton production in the pelagic zone (Fig. 8.3). However, there are many cold-water algal species, especially diatoms, that live in association with the Baltic Sea ice in winter and early spring (*cf.* Table 9.2).

8.2.5 Spatial patterns in the Gulf of Bothnia

Phytoplankton communities vary between the different basins of the Baltic Sea, as well as along gradients from the coasts to the open sea. Even though phytoplankton community composition is expected to be mainly shaped by local chemical, physiological and biological factors, large-scale interrelations also exist (Ptacnik et al. 2010).

In the Bothnian Bay, salinity and phosphate concentrations are low (*cf.* Figs. 2.15 and 2.23). Phosphorus is the limiting nutrient for primary production while nitrogen is usually limiting in other parts of the Baltic Sea (Tamminen and Andersen 2007). As a result of the low phosphate concentrations, biomass is generally low in the Bothnian Bay and diatoms and freshwater chlorophytes are relatively high in abundance (Fig. 8.6a) due to low salinity, silicate-rich water and weak vertical water stratification. Diatoms (especially the chain-forming species *Chaetoceros holsaticus*, *Chaetoceros wighamii*, *Pauliella taeniata* and *Thalassiosira levanderi*) and dinoflagellates (especially the chain-forming species *Peridiniella catenata*, Fig. 8.7a), dominate the spring bloom and there is also an autumn peak of diatoms (Fig. 8.6). During summer, the chlorophyte *Monoraphidium contortum*, as well as chrysophytes, cryptophytes and picocyanobacteria are abundant.

In the Bothnian Sea, the spring bloom (Fig. 8.6b) is dominated by the diatom *Thalassiosira baltica* (*cf.* Box Fig. 2.7a, b) and the dinoflagellate *Peridiniella catenata* while the photosynthetic ciliate *Mesodinium rubrum* (syn. *Myrionecta rubra*, Fig. 8.8) increases in abundance during and after the spring bloom. Haptophytes (syn. prymnesiophytes), filamentous cyanobacteria and picocyanobacteria are abundant during the summer.

In the Archipelago Sea, the combined nitrogen and phosphorus limitation in summer has, since the late 1990s, shifted towards nitrogen limitation. Consequently, the late-summer cyanobacterial biomass has increased. In spring, the diatom genera *Chaetoceros*, *Pauliella*, *Skeletonema* and *Thalassiosira*, and the dinoflagellate genera *Gymnodinium*,

Fig. 8.5 Summer blooms of filamentous cyanobacteria dominated by *Nodularia spumigena*. (**a**) Surface bloom in the open Baltic Sea. (**b**) The surface blooms often reach coastal areas with winds and currents. **c** Decaying cyanobacterial blooms in coastal waters may be toxic and do not encourage bathing in the sea. Photo: (**a**) © Finnish Coast Guard, (**b**) © Anna Ulanova, (**c**) © Gary Wife

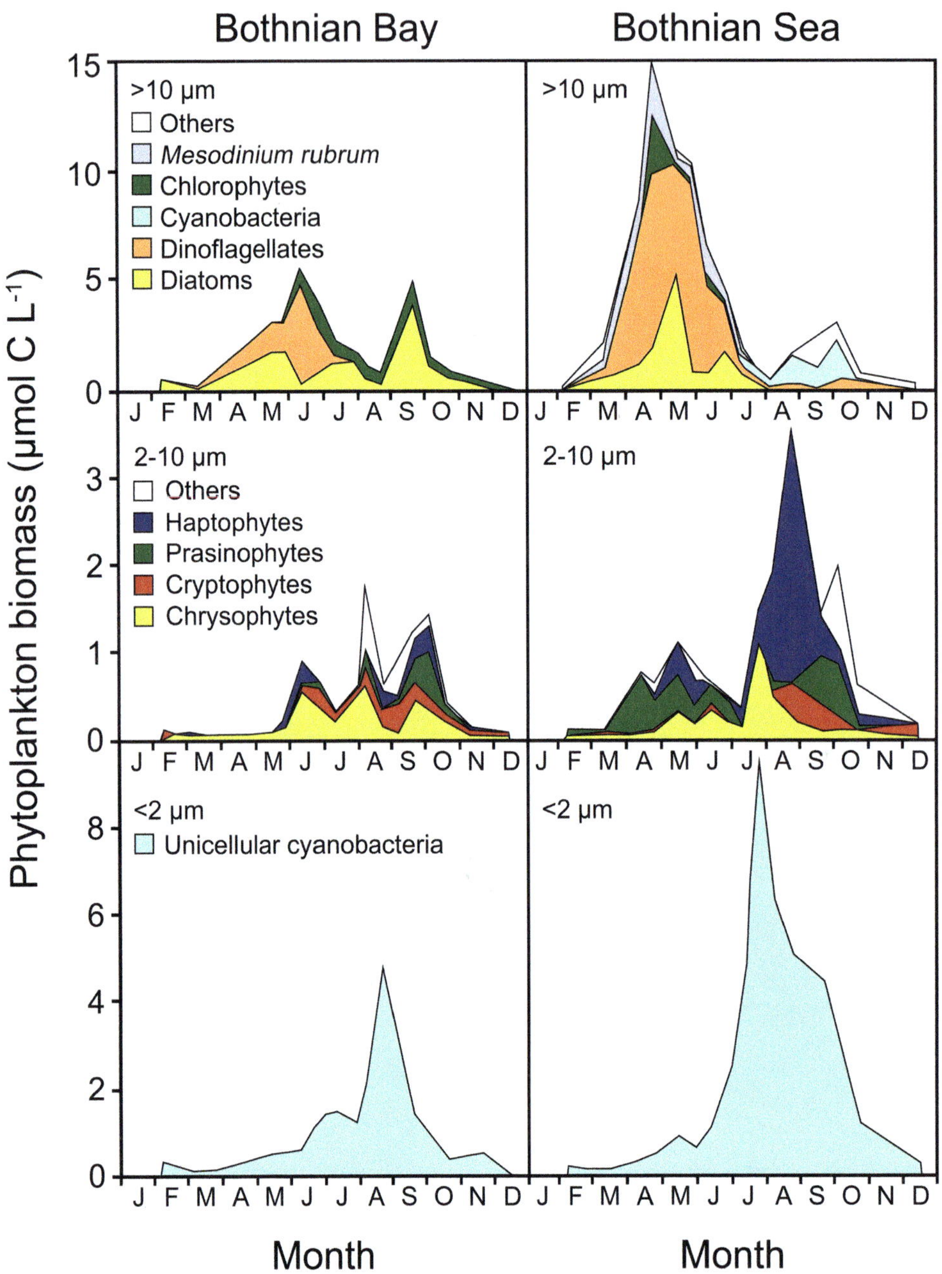

Fig. 8.6 Seasonal succession of phytoplankton in the Gulf of Bothnia during 1991. (**a**) Station F9 in the Bothnian Bay. (**b**) Station US5B in the Bothnian Sea. The phytoplankton organisms were grouped into different size classes (>10 µm, 2–10 µm, <2 µm), and within these size classes into taxonomic groups. Figure modified from Andersson et al. (1996)

Peridiniella and *Protoperidinium* are well represented. *Mesodinium rubrum* peaks after the spring bloom, and the summer minimum is dominated by chrysophytes, cryptophytes, haptophytes, and the prasinophyte *Pyramimonas* (Lagus 2009). Mixotrophic dinoflagellates, especially *Dinophysis acuminata* and *Dinophysis norvegica* (Fig. 8.7b) contribute to the biomass in summer, and the late-summer cyanobacterial blooms may be followed by blooms of the dinoflagellates *Heterocapsa triquetra* and *Prorocentrum cordatum* (Fig. 8.7c, d). In the shallow coastal waters around the Åland archipelago, the dinoflagellates *Alexandrium ostenfeldii* and *Heterocapsa triquetra* may form late-summer blooms (Lindholm and Nummelin 1999; Kremp et al. 2009).

8.2.6 Spatial patterns in the central and southern Baltic Sea

In the Gulf of Finland and the northern and central Baltic Sea proper, the dinoflagellates *Biecheleria baltica*,

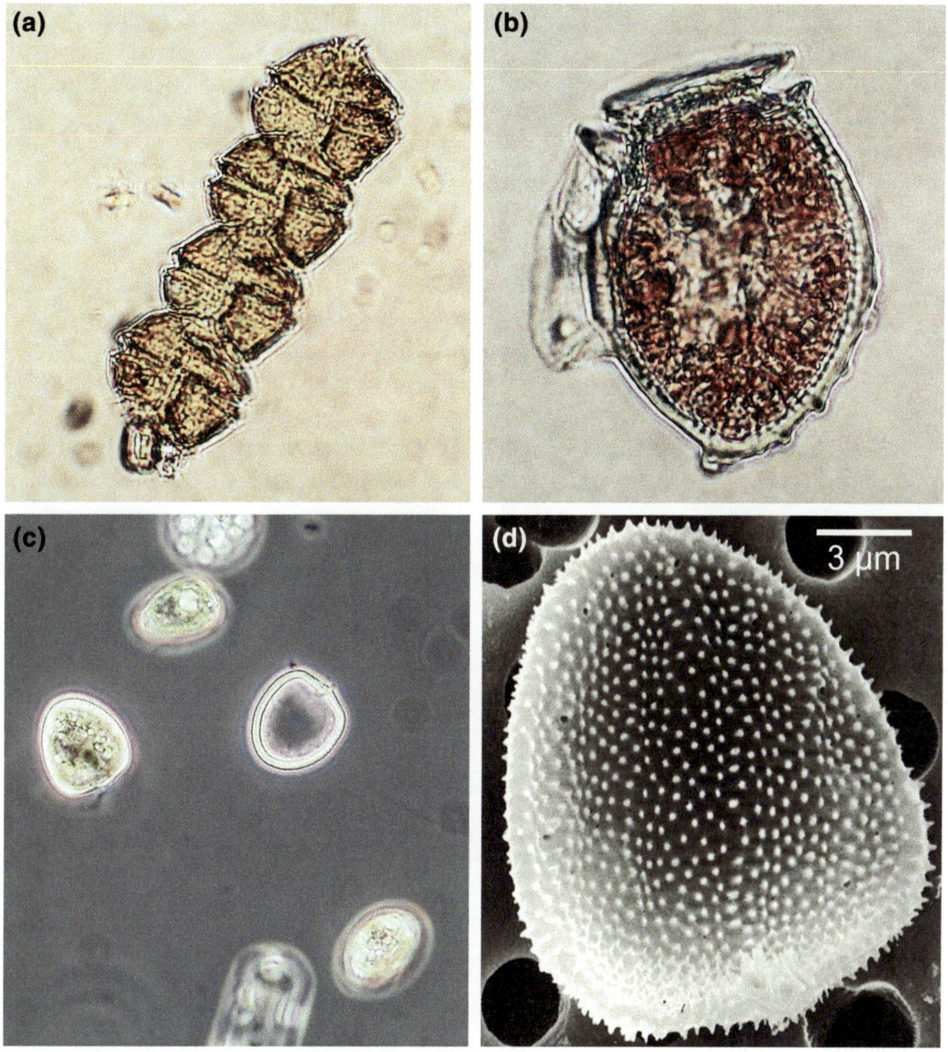

Fig. 8.7 Three abundant dinoflagellate species in the Baltic Sea. (**a**) A colony of four cells of the chain-forming Arctic-Boreal species *Peridiniella catenata* with a cell width of 22 µm. (**b**) The Arctic-Boreal species *Dinophysis norvegica* with a cell width of 45 µm. (**c**) The cryptogenic species *Prorocentrum cordatum* (syn. *Prorocentrum minimum*) with a cell width of 15 µm. (**d**) Scanning electron micrograph of a *Prorocentrum cordatum* cellulose thecal plate covered by minute spines (papillae). One cell is covered with two thecal plates. Photo: (**a**, **b**) © Pauline Snoeijs-Leijonmalm, (**c**) © Regina Hansen, (**d**) © Regina Hansen and Rainer Bahlo

Gymnodinium corollarium and *Scrippsiella* sp. are abundant spring-bloom species while *Pauliella taeniata* is the dominant diatom (Wasmund et al. 2011). In the Gulf of Riga, the spring bloom is dominated by *Peridiniella catenata* and diatoms. Late-summer communities in the Gulf of Finland, Gulf of Riga, and the northern and central Baltic Sea proper are characterised by cyanobacterial blooms dominated by *Aphanizomenon flos-aquae* and *Nodularia spumigena*, accompanied by colonial cyanobacteria and small flagellates (Suikkanen et al. 2007).

In the southern Baltic Sea proper, as well as in the Belt Sea, diatoms dominate. In spring, the genera *Chaetoceros*, *Coscinodiscus*, *Rhizosolenia*, *Skeletonema* and *Thalassiosira* are usually abundant. In summer, it is the genera *Actinocyclus*, *Cerataulina*, *Dactyliosolen*, *Proboscia* and *Pseudonitzschia*, and in autumn *Coscinodiscus granii* (Fig. 8.4c), which are abundant (Wasmund et al. 2008, 2011; Henriksen 2009). Also the dinoflagellates *Ceratium* and *Prorocentrum* may form blooms in these areas in autumn. Early-summer blooms of dictyochophytes have been reported from the Mecklenburg Bay (Wasmund et al. 2011). The main representative is *Dictyocha speculum*, which occurs mostly in its naked form (Jochem and Babenerd 1989).

8.2.7 Temporal oscillations

In many cases, even long-term changes in the Baltic Sea phytoplankton community composition tend to represent

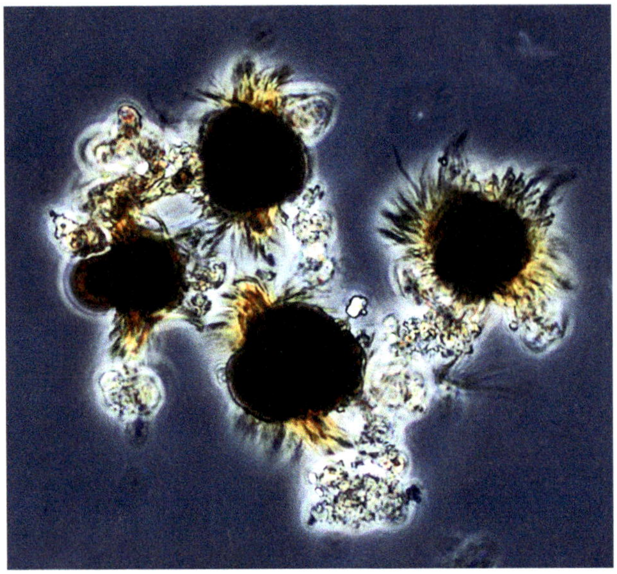

Fig. 8.8 Four individuals of the mixotrophic ciliate species *Mesodinium rubrum* (syn. *Myrionecta rubra*), which is very common in the phytoplankton communities of the Baltic Sea. The cells are filled with cryptophyte-type chloroplasts through kleptoplastidy. The cell size is highly variable, 10–65 μm in diameter in the Baltic Sea (Olenina et al. 2006). Photo: © Helena Höglander

basin-specific oscillations rather than directional trends (Jaanus et al. 2011). In addition, now and then the Baltic Sea is the scene of unusual phytoplankton events: for example, the occurrence of an extensive bloom of the haptophyte *Prymnesium polylepis* in autumn-spring 2007–2008 (Majaneva et al. 2012). However, no measurable negative responses to *Prymnesium polylepis* were observed for either the total phytoplankton stocks or the zooplankton community (Gorokhova et al. 2014).

The only negative response regarding other plankton, possibly associated with *Prymnesium polylepis* occurrence, was a significantly lower abundance of dinoflagellates both during and after the bloom in 2008. Moreover, contrary to the expected negative effects, there were significantly higher total phytoplankton abundances as well as significantly higher winter abundances and winter-spring RNA:DNA ratios in dominant zooplankton species in 2008. This indicated that the *Prymnesium polylepis* bloom coincided with favourable feeding conditions for zooplankton. However, in the benthic zone negative effects on the condition of the blue mussel *Mytilus trossulus* in 2008 was ascribed to *Prymnesium polylepis* (Larsson et al. 2014a, b).

Short-term, sudden changes in the phytoplankton community can occur as a result of upwelling events, which are common along the coasts of the Baltic Sea (Lehmann and Myrberg 2008). During an upwelling event, cold nutrient-rich water can break through the summer thermocline and cause changes in species composition and biomass (Laanemets et al. 2004; Smayda and Trainer 2010). Typically, the lowest chlorophyll *a* concentrations are recorded in the upwelled water, especially at the upwelling centres, and the highest concentrations about two weeks after the upwelling peak (Uiboupin et al. 2012).

8.2.8 Temporal trends

Recent studies show that considerable changes in the phytoplankton communities have taken place during the past 40–50 years. For example, in the Baltic Sea proper the abundance and intensity of cyanobacterial blooms have increased significantly (Funkey et al. 2014). In the Gulf of Finland, the proportion of dinoflagellates from the *Biecheleria baltica* complex (the three species: *Biecheleria baltica*, *Gymnodinium corollarium* and *Scrippsiella hangoei*) has increased in the pelagic spring bloom at the cost of diatoms and the dinoflagellate *Peridiniella catenata* (Fig. 8.9a). *Peridiniella catenata* is a typical cold-water species, which is common in the phytoplankton of the North-Atlantic and Arctic regions. The same trends for the *Biecheleria baltica* complex (positive) and diatoms (negative) with time were found in the Baltic Sea proper, but not for *Peridiniella catenata* (Fig. 8.9). In the northern Baltic Sea proper and the Gulf of Finland, phytoplankton biomass (chlorophyll *a* concentration) has increased (Suikkanen et al. 2013) and phytoplankton diversity has increased during the same time (Olli et al. 2014).

Analyses of long-term data have revealed changes in phytoplankton community structure. For example, in the northern Baltic Sea proper and the Gulf of Finland communities have shifted from a more autotrophic species composition towards a more mixotrophic one, with smaller-sized organisms (Suikkanen et al. 2013). This change was correlated with increased late-summer water temperature, increased dissolved nutrient concentrations and decreased salinity. This shift in community structure implies an increase in "low-quality" food for the micro- and mesozooplankton in the northern Baltic Sea proper and the Gulf of Finland.

In the southern Baltic Sea proper, increases in spring phytoplankton biomass (chlorophyll *a* concentration) have been recorded (Wasmund et al. 2011). In the Kiel Bay in the southwestern Belt Sea, the total phytoplankton biomass has roughly doubled, the summer and autumn blooms of *Chaetoceros* and *Skeletonema* have been partly replaced by other diatoms (*Cerataulina pelagica*, *Dactyliosolen fragilissimus*, *Proboscia alata* and *Pseudonitzschia* spp.) and some new bloom-forming, potentially toxic species (*Dictyocha speculum*, *Prorocentrum cordatum*, *Pseudonitzschia* spp.) have appeared during the last 100 years (Wasmund et al. 2008).

At the moment, the potentially toxic dinoflagellate *Prorocentrum cordatum* is the only known non-native

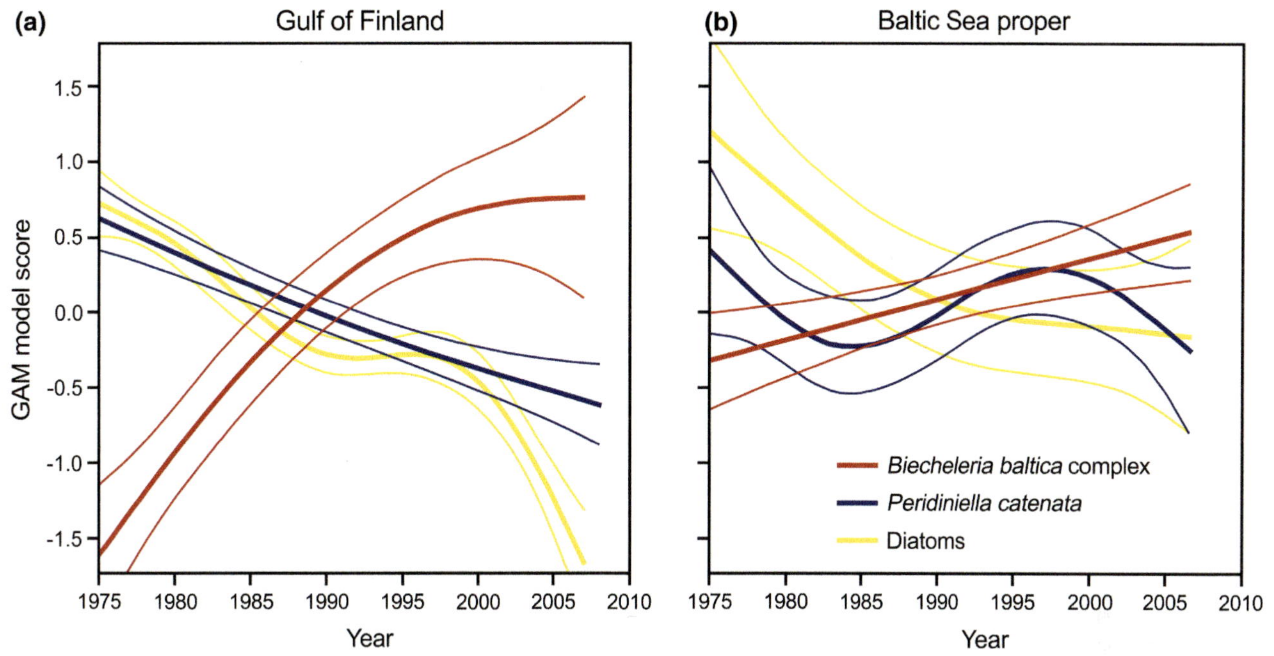

Fig. 8.9 Long-term trends of phytoplankton groups in (**a**) the Helsinki archipelago in the Gulf of Finland (1975–2008) and (**b**) Station B1 on the Swedish east coast in the Baltic Sea proper (1983–2004). The long-term trend curves were estimated using generalised additive model (GAM) analyses after accounting for all significant weather variables. Thick lines show mean values and thin lines denote double standard errors. *Peridiniella catenata* is a chain-forming dinoflagellate species (Fig. 8.7a) and the *Biecheleria baltica* complex refers to a species complex of dinoflagellates including *Biecheleria baltica*, *Gymnodinium corollarium* (Box Fig. 8.3f) and *Scrippsiella hangoei*. All positive trends with time for the *Biecheleria baltica* complex, and negative trends with time for diatoms, are significant, but for *Peridiniella catenata* there is only a significant negative trend in the Gulf of Finland. Figure modified from Klais et al. (2013)

phytoplankton species that has been spreading northwards in the Baltic Sea since the early 1980s (Hajdu et al. 2000, 2005). This cryptogenic species has successfully established itself in the Baltic Sea and is now a dominant summer species in the southern Baltic Sea proper, and reaches into the low-salinity waters of the Baltic Sea as far as the central Gulf of Finland. In the 1990s, *Prorocentrum cordatum* developed several coastal blooms in the eastern and northern Baltic Sea proper, although these blooms occurred irregularly between years.

8.2.9 Causes of the on-going changes

The long-term changes in the phytoplankton communities of the Baltic Sea may reflect both hydrographical variability and the human-induced eutrophication process (Wasmund et al. 2011). For example, the increases in the summer cyanobacterial blooms have been attributed to eutrophication (Funkey et al. 2014). Increased phytoplankton biomass production as a result of increased access to dissolved inorganic nutrients is a reversible process. This is evident along the Danish coasts, where the phytoplankton biomass has decreased since the late 1970s in parallel with a decrease in the nitrogen supply and an increase in water temperature (Henriksen 2009). Increased grazing pressure with increased water temperature has been proposed as one of the processes underlying this reduction in phytoplankton biomass. The diatom and dinoflagellate species that were dominant in spring and autumn net tows ~100 years ago are still among the dominant species along the Danish coasts. However, they are nowadays accompanied by several other species that were not present in samples collected around 1900.

The increase in the proportion of dinoflagellates during the spring bloom in many areas of the Baltic Sea has been linked primarily to a shift in weather and hydrographical conditions (Wasmund et al. 2008; Klais et al. 2011, 2013). Thick (>30 cm) and long-lasting ice cover favours diatom-dominated spring blooms. However, mild winters, with storms and a thin (10–20 cm) ice cover, support a bloom of a complex of at least three medium-sized dinoflagellate species (*Biecheleria baltica*, *Gymnodinium corollarium* and *Scrippsiella hangoei*). This is also a reversible and oscillating process: after cold winters diatoms dominate the spring bloom and after warm winters dinoflagellates dominate. However, since 1980 the ice cover in the Baltic Sea proper has been declining, which is reflected in the community composition of the spring bloom, and with continued global warming this trend may proceed (Klais et al. 2013).

The phytoplankton communities in the Baltic Sea are not the same as they were 50 years ago. In an analysis of ecosystem-wide monitoring data, including 7272 summer samples and 815 phytoplankton taxa for the years 1966–2008, Olli et al. (2011) found a significant, gradual temporal change in phytoplankton community composition in most parts of the Baltic Sea. The observed changes do not depend on the occurrence or abundance pattern of a particular species or group but reflect an overall shift in community composition. Although these changes are only weakly associated with known eutrophication gradients, it is clear that the summer Baltic Sea phytoplankton communities are not in a steady state equilibrium, but experience a directional change.

8.3 Prokaryotes

8.3.1 Available methodology affects our scientific concepts

Although bacteria are the most important organisms for sustaining life on our planet, microbiology itself is a relatively young scientific discipline. Practically everything started with Antonie van Leeuwenhoek's first ground-breaking microscopic observations and the discovery of bacteria in the late 17th century. In principle, the "traditional" microbial ecology was based on microscopic observations, isolation and cultivation of microorganisms, or measurement of their general activity, and this approach was prevalent for nearly 300 years.

In the sea, bacteria were originally regarded mainly as decomposers and important remineralisers of dead organic matter, and not as a part of the pelagic food web. This view changed in the 1970s, mainly due to the application of a range of new techniques, such as the use of radioactive isotopes and fluorescence microscopy, which revealed a better picture of the functional role of bacteria in the pelagic carbon cycle. Although the lack of information on the taxonomic composition of the planktonic bacteria meant that they were still regarded as a "black box"-type compartment of the pelagic food web, it became clear that bacteria account for a significant proportion of the planktonic biomass, and that a major part of the pelagic carbon flux is transferred through the microbial food web rather than through the classical pelagic food web from phytoplankton to zooplankton to fish (Azam et al. 1983, Fig. 8.1).

The developments in molecular and gene-based methods during the last 20–30 years have allowed for fundamental cultivation-independent diversity analyses of microbial communities. This has been a major step forward because appropriate techniques to culture a vast majority of marine bacteria still do not exist. Thus, a level of information similar to that available for higher organisms for more than 100 years is now accessible for microorganisms as well. Identification of prokaryotic microorganisms by sequencing, and phylogenetic analysis of their 16S rRNA gene, makes it possible to determine spatial and temporal variations in composition (Herlemann et al. 2011), identify key players in the microbial realm and their specific population dynamics, and link the fate of specific prokaryotic taxa to some controlling mechanisms (*e.g.* resource availability, predation). Therefore, a whole range of ecological questions regarding prokaryotes can be addressed at present, just like it has been possible, for a long time, with respect to larger eukaryotes. However, despite this enormous methodological leap forward, the current knowledge of the bacterioplankton in the Baltic Sea still lags far behind the knowledge of the phyto- and zooplankton, as these novel techniques have become available only recently and because they are costly and technically demanding.

Still more challenging than pure diversity assessments is linking microbial identities to their specific activities and functions. The so-called "omics techniques", in combination with high-throughput sequencing technologies, are becoming increasingly important. These techniques allow the simultaneous assessment of the microorganisms' identity, their proportion in the community, their physiological status, and their biogeochemical functions. It is conceivable that these approaches will again lead to major new insights and possibly to alteration of the current understanding of the ecology of planktonic microbes.

8.3.2 The development of Baltic Sea microbiology

Microbiological investigations in the Baltic Sea started at the end of the 19th century with the description of halophilic bacteria involved in the recycling of organic material (Rheinheimer 1984). As in other marine areas, microbiological studies relied on the isolation of bacteria and their culturing on agar plates. When the "microbial loop" concept was first introduced in the 1970s (Azam et al. 1983), Gerhard Rheinheimer's marine microbiology group at the Marine Science Institute at Kiel University (Germany) played an important role in the development, modification and application of techniques for the assessment of heterotrophic activities by measuring the uptake of radiolabelled organic substrates, for the enumeration of nucleoid-stained bacterioplankton using epifluorescence microscopy, and for the determination of the abundance of active bacteria by microautoradiography. These techniques were applied mainly to the waters of the Baltic Sea (Rheinheimer 1977).

An excellent and detailed overview of the whole history of the Baltic Sea microbiology, spanning the period from

1894 to the early 1980s was published by Rheinheimer (1984). It is interesting to read his claim that, close to the end of the 20th century, knowledge of Baltic Sea microbiology was – despite the fact that the Baltic Sea was amongst the best microbially investigated areas in the world at that time – still restricted "…to small areas usually in the vicinity of institutes with marine microbiological study groups such as there are at… Kiel, Aarhus, Göteborg, and Tvärminne". To some extent, this still remains the case: not all areas and subsystems of the Baltic Sea have been equally well studied with respect to microbiology.

In addition to areas in the vicinity of marine research institutes, the focus was placed on the microbiology of locations featuring particular gradients, such as the Gotland deep and the Landsort deep with their oxic/anoxic interfaces. Overall, current knowledge of the taxonomic composition of the bacterioplankton in the Baltic Sea still lags behind that in the major oceans for which there are data on the most abundant prokaryotes. However, as molecular techniques are gaining importance in Baltic Sea science, this situation is now rapidly changing.

8.3.3 Bacterioplankton – not only bacteria

The current Bacteria and Archaea are descendants of the oldest life forms on Earth. As their cells do not possess a nucleus, members of the two domains are referred to as "prokaryotes". For about two decades it has been known that planktonic prokaryotes also comprise, in addition to members of the Bacteria domain, representatives of the Archaea, previously known only from extreme environments (Fuhrman 2002). Mesophilic members of the Archaea are abundant in marine systems, especially in the deeper strata of the ocean (mesopelagial), and have also been recorded in the Baltic Sea.

The Archaea (formerly known as the Archaebacteria) differ considerably from the Bacteria in both phylogeny and biochemical cell composition. The functional and ecological role of the Archaea in the ocean is not fully understood yet, but there is evidence of an important role in chemoautotrophy (CO_2 fixation) and energy generation by oxidation of ammonium. This applies also to the oxygen-poor ("suboxic") zone of the deep basins of the Baltic Sea where the Archaea (more specifically members of the order Thaumarchaeota) are probably the major nitrifiers (Labrenz et al. 2010; Berg et al. 2015). The archaeal contribution to prokaryotic abundance in the Baltic Sea surface waters is poorly known at present.

Most of the ecologically focused marine field studies in which the dynamics of bacterioplankton have been examined simply refer to all the cells enumerated by epifluorescence microscopy as "bacteria" or "bacterioplankton". As bacterial and archaeal cells cannot be distinguished by microscopy, it would be more correct to replace the term "bacteria" with "prokaryotes" when no further information on their composition is available. However, for the sake of consistency and to maintain links with other studies, the terms "bacterioplankton" or "bacteria" in an ecological/functional context will be used hereafter because the Archaea are probably quantitatively more important only in certain specific habitats such as the deep ocean, redox gradients and sediments. Nevertheless, it is worth keeping in mind that suspended prokaryotes potentially encompass cells belonging to both domains, the Archaea and the Bacteria.

8.3.4 Bacterioplankton – more than heterotrophs

The bacterioplankton contributes a significant biomass to pelagic systems and strongly dominates the heterotrophic activity. The usual functional perception of bacterioplankton is that of heterotrophic decomposers of organic matter. This is indeed the case for the majority of the prokaryotes, but some important exceptions exist. Prokaryotes can be autotrophic, mixotrophic or heterotrophic and play diverse, fundamental roles in energy and carbon fluxes and the cycling of elements in aquatic ecosystems. Being small, the prokaryotes have high surface to volume ratios and thus high specific activity, whereby they are able to outcompete all other organisms in the uptake of dissolved compounds.

Cyanobacteria are primary producers that belong to the domain Bacteria. Large filamentous cyanobacteria belonging to the genera *Aphanizomenon*, *Dolichospermum* and *Nodularia* are conspicuous components of the Baltic Sea phytoplankton communities. Chroococcid cyanobacteria, *e.g.* the genus *Synechococcus*, which are functionally grouped together with heterotrophic bacteria into the picoplankton (0.2–2 µm), can be quantified by their autofluorescence and distinguished by their pigments (Larsson et al. 2014a, b).

Several prokaryotic groups are mainly, but not entirely, heterotrophic. Many bacteria in surface waters are known to be capable of generating energy using light. A light-generated proton gradient for synthesising ATP is produced either with the help of bacteriochlorophyll (organisms with this ability are termed aerobic anoxygenic phototrophs, AAPs) or with the pigment proteorhodopsin. However, these bacteria are not considered autotrophs as they obtain their carbon from dissolved organic carbon (DOC) rather than from dissolved inorganic carbon (DIC). There is evidence that, in certain situations, additional phototrophic ATP generation offers a competitive advantage. This combination of heterotrophic and phototrophic metabolic pathways, found among several bacterial lineages, is called "photoheterotrophy" and thus constitutes a form of mixotrophy.

Both AAPs and proteorhodopsin-containing bacteria have been identified in the surface waters of the Baltic Sea (Atamna-Ismaeel et al. 2008; Salka et al. 2014).

In some locations, particularly around oxic/anoxic interfaces, prokaryotes, which gain energy by oxidising reduced inorganic compounds such as ammonium, sulphides, reduced forms of iron and manganese (*cf.* Table 3.1), are abundant. They use chemical energy for CO_2 fixation, as do photosynthetic organisms, but without the need for light. In addition to the CO_2-fixing pathway of green plants via the Ribulose-1, 5-Bisphosphate Carboxylase/Oxygenase (RuBisCO)-catalysed Calvin cycle, which may represent the most important autotrophic CO_2 fixation pathway, other CO_2 fixing mechanisms have been discovered as well (Hügler and Sievert 2011). The organisms that employ these mechanisms are called chemolithoautotrophs and occur among both the Bacteria and the Archaea. Since some of them can also utilise certain organic molecules for energy generation and as a carbon source, they are also potentially mixotrophs. In the water column of the Baltic Sea proper, chemolithoautotrophic prokaryotes can gain considerable importance in the oxic/anoxic interface of the deep basins (*cf.* Sect. 8.3.12).

8.3.5 Aggregates as "hotspots" of bacterial activity

The term "bacterioplankton" brings to mind an assemblage of solitary suspended cells. For the majority of cells in the pelagic environment this is a true image, but a variable portion of the prokaryotic cells is attached to suspended particles produced mainly in the surface layer by various mechanisms (*e.g.* phytoplankton bloom collapse, zooplankton faeces).

At certain times of the seasonal succession, particle-attached microorganisms play a significant role, and suspended aggregates ("marine snow") constitute "hotspots" of microbial abundance and activity (Simon et al. 2002). Aggregates offer habitats different from the open water. They resemble suspended biofilms, which support high concentrations of organic substrates, prokaryotes and protists and have a different microbial community composition than the water-suspended bacterioplankton. Polymer-degrading taxa, such as some groups within the Bacteroidetes, are more abundant on particles and use exoenzymes to hydrolyse different carbon compounds, thereby releasing dissolved substrates into the environment. When analysing pelagic prokaryotic communities, both particle-attached and free-living bacteria have to be taken into account. In fact, most sampling protocols involve the collection of both freely suspended single cells and particle-attached microbes that are analysed together, unless the cells in a water sample are fractionated into different size classes by appropriate filters or when aggregates are sampled and analysed separately, *e.g.* by collecting them with syringes.

In the Baltic Sea, only a few *in situ* studies (*e.g.* Schumann et al. 2003 for coastal waters) have been conducted in which the quantitative contribution of aggregates, their structure and particle-associated bacteria have been investigated in more detail. In the Belt Sea, 1–23 % of the total bacterioplankton was found to be attached to detrital particles. High concentrations of aggregates can also be found in the fluffy layer above the sediment (*cf.* Sect. 13.3.7). The agglomeration of filamentous cyanobacteria that floats on the water surface during calm summer days also forms a microhabitat with a high substrate supply, many particle-associated microbes and high bacterial activity.

8.3.6 Specific features that influence the Baltic Sea bacterioplankton

The general patterns of occurrence and the regulating mechanisms of bacterioplankton in the Baltic Sea are principally similar to those in marine and limnic pelagic systems. However, the Baltic Sea environment shows a number of specific features, which are assumed to also affect prokaryotic communities. These features include:

1. river inflow-borne supply of freshwater and terrestrial bacteria
2. river inflow-borne supply of dissolved and particulate organic material and lateral transport of particulates into the deeper basins
3. large cyanobacterial blooms in summer
4. stronger influence of a winter ice cover in the northern subbasins
5. temperature control of bacterial activities during winter-spring
6. influence of the Baltic Sea-specific pelagic food web structure
7. a brackish-water environment and horizontal and vertical salinity gradients
8. the development of oxygen-poor and anoxic deep waters.

8.3.7 Bacterial growth requires organic and inorganic nutrients

Bacteria process the bioavailable part of particulate detritus and dissolved organic matter (DOM), either produced *in situ* by primary producers or imported from terrestrial sources and wetlands, and respire most of it to CO_2. The organic nitrogen and phosphorus associated with the substrates may become remineralised to ammonia and phosphate, depending on their organic matter content and the bacterial demand.

Much of the organic matter decomposition takes place in the surface water layer, but decomposition also occurs on sinking particles prior to sedimentation. Generally, the DOM imported from terrestrial sources (*e.g.* plant material) is largely refractory, with a high level of humic acids and fulvic acids, whereas the DOM originating from phytoplankton (via direct exudation, lysis or grazing) is a substrate that is much more readily utilised. Terrestrial DOM has been suggested to be a major driver of bacterial production in the Gulf of Bothnia (Sandberg et al. 2004; Wikner and Andersson 2012; Andersson et al. 2015), but there are also studies pointing to a rather low degradability of terrestrially derived DOM (*e.g.* Herlemann et al. 2014).

The reliance of bacterioplankton distribution on algal-derived substrates is apparent in the vertical profiles of and seasonal changes in bacterioplankton abundance and productivity, which are generally correlated with phytoplankton biomass. The highest levels of bacterial abundances, production and all the activity parameters are generally found in the photic zone, and decline with depth. In addition, the seasonal variability in bacterioplankton generally follows that of the phytoplankton, including blooms (Kuosa and Kivi 1989; Samuelsson et al. 2006). As documented in a number of empirical studies (*e.g.* Cole et al. 1988), bacterial parameters (abundance, biomass, production) and phytoplankton biomass and production are positively correlated across different aquatic systems of differing productivities, particularly when annual mean values are considered.

Growth and abundance of planktonic prokaryotes is controlled by the availability of resources (bottom-up regulation) and predation (top-down regulation). Although bacterial growth is generally limited by the supply of labile organic carbon, numerous studies from marine and freshwater systems have demonstrated that inorganic nutrient limitation by nitrogen or phosphorus may occur as well. Experiments in which the response to different combinations of nutrients is examined have also been conducted in the Baltic Sea. Although temporal and spatial changes in the degree of limitation by inorganic nutrients do occur, it is obvious that labile organic carbon alone is often insufficient for a strong bacterial growth, and that additional inorganic nitrogen and phosphorus are required (Zweifel et al. 1993; Lignell et al. 2008).

8.3.8 Temperature as a growth-limiting factor for bacteria

Temperature is an important physical factor that can limit bacterial growth despite an adequate substrate supply. At lower temperatures, bacterial growth rate is closely correlated with temperature. The temperature range at which bacterial growth is most strongly influenced seems to vary between systems. Temperature up-shift experiments with the Baltic Sea water in bottles and indoor mesocosms revealed that warming within the 2–10 °C range accelerated bacterial development and increased bacterial production relative to primary production (Autio 1998; Wohlers et al. 2009). This has implications for the functional role of the bacterioplankton in the pelagic carbon cycle during the era of global warming, particularly when winter water temperature increases.

Despite the general validity of the temperature limitation of their growth, prokaryotes also inhabit the sea ice that covers the northern Baltic Sea during winter (*cf.* Fig. 2.19). They live in brine channel systems within the sea ice where concentrations of organic and inorganic nutrients may be higher than in the seawater. The sea-ice bacteria living in the brine channels are adapted not only to low temperatures, but also to fluctuating salinity. This habitat of the Baltic Sea supports cold-adapted (psychrophilic) bacterial taxa related to those found in polar regions (Granskog et al. 2006).

8.3.9 Top-down regulation of bacterioplankton

The top-down factors that affect the bacterioplankton include grazing by heterotrophic nanoflagellates (2–20 µm), which constitute a highly diverse group of protists (Jürgens and Massana 2008). Field and experimental studies (the latter in mesocosms in which food web structure or nutrient supply was manipulated) have established the important role that heterotrophic nanoflagellates play as bacterial consumers in the Baltic Sea (Kuosa and Kivi 1989; Uitto et al. 1997). Other potential consumers of bacterioplankton are ciliates which may become temporarily abundant, and some groups of metazoan filter feeders (cladocerans, *e.g. Bosmina*). Viral lysis is another major mortality-causing factor for bacteria. However, this process is more difficult to assess than grazing, and therefore quantitative data are rare.

Also, indirect effects of the mesozooplankton, *e.g.* via predation on protozoans, can affect the bacterioplankton (Kivi et al. 1996; Zöllner et al. 2009). In mescocosm experiments where the copepod dominance in the zooplankton was manipulated, mechanisms of predatory trophic cascades from the mesozooplankton to the level of bacterial activity and diversity have been demonstrated (Fig. 8.10). Selectively feeding copepods exert a strong top-down control on planktonic ciliates. This in turn results in some release of autotrophic and heterotrophic nanoplankton (including the most important bacterivores) from predation pressure, and the nanoplankton increases in abundance.

This three-tier trophic cascade is a truly community-wide one. The effect on the bacterioplankton proved to be more complex, and the increase of heterotrophic nanoflagellates

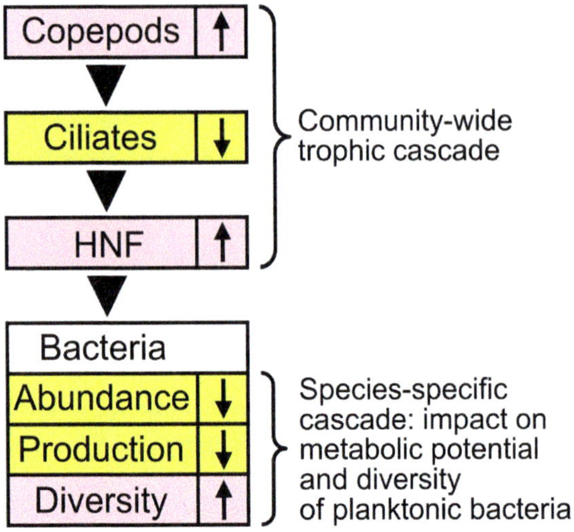

Fig. 8.10 Conceptual model of a copepod-mediated trophic cascade as observed in mesocosm experiments: trophic cascade dampening and differential effects on bacteria are demonstrated. The arrows in boxes indicate the positive or negative responses of the functional groups or parameters. HNF = Heterotrophic nanoflagellates. Figure modified from Zöllner et al. (2009)

abundance produced only a slight reduction of bacterial abundance, possibly due to the development of predation resistance. However, bacterial community composition and activity were observed to change substantially, providing evidence of species-specific cascading effects. Therefore, it seems that predatory cascades generally do not affect the bacterioplankton biomass as a whole but impact its community composition and activity.

8.3.10 Salinity effects on microbial growth

As a physical determinant, salinity has biological relevance because of its influence on ambient medium density (together with temperature and hydrostatic pressure), which necessitates osmoregulation by organisms (*cf.* Fig. 7.7). Distributions of phytoplankton, macrophytobenthos, macrozoobenthos and fish in aquatic habitats are known to be salinity-dependent (*cf.* Fig. 4.18). However, the impact of salinity on the growth and distribution of bacteria is less known, and the current knowledge stems mainly from studies in estuaries, areas of a highly complex dynamics related to shifts in several physico-chemical factors.

Although statistical analyses identified salinity as an important constraint for microbial activity among 33 physical, chemical and biological parameters, none of the individual parameters was a major controlling factor in shallow brackish-water systems such as the Kiel Bay and Kiel Fjord in the southwestern Belt Sea (Bölter et al. 1981). Also, in a survey of general bacterial parameters in the surface water layer along a transect from the Kiel Bay to the Bothnian Bay, it was found that differences in biomass and bacterial activity were rather small between sampling sites (Gocke and Rheinheimer 1991). Despite substantially differing environmental conditions between the different subregions of the Baltic Sea, bacterial numbers fluctuated between 3 and 4 million cells mL^{-1}, whereas activity parameters were somewhat more variable.

However, incubation experiments in which natural Baltic Sea microbial communities were exposed to different salinities and water masses showed that even moderate changes in salinity can affect bacterial functional performance such as biomass production, growth efficiency and growth rates (Langenheder et al. 2003). Growth rates increased, but growth efficiency (biomass yield per DOC utilised) decreased when bacteria were confronted with a shift in salinity, and more of the utilised carbon was respired. Also during these experiments, bacterial community composition changed with salinity. Strains belonging to the α- and β-Proteobacteria, Actinobacteria and γ-Proteobacteria other than the genus *Pseudomonas*, showed higher relative abundances under freshwater conditions, whereas strains of the genus *Pseudomonas* and the *Cytophaga-Flavobacterium-Bacteroides* group were favoured by estuarine conditions.

The ratio of colony-forming to non-colony-forming bacteria was found to be ~10 to 20 times higher in the brackish Baltic Sea than in the Skagerrak (Simu et al. 2005). It is not known yet what causes this difference, but the higher ratio in the Baltic Sea emphasises the peculiarity of the brackish settings. The two sea areas differ in their levels of dissolved organic carbon (lower in the Skagerrak), bacterial production (lower in the Skagerrak), and salinity (higher in the Skagerrak).

8.3.11 Salinity effects on community composition

The open Baltic Sea, with its relatively stable distinct horizontal and vertical salinity gradients (being disturbed only by MBIs, *cf.* Box 2.1), provides an ideal system for studying the effects of salinity on the community composition of autochthonous brackish bacterioplankton. Salinity effects on microbial community composition in the Baltic Sea were already demonstrated in early culture experiments. Saprophytic bacteria were frequently isolated from samples collected in the western part, whereas brackish-water species and salt-tolerant freshwater species were more frequent in the central and eastern parts of the Baltic Sea Area (Rheinheimer 1974). Twenty years later, these observations were supported by 16S rRNA sequencing, which demonstrated the lack of typical marine bacteria among isolates originating from the low-salinity northern Baltic Sea (Hagström et al. 2000).

The general importance of freshwater bacteria in the Baltic Sea was demonstrated by cultivation-dependent, as well as by cultivation-independent, studies. For example, typical freshwater phylotypes within the Actinobacteria, Verrucomicrobia and β-Proteobacteria are abundant in the Landsort deep off the Swedish coast (Riemann et al. 2008), and high counts of typical freshwater Actinobacteria were recorded along two surface transects in the Gulf of Bothnia, together with a change in the Actinobacteria composition in relation to salinity (Holmfeldt et al. 2009). Freshwater bacteria can successfully migrate into the brackish Baltic Sea where they gain a selective advantage when riverine DOC is the main carbon source (Kisand et al. 2005).

At the Skagerrak-Kattegat front which separates the Baltic-Sea-influenced water from pure North Sea water (*cf.* Sect. 2.2.2), it was found that some bacteria related to the *Verrucomicrobium* clade and Bacteroidetes, as well as prosthecate bacteria (*e.g. Hyphomonas*), showed distinct distribution patterns on each side of the front, and also with depth (Fig. 8.11). A more comprehensive view of the salinity-related Baltic Sea bacterial biogeography was recently produced by the analysis of more than 200 samples from horizontal and vertical salinity gradients using high-throughput 454 pyrosequencing of the 16S rRNA gene (Herlemann et al. 2011). Hierarchical clustering of the surface samples separated marine-brackish, brackish, and brackish-freshwater bacterial communities (Fig. 8.12). The

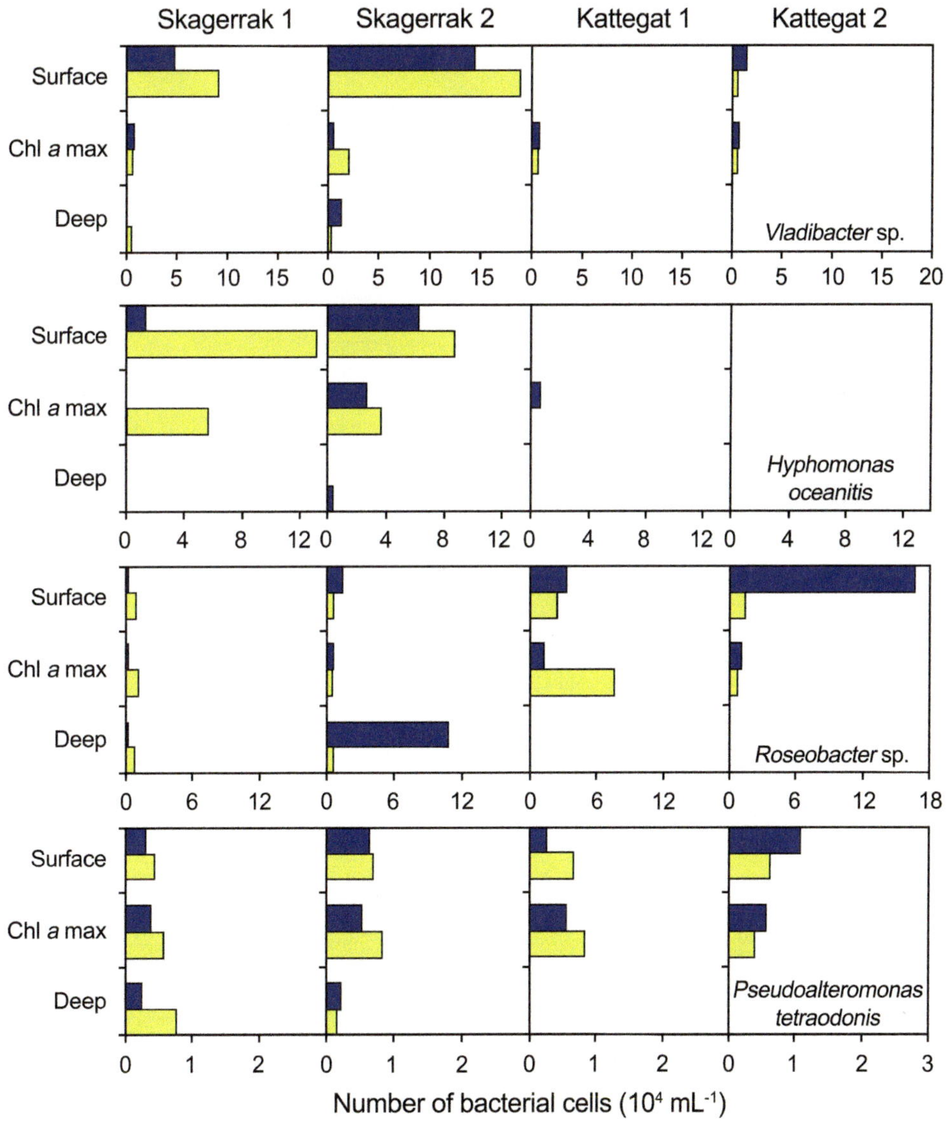

Fig. 8.11 Spatial distribution of numerically abundant marine bacteria at either side of the Skagerrak-Kattegat front which separates Baltic-Sea-influenced water from pure North Sea water. Blue and yellow columns indicate morning and noon samples, respectively. The abundance of specific bacteria was determined by whole-genome hybridisation. Figure modified from Pinhassi et al. (2003)

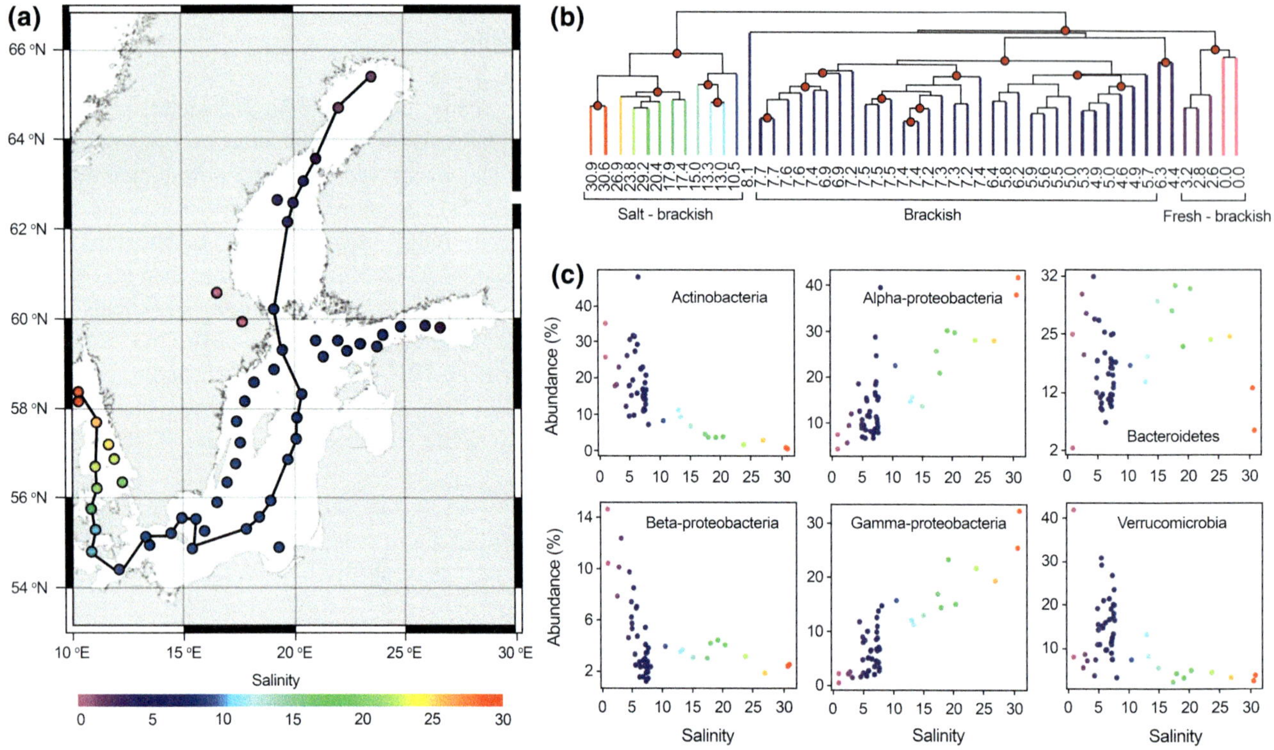

Fig. 8.12 Bacterial communities of surface samples (taken in summer 2008) along the Baltic Sea salinity gradient from the Skagerrak to the Bothnian Bay and the Gulf of Finland. (**a**) Map of the Baltic Sea Area with sampling stations colour-coded according to the measured surface-water salinity. (**b**) Hierarchical clustering based on bacterial community composition similarities. Nodes supported by high bootstrap values (>90 %) are marked with red circles. The samples cluster in three salinity ranges: freshwater-brackish (0–3.2), brackish (4.6–7.7), and marine-brackish (10.5–30.9). Three samples did not show a clear affiliation to any cluster (8.1, 6.3 and 4.4). The samples with salinities of 6.3 and 4.4 were sampled closest to the two largest cities in the region, Stockholm and Sankt-Petersburg, respectively, and may have been influenced by anthropogenic emissions of nutrients and/or chemical pollutants. (**c**) Relative abundance *versus* salinity for abundant bacterial phyla and proteobacterial classes. Figure reprinted from Herlemann et al. (2011) with permission from The ISME Journal

surface water communities were dominated by members of the Actinobacteria, α-Proteobacteria, β-Proteobacteria, γ-Proteobacteria, Bacteroidetes, Verrucomicrobia and Cyanobacteria. On this broad phylogenetic level, the relative abundance of distinct α-Proteobacteria and γ-Proteobacteria increased with salinity, whereas an opposite trend was exhibited by the Actinobacteria and β-Proteobacteria.

The pattern revealed in this study along the large-scale Baltic Sea gradient is consistent with distributions known from estuaries. Moreover, a *Verrucomicrobium* was identified as an abundant organism under brackish conditions (Herlemann et al. 2011). This organism was represented by a single dominant OTU (operational taxonomic unit) and was phylogenetically affiliated with the Spartobacteriaceae. Given its high abundance over a range of salinities and depths, this verrucomicrobial OTU probably represents one of the most abundant organisms in the Baltic Sea. The reconstruction of its genome by metagenomics revealed many genes potentially involved in the processing of polysaccharides, which are produced by phytoplankton (Herlemann et al. 2013).

8.3.12 Biogeochemical roles of prokaryotes in pelagic redoxclines

In the deep anoxic basins of the Baltic Sea proper, the pelagic redoxcline or chemocline (the transition layer between the oxic and anoxic water column) is the site of biogeochemical processes, which are mostly related to the cycles of oxygen, sulphur, nitrogen and carbon but also to the cycling of metals (*cf.* Fig. 3.17). Most biogeochemical transformations in the redoxcline are controlled by prokaryotes, and the knowledge of the relationships between microbial functions, identity and microbe-mediated processes is increasing. As chemolithoautotrophic bacteria and archaea dominate the prokaryotic abundance and production in the redoxcline, their quantitative and functional importance is evident (Jost et al. 2010). In fact, the chemolithoautotrophic production constitutes the base of the microbial food web, including different protistan groups, at the oxic/anoxic interface (Glaubitz et al. 2009; Anderson et al. 2012).

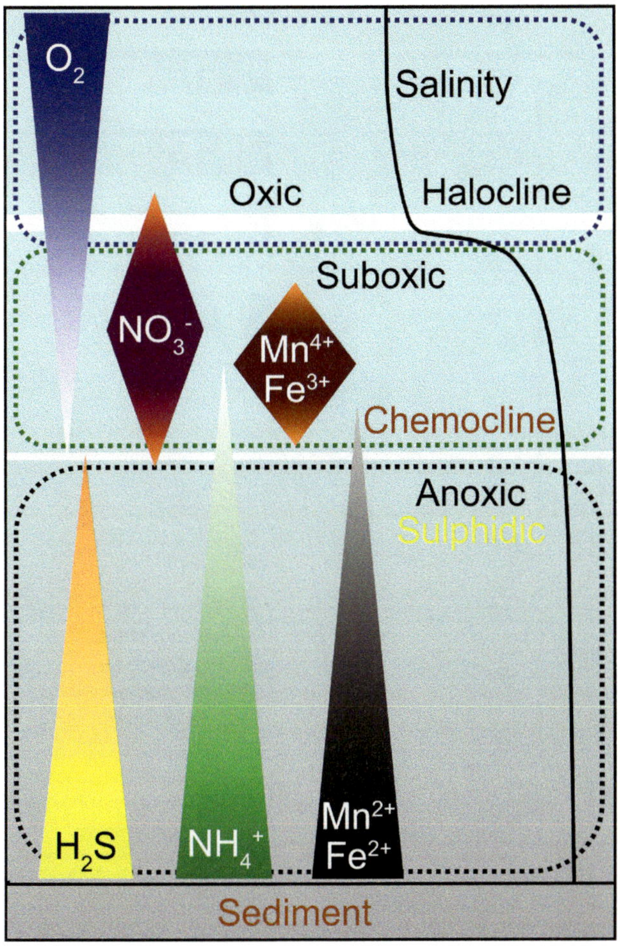

Fig. 8.13 Chemical zonation of the water column in the deep basins of the Baltic Sea proper. At the oxic/anoxic interface (chemocline or redoxcline), different functional groups of chemolithoautotrophic prokaryotes mediate the oxidation of reduced inorganic substrates (H_2S, NH_4^+, Mn^{2+}, Fe^{2+}). Figure: © Klaus Jürgens and Matthias Labrenz

Pelagic redoxclines show steep chemical gradients, intensive mineral particle formation, high dark CO_2 fixation rates and active biogeochemical processes connecting different elemental cycles (Fig. 8.13). These biogeochemical processes have many similarities to those in other marine hypoxic systems (<2 mL O_2 L^{-1}), particularly to fully anoxic and sulphidic systems as found *e.g.* in the Black Sea, the Cariaco basin (Venezuela), and in many fjords. Similarities in microbial community composition and biogeochemistry can also be found in the oceanic oxygen minimum zones (OMZs) near upwelling areas (Wright et al. 2012), as well as in sediments (*cf.* Fig. 3.16).

A detailed survey of the depth distribution of bacterial communities in the Baltic Sea was conducted by using the 454 pyrosequencing approach (Herlemann et al. 2011). In oxygen-sulphide transition layers, the bacterial community composition shifts drastically towards a community dominated by ε-Proteobacteria (Fig. 8.14a). In the deeper, purely sulphidic zones, the sulphate-reducing δ-Proteobacteria become a dominant group. No such shift in bacterial community composition occurs where oxygen is present throughout the water column, *e.g.* in the Åland deep (Fig. 8.14b).

Only a limited number of prokaryotic key players mediate processes within the nitrogen and sulphur cycles. The co-occurrence of high denitrification and dark CO_2 fixation (Brettar and Rheinheimer 1991) indicates that chemolithoautotrophic oxidation of sulphur compounds coupled with nitrate reduction could be important in the pelagic redoxcline. The ε-Proteobacteria of the genus *Sulfurimonas* are responsible for the major share of chemoautotrophic production around the oxic/anoxic interface (Grote et al. 2008). This bacterial group combines denitrification with the oxidation of reduced sulphur compounds, and its activity is largely responsible for detoxification of sulphide before it enters oxic waters. It is also responsible for the major nitrogen loss in the water column of the Baltic Sea and in other systems with a nitrate/sulphide interface. Detailed genomic and physiological studies of a representative of this group, *Sulfurimonas gotlandica* strain GD1, revealed an enormous metabolic flexibility and specific adaptations to survival in pelagic redoxclines (Grote et al. 2012; Labrenz et al. 2013). Recently, it was also found that members of the γ-proteobacterial SUP05 cluster, known already from other marine oxygen-deficient systems, could play a role in autotrophic denitrification around the oxic/anoxic interface (Glaubitz et al. 2013).

As in the case of autotrophic denitrification, aerobic ammonium oxidation was also found to be catalysed by a microbial community that is much less diverse than previously thought. These nitrifying microbes belong mainly to the Thaumarchaea, related to the only isolate available hitherto, *Candidatus Nitrosopumilus maritimus*, and are active mainly in the oxygen-poor zone above the oxic/anoxic interface (Labrenz et al. 2010; Berg et al. 2015). The relatively low diversity of microbial communities in the redoxclines of the Baltic Sea proper makes those communities ideal models for studies aimed at a better understanding of the regulation and function of microbe-dominated hypoxic systems.

8.4 Heterotrophic protists

8.4.1 Protozoa in the pelagic food web

Heterotrophic protists (also known as "Protozoa") can be defined functionally as simple phagotrophic "animals". They are common in pelagic systems and a large proportion of the organic carbon in bacteria-based food webs passes through

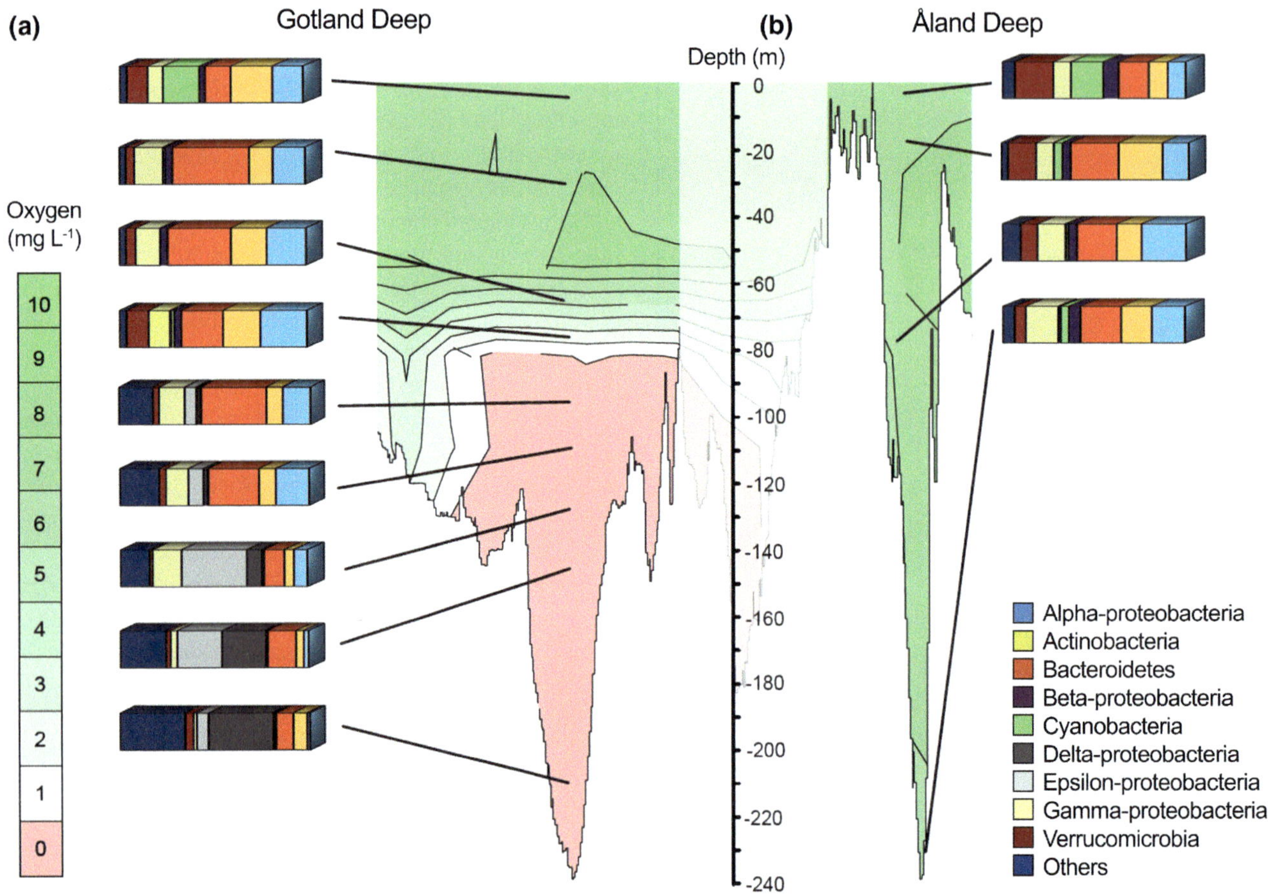

Fig. 8.14 Bacterial community composition along depth profiles of the Gotland deep and the Åland deep as determined by 16S rRNA 454 barcode pyrosequencing. Samples were taken in summer 2008. Note that the bacterial composition shifts at broad phylogenetic levels below the oxygen chemocline in the Gotland deep, but does not change considerably in the deep water layers of the Åland deep, which was not anoxic. Figure reprinted from Herlemann et al. (2011) with permission from the ISME Journal

this protistan level (Berglund et al. 2007, Fig. 8.1). Although purely bacteria-based food webs are absent from pelagic systems, in areas with a high content of humic substances, *e.g.* in the Bothnian Bay, a relatively large part of the food web can be based on bacterial production (Samuelsson et al. 2006), hence the importance of heterotrophic protists there. In pelagic microbial food webs, these protists often constitute the second and third trophic levels (Samuelsson et al. 2002). They engulf prey particles such as bacteria and other protists (including small phytoplankton) and are classified as interception feeders, filter feeders or raptorial feeders (Fenchel 1987). Many protists are omnivores feeding on different trophic levels below theirs and on varying types of organisms.

This diverse group of unicellular organisms is difficult to identify under the microscope due to their lack of clear cellular characters. Protists are renowned for having entered into permanent symbiotic relationships with their engulfed prey. According to the "endosymbiotic theory" (symbiogenesis), which is an evolutionary theory that explains the origin of eukaryotic cells from prokaryotes (Stewart and Mattox 1980), several key organelles of eukaryotes (plastids, mitochondria) have been formed through symbiosis between separate unicellular organisms. For example, the eukaryotic algae have evolved from a phagotrophic protist engulfing a photosynthetic prokaryote, after which the latter remained an endosymbiont of the former. In natural systems, including the Baltic Sea, symbiosis between protists and photosynthetic organisms is common. Some protists support photosynthetic symbionts, *e.g.* the ciliate *Mesodinium rubrum* (Fig. 8.8), whose cells are filled with cryptophyte-type chloroplasts (Johnson et al. 2007; Myung et al. 2011). This ciliate retains plastids from the engulfed cryptophyte cells to acquire and maintain its photosynthetic capacity, so it steals plastids (kleptoplastidy). In the Baltic Sea, *Mesodinium rubrum* occurs from north to south, and is often dominant in the protistan and phytoplankton communities (Andersson et al. 1996).

Often, the two major trophic levels occupied by heterotrophic protists are small-sized nanoflagellates (<10 µm) and larger ciliates (20–60 µm) (Berglund et al. 2007). The protist nanoflagellates are often subdivided into choanoflagellates (unicellular and colonial filter-feeding protozoans, which are the closest known relatives of metazoans, King et al. 2008) and "other heterotrophic nanoflagellates". The larger ciliates consume the smaller nanoflagellates, but a group of choanoflagellates have a lorica (a siliceous exoskeleton, Leadbeater et al. 2009), which prevents them from being consumed by ciliates. In the Baltic Sea, these choanoflagellates are instead eaten by the next trophic level up, the mesozooplankton, which also consume ciliates. In experimental systems it has been shown that choanoflagellates, especially the loricated forms, are promoted by the presence of biosurfaces provided by phytoplankton cells (Samuelsson et al. 2006). In contrast, other heterotrophic flagellates are not affected by the presence of phytoplankton surfaces. In the Baltic Sea, heterotrophic nanoflagellates have been shown to be common in coastal as well as offshore areas, and are major consumers of the picoplankton (Uitto et al. 1997).

Many dinoflagellates have chloroplasts, and are thus photosynthetic, while others are mixotrophic or have photosynthetic endosymbionts; still others are purely heterotrophic (*cf.* Table 4.4). On account of their phagotrophy, the two latter groups can be considered protozoans. However, traditional studies of non-autotrophic protistan communities tended to disregard dinoflagellates and treated them as members of the phytoplankton.

In the Gulf of Bothnia, relatively small flagellates and ciliates dominate the protistan community, while further south (in the northern Baltic Sea proper) larger-sized flagellates and ciliates make up a larger proportion of the protistan biomass (Fig. 8.15). This difference in community size structure may partly be a reflection of the differences in nutrient concentrations from north to south in the Baltic Sea (*cf.* Fig. 2.23). In the nutrient-poor north, smaller organisms are promoted, while larger organisms are favoured in the more productive south.

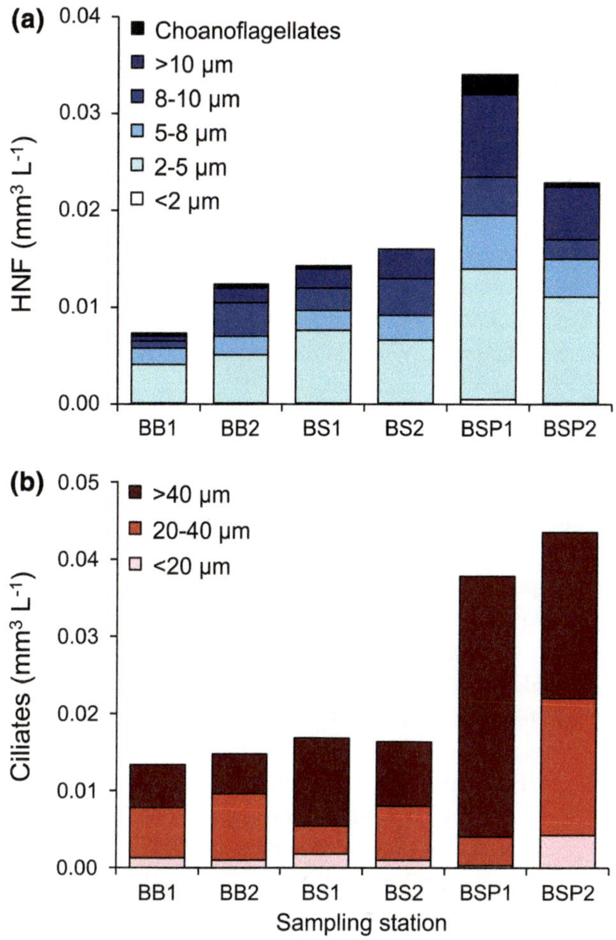

Fig. 8.15 Annual average biovolumes of different size groups of heterotrophic protists at offshore stations in the Bothnian Bay (BB), Bothnian Sea (BS) and the northern Baltic Sea proper (BSP). (**a**) heterotrophic nanoflagellates (HNF). (**b**) Ciliates. Figure modified from Samuelsson et al. (2006)

8.4.2 Heterotrophic nanoflagellates – the smallest protists

Common among the small heterotrophic nanoflagellates are Chrysophyceae (*cf.* Table 4.4), with especially the genera *Paraphysomonas* and *Spumella* occurring frequently (Samuelsson et al. 2006). In the Bothnian Bay, chrysophyceans dominate, while this group contributes with less than 10 % of the total biomass of heterotrophic nanoflagellates in the Baltic Sea proper. Several other taxa of small flagellates, *e.g.* *Goniomonas* sp. (a cryptophyte), have been shown to form short-lived peaks in the Bothnian Bay. Two larger cryptophyte flagellates, with their maxima during the spring bloom, are *Katablepharis* cf. *remigera*, found throughout the Baltic Sea, and *Leucocryptos marina*, which mainly occurs in the Baltic Sea proper.

The taxonomic composition of loricated choanoflagellates differs between spring and autumn and changes markedly from north to south. In spring, *Diaphanoeca grandis* dominates in the north, while *Calliacantha natans* and *Calliacantha simplex* are dominants in the northern Baltic Sea proper. In autumn, *Cosmoeca* sp. and *Acanthocorbis* sp. dominate in the Bothnian Bay, while *Acanthocorbis* cf. *apoda* and *Acanthocorbis* cf. *asymmetrica* as well as *Stephanoeca* sp. dominate in the Bothnian Sea. Among the non-loricated choanoflagellates, *Monosiga* sp. is common (Samuelsson et al. 2006). Some choanoflagellates (*e.g.* species within the genus *Codosiga*) are adapted to oxygen-poor environments and consume bacteria in the hypoxic areas of the Baltic Sea (Wylezich et al. 2012). The application of 18S rRNA gene-based techniques have revealed that some flagellate groups, especially jacobids and symbiontids, inhabit even the anoxic and sulphidic waters of the deep basins in the Baltic Sea (Weber et al. 2014).

8.4.3 Ciliates – a highly diverse group

The ciliates of the Baltic Sea are highly diverse, and their taxonomic composition varies spatially (Setälä and Kivi 2003; Johansson et al. 2004). Ciliates occurring in the open Baltic Sea include the autotrophic species *Mesodinium rubrum* and heterotrophic species of the orders Choreotrichida (*Helicostomella*, *Lohmaniella*, *Strobilidium*, *Tintinnidium*, *Tintinnopsis*), Euplotida (*Euplotes*), Haptorida (*Askenasia*, *Didinium*), Oligotrichida (*Strombidium*), Prostomatea (*Balanion*, *Holophrya*) and Sessilida (*Vorticella*, Vaginicolidae) (Johansson et al. 2004).

Despite this high diversity, a single order, the oligotrichids, are usually dominant. In the open northern Baltic Sea proper, species of the oligotrichid genus *Strombidium* and the choreotrichid genus *Strobilidium* often account for 50 % of the ciliate biomass (Johansson et al. 2004). It is only during spring that the biomass of haptorid species (mainly *Askenasia* cf. *stellaris* and *Didinium gargantua*) is higher than that of choreotrichids or oligotrichids. The prostomes (*e.g.* *Balanion comatum* and *Holophrya* spp.) usually account for up to ~10 % of the ciliate biomass. Dominants in coastal areas are different from the species dominating offshore. For example, large ciliates, such as members of the genera *Didinium*, *Euplotes*, *Lacrymaria* and *Tintinnopsis*, are common in the Gdańsk Bay (Witek 1998).

The upper water layers in coastal and offshore areas are normally oxygenated. However, in eutrophic waters, oxygen levels can be low, *e.g.* in semi-enclosed bays and in the deeper parts of the Baltic Sea proper. The protistan community occurring in such habitats differs from that found in oxygenated water (Fenchel et al. 1990; Setälä and Kivi 2003). While the oxic upper water layer may support bicoecids, choanoflagellates and chrysomonads, diplomonad flagellates are common in the deeper anoxic water.

Oxyclines, narrow water layers characterised by a steep reduction of the oxygen concentration, show peaks of bacteria, heterotrophic nanoflagellates and ciliates. The dominant ciliates in such environments are usually tintinnids and other oligotrichs, in addition to gymnostome ciliates. One characteristic ciliate group inhabiting the oxic/anoxic interface in the Baltic Sea consists of species of the genus *Metacystis* (Weber et al. 2014), none of which has been cultivated to date.

Ciliates become the major bacterivores with increasing hypoxia in the waters of the Baltic Sea, whereas the importance of nanoflagellates as bacterial grazers decreases (Anderson et al. 2012). The underlying anoxic systems support a considerable ciliate diversity, for example within the order Trichostomatida. Many of the species present in the anoxic zone harbour prokaryotic endosymbionts, such as *e.g.* methanogenic bacteria (Fenchel et al. 1990). However, the biomass ratio between phagotrophs and their bacterial prey is still relatively low in this zone, which can be explained by the low growth efficiency of anaerobic eukaryotes.

8.4.4 Dinoflagellates – overlooked protists

Heterotrophic dinoflagellates such as *Protoperidinium*, *Gyrodinium* and *Gymnodinium* are abundant in the pelagic communities, especially in nutrient-rich coastal areas such as the Gdańsk Bay (Bralewska and Witek 1995). In such areas, heterotrophic dinoflagellates can account for 50 % of the zooplankton biomass (Rychert 2011). These dinoflagellates are mostly included in the phytoplankton because they are difficult to distinguish from autotrophic forms with the Utermöhl technique. With epifluorescence microscopy it is possible to observe if cells contain chloroplasts, but chloroplasts should not be confused with ingested cells. Furthermore, many dinoflagellates (*e.g.* *Dinophysis*) contain cryptophyte endosymbionts, which fluoresce orange with epifluorescence, similarly to chloroplast-containing cells.

8.4.5 Temporal variations in the protistan community

During the spring phytoplankton bloom, choanoflagellates peak in all three of the Baltic Sea's largest basins: the Baltic Sea proper, the Bothnian Sea and the Bothnian Bay (Fig. 8.16). This is consistent with the enhanced availability of bio-surfaces (phytoplankton) in the water during this time of the year. Choanoflagellates also have an autumn peak, which coincides with the small autumn phytoplankton bloom. Other heterotrophic nanoflagellates (HNF) peak during the summer, concurrently with, or shortly after, the peak of heterotrophic bacteria, which can be interpreted as a predator-prey interaction. Ciliates peak during the spring bloom, probably because they feed on the phytoplankton. Another ciliate peak is observed shortly after the summer peaks of bacteria and HNF.

Taken together, the microbial communities in all subregions of the Baltic Sea seem to be involved in predator-prey interactions. The temporal variations are, in general, more extensive in the Baltic Sea proper than in the Gulf of Bothnia, probably as a result of higher nutrient availability there, which allows for larger changes in different plankton groups. The size structure within the ciliate community follows that of the phytoplankton community, with larger cells prevailing in spring and smaller cells being more abundant in summer (Witek 1998; Johansson et al. 2004).

Generally, there is an inverse relationship between ciliates and mesozooplankton. When the copepod biomass increases that of larger ciliates decreases, indicating a

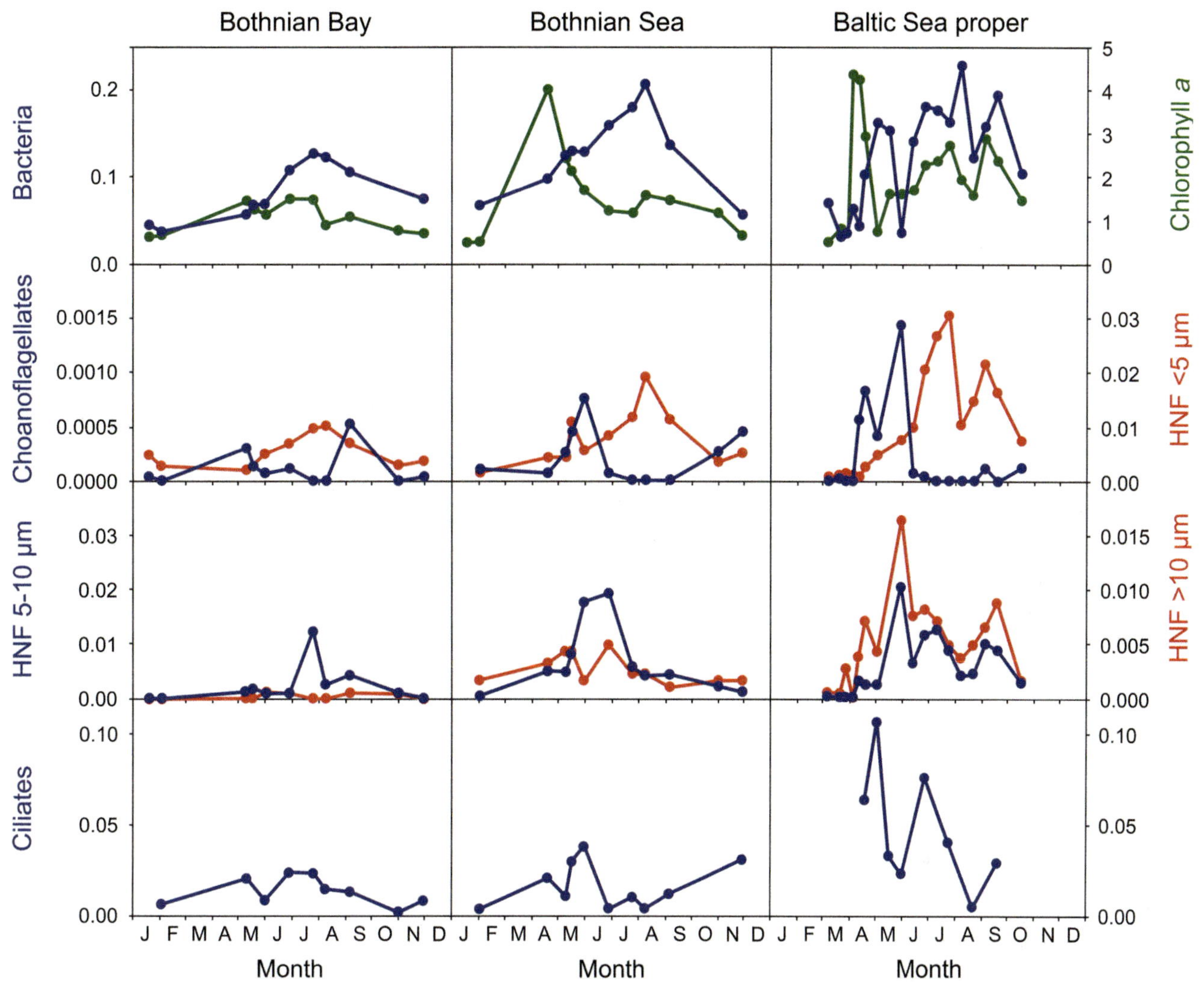

Fig. 8.16 Seasonal dynamics of chlorophyll *a* concentration (in µg L^{-1}), and the biovolumes of bacteria, flagellates and ciliates (in mm^3 L^{-1}) in the Bothnian Bay, the Bothnian Sea and the Baltic Sea proper. HNF = Heterotrophic nanoflagellates other than choanoflagellates. Figure modified from Samuelsson et al. (2006)

predator-prey relationship (Smetacek 1981; Johansson et al. 2004). Despite an increase in the potential for ciliate production, the summer increase in copepod biomass is accompanied by a decrease in the total ciliate biomass. This clearly indicates a mesozooplankton predation control over the ciliates.

The potentially mixotrophic flagellates, *e.g.* haptophytes, chrysophyceans and dinophytes, are common in the nutrient-poor summer water in the pelagic zone of the Baltic Sea (Andersson et al. 1996; Hajdu et al. 1996). In an experimental system, some of these flagellates were shown to employ photosynthesis as a survival mechanism, and to become dominant in nutrient-depleted water (Andersson et al. 1989; Samuelsson et al. 2002). However, in nutrient-rich water, they are outcompeted by other phytoplankton or by heterotrophic species (Samuelsson et al. 2002; Andersson et al. 2006).

8.5 Variability at the base of the food web

8.5.1 Aquatic food webs – different pathways

Autotrophic phytoplankton and heterotrophic bacteria constitute the "base" of the pelagic food web. They are diffusion feeders and take up dissolved substances through their cell membranes. The autotrophic phytoplankton are primary producers that use solar radiation as their energy source and assimilate inorganic carbon (carbon dioxide) from the water to produce carbon biomass. The energy and carbon source for heterotrophic bacteria is organic carbon, which can be either autochthonous (originating from the phytoplankton) or allochthonous (originating from outside the system, *e.g.* from riverine runoff). Due to their large surface-to-volume ratio, bacteria are more efficient in absorbing nutrients at low nutrient concentrations compared to phytoplankton, and may therefore outcompete phytoplankton if an allochthonous carbon supply is available.

The carbon produced at the base of the food web is transferred through a number of trophic levels before reaching the top consumers. The phytoplankton are grazed upon by the zooplankton, which in turn are consumed by zooplanktivorous fish. However, bacteria are too small to be readily eaten by many marine zooplankton organisms. Instead, heterotrophic flagellates and ciliates prey on bacteria, forming a link between bacteria and the zooplankton (Azam et al. 1983). Hence, the energy produced by bacteria is transferred through more trophic levels before reaching the top consumers in the ecosystem than the energy produced by the phytoplankton (Fig. 8.1) This inevitably means a larger loss of energy in bacteria-based food webs, because at each trophic level ~70 % of the consumed energy is lost (Straile 1997; Berglund et al. 2007).

Phytoplankton-based food webs are generally named the "classical pelagic food webs" or "grazing food webs", while bacteria-based food webs are called the "microbial food webs" or "microbial loops" (Azam et al. 1983; Legendre and Rassoulzadegan 1995). The classical pelagic food web is known to dominate in nutrient-rich waters, while microbial food webs dominate in nutrient-poor waters or in systems that are highly influenced by allochthonous organic matter. Thus, the pelagic food web has quite different structures and pathways depending on nutrient levels and other drivers in the system.

8.5.2 Food web efficiency – a measure of food web functioning

At each trophic level, a significant part of the consumed energy (carbon) is lost due to *e.g.* respiration, excretion and "sloppy" feeding (Azam et al. 1983; Straile 1997). The ratio between production by the top trophic level and the basal trophic level, *i.e.* the food web efficiency (FWE), can be calculated and used as a measure of the overall system efficiency (Rand and Stewart 1998). Accordingly, a more bacteria-based system will have lower FWE than a more phytoplankton-based system.

The gross growth efficiency is defined as the growth of an organism divided by the food ingestion (Fenchel 1987). In ecosystem studies, this concept can be applied to each trophic level and is often called the "trophic transfer efficiency". The trophic transfer efficiency is usually in the range of ~15–35 % (Welch 1968; Straile 1997), and depends on *e.g.* food availability. A higher food availability means a lower assimilation efficiency and thus a low growth efficiency (Welch 1968), which in turn leads to a decreased FWE.

In aquatic food webs where edible phytoplankton constitute the base, zooplankton forms the intermediate level and planktivorous fish are the highest trophic level. If the trophic transfer efficiency is 25 %, the FWE would be 6.25 % $[(0.25)^2]$. If bacteria constitute the base of the food web, an additional intermediate trophic level is established (protists), and the FWE will end up being 1.6 % $[(0.25)^3]$. However, since bacterial production and primary production generally co-occur in natural systems, the FWE in systems with planktivorous fish as the highest trophic level might be somewhere in between 1 % and 6 %.

Many other factors may affect the FWE, *e.g.* the edibility of the basal producers. In nutrient-rich systems, the phytoplankton community is often dominated by inedible or poorly edible forms, *e.g.* filamentous cyanobacteria or green algae, which results in a very low FWE (Andersson et al. 2013). On the contrary, the occurrence of omnivory, *i.e.* feeding by an organism on different trophic levels (Sprules and Bowerman 1988; Burns 1989; Thompson et al. 2007), would increase the FWE. Taken together, there are several factors which may govern food web functioning. Understanding the regulation of food web efficiency in an aquatic ecosystem, *e.g.* the Baltic Sea, enables assessments and predictions of how environmental change will affect ecosystem functioning.

8.5.3 Plankton production and food web functioning

In the pelagic zone of the Baltic Sea, the production at the base of the food web exhibits large differences from north to south. In the offshore waters of the Bothnian Bay, the Bothnian Sea and the Baltic Sea proper, the bacterial production equals 67 %, 38 % and 5 % of the primary production, respectively (Table 8.1). Annual phytoplankton primary production shows a 10-fold increase from north to south, while bacterial production is more stable along the geographical gradient. The reason for the relatively high importance of bacteria in the north is probably the fueling of bacteria with allochtonous organic carbon arriving via the rivers, in combination with low phosphorus availability (Sandberg et al. 2004; Andersson et al. 2015), which facilitates bacterial competition for inorganic nutrients. Furthermore, the high concentrations of coloured dissolved organic matter (CDOM, *cf.* Sect. 15.2.6) reduce the light in the water, causing primary production to decrease. Accordingly, the Bothnian Bay has been shown to be net-heterotrophic, emitting carbon dioxide, while further south heterotrophy is balanced by autotrophy (Algesten et al. 2006).

Since bacteria have an average cell size of ~0.5 µm (*cf.* Box Fig. 4.1), and the phytoplankton communities in the Bothnian Sea and the Baltic Sea proper consist of organisms with an average cell size of >20 µm, the "basal production" is anticipated to enter the food web at a lower level in the

Table 8.1 Annual average primary and bacterial production, the hypothetical food web efficiency (FWE) and number of trophic levels from basal producers to the mesozooplankton in the three largest basins of the Baltic Sea: the Bothnian Bay, the Bothnian Sea and the Baltic Sea proper. Data from Samuelsson et al. (2006) and Larsson et al. (2010)

	Bothnian Bay	Bothnian Sea	Baltic Sea proper
Primary production (mol C m^{-2} $year^{-1}$)	1.5	3.9	14.3
Bacterial production (mol C m^{-2} $year^{-1}$)	1.0	1.5	0.7
FWE (%)	17	20	24
Number of trophic levels	1.6	1.4	1.0

northern basins. In a size-structured food web, there is at least one trophic level between ~0.5 and 40 µm, comprising protists, heterotrophic nanoflagellates or ciliates. In line with this, HNF <5 µm are relatively more abundant (among HNF) in the north, while HNF >10 µm are relatively more abundant in the south (Fig. 8.16). Furthermore, small ciliates (<40 µm) dominate in the north, while larger ciliates (>40 µm) are abundant in the south. Accordingly, both heterotrophic nanoflagellates and ciliates have been shown to increase in biovolume from north to south in the Baltic Sea, in accordance with the spatial pattern of primary production (Samuelsson et al. 2006).

The pelagic food web is thus of a "microbial food web" type in the northern Baltic Sea, with small cells at the base followed by high abundances of small protozoan cells, while in the south the basal producers are constituted by larger cells giving rise to larger protozoan cells and possibly fewer trophic steps. By assuming full edibility of the basal producers, food web structures as presented in Fig. 8.1 and a trophic transfer efficiency of 25 % (Welch 1968; Straile 1997), it is possible to calculate a hypothetical FWE from the basal producers to mesozooplankton in the three largest basins of the Baltic Sea. The FWE would increase from 17 % to 24 % along the north-south gradient (Table 8.1). This is caused by a 0.6-step longer food web in the north, which leads to higher energy loss. However, this is a theoretical calculation which may be contradicted by many factors such as varying edibility and trophic transfer efficiency. Nevertheless, the calculation gives an indication of the differences in food web functioning between different areas of the Baltic Sea.

8.6 Mesozooplankton

8.6.1 Major components

Mesozooplankton constitutes an important link between the lower and higher trophic levels. This group comprises the size range of 0.2–2 mm and consists mainly of rotifers and small crustaceans. Zooplankton organisms with adult stages larger than 2 mm, *e.g.* mysids, are included in the macrozooplankton. Both the meso- and macrozooplankton are important grazers and predators of phytoplankton, heterotrophic flagellates and microzooplankton, and transfer energy to the next trophic level up, *i.e.* planktivorous fish. The meso- and macrozooplankton include organisms with different spatial and seasonal distributions, life strategies, feeding habits and interactions with predators.

The dominant mesozooplankton groups are rotifers (Rotifera), water fleas (Cladocera) and copepods (Copepoda), including their larval stages (Fig. 8.17a–c). Taxonomically, rotifers as a group have a phylum status, while cladocerans and copepods are crustaceans and belong to the phylum Arthropoda. In addition to these groups, planktonic larvae of bivalves (Fig. 8.17d), gastropods and polychaetes, collectively termed the "meroplankton", may at times be abundant members of the mesozooplankton. The larvae of *Cerastoderma glaucum*, *Macoma balthica*, *Mya arenaria* and *Mytilus trossulus* may especially, for short periods and often in late summer, be very abundant in zooplankton samples. Another common component of the meroplankton consists of the planktonic larvae of the bay barnacle *Amphibalanus improvisus*. In addition, benthic ostracods may also occasionally enter the water column.

Generally, the diversity of the mesozooplankton in the pelagic zone of the open Baltic Sea is low. For example, in the northern Baltic Sea, usually three to five species account for 80–90 % of the biomass of the mesozooplankton communities throughout the year (Viitasalo 1992a). Zooplankton organisms are traditionally sampled with with plankton nets (Box 8.2).

8.6.2 Rotifers

The name "rotifers" derives from the Latin word for "wheel-bearers" and refers to the corona, a ring of moving cilia around the mouth (Fig. 8.18a, b). The corona propels the animal forward and allows it to grab various food particles. The most abundant rotifer species in the pelagic zone of the Baltic Sea are members of the genera *Synchaeta* and *Keratella*. Low-salinity coastal areas support more rotifer species than the open Baltic Sea (Telesh and Heerkloss 2004). Rotifers feed on bacteria, small phytoplankton cells and protists as well as on detritus. In early summer, rotifers,

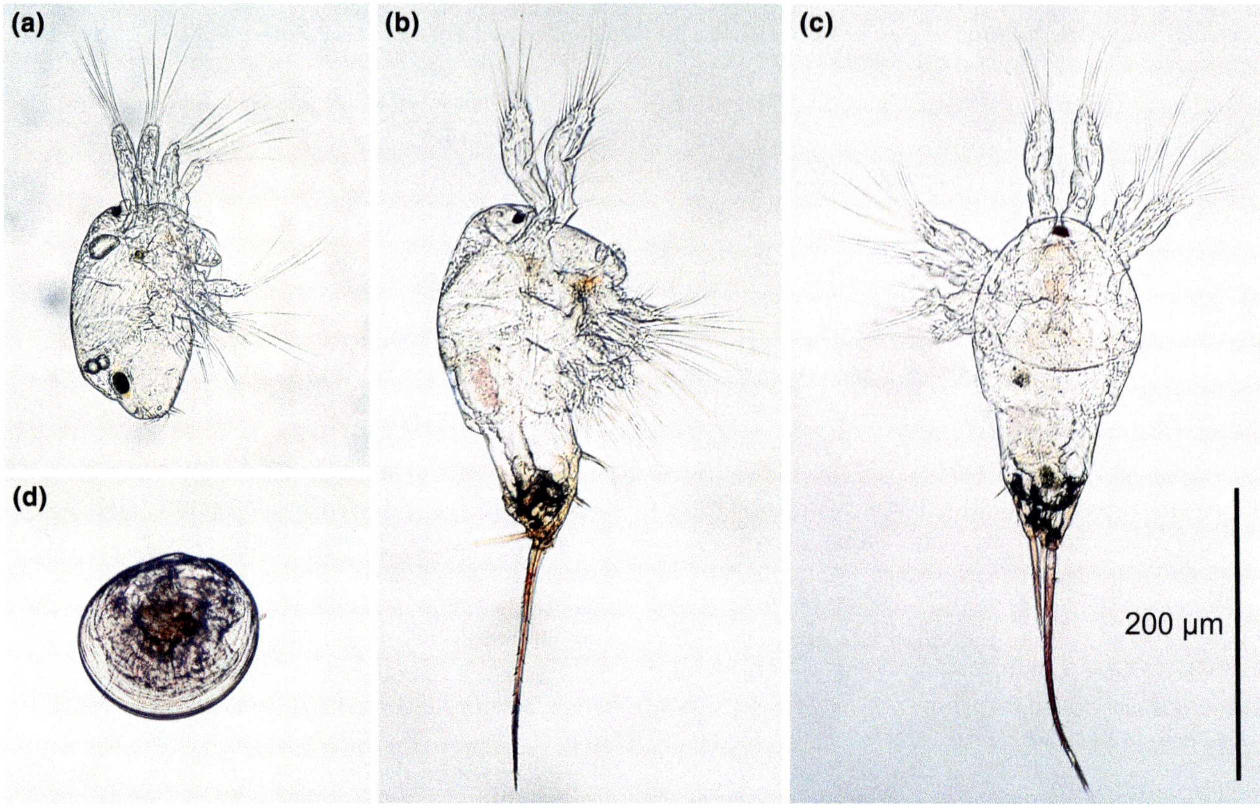

Fig. 8.17 Light micrographs of some larval stages in the zooplankton of the Baltic Sea (Hanöbukten, Baltic Sea proper, November). (**a**) Lateral views of a copepod nauplius, which may be difficult to identify with regard to species or even genus. (**b**, **c**) Lateral and dorsal views of a nauplius of the copepod genus *Temora*. (**d**) Larva of a bivalve mollusc. The scale for all images is given in (**c**). Photo: © Pauline Snoeijs-Leijonmalm

especially *Synchaeta baltica*, may dominate the mesozooplankton communities and contribute more than 60 % of the mesozooplankton biomass in the open northern Baltic Sea proper (Johansson et al. 2004).

Rotifers seem to be underrepresented in the Baltic Sea zooplankton studies. This is possibly due to their transparency, which makes them difficult to detect in water. However, in the northern Baltic Sea proper the biomass of *Synchaeta* spp. was found to be positively correlated with the phytoplankton biomass in autumn (Johansson 1983), although their low energy content probably makes them a less attractive food for planktivorous fish.

8.6.3 Cladocerans

The cladocerans of the open Baltic Sea consist mainly of two groups: bosminids and polyphemoids. The bosminids contain one single, endemic species, *Bosmina longispina maritima* (syn. *Eubosmina coregoni maritima*, Fig. 8.18d). This is a relatively small mesozooplankton species, with an average body volume of 0.016 mm^3, compared to 0.040–0.071 mm^3 for different copepod species (Snoeijs and Häubner 2014). In summer, *Bosmina longispina maritima* may occur in abundances as high as several hundred individuals per litre. In some years, masses of *Bosmina longispina maritima* may form peculiar surface "blooms" with millions of dying individuals stuck to the water surface. The reason for this die-off is not known, but the aggregation of the moribund individuals at the water surface may be partly explained by the hydrophobicity of the bosminid carapace, whereby they adhere tightly to the surface-water film. Bosminids feed by filtering out food from the feeding current that they create. Their filter feeding is selective, since they select and reject some particles they catch, apparently guided by mechanical and biochemical cues.

The polyphemoids received their name from Polyphemus, the famous Cyclops of the Greek mythology, whose only eye was speared by Odysseus. Polyphemoids have a conspicuous ball-shaped compound eye, which is probably useful in visual orientation. The Baltic Sea polyphemoids are dominated by four species: the abundant, small-sized *Pleopis polyphemoides*, *Podon intermedius* and *Podon leuckartii* (both very similar to *Pleopis polyphemoides*, but larger), and *Evadne nordmanni* (Fig. 8.18c). The latter species is distinguished from other three species by its cone-shaped egg

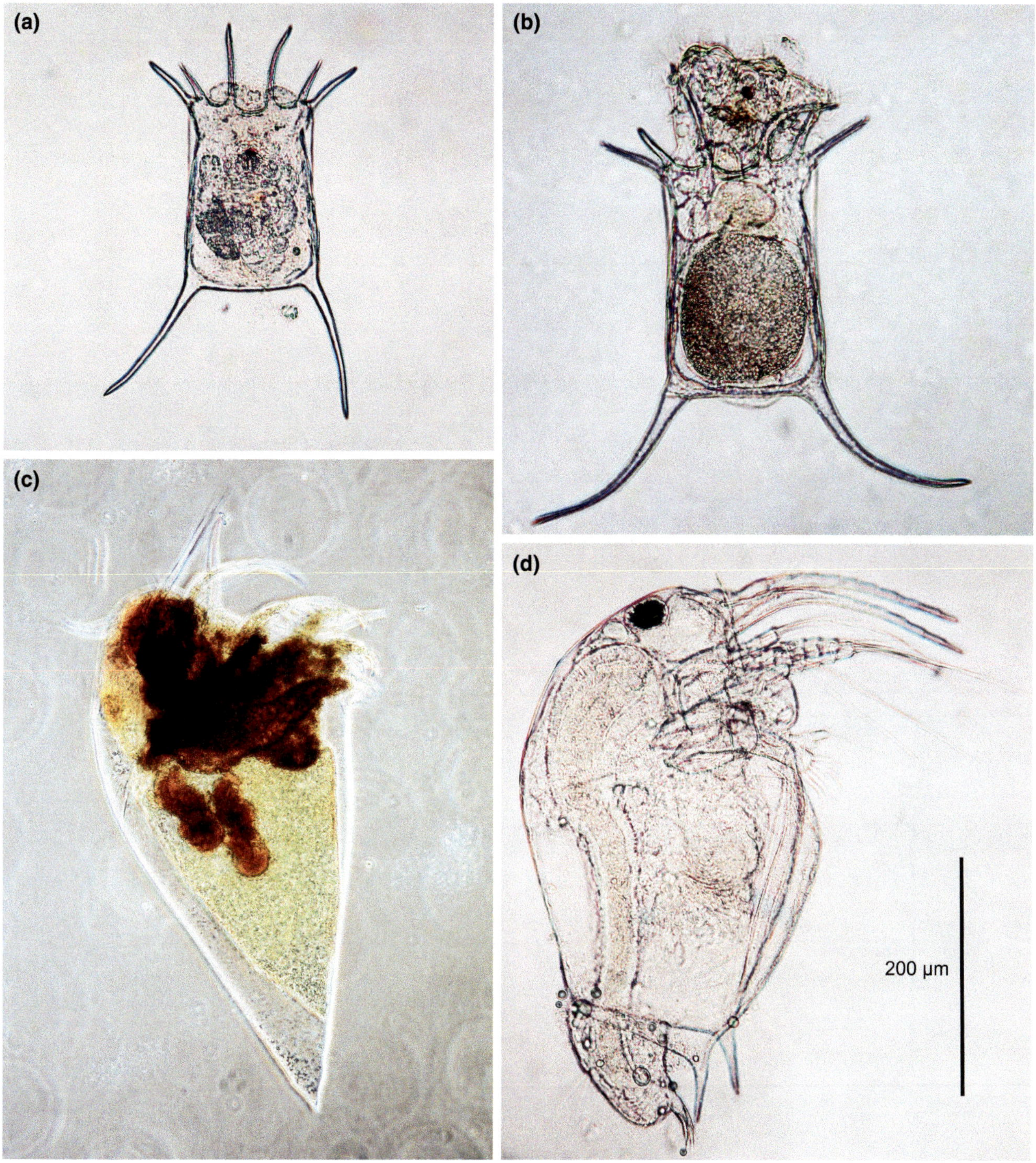

Fig. 8.18 Light micrographs of some typical zooplankton organisms of the Baltic Sea (Hanöbukten, Baltic Sea proper, November). (**a**) The rotifer *Keratella quadrata platei*. (**b**) *Keratella quadrata platei* with a clearly visible corona. (**c**) The cladoceran *Evadne nordmanni*. (**d**) The cladoceran *Bosmina longispina maritima*. The scale for all images is given in (d). Photo: © Pauline Snoeijs-Leijonmalm

sac that may contain up to ~10 parthenogenetically produced embryos. All polyphemoids feed by grabbing particles one by one, and they apparently hunt for smaller plankton. However, they are poor swimmers and are probably able to catch only relatively small and slow prey. Because of their poor escape abilities, dark-coloured eye and large egg sac, polyphemoids are a preferred food of planktivorous fish in the Baltic Sea (Flinkman et al. 1992).

During the 1990s, two new polyphemoids, *Cercopagis pengoi* (first record in 1992, *cf.* Box 5.5) and *Evadne anonyx*

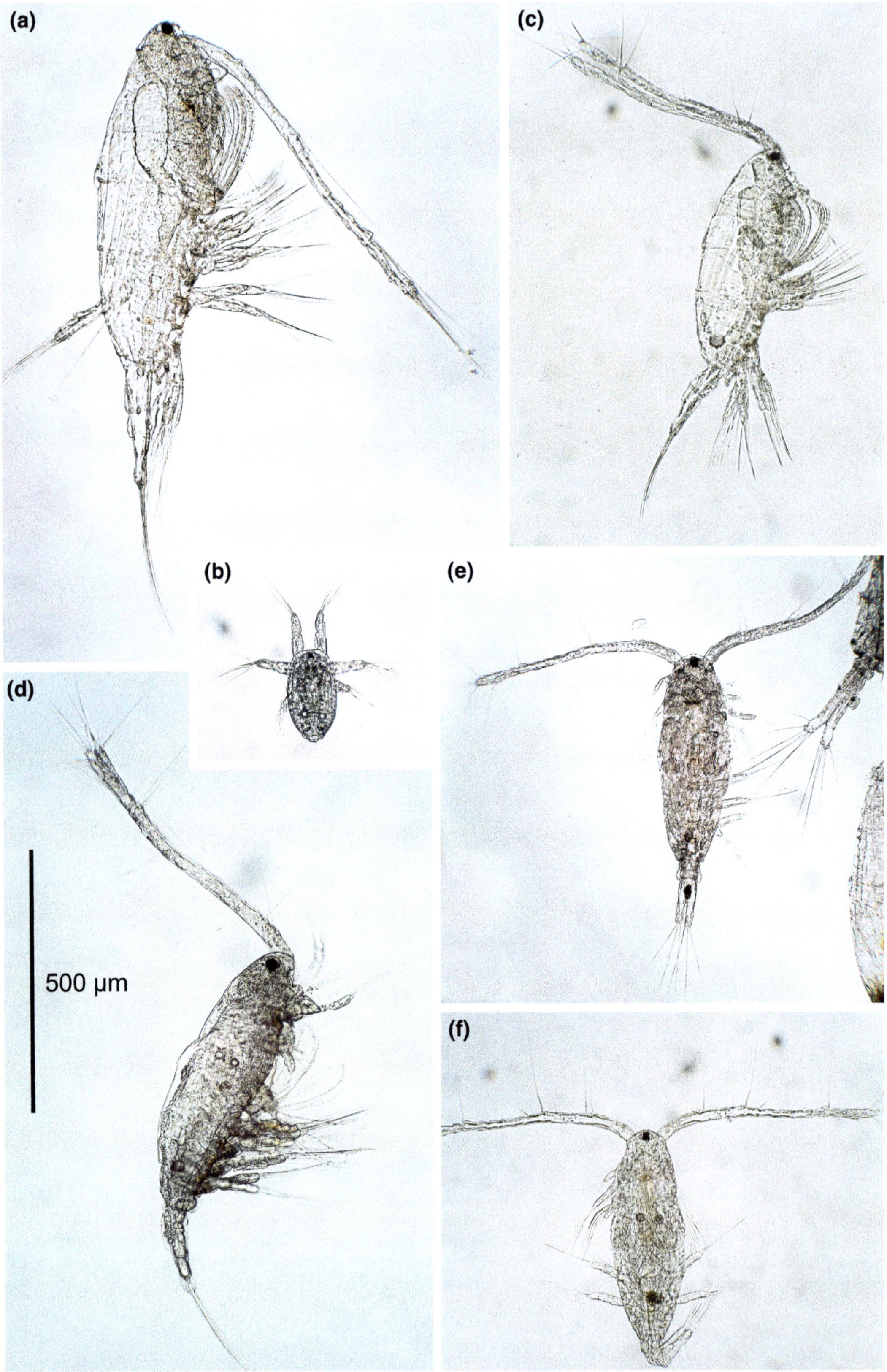

Fig. 8.19 Light micrographs of calanoid copepods of the genus *Acartia* in the pelagic zone of the Baltic Sea (Hanöbukten, Baltic Sea proper, November). (**a**) Adult female of *Acartia longiremis*. (**b**) Nauplius of *Acartia* sp. (**c–f**) Copepodite stages C3–C5 of *Acartia longiremis*. The scale for all images is given in (**d**). Photo: © Pauline Snoeijs-Leijonmalm

(first record in 1999), have spread to the Baltic Sea from the Ponto-Caspian region. *Cercopagis pengoi* is 1–2 mm long, but it has an elongated tail-spine that may be up to 15 mm in length. The spine ends with a hook at the end, which explains the organism's common name "the fish-hook water flea". Because of the hook, *Cercopagis pengoi* individuals are easily entangled with each other in zooplankton samples. In the sea, they may occur at such high densities that they clog fishing nets, and are therefore considered a nuisance species. *Cercopagis pengoi* is a voracious predator of smaller zooplankton, and is nowadays an important food item for planktivorous fish in the Baltic Sea (Gorokhova et al. 2004; Ojaveer et al. 2004).

In addition to bosminids and polyphemoids, the pelagic cladoceran fauna of the Baltic Sea also contains daphnids. Although they usually occur in low abundances, daphnids may at times – because of their large body size – significantly contribute to the mesozooplankton biomass in coastal areas (Viitasalo 1992a). It is not known why this originally freshwater taxon may be also present in the open Baltic Sea. One explanation may be that surface-water currents transport the daphnids offshore from low-salinity areas such as river mouths and estuaries.

8.6.4 Copepods

The Copepoda, meaning "oar-footed" in Latin, are the most ubiquitous group of mesozooplankton in the Baltic Sea. The most common species belong to the order Calanoida (calanoid copepods). Although occurring throughout the year, they develop the highest biomass in mid- and late summer. In the northern Baltic Sea, *Acartia* spp. (Fig. 8.19) and *Eurytemora affinis* often dominate the copepod community throughout the growing season, whereas the more marine species *Centropages hamatus*, *Pseudocalanus acuspes* and *Temora longicornis* (Fig. 8.20) are more abundant later in the season and in deeper and more saline waters (Hernroth and Ackefors 1979; Viitasalo 1992a).

Zooplankton samples also occasionally contain representatives of the copepod orders Cyclopoida and Harpacticoida. The latter are predominantly benthic species, but some of them are capable of active swimming and may enter the water column. Cyclopoids are a freshwater group, but certain predatory cyclopoids, *e.g. Oithona similis*, may be abundant at times in the southernmost Baltic Sea and the Kattegat (Maar et al. 2006). In addition, the deep cool basins of the northern Baltic Sea support *Limnocalanus macrurus*, a glacial relict. This species resembles the marine *Calanus* spp. in appearance, and with its 2–3 mm long body is by far the largest copepod in the Baltic Sea. It is assumed to feed actively on other copepod species, but may at times switch to herbivorous feeding on diatoms (Dahlgren et al. 2012). Because of its large body size and abundant lipid reserves, *Limnocalanus macrurus* is an important prey for clupeids in the Gulf of Bothnia (Sandström 1980).

Most copepods can consume a variety of food items, but their feeding is selective. They can detect moving prey by responding to its hydromechanical signals, and select prey according to its size, shape and taste. Certain copepod species, such as *Acartia tonsa*, feed on small items, *e.g.* flagellates, by creating a feeding current, but may switch to "ambush feeding" (waiting in a concealed position and then rapidly executing a surprise attack) when seeking motile prey such as ciliates (Kiørboe et al. 1996). When employing the latter feeding mode, they sink motionless through the water, sensing hydrodynamic signals created by their prey with the sensory hairs covering their antennae, then attack the prey with a sudden jump. In addition, they can sense approaching predators such as planktivorous fish and mysids by the hydrodynamic signals that these animals create, and escape at speeds that may exceed 100 times their body length per second. Compared to a 1.8 m tall human, such an escape speed would correspond to 650 km h^{-1} (Viitasalo et al. 1998).

8.6.5 Sexual and asexual reproduction in rotifers and cladocerans

During their season of peak abundance, both rotifers and cladocerans reproduce parthenogenetically. Asexual reproduction is efficient, *e.g.* cladoceran embryos may be seen developing their own embryos when they are still in their mother's egg sac, which is a phenomenon called paedogenesis that ensures an extremely rapid population increase. Sexual reproduction intensifies when living conditions start to deteriorate in late summer. The transition from asexual to sexual reproduction is triggered by temperature, food quantity and quality as well as by photoperiod (Miyashita et al. 2011). The sexually produced resting eggs sink to the bottom and overwinter there. At the seafloor the eggs may become buried in the sediment and remain dormant for a long period of time. For example, sediment-buried eggs of *Bosmina*, *Evadne* and *Podon/Pleopis* have been observed to hatch after 8–10 years of dormancy (Viitasalo and Katajisto 1994).

8.6.6 Life-cycle strategies of copepods

All copepods reproduce sexually. While a small number of copepods overwinter as plankton in the water column, most calanoid species in the Baltic Sea produce overwintering

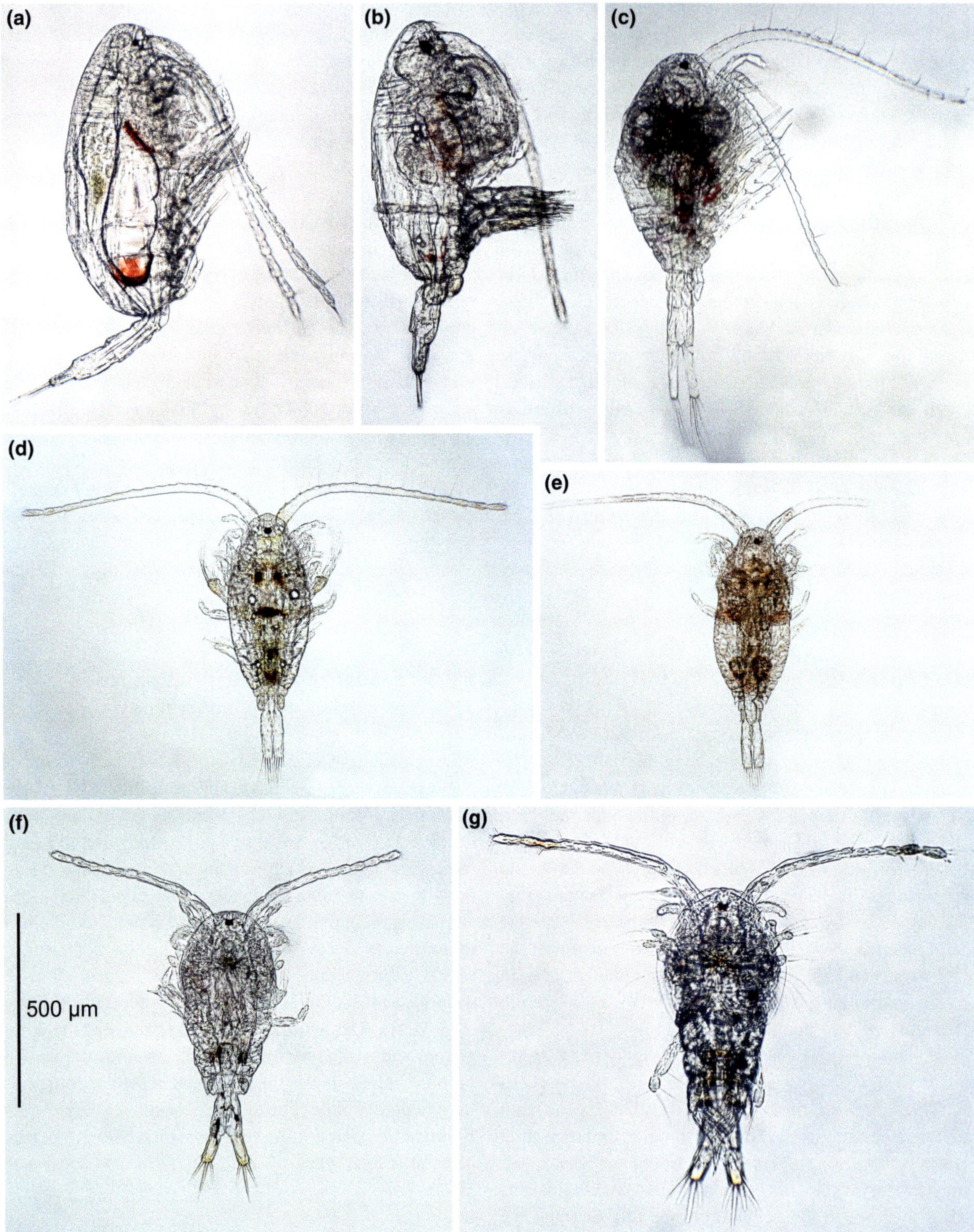

Fig. 8.20 Light micrographs of some dominant marine calanoid copepod species in the pelagic zone of the Baltic Sea (Hanöbukten, Baltic Sea proper, November). Generally, these species are more abundant later in the season and in deeper and more saline waters than *Acartia* spp. and *Eurytemora* spp. (**a**) *Pseudocalanus acuspes*, copepodite stage C4. (**b**) *Centropages hamatus*, copepodite stage C4. (**c**) *Temora longicornis*, copepodite stage C4. (**d**, **e**) *Centropages hamatus*, copepodite stage C2. (**f**) *Temora longicornis*, copepodite stage C1. (**g**) *Temora longicornis*, copepodite stage C2. The scale for all images is given in (**f**). Photo: © Pauline Snoeijs-Leijonmalm

resting eggs from which the population is replenished in the early summer (Viitasalo 1992b). Different calanoid groups show differences in the egg spawning and protection strategies which probably stem from the copepod evolutionary history. According to Marcotte (1982), calanoid copepods evolved from a near-benthic harpacticoid-like ancestor. Harpacticoid copepod females carry their eggs in egg sacs which protect the eggs from being eaten by small invertebrate predators. Large egg sacs, however, draw the attention of visual predators, and spawning the eggs freely in the water has probably evolved as an adaptation to a truly pelagic existence in areas where fish predation on egg-carrying females is intense (Webb and Weaver 1988). In shallower waters, free-spawned eggs are, however, in danger of sinking to the bottom and getting buried in the sediment.

Unlike rotifers and cladocerans, most calanoids do not produce true diapausing eggs that require a cooling period before development can be resumed. However, in the calanoid superfamily Centropagoidea, the members of which dwell in near-shore areas, resistance to adverse environmental conditions has probably evolved as an adaptation to coastal conditions where eggs are likely to sink to the bottom. If buried under a deep layer of sediment, calanoid eggs can remain viable for up to 19 years (Katajisto 1996), but hatch soon after the external conditions become suitable. The egg viability and hatching success depend on the depth of burial, on processes affecting the location of the eggs in the sediment (bioturbation, bottom currents) and on oxygen conditions in the sediment and in the water (Katajisto 2004). Mesozooplankton eggs are also resistant to adverse conditions in environments other than sediments, *e.g.* cladoceran and copepod eggs have been observed to survive the passage through the guts of planktivorous fish (Flinkman et al. 1994).

8.6.7 Hydrography sets geographical limits

The main factors that affect the geographical distribution and seasonal variability of the zooplankton include temperature, salinity, food availability and predation. Annual variations in temperature and light conditions modify the growth rates, production and seasonal succession of all primary and secondary producers in the Baltic Sea. Increasing temperature favours rapid parthenogenetic reproduction of rotifers and cladocerans, and these taxa peak in early summer to mid-summer. According to the classic concept of Remane (1934), only truly estuarine species and species with a wide salinity tolerance can cope with salinity 5–7 (*cf.* Fig. 4.21). In accordance with these early observations, salinity determines the geographical ranges for zooplankton of both marine and freshwater origin in the Baltic Sea (*cf.* Table 4.5). In addition, salinity varies both vertically and seasonally in the water column. Species with a preference for higher salinity are most abundant in deeper waters and later in the growing season, when wind-induced mixing increases the salinity of the surface layer.

8.6.8 Temperature and food availability affect seasonal succession

While the timing of the different zooplankton species' peak abundances changes from year to year, the basic seasonal successional pattern of the main mesozooplankton groups remains fairly stable (Fig. 8.21). It starts with peaks of rotifers in spring and early summer, followed by increased abundances of small cladocerans (*e.g. Bosmina*) and copepods (*e.g. Acartia*) around mid-summer, while marine copepods (*e.g. Pseudocalanus* and *Temora*) peak in late summer and autumn (Hernroth and Ackefors 1979; Viitasalo 1992a).

Small numbers of adult and subadult copepods survive through the winter. Certain species, *e.g. Pseudocalanus acuspes* and *Limnocalanus macrurus*, can survive on their lipid reserves, whereas species that store less fat, such as *Acartia* spp., probably rely on ice algae as a food source. These adults thus give rise to the population increases during the following spring. In the northern Baltic Sea, nauplii of *Acartia* spp. (Fig. 8.19b) appear in the plankton immediately after the melting of the sea ice.

Rotifers constitute the most abundant zooplankton group in the transition period from spring to summer. Their average abundances in the water column of the northern Baltic Sea may, at times, exceed 50 individuals per litre (Viitasalo 1992a). In summer, the mesozooplankton communities can be subdivided into a surface-dwelling part, which consists of rotifers, cladocerans and copepod nauplii, and a migrating part consisting of the older copepodite stages of copepods and cladocerans. The latter group, particularly *Bosmina longispina maritima*, can be very abundant in late summer, with water column averages exceeding 100 individuals per litre. In autumn, the abundances of rotifers and cladocerans decline, and the mesozooplankton communities consist mainly of calanoid copepods, both in the northern (Viitasalo 1992a; Ojaveer et al. 1998) and southern Baltic Sea (Möllmann et al. 2000; Aleksandrov et al. 2009).

The importance of food for the distribution and abundance of zooplankton is difficult to distinguish from the effects of other seasonally varying factors. However, the positive correlations between long-term changes in the body size of certain copepods and temperature suggest food abundance as a factor affecting the zooplankton dynamics (Viitasalo et al. 1995, 2015).

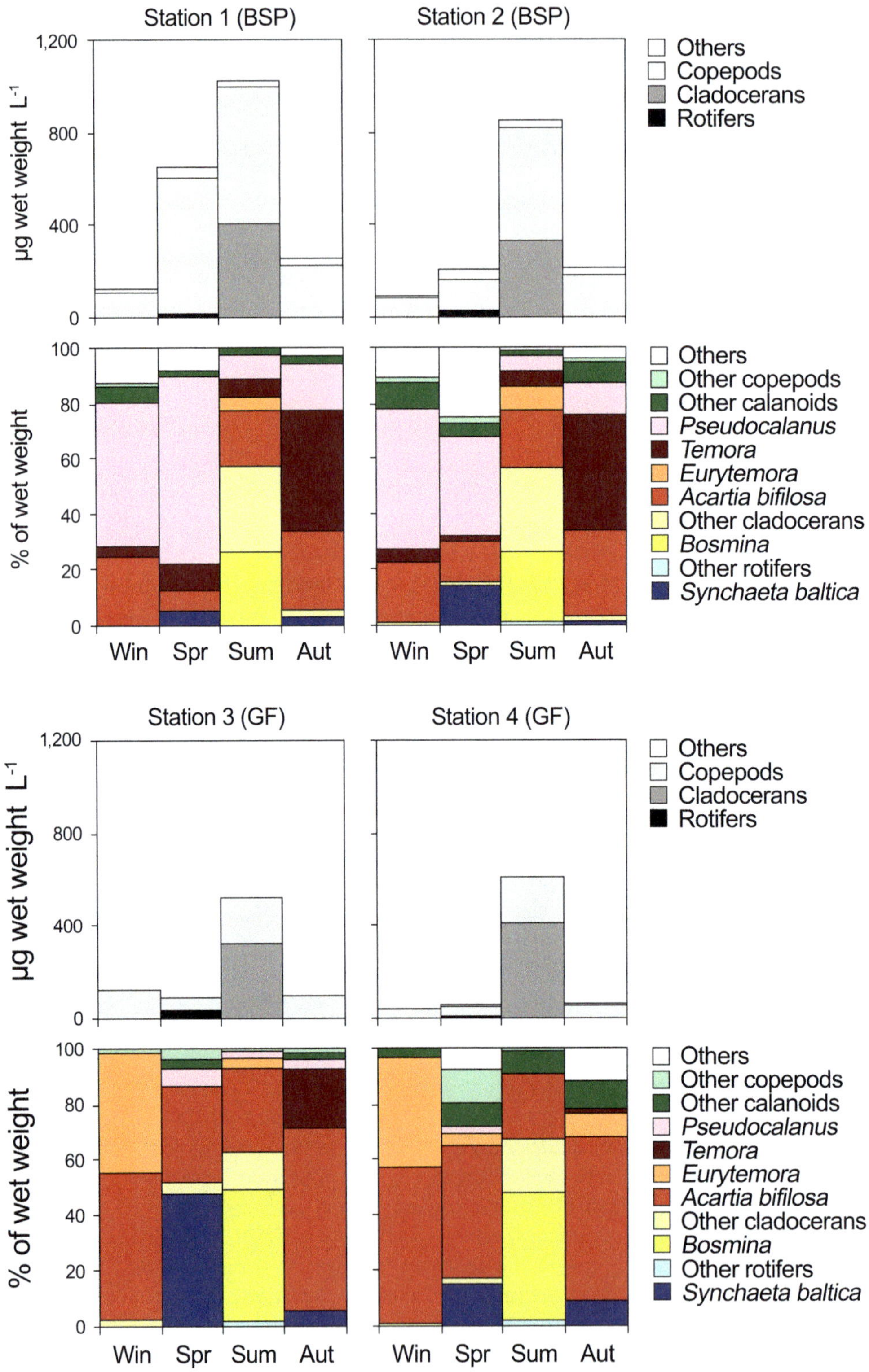

Fig. 8.21 The average seasonal development of zooplankton biomass and community composition for the years 1986–1990 at four offshore stations in winter (January-April), spring (May-June), summer (July-September), and autumn (October-December). The samples were taken with a 100 µm WP-2 closing net from 25 m of water depth to the surface. The sampling stations follow a gradient: Station 1 is in the northern Baltic Sea proper (BSP), Station 2 is in the northern Baltic Sea proper near the entrance to the Gulf of Finland (GF), Station 3 is in the Gulf of Finland, and Station 4 is in the Gulf of Finland east of Station 3. Figure modified from Viitasalo (1992a)

8.7 Macrozooplankton

8.7.1 Mysids

Mysids are common pelagic and benthic organisms in the Baltic Sea, both in the sublittoral zone and the open sea. The open sea species, *Mysis mixta* and *Mysis relicta* are ~1 cm long crustaceans that remain close to the seafloor during the day and ascend the water column during the night. They are omnivorous, feeding on detritus along with living and dead particles on and in the sediment during the day and on the phytoplankton and zooplankton during the night (Viherluoto et al. 2000). Like copepods, they can create a feeding current and switch to grabbing more evasive prey when necessary (Viitasalo and Rautio 1998). Mysids constitute important food items for clupeids, particularly for larger individuals of the Atlantic herring *Clupea harengus* (Aneer 1980).

8.7.2 Jellyfish

The jellyfish are the largest truly planktonic species in the Baltic Sea Area: the lion's mane jellyfish *Cyanea capillata* (*cf.* Fig. 4.19a) can grow to a diameter of 60 cm. *Cyanea capillata* occurs in the Kattegat and the Belt Sea, while the common jellyfish *Aurelia aurita* (Fig. 8.22) is distributed almost throughout the entire Baltic Sea. While *Aurelia aurita* individuals can grow to a diameter of 30 cm and may weigh 1 kg in the southern Baltic Sea, the largest individuals found in the north are usually <20 cm in diameter.

In the southern Baltic Sea, the peak jellyfish abundance occurs in August, somewhat earlier than in the north. The life cycle of *Aurelia aurita* involves pelagic sexual and benthic asexual reproduction. The sexually produced eggs develop into planula larvae that attach to hard substrates such as rocks and boulders, and develop into the polyp stage. The polyp reproduces asexually, sometimes growing an entire colony of polyps. Mature polyps start "budding" planktonic ephyrae that gradually grow into adult jellyfish, which again reproduce sexually.

Fig. 8.22 Mass occurrence of the common jellyfish *Aurelia aurita* in the Baltic Sea proper. Photo: © Hans Kautsky

Aurelia aurita occurs often in a relatively thin layer close to the surface, and can accumulate in masses in semi-enclosed bays. The summer water column median abundances in the Kiel Bay (southwestern Belt Sea) have been estimated to range from 0.2 to 16 individuals per 100 m^3. Schneider and Behrends (1994) calculated that *Aurelia aurita* may consume two-thirds of the daily zooplankton production during their "blooms", while Barz and Hirche (2005) found no significant effect on the mesozooplankton. The prey species of *Aurelia aurita* vary according to the degree to which they are affected by jellyfish predation (Behrends and Schneider 1995), possibly because the more evasive copepods can avoid predation, while slow swimmers, such as bivalve larvae, cannot.

8.7.3 Ctenophores

While the ctenophore *Pleurobrachia pileus* (sea gooseberry) is thought to be common in the Baltic Sea Area, the American comb jelly *Mnemiopsis leidyi* made its first appearance here in autumn 2006 (*cf.* Box 5.10). The introduction of *Mnemiopsis leidyi* may have consequences for the ecosystem in the southern part of the Baltic Sea. This species is capable of appearing in masses on account of its extremely efficient reproduction: it is hermaphroditic (*i.e.* capable of self-fertilisation), produces several thousand eggs per day, can reproduce by division, and can rapidly grow to a length of 10 cm at favourable temperatures and food conditions. It is a voracious predator of zooplankton, fish eggs and even fish larvae.

As early as in 2007, small (1–2 mm) ctenophores found in the northern Baltic Sea were identified as *Mnemiopsis leidyi*. Subsequent molecular analyses demonstrated, however, that those individuals belonged to an entirely different species, the Arctic comb jelly *Mertensia ovum* (Gorokhova et al. 2009). This finding raised doubts as to earlier ctenophore species records in the northern Baltic Sea in general. It is possible that at least some *Pleurobrachia pileus* have been misidentified, and *Mertensia ovum* has been present in the Baltic Sea since much earlier than previously thought. Thus far, all the specimens found in the northern Baltic Sea have been positively identified as *Mertensia ovum*.

8.8 Pelagic vertebrates

8.8.1 Fish

The pelagic fish community of the Baltic Sea is extremely species-poor. It consists mainly of only three marine fish species: the Atlantic herring *Clupea harengus*, the European sprat *Sprattus sprattus* and the Atlantic cod *Gadus morhua* (Fig. 8.23). The cod is the only abundant pelagic piscivorous fish in the Baltic Sea and much of its diet consists of herring and sprat. The Atlantic salmon *Salmo salar* occurs in the pelagic zone of the Baltic Sea as well, but wild stocks are rare and its biomass is much lower than that of cod. During the last 40 years the dynamics of the fish stocks of the Baltic Sea have, besides natural variation through *e.g.* climatic fluctuations, been strongly affected by high recruitment, eutrophication and fishing (*cf.* Fig. 18.13).

In the Baltic Sea, the Atlantic herring *Clupea harengus* is smaller and less fatty than its Atlantic conspecific. It seems to be physiologically well adapted to the low salinities of the Baltic Sea and thrives in low-salinity areas such as the Bothnian Sea. Previously, the Baltic herring has been considered a subspecies on its own, but genetic differentiation between the Baltic and Atlantic populations is small (*cf.* Sect. 6.3.4).

The European sprat *Sprattus sprattus*, which is slightly smaller than the herring, is relatively more marine, and its distribution does not extend as far north and east as that of the herring. There are also differences in the feeding habits of these two clupeids. While sprat is considered to be an obligatory planktivore and mainly feeds on the mesozooplankton, adult herring also utilises other food items such as benthic animals (Ojaveer et al. 1981). The largest herring individuals may also be piscivorous, as shoals of very large and fast-growing "giant" herring have been observed to feed on smaller fish in the Baltic Sea.

The three-spined stickleback *Gasterosteus aculeatus*, an abundant fish that usually lives amongst the macrophytes in the sublittoral zone, may also form large pelagic shoals in the open sea near the edge of the coastal zone. In the open sea, these stickleback shoals can be important predators of smaller zooplankton such as *Bosmina longispina maritima*.

8.8.2 Birds

A few birds feed mainly on fish in the pelagic zone of the Baltic Sea. The best-known species are the common guillemot *Uria aalge* (*cf.* Fig. 16.2) and the razorbill *Alca torda*, which feed predominantly on the clupeids sprat and herring, both when they bring food to their young in summer and when they stay in the open sea far from land in winter. These two auk species are excellent swimmers and dive regularly down to 25 m or more, even down to 100 m, when catching schooling fish.

During autumn and spring migration, as well as in mid-winter, both the black-throated diver *Gavia arctica* (Fig. 8.24d) and the red-throated diver *Gavia stellata* (Fig. 8.24c) are regularly recorded as fishing in the pelagic zone of the Baltic Sea. In winter, both divers feed entirely on fish. Also, other species of waterbirds, such as grebes, red-breasted mergansers *Mergus serrator* (Fig. 8.24a) and great cormorants *Phalacrocorax carbo sinensis* (Fig. 8.24b) are observed to fish in the pelagic zone during winter, although not exclusively so.

Fig. 8.23 The three dominant fish species in the pelagic zone of the Baltic Sea. (**a**) The Atlantic herring *Clupea harengus* and the European sprat *Sprattus sprattus*. (**b**) The Atlantic cod *Gadus morhua*. Photo: © Pauline Snoeijs-Leijonmalm

Fig. 8.24 Examples of bird species that can be observed as fishing in the pelagic zone of the Baltic Sea in winter. (**a**) The red-breasted merganser *Mergus serrator*. (**b**) The great cormorant *Phalacrocorax carbo sinensis*. (**c**) The red-throated diver *Gavia stellata*. (**d**) The black-throated diver *Gavia arctica*. Photo: (**a**, **c**, **d**) © Bo Tallmark, (**b**) © Stanislaw Węsławski

8.8.3 Mammals

The four marine mammals of the Baltic Sea: the harbour seal *Phoca vitulina*, the grey seal *Halichoerus grypus*, the ringed seal *Pusa hispida* (*cf.* Box 4.13) and the harbour porpoise *Phocoena phocoena* (*cf.* Fig. 4.15) are, together with birds, the natural fish-consuming top predators in the pelagic food web. Stomach content analyses have shown that the grey seal mainly consumes Atlantic herring *Clupea harengus*, followed by European sprat *Sprattus sprattus*, in the Baltic Sea proper and by common whitefish *Coregonus lavaretus* in the Gulf of Bothnia (Lundström et al. 2010). However, the remains of altogether 16 fish taxa were detected in grey seal stomachs, including Atlantic cod *Gadus morhua* and Atlantic salmon *Salmo salar*. Similarly, the other three mammals feed on the available pelagic fish in the different subregions of the Baltic Sea, thus mainly on herring, sprat and cod.

In the early 20th century, the four pelagic mammals were the dominant top predators, and they consumed a large part of the fish production of the Baltic Sea. However, seal populations declined by ~95 % during the last century (Harding and Härkönen 1999), first due to hunting (1900–1940) and later due to toxic compounds (1965–1975). The harbour porpoise was also radically reduced by *e.g.* hunting and incidental catches in gill nets (MacKenzie et al. 2002). The virtual elimination of marine mammals caused a shift from seal to cod domination and can be seen as the first human-induced regime shift in the Baltic Sea (Österblom et al. 2007). Humans have, through predation, pollution and competition, replaced the seals as the major top consumers of the Baltic Sea food web (Folke et al. 1991).

8.9 Trophic interactions in the pelagic food web

8.9.1 Zooplankton do not feed only on phytoplankton

Mesozooplankton occupies a central functional position within the pelagic food web of the Baltic Sea, where the

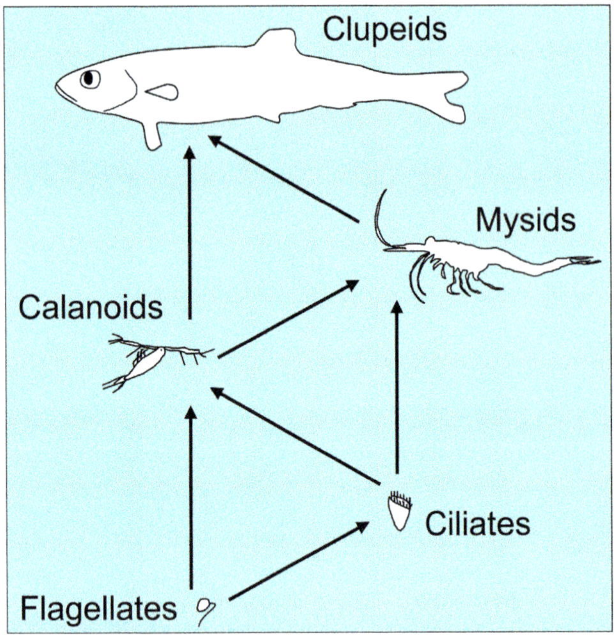

Fig. 8.25 Intraguild predatory interactions in the pelagic zone of the Baltic Sea. Figure: © Markku Viitasalo

energy-flow pathways from the autotrophic phytoplankton and heterotrophic producers of the microbial loop cross (Fig. 8.25). However, in the northern Baltic Sea the calanoid copepods consume less than 5 % of the net primary production, due to the seasonal "mismatch" with the spring phytoplankton bloom (Lignell et al. 1993). Thus, the traditional classical pelagic food web (diatom-copepod-fish) concept (Fig. 8.1) is not fully appropriate in the Baltic Sea, at least in its northern parts. Instead, copepods and other mesozooplankton groups serve as an important channel of energy derived from dissolved organic carbon. The mesozooplankton channels this energy from the microbial loop (microbes, flagellates and protists) to fish.

Copepods function as consumers at several trophic levels, as they eat both phytoplankton/flagellates and their predators (ciliates). This kind of "intraguild" relationship, and opportunistic switching between the most abundant prey, means that copepods provide a stabilising function in the food web (Gismervik and Andersen 1997). In fact, intraguild interactions are typical of the low-diversity food web of the Baltic Sea: similar "triangles" exist between mysids, copepods and ciliates, and between clupeids, mysids and copepods (Fig. 8.25).

8.9.2 Zooplankton contribute to nutrient recycling

Mesozooplankton contributes to the recycling of nutrients and organic matter in the Baltic Sea. Stoichiometric analyses have shown that the Baltic mesozooplankton need more phosphorus than normally available in their food. This makes them a phosphorus sink, just like the zooplankton in freshwater systems (Pertola et al. 2002).

The rate at which different mesozooplankton groups recycle their food differs: copepods encapsulate their egesta in small pellets that tend to sink to the bottom, while cladocerans are said to have constant diarrhoea, thus releasing their stomach contents and associated nutrients rapidly into the photic zone. On the other hand, the few detailed investigations into the sinking rates of copepod faecal pellets showed that most of the pellets seem to be broken down and recycled rapidly, and only a small fraction of all the pellets produced actually reaches the sediment (Viitasalo et al. 1999).

8.9.3 Interactions with cyanobacteria

The toxins that can be produced by the massive cyanobacterial blooms that occur in the open Baltic Sea (Fig. 8.5) may be expected to threaten the pelagic food web. Nodularin, the major toxin produced by *Nodularia spumigena*, is a hepatotoxin acutely harmful to vertebrates (*cf.* Box 16.4). Consequences of cyanobacterial blooms for the Baltic Sea zooplankton have been thoroughly investigated. The results show that the filamentous cyanobacteria are low-quality food for copepods and do not cause a drastic decline in the zooplankton abundance. In fact, it appears that the copepod growth may also remain high during a heavy cyanobacterial bloom, apparently because the selectively feeding copepods find ample food among the microbial loop organisms (flagellates and ciliates), which thrive in the decaying bloom (Koski et al. 2002).

8.9.4 Predator avoidance

The "predatory cycle" includes search, detection, pursuit, attack, capture (or escape), handling and feeding (or rejection). A prey organism can avoid predation in various ways at every stage of its life cycle. Transparent prey can go unnoticed, small prey can remain unattacked, agile swimmers can avoid being caught, and spiny or unappetising prey can be rejected. Naturally, the performance of the predator also influences the outcome of the hunt: fast, perceptive, agile and hydrodynamically inconspicuous predators catch prey more often than slow and clumsy ones do (Viitasalo et al. 2001).

Many morphological adaptations in the plankton are, at least partly, predation defences. The spiny armour of the rotifer *Keratella* (Fig. 8.18a, b) most likely reduces the probability of it being caught by copepods, and the enormously elongated tail spine of *Cercopagis pengoi* makes the capture and handling of this species by smaller planktivorous fish difficult. Moreover, the zooplankton escape ability varies from species to species. The smaller cladocerans such as *Bosmina* and *Pleopis* are relatively slow swimmers with almost no escape ability if hunted by larger planktivores. As for copepods, *Acartia* and *Eurytemora* have been shown to be more agile in escaping suction from a pipette (simulated predation) than the more sluggish *Pseudocalanus* and *Temora* (Viitasalo et al. 2001). The sluggishness and the visual conspicuousness of the latter two species make them vulnerable to predation by planktivores.

Many Baltic Sea zooplankton taxa perform a distinct diurnal vertical migration (Burris 1980). They ascend to feed at night, and descend to deeper layers during daylight hours. While several (*e.g.* energetic) benefits have been evoked to explain this behaviour, the most common explanation refers to avoidance of visual predation by planktivorous fish. This has been demonstrated in the Baltic Sea for *Eurytemora affinis* by comparing the vertical migration activity of the visually most conspicuous egg-carrying females and smaller individuals, and by field and experimental studies where both herring and three-spined sticklebacks were found to prefer ovigerous females over non-ovigerous ones (Vuorinen et al. 1983; Flinkman et al. 1992). In addition to copepods, distinct diurnal migration is also typical of mysids, which also ascend to the surface waters at night.

8.9.5 Interactions between copepods, clupeids and cod

Mesozooplankton populations can be affected by planktivorous fish. In autumn, the quantity of copepods eaten by clupeids and mysids exceeds the copepod production, and thus predation is probably the main reason for the autumnal population decline of copepods in the northern Baltic Sea (Rudstam et al. 1992). As the clupeid populations are affected by the foraging intensity of their main predator, cod, environmental drivers affecting cod may also influence zooplankton due to food web effects "cascading down the food web".

The cod population is highly sensitive to variations in the basic hydrography of the Baltic Sea, particularly the salinity stratification and oxygen conditions. During the past 20 years, the Baltic cod stocks have declined because of overfishing, anoxia and decreased salinity. Fewer cod has caused an increase in sprat populations, which compete with herring for food. Being a marine species, the cod is adapted to marine waters where, after spawning, cod eggs float freely in the water. The Baltic Sea salinity is much lower and eggs tend to sink to a depth at which they have neutral buoyancy at salinity ~ 11 (Westin and Nissling 1991). Thus, the conditions for cod reproduction are generally more favourable in the Bornholm deep than in the eastern spawning areas (Gdańsk deep, Gotland deep), as these latter areas are less frequently influenced by water influx from the Kattegat (Wieland et al. 1994, 2000).

During the period of declining Baltic Sea salinity since 1977, interrupted only by the MBIs in 1993, 2003 and 2014 (*cf.* Fig. 2.13a), the cod reproduction failed in the northern part of the sea and resulted in a population crash. This led to an increase in the sprat population. Meanwhile, herring populations have not increased as much, and between 1982 and 1992, the weight of herring at any given age decreased by 50 % (Flinkman et al. 1998). This was probably due to both "bottom-up" and "top-down" effects. The herring food supply was probably reduced as a result of changes in the zooplankton community composition: as salinity declined, abundances of marine copepods preferred by herring were reduced as well, and the herring selectively feeds on larger mesozooplankton (Fig. 8.26). This diminished the energy supply to the herring, whereas that of sprat increased, since sprat feed more on other zooplankton taxa, which had not decreased in abundance (Rönkkönen et al. 2003). Sprat is assumed to be a stronger competitor for planktonic food than herring, and therefore the resource competition with sprat worsened the situation for the herring (Möllmann et al. 2005, Casini et al. 2006).

8.10 Pelagic-benthic coupling

The pelagic primary producers generate a large amount of organic matter that is not fully consumed within the pelagic zone. The unused part sinks in the water column to eventually reach the seafloor. This sedimented material is the main source of energy for benthic organisms at larger depths, such as polychaetes, bivalves and benthic crustaceans (*cf.* Sect. 10.9.1). Most of this vertical transport of material takes place at the end and immediately after the phytoplankton spring bloom. In contrast, during the summer a larger part of the organic matter is produced by small-cell organisms, such as flagellates, and is recycled within the surface layer (Heiskanen 1998).

The sedimentation rate (sinking rate) of particles depends on phytoplankton composition (*cf.* Fig. 3.12) as well as on the stability of the water column (whether it is stratified or mixed). Once the sedimenting organic material reaches the seafloor, it becomes involved in benthic biogeochemical processes, which decide whether the sediment acts as a sink or a source of nutrients to the overlying pelagic habitat and its community. After the organic matter has reached the bottom, benthic organisms consume, digest and excrete the organic matter, and thus release nutrients and other chemical compounds back into the pore water (Karlson et al. 2007). In addition, burrowing animals aerate the sediment, thus affecting chemical processes within the sediment

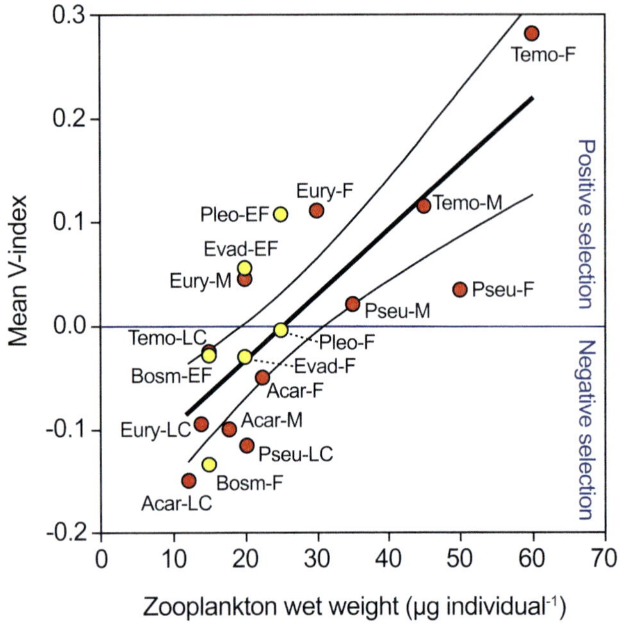

Fig. 8.26 A prey selection index (V-index) for the Atlantic herring *Clupea harengus* related to the body weight of the major mesozooplankton organisms in the northern Baltic Sea proper and western Gulf of Finland. The stomach contents of herrings (total body length 15–17 cm, age 2–6 years) were compared with zooplankton samples taken in concert with fish sampling using a 100 µm WP-2 closing net from 25 m of water depth to the surface. The V-index (Pearre 1982) is calculated using the average abundance percentages for each mesozooplankton size group in herring stomachs and in the respective plankton samples. The V-index is a useful measure of prey selection because, in contrast to many other indices, it is zero for no selection and is statistically testable. Values above the horizontal blue line indicate positive selection, and values below this line indicate negative selection. The thick black line is a significant linear regression line and thin black lines are the 95 % confidence limits of the regression analysis. Red dots represent copepods: Acar = *Acartia* sp., Eury = *Eurytemora affinis*, Pseu = *Pseudocalanus acuspes*, Temo = *Temora longicornis*. Yellow dots represent cladocerans: Bosm = *Bosmina longispina maritima*, Evad = *Evadne nordmanni*, Pleo = *Pleopis polyphemoides*. LC = large copepodite, EF = egg-carrying female, F = non-egg-carrying adult female, M = adult male. Figure modified from Viitasalo et al. (2001)

(*cf.* Sect. 10.10). Under oxic conditions, phosphorus is bound to iron compounds, which prevents its release into the overlying water. A healthy community of benthic animals thus affects the pelagic ecosystem by reducing the "vicious circle" of internal nutrient loading (Vahtera et al. 2007).

Another type of pelagic-benthic coupling involves sedimentation, burial and hatching of zooplankton resting eggs and phytoplankton cysts (Fig. 8.27). They form a propagule reserve for the future generations and may either remain buried or, under suitable conditions, hatch and initiate a new generation (*cf.* Sects. 8.6.5 and 8.6.6).

Dinoflagellate cysts also tend to sink to the bottom and may germinate later. This process, it has been suggested, is favoured by climate warming. For example, the dinoflagellate *Biecheleria baltica* has expanded its range in recent decades, which has been attributed to the species' efficient benthic cyst production (Olli and Trunov 2010). Given the rise in deep-water temperatures in the Baltic Sea proper, the germination of *Biecheleria baltica* benthic cysts may have been enhanced and contributed to the success of this species in the Baltic Sea (Kremp et al. 2008).

8.11 Climate-change effects on the pelagic food web

8.11.1 Climate change scenario for the Baltic Sea

It is predicted that global warming will raise the average surface (upper 100 m) temperature of the ocean by up to ~2 °C during the 21st century (Stocker et al. 2013). For the northern Baltic Sea, the predicted temperature increase is larger than for the southern Baltic Sea because of the decease in albedo (the fraction of solar energy reflected from the Earth back to space) when the ice season shortens. According to regional climate projections, the air temperature over the Baltic Sea will increase by 1–6 °C during the 21st century, the largest changes (up to 6 °C) being predicted for the winter and the northernmost parts (Christensen et al. 2015). Consequently, the annual average of the sea surface

Fig. 8.27 The silicified resting stages (stomatocysts) of chrysophytes (heterokont algae with a long "flimmer" flagellum and a short "whiplash" flagellum) possess a plugged pore. The scale and bristle morphology is species-specific. Like the silica frustules of diatoms, stomatocysts are often preserved in the sediment, and can be used as palaeo-environmental indicators. (**a**) Scanning election micrograph of seven cysts with the pore visible in the middle one. (**b**) Light micrograph of one stomatocyst. (**c**) Scanning electron micrograph of one stomatocyst. Photo: (**a**, **c**) © Regina Hansen and Rainer Bahlo, (**b**) © Regina Hansen

temperature will rise by 2–3 °C. However, the seasonal and spatial variations are large. For example, according to one model's prediction, the sea surface temperature of the Bothnian Bay will increase by 4.4 °C (Meier 2015).

In addition, salinity is expected to decrease because of increasing rainfall and thus a higher freshwater runoff entering the sea. The surface layer salinity of the Baltic Sea proper is predicted to decrease from 7 to 5 depending on changes in the precipitation-evaporation balance (Meier 2015).

Changes in water temperature and salinity will affect the structure, diversity and succession of the pelagic food web as well as the overall carbon fluxes.

8.11.2 Changes at the base of the food web

Increased temperature will probably produce different effects on the timing of the spring bloom and on heterotrophic consumers, which might result in a mismatch scenario and disturbed energy flows in the system (Cushing 1990). Climate-induced reduction of ice cover may lead to an earlier onset of the spring phytoplankton bloom by up to one month in the northern Baltic Sea (Eilola et al. 2013). Experiments have shown that phytoplankton-zooplankton interactions will change due to warming (Lewandowska and Sommer 2010) and a tighter coupling between the spring phytoplankton bloom and decomposing heterotrophic bacteria might ensue (Wohlers et al. 2009). This suggests that organic matter cycling within the microbial food web will be accelerated at higher temperatures, and less organic matter will be available for higher trophic levels or for export to deeper waters.

The effects of a climate-induced increase in the river inflow might be particularly strong in the north of the Baltic Sea, depending on changes in the precipitation-evaporation balance. Here the riverine water contains very high concentrations of humic substances, which may serve as a nutrient and energy source for bacterial growth but reduce phytoplankton growth by decreasing light penetration. Under such conditions, bacteria would be better competitors for inorganic nutrients than phytoplankton. This may potentially lead to an increasing importance of bacteria as basal producers, while the proportion of phytoplankton-based production will decrease (Wikner and Andersson 2012; Lefébure et al. 2013; Andersson et al. 2015). It may be speculated that such changes would lead to reduced fish production due to the inclusion of more trophic levels into the food web.

The increased riverine runoff of nutrients and higher temperatures may intensify internal nutrient cycling (Meier et al. 2011), which may increase both primary production and oxygen consumption. This will lead to increased phosphorus release from sediments as well as reduced denitrification efficiency (Meier et al. 2012). Climate warming may therefore lead to earlier and more frequent cyanobacterial blooms, as already observed from surface accumulations of cyanobacteria (Kahru and Elmgren 2014). Diazotrophic cyanobacteria will supply the ecosystem with available nitrogen, but since they are of poor food quality for consumers, the efficiency of energy transfer to higher trophic levels may be reduced. Still, increased nitrogen fixation can be a subsidy for the marine system.

8.11.3 Changes in zooplankton communities

Any climate-induced change in the basic hydrography of the Baltic Sea will affect zooplankton communities and zooplankton-predator relationships. Temperature variations are particularly important for the species dwelling above the thermocline, such as cladocerans. Rising temperature increases the metabolic rates of organisms and thus speeds up reproduction and rapidly increases population growth rates of parthenogenetically reproducing species. Increased water temperature in the Baltic Sea may increase the growth rates of fish larvae. This was shown in *e.g.* herring larvae and may have been caused either by a direct temperature effect (changes in metabolic rate) or by the indirect effect of changes in food availability (Hakala et al. 2003).

The same basic environmental drivers affect the seasonal variability and geographic distribution of the mesozooplankton and the long-term development of their populations. Because many zooplankton species originate either from freshwater or marine environments, relatively small changes in salinity may induce large shifts in their geographical distributions. For example, during the "oceanisation" of the Baltic Sea (Segerstråle 1969) that took place between 1936 and 1954, various marine copepods, such as *Pseudocalanus* and *Temora*, spread hundreds of km northwards, whereas species preferring low salinities, particularly *Limnocalanus macrurus*, retreated. Later on (from the 1970s), a reduction in the salinity of the Gulf of Finland was accompanied by a decline in the *Pseudocalanus* populations in the northern Baltic Sea (Lumberg and Ojaveer 1991).

8.11.4 Changes at higher trophic levels

Climate change may increase fish production in some high-latitude regions, like the Baltic Sea, because of warmer water and decreased ice cover (Brander 2007). Also, the expected salinity decline in the Baltic Sea will probably result in a decrease in abundance and habitat occupied by marine species (MacKenzie et al. 2007). In contrast, freshwater species, particularly those whose growth or survival are enhanced by higher temperature, will increase. On the other hand, such changes will only become evident if other factors, such as fisheries and eutrophication, will not counteract them.

In the pelagic zone of the Baltic Sea, the cod is expected to be negatively affected by climate change while a gradual increase in temperature will have a positive impact on the productivity of sprat and herring stocks (MacKenzie and Köster 2004; MacKenzie et al. 2007; Bartolino et al. 2014). Higher temperature supports sprat recruitment directly via increased egg survival, and indirectly via improved food conditions (notably access to *Acartia* spp.). Higher temperature also promotes herring recruitment and may improve its feeding conditions (Rajasilta et al. 2014).

Reproduction of the eastern Baltic cod population will probably decline if salinities fall and temperatures rise (MacKenzie et al. 2007). The eggs of the eastern Baltic cod population survive only in water layers with oxygen concentrations >2 mL L^{-1}. Because of their specific gravity, cod eggs sink to deep water layers, which nowadays mostly are anoxic (Wieland et al. 1994). Moreover, higher water temperature will increase oxygen consumption rates in the deeper parts of the Baltic Sea where cod eggs live, thereby further reducing the size of cod spawning habitats (MacKenzie et al. 2007). In contrast, sprat eggs float at shallower depths than cod eggs (Nissling et al. 2003), and consequently their survival is less affected by deep-water anoxia. Furthermore, while cod as adults are voracious predators on sprat, sprat actually prey upon cod eggs. This also decreases cod recruitment during periods when sprat is abundant.

It has been speculated that seals can have contributed to the decline of the cod in the Baltic Sea. Modelling efforts have, however, shown that seal has a much lower impact on cod than fisheries and salinity. If fisheries were following management plans, the eastern Baltic cod population would be higher than today, even if salinity remained stable and seal predation would increase (MacKenzie et al. 2011). With similar seal and fishing levels, but with 15 % lower salinity, the Baltic Sea will still be able to support a cod population that can sustain a fishery. Hence, simultaneous recovery of seal and cod populations is possible, but the end result depends on how climate change affects cod recruitment (MacKenzie et al. 2011).

8.11.5 Effects of climate change are difficult to predict

The climate-induced changes in the hydrography of the Baltic Sea are likely to be accompanied by shifts in the geographical distribution and population sizes of species (Vuorinen et al. 2015). If the projected changes do take place, those species that can cope with lower salinity and higher temperature, such as cladocerans and certain euryhaline or limnic copepod species, will probably become more abundant than they are today. On the other hand, a temperature increase may induce unexpected population effects. Increasing temperature favours both jellyfish (Mills 2001) and non-indigenous species originating from warmer sea areas, such as the Ponto-Caspian *Cercopagis pengoi* (Leppäkoski and Olenin 2000). It is difficult to predict how such changes may influence the food web in the future. For example, studies employing stable isotopes have shown that the establishment of *Cercopagis pengoi* in the Baltic Sea added one trophic level to the food web extending from primary producers to fish (Gorokhova et al. 2005).

Hänninen et al. (2000) suggested that the bottom-up connection between climate, zooplankton and herring makes it possible to predict both zooplankton variations and herring condition from climatic variations several years ahead. However, because of confounding top-down effects (*e.g.* human influence, non-indigenous species) and other complex food web interactions, the effects of climate change on the pelagic ecosystem are, in fact, difficult to predict (Viitasalo et al. 2015). If the Baltic Sea salinity decreases and anoxia persists, cod reproduction will stay low and the pelagic ecosystem will probably remain dominated by sprat. Because humans will then be the major harvesters of clupeids in the Baltic Sea, the importance of fisheries as a regulator of the pelagic ecosystem of the Baltic Sea will increase.

Review questions

1. How do the gradients in salinity and oxygen influence the composition of the microbial communities in the Baltic Sea?
2. What are the major groups of phytoplankton in the Baltic Sea and what spatial and temporal variations do they exhibit?
3. What are the most common protistan groups in pelagic environments and what are their nutritional characteristics and feeding mechanisms?
4. How does the pelagic food web structure differ between the north and south of the Baltic Sea? Describe the geographical variations and explain the underlying mechanisms controlling the differences.
5. How might climate change influence trophic interactions between zooplankton and fish?

Discussion questions

1. What are the major similarities and differences between aquatic and terrestrial food webs?
2. What roles do prokaryotes play in the pelagic food web?
3. How does phosphorus release from sediments affect pelagic food webs from basal producers to fish?

4. What are the major anthropogenic pressures on the pelagic food web in the different basins of the Baltic Sea?
5. What measures would you take to improve the environmental conditions in the pelagic zone of the Baltic Sea? Include in your discussion the oxygen situation, eutrophication, biodiversity, inflow of chemical pollutants and the biological/ecosystem capacity to survive change.

References

Aleksandrov SV, Zhigalova NN, Zezera AS (2009) Long-term dynamics of zooplankton in the southeastern Baltic Sea. Russian Journal of Marine Biology 35:296–304

Algesten G, Brydsten L, Jonsson A, Kortelainen P, Lövgren S et al (2006) Organic carbon budget for the Gulf of Bothnia. Journal of Marine Systems 63:155–161

Anderson R, Winter C, Jürgens K (2012) Protist grazing and viral lysis as prokaryotic mortality factors at Baltic Sea oxic-anoxic interfaces. Marine Ecology Progress Series 467:1–14

Andersson A, Falk S, Samuelsson G, Hagström A (1989) Nutritional characteristics of a mixotrophic nanoflagellate, *Ochromonas* sp. Microbial Ecology 17:251–262

Andersson A, Hajdu S, Haecky P, Kuparinen J, Wikner J (1996) Succession and growth of phytoplankton in the Gulf of Bothnia (Baltic Sea). Marine Biology 126:791–801

Andersson A, Samuelsson K, Haecky P, Albertsson J (2006) Changes in the pelagic microbial food web due to artificial eutrophication. Aquatic Ecology 40:99–313

Andersson A, Jurgensone I, Rowe OF, Simonelli P, Bignert A et al (2013) Can humic water discharge counteract eutrophication in coastal waters? PLoS ONE 8(4):e61293

Andersson A, Meier HEM, Ripszam M, Rowe O, Wikner J et al (2015) Projected future climate change and Baltic Sea ecosystem management. Ambio 44(Supplement):S345–S356

Aneer G (1980) Estimates of feeding pressure on pelagic and benthic organisms by Baltic herring (*Clupea harengus* v. *membras* L.). Ophelia 1:65–275

Atamna-Ismaeel N, Sabeh G, Sharon I, Witzel KP, Labrenz M et al (2008) Widespread distribution of proteorhodopsins in freshwater and brackish ecosystems. The ISME Journal 2:656–662

Autio R (1998) Response of seasonally cold-water bacterioplankton to temperature and substrate treatments. Estuarine, Coastal and Shelf Science 46:465–474

Azam F, Fenchel T, Field JG, Gray JS, Meyer-Reil LA, Thingstad F (1983) The ecological role of water-column microbes in the sea. Marine Ecology Progress Series 10:257–263

Backer J, Biemans C, van Doorn J, Krab K, Reinders W et al (2014) Van Leeuwenhoek: groots in het kleine. Veen Media, Amsterdam, 160 pp [in Dutch]

Bartolino V, Margonski P, Lindegren M, Linderholm H, Cardinale et al (2014) Forecasting fish stock dynamics under climate change: Baltic herring (*Clupea harengus*) as a case study. Fisheries Oceanography 23:258–269

Barz K, Hirche HJ (2005) Seasonal development of scyphozoan medusae and the predatory impact of *Aurelia aurita* on the zooplankton community in the Bornholm basin (central Baltic Sea). Marine Biology 147:465–476

Behrends G, Schneider G (1995) Impact of *Aurelia aurita* medusae (Cnidaria, Scyphozoa) on the standing stock and community composition of mesozooplankton in the Kiel Bight (western Baltic Sea). Marine Ecology Progress Series 127:39–45

Berg C, Vandieken V, Thamdrup B, Jürgens K (2015) Significance of archaeal nitrification in hypoxic waters of the Baltic Sea. The ISME Journal 9:1319–1332

Berglund J, Muren U, Båmstedt U, Andersson A (2007) Efficiency of a phytoplankton-based and a bacteria-based food web in a pelagic marine system. Limnology and Oceanography 52:121–131

Bölter M, Meyer-Reil LA, Dawson R, Liebezeit G, Wolter K, Szwerinski H (1981) Structure analysis of shallow water ecosystems: interaction of microbiological, chemical and physical characteristics measured in the overlying waters of sandy beach sediments. Estuarine, Coastal and Shelf Science 13:579–589

Bralewska JM, Witek Z (1995) Heterotrophic dinoflagellates in the ecosystem of the Gulf of Gdańsk. Marine Ecology Progress Series 117:241–248

Brander KM (2007) Global fish production and climate change. Proceedings of the National Academy of Sciences of the USA 104:19709–19714

Brettar I, Rheinheimer G (1991) Denitrification in the central Baltic: evidence for H_2S-oxidation as motor of denitrification at the oxic-anoxic interface. Marine Ecology Progress Series 77: 157–169

Burns TP (1989) Lindeman's contradiction and the trophic structure of ecosystems. Ecology 70:1355–1362

Burris JE (1980) Vertical migration of zooplankton in the Gulf of Finland. American Midland Naturalist 103:316–322

Casini M, Cardinale M, Hjelm J (2006) Inter-annual variation in herring, *Clupea harengus*, and sprat, *Sprattus sprattus*, condition in the central Baltic Sea: what gives the tune? Oikos 112:638–650

Christensen OB, Kjellström E, Zorita E (2015) Projected change – atmosphere. In: BACC Author Team (ed) Second assessment of climate change for the Baltic Sea basin. Regional Climate Studies. Springer, Berlin, pp 217–233

Cole JJ, Findley S, Pace ML (1988) Bacterial production in fresh- and saltwater ecosystems: a cross-system overview. Marine Ecology Progress Series 43:1–10

Cushing DH (1990) Plankton production and year-class strength in fish populations: an update of the Match/Mismatch hypothesis. Advances in Marine Biology 26:249–293

Dahlgren K, Olsen BR, Troedsson C, Båmstedt U (2012) Seasonal variation in wax ester concentration and gut content in a Baltic Sea copepod [*Limnocalanus macrurus* (Sars 1863)]. Journal of Plankton Research 34:286–297

Eilola K, Mårtensson S, Meier HEM (2013) Modeling the impact of reduced sea ice cover in future climate on the Baltic Sea biogeochemistry. Geophysical Reseach Letters 40:1–6

Fenchel T (1987) Ecology of protozoa: the biology of free-living phagotrophic protists. Springer, Berlin 197 pp

Fenchel T, Kristensen LD, Rasmussen L (1990) Water column anoxia: vertical zonation of planktonic protozoa. Marine Ecology Progress Series 62:1–10

Flinkman J, Vuorinen I, Aro E (1992) Planktivorous Baltic herring (*Clupea harengus*) prey selectively on reproducing copepods and cladocerans. Canadian Journal of Fisheries and Aquatic Sciences 49:73–77

Flinkman J, Vuorinen I, Christiansen M (1994) Calanoid copepod eggs survive passage through fish digestive tracts. ICES Journal of Marine Science 51:127–129

Flinkman J, Aro E, Vuorinen I, Viitasalo M (1998) Changes in the northern Baltic zooplankton and herring nutrition from 1980s to 1990s: top-down and bottom-up processes at work. Marine Ecology Progress Series 165:127–136

Folke C, Hammar M, Jansson AM (1991) Life-support value of ecosystems: a case study of the Baltic Sea region. Ecological Economics 3:123–137

Fuhrman JA (2002) Community structure and function in prokaryotic marine plankton. Antonie van Leeuwenhoek 81:521–527

Funkey CP, Conley DJ, Reuss NS, Humborg C, Jilbert T, Slomp CP (2014) Hypoxia sustains Cyanobacteria blooms in the Baltic Sea. Environmental Science and Technology 48:2598–2602

Gismervik I, Andersen T (1997) Prey switching by *Acartia clausi*: experimental evidence and implications of intraguild predation assessed by a model. Marine Ecology Progress Series 157: 247–259

Glaubitz S, Lueders T, Abraham WR, Jost G, Jürgens K, Labrenz M (2009) ^{13}C-isotope analyses reveal that chemolithoautotrophic Gamma- and Epsilonproteobacteria feed a microbial food web in a pelagic redoxcline of the central Baltic Sea. Environmental Microbiology 11:326–337

Glaubitz S, Kießlich K, Meeske C, Labrenz M, Jürgens K (2013) SUP05 dominates the gammaproteobacterial sulfur oxidizer assemblages in pelagic redoxclines of the central Baltic and Black Seas. Applied and Environment Microbiology 79:2767–2776

Gocke K, Rheinheimer G (1991) A synoptic survey on bacterial numbers, biomass and activity along the middle line of the Baltic Sea. Kieler Meeresforschungen Sonderheft 8:1–7

Gorokhova E, Fagerberg T, Hansson S (2004) Predation by herring (*Clupea harengus*) and sprat (*Sprattus sprattus*) on *Cercopagis pengoi* in a western Baltic Sea bay. ICES Journal of Marine Science 61:959–965

Gorokhova E, Hansson S, Höglander H, Andersen CM (2005) Stable isotopes show food web changes after invasion by the predatory cladoceran *Cercopagis pengoi* in a Baltic Sea bay. Oecologia 143:25–259

Gorokhova E, Lehtiniemi M, Viitasalo-Frösen S, Haddock SHD (2009) Molecular evidence for the occurrence of ctenophore *Mertensia ovum* in the northern Baltic Sea and implications for the status of the *Mnemiopsis leidyi* invasion. Limnology and Oceanography 54:2025–2033

Gorokhova E, Hajdu S, Larsson U (2014) Responses of phyto- and zooplankton communities to *Prymnesium polylepis* (Prymnesiales) bloom in the Baltic Sea. PLoS ONE 9(11):e112985

Granskog M, Kaartokallio H, Kuosa H, Thomas DN, Vainio J (2006) Sea ice in the Baltic Sea – a review. Estuarine, Coastal and Shelf Science 70:145–160

Grote J, Jost G, Labrenz M, Herndl GJ, Jürgens K (2008) Epsilonproteobacteria represent the major portion of chemoautotrophic bacteria in sulfidic waters of pelagic redoxclines of the Baltic and Black Seas. Applied and Environment Microbiology 74:7546–7551

Grote J, Schott T, Bruckner CG, Glöckner FO, Jost G et al (2012) Genome and physiology of a model Epsilonproteobacterium responsible for sulfide detoxification in marine oxygen depletion zones. Proceedings of the National Academy of Sciences of the USA 109:506–510

Hagström Å, Pinhassi J, Zweifel UL (2000) Biogeographical diversity among marine bacterioplankton. Aquatic Microbial Ecology 21:231–244

Hajdu S, Larsson U, Moestrup O (1996) Seasonal dynamics of *Chrysochromulina* species (Prymnesiophyceae) in a coastal area and a nutrient-enriched inlet of the northern Baltic proper. Botanica Marina 39:281–295

Hajdu S, Edler L, Olenina I, Witek B (2000) Spreading and establishment of the potentially toxic dinoflagellate *Prorocentrum minimum* in the Baltic Sea. International Review of Hydrobiology 85:561–575

Hajdu S, Pertola S, Kuosa H (2005) *Prorocentrum minimum* (Dinophyceae) in the Baltic Sea: morphology, occurrence – a review. Harmful Algae 4:471–480

Hakala T, Viitasalo M, Rita H, Aro E, Flinkman J, Vuorinen I (2003) Temporal and spatial variability in the growth rates of Baltic herring (*Clupea harengus membras* L.) larvae during summer. Marine Biology 142:25–33

Halinen K, Jokela J, Fewer DP, Wahlsten M, Sivonen K (2007) Direct evidence for production of microcystins by *Anabaena* strains from the Baltic Sea. Applied and Environment Microbiology 73:6543–6550

Hällfors G (2004) Checklist of Baltic Sea phytoplankton species. Baltic Sea Environment Proceedings 95:1–208

Hänninen J, Vuorinen I, Hjelt P (2000) Climatic factors in the Atlantic control the oceanographic and ecological changes in the Baltic Sea. Limnology and Oceanography 45:703–710

Harding KC, Härkönen T (1999) Development in the Baltic grey seal (*Halichoerus grypus*) and ringed seal (*Phoca hispida*) populations during the 20th century. Ambio 28:619–627

Heiskanen AS (1998) Factors governing sedimentation and pelagic nutrient cycles in the northern Baltic Sea. University of Helsinki, Tammer-Paino Oy, Tampere [PhD Thesis]

HELCOM (2015) Manual for marine monitoring in the COMBINE programme of HELCOM. HELCOM, Helsinki, 413 pp. http://www.helcom.fi

Henriksen P (2009) Long-term changes in phytoplankton in the Kattegat, the Belt Sea, the Sound and the western Baltic Sea. Journal of Sea Research 61:114–123

Hensen V (1887) Über die Bestimmung des Planktons oder des im Meer treibenden Materials an Pflanzen und Thieren. Bericht der Kommission zur wissenschaftlichen Untersuchung der deutschen Meere 5:1–108 [in German]

Herlemann DPR, Labrenz M, Jürgens K, Bertilsson S, Waniek JJ, Andersson AF (2011) Transitions in bacterial communities along the 2,000 km salinity gradient of the Baltic Sea. The ISME Journal 5:1571–1579

Herlemann DPR, Lundin D, Labrenz M, Jürgens K, Zheng Z et al (2013) Metagenomic *de novo* assembly of an aquatic representative of the Verrucomicrobia class Spartobacteria. mBio 4(3):e00569-12

Herlemann DPR, Manecki M, Meeske C, Pollehne F, Labrenz M et al (2014) Uncoupling of bacterial and terrigenous dissolved organic matter dynamics in decomposition experiments. PLoS ONE 9(4): e93945

Hernroth L, Ackefors H (1979) The zooplankton of the Baltic Proper. A long-term investigation of the fauna, its biology and ecology. Report of the Fishery Board of Sweden, Institute of Marine Research 2:1–59

Holmfeldt K, Dziallas C, Titelman J, Pohlmann K, Grossart HP, Riemann L (2009) Diversity and abundance of freshwater actinobacteria along environmental gradients in the brackish northern Baltic Sea. Environmental Microbiology 11:2042–2054

Hügler M, Sievert SM (2011) Beyond the Calvin cycle: autotrophic carbon fixation in the ocean. Annual Review of Marine Sciences 3:261–289

Jaanus A, Andersson A, Olenina I, Törning K, Kaljurand K (2011) Changes in phytoplankton communities along a north-south gradient in the Baltic Sea between 1990 and 2008. Boreal Environment Research 16(supplement A):191–208

Jochem F, Babenerd B (1989) Naked *Dictyocha speculum* – a new type of phytoplankton bloom in the western Baltic. Marine Biology 103:373–379

Johansson S (1983) Annual dynamics and production of rotifers in a eutrophication gradient in the Baltic Sea. Hydrobiologia 14:335–340

Johansson M, Gorokhova E, Larsson U (2004) Annual variability in ciliate community structure, potential prey and predators in the open northern Baltic Sea proper. Journal of Plankton Research 26:67–80

Johnson MD, Oldach D, Delwiche CF, Stoecker DK (2007) Retention of transcriptionally active cryptophyte nuclei by the ciliate *Myrionecta rubra*. Nature 445:426–428

Jost G, Martens-Habbena W, Pollehne F, Schnetger B, Labrenz M (2010) Anaerobic sulfur oxidation in the absence of nitrate dominates microbial chemoautotrophy beneath the pelagic

chemocline of the Eastern Gotland basin, Baltic Sea. FEMS Microbiology Ecology 71:226–236
Jürgens K, Massana R (2008) Protistan grazing on marine bacterioplankton. In: Kirchman DL (ed) Microbial ecology of the oceans, 2nd edn. Wiley, New York, pp 383–441
Kahru M, Elmgren R (2014) Multi-decadal time series of satellite-detected accumulations of cyanobacteria in the Baltic Sea. Biogeosciences 11:3619–3633
Kaitala S, Hällfors S, Maunula P (2011) Phytoplankton biomass and species succession. HELCOM Baltic Sea Environment Fact Sheet. http://www.helcom.fi
Karlson K, Bonsdorff E, Rosenberg R (2007) The impact of benthic macrofauna for nutrient fluxes from Baltic Sea sediments. Ambio 36:161–167
Katajisto T (1996) Copepod eggs survive a decade in the sediments of the Baltic Sea. Hydrobiologia 320:153–159
Katajisto T (2004) Effects of anoxia and hypoxia on the dormancy and survival of subitaneous eggs of *Acartia bifilosa* (Copepoda: Calanoida). Marine Biology 145:751–757
King N, Westbrook MJ, Young SL, Kuo A, Abedin M et al (2008) The genome of the choanoflagellate *Monosiga brevicollis* and the origin of metazoans. Nature 451:783–788
Kiørboe T, Saiz E, Viitasalo M (1996) Prey switching behaviours in the planktonic copepod *Acartia tonsa*. Marine Ecology Progress Series 143:65–75
Kisand V, Andersson N, Wikner J (2005) Bacterial freshwater species successfully immigrate to the brackish water environment in the northern Baltic. Limnology and Oceanography 50:945–956
Kivi K, Kuosa H, Tanskanen S (1996) An experimental study on the role of crustacean and microprotozoan grazers in the planktonic food web. Marine Ecology Progress Series 136:59–68
Klais R, Tamminen T, Kremp A, Spilling K, Olli K (2011) Decadal-scale changes of dinoflagellates and diatoms in the anomalous Baltic Sea spring bloom. PLoS ONE 6(6):e21567
Klais R, Tamminen T, Kremp A, Spilling K, Woong An B et al (2013) Spring phytoplankton communities shaped by interannual weather variability and dispersal limitation: mechanisms of climate change effects on key coastal primary producers. Limnology and Oceanography 58:753–762
Koski M, Schmidt K, Engström-Öst J, Viitasalo M, Jónasdóttir SH et al (2002) Calanoid copepods feed and produce eggs in the presence of toxic cyanobacteria *Nodularia spumigena*. Limnology and Oceanography 47:878–885
Kremp A, Tamminen T, Spilling K (2008) Dinoflagellate bloom formation in natural assemblages with diatoms: nutrient competition and growth strategies in Baltic spring phytoplankton. Aquatic Microbial Ecology 50:181–196
Kremp A, Lindholm T, Dreßler N, Erler K, Gerdts G et al (2009) Bloom forming *Alexandrium ostenfeldii* (Dinophyceae) in shallow waters of the Åland archipelago, Northern Baltic Sea. Harmful Algae 8:318–328
Kuosa H, Kivi K (1989) Bacteria and heterotrophic flagellates in the pelagic carbon cycle in the northern Baltic Sea. Marine Ecology Progress Series 53:93–100
Laamanen M, Kuosa H (2005) Annual variability of biomass and heterocysts of the N_2-fixing cyanobacterium *Aphanizomenon flos-aquae* in the Baltic Sea with reference to *Anabaena* spp. and *Nodularia spumigena*. Boreal Environment Research 10:19–30
Laanemets J, Kononen K, Pavelson J, Poutanen EL (2004) Vertical location of seasonal nutriclines in the western Gulf of Finland. Journal of Marine Systems 52:1–13
Labrenz M, Sintes E, Toetzke F, Zumsteg A, Herndl GJ et al (2010) Relevance of a crenarchaeotal subcluster related to *Candidatus Nitrosopumilus maritimus* to ammonia oxidation in the suboxic zone of the central Baltic Sea. The ISME Journal 4:1496–1508
Labrenz M, Grote J, Mammitzsch K, Boschker HTS, Laue M et al (2013) *Sulfurimonas gotlandica* sp. nov., a chemoautotrophic and psychrotolerant Epsilonproteobacterium isolated from a pelagic redoxcline, and an emended description of the genus *Sulfurimonas*. International Journal of Systematic and Evolutionary Microbiology 63:4141–4148
Lagus A (2009) Role of nutrients in regulation of phytoplankton community in the Archipelago Sea, northern Baltic Sea. Annales Universitatis Turkuensis, Ser. AII 239:1–56 [PhD Thesis]
Langenheder S, Kisand V, Wikner J, Tranvik LJ (2003) Salinity as a structuring factor for the composition and performance of bacterioplankton degrading riverine DOC. FEMS Microbiology Ecology 45:189–202
Larsson U, Nyberg S, Andreasson K, Lindahl O, Wikner J (2010) Phytoplankton production – measurements with problems. Havet – om miljötillståndet i svenska havsområden 2010:26–29 [in Swedish]
Larsson J, Celepli N, Ininbergs K, Dupont CL, Yooseph S et al (2014a) Picocyanobacteria containing a novel pigment gene cluster dominate the brackish-water Baltic Sea. The ISME Journal 8:1892–1903
Larsson K, Hajdu S, Kilpi M, Larsson R, Leito A et al (2014b) Effects of an extensive *Prymnesium polylepis* bloom on breeding eiders in the Baltic Sea. Journal of Sea Research 88:21–28
Leadbeater BSC, Yu Q, Kent J, Stekel D (2009) Three-dimensional images of choanoflagellate loricae. Proceedings of the Royal Society B 276:3–11
Lefébure R, Degerman R, Andersson A, Larsson S, Eriksson LO et al (2013) Impacts of elevated terrestrial nutrient loads and temperature on pelagic food-web efficiency and fish production. Global Change Biology 19:1358–1372
Legendre L, Rassoulzadegan F (1995) Plankton and nutrient dynamics in marine waters. Ophelia 41:153–172
Lehmann A, Myrberg K (2008) Upwelling in the Baltic Sea. Journal of Marine Systems 74:S3–S12
Lehtimäki J, Moisander P, Sivonen K, Kononen K (1997) Growth, nitrogen fixation, and nodularin production by two Baltic Sea cyanobacteria. Applied and Environment Microbiology 63:1647–1656
Leppäkoski E, Olenin S (2000) Non-native species and rates of spread: lessons from the brackish Baltic Sea. Biological Invasions 2:151–163
Lewandowska AM, Sommer U (2010) Climate change and the spring bloom: a mesocosm study on the influence of light and temperature on phytoplankton and mesozooplankton. Marine Ecology Progress Series 405:101–111
Lignell R, Heiskanen AS, Kuosa H, Gundersen K, Kuuppo-Leinikki P et al (1993) Fate of phytoplankton spring bloom: sedimentation and carbon flow in the planktonic food web in the northern Baltic. Marine Ecology Progress Series 94:239–252
Lignell R, Hoikkala L, Lahtinen T (2008) Effects of inorganic nutrients, glucose and solar radiation on bacterial growth and exploitation of dissolved organic carbon and nitrogen in the northern Baltic Sea. Aquatic Microbial Ecology 51:209–221
Lindholm T, Nummelin C (1999) Red tide of the dinoflagellate *Heterocapsa triquetra* (Dinophyta) in a ferry-mixed coastal inlet. Hydrobiologia 393:245–251
Lohmann H (1908) Untersuchungen zur Feststellung des vollständigen Gehaltes des Meeres an Plankton. Wissenschaftliche Meeresuntersuchung Abteilung Kiel N.F. 10:131–370 [in German]
Lumberg A, Ojaveer E (1991) On the environment and zooplankton dynamics in the Gulf of Finland in 1961–1990. Proceedings of the Estonian Academy of Sciences, Biology and Ecology 1(N3): 131–140
Lundström K, Hjerne O, Lunneryd SG, Karlsson O (2010) Understanding the diet composition of marine mammals: grey seals

(*Halichoerus grypus*) in the Baltic Sea. ICES Journal of Marine Science 67:1230–1239
Maar M, Visser AW, Nielsen TG, Stips A, Saito H (2006) Turbulence and feeding behaviour affect the vertical distributions of *Oithona similis* and *Microsetella norwegica*. Marine Ecology Progress Series 313:157–172
MacKenzie BR, Köster FW (2004) Fish production and climate: sprat in the Baltic Sea. Ecology 85:784–794
MacKenzie BR, Alheit J, Conley DJ, Holm P, Kinze CK (2002) Ecological hypotheses for a historical reconstruction of upper trophic level biomass in the Baltic Sea and Skagerrak. Canadian Journal of Fisheries and Aquatic Sciences 59:173–190
MacKenzie BR, Gislason H, Möllmann C, Köster FW (2007) Impact of 21st century climate change on the Baltic Sea fish community and fisheries. Global Change Biology 13:1–20
MacKenzie BR, Eero M, Ojaveer H (2011) Could seals prevent cod recovery in the Baltic Sea? PLoS ONE 6(5):e18998
Majaneva M, Rintala JM, Hajdu S, Hällfors S, Hällfors G et al (2012) The extensive bloom of alternate-stage *Prymnesium polylepis* (Haptophyta) in the Baltic Sea during autumn–spring 2007–2008. European Journal of Phycology 47:310–320
Marcotte BM (1982) Evolution within the Crustacea. Part 2: Copepoda. In: Abele LG (ed) The biology of Crustacea, vol 1. Academic Press, New York, pp 185–197
Meier HEM (2015) Projected change – marine physics. In: BACC Author Team (ed) Second assessment of climate change for the Baltic Sea basin. Regional Climate Studies. Springer, Berlin, pp 243–252
Meier HEM, Andersson HC, Eilola K, Gustafsson BG, Kuznetsov I et al (2011) Hypoxia in future climates – a model ensemble study for the Baltic Sea. Geophysical Reseach Letters 38:L24608
Meier HEM, Müller-Karulis B, Andersson HC, Dieterich C, Eilola K et al (2012) Impact of climate change on ecological quality indicators and biogeochemical fluxes in the Baltic Sea: A multi-model ensemble study. Ambio 41:558–573
Mills CE (2001) Jellyfish blooms: are populations increasing globally in response to changing ocean conditions? Hydrobiologia 451:55–68
Miyashita LK, Gaeta SA, Lopes RM (2011) Life cycle and reproductive traits of marine podonids (Cladocera, Onychopoda) in a coastal subtropical area. Journal of Plankton Research 33:779–792
Möllmann C, Kornilovs G, Sidrevics L (2000) Long-term dynamics of main mesozooplankton species in the central Baltic Sea. Journal of Plankton Research 22:2015–2038
Möllmann C, Kornilovs G, Fetter M, Köster FW (2005) Climate, zooplankton and pelagic fish growth in the central Baltic Sea. ICES Journal of Marine Science 62:1270–1280
Muro-Pastor AM, Hess WR (2012) Heterocyst differentiation: from single mutants to global approaches. Trends in Microbiology 20:549–557
Myung G, Hyung SK, Jong SP, Myung GP, Wonho Y (2011) Population growth and plastic type of *Myrionecta rubra* depend on the kinds of available cryptomonad prey. Harmful Algae 10: 536–541
Nissling A, Müller A, Hinrichsen HH (2003) Specific gravity and vertical distribution of sprat (*Sprattus sprattus*) eggs in the Baltic Sea. Journal of Fish Biology 63:280–299
Ojaveer E, Lindroth A, Bagge O, Lehtonen H, Toivonen J (1981) Fish and Fisheries. In: Voipio A (ed) The Baltic Sea. Elsevier Oceanography Series, vol 30, pp 275–350
Ojaveer E, Lumberg A, Ojaveer H (1998) Highlights of zooplankton dynamics in Estonian waters (Baltic Sea). ICES Journal of Marine Science 55:748–755
Ojaveer H, Simm M, Lankov A (2004) Population dynamics and ecological impact of the non-indigenous *Cercopagis pengoi* in the Gulf of Riga (Baltic Sea). Hydrobiologia 522:261–269
Olenina I, Hajdu S, Edler L, Andersson A, Wasmund N et al (2006) Biovolumes and size-classes of phytoplankton in the Baltic Sea. Baltic Sea Environment Proceedings 106:1–144
Olli K, Trunov K (2010) Abundance and distribution of vernal bloom dinoflagellate cysts in the Gulf of Finland and Gulf of Riga (the Baltic Sea). Deep-Sea Research II 57:235–242
Olli K, Klais R, Tamminen T, Ptacnik R, Andersen T (2011) Long term changes in the Baltic Sea phytoplankton community. Boreal Environment Research 16:3–14
Olli K, Ptacnik R, Andersen T, Trikk O, Klais R et al (2014) Against the tide: recent diversity increase enhances resource use in a coastal system. Limnology and Oceanography 59:267–274
Österblom H, Hansson S, Larsson U, Hjerne O, Wulff F et al (2007) Human-induced trophic cascades and ecological regime shifts in the Baltic Sea. Ecosystems 10:877–889
Paxinos R, Mitchell JG (2000) A rapid Utermöhl method for estimating algal numbers. Journal of Plankton Research 22:2255–2262
Pearre S (1982) Estimating prey preference by predators: uses of various indices, and a proposal of another based on χ^2. Canadian Journal of Fisheries and Aquatic Sciences 39:914–923
Pertola S, Koski M, Viitasalo M (2002) Stoichiometry of mesozooplankton in N- and P-limited areas of the Baltic Sea. Marine Biology 140:425–434
Pinhassi J, Winding A, Binnerup SJ, Zweifel UL, Riemann B, Hagström Å (2003) Spatial variability in bacterioplankton community composition at the Skagerrak-Kattegat front. Marine Ecology Progress Series 255:1–13
Ptacnik R, Andersen T, Brettum P, Lepistö L, Willén E (2010) Regional species pools control community saturation in lake phytoplankton. Proceedings of the Royal Society B 277:3755–3764
Rajasilta M, Hänninen J, Vuorinen I (2014) Decreasing salinity improves the feeding conditions of the Baltic herring (*Clupea harengus membras*) during spring in the Bothnian Sea, northern Baltic. ICES Journal of Marine Science 71:1148–1152
Rand PS, Stewart DJ (1998) Prey fish exploitation, salmonine production, and pelagic food web efficiency in Lake Ontario. Canadian Journal of Fisheries and Aquatic Sciences 55:318–327
Rantajärvi E, Olsonen R, Hällfors S, Leppänen JM, Raateoja M (1998) Effect of sampling frequency on detection of natural variability in phytoplankton: unattended high-frequency measurements on board ferries in the Baltic Sea. ICES Journal of Marine Science 55: 697–704
Remane A (1934) Die Brackwasserfauna. Verhandlungen der Deutschen Zoologischen Gesellschaft 36:34–74 [in German]
Repka S, Meyerhöfer M, von Bröckel K, Sivonen K (2004) Associations of cyanobacterial toxin, nodularin, with environmental factors and zooplankton in the Baltic Sea. Microbial Ecology 47:350–358
Rheinheimer G (1974) Bakterien und Pilze. In: Magaard L, Rheinheimer G (eds) Meereskunde der Ostsee. Springer, Berlin, pp 161–170 [in German]
Rheinheimer G (ed) (1977) Microbial ecology of a brackish water environment. Ecological Studies, vol 25. Springer, Berlin, 296 pp
Rheinheimer G (1984) Bacterial ecology of the North and Baltic seas. Botanica Marina 27:277–299
Riemann L, Leitet C, Pommier T, Simu K, Holmfeldt K et al (2008) The native bacterioplankton community in the central Baltic Sea is influenced by freshwater bacterial species. Applied and Environment Microbiology 74:503–515
Rönkkönen S, Ojaveer E, Raid T, Viitasalo M (2003) Long-term changes in the Baltic herring growth. Canadian Journal of Fisheries and Aquatic Sciences 61:219–229
Rudstam LG, Hansson S, Johansson S, Larsson U (1992) Dynamics of planktivory in a coastal area of the northern Baltic Sea. Marine Ecology Progress Series 80:159–173

Rychert K (2011) Communities of heterotrophic protists (Protozoa) in the near-bottom zone of the Gdańsk basin. Oceanological and Hydrobiological Studies 40:67–73

Salka I, Wurzbacher C, Garcia SL, Labrenz M, Jürgens K, Grossart HP (2014) Distribution of acI-actinorhodopsin genes in Baltic Sea salinity gradients indicates adaptation of facultative freshwater photoheterotrophs to brackish waters. Environmental Microbiology 16:586–597

Samuelsson K, Berglund J, Haecky P, Andersson A (2002) Structural changes in an aquatic microbial food web caused by inorganic nutrient addition. Aquatic Microbial Ecology 29:29–38

Samuelsson K, Berglund J, Andersson A (2006) Factors controlling the heterotrophic flagellate and ciliate community along a brackish water primary production gradient. Journal of Plankton Research 28:345–359

Sandberg J, Andersson A, Johansson S, Wikner J (2004) Pelagic food web structure and carbon budget in the northern Baltic Sea: potential importance of terrigenous carbon. Marine Ecology Progress Series 268:13–29

Sandström O (1980) Selective feeding by Baltic herring. Hydrobiologia 69:199–207

Schneider G, Behrends G (1994) Population dynamics and the trophic role of *Aurelia aurita* medusae in the Kiel Bight and western Baltic. ICES Journal of Marine Science 51:359–367

Schumann R, Rieling T, Görs S, Hammer A, Selig U, Schiewer U (2003) Viability of bacteria from different aquatic habitats I. Environmental conditions and productivity. Aquatic Microbial Ecology 32:121–135

Segerstråle SG (1969) Biological fluctuations in the Baltic Sea. Progress in Oceanography 5:169–184

Seppälä J, Ylöstalo P, Kaitala S, Hällfors S, Raateoja M, Maunula P (2007) Ship-of-opportunity based phycocyanin fluorescence monitoring of the filamentous cyanobacteria bloom dynamics in the Baltic Sea. Estuarine, Coastal and Shelf Science 73:489–500

Setälä O, Kivi K (2003) Planktonic ciliates in the Baltic Sea in summer. Distribution, species association and estimated grazing impact. Aquatic Microbial Ecology 32:287–297

Simon M, Grossart HP, Schweitzer B, Ploug H (2002) Microbial ecology of organic aggregates in aquatic ecosystems. Aquatic Microbial Ecology 28:175–211

Simu K, Holmfeldt K, Zweifel UL, Hagström Å (2005) Culturability and coexistence of colony-forming and single-cell marine bacterioplankton. Applied and Environment Microbiology 71:4793–4800

Sivonen K, Kononen K, Carmichael WW, Dahlem AM, Rinehart KL et al (1989) Occurrence of the hepatotoxic cyanobacterium *Nodularia spumigena* in the Baltic Sea and structure of the toxin. Applied and Environment Microbiology 55:1990–1995

Sivonen K, Niemelä SI, Niemi RM, Lepistö L, Luoma TH, Räsänen LA (1990) Toxic cyanobacteria (blue-green algae) in Finnish fresh and coastal waters. Hydrobiologia 190:267–275

Smayda TJ, Trainer VL (2010) Dinoflagellate blooms in upwelling systems: seeding, variability, and contrasts with diatom bloom behaviour. Progress in Oceanography 85:92–107

Smetacek V (1981) The annual cycle of protozooplankton in the Kiel Bight. Marine Biology 63:1–11

Snoeijs P, Häubner N (2014) Astaxanthin dynamics in Baltic Sea mesozooplankton communities. Journal of Sea Research 85: 131–143

Sprules W, Bowerman J (1988) Omnivory and food web lengths in zooplankton food webs. Ecology 69:418–426

Stewart KD, Mattox KR (1980) Phylogeny of phytoflagellates. In: Cox ER (ed) Phytoflagellates. Developments in Marine Biology, vol 2. Elsevier, Amsterdam, pp 433–462

Stocker TF, Qin D, Plattner GK, Tignor M, Allen SK et al (eds) (2013) The physical science basis – contribution of working group I to the fifth assessment report of the intergovernmental panel on climate change. Cambridge University Press, Cambridge 1535 pp

Straile D (1997) Gross growth efficiencies of protozoan and metazoan zooplankton and their dependence on food concentration, predator-prey weight ratio, and taxonomic group. Limnology and Oceanography 42:1375–1385

Suikkanen S, Laamanen M, Huttunen M (2007) Long-term changes in summer phytoplankton communities of the open northern Baltic Sea. Estuarine, Coastal and Shelf Science 71:580–592

Suikkanen S, Pulina S, Engström-Öst J, Lehtiniemi M, Lehtinen S, Brutemark A (2013) Climate change and eutrophication induced shifts in northern summer plankton communities. PLoS ONE 8(6): e66475

Tamminen T, Andersen T (2007) Seasonal phytoplankton nutrient limitation patterns as revealed by bioassays over Baltic Sea gradients of salinity and eutrophication. Marine Ecology Progress Series 340:121–138

Telesh I, Heerkloss R (2004) Atlas of Estuarine Zooplankton of the Southern and Eastern Baltic Sea. Part II: Crustacea. Verlag Dr. Kovač, Hamburg, p 118

Thompson RM, Hemberg M, Strazomski BM, Shurin JB (2007) Trophic levels and trophic tangels: the prevalence of omnivory in real food webs. Ecology 88:612–617

Uiboupin R, Laanemets J, Sipelgas L, Raag L, Lips I, Buhhalko N (2012) Monitoring the effect of upwelling on the chlorophyll *a* distribution in the Gulf of Finland (Baltic sea) using remote sensing and *in situ* data. Oceanologia 54:395–419

Uitto A, Heiskanen AS, Lignell R, Autio R, Pajuniemi R (1997) Summer dynamics of the coastal planktonic food web in the northern Baltic Sea. Marine Ecology Progress Series 151: 27–41

Utermöhl H (1931) Neue Wege in der quantitativen Erfassung des Planktons (mit besonderer Berücksichtigung des Ultraplanktons). Verhandlungen der Internationale Vereinigung für Theoretische und Angewandte Limnologie 5:567–595 [in German]

Utermöhl H (1958) Zur Vervollkommnung der quantitativen Phytoplankton-Methodik. Mitteilungen der Internationale Vereinigung für Theoretische und Angewandte Limnologie 9:1–38 [in German]

Vahtera E, Conley DJ, Gustafsson BG, Kuosa H, Pitkänen H et al (2007) Internal ecosystem feedbacks enhance nitrogen-fixing cyanobacteria blooms and complicate management in the Baltic Sea. Ambio 36:186–194

Viherluoto M, Kuosa H, Flinkman J, Viitasalo M (2000) Food utilisation of pelagic mysids, *Mysis mixta* and *M. relicta* during their growing season in the northern Baltic Sea. Marine Biology 136:553–559

Viitasalo M (1992a) Mesozooplankton of the Gulf of Finland and northern Baltic Proper – a review of monitoring data. Ophelia 35:147–168

Viitasalo M (1992b) Calanoid resting eggs in the Baltic Sea: implications for the population dynamics of *Acartia bifilosa* (Copepoda). Marine Biology 114:397–405

Viitasalo M, Katajisto T (1994) Mesozooplankton resting eggs in the Baltic Sea – identification and vertical distribution in laminated and mixed sediments. Marine Biology 120:455–466

Viitasalo M, Rautio M (1998) Zooplanktivory by *Praunus flexuosus* (Crustacea: Mysidacea): functional responses and prey selection in relation to prey escape responses. Marine Ecology Progress Series 174:77–87

Viitasalo M, Koski M, Pellikka K, Johansson S (1995) Seasonal and long-term variations in the body size of planktonic copepods in the northern Baltic Sea. Marine Biology 123:241–250

Viitasalo M, Kiørboe T, Flinkman J, Pedersen LW, Visser AW (1998) Predation vulnerability of planktonic copepods: consequences of

predator foraging strategies and prey sensory abilities. Marine Ecology Progress Series 175:129–145
Viitasalo M, Rosenberg M, Heiskanen AS, Koski M (1999) Sedimentation of copepod fecal material in the coastal northern Baltic Sea: where did all the pellets go? Limnology and Oceanography 44:1388–1399
Viitasalo M, Flinkman J, Viherluoto M (2001) Zooplanktivory in the Baltic Sea: a comparison of prey selectivity by *Clupea harengus* and *Mysis mixta*, with reference to prey escape reactions. Marine Ecology Progress Series 216:191–200
Viitasalo M, Blenckner T, Gårdmark A, Kaartokallio H, Kautsky L et al (2015) Environmental impacts – marine ecosystems. In BACC Author Team (ed) Second assessment of climate change for the Baltic Sea basin. Regional Climate Studies. Springer, Berlin, pp 363–380
Vuorinen I, Rajasilta M, Salo J (1983) Selective predation and habitat shift in a copepod species – support for the predation hypothesis. Oecologia 59:62–64
Vuorinen I, Hänninen J, Rajasilta M, Laine P, Eklund J et al (2015) Scenario simulations of future salinity and ecological consequences in the Baltic Sea and adjacent North Sea areas – implications for environmental monitoring. Ecological Indicators 50:196–205
Wasmund N, Göbel J, von Bodungen B (2008) 100-years-changes in the phytoplankton community of Kiel Bight (Baltic Sea). Journal of Marine Systems 73:300–322
Wasmund N, Tuimala J, Suikkanen S, Vandepitte L, Kraberg A (2011) Long-term trends in phytoplankton composition in the western and central Baltic Sea. Journal of Marine Systems 87:145–159
Webb DG, Weaver AJ (1988) Predation and the evolution of free spawning in marine calanoid copepods. Oikos 51:189–192
Weber F, Anderson R, Foissner W, Mylnikov AP, Jürgens K (2014) Morphological and molecular approaches reveal highly stratified protist communities along Baltic Sea pelagic redox gradients. Aquatic Microbial Ecology 73:1–16
Welch HE (1968) Relationship between assimilation efficiencies and growth efficiencies for aquatic consumers. Ecology 49:755–759
Westin L, Nissling (1991) Effects of salinity on spermatozoa motility, percentage of fertilized eggs and egg development of Baltic cod (*Gadus morhua*), and implications for cod stock fluctuations in the Baltic. Marine Biology 108:5–9
Wieland K, Waller U, Schnack D (1994) Development of Baltic cod eggs at different levels of temperature and oxygen content. Dana 10:163–177
Wieland K, Jarre-Teichmann A, Horbowa K (2000) Changes in the timing of spawning of Baltic cod: possible causes and implications for recruitment. ICES Journal of Marine Science 57: 452–464
Wikner J, Andersson A (2012) Increased freshwater discharge shifts the trophic balance in the coastal zone of the northern Baltic Sea. Global Change Biology 18:2509–2519
Witek M (1998) Annual changes of abundance and biomass of planktonic ciliates in the Gdańsk basin, southern Baltic. International Review of Hydrobiology 83:163–182
Wohlers J, Engel A, Zöllner E, Breithaupt P, Jürgens K et al (2009) Changes in biogenic carbon flow in response to sea surface warming. Proceedings of the National Academy of Sciences of the USA 106:7067–7072
Wright JJ, Konwar KM, Hallam SJ (2012) Microbial ecology of expanding oxygen minimum zones. Nature Reviews Microbiology 10:381–394
Wylezich C, Karpov SA, Mylnikov AP, Anderson R, Jürgens K (2012) Ecologically relevant choanoflagellates collected from hypoxic water masses of the Baltic Sea have untypically mitochondrial cristae. BMC Microbiology 12:271, 13 pp
Zöllner E, Hoppe HG, Sommer U, Jürgens K (2009) Effect of zooplankton-mediated trophic cascades on marine microbial food web components (bacteria, nanoflagellates, ciliates). Limnology and Oceanography 54:262–275
Zweifel UL, Norrman B, Hagström Å (1993) Consumption of dissolved organic carbon by marine bacteria and demand for inorganic nutrients. Marine Ecology Progress Series 101:23–32

Life associated with Baltic Sea ice

9

David N. Thomas, Hermanni Kaartokallio, Letizia Tedesco, Markus Majaneva, Jonna Piiparinen, Eeva Eronen-Rasimus, Janne-Markus Rintala, Harri Kuosa, Jaanika Blomster, Jouni Vainio, and Mats A. Granskog

Abstract

1. The formation of sea ice impacts directly on the physical dynamics of water masses (*e.g.* wind stress at the sea surface) and air-sea exchange processes (*e.g.* vertical heat fluxes).
2. The annual cycle of formation, consolidation and melting of sea ice has a major influence on the ecology of both the benthic and pelagic components of the Baltic Sea ecosystem.
3. There is considerable inter-annual variation in the extent of sea ice in the Baltic Sea and thus in the size of the habitat for sympagic (ice-associated) microbial and metazoan communities as well as for larger organisms living on the ice, notably the ringed seal.
4. There is a pronounced gradient in ice characteristics, from more saline ice in the south of the Baltic Sea to freshwater ice in the north. The former is more porous and supports more ice-associated biology than the latter.
5. The Baltic sympagic communities consist mainly of prokaryotic and eukaryotic microbes (bacteria, diatoms, dinoflagellates, flagellates), ciliates and rotifers. These communities are recruited from the plankton when the ice forms, followed by an ice-adapted successional pattern with an expansion of substrate-bound pennate diatoms, which does not occur in the seawater beneath the ice.
6. The sea-ice food webs inside the ice are truncated compared to the open-water food webs because organisms larger than the upper size limit of the brine channels are lacking in the internal sympagic communities.
7. Global climate change decreases the extension and thickness of the sea ice as well as the length of the ice season, and therefore the seasonal effects that sea ice has on the Baltic Sea winter-spring ecosystem dynamics.

Keywords
Baltic Sea gradient • Brine channels • Climate change • Productivity • Sea ice habitat • Sympagic communities • Truncated food web

D.N. Thomas (✉)
School of Ocean Sciences, Bangor University, Menai Bridge, Anglesey, LL59 5AB, UK
e-mail: d.thomas@bangor.ac.uk

H. Kaartokallio · L. Tedesco · J. Piiparinen · E. Eronen-Rasimus · H. Kuosa
Marine Research Centre Finnish Environment Institute, P.O. Box 140, 00251 Helsinki, Finland
e-mail: hermanni.kaartokallio@ymparisto.fi

L. Tedesco
e-mail: letizia.tedesco@ymparisto.fi

J. Piiparinen
e-mail: jonna.piiparinen@ymparisto.fi

E. Eronen-Rasimus
e-mail: eeva.eronen-rasimus@ymparisto.fi

H. Kuosa
e-mail: harri.kuosa@ymparisto.fi

M. Majaneva
Department of Natural History, Norwegian University of Science and Technology, 7491 Trondheim, Norway
e-mail: markus.majaneva@ntnu.no

P. Snoeijs-Leijonmalm et al. (eds.), *Biological Oceanography of the Baltic Sea*,
DOI 10.1007/978-94-007-0668-2_9

Fig. 9.1 Pack ice in the Baltic Sea and in the Arctic Ocean. On the surface it all looks the same, but underneath the snow the systems are quite different. (**a**) The Finnish research vessel Aranda in the Baltic Sea sea ice. (**b**) The Swedish ice-breaker Oden in the Arctic Ocean sea ice. Photo: (**a**) © Ilkka Lastumäki, (**b**) © Pauline Snoeijs-Leijonmalm

9.1 Sea ice is an integral part of the Baltic Sea ecosystem

9.1.1 Ice shapes the ecology of the Baltic Sea

One of the major drivers of the seasonal dynamics in the ecology of the Baltic Sea is the formation of sea ice (*cf.* Figs. 2.17–2.19). Vast areas of the sea are covered each winter by an ice layer that restricts the penetration of light, reduces water mixing by winds and changes heat and momentum flux (Leppäranta and Myrberg 2009). These physical changes strongly impact the biology living under the ice since water circulation and mixing are so clearly altered. A good example is that river plumes extend under the ice to a much larger distance and with higher stability than in ice-free conditions. Under-ice river plumes not only alter the mixing properties of the waters, they also result in changed ice growth dynamics, and ice-associated (sympagic) biological communities, with the underside of the ice being encased, in the extreme case, with a frozen freshwater layer (Granskog et al. 2006).

Baltic Sea ice reaches its maximum extent in February-March with a maximum thickness of ~0.8 m for landfast ice and 0.3–0.5 m for offshore ice, even during mild winters in the Bothnian Bay (Vihma and Haapala 2009). Further to the south, the thickness of the ice decreases. Many shallow coastal benthic habitats of the northern Baltic Sea are disturbed by ice scouring (*cf.* Sect. 11.2.5). The ice itself forms a temporary habitat for many plankton organisms: here they undergo a "lifestyle" shift from a viscous, sparsely populated liquid habitat to one comprised of a semi-solid matrix with liquid inclusions and dense communities of organisms (Thomas and Dieckmann 2010).

One of the most striking features of a winter Baltic Sea scene is how similar the sea-ice surface looks to the ice-covered Arctic and Southern Oceans (Fig. 9.1). However, while there are similarities between the biology associated with ice formed from the Baltic Sea water and that associated with thicker sea ice in the Arctic and Southern Oceans, findings from the frozen polar oceans cannot necessarily be transcribed to the frozen Baltic Sea and *vice versa*. This is mainly a result of the parent waters' different salinities where the ice is formed.

9.1.2 Sea ice and biogeochemical cycling

Sympagic communities, consisting of the organisms living within and on the sea ice, as well as in the seawater immediately below the ice, have been a source of study since the early days of polar exploration, and it is clear that they have a key role in the ecology of ice-covered waters (Thomas and Dieckmann 2002, 2010). This is also true of the Baltic Sea ice habitat. In the early days of sea-ice biological

J.-M. Rintala · J. Blomster
Department of Environmental Sciences, University of Helsinki, P.O. Box 65, 00014 Turku, Finland
e-mail: janne.rintala@helsinki.fi

J. Blomster
e-mail: jaanika.blomster@helsinki.fi

J. Vainio
Finnish Meteorological Institute, P.O. Box 503, 00101 Helsinki, Finland
e-mail: jouni.vainio@fmi.fi

M.A. Granskog
Norwegian Polar Institute, Fram Centre, Hjalmar Johansens Gatan 14, 9296 Tromsø, Norway
e-mail: mats.granskog@npolar.no

research, the scientists' curiosity about the identity of the ice-associated organisms was probably the primary driver for many of the studies.

The ice itself has been considered a relatively impervious barrier between air and water. Nowadays we know that the ice compartment significantly contributes to biogeochemical cycling since biological and chemical processes within the ice and at the ice peripheries can considerably influence gas fluxes to and from the atmosphere as well as matter fluxes between the ice and the surface waters (Rysgaard et al. 2011).

The ice cover can also modify nutrient fluxes and heavy metal concentrations as well as those of other chemical pollutants. Chemical compounds can accumulate in snow and ice, following atmospheric deposition, and/or by organism uptake or regeneration within ice, followed by subsequent pulsed releases into the water column during spring melt (Granskog and Kaartokallio 2004). For example, for the Bothnian Bay it has been estimated that 5 % of the total annual flux of nitrogen and phosphorus and 20–40 % of the total annual flux of lead and cadmium may be deposited on to the ice fields from the atmosphere (Granskog et al. 2006).

9.2 Characteristics of Baltic Sea ice

9.2.1 Seasonal and spatial variation of the ice cover

Annually, sea ice covers an average of ~44 % of the Baltic Sea water surface area and the median maximum ice extent for the period 1961–2010 was 186,000 km^2 (Fig. 9.2, Table 9.1). However, the interannual variability in the maximum ice extent is large as it ranges from 10 % to 100 % of the Baltic Sea water surface area (Vihma and Haapala 2009).

Ice formation begins along the coasts in the northernmost Bothnian Bay and in the easternmost Bothnian Sea, usually in November. Next to freeze is the shallow and narrow Norra Kvarken sill between the Bothnian Bay and the Bothnian

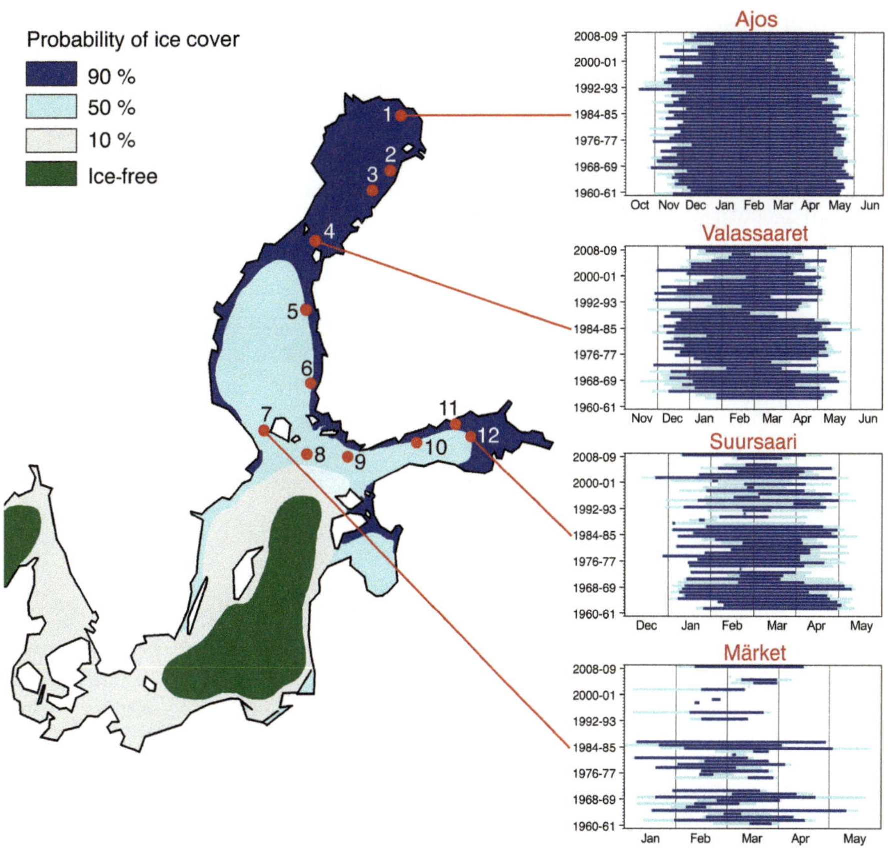

Fig. 9.2 Map of the Baltic Sea, showing the probability of sea-ice coverage. The locations 1–12 denote the measuring stations for the ice statistics presented in Table 9.1. The annual duration of the sea-ice coverage for the time period 1961–2010 are plotted for Station 1 (Ajos), Station 4 (Valassaaret), Station 12 (Suursaari, Gogland in Russian) and Station 7 (Märket). The left-hand side of each annual bar indicates the first appearance of ice and the right-hand side indicates the final disappearance of ice. The dark-blue regions of each bar represent the period of permanent ice cover and the light-blue regions represent the period of temporary ice cover. The absence of a bar indicates no ice cover at the station. Figure based on ice data from the Finnish Ice Service. Figure: © Jouni Vainio

Table 9.1 Median dates of appearance, disappearance, number of sea-ice days per year and number of ice-free years at 12 measuring stations in the Baltic Sea for the time period 1961–2010. The locations of the measuring stations are shown in Fig. 9.2. Data from the Finnish Ice Service

Station number	Station name	First appearance of ice	Start of permanent ice cover	End of permanent ice cover	Final disappearance of ice	Number of sea-ice days	Number of ice-free years
1	Kemi, Ajos	12 November	21 November	11 May	17 May	181	0
2	Raahe, Lapaluoto	24 November	5 December	30 April	7 May	161	0
3	Ulkokalla, off side	2 January	20 January	29 April	15 May	120	0
4	Valassaaret	16 December	27 December	20 April	6 May	133	0
5	Sälgrund	16 December	3 January	8 April	14 April	104	0
6	Raumanmatala	17 January	8 February	22 March	2 April	44	8
7	Märket	7 February	13 February	25 March	27 March	20	17
8	Utö	2 February	8 February	25 March	4 April	25	14
9	Bengtskär	31 January	8 February	28 March	7 April	34	13
10	Helsingin matala	23 January	4 February	31 March	13 April	58	6
11	Orrengrund	1 January	20 January	16 April	25 April	96	1
12	Suursaari, Gogland	20 January	26 January	7 April	25 April	90	1

Sea, followed by the coastal areas of the Bothnian Sea. In average winters, the ice covers the Gulf of Bothnia, the Archipelago Sea, the Gulf of Finland, the Gulf of Riga and the northern part of the Baltic Sea proper. The areas along the Polish and German coasts can also freeze over and in severe winters the southern Baltic Sea proper and the Belt Sea are also covered with ice.

Locally, ice formation usually starts within sheltered bays and around skerries (small rocky islands) and the ice edge moves outward from the coasts as the winter progresses (Fig. 9.3). Ice forms first at the inner skerries and bays where the water is often fresher and shallower than in the open Baltic Sea (and thus has a lower heat content), and where the ice cover can be anchored to islands and shoals. The landfast ice cover usually extends to the outer skerries, where the water depth is typically between 5 and 15 m. Along with increasing solar radiation in spring, the ice begins to melt starting from the south. The northern Baltic Sea proper is normally open by the beginning of April. By early May the sea ice is present only in the Bothnian Bay, and by early June it has usually also completely melted in the far north of the Baltic Sea.

In the early stages of ice formation when there is significant turbulence in the water, the first stages of ice formation may involve the formation of “pancake ice”. The water movement induces the ice crystals to coalesce into small discs (“pancakes”) that become progressively larger. These can freeze together to form closed ice sheets. Under more quiescent conditions, more typical of coastal waters, the ice cover tends to be a rather more uniformly flat sheet

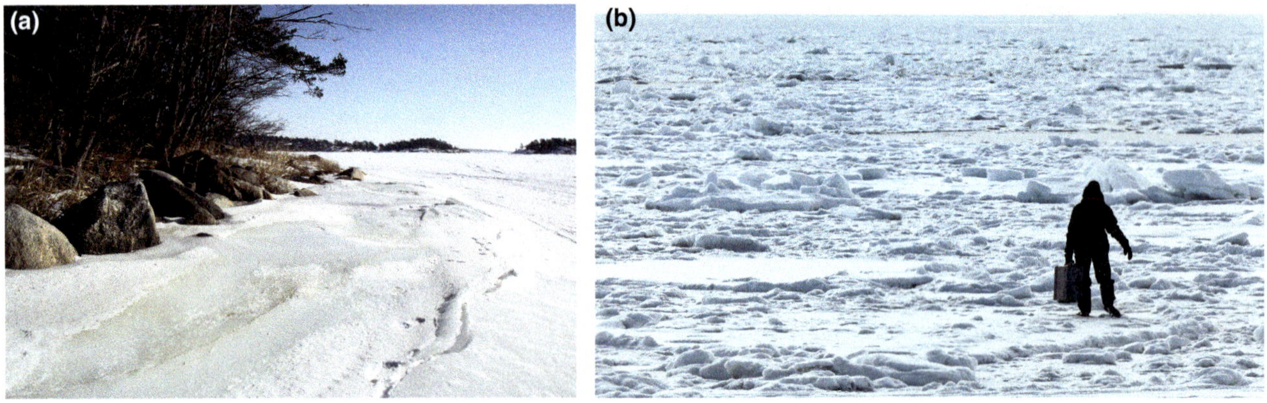

Fig. 9.3 Two different types of sea ice in the Baltic Sea. (**a**) Land-fast sea ice. (**b**) Pack ice. Photo: (**a**) © David N. Thomas, (**b**) © Ilkka Lastumäki

ice that is cracked and fissured through larger scale water movements.

9.2.2 Ice formation

Ice formation in brackish waters with salinity below 24.7, such as the Baltic Sea water, resembles more the ice formation in freshwater lakes than that in the oceans. When brackish water cools down to the freezing point (which is determined by salinity and pressure), the temperature of the maximum density of fresh and brackish waters is reached prior to this (*cf.* Sect. 2.4.4, Fig. 2.17b). Consequently, vertical convection ceases.

Moreover, the physical processes at the ice/water interface that result in the salinity distribution in the solidifying sea-ice matrix may differ, partly because thermal convection at the ice/water interface is restricted as the result of the low salinity of the Baltic Sea water (Granskog et al. 2006). Even though the northern Baltic Sea has low surface-water salinity (*cf.* Fig. 2.15), at salinities higher than ~0.6 in the parent water (the water mass ice forms from), the ice structure resembles that of sea ice. Close to the mouths of rivers where large volumes of freshwater enter the coastal waters the ice often has typical freshwater ice properties.

9.2.3 Salinity and temperature define ice properties

The structure of sea ice differs from that of freshwater ice as some of the salts and other dissolved constituents of the parent water, which are not incorporated into the crystal lattices, are entrapped between the ice crystals and form a hypersaline liquid solution referred to as "brine". Brine inclusions within sea ice form pockets and interconnected channels, and the brine concentration and corresponding brine volume are directly proportional to the ice temperature (Thomas and Dieckmann 2010).

In addition, brines move inside the ice sheet and across the ice/water interface due to gravity drainage and thermodynamic processes. For the establishment of a sympagic community, and the flux of water and/or gas through the ice or across the ice/water interface, the most important physical parameters are the ice porosity (*i.e.* the total volume of brine inclusions) and the permeability (the ability to transport fluid) of the ice. Sea ice is generally more porous than freshwater ice, the ice porosity being affected by the interaction of temperature, brine salinity and brine volume.

According to the "Law of fives", sea ice with a temperature of −5 °C and a bulk sea-ice salinity (melted sea ice including brine) of 5 has a brine volume of 5 % (Golden 2009), which makes it permeable because brine pockets and inclusions are then assumed to be interconnected. However, for the Baltic Sea the "Law of fives" obviously does not hold true since the bulk sea-ice salinity does not reach 5, except in the very southernmost parts (Granskog et al. 2006). However, it can be calculated that for ice with a bulk salinity of 1, a temperature of −1 °C is needed for brine volumes to be large enough for Baltic Sea ice to become permeable ("Law of ones", Leppäranta and Manninen 1988).

For this reason, much of the biology of Baltic Sea ice is restricted to the bottommost part of ice floes, where dense accumulations of organisms give the ice a distinctive brownish colouration in the permeable layer (Fig. 9.4). The rest of the ice, the "internal ice habitat" without visible colouration, has extremely low biological activity because it contains brine of too low a salinity to allow for the interconnection of fluids and permeability. The extent of the permeable layer varies in the different basins of the Baltic Sea as a function of brine salinity and temperature. Thus, as brine salinity and temperature decrease northwards, the permeable layer of the ice becomes narrower towards the north of the Baltic Sea (Fig. 9.5).

9.2.4 Desalination

Sea ice contains a fraction of salts entrapped in so-called "brine channels" and pockets between ice crystals. Brine channels form an interconnected system that allows brine movement inside the ice as well as brine transport from the ice to the underlying water. Since the major part of the sea ice volume

Fig. 9.4 Sample from the lower part of the Baltic Sea ice cover, showing the brown colour of the bottom layer and the brine drainage holes (diameter ≤1 cm) to the water column. Photo: © Jari Uusikivi

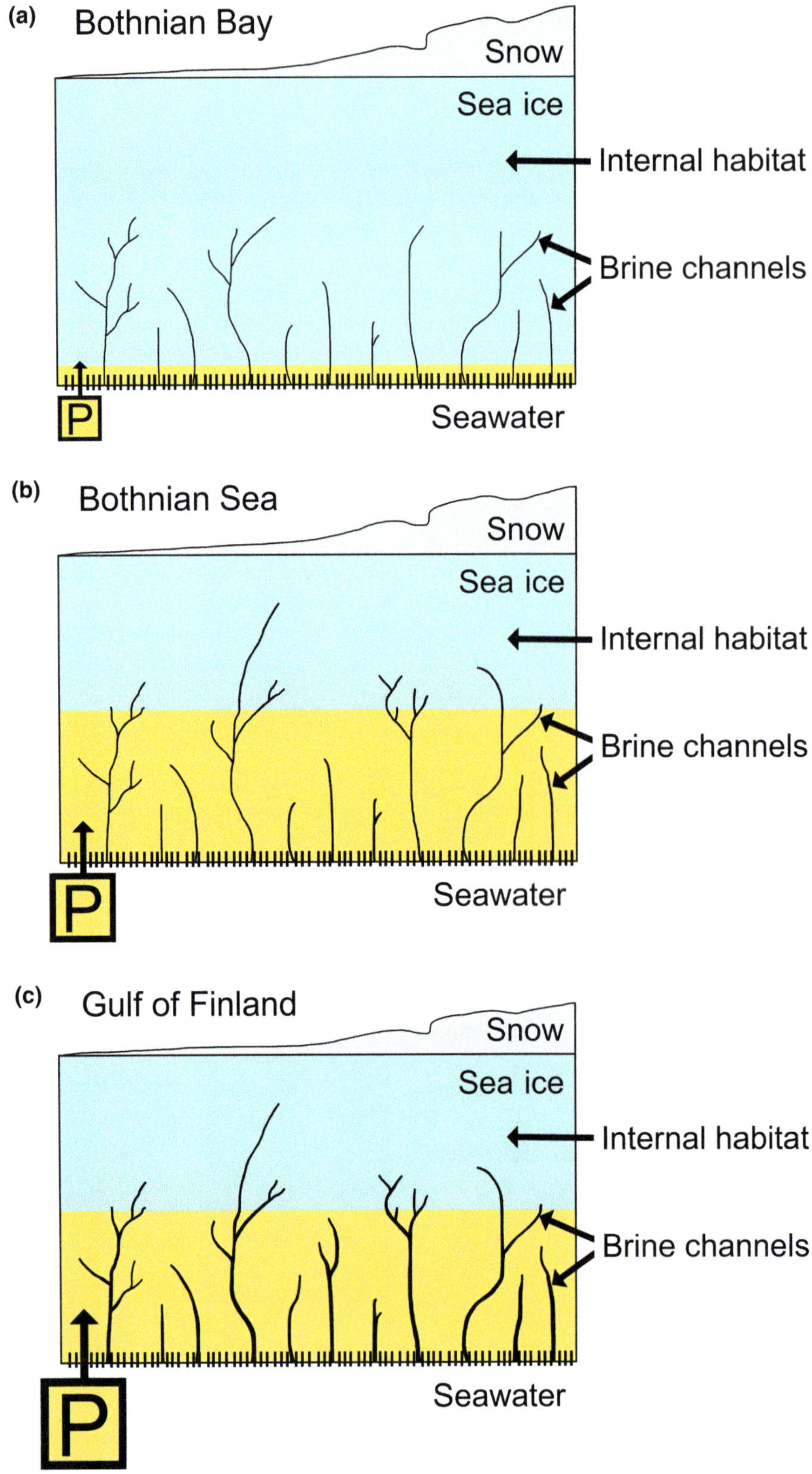

Fig. 9.5 Schematic presentation of sea ice in three different sea areas of the northern Baltic Sea: (**a**) the Bothnian Bay, (**b**) the Bothnian Sea, (**c**) the western Gulf of Finland. Note the increasing size of the brine channels from (**a**) to (**c**). The figures show the snow layer (grey), the internal sea-ice layer with no or very low biological activity due to low ice porosity (blue) and the Biologically Active Layer in the lower part of the sea ice (yellow). P indicates the direction and size of the phosphorus flux in the different sea areas. Figure modified from Kuparinen et al. (2007)

consists of pure ice crystals, and all dissolved constituents, including salts, are in the brine fraction, the transport of saline brines out of an ice sheet causes desalination of the ice.

Ice salinity is a dynamic variable, governed by initial brine entrapment during ice formation and subsequent desalination processes. The initial brine entrapment depends on the salinity of the parent water and the growth velocity of the ice, with faster ice growth at lower temperatures allowing for more brine entrapment between the crystal lattice at the ice bottom and thus higher bulk ice salinity. The initial salt entrapment into growing ice has been measured to be smaller in the Baltic Sea than in marine sea ice (Uusikivi 2013). Once the ice is formed, several processes cause brine loss and thereby desalination.

In general, the desalination processes can be subdivided into those in colder ice during ice growth and those in warmer ice with increased ice porosity (Thomas and Dieckmann 2010). Cold-ice processes are related to pressure build-up within the brine channel system when the ice temperature changes, which can lead to brine expulsion from the upper surface of the ice, and the growth of frost flowers, or at the ice/water interface, brine loss to the underlying water. Other processes are based on gravity drainage, *i.e.* cold saline brines flowing downwards in the ice due to their higher density and brine channel flushing due to the pressure built up melt water accretion in the upper surface of the ice (Granskog et al. 2003a).

Warming of the sea ice leads to melting at the brine channel walls, enlargement of the brine channels and coalescence of isolated brine inclusions (Meese 1989; Weeks 1998). Ice warming thus increases both the ice porosity and the connectivity between the ice and the underlying seawater. The most efficient desalination processes are related to changes in ice porosity. Since the porosity of Baltic Sea ice is generally lower than that of marine sea ice due to the low salinity of the parent water (Meiners et al. 2002), it can be expected that desalination processes function differently. Quantitative information on desalination processes in Baltic Sea ice is virtually non-existent (Granskog et al. 2010), but measurements made by Uusikivi et al. (2006) suggest that the salinity fluxes from ice to seawater in the Baltic Sea are small compared to those from the sea ice formed in ocean waters.

9.3 Baltic sea ice as a habitat

9.3.1 Brine channels

For sympagic organisms, there are several habitats within and associated with the sea ice in which they thrive. Within the ice sheet, the brine channel system is the primary habitat of the ice biota (Fig. 9.5). This is a semi-enclosed system, consisting of partially interconnected small pockets and elongated vertical channels that form at ice-crystal junctions when the ice sheet grows. In the bottommost permeable layer, the brine channels are open to the underlying water and enable the movement of motile organisms into the ice. Not only the brine, but also the brine channel ice surfaces, are colonised by sympagic organisms, and many of the biological and chemical interactions in the ice system may be more like those found in aquatic biofilms than in pelagic systems.

The brine channel habitat is characterised by steep vertical gradients in temperature and salinity that can change on a diel scale following temperature changes. Salinity, pH, dissolved inorganic nutrients and dissolved organic matter within the brine channels change over seasonal scales following the succession of the sympagic communities. Also, the light field inside the ice may rapidly vary as a result of changes in the incoming solar radiation in combination with snow cover thickness. Inside the brine channels all biomass is confined to a space of 2–10 % of the total ice volume, and the brine channel habitat is typically densely packed with organisms compared to the underlying water. The maximum chlorophyll *a* (Chl *a*) concentrations, which may be regarded as a measure of the phototrophic biomass, can reach values of 800–2,000 μg Chl *a* L^{-1} in the brine fraction (Granskog et al. 2006). The brine channel diameter sets an approximate upper limit of 0.2 mm for organism body size (Weissenberger et al. 1992), although channels can be larger during the melting phase.

9.3.2 Spatial variability of ice properties

The other main sea ice habitat, besides brine channels, is the upper ice surface habitat, which consists of slush layers at the ice/snow interface and meltwater ponds on the ice. In addition, ice-bottom habitats occur at the ice/water interfaces, and these consist of a porous skeletal ice layer associated with relatively stable water layers immediately beneath the ice.

The spatial variability in sea-ice properties is controlled mainly by the bulk sea-ice salinity in the different parts of the Baltic Sea since the bulk salinity regulates the ice porosity and habitable space within the ice (larger space with higher salinity). Thus, the variability in ice properties on a subregional scale is largely controlled by the Baltic Sea salinity gradient (*cf.* Fig. 2.15, Meiners et al. 2002; Granskog et al. 2003a).

The regional- and local-scale variation in sea-ice properties in coastal areas is mainly caused by onshore-offshore gradients in salinity created by river water inflows (Granskog et al. 2005; Steffens et al. 2006; Piiparinen et al. 2010). However, even on small scales (tens of metres) the variability in ice properties can be as large as that on local or regional scales (Steffens et al. 2006). The bulk salinity of the ice bottom typically reflects the inshore-offshore salinity gradients in the ice parent water. It has been suggested that this controls the amount and distribution of sea-ice phototrophic biomass (as Chl *a*) and the composition of the sympagic communities in the Baltic Sea (Kaartokallio et al. 2007; Piiparinen et al. 2010), as well as in other non-polar sea ice-covered areas (Thomas and Dieckmann 2010).

9.3.3 Landfast ice *versus* pack ice

The fundamental differences between landfast ice and pack ice (Fig. 9.3) are the location of formation (coast *versus* open sea) and the dynamics (drift/ridging/rafting *versus* stability). This is expected to result in different sympagic communities in these two classes of ice. As drifting pack ice is transported by wind and currents, the sympagic community represents the species composition at the site of ice formation rather than a community typical of the area to where the ice has drifted.

Thus, at the same latitude, the landfast ice, which remains more or less at the same spot from ice formation to spring thaw, may differ greatly in biological properties from the pack ice in the same area. In the Bothnian Bay, the low-biomass early stages of sympagic communities were found to be similar in both types of ice (Piiparinen et al. 2010; Rintala et al. 2010b). However, with the advancing ice season, the communities start to deviate, especially in terms of chlorophyte, ciliate, and rotifer biomass. Chlorophytes (*e.g. Chlamydomonas* sp. and *Dictyosphaerium* sp.) show decreasing trends from landfast to pack ice, while ciliates (*e.g. Lacrymaria rostrata* and *Strombidium* sp.) and rotifers (mainly *Synchaeta* cf. *littoralis*) show the opposite trend (Meiners et al. 2002; Piiparinen et al. 2010; Rintala et al. 2010b).

9.3.4 High amounts of snow-ice are typical of the Baltic Sea

As a consequence of the relatively heavy snow load on Baltic Sea ice, the weight of the snow frequently submerges the ice surface. Thus, the seawater floods the upper ice surface and "snow-ice" is formed when the flooded layers freeze (Granskog et al. 2003b). The superimposed ice is the ice formed from the freezing of the snow melt during melt-freeze cycles brought about by short-term warm weather events (Fig. 9.6), and is a favourable habitat for sympagic communities. This habitat type is especially important in the low-salinity

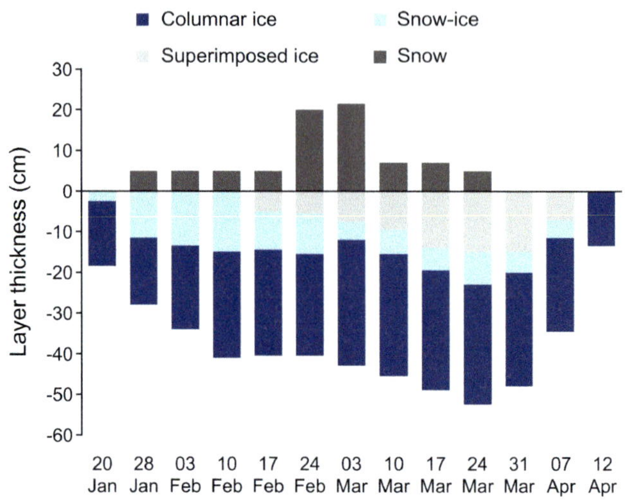

Fig. 9.6 The development of an ice cover at Santala Bay (Gulf of Finland) during winter 1999. The snow/granular ice interface is shown as the reference level (0 cm). Measurements of $\delta^{18}O$ showed that the snow-ice layer consisted of a mixture of ice and seawater while the superimposed ice was formed from melted snow alone. Figure modified from Kawamura et al. (2001)

Bothnian Bay, where the low brine volumes inside the ice greatly constrain the habitability of the ice (Fig. 9.5).

In the Bothnian Bay, the nutrient-rich and well-illuminated snow-ice habitat is primarily occupied by chain-forming centric diatoms, with *Melosira arctica* (Fig. 9.7) as the dominant species (Piiparinen et al. 2010; Rintala et al. 2010b). A dominance of centric diatoms in the ice-surface layer has also been observed in the Gulf of Finland, but here the dominant species is *Chaetoceros wighamii* (Kaartokallio et al. 2007). The contribution of centric diatoms to the total sympagic biomass generally decreases from the Bothnian Bay to the Gulf of Finland, possibly due to the decrease of snow-ice (Rintala et al. 2010b).

9.3.5 Rafting and ridging

Currents or winds often push undeformed ice, pancake ice and larger ice floes around so that they slide over each other, a process known as rafting. Thicker sea ice may fracture and

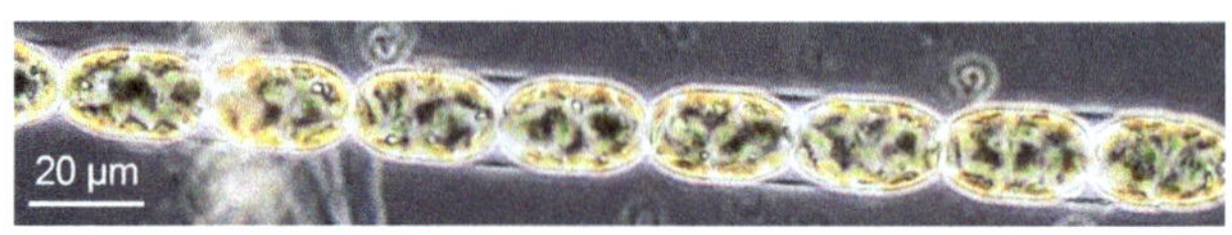

Fig. 9.7 The chain-forming diatom *Melosira arctica* is a typical ice-associated alga which occurs both in the Baltic Sea and in the Arctic Ocean. Photo: © Regina Hansen

pile up under the influence of strong winds and pressure in the ice, forming ridges on the ice surface. The formation of pressure ridges is a common phenomenon in the pack ice of the Baltic Sea and they are typically 3–5 m thick (Kankaanpää 1997). However, freely floating ridges of up to 25 m thick have been observed (Haapala et al. 2015).

In addition to an increase in ice thickness and changes in the ice structure profiles, ridges often increase the sympagic biomass in the ice. Peaks of chlorophyll *a* and other biomass estimates in the centre of rafted sea ice are typical signs of rafting. These biomass peaks originate from the bottom-ice communities of the overlying ice floe and/or the surface-ice communities of the underlying ice floe (Rintala et al. 2010b). Occasionally the dynamic forces in a pack ice field may flip ice floes, resulting in the bottom-ice communities becoming effective ice-surface communities and *vice versa*. Thus, dynamic events in the pack ice may subject sympagic communities to changed physico-chemical conditions under which *e.g.* irradiation and/or nutrients reaching the organisms may be reduced or increased.

9.4 Productivity in Baltic Sea ice

9.4.1 Factors that regulate growth of sympagic organisms

The key factors regulating the growth and succession of autotrophic sympagic organisms are light, salinity and nutrients, while for heterotrophic organisms the key factors are salinity and organic matter. In general terms, the main nutrient supply in the internal sea ice habitat consists of the initial nutrients incorporated when the ice once formed. In older ice brine channels, exchange can result in the transport of nutrients across the ice/water interface. Concentrations of nitrogenous inorganic nutrients in melted sea ice are typically higher than those in the under-ice water, whereas dissolved inorganic phosphate concentrations show an opposite trend (Kuparinen et al. 2007).

The snow cover on the ice accumulates nutrients carried by precipitation, and these nutrients can be transported down into the ice sheet through snow-melting during warm weather (Granskog et al. 2003a; Granskog and Kaartokallio 2004). The recycling of nutrients from allochthonous (transported into the system) and autochthonous (produced within the system) biomass in the sea ice, through decomposition and nutrient regeneration carried out by sympagic heterotrophs, can also be an important source of nutrient supply inside the ice. Here the main actors are heterotrophic bacteria, which presumably degrade both particulate and dissolved organic matter and regenerate nutrients, as in other aquatic systems. However, phagotrophic protists are also likely to be key nutrient regenerators in the sea-ice environment (Kaartokallio 2004).

9.4.2 Autotrophic biomass and primary production

About 85 % of the sympagic communities' total biomass in the Baltic Sea consist of microalgae (Figs. 9.8 and 9.9).

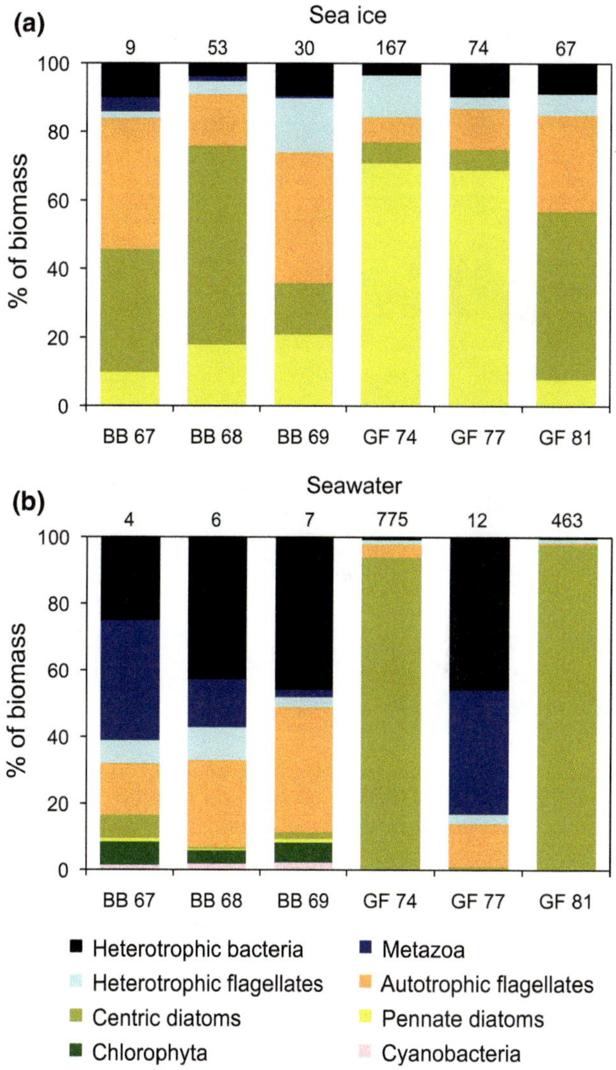

Fig. 9.8 Community composition in sea ice and seawater based on the examination of integrated samples taken at three ice stations in the Bothnian Bay (BB) and three ice stations in the Gulf of Finland (GF) in March 2000. (**a**) Community composition in 23–30 cm thick sea ice after melting in the laboratory. (**b**) Community composition in the water column immediately under the ice (at a 0–10 m water depth). The numbers above the bars denote the total biomass in μg C L^{-1}. Figure modified from Meiners et al. (2002)

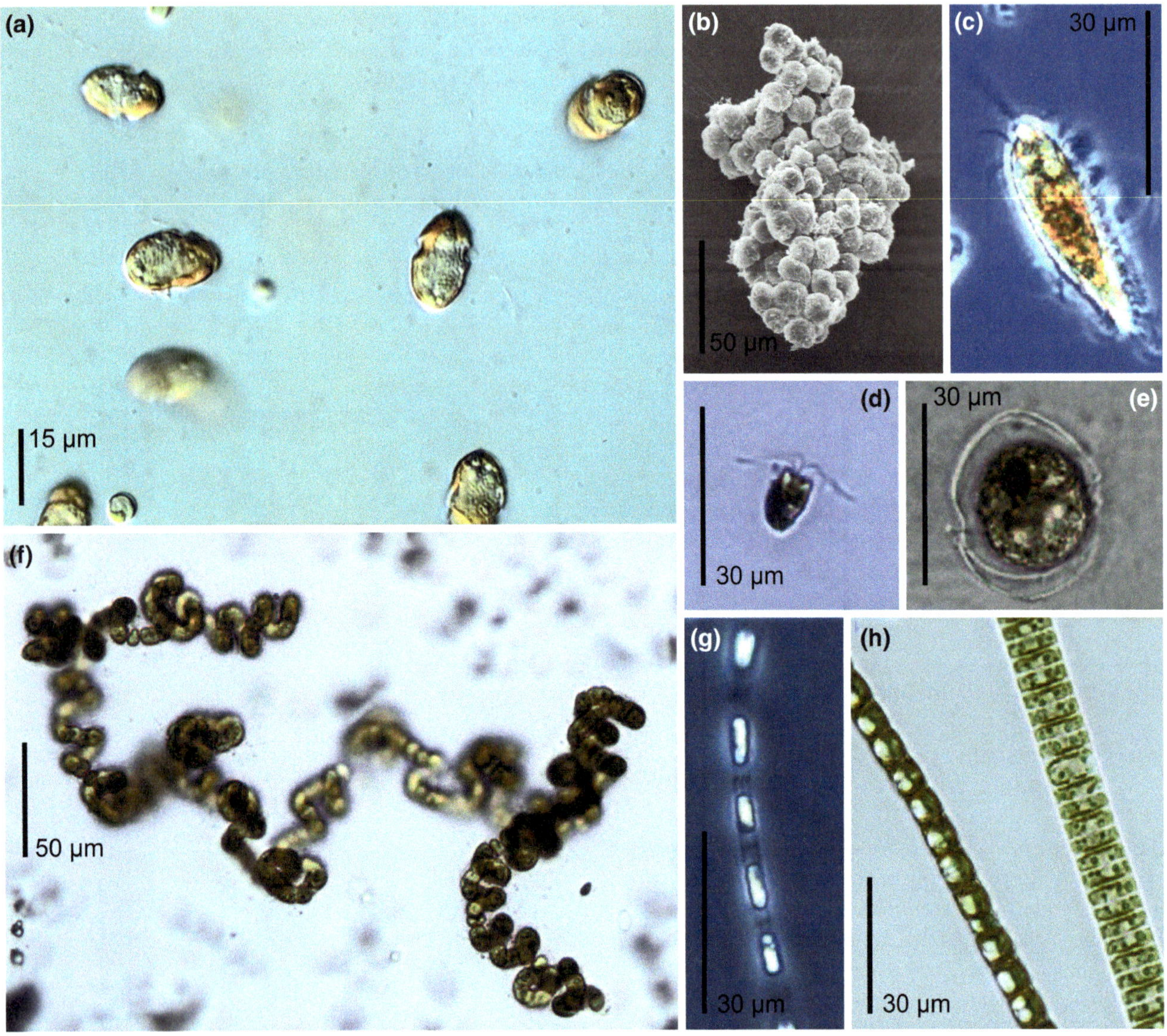

Fig. 9.9 Examples of microorganisms found in sea ice: (**a**) the dinoflagellate *Heterocapsa arctica* subsp. *frigida*, (**b**) the dinoflagellate *Scrippsiella hangoei*, (**c**) a ciliate, (**d**) the prasinophyte *Pyramimonas* sp., (**e**) the dinoflagellate *Scrippsiella hangoei* during its transformation into a pellicular cyst when the cell has already shredded the flagellate and is withdrawn from its theca, (**f**) the cyanobacterium *Nodularia spumigena*, not a sympagic species (**g**) a diatom chain of *Skeletonema* sp. cells, (**h**) two diatom chains of *Pauliella taeniata* cells seen from different angles. All images show organisms from Baltic Sea ice samples or from cell cultures isolated from Baltic Sea ice samples. All images were taken under light microscopy, except for (**b**), which was taken with scanning electron microscopy (SEM). Photo: © Janne-Markus Rintala

Autotrophic growth in Baltic Sea ice is sequentially light-limited and nutrient-limited as winter progresses, as in polar sympagic communities. Phosphorus is thought to be the most important single limiting nutrient (Haecky et al. 1998; Kuosa and Kaartokallio 2006).

The *in situ* level of primary production in sea ice is difficult to quantify due to the lack of suitable methods (Box 9.1). Haecky and Andersson (1999) estimated that the ice-algal production ($\sim$0.1 g C m^{-2}) accounted for only $\sim$1 % of the total annual open-sea production (ice and pelagic) and $\sim$10 % of the total open-sea production during the ice-cover season. However, photosynthetic parameters measured in Baltic Sea ice point to a highly variable, but at times very active, primary productivity in all parts of the Baltic Sea (Piiparinen et al. 2010; Rintala et al. 2010b; Piiparinen and Kuosa 2011).

The average biomass of the phototrophic part of the sympagic communities during the ice season is generally higher in the Gulf of Finland than in the Gulf of Bothnia. Typical chlorophyll *a* concentrations in Gulf of Finland ice are up to ~5.5 mg m^{-2}, but only up to ~2.2 mg m^{-2} in Gulf of Bothnia ice (Granskog et al. 2006). The under-ice water has typically a lower phototrophic biomass than the ice, although a high phototrophic biomass (>200–300 µg Chl *a* L^{-1}) can occur during bloom conditions under the ice just before ice break-up (*e.g.* Stations GF 74 and GF 81 in Fig. 9.8b). Pressure ridges may also be hotspots of phototrophic biomass accumulation as the biomass in the keel ice blocks and interstitial water has been reported to be high (up to 50 µg Chl *a* L^{-1}, Kuparinen et al. 2007), even in comparison with intense phytoplankton blooms in open Baltic Sea waters.

In the Bothnian Bay, the maximum phototrophic biomass is restricted to the ice bottom in both landfast ice and pack ice (Kuparinen et al. 2007, Rintala et al. 2010b). Also, in the Gulf of Finland, the maximum phototrophic biomass is typically found in the lowermost ice section where phosphorus accumulates (Granskog et al. 2005). However, in the Norra Kvarken area between the Bothnian Bay and the Bothnian Sea, both the major nutrient regeneration activity and the major ice phototrophic spring bloom were found in the interior ice layers (Norrman and Andersson 1994; Haecky et al. 1998), although ice-bottom maxima may also occur in this region (Haecky and Andersson 1999; Kaartokallio 2004).

Both surface and interior phototrophic biomass maxima have also been found in coastal ice in the Gulf of Bothnia and the Gulf of Finland (Granskog et al. 2005; Rintala et al. 2006; Kaartokallio et al. 2007; Piiparinen et al. 2010). As the research progresses, evidence regarding the occurrence of surface biomass maxima is accumulating (Piiparinen et al. 2010; Rintala et al. 2010b). Those combined observations show vertical biomass distribution patterns in the Baltic Sea, whereby the Bothnian Bay ice displays blooms that are restricted to the ice bottom and generally lower phototrophic biomass. More to the south, the location of the sympagic biomass maximum is variable, and the biomass is generally higher. It is probable that the availability of nutrients and their transport from the underlying water to the ice, as well as nutrient regeneration inside the ice, are decisive factors for biomass development when habitat space is available.

9.4.3 Bacterial biomass and activity

About 8 % of the total biomass of the sympagic communities in the Baltic Sea consist of bacteria (Fig. 9.8a). Heterotrophic bacteria in Baltic Sea ice also show vertical biomass distribution patterns that are analogous to those of the phototrophic part of the sympagic communities. In general, bacterial biomass and production in the ice are lowest in the Bothnian Bay, whereas the sea ice in the Norra Kvarken area and Gulf of Finland shows a higher bacterial biomass and higher production rates (Kuparinen et al. 2007).

Bacterial biomass in the sea ice of the Norra Kvarken area is in the range of 1–8 µg C L^{-1}. In the Gulf of Finland it is 4–10 µg C L^{-1} while production rates are 0.001–0.6 µg C L^{-1} h^{-1} and 0.03–1.1 µg C L^{-1} h^{-1}, respectively (Kuparinen et al. 2007). The maximum bacterial production rates in the sympagic communities are similar to the maximum values measured in the Baltic Sea surface waters during plankton blooms at other times of the year (Kuparinen et al. 2007, 2011). Short turn-over times and high per-cell activity in sympagic bacteria, compared to the open-water bacteria, imply that sympagic bacteria have a high capacity to process organic carbon and to regenerate nutrients (Kuparinen et al. 2007).

9.4.4 Dissolved organic matter (DOM)

Bacterial growth in Baltic Sea ice is sequentially limited by nutrients and substrate as winter progresses (Kuosa and Kaartokallio 2006). This suggests that the bacteria depend on the production of autochthonous DOM by phototrophs. As opposed to polar sea ice, DOM concentrations in Baltic Sea ice are lower than in underlying waters because of the generally high concentrations of terrestrially-derived DOM in Baltic Sea water (*cf.* Sect. 15.2.6). However, the DOM concentrations in Baltic Sea ice are still higher than those in Arctic sea ice. The high DOM loading of Baltic Sea waters and ice result in the ice having quite different chemical and optical characteristics compared to those known from polar oceans (Granskog et al. 2006).

Inside Baltic Sea ice, the DOM is thought to originate both from material incorporated into the ice during its formation and from autochthonous matter produced by the organisms inhabiting the ice, the latter largely comprising carbohydrate-rich polysaccharides (Underwood et al. 2013; Krembs et al. 2011). Thus, the DOM in Baltic Sea ice has a complex origin, being partially terrestrial and partially produced within the ice (Stedmon et al. 2007). The complex origin of the ice DOM may lead to uncoupled dynamics of DOC (dissolved organic carbon), DON (dissolved organic nitrogen) and DOP (dissolved organic phosphorus), as shown by *e.g.* the lack of significant correlations between DOM and bacterial parameters in Baltic Sea ice (Kaartokallio et al. 2007).

Box 9.1: Sampling of Sympagic Communities

Ice coring

Sampling of sea ice presents some intriguing difficulties because, in order to investigate the organisms living in the ice itself, the habitat has to be destroyed (Box Fig. 9.1). This means melting the ice. Sampling the water column under the ice is not quite as problematic, but it is nonetheless not a trivial undertaking if the ice is thick. The polar, lake and Baltic Sea ice is sampled mainly by using ice corers that are power-driven, or on thin ice (<50 cm), hand-held. Most commercially available corers produce cores with diameters around 10 cm (Box Fig. 9.1a). During the ice core removal, the core is compromised in that the brine it contains is drained out. There is no solution to this problem, and draining is largest when the ice is warm and brine channels are wide and strongly interconnected. Other problems involve the warming up and melting of the ice core and a change in the light environment. Thus, the highest priority is to sample the ice before these environmental changes induce alterations to the biology and/or chemistry of the ice that are too significant. Traditionally, the retrieved ice core is sectioned (using ice saws) into ice horizons (slices typically 2–10 cm wide) as soon as possible after retrieval. Each of the sections is immediately placed in a clean plastic container that can be closed for safe return to the laboratory (Box Fig. 9.1e). Usually, replicate cores are taken at the same sampling site, since there is often not enough volume of ice in one core to perform the standard suite of measurements that are typically carried out.

Studies under the ice

The ice core collection leaves a convenient hole in the ice through which water-sampling bottles can be operated for the purpose of taking under-ice seawater samples (Box Fig. 9.1b). The hole can be also used to deploy other equipment such as small CTDs (to measure conductivity, temperature and water depth), video/camera equipment (to examine the underside of the ice cover) and light sensors (to measure the amount and quality of the light penetrating through the ice and snow covers).

Processing of ice cores

There has been much debate over the years concerning how to best treat these sections for subsequent analyses. Rintala et al. (2014) have shown that probably the best method for most usual biological and chemical analyses of Baltic Sea ice is to melt the ice sections as rapidly as possible at room temperature, while gently shaking the sample to keep it cool. If the sample is subsampled just as the last ice melts, the temperature of the whole sample should still be only just above the freezing point. The standard suite of measurements that are typically carried out include measurements of ice crystal structure, salinity, temperature, chlorophyll *a*, dissolved inorganic and organic constituents, bacterial, algal and other protists' enumeration and activities. It is of key importance to protect the cores from contamination by seawater and/or by material from clothing and hands. Therefore, the cores should be handled as little as possible.

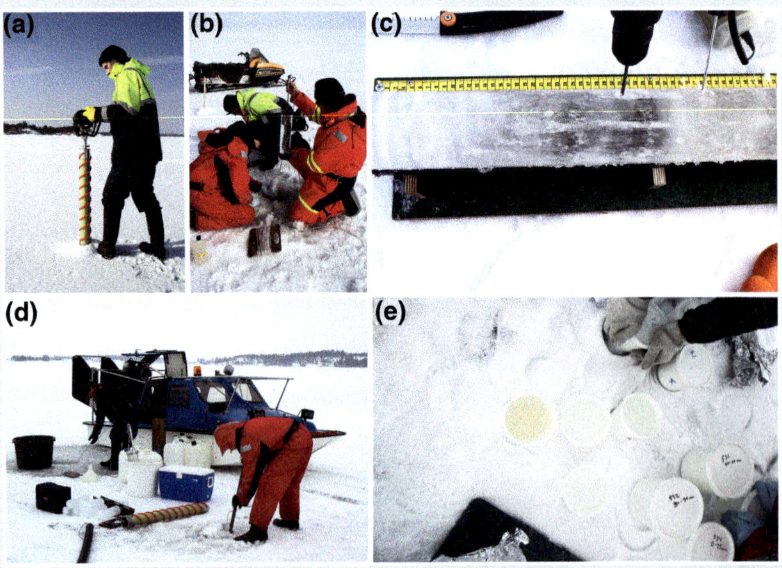

Box Fig. 9.1 Sampling of ice and seawater in the northern Baltic Sea. (**a**) A scientist operating an ice corer to make a hole in the ice and retrieve an ice core. (**b**) Sampling of under-ice seawater through the hole made by the ice corer. (**c**) Treatment of an ice core in the field. Holes are drilled and the temperature inside the ice is measured. (**d**) Salinity and other parameters are measured in the water under the ice. (**e**) The ice core is subdivided into slices and each slice is put into a separate container. Photo: © David N. Thomas

9.5 Diversity of sympagic organisms

9.5.1 Composition of sympagic communities

Our knowledge of the taxonomic composition of sympagic communities has greatly increased over the last 20 years and new species continue to be described (Rintala et al. 2010a). This progress has been achieved by microscopy and conventional cell counts (Figs. 9.8 and 9.9), but also by molecular techniques (Fig. 9.10).

In general, molecular studies on sea-ice eukaryotic diversity in the Baltic Sea provide results similar to morphological studies, although the diversity resolved with molecular tools is generally higher than that produced by studies based on morphology. Several organism groups, such as heterotrophic flagellates, may be better represented in molecular studies. The reverse is true for other groups, such as haptophytes and diatoms, which are generally underrepresented in environmental molecular studies.

Based on molecular studies of Baltic Sea ice, the most common phototrophic groups (in the order of decreasing species richness) are diatoms, dinoflagellates, chlorophytes, cryptophytes, pelagophytes, haptophytes, synurophytes, bolidophytes, dictyochophytes, eustigmatophytes and chrysophytes. The most species-rich groups of heterotrophic protists in Baltic Sea ice are ciliates, cercozoans, dinoflagellates, choanoflagellates, novel unnamed stramenopiles, chrysophytes, labyrinthulids and *Telonema* spp. (Majaneva et al. 2012).

Many of the frequently observed phototrophic sympagic taxa in the Baltic Sea, such as *Melosira arctica*, *Pauliella taeniata* and *Peridiniella catenata*, are also commonly found in association with Arctic sea ice. As in the Arctic Ocean, the ice near river mouths in the Baltic Sea has a clear freshwater component with abundant chlorophytes. However, not all organisms found in the sea ice are "truly" sympagic. For example, some colonial filamentous cyanobacteria typical of Baltic Sea summer blooms, such as *Nodularia spumigena* (Fig. 9.9f), may survive in Baltic Sea ice, possibly overwintering there (Rintala et al. 2010b).

9.5.2 Bacteria and viruses

Heterotrophic bacteria are the most abundant prokaryote group in the sea ice. They contribute 4–11 % (average 8 %) to the total biomass of the Baltic Sea sympagic communities

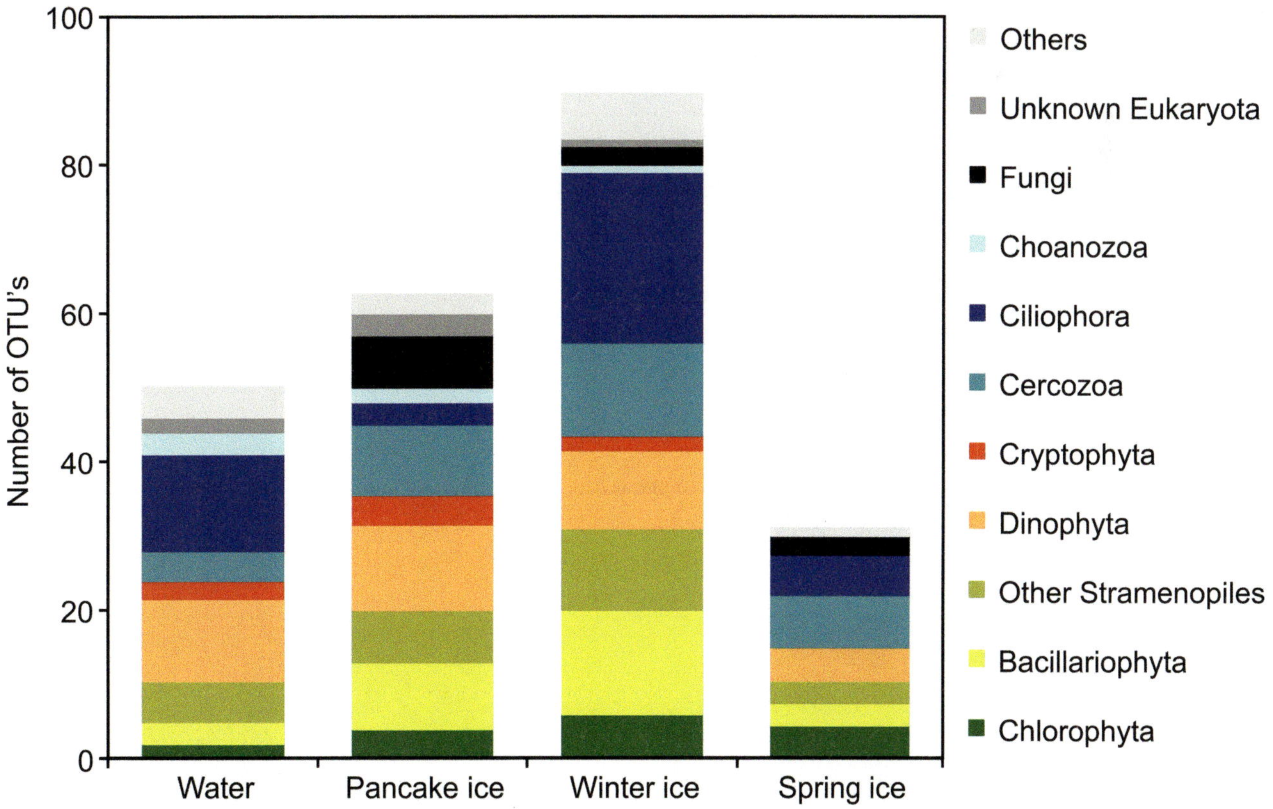

Fig. 9.10 Succession of eukaryotic community composition based on sequencing of the 18S rRNA gene from integrated samples of the Gulf of Finland and the Gulf of Bothnia. OTU = operational taxonomic unit. Figure modified from Majaneva et al. (2012)

Table 9.2 Examples of eukaryotes found in sympagic communities in the Baltic Sea. Data updated from Granskog et al. (2006). Citations: 1 = Hällfors and Niemi (1974), 2 = Huttunen and Niemi (1986), 3 = Norrman and Andersson (1994), 4 = Larsen et al. (1995), 5 = Ikävalko and Thomsen (1997), 6 = Haecky et al. (1998), 7 = Meiners et al. (2002), 8 = Kaartokallio (2004), 9 = Werner and Auel (2004), 10 = Piiparinen et al. (2010), 11 = Rintala et al. (2010a), 12 = Rintala et al. (2010b), 13 = Majaneva et al. (2012)

Species	Taxonomic group	Citations
Autotrophs		
Chaetoceros neogracilis	Heterokontophyta (centric diatom)	13
Chaetoceros wighamii	Heterokontophyta (centric diatom)	5, 6, 10, 12
Melosira arctica	Heterokontophyta (centric diatom)	2, 3, 5, 6, 7, 10
Skeletonema grevillei	Heterokontophyta (centric diatom)	13
Thalassiosira hyperborea	Heterokontophyta (centric diatom)	7
Fragillariopsis cylindrus	Heterokontophyta (pennate diatom)	7
Navicula pelagica	Heterokontophyta (pennate diatom)	3, 5
Navicula vanhoeffenii	Heterokontophyta (pennate diatom)	2, 3, 5, 10, 12
Nitzschia frigida	Heterokontophyta (pennate diatom)	2, 3, 5, 6, 7, 12
Pauliella taeniata	Heterokontophyta (pennate diatom)	2, 3, 6, 10, 7, 12
Nannochloropsis limnetica	Heterokontophyta (eustigmatophyte)	13
Paraphysomonas spp.	Heterokontophyta (chrysophyte)	5
Chrysochromulina birgeri	Haptophyta (prymnesiophyte)	1
Biecheleria baltica	Dinophyta	13
Gyrodinium fusiforme	Dinophyta	13
Heterocapsa arctica subsp. *frigida*	Dinophyta	11
Pentapharsodinium tyrrhenicum	Dinophyta	13
Peridiniella catenata	Dinophyta	2, 3, 5, 7, 10, 12
Polarella glacialis	Dinophyta	13
Scrippsiella hangoei	Dinophyta	4, 10, 12, 13
Chlamydomonas raudensis	Chlorophyta	13
Dictyosphaerium sp.	Chlorophyta	10
Monoraphidium contortum	Chlorophyta	5, 7, 10, 12
Oocystis heteromucosa	Chlorophyta	13
Ostreococcus tauri	Chlorophyta	13
Planktonema lauterbornii	Chlorophyta	10
Polytoma papillatum	Chlorophyta	5
Pyramimonas gelidicola	Chlorophyta	13
Heterotrophs		
Diaphanoeca sphaerica	Choanozoa	5
Savillea micropora	Choanozoa	5
Cryothecomonas aestivalis	Protozoa *incertae sedis*	13
Cryothecomonas armigera	Protozoa *incertae sedis*	5, 7
Cryothecomonas longipes	Protozoa *incertae sedis*	13
Quadricilia rotundata	Protozoa *incertae sedis*	5
Bursaria sp.	Ciliophora	8
Lacrymaria rostrata	Ciliophora	12, 13
Mesodinium rubrum	Ciliophora	8, 12, 13
Strombidium spp.	Ciliophora	5, 8, 12
Keratella spp.	Rotifera	7, 9

(continued)

Table 9.2 (continued)

Species	Taxonomic group	Citations
Synchaeta baltica	Rotifera	3, 7
Synchaeta cf. *littoralis*	Rotifera	8, 9, 12
Acartia bifilosa	Arthropoda (Copepoda)	8, 9
Fungi		
Eurotium rubrum	Ascomycota	13
Debaryomyces hansenii	Ascomycota	13
Rhodotorula mucilaginosa	Basidiomycota	13
Sclerotium sp.	Basidiomycota	13
Graphiola phoenicis	Basidiomycota	13

(Fig. 9.8, Meiners et al. 2002). Picocyanobacterial cells may be abundant, but never exceed 0.1 % of the total sympagic biomass. Studies that focus on Baltic Sea ice bacteria have been performed in the Gulf of Bothnia, the Gulf of Finland and in the Kiel Bay (southwestern Belt Sea). In general, the bacterial diversity found in Baltic Sea ice is fairly similar to that of the sea ice in both the Arctic Ocean and the Southern Ocean. This suggests that the factors shaping the communities are the same despite the different geographical locations.

The bacteria in the sympagic communities develop from bacteria in the parent water as a result of physical, chemical and biological processes. At the beginning of winter, the bacteria seem to go through an adaptive phase. When sea ice forms, the bacterial numbers are relatively high and cells are dividing, but activities are lower than in the under-ice water. Another important phase change for the sympagic bacteria is a peak in bacterial production after an ice-algal bloom (Kaartokallio 2004; Kaartokallio et al. 2008).

Bacteria from α-Proteobacteria (*e.g.* the genus *Loktanella*), β-Proteobacteria (*e.g.* the family Comamonadaceae), and γ-Proteobacteria (*e.g.* the genera *Colwellia*, *Psychromonas* and *Shewanella*), Bacteroidetes (*e.g.* the genus *Flavobacterium*) and Actinobacteria have been recorded in Baltic Sea ice. The same bacterial classes and phyla occur in Baltic Sea water. The bacterial components of the sympagic communities apparently have successional patterns resulting from exchange processes at the ice/water interface, the maturity of the ice and the availability of substrate for ice algae (Kaartokallio et al. 2008; Eronen-Rasimus et al. 2015).

At the early stages of the ice formation (nilas and pancake ice) the sympagic bacteria in the Gulf of Bothnia drift ice are reminiscent of those in the parent water with a dominance of Actinobacteria and α-Proteobacteria, whereas the older columnar ice supports typical sympagic communities with a dominance of Flavobacteriia and γ-Proteobacteria, which is similar to polar sea ice (Eronen-Rasimus et al. 2015). Dominant taxa in older columnar ice are known to be able to efficiently utilise high substrate concentrations, *e.g.* in conjunction with ice-algal blooms. The maturity of ice as a structuring factor regarding the bacterial component of the sympagic communities may thus be related to increased supply of autochthonous organic matter as a substrate rather than to the time elapsed from ice formation *per se*.

In a recent study, bacteria isolated from Baltic Sea ice were shown to affiliate with *Flavobacterium gelidilacus*, *Shewanella baltica* and *Shewanella frigidimarina* (Luhtanen et al. 2014). *Flavobacterium* and *Shewanella* are common sea ice bacterial genera (Thomas and Dieckmann 2010). In the isolated bacterial strains, a total of seven bacteria-infecting viruses (bacteriophages) were found. These viruses represented the families *Siphoviridae* and *Myoviridae* with hosts belonging to the classes Flavobacteriia and γ-Proteobacteria (Luhtanen et al. 2014). The ecological significance of these viruses is still unknown, but since the host organisms are common sea-ice bacteria, the viruses can potentially modify the sympagic community dynamics in Baltic Sea ice.

9.5.3 Diatoms

In their recent synthesis of sympagic organisms from Arctic sea ice, Poulin et al. (2011) found 71 % of the 1,027 Arctic eukaryote taxa to be diatoms. Diatoms are also the dominant primary producers in and under Baltic Sea ice. Both pennate and centric diatoms usually dominate the sympagic community biomass with 10–71 % (average 33 %) and 6–57 % (average 29 %) of the total biomass, respectively (Fig. 9.8). The pennates *Pauliella taeniata* and *Nitzschia* spp., and the centrics *Melosira arctica* and *Chaetoceros* spp., are the most typical sea-ice diatoms in Baltic Sea ice (Table 9.2, Piiparinen et al. 2010; Rintala et al. 2010b; Majaneva et al. 2012). *Pauliella taeniata* and *Melosira arctica* are examples of glacial relicts, which today occur in association with sea ice both in the Baltic Sea and in the Arctic Ocean.

Box 9.2: Succession in sympagic communities

Spational and seasonal changes of the sea ice habitat

Ice formation, ice growth, ice melting and community succession follow seasonal changes in solar angle and air temperature (Box Fig. 9.2). The timing of these processes depends on latitude and weather conditions (*cf.* Fig. 9.2). The coldest months are January and February. Sea ice formation begins in January and the ice cover continues to grow downwards through February and reaches its maximum thickness in March. Ice melting starts in early April and ice breaks up and finally disappears between mid-April and May. Low-saline river waters and ice-melt water are fresher and therefore lighter than the brackish under-ice water (*cf.* Sect. 1.2.3), and can form distinct freshwater layers under the ice (Box Fig. 9.2). Snow accumulates on top of the ice and its weight can submerge the ice, leading to snow-ice formation as the slush layer freezes (*cf.* Fig. 9.6). Superimposed ice is ice formed from the freezing of melted snow during melt-freeze cycles brought about by short-term warm weather events, *e.g.* when a diminishing snowpack enhances the amount of solar radiation energy that is able to pass into the ice.

Three successional stages in the sympagic communities are distinguished:

Phase A is a low-productivity winter stage with only a low amount of solar radiation entering the ice; accordingly, biomass is low. The dominance of rotifers early in the season in thin young ice also occurs in the Arctic (Friedrich and De Smet 2000). This is possibly due to the effective entrapment of pelagic rotifers (or their eggs) from the parent water mass during ice formation, or their active migration into the ice from under-ice water. Rotifers are known to effectively utilise detritus as a food source, which explains their growth in this low-productivity phase.

Phase B starts when the amount of light increases after the mid-winter minimum and the ice-algal bloom period with high autotrophic productivity and biomass formation sets in. This phase is characterised by a strong dominance of autotrophic organisms in the total organism standing stocks. Diatoms and dinoflagellates dominate the biomass.

Phase C is the post-bloom phase, which starts when the ice-algal bloom is terminated. This phase is characterised by high heterotrophic productivity, with rotifers and heterotrophic bacteria being the most important organism groups in terms of biomass.

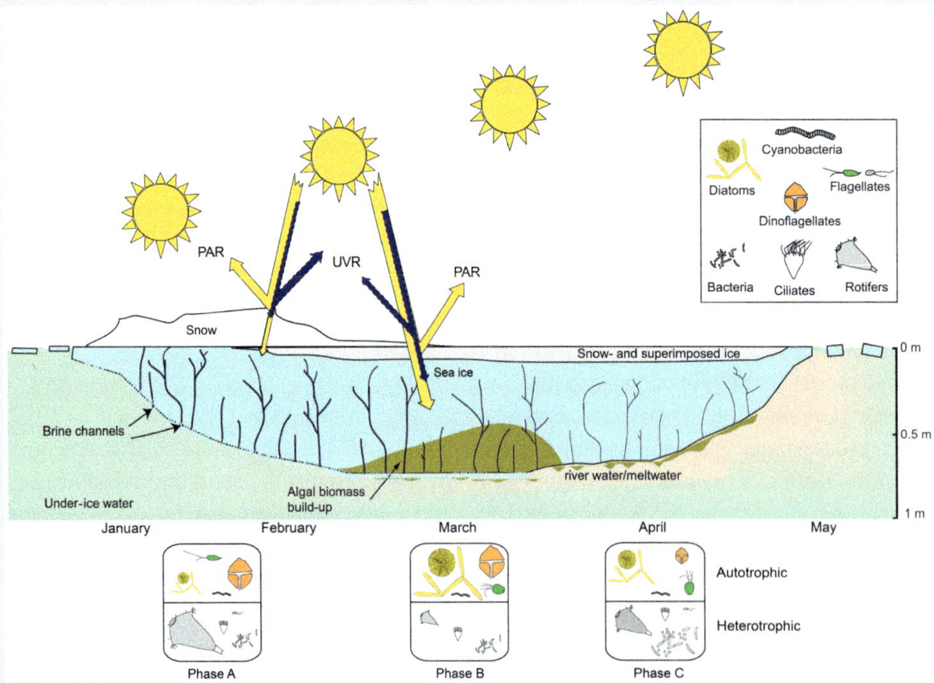

Box Fig. 9.2 A schematic presentation showing the major successional stages of sympagic communities during the ice season in the Baltic Sea. The size of the symbols denoting a group of organisms (diatoms, dinoflagellates, bacteria, ciliates and rotifers) is roughly proportional to the biomass of those groups within each successional stage. The total biomass of the ice organisms also varies between phases, with Phase A having the lowest and Phase B the highest biomass. PAR = photosynthetically active radiation, UVR = ultraviolet radiation. Figure: © Hermanni Kaartokallio

When morphological studies of the same sea-ice samples were compared to the molecular data, the molecular analyses showed a lower relative diatom richness compared to morphological analyses (Majaneva et al. 2012). It is likely that diatoms are better represented in light-microscopy studies because they generally have a larger cell size and more distinguishable morphological characters compared to other sympagic eukaryotes.

A striking difference between the community composition in the sea ice and that of the phytoplankton under the sea ice is the enormous difference in the abundances of raphe-bearing pennate diatoms such as *Navicula* and *Nitzschia* species (Table 9.2). This group is usually nearly absent in the phytoplankton, but may completely dominate the biomass during the sympagic community bloom period inside the ice (Box 9.2, Fig. 9.8).

Diatoms are not able to move in water (except for flagellated male gametes), but pennate species that possess a raphe can glide over a surface (*e.g.* ice) at speeds >10 μm s^{-1} (Cohn and Witzell 1996). Motility is even an advantage inside the ice as the cells have the ability to actively expose themselves to optimal light conditions, *i.e.* they can move towards the light and they can also move away from supersaturating irradiance to avoid oxidative damage. Another advantage of motility is that the cells can move to more nutrient-rich microhabitats when nutrients are limiting growth.

Non-motile pennate chain-forming diatoms (*e.g. Pauliella taeniata*) are more confined to the ice/water interface, using the ice for attachment of the chains, while the centric chain-forming diatoms *Chaetoceros wighamii* and *Melosira arctica* are dominant in surface layers composed of snow-ice (Kaartokallio et al. 2007; Piiparinen et al. 2010; Rintala et al. 2010b).

9.5.4 Flagellates

When diatoms do not dominate the sympagic community biomass, flagellates (especially dinoflagellates) usually become dominants in Baltic Sea ice (Kuosa and Kaartokallio 2006; Kaartokallio et al. 2007). Autotrophic flagellates contribute 8–38 % (average 23 %), and heterotrophic flagellates 2–17 % (average 7 %) to the total biomass (Meiners et al. 2002). Besides dinoflagellates, the flagellate group includes bolido-, chryso-, crypto-, dictyocho-, dino-, eustigmato-, hapto-, pelago- and prasinophytes and Cercozoa (Table 9.2, Spilling 2007; Piiparinen et al. 2010; Rintala et al. 2010b).

The smaller flagellates are usually well-represented in molecular data but in light-microscopy studies they are often combined into a generic group of "auto- and heterotrophic flagellates" due to their small size and general lack of distinct visible features. Among the Cercozoa (Rhizaria), especially abundant in Baltic Sea ice, are members of the genera *Cryothecomonas* and *Protaspis* (Ikävalko 1998; Majaneva et al. 2012).

Some phototrophic flagellates are capable of mixotrophic modes of nutrition, meaning that they can combine autotrophy and heterotrophy both by taking up inorganic carbon, nitrogen and phosphorus directly from the water and by utilising organic nutrients, *e.g.* via phagotrophy (both phagocytosis and pinocytosis). Extracellular enzymes have been successfully used to detect the occurrence of mixotrophy in aquatic organisms and these enzymes have also been found in the sympagic communities of the Baltic Sea (Rintala 2009). The capability of mixotrophy provides any organism with a competitive advantage over strict autotrophs and heterotrophs, *e.g.* when sudden changes in snow cover have a large impact on the incoming solar radiation.

Some flagellate species possess unique survival strategies in the form of resting stages, which can enable a population to survive unfavourable environmental conditions. An example of the successful use of a resting stage in sympagic communities is the formation of cysts of the dinoflagellate *Scrippsiella hangoei* in Baltic Sea ice (*cf.* Fig. 9.9e, Rintala et al. 2007). It is evident that encystment provides *Scrippsiella hangoei* populations with a possibility to survive prolonged periods of low irradiance, which are common in the sea ice (especially when it is snow-covered).

9.5.5 Ciliates

The main heterotrophic protists in Baltic Sea ice, besides heterotrophic flagellates, are ciliates of various cell sizes. Ciliates seem to be more important in Baltic Sea ice than in other non-polar ice-covered areas (Thomas and Dieckmann 2010), probably reflecting the paucity of metazoans. The present data are still limited, but the ciliate component of the Baltic Sea sympagic communities seems to be dominated by species of the genus *Strombidium*. Molecular analyses have confirmed the dominance of Strombidiidae and also identified the genus *Lacrymaria* and several unidentified ciliates as common (Majaneva et al. 2012).

As in the Arctic sea ice, most of the ciliates are relatively small (20–80 μm), which implies their possible role as grazers of small particles, including bacteria. Mixotrophic ciliate species have not been recorded in the Baltic Sea sympagic communities, except for *Mesodinium rubrum* (Kaartokallio et al. 2007; Rintala et al. 2010b). A notable feature of the ice-associated ciliate fauna is the growth of large ciliates (*Bursaria* sp.) under the ice (Kaartokallio 2004, Kaartokallio et al. 2007; Rintala et al. 2010b). These large species obviously graze on dinoflagellates, which is otherwise not common for Baltic Sea ciliates.

9.5.6 Metazoans

In contrast to polar areas where a diverse metazoan fauna occurs in sea ice (Thomas and Dieckmann 2010), the metazoans represented in Baltic Sea ice are limited to rotifers and the nauplii stages of copepods (Meiners et al. 2002; Werner and Auel 2004). Metazoans contribute less than 4.3 % (average 1.0 %) to the total biomass of the sympagic communities (Meiners et al. 2002). Typical taxa encountered in Baltic Sea ice are the rotifers *Keratella* spp. and *Synchaeta* spp., and nauplii of the copepod *Acartia bifilosa* (Table 9.2).

Rotifer biomass can make up as much as 30 % of the total sympagic community biomass during the low-productive winter period, but is reported to be in the range of 1–7 % during ice-algal bloom periods (Meiners et al. 2002; Granskog et al. 2006). It is possible that the sea ice in the Baltic Sea functions as an overwintering habitat and feeding ground for rotifers and copepod nauplii, as well as a reservoir for rotifer resting eggs (Granskog et al. 2006). The underside of the ice is also an important feeding ground in winter for calanoid copepod species dominant in the Baltic Sea, such as *Acartia bifilosa*, and populations of this species have been shown to reproduce, grow and develop under the ice cover (Werner and Auel 2004).

9.5.7 Fungi and parasitic protists

Generally, fungi are an important group in the recycling of organic matter, and they occur in high abundances in both Baltic Sea water and ice. The presence of sympagic microscopic fungi, consisting of unicellular Ascomycota (*e.g. Eurotium* molds and *Debaryomyces* yeasts) and Basidiomycota (*e.g. Rhodotorula* yeasts) associated with Baltic Sea ice, has been confirmed with molecular techniques (Fig. 9.10, Majaneva et al. 2012). Also chytrids and *Graphiola* and *Sclerotium* species have been found within the sea ice, suggesting that besides being saprophytic, some fungi in the ice may be parasitic. Other parasites in Baltic Sea ice include members of the endosymbiont dinoflagellate order Syndiniales (Majaneva et al. 2012). Parasites of protists and animals are good examples of organisms that are generally easier to detect with molecular techniques than through direct microscopic observation.

9.6 Sea-ice food webs

9.6.1 The food webs in Baltic Sea ice are truncated

Due to space limitation in the brine channels, the internal sea-ice food webs are truncated compared to the open water food webs. The reason for the occurrence of these truncated food webs is that organisms larger than the upper size limit of the brine channels are lacking from the sympagic communities. In Baltic Sea ice, where low ice porosity and small brine channel diameter severely restrict the upper size of organisms in brine channels, the largest metazoan animals are occasional copepod nauplii. The absence, or low numbers, of metazoans simplifies the food webs by lowering the number of trophic interactions.

However, describing sea-ice food webs and their function is challenging due to sampling difficulties (Box 9.1), the complex dynamics of the physical environment exerting control over biological communities (Zhou et al. 2014) and an open boundary with the underlying water food webs that allows the exchange of matter and ecological interactions across it. Different "short circuits" in the organic matter and energy flows are suggested to be typical of microbial food webs inside the sea ice. These include herbivory by ciliates and flagellates, ciliate bacterivory and the direct utilisation of DOM by heterotrophic flagellates. Of these, at least direct utilisation of DOM by flagellates and ciliate grazing over several size classes has been suggested to be functional in the Baltic Sea sympagic communities (Haecky and Andersson 1999; Kaartokallio 2004).

The ice food web is characterised by the importance of heterotrophic bacteria as they are able to recycle DOM and serve as prey for both flagellates and small ciliates (Kaartokallio 2004). In Baltic Sea ice, DOM comprises both allochthonous (originating from parent seawater) and autochthonous components (produced by ice algae) (Stedmon et al. 2007). As in polar sea ice, they can be considered a major link between primary and secondary producers (Gradinger et al. 1992). Bacteria are able to directly utilise DOM for their growth and benefit from the generally high substrate availability in the sea ice environment. Ice bacteria can then serve as a food source for flagellates and small ciliates capable of ingesting bacteria-sized prey (Kaartokallio 2004).

The sparse data available suggest that there are differences in the sea-ice food web structure and nutrient dynamics between the subregions of the Baltic Sea. Auto- and heterotrophic flagellates seem to be more dominant in the ice food webs in the Gulf of Bothnia compared with those in the Gulf of Finland (Haecky and Andersson 1999; Meiners et al. 2002; Kaartokallio 2004). The parent-water nutrient concentrations affect the amount of nutrients entrapped during ice formation. Nutrients entrapped in ice, and the subsequent regeneration of this pool, are probably more important in the Gulf of Bothnia than in the Gulf of Finland (Haecky and Andersson 1999; Kaartokallio 2004; Granskog et al. 2005; Piiparinen et al. 2010). In the Gulf of Finland, phosphorus accumulates in the lower ice layers typically supporting the maximum algal biomass (Granskog et al. 2005), whereas in the Gulf of Bothnia the ice algal spring bloom is located in the interior ice layers (Norrman and Andersson 1994; Haecky et al. 1998) with the highest nutrient regeneration activity (Kaartokallio 2001).

9.6.2 Seasonal succession of autotrophy and heterotrophy

The typical successional pattern of the sympagic communities in Baltic Sea ice is analogous to that in polar sea ice. The initial colonisation during sea ice formation is followed by a low-productivity mid-winter stage, a bloom of the sea ice phototrophs and finally a heterotrophy-dominated stage late in the season (Box 9.2). Biomass accumulation of the sea ice phototrophs generally follows the seasonal increase in solar radiation beginning at the transition of winter and spring and lasts until the onset of ice melt (Norrman and Andersson 1994; Haecky and Andersson 1999). The relative importance of heterotrophic and autotrophic processes in the sea ice changes during ice season progress, and is driven by changes in incoming solar radiation. The growth of sea ice algae and bacteria is sequentially limited by light, nutrients and substrate (for bacteria) as the ice season progresses (Haecky et al. 1998; Kuosa and Kaartokallio 2006).

Phosphorus is probably the most important single limiting nutrient for ice-associated algae (Haecky and Andersson 1999; Granskog et al. 2005; Kuosa and Kaartokallio 2006). The spring ice phototrophic blooms typically occur in March in the Gulf of Finland and in March-April in the Gulf of Bothnia. In the Gulf of Finland, the occurrence of another, minor phototrophic biomass maximum during a low-light period in January under snow-free ice has also been reported (Kaartokallio 2004). The significance of heterotrophic processes in carbon cycling increases during late-bloom and post-bloom situations towards the end of the sea ice season.

9.6.3 Under-ice microalgal blooms

Under-ice microalgal blooms starting at the ice/water interface before ice-breakup, and facilitated by a stable melting-water layer under the ice, were reported from the southwestern coast of Finland (Spilling 2007). These blooms are assumed to contribute to the onset of the major phytoplankton spring bloom after ice break-up. In the Gulf of Finland these blooms are dominated by phototrophic dinoflagellates (Spilling 2007). Dense under-ice algal blooms dominated by dinoflagellates may also occur due to river plumes under the ice over the whole ice season (Larsen et al. 1995).

Quite specific circumstances are required to produce these blooms: there has to be a layer of low-saline water under the ice and the ice should be free of snow. The blooms can be very patchy and are concentrated to the upper few cm of the water column due to the active movement of the dinoflagellates, or as a result of very shallow vertical salinity gradients. A large haptophyte, *Chrysochromulina birgeri*, can also form under-ice blooms similar to those of the dinoflagellates. The distribution of these blooms, their effects on winter productivity and implications for the survival of metazoans are not yet fully understood (Spilling 2007).

9.6.4 Do sea-ice diatoms "seed" the pelagic spring bloom?

Most of the sympagic diatoms sink rapidly, almost immediately after the ice melt, but there is a difference in the dominant pelagic spring-bloom diatom species, depending on whether or not the spring bloom is formed after an ice-free or an ice-covered sea in winter (Haecky et al. 1998). The spring bloom of ice-free locations is dominated by centric diatoms such as *Skeletonema marinoi* and *Thalassiosira baltica*, whereas the blooms following ice-covered areas are dominated by *Pauliella taeniata*, *Chaetoceros wighamii* and *Nitzschia frigida*. The latter two species are thought to be introduced into the pelagic zone from the ice, but *Pauliella taeniata* could also originate from sediments (Piiparinen et al. 2010).

9.7 Ice and light

9.7.1 Ice optical properties

Light is the key controlling factor in determining the seasonal development of the sympagic phototrophs. Along with the seasonal development of incoming solar radiation due to changing solar zenith and day length, the light regime in ice is strongly dependent on the surface characteristics, which affect the albedo (the fraction of solar energy reflected from the Earth back to space) and the attenuation coefficients (*i.e.* the attenuation of light with depth by absorption and scattering). For example, dry surfaces (including ice) efficiently reflect solar radiation, whereas wet surfaces retain more radiation than they reflect (Rasmus et al. 2002; Ehn et al. 2004).

Due to the highly scattering nature of snow and the inclusion of variable amounts of absorptive impurities (*e.g.* soot, dust), the attenuation coefficients are generally higher for snow than for bare ice. Thus, the presence of a snow cover may greatly reduce the amount of light reaching the sea ice, depending on the thickness of the snow cover as well as on other properties of snow. Similarly to snow, the transmission of light in the sea ice is also changed by scattering (air bubbles, brine pockets), and by absorption by the biota, other particles and DOM.

Ultraviolet radiation (UVR, 280–400 nm) attenuates faster than photosynthetically active radiation (PAR, 400-700 nm) and thus most of the transmitted light is in the PAR band (Rasmus et al. 2002; Ehn et al. 2004). Some characteristics of Baltic Sea ice optical properties are different from those of Arctic sea ice. In the Baltic Sea, the typically large snow-ice fraction in the surface ice enhances scattering, while large amounts of atmospheric fallout, DOM and POM reduce the transmittance by absorption at shorter wavelengts. Therefore the maximum transmittance wavelength is shifted from 500–550 nm in the Arctic Ocean to 562–570 nm in the Baltic Sea (Ehn et al. 2004; Uusikivi et al. 2010).

As in all aquatic systems, the ability of phototrophs to increase their cellular Chl *a* content in response to low irradiances, and to decrease it at high irradiances (photoacclimation) is a common phenomenon in Baltic Sea ice phototrophs (Rintala et al. 2006; Piiparinen and Kuosa 2011). The photosynthetic components of the Baltic Sea sympagic communities are not as strongly dark-adapted as their polar counterparts, possibly due to the relatively thin snow and ice cover, the rapid changes in snow-cover thickness, and the shorter period of low irradiance in mid-winter (Piiparinen et al. 2010; Rintala et al. 2010b).

9.7.2 Ultraviolet radiation

Sympagic organisms are potentially more susceptible to UVR than planktonic organisms due to limited vertical movement inside the ice and the occasionally high O_2 concentrations in the brine (mainly from photosynthesis), which induces the formation of reactive oxygen species (ROS) in the presence of UVR. High-energy UV photons impair cell's normal functions, either directly or indirectly through formation of ROS, by promoting damage in DNA, RNA, proteins and membranes.

However, organisms can partially cope with the harmful effect of UVR by generic DNA repair and by the synthesis of UV-absorbing compounds, pigments and antioxidants in the cell. One specific group of UV-screening compounds, found in many algae and cyanobacteria, are the mycosporine-like amino acids (MAAs), which absorb between 309 and 362 nm (Uusikivi et al. 2010; Piiparinen et al. 2015). The stronger attenuation of shorter wavelengths in ice results in UV effects being concentrated mostly in the sympagic communities in the top 10 cm of sea ice. The high concentrations of the MAAs palythine and shinorine reported from the surface layer of snow-free ice indicate that the sympagic organisms in the Baltic Sea need to protect themselves against UVR (Uusikivi et al. 2010).

The few existing studies on the effects of UVA (315–400 nm) on bacteria and phototrophs in Baltic Sea ice indicate that UVR (which in sea ice consists mainly of UVA) shapes the vertical distribution of organisms in the ice column to some degree (Piiparinen and Kuosa 2011). This seems to apply especially to chlorophytes and pennate

Fig. 9.11 Set-up of an UVA experiment on Baltic Sea ice. One of the tent-shaped frames is covered with UVA opaque foil (PAR treatment) and the other is covered with UVA transmitting foil (PAR + UVA treatment). Photo: © Jari Uusikivi

diatoms, which increased in biomass in the surface layers of snow-free ice when UVA was experimentally filtered off (Fig. 9.11), whereas exposure to UVA had the opposite effect on these two algal groups. On the other hand, centric diatoms showed sensitivity to both PAR and PAR + UVA and concentrated in deeper ice layers under these exposure regimes.

Bacterial production is closely linked to changes in algal biomass and species composition, indicating that the UV-effects may extend throughout the sea-ice food web. The bacteria in the surface ice are also affected by UVA: α-, β-, and γ-Proteobacteria are sensitive to UVA whereas the Flavobacteriia seem to be UVA-resilient. When snow covers the sea ice it provides efficient protection against UVR for sympagic organisms, but at the same time it reduces the PAR intensity.

9.8 Modelling the Baltic Sea ice system

9.8.1 Climate change in the Baltic Sea Area

The global air temperature is increasing (*cf.* Fig. 2.28; Stocker et al. 2013), and the Baltic Sea Area is no exception. The importance of sea ice for the functioning of the Baltic Sea ecosystem needs to be well understood since the sea ice is expected to disappear from most of the basins in the near future. Over the past century the air temperature in the Baltic Sea Area has increased by 0.7 °C, which exceeds the global average of 0.5 °C (BACC Author Team 2015). The temperature trend in the Baltic Sea surface waters over the time period of 1982–2010 indicates a warming rate of 0.063–0.078 °C per year (Baker-Austin et al. 2013). The increasing temperature shrinks the extension of Baltic Sea ice, as already observed *e.g.* at the ice monitoring stations Märket and Suursaari (Fig. 9.2).

The increase of both air temperature and precipitation is predicted to continue over the coming decades. A projection of the B2 SRES climate change scenario, a mild IPCC climate change scenario (Stocker et al. 2013), to the northeastern Baltic Sea, shows an increase in the air temperature by ~3 °C in spring and autumn, ~4 °C in summer and ~4.5 °C in winter in the time slice 2071–2090 (BACC Author Team 2015). The same scenario predicts an increase of precipitation in the northeastern Baltic Sea by ~5 % in summer, ~ 10 % in autumn, ~15 % in spring and ~ 20 % in winter.

Therefore, the largest changes, both in temperature and in precipitation, are projected to occur in winter, whereby the ice cover will be reduced, both in time and space. The ice-covered area in the Baltic Sea is estimated to shrink by ~45,000 km^2 for each 1°C increase in the average temperature. With an increase of 1 °C only the northernmost part of the Gulf of Bothnia and the easternmost part of the Gulf of Finland would freeze during mild winters and the ice would be thinner and more easily movable. With an air temperature increase of ~4 °C in winter, the Baltic Sea would become completely ice-free (Omstedt and Hansson 2006).

9.8.2 Sea ice is a challenge for modellers

To understand and predict the consequences of global climate change and other non-climate stressors, the development of ecosystem modelling applications that can act as a decision support system (tool) for policy makers has become a major task for the scientific community. Assessing the qualitative and quantitative role of the sea ice is a challenge facing ecosystem modellers of polar and subpolar regions. For example, how will the absence of an ice cover affect primary production, and ultimately fish production, in these regions?

Very few studies have dealt with modelling of the sea-ice habitat (Tedesco and Vichi 2014), and modelling of sea ice biota is a challenge, firstly because there is a scarcity of observational data and secondly because of the complexity of the system. *In situ* observations of spatial and temporal variability of the sea-ice properties and processes are needed for model development and evaluation, but usually only point data with small spatial resolution are available. Thus, while the principal biological and ecological processes that characterise a certain sea-ice area may be known, the variability of the biogeochemical properties remains still largely unknown, which lowers the reliability of a model.

Compared to ocean biogeochemical models, those developed for sea-ice biogeochemistry are more complex as they include also ice physics, specific light parameterisations and specific fluxes at the ice/water interface. Sea-ice models may differ in complexity in terms of resolution and biology. A single-layer model of a preset ice thickness has a lower resolution than a multi-layer model with several sea-ice layers. The number and type of biological tracers, functional groups and ecological and physiological processes affect model complexity. For example, a silica-based model representing only one group of primary producers (diatoms) is less complex than a multi-nutrient-based model with several living components.

9.8.3 A biogeochemical model of Baltic Sea ice

A sea ice biogeochemical model considering the specific physical, chemical and biological characteristics of Baltic Sea ice has been developed by Tedesco et al. (2009, 2010). In the seasonal ice of the Baltic Sea, the bottommost part usually is the most productive layer, and the concept of the "Biologically Active Layer" (Fig. 9.5) is suitable to represent the vertical resolution of the sea ice biogeochemical model.

The model includes the state variables: nitrogen, phosphorus, silica, chlorophyll *a* and carbon in a flexible stoichiometry framework as well as two functional groups of algae (adapted diatoms and surviving sea ice algae), sea ice bacteria, sea-ice fauna, sea-ice organic matter (dissolved and particulate) and gases such as oxygen and carbon dioxide (Fig. 9.12). The physical part of the model considers only two sea-ice layers, the Biologically Active Layer and the rest of the sea ice with no, or very low, biological activity. Included are several snow layers that describe the different characteristics of the upper layers, such as freshly deposited snow, compacted snow, snow-ice or superimposed ice (Tedesco et al. 2009, 2010).

9.8.4 A climate change scenario for the Baltic Sea

The Baltic Sea ice biogeochemical model of Tedesco et al. (2009) was first used to simulate the ice season 1999–2000 at Santala Bay in the Gulf of Finland (Tedesco et al. 2010). The Biologically Active Layer showed a dynamic thickness and biological production during the ice season that compared well with the available physical and biochemical observations (Fig. 9.13). In particular, the model was able to reproduce a rather mild winter during which the sympagic community appeared to be very active throughout most of the ice season.

When the B2 SRES climate change scenario (Stocker et al. 2013) was applied to Santala Bay in the Baltic Sea ice biogeochemical model of Tedesco et al. (2010), only six days during the whole of winter were projected to be below

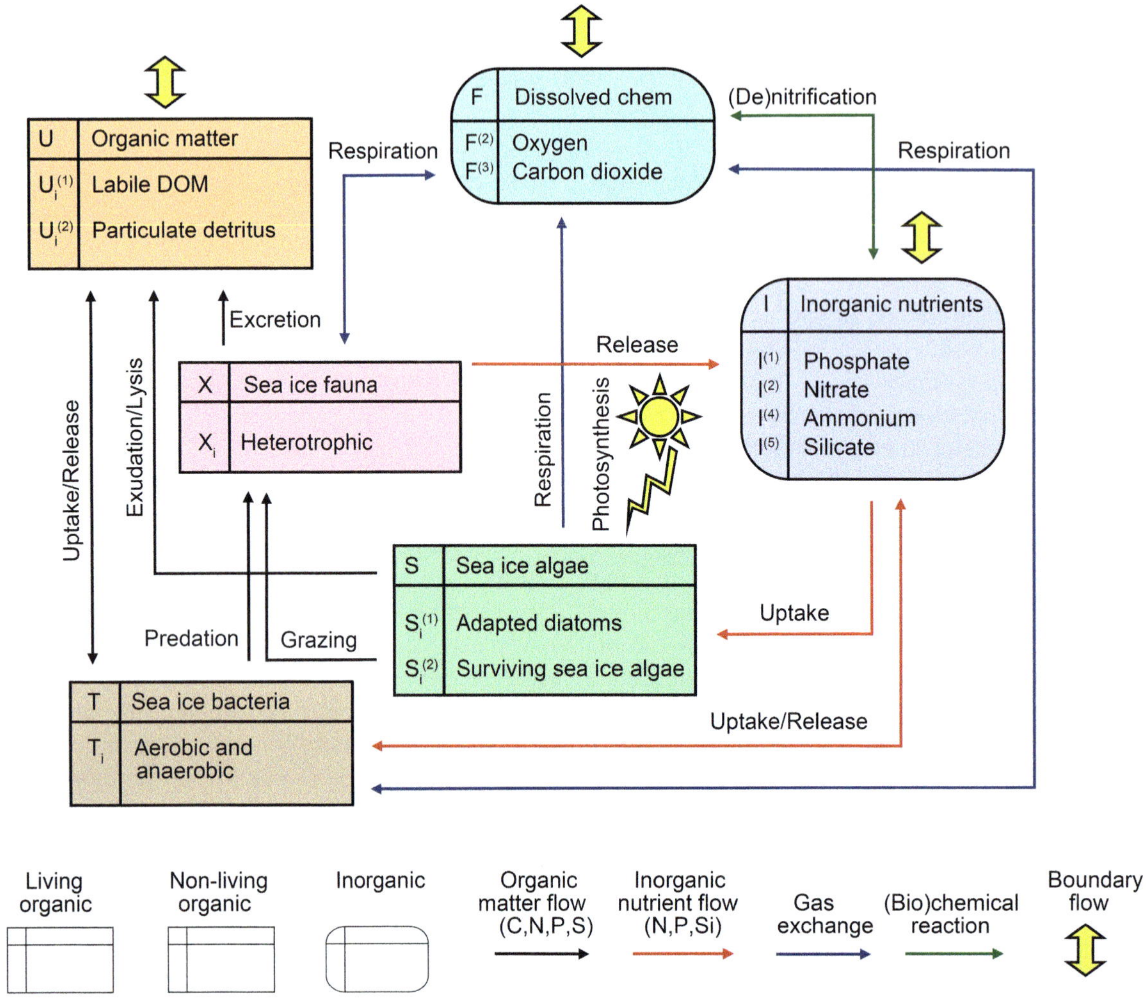

Fig. 9.12 Scheme of the state variables and interactions used in the sea-ice biogeochemical model for the Baltic Sea. Figure: © Letizia Tedesco

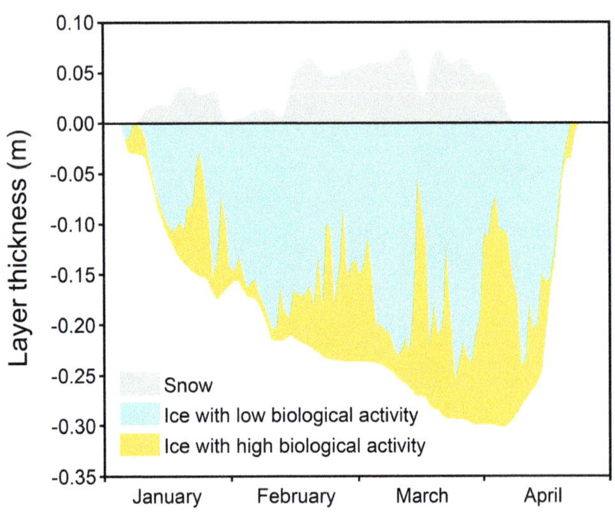

Fig. 9.13 Model simulation of the thickness of the snow layer (grey), the internal sea-ice layer with no or very low biological activity (blue) and the Biologically Active Layer in the sea ice (yellow) during the ice season 1999–2000 at Santala Bay in the Gulf of Finland. Figure modified from Tedesco et al. (2010)

the freezing point of the brackish Baltic Sea water. In comparison, the ice season of 1999–2000 had 138 days below the freezing point at Santala Bay. Six days is not enough for the sea ice to become firm and persistent.

Without an ice cover, the pelagic phytoplankton bloom was projected to occur later than in the reference simulation of 1999–2000 (Fig. 9.14). Furthermore, the biomass of the pelagic spring bloom following an ice-free winter was projected to be larger with community composition to be dominated by flagellates instead of diatoms. This flagellate bloom consists mainly of dinoflagellates, which are more efficient in warmer waters and currently already compete with diatoms in large areas of the Baltic Sea during the pelagic spring bloom (Spilling 2007). The change in the biomass and composition of the phytoplankton bloom will presumably affect the composition and timing of the zooplankton peak, and bacterial dynamics are expected to change as well. In general, the whole food web is projected to change, ultimately impacting top predator populations.

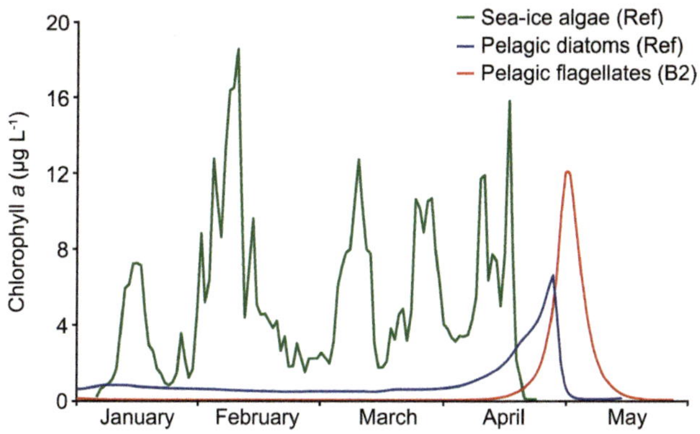

Fig. 9.14 Modelled sea-ice and pelagic chlorophyll *a* concentrations during the ice season and the pelagic spring bloom in April-May at Santala Bay (Gulf of Finland), showing two simulations. Ref = winter 1999–2000 with 138 days below the freezing point of the brackish Baltic Sea water and an ice cover, yielding a pelagic bloom dominated by diatoms. B2 = IPCC climate change scenario B2 SRES with only six days below the freezing point of the brackish Baltic Sea water and no ice cover, yielding a pelagic bloom dominated by flagellates. Figure: © Letizia Tedesco

Fig. 9.15 The ringed seal *Pusa hispida* is the only ice-obligate marine mammal in the Baltic Sea. This species occurs both in the Baltic Sea and in the Arctic Ocean. Photo: © Charlotta Moraeus/Azote

More extensive models, *i.e.* encompassing the entire ice-covered Baltic Sea and its biology, are needed. Their construction requires more field observations and the development of new parameterisations for sea ice physical, chemical and biological processes at large spatial scales. Large-scale models should also include the biogeochemical importance of ice movement as thinner ice is more readily deformed by winds, which may result in increased roughness of the ice and rafting and ridging among ice floes (Stirling et al. 2008; Löptien et al. 2013). Currently, *e.g.* pressure ridges are missing in all sea-ice models existing worldwide, and this most likely leads to an underestimation of the sea ice biological production.

9.9 Life on the Baltic Sea ice

In the Arctic Ocean, four ice-obligate marine mammals occur: the ringed seal *Pusa hispida*, the walrus *Odobenus rosmarus*, the bearded seal *Erignathus barbatus* and the polar bear *Ursus maritimus* (Moore and Huntington 2008). In the Baltic Sea there is only one, the ringed seal (Fig. 9.15, *cf.* Box 4.13). In the Baltic Sea, the ringed seal occupies the seasonal sea ice from the time of ice formation until ice break-up, including the breeding season. One of the critical periods in the ringed seals' life is the "subnivean period" in lairs under the snow.

Already in the 1950s it was hypothesised that the Archipelago Sea subpopulation of the ringed seal had declined due to the mild winters in the 1930s; later findings confirmed the correlation between high pup mortality and restricted ice coverage (Meier et al. 2004). This is because successful breeding depends on the seals' ability to carve lairs in the snow and the subsequent (about a month long) period in March and early April during which pup survival depends on stable ice conditions, as the pups still live under the protection of lairs.

Because of the projected climate change, in the future the breeding of all the southern subpopulations of the Baltic ringed seal will only be possible in exceptionally cold years. Only the Bothnian Bay subpopulation is likely to survive since the present climate change scenarios still predict an annual two-month long ice-cover period here, which is the minimum time required to sustain a ringed seal population (Meier et al. 2004).

Review questions

1. What are the characteristics of the three main habitats of sympagic communities?
2. What is the major driver of the seasonal succession in sympagic communities?
3. What limits primary production in Baltic Sea ice?

4. How do the main Baltic Sea ice habitats differ from those in the Arctic Ocean, and what are the biological consequences of these differences?
5. Which factors affect the transmission of light in sea ice?

Discussion questions

1. What impacts on Baltic Sea ice are predicted by climate change scenarios and what consequences of those impacts can you envisage for the Baltic Sea ecology?
2. Why is modelling an important tool in studying sea ice?
3. Why are diatoms the dominating algal group in sea ice?
4. Can the ice cover affect the productivity of the Baltic Sea in other seasons?
5. Why are bacteria in sea ice able to grow almost as fast as they do in the open water in summer?

References

BACC Author Team (2015) Second assessment of climate change for the Baltic Sea basin. Regional climate studies. Springer, Berlin 501 pp

Baker-Austin C, Trinanes JA, Taylor NGH, Hartnell R, Siitonen A, Martinez-Urtaza J (2013) Emerging *Vibrio* risk at high latitudes in response to ocean warming. Nature Climate Change 3:73–77

Cohn SA, Witzell RE (1996) Ecological considerations of diatom cell motility. I. Characterization of motility and adhesion in four diatom species. Journal of Phycology 32:928–939

Ehn J, Granskog MA, Reinart A, Erm A (2004) Optical properties of melting landfast sea ice and underlying seawater in Santala Bay, Gulf of Finland. Journal of Geophysical Research 109 (C09003):1–12

Eronen-Rasimus E, Lyra C, Rintala JM, Jürgens K, Ikonen V, Kaartokallio H (2015) Ice formation and growth shape bacterial community structure in Baltic Sea drift ice. FEMS Microbiology Ecology 91:13. doi: 10.1093/femsec/fiu022

Friedrich C, De Smet WH (2000) The rotifer fauna of Arctic sea ice from the Barents Sea, Laptev Sea and Greenland Sea. Hydrobiologia 432:73–89

Golden KM (2009) Climate change and the mathematics of transport in sea ice. Notices of the American Mathematical Society 56:562–584

Gradinger R, Spindler M, Weissenberger J (1992) On the structure and development of Arctic pack ice communities in Fram Strait: a multivariate approach. Polar Biology 12:727–733

Granskog MA, Kaartokallio H (2004) An estimation of the potential fluxes of nitrogen, phosphorus, cadmium and lead from sea ice and snow in the northern Baltic Sea. Water, Air and Soil Pollution 154:331–347

Granskog MA, Kaartokallio H, Kuosa H (2010) Sea ice in non-polar regions. In: Thomas DN, Dieckmann GS (eds) Sea ice, 2nd edn. Wiley-Blackwell Publishing, Oxford, pp 531–577

Granskog MA, Kaartokallio H, Kuosa H, Thomas DN, Vainio J (2006) Sea ice in the Baltic Sea: a review. Estuarine, Coastal and Shelf Science 70:145–160

Granskog MA, Kaartokallio H, Shirasawa K (2003a) Nutrient status of Baltic Sea ice: evidence for control by snow-ice formation, ice permeability and ice algae. Journal of Geophysical Research 108:3253–3256

Granskog MA, Kaartokallio H, Thomas DN, Kuosa H (2005) The influence of freshwater inflow on the inorganic nutrient and dissolved organic matter within coastal sea ice and underlying waters in the Gulf of Finland (Baltic Sea). Estuarine, Coastal and Shelf Science 65:109–122

Granskog MA, Martma TA, Vaikmäe RA (2003b) Development, structure and composition of landfast sea ice in the northern Baltic Sea. Journal of Glaciology 49:139–148

Haecky P, Andersson A (1999) Primary and bacterial production in sea ice in the northern Baltic Sea. Aquatic Microbial Ecology 20:107–118

Haecky P, Jonsson S, Andersson A (1998) Influence of sea ice on the composition of the spring phytoplankton bloom in the northern Baltic Sea. Polar Biology 20:1–8

Hällfors G, Niemi Å (1974) A *Chrysochromulina* (Haptophyceae) bloom under the ice in the Tvärminne archipelago, southern coast of Finland. Memoranda Societatis pro Fauna et Flora Fennica 50:89–104

Haapala JJ, Ronkaonen L, Schmelzer N, Sztobryn M (2015) Recent change—sea ice. BACC Author Team. Second assessment of climate change for the Baltic Sea basin. Regional Climate Studies, Springer, Berlin, pp 145–153

Huttunen M, Niemi Å (1986) Sea-ice algae in the northern Baltic Sea. Memoranda Societatis pro Fauna et Flora Fennica 62:58–62

Ikävalko J (1998) Further observations on flagellates within sea ice in northern Bothnian Bay, the Baltic Sea. Polar Biology 19:323–329

Ikävalko J, Thomsen HA (1997) The Baltic Sea ice biota (March 1994): a study of the protistan community. European Journal of Protistology 33:229–243

Kaartokallio H (2001) Evidence for active microbial nitrogen transformations in sea ice (Gulf of Bothnia, Baltic Sea) in midwinter. Polar Biology 24:21–28

Kaartokallio H (2004) Food-web components, and physical and chemical properties of Baltic Sea ice. Marine Ecology Progress Series 273:49–63

Kaartokallio H, Kuosa H, Thomas DN, Granskog MA, Kivi K (2007) Changes in biomass, composition and activity of organism assemblages along a salinity gradient in sea ice subjected to river discharge. Polar Biology 30:186–197

Kaartokallio H, Tuomainen J, Kuosa H, Kuparinen J, Martikainen P, Servomaa K (2008) Succession of sea-ice bacterial communities in the Baltic Sea fast ice. Polar Biology 31:783–793

Kankaanpää P (1997) Distribution, morphology and structure of sea ice pressure ridges in the Baltic Sea. Fennia 175:139–240

Kawamura T, Shirasawa K, Ishikawa N, Lindfors A, Rasmus K et al (2001) Time-series observations of the structure and properties of brackish ice in the Gulf of Finland. Annals of Glaciology 33:1–4

Krembs C, Eicken H, Deming JW (2011) Exopolymer alteration of physical properties of sea ice and implications for ice habitability, biogeochemistry and persistence in a warming climate. Proceedings of the National Academy of Sciences of the USA 108:3653–3658

Kuosa H, Kaartokallio H (2006) Experimental evidence on substrate and nutrient limitation of Baltic Sea sea-ice algae and bacteria. Hydrobiologia 554:1–10

Kuparinen J, Autio R, Kaartokallio H (2011) Sea ice bacteria have high growth rate, growth efficiency and preference for inorganic nitrogen sources in the Baltic Sea. Polar Biology 34:1361–1373

Kuparinen J, Kuosa H, Andersson A, Autio R, Granskog MA et al (2007) Role of sea-ice biota in nutrient and organic material cycles in the northern Baltic Sea. AMBIO 36:149–154

Larsen J, Kuosa H, Ikävalko J, Kivi K, Hällfors S (1995) A redescription of *Scrippsiella hangoei* (Schiller) comb. nov.—a “red tide” dinoflagellate from the northern Baltic. Phycologia 34:135–144

Leppäranta M, Manninen T (1988) The brine and gas content of sea ice with attention to low salinities and high temperatures. Finnish Institute for Marine Research, Helsinki, Report 2:1–14

Leppäranta M, Myrberg K (2009) Physical oceanography of the Baltic Sea. Springer, Berlin 378 pp
Löptien U, Mårtensson S, Meier HEM, Höglund A (2013) Long-term characteristics of simulated ice deformation in the Baltic Sea (1962-2007). Journal of Geophysical Research: Oceans 118:801–815
Luhtanen AM, Eronen-Rasimus E, Kaartokallio H, Rintala JM, Autio R, Roine E (2014) Isolation and characterization of phage-host systems from the Baltic Sea ice. Extremophiles 18:121–130
Majaneva M, Rintala JM, Piisilä M, Fewer DP, Blomster J (2012) Comparison of wintertime protists and fungi from sea ice and open water in the Baltic Sea. Polar Biology 35:875–889
Meese DA (1989) The chemical and structural properties of sea ice in the southern Beaufort Sea. Cold Regions Research and Engineering Laboratory Monograph 1989 25:1–144
Meier HEM, Döscher R, Halkka A (2004) Simulated distributions of Baltic Sea ice in warming climate and consequences for the winter habitat of the Baltic ringed seal. AMBIO 33:249–256
Meiners K, Fehling J, Granskog MA, Spindler M (2002) Abundance, biomass and composition of biota in Baltic Sea ice and underlying water (March 2000). Polar Biology 25:761–770
Moore SE, Huntington HP (2008) Arctic marine mammals and climate change: impacts and resilience. Ecological Applications 18:157–165
Norrman B, Andersson A (1994) Development of ice biota in a temperate sea area (Gulf of Bothnia). Polar Biology 14:531–537
Omstedt A, Hansson D (2006) The Baltic Sea ocean climate system memory and response to changes in the water and heat balance components. Continental Shelf Research 26:236–251
Piiparinen J, Enberg S, Rintala JM, Sommaruga R, Majaneva M et al (2015) The contribution of mycosporine-like amino acids, chromophoric dissolved organic matter and particles to the UV protection of sea-ice organisms in the Baltic Sea. Photochemical and Photobiological Sciences 14:1025–1038
Piiparinen J, Kuosa H (2011) Impact of UVA radiation on algae and bacteria in Baltic Sea ice. Aquatic Microbial Ecology 63:75–87
Piiparinen J, Kuosa H, Rintala JM (2010) Winter-time ecology in Bothnian Bay, Baltic Sea: nutrients and algae in fast-ice. Polar Biology 33:1445–1461
Poulin M, Daugbjerg N, Gradinger R, Ilyash L, Ratkova T, von Quillfeldt C (2011) The Pan-Arctic biodiversity of marine pelagic and sea-ice unicellular eukaryotes: a first-attempt assessment. Marine Biodiversity 41:13–28
Rasmus K, Ehn J, Granskog MA, Kärkäs E, Leppäranta M et al (2002) Optical measurements of sea ice in the Gulf of Finland. Nordic Hydrology 33:207–226
Rintala (2009) A systematic-ecological approach to Baltic Sea ice studies of algae and protists. University of Helsinki, Walter and Andrée de Nottbeck Foundation Scientific Reports 34:1–48 [PhD Thesis]
Rintala JM, Hällfors H, Hällfors S, Hällfors G, Majanerva M, Blomster J (2010a) *Heterocapsa arctica* subsp. *frigida* subsp. nov. (Peridiniales, Dinophyceae)—description of a new dinoflagellate and its occurrence in the Baltic Sea. Journal of Phycology 46:751–762
Rintala JM, Piiparinen J, Ehn J, Autio R, Kuosa H (2006) Changes in phytoplankton biomass and nutrient quantities in sea ice as responses to light/dark manipulations during different phases of the Baltic winter 2003. Hydrobiologia 554:11–24
Rintala JM, Spilling K, Blomster J (2007) Temporary cyst enables long term dark survival of Scrippsiella hangoei (Dinophyceae). Marine Biology 152:57–62
Rintala JM, Piiparinen J, Uusikivi J (2010b) Drift-ice and under-ice water communities in the Gulf of Bothnia (Baltic Sea). Polar Biology 33:179–191
Rintala JM, Piiparinen J, Blomster J, Majaneva M, Müller et al (2014) Fast direct melting of brackish sea-ice samples results in biologically more accurate results than slow buffered melting. Polar Biology 37:1811–1822
Rysgaard S, Bendtsen J, Delille B, Dieckmann GS, Glud RN et al (2011) Sea ice contribution to the air-sea CO_2 exchange in the Arctic and Southern Oceans. Tellus B 63:823–830
Spilling K (2007) Dense sub-ice bloom of dinoflagellates in the Baltic Sea, potentially limited by high pH. Journal of Plankton Research 29:895–901
Stedmon CA, Thomas DN, Granskog M, Kaartokallio H, Papadimitriou S, Kuosa H (2007) The characteristics of dissolved organic matter in Baltic coastal sea ice: allochthonous or autochthonous origins? Environmental Science and Technology 41:7273–7279
Steffens M, Granskog MA, Kaartokallio H, Kuosa H, Luodekari K et al (2006) Spatial variation of biogeochemical properties of landfast sea ice in the Gulf of Bothnia (Baltic Sea). Annals of Glaciology 44:80–87
Stirling I, Richardson E, Thiemann GW, Derocher AE (2008) Unusual predation attempts of polar bears on ringed seals in the southern Beaufort Sea: possible significance of changing spring ice conditions. Arctic 61:14–22
Stocker TF, Qin D, Plattner GK, Tignor M, Allen SK et al (eds) (2013) The physical science basis—contribution of Working Group I to the Fifth Assessment Report of the Intergovernmental Panel on Climate Change. Cambridge University Press, Cambridge 1535 pp
Tedesco L, Vichi M (2014) Sea ice biogeochemistry: a guide for modellers. PLoS ONE 9(2):e89217
Tedesco L, Vichi M, Haapala J, Stipa T (2009) An enhanced sea ice thermodynamic model applied to the Baltic Sea. Boreal Environment Research 14:68–80
Tedesco L, Vichi M, Haapala J, Stipa T (2010) A dynamic biologically-active layer for numerical studies of the sea ice ecosystem. Ocean Modelling 35:89–104
Thomas DN, Dieckmann GS (2002) Antarctic sea ice—a habitat for extremophiles. Science 295:641–644
Thomas DN, Dieckmann GS (eds) (2010) Sea ice, 2nd edn. Wiley-Blackwell Publishing, Oxford 621 pp
Underwood GJC, Salam SN, Michel C, Niemi A, Norman L et al (2013) Broad-scale predictability of carbohydrates and exopolymers in Antarctic and Arctic sea ice. Proceedings of the National Academy of Sciences of the USA 110:15734–15739
Uusikivi J (2013) On optical and physical properties of sea ice in the Baltic Sea. Report Series in Geophysics 73:1–49
Uusikivi J, Ehn J, Granskog MA (2006) Direct measurements of turbulent momentum, heat and salt fluxes under landfast ice in the Baltic Sea. Annals of Glaciology 44:42–46
Uusikivi J, Vähätalo AV, Granskog MA, Sommaruga R (2010) Contribution of mycosporine-like amino acids and colored dissolved and particulate matter to sea ice optical properties and ultraviolet attenuation. Limnology and Oceanography 55:703–713
Vihma T, Haapala J (2009) Geophysics of sea ice in the Baltic Sea: a review. Progress in Oceanography 80:129–148
Weeks WF (1998) Growth conditions and the structure and properties of sea ice. In: Leppäranta M (ed) Physics of ice-covered seas. Department of Geophysics, University of Helsinki, Helsinki, pp 25–104
Weissenberger J, Dieckmann GS, Gradinger R, Spindler M (1992) Sea ice: a cast technique to examine and analyse brine pockets and channel structure. Limnology and Oceanography 37:179–183
Werner I, Auel H (2004) Environmental conditions and overwintering strategies of planktonic metazoans in and below coastal fast ice in the Gulf of Finland (Baltic Sea). Sarsia 89:102–116
Zhou J, Delille B, Kaartokallio H, Kattner G, Kuosa H et al (2014) Physical and bacterial controls on inorganic nutrients and dissolved organic carbon during a sea ice growth and decay experiment. Marine Chemistry 166:59–69

Deep soft seabeds

10

Urszula Janas, Erik Bonsdorff, Jan Warzocha, and Teresa Radziejewska

Abstract

1. The deep soft seabeds of the Baltic Sea Area offer a wide range of ecological niches for invertebrates (zoobenthos), from the high-diversity marine regions characterised by large and long-lived organisms in the Skagerrak to the species-poor, almost limnic, systems in the inner reaches of the Bothnian Bay and the Gulf of Finland.
2. The zoobenthos processes nutrients and organic matter in the sediments, oxygenates the sediments through bioturbation and bioventilation, affects nutrient fluxes at the sediment/water interface, and acts as a link in both bottom-up and top-down control of the entire Baltic Sea ecosystem.
3. The steep spatial and seasonal gradients of the Baltic Sea structure the zoobenthic communities and shape their functional roles in the food webs as well as in benthic-pelagic coupling.
4. Eutrophication and widespread hypoxia and anoxia are major factors that shape the taxonomic composition, functionality and successional patterns of the zoobenthic communities.
5. As the zoobenthos still recovers from the last glaciation through an on-going succession, there are plenty of vacant niches available in the Baltic Sea for the introduction and establishment of non-indigenous species, and these species may have profound impacts on the whole ecosystem.
6. Due to the sensitivity of the zoobenthos to environmental change and its relative longevity, zoobenthos abundance, biomass and community composition are used as indicators of ecosystem health.
7. Modern science combines field surveys with experiments and advanced mathematical modelling, linking physical and chemical drivers with food web processes. Flux measurements and broad functional analyses, in combination with molecular studies, provide information on processes that reshape our understanding of ecosystem functioning.

Keywords

Baltic Sea gradient • Benthic food web • Benthic-pelagic coupling • Bioturbation • Human impacts • Hypoxia • Organic matter mineralisation • Zoobenthic communities

U. Janas (✉)
Institute of Oceanography, University of Gdańsk,
Al. M. Piłsudskiego 46, 81378 Gdynia, Poland
e-mail: oceuj@ug.edu.pl

E. Bonsdorff
Department of Biosciences, Environmental and Marine Biology,
Åbo Akademi University, Artillerigatan 6, 20520
Turku, Finland
e-mail: ebonsdor@abo.fi

J. Warzocha
Department of Oceanography and Marine Ecology,
National Marine Fisheries Research Institute, Kollataja 1,
81332 Gdynia, Poland
e-mail: janw@mir.gdynia.pl

T. Radziejewska
Faculty of Geosciences, ul. Mickiewicza 16A,
70383 Szczecin, Poland
e-mail: tera@univ.szczecin.pl

P. Snoeijs-Leijonmalm et al. (eds.), *Biological Oceanography of the Baltic Sea*,
DOI 10.1007/978-94-007-0668-2_10

10.1 Environmental conditions in deep water

10.1.1 The deep soft seabeds of the Baltic Sea

The deep soft seabeds of the Baltic Sea are the areas of the seafloor that are covered by fine-grained sediments (fine sand, silt, clay) and located at depths below the photic zone and often also below the halocline. In the transition zone (Belt Sea and Kattegat) the halocline is found at a ~15–20 m water depth, but inside the Baltic Sea it occurs much deeper, down to an ~80 m water depth in the Gotland deep (*cf.* Table 2.6). Thus, the term "deep soft seabeds" usually refers to the bottoms of the deep basins ("deeps") of the Baltic Sea, but it is also used for shallower areas where haloclines are weaker, *e.g.* in the Gulf of Riga where the deep water from the Baltic Sea proper cannot enter.

The environmental conditions created by the halocline are of paramount importance for the life in and on the deep soft seabeds of the Baltic Sea. The salinity of the below-halocline water decreases along the large-scale Baltic Sea gradient, from >15 in the transition zone to ~8–12 in the Baltic Sea proper and 4–5 in the Bothnian Bay (*cf.* Fig. 2.15). Occasional large inflows of Kattegat water ("major Baltic inflows", *cf.* Sect. 2.3.9) may temporarily increase the below-halocline salinity of the Baltic Sea proper. The water temperature below the halocline is relatively stable throughout the Baltic Sea proper (4–6 °C), although the occasional large inflows may slightly raise the bottom-water temperature as well (Voipio 1981) since the temperature of the inflowing deep water of the transition zone is 6–8 °C (*cf.* Fig. 2.16c). In the Bothnian Bay the temperature of the deep water is only 2–4 °C.

There is also a slight seasonal variability in the bottom water temperature with a time lag to the surface temperature where the difference between the surface water and the deep water depends on water depth. For example, the surface/bottom temperature in the shallow Belt Sea is 2–4/3–5 °C in January and 16–17/8–14 °C in July, while in the deeper Bornholm Sea this is 4/5–8 °C in January and 16–17/5–6 °C in July (Leppäranta and Myrberg 2009).

10.1.2 Oxygen conditions in deep water

The stable vertical stratification in the Baltic Sea generates a zonal distribution of the zoobenthos along the depth gradient. This zonation is strengthened by hypoxia (<2 mL O_2 L^{-1}) (Fig. 10.1). Oxygen deficiency results from a complex interplay of climatological (atmospheric circulation, prevailing winds inducing saltwater inflows), oceanographic (density currents accompanying the inflows, vertical mixing or lack thereof), and anthropogenic impacts associated with eutrophication.

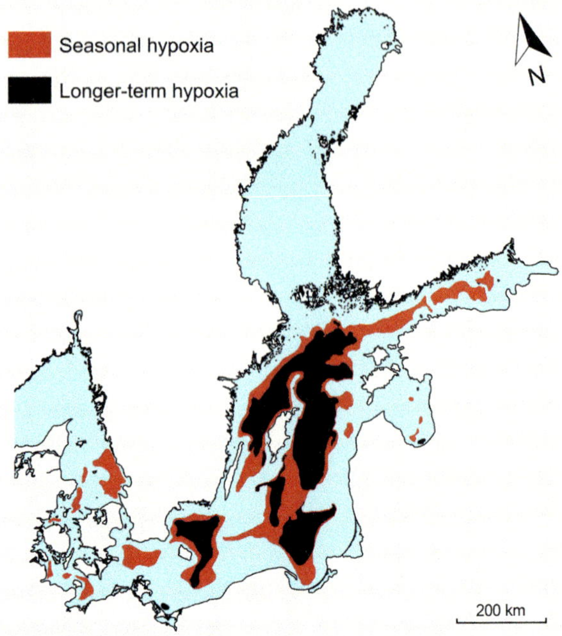

Fig. 10.1 Extent of seasonal and longer-term near-bottom hypoxia (<2 mL O_2 L^{-1}) in the time period 2001–2006. Figure modified from HELCOM (2009b)

Since the 1970s, major Baltic inflows have occurred less frequently and were usually of low intensity (*cf.* Fig. 2.13a). Consequently, the vertical salinity difference between the upper and bottom layers has decreased. Oxygen conditions have deteriorated markedly after 1999. While large areas of the Bornholm, Gdańsk and Gotland deeps as well as the Gulfs of Finland and Riga were before hypoxic with only very small anoxic areas, the extent of both hypoxia and anoxia observed after 1999 increased to an unprecedented magnitude (Fig. 10.1). In the Baltic Sea proper, the seabed area affected by hypoxia increased from 22 to 28 %, while the anoxic bottom area increased from 5 to 15 % (Conley et al. 2009; Hansson et al. 2011).

10.1.3 Anthropogenic impacts or natural conditions?

In a traditional view, the oxygen deficiency in the below-halocline water was closely associated with eutrophication, a process that has accelerated in the Baltic Sea since the mid-20th century. However, excessive nutrient loads-the main factor underlying increased production-have decreased since the 1980s (*cf.* Box 2.2) because of management efforts by the countries surrounding the Baltic Sea (*cf.* Sect. 18.8). In contrast, the severity of hypoxia has increased during this time period. This suggests that climatic-oceanographic factors, such as changes in the wind regime and the frequency and nature of inflows, weaker vertical mixing, and permanent stratification, may have gained the upper hand when it comes to hypoxia. However, many of the changes in the zoobenthic communities during the last decades can still be ascribed to eutrophication (HELCOM 2009b).

In addition to the oxygen conditions, the distribution of the zoobenthos in and on the deep soft seabeds of the Baltic Sea is affected by the nature of the sedimentary environment itself. While sand dominates down to 60–80 m in the southern Baltic Sea proper, the bottoms at larger depths feature a more fine-grained sedimentary cover (silty sands, sandy silts, silts, and clays) and are referred to as the "muddy bottom" (*cf.* Fig. 2.6). Local anthropogenic activities such as dredging, trawling and the proximity of pollution sources may also modify the environmental conditions on the deep soft seabed.

Both natural gradients and anthropogenic stressors (eutrophication, chemical contamination by hazardous substances, fishing pressure, introductions of non-indigenous species) influence the structure and functioning of the zoobenthic communities in the Baltic Sea. The combined impacts of natural drivers, anthropogenic stressors, and climate change-induced shifts in temperature and salinity on the zoobenthos of the deep soft seabeds are difficult to predict as the severity of many stressors' impact is dependent on the physical environment (Leppäkoski and Bonsdorff 1989). This poses future challenges for science and society that need to be faced by accurate knowledge of the spatial and temporal patterns exhibited by the zoobenthic communities of the deep soft seabeds. The ecological status ("health") of the benthic system must be classified in order to understand the changes and differences in the system, and also as a basis for management under *e.g.* the European Marine Strategy (Villnäs et al. 2015).

10.2 Classification and characteristics of the fauna

10.2.1 Macrofauna and meiofauna

In addition to heterotrophic microorganisms, the deep soft seabeds of the Baltic Sea are inhabited by the zoobenthos. Sampling of the deep bottom fauna to estimate abundance, biomass and community composition takes place from ships (Box 10.1). Recently, methods for *in situ* studies have also been developed (Box 10.2).

For methodological reasons, the zoobenthos is subdivided into two main size groups: the smaller organisms are referred to as the meiofauna (or meiobenthos) and the larger ones as the macrofauna (or macrozoobenthos). Fairly often the macrozoobenthos is simply called "benthos" or "macrobenthos", but this is undesirable because it may create confusion with the phytobenthos. For the meiobenthos there is no such problem because the "meio" category is not used for algae or vascular plants (*cf.* Table 4.1).

The most common operational borderline between the meio- and macrofauna is the 1 mm mesh size of a sieve used to separate the animals from the sediment, although a limit of 0.5 mm mesh size is sometimes used as well. The lower limit of meiofauna body size is defined as 0.044 or 0.063 mm (*cf.* Table 4.1). In addition to being operationally defined, the meiobenthos has a faunistic meaning as well. The meiofauna comprises many representatives of orders, classes, and even phyla (*cf.* Box 4.8) that are not found among the macrofauna.

The macrozoobenthos of deep soft seabeds has for a long time been the primary focus of biological studies in the Baltic Sea (Petersen and Boysen-Jensen 1911; Petersen 1913; Demel and Mulicki 1954; Segerstråle 1957a) and remains one of the best-studied ecosystem components of the Baltic Sea to date. The deep soft-bottom macrofauna is composed of species showing different levels of mobility:

1. Sessile species, the adult stages of which are usually permanently attached to a substrate, *e.g.* the blue mussel *Mytilus trossulus* (*cf.* Fig. 4.29)
2. Vagile species, able to move for a short distance, such as the Baltic clam *Macoma balthica* (Fig. 10.2b) or species with nocturnal swimming activity such as several amphipods, *e.g. Monoporeia affinis* (*cf.* Fig. 4.30).
3. Highly mobile species, such as mysids (*cf.* Box Fig. 6.1).
4. Species the adult stages of which are actively swimming at biologically high Reynolds numbers (10^3–10^9) (*cf.* Fig. 1.7), *e.g.* the fish swimming immediately above the sediment also belong to the deep soft seabed community, linking the benthic and pelagic food webs.

10.2.2 Characteristics of the Baltic Sea zoobenthos

Compared to the zoobenthos of fully marine areas, the fauna inhabiting the deep soft seabeds of the Baltic Sea consists of a lower number of species. The Baltic Sea species are of marine origin, freshwater origin or migratory species alternating between marine and fresh waters. Most marine species in the Baltic Sea have Atlantic origin, but they may also be cosmopolitan (*cf.* Sect. 4.2). In contrast to the shallow, near-shore areas of the Baltic Sea, non-indigenous species are rather rare in deep soft-bottom habitats. On the other hand, characteristic of deep soft seabeds are some glacial relicts, *e.g.* the crustaceans *Monoporeia affinis*, *Mysis* spp. and *Saduria entomon* (Fig. 10.2c) and the priapulid *Halicryptus spinulosus*

Box 10.1: Sampling of zoobenthic communities

Standardised methods

When sampling and analysing the deep soft-bottom zoobenthic communities, it is necessary to use standardised methods to make comparisons possible (Box Figs. 10.1, 10.2 and 10.3). Thus, it is important to consider the type of sampling gear (grab, core sampler or dredge) and the mesh size of the sieve (*e.g.* a limit of 0.5 or 1 mm between meio- and macrofauna). Different sieves are used for macrofauna and meiofauna, and in many cases also for offshore monitoring and coastal research.

Box Fig. 10.1 Different types of soft-bottom sampling equipment, (**a**) Van Veen grab, (**b**) Ekman-Birge grab, (**c**) Box corer. Photo: (**a**) © Rafal Michon, (**b**) © Tore Lindholm, (**c**) © Erik Bonsdorff

Box Fig. 10.2 The sieving of samples to separate the macrofauna from the bottom sediments. (**a**) A Kattegat sample from deep marine sediments is sieved over a coarse mesh. (**b**) A Baltic Sea sample from coastal sediments is sieved over a 0.5 mm mesh. The Kattegat sample is dominated by polychaetes (*e.g. Heteromastus*) and echinoderms (brittle stars *Amphiura* and sea urchins *Brissopsis*) while the Baltic Sea sample is dominated by the bivalve *Macoma balthica*. Photo: (**a**) © Erik Bonsdorff, (**b**) © Benjamin Weigel

Box Fig. 10.3 To identify the benthic in- and epifaunal species, such as oligochaetes, polychaetes, crustaceans and molluscs, it is necessary to use a stereomicroscope with a zoom range of 0.75–11.25 magnification. Photo: © Claes Björkholm

Sampling strategy and design

It is vital to consider appropriate sampling strategy and design, *i.e.* monitoring programmes should be designed in a way so that even unexpected changes or threats can be detected. When carrying out specific case studies, it is necessary to distinguish control from impact as well as situations before, during and after the impact (be it chemical, physical or climate-driven). To test specific hypotheses, it may be necessary to conduct experiments, for which methods may need to be designed to address specific questions. Experiments can be carried out in the field or under more controlled conditions in the laboratory.

Data collection

In the laboratory, aquaria allow for rigid replication, whereas experimenting in the field is often restricted in space and time. On the other hand, field experiments may yield information collected under more realistic and complex conditions with natural combinations of environmental (abiotic and biotic) factors. For example, measurements of fluxes in and out of the sediment, or of recruitment to defaunated areas, may require field manipulations rather than aquaria incubations (Villnäs et al. 2012). Finally, the collected data must allow for rigid numerical analysis using appropriate statistical tools and modelling. Many of these aspects have been treated in methodological textbooks applicable to the Baltic Sea benthic systems (Rumohr 1999; Eleftheriou and McIntyre 2005; Gray and Elliott 2009). However, for the numerical treatment of data, new techniques and programmes are constantly being developed.

Box 10.2: *In situ* sediment profiles

A method to study the sediment, the organisms, and the structures they create in the sediment (voids, tubes, burrows) *in situ* is based on photographic sediment profile imagery (SPI or REMOTS, using a sediment profile camera (Box Fig. 10.4) that provides images of the sediment structure and conditions (Box Fig. 10.5). The method was developed for the study of animal-sediment relationships during various stages of colonisation and succession along gradients of environmental stress (Rhoads and Germano 1982, 1986). Since then it has become an important and reliable tool to characterise the soft-bottom habitats and the infauna in a wide array of environments and conditions (Bonsdorff et al. 1996; Rumohr and Karakassis 1999; Nilsson and Rosenberg 2000; Smith et al. 2003). *In situ* sediment profiling has become a tool for assessing ecosystem health under the EU Water Framework Directive (Rosenberg et al. 2009). Several variations and developments of the method have been developed, including elaborate 3D X-ray images of deep sediment cores to study post-glacial subfossilised biogenic structures such as burrows and tubes (Virtasalo et al. 2006, 2011a, b).

Box Fig. 10.4 A SPI/REMOTS-camera ready for deployment. Photo: © Marine Monitoring (Lysekil, Sweden)

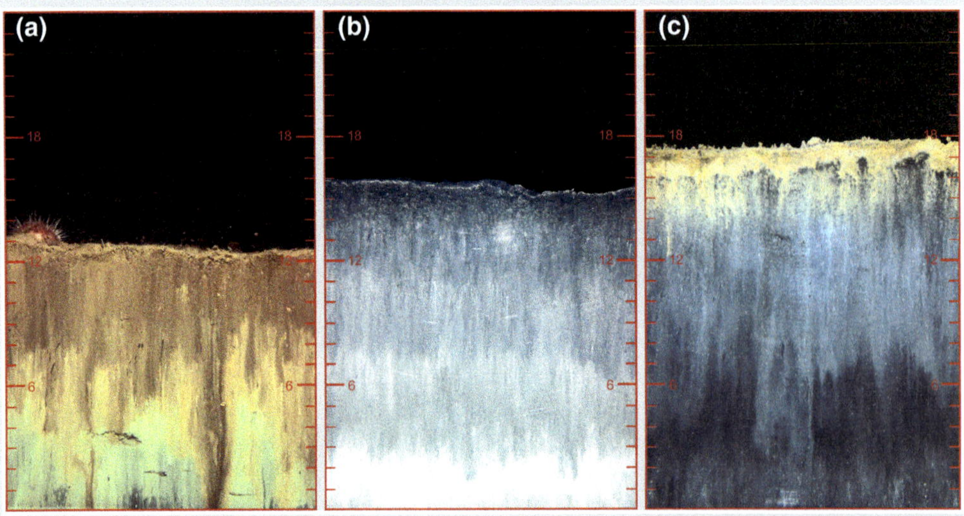

Box Fig. 10.5 Three different sedimentary conditions recorded with an SPI/REMOTS camera. (**a**) Well-oxygenated deep sediment. (**b**) Anoxic sediment. (**c**) Sediment during recovery from anoxia, showing numerous polychaete tubes sticking out of the sediment. The scale is given in cm. Photo: © Marine Monitoring (Lysekil, Sweden)

Fig. 10.2 Examples of invertebrates living in the deep soft seabeds of the central Baltic Sea proper. (**a**) The polychaete *Bylgides sarsi*. (**b**) The bivalve *Macoma balthica*. (**c**) The isopod *Saduria entomon*. (**d**) The amphipod *Pontoporeia femorata*. (**e**) The priapulid *Halicryptus spinulosus*. Photo: © Piotr Wysocki

(Fig. 10.2e), which immigrated to the Baltic Sea during the Baltic Ice Lake or Yoldia Sea stages (*cf.* Sect. 2.5). These glacial relicts are also found in some large lakes in Fenno-Scandia and northeastern Russia (Audzijonyte and Väinölä 2005).

As the environmental conditions changed during the geological development of the Baltic Sea, some of the less stress-tolerant species died out, while those with wider tolerance limits (often called "opportunistic" species) survived (Segerstråle 1957a). Their opportunistic traits allowed some animals to extend their distributional (depth) range in the Baltic Sea. Thus, the deep Baltic Sea bottoms often support species that are common in shallow areas of fully marine seas, an effect known as the "brackish-water submergence" (*cf.* Sect. 11.2.2).

In many ways, the zoobenthos of the Baltic Sea, restricted by the steep environmental gradients (salinity, temperature, oxygen), is still in the process of post-glacial succession, *i.e.* in large parts of the Baltic Sea the zoobenthic communities have not reached a "climax" or "mature" state yet (Rumohr et al. 1996). This in turn may partly explain why the Baltic Sea zoobenthic communities are so sensitive to unexpected or sudden changes in the environment. Many species found in the Baltic Sea live on the edges of their physical and ecological tolerance limits. Should conditions suddenly change, there would be no natural replacement for a possible (local, regional or ecosystem-wide) loss of species or ecological functions. Such perturbations in species composition may have a profound impact on the functioning of entire zoobenthic communities. Moreover, the specific conditions prevailing in the Baltic Sea have also affected the zoobenthic organisms themselves, including their morphology, physiology (*cf.* Sect. 7.1) and genetic composition (*cf.* Sect. 6.3). For example, most marine species living in the Baltic Sea grow to a much smaller body size (*cf.* Fig. 4.5a), and sometimes also display reduced reproductive capacity, compared with their conspecifics living in marine conditions.

10.3 Meiobenthos

10.3.1 The Baltic Sea meiobenthos is poorly known

The meiofaunal component of the zoobenthic communities on the Baltic Sea deep soft seabeds is rather poorly known. Spatially resolved faunistic surveys are mainly restricted to a few areas in the Stockholm archipelago and the southern Baltic Sea. In most cases, the taxonomic resolution of the meiofauna is fairly coarse: analyses based on morphological characteristics can usually only be carried out for higher-ranked taxa such as family, class or phylum (Ojaveer et al. 2010). Representatives of only very few groups of the Baltic meiobenthos have been identified to the species level, *e.g.* the harpacticoid copepods of the Słupsk channel (Drzycimski 2000).

10.3.2 The composition of the meiobenthos is depth-dependent

The limited existing evidence shows that the structure of the meiobenthos changes with water depth. Although dominated by free-living nematodes (60–90 % and more of the total abundance) throughout the Baltic Sea, the meiobenthos is most diverse on sandy sediments at the near-halocline depths. There, the ever-present nematodes are accompanied by gastrotrichs, halacarids, harpacticoid copepods, kinorhynchs, ostracods, turbellarians, and juvenile forms of the macrozoobenthos. Many of these juvenile forms are restricted to certain seasons; particularly in late spring-summer, young amphipods, bivalves, gastropods and polychaetes may become numerically quite important in the meiobenthos.

With increasing depth, as the sand becomes enriched with finer silt particles to grade into muddy sand, sandy mud, and finally mud, there is an increasing dominance of nematodes in the meiobenthos. The overall density of the meiofauna may increase with depth as well, provided the near-bottom water is not hypoxic (Pallo et al. 1998). At still larger depths, in the deepest, muddy parts of the Baltic Sea (experiencing hypoxia and anoxia and devoid of macrozoobenthos), the meiobenthos may still be represented by sparsely occurring nematodes (Elmgren et al. 1984; Radziejewska 1989).

10.4 Macrozoobenthos

10.4.1 Colonisation of the Baltic Sea

In the Belt Sea, west of the Darß sill where salinity is higher and more fluctuating (*cf.* Fig. 4.2), many typical marine species still occur, *e.g.* the bivalves *Corbula gibba* and *Scrobicularia plana*, which are absent in the Baltic Sea east of the sill. Slightly more euryhaline marine species, *e.g.* the bivalve *Astarte borealis*, the polychaete *Travisia forbesii* and the crustacean *Nymphon grossipes*, reach the limits of their distributions in the southern Baltic Sea, while others may penetrate into the gulfs. *Pontoporeia femorata* is an example of a species that, despite its marine origin, has colonised the entire Baltic Sea all the way up to the Bothnian Sea.

In the Gulf of Bothnia, the zoobenthos contains a large proportion of freshwater species. Here, the bottom-water temperature may for some species be a stronger barrier than salinity. The macrozoobenthos of the Gulf of Bothnia is dominated by the cold-water crustaceans *Monoporeia affinis*, *Pontoporeia femorata*, and *Saduria entomon*. With their preferences towards cold and oligohaline waters, these species are present also underneath the layer of cold winter water above the halocline in the Baltic Sea proper.

Fig. 10.3 *Marenzelleria* is an example of a highly successful non-indigenous polychaete genus in the Baltic Sea. At least three species (*Marenzelleria arctia*, *Marenzelleria neglecta* and *Marenzelleria viridis*) have colonised the Baltic Sea since the 1980s (*cf.* Box 5.3). Their overlapping ranges now cover the entire Baltic Sea, from south to north and from shallow to deep soft seabeds. Photo: © Jan-Erik Bruun

That the post-glacial immigration of species to the Baltic Sea is an on-going process (Segerstråle 1957a, b; Rumohr et al. 1996; Bonsdorff 2006) is exemplified by the recent colonisation of the entire Baltic Sea by polychaetes of the non-indigenous genus *Marenzelleria* (Fig. 10.3; *cf.* Box 5.3). The invasion history of *Marenzelleria* spp. is an intriguing case of unexpected, fundamental changes in the structure and function of the ecosystem (Blank et al. 2008; Norkko et al. 2012; Rousi et al. 2013).

Like other areas with low salinity and low biodiversity, the Baltic Sea is regarded as particularly vulnerable to colonisation by non-indigenous, accidentally introduced species (*cf.* Sect. 5.1). This vulnerability is thought to be enhanced by the high anthropogenic stress on the Baltic Sea ecosystem, which leads to reduced population sizes of some native species and thus opens a window of opportunity for non-indigenous species (*e.g. Marenzelleria* spp.). The vulnerability is further emphasised by the fact that the entire Baltic Sea ecosystem is young (*cf.* Sect. 2.5.1) and still undergoing post-glacial colonisation (Rumohr et al. 1996; Bonsdorff 2006).

10.4.2 Depth-dependent distributions

The deep soft seabed macrofauna of the Baltic Sea consists mainly of the classes Bivalvia, Crustacea and Polychaeta.

Species representing the Anthozoa and Echinodermata are restricted to the westernmost part of the Baltic Sea and the transition zone. The spatial distribution of the macrozoobenthos reflects vertical (bathymetric) and horizontal (environmental) gradients as well as the zoogeographic origins and adaptive abilities of individual species (Bonsdorff 2006).

The depth gradient in the Baltic Sea proper is associated with increasing salinity below the halocline, whereby the sediment is more fine-grained, the temperature decreases and becomes more stable, and oxygen conditions frequently deteriorate. The proportion of marine species in the zoobenthic communities, *e.g.* that of the polychaetes *Scoloplos armiger* and *Terebellides stroemii* or the crustacean *Pontoporeia femorata* (Fig. 10.2d), increases with depth. Some opportunistic species display submergence, the most typical one being the bivalve *Macoma balthica*, which occurs in the entire Baltic Sea and throughout the whole range of depths.

10.4.3 Diversity along the large-scale Baltic Sea gradient

The composition and distribution of macrofaunal species along the depth gradient is modified by the horizontal gradients of salinity and temperature, which extend from the transition zone through the Baltic Sea proper to the Gulf of Bothnia and the Gulf of Finland. The species that have colonised the Baltic Sea differ in their salinity tolerance. Hence, those species least resistant to reduced salinity occur only in the transition zone where the salinity is higher. The number of marine macrofauna species diminishes with increasing distance from the transition zone.

This trend is illustrated by the census of macroscopic species (those visible to the naked eye) along the salinity gradient: species richness drops dramatically from >1600 in the North Sea and ~500 at the entrance of the Baltic Sea to <20 in the central and northern Baltic Sea proper and ~20 in the Gulf of Bothnia and the Gulf of Finland (Fig. 10.4). In the innermost reaches of the system (but outside the deep soft-bottom realm), the number of aquatic invertebrate species increases again on account of an increasing proportion of freshwater organisms, such as some bivalves and gastropods, and typical freshwater groups such as oligochaetes and insect larvae (Bonsdorff 2006; Villnäs and Norkko 2011; Zettler et al. 2014; Gogina et al. 2016).

Although the number of zoobenthic species recorded in and on deep soft seabeds is far higher in the Belt Sea than in the eastern Baltic Sea proper (Fig. 10.4), the differences in the number of species recorded at a specific station per sampling occasion (usually three samples are taken with an 0.1 m^2 van Veen grab, Box 10.1a) are not so large (Villnäs and Norkko 2011). In the Arkona Sea, a total species pool

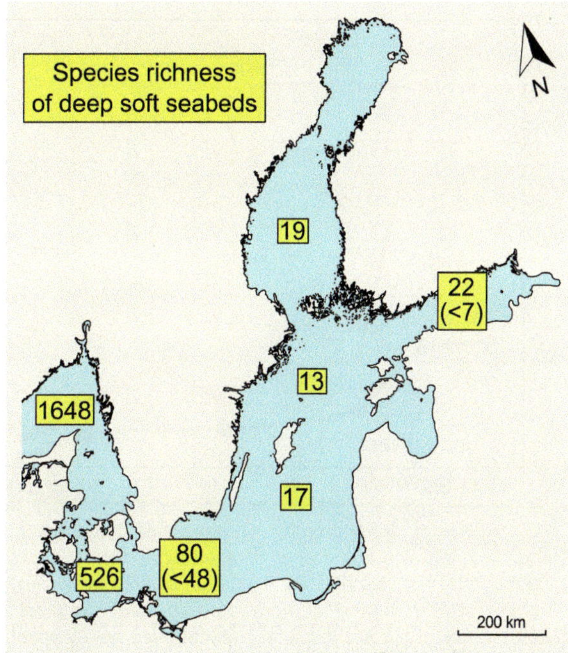

Fig. 10.4 Species richness of the macroscopic fauna (visible to the naked eye) along the large-scale Baltic Sea gradient on soft bottoms deeper than 20 m. The numbers in brackets refer to data from below the halocline in areas with annual to semi-annual hypoxia. Figure modified from Bonsdorff (2006)

(γ-diversity) of 78 species was recorded, but no more than 23 species were encountered on any sampling occasion (α-diversity) in the time period 1964–2006 (Fig. 10.5). The difference between these two properties describing species richness continues to decrease northwards in the Baltic Sea because of a diminishing number of rare species. Therefore, to record the maximum regional species richness in the northern Baltic Sea, it is usually sufficient to sample only a few stations, whereas a much higher sampling effort is required in the western part of the Baltic Sea.

10.4.4 Patterns of production

Macrozoobenthic biomass, a rough measure of production, varies along the bathymetric and horizontal gradients of the Baltic Sea. Typically, the biomass decreases with depth, from high values on the order of magnitude of thousands of grams wet weight (ww) m^{-2} at shallow sites (mainly mussel beds) to only a few grams in deep-water sediments (or a complete lack of fauna as a consequence of severe hypoxia or anoxia). The higher biomasses in shallow water can be explained by a higher availability of food from primary production and a higher proportion of filter feeders in the zoobenthos.

The macrozoobenthic biomass is also observed to drop along the Baltic Sea gradient from the Belt Sea (several hundreds g ww m^{-2}) to the Baltic Sea proper

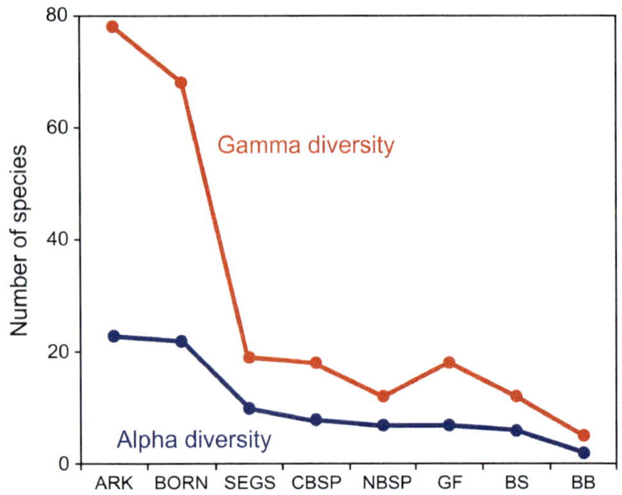

Fig. 10.5 α-diversity and γ-diversity of zoobenthic communities on the Baltic Sea soft bottoms deeper than 40 m in the time period 1964–2006. α-diversity describes the maximum number of species per station observed on any sampling occasion. γ-diversity describes the maximum number of species per subregion observed on all sampling occasions combined. Subregions: ARK = Arkona Sea, BORN = Bornholm Sea, SEGS = southeastern Gotland Sea, CBSP = central Baltic Sea proper, NBSP = northern Baltic Sea proper, GF = Gulf of Finland, BS = Bothnian Sea, BB = Bothnian Bay. Figure modified from Villnäs and Norkko (2011)

(~100 g ww m^{-2}) and further to the Gulf of Bothnia (a few g ww m^{-2}). To a certain extent, this trend can be explained by similar horizontal changes in primary production (*cf.* Figs. 2.25 and 11.37) and the lack of large suspension-feeding mussel populations in the Gulf of Bothnia (Bonsdorff and Pearson 1999). However, in the eastern Baltic Sea proper the biomass of the patchily distributed bivalves may locally exceed levels typical of the Belt Sea.

10.5 Benthivorous fish

In addition to benthic invertebrates, fish also live on the deep soft seabed of the Baltic Sea or migrate there to feed. The number of fish species in the Baltic Sea decreases along the Baltic Sea gradient from ~200 in the transition zone to ~70 in the northern Baltic Sea (*cf.* Fig. 4.10d). The deep soft seabeds of the Baltic Sea are inhabited by demersal fish of various origins, mainly marine species and glacial relicts, but close to the coast in the northern parts of the Baltic Sea some freshwater fish species are also represented.

The freshwater fish feeding on the zoobenthos are dominated by cyprinids, *e.g.* the roach *Rutilus rutilus*, the bream *Abramis brama* and the whitefish *Coregonus* sp. Marine species such as the Atlantic cod *Gadus morhua*, the European flounder *Platichthys flesus*, the European plaice *Pleuronectes platessa* and turbot *Scophthalmus maximus*, are more abundant in the southern Baltic Sea.

Glacial relict fish, *e.g.* the fourhorn sculpin *Myoxocephalus quadricornis*, the lumpsucker *Cyclopterus lumpus* (Fig. 10.6a) and the common seasnail *Liparis liparis* (Fig. 10.6c) are present in relatively low abundances and occur in deeper areas as they require cold and well-oxygenated water (HELCOM 2009a). The snakeblenny *Lumpenus lampretaeformis* is probably a glacial relict from the Yoldia Sea stage (*cf.* Sect. 2.5.6) and occurs in the Baltic Sea and the northern Atlantic Ocean (Segerstråle 1957a, b). This elongated eel-like benthic fish is reproductive in the entire Baltic Sea (HELCOM 2012) and lives in Y-shaped tubes in muddy bottoms at a >30 m water depth. The snakeblenny feeds on meiofauna and small macrozoobenthic crustaceans such as *Pontoporeia femorata*. Together with the isopod *Saduria entomon* the snakeblenny is an important prey species for deep-water cod. Another cold-water species, the viviparous eelpout *Zoarces viviparus* (Fig. 10.6b), can also inhabit coastal areas, but females require low temperature to give birth to the young, and therefore the species spawns during the cold season. *Zoarces viviparus* is the only viviparous fish species that occurs in the Baltic Sea.

10.6 Zoobenthic communities

10.6.1 Characteristics of Baltic Sea zoobenthic communities

Although Karl-August Möbius coined the concept "biocenosis" to denote a "living community" in his 1877 work "The oyster and oyster culture" (Möbius 1877; Glaubrech 2008), detailed research on zoobenthic communities worldwide was initiated in the Baltic Sea by Johannes Petersen in the early 1900s (Petersen and Boysen-Jensen 1911; Petersen 1913). The community concept in a broad sense (*cf.* Sect. 4.1.2) facilitates the understanding of structural relationships between groups of organisms, although it may also denote a supra-organismal unit the components of which are connected by strong biological interactions (Mills 1969). At present, communities are viewed in a more complex way, as it has become obvious that sets of local communities are linked on various spatial and temporal scales. Such sets are termed "metacommunities" and exhibit various traits such as resilience, connectivity, recruitment and succession (Leibold et al. 2004).

Petersen (1913), and those following him, defined a community as "a regularly occurring combination of certain species, normally present in quite large numbers". Although interrelationships between the component species were regarded as the principal community-structuring factors, Petersen placed a particular emphasis on the dominant species, and thus described almost the entire Baltic Sea as "the *Macoma balthica* community". Later studies showed that the

Fig. 10.6 Examples of fish species found in deep areas of the Baltic Sea and showing preference towards low temperature. (**a**) The lumpsucker *Cyclopterus lumpus*. (**b**) The viviparous eelpout *Zoarces viviparus*. (**c**) The common seasnail *Liparis liparis*. Photo: © Piotr Wysocki

boundaries observed to exist between areas inhabited by specific groups of species reflect changes in environmental conditions rather than altered biological interrelationships.

A characteristic feature of the macrozoobenthos in the central Baltic Sea proper is a high similarity between communities, although variability can be high depending on variability in physical and chemical conditions in time and space (Bonsdorff and Pearson 1999; Laine 2003; Villnäs and Norkko 2011). The structure and distribution of the macrozoobenthic communities reflect a combination of the main environmental drivers (oxygen, salinity, temperature, sediment type) and the environmental preferences of each species. For example, the Baltic Sea proper is dominated by communities consisting of marine species, mainly *Macoma balthica* (in the Baltic Sea a mixture of Pacific and Atlantic genes, *cf.* Box 6.4), whereas the northern Baltic Sea supports communities in which cold-water species such as *Monoporeia affinis*, *Pontoporeia femorata* and *Saduria entomon* are prevalent.

10.6.2 Succession of Baltic Sea zoobenthic communities

Based on a theoretical model of zoobenthic succession in fully marine seas (Pearson and Rosenberg 1978), Rumohr et al. (1996) described community changes along a gradient of organic enrichment (leading to hypoxia, anoxia, H_2S). The successional stages of the Baltic Sea zoobenthos were analysed along a south-north transect from the Belt Sea to the Bothnian Sea and with increasing water depth. The model predicts five successional stages ranging from diverse communities to dead bottoms (Fig. 10.7):

I. Climax communities, which are dominated by deep-burrowing and long-lived *K*-selected species (*cf.* Box 4.14), echinoderms in particular, as well as large polychaetes and epifaunal crustaceans. A distinct redox potential discontinuity (RPD, *cf.* Fig. 12.6) layer is found deep in the sediment. This stage is found only in the Belt Sea.

II. Communities showing the first signs of stress, which can be due to *e.g.* eutrophication or physical disturbance through intensive trawling. The RPD layer is located shallow in the sediment. Characteristic are spatial and temporal fluctuations in the biota. Molluscs and long-lived polychaetes dominate the community and biomass production shows an overall increase. This stage can be regarded as the "basic zoobenthic community" in the present-day Baltic Sea.

III. Communities showing additional stress, usually caused by organic enrichment, oxygen deficiency and deteriorated environmental conditions. The RPD layer lies at the sediment/water interface. This successional stage is characterised by low biodiversity, high abundance, high dominance and low biomass of the zoobenthos dominated by small polychaetes (*e.g.* *Capitella capitata*, *Polydora* spp., *Pygospio elegans*) or abundant populations of bivalves (*Abra alba*, *Kurtiella bidentata*, *Macoma balthica*) or amphipods (*Monoporeia affinis*, *Pontoporeia femorata*) living close to or on the sediment surface. Species may disappear as a result of hypoxia or anoxia, but recovery is rapid because *r*-selected species dominate.

Box 10.3: Biological traits analysis (BTA)

Traits

The most recent development in our understanding of the functioning of the benthic ecosystem comes from the so-called "biological traits analysis" (BTA). BTA simultaneously analyses changes in multiple biological traits (divided into modalities) reflecting different functionalities of organisms by putting together "traditional" knowledge on species morphology (*e.g.* the structure of the feeding apparatus), biology (*e.g.* reproductive capacity), physiology (*e.g.* metabolic rate) and behaviour (*e.g.* movement, feeding behaviour) (Villnäs and Norkko 2011; Törnroos and Bonsdorff 2012; Villnäs et al. 2012; Törnroos et al. 2015).

BTA as a predictive tool for functional characteristics of communities

BTA provide a basis on which the specific capacity and role of individual species and communities in the sedimentary environment can be tested experimentally (Karlson et al. 2007; Norling et al. 2007; Josefson et al. 2012), and adequate modelling tools can be developed (Norkko et al. 2012; Timmermann et al. 2012). By combining detailed species-specific knowledge with long-term data on community composition and abundance (*cf.* Fig. 10.9), our capacity to understand changes in the benthic communities develops into a predictive tool for the benefit of the Baltic Sea environment. Analysing large-scale patterns of the functional traits allows us to visualise not only the spatio-temporal changes in species composition, abundance and/or biomass, but also to comprehend the functional characteristics of the biota across the large-scale Baltic Sea gradient (*cf.* Fig. 10.10).

IV. Communities showing severe stress, characterised by long periods of severe hypoxia or anoxia. *Beggiatoa* spp. mats form on the sediment surface and the RPD layer is at the sediment/water interface. The macrozoobenthos is generally absent, although some species, *e.g. Bylgides sarsi* (Fig. 10.2a) may feed on the sediment surface.
V. In the last successional stage there is a complete absence of the zoobenthos as a result of long-term (years, decades) hypoxia or anoxia. The sediment is laminated due to the absence of bioturbation. The RPD layer may be absent, and the boundary between good and bad oxygen conditions may lie in the water column at some distance above the sediment.

10.6.3 Functionality of zoobenthic communities

It is commonly recognised that the structure of the zoobenthic communities depends largely on sediment parameters. Thus, a change in sediment type with depth causes a shift in community composition and consequently a shift in structural characteristics. There are a number of functional traits that can be determined, *e.g.* the mode of movement and feeding habits of the constituent species, and species can be combined into functional groups (Bonsdorff and Pearson 1999).

A further development of the functional group concept is the use of biological traits (*cf.* Sect. 4.7). Biological traits analysis (BTA; Box 10.3) assigns various functional properties (biological trait categories) to each species, which then provides an estimate of the functional diversity of a community (Villnäs and Norkko 2011; Törnroos and Bonsdorff 2012; Törnroos et al. 2015).

The zoobenthos in deep soft-bottom areas is dominated by infauna (organisms living below the sediment surface), such as *Macoma balthica*, *Marenzelleria* spp. *Monoporeia affinis*, *Pontoporeia femorata*, *Priapulus caudatus*, *Scoloplos armiger* and *Terebellides stroemii*. The dominance of the infauna increases with depth while the number of mobile epibenthic species (*e.g. Bylgides sarsi* and *Saduria entomon*) decreases.

Changes in the functional structure of zoobenthic communities in deep-water areas are also apparent if we follow the Baltic Sea gradient from the Kattegat-Skagerrak area to the Gulf of Bothnia. Along the gradient the number of functional groups drops from 20 down to only four (Fig. 10.8). Only one group, mobile predators, is present in all the regions of the Baltic Sea. A characteristic feature is the rapid disappearance of deep-burrowing detritivores and sedentary organisms; consequently, the relative importance of mobile fauna moving or feeding on the sediment surface increases.

10.6.4 Gradients of functionality

As the number of species is reduced with decreasing salinity (Fig. 10.9), important species-specific traits and community functions are being lost. An example is furnished by the deep-burrowing thallassinidean shrimp *Calocaris*

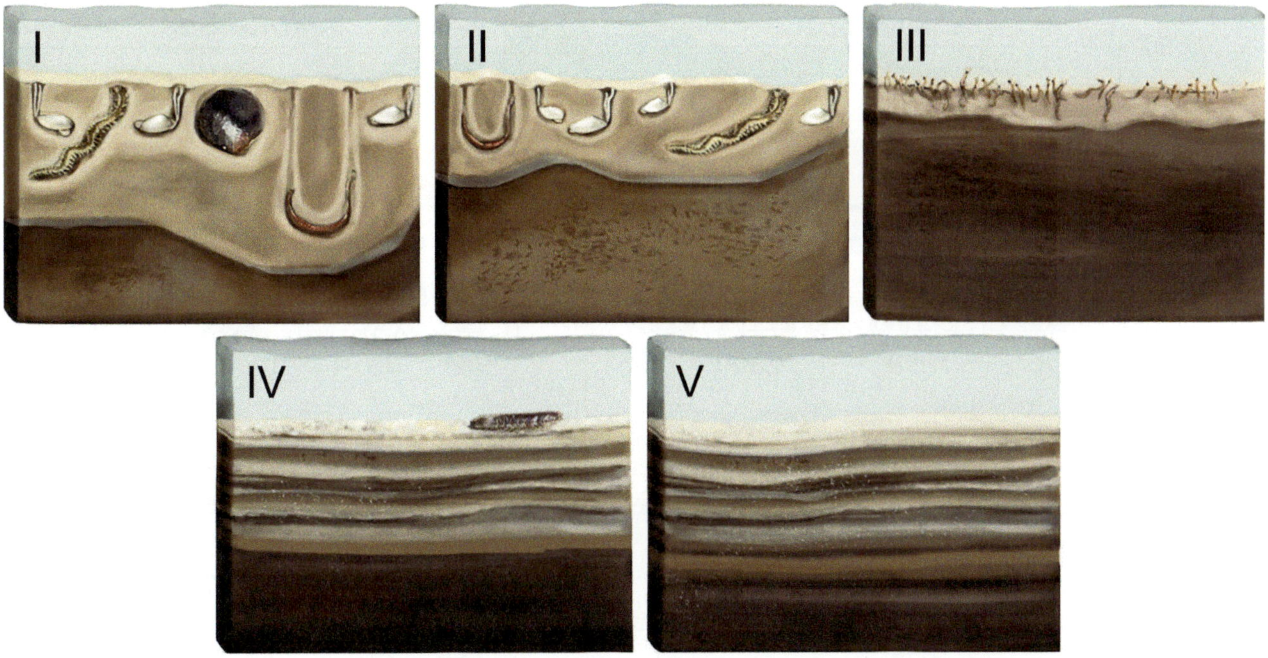

Fig. 10.7 A schematic illustration of the zoobenthic succession in a gradient along a south-north transect from the Belt Sea to the Bothnian Sea with increasing water depth. I = Climax community with deep-burrowing long-lived species. II = Community showing the first signs of stress, dominated by molluscs and large polychaetes. III = Community showing additional stress, dominated by small polychaetes. IV = Community under severe stress, with *Beggiatoa* spp. mats and few *Bylgides sarsi*. V = The last successional stage: laminated sediments and absence of zoobenthos. Figure based on information in Rumohr et al. (1996). Figure: © Małgorzata Piłka

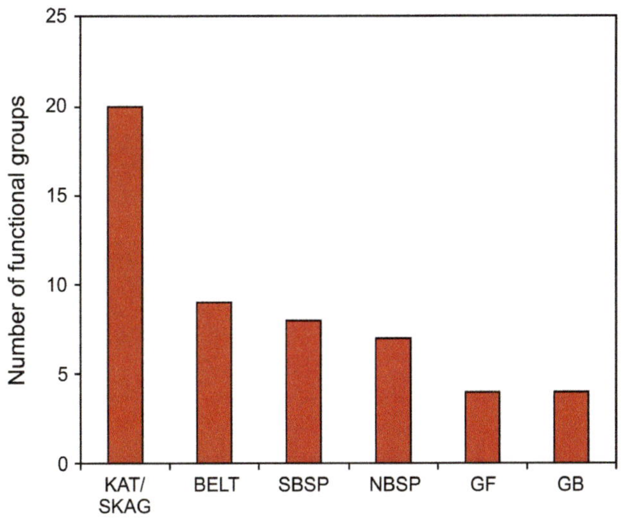

Fig. 10.8 Number of macrozoobenthic functional groups, defined by mobility, feeding apparatus and feeding type of the species, along a south-north transect through the deep (subhalocline) soft bottoms of the Baltic Sea. Subregions: KAT/SKAG = Kattegat/Skagerrak, BELT = Belt Sea, SBSP = southern Baltic Sea proper, NBSP = northern Baltic Sea proper, GF = Gulf of Finland, GB = Gulf of Bothnia (Bothnian Sea and Bothnian Bay). Figure modified from Bonsdorff and Pearson (1999)

macandreae which lives in the Skagerrak and Kattegat. As a very active bioturbator, this shrimp is capable of affecting oxygen and nutrient fluxes and organic matter mineralisation more than any other species (Norling et al. 2007). However, *Calocaris macandreae* is absent in the Baltic Sea proper.

The number of species per functional group is about five in the Kattegat and Skagerrak, and drops sharply to below two throughout the Baltic Sea (Bonsdorff and Pearson 1999). This underlines the sensitivity of the Baltic Sea deep soft-bottom community to stress, *e.g.* if the loss of one species' function is not compensated for by another species. Along the large-scale Baltic Sea gradient, it is not only species composition that changes (Fig. 10.9), but community functioning as well, with fundamental consequences for the ecosystem (Fig. 10.10).

The living conditions of the zoobenthos gradually change with time due to climate change, eutrophication, hypoxia and anoxia, the success of non-indigenous species in establishing themselves and altered food web cascades, both bottom-up through increased sedimentation of food particles and top-down through altered fish communities and hence predation pressure. In this context it is essential to realise that reference conditions change as well ("shifting baselines" *cf.* Fig. 17.1; Villnäs and Norkko 2011). Thus, our

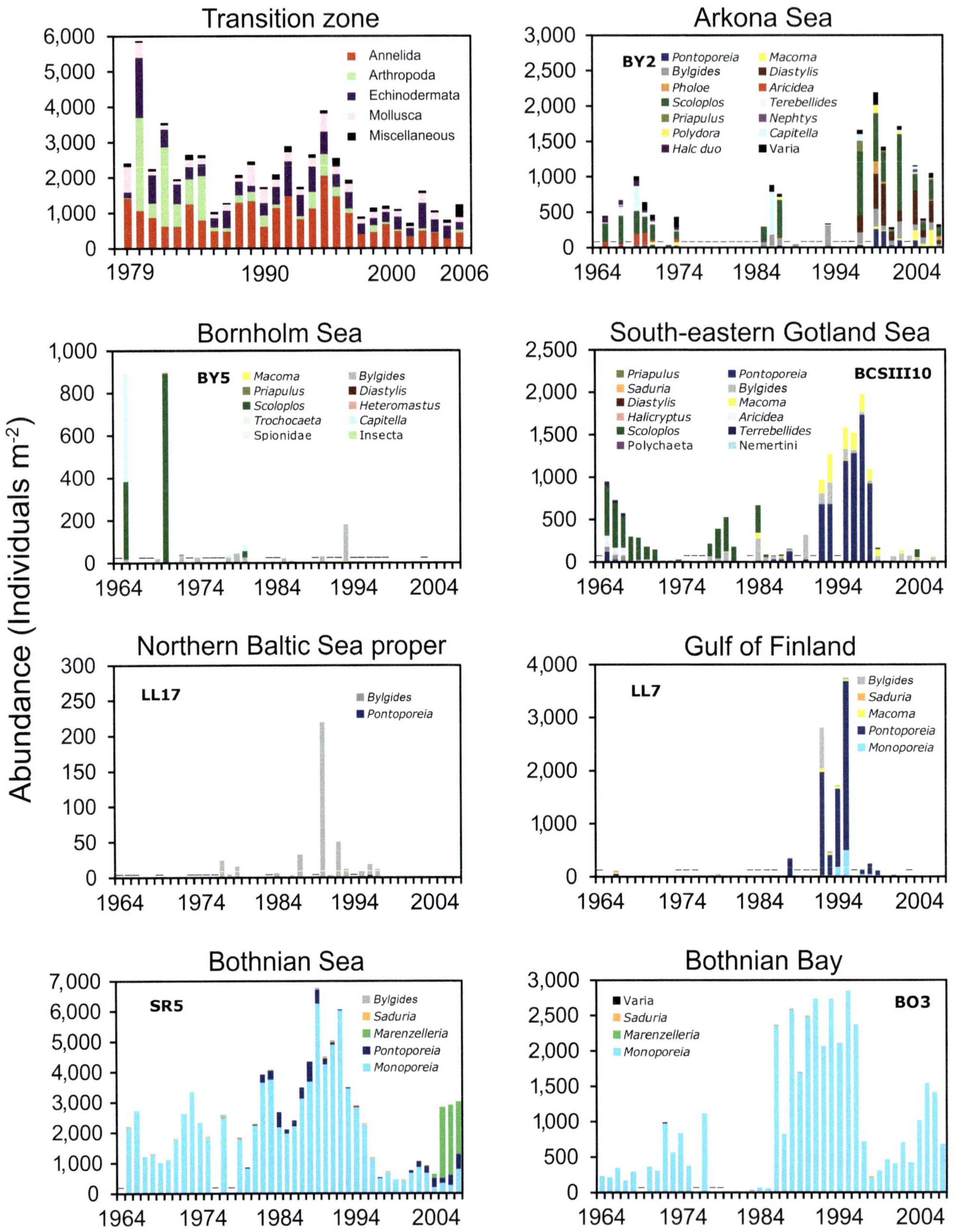

Fig. 10.9 Long-term changes in abundance and community composition of benthic invertebrates at depths >40 m in the transition zone (Belt Sea and Kattegat) and seven subregions of the Baltic Sea. Note the different scales on the x- and y-axes and the difference in the grouping of taxa between the transition zone and the Baltic Sea. Data from Norkko and Jaale (2008) and A. Josefson (unpublished). Dashes indicate that no data are available. Figure with small modifications reprinted from HELCOM (2009b) with permission from Alf Norkko and HELCOM

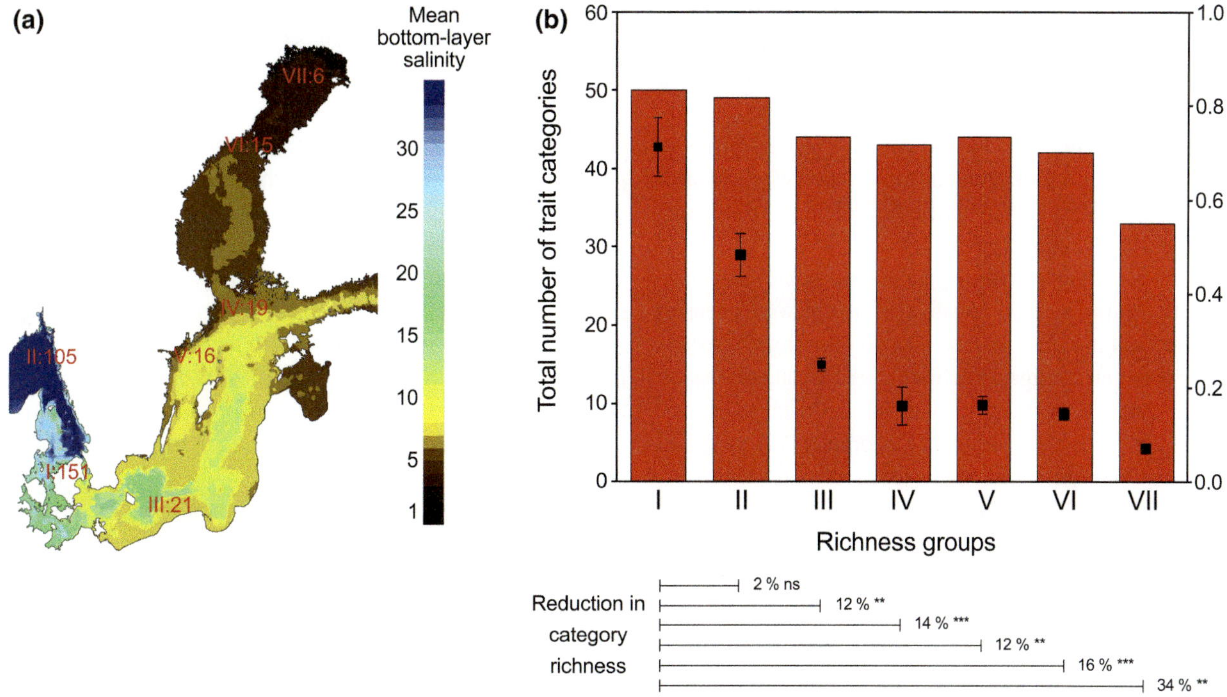

Fig. 10.10 An example of the functional diversity of soft-bottom fauna along the large-scale Baltic Sea gradient of taxonomic richness. (**a**) Map of the Baltic Sea Area with the locations of seven defined richness group regions (I–VII) with the total number of taxa, varying from 151 in richness group I to 6 in richness group VII. (**b**) The number of functional trait categories (left-hand y-axis, red bars) and the functional diversity-index $FD_{Petchey\ \&\ Gaston}$ (right-hand y-axis, black squares with SE) for each richness group. The defined richness group regions are (I) Belt Sea and Kattegat, (II) the Swedish Skagerrak coast, (III) the Polish coast, (IV) the Stockholm archipelago, (V) the Askö area, (VI) the Norra Kvarken area, including stations from both Swedish and Finnish waters, and (VII) the innermost part of the Bothnian Bay. Note that the 34 % reduction in category richness between the regions with highest species richness (I) and lowest species richness (VII) means that still 66 % of all trait categories were represented in both I and VII. However, the reduction in FD between the richness groups along the gradient was steeper. Figure modified from Törnroos et al. (2015)

understanding of the benthic system must change from the traditional static view to a more flexible conceptual framework (Galil et al. 2014; Katsanevakis et al. 2014).

10.7 Seasonal variability

10.7.1 Reproduction seasons

Although seasonality on a deep soft seabed is not as apparent as in the coastal zone, abundances of individual species vary significantly from season to season as a result of seasonal reproduction with distinct cohorts (age classes). Most species produce offspring in the spring, *e.g.* the crustaceans *Monoporeia affinis* and *Pontoporeia femorata*. For other species the reproductive period extends over several months (*e.g. Mytilus trossulus*) or even takes the whole year (*e.g. Saduria entomon*). Many species (molluscs, polychaetes) have pelagic larval stages, and when these begin to settle on the bottom as members of the so-called “temporary meiobenthos”, the species may occur in vast numbers of up to several hundred thousand individuals m^{-2}.

The macrozoobenthos usually consists of relatively long-lived species, with life cycles from several years (*e.g. Macoma balthica*) up to decades (*e.g. Arctica islandica*). Hence their biomass, consisting largely of adult individuals, does not normally exhibit such dramatic seasonal variations as the abundance, provided the environmental conditions stay within their natural limits.

10.7.2 Seasonality of detritus

The macrozoobenthos of deep soft seabeds relies to a large extent on the amount of organic material settling from the water column. This organic material consists mainly of phytodetritus (remains of phytoplankton blooms), which may be taken up directly (primarily by suspension feeders) or be incorporated in the sediment. The latter material undergoes diagenetic and microbial transformations and increases the organic matter sediment reserve to be utilised later (primarily by deposit feeders).

The amount and quality of the phytodetritus reaching the seafloor vary seasonally, depending on the timing, composition and magnitude of phytoplankton blooms. Thus, the benthic suspension feeders and deposit feeders must have efficient energy storage systems to be able to cope with long periods of poor food availability. Some species, such as the facultative suspension-surface deposit feeder *Macoma balthica*, accumulate glycogen and lipid reserves soon after the spawning seasons in the spring and autumn (Bonsdorff and Wenne 1989; Janas et al. 2007).

Even in deep water with a year-round stable low temperature, the animals grow the fastest in the warm season when more food is available. The growth of mollusc shells is shown by concentric rings, which are the external expression of growth rate. In winter, growth in both mussels and crustaceans is arrested, and their locomotor activity and metabolic activity are significantly reduced.

At the onset of the cold season, the deep soft-bottom areas experience the arrival of abundant nectobenthos. During the warm season, nectobenthic organisms, *e.g.* the brown shrimp *Crangon crangon*, the native prawn *Palaemon adspersus* and the non-indigenous prawn *Palaemon elegans*, stay in the coastal area, but in winter they migrate to the more thermally stable deeper zones.

10.8 Temporal trends

10.8.1 Prolonged periods of anoxia lead to regime shifts

The most dramatic changes, leading to a significant reduction or even complete elimination of the macrofauna, are observed on the seabed beneath the halocline. At first, these changes were not associated with eutrophication, which used to be regarded as a problem affecting only coastal waters and which generally increased the biomass. It was not thought then that anthropogenic activities could affect the entire Baltic Sea (Elmgren 2001). However, the specific impacts of eutrophication also became gradually acknowledged for the zoobenthos (Cederwall and Elmgren 1980, 1990; Perus and Bonsdorff 2004).

Fonselius (1969) found that the deep-water oxygen deficiencies in the Baltic Sea proper in the 1960s were substantially larger than those recorded before the 1960s and suggested that they could be linked to anthropogenic activities. Since then it has been demonstrated that the additional stress generated by eutrophication has modified the distribution of the zoobenthos in many regions of the Baltic Sea (HELCOM 1990). This process is usually described as a continuum of changes in the zoobenthos along an enrichment gradient, in which an undisturbed, species-rich community is first transformed into a species-poor but highly productive community and finally into total lack of macrofauna (equivalent to Rumohr et al.'s 1996 model stage V) through prolonged periods of anoxia (Fig. 10.7e).

In accordance with the Pearson and Rosenberg (1978) and Rumohr et al. (1996) models, this process is theoretically reversible as long as conditions improve. However, the physical and chemical changes of the sediment properties can continue to affect biological processes for a long time, even if other factors do improve. This means that the resistance (and inertia) of the system is important, in spite of its resilience (recovery potential), and several ecological thresholds and alternative stable states may affect the potential recovery. Hence, the concept of "regime shifts" is a consequence of non-linear impacts of various drivers that affect the ecosystem simultaneously. Consequently, recolonisation of defaunated areas may not restore a zoobenthic community to the state it was in before the onset of the perturbing factors.

10.8.2 Regime shifts below the halocline

Studies carried out in the southern Baltic Sea before 1950 showed that the entire deep-water bottom area supported macrozoobenthic communities and were dominated by polychaetes and bivalves (Demel and Mulicki 1954). Subsequently, changes in the structures of these communities were observed (Żmudziński 1977), during which some bivalve species (*Astarte* spp., *Macoma calcarea*) were reduced in abundance or disappeared altogether, to be replaced by polychaetes. During the last 50 years, significant changes have been documented in many areas of the Baltic Sea (Villnäs and Norkko 2011). The shift from a marine to a brackish community over time was demonstrated for the southeastern Gotland Sea and cyclic changes in species abundances were revealed in the Bothnian Sea and Bothnian Bay (Fig. 10.9).

The permanent absence of macrozoobenthos due to hypoxic and anoxic conditions was first discovered in the Bornholm deep in 1964–65. The same was found at the bottom of the Gotland and Gdańsk deeps in the early 1970s. In later years, the seabed areas with heavily reduced macrofauna or devoid of benthic invertebrates increased in spatial extent, even including areas at shallower depths (60–80 m) in the Baltic Sea proper (Villnäs and Norkko 2011).

Laminated sediments, indicative of the absence of burrowing animals, are now estimated to cover one-third of the total sedimentary area, which represents a 3.5-fold increase since the 1960s (Conley et al. 2009; Villnäs and Norkko 2011). With the on-going climate change, there is an evident risk of increased decomposition and increased oxygen consumption with higher temperature, so that the zoobenthic communities below the halocline may become even more impoverished during the 21st century.

10.8.3 Community changes above the halocline

The macrozoobenthos in the above-halocline soft-bottom areas has generally shown a biomass increase during the last decades, which is ascribed to a larger availability of food resulting from elevated levels of primary production caused by eutrophication (HELCOM 2009b). The overfishing of predatory demersal fish has also been considered as a possible cause of the increased macrozoobenthic biomass. However, also in the Bothnian Sea, beyond the distributional range of benthivorous flatfish and cod, the macrozoobenthos has increased. This suggest the gradual spreading of eutrophication effects throughout the entire Baltic Sea (HELCOM 2009a, b).

The near-bottom water in shallow areas where there is no halocline is also susceptible to oxygen depletion, and hypoxic events in coastal regions of the Baltic Sea have become increasingly common during the last decades (Conley et al. 2011). In areas where the bottom water is seasonally or episodically hypoxic, macrozoobenthic communities are impoverished and consist of hypoxia-tolerant species adapted to unfavourable oxygen conditions. Examples of such species are *Macoma balthica* and *Marenzelleria* spp. with rapid turnover time and widespread larval recruitment. With the gradual escalation of climate change, eutrophication and hypoxia, the soft-bottom zoobenthic communities will change even in coastal waters, and the links to other trophic levels (*e.g.* fish) may alter ecosystem functioning (Snickars et al. 2015; Weigel et al. 2015).

The largest changes in both species composition and community structure have been observed in recent years in the Bothnian Sea and the Gulf of Finland. In the 1990s, hypoxia and anoxia in the Gulf of Finland was recorded below a 20–30 m water depth, most probably as a result of density stratification in combination with escalating eutrophication. Abundant populations of the amphipods *Monoporeia affinis* and *Pontoporeia femorata*, typical of the deep areas of the eastern Gulf of Finland and Bothnian Sea, were depleted and did not fully recover, and were replaced by *Marenzelleria* spp., primarily by *Marenzelleria arctia*. During the last ten years, a massive increase of the polychaete's population has led to an increase in the overall macrozoobenthic biomass, which is now completely dominated by *Marenzelleria* (Maximov 2011; Kauppi et al. 2015; Maximov et al. 2015).

In summary, due in part to continuous eutrophication and in part to other large-scale changes such as ecosystem-wide regime shifts (Österblom et al. 2007), and a gradual climate change in the Baltic Sea, the entire system has changed. It is notable that the soft-bottom communities above the halocline have generally increased in biomass, the increase being in part due to the increased abundances of the bivalve *Macoma balthica* and the polychaetes *Marenzelleria* spp.

10.9 The deep soft-bottom food web

10.9.1 The base of the food web

The sediment-dwelling fauna is highly dependent on the energy produced in the photic zone, which is supplied to the soft-bottom habitats as organic matter. The organic material is delivered to the deep seabed via sedimentation (vertical flux) or horizontal transport (lateral advection). In some places, *e.g.* on the slopes of the deeps, the lateral advection may be equally important or even more important than the vertical flux (Graf 1992). The meiobenthic and macrozoobenthic organisms consume the sedimented phytoplankton, phytodetritus and/or the bacteria associated with it. The meio- and macrozoobenthos in turn provide food for higher trophic levels, *i.e.* predatory invertebrates and fish (Fig. 10.11). In addition, bacteria break down the detritus; the resultant compounds enter the microbial loop and become recycled into inorganic nutrients.

Benthic animals represent different feeding types. Bivalves are mainly suspension feeders (*e.g. Mytilus trossulus*) or facultative suspension-surface deposit feeders (*e.g. Macoma balthica*), their major food item being phytoplankton, especially diatoms. Polychaetes and crustaceans are deposit feeders, predators or (very often) omnivores (*e.g. Saduria entomon*). Some opportunistic species can alter their diet depending on food availability. For example, the polychaete *Hediste diversicolor* can use more than one feeding mode and switches between them depending on the prevalent conditions: it can be a suspension feeder, a deposit feeder, an opportunistic predator, an herbivore or a scavenger.

10.9.2 Benthic consumers

The most important consumers of the macrozoobenthos in the deep soft-bottom habitats of the Baltic Sea include cod, flounder and the viviparous eelpout *Zoarces viviparus*. Cod and the Atlantic herring *Clupea harengus* rely on epibenthic crustaceans such as mysids. During their winter migrations to the deep soft-bottom habitats, prawns (*e.g. Palaemon elegans*) are an easy catch for predators such as cod. In addition, pelagic fish such as herring and smelt (*Osmerus eperlanus*) forage on the nectobenthic polychaete *Bylgides sarsi* and the amphipods *Monoporeia affinis* and *Pontoporeia femorata*, which migrate up and down in the water column at night.

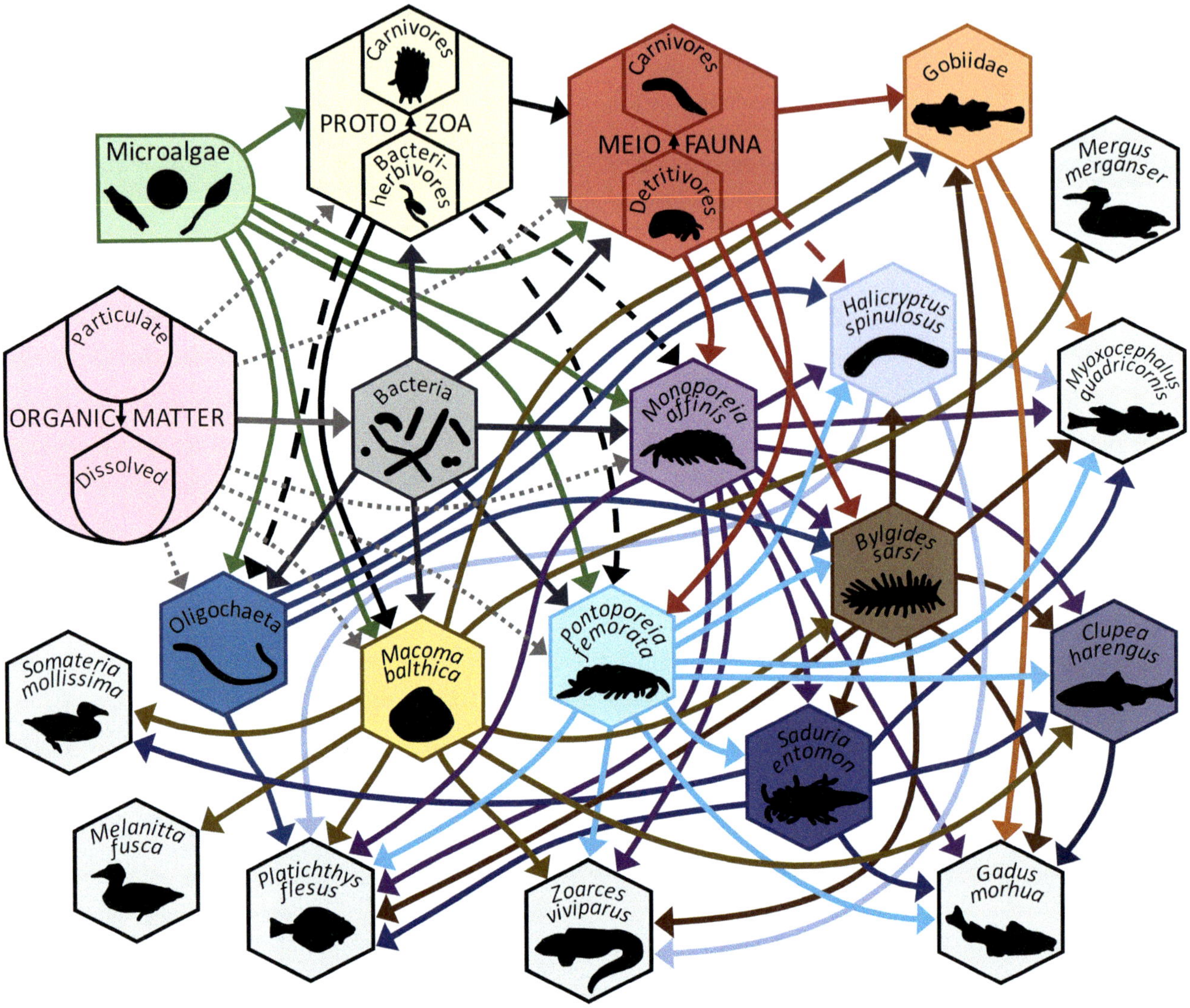

Fig. 10.11 The classic view of the position of the zoobenthos in the Baltic Sea food web, showing some of the key links and energy transfer pathways from the organic matter (food source based on primary production) through bacteria, protozoans, meio- and macrofauna to fish and birds. It is known at present that the organic matter plays a significant role in the food web, and research involving the use of stable isotopes of carbon and nitrogen has shown that several species change their position in the food web depending on the local conditions. Figure modified from Ankar (1977)

10.9.3 Benthic-pelagic coupling

Benthic invertebrates are an integral part of the Baltic Sea ecosystem: being consumed by fish, they ensure the energy flow from the benthic to the pelagic zone via a set of interactions known as "benthic-pelagic coupling". The zoobenthic organisms are a good source of food for higher trophic levels in terms of energy content and essential nutrients such as proteins, polyunsaturated fatty acids (PUFAs) and antioxidants (Brey et al. 2010; Røjbek et al. 2012; Box 10.4). Recently, analyses of stable isotopes of nitrogen and carbon have illuminated both the trophic position of organisms in the food web and the efficiency of transfer of carbon up the food web (Nordström et al. 2009). Furthermore, network analysis and food web models have proven to be promising tools to increase our understanding of the patterns and processes in complex food web interactions (Tomczak et al. 2013).

10.10 Bioturbation

10.10.1 Bioturbation oxygenates sediments

In sediments devoid of macrofauna, the presence of oxygen is limited to the surface layer, usually to the uppermost few mm. Therefore, burrowing animals that improve oxygen conditions are crucial for biogeochemical processes within the sediments and for ecosystem functioning in general. The soft-bottom benthic fauna penetrating the sediment often constructs tubes and burrows, thereby creating a three-dimensional sediment matrix with a mosaic of microhabitats. In the Baltic Sea, such fauna may live down to 15 cm beneath the sediment surface (*e.g.* the polychaete *Hediste diversicolor*), or even as far down as 35 cm (*e.g. Marenzelleria* spp.).

The deep-penetrating animals ventilate the sediment by drawing oxygenated water into it, thus creating an oxic and oxidised zone around their burrows. This produces conditions favourable for the development of bacteria and meiofauna, and changes the vertical distribution and composition of the zoobenthic communities. The oxidised compounds (electron acceptors) in the lining of tubes and burrows, and the halo around them, may be colonised by aerobic bacteria that are active in organic matter decomposition and in the natural bioremediation of *e.g.* oil-polluted sediments.

Organic matter mineralisation is enhanced mainly by the oxygenation of the subsurface sediment effected by macrozoobenthos via irrigation, but also by the redistribution of particles as a result of macrofaunal feeding and burrowing activity (Fig. 10.12). This is especially important for aged detritus which, under oxic conditions, is degraded up to ten times faster than under anoxia. Vertical transport of the sediment, activated by the macro- and meiofauna, enhances biodegradation of sediment-bound organic contaminants. Moreover, the fauna also produces digestive solubilisers, which improve the bioavailability of hydrocarbons to hydrocarbon-degrading bacteria.

10.10.2 Nutrient fluxes at the sediment/water interface

Generally, the activities of the benthic macrofauna affect the sediment biogeochemistry to a high extent and often stimulate processes that counteract eutrophication, *e.g.* increased phosphorus retention of the sediment. The degree of faunal impact on such processes varies depending on faunal density and functional group composition (Karlson et al. 2007).

Bioturbation performed by the zoobenthos exerts significant indirect effects on the form and amount of nutrients that are released to the overlying water column. For example, under anoxic conditions, insoluble Fe(III) is reduced to soluble Fe(II) ions and phosphorus is released into the pore water and diffuses upwards into the water column above the sediment (Fig. 10.12a). Under oxic conditions, phosphorus is bound to iron(III) oxyhydroxides, which decreases the amount of phosphorus released to the pore water during organic matter degradation (Fig. 10.12b). Thus, improvement of the sediment oxic conditions through bioturbation can decrease the phosphate flux from the sediment to the water column.

Bioturbated sediments support a mosaic of oxic and anoxic microenvironments around biogenic structures, which form sites amenable to coupled nitrification/denitrification (Karlson et al. 2007). Nitrification depends on the availability of oxygen (*cf.* Sect. 3.6.3) and is thus inhibited by hypoxia, which results in ammonium efflux

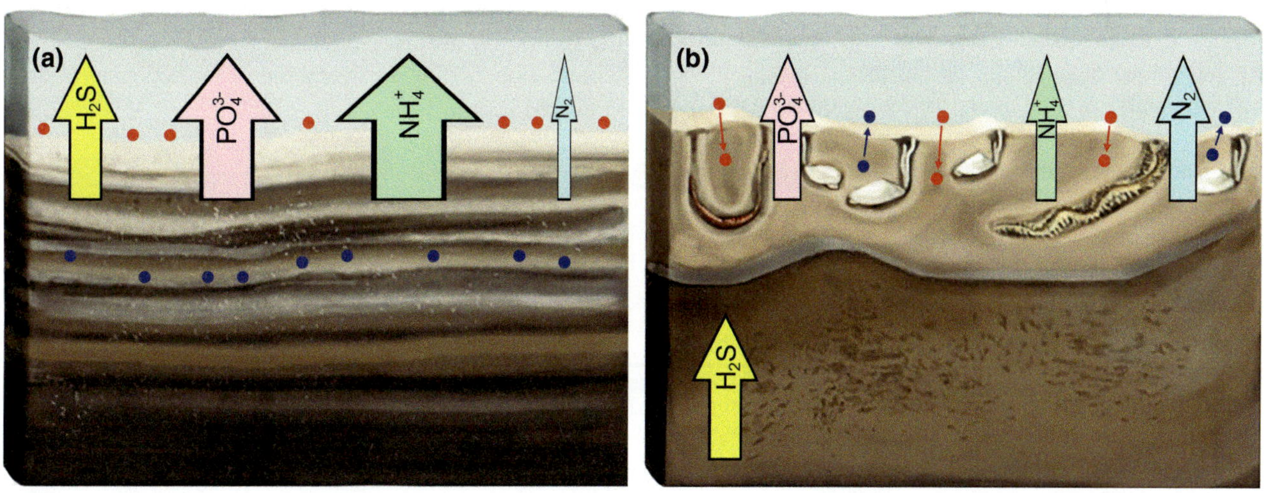

Fig. 10.12 A schematic illustration of the effects of the benthic fauna on processes at the sediment/water interface. (**a**) Sediment devoid of fauna. (**b**) Bioturbated sediment populated by fauna. The effects of faunal activity are as follows: (1) increased redistribution of particles into (red dots) and from (blue dots) the sediment, (2) oxygenation of the sediment and reduction of hydrogen sulphide and phosphate outflux (in oxygenated sediments, phosphates are accumulated in the sediment bound to iron minerals), (3) a general decrease of the ammonium outflux and an increase in the dinitrogen gas efflux due to coupled nitrification/denitrification in bioturbated sediments. Figure based on information in Karlson et al. (2007) and Norkko et al. (2012). Figure: © Małgorzata Piłka

Box 10.4: Prey caloric value

Lipids, carbohydrates and proteins

The value of organisms as food for others can be assessed by determining their energy content (the number of energy units per gram body mass or the number of energy units per individual). The energy content depends on the body biochemical composition; it is primarily strongly correlated with lipid content. Lipids are energy-rich compounds, and their energy content exceeds that of carbohydrates and proteins. Organisms that differ in their feeding and mobility modes are known to differ in their energy content (Rumohr et al. 1987; Brey et al. 1988, 2010). In general, carnivores are more protein-rich, contain less carbohydrates in dry weight (dw), and show a higher energy content than herbivores, whereas omnivores are intermediate in terms of their energy content.

Variability of energy content between and within species

The energy content of sessile organisms is lower than that of active swimmers. For example, the energy content of the large isopod *Saduria entomon*, the more active non-indigenous prawn *Palaemon elegans*, and the active swimmer mysid *Mysis mixta*, average 10.5 J mg^{-1} dw ($\sim$1,900 J $individual^{-1}$), 16.5 J mg^{-1} dw ($\sim$1,537 J $individual^{-1}$), and 24.8 J mg^{-1} dw ($\sim$260 J $individual^{-1}$), respectively (Szaniawska 1991; Janas and Bruska 2010). On account of high glycogen and lipid contents, the energy content of the Baltic clam *Macoma balthica*, the most abundant bivalve in the deep soft seabed of the Baltic Sea, is high, averaging 22 J mg^{-1} dw (soft tissues, without shell) ($\sim$1,370 J $individual^{-1}$). The variation of the energy content within species is governed by endogenous factors such as sex or life cycle stage, as well as by exogenous factors such as food availability, temperature and oxygen status of the environment.

from the sediment (Fig. 10.12). Since denitrification depends on nitrate produced by nitrification, denitrification (*cf.* Sect. 3.6.4) is also impeded, and less elemental nitrogen (N_2) is released from the sediment (Jørgensen 1996).

10.10.3 The effect of bioturbation is species-dependent

Effects of the zoobenthos on processes in the sediment depend, among other things, on the species composition of the community, as well as on the total biomass and physiological status of the animals. For example, more effective bioturbators such as the amphipods *Monoporeia affinis* and *Pontoporeia femorata* or the polychaete *Hediste diversicolor* may have a stronger impact on biochemical processes than bivalves such as *Macoma balthica*. The amphipods are highly mobile and the polychaete extensively ventilates its burrows, whereas the bivalves are less mobile in the sediment and pump water through their inhalant siphons.

The impact of *Marenzelleria* spp. on the sediment/water interface is not fully understood yet. These polychaete worms burrow deeper than most native species do, and seem to affect sediment biogeochemistry both through bioventilation and bioirrigation. Short-term laboratory experiments have revealed a negative effect in the form of increased hydrogen sulphide and nutrient concentrations in the water above the sediment. This can be explained by the percolation of water from deeper sediments to the sediment surface through the activities of *Marenzelleria* (Quintana et al. 2011).

However, a model study demonstrated that in the long run (years), under a high but natural abundance of *Marenzelleria* (>3,000 individuals m^{-2}), which allows a sufficient amount of oxygen to be drawn into the sediment, the polychaete can increase phosphorus retention in the sediment (Norkko et al. 2012). This might have a potentially positive ecosystem effect by reducing the near-bottom water phosphorous contribution to eutrophication. Repeated small-scale disturbances (*e.g.* hypoxia, currents, sediment transport) may also play fundamental roles for biodiversity, bioturbation, and fluxes in and out of the sediment (Villnäs et al. 2013).

10.11 Hypoxia and anoxia

10.11.1 Eutrophication increases hypoxia

Increased food availability, as a result of higher nutrient loading and organic enrichment at the early stage of eutrophication, generally increases the zoobenthic biomass. At a later eutrophication stage, when hypoxia and anoxia develop, impoverishment or complete loss of benthic macrofauna is observed (Pearson and Rosenberg 1978). Hypoxia and anoxia are two of the most important structuring factors for the zoobenthic communities of the Baltic Sea. In addition to salinity and temperature, they operate on a large scale and produce substantial adverse consequences. Changes in biogeochemical processes as a result of decreases and loss of bioturbating animals from the sediment may have consequences for the entire ecosystem. The internal nutrient loading may increase through the denitrification rate being reduced and the elevated release of phosphorus from sediments (*cf.* Sects. 10.10.2 and 10.10.3). A change in the ratio of dissolved nutrients in the water column may in turn intensify algal blooms and thus exacerbate the problems of eutrophication and hypoxia.

10.11.2 Species-specific sensitivity to hypoxia

The loss of macrofaunal biomass caused by hypoxia and anoxia in the Baltic Sea has been estimated at 1.4 million tonnes (Karlson et al. 2002). Such a huge loss may have profound consequences for both fish stocks and biogeochemical processes. Losses may be gradual or occur as a mass mortality event. One of the most severe, well-documented instances of mass mortality of the benthic fauna caused by oxygen depletion was observed in the Kiel Bay (Belt Sea) in the late summer of 1981, when ~30,000 tonnes of macrofauna (97 % of the standing stock) perished below the halocline (Weigelt and Rumohr 1986). Of the 60 zoobenthos species reported from the area, only a few of the most tolerant species survived the mass mortality event: the clams *Arctica islandica*, *Astarte* spp. and *Corbula gibba*, and the priapulid *Halicryptus spinulosus*.

Zoobenthic species have developed diverse strategies to cope with adverse oxygen conditions. The key factor for long-term survival under anoxia is the metabolic rate reduction. The highly tolerant bivalves *Astarte borealis* and *Arctica islandica* reduce their metabolic rates to less than 1 % of the aerobic rate after 20 days of anoxic incubation, consuming only about half their glycogen reserves during that time. It is worth mentioning that the ocean quahog clam *Arctica islandica* is an extremely slow-growing and long-lived invertebrate, with a maximum recorded age of 507 years in the northern Atlantic Ocean: indeed, it is the longest-lived non-colonial animal whose age at death can be accurately determined (Butler et al. 2013).

Bottom-dwelling organisms such as bivalves and polychaetes are usually more tolerant than the mobile fauna (Diaz and Rosenberg 1995). In general, the larger the energy reserves and ability to reduce metabolic rate, the better the chances of survival. The most tolerant bivalves are able to survive even two months without oxygen, for example *Astarte borealis* (Lt_{50} = 69 days at 10 °C) and *Arctica islandica* (Lt_{50} = 52 days), Lt_{50} being the time after which 50 % of individuals have perished.

Except for the isopod *Saduria entomon*, Baltic crustaceans are relatively sensitive to oxygen deficiency. Like the bivalve *Macoma balthica*, *Saduria entomon* can survive for several days under anoxia (Normant and Szaniawska 2000), whereas very sensitive species living in the coastal area (*e.g.* the brown shrimp *Crangon crangon*) survive anoxia for only several hours (Hagerman and Szaniawska 1986). The mortality of *Crangon crangon* dramatically increases at oxygen concentration <2 mL O_2 L^{-1} and is enhanced by hydrogen sulphide, high temperature or unfavourable salinity.

10.11.3 Hypoxia and reproduction

Reproduction may be impaired already at oxygen concentrations that do not or only slightly affect adult behaviour. For example, exposure of the amphipod *Monoporeia affinis* to low oxygen levels (1.4–3.1 mL O_2 L^{-1}) was observed to reduce the fertility and to increase the dead brood rate (HELCOM 2010). The effect was similar to, or even more pronounced than, the effect of harbour sediment highly contaminated by heavy metals, PAHs and PCBs (*cf.* Table 16.1) and illustrates the severity of even a slight reduction of oxygen content in the water body (Conley et al. 2011).

The oxygen conditions in the water beneath the halocline are extremely important for the reproduction of the eastern Baltic cod which spawn in the Bornholm Sea, the Gdańsk deep, and the Gotland Sea. The spawning success depends on water quality, as the eggs will only float and survive in fairly saline (>11) and oxygen-rich (>2 mL O_2 L^{-1}) water. Such conditions are dependent on the large marine inflows, on the one hand, and on eutrophication resulting in faster rates of organic matter sedimentation and mineralisation in deep waters, leading to oxygen deficiency, on the other. When the high amount of organic matter accumulated in the sediment is mineralised and the water under the halocline stagnates, H_2S migrates up the water column. Not only does this deteriorate the oxygen conditions in the water, but may adversely influence the survival and development of cod eggs and larvae.

Since the mid-1960s, the volume of water masses with salinity and oxygen conditions suitable for cod reproduction has been steadily declining (ICES 2007). In 1986–2007, cod reproduction was restricted mainly to the Bornholm Sea, while suitable reproduction conditions in the Gdańsk deep and in the Gotland Sea occurred only occasionally (after inflows in 1993, 2003 and 2014 *cf.* Fig. 2.13a). After the maximum cod recruitment (800 million individuals) in the mid-1970s, the annual population size decreased drastically and was estimated to have reached 50 million by 1990 (Karlson et al. 2002). The past few years witness an improvement in cod reproduction, but the reasons are not fully understood.

10.11.4 Behavioural adaptations to hypoxia

The changes observed at the community level, and the subsequent changes in ecosystem functioning, are a consequence of direct impacts of stressors such as hypoxia, anoxia and the presence of hydrogen sulphide on the behaviour and life

Fig. 10.13 Like many other bivalves, the Baltic clam *Macoma balthica* responds to hypoxia by emerging on to the sediment surface and extending their siphons into the water column where more oxygen may be found. Photo: © Urszula Janas

processes of the affected organisms and populations. The mobile fauna migrates away from areas affected by the stressors, while the less mobile fauna shows impaired locomotor activity, bioventilation rate, metabolic rate, and feeding intensity. Bivalves react to hypoxia by stretching their siphons into the water column and emerging on to the sediment surface (Fig. 10.13). While more oxygen may be found in the water column, the clams become easier targets for predation or siphon cropping, *e.g.* by flatfish. In contrast, crustaceans and polychaetes increase their bioventilation rates.

An example of decreased feeding intensity and behavioural adaptation is the effect of oxygen conditions on the predator-prey interaction between the large isopod *Saduria entomon* and its prey amphipods *Monoporeia affinis* and *Pontoporeia femorata* (Johansson 1999). The predation success (feeding intensity) of *Saduria entomon* declines at oxygen concentrations of $\sim$3.2 mL O_2 L^{-1} because the amphipods change their behaviour by remaining inactive in the sediment for longer periods. When the hunting ability of *Saduria entomon* itself is reduced at still lower oxygen levels of $\sim$0.5 mL O_2 L^{-1}, the amphipods come to the surface and obtain the oxygen they need to survive without markedly increasing their risk of being eaten by the predator.

10.11.5 Hypoxia-induced changes in metabolism

Under hypoxia, the less mobile fauna uses aerobic metabolism for as long as possible. For example, the polychaete *Scoloplos armiger* can maintain a fully aerobic metabolism down to $\sim$0.8 mL O_2 L^{-1}, and even at concentrations of $\sim$0.4 mL O_2 L^{-1} a partially aerobic metabolism is retained. In addition, many hypoxia-exposed crustaceans (*e.g. Saduria entomon*) can increase their haemocyanin production, which enables them to transport as much oxygen as during normal oxygen conditions.

Anaerobic metabolism comes into play if there is not enough oxygen for aerobic metabolism to function. During anaerobic metabolism, animals use a metabolic storage substrate such as glycogen. Large glycogen reserves of up to 20 % dry weight in *Macoma balthica* or 12 % dry weight in *Astarte borealis* are essential for short-term anaerobic metabolism. Anaerobic metabolism results in accumulations of large amounts of end products which differ among taxonomic groups and species. Under anaerobic conditions, a bivalve such as *Macoma balthica* produces mainly succinate, whereas the isopod *Saduria entomon* accumulates lactate when it is active under short-time anoxia or alanine if inactive under long-term anoxia (Hagerman 1998). This shows the versatility of adaptations in the benthic animals to cope with environmental stress.

10.11.6 Metabolic disposal of toxic by-products

Burrowing animals need to get rid of the toxic hydrogen sulphide that diffuses from the deeper layers of sediment. During hypoxia, there may be not enough oxygen to oxidise the hydrogen sulphide in the interstitial water. Besides the strategies used under hypoxic and anoxic conditions, such as escape or switch to anaerobic metabolism, benthic animals have the option of preventing H_2S from entering the body (Hagerman 1998). Bivalves were observed to keep H_2S out by closing their valves, while in the priapulid *Halicryptus spinulosus* H_2S binding to iron was reported. Once in the animal body, hydrogen sulphide must be detoxified (oxidised) to a less harmful substance, mainly thiosulphate or sulphite.

Mitochondrial oxidation of hydrogen sulphide has been demonstrated in *e.g. Halicryptus spinulosus* and *Saduria entomon*. However, this process is feasible only when tissue sulphide concentrations are low (at the micromolar level) and in the presence of oxygen. When the tissue sulphide concentrations are higher, the respiratory chain becomes completely inhibited. Anaerobic metabolism has been shown to be initiated at certain sulphide concentrations even if oxygen is present at concentrations that would normally permit aerobic metabolism. Thus, anaerobic metabolism is used as a defence against both hypoxic or anoxic conditions and high hydrogen sulphide concentrations.

Even with these adaptations, animals can survive under hypoxic or anoxic conditions, or high hydrogen sulphide concentrations for only a very limited period of time because of finite energy reserves, a low amount of energy produced during anaerobic metabolism, and the accumulation of toxic products.

10.11.7 Limitations of our knowledge

Our knowledge about responses of benthic invertebrates to hypoxia is insufficient. It is based on observations in the field and on laboratory experiments, but the results are sometimes contradictory. This is due *e.g.* to the behaviour of animals in the wild, which is difficult or impossible to induce under laboratory conditions. New ecophysiological approaches, *e.g.* the integration of the various "omics techniques" into ecophysiology (bulk gene expression and protein production), offer promising tools to increase our understanding of how hypoxia affects the physiology of invertebrates from the molecular level all the way up to the whole organism (Spicer 2014).

10.12 Hazardous substances

10.12.1 Types of hazardous substances in sediments

The soft-bottom sediments in the Baltic Sea have accumulated high concentrations of hazardous substances which may potentially influence the benthic fauna. Notorious toxic substances include pesticides, polycyclic aromatic hydrocarbons (PAHs), radionuclides, heavy metals, the organotin compound TBT (*cf.* Table 16.1) and some of the military waste materials deposited on the seafloor after World War II (HELCOM 2010).

In 1945–48, enormous quantities of conventional ammunition, gas shells and blister gases encased in metal containers were dumped in the Baltic Sea Area. This took place mainly in three regions: the Lillebælt, the Bornholm deep and the Gotland deep. However, during the transport to these dump sites part of the chemical arsenal was already thrown overboard and never reached its final destination (Andrulewicz 2007). An estimated 40–60 kilotonnes of chemical weapons, (*cf.* Sect. 16.1.12), are still present at the Baltic Sea bottom. The largest environmental hazards are posed by mustard gas and arsenic. The chemical weaponry stored on the Baltic Sea bottom does not seem to cause widespread ecological damage today, the impact being rather local. However, the hazard may be augmented by corrosion the metal containers, *e.g.* by direct disturbance at the sea bottom. It is also possible that arsenic will bioaccumulate in marine organisms, but biomagnification of arsenic in the Baltic Sea food web has not yet been conclusively demonstrated.

The number of new chemicals synthesised and marketed worldwide is increasing exponentially (*cf.* Sect. 16.1). The registry database of the Chemical Abstracts Service (CAS), a division of the American Chemical Society and the world's authority for chemical information, contains at present more than 33 million organic and inorganic substances, some 4,000 new substances being added to the list daily. Most of these compounds find their way to the seas and oceans where they react with other compounds, undergo chemical, physical and biological changes, and can affect marine life.

10.12.2 Accumulation of hazardous substances in sediments

As a consequence of physical adsorption and chemical bonding, chemical pollutants accumulate in sediments in concentrations far exceeding those in the overlying water. Often referred to as a "pollutant sink", sediments may be toxic due to the presence of both water-soluble and sediment-bound hazardous substances. Sediment toxicity is often associated with the fine sediment fraction, the amounts of which increase with water depth. The seabed of the southern and southeastern Baltic Sea is predominantly sandy with a low organic content, but in many other areas the sediments are fine-grained and rich in organic matter, and thus have a larger capacity to accumulate hazardous substances (HELCOM 2010).

Moreover, anoxic conditions in the deeper layers of many parts of the seabed adversely affect organisms, which inhibits the oxidative decomposition of substances, such as petroleum hydrocarbons, and immobilises metals. Thus, chemical pollutants may persist in sediments for a long time. Their conversions are long-drawn processes and their reaction products are released into the water column only slowly, *e.g.* as a result of erosion or bioturbation. Benthic organisms are thus exposed to these substances for a very long time.

10.12.3 Effects of hazardous substances in soft-bottom organisms

Hazardous substances may be absorbed by zoobenthic animals directly through the body surface or gills, or may enter an animal when it feeds on contaminated prey or organic matter. An organism often absorbs a contaminant at a rate faster than that at which the substance is lost, an effect known as "bioaccumulation". Concentrations of some chemicals such as methyl mercury increase with transfer through the food web to higher trophic levels, an effect known as "biomagnification".

Effects of chemical pollutants are manifested at different levels of biological organisation of the body, *i.e.* molecules, cells, tissues or organs. This situation may in turn elicit physiological and behavioural changes, and in consequence affect populations and the structure and functioning of whole communities. For example, shell and skeletal deformations, tumours, histological and cytogenetic abnormalities or pollution-induced diseases have been reported from fish and invertebrates in the Baltic Sea (Leppäkoski and Bonsdorff

1989; HELCOM 2010). In addition, contaminated benthic fauna showed decreased swimming or burrowing activity, which could lead to increased vulnerability to predation pressure.

Reproductive disorders have been observed in the amphipod *Monoporeia affinis* in the vicinity of an aluminium smelter and pulp mills in the northern Baltic Sea proper and the Bothnian Sea (Sundelin and Eriksson 1998; *cf.* Sect. 16.2.5). Some reproductive disorders were assumed to be due to other factors such as hypoxia, temperature or food deficiency. However, on account of its susceptibility to pollution effects, *Monoporeia affinis* is used in long-term and laboratory sediment toxicity monitoring programmes (Sundelin et al. 2008).

Baltic Sea organisms may be particularly vulnerable to hazardous substances, as the steep environmental gradients co-occur with gradients in toxicity for many chemical pollutants. In many cases (*e.g.* heavy metals), toxicity is highly dependent on temperature and salinity. Adding to this is the gradual shift in the ecosystem driven by climate change and increased hypoxia and anoxia, which may affect the bioavailability of chemical pollutants buried in the sediments for decades (Leppäkoski and Bonsdorff 1989).

It could be argued that the naturally stressed ecosystem of the Baltic Sea, supporting many eurytolerant species, should be more resistant to additional stressors than ecosystems not exposed to severe natural environmental stressors (Leppäkoski and Bonsdorff 1989). However, many of the substances produced by man and reaching the seas are completely new and foreign to the biota (xenobiotics), and organisms that have never come into contact with such chemicals have not evolved any defence mechanisms against them. Still other compounds, such as endocrine disrupting chemicals, can mimic an animal's own hormones and may induce negative health effects. The presence of hazardous substances is an additional factor that may directly or indirectly affect the already disturbed deep soft-bottom zoobenthic communities subject to hypoxic stress.

Review questions

1. Which of the deep soft-bottom macrozoobenthic species in the Baltic Sea are most threatened by humans?
2. What are the basic zoobenthic communities found in the various regions of the Baltic Sea and how do they change over time and under stressful conditions?
3. What are the main drivers controlling the structure and functioning of the zoobenthic communities?
4. Are changes in the zoobenthic communities reversible, *i.e.* are the zoobenthic communities resilient in relation to environmental stress, and is this resilience scale-dependent?
5. How long can benthic animals withstand adverse oxygen conditions? What kind of adaptations have they evolved?

Discussion questions

1. Can the zoobenthos be replaced by other food items in fish nutrition in the Baltic Sea and what would be the result of this?
2. What will we gain when the oxygen conditions in the Baltic Sea improve as a result of reduced external nutrient loads and a decline in primary production?
3. How can the zoobenthos play a role in reducing the effects of the internal nutrient loading of the Baltic Sea, especially in relation to the large-scale hypoxia?
4. A rapid increase in the anthropogenic use of offshore areas such as pipelines for gas and oil and offshore wind farms is observed. In what way could such activities be dangerous for the deep soft-bottom communities and demersal fish?
5. What effects in the deep soft-bottom macrozoobenthic communities can be potentially associated with climate change?

References

Andrulewicz E (2007) Chemical weapons dumped in the Baltic Sea. In: Gonenc IE, Koutitonsky VG, Rashleigh B, Ambrose RB Jr, Wolflin JP (eds) Assessment of the fate and effects of toxic agents on water resources. NATO Security through Science Series. Springer, Dordrecht, pp 299–319

Ankar S (1977) The soft bottom ecosystem of the northern Baltic Proper with special reference to the macrofauna. Contributions from the Askö Laboratory, vol 19. Stockholm University, Sweden, pp 1–161

Audzijonyte A, Väinölä R (2005) Diversity and distributions of circumpolar fresh- and brackish-water *Mysis* (Crustacea: Mysida): descriptions of *M. relicta* Lovén, 1862, *M. salemaai* n. sp., *M. segerstralei* n. sp., based on molecular and morphological characters. Hydrobiologia 544:89–141

Blank M, Laine AO, Jürss K, Bastrop R (2008) Molecular identification key based on PCR/RFLP for three polychaete sibling species of the genus *Marenzelleria*, and the species' current distribution in the Baltic Sea. Helgoland Marine Research 62:129–141

Bonsdorff E (2006) Zoobenthic diversity-gradients in the Baltic Sea: continuous post-glacial succession in a stressed ecosystem. Journal of Experimental Marine Biology and Ecology 330:383–391

Bonsdorff E, Diaz RJ, Rosenberg R, Norkko A, Cutter G (1996) Characterization of soft-bottom benthic habitats of the Åland Islands, northern Baltic Sea. Marine Ecology Progress Series 142:235–245

Bonsdorff E, Pearson TH (1999) Variation in the sublittoral macrozoobenthos of the Baltic Sea along environmental gradients: a functional-group approach. Australian Journal of Ecology 24:312–326

Bonsdorff E, Wenne R (1989) A comparison of condition indices of *Macoma balthica* (L.) from the northern and southern Baltic Sea. Netherlands Journal of Sea Research 23:45–55

Brey T, Müller-Wiegemann C, Zittier Z, Hagen W (2010) Body composition in aquatic organisms – a global data bank of relationships between mass, elemental composition and energy content. Journal of Sea Research 64:334–340

Brey T, Rumohr H, Ankar S (1988) Energy content of macrobenthic invertebrates: general conversion factors from weight to

energy. Journal of Experimental Marine Biology and Ecology 117:271–278
Butler PG, Wanamaker AD, Scourse JD, Richardson CA, Reynolds DJ (2013) Variability of marine climate on the North Icelandic Shelf in a 1357-year proxy archive based on growth increments in the bivalve *Arctica islandica*. Palaeogeography, Palaeoclimatology, Palaeoecology 373:141–151
Cederwall H, Elmgren R (1980) Biomass increase of benthic macrofauna demonstrates eutrophication of the Baltic Sea. Ophelia Supplement 1:287–304
Cederwall H, Elmgren R (1990) Biological effects of eutrophication of the Baltic Sea, particularly the coastal zone. AMBIO 19:142–151
Conley DJ, Björck S, Bonsdorff E, Carstensen J, Destouni G et al (2009) Hypoxia-related processes in the Baltic Sea. Environmental Science and Technology 43:3412–3420
Conley DJ, Carstensen J, Aigars J, Axe P, Bonsdorff E et al (2011) Hypoxia is increasing in the coastal zone of the Baltic Sea. Environmental Science and Technology 45:6777–6783
Demel K, Mulicki Z (1954) Quantitative investigations on the biological bottom productivity in the southern Baltic. Prace Morskiego Instytutu Rybackiego Gdynia 7:75–126
Diaz RJ, Rosenberg R (1995) Marine benthic hypoxia: a review of its ecological effects and the behavioural responses of benthic macrofauna. Oceanography and Marine Biology – An Annual Review 33:245–303
Drzycimski I (2000) The Słupsk Furrow as a marine protected area in the Baltic. Oceanological Studies 29:33–42
Eleftheriou A, McIntyre A (eds) (2005) Methods for the study of marine benthos, 3rd edn. Blackwell Publishing, Oxford 418 pp
Elmgren R (2001) Understanding human impact on the Baltic ecosystem: changing views in recent decades. AMBIO 30:222–231
Elmgren R, Rosenberg R, Andersin AB, Evans S, Kangas P et al (1984) Benthic macro- and meiofauna in the Gulf of Bothnia (northern Baltic). Finnish Marine Research 250:3–18
Fonselius SH (1969) Hydrography of the Baltic deep basins III. Fishery Board of Sweden. Series Hydrography. Report 23:1–97
Galil BS, Marchini A, Occhipinti-Ambrogi A, Minchin D, Narščius A et al (2014) International arrivals: widespread bioinvasions in European seas. Ethology, Ecology and Evolution 26:152–171
Glaubrecht M (2008) Homage to Karl August Möbius (1825-1908) and his contributions to biology: zoologist, ecologist, and director at the "Museum für Naturkunde" in Berlin. Zoosystematisc and Evolution 84:9–30
Gogina M, Nygård H, Blomqvist M, Daunys D, Josefson AB et al (2016) The Baltic Sea scale inventory of benthic faunal communities. ICES Journal of Marine Science 73:1196–1213
Graf G (1992) Benthic-pelagic coupling: a benthic view. Oceanography and Marine Biology – An Annual Review 30:149–190
Gray JS, Elliott M (2009) Ecology of marine sediments: from science to management. Oxford University Press, Oxford 225 pp
Hagerman L (1998) Physiological flexibility; a necessity for life in anoxic and sulphidic habitats. Hydrobiologia 375/376:241–254
Hagerman L, Szaniawska A (1986) Behaviour, tolerance and anaerobic metabolism under hypoxia in the brackish-water shrimp *Crangon crangon*. Marine Ecology Progress Series 34:125–132
Hansson M, Andersson L, Axe P (2011) Areal extent and volume of anoxia and hypoxia in the Baltic Sea, 1960–2011. Swedish Meteorological and Hydrological Institute, Norrköping, Report Oceanography 42:1–63
HELCOM (1990) Second periodic assessment of the state of the marine environment of the Baltic Sea, 1984–1988. Baltic Sea Environment Proceedings 35B:1–432
HELCOM (2009a) Biodiversity in the Baltic Sea – an integrated thematic assessment on biodiversity and nature conservation in the Baltic Sea. Baltic Sea Environment Proceedings 116B:1–188
HELCOM (2009b) Eutrophication in the Baltic Sea – an integrated thematic assessment of the effects of nutrient enrichment and eutrophication in the Baltic Sea region. Baltic Sea Environment Proceedings 115B:1–148
HELCOM (2010) Hazardous substances in the Baltic Sea – an integrated thematic assessment of hazardous substances in the Baltic Sea. Baltic Sea Environment Proceedings 120B:1–116
HELCOM (2012) Checklist of Baltic Sea macro-species. Baltic Sea Environment Proceedings 130:1–203
ICES (2007) Report of the study group on Baltic Sea productivity (SGPROD), 23–26 January 2007, Gdynia, Poland. ICES CM 2007/BCC:02, 70 pp
Janas U, Bruska O (2010) Energy values and energy resources of two prawns in Baltic coastal waters: the indigenous *Palaemon adspersus* and the non-indigenous *Palaemon elegans*. Oceanologia 52:281–297
Janas U, Nowodworska E, Bezdzietny E (2007) Fitness and chemical composition of the Baltic clam *Macoma balthica* (Linnaeus, 1758) from sulphidic habitats in the Gulf of Gdańsk (southern Baltic). Thermochimica Acta 458:112–117
Johansson B (1999) Influence of oxygen levels on the predatory behaviour of the isopod *Saduria entomon*. Marine and Freshwater Behaviour and Physiology 32:223–238
Jørgensen BB (1996) Material flux in the sediment. In: Jørgensen BB, Richardson K (eds) Eutrophication in coastal and marine ecosystems. American Geophysical Union, Washington, DC, pp 115–135
Josefson AB, Norkko J, Norkko A (2012) Burial and decomposition of plant pigments in surface sediments of the Baltic Sea: role of oxygen and benthic fauna. Marine Ecology Progress Series 455:33–49
Karlson K, Bonsdorff E, Rosenberg R (2007) The impact of benthic macrofauna for nutrient fluxes from Baltic Sea sediment. AMBIO 36:161–167
Karlson K, Rosenberg R, Bonsdorff E (2002) Temporal and spatial large-scale effects of eutrophication and oxygen deficiency on benthic fauna in Scandinavian and Baltic waters – a review. Oceanography and Marine Biology – An Annual Review 40:427–489
Katsanevakis S, Wallentinus I, Zenetos A, Leppäkoski E, Cinar ME et al (2014) Impacts of invasive alien marine species on ecosystem services and biodiversity: a pan-European review. Aquatic Invasions 9:391–423
Kauppi L, Norkko A, Norkko J (2015) Large-scale species invasion into a low-diversity system: spatial and temporal distribution of the invasive polychaetes *Marenzelleria* spp. in the Baltic Sea. Biological Invasions 17:2055–2074
Laine AO (2003) Distribution of soft-bottom macrofauna in the deep open Baltic Sea in relation to environmental variability. Estuarine and Coastal Shelf Sciences 57:87–97
Leibold MA, Holyoak M, Mouquet N, Amarasekare P, Chase JM et al (2004) The metacommunity concept: a framework for multi-scale community ecology. Ecology Letters 7:601–613
Leppäkoski E, Bonsdorff E (1989) Ecosystem variability and gradients. Examples from the Baltic Sea as a background for hazard assessment. In: Landner L (ed) Chemicals in the aquatic environment – advanced hazard assessment. Springer Series in Environmental Management, Springer, Berlin, pp 6–58
Leppäranta M, Myrberg K (2009) Physical oceanography of the Baltic Sea. Springer, Berlin, 378 pp
Maximov A (2011) Large-scale invasion of *Marenzelleria* spp. (Polychaeta, Spionidae) in the eastern Gulf of Finland, Baltic Sea. Russian Journal of Biological Invasions 2:11–19
Maximov A, Bonsdorff B, Eremina T, Kauppi L, Norkko A, Norkko J (2015) Context-dependent consequences of *Marenzelleria* spp. (Spionidae: Polychaeta) invasion for nutrient cycling in the northern Baltic Sea. Oceanologia 57:342–348

Mills EL (1969) The community concept in marine zoology, with comments on continua and instability in some marine communities, a review. Journal of Fisheries Research Board of Canada 26:1415–1428
Möbius KA (1877) Die Auster und die Austernwirtschaft. Wiegandt, Hempel and Parey, Berlin [in German]
Nilsson HC, Rosenberg R (2000) Succession in marine benthic habitats and fauna in response to oxygen deficiency: analysed by sediment profile-imaging and by grab samples. Marine Ecology Progress Series 197:139–149
Nordström M, Aarnio K, Bonsdorff E (2009) Temporal variability of a benthic food web: patterns and processes in a low-diversity system. Marine Ecology Progress Series 378:13–26
Norkko A, Jaale M (2008) Trends in soft sediment macrozoobenthic communities in the open sea areas of the Baltic Sea. MERI – Finnish Institute of Marine Research 62:73–80
Norkko J, Reed DC, Timmermann K, Norkko A, Gustafsson BG et al (2012) A welcome can of worms? – hypoxia mitigation by an invasive species. Global Change Biology 18:422–434
Norling K, Rosenberg R, Hulth S, Gremare A, Bonsdorff E (2007) Importance of functional biodiversity and species-specific traits of benthic fauna for ecosystem functions in marine sediment. Marine Ecology Progress Series 332:11–23
Normant M, Szaniawska A (2000) Behaviour, survival and glycogen utilisation in the Baltic isopod *Saduria entomon* exposed to long-term oxygen depletion. Marine and Freshwater Behaviour and Physiology 33:201–211
Ojaveer H, Jaanus A, MacKenzie BR, Martin G, Olenin S et al (2010) Status of biodiversity in the Baltic Sea. PLoS ONE 5(9):e12467
Österblom H, Hansson S, Larsson U, Hjerne O, Wulff F et al (2007) Human-induced trophic cascades and ecological regime shifts in the Baltic Sea. Ecosystems 10:877–889
Pallo P, Widbom B, Olafsson E (1998) A quantitative survey of the benthic meiofauna in the Gulf of Riga (eastern Baltic Sea), with special reference to the structure of nematode assemblages. Ophelia 49:117–139
Pearson T, Rosenberg R (1978) Macrobenthic succession in relation to organic enrichment and pollution of the marine environment. Oceanography and Marine Biology – An Annual Review 16:229–311
Perus J, Bonsdorff E (2004) Long-term changes in macrozoobenthos in the Åland archipelago, northern Baltic Sea. Journal of Sea Research 52:45–56
Petersen CGJ (1913) Valuation of the sea II. The animal communities of the sea bottom and their importance for marine zoogeography. Report of the Danish Biological Station to Board of Agriculture, Copenhagen 21:1–44
Petersen CGJ, Boysen-Jensen P (1911) Valuation of the sea. Animal life of the sea bottom, its food and quantity. Report of the Danish Biological Station to Board of Agriculture, Copenhagen 10:1–76
Quintana CO, Hansen T, Delefosse M, Banta G, Kristensen E (2011) Burrow ventilation and associated irrigation by polychaete *Marenzelleria viridis*. Journal of Experimental Marine Biology and Ecology 397:179–187
Radziejewska T (1989) Large-scale spatial variability in the southern Baltic meiobenthos distribution as influenced by environmental factors. In: Klekowski Z, Styczyńska-Jurewicz E (eds) Proceedings of the 21st EMBS meeting, Gdańsk, 14–19 September 1986. Polish Academy of Sciences, Institute of Oceanology, Ossolineum, pp 403–412
Rhoads DC, Germano JD (1982) Characterization of organism-sediment relations using sediment profile imaging: an efficient method of remote ecological monitoring of the seafloor (REMOTSTM System). Marine Ecology Progress Series 8:115–128
Rhoads DC, Germano JD (1986) Interpreting long-term changes in benthic community structure: a new protocol. Hydrobiologia 142:291–308
Røjbek MC, Jacobsen C, Tomkiewicz J, Støttrup JG (2012) Linking lipid dynamics with the reproductive cycle in Baltic cod *Gadus morhua*. Marine Ecology Progress Series 471:215–234
Rosenberg R, Magnusson M, Nilsson HC (2009) Temporal and spatial changes in marine benthic habitats in relation to the EU Water Framework Directive: the use of sediment profile imagery. Marine Pollution Bulletin 58:565–572
Rousi H, Laine AO, Peltonen H, Kangas P, Andersin AB et al (2013) Long-term changes in coastal zoobenthos in the northern Baltic Sea: the role of abiotic environmental factors. ICES Journal of Marine Science 70:440–451
Rumohr H (1999) Soft bottom macrofauna: collection, treatment, and quality assurance of samples. ICES Techniques in Marine Environmental Sciences 27:1–26
Rumohr H, Bonsdorff E, Pearson TH (1996) Zoobenthic succession in Baltic sedimentary habitats. Archive of Fishery and Marine Research 44:179–214
Rumohr H, Brey T, Ankar S (1987) A compilation of biometric conversion factors for benthic invertebrates of the Baltic Sea. The Baltic Marine Biologists, Publication 9:1–55
Rumohr H, Karakassis I (1999) Comparison of multivariate patterns: different taxonomic levels in macrofaunal analysis *versus* sediment profiling imagery (SPI). Marine Ecology Progress Series 190: 125–132
Segerstråle SG (1957a) Baltic Sea. Geological Society of America Memoirs 67:751–800
Segerstråle SG (1957b) On the immigration of the glacial relicts of northern Europe, with remarks on their prehistory. Societas Scientiarum Fennica, Commentationes Biologicae 16:1–117
Smith CJ, Rumohr H, Karakassis I, Papadopoulou KN (2003) Analysing the impact of bottom trawls on sedimentary seabeds with sediment profile imagery. Journal of Experimental Marine Biology and Ecology 285/286:479–496
Snickars M, Weigel B, Bonsdorff E (2015) Impact of eutrophication and climate change on fish and zoobenthos in coastal waters of the Baltic Sea. Marine Biology 162:141–151
Spicer JI (2014) What can an ecophysiological approach tell us about the physiological responses of marine invertebrates to hypoxia? Journal of Experimental Biology 217:46–56
Sundelin B, Eriksson AK (1998) Malformations in embryos of the deposit-feeding amphipod *Monoporeia affinis* in the Baltic Sea. Marine Ecology Progress Series 171:165–180
Sundelin B, Eriksson-Wiklund AK, Ford A (2008) The use of embryo aberrations in amphipod crustaceans for measuring effects of environmental stressors. ICES Techniques in Marine Environmental Sciences 41:1–23 [http://www.ices.dk]
Szaniawska A (1991) Gospodarka energetyczna bezkręgowców bentosowych występujących w Zatoce Gdańskiej. University of Gdańsk, 121 pp [in Polish]
Timmermann K, Norkko J, Janas U, Norkko A, Gustafsson BG, Bonsdorff E (2012) Modelling macrofaunal biomass in relation to hypoxia and nutrient loading. Journal of Marine Systems 105–108:60–69
Tomczak MT, Heymans JJ, Yletyinen J, Niiranen S, Otto SA, Blenckner T (2013) Ecological network indicators of ecosystem status and change in the Baltic Sea. PLoS ONE 8(10):e75439
Törnroos A, Bonsdorff E, Bremner J, Blomqvist M, Josefson AB et al (2015) Marine benthic ecological functioning over decreasing taxonomic richness. Journal of Sea Research 98:49–56
Törnroos AM, Bonsdorff E (2012) Developing the multitrait concept for functional diversity: lessons from a system rich in functions but poor in species. Ecological Applications 22:2221–2236
Villnäs A, Hewitt J, Norkko A (2015) Evaluating the performance of benthic multi-metric indices across broad-scale environmental gradients. Ecological Indicators 58:382–391

Villnäs A, Norkko A (2011) Benthic diversity gradients and shifting baselines: implications for assessing environmental status. Ecological Applications 21:2172–2186

Villnäs A, Norkko J, Hietanen S, Josefson AB, Lukkari K, Norkko A (2013) The role of recurrent disturbances for ecosystem multifunctionality. Ecology 94:2275–2287

Villnäs A, Norkko J, Lukkari K, Hewitt J, Norkko A (2012) Consequences of increasing hypoxic disturbance on benthic communities and ecosystem functioning. PLoS ONE 7(10):e44920

Virtasalo JJ, Bonsdorff E, Moros M, Kabel K, Kotilainen AT et al (2011a) Ichnological trends along an open-water transect across a large marginal-marine epicontinental basin, the modern Baltic Sea. Sedimentary Geology 241:40–51

Virtasalo JJ, Kotilainen AT, Gingras MK (2006) Trace fossils as indicators of environmental change in Holocene sediments of the Archipelago Sea, northern Baltic Sea. Paleogeography, Paleoclimatology, Paleoecology 240:453–467

Virtasalo JJ, Leipe T, Moros M, Kotilainen AT (2011b) Physicochemical and biological influences on sedimentary-fabric formation in a salinity and oxygen-restricted semi-enclosed sea: Gotland deep, Baltic Sea. Sedimentology 58:352–375

Voipio A (1981) The Baltic Sea. Elsevier, Amsterdam, 418 pp

Weigel B, Andersson HC, Meier HEM, Blenckner T, Snickars M, Bonsdorff E (2015) Long-term progression and drivers of coastal zoobenthos in a changing system. Marine Ecology Progress Series 528:141–159

Weigelt M, Rumohr H (1986) Effects of wide-range oxygen depletion on benthic fauna and demersal fish in Kiel Bay 1981–1983. Meeresforschung 31:124–136

Zettler M, Karlsson A, Kontula T, Gruszka P, Laine AO et al (2014) Biodiversity gradient in the Baltic Sea: a comprehensive inventory of macrozoobenthos data. Helgoland Marine Research 68:49–57

Żmudziński L (1977) The Baltic deserts. Annales Biologiques, Copenhagen 32:50–51

The phytobenthic zone

11

Hans Kautsky, Georg Martin, and Pauline Snoeijs-Leijonmalm

Abstract

1. Phytobenthic communities consist of macrophytes (macroalgae, vascular plants and mosses) with their accompanying fauna and microorganisms.
2. The phytobenthic communities occur in the photic zone, which in the Baltic Sea extends from the water surface down to a ~20 m water depth, but in turbid coastal waters only down to ~5 m.
3. The type of vegetation is determined by the available substrate, which is a result of geography and geology in combination with currents. Most macroalgae grow attached to hard substrates whereas vascular plants and charophytes grow on sandy or soft (silt and mud) substrates.
4. Generally, the coastal areas of the Baltic Sea consist of mixed substrates with an intermingled vegetation of vascular plants and algae. In the northern Baltic Sea hard substrates dominate in the outer archipelagos, and in the southeastern Baltic Sea sandy and muddy substrates dominate.
5. Luxuriant stands of macrophytes provide food, shelter and spawning habitats for the associated sessile and mobile micro-, meio- and macrofauna, including fish.
6. On an ecosystem-wide scale, the phytobenthic communities vary along the large-scale Baltic Sea gradient. Biomass decreases with lower salinity and colder climate, while the proportion of freshwater species increases.
7. On a local scale, the phytobenthic communities are mainly, directly or indirectly, shaped by water movement (*e.g.* by the occurrence of sandy beaches and rocky shores) and winter ice cover. Light and substrate availability give rise to typical depth zonation patterns, ending with soft-substrate communities deepest down.
8. On a small scale (patches), phytobenthic community structure and composition is influenced by microhabitat structure and biotic interactions.
9. The phytobenthic communities in the brackish Baltic Sea are more sensitive to disturbance than their marine counterparts due to low diversity, physiological stress and the loss of sexual reproduction when species approach their salinity limit.

Keywords

Baltic Sea gradient • Community structure • Food-web interactions • Habitat-forming macrophytes • Human impacts • Phytobenthic communities • Productivity

H. Kautsky · P. Snoeijs-Leijonmalm (✉)
Department of Ecology, Environment and Plant Sciences, Stockholm University, 10691 Stockholm, Sweden
e-mail: pauline.snoeijs-leijonmalm@su.se

H. Kautsky
e-mail: hans.kautsky@su.se

G. Martin
Estonian Marine Institute, University of Tartu, 14 Mäealuse Street, 12618 Tallinn, Estonia
e-mail: georg.martin@ut.ee

P. Snoeijs-Leijonmalm et al. (eds.), *Biological Oceanography of the Baltic Sea*,
DOI 10.1007/978-94-007-0668-2_11

11.1 The substrate shapes the vegetation

11.1.1 Phytobenthic communities

The phytobenthic zone comprises the photic zone that contains vegetation growing on hard, sandy and soft substrates (Fig. 11.1). Usually, the coastal areas of the Baltic Sea show mixes of these substrates ("mixed substrates") on a local scale. Phytobenthic communities include all organisms (*e.g.* bacteria, protists and fauna) associated with the vegetation of macroalgae, mosses and vascular plants that forms the three-dimensional structure on the substrate.

Quantitative sampling of phytobenthic communities growing on hard or mixed substrates is impossible without SCUBA diving (Boxes 11.1 and 11.2). Phytobenthic investigations in the Baltic Sea Area include also the *Mytilus trossulus*-dominated belts at a ~20–30 m water depth, just below the photic zone, even if macrophytes are basically absent here. These belts are a continuation of the presence of *Mytilus trossulus* in the phytobenthic communities of the photic zone and are also preferably studied by SCUBA diving. Zoobenthic communities on soft substrates are, on the contrary, sampled usually from a ship using cores or other sampling devices (*cf.* Box 10.1).

Substrates in the photic zone that are highly unstable on a short time scale, such as constantly moving sand or gravel on exposed coasts, lack any macroscopic vegetation. This is especially the case along the southern and southeastern coasts of the Baltic Sea proper where immense amounts of sand originating from land are delivered from glacial deposits and by riverine runoff to the sea (*cf.* Fig. 2.6). However, whenever even a small patch of hard substrate occurs in a sand-dominated environment, macroalgae attach and the biodiversity of the area increases (Fig. 11.2c).

11.1.2 Algae and vascular plants grow on different substrates

The phytobenthic communities on hard substrates are dominated by attached macroalgae (Fig. 11.2a, c), but aquatic mosses, too, may attach to hard substrates. Rooted vascular plants increase proportionally with decreasing substrate grain size (*cf.* Table 2.3), and on sandy and soft substrates (silt and

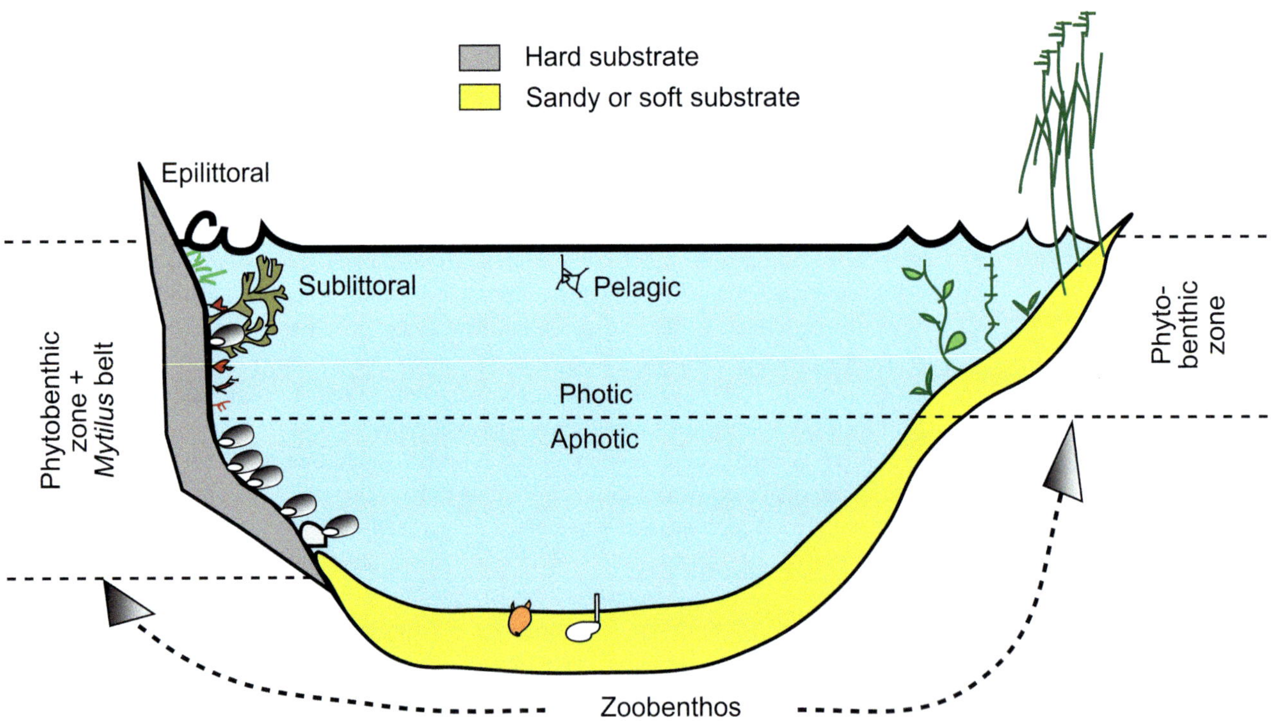

Fig. 11.1 Schematic summary of habitat variability in the phytobenthic zone, which comprises the photic zone that contains vegetation on hard, sandy and soft substrates. The *Mytilus trossulus* belt, which stretches down to a ~30 m water depth into the aphotic zone, is usually included in phytobenthic studies in the Baltic Sea Area. Figure: © Hans Kautsky

Fig. 11.2 Phytobenthic communities on different substrates in the Baltic Sea. (**a**) A *Fucus*-dominated vegetation on hard substrates. (**b**) A *Zostera marina*-dominated vegetation on sand and gravel to the right, red algae on stones to the left. (**c**) A *Fucus*-dominated vegetation on a boulder, surrounded by sand on which a macroscopic vegetation is lacking because of substrate instability. (**d**) A *Potamogeton perfoliatus*-dominated vegetation with the brown filamentous alga *Pylaiella littoralis* as epiphyte on soft substrate. (**a**, **c**, **d**) Höga Kusten, Bothnian Sea, (**b**) Askö, Baltic Sea proper. Photo: © Hans Kautsky

mud) they dominate the vegetation (Fig. 11.2b, d). One group of algae, the charophytes, grow on sandy and soft substrates as well. They are anchored with root-like filaments (rhizoids) that can be used for nutrient uptake (Vermeer et al. 2003). The vegetation on sandy and soft substrates stabilises the substrate: vascular plants achieve this by their rhizomes and roots and the charophytes by their rhizoids.

A mosaic of different substrates ("mixed substrates") yields the highest biodiversity of macrophytes and associated fauna since algae, mosses, vascular plants, epifauna and infauna occur side by side (Fig. 11.3). At some highly exposed sites, wind-induced wave action may affect the whole photic zone. However, there usually is a depth gradient with boulders and stones at the water line, because the influence of wind-induced wave action is strongest at the sea surface, and finer material deeper down, with silt and mud in the deepest parts. In most parts of the Baltic Sea, the seafloor in the lower part of the photic zone is covered by soft substrates.

The grain size of the substrate that occurs at a given site is to a large extent determined by sorting through water movement (Box 11.3). At sites continuously exposed to strong wave action (except for areas with sand deposition), only bedrock and large boulders are left as the finer grain sizes are continuously washed away, and the rock is covered by macroalgae. The substrate gets finer in concert with a decrease in wave exposure or current velocity: from boulders to stones (cobbles and pebbles), gravel, sand, silt, and finally to mud (*cf.* Table 2.3). Thereby, the proportion of rooted vascular plants and charophytes in the phytobenthic community increases.

Box 11.1: Methods for studying phytobenthic communities

Different methods

Depending on the scope of the study, indirect or direct methods are used to investigate phytobenthic communities (Box Fig. 11.1). The indirect methods include satellite imagery, aerial photography, laser scanning, multibeam and echosounding detection, which can cover large geographical areas, but the resolution of the community descriptors is poor. These indirect methods yield signals that need to be interpreted and verified through sea-truthing. The observations from space and air also depend on weather conditions (*e.g.* cloudiness) and water turbidity, and biological parameters can usually only be mapped down to a few metres water depth in the dark coastal waters of the Baltic Sea. At the other end of the scale we find direct observation by divers. In many cases, scientific diving is the most reliable method to study the phytobenthic zone because divers can easily take a closer look or remove obscuring objects. Quantitative sampling during diving is the most elaborate method, but achieves the highest resolution of biomass and community composition. In between, there are various other methods, some of which also include diving or more complicated equipment such as underwater vehicles and video equipment hanging from boats. Each method is a trade-off between the area covered, data resolution, time spent to collect the data and information achieved (Box Fig. 11.1).

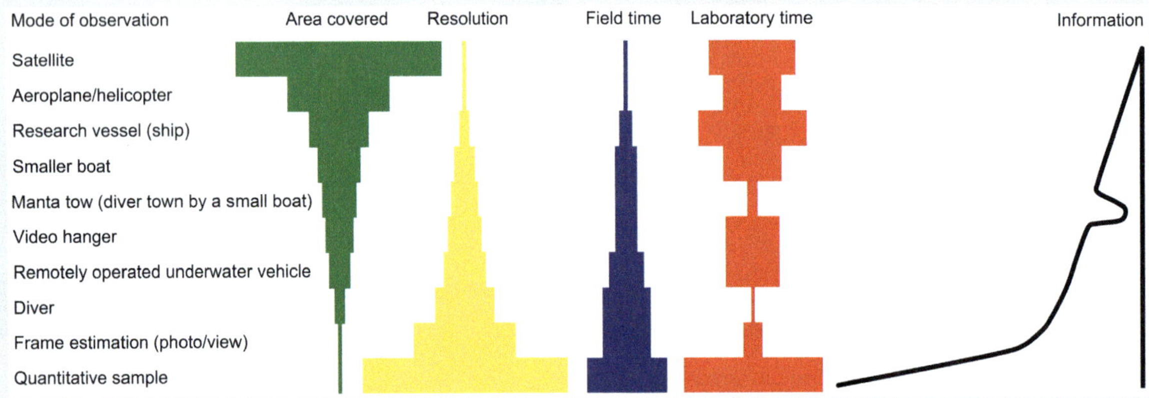

Box Fig. 11.1 Comparison of the different methods used to study the phytobenthic zone. Figure modified from Kautsky (2013)

Box Fig. 11.2 SCUBA diving to a maximum water depth of 30 m is used for quantitative biological studies in the phytobenthic zone. (**a**) Boat with diving flag and diver making notes. (**b**, **c**) Diver working along a transect in the *Fucus* belt of the Baltic Sea. Photo: © Hans Kautsky

Box Fig. 11.3 A diver using a "Kautsky frame" (recommended by the HELCOM COMBINE Guidelines, http://www.helcom.fi), which is a square with three metal sides and a sampling bag attached to the fourth side of the square (Kautsky 2013). A scraper is used to scrape the phytobenthic communities from the rock into the sampling bag. Photo: © Hans Kautsky

Diving

Marine biologists use diving as a tool to perform observations or experiments below the water surface (Box Fig. 11.2). The diving procedure should be kept as simple as possible so that the diver can focus primarily on the work to be carried out. There should never be any violation of safety rules. These rules are established in national safety regulations for scientific diving and must be followed in the respective countries when diving. Diving may be performed with SCUBA (self-contained underwater breathing apparatus) or with air from the surface (using a hose). The latter method gives the diver unlimited time under water, but the diver is spatially limited by the length of the hose. It is recommended to wear a dry suit with clothes under while SCUBA diving in the cold waters of the Baltic Sea. Since the Baltic Sea underwater environment is also relatively dark, it is also recommended to wear a bright-coloured diving suit or at least bright-coloured flasks, fins and gloves. Divers completely dressed in black are difficult to observe under water, which may be fatal in case of an accident. When using SCUBA diving, field work is limited by depth and time spent under water. For security reasons the maximum depth should not exceed 30 m unless the diver is especially trained for deep-water diving. The work under water should be carried out as easily as possible by not carrying more air and equipment than necessary and by good ahead-planning of the dive, *e.g.* planning to not stay too long in deep water in order to reduce the decompression time on the way back to the surface. The recommended equipment is 2·4·300 atm air tanks (2400 L). Technical diving increases exposure time, but at the cost of heavier equipment and a more complicated handling of the equipment, *e.g.* gas controls. This means that one has more details to keep in mind, which is less secure. Never dive alone unless you have contact to the surface by two-way communication and a security diver at the surface.

Quantitative sampling and depth distributions

The recording of phytobenthic community parameters can be performed along transect lines, with or without frames placed on the seafloor (Kautsky 2013). Estimations of the substrate type and species composition, and their cover on the substrate, requires a botanically skilled diver. A new diver has to practice under surveillance before good work can be performed. Free estimates of overall vegetation coverage in an area are made faster than cover estimates of species within a number of replicate frames (Box Fig. 11.3). However, cover estimates always have a degree of subjectiveness and destructive quantitative sampling is the best method to obtain accurate data on species composition and biomass. Drawbacks are the limited area sampled and the need for many replicate samples to reflect an entire area because the phytobenthic vegetation is usually patchy. Such samples take time (*i.e.* are expensive) to process. The lower limits of occurrence of many species, including the key species *Fucus vesiculosus* and *Zostera marina*, were in earlier years recorded by dredging, *i.e.* without knowing if the species was actually growing at the dredged depth or had drifted there. It was not until SCUBA diving was introduced that reliable data of depth distributions could be collected.

Box 11.2: The early days of scientific diving in the Baltic Sea

Marianne Hielm Pedersén

A pioneer diving ecologist in the Baltic Sea

In 1938, Mats Wærn, a PhD student at Uppsala University (Sweden) became a pioneer in the systematic use of diving and underwater photography for collecting scientific data in the phytobenthic zone of the Baltic Sea (Wallentinus et al. 1992). Until then, dredging had been the only way to study submarine organisms. However, sampling with a dredge mixes organisms from different substrates and depths, so that their exact habitats cannot be determined, and biomass cannot be measured in a reliable way. Mats Wærn was interested in how the macroalgal vegetation on rocky shores changes along environmental gradients, *e.g.* salinity, exposure to wave action and water depth. He was also in this respect a pioneer because at that time ecology was a young science and most phycologists were taxonomists. Mats

Box Fig. 11.4 Mats Wærn and his team in the 1940s. During his diving expeditions he was assisted by six persons, one diving chief, two persons pumping down air, one person to keep records of the underwater observations transmitted by the diver using a telephone, and two others in a small rowing boat to assist with the camera and other equipment. The copper helmet was put on after the pumping had started and the diver was lowered to the seafloor, and taken up after the dive, with the help of a rope. Anecdotically: before young Mats was allowed to dive, his mother tested walking on the seafloor in a heavy diving suit and found it safe enough for her son to practice. (**a**) Mats in the thick clothing he wore under his diving suit. (**b**) Mats in his heavy diving suit with the rope around his chest. (**c**) Nils Quennerstedt (also a PhD student at Uppsala University at that time) preparing for a dive. (**d**) Mats going down, note the rope, air hose and telephone line. Photo:

Wærn's herbarium, kept at Uppsala University (Sweden), is an "ecological herbarium", *i.e.* he mounted whole communities on the herbarium sheets (*cf.* Fig. 11.28a), including animals. His doctoral thesis "Rocky-shore algae in the Öregrund archipelago" (Wærn 1952) is a rich source of both ecological and taxonomic information on the macroalgae of the Baltic Sea.

A heavy diving suit

Mats Wærn still used a heavy diving suit (Box Fig. 11.4) because the aqua-lung, the first open-circuit, self-contained underwater breathing apparatus ("SCUBA") had not been invented until 1943, by the engineer Émile Gagnan and the naval lieutenant, explorer and filmmaker Jacques-Yves Cousteau. While a SCUBA diver floats in the water, a diver with a heavy diving suit walks around on the seafloor. This has both advantages and disadvantages: walking on the seafloor may be more stable, like walking on land, and diving time is much longer (Mats Wærn's dives usually lasted for 4–5 hours); however, it is easy to stir up sediments and destroy one's view. The diving suit Mats Wærn used was the so-called "German double suit" with the lead weights around the waist, which was safer for a marine biologist compared to the alternative in the 1940s, an "English heavy suit" with the lead weights hanging on the chest. Since the diver had to crawl on his knees when sampling the algal vegetation there was a more significant risk with the English suit that the diver would come up with his legs first if he should lose his balance and allow air into the lower part of the diving suit.

Underwater photography

Mats Wærn designed a water-proof camera housing for photographing the *in situ* phytobenthic communities (Box Fig. 11.5). This enabled him to use the camera under water and even to adjust the distance and wind the film, the lock on the camera housing having been tightened with 16 winged nuts. However, he could not set the aperture and exposure time under water. Therefore, Mats first measured the aperture and exposure time with a Sixtus light meter in a glass jar and reported the data to his assistants in the boat by telephone. Then the assistants prepared the camera, tightened the 16 winged nuts, and lowered the camera down into the water with a rope. When Mats received the camera he waited until the light meter gave the same values as before and then took the photograph. The phytobenthic communities were recorded following a standard procedure. When Mats found a suitable place he ordered "weight down" and the water depth was determined with a rope and a weight at its lower end. Then he made a general survey of the type of seafloor and all vegetation in sight, usually within a distance of 2–8 m, and estimated the degree of coverage for the different algal species on a 1–5 ordinal scale. All observations were directly reported by telephone to the record-keeper in the boat. A frame was placed on the rock and the cover in the frame was estimated, after which all algae within the frame were scraped off from the rock surface, collected in a bag, and sent up for later detailed analyses in the laboratory, including microscopy and herbarium mounting.

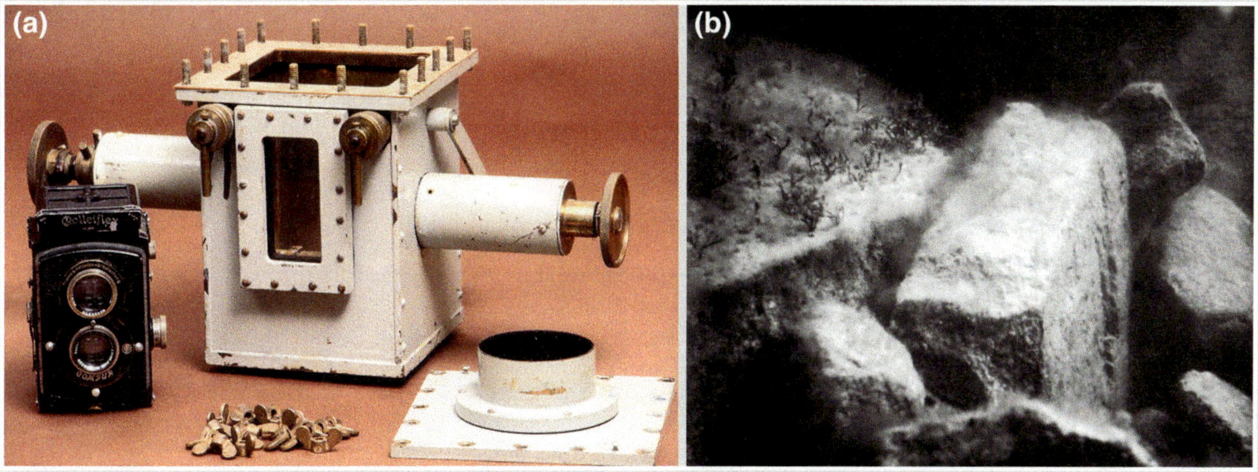

Box Fig. 11.5 Early underwater photography in the Baltic Sea. (**a**) Mats Wærn's Rolliflex camera with the water-proof housing. (**b**) A photograph taken with this camera, showing the lower limit of the *Fucus* vegetation (in the upper left part of the photograph) and *Battersia arctica* on the vertical surface (in the right part of the photograph) at a water depth of 10 m at an exposed site at Halsaren in the Södra Kvarken area (Sweden). The photograph was taken on 17 July 1944 at 4 p.m. with aperture 3.5 and exposure time 1/2 seconds. Photo: © Uppsala University

Box 11.3: Water movement determines the type of substrate

Water movement determines the type of substrate along the coasts because sediment grains are sorted by grain size in water movement. For example, when the waves lose energy by friction, first stones, then gravel and finally sand are deposited on a beach (Box Fig. 11.6a).

Along sandy coasts, a constant transport of sand takes place. Depending on the local water movement, sand is either deposited along the shoreline or an aberration of the shore occurs. On sandy bottoms, ripple marks are shaped by the sorting of grain size (Box Fig. 11.6b), and can be observed from the water surface down to a water depth of >30 m at offshore sites.

On unstable sandy beaches the sand is constantly moved by wave action. When the sand is washed away by a storm, larger stones and boulders are left and are subsequently covered by a new sand layer. This is particularly visible after storms when the shoreline can look entirely different than before the storm, *e.g.* a beach can suddenly be covered by stones when storm waves wash away the sand that completely covered the stones before the storm (Box Fig. 11.6c). This also occurs below the water surface and, especially in areas where the waves break, temporary stone ridges can appear.

A steady stream of large waves leaves only bedrock and boulders on an exposed beach. Often foam is created in this process (Box Fig. 11.6d), which is a natural phenomenon; decaying organic matter allows the seawater to mix with air and foam can accumulate where waves break on the coast.

Box Fig. 11.6 Water movement determines the type of substrate along the coasts. (**a**) The waves are sorting sediment by grain size. (**b**) Ripple marks shaped by sorting according to grain size. (**c**) An unstable sandy beach covered by stones and boulders after a storm. (**d**) A steady stream of large waves leaves only bedrock and boulders on an exposed beach. Photo: (**a, d**) © Pauline Snoeijs-Leijonmalm, (**b, c**) © Hendrik Schubert

Fig. 11.3 The highest diversity in the phytobenthic communities of the Baltic Sea is found on mixed substrates. Here, patches of *Fucus vesiculosus*-dominated vegetation on hard substrate grow mixed with patches of *Stuckenia pectinata*-dominated vegetation on soft substrate at Askö, Baltic Sea proper. Photo: © Hans Kautsky

Coasts with bedrock and boulder fields dominate in the western and northeastern Baltic Sea proper, as well as in parts of the Gulf of Finland and the Gulf of Bothnia (*cf.* Fig. 2.8). On coasts directly exposed to the open sea, a macroalgal vegetation can be found on the bedrock and large boulders throughout the whole photic zone (Fig. 11.4). In boulder fields, which are more common in the Baltic Sea than exposed bedrock, the boulders often occur mixed with patches of smaller-grained and soft substrate. Therefore, boulder fields usually support a "mixed vegetation" (consisting of both macoalgae and vascular plants) that is dominated, depending on the relative amounts of the different substrate types, by attached algae or rooted vascular plants and/or charophytes (Fig. 11.5).

11.1.3 The vegetation along a substrate gradient

The typical archipelago coasts of the Baltic Sea are characterised by gradients in exposure to wave action. Water circulation strengthens towards the outer archipelago (Fig. 11.6). This increases the transport of particles and decreases sedimentation so that bedrock is mainly found at the outer edges of the archipelago while boulders tend to dominate the outer and middle parts, and soft substrates the inner parts. At the outer skerries, soft substrates start at a ~25 m water depth or deeper, but in the innermost archipelago area soft substrates with submerged vascular plants may start already at an 0.1 m water depth – if not replaced by a reed belt.

Fig. 11.4 Bedrock is dominated by algal communities. (**a**) The *Cladophora* belt and the *Fucus* belt on bedrock at Askö in the Baltic Sea proper. (**b**) A *Pylaiella littoralis*-dominated vegetation on bedrock and boulders at Höga Kusten in the Bothnian Sea. Photo: © Hans Kautsky

From the inner archipelago to the outer archipelago, the phytobenthic vegetation shifts in concert with the available substrate (Fig. 11.6), from a vegetation dominated by rooted plants (Fig. 11.7a) via a mixture of algae and vascular plants (Fig. 11.3) to algae-dominated communities (Fig. 11.4). Common on soft substrates are also communities that consist of entangled loose-lying vascular plants and/or algae (Fig. 11.7b).

Along an archipelago gradient, salinity-dependent community changes may occur in the case of freshwater runoff causing a horizontal salinity gradient. Furthermore, archipelago gradients are often accompanied by nutrient gradients because the influence of land-derived nitrogen and phosphorus is usually larger in the inner parts than in the outer parts of an archipelago.

Fig. 11.5 Boulder fields with algal and mixed vegetation. (**a**) Boulders at a 1–2 m water depth with a zonation of *Cladophora glomerata*, *Fucus vesiculosus*, *Ceramium tenuicorne* and *Mytilus trossulus*. (**b**) A biologist recording the vegetation on boulders covered by *Fucus vesiculosus* at a ~2 m water depth. (**c**) Boulders at a ~2 m water depth with *Fucus vesiculosus* surrounded by patches of smaller-grained stones (unstable substrate) without vegetation. (**d**) Boulders at a ~0.5 m water depth with filamentous algae surrounded by patches of soft bottom with *Stuckenia filiformis* (syn. *Potamogeton filiformis*). (**a**, **b**, **c**) Askö, Baltic Sea proper, middle archipelago, (**d**) Höga Kusten, Bothnian Sea, inner archipelago. Photo: © Hans Kautsky

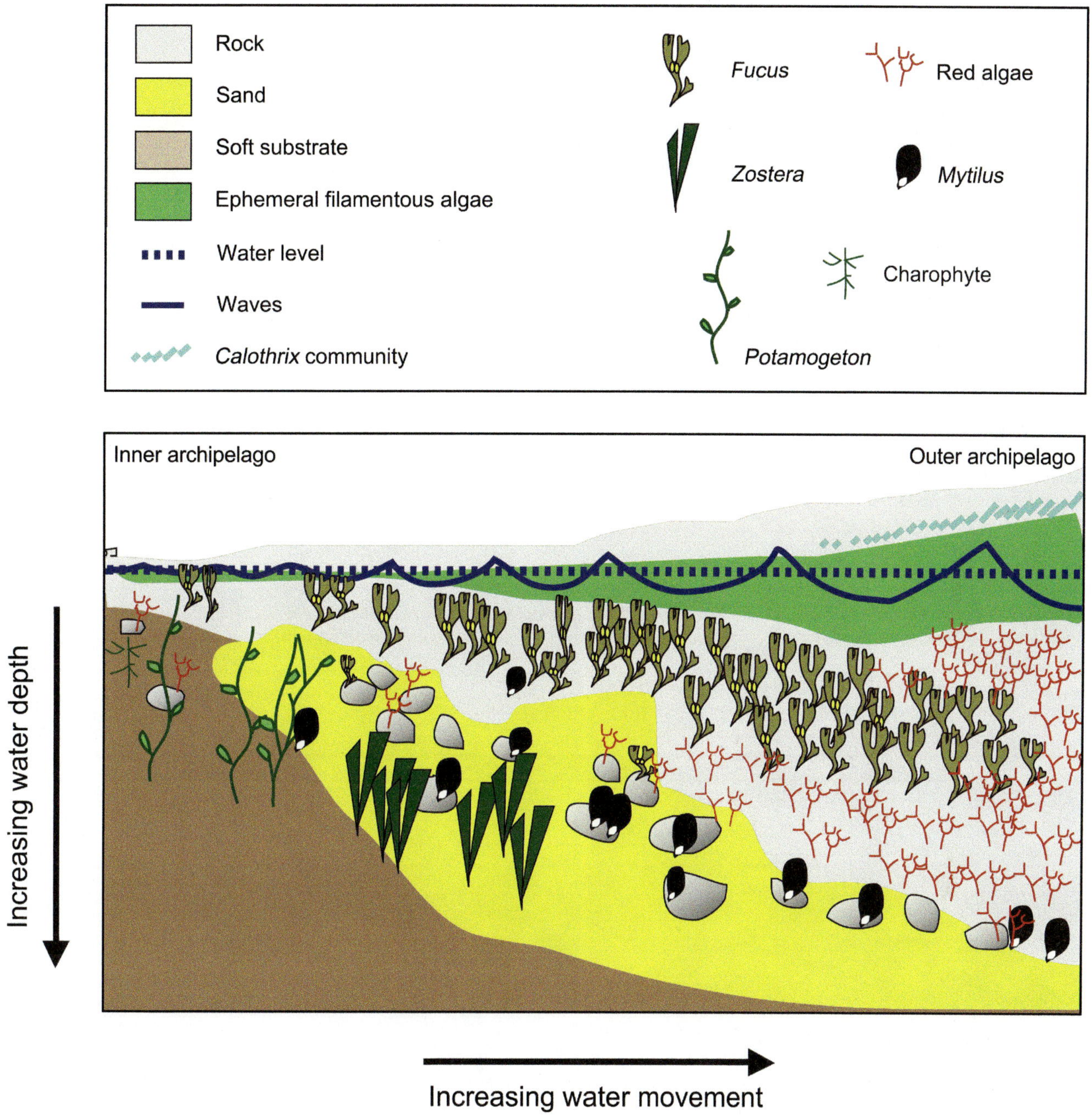

Fig. 11.6 Schematic summary of the relationship between exposure to waves, water depth, substrate and the type of phytobenthic vegetation along an archipelago gradient. Figure: © Hans Kautsky

11.1.4 Different rock types

Most of the rock on which phytobenthic communities grow in the Baltic Sea Area is hard, acidic igneous rock (granite) or metamorphic rock (gneiss). However, porous sedimentary calcareous rock (limestone) occurs in a belt from the southwestern Gulf of Finland, the Estonian islands of Saaremaa and Hiiumaa and Swedish islands of Gotland and Öland to the Danish island of Bornholm. Parts of these limestone coasts consist of klints (*cf.* Fig. 2.8), which are stepwise deeper terraces that follow the geological stratification. The uppermost klint terrace is often sheltered as waves break at its edge some distance from the shore, and it usually has a luxuriant growth of attached vegetation due to high insolation and good water exchange (Fig. 11.8). The klint coasts on the eastern sides of the islands of Saaremaa and Gotland often do not reach a water depth of 10 m until several km away from the coast. This allows phytobenthic vegetation to grow far off from the coastline, forming extensive areas with high primary production.

Fig. 11.7 Soft substrates (silt and mud) are dominated by rooted vascular plant communities or more or less loose-lying communities of vascular plants and/or algae. (**a**) A biologist studying a phytobenthic community dominated by *Potamogeton perfoliatus* and *Stuckenia pectinata* in Norafjärden, Bothnian Sea. (**b**) A phytobenthic community of loose-lying *Ceratophyllum demersum* entangled with loose-lying *Fucus vesiculosus* on soft substrate at Askö, Baltic Sea proper. Photo: © Hans Kautsky

Fig. 11.8 Limestone cliffs and their submerged vegetation. (**a**) Limestone rauk coast at Gotland in the Baltic Sea proper, with *Fucus vesiculosus* in the small lagoons. (**b**) *Fucus vesiculosus*-dominated vegetation at a wave-exposed site with a small patch of *Ruppia spiralis*. (**c**) A limestone cliff exposed to strong wave action on the eastern coast of Gotland with an unstable substrate of rounded boulders and stones at a 2–3 m water depth. More stable boulders support a luxuriant growth of the annual brown algae *Chorda filum* and *Pylaiella littoralis*. The surrounding rock is covered by the red alga *Ceramium tenuicorne*. Photo: © Hans Kautsky

Limestone shows large variability in quality as the substrate type depends on the geological origin of the exposed layer, which determines the type of vegetation. However, it is the grain size and stability, not the chemistry, of the substrate that affects the species composition of the phytobenthic communities. A typical vegetation of the first klint consists of, *inter alia*, the brown macroalgae *Fucus vesiculosus*, *Chorda filum* and *Dictyosiphon foeniculaceus* and colonies of the cyanobacterium *Rivularia atra*. Also vascular plants, such as *Ruppia maritima*, *Ruppia cirrhosa* (syn. *Ruppia spiralis*), *Stuckenia pectinata* (syn. *Potamogeton pectinatus*) and *Zannichellia palustris*, as well as charophytes, grow here in limestone gravel. Limestone with substantial amounts of clay incorporated ("marl") is rather unstable and, therefore, a poor substrate for algae with large thalli such as *Fucus vesiculosus*, but it creates an excellent habitat for filamentous algae.

11.1.5 Substrates far away from coasts

Boulder fields deposited by the glacial ice have created submerged offshore stone reefs in the Baltic Sea Area, *e.g.* the Słupsk bank off the Polish coast, the Adlergrund in the Arkona Sea and Lilla Middelgrund, Stora Middelgrund, Morups bank and Fladen in the Kattegat (*cf.* Fig. 2.2; Pedersén and Snoeijs 2001; Andrulewicz et al. 2004; Zettler and Gosselck 2006). At these reefs, continuous currents prevent sedimentation so that macroalgae can attach, but the boulders and algae also create refuges where benthic organisms are protected from mechanical disturbance by the same currents.

Compared to the sand-dominated coasts of Poland and Germany, the Słupsk bank and the Adlergrund host communities with rich growth and high diversity of macroalgae, invertebrates and fish (Andrulewicz et al. 2004). Since these offshore stone reefs are unaffected by direct land-runoff and sedimentation, but are subject to strong currents, the algae are remarkably clean and healthy and may, if hard substrate is available, penetrate deeper than at most coastal sites. In the shallow Kattegat, the deep-water currents of the continuous inflow of saline water from the Skagerrak into the Baltic Sea (*cf.* Sect. 2.3.7) pass the offshore stone reefs below the halocline, which is probably an additional reason for the extremely high diversity and good condition of the algae at the offshore stone reefs in the Kattegat (Fig. 11.9).

Other shallow offshore areas in the Baltic Sea proper and the Gulf of Bothnia, *e.g.* the Hoburgs bank, Södra Midsjö bank and Norra Midsjö bank (*cf.* Fig. 2.2), are dominated by relatively unstable wave-sorted till deposits, although attached specimens of the red alga *Coccotylus truncatus* can be observed here at a ~32 m water depth. In contrast to the offshore stone reefs in the Kattegat, the Hoburgs bank and the Midsjö banks are low in biodiversity. Due to the rather unstable substrate, perennial species do not persist and annual species dominate. However, *Mytilus* beds with >3 kg dry weight m^{-2} can be found at the Hoburgs bank and the Midsjö banks (Kautsky 1984), and are a sign of the high productivity of these areas.

Of still lower diversity, due to continuous substrate instability, are the submerged sand banks in the Baltic Sea proper (*e.g.* the Odra bank) and the Bothnian Bay. Such sand banks lack attached vegetation and only a few invertebrate species can cope with the extreme environmental conditions

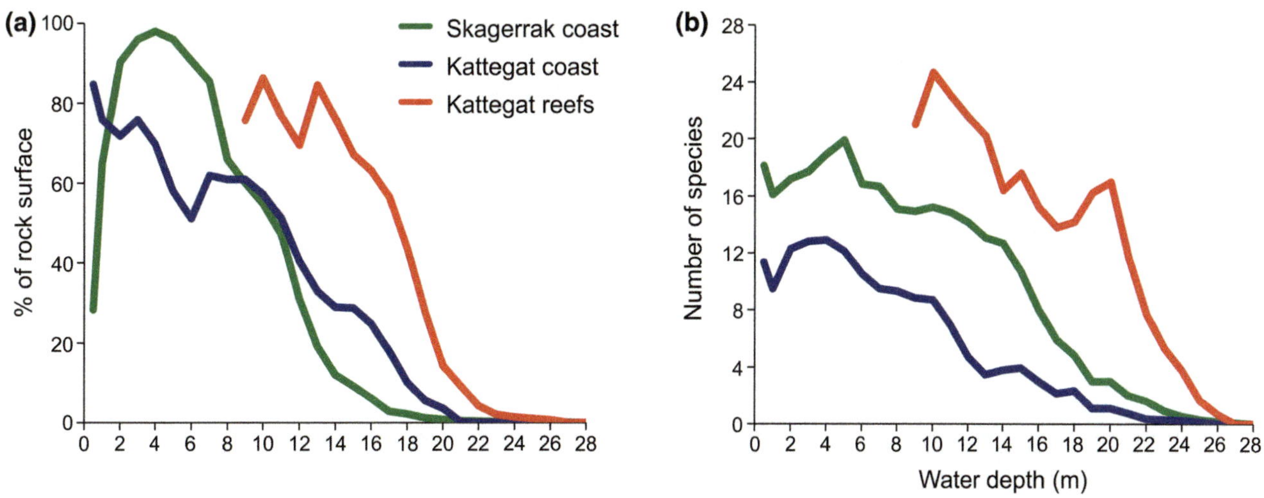

Fig. 11.9 Comparisons between the phytobenthic communities at offshore stone reefs in the Kattegat and those at wave-exposed coastal sites in the Kattegat and the Skagerrak. (**a**) Vegetation coverage. (**b**) Species richness. The higher diversity at the coastal sites in the Skagerrak is due mainly to the higher salinity. The higher diversity and cover at the reefs are caused by a combination of low sedimentation and higher salinity below the halocline (in the Kattegat situated at a 15–20 m water depth). The graphs represent means of 15 diving transects at coastal sites in the Skagerrak, 23 transects at coastal sites in the Kattegat and 12 transects at the offshore stone reefs in the Kattegat. Figure modified from Pedersén and Snoeijs (2001)

Box 11.4: Commercial harvest of macroalgae

The target compounds: polysaccharides

Polysaccharides are long molecules in which there are spaces that can be filled by water. This water-binding capacity is utilised when polysaccharides from algae are employed as additives to stabilise, thicken and smoothen human foods such as ice-cream, puddings, weight-loss products and as an alternative to gelatin for vegetarians (additives numbered E400-407a in the European Union classification). The algae use these polysaccharides to strengthen their cell walls as a protection against damage from being repeatedly thrown against the rocks by strong waves and to resist desiccation in the intertidal zone. The polysaccharides agar and carrageenan are extracted from red algae, and alginates are derived from brown algae. While no negative health effects have been reported for agar and alginates, the use of carrageenan in human food has been debated since the 1970s, especially in connection with colon cancer and diabetes (Tobacman 2001; Yang et al. 2012; Bhattacharyya et al. 2015), although food authorities in Europe and the USA consider it safe.

The algae harvested

Norway is one of the world's largest producers of alginates from Laminariaceae harvested in the wild along the Atlantic coast. In tropical countries, the red algae *Gracilaria* spp. (agar), *Eucheuma denticulatum* (carrageenan) and *Kappaphycus alvarezii* (carrageenan) are cultivated in large amounts, but in the Baltic Sea Area wild *Furcellaria lumbricalis* (Box Fig. 11.7) is harvested to obtain a product called "furcellaran", "Danish agar" or "Baltic agar", which is a form of carrageenan. The majority of the global *Furcellaria lumbricalis* harvest comes from Denmark and Canada, but the eastern coast of the Baltic Sea proper also hosts a large loose-lying *Furcellaria lumbricalis*-dominated community, which is commercially harvested in Estonia (Martin et al. 2006a, b; Bučas et al. 2009). In the early 1900s, large quantities of loose-lying *Furcellaria lumbricalis* occurred in the Kattegat (Austin 1959), but because of overexploitation this population has declined. *Furcellaria lumbricalis* was previously included in the HELCOM list of threatened or declining species in the Baltic Sea Area (HELCOM 2007), but is not on this list anymore (HELCOM 2013a). However, it is necessary to carefully monitor its standing stock and to decide if and how much can be sustainably harvested each year.

Box Fig. 11.7 *Furcellaria lumbricalis* is common along the northern European and Canadian marine coasts where it can be 30 cm in diameter. It is the largest red alga with a wide distribution in the Baltic Sea, although it is smaller in thallus size than on fully marine coasts. Together with other red algal species *Furcellaria lumbricalis* forms the red algal belt of the Baltic Sea proper. *Furcellaria lumbricalis* is a key species with an important structural role in the ecosystem, either attached to hard substrates or in loose-lying algal aggregates. Photo: © Hans Kautsky of high exposure to currents and low salinity due to runoff from local rivers. However, some species of suspension-feeding invertebrates can reach high densities at the Odra bank, *e.g.* the bristle worm *Pygospio elegans*, the amphipod *Bathyporeia pilosa*, the brown shrimp *Crangon crangon*, and the bivalves *Cerastoderma glaucum*, *Macoma balthica*, *Mya arenaria* and *Mytilus trossulus* (Zettler and Gosselck 2006). These high concentrations of invertebrates attract fish and waterbirds, as well as fisheries, which creates conflicting interests (Sonntag et al. 2012).

11.1.6 Macrophyte vegetation without a substrate

Macrophytes may be ripped off their substrate by wave forces and transported away with the currents. This is an important mechanism of vegetative dispersal for algae, and even for vascular plants, as they may find a new substrate elsewhere and grow further, often by first getting entangled in the attached vegetation. However, it is also common for the ripped-off macrophytes to end up on the shore and die (Fig. 11.10).

Some algae clearly have two different life forms: they occur both attached to a hard substrate or as loose-lying balls moving around with the prevailing currents. Examples of such species are the marine red alga *Furcellaria lumbricalis* (Box 11.4) in the Kattegat and the southeastern Baltic Sea proper and the freshwater green alga *Aegagropila linnaei* (syn. *Cladophora aegagropila*) in the Gulf of Bothnia.

In nutrient-rich places with weak water exchange, loose-lying algae may form aggregates called "algal mats". These algal mats may be lying on the seafloor and/or float in the water column and form species-rich communities, often with macroalgae as a basis and accompanied by bacteria, cyanobacteria, protists and fauna. In nutrient-rich sheltered bays algal mats may be a natural phenomenon in summer, but the occurrence and abundance of algal mats increase with eutrophication and may become a nuisance, *e.g.* by preventing growth of an attached vegetation or by clogging of fishing nets. Furthermore, thick layers of decomposing algal mats can cause shallow-water anoxia, kill the benthic fauna, lead to the formation of hydrogen sulphide and the release of nutrients from the sediment in the same way as in the deep soft-bottom system of the Baltic Sea (*cf.* Fig. 10.12).

11.2 Vegetation changes along the large-scale Baltic Sea gradient

11.2.1 Loss of community diversity

Salinity is the most important factor for the macrophyte species distributions in the Baltic Sea on an ecosystem-wide scale (*cf.* Figs. 4.10a and 4.18b) and is based on the species' salinity tolerances. The loss of marine macroalgal species along the Baltic Sea salinity gradient governs the structure and composition of phytobenthic communities to a large extent. Most phytobenthic vascular plant and charophyte species in the Baltic Sea are of freshwater origin, but, given their different substrate requirements, they cannot compensate for the loss of macroalgal species on hard substrates.

The changes in community composition from the Baltic Sea proper to the Bothnian Sea and further to the Bothnian Bay, are more stepwise than gradual, *i.e.* community composition changes radically at the sill areas of Södra Kvarken and Norra Kvarken because salinity conditions are relatively stable within each of the three basins, but there is a salinity gradient between them (*cf.* Fig. 4.2). Since there is no sill between the Gulf of Finland and the Baltic Sea proper the changes in community composition with salinity are more gradual here, but may be overshadowed by vegetation changes caused by the heavy eutrophication of the inner Gulf of Finland.

11.2.2 The "downward process" of marine macroalgae

When following the salinity gradient from the Skagerrak via the Kattegat, the Belt Sea and into the Baltic Sea, many marine algae that on the Atlantic and North Sea coasts are found in the intertidal and upper sublittoral zones occur in increasingly deeper water. For example, the mean upper depth limit of the relatively euryhaline species *Saccharina latissima* (syn. *Laminaria saccharina*) changes from a 1.5 m water depth in the Skagerrak to a 4 m depth in the Kattegat (Pedersén and Snoeijs 2001). Simultaneously, the less euryhaline species *Halidrys siliquosa* extends its depth range from 1 to 9 m and *Corallina officinalis* from 2 to 12.5 m.

The three canopy-forming brown algae: *Fucus vesiculosus*, *Fucus serratus* and *Saccharina latissima* differ in how far they penetrate into the Baltic Sea (Fig. 11.11). Of these three species, *Fucus vesiculosus* is best adapted to the low salinity of the Baltic Sea and *Saccharina latissima* least. At the inner Baltic distributional limit of *Saccharina latissima* in the southern Baltic Sea proper (the island of Bornholm, Denmark), its upper limit lies at the water depth of $\sim$20 m (Wærn 1965).

This phenomenon of successive downward dislocation with lower salinity is called "brackish-water submergence" or, as defined for the vegetation in the Baltic Sea, "the downward process" (Svedelius 1901; Wærn 1965). However, in the case of the large-scale Baltic Sea gradient, this process is more related to the horizontal salinity gradient than to a vertical salinity gradient in the water column. Initially, in the Skagerrak and Kattegat, the downward

Fig. 11.10 Algae washed ashore. (**a**) Red algae at the island of Gräsö in the southern Bothnian Sea. (**b**) The common gull *Larus canus* looking for something to eat among the algal debris in the Ekenäs archipelago in the Gulf of Finland. (**c**) Brown and red algae at the island of Gotland in the Baltic Sea proper. Photo: (**a**, **b**) © Pauline Snoeijs-Leijonmalm, (**c**) © Hans Kautsky

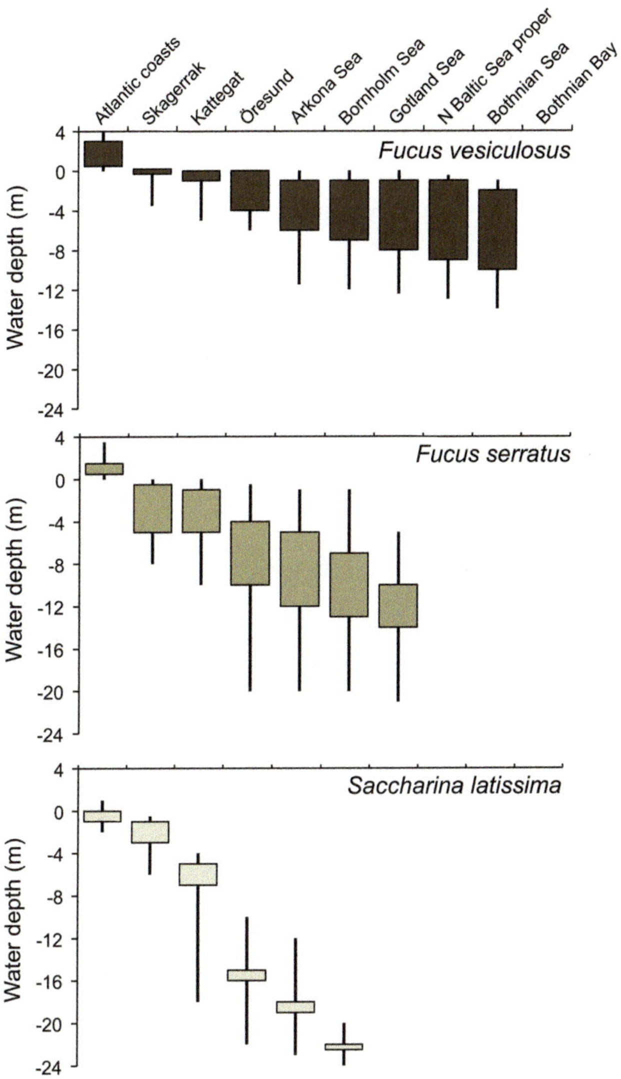

Fig. 11.11 Schematic summary of the downward process of the canopy-forming brown algae *Fucus vesiculosus*, *Fucus serratus* and *Saccharina latissima* along the large-scale Baltic Sea gradient. In the Bothnian Sea "*Fucus vesiculosus*" also includes *Fucus radicans*. The "zero" line is set at the waterline. Figure based on depth-distribution data from Wærn (1952, 1965), von Wachenfeldt (1975) and Pedersén and Snoeijs (2001)

process is obviously in some way, directly or indirectly, related to the reduction in surface-water salinity with water depth, as well as to the reduction of the tides towards the Baltic Sea (*cf.* Sect. 2.3.3). The Atlantic intertidal species *Fucus vesiculosus* and *Fucus serratus* live permanently submerged in the Baltic Sea and broaden their vertical distribution from 1–2 m to 15–20 m (Fig. 11.11). On the Atlantic coasts each of these two *Fucus* species is restricted to a specific narrow depth interval through competition with the species above and below them. Already in the Skagerrak they live submerged and extend deeper than many of their competitors, which are weakened by the low salinity.

11.2.3 The "downward process" inside the Baltic Sea

Inside the Baltic Sea the halocline is located far below the photic zone (*cf.* Fig. 2.15), and the downward process is driven by the tolerances of the algal species to low salinity, not by a vertical salinity gradient. From the Kattegat to the Bornholm Sea the depth extensions of *Fucus vesiculosus* and *Fucus serratus* maximise, but north of the Kalmarsund area and at the southern tip of the island of Gotland *Fucus serratus* has its northern limit.

Along the large-scale Baltic Sea gradient, *Fucus vesiculosus* seems to "press down" *Fucus serratus*, which in turn "presses down" *Saccharina latissima* and other species through competition (Fig. 11.11). This may partly be explained by a reduced competition when species diversity decreases successively along the Baltic salinity gradient (Torn et al. 2006a). However, at sites where *Fucus vesiculosus* and *Fucus serratus* occur together, *Fucus vesiculosus* grows from an ~0.5–1 m water depth and is joined by *Fucus serratus* at a ~3 m depth. The two species then grow together in approximately equal proportions down to the lower limit of *Fucus vesiculosus* at a 10–12 m water depth, and thereafter *Fucus serratus* is found deepest, usually a few metres deeper than *Fucus vesiculosus*. As they grow together in a broad belt, competition between these two *Fucus* species does not seem to be strong in the Baltic Sea. It is possible that *Fucus serratus* is better adapted to low irradiance, but this has not been proven.

11.2.4 Depth penetration – set by CDOM or eutrophication?

The attached vegetation in the Kattegat generally penetrates deeper (on average down to ~25 m) than it does in the Baltic Sea (on average down to 10–20 m) (Pekkari 1965; Wærn 1965; Kautsky 1995a; Pedersén and Snoeijs 2001). Exceptions are single specimens of *e.g.* the brown alga *Battersia arctica* (syn. *Sphacelaria arctica*) and some species of crustose algae, which may penetrate 5–10 m deeper than the lower limit of most other primary producers in the Baltic Sea.

It has been suggested that high concentrations of coloured dissolved organic matter (CDOM, *cf.* Sect. 15.2.6), which gives the water a yellowish-brown colour, limit the depth extension of attached growth of benthic primary producers

Box 11.5: The measurement of water movement

Different methods are used to measure or predict the forces of the wave action experienced by primary producers and animals in the phytobenthic communities at coastal sites. The method chosen depends on the focus of the study. The methods most frequently applied today involve *in situ* sensors and GIS-based modelling.

***In situ* sensors** perform direct measurements, thus providing the best information on the degree of water movement to which the organisms living at that spot are exposed. An abundance of instruments has been described in literature, from highly complicated expensive constructions to pieces of chalk that slowly dissolve in the water. Today, small cheap sensors are available that can be applied in the field. However, when employing instruments in the field, it is of course necessary to ensure that their mere presence does not alter the currents.

The Baardseth index is a simple cartographic measure of the "openness" of a coastal site (Baardseth 1970). It can be determined by placing the centre of a transparent circular disc with a radius of 7.5 km and subdivided into 40 sectors (each with an angle of 9°) at the study site on a 1:50,000 nautical chart (Ruuskanen et al. 1999). The Baardseth index represents the number of sectors without skerries, islands or mainland. Small solitary rocks at the periphery of the disc may be ignored. The Baardseth index varies between 0 and 40, where 0 indicates the extreme shelter and 40 denotes the maximum exposure.

The effective fetch is a cartographic measure that describes the average distance within which a wave can collect energy before it meets a coastal study site. This measure was originally designed for lakes, and is based on 15 measurements of the distance between the study site and the nearest skerry, island or mainland (Håkansson 1977). The central radius of a 90° fan shape is positioned from the study site towards the main wind direction (Box Fig. 11.8) and the distance (χ_i in km) from the study site to land is measured for each of the deviation angles from the central radius (γ_i) of −42°, −36°, −30°, −24°, −18°, −12°, −6°, 0°, 6°, 12°, 18°, 24°, 30°, 36° and 42°. The effective fetch (L_f) is then calculated as $L_f = (\sum \chi_i \cos|\gamma_i|)/(\sum \cos|\gamma_i|)$. The value of the effective fetch depends on the measured distances and is not restricted by a maximum value (if not defined). Additional calculations based on an effective fetch can be made by including *e.g.* wind speed and water depth.

GIS-based wave exposure models have been developed to provide estimations of water movement more precisely than the Baardseth index or the effective fetch (Ekebom et al. 2003; Isæus 2004). In the geographic information system (GIS), detailed wind direction, wind speed and water movement data from nearby meteorological stations are included in calculations for each angular section, and parameters such as wave height can also be estimated.

Microhabitats may have exposure conditions that differ substantially from cartographic measures and GIS-based models. Correct measurement of water movement in defined microhabitats can be obtained only with *in situ* sensors. It should be kept in mind that there are major differences in water movement on a small scale, *e.g.* between the top and the side of a boulder.

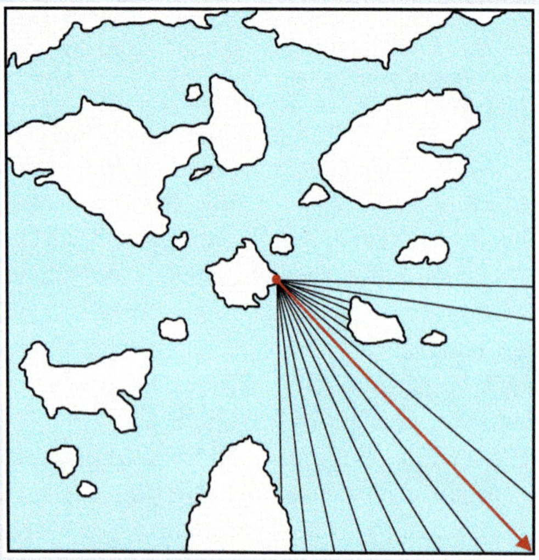

Box Fig. 11.8 The principle of calculating the effective fetch for a defined sampling station (red dot). The red arrow denotes the main wind direction. The black lines indicate the directions of the lines (each with a defined deviation angle). The lengths between the sampling station and the nearest land for all of these lines are used in the calculations. Figure: © Pauline Snoeijs-Leijonmalm as it does for phytoplankton (*cf.* Sect. 2.4.6). However, although CDOM does change light conditions in the water, some well-studied species (*e.g. Fucus vesiculosus*) grow generally deeper down in the CDOM-richer Bothnian Sea than in the Baltic Sea proper (Fig. 11.11). This is probably related to the lower degree of eutrophication (lower phytoplankton biomass) in the Bothnian Sea compared to the Baltic Sea proper.

The opposite is found in the eastern Gulf of Finland, where the lower limit of the vegetation is usually much shallower than in the Baltic Sea proper due to heavy eutrophication. Today the high phosphorus input from the city of Sankt-Petersburg has decreased, but there is still a high nitrogen input from Lake Ladoga via the Neva river. Eutrophication usually increases the turbidity of the water column by increasing phytoplankton growth, which decreases light availability on the seafloor (Kautsky et al. 1986). At the same time sedimentation increases and decreases the available area of hard substrate so that algae cannot attach with their holdfasts anymore. However, sedimentation is usually of lower importance than shading by phytoplankton because wind-induced waves do not change with eutrophication at a given site, *i.e.* at exposed sites the sedimented material is removed by water movement and at sheltered sites soft bottoms already occur.

11.2.5 Shading and scouring by ice

Further to the north of the Baltic Sea, in the Bothnian Bay, another factor that is thought to limit the depth distribution of the vegetation, probably much more important than the high CDOM concentrations here, is the ice cover that shades the seafloor for up to six months per year. Furthermore, at the northernmost latitudes of the Baltic Sea (64–66 °N) the lower declination of the sun causes a larger portion of the insolation to reflect due to higher albedo, despite long summer days with almost midnight sun. Thus, less light energy can penetrate deeper down into the water column in the Bothnian Bay than in the rest of the Baltic Sea, and the lower limit of the vegetation here is only $\sim$10 m.

The scouring of ice can mechanically disturb the phytobenthic communities close to the water surface, usually down to a $\sim$1 m water depth but occasionally deeper. The largest damage to the attached vegetation by ice scour is caused at wave-exposed open sites during ice break-up in spring. The influence of ice scouring is strongest in the northern Gulf of Bothnia and the eastern Gulf of Finland, where an ice cover occurs every year and lasts longer than in the south of the Baltic Sea (up to half a year in the Bothnian Bay, *cf.* Sect. 2.4.4). The most dramatic effects of ice scouring may be observed on some coasts in the Norra Kvarken area facing the Bothnian Bay, where land vegetation (including lichens on the rock) is destroyed for several metres high up on the shore (Fig. 11.12a). In this area, 17-m high ice piles from ridging have been observed on land, as well as substantial damage to the sublittoral phytobenthic communities (Fig. 11.12b, c).

11.3 Factors structuring the vegetation on local scales

11.3.1 Water movement

Besides its indirect effect on community composition by substrate sorting (*cf.* Sect. 11.1), water movement has a significant effect on macrophyte growth, both by its physical drag force and by its transport of nutrients and metabolites. At low to moderate velocities, currents typically stimulate both the biomass and diversity of the phytobenthic communities, but at higher velocities the macrophyte growth is reduced. For example, the tallest and widest *Fucus vesiculosus* thalli are found at the sheltered end of an exposure gradient and the smallest and narrowest thalli at the most exposed end (Ruuskanen et al. 1999).

The actual water movement at a given site (microcurrents, *e.g.* around a boulder) can be measured with *in situ* sensors (Box 11.5). Also, several exposure indices based on simple cartographic models, such as the Baardseth index and the effective fetch, have been designed to estimate the amount of water movement that is experienced by the phytobenthic communities. However, more detailed modelling, based on geography, wind and current data in the geographic information system (GIS-based), is often used today.

Extreme wave exposure can even turn boulders around at a substantial water depth and restrict macroalgal growth (Fig. 11.4b). On strongly wave-exposed bedrock, the algal vegetation is kept short by the constant mechanical disturbance and/or can only attach in sheltered microenvironments such as rock crevices (Fig. 11.13a–c). With less water movement, *e.g.* at the lee side of an island, or within archipelagos, conditions are more sheltered and the vegetation can grow higher and form more diverse phytobenthic communities.

Larger algae may utilise the forces of waves to control their filamentous algal competitors by the so-called "whiplash effect" (Dayton 1975; Kiirikki 1996a). For example, especially when the *Fucus vesiculosus* vegetation is sparse, filamentous epiphytes and undergrowth are continuously removed by the *Fucus* tufts as they are swept around by waves (Fig. 11.13d).

11.3.2 Light and substrate availability

In an archipelago area, the depth penetration of attached macroalgae usually increases from the inner archipelago to

Fig. 11.12 Effects of the annual ice scouring on the coast of Holmöarna at the Norra Kvarken sill between the Bothnian Sea and the Bothnian Bay. (**a**) Enormous amounts of ice have been pressed on to the cliffs and have thereby scraped off the lichens several metres up on the shore. Only bare rock with some terrestrial vegetation in deep crevices is left. (**b**) A biologist documenting the effects of ice scouring on the *Fucus* vegetation. (**c**) Ice scouring has ripped off older *Fucus* plants and new recruits have started to grow. Photo: © Hans Kautsky

the islands furthest away from the coastline (Fig. 11.6). The reason for this is twofold. Firstly, an increase in wind-induced currents towards the outer archipelago increases the amount of suitable (hard) substrate deeper down. Secondly, the influence from terrestrial runoff, and thus the turbidity of the water, decreases.

The algal depth penetration at the outer archipelago is usually set by light availability. When hard substrate is available down to the deepest part the photic zone, the algae become smaller and darker with water depth through slow growth and high pigment concentrations proportional to the increasing light stress (Fig. 11.14a). In the case of a sudden disappearance of the algal vegetation with water depth, this is usually due to the lack of suitable substrate deeper down (Fig. 11.14b). With the decrease in water movement towards the inner archipelago, the hard substrate deeper down turns into mixed and soft substrates closer and closer to the water surface, limiting the algal depth penetration. It is not the light but the lack of suitable substrate deeper down that limits the depth penetration of the macroalgae, and they are replaced by rooted plants.

Sublittoral light quality depends on water depth. In coastal waters the maximum transmittance occurs at 500–575 nm (Jerlov 1976), *i.e.* in the green part of the energy spectrum, which means that green light penetrates deepest down in the water column. This has been used to explain why red algae occur deepest on *e.g.* Atlantic coasts according to the "theory of complementary chromatic adaptation", and even to explain why red algae have evolved red pigments in the form of phycobilins. However, the combined physiological evidence supports the notion that the changes in pigment composition that are observed with increasing depth in marine algae are largely adaptations to low irradiance, and not to the spectral composition of underwater light (Dring 1981). For example, increases in phycoerythrin concentrations in red algae are responses to low irradiances, and not to green wavelengths, of light. More important for algal zonation is that the photosynthetic apparatus of the species growing deeper down is more sensitive to low irradiance (Johansson and Snoeijs 2002) and that a thin thallus is essential for the growth and survival of marine macroalgae at great depths (Markager and Sand-Jensen 1992).

11.4 Microhabitats and biotic interactions

11.4.1 Vegetation layers

Like a terrestrial forest, the attached aquatic vegetation can be subdivided into several vertical layers that create

Fig. 11.13 Mechanical disturbances at sites exposed to strong wave action. (**a**) The *Cladophora glomerata* vegetation on the bedrock is kept short by the continuous mechanical forces on the algal thalli. (**b**) *Cladophora glomerata* can only attach in deep crevices in the bedrock. (**c**) The thallus of *Fucus vesiculosus* is continuously ripped off and only basal parts of the alga are left. Here the basal parts of *Fucus vesiculosus* are accompanied by *Ceramium tenuicorne* and *Mytilus trossulus*. (**d**) A *Fucus vesiculosus* tuft controlling the growth of filamentous algae (epiphytes and undergrowth) by the whiplash effect. Photo: (**a–c**) © Hans Kautsky, (**d**) © Svante Pekkari

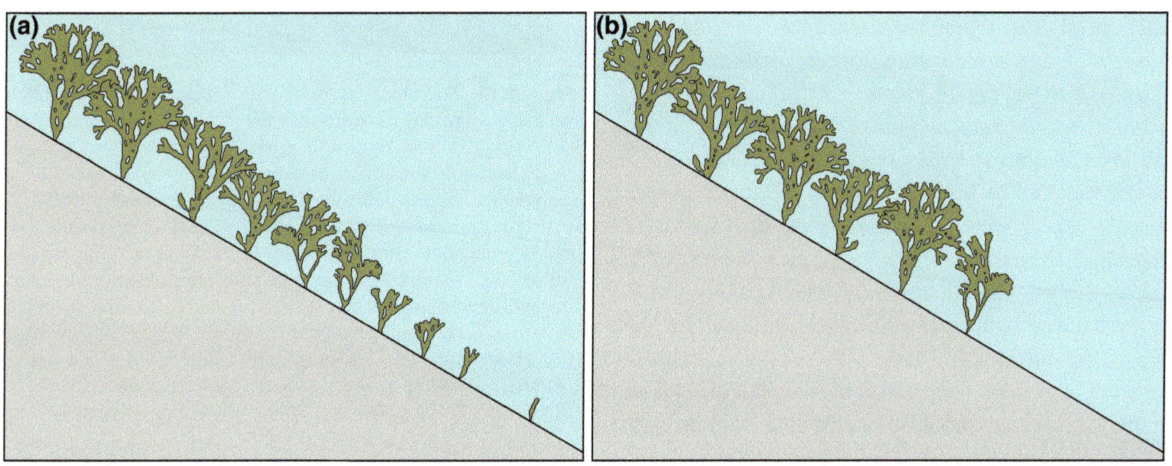

Fig. 11.14 The lower depth limit of *Fucus vesiculosus* can be set by light or by substrate. (**a**) Light penetration through the water column gradually decreases with increasing water depth, which produces a gradual decrease in the thallus size of *Fucus vesiculosus* and dwarf morphology of the lowermost individuals. (**b**) Lack of substrate below a certain depth, often caused by sedimentation, produces an abrupt elimination of the *Fucus vesiculosus* vegetation with no decrease in thallus size. Figure modified from Kiirikki (1996b)

microhabitats for other species in the phytobenthic communities (Fig. 11.15a). The canopy layer consists of large macrophytes such as the bladderwrack *Fucus vesiculosus* and the common eelgrass *Zostera marina*, with no other vegetation above them. The bush layer consists of smaller macrophytes such as *Ceramium tenuicorne*, *Furcellaria lumbricalis* and *Zannichellia palustris*, and the ground layer contains the species that are barely elevated above the substrate, such as crustose algae and *Rivularia atra*.

Each of these three layers supports epiphytes, *i.e.* all species (including animals) attached to the macrophytes. The different vegetation layers are not exactly defined by the macrophyte species in them and the same macrophyte species can create different microhabitat types. For example, the same macrophyte species can belong to different vegetation layers depending on the age of the algal tuft: an attached *Fucus* germling occurs in the ground layer, a first-year *Fucus* in the bush layer and an adult *Fucus* in the canopy layer. Filamentous algae such as *Pylaiella littoralis* and *Ceramium tenuicorne* can occur in the bush layer as well as epiphytically.

11.4.2 Diatoms: an integral part of all vegetation layers

Hundreds of diatom species occur in the sublittoral zone of the Baltic Sea (*cf.* Box Fig. 4.7) and form an important food source for invertebrates, *e.g.* for deposit-feeding snails, in the phytobenthic communities. The diatoms may belong to the inflora (*e.g. Campylodiscus clypeus*, *Martyana atomus*, *Surirella brebissonii*), ground layer (*e.g. Mastogloia smithii*, *Navicula perminuta*, *Nitzschia inconspicua*), epiphytes (*e.g. Cocconeis pediculus*, *Gomphonema olivaceum*, *Rhoicospenia abbreviata*), and some colony-forming diatoms can even belong to the bush layer (*e.g. Berkeleya rutilans*, *Encyonema silesiacum*, *Navicula ramosissima*) (Svensson et al. 2014).

Bush-layer diatom species often live in gelatinous tubes and the colonies can be several dm high, especially in spring (Snoeijs 1990a). At first sight, they look like brown filamentous algae, but when one tries to pick them they fall apart. In the early days of diatom research all colony-forming species were united in the genus *Schizonema* (Greek for "split thread", *cf.* Agardh 1824, 1830), but today they belong to many genera, *e.g. Berkeleya*, *Encyonema* and *Parlibellus*.

Another aspect that should be taken into account is that the biomass of epiphytic microalgae on macrophytes can be extremely high, especially on thin filamentous algae. For example, >95 % of the biomass of a "*Pylaiella littoralis*" sample can in fact consist of diatom cells (Snoeijs 1995). In such a case it would perhaps be fair to consider the algal tuft as a diatom colony instead of a macroalga.

Fig. 11.15 Microhabitats in the phytobenthic zone. (**a**) Schematic summary, showing the different microhabitats: A = canopy layer, B = epiphytes, C = bush layer, D = ground layer, E = inflora and infauna in sandy and soft substrates. Free-living among the vegetation are *e.g.* gammarids, mysids and fish. (**b**) On larger boulders a distinct vertical microzonation pattern may be observed with algae on the top of the boulder, the blue mussel *Mytilus trossulus* (black) below the algae, followed by the balanoid *Amphibalanus improvisus* (white), and in the lowest, darkest places the hydrozoan *Cordylophora caspia* and the bryozoan *Einhornia crustulenta* are found. Besides this microzonation, the photograph also shows a large number of *Fucus vesiculosus* recruits, only a few of which will survive to reproduce after 5–10 years. Figure: (**a**) © Hans Kautsky, Photo: (**b**) © Hans Kautsky

11.4.3 Fauna abundance and species composition

Micro-, meio- and macrofauna organisms seek food, shelter and spawning habitats in the macrophyte vegetation. Some

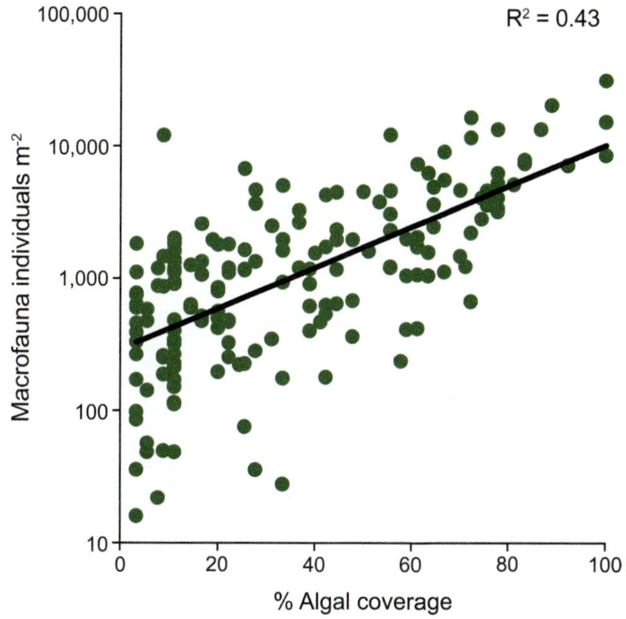

Fig. 11.16 The relationship between epilithic algal coverage and the density of the associated macrofauna (>1 mm) in the *Cladophora* belt at Forsmark in the southern Bothnian Sea (linear regression analysis). With an increase in algal coverage from 1 % to 100 %, the macrofaunal density increases from an average of ~300 individuals m^{-2} to an average of ~10,000 individuals m^{-2}. The data shown in this figure represent 11 sampling sites on 18 sampling occasions throughout one annual cycle. Note the logarithmic scale on the y-axis. Figure based on data in Snoeijs (1989)

animals stay in one of the vegetation layers, but many free-living animals, *e.g.* snails, gammarids, isopods, mysids and fish, move around between the different layers. Below the ground layer there is often a rich inflora and infauna, *i.e.* species living inside sandy or soft substrate below the seafloor, which also belong to the phytobenthic communities.

The number of animals associated with the vegetation is usually directly proportional to the abundance of the vegetation (Fig. 11.16). The species composition of the animals usually depends on the vegetation type. For example, in the western Gulf of Finland different *Gammarus* species prefer different microhabitats in the phytobenthic zone following the algal depth zonation. *Gammarus zaddachi* (Fig. 11.17a) lives mainly at a 0–1 m water depth between filamentous algae and in sheltered locations also in the *Fucus* vegetation, whereas *Gammarus oceanicus* (Fig. 11.17b) lives mainly at a 1–4 m water depth in the *Fucus* belt, and *Gammarus salinus* (Fig. 11.17c) occurs deeper than 4 m associated with red algae and *Mytilus trossulus*. A fourth gammarid abundant in the Baltic Sea, *Gammarus duebeni*, is a typical rock-pool species (Korpinen and Westerbom 2010).

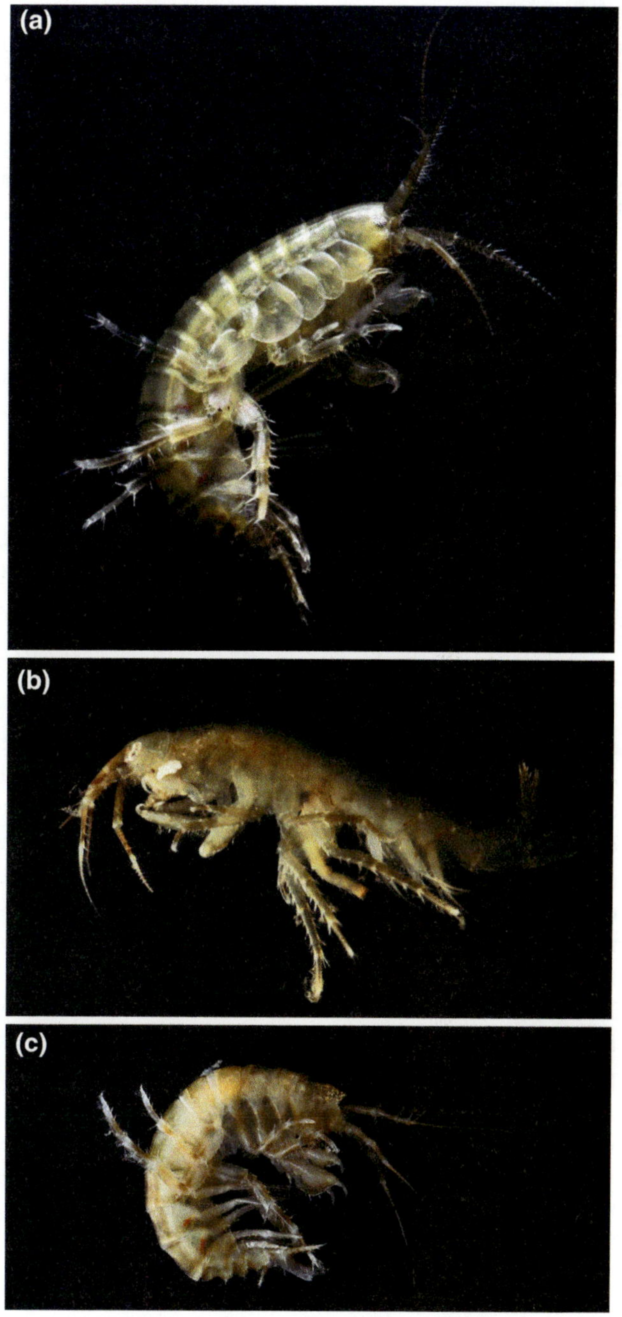

Fig. 11.17 The three native *Gammarus* species that are common in the phytobenthic zone of the Baltic Sea. (**a**) *Gammarus zaddachi* lives mainly at a 0–1 m water depth. (**b**) *Gammarus oceanicus* lives mainly at a 1–4 m water depth. (**c**) *Gammarus salinus* occurs deeper than 4 m. Photo: © Maria Włodarska-Kowalczuk and Piotr Bałazy

11.4.4 Biotic interactions

Biotic processes such as predation and competition operate on smaller scales than environmental drivers such as salinity or currents, *i.e.* at a patch or community scale. In comparison with fully marine systems, biotic interactions seem to be of lower importance in determining the community composition in the phytobenthic zone of the Baltic Sea (Kautsky and van der Maarel 1990; Nyström-Sandman et al. 2012). This is related to the loss of macroscopic species diversity along the large-scale Baltic Sea gradient (*cf.* Fig. 4.10). Thus, the current increase of the diversity of the Baltic Sea through introductions of non-indigenous species (*cf.* Sect. 5.1) has the potential to increase biotic interactions.

The absence of large predators such as starfish (*Asterias* spp.) and larger crabs (*Carcinus* spp.) in the major part of the Baltic Sea is probably the reason why *Mytilus trossulus* can become so abundant (Kautsky 1981). Also, major groups of large marine herbivores such as sea urchins, *Littorina* spp. and *Patella* spp. occur up to the Arkona Sea at most, which decreases the grazing pressure on the aquatic vegetation. However, the Baltic Sea hosts abundant populations of mesoherbivores that may regulate the density and species composition of the vegetation, especially at the early stage of algal colonisation.

An example of competition for space in the Baltic Sea is the distinct vertical microzonation that may be observed on larger boulders with macroalgae on top of the boulder, *Mytilus trossulus* on the side below the algae, the barnacle *Amphibalanus improvisus* below the blue mussel, and lowest down in crevices of the rock the hydrozoan *Cordylophora caspia* and the bryozoan *Einhornia crustulenta* (syn. *Electra crustulenta*) (Figs. 11.15b and 11.18). However, strong competition for space between different *Fucus* species as documented in the intertidal zone of Atlantic coasts (Hawkins and Hartnoll 1985) does not seem to take place in the Baltic Sea, where >10 m wide belts of mixed *Fucus serratus* and *Fucus vesiculosus* occur in the southwestern Baltic Sea proper and >10 m wide belts of mixed *Fucus radicans* and *Fucus vesiculosus* in the Bothnian Sea. On the other hand, *Fucus* recruits are sensitive to both intraspecific and interspecific competition. Out of a thousand established individuals only one or two will survive to reproduce 5–10 years later (Fig. 11.15b).

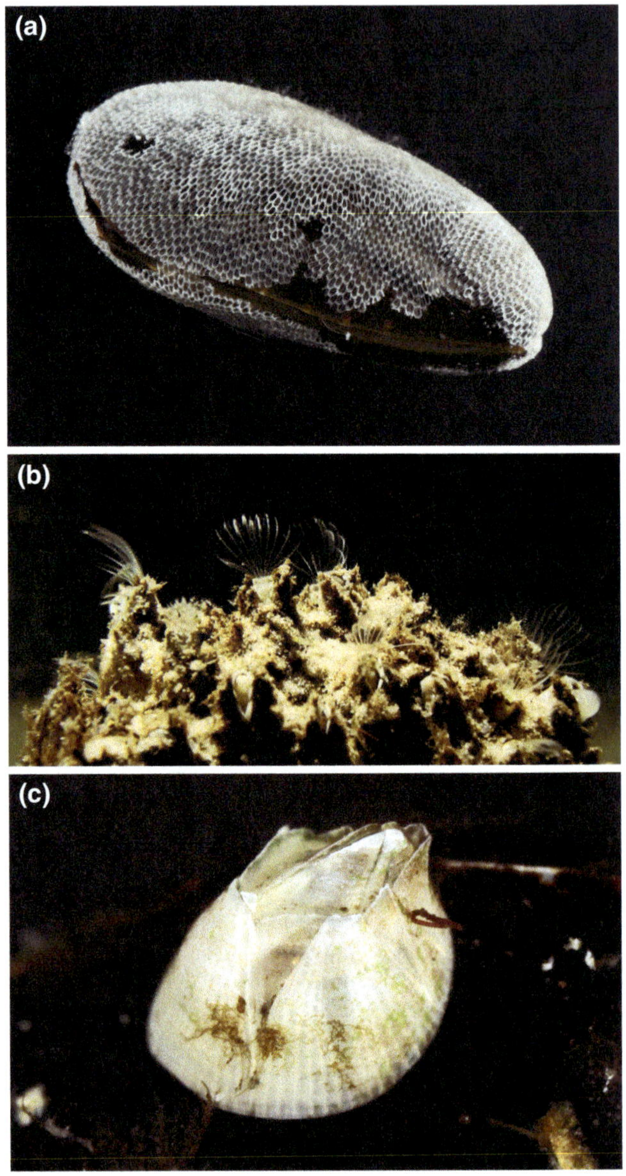

Fig. 11.18 Sessile filter-feeding animals that are common in the phytobenthic zone of the Baltic Sea. (**a**) The bryozoan *Einhornia crustulenta*. (**b**) The barnacle *Amphibalanus improvisus*. (**c**) A closed specimen of *Amphibalanus improvisus*. Photo: (**a**, **c**) © Maria Włodarska-Kowalczuk and Piotr Bałazy, (**b**) Piotr Bałazy

11.4.5 Experimental studies reveal biotic interactions

In general, biotic interactions in the phytobenthic zone are difficult to identify by field observations, and in most cases an experimental approach is necessary. Laboratory experiments neatly show the sensitivity of species and interaction mechanisms when factors such as population density, species richness or grazing pressure are varied under controlled conditions at different temperatures, nutrient concentrations and/or concentrations of hazardous substances. While the results obtained in such experiments often show clear cause-effect relationships, it is often more complicated to evaluate their relevance and/or importance for processes observed in nature. In the field there are so many factors that simultaneously influence biotic interactions in a community that they cannot be addressed in a single experimental set-up.

A compromise is to perform experimental manipulations of biotic factors (in combination with variation in environmental drivers) directly in the field or in mesocosms, *i.e.* in enclosures of a part of the ecosystem. In the phytobenthic zone, such field experiments are especially important for understanding recruitment mechanisms and interactions between the primary producers and the fauna. Field experiments have elucidated several crucial mechanisms that can regulate the abundance and species composition of the primary producers, such as recruitment of algae, nutrient recycling, food preference of grazers and trophic cascades governed by top predator abundances (Eriksson and Johansson 2003; Worm et al. 1999, 2001; Eriksson et al. 2006, 2009).

However, in a meta-analysis including a large number of the same type of experiments, the biotic interactions may not be so clear as in single experiments, or may even be contradictory (Lyons et al. 2014; Thomsen and Wernberg 2015). One of the reasons for such discrepancies is that no field experiments are carried out under exactly the same conditions because environmental conditions, species, life stages of species and/or genetic diversity within species, etc. may differ between experiments. To draw wider conclusions, *e.g.* for ecosystem management, it is necessary that support for the processes discovered in experiments is observed in nature. Thus, a combination of experiments and field observations is always recommended.

11.5 Biological traits and ecological strategies

11.5.1 Functional traits

To predict and assess community shifts and their consequences, ecologists are increasingly investigating how the functional traits of primary producers determine their relative fitness along environmental gradients (*cf.* Sect. 4.7). The trait-based approach in ecology offers the opportunity to tackle the complexity of species-rich communities by constructing simple taxon-independent models of community structure and community dynamics in relation to the environment. By defining species according to their form and function, communities with different species can be compared through the "common currency" of their traits. For macrophytes such traits include *e.g.* growth form, body size, life span, reproductive strategy, season, reproductive performance and fecundity.

11.5.2 Growth forms of macroalgae

Growth form groups of marine macroalgae are often used to functionally relate the vegetation to environmental change (Littler and Littler 1984). In such models, a wide range of ecological and physiological functions are assumed to be correlated with different categories of thallus morphology, *e.g.* crustose, filamentous, coarsely branched, foliose, and this occurs often in combination with thallus thickness, *e.g.* uniseriate (consisting of rows of single cells) filamentous, thin foliose, thick leathery foliose, etc.

Such classifications have been found appropriate in many cases, but not all, and measurements based on specific ecological functions, *e.g.* photosynthetic rate, nutrient uptake rate (Box 11.6) or resistance to disturbances (including *e.g.* grazing), rather than morphology grouping, may be more reliable (Padilla and Allen 2000). Measurements of variables such as photosynthesis or nutrient uptake also provide the advantage of interval-scale data, which are more rigorous in statistical analyses than nominal-scale data (using categories). On the other hand, it has to be taken into account that such measurements are often made in the laboratory from isolated thallus parts under quite different conditions than those in the field.

11.5.3 Life-history strategies

Traits such as body size, life span and reproductive strategy are often clustered and incorporated in theories to explain ecological patterns, *e.g.* "*r*/*K* selection" (*cf.* Box 4.14) and "life-history evolution" (Stearns 1989). Thin filamentous algae are mostly short-lived species categorised as *r*-strategists or "opportunists" with an exponential growth curve. Such species can quickly colonise new areas when the opportunity arises. In comparison, slow-growing, long-lived perennial species with larger thalli, which are categorised as *K*-strategists, remain at their site once they are established. These species are good competitors for space. Typical *r*-strategists in the Baltic Sea are ephemeral algae such as *Ulva linza* (syn. *Enteromorpha ahlneriana*) and *Spirogyra* sp., which can appear one week and disappear the next. Fast-growing annual filamentous algae that usually stay attached for longer periods during the growing season, *e.g.* *Ceramium tenuicorne*, *Cladophora glomerata*, *Polysiphonia fucoides* and *Pylaiella littoralis*, are still more *r*- than *K*-selected. Included in this group of "annual" algae are also "pseudo-annuals", *i.e.* they have seasonal growth, but can hibernate as dark, rigid basal parts of up to a few cm in length (*e.g.* *Cladophora glomerata*) and start to grow from there when conditions become favourable again. Typical *K*-strategists in the Baltic Sea, with slower growth rates and thicker and more complex thalli, are perennial species such as the brown algae *Fucus vesiculosus*, *Fucus serratus*, *Fucus radicans*, the red algae *Coccotylus truncatus*, *Furcellaria lumbricalis*, *Phyllophora pseudoceranoides*, *Rhodomela confervoides* and the vascular plants *Zostera marina* and *Ruppia maritima*.

An increase in the proportion between annual and perennial algae and plants may indicate a disturbance that benefits an *r*-strategy over a *K*-strategy. An *r*-strategist usually profits the most from growth-enhancing

Box 11.6: Photosynthetic rates and nutrient uptake rates

Algal morphology and surface:volume ratio

Photosynthetic rate, nutrient uptake rate and growth rate are roughly related to algal growth form and life-history strategies, being lowest in slow-growing thick leathery algae (Box Fig. 11.9a, b) and highest in fast-growing thin filamentous algae (Box Fig. 11.9a, c). The differences in the rates between species are based on the relative cell surface area (surface:volume ratio) that has direct contact with the environment. For example, the maximum photosynthetic rates (P_{max}), as well as most of the maximum nutrient uptake rates (V_{max}) of the finely branched algae *Ceramium tenuicorne* (red) and *Dictyosiphon foeniculaceus* (brown) are higher than those of *Furcellaria lumbricalis* (red) and *Fucus vesiculosus* (brown) with coarser thallus morphology. However, when comparing the rates at which different algal species take up nitrate, ammonia and phosphate from the water (Wallentinus 1984), the red algae *Ceramium tenuicorne* and *Furcellaria lumbricalis* (with different morphologies) have lower phosphate and nitrate uptake rates than brown and green algae with comparable morphologies. On the other hand, the large differences in the ammonia uptake rate and photosynthetic rate between *Ceramium tenuicorne* and *Furcellaria lumbricalis* do reflect their respective morphologies.

The efficiency of photosynthesis

Uniseriate algae (consisting of rows of single cells) are expected to have the highest photosynthetic rates because they have the highest surface:volume ratio, but this is not necessarily so. In natural communities (Box Fig. 11.10a), many factors other than mere morphology are involved, including thallus self-shading, the abundance of microscopic

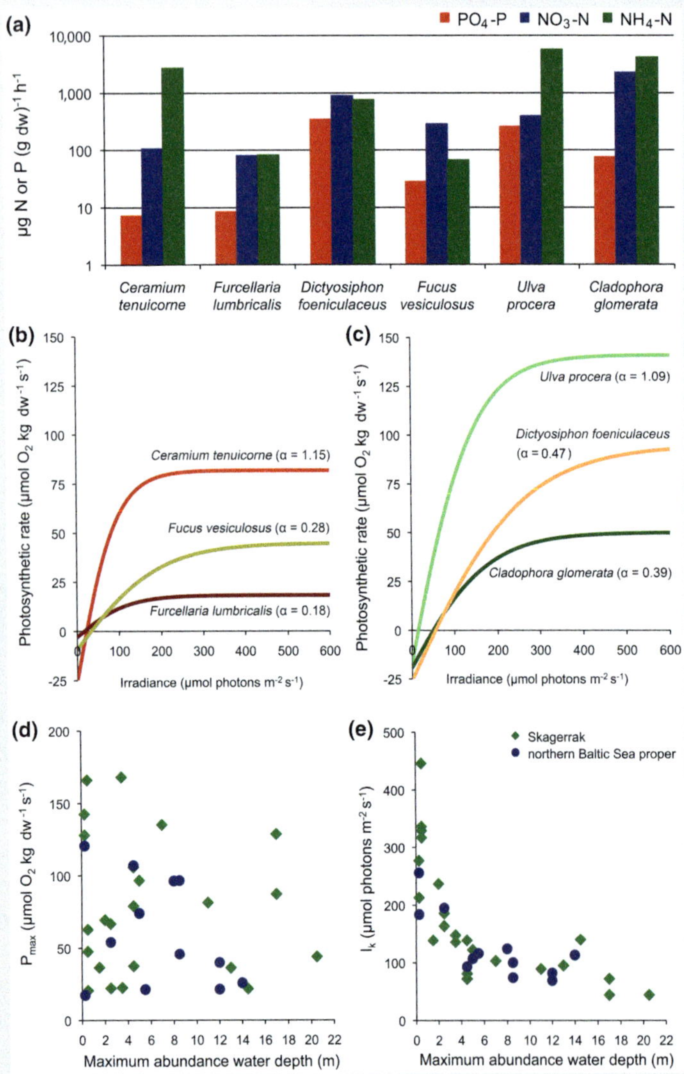

Box Fig. 11.9 Nutrient uptake rates and photosynthetic properties of different macroalgal species. (**a**) Maximum uptake rates (V_{max}) of the inorganic nutrients phosphate, nitrate and ammonia for the two red algae *Ceramium tenuicorne* (finely branched) and *Furcellaria lumbricalis* (coarsely branched), the two brown algae *Dictyosiphon foeniculaceus* (finely branched) and *Fucus vesiculosus* (flat thallus, young parts were used) and the two green algae *Ulva linza* (tube-shaped thin filaments) and *Cladophora glomerata* (uniseriate filaments) in the Baltic Sea proper at Askö (Sweden). Note that a logarithmic scale is used on the y-axis. (**b, c**) Photosynthesis-irradiance curves based on oxygen evolution measurements for the same six species as in (**a**), also from Askö. *Ulva linza* has the the highest photosynthetic rate at saturating light (P_{max}) and *Furcellaria lumbricalis* the lowest. α = initial slope at limiting irradiance levels [in µmol O_2 kg dry weight^{-1} (µmol photons m^{-2})$^{-1}$]. (**d, e**) P_{max} and the light saturation parameter I_k (= P_{max}/α) for 23 red and brown algal species in the Skagerrak at Fiskebäckskil (Sweden) and 12 red and brown algal species in the Baltic Sea proper at Askö. Figure (**a**) based on data in Wallentinus (1984), (**b–e**) based on data in Johansson and Snoeijs (2002) and Choo et al. (2005)

epiphytes and the efficiency of the photosynthetic apparatus. For example, the P_{max} and the initial slope at limiting irradiance levels (α) of the green alga *Cladophora glomerata* (uniseriate branched, Box Fig. 11.10b) is only slightly higher than that of young *Fucus vesiculosus*. Another example is the photosynthetic rate of the thin filamentous *Ulva linza* (Hayden et al. 2003), which is much higher than that of *Cladophora glomerata*, both when normalised to dry weight and when normalised to chlorophyll *a* content (Choo et al. 2005). These two green algal species often co-exist in the upper littoral zone of the Baltic Sea and, macroscopically, *Ulva linza* is easily confused with *Cladophora glomerata*. The older parts of *Cladophora glomerata* are usually completely overgrown by diatoms and the cyanobacterium *Heteroleibleinia* cf. *kützingii* (Box Fig. 11.10c) while *Ulva* is remarkably devoid of epiphytes (Box Fig. 11.10d, e), except when it is decaying. This is attributed to the release of large amounts of hydrogen peroxide as part of the alga's defence mechanism against oxidative stress in *Ulva linza* (Choo et al. 2005) and intercalary growth instead of apical growth. The low degree of shading by epiphytes of the *Ulva* species may partly explain their high photosynthetic rates.

Water depth

The P_{max} of the macroalgae in the Baltic Sea Area and the Skagerrak is independent of the water depth at which they are attached (Box Fig. 11.9d). However, the light saturation parameter I_k shows that the species growing deepest down in the sublittoral zone are more sensitive to light (Box Fig. 11.9e), *i.e.* their photosynthesis is more efficient at low irradiance levels compared with other species growing higher up in the sublittoral zone (Johansson and Snoeijs 2002). This is a physiological adaptation to low-light conditions that has nothing to do with algal morphology.

Box Fig. 11.10 Common green algae in the upper sublittoral zone of the Baltic Sea. (**a**) Bright green *Ulva linza* at the waterline, with *Cladophora glomerata* (coloured brownish-green by high abundances of microscopic epiphytes) and deeper down the tubular alga *Ulva intestinalis* with bright green lower parts and light green upper parts (from which spores have been released) in the Baltic Sea proper at Kråkelund (Sweden). Above the vegetation, the jellyfish *Aurelia aurita* and a three-spined stickleback *Gasterosteus aculeatus* are seen. (**b**) Light micrograph of *Cladophora glomerata* cells with the epiphytic diatoms *Brebissonia lanceolata* and *Gomphonema olivaceum*. (**c**) Light micrograph of old *Cladophora glomerata* cells completely overgrown with epiphytes, mainly the diatom *Rhoicosphenia abbreviata* (*cf.* Fig. 12.8) and the filamentous cyanobacterium *Heteroleibleinia* cf. *kützingii*. (**d**) Light micrograph showing cells of the branched tube-shaped alga *Ulva linza*, which are arranged in rows and have one pyrenoid. (**e**) Light micrograph showing cells of the unbranched tube-shaped alga *Ulva intestinalis*, which are not arranged in rows and have one pyrenoid. Photo: © Pauline Snoeijs-Leijonmalm environmental conditions that lead to an increase in the rate of supply of organic matter in an ecosystem (Nixon 1995), such as nutrient emissions, increased water temperature or the natural aging of a water body. Natural disturbances that favour *r*-strategists in the Baltic Sea are *e.g.* the irregular water level fluctuations following the frequency and intensity of weather disturbances in the upper sublittoral (*cf.* Sect. 2.3.4) and the harsh climate of the Bothnian Bay. Additionally, an unstable substrate (but not so unstable to totally prevent vegetation) usually benefits *r*-strategists that can recolonise fast.

11.5.4 Eutrophication indices

The ratio of annual to perennial macroalgae, as well as the ratio of filamentous algae to *Zostera marina*, can be used to indicate eutrophication because increased nutrient concentrations generally favour the growth of filamentous ephemeral algae (Korpinen et al. 2007). However, such indices should be used with great care because the abundance of filamentous algae depends not only on nutrient concentrations, but also on other factors such as salinity, grazing, season, sea level fluctuations and duration, thickness and ice-type of the winter ice cover. For example, in a broad salinity range, the fraction of opportunistic algae responds predominantly to salinity and not to eutrophication (Fig. 11.19). Thus, if a eutrophication gradient is accompanied by a salinity gradient, an index based on the proportion of filamentous algae is useless.

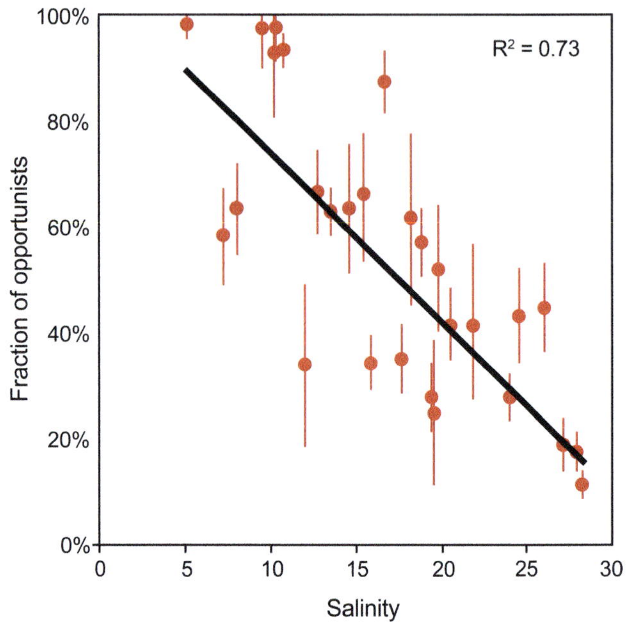

Fig. 11.19 Fraction of opportunistic algae as a function of salinity (linear regression analysis) in 28 brackish-water areas in the Bornholm Sea, Belt Sea, Kattegat, Limfjorden, Nissum Fjord and Ringkøbing Fjord. The red dots represent area-specific values, and error bars represent 95 % confidence intervals. Data from the National Danish monitoring programme. Figure modified from Krause-Jensen et al. (2007)

11.6 The epilittoral zone

11.6.1 Rock covered by lichens and cyanobacteria

The epilittoral zone of the Baltic Sea is the part of the coast that is influenced by waves and sea spray. The drier part of this zone is inhabited by lichens, which grow in distinct zonation patterns determined by the amount of sea spray they receive. The epilittoral lichen vegetation of the Baltic Sea Area is basically the same as that along other sea coasts in northern Europe (Ferry and Sheard 1969). The black tar lichen *Verrucaria maura* grows just above the waterline and orange sea lichen species of the genus *Caloplaca* are found slightly higher up on the shore (Fig. 11.20b, c). Species belonging to the green algal genus *Prasiola* can be found growing in shaded rock crevices some metres above the water line.

The black *Verrucaria maura* layer may be confused with the dark-green to black microbial layer dominated by the cyanobacterium *Calothrix scopulorum* (Fig. 11.20a). The two layers can be roughly distinguished by pure drastic experience: *Calothrix scopulorum* grows on sun-exposed parts and is extremely slippery, causing many unwanted baths, whereas *Verrucaria maura* is found on shaded parts of the shore and is not so slippery.

Common microbial species growing mixed with *Calothrix scopulorum* are the cyanobacteria *Gloeocapsopsis crepidinum* (syn. *Gloeocapsa crepidinum*) and *Phormidium* spp., and at the lower end of the layer (closest to the waterline) also *Rivularia atra* (Snoeijs and Prentice 1989). In contrast to *Rivularia atra*, *Calothrix scopulorum* does not live permanently submerged, except at a sudden high water level. Benthic diatoms typically occur associated with cyanobacteria (Snoeijs and Wakuru-Murasi 2004). At a low water level, this cyanobacteria-diatom layer dries out to form a distinct white crust, which mainly consists of the silica frustules of the diatoms. Also, dry *Cladophora glomerata* remnants at low water may be white from the silica frustules of its epiphytic diatoms.

11.6.2 Rock pools

Rock pools occur in depressions in the bedrock. These patchy habitats are characterised by a low temporal stability because daily temperature and salinity fluctuations (by evaporation or precipitation) may be large and smaller rock pools easily dry out during sunny days. Rock pools receive

Fig. 11.20 Typical cyanobacteria and lichens in the epilittoral zone of the Baltic Sea. (**a**) At sites exposed to medium to strong wave action, a microbial layer dominated by the cyanobacterium *Calothrix scopulorum* (dark green - blackish) occurs just above the upper *Cladophora glomerata* tufts (light green). (**b**) The *Calothrix scopulorum* layer grows on sun-exposed parts of the shore closest to the waterline while the black tar lichen *Verrucaria maura* also grows close to the waterline, but on shaded parts of the shore (here on the front rock facing the photographer). Orange sea lichen species of the genus *Caloplaca* are found higher up on the shore. (**c**) Close-up of *Caloplaca* sp. Photo: © Pauline Snoeijs-Leijonmalm

water from surf, large waves and/or precipitation and can vary widely in salinity, temperature and nutrient concentrations depending on weather conditions, placement on the shore and pool size. Because of the large environmental variability between rock pools, they are inhabited by many different types of organisms, including microalgae, macroalgae, mosses, macrophytes and associated invertebrates (Fig. 11.21; Ganning 1971; Hällfors 1984). Within a rock pool, biological diversity is generally low (lowest in small pools) and with a simple food web structure.

When conditions deteriorate, rock pool organisms either have to escape from the habitat or survive the unfavourable

Fig. 11.21 Rock pools with different types of vegetation. (**a**) The vascular plant *Callitriche* sp. (**b**) Green filamentous algae. (**c**) Cysts of the green microalga *Haematococcus pluvialis*, which are coloured red by the carotenoid astaxanthin. Photo: (**a**) © Hans Kautsky, (**b**, **c**) © Pauline Snoeijs-Leijonmalm

period in a dormant state to stay alive. For example, the green alga *Haematococcus pluvialis* survives nutrient starvation and desiccation as red cysts filled with the antioxidant carotenoid astaxanthin (Fig. 11.21c) and cladocerans of the genus *Daphnia* can survive as dormant eggs. The production of dormant eggs in *Daphnia* spp. is induced by stimuli associated with deteriorated growth conditions (Pauwels 2007). The two species most commonly found in Baltic Sea rock pools, the green alga *Ulva intestinalis* (syn. *Enteromorpha intestinalis*) and the amphipod *Gammarus duebeni*, both display extremely wide environmental tolerances (Ganning 1971).

11.7 The *Cladophora* belt

11.7.1 *The green alga Cladophora glomerata*

The general lack of perennial vegetation down to a water depth of ~0.5–1 m found everywhere in the Baltic Sea is caused by the irregular water level fluctuations due to weather conditions (*cf.* Sect. 2.3.4) and not primarily by ice scouring. In this zone, the freshwater green filamentous alga *Cladophora glomerata* is an imperative habitat builder (Jansson 1974; Salovius and Kraufvelin 2004). During prolonged periods of high atmospheric pressure, often in combination with off-land wind, the upper ~0.5–1 m of the sublittoral desiccates and the algae in the *Cladophora* belt dry out and die (Fig. 11.22). Ephemeral algae (*Cladophora glomerata* and others) readily recolonise the upper sublittoral zone when the water level increases again or can start growing again from surviving basal cells of the old tufts. Perennial species are slower colonisers and/or are confined to colonisation in a specific season. The low water level occurs irregularly, and the few perennial specimens that may settle close to the water surface will sooner or later desiccate and disappear again.

Cladophora glomerata is widely distributed in the entire Baltic Sea, and globally it is perhaps the most common attached alga in temperate fresh and brackish waters (Zulkifly et al. 2013). Recruitment is mainly through diploid biflagellate zoospores that directly germinate into diploid filaments. Each *Cladophora glomerata* tuft has usually

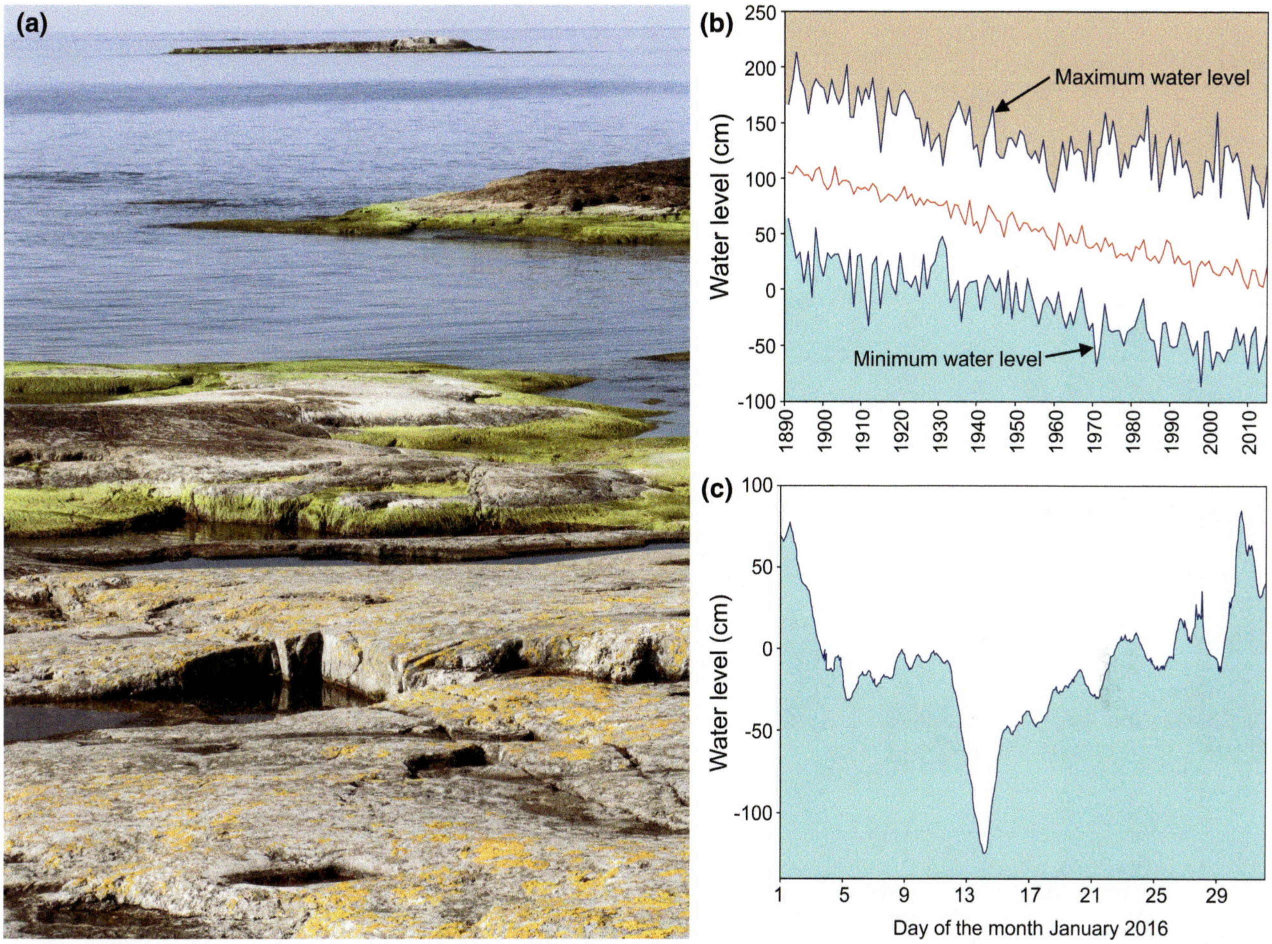

Fig. 11.22 Irregular changes in water level desiccate the algal vegetation in the upper littoral zone of the Baltic Sea. (**a**) A desiccated *Cladophora* belt at low water level at Askö, Baltic Sea proper. (**b**) The yearly minimum and maximum water levels at Ratan (63°59′ N, 20°54′ E, just north of Norra Kvarken) for the years 1891–2015 using the Swedish National RH2000 System. The amplitude between the minimum and maximum water levels is ~1.5 m. The red line is the yearly average water level. The downward trend with time is caused by the land uplift in the area (*cf.* Fig. 2.26b). (**c**) The daily mean water level at Haparanda (65°48′ N, 24°08′ E, northernmost Bothnian Bay). The amplitude between the minimum and maximum water levels is ~2.0 m and can change by more than 1.0 m from day to day. Photo: (**a**) © Hans Kautsky. Figures (**b**) and (**c**) based on data from the Swedish Hydrological and Meteorological Institute, SMHI (http://www.smhi.se)

grown from one zoospore (Figs. 11.20a and 11.23c). These zoospores are released throughout the growing season, from spring to autumn, and enable fast recruitment (Snoeijs and Prentice 1989; Hillebrand et al. 2010). Several generations of new asexual recruitment of *Cladophora glomerata* are often observed in horizontal rows above each other, reflecting past water level changes. In the low-salinity Bothnian Bay, a species related to *Cladophora*, *Aegagropila linnaei*, becomes belt-forming, but also here *Cladophora glomerata* dominates the upper sublittoral zone.

11.7.2 The *Cladophora* belt of the Baltic Sea proper

Like in the Skagerrak, the upper sublittoral zones of the Kattegat and Belt Sea are still inhabited by a *Fucus*-dominated vegetation. However, in the Baltic Sea proper the sublittoral zone between a 0 and 0.5–1 m water depth is dominated by filamentous algae. This zone is highly dynamic due to the irregular water level fluctuations typical of the Baltic Sea. *Cladophora glomerata* starts to colonise here in spring (April-May) and attains its full growth in early summer (June). Usually this species totally dominates the upper 0.5–1 m until September-October (Fig. 11.23), and this is the reason why this dynamic vegetation belt is called the "*Cladophora* belt" (Jansson 1974; Snoeijs 1990b)

The *Cladophora* belt shows a distinct seasonality where overwintering red algae are replaced sequentially by brown and green algae in spring and then finally by a pure green algal belt in summer. This cycle is partly triggered by repeatedly occurring longer low-water periods when this zone is desiccated, killing the standing population. When the water returns, the vegetation is replaced by the next

Fig. 11.23 Different appearance of the *Cladophora* belt of the Baltic Sea. (**a**) Down to a ~0.5–1 m water depth the rock is covered by filamentous algae; the green alga *Cladophora glomerata* dominates during summer often (like in this photograph) accompanied by *Ulva* spp. (**b**) In autumn *Cladophora glomerata* is replaced by the red alga *Ceramium tenuicorne*. (**c**) Exposure to strong wave action may hamper the growth of *Cladophora glomerata* as the tufts partly detach when they grow too large. (**d**) Filamentous algae are preferred food for grazers such as the snail *Theodoxus fluviatilis*, which can denude small patches of the substrate from vegetation. (**e**) *Cladophora glomerata* growing on the upper part of a boulder near the water level just above the *Fucus* belt. (**f**) *Cladophora glomerata* accompanied by the brown alga *Scytosiphon lomentaria*. Photo: © Hans Kautsky

"colour". In summer, other species common in the *Cladophora* belt of the Baltic Sea proper are the brown filamentous algae *Dictyosiphon foeniculaceus* and *Pylaiella littoralis*, different species of the green algal genus *Ulva*, as well as colonies of the cyanobacterium *Rivularia atra* (Fig. 11.24). In late summer-autumn, the red alga *Ceramium tenuicorne* (Fig. 11.24a), sometimes together with the brown algae *Scytosiphon lomentaria* (Fig. 11.23f) and *Spongomorpha aeruginosa*, colonises the upper sublittoral when *Cladophora glomerata* declines.

Ceramium tenuicorne persists through winter and dominates the upper sublittoral until spring. In early spring, it is joined by the green leaf-like alga *Ulvopsis grevillei*, green unbranched filamentous algae such as *Ulothrix* spp. and *Urospora penicilliformis*, and sometimes also by the red unbranched filamentous alga *Bangia atropurpurea* at the waterline. Later in spring, *Acrosiphonia centralis* (green) and *Pylaiella littoralis* (brown), followed by *Cladophora glomerata*, colonise. Also in late spring the brown algae *Dictyosiphon chordaria*, *Eudesme virescens*, *Halosiphon tomentosus* and *Scytosiphon lomentaria* can be found around a ~0.5 m water depth. *Bangia atropurpurea*, *Ulothrix* spp. and *Urospora penicilliformis* may reoccur in autumn.

11.7.3 The *Cladophora* belt of the Gulf of Bothnia

As in the Baltic Sea proper, the upper sublittoral of the Bothnian Sea is dominated by *Cladophora glomerata* from spring to autumn, mixed with *Pylaiella littoralis* (spring), *Ulva intestinalis* (spring-autumn), *Ulva flexuosa* (syn. *Enteromorpha flexuosa*, summer), *Ulva linza* (summer-autumn) and *Ceramium tenuicorne* (autumn). In the Bothnian Sea, *Cladophora glomerata* often extends deeper down than ~0.5 m because, due to ice scouring, the *Fucus* belt usually starts deeper down than in the Baltic Sea proper (Fig. 11.11).

In the Bothnian Bay *Fucus* spp. are absent. *Cladophora glomerata* totally dominates the upper sublittoral in summer, and below this *Aegagropila linnaei* forms a perennial vegetation all the way down to the lower depth limit of the vegetation at the water depth of ~10 m (Fig. 11.25). The green algae are usually completely overgrown by epiphytic diatoms, which gives them a yellowish appearance (Fig. 11.25c, d). An eye-catching species that occurs scattered in this vegetation type is the aquatic moss *Fontinalis dalecarlica* (*cf.* Fig. 11.31e). In the uppermost sublittoral zone, *Cladophora glomerata* is often accompanied by other green filamentous freshwater species such as *Ulothrix zonata*.

Fig. 11.24 In autumn the red alga *Ceramium tenuicorne* dominates the *Cladophora* belt. *Ceramium tenuicorne* is often associated with colonies of the cyanobacterium *Rivularia atra*. (**a**) A boulder with still some *Cladophora glomerata* on the top and *Ceramium tenuicorne* with *Rivularia atra* lower down. (**b**) Colonies of the cyanobacterium *Rivularia atra*, of the diameter up to ~6 mm, on a stone picked up from the *Cladophora* belt. Photo: (**a**) © Hans Kautsky, (**b**) © Pauline Snoeijs-Leijonmalm

11.8 The *Fucus* belt

11.8.1 Four *Fucus* species occur in the Baltic Sea Area

Globally, the bladderwrack *Fucus vesiculosus* is a widely distributed intertidal marine species in the northern hemisphere. In the Baltic Sea it lives permanently submerged and can become almost one metre high (Fig. 11.26). *Fucus vesiculosus* is considered the structurally most important phytobenthic alga in the Baltic Sea because it is the most widely distributed species with a larger thallus size. It is only

Fig. 11.25 The vegetation of the Bothnian Bay. (**a**) An *Aegagropila linnaei*-dominated vegetation covered with diatoms in the northernmost bay of the Baltic Sea (Töreviken) at a ~3 m water depth. (**b**) A biologist working in a sublittoral environment typical of the Bothnian Bay. (**c**) A *Cladophora glomerata*-dominated vegetation covered with diatoms at a ~3 m water depth. (**d**) A close-up of epiphytic diatoms completely covering filamentous algae. Photo: (**a–c**) © Hans Kautsky, (**d**) © Svante Pekkari

absent from the Bothnian Bay and coastal areas elsewhere in the Baltic Sea with salinity below ~4. There are occasional reports of isolated and sparse populations or single individuals of *Fucus vesiculosus* found at salinities down to 2, but they concern the algae that grew within the radius <1 m around outlets of pipes releasing untreated sewage, which locally increased water conductivity (Wærn 1952; Pekkari 1965).

In the Bothnian Sea a "dwarf form" of *Fucus vesiculosus* has long been recognised (Wærn 1952; Kautsky et al. 1992). This form was recently described as a species of its own, *Fucus radicans* (*cf.* Fig. 6.2; Bergström et al. 2005). *Fucus vesiculosus* and *Fucus radicans* occur together in the Bothnian Sea and at the island of Saaremaa (Estonia). However, as *Fucus vesiculosus* decreases in size with decreasing salinity, it may be difficult to distinguish between the two species if they do not grow side by side (Fig. 11.26b).

Two other *Fucus* species occur in the Baltic Sea Area. *Fucus serratus* (*cf.* Fig. 4.27a) is belt-forming from the Skagerrak up to the Gotland Sea (Fig. 11.11). The non-indigenous species *Fucus evanescens* (an Arctic species) has been introduced to the Kattegat and the Belt Sea, where it has been observed to grow close to the water surface. Its occurrence is largely confined to harbour areas (being transported there by ships) and low salinity seems to restrict its further distribution into the Baltic Sea (Wennberg 1992; Wikström et al. 2002).

11.8.2 The key species *Fucus vesiculosus*

Fucus belts (Fig. 11.26) have high biomass and play an important structuring role in the phytobenthic zone of the Baltic Sea, except for the Bothnian Bay. *Fucus vesiculosus* creates habitats for species-rich communities of epiphytes and invertebrates, as well as for the recruitment of fish (Kautsky et al. 1992). In coastal areas of the Baltic Sea, fluctuations in the distribution and abundance of *Fucus vesiculosus* influence the ecosystem at all trophic levels. A species with these characteristics is called a "key species", an "engineering species", or simply a "dominating species".

Fig. 11.26 Different appearance of the *Fucus* belt of the Baltic Sea. (**a**) A dense *Fucus* vegetation. (**b**) Co-occurrence of *Fucus vesiculosus* (to the left) and *Fucus radicans* (to the right). (**c**) *Fucus vesiculosus* with *Dictyosiphon foeniculaceus* as an epiphyte. (**a**, **b**) Höga Kusten, Bothnian Sea, (**c**) Askö, Baltic Sea proper. Photo: © Hans Kautsky

All of these terms indicate the importance of *Fucus vesiculosus* for the structure of the ecosystem. *Fucus vesiculosus* is not a "keystone species", which by definition is a species that, relative to its abundance, has a disproportionately large effect on community structure, usually a predator (Paine 1966, 1995).

Although the habitat requirements of *Fucus vesiculosus* with respect to salinity are fulfilled almost everywhere in the Baltic Sea, rich *Fucus* belts grow only on stable substrates at sites not exposed to strong wave action. *Fucus vesiculosus* has air bladders that keep the thalli upright in the water towards the light. With increasing wave exposure, the air bladders decrease in size and are absent at the most exposed sites where *Fucus vesiculosus* can grow. At high wave exposure there is no need to keep the thalli upright and bladders increase the drag force of waves with the risk of damaging and ripping off the algae. Deeper down in the sublittoral zone at exposed sites, where the forces of the waves are smaller, bladders still occur. Furthermore, detached *Fucus vesiculosus* is kept floating by means of the air bladders, and this is thought an advantage for long-distance dispersal of the species (van den Hoek 1987).

11.8.3 The *Fucus-Ascophyllum* belt of the Kattegat

The Atlantic intertidal species *Fucus vesiculosus* lives permanently submerged already in the Kattegat. It grows just below the water surface, with *Fucus spiralis* above it and *Ascophyllum nodosum* and *Fucus serratus* below it. These four perennial species form the *Fucus-Ascophyllum* belt in the Kattegat, together with many annual and pseudo-annual species that have seasonal abundance maxima in this belt. For example, in the end of February *Dumontia contorta* and *Ulvopsis grevillei* occur just beneath the water surface. Towards summer they give place to *Nemalion multifidum* in July and August, followed by *Polysiphonia fibrillosa* from October to February. *Fucus serratus* stretches from a 0.5–1 m down to ~4–5 m water depth, often accompanied by *Chondrus crispus* and *Chorda filum*. Filamentous algae such as *Ceramium* spp., *Cladophora* spp., *Dictyosiphon foeniculaceus*, *Ectocarpus* spp., *Pylaiella littoralis* and *Ulva* spp. start to grow in spring, epiphytic on *Fucus* or directly on the rock.

If an ice cover has been present in the Kattegat during the winter, the upper perennial algae may have been ripped off and the bare rock left is often colonised by *Osmundea truncata* and *Scytosiphon lomentaria* in spring. When they die off in May, regeneration of *Fucus* and *Ascophyllum nodosum* starts by colonisation of germlings in the gaps where the perennial vegetation was damaged by the ice.

11.8.4 The *Fucus* belt of the Baltic Sea proper

In the southern Baltic Sea proper, the sublittoral zone between a 0.5–1 m and 4–5 m water depth is characterised by belt-forming *Fucus vesiculosus* (Fig. 11.26), except for sheltered sites and sites heavily exposed to wave action. Up to a line between just north of the island of Öland (Sweden)

and the southern part of the island of Gotland (Sweden), the *Fucus* belts also contain *Fucus serratus* (Fig. 11.11). In the northern Baltic Sea proper *Fucus serratus* does not occur anymore and *Fucus vesiculosus* is the only *Fucus* species left here.

The lower depth limit of *Fucus vesiculosus* in the northern Baltic Sea proper varies between 4 and 9 m, but it has been found attached down to a ~14 m water depth. In archipelago areas, the lower depth limit extends deeper towards more exposed and cleaner sites in the outer archipelago. Typical epiphytes on *Fucus vesiculosus* are filamentous algae such as *Ceramium tenuicorne*, *Elachista fucicola* (obligate on *Fucus* spp.), *Pylaiella littoralis* (especially in spring and early summer), *Ectocarpus siliculosus* (in summer), *Dictyosiphon foeniculaceus* (in late summer).

The *Fucus vesiculosus* thalli support a high diversity of animals living on and in between them. Typical attached faunal species are *Amphibalanus improvisus* (Fig. 11.18b, c), *Einhornia crustulenta* (Fig. 11.18a) and *Mytilus trossulus* (*cf.* Fig. 4.29). High abundances of gammarids and isopods are also common, as well as grazing snails such as *Theodoxus fluviatilis*, *Ecrobia ventrosa* (syn. *Hydrobia ventrosa*) and *Peringia ulvae* (syn. *Hydrobia ulvae*). Below the *Fucus vesiculosus* canopy, an understory bush vegetation of *Ceramium tenuicorne*, *Cladophora rupestris*, *Furcellaria lumbricalis*, *Polysiphonia fucoides* and *Stictyosiphon tortilis* is found. *Chorda filum* usually occurs as single threads, but may occasionally form dense bundles on less stable substrates such as small stones and gravel (Fig. 11.8c).

11.8.5 The *Fucus* belt of the Bothnian Sea

In the low salinity (~5) of the Bothnian Sea, *Fucus vesiculosus* continues to grow in dense stands until close to its northern limit at Norra Kvarken (at salinity ~4). *Fucus vesiculosus* is in the Bothnian Sea accompanied by *Fucus radicans* (Fig. 11.26b), which has its northern limit of occurrence only ~20 km north of that of *Fucus vesiculosus*. Furthermore, detached, up to football-sized floating bundles of *Fucus* can cover large areas of the shallow sandy sublittoral zone of the western Bothnian Sea.

In the Bothnian Sea, the *Fucus* belt usually starts deeper down than in the Baltic Sea proper due to ice scouring. However, in protected places, *e.g.* behind large boulders, the two *Fucus* species can be found already at a ~0.5 m water depth. Like in the *Cladophora* belt, the aquatic moss *Fontinalis dalecarlica* (*cf.* Fig. 11.31e) is a conspicuous species in the Bothnian Sea *Fucus* belt. The distribution of this moss in the Baltic Sea stretches from the Gräsö area in the southern Bothnian Sea up to the northernmost Bothnian Bay. In the Södra Kvarken area, *Fontinalis dalecarlica* can be used as an indicator of Bothnian Sea water, *e.g.* areas of the Åland Sea that are affected by Bothnian Sea water may be classified as belonging to "the *Fontinalis* district" due to the occurrence of this moss (Wærn 1952).

Other species in the *Fucus* belt of the Bothnian Sea are the green algae *Aegagropila linnaei* and *Cladophora glomerata*, the brown algae *Chorda filum* and *Pylaiella littoralis*, and the red algae *Ceramium tenuicorne*, *Coccotylus truncatus*, *Furcellaria lumbricalis*, *Phyllophora pseudoceranoides* and *Polysiphonia fucoides*. At the lower end of the *Fucus* belt, in the whiplash zone where the *Fucus* vegetation is less dense, *Cladophora rupestris* and *Rhodochorton purpureum* are frequently found, together with the belt-forming species *Battersia arctica*.

11.9 The red algal belt

11.9.1 The red alga *Furcellaria lumbricalis*

The marine red alga *Furcellaria lumbricalis*, with a rather stable, leathery thallus (Fig. 11.27), is the largest red alga with a wide distribution in the Baltic Sea. It occurs up to Norra Kvarken at the islands of Holmöarna (Sweden) and the Vaasa archipelago (Finland). *Furcellaria lumbricalis* is belt-forming in two forms, either attached to a stable hard substrate in the lower part of phytobenthic zone (Kautsky 1995a) or in loose-lying aggregates at a 4–10 m water depth (Martin et al. 2006a; Bučas et al. 2009).

Together with other red algal species, *Furcellaria lumbricalis* forms the "red algal belt" of the Baltic Sea. Its morphology is governed by salinity and light and the maximum size of the *Furcellaria lumbricalis* tufts in the Baltic Sea is ~12 cm in diameter, which is much smaller than on fully marine coasts. At its lower depth limit (stressed by low irradiance), and at its northern limit of occurrence in the Baltic Sea (stressed by low salinity), *Furcellaria lumbricalis* tufts are only a few cm in diameter, almost black in colour and with thin branches (Box 11.7).

11.9.2 The red algal belt of the Baltic Sea

Many marine red algae cannot survive the low salinity of the Baltic Sea. Species common at Atlantic coasts, such as *Corallina officinalis*, *Dilsea carnosa*, *Odonthalia dentata*, *Osmundea truncata*, *Phyllophora crispa*, *Polysiphonia brodiei*, *Porphyra umbilicalis*, *Pterothamnion plumula* and *Scagelothamnion pusillum* (syn. *Antithamnion boreale*) are still found in the lower sublittoral of the Belt Sea, but do not enter the Baltic Sea (HELCOM 2012). Some others, *e.g.* *Brongniartella byssoides*, *Palmaria palmata*, *Plumaria plumosa* and *Spermothamnion repens* occur in the Arkona Sea but do not penetrate further into the Baltic Sea.

Fig. 11.27 Different appearance of the red algal belt of the Baltic Sea. (**a**) *Furcellaria lumbricalis* with *Ceramium tenuicorne* as an epiphyte, growing together with *Polysiphonia fucoides* and the animals *Mytilus trossulus*, *Amphibalanus improvisus* and *Einhornia crustulenta* at a ~4 m water depth. (**b**) *Furcellaria lumbricalis*, *Polysiphonia fucoides* and *Mytilus trossulus* at a ~7 m water depth; the green algae are patches of *Spirogyra* sp. (**c**) A loose-lying *Mytilus trossulus* community with intertwined *Furcellaria lumbricalis*, *Coccotylus truncatus* and other algae at a ~10 m water depth. (**a–c**) Askö, Baltic Sea proper. Photo: © Hans Kautsky

Consequently, the distinct "red algal belt" of the Baltic Sea proper (Fig. 11.27) consists of only nine commonly occurring species. The dominant species, *i.e.* that with the highest biomass, is usually *Furcellaria lumbricalis*. It is accompanied by *Ceramium tenuicorne*, *Ceramium virgatum* (syn. *Ceramium rubrum*), *Coccotylus truncatus*, *Phyllophora pseudoceranoides*, *Polysiphonia fibrillosa*, *Polysiphonia fucoides* and *Rhodomela confervoides*, and rarely also by *Ahnfeltia plicata*. The red algae successively disappear when the *Battersia arctica* belt takes over in the northern Baltic Sea proper.

In the Bothnian Sea some red algae, *e.g. Furcellaria lumbricalis* and *Coccotylus truncatus*, may still grow in a narrow belt below the *Fucus* belt. However, they usually occur here in the lower part of the *Fucus* belt or the upper part of the *Battersia* belt and not in a red algal belt of their own. All red algal species have smaller thalli in the Bothnian Sea than in the Baltic Sea proper.

The only frequently occurring marine species left in the Bothnian Bay is *Ceramium tenuicorne*, but it has to be searched for since it is small and often overgrown by epiphytic diatoms. The marine crustose alga *Hildenbrandia rubra*, which still occurs in the Bothnian Sea, is in the Bothnian Bay replaced by its freshwater relative *Hildenbrandia rivularis* (Pekkari 1965). The latter species occurs on hard substrates from the low water line down to the water depth of ~10 m, and it may, especially in deeper water, become well developed and up to 2–3 dm^2 large (Forsberg and Pekkari 1999).

11.9.3 Red algae and *Mytilus*

The red algal belt of the Baltic Sea proper is often associated with *Mytilus trossulus*. Aggregates of *Furcellaria lumbricalis* are attached to the blue mussel shells or entangled in the byssus treads. Other red algae that are habitually associated with these *Furcellaria-Mytilus* aggregates (Fig. 11.27c) are *Coccotylus truncatus*, *Phyllophora pseudoceranoides* and *Rhodomela confervoides*. This community type forms distinct high-diversity patches on flat, soft substrates. For example, in the northern Baltic Sea proper a total of 24 associated macrofaunal species were identified in such patches, 11 of which were not present on the bare sediment outside the patches (Norling and Kautsky 2008).

In the eastern Baltic Sea proper, small areas densely vegetated with red algae are valuable marine biodiversity hotspots surrounded by vast areas with sandy substrate. The Estonian coast hosts a large loose-lying *Furcellaria lumbricalis*-dominated community, which is commercially harvested (Box 11.4). *Furcellaria lumbricalis*, mixed with *Coccotylus truncatus*, covers up to 120 km^2 of the seafloor with more than 140 kilotonnes of wet weight biomass in Kassari Bay between the islands of Saaremaa and Hiiumaa (Martin et al. 2006a, b). Also along the coasts of Latvia and Lithuania a loose-lying *Furcellaria lumbricalis* dominated vegetation is abundant (Bučas et al. 2009).

Box 11.7: Norra Kvarken: the northern distributional limit of marine algae

In the Bothnian Sea with salinity ~5, most marine algae that occur in the northern Baltic Sea proper are still found, but their thallus size tends to be smaller. Along the Bothnian Sea coast some marine species reach the northern limit of their distribution. However, the bulk disappearance of marine species occurs in the Norra Kvarken area at salinity ~4. Within a few tens of km almost all marine species are gone and freshwater species take over completely (Box Fig. 11.11).

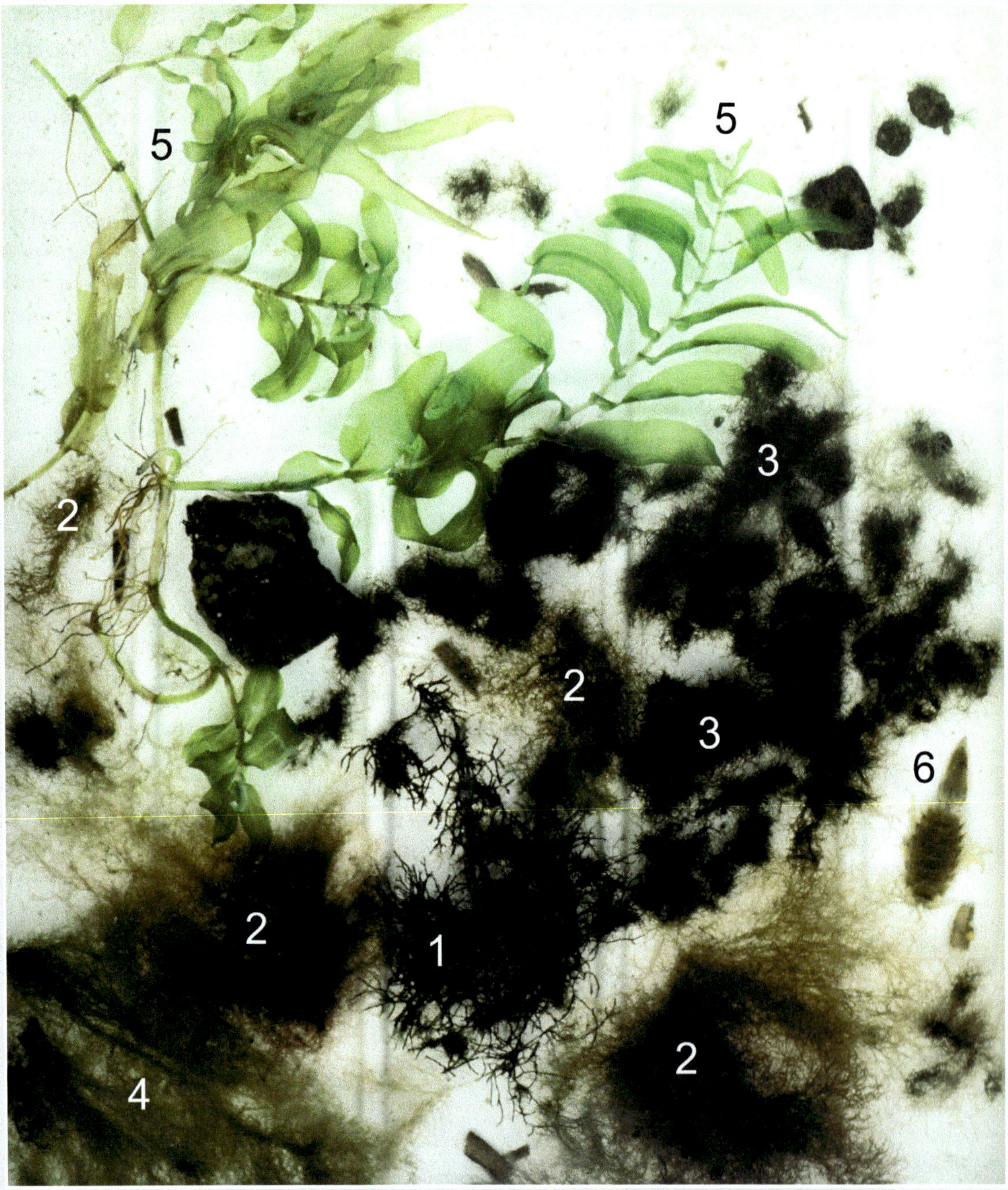

Box Fig. 11.11 Marine macroalgae reach the northern limit of their distribution in the Baltic Sea at the islands of Holmöarna in the Norra Kvarken area as shown by a sample taken in this area including *Furcellaria lumbricalis* (1) with entangled *Coccotylus truncatus*, *Ceramium tenuicorne* (2) and *Battersia arctica* (3). They occur together with the freshwater green alga *Cladophora glomerata* (4), the freshwater vascular plant *Potamogeton perfoliatus* (5) and the isopod *Saduria entomon* (6). Photo: © Hans Kautsky

11.10 The *Battersia* belt

11.10.1 The brown alga *Battersia arctica*

The marine brown alga *Battersia arctica* is an Arctic species that has not been reported on coasts south of northern Norway, except in the Baltic Sea Area. *Battersia* occurs from the Kattegat to the Norra Kvarken area, down to the salinity of ~3–4 (Wærn 1952, 1965). Tufts of this slow-growing perennial species are usually 2–5 cm high, but may grow up to 8 cm (Fig. 11.28a).

Battersia arctica has a rather stiff thallus, which is probably the reason why this alga is able to cope with moderate amounts of sedimentation. Especially the deepest-growing *Battersia arctica* specimens may be periodically almost completely covered with sediment with only the upper shoots protruding above the sediment surface (Eriksson et al. 1998).

11.10.2 The *Battersia* belt of the Baltic Sea

The *Battersia* belt, completely dominated by *Battersia arctica*, forms the deep-water vegetation on hard substrates below the red algal belt in the Baltic Sea proper and immediately below the *Fucus* belt in the Bothnian Sea (Fig. 11.28). Only some crustose algae, *e.g. Hildenbrandia rubra* (red) and *Pseudolithoderma rosenvingei* (brown), can penetrate deeper down in the sublittoral zone than *Battersia arctica*.

In the Bothnian Sea, a vegetation dominated by *Battersia arctica* may occur below ~10 m when substrate is available (Wærn 1952, 1965). For example, in the Öregrund archipelago (southern Bothnian Sea), *Battersia arctica* is the most common alga, covering more than half of the phytobenthic zone (Wærn 1945, 1952). However, population size seems to vary over time (Kautsky et al. 1986, Eriksson et al. 1998).

11.11 *Zostera* meadows

11.11.1 The distribution of *Zostera marina*

The only truly marine vascular plant with a wide distribution in the Baltic Sea is the seagrass *Zostera marina* (common eelgrass). This is the most common marine vascular plant in the cooler coastal waters of the northern hemisphere. The plants are anchored by roots and rhizomes in soft or sandy substrates. Contrary to the more marine areas in the Skagerrak and Kattegat, where it mainly grows on soft substrates, *Zostera marina* seems to prefer sand and sand mixed with gravel in the Baltic Sea. A smaller relative of

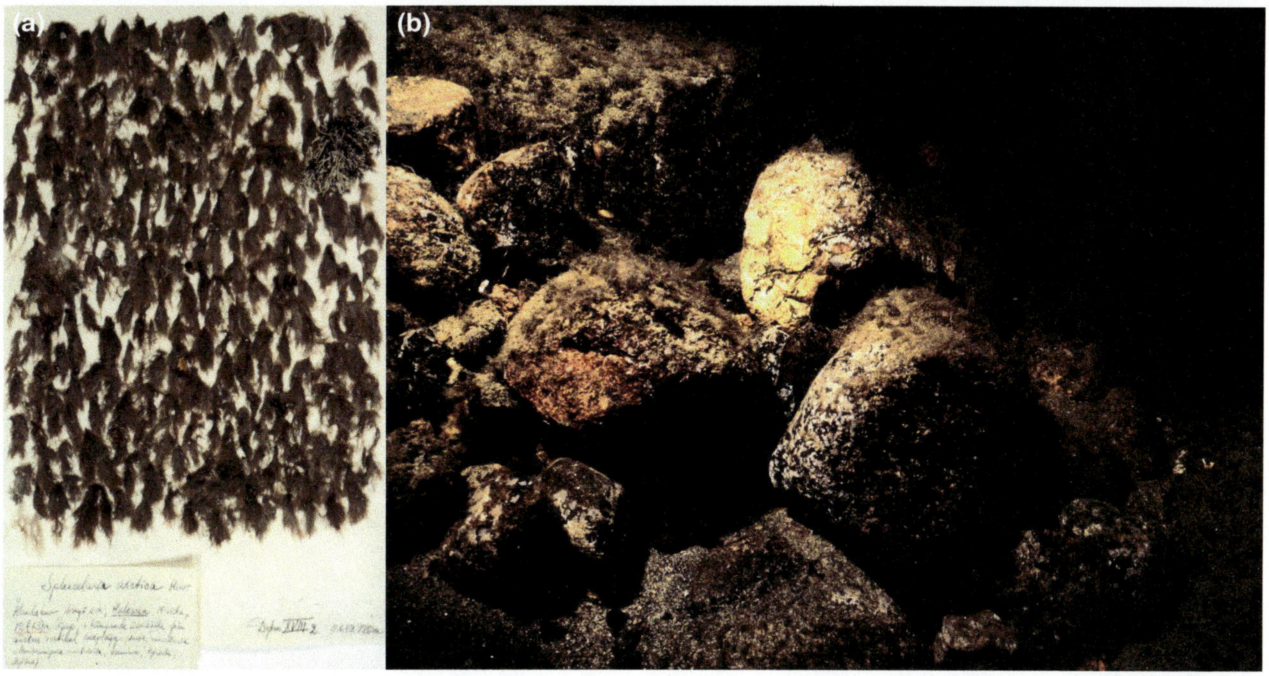

Fig. 11.28 The *Battersia* belt of the Baltic Sea. (**a**) A herbarium sheet of material sampled from an almost vertical bedrock cliff at a ~12–13 m water depth during a dive by Mats Wærn at Halsaren (Åland Sea) on 11 June 1943. The dominant species on the herbarium sheet is *Battersia arctica* (syn. *Sphacelaria arctica*), accompanied by *Ceramium tenuicorne*, *Furcellaria lumbricalis* (with epiphytic *Einhornia crustulenta*), *Pylaiella littoralis* and *Mytilus trossulus*. (**b**) *Battersia arctica* growing on boulders at a ~15 m water depth at Höga Kusten in the Bothnian Sea. Photo: (**a**) © Marianne Hielm Pedersén, (**b**) © Hans Kautsky

Zostera marina, the dwarf eelgrass *Zostera noltei*, which is widely distributed along the European Atlantic coasts, the Mediterranean Sea and the Black Sea, is rare in the Baltic Sea Area and restricted to the Belt Sea west of the Darß sill.

The range of *Zostera marina* in the Baltic Sea Area extends from the Kattegat to a line from the northern Stockholm archipelago to Rauma in southwestern Finland, with its main distribution along the ice-marginal Salpausselkä end moraines (Boström et al. 2006). The lower limit of its salinity distribution in the Baltic Sea is ~5 (Boström et al. 2014). The total areal cover of *Zostera marina* in the Baltic Sea Area is estimated at a minimum of 1,222 km^2, ~74 % of which is situated in the Kattegat and the Belt Sea and ~26 % in the Baltic Sea proper (Boström et al. 2014). This estimate is largely influenced by Danish cover data, which range from 673 to 1,345 km^2, assuming the present covers constituting 10 % to 20 % of the historical distribution, respectively. Thus, when using the more optimistic estimate for Denmark, the total eelgrass area in the Baltic Sea Area would be almost 1,900 km^2.

The most luxuriant *Zostera* meadows in the Baltic Sea Area, with plants over half a metre high, are found in the Belt Sea, as well as on the west coast of the island of Öland and around the island of Gotland in the Baltic Sea proper. In these areas sandy substrates prevail, and almost pure stands of *Zostera marina* may form extensive underwater meadows of several km^2 in size (Fig. 11.29a, b).

Shoot density at the time of maximum above-ground biomass is highly variable (72–3,948 shoots m^{-2}) but does not show significant differences between the different regions in the Baltic Sea Area (Boström et al. 2014). Although there is no significant relationship between salinity and shoot density within the salinity range 5–26, the shoot density is generally highest in shallow water and decreases exponentially with depth. However, the *Zostera* meadows in the northern Baltic Sea proper tend to develop smaller shoots, and the above-ground biomass is lower (range 37–73 g dry weight m^{-2}) than in the Skagerrak and the Kattegat - Belt Sea area (average 281 g dry weight m^{-2}), where shoots are generally larger (Boström et al. 2014).

Fig. 11.29 *Zostera* meadows in the Baltic Sea. (**a**) A biologist (the late Bengt-Owe Jansson) working in a dense *Zostera* meadow in the Baltic Sea proper at the northeastern coast of the island of Gotland (Baltic Sea proper) at a ~4 m water depth. (**b**) *Zostera marina* on a substrate of sand, gravel and mollusc shells. (**c**) *Zostera marina* in a vegetation on mixed substrates, growing together with the brown alga *Pylaiella littoralis* and the hydrozoan *Cordylophora caspia*. (**d**) *Zostera marina* growing together with other vascular plants at the northern limit of its extension a few km north of Stockholm (northern Baltic Sea proper). Photo: © Hans Kautsky

11.11.2 The key species *Zostera marina*

Similarly to *Fucus vesiculosus* on hard substrates, *Zostera marina* plays critical structural and functional roles in the Baltic Sea ecosystem on sandy substrates, and is therefore considered a key species within its distributional range. On the otherwise species-poor sandy substrates, *Zostera marina* communities have high biomass with a positive effect on biodiversity by creating habitats for epiphytes, invertebrates and fish reproduction. In addition, the complex rhizome and root system of *Zostera marina* stabilises the substrate and facilitates the existence of diverse infaunal communities (Boström et al. 2002).

The higher-salinity areas in the Kattegat and the Belt Sea support monospecific meadows, with higher productivity (3–10 g dry weight m^{-2} d^{-1}) and higher genetic connectivity. In the Baltic Sea proper, *Zostera marina* productivity is generally low (<2 g dry weight m^{-2} d^{-1}) and meadows are isolated and genetically impoverished (Boström et al. 2014). Close to the limit of its salinity tolerance in the Baltic Sea, *Zostera marina* mainly grows vegetatively; flowering shoots are rare, and seeds do not ripen during the season. As a consequence, a *Zostera* meadow may consist of a single genotype (individual) and these mega-clones can reach an age of >1,000 years (Reusch et al. 1999).

A large proportion of the total faunal biomass associated with the *Zostera marina* vegetation in the Skagerrak and the northern Belt Sea consists of fish, ~70 % and ~40 %, respectively (Fig. 11.30). In the southern Belt Sea, fish make up as little as a few % at most of the total faunal biomass, and in the Baltic Sea even less. Omnivorous crustaceans decrease in relative abundance along the large-scale Baltic Sea gradient northward and are replaced by grazing crustaceans and gastropods. This dominance of grazers within the communities in the northern Baltic Sea proper appears to constitute an important buffer against epiphytic overgrowth of the *Zostera marina* plants.

11.11.3 *Zostera* meadows support high diversity

In the Baltic Sea proper, the epifaunal diversity in the *Zostera* meadows seems to be as high as in the *Fucus*-dominated vegetation (Kautsky and van der Maarel 1990). However, as a diverse infaunal community thrives in the organic-rich sediments among the seagrass roots, the total invertebrate diversity of *Zostera* meadows is usually higher than on rocky coasts. Except for pipefish (*Nerophis ophidion* and *Sygnathus typhle*), no associated species seem to be specific for the *Zostera* meadows in the Baltic Sea proper, probably because most species in the Baltic Sea proper are generalists rather than specialists.

The highest diversity of the *Zostera marina* communities in the Baltic Sea proper is found in areas with mixed substrates. The plants are often rooted next to patches of hard substrate with attached algae (Fig. 11.29c) and may provide a secondary hard substrate for epilithic species that already

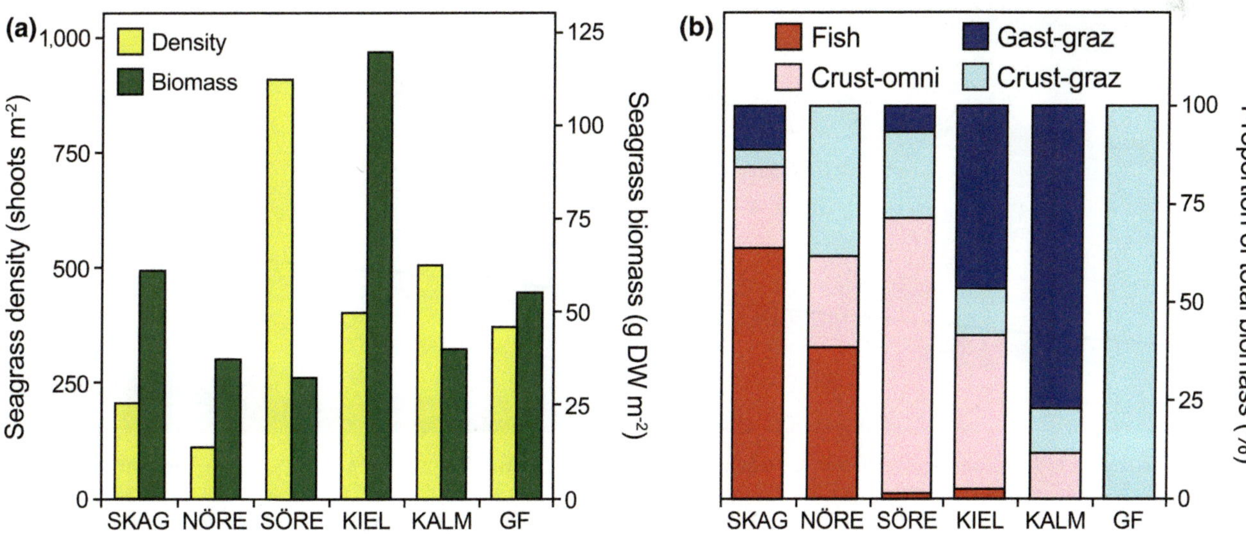

Fig. 11.30 The structure of *Zostera* meadows and their main epifaunal groups in six regions of the Baltic Sea Area. (**a**) Habitat complexity (expressed as shoot density) and the biomass of *Zostera marina* from which the epifauna was sampled. Note that this graph is not representative for shoot density or biomass differences between the six regions. (**b**) Relative biomass contribution (% of wet weight) of different groups of macrofauna to the total biomass of epifauna and fish. Crust-omni = omnivorous crustaceans, Crust-graz = grazing crustaceans, Gast-graz = grazing gastropods. SKAG = Skagerrak, NÖRE = northern Öresund (transition zone), SÖRE = southern Öresund (transition zone), KIEL = Kiel Bay (transition zone), KALM = Kalmarsund (Baltic Sea proper), GF = Gulf of Finland. Figure modified from Boström et al. (2014)

Fig. 11.31 Examples of freshwater species in the vegetation of the Baltic Sea. (**a**) The charophyte *Chara aspera*. (**b**) The charophyte *Chara tomentosa*. (**c**) The vascular plant *Stuckenia pectinata*. (**d**) *Najas marina* with epiphytic green colonies of the colonial ciliate *Ophrydium versatile*, which congregates in large mucilaginous masses. (**e**) The epilithic moss *Fontinalis dalecarlica* growing together with *Fucus*. (**f**) The vascular plant *Nuphar lutea* growing in the mouth of the Råneå river in the Bothnian Bay. The water is yellowish from high concentrations of coloured dissolved organic matter (CDOM). Photo: © Hans Kautsky

occur in the area (Boström and Bonsdorff 1997; Möller et al. 2014). Additionally, in the Baltic Sea proper *Zostera marina* often grows in mixed stands with other vascular plants such as *Myriophyllum spicatum*, *Potamogeton perfoliatus*, *Ruppia maritima*, *Stuckenia pectinata* and *Zannichellia palustris*, which also increases community diversity. The growth of *Zostera marina* seems to be positively influenced by these other vascular plants as it produces more shoots and biomass in mixed stands (Salo et al. 2009; Gustafsson and Boström 2010, 2013; Fig. 11.29d).

11.12 Freshwater influences in the vegetation

11.12.1 Vascular plants in the Baltic Sea proper

A major difference in the vegetation on mixed, sandy and soft substrates between the Kattegat area and the Baltic Sea proper is the higher abundance of freshwater vascular plant and charophyte species in the latter (Fig. 11.31). The vegetation of the Baltic Sea proper often consists of mixed stands of *Zostera marina*, the brackish-water species *Ruppia cirrhosa*, and freshwater vascular plants such as *Myriophyllum spicatum*, *Potamogeton perfoliatus*, *Stuckenia pectinata* and *Zannichellia palustris* (Luther 1951a, b; Wallentinus 1979; Kautsky 1989). In the Kattegat-Belt Sea area these freshwater vascular plants are restricted to bays with a strong freshwater influence.

11.12.2 Vascular plants in the Bothnian Sea

The dominating coast type in the Bothnian Sea, except for the rocky coasts of Höga Kusten (Sweden), is a flat coast with mixed, sandy and soft substrates. Flushed till from the last glaciation characterises the landscape below the sea surface (seascape) in the Bothnian Sea and this type of substrate supports abundant *Myriophyllum spicatum*, *Potamogeton perfoliatus*, *Stuckenia pectinata*, and *Zannichellia palustris*. *Zostera marina* and *Ruppia cirrhosa* are absent, but the brackish-water species *Ruppia maritima* (which despite its Latin name is not a marine species) still occurs in the southern half of the Bothnian Sea, up to Höga Kusten. A number of additional freshwater vascular plants emerge in the Bothnian Sea, *e.g. Callitriche* spp., *Isoetes echinospora*, *Subularia aquatica*, as well as the submerged aquatic bryophytes *Drepanocladus* spp. and *Fontinalis dalecarlica* (Fig. 11.31e).

11.12.3 Vascular plants in the Bothnian Bay

Exposed bedrock is rare in the Bothnian Bay and mixed, sandy and soft substrates dominate the shallow coastline. The long winter period makes it difficult for perennial species to survive. However, sites sheltered from ice scouring may host the luxuriant growth of vascular plants, with *Potamogeton perfoliatus* and *Stuckenia pectinata* as dominants, accompanied by *Myriophyllum spicatum*, *Ranunculus palustris* and others. Some additional freshwater vascular plants that occur in the phytobenthic vegetation here, but not in the Bothnian Sea, are *Limosella aquatica* and the red-listed species *Alisma wahlenbergii* (HELCOM 2013a), as well as some mosses (*Fissidens fontanus*, *Fontinalis* spp.). *Limosella aquatica* and *Alisma wahlenbergii* seem to be even more abundant in the Bothnian Bay than in freshwater, perhaps due to lower competition.

Obligate freshwater species such as *Nuphar lutea* (Fig. 11.31f) occur only in almost pure freshwater in river mouths. The more wave-exposed coasts in the Bothnian Bay are often characterised by extensive sand fields, which are mainly inhabited by benthic diatoms and meiofauna. On boulders and stones, short tufts of *Aegagropila linnaei*, benthic diatoms and freshwater bivalves (*Anodonta* sp., *Pisidium* spp., *Sphaerium* sp.) and freshwater snails such as *Radix balthica* (syn. *Radix ovata*), *Radix labiata* (syn. *Radix peregra*), *Theodoxus fluviatilis* and *Valvata piscinalis*, may occur.

11.12.4 Charophytes

Charophytes may occur in high abundances and cover large areas on sandy and soft bottoms in shallow sheltered bays (Fig. 11.31a, b). The distribution of charophytes in the Baltic Sea is mainly governed by salinity, water depth, sediment type and exposure to wave action (Schubert and Blindow 2003; Torn and Martin 2004). The most widely distributed *Chara* species in the Baltic Sea are *Chara aspera* and *Chara baltica* (Torn 2008), but *Chara tomentosa* may also form high-biomass nearly monospecific aggregations locally in the northeastern Baltic Sea proper. Here the primary productivity of *Chara tomentosa* can be as high as ~30 μmol O_2 (kg dry weight)$^{-1}$ s^{-1} in July (Torn et al. 2006b), which is similar to that of *Cladophora glomerata* and *Fucus vesiculosus* (Box 11.6).

In recent decades, the number of species, distribution area and biomass of charophytes have significantly declined in the Baltic Sea. Declines have been described for several

Fig. 11.32 *Mytilus trossulus* at Askö, Baltic Sea proper. (**a**) A dense patch of *Mytilus trossulus* in the *Fucus* belt, growing together with *Ceramium tenuicorne* and *Einhornia crustulenta* at a ~2 m water depth. (**b**) *Mytilus trossulus* growing together with short tufts of *Ceramium tenuicorne* at a ~2 m water depth at an exposed site where *Mytilus trossulus* can only stay attached in the crevices of the rock. (**c**) *Mytilus trossulus* on bedrock cliffs at a ~4 m water depth. Here *Mytilus trossulus* can attach to vertical surfaces only because the horizontal surfaces hold too much silt for the byssus to anchor. (**d**) Loose-lying *Mytilus trossulus* aggregates on sandy-soft substrate at a ~7 m water depth. (**e**) *Mytilus trossulus* attaches to any hard substrate, including beer cans. (**f**) *Mytilus trossulus* covered by an algal mat of decaying *Spirogyra* sp. at a ~4 m water depth. The decay process creates hypoxic conditions, but *Mytilus trossulus* individuals pump oxygenated seawater to the seafloor through the craters in the algal mat. A bright green branch of *Callitriche hermaphroditica* grows on the algal mat. Photo: © Hans Kautsky

subregions of the Baltic Sea, *e.g.* the coastal waters of Schleswig-Holstein (Germany), the Swedish west coast and the Hanko Peninsula in southwestern Finland (Blindow 2000; Schubert and Blindow 2003; Munsterhjelm 2005). Such declines are related to increased anthropogenic pressure (Torn 2008), and they may occur in other (less investigated) areas of the Baltic Sea as well. Charophytes are especially sensitive to mechanical stress (habitat destruction) as well as to eutrophication (Yousef et al. 2001; Schubert and Blindow 2003; Munsterhjelm 2005).

11.13 The role of the fauna in the phytobenthic zone

11.13.1 The key species *Mytilus trossulus*

The blue mussel *Mytilus trossulus* is a highly abundant habitat-forming key species in the Baltic Sea proper from just below the water surface down to a ~20 m water depth, but it can live at greater depth as well (Fig. 11.32). A biomass of ~300 g dry weight m^{-2} is common along the western coast of the Baltic Sea proper where 80–90 % of the total animal biomass in the sublittoral zone consists of *Mytilus trossulus* (Jansson and Kautsky 1977). In the Baltic Sea it is found up to the northernmost part of the Bothnian Sea with salinity ~4, but at this low salinity it seldom reaches high biomass. In the Gulf of Finland *Mytilus trossulus* extends to the central parts, down to salinity ~4.5 (Westerbom et al. 2002).

The energy demand for coping with osmotic stress makes the Baltic Sea blue mussel *Mytilus trossulus* smaller in body size (*cf.* Fig. 4.5b), with thinner shells (*i.e.* more easily predated) and byssus weaker (*i.e.* easily detached from the substrate) than its North Sea counterpart (Tedengren and Kautsky 1986; Tedengren et al. 1990). In the Åland Sea the shells are so thin that the mussel is an easy prey for fish and waterbirds. Therefore, *Mytilus trossulus* is usually rare in the entire Bothnian Sea, and scattered individuals are mainly found sheltered under boulders and stones where predators cannot reach them.

The high biomass of *Mytilus trossulus* in the Baltic Sea proper is also an important food source for fish and waterbirds, but here predators usually have little effect on the *Mytilus trossulus* population (Kautsky 1981). This is in contrast to more marine habitats where *Mytilus edulis* is constantly preyed upon by benthic predators (*cf.* Fig. 4.32) and the high biomass of the mussel is restricted to places where these benthic predators cannot reach them, *e.g.* high up on piers exposed to strong wave action.

Mytilus trossulus is a filter feeder that consumes both plankton and suspensions of benthic microorganisms directly, and is an important link between the benthic and pelagic components of the Baltic Sea ecosystem in channelling the matter and energy flows. In nearly all habitat types, pelagic feeding by *Mytilus trossulus* exceeds that of benthic feeding (Lauringson et al. 2009). One blue mussel individual is estimated to filter 5 L of seawater per hour and, theoretically, the *Mytilus trossulus* population of the entire Baltic Sea could recirculate the entire water column four times per year (Kautsky and Wallentinus 1980; Kautsky and Evans 1987). The recirculation of nitrogen and phosphorus from pelagic production by *Mytilus trossulus* is sufficient to support the phytobenthic system and also to export back nutrients to the pelagic system.

11.13.2 Substrate, wave exposure and the distribution of *Mytilus*

The lower depth limit of the blue mussel *Mytilus trossulus* in the sublittoral zone is usually set by the lack of suitable substrate (Fig. 11.32c). Close to the water surface the mussel attaches to any hard substrate, including *Fucus vesiculosus*. Where there are fewer algae, the competition for space between algae and the bivalve is lower and *Mytilus trossulus* can dominate the substrate. Therefore, the mussel may form belts that cover 100 % of the hard substrate below the algal belts or on vertical rock surfaces. On gently sloping seafloors, *Mytilus trossulus* is commonly found in clusters lying scattered on the substrate, which may be everything from gravel to soft substrate, often associated with red algae (Figs. 11.27c and 11.32d)

Water circulation is of major importance for food supply to *Mytilus trossulus*, *e.g.* an extremely high biomass of >3.5 kg dry weight m^{-2} (including shells) has been recorded at blue mussel beds on the strongly exposed offshore Sandö bank (Sweden) at an ~18 m water depth (Kautsky 1984). Along a wave-exposure gradient in the Archipelago Sea (Finland), the lowest densities and biomass of *Mytilus trossulus* were found in the innermost archipelago (Vuorinen et al. 2002). Most of the populations in the inner archipelago consisted of small, 4–6 mm long individuals. In the middle and outer archipelago areas they were twice that size, and the average growth rate was highest in the middle archipelago. Similar observations were made in the Ekenäs archipelago (Gulf of Finland) where the *Mytilus trossulus* density also increased steadily with increasing wave exposure towards the outer archipelago, but biomass was highest at intermediate exposure (Fig. 11.33, Westerbom and Jattu 2006). This suggests that *Mytilus trossulus* may be sensitive to intraspecific competition towards the outer archipelago since density increases but biomass decreases (*i.e.* body size decreases).

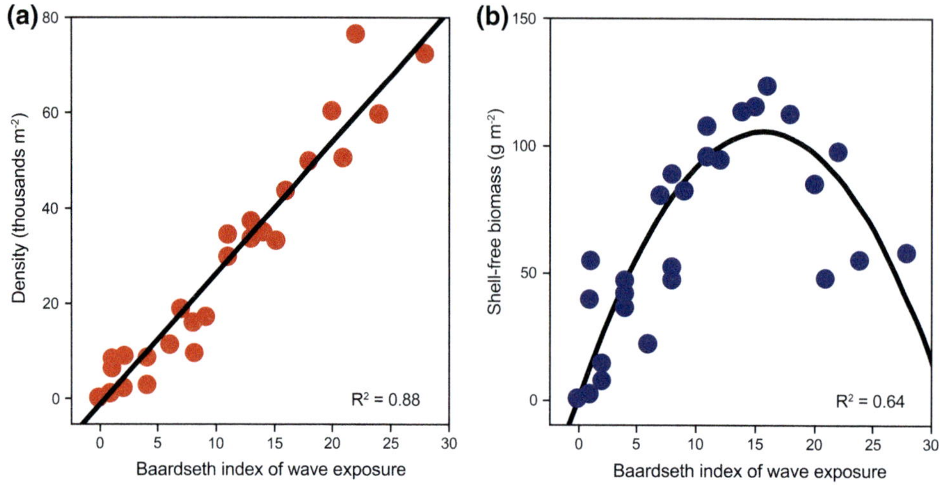

Fig. 11.33 Density and biomass of *Mytilus trossulus* in relation to wave exposure (Baardseth index, *cf.* Box 11.5) at a water depth of 8 m for 30 sampling stations near Tvärminne close to the entrance to the Gulf of Finland. (**a**) Relationship between the mean density of *Mytilus trossulus* per station and wave exposure. (**b**) Relationship between the mean shell-free biomass per station and wave exposure. Figure modified from Westerbom and Jattu (2006)

11.13.3 Invertebrate grazers

Grazers can have a direct negative grazing effect on macrophytes, but in the Baltic Sea it is rare for grazers to crop down entire phytobenthic communities because the grazer populations are usually not abundant enough. Rather it seems that grazing mainly keeps the growth of algal epiphytes down to the advantage of the phytobenthic communities' key species (Boström et al. 2014). Dominant perennial algae in the Baltic Sea, *e.g. Fucus vesiculosus*, *Fucus serratus*, *Furcellaria lumbricalis* and *Phyllophora pseudoceranoides*, have been shown to use chemical defences against grazing by the isopod *Idotea balthica* (Rodhe et al. 2004; Rodhe and Wahl 2008).

Typical grazers in the Baltic Sea are snails (*e.g. Bithynia tentaculata*, *Radix labiata* and *Theodoxus fluviatilis*, Fig. 11.34a), which mainly consume diatoms and filamentous algae. However, it has been demonstrated that the radula of the snails can also damage the thallus of robust algae, including the stiff leathery thallus of *Fucus vesiculosus*. Also, isopods (*e.g. Idotea* spp., Fig. 11.34b, c) and amphipods (*e.g. Gammarus* spp., Fig. 11.17) are grazers on diatoms, macroalgae and/or vascular plants. An exception is the large isopod *Saduria entomon*, which is basically a detritivore but also a carnivore.

Species of the marine genus *Idotea* can attain impressive feeding rates on a range of macroalgae and vascular plants (Vesakoski et al. 2008; Leidenberger et al. 2012). The three *Idotea* species that occur in the Baltic Sea exhibit a habitat segregation according to their lower salinity limit and preference for vegetation type and exposure to wave action. *Idotea balthica* (Fig. 11.34b) occurs down to salinity ~ 2.7 and is often dominant in the *Fucus* belt at varying exposure, *Idotea chelipes* (Fig. 11.34c) occurs down to salinity ~ 3.2 and is often dominant in *Zostera* meadows at varying exposure while *Idotea granulosa* occurs down to salinity ~ 4.6 and is restricted to fully exposed coasts (Leidenberger et al. 2012).

11.13.4 Grazing in the *Fucus* belt

An early model of the general decline of *Fucus* in the Baltic Sea in the 1970s suggested that filamentous algal epiphytes on *Fucus* would benefit from eutrophication and that these epiphytes would outcompete *Fucus* in the competition for light and nutrients (Kangas et al. 1982). Furthermore, the filamentous algae were thought to increase the abundance of grazers (notably *Idotea balthica*), which then also crop down the *Fucus* vegetation.

In the early 1990s, *Fucus* disappeared from the deeper parts of the sublittoral and also from whole depth transects in the Kalmarsund area (Sweden) in the Baltic Sea proper, which coincided with a mass occurrence of *Idotea balthica* (Engkvist et al. 2000). Experimental studies demonstrated the isopod's voracious feeding on *Fucus* and showed that 800 individuals, but not 600 or fewer, of *Idotea balthica* kg^{-1} of *Fucus* wet weight are able to graze down the *Fucus* vegetation (Engkvist et al. 2000; Svensson et al. 2004). However, such high abundances of *Idotea balthica* are extremely rare in nature (Leidenberger et al. 2012).

Furthermore, multiple choice experiments showed that *Idotea balthica* in fact prefers the green filamentous algae

Fig. 11.34 Three species of grazers that are common in the phytobenthic zone of the Baltic Sea. (**a**) The snail *Theodoxus fluviatilis* (**b**) The isopod *Idotea balthica*, see also Fig. 6.1. (**c**) The isopod *Idotea chelipes*. Photo: © Maria Włodarska-Kowalczuk and Piotr Bałazy

Cladophora glomerata and *Ulva intestinalis* above the leathery thallus of *Fucus vesiculosus* (Goecker and Kåll 2003). Other experiments confirmed that *Idotea balthica* prefers *Cladophora glomerata* above *Fucus vesiculosus* as a food item, but that vascular plants are even more appealing. The preferred food items of *Idotea balthica* can be arranged in the following series: *Stuckenia pectinata* > *Zannichellia palustris* > *Zostera marina* > *Cladophora glomerata* > *Fucus vesiculosus*, while this isopod's habitat preference order is *Fucus vesiculosus* > *Stuckenia pectinata* > *Zostera marina* > *Zannichellia palustris* > *Cladophora glomerata* (Vesakoski et al. 2008).

Other experiments have shown that moderate abundances of grazers are beneficial for the growth of *Fucus vesiculosus* during nutrient enrichment since they crop down filamentous epiphytes, which lowers shading by epiphytes and decreases competition for nutrients (Worm et al. 1999). Furthermore, it was shown that grazers may even enhance *Fucus vesiculosus* recruitment by selective consumption of filamentous algae (Worm et al. 1999, 2001). Thus, the effect of grazers on *Fucus vesiculosus* in eutrophic environments may be negative or positive, depending on the local characteristics, the life stage of *Fucus vesiculosus* and the magnitude of grazing.

11.13.5 Bottom-up and top-down control of grazers

Theoretically, eutrophication would support higher grazer densities by increasing the availability and quality of algal food to the grazers (Hemmi and Jormalainen 2002). This would lead to an increasing density of grazers such as snails, isopods and amphipods (bottom-up control). The abundance of epiphytic diatoms and filamentous algae in the phytobenthic community would decrease to a point where they switch to feed on *Fucus*. This would then damage the *Fucus* thallus so that it is more easily torn away from its substrate by wave action and there would be a loss of *Fucus* biomass.

The top-down approach postulates that when a piscivorous fish (*e.g.* the Atlantic cod *Gadus morhua*) decreases in abundance, the population sizes of its prey (invertivorous fish) will increase and the population sizes of the invertebrate grazers (snails, amphipods, isopods) will decrease, which leaves the filamentous alga to grow freely with low grazing pressure. Experimental evidence for this mechanism has been presented for the Baltic Sea Area (Moksnes et al. 2008; Baden et al. 2010; Eriksson et al. 2009, 2011).

Both the bottom-up and top-down mechanisms cannot be confirmed by long-term field observations in the Askö area in the Baltic Sea proper. While eutrophication decreased and the cod collapsed in this area between 1993 and 2012, there were increasing trends in both grazer biomass and *Fucus* biomass, whereas filamentous algae did not show any particular trend (Fig. 11.35a–c). This suggests that the grazer pressure in the Askö area is not high enough to damage the *Fucus* vegetation.

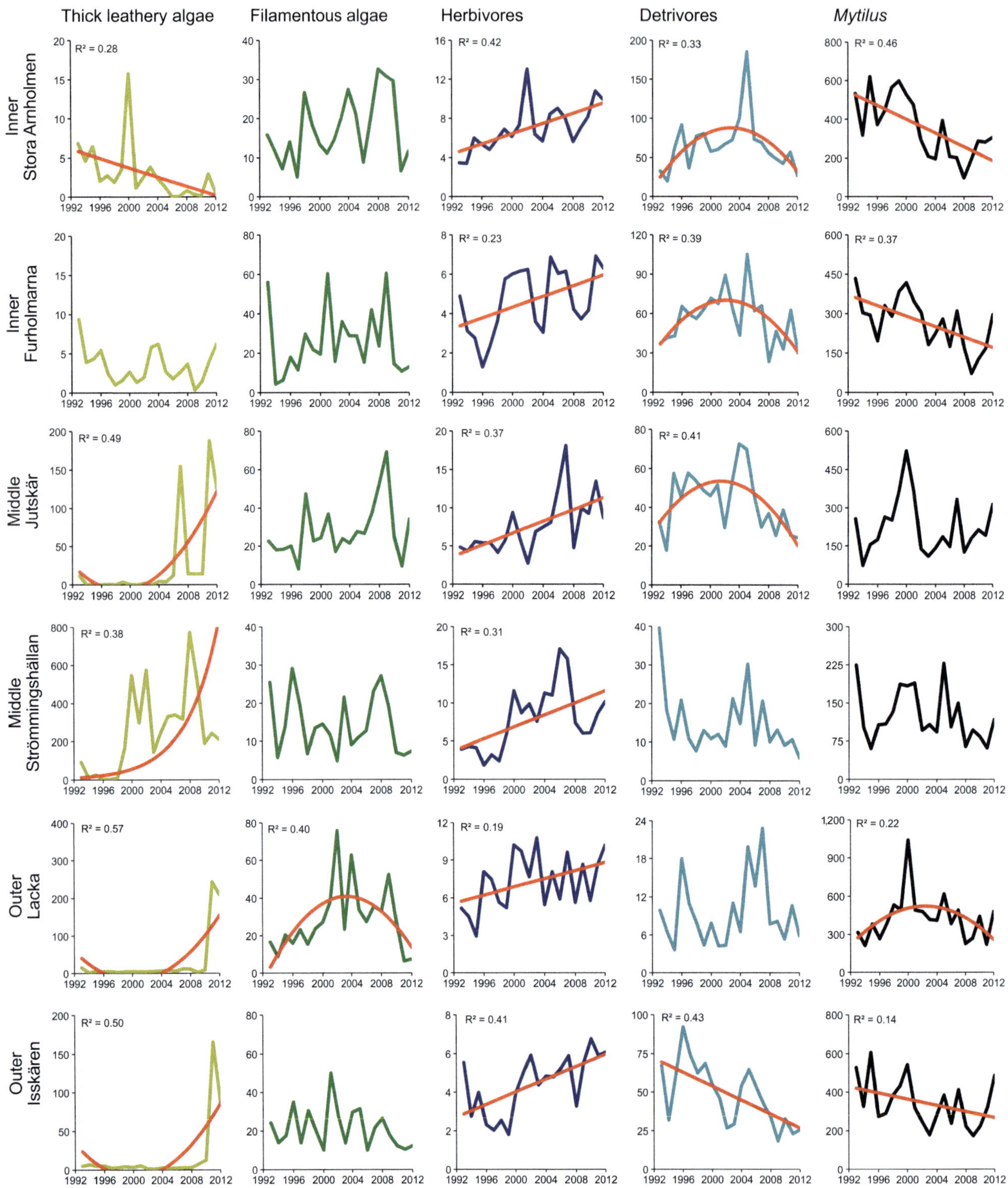

Fig. 11.35 Long-term field observations (1993–2012) on the phytobenthic communities, including the associated macrofauna, in the Askö area (Baltic Sea proper) at two stations in the inner archipelago (Stora Arnholmen, Furholmarna), two stations in the middle archipelago (Jutskär, Strömmingshällan) and two stations in the outer archipelago (Lacka, Isskären). This is shown as the mean dry weight m^{-2} per station for thick leathery algae (mainly *Fucus*), filamentous algae, herbivores, detritivores and *Mytilus trossulus* (including shells) at each station. Figure based on data in Kautsky et al. (2011) and additional unpublished data from the Swedish monitoring programme for phytobenthos

11.13.6 Invertebrate carnivores and omnivores

There are many invertebrate carnivorous species that pick their prey within the phytobenthic communities. Many of these species are not obligate carnivores as they can also eat algae and plants, and are thus in fact omnivores. Some of them eat invertebrate prey as well as detritus. The polychaete *Hediste diversicolor* and several insect larvae (*e.g.* the Polycentropodidae, Tanypodinae, Zygoptera) are carnivores, but many insect larvae are herbivores, and some even breed and harvest their own food (*e.g.* some tube-dwelling Chironomidae and Trichoptera larvae, Hasselrot 1993). None of the carnivorous or omnivorous invertebrate species in the phytobenthic zone of the brackish Baltic Sea occur in abundances high enough to alter the community composition like large predators such as starfish (*Asterias* spp.) and larger crabs (*Carcinus* spp.) are able to do in marine areas (*cf.* Fig. 4.32).

11.13.7 Invertebrate detritivores

Most of the energy fixed in the phytobenthic zone is not consumed directly. Most primary producers are first decomposed and then consumed as detritus (dead organic matter). Decomposition rates can be so high that hypoxic (<2 mL O_2 L^{-1}) and anoxic conditions occur as patches in the phytobenthic zone. This is often shown as white patches of bacteria belonging to the genus *Beggiatoa*, which oxidise hydrogen sulphide (H_2S) as an energy source, or as a purplish to wine-red or blue-green layer of the cyanobacterium *Spirulina*, which covers patches of decaying primary producers on the seafloor.

The detritivores, which recirculate the nutrients, have a key function in the ecosystem. The detritivores are scavengers, deposit feeders and suspension feeders and are represented in most invertebrate groups. Abundant detritivores in the phytobenthic communities of the Baltic Sea are the bivalves *Cerastoderma glaucum* and *Macoma balthica*, which filter suspended matter from the water or consume the organic matter from the seafloor.

11.13.8 Fish

Small fish species and juvenile fish seek shelter from predators and find their food in the phytobenthic communities (Aneer 1985; Kautsky et al. 1992; Rönnbäck et al. 2007). The two perhaps most common benthic fish in the shallow waters of the Baltic Sea, the sand goby *Pomatoschistus minutus* and the common goby *Pomatoschistus microps*, are difficult to observe without diving (Nellbring 1985). However, other common inhabitants of the phytobenthic zone, the three-spined stickleback *Gasterosteus aculeatus* (Fig. 12.15a) and the common minnow *Phoxinus phoxinus*, can be easily observed swimming among the vegetation close to the water surface.

One of the dominant pelagic fish in the Baltic Sea, the Atlantic herring *Clupea harengus*, has its spawning habitats among the macrophytes (*cf.* Fig. 4.28), while the Atlantic cod *Gadus morhua* seeks its prey fish here. Freshwater piscivores such as European perch *Perca fluviatilis* and northern pike *Esox lucius* use the shallow bays of the Baltic Sea as spawning habitats because they are rich in vegetation and warm up quickly by the sun during the day (*cf.* Box 4.9). Restoration of these habitats is nowadays being carried out to stop the decline of the benthic fish populations of the Baltic Sea proper.

In open coastal waters of the southwestern Baltic Sea proper, the coastal predators, the perch and the pike, have decreased markedly in abundance while small fish species such as sticklebacks have increased (Ljunggren et al. 2010). One hypothesis claims that the now numerous sticklebacks induced the decline of their predators by eating their eggs (Nilsson et al. 2004; Nilsson 2006), but this was based on the stomach content of the sticklebacks and not so much on their actual numbers in field. An alternative hypothesis proposes that the decline of the perch and pike in the phytobenthic zone are caused by a trophic cascade in the Baltic Sea offshore pelagic system where the European sprat *Sprattus sprattus* has caused a decline in the mesozooplankton populations (*cf.* Sect. 8.9.5) and thus limiting food for young recruits of perch and pike (Ljunggren et al. 2010). Thus, the dramatic change in the offshore system may have propagated to the coast.

The ground-dwelling fish all have their microhabitat preferences, and in general their diversity increases with the heterogeneity of the substrate (Fig. 11.36). These fish species are usually stationary and may reflect the local environmental conditions. For example, the viviparous eelpout *Zoarces viviparus* is used in environmental monitoring along the Baltic Sea coast as a bioindicator (*cf.* Sect. 14.6.2). The male black goby *Gobius niger* is easily observed on mixed substrates where it defends its nest below a boulder. The well-camouflaged European bullhead *Cottus gobio* is more difficult to see as it lies still among stones and algae, waiting for its prey to swim just in front of its large mouth. The pipefish *Syngnathus typhle* floats around in an upright position and sucks in small animals or nibbles off the siphons of hydrobiid snails. Small schools of the common

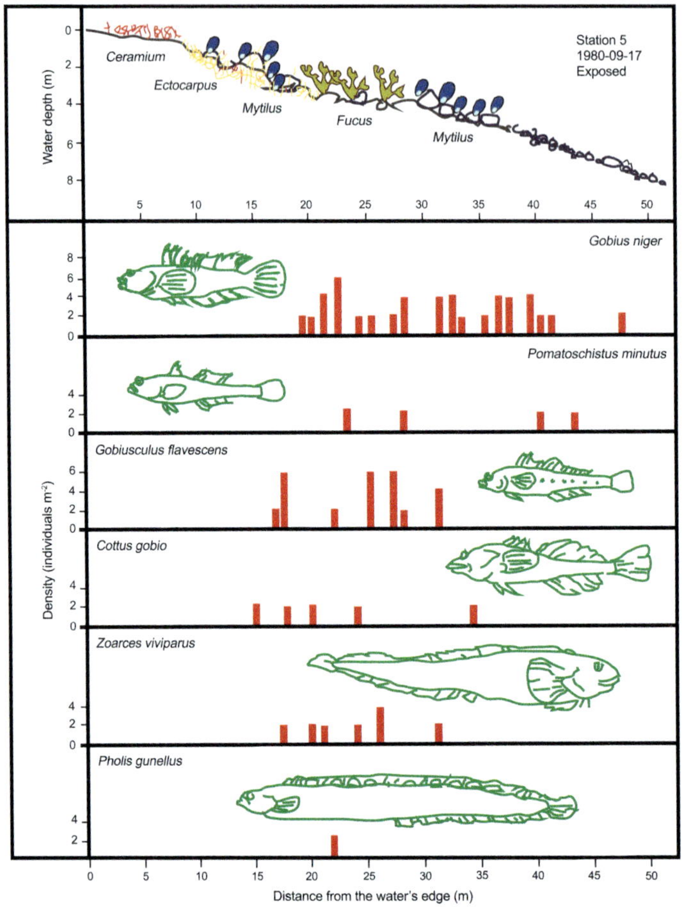

Fig. 11.36 Spatial distribution of demersal fish in the phytobenthic zone of the Askö area in the Baltic Sea proper. Figure modified from Jansson et al. (1985)

minnow *Phoxinus phoxinus* hunt for large schools of mysids that live on invertebrate prey found within the *Fucus* belt.

11.13.9 Birds

Several species of diving waterbirds feed in the phytobenthic zone either close to the coast or at shallow offshore banks far from land. Some species of diving ducks and especially the sea ducks feed to a large extent on mussels, clams and other benthic fauna. For example, in winter a single common eider *Somateria mollissima* may consume up to two kilogram wet weight blue mussels per day and the smaller long-tailed duck *Clangula hyemalis* (*cf.* Box Fig. 11.12) may consume one kilogram or more blue mussels per day. In total, two to three million sea ducks, including Arctic-breeding long-tailed ducks, common scoters *Melanitta nigra* and velvet scoters *Melanitta fusca* as well as the Baltic-breeding common eiders and velvet scoters spend the non-breeding season in the Baltic Sea (Skov et al. 2011). Marine areas with dense populations of blue mussels of high quality and of right size are therefore needed to support the European sea duck populations (Box 11.8).

Fish-eating waterbirds such as cormorants and grebes are often found foraging in the phytobenthic zone. These bird species catch both pelagic and benthic fish and in some cases also benthic invertebrates. Among the auks, the feeding behaviour of black guillemot *Cepphus grylle* differs from the more offshore and pelagic feeding common guillemot *Uria aalge* (*cf.* Fig. 16.2) and razorbill *Alca torda*. In winter, the black guillemot mainly occurs at offshore banks and at coastal areas which indicate that they primarily search for fish and invertebrates in the benthic zone.

11.14 Biomass and productivity in the phytobenthic zone

11.14.1 Macrophyte and macrofauna biomass

The average biomass in the phytobenthic zone, both that of the macrophytes and of the associated fauna, is about the same in the Baltic Sea proper as in the Kattegat, ~50–300 g dry weight m^{-2} for both groups (Fig. 11.37). However, extremely high biomasses of ~2 kg dry weight of primary producers m^{-2} and ~6 kg dry weight of animals including shells m^{-2}, such as those recorded *e.g.* in the marine Gullmarsfjorden in the Skagerrak, are not found in the Baltic Sea.

In the Bothnian Sea the average biomass is less than one-third of that in the Baltic Sea proper. In the Bothnian Bay the average biomass becomes extremely low, ~100 times lower than in the Baltic Sea proper, even if these data were collected in patches sheltered from ice; thus, in practice, the average biomass is even lower. The large differences in the biomass of the primary producers between the three major basins of the Baltic Sea are also reflected in the biomass of the animals (Figs. 11.37 and 11.38).

The reasons for the drastic decrease of primary production in the phytobenthic zone from the Baltic Sea proper to the Bothnian Bay are, in the first place, a lack of nutrients, and secondly, the increasingly harsh climatic conditions towards the north with ice scouring, shading by ice and a short growing season. The Bothnian Bay could be considered ultra-oligotrophic because of its low phosphate concentrations in the water. Also, while the Bothnian Sea is oligotrophic, the Baltic Sea proper has higher nutrient concentrations and is classified as mesotrophic (*cf.* Figs. 3.24 and 3.25).

11.14.2 Microphytobenthic biomass

A diatom spring bloom does not only occur in the pelagic zone (*cf.* Sect. 8.2.4), but also in the phytobenthic zone. In the upper sublittoral zone, before *Cladophora glomerata* colonises, these diatom blooms may consist of dense 10–30 cm high colonies of tube-dwelling diatoms such as *Berkeleya rutilans* and *Navicula ramosissima* (Snoeijs and Kautsky 1989) and biomass can be as high as 570 g dry weight m^{-2} after a warm winter with little ice cover in the Bothnian Sea (Snoeijs 1990a). This is higher than the average biomass in the entire phytobenthic zone in summer (*cf.* Sect. 11.14.1).

After an average winter, the biomass of diatoms and filamentous algae (mainly *Pylaiella littoralis*) on rocks in the upper sublittoral zone (0.2–0.7 m of water depth) during the benthic spring bloom of diatoms in April-May (about three weeks after ice break) is ~100–200 g dry weight m^{-2} along the Baltic Sea gradient from the Öresund to the northern

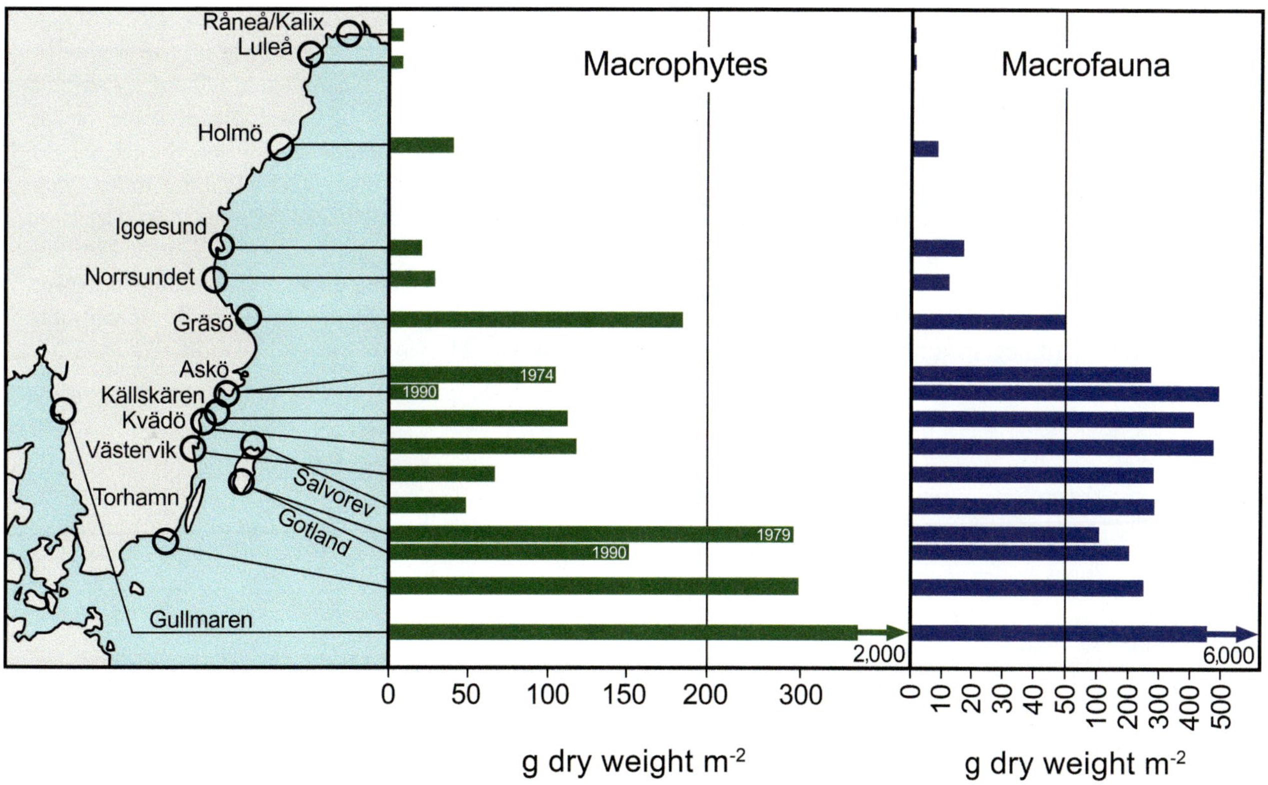

Fig. 11.37 The average total biomasses of macrophytes and associated macrofauna in the phytobenthic zone along the large-scale Baltic Sea gradient. Variation between years can be large as shown for Askö (1974 and 1990) and Gotland (1979 and 1990). In the Skagerak (Gullmarsfjorden) biomass is much higher than in the Baltic Sea for both macrophytes and macrofauna. The fauna dry weight includes shells. Note the change of scale above 200 g and 50 g of macrophytes and macrofauna, respectively. Figure modified from Kautsky (1995a)

Box 11.8: The Baltic Sea is an important wintering area for waterbirds

Kjell Larsson

Waterbirds rely on marine food all year round

Waterbird populations are fascinating elements of the Baltic Sea ecosystem. In summer, several species of diving and dabbling ducks, gulls, terns, auks, grebes, mergansers, other waterfowl as well as waders breed along the Baltic Sea coast. However, the Baltic Sea is not just an important breeding area. Several million waterbirds are observed in the Baltic Sea in the non-breeding season when Arctic breeding species have returned from the north, and birds that mainly breed in inland freshwater lakes have moved to the coast.

Many waterbirds feed on fish or benthic fauna

About 30 waterbird species in the Baltic Sea region are completely or to a very large extent dependent on fish, mussels or other marine invertebrates as food in winter and spring. Auks, cormorants, divers, grebes, mergansers, terns and some species of gulls are specialised fish predators. Auks such as the common guillemot *Uria aalge* and the razorbill *Alca torda* and the wintering black- and red-throated divers *Gavia arctica* and *Gavia stellata* usually forage in the pelagic zone far from land. Auks may dive to 25 m depth or more, and occasionally even down to 100 m, in search for schooling fish such as sprat and herring. Great cormorants *Phalacrocorax carbo sinensis* predate on a whole range of fish species, usually in more shallow coastal waters, but can easily dive to 10 m depth. Gulls and terns, on the other hand, catch fish that swim close to the surface. Wintering diving ducks such as the tufted duck *Aythya fuligula*, the common goldeneye *Bucephala clangula* and the greater scaup *Aythya marila* and sea ducks such as the common eider *Somateria mollissima* (*cf.* Box 4.10), the common scoter *Melanitta nigra*, the velvet scoter *Melanitta fusca* and the long-tailed duck *Clangula hyemalis* (Box Fig. 11.12) feed to a very large extent on benthic fauna, especially on mussels and clams but also on crustaceans and other invertebrates.

Wintering areas

The major part of the European long-tailed duck population winter at offshore banks far from land in the central and southern Baltic Sea. In mid-winter the most important sites are the Hoburgs bank, Södra Midsjö bank, Norra Midsjö bank, Odra bank, Adlergrund and Słupsk bank (*cf.* Fig. 2.2). There the long-tailed ducks regularly dive down to 10–25 m depth to feed mainly on *Mytilus trossulus* but also to some extent on *Macoma balthica* and *Saduria entomon*. Wintering common eiders and common and velvet scoters prefer offshore *Mytilus* beds and coastal waters in the southern and southwestern parts of the Baltic Sea, for example at the Odra bank and in Danish waters. In general, the common eider and scoters prefer somewhat shallower areas than the long-tailed duck. The wintering tufted duck, common goldeneye and greater scaup feed mainly on small bivalves and invertebrates in coastal waters less than 10 m deep. In late March and April, the Baltic common eider population starts migrating northward to their main breeding areas in Sweden and Finland. In late April and the beginning of May, the long-tailed ducks and common and velvet scoters aggregate in large flocks in the Gulf of Riga and Gulf of Finland before they leave the Baltic Sea in mid-May for their northward migration to their Arctic breeding grounds (Box Fig. 11.12a).

Box Fig. 11.12 The long-tailed duck *Clangula hyemalis*. (**a**) Long-tailed ducks aggregate in large flocks in April and May in the Gulf of Riga and Gulf of Finland before they leave the Baltic Sea for northward migration to their Arctic breeding grounds. (**b**) Long-tailed ducks in winter: four adult males and one female. Photo: © Kjell Larsson

A link between phytoplankton and sea ducks

Sea ducks swallow blue mussels whole, but since it is only the soft parts that is of nutritional value they must consume very large quantities of mussels each day, as well as get rid of large quantities of crushed shells, to maintain their energy balance. Mussels feed by filtering plankton from the water. Because different phytoplankton groups have different energetic and nutritious qualities, the concentrations and species composition of the phytoplankton may affect the growth and condition of the mussels and thus their quality as a food for sea ducks. Large-scale changes at the lower trophic levels in the Baltic Sea may thus affect the population dynamics of sea ducks (Box Fig. 11.13). Sea duck populations may also be affected by large-scale anthropogenic pressures such as habitat destruction, oil spills, bycatch and hunting, as well as by predation and diseases.

Trends in numbers differ between species

The trends in numbers of individuals differ greatly between different waterbird species (Durinck et al. 1994; Skov et al. 2011). Recent surveys of wintering waterbirds in the Baltic Sea indicate that populations of herbivorous waterbirds, for example populations of the mute swan *Cygnus olor*, the mallard *Anas platyrhynchos* and the Eurasian coot *Fulica atra*, are stable or increasing. The trends of fish-eating waterbirds also differ between species. Cormorants and common guillemots have increased in numbers, although the number of breeding cormorants in the Baltic Sea has levelled off in recent years. The trends of fish-eating grebes and divers are more uncertain. A number of surveys have shown that sea duck populations with offshore distribution in winter, especially populations of the long-tailed duck, the common eider and the velvet scoter, have decreased by 50 % or more since the beginning of the 1990s. The rapid declines can most likely be explained by a combination factors including: (1) ecosystem changes affecting the quality and quantity of food resources in the Baltic Sea during the non-breeding season, (2) changing predation pressures in the Baltic region in summer due to the return of the white-tailed eagle *Haliaeetus albicilla* and climate changes affecting the lemming-waterbird-predator relationships in the Arctic, (3) elevated mortality rates because of recurrent oil spills at important wintering sites, and (4) elevated mortality rates because of by-catches of birds in fishing nets and by hunting (Larsson and Tydén 2005; Skov et al. 2011; Bellebaum et al. 2013; Kilpi et al. 2015).

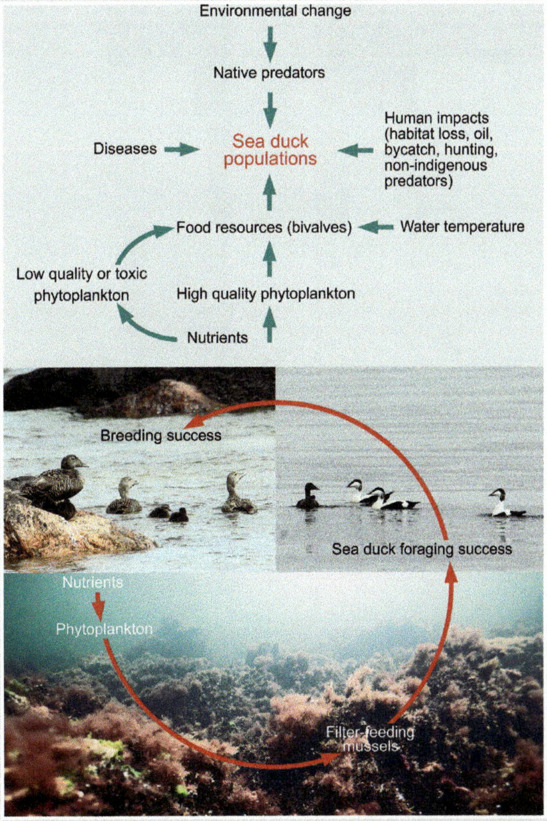

Box Fig. 11.13 Schematic view of how changes at lower trophic levels in the Baltic Sea may affect the population dynamics of sea ducks. Photo: birds © Kjell Larsson, seafloor © Hans Kautsky

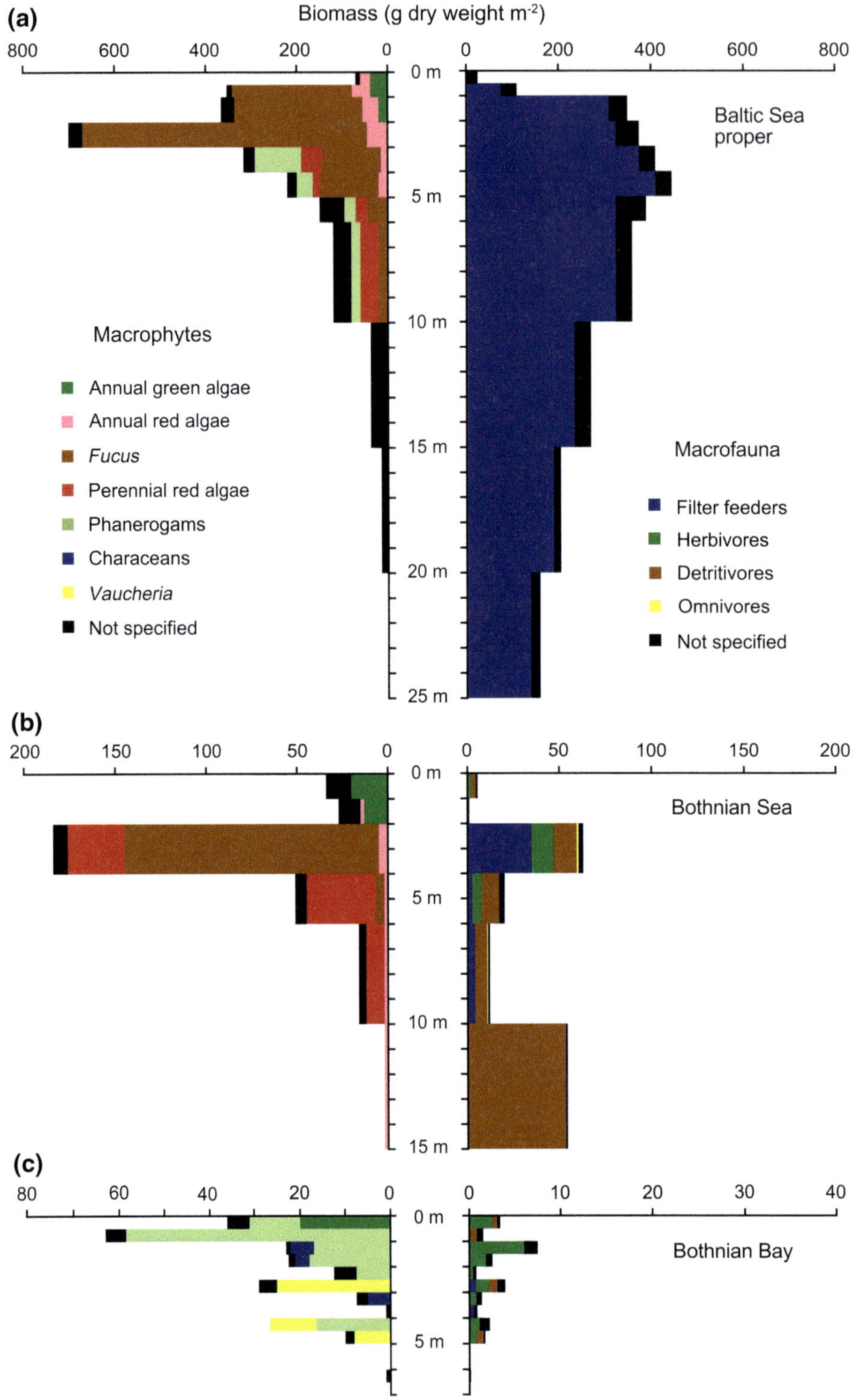

Fig. 11.38 Comparison of the biomass of macrophytes and associated macrofauna in the phytobenthic zone in different parts of the Baltic Sea. (**a**) The Baltic Sea proper at Askö. (**b**) The southern Bothnian Sea at Gräsö. (**c**) The Bothnian Bay at Luleå. The category "Not specified" macrophytes contains mainly filamentous brown algae, both annuals (*e.g. Pylaiella littoralis*) and the perennial *Battersia arctica*. In (**a**) all filter feeders are *Mytilus trossulus*, and all other animals are in the category "Not specified" fauna. In (**c**) the vascular plants are mainly *Isoetes lacustris* at a water depth of 1.0–1.5 m and *Potamogeton perfoliatus* in deeper water, while the herbivores consist mainly of the freshwater snails *Gyraulus acronicus*, *Radix labiata*, *Theodoxus fluviatilis* and *Valvata piscinalis*, the filter feeder is the freshwater clam *Pisidium* sp. and the detritivores are oligochaetes and *Saduria entomon*. Figure modified from Kautsky (1995a)

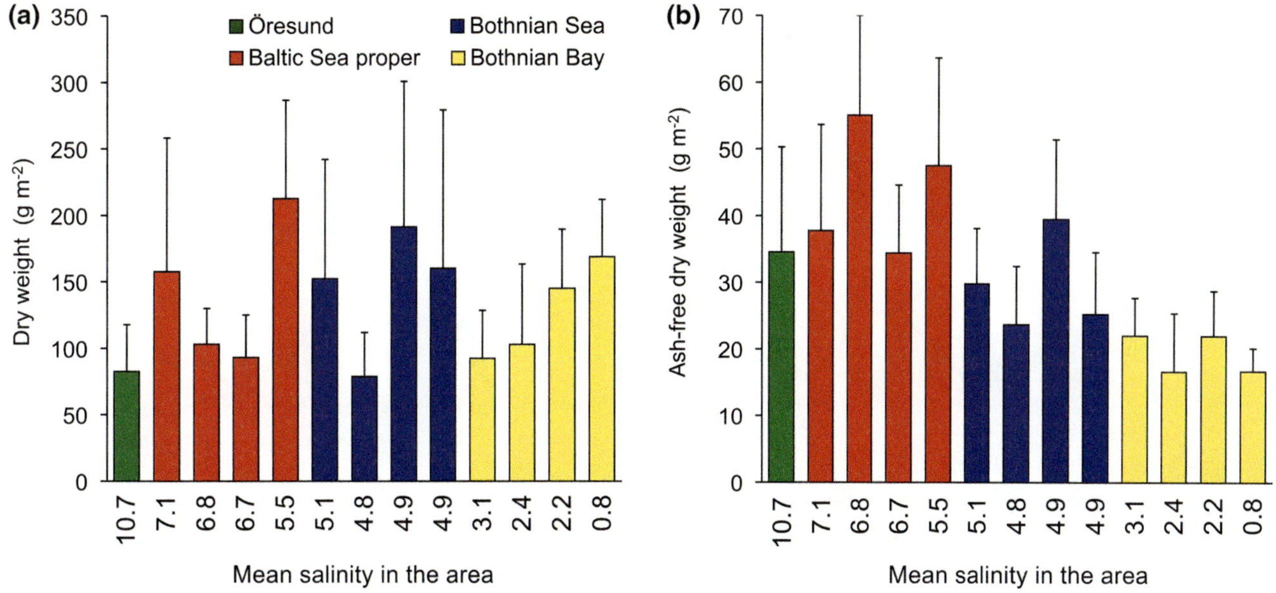

Fig. 11.39 Average biomass of macro- and microphytobenthos in 13 areas along the large-scale Baltic Sea gradient in the upper sublittoral zone at a water depth of 0.2–0.7 m during the benthic spring bloom of diatoms in April-May 1990 and 1991. Within each area, the sampling sites were located along a gradient from the inner to the outer archipelago. Altogether, 135 sampling stations are included. All macroalgae in the samples were filamentous algae, mainly *Pylaiella littoralis*. (**a**) Dry weight of macro- and microphytobenthos. (**b**) Ash-free dry weight of the same samples (without diatom silica frustules and other inorganic materials). Figure modified from Ulanova et al. (2009)

Bothnian Bay (Fig. 11.39). Ash-free dry weight, on the other hand, shows a decreasing trend towards the north, with ~45 g m^{-2} in the Baltic Sea proper, ~30 g m^{-2} in the Bothnian Sea, and ~20 g m^{-2} in the Bothnian Bay (Fig. 11.39). This suggests that, at least in spring, there is an increase in the relative proportion of diatoms towards the north because the silica frustules (as well as other inorganic materials) are not included in the ash-free dry weight.

11.14.3 Biomass and substrate availability

Differences in biomass and composition along the large-scale Baltic Sea gradient are explained not only by salinity and nutrient concentrations but also by substrate availability. For example, a steep coast dominated by hard substrate is inhabited by macroalgae and supports a higher biomass than a gently sloping coast with a rooted vegetation on sand and gravel. Hard substrates are rarer towards the north because of the geological conditions and the flat topography.

The coasts of the Bothnian Bay are characterised by scattered boulders, stones and gravel, intermingled with sand and soft substrates. In the more wave-exposed areas, vast sandflats with no vegetation prevail. Wherever hard substrate occurs in the Bothnian Bay, it is usually covered with a low biomass of filamentous algae and diatoms since no larger marine algae can live in the low salinity. However, in places sheltered from ice scouring, extensive growth of annual freshwater vascular plants and charophytes may occur, and biomass can be as high as in similar habitats in the Bothnian Sea.

11.14.4 Productivity in the phytobenthic zone

Over the year, the primary productivity (biomass production per unit time) in the coastal zone of the Baltic Sea proper is about three times higher in the water column than in the phytobenthic communities (Fig. 11.40a). However, in the Bothnian Sea the primary productivity in the phytobenthic communities is higher than that in the water column in summer, and the yearly primary production is similar in the pelagic and the phytobenthic zone in coastal waters (Fig. 11.40b). Both macrophyte and animal respiration in the phytobenthic zone of the Bothnian Sea are much lower than in the Baltic Sea proper, which is explained by the absence of high *Mytilus trossulus* biomass. In the Baltic Sea proper, *Mytilus trossulus* contributes with 80–90 % of the total animal respiration, but in the Bothnian Sea this is <1 %.

11.14.5 Energy flows in the phytobenthic zone

A holistic approach to understanding the ecosystem functioning is to model the energy flow between different

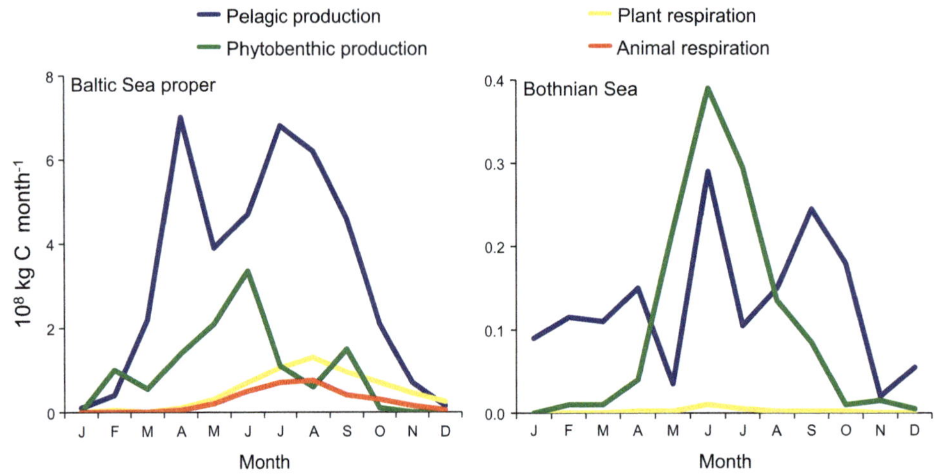

Fig. 11.40 Comparison of the net primary productivity by the macro- and microphytobenthos at a water depth of 0–30 m calculated for the coastal zones of the whole Baltic Sea proper and the whole Bothnian Sea. In the Bothnian Sea animal respiration is negligible. Note the different scales on the y-axes (twenty times lower in the Bothnian Sea). Figure based on data in Kautsky (1995b)

ecosystem compartments (Jansson 1978; Kautsky and Kautsky 1995). Such models can help us to determine whether a studied mechanism (*e.g.* the possibility for *Idotea* spp. to graze down the *Fucus* vegetation) is realistic or not. Simple energy-flow measurements based on biomass and food demand and the food preference of single species can depict mismatches, missing parts and efficiency of the ecosystem. By answering questions such as "can the measured primary production support all the organisms found?" and "are there alternative energy sources?" different aspects of ecosystem functioning can be disclosed and also make it possible to predict changes when conditions are altered, *e.g.* through eutrophication, climate change or overfishing.

The pelagic and phytobenthic energy flows differ between the Bothnian Bay, Bothnian Sea and Baltic Sea proper (Fig. 11.41). The standing stock of benthic and pelagic primary producers is about four times higher in the Bothnian Sea than in the Bothnian Bay. In the Baltic Sea proper, the pelagic standing stock is again about four times higher than in the Bothnian Sea, but the pelagic primary production has increased more than ten-fold (from ~2,240 to ~39,300 10^5 kg C $year^{-1}$). The phytobenthic biomass and primary production in the Baltic Sea proper are about ten times and almost eight times those in the Bothnian Sea, respectively.

Of the total annual benthic and pelagic primary production in the coastal zone of the Baltic Sea (with the border to open water set at the 25 m depth isoline), the phytobenthos production is estimated to contribute 12 % to the total coastal production in the Baltic Sea proper, 23 % in the Bothnian Sea and 50 % in the Bothnian Bay (Kautsky and Kautsky 1995). According to these estimates, the phytobenthic zone accounts for a significant part of the total primary production and carbon turnover in the coastal areas of the Baltic Sea. Also, benthic-pelagic coupling seems to be strong.

11.15 Anthropogenic impacts

11.15.1 Sensitivity to disturbances

Since attached organisms cannot escape when environmental conditions become detrimental (unless they die), they reflect what actually happens at a specific site and are therefore excellent indicators of environmental change. The long water residence time of the Baltic Sea, with ~85 million people living in its drainage area, makes the Baltic Sea one of the most polluted and eutrophicated seas in the world. Much is done to reduce the anthropogenic impacts (*cf.* Sects. 17.8 and 18.5), and deterioration as well as recovery of the environment in the Baltic Sea are closely monitored.

In general, the low-salinity phytobenthic communities in the Baltic Sea are more sensitive to disturbance than their marine counterparts since many species are already stressed by low salinity. In addition, many macrophytes lose the ability to reproduce sexually when they approach their salinity limit, and for this reason re-establishment after a disturbance may be more difficult than under fully marine conditions. When one of the habitat-forming key species, *e.g. Fucus vesiculosus*, *Zostera marina* or *Mytilus trossulus*, disappears from an area, this has a major negative effect on the pelagic system as well. For example, the phytobenthic communities act as herring spawning habitats (*cf.* Fig. 4.28) and the *Mytilus* beds recirculate nutrients.

A macrophyte vegetation also reduces water movement, both within and adjacent to the vegetation itself, resulting in increased sedimentation and reduced turbidity in the water. Reduced turbidity increases light availability for macrophytes, increasing their growth. Additionally, macrophytes affect the distribution, composition and grain size of sediments and reduce erosion. Therefore, phytobenthic

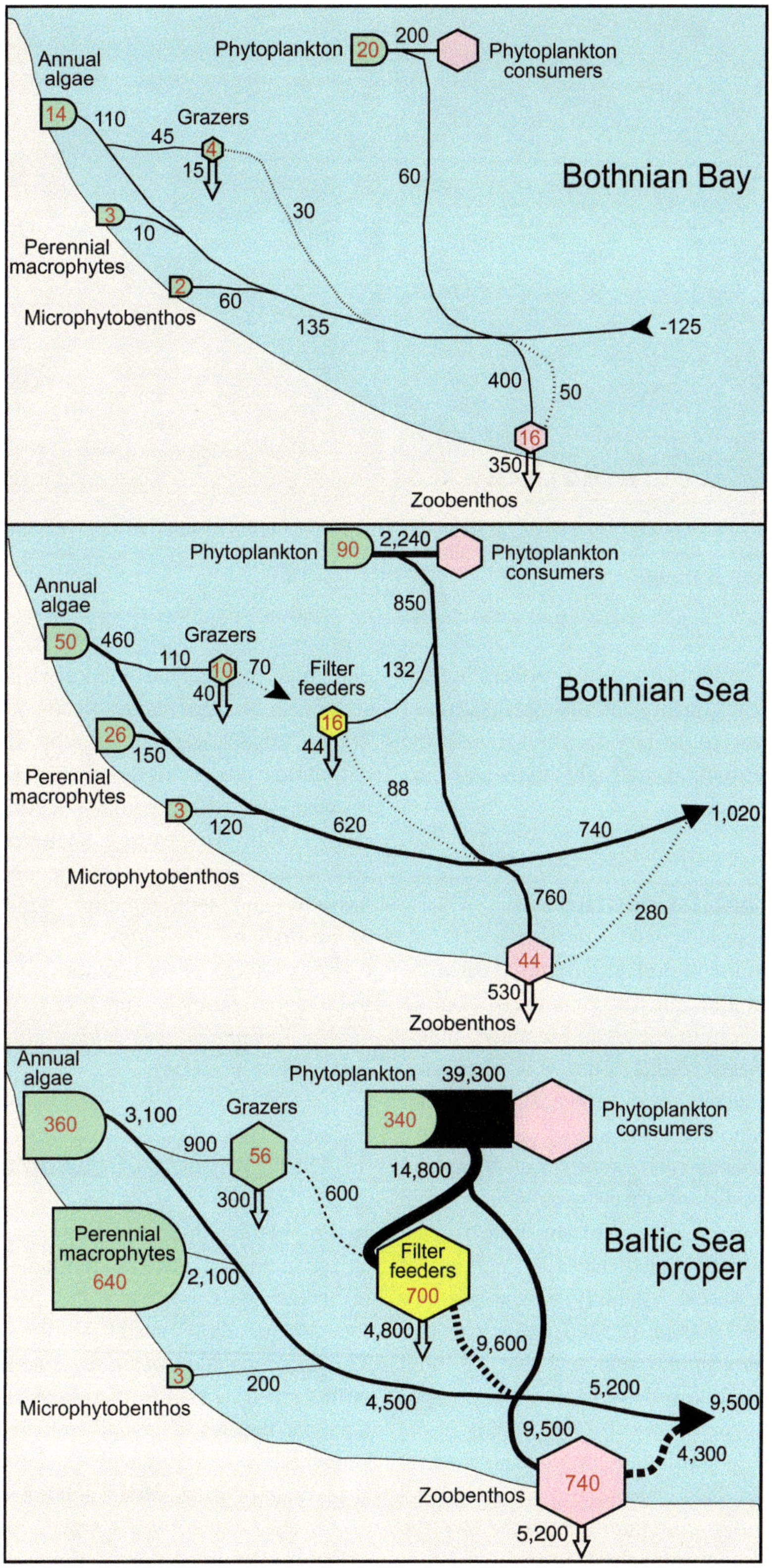

Fig. 11.41 Standing stocks and flows of carbon through the Baltic Sea coastal ecosystem in the three largest subbasins, the Bothnian Bay, the Bothnian Sea and the Baltic Sea proper. Bullet-shaped symbols are primary producers and hexagons are consumers. Phytoplankton consumers include both zooplankton and the microbial loop. The zoobenthos consists mainly of detritivores. Numbers in red show standing stocks in 10^5 kg C, and numbers in black show flows of carbon in 10^5 kg C year^{-1}. Solid lines show primary produced matter and dotted lines show secondary production and faeces. Open arrows pointing down from storages (the boxes) indicate respiration. Filled black arrows indicate export to the open waters and deeper benthos or, as in Bothnian Bay, the need of import of energy input from the open water and deeper benthos. Figure modified from Kautsky and Kautsky (1995)

Fig. 11.42 Impacts of ferry traffic on coastal habitats in the Baltic Sea. (**a**) *Cladophora glomerata* grows higher up on the bedrock than normal because of the regular occurrence of large waves. (**b**) smaller-grained substrates are swept away by the large waves, causing severe erosion. Photo: (**a**) © Michael Borgiel, (**b**) © Hans Kautsky

communities provide important ecosystem services, including the improvement of water quality and the stabilisation of sediments. The major threats to the phytobenthic zone of the Baltic Sea are habitat destruction, eutrophication and hazardous substances.

11.15.2 Impacts of habitat destruction

Direct physical damage to the littoral zone impoverishes the phytobenthic communities through fractionation of populations and destruction of recruitment areas for algae, plants and animals. Such damage results from *e.g.* dredging, locating marinas in sheltered bays, increasing harbour facilities by land reclamation, destruction of salt marshes, building offshore constructions such as windmill farms and installing pipelines. Stone-fishing (harvesting stones from the seafloor) decreases the available hard substrate for the settlement of algae. Sand extraction reduces the amount of the substrate, but also induces instability of the seafloor which may cause coastal landslides. Sand extraction and pipelines may create holes in the seafloor, that fill with decaying organic material followed by bacterial decomposition and hypoxia.

Waves induced by ship traffic not only cause coastal erosion (Fig. 11.42b), but also change the structure and composition of phytobenthic communities. Especially the high-speed ferries that create "artificial tides" of up to 1.8 m when they displace a water volume of the size of a small lake, affect the vegetation along their routes in the archipelagos of the Baltic Sea (Östman and Rönnberg 1991; Kurennoy et al. 2009). The regular strong water circulation induced by the ship traffic may have a local positive effect on perennial macroalgae by keeping the seafloor free from sediments and drift algae (Roos et al. 2003; Eriksson et al. 2004). Algal diversity is not affected, but *Cladophora glomerata* grows higher up on the bedrock than normal because of the regular occurrence of large waves (Fig. 11.42a). However, at sheltered sites and in inlets along ferry routes, perennial algae decline as a result of increased turbidity and sedimentation, which seems to favour *e.g. Myriophyllum spicatum* and *Ceratophyllum demersum* (species typical of turbid, nutrient-rich muddy habitats) and to disfavour exposure-sensitive mud-thriving species such as *Chara tomentosa* and *Najas marina* (Roos et al. 2003; Eriksson et al. 2004).

11.15.3 Impacts of eutrophication

In the phytobenthic zone, eutrophication is often manifested as an increased growth of filamentous algae, which may occur as mats covering the perennial vegetation or as free-floating algal mats at the water surface (Fig. 11.43a, b). The decomposition of these algae may cause hypoxia at the seafloor (Fig. 11.43c). Increased abundances of opportunistic algal species, especially along the eastern coast of the Baltic Sea proper, as well as the occurrence of algal mats consisting of filamentous algae such as *Cladophora glomerata*, *Pylaiella littoralis* and *Ectocarpus siliculosus* in the outer Archipelago Sea (Finland), may be caused by eutrophication (Vahteri et al. 2000; Korpinen et al. 2007). Eutrophication is also manifested as increased growth of phytoplankton; as a result, less light can penetrate through the water column to the phytobenthic communities (Sandén and Håkansson 1996).

Fig. 11.43 Signs of eutrophication. (**a**) A floating algal mat at a sheltered site influenced by excess nutrients. (**b**) Overgrowth of perennial algae by filamentous algae. (**c**) White sulphur bacteria (*Beggiatoa*) at the seafloor indicate hypoxia. Photo: (**a**) © Pauline Snoeijs-Leijonmalm, (**b**, **c**) © Hans Kautsky

The significance of eutrophication and substrate type for the distribution of phytobenthic communities is clearly observed in an example from the Gulf of Riga (Fig. 11.44). The station just north of the city of Riga and the two stations closest to the city of Pärnu were strongly influenced by city effluents, and biomass was low due to the absence of perennial algae, notably *Fucus vesiculosus* and *Furcellaria lumbricalis*. The biomass was also very low at two unpolluted stations, but here the reason was the lack of hard substrate. The station just north of Riga, where the Daugava river enters the Gulf of Riga, was totally dominated by *Cladophora glomerata* and omnivores (*Gammarus* spp.). The relative amount of green algae represented in the samples decreased with increasing distance from the cities and may be an indicator of eutrophication.

The western and eastern coasts of the Gulf of Riga furthest away from the cities, as well as the northern coast of the island of Saaremaa, support luxuriant phytobenthic communities of composition and biomass comparable to those found elsewhere in less eutrophied areas of the Baltic Sea proper. In comparison to the macrophyte biomass, the animal biomass was much lower and more evenly distributed in the Gulf of Riga with the filter feeders *Amphibalanus improvisus* (Fig. 11.18b, c) and *Mytilus trossulus* (Fig. 11.32). Moreover, at salinity <4 in the reach of the Daugava river, the non-indigenous freshwater species *Dreissena polymorpha* (*cf.* Box 5.4) and the detritivore *Macoma balthica* (*cf.* Box 13.5) were dominant.

11.15.4 Impacts of hazardous substances

Contrary to the nutrients nitrogen and phosphorus, which cause eutrophication, hazardous substances are detrimental or directly toxic to organisms. The same hazardous substance can be more toxic to one group of species than to another. For example, chlorine, which previously was used for bleaching in pulp mills, affects mainly brown algae. Thus, hazardous substances can change the competitive balance within a phytobenthic community and alter its species composition and structure.

Especially in the 1960s–1980s, pulp mills along the coasts of the northern Baltic Sea were severe and complex pollution sources causing increased turbidity (decreased light penetration), increased levels of hazardous substances and eutrophication in the coastal environment. In the direct vicinity of pulp mill effluents, the biomass and species richness of the phytobenthic communities were low, with green filamentous algae and filter-feeding animals dominating (Fig. 11.45). With improving conditions further away from the effluents, species richness increased and reached a maximum at an intermediate distance from the pollution source. Here the species were relatively evenly distributed among the functional groups. In the least polluted areas farther away from the pollution source, species richness and evenness among functional groups was lower than in the intermediate area. This pattern is well in

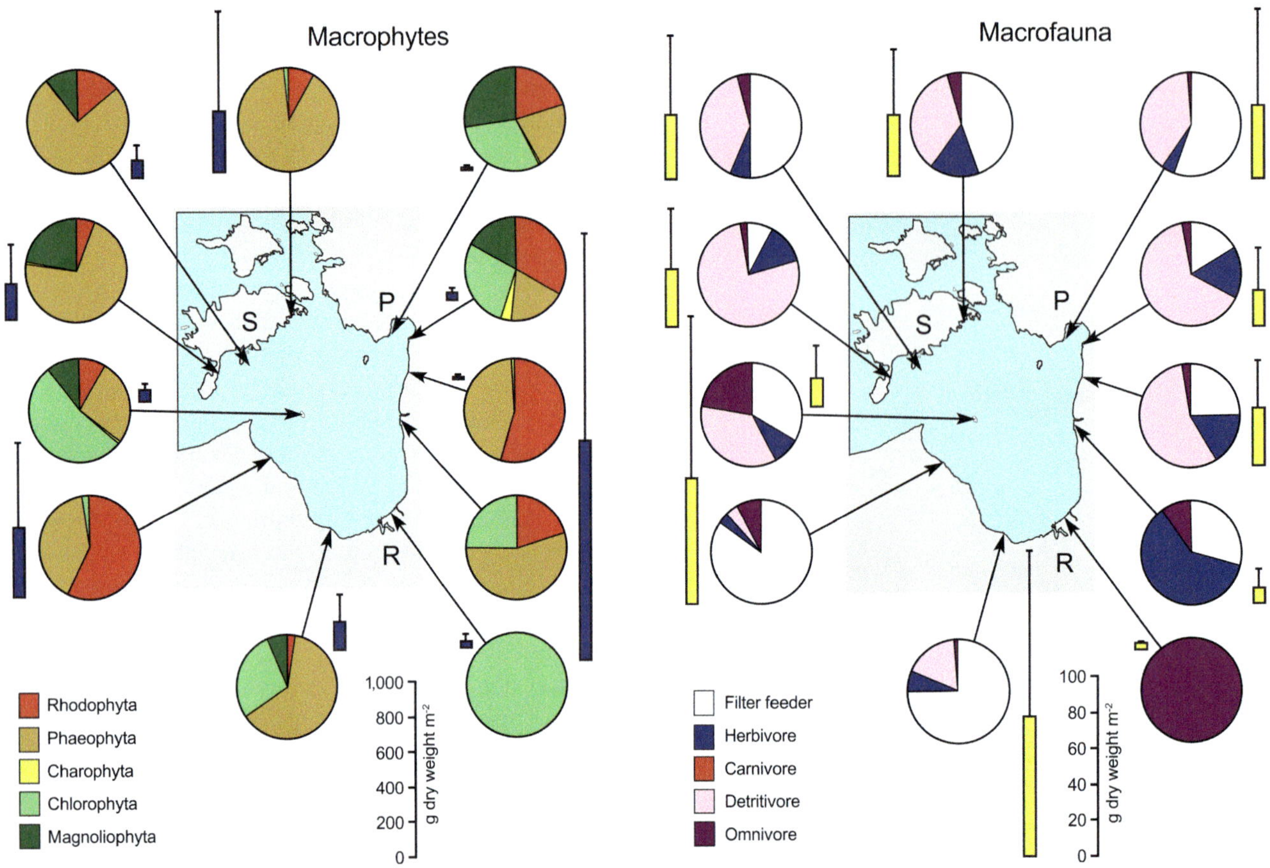

Fig. 11.44 The composition of macrophytes and associated macrofauna in the phytobenthic zone of 11 diving transects in the Gulf of Riga in 1995–1996. (**a**) Macrophytes subdivided into taxonomic groups, in % of the average biomass in each transect. (**b**) Animals subdivided into feeding groups, in % of the average biomass in each transect. The vertical bars show the average biomass in each transect. P = the city of Pärnu (Estonia), R = the city of Riga (Latvia), where the Daugava river enters the Gulf of Riga, S = the island of Saaremaa. Figure modified from Kautsky et al. (1999)

accordance with the "intermediate disturbance hypothesis" (*cf.* Box 4.16).

11.15.5 Climate change in the Baltic Sea Area

Regional climate scenarios for the brackish Baltic Sea Area predict, besides an increase in the sea surface temperature, a decrease in salinity with a changed water budget (*cf.* Fig. 2.12), including increased terrestrial runoff due to precipitation (HELCOM 2013b). Signs of the on-going climate change have already been observed in the temperature record. Historical data on the maximum summer surface-water temperature in the Baltic Sea show an increase of ~1.3 °C between 1861–1900 and 1985–2005 (Mackenzie and Schiedek 2007).

The changing environment created by climate change will alter the composition of the phytobenthic communities in the Baltic Sea Area. Critical questions to be answered are how communities will reorganise at a higher temperature and lower salinity and what the consequences of these community changes will be for ecosystem functioning. It is not possible to project recorded changes along the large-scale Baltic Sea gradient directly on climate scenarios because along this gradient the temperature and salinity decrease in concert while with climate change the temperature will increase but the salinity is expected to decrease.

11.15.6 Experiences from artificially raised water temperature

Studies in cooling-water discharges from nuclear power plants into the Baltic Sea, *i.e.* artificially heated natural environments, have identified the following changes that occur in the phytobenthic communities at higher water temperature and the absence of a winter ice cover: extension of the growing season, increased primary production, species shifts towards a higher proportion of opportunistic species, and increased temporal β-diversity (species turnover) (Snoeijs and Prentice 1989; Snoeijs 1989; Ilus 2009; Hillebrand et al. 2010).

Reduction in body size has been proposed as a universal response of organisms to both warming and decreased salinity. For example, the average cell volume was found to decrease linearly with ~2.2 % per °C in species-rich diatom communities and with ~2.5 % per °C within protistan species (Atkinson et al. 2003; Svensson et al. 2014). These observations confirm the "temperature-size rule" (Atkinson 1994; Foster et al. 2013).

While such linear relationships may be found in large pooled data sets, the conditions in the field (including factors other than temperature) are higly variable on a temporal scale. For diatoms in the phytobenthic zone of the Baltic Sea it was suggested that climate change in this century may polarise seasonality by creating two new ecological niches, one with elevated temperature at high nutrient concentrations in the cold season (increasing cell size) and the other one with elevated temperature at low nutrient concentrations in the warm season (decreasing cell size) (Svensson et al. 2014). Thus, higher temperature in summer and lower salinity by a changed water budget in the Baltic Sea are expected to decrease the average cell size of the microphytobenthos, which is likely to affect the transfer of energy to higher trophic levels.

11.15.7 Impacts of climate change

Species shifts in phytobenthic communities at elevated temperature include *e.g.* increases in the green filamentous algae *Cladophora glomerata* and *Ulva intestinalis* and decreases in perennial algae, *e.g. Fucus radicans* and *Fucus vesiculosus* disappear (Snoeijs and Prentice 1989). Parallel changes occur in the fauna, *e.g.* snails that can reproduce all year round (*Potamopyrgus antipodarum* and *Theodoxus fluviatilis*) increase in abundance at the expense of snails that hatch only during early summer (*Bithynia tentaculata* and *Radix labiata*) (Snoeijs 1989). Noteworthy is also the increase in non-indigenous species in cooling-water discharge areas, which may cause changes in food web interactions (*cf.* Box 5.9).

The projected decrease in salinity in the Baltic Sea as a result of climate change will clearly affect the key species *Fucus vesiculosus*, *Fucus radicans* and *Mytilus trossulus* at their northern limit of distribution, which may move 400 km to the south, from the Norra Kvarken area to the Södra Kvarken area (Strandmark et al. 2015). In this case, marine algae will be replaced by freshwater species in the Bothnian Sea, the large algal belts will disappear, and consequently the diversity of associated invertebrates will decrease.

Similarly affected will be the marine vascular plant *Zostera marina*, which has its northern salinity limit in the Stockholm archipelago and the southwestern coast of Finland (Boström et al. 2002). If salinity decreases it will probably be found only on the southern coasts of the Baltic Sea, which would have fundamental consequences for many associated species, including the coastal fish species that use seagrass meadows as hatching and nursery habitats. Altogether, climate change may reverse the recent improvements in the phytobenthic zone that have been achieved through the eutrophication decrease (oligotrophication) (Fig. 11.35).

11.15.8 Impacts of acidification

The global increase in the partial pressure of atmospheric CO_2 decreases the pH of marine surface waters worldwide, a process known as "ocean acidification". The pH decreases because CO_2 reacts with water to form hydrogen carbonate (bicarbonate) and hydrogen ions (*cf.* Sect. 1.3.6). The pH shift changes the equilibrium between hydrogen carbonate and carbonate, thereby depleting the available carbonate pool, which increases the solubility of $CaCO_3$ (*cf.* Sect. 1.3.7).

The ocean surface-water pH is expected to have dropped by an average 0.5 pH units at the end of the present century (Caldeira and Wickett 2003; Raven et al. 2005). For the Baltic Sea it has been estimated that this process will result in a reduction by 0.2–0.4 pH units by the year 2100 (Havenhand 2012). Coastal pH is more variable and difficult to predict than that of the open sea through complicated local effects of *e.g.* runoff (Andersson et al. 2005). The pH also varies on diurnal and seasonal scales, depending on photosynthetic carbon uptake, especially in shallow sheltered bays (Saderne et al. 2013).

Acidification changes the rates of some important biological processes, *e.g.* it stimulates photosynthetic carbon uptake and counteracts calcification. Artificially elevated CO_2 concentrations in Baltic Sea water were shown to increase the photosynthetic carbon uptake of *Chara aspera*, *Chara horrida* and *Chara tomentosa* (Pajusalu et al. 2015), as well as that of *Furcellaria lumbricalis* and *Ulva intestinalis*, but not of *Fucus vesiculosus* (Pajusalu et al. 2013). This suggests that the natural content of CO_2 in the water of the Baltic Sea may limit the primary production of macrophytes today.

The observed positive effect of acidification on photosynthetic activity seems to be stronger in species with fast growth rates, such as *Ulva intestinalis*, than in species with low growth rates, such as *Fucus vesiculosus* (Fig. 11.46), which would lead to the conclusion that the predicted acidification in the Baltic Sea may cause shifts in the species composition of macrophyte communities towards a larger proportion of opportunistic filamentous algae. However, the majority of algae possess carbon-concentrating mechanisms which enable them to take up hydrogen carbonate as well as CO_2 (Raven et al. 2012), and it is uncertain how an 0.2–0.4 pH unit reduction, in combination with changes in other environmental drivers, would affect a macrophyte vegetation in the field.

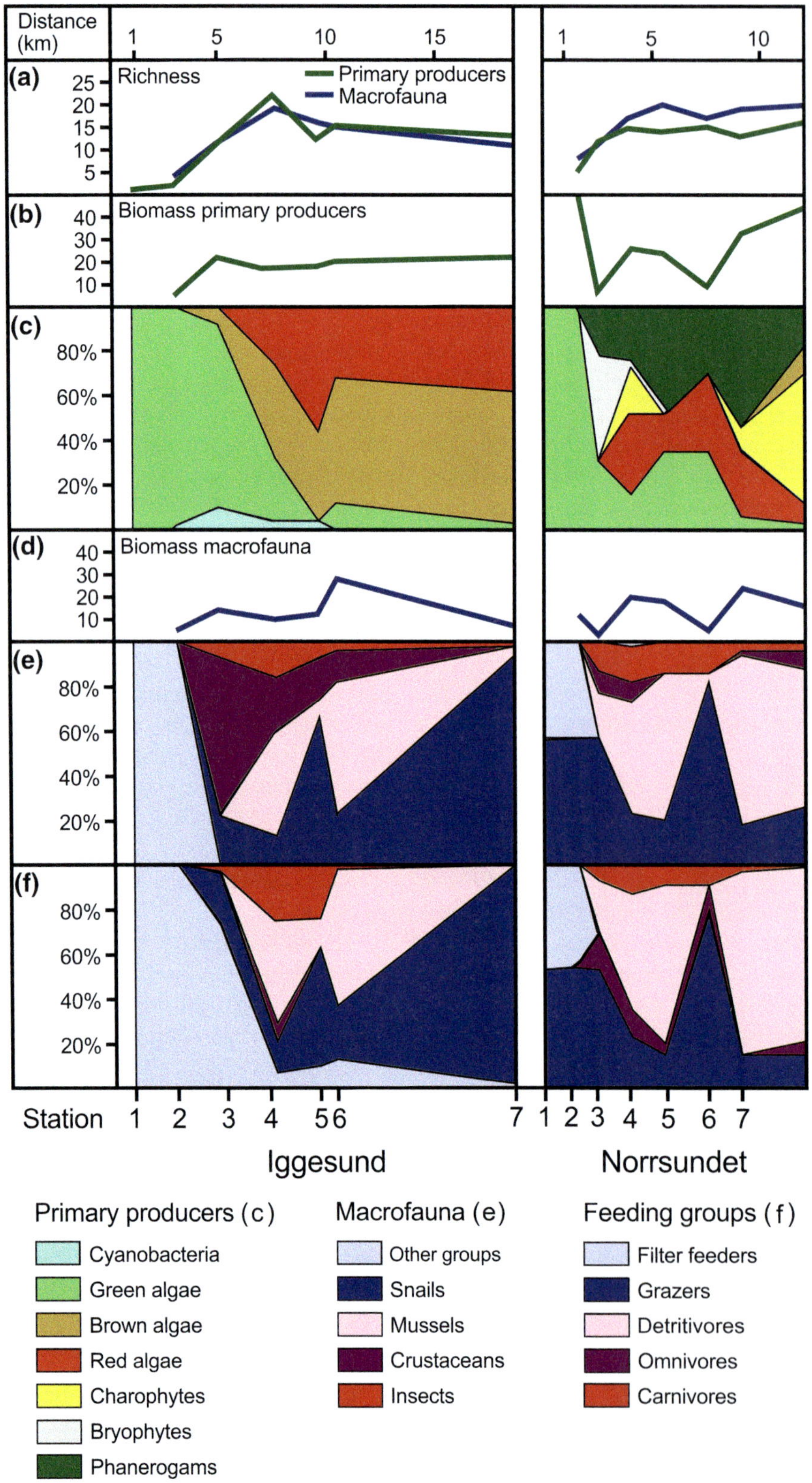

Fig. 11.45 Distribution of primary producers (macrophytes and cyanobacteria) and associated macrofauna in the phytobenthic zone along pollution gradients outside two pulp mills in the Bothnian Sea (Iggesund and Norrsundet) in the 1980s. (**a**) Species richness of primary producers and macrofauna. (**b**) mean biomass of primary producers. (**c**) % of total biomass of macrophytes + cyanobacteria. (**d**) = mean biomass of the macrofauna. (**e**) % of total biomass of the macrofauna. (**f**) % of total biomass of macrofaunal feeding groups. Numbers 1–7 indicate the sampling stations at each pulp mill, and the distance denotes the distance between the pulp mill outlet and the sampling station. Figure modified from Kautsky (1992)

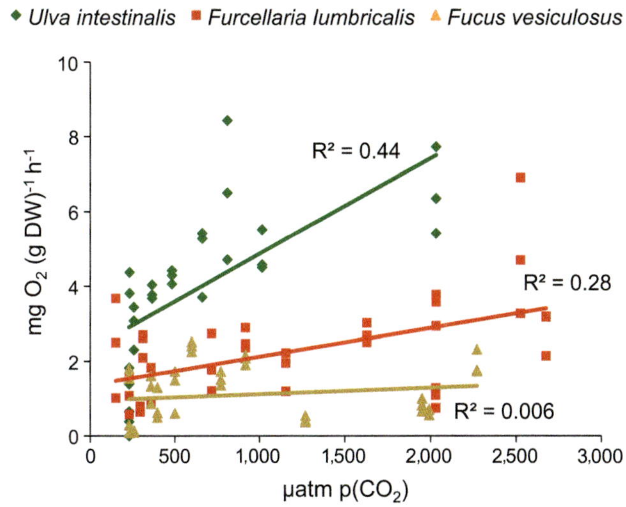

Fig. 11.46 Net primary production rates (expressed as oxygen evolution) of the three macroalgal species *Ulva intestinalis*, *Fucus vesiculosus*, and *Furcellaria lumbricalis* at different CO_2 concentrations in short-term incubation experiments. Figure modified from Pajusalu et al. (2013)

11.16 Long-term trends

11.16.1 Long-term trends in the *Fucus* belt

The external nutrient inputs into the Baltic Sea as a whole have dramatically increased since ~1950, having peaked in the 1980s and significantly decreased after that (*cf.* Box 2.2). These changes in nutrient inputs caused changes in the *Fucus vesiculosus* vegetation. The classical work in the Södra Kvarken area by Mats Wærn in the 1940s provides a baseline for the depth distribution of *Fucus vesiculosus* in unpolluted water (Wærn 1952). In 1943/44, the maximum development of the *Fucus* belt was found at a 5–6 m water depth and the lower depth limit of *Fucus vesiculosus* was 11.5 m. In 1984, the maximum development occurred at a 3–4 m water depth, and the lower depth limit was 8.5 m, in the same diving transects (Kautsky et al. 1986).

Both in 1943/44 and 1984 the decrease in *Fucus vesiculosus* coverage with water depth towards its lower depth limit could be approximately fitted to an exponentially decreasing light attenuation curve. These results indicate that the decreased water transparency in 1984 was due to a 40–50 % increases in summer chlorophyll *a* and nutrient concentrations in the offshore surface water of the Baltic Sea since the 1940s. Since the 1980s, *Fucus vesiculosus* has been expanding its depth distribution towards greater water depth, and in 2006 at almost all stations revisited the bladderwrack went as deep down as in the 1940s (Fig. 11.47), which suggests that the light conditions in the water column have improved. Also, in the Askö area of the Baltic Sea proper, the lower depth limit of *Fucus vesiculosus* has increased, from ~6 m in the 1970s to ~9 m in the 2010s, in the same diving transects.

11.16.2 Long-term trends in *Zostera* meadows

In a longer time perspective, the areas covered by the common eelgrass *Zostera marina* have decreased in the Skagerrak, Kattegat, Belt Seas and Baltic Sea proper (Boström et al. 2014). In the early 1930s, a fungal infection, the "wasting disease", caused by the slime mold *Labyrinthula zosterae*, eradicated a large proportion of the populations in the Kattegat, Belt Sea and the southern Baltic Sea, similarly to many other areas in the northern hemisphere (Rasmussen 1977). This event highlighted the importance of *Zostera marina* in protecting seashores from erosion and as a habitat for fish recruitment (Pihl et al. 2006).

Only in the 1950s and 1960s did the *Zostera marina* populations in the Baltic Sea Area start to recover after the fungal infection; however, since the 1980s declines in eelgrass depth limits and areal cover have been documented again, particularly in regions experiencing high anthropogenic pressure. Although the distribution may vary inter-annually by ~5 %, a large-scale long-term decline has been confirmed (Baden et al. 2003; Nyqvist et al. 2009). For example, the areal extent of eelgrass along the Swedish Skagerrak coast has declined by 60 % since the mid-1980s. In well-investigated areas such as the Puck Bay (Poland), the disappearance of *Zostera marina* was clearly related to

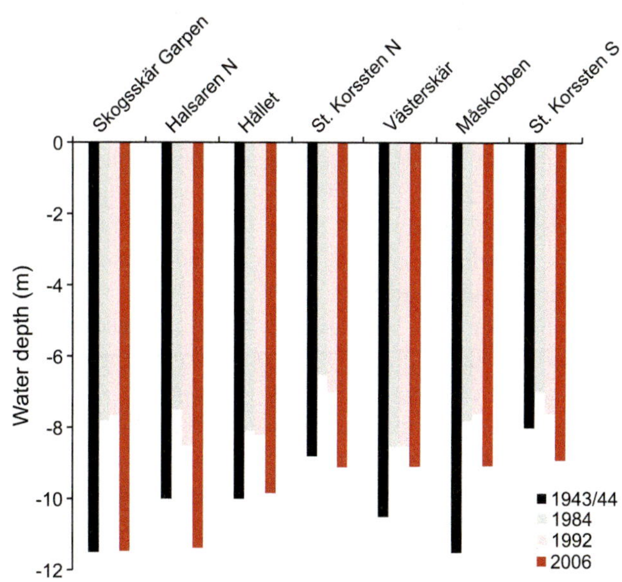

Fig. 11.47 The lower depth limit of *Fucus vesiculosus* at seven diving transects in the Södra Kvarken area. All transects were visited four times. Figure based on data for 1943/44 in Wærn (1952), for 1984 in Kautsky et al. (1986) and and additional unpublished data for 1992 and 2006 from the Swedish monitoring programme for phytobenthos

heavy eutrophication of the bay. Since the 1990s conditions have been improving, but the *Zostera marina* vegetation of the Puck Bay recovers slowly.

The failure of eelgrass to re-establish itself in affected areas, despite nutrient reductions and improved water quality, signals complex recovery trajectories and calls for large conservation efforts to protect the existing meadows (Boström et al. 2014). Even if the aboveground biomass in a *Zostera* meadow may be low in certain years with adverse growing conditions, new shoots will emerge in the next growing season if the rhizomes remain. However, once *Zostera marina* plants have disappeared, including their rhizomes that stabilise the substrate, it is difficult for new plants to re-establish healthy communities because of substrate instability. Close to the limit of its salinity tolerance in the Baltic Sea, where *Zostera marina* mainly grows vegetatively (Reusch et al. 1999), it may take a very long time (10–50 years or more) before the species re-establishes because vegetative (clonal) growth is slow (only 10–20 cm per year).

11.16.3 Long-term trends in *Mytilus*

In some areas of the Baltic Sea, a decreasing trend in the *Mytilus trossulus* biomass has been observed since the 1990s. This may be caused by decreased food availability (plankton) due to the eutrophication decrease in coastal waters. For example, in the Askö area in the Baltic Sea proper, the phytoplankton spring bloom has been almost halved between the years 1980 and 2012, and since the early 1990s *Mytilus* has decreased as well (Fig. 11.35). These changes are especially obvious in the inner parts of the archipelago, but there are also signs of a decrease in the outer archipelago. Detritivores should also decrease with a eutrophication decrease, and since the turn of the century they actually have decreased in the Askö area (Fig. 11.35).

Considerable fluctuations in the densities and biomass of *Mytilus trossulus* recorded at different locations in the Baltic Sea are difficult to explain by variation in environmental conditions such as salinity or nutrient concentrations (Westerbom and Jattu 2006). A virus disease or periodic oxygen deficiency may have been involved, especially in areas with large eutrophication problems. A decline of the *Mytilus trossulus* population in the Askö area in 1994, which also was observed in most parts of the Baltic Sea proper, was most probably due to unusually high temperatures down to the thermocline (at a ~20 m water depth) for a long period in summer and a simultaneous low pelagic primary production (Axén 1999). Due to the lack of food, the mussels probably respired themselves to death.

Finally, especially in the southern and eastern Baltic Sea proper, non-indigenous species may increase the predation pressure on *Mytilus trossulus* to an extent resulting in a drastic population decline. For example, *Mytilus trossulus* seems to be a favourite food item for the Chinese mitten crab *Eriocheir sinensis* (*cf.* Box 5.8) and the round goby *Neogobius melanostomus* (*cf.* Box 5.6), two non-indigenous species that are currently spreading in the Baltic Sea (Karlson et al. 2007; Wójcik et al. 2015).

Review questions

1. How does the vegetation reflect the type of substrate in an archipelago of the Baltic Sea? How does this relate to water movement?
2. How can morphology groups of macroalgae be used in ecology?
3. How would you distinguish between *Calothrix scopulorum* and *Verrucaria maura* macroscopically and microscopically?
4. Describe how the algal zonation on hard substrates differs between the Kattegat, Baltic Sea proper, Bothnian Sea and Bothnian Bay. How are these differences explained?
5. Which key species of the Baltic Sea ecosystem live in the phytobenthic zone? Why are they key species?

Discussion questions

1. Why are biotic interactions less important in the phytobenthic zone of the Baltic Sea than in the phytobenthic zone of Atlantic coasts?
2. Why is the phytobenthos important for the fish stocks of the Baltic Sea?
3. Why is the sequence "algae – *Amphibalanus improvisus* – *Mytilus trossulus* – *Cordylophora caspia* and *Einhornia crustulenta*" (and not in another order) in the microzonation shown in Fig. 11.15b?
4. What is the best way to study phytobenthic communities?
5. What is the best way to protect phytobenthic communities?

References

Agardh CA (1824) Systema algarum. Lundae Literis Berlingianis, 312 pp [in Latin]
Agardh CA (1830) Conspectus criticus diatomacearum, Part 1. Lundae Literis Berlingianus, 16 pp [in Latin]
Andersson AJ, Mackenzie FT, Lerman A (2005) Coastal ocean and carbonate systems in the high CO_2 world of the anthropocene. American Journal of Science 305:875–918
Andrulewicz E, Kruk-Dowgiałło L, Osowiecki A (2004) Phytobenthos and macrozoobenthos of the Słupsk bank stony reefs, Baltic Sea. Hydrobiologia 514:163–170
Aneer G (1985) Some speculations about the Baltic herring in connection with the eutrophication of the Baltic Sea. Canadian Journal of Fisheries and Aquatic Sciences 42. Supplement 1:83–90

Atkinson D (1994) Temperature and organism size – a biological law for ectotherms? Advances in Ecological Research 3:1–58

Atkinson D, Ciotti BJ, Montagnes DJS (2003) Protists decrease in size linearly with temperature: *ca.* 2.5 % degrees C^{-1}. Philosophical Transactions of the Royal Society B 270:2605–2611

Austin A (1959) Observations on *Furcellaria fastigiata* (L.) Lam. forma *aegagropila* Reinke in Danish waters together with a note on other unattached algae. Hydrobiologia 14:255–277

Axén J (1999) A STELLA®-simulation. Predicting population fluctuations of Baltic Sea blue mussels, *Mytilus edulis* L. Department of Systems Ecology, Stockholm University 1999,17:1–24 [MSc Thesis]

Baardseth E (1970) A square-scanning, two-stage sampling method of estimating seaweed quantities. Reports of the Norwegian Institute for Seaweed Research 33:1–41

Baden SP, Boström C, Arponen H, Tobiasson S, Moksnes PO (2010) Relative importance of trophic interactions and nutrient enrichment in seagrass ecosystems: a broadscale field experiment in the Baltic-Skagerrak area. Limnology and Oceanography 55:1435–1448

Baden SP, Gullström M, Lundén B, Pihl L, Rosenberg R (2003) Vanishing seagrass (*Zostera marina* L.) in Swedish coastal waters. AMBIO 32:374–377

Bellebaum J, Schirmeister B, Sonntag N, Garthe S (2013) Decreasing but still high: bycatch of seabirds in gillnet fisheries along the German Baltic coast. Aquatic Conservation: Marine and Freshwater Ecosystems 23:210–221

Bergström L, Tatarenkov A, Johannesson K, Jönsson RB, Kautsky L (2005) Genetic and morphological identification of *Fucus radicans* sp. nov. (Fucales, Phaeophyceae) in the brackish Baltic Sea. Journal of Phycology 41:1025–1038

Bhattacharyya S, Feferman L, Unterman T, Tobacman JK (2015) Exposure to common food additive carrageenan alone leads to fasting hyperglycemia and in combination with high fat diet exacerbates glucose intolerance and hyperlipidemia without effect on weight. Journal of Diabetes Research 2015. Article ID 513429:1–13

Blindow I (2000) Distribution of charophytes along the Swedish Coast in relation to salinity and eutrophication. Internationale Revue für Hydrobiologie 85:707–717

Boström C, Baden S, Bockelmann AC, Dromph K, Fredriksen S et al (2014) Distribution, structure and function of Nordic eelgrass (*Zostera marina*) ecosystems: implications for coastal management and conservation. Aquatic Conservation: Marine and Freshwater Ecosystems 24:410–434

Boström C, Bonsdorff E (1997) Community structure and spatial variation of benthic invertebrates associated with *Zostera marina* (L.) beds in the northern Baltic Sea. Journal of Sea Research 37:153–166

Boström C, Bonsdorff E, Kangas P, Norkko A (2002) Long-term changes in a brackish water *Zostera marina* community indicate effects of eutrophication. Estuarine Coastal Shelf Science 55: 795–804

Boström C, O'Brien K, Roos C, Ekebom J (2006) Environmental variables explaining structural and functional diversity of seagrass macrofauna in an archipelago landscape. Journal of Experimental Marine Biology and Ecology 335:52–73

Bučas M, Daunys D, Olenin S (2009) Recent distribution and stock assessment of the red alga *Furcellaria lumbricalis* on an exposed Baltic Sea coast: combined use of field survey and modelling methods. Oceanologia 51:341–359

Caldeira K, Wickett ME (2003) Oceanography: anthropogenic carbon and ocean pH. Nature 425:365

Choo KS, Nilsson J, Pedersén M, Snoeijs P (2005) Photosynthesis, carbon uptake and antioxidant defence in two coexisting filamentous green algae under different stress conditions. Marine Ecology Progress Series 292:127–138

Dayton PK (1975) Experimental evaluation of ecological dominance in a rocky intertidal algal community. Ecological Monographs 48:137–159

Dring M (1981) Chromatic adaptation of photosynthesis in benthic marine algae: an examination of its ecological significance using a theoretical model. Limnology and Oceanography 26:271–284

Durinck J, Skov H, Jensen FP, Pihl S (1994) Important marine areas for wintering birds in the Baltic Sea. EU DG XI research contract no. 2242/90-09-01. Ornis Consult Report, 110 pp

Ekebom J, Laihonen P, Suominen T (2003) A GIS-based stepwise procedure for assessing physical exposure in fragmented archipelagos. Estuarine, Coastal and Shelf Science 57:887–898

Engkvist R, Malm T, Tobiasson S (2000) Density-dependent grazing effects of the isopod *Idotea balthica* Pallas on *Fucus vesiculosus* L in the Baltic Sea. Aquatic Ecology 34:253–260

Eriksson BK, Johansson G (2003) Sedimentation reduces recruitment success of *Fucus vesiculosus* (Phaeophyceae) in the Baltic Sea. European Journal of Phycology 38:217–222

Eriksson BK, Johansson G, Snoeijs (1998) Long-term changes in the sublittoral zonation of brown algae in the southern Bothnian Sea. European Journal of Phycology 33:241–249

Eriksson BK, Ljunggren L, Sandström A, Johansson G, Mattila J, Rubach A, Råberg S, Snickars M (2009) Declines in predatory fish promote bloom-forming macroalgae. Ecological Applications 19:1975–1988

Eriksson BK, Rubach A, Hillebrand H (2006) Community dominance by a canopy species controls the relationship between macroalgal production and species richness. Limnology and Oceanography 51:1813–1818

Eriksson BK, Sandström A, Isæus M, Schreiber H, Karås P (2004) Effects of boating activities on aquatic vegetation in the Stockholm archipelago, Baltic Sea. Estuarine, Coastal and Shelf Science 61:339–349

Eriksson BK, Sieben K, Eklöf J, Ljunggren L, Olsson J, Casini M, Bergström U (2011) Effects of altered offshore food webs on coastal ecosystems emphasize the need for cross-ecosystem management. AMBIO 40:786–797

Ferry BW, Sheard JW (1969) Zonation of supralittoral lichens on rocky shores around the Dale peninsula, Pembrokeshire. Field Studies 3:41–67

Forsberg Å, Pekkari S (1999) Undersökning av undervattensvegetation och vattenkemi i nordligaste Bottenviken. Länsstyrelsen i Norrbottens län, Rapportserie 3/1999. [in Swedish]

Foster J, Hirst AG, Esteban (2013) Achieving temperature-size changes in a unicellular organism. The ISME Journal 7:28–36

Ganning B (1971) Studies on chemical, physical and biological conditions in Swedish rockpool ecosystems. Ophelia 9:51–105

Goecker ME, Kåll SE (2003) Grazing preferences of marine isopods and amphipods on three prominent algal species of the Baltic Sea. Journal of Sea Research 50:309–314

Gustafsson C, Boström C (2010) Biodiversity influences ecosystem functioning in aquatic angiosperm communities. Oikos 120: 1037–1046

Gustafsson C, Boström C (2013) Influence of neighboring plants on shading stress resistance and recovery of eelgrass, *Zostera marina* L. PLoS ONE 8(5):e64064

Håkansson L (1977) The influence of wind, fetch, and water depth on the distribution of sediments in Lake Vänern, Sweden. Canadian Journal of Earth Sciences 14:397–412

Hällfors G (1984) Filamentous rock-pool algae in the Tvärminne archipelago, S. coast of Finland. Acta Botanica Fennica 126:1–111

Hasselrot AT (1993) Insight into a psychomyiid life: towards the understanding of the ecology of the caddis fly *Tinodes waeneri* L. (Trichoptera, Psychomyiidae). Acta Universitatis Upsaliensis: Comprehensive summaries of Uppsala dissertations from the Faculty of Science and Technology, Uppsala University 13:1–24

Havenhand JN (2012) How will ocean acidification affect Baltic Sea ecosystems? An assessment of plausible impacts on key functional groups. AMBIO 41:637–644
Hawkins SJ, Hartnoll RG (1985) Factors determining the upper limits of intertidal canopy-forming algae. Marine Ecology Progress Series 20:265–271
Hayden HS, Blomster J, Maggs CA, Silva PC, Stanhope MJ, Waaland JR (2003) Linnaeus was right all along: *Ulva* and *Enteromorpha* are not distinct genera. European Journal of Phycology 38:277–294
HELCOM (2007) HELCOM list of threatened and/or declining species and biotopes/habitats in the Baltic Sea Area. Baltic Sea Environment Proceedings 113:1–17
HELCOM (2012) Checklist of Baltic Sea macro-species. Baltic Sea Environment Proceedings 130:1–203
HELCOM (2013a) HELCOM red list of Baltic Sea species in danger of becoming extinct. Baltic Sea Environment Proceedings 140:1–106
HELCOM (2013b) Climate change in the Baltic Sea Area: HELCOM thematic assessment in 2013. Baltic Sea Environment Proceedings 137:1–66
Hemmi A, Jormalainen V (2002) Nutrient enhancement increases performance of a marine herbivore via quality of its food alga. Ecology 83:1052–1064
Hillebrand H, Soininen J, Snoeijs P (2010) Warming leads to higher species turnover in a coastal ecosystem. Global Change Biology 16:1181–1193
Ilus E (2009) Environmental effects of thermal and radioactive discharges from nuclear power plants in the boreal brackish-water conditions of the northern Baltic Sea. STUK – Radiation and Nuclear Safety Authority, Helsinki, Finland, 380 pp
Isaeus M (2004) Factors structuring *Fucus* communities at open and complex coastlines in the Baltic Sea. Stockholm University, 24 pp [PhD Thesis]
Jansson AM (1974) Community structure, modelling and simulation of the *Cladophora* ecosystem in the Baltic Sea. Contributions from the Askö Laboratory, Stockholm University 5:1–130
Jansson AM, Kautsky N (1977) Quantitative survey of hard bottom communities in a Baltic archipelago. In: Keegan BF, O'Ceidigh P, Boaden PJS (eds) Biology of benthic organisms. Pergamon Press, Oxford, pp 359–366
Jansson BO (1978) The Baltic – a systems analysis of a semi-enclosed sea. In: Charnock H, Deacon G (eds) Advances in Oceanography. Plenum Press, New York NY, pp 131–183
Jansson BO, Aneer G, Nellbring S (1985) Spatial and temporal distribution of demersal fish fauna in a Baltic archipelago as estimated by SCUBA-census. Marine Ecology Progress Series 23:31–43
Jerlov NG (1976) Marine optics. Elsevier, New York NY, 231 pp
Johansson G, Snoeijs P (2002) Macroalgal photosynthetic responses to light in relation to thallus morphology and depth zonation. Marine Ecology Progress Series 244:63–72
Kangas Autio H, Hällfors G, Luther H, Niemi Å, Salemaa H (1982) A general model of the decline of *Fucus vesiculosus* at Tvärminne, south coast of Finland in 1977–1981. Acta Botanica Fennica 188:1–27
Karlson AML, Almqvist G, Skóra KE, Appelberg M (2007) Indications of competition between non-indigenous round goby and native flounder in the Baltic Sea. ICES Journal of Marine Science 64:479–486
Kautsky H (1984) Inventering av de grunda, vegetationstäckta bottnarna inom det planerade marina naturreservatet Salvorev, Sandö bank och Kopparstenarna, maj-juni 1983. Länsstyrelsen i Gotlands Län, 65 pp [in Swedish]
Kautsky H (1989) Quantitative distribution of plant and animal communities of the phytobenthic zone in the Baltic Sea. Contributions from the Askö Laboratory, Stockholm University 35:1–80
Kautsky H (1992) The impact of pulp mill effluents on phytobenthic communities of the Baltic Sea. AMBIO 21:308–313
Kautsky H (1995a) Quantitative distribution of sublittoral plant and animal communities in the Baltic Sea gradient. In: Eleftheriou A, Ansell AA, Smith CJ (eds) Biology and ecology of shallow coastal waters. Olsen and Olsen, Fredensborg, pp 23–30
Kautsky H (2013) Phytobenthos techniques. In: Eleftheriou A (ed) Methods for the study of marine benthos, 4th edn. John Wiley and Sons, Hoboken, NJ, pp 427–465
Kautsky H, Kautsky L, Kautsky N, Kautsky U, Lindblad C (1992) Studies on the *Fucus vesiculosus* community in the Baltic Sea. Acta Phytogeographica Suecica 78:33–48
Kautsky H, Martin G, Mäkinen A, Borgiel M, Vahteri P, Rissanen J (1999) Structure of phytobenthic and associated animal communities in the Gulf of Riga. Hydrobiologia 393:191–200
Kautsky H, Tobiasson S, Karlsson J (2011) Komplexa samband på bottnarna. Havet – om miljötillståndet i svenska havsområden 2011:36–41 [in Swedish]
Kautsky H, van der Maarel E (1990) Multivariate approaches to the variation in phytobenthic communities and environmental vectors in the Baltic Sea. Marine Ecology Progress Series 60:169–184
Kautsky N (1981) On the trophic role of the blue mussel (*Mytilus edulis*) in a Baltic coastal ecosystem and the fate of the organic matter produced by the mussels. Kieler Meeresforschung, Sonderhäft 5:454–461
Kautsky N, Evans S (1987) Role of biodeposition by *Mytilus edulis* in the circulation of matter and nutrients in a Baltic coastal ecosystem. Marine Ecology Progress Series 38:201–212
Kautsky N, Kautsky H, Kautsky U, Wærn M (1986) Decreased depth penetration of *Fucus vesiculosus* L. since the 1940s indicates eutrophication of the Baltic Sea. Marine Ecology Progress Series 28:1–8
Kautsky N, Wallentinus I (1980) Nutrient release from a Baltic *Mytilus*-red algal community and its role in benthic and pelagic productivity. Ophelia, Supplement 1:17–30
Kautsky U (1995b) Ecosystem processes in coastal areas of the Baltic Sea. Stockholm University, 25 pp [PhD Thesis]
Kautsky U, Kautsky H (1995) Coastal production in the Baltic Sea. In: Eleftheriou A, Ansell AA, Smith CJ (eds) Biology and ecology of shallow coastal waters. Olsen and Olsen, Fredensborg, pp 31–38
Kilpi M, Lorentsen SH, Petersen IK, Einarsson A (2014) Trends and drivers of change in diving ducks. TemaNord 2015: 516. Nordic Council of Ministers, Copenhagen, p 56
Kiirikki M (1996a) Experimental evidence that *Fucus vesiculosus* (Phaeophyta) controls filamentous algae by means of the whiplash effect. European Journal of Phycology 31:61–66
Kiirikki M (1996b) Mechanisms affecting macroalgal zonation in the northern Baltic Sea. European Journal of Phycology 31:225–232
Korpinen S, Honkanen T, Vesakoski O, Hemmi A, Koivikko R (2007) Macroalgal communities face the challenge of changing biotic interactions: review with focus on the Baltic Sea. AMBIO 36: 203–211
Korpinen S, Westerbom M (2010) Microhabitat segregation of the amphipod genus *Gammarus* (Crustacea: Amphipoda) in the northern Baltic Sea. Marine Biology 157:361–370
Krause-Jensen D, Carstensen J, Dahl K (2007) Total and opportunistic algal cover in relation to environmental variables. Marine Pollution Bulletin 55:115–125
Kurennoy D, Soomere T, Parnell KE (2009) Variability in the properties of wakes generated by high-speed ferries. Journal of Coastal Research SI 56:519–523
Larsson K, Tydén (2005) Effekter av oljeutsläpp på övervintrande alfågel *Clangula hyemalis* vid Hoburgs bank i centrala Östersjön mellan 1996/97 och 2003/04. Ornis Svecica 15:161–171 [in Swedish]

Lauringson V, Kotta J, Orav-Kotta H, Kotta I, Herkül K, Põllumäe A (2009) Comparison of benthic and pelagic suspension feeding in shallow water habitats of the northeastern Baltic Sea. Marine Ecology 30:43–55

Leidenberger S, Harding K, Jonsson PR (2012) Ecology and distribution of the isopod genus *Idotea* in the Baltic Sea: key species in a changing environment. Journal of Crustacean Biology 32:359–381

Littler NN, Littler DS (1984) Relationships between macroalgal functional form groups and substrata stability in a subtropical rocky-intertidal system. Journal of Experimental Marine Biology and Ecology 74:13–34

Ljunggren L, Sandström A, Bergström U, Mattila J, Lappalainen A et al (2010) Recruitment failure of coastal predatory fish in the Baltic Sea coincident with an offshore ecosystem regime shift. ICES Journal of Marine Science 67:1587–1595

Luther H (1951a) Verbreitung und Ökologie der höheren Wasserpflanzen im Brackwasser der Ekenäs-Gegend in Südfinnland. I. Algemeiner Teil. Acta Botanica Fennica 49:1–231 [in German]

Luther H (1951b) Verbreitung und Ökologie der höheren Wasserpflanzen im Brackwasser der Ekenäs-Gegend in Südfinnland. II. Spezieller Teil. Acta Botanica Fennica 50:1–370 [in German]

Lyons D, Arvanitidis C, Blight A, Chatzinikolaou E, Guy-Haim T et al (2014) Macroalgal blooms alter community structure and primary productivity in marine ecosystems. Global Change Biology 20:2712–2724

Mackenzie BR, Schiedek D (2007) Daily ocean monitoring since the 1860s shows record warming of northern European seas. Global Change Biology 13:1335–1347

Markager S, Sand-Jensen K (1992) Light requirements and depth zonation of marine macroalgae. Marine Ecology Progress Series 88:83–92

Martin G, Paalme T, Torn K (2006a) Growth and production rates of loose-lying and attached forms of the red algae *Furcellaria lumbricalis* and *Coccotylus truncatus* in Kassari Bay, the West Estonian Archipelago Sea. Hydrobiologia 554:107–115

Martin G, Paalme T, Torn K (2006b) Seasonality pattern of biomass accumulation in drifting *Furcellaria lumbricalis* community in waters of the West Estonian archipelago, Baltic Sea. Journal of Applied Phycology 18:557–563

Moksnes PO, Gullström M, Tryman K, Baden SP (2008) Trophic cascades in a temperate seagrass community. Oikos 117:763–777

Möller T, Kotta J, Martin G (2014) Spatiotemporal variability in the eelgrass *Zostera marina* L. in the northeastern Baltic Sea: canopy structure and associated macrophyte and invertebrate communities. Estonian Journal of Ecology 63:90–108

Munsterhjelm R (2005) Natural succession and human-induced changes in the soft-bottom macrovegetation of shallow brackish bays on the southern coast of Finland. Walter and Andrée de Nottbeck Foundation Scientific Reports 26:1–53

Nellbring S (1985) Abundance, biomass, and seasonal variation of fish on shallow soft bottoms in the Askö area, northern Baltic Proper. Sarsia 70:217–225

Nilsson J, Andersson J, Karås P, Sandström O (2004) Recruitment failure and decreasing catches of perch (*Perca fluviatilis* L.) and pike (*Esox lucius* L.) in the coastal waters of south-east Sweden. Boreal Environment Research 9:295–306

Nilsson J (2006) Predation of northern pike (*Esox lucius* L.) eggs: a possible cause of regionally poor recruitment in the Baltic Sea. Hydrobiologia 553:161–169

Nixon SW (1995) Coastal marine eutrophication: a definition, social causes, and future concerns. Ophelia 41:199–219

Norling P, Kautsky N (2008) Patches of the mussel *Mytilus* sp. are islands of high biodiversity in subtidal sediment habitats in the Baltic Sea. Aquatic Biology 4:75–87

Nyqvist A, André C, Gullström M, Baden SP, Åberg P (2009) Dynamics of seagrass meadows on the Swedish Skagerrak coast. AMBIO 38:85–88

Nyström-Sandman A, Wikström SA, Blomqvist M, Kautsky H, Isæus M (2012) Scale-dependent influence of environmental variables on species distribution: a case study on five coastal benthic species in the Baltic Sea. Ecography 35:1–10

Östman M, Rönnberg O (1991) Effects of ships' waves on rockpools in the Åland archipelago, northern Baltic Sea. Sarsia 76:125–132

Padilla DK, Allen BJ (2000) Paradigm lost: reconsidering functional form and group hypotheses in marine ecology. Journal of Experimental Marine Biology and Ecology 250:207–221

Paine RT (1966) Food web complexity and species diversity. American Naturalist 100:65–75

Paine RT (1995) A conversation on refining the concept of "keystone species". Conservation Biology 9:962–964

Pajusalu L, Martin G, Põllumäe A, Paalme T (2013) Results of laboratory and field experiments of the direct effect of increasing CO_2 on net primary production of macroalgal species in the brackish water ecosystems. Proceedings of the Estonian Academy of Sciences, Biology and Ecology 62:148–154

Pajusalu L, Martin G, Põllumäe A, Torn K, Paalme T (2015) Direct effects of increased CO_2 concentrations in seawater on the net primary production of charophytes in a shallow, coastal, brackish-water ecosystem. Boreal Environment Research 20:413–422

Pauwels K (2007) Biochemical adaptation for dormancy in subitaneous and dormant eggs of *Daphnia magna*. Hydrobiologia 594:91–96

Pedersén M, Snoeijs P (2001) Patterns of macroalgal diversity, community composition and long-term changes along the Swedish west coast. Hydrobiologia 459:83–102, Erratum (2002) Hydrobiologia 468:233

Pekkari S (1965) Notes on aquatic vegetation. Acta Phytogeographica Suecica 50:209–214

Pihl L, Baden S, Kautsky N, Rönnbäck P, Söderqvist T, Troell M, Wennhage H (2006) Shift in fish assemblage structure due to loss of seagrass *Zostera marina* habitats in Sweden. Estuarine, Coastal and Shelf Science 67:123–132

Rasmussen E (1977) The wasting disease of eelgrass *Zostera marina* and its effects on environmental factors and fauna. In: McRoy CP, Helfferich C (eds) Seagrass ecosystems – a scientific perspective. Marcel Dekker, New York NY, pp 1–51

Raven JA, Giordano M, Beardall J, Maberly SC (2012) Algal evolution in relation to atmospheric CO_2: carboxylases, carbon-concentrating mechanisms and carbon oxidation cycles. Philosophical Transactions of the Royal Society B 367:493–507

Raven J, Caldeira K, Elderfield H, Hoegh-Guldberg O, Liss P et al (2005) Ocean acidification due to increasing atmospheric carbon dioxide. The Royal Society Policy Document 12(05):1–60

Reusch TBH, Boström C, Stam WT, Olsen JL (1999) An ancient eelgrass clone in the Baltic. Marine Ecology Progress Series 183:301–304

Rodhe S, Molis M, Wahl M (2004) Regulation of anti-herbivore defence by *Fucus vesiculosus* in response to various cues. Journal of Ecology 92:1011–1018

Rodhe S, Wahl M (2008) Antifeeding defense in Baltic macroalgae: induction by direct grazing *versus* waterborne cues. Journal of Phycology 44:85–90

Rönnbäck P, Kautsky N, Pihl L, Troell M, Söderqvist T, Wennhage H (2007) Ecosystem goods and services from Swedish coastal habitats: identification, valuation, and implications of ecosystem shifts. AMBIO 36:534–544

Roos C, Rönnberg O, Berglund J, Alm A (2003) Long-term changes in macroalgal communities along ferry routes in a northern Baltic archipelago. Nordic Journal of Botany 23:247–259

Ruuskanen A, Bäck S, Reitalu T (1999) A comparison of two cartographic exposure methods using *Fucus vesiculosus* as an indicator. Marine Biology 134:139–145

Saderne V, Fietzek P, Herman PMJ (2013) Extreme variations of pCO_2 and pH in a macrophyte meadow of the Baltic Sea in summer: evidence of the effect of photosynthesis and local upwelling. PLoS ONE 8(4):e62689

Salo T, Gustafsson C, Boström C (2009) Effects of plant diversity on primary production and species interactions in a brackish water seagrass community. Marine Ecology Progress Series 396: 261–272

Salovius S, Kraufvelin P (2004) The filamentous green alga *Cladophora glomerata* as a habitat for littoral macrofauna in the northern Baltic Sea. Ophelia 58:65–78

Sandén P, Håkansson B (1996) Long-term trends in Secchi depth in the Baltic Sea. Limnology and Oceanography 41:346–351

Schubert H, Blindow I (eds) (2003) Charophytes of the Baltic Sea. Koeltz, Königstein 332 pp

Skov H, Heinänen S, Žydelis R, Bellebaum J, Bzoma S, et al. (2011) Waterbird populations and pressures in the Baltic Sea. TemaNord 2011:550. Nordic Council of Ministers, Copenhagen, 201 pp

Snoeijs P (1989) Effects of increasing water temperatures and flow rates on epilithic fauna in a cooling-water discharge basin. Journal of Applied Ecology 26:935–956

Snoeijs P (1990a) Effects of temperature on spring bloom dynamics of epilithic diatom communities in the Gulf of Bothnia. Journal of Vegetation Science 1:599–608

Snoeijs P (1990b) Marine and brackish waters. Acta Phytogeographica Suecica 84:187–212

Snoeijs P (1995) Effects of salinity on epiphytic diatom communities on *Pylaiella littoralis* (Phaeophyceae) in the Baltic Sea. Ecoscience 2:382–394

Snoeijs P, Kautsky U (1989) Effects of ice-break on the structure and dynamics of a benthic diatom community in the northern Baltic Sea. Botanica Marina 32:547–562

Snoeijs P, Prentice IC (1989) Effects of cooling water discharge on the structure and dynamics of epilithic algal communities in the northern Baltic. Hydrobiologia 184:99–123

Snoeijs P, Wakuru-Murasi L (2004) Symbiosis between diatoms and cyanobacterial colonies. Vie Milieu 54:163–170

Sonntag N, Schwemmer H, Fock HO, Bellebaum J, Garthe S (2012) Seabirds, set-nets, and conservation management: assessment of conflict potential and vulnerability of birds to bycatch in gillnets. ICES Journal of Marine Science 69:578–589

Stearns SC (1989) Trade-offs in life-history evolution. Functional Ecology 3:259–268

Strandmark A, Bring A, Cousins SAO, Destouni G, Kautsky H et al (2015) Climate change effects on the Baltic Sea borderland between land and sea. AMBIO 44(Supplement):S28–S38

Svedelius N (1901) Studier öfver Östersjons hafsalgflora. Uppsala University, 140 pp [PhD Thesis, in Swedish]

Svensson F, Norberg J, Snoeijs P (2014) Diatom cell size, coloniality and motility: trade-offs between temperature, salinity and nutrient supply with climate change. PLoS ONE 9(10):e109993

Svensson PA, Malm T, Engkvist R (2004) Distribution and host plant preference of *Idotea balthica* (Pallas) (Crustacea: Isopoda) on shallow rocky shores in the central Baltic Sea. Sarsia 89:1–7

Tedengren M, André C, Johannesson K, Kautsky N (1990) Genotypic and phenotypic differences between Baltic and North Sea populations of *Mytilus edulis* evaluated through reciprocal transplantations, III Physiology. Marine Ecology Progress Series 59:221–227

Tedengren M, Kautsky N (1986) Comparative study of the physiology and its probable effect on size in blue mussels (*Mytilus edulis* L.) from the North Sea and the Northern Baltic Proper. Ophelia 25:147–155

Thomsen MS, Wernberg T (2015) The devil in the detail: harmful seaweeds are not harmful to everyone. Global Change Biology 21:1381–1382

Tobacman J (2001) Review of harmful gastrointestinal effects of carrageenan in animal experiments. Environmental Health Perspectives 109:983–994

Torn K (2008) Distribution and ecology of charophytes in the Baltic Sea. Dissertationes Biologicae Universitatis Tartuensis 143:1–31

Torn K, Krause-Jensen D, Martin G (2006a) Present and past depth distribution of bladderwrack (*Fucus vesiculosus*) in the Baltic Sea. Aquatic Botany 84:53–62

Torn K, Martin G (2004) Environmental factors affecting the distribution of charophyte species in Estonian coastal waters, Baltic Sea. Proceedings of the Estonian Academy of Sciences, Biology and Ecology 53:251–259

Torn K, Martin G, Paalme T (2006b) Seasonal changes in biomass, elongation growth and primary production rate of *Chara tomentosa* in the NE Baltic Sea. Annales Botanici Fennici 43:276–283

Ulanova A, Busse S, Snoeijs P (2009) Coastal diatom-environment relationships in the brackish Baltic Sea. Journal of Phycology 45:54–68

Vahteri P, Mäkinen A, Salovius S, Vuorinen I (2000) Are drifting algal mats conquering the bottom of the Archipelago Sea, SW Finland? AMBIO 29:338–343

van den Hoek C (1987) The possible significance of long-range dispersal for the biogeography of seaweeds. Helgoländer Meeresuntersuchungen 41:261–272

Vermeer CP, Escher M, Portielje R, de Klein JJM (2003) Nitrogen uptake and translocation by *Chara*. Aquatic Botany 76:245–258

Vesakoski O, Boström C, Ramsay T, Jormalainen V (2008) Sexual and local divergence in host exploitation in the marine herbivore *Idotea balthica* (Isopoda). Journal of Experimental Marine Biology and Ecology 367:118–126

von Wachenfeldt T (1975) Marine benthic algae and the environment in the Öresund I-III. University of Lund, 328 pp [PhD Thesis]

Vuorinen I, Antsulevich AE, Maximovich NV (2002) Spatial distribution and growth of the common mussel *Mytilus edulis* L. in the archipelago of SW Finland, northern Baltic Sea. Boreal Environment Research 7:41–52

Wærn M (1945) Remarks on some Swedish Sphacelariaceae. Svensk Botanisk Tidskrift 39:396–420

Wærn M (1952) Rocky-shore algae in the Öregrund archipelago. Acta Phytogeographica Suecica 30:1–298

Wærn M (1965) A vista on the marine vegetation. Acta Phytogeographica Suecica 50:13–27

Wallentinus I (1979) Environmental influence on benthic macrovegetation in the Trosa-Askö area, northern Baltic Proper. II. The ecology of macroalgae and submersed phanerogams. Contributions from the Askö Laboratory, Stockholm University 25:1–210

Wallentinus I (1984) Comparisons of nutrient uptake rates for Baltic macroalgae with different thallus morphologies. Marine Biology 80:215–225

Wallentinus I, Pedersén M, Snoeijs P (1992) A tribute to Mats Wærn – the nestor of modern Swedish phycology. Acta Phytogeographica Suecica 78:1–9

Wennberg T (1992) Colonization and succession of macroalgae on a breakwater in Laholm Bay, a eutrophicated brackish water area (SW Sweden). Acta Phytogeographica Suecica 78:65–77

Westerbom M, Jattu S (2006) Effects of wave exposure on the sublittoral distribution of blue mussels *Mytilus edulis* in a heterogeneous archipelago. Marine Ecology Progress Series 306:191–200

Westerbom M, Kilpi M, Mustonen O (2002) Blue mussels, *Mytilus edulis*, at the edge of the range: population structure, growth and biomass along a salinity gradient in the northeastern Baltic Sea. Marine Biology 140:991–999

Wikström SA, von Wachenfeldt T, Kautsky L (2002) Establishment of the exotic species *Fucus evanescens* C. Ag. (Phaeophyceae) in Öresund, southern Sweden. Botanica Marina 45:510–517

Wójcik D, Normant M, Dmochowska B, Fowler A (2015) Impact of Chinese mitten crab *Eriocheir sinensis* on blue mussel *Mytilus edulis trossulus* – laboratory studies of claw strength, handling behavior, consumption rate, and size selective predation. Oceanologia 57:263–270

Worm B, Lotze HK, Bostrom C, Engkvist R, Labanauskas V, Sommer U (1999) Marine diversity shift linked to interactions among grazers, nutrients and propagule banks. Marine Ecology Progress Series 185:309–314

Worm B, Lotze HK, Sommer U (2001) Algal propagule banks modify competition, consumer and resource control on Baltic rocky shores. Oecologia 128:281–293

Yang B, Bhattacharyya S, Linhardt R, Tobacman J (2012) Exposure to common food additive carrageenan leads to reduced sulfatase activity and increase in sulfated glycosaminoglycans in human epithelial cells. Biochimie 94:1309–1316

Yousef MAM, Schubert H, von Nordheim H (eds) (2001) Charophytes in the Baltic Sea: threats and conservation. Schriftenreihe für Landschaftspflege und Naturschutz 72:1–44

Zettler ML, Gosselck F (2006) Benthic assessment of marine areas of particular ecological importance within the German Baltic Sea EEZ. In: von Nordheim H, Boedeker D, Krause JC (eds) Progress in marine conservation in Europe – NATURA 2000 sites in German offshore waters. Springer, Berlin, pp 141–156

Zulkifly SB, Graham JM, Young EB, Mayer RJ, Piotrowski MJ et al (2013) The genus *Cladophora* Kützing (Ulvophyceae) as a globally distributed ecological engineer. Journal of Phycology 49:1–17

Sandy coasts 12

Teresa Radziejewska, Jonne Kotta, and Lech Kotwicki

Abstract

1. Sandy coasts, including the epilittoral part of sandy beaches and the shallow sandy sublittoral, are particularly extensive in the southern and southeastern part of the Baltic Sea.
2. In the Baltic Sea ecosystem, sandy coasts function as biocatalytic filters by decomposing organic matter (including detritus) most of which originates directly or indirectly (*e.g.* via waterbirds) from the sea.
3. Sandy coasts are unstable, erodable environments which change in time and space due to *e.g.* erosion in winter and deposition of sand on the beaches in summer, and to the constant shifting of the substrate by winds and currents.
4. The sandy epilittoral and shallow sublittoral habitats support a variety of life forms, from microbes to birds, and are the space in which diverse processes involved in energy flow and matter cycling operate at different temporal and spatial scales.
5. The sandy coast food webs are partly based on the direct input of solar energy and nutrients used by primary producers (phytoplankton, microphytobenthos, macrophytes) whose production is subsequently utilised by invertebrates (meiobenthos, macrozoobenthos), fish and birds.
6. Another part of the sandy coast food webs is based on the input of organic material in the form of detritus, a source of energy for microbial communities consisting of bacteria, fungi, yeasts and actinomycetes as well as of heterotrophic protists living attached to sand grains and in the interstices.
7. Birds collect invertebrate prey from the sand on the beach or from the shallow sublittoral and contribute to the organic matter pool of the sandy habitat.
8. The sandy coasts of the Baltic Sea experience heavy anthropogenic pressure which primarily involves tourism and recreation, but also effects of eutrophication, establishment of non-indigenous species, sand extraction and dredging, fishing, infrastructure and shore defence constructions.

Keywords

Community structure • Detritus-based energy flow • Food-web interactions • Human impacts • Productivity • Sandy habitats

T. Radziejewska (✉)
Faculty of Geosciences, University of Szczecin,
ul. Mickiewicza 16a, 70383 Szczecin, Poland
e-mail: tera@univ.szczecin.pl

J. Kotta
Estonian Marine Institute, University of Tartu,
14 Mäealuse Street, 12618 Tallinn, Estonia
e-mail: jonne.kotta@ut.ee

L. Kotwicki
Institute of Oceanology, Polish Academy of Sciences,
ul. Powstańców Warszawy 55, 81712 Sopot, Poland
e-mail: lechk@iopan.gda.pl

P. Snoeijs-Leijonmalm et al. (eds.), *Biological Oceanography of the Baltic Sea*,
DOI 10.1007/978-94-007-0668-2_12

12.1 Shared by humans and other organisms

12.1.1 Sandy coasts are used by humans

Sandy coasts include the epilittoral part of sandy beaches and the shallow sandy sublittoral (*cf.* Fig. 11.1). A "sandy beach" is an expression that, to many a reader, conjures images of sunny holidays, sunbathing, swimming, having good time (Fig. 12.1a). The economic value of the sandy coasts of the Baltic Sea for the tourism industry is immense. To most beach visitors, the beach is just a stretch of sand-covered coast, devoid of any particular forms of (non-human) life save an occasional live bivalve found among shell debris at the waterline, and a seagull or two soaring above. To a marine biologist, a sandy beach and its extension, the shallow sandy sublittoral, are extremely interesting habitats which literally teem with life!

Like many popular beaches worldwide, the sandy coasts of the Baltic Sea experience numerous pressures. In addition to natural geological and climatic events, they are impacted by diverse anthropogenic activities associated with tourism and recreation, coastal defence construction, infrastructure development, etc. (Fig. 12.1). Periodically, the appearance of the beaches reflects the severe eutrophication of the Baltic Sea as they intercept masses of decaying algae and other debris (Fig. 12.2). These are usually the remains of large phytoplankton blooms and macrophytes wrecked in tangled smelly bundles or mats on the coast.

In addition to human uses, the sandy coasts provide important ecosystem services. They constitute a major link

Fig. 12.1 Examples of human uses of the large sandy shores in the southern Baltic Sea proper. (**a**) Bathing and kiting. (**b**) Fishing and industry. (**c**) Harbour constructions. Photo: © Hendrik Schubert

Fig. 12.2 Debris from the sea wrecked on the sandy shores of the southern Baltic Sea proper. (**a**) Green algae (*Cladophora glomerata* and *Ulva* sp.) at Świnoujście, Poland, (**b**) A mixture of filamentous algae at Kakumäe, Estonia, (**c**) The red alga *Furcellaria lumbricalis* and the round goby *Neogobius melanostomus*. (**d**) A pelagic bloom of cyanobacteria stranded on the beach. Photo: (**a**) © Teresa Radziejewska, (**b**) © Tiina Paalme, (**c**, **d**) © Sergej Olenin

in the Baltic Sea ecosystem as they function as biocatalytic filters by decomposing organic matter (including detritus) most of which originates directly or indirectly (*e.g.* via waterbirds) from the sea. Some inshore areas of the sandy sublittoral feature seagrass meadows (*cf.* Sect. 11.11). These habitats support a large diversity of autotrophic and heterotrophic life forms and function as nurseries for fish.

12.1.2 Sandy coasts are inhabited by biological communities

All species that inhabit sandy substrates have evolved tolerances to the ambient environmental conditions. The realised community structure integrates the outcome of behavioural choices made by individuals, the demographic performance of populations in different habitats, and intra- and inter-specific interactions. Biological patterns are often scale-dependent, and by describing spatial patterns on a range of scales we can capture essential elements of heterogeneity and match patterns with processes (Bergström et al. 2002).

The formation and development of sandy bottom communities are governed by numerous processes operating at various spatial and temporal scales. It is plausible that large-scale environmental drivers, *e.g.* climate-driven changes in water temperature, salinity, sea level or the intensity of ice scouring, can synchronise community changes over wide geographical areas and define broad patterns in sandy habitats. Within these broad patterns, smaller-scale processes, *e.g.* responses to oxygen level or organic enrichment operate at a lower intensity in order to modify distributions, abundances and functioning of the communities. As shown for the macrozoobenthos on intertidal sandflats outside of the Baltic Sea (Hewitt and Thrush 2009), the macrofaunal

Box 12.1: The mapping of sandy coasts

For conservation purposes, including the designation of coastal areas deserving protection, it is necessary that different shore biotopes be mapped. The spatial extent of different habitats within a sandy coast can be visualised using various methods and techniques. Among them, remote sensing is a tool that meets the requirements of large-scale mapping. It provides evidence of environmental changes and allows the deciphering of the processes responsible for these changes. A large variety of remote sensing instruments, from aircraft (Box Fig. 12.1) to satellites, is used to monitor sandy habitats (Kenny et al. 2003). Although remote sensing cannot capture the species composition of the organisms living in and on a sandy seabed, it has a significant advantage over traditional techniques (ships, diving) along the sandy coasts of the southern Baltic Sea proper, as it is spatially comprehensive. An assessment of the status of sandy habitats often does not require the distinction of species because variables such as the total macrophyte cover and/or different macrophyte growth forms are sufficient for such purposes. Nevertheless, the mapping of sandy habitats is still a considerable challenge in the Baltic Sea, both when using remote sensing and traditional techniques, because of the high turbidity of the water (Box Fig. 12.2).

Box Fig. 12.1 Aerial view of a typical beach in the southern Baltic Sea proper. The underwater bars parallel to the coast are visible as are signs of beach erosion in the form of arch-like incisions of the shoreline. Photo: © Piotr Domaradzki

Box Fig. 12.2 Differences in water turbidity just below the surface in different types of water bodies. (**a**) An oligotrophic clear-water lake. (**b**) A North-Atlantic kelp forest. (**c**) A coastal habitat in the southern Baltic Sea in summer. Photo: © Hendrik Schubert

responses to broad-scale climatic factors (*e.g.* variables related to the El Niño Southern Oscillation) depend on interactions between those factors and smaller-scale environmental variability induced by *e.g.* wind-generated wave exposure or turbidity, and the responses were not always consistent across sites or species during the 13-year period of study. It would be interesting to find out whether this inconsistency applies to the virtually non-tidal Baltic Sea as well. The knowledge of such variability may control our ability to predict the effects of environmental changes on communities inhabiting sandy habitats and model their spatio-temporal distribution.

12.2 Spatial extent of sandy coasts in the Baltic Sea Area

12.2.1 Geography

Sandy bottoms form a distinct zone of sedimentation in the Baltic Sea. Particularly spectacular is the almost unbroken stretch of sandy sediments from the northern coast of the Jylland Peninsula in Denmark (at the Kattegat-Skagerrak boundary) along the southern and eastern parts of the Baltic Sea coast to the Gulf of Riga (*cf.* Fig. 2.6). It is a zone of coastal sand accumulation that has been going on in the Baltic Sea during the Quaternary (Voipio 1981), especially in the southern and southeastern parts of the Baltic Sea proper and the Belt Sea. Large areas with sandy deposits also occur in some areas off the coast of the southern tip of Sweden, the islands of Gotland and Öland, and in some parts of the Gulfs of Bothnia and Finland. Beside these large coherent areas, smaller-sized sections of sandy bottoms occur all along the Baltic Sea coasts (Fig. 12.3).

12.2.2 A variety of habitats

Sandy coasts encompass a large variety of habitats, from plain beaches and underwater "desert plains" to species-rich and complex macrophyte meadows (Box 12.1). The complexity of sandy coasts increases northwards along the eastern coast of the Baltic Sea proper. The most diverse macrophyte communities are found *e.g.* along the Estonian coast. Locally, unvegetated substrates vary from plain sand to a mixture of silt, sand, pebble and rocks of different sizes.

Fig. 12.3 A small sandy beach on the Finnish archipelago coast. Photo: © Maria Laamanen

The vegetation on sandy substrates varies from sparsely occurring macrophytes to densely overgrown meadows. The vegetation may also vary from monospecies patches of different macrophytes to a mixture of vascular plants, ephemeral algae and perennial algae of different densities, dependent on the available substrate and exposure (*cf.* Sect. 11.1).

12.3 The sandy coastal environment

12.3.1 The physical environment

The sandy coast acts as an interface between sand, water and air. The beaches are formed by the deposition of particles brought in by currents from other areas. A part of the material transported results from shore erosion, but the major proportion originates from land and is delivered with the riverine runoff to the sea. The sandy coasts of the Baltic Sea are built primarily of quartz sands.

The sandy coasts are characterised by their slope, a product of wind patterns, storm events and the accompanying sand movement. The slope may vary over short distances and with the season. Autumn and winter storms tend to wash sediments down the slope, while under calm summer conditions sediments are washed up the slope. Accordingly, beaches are usually wider in summer than in the rough seasons. Wave action shapes the beach slope as well, which generally implies that the higher the wave energy, the stronger the water transport when the waves break and the flatter the beach.

As the Baltic Sea is a microtidal to non-tidal area (*cf.* Sect. 2.3.3), tides play no role in shaping the sandy coasts. Sand is typically deposited on beaches as the waves break on the shoreline and their energy dissipates. Particulates suspended in the wave are deposited on the beach and then dragged down its face again in the wave's backwash. Since the backwash energy tends to be much lower than the initial energy of the wave, there is typically a net onshore transport of sand.

Sandy substrates are unstable, erodable environments. Currents may be deflected by bottom topography to create horizontal pressure gradients at the sediment surface. This leads to advective flow of water through the sediment. Water movement also transports sand close to the seabed (a process known as the "bedload transport") and in suspension in the water column (McLachlan and Brown 2006). Signs of sand movements include, in the underwater part of the coastal area, ripple marks on the seabed (Fig. 12.4), sand accretion (a steep shore profile), and sand erosion (a flattened sand

Fig. 12.4 Ripple marks on a sandy seabed in the shallow sublittoral. (**a**) Bottom devoid of vegetation and with little debris. (**b**) Bottom with sorted sandy sediments, some debris and a large number of mysids. Photo: (**a**) © Christoffer Boström, (**b**) © Martynas Bučas

surface). Generally, sandy beaches experience cycles of erosion (primarily during storms) and accretion (during calm periods). Beach morphology responds dynamically to these cycles, and the beach protects the inland part of the coast in the process.

Morphodynamically, beaches are classified as: (1) dissipative beaches (wave energy dissipates before reaching the beach so that the beach is flat, depositional, and the fine sand-covered bottom features bars parallel to the shoreline), (2) intermediate beaches, and (3) reflective beaches (the wave energy is reflected from the beach face and the sand is coarse). The morphodynamics of a beach depends to some extent on its exposure to wind and wave action. Thus, beaches may be sheltered or exposed (McLachlan and Brown 2006).

12.3.2 The structure of the sandy habitat

The sandy coastal habitat is called the arenal or psammal, derived from the Latin word "arena" and the Greek word "psammos", both meaning "sand". Depending on the moisture content in the sand (and the proximity to the water line), the arenal represents the eupsammal (the upper part of the beach, a deep-lying water table, frequent freshwater seepage), the hygropsammal (the lower part of the beach intermittently covered by swash) and the hydropsammal (sand surface permanently covered by the water). These zones differ in their (micro-) habitat conditions, and their physical and chemical regimes affect the distribution of the sand inflora and infauna (called the "psammon"), particularly the smallest organisms living on and in-between the sand grains (Fig. 12.5).

Sands are classified by their grain size into *e.g.* coarse, medium or fine sand (*cf.* Table 2.3). The sand grains in the sediment are separated by spaces called the interstices that may account for 25–40 % of the total sediment volume. The interstices may be filled with water, gases, detritus and/or organisms (Fig. 12.5). The volume of the pores in the sand is called the sand porosity, a property used as a measure of the moisture-holding capacity of sand. Generally, the finer the sand, the higher its porosity, despite the decreasing size of the individual pore spaces (McLachlan and Brown 2006).

Another key feature of sandy sediments is their permeability, a measure of water flow rate through the interstices, which is important for the transport of *e.g.* oxygen into the sediment; generally, the finer the sand, the lower its permeability. Clogging of the interstices by organic material may temporarily reduce sediment permeability.

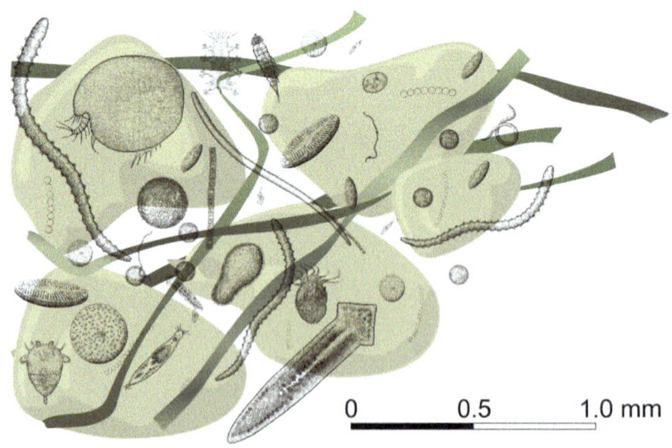

Fig. 12.5 Sand grains are separated by tiny voids called the interstitial spaces. These spaces are filled with water and support specialised organisms termed the "interstitial flora and fauna". The organisms shown schematically here include diatoms, ciliates, turbellarians, rotifers, gastrotrichs, oligochaetes, tardigrades, kinorhynchs and microcrustaceans. Green strands symbolise plant debris (e.g. seagrass blades) that may be buried in the sand. Figure: © Lech Kotwicki

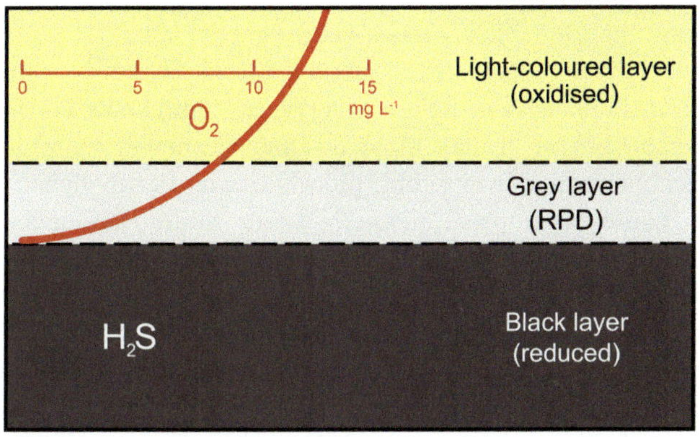

Fig. 12.6 Along a vertical sediment profile, the oxygen concentration (indicated in red) decreases to the point of oxygen depletion (hypoxia) in the redox potential discontinuity (RPD) layer, which is marked by a grey colour of the sand. Below the RPD layer there is a total lack of oxygen (anoxia), which is marked by a black colour of the sand due to the presence of hydrogen sulphide. Figure: © Halina Dworczak

12.3.3 Chemistry of the sandy habitat

The physics (*e.g.* temperature) and chemistry (*e.g.* salinity and the concentrations of oxygen, nutrients and organic compounds) of the interstitial water define the individual microhabitats within the psammic environment. In the sandy habitats of the Baltic Sea, the temperature changes seasonally and depends on the temperature of the air (primarily in the eupsammal) and of the seawater (in the hygro- and hydropsammal). During particularly severe winters, the interstitial water may freeze.

Temperature changes affect the rates of chemical processes on the sand surface and within the sandy sediments. Those processes are closely associated with the input of organic material supplied as dissolved organic matter (DOM) and particulate organic matter (POM) (Kotwicki et al. 2005b, c). Large organic matter inputs, coupled with low permeability of sand (dissipative beaches), result in sharp oxygen concentration gradients with depth in the sand (Fig. 12.6). Oxygen concentration decreases to the point of oxygen depletion (hypoxia) in the so-called "redox potential discontinuity" (RPD) layer, which is marked by a grey colour of the sand. The lack of oxygen (anoxia) below the RPD layer is manifested as a black colouration of the sand due to the presence of hydrogen sulphide (or, more precisely, of metal sulphides).

Both forms of organic matter in the sandy sediments (DOM and POM) undergo mineralisation mediated by microorganisms. The resultant nutrients, primarily nitrates and ammonia (McLachlan and Brown 2006), are stored in the microbial biomass and/or are adsorbed onto sand grains. The eutrophic water of the Baltic Sea proper and the Gulfs of Finland and Riga may supply additional nutrients to the coastal environment and enhance the growth of cyanobacteria, microalgae and macrophytes.

The net result of the combination of different environmental drivers is the presence of a variety of habitats, discernible on various scales, from (1) macrohabitats, *e.g.* different beach types, to (2) mesohabitats, *e.g.* shell accumulations (Fig. 12.7), debris-covered sand (Fig. 12.2) and clean sand within a beach, to (3) microhabitats, *e.g.* oxygen concentration-dependent vertical zonation within the sand.

Fig. 12.7 Some beach stretches feature copious accumulation of shell debris, as shown here on a beach in the southern Baltic Sea at Świnoujście, Poland. The presence of such shell debris is in turn conducive to heavy accumulation of coarse and fine detritus. Photo: © Teresa Radziejewska

12.4 Heterotrophic microbial communities

12.4.1 Microbial biomass and abundance

Broadly speaking, the sandy coast food webs involve two basic, interlinked energy flow and matter cycling pathways. One is based on a direct input of nutrients and solar energy, utilised by primary producers (microphytobenthos, phytoplankton and macrophytes) during photosynthesis. The net result is growth of the primary producers and their subsequent utilisation by primary consumers among the meiobenthos and macrozoobenthos. The other pathway relies on DOM and POM (the latter termed "detritus") supplied by the sea, the land and the atmosphere. Autochthonous primary producers and their consumers on the sandy coast itself contribute to the detrital pool as well. The detrital pool is a source of energy for a rich, albeit incompletely known, microbial community consisting of bacteria, fungi, yeasts and actinomycetes as well as protists. They live attached to sand grains and in the interstices.

Bacteria are by far the best known component of the heterotrophic microbial communities on the sandy coasts of the transition zone (Belt Sea and Kattegat) and the southern Baltic Sea (Meyer-Reil et al. 1978; Sundbäck et al. 1996; Podgórska et al. 2008). The bacterial abundances range within $0.03–2.7 \times 10^9$ cells g^{-1} dry sediment and the biomass ranges between 5.6 and 14.6 µg C g^{-1} dry sediment. On the beaches of the Kiel Fjord and the Kiel Bay (southwestern Belt Sea), the bacterial biomass was estimated to be four times that of the meiofauna (Meyer-Reil and Faubel 1980).

On a southern Baltic Sea beach on the Polish coast, studied by Podgórska et al. (2008), bacterial communities showed both vertical and cross-shore zonation in their distribution. Bacterial genera such as *Acinetobacter*, *Bacillus*, *Escherichia*, *Pseudomonas* and *Serratia* were found concentrated at the sand surface whereas *Micrococcus*, *Photobacterium* and *Vibrio* were more abundant deeper (5–10 cm) in the sediment. The highest bacterial abundances and biomass were recorded at the water line (Podgórska et al. 2008).

12.4.2 Microbial decomposition processes

On account of all the processes mediated by microbial communities, the sandy coastal environment, and its interstitial system in particular, acts as a natural biological nutrient filter. The many different strains of bacteria occurring on the sandy coasts differ in their physiology and activity, as shown by differences in the substances they utilise (decompose) and the rate of the decomposition process. Decomposition of organic matter proceeds under both oxic and anoxic conditions, the former being much faster than the latter.

As a source of energy, bacteria use primarily amino acids (from proteins) and lipids (Mudryk and Podgórska 2005). Some bacteria, mainly those dwelling deeper in the sediment, are active in anaerobic processes and reduce sulphates to hydrogen sulphide (*Desulfovibrio*), while *Thiobacillus* oxidises hydrogen sulphide back to sulphates. Other bacteria are responsible for oxidation of ammonia to nitrogen dioxide and further to nitrates in the aerobic nitrification process. The nitrates are subsequently reduced to ammonium and nitrogen gas (denitrification). Diazotrophic (nitrogen-fixing) bacteria (cyanobacteria and heterotrophs) convert nitrogen gas into nitrate (nitrification). Thus, bacteria play an extremely important role in the decomposition (mineralisation) of numerous organic compounds and in nutrient recycling on the sandy coast.

Bacteria are also known to be involved in decomposition and inactivation of chemical pollutants (petroleum products, heavy metals) in the sandy coastal environment and are therefore important for beach cleaning. The bacterial genera *Acinetobacter* and *Micrococcus*, which dominate the microbial communities on the sandy coasts of the southern Baltic Sea (Mudryk and Podgórska 2005), seem to be particularly abundant on polluted beaches.

Other heterotrophic microbes (fungi, yeasts, actinomycetes) of sandy coastal environments of the Baltic Sea have been studied much less extensively than bacteria. On the southern Baltic Sea beaches, the respective abundances of fungi, yeasts and actinomycetes were found to range within 70–22,500, 0–2,400 and 0–1,330 g^{-1} dry sediment (Mudryk and Podgórska 2007). These organisms occur most abundantly near dunes and are involved in the colonisation and aggregation of dune sand.

12.5 Primary producers

12.5.1 Microphytes

The sandy beaches and the shallow sandy sublittoral of the Baltic Sea support photosynthetic organisms representing different levels of organisation and modes of life. This includes microbenthic algae (*e.g.* diatoms and cyanobacteria), macroalgae (*e.g.* unattached or epiphytic green algae such as *Ulva* spp. and *Cladophora glomerata* and brown algae such as *Pylaiella littoralis*) and vascular plants (*e.g.* the common eelgrass *Zostera marina*).

An additional photosynthetic component is the coastal phytoplankton in the water column above the shallow sandy substrate, which, depending on the season, is dominated by

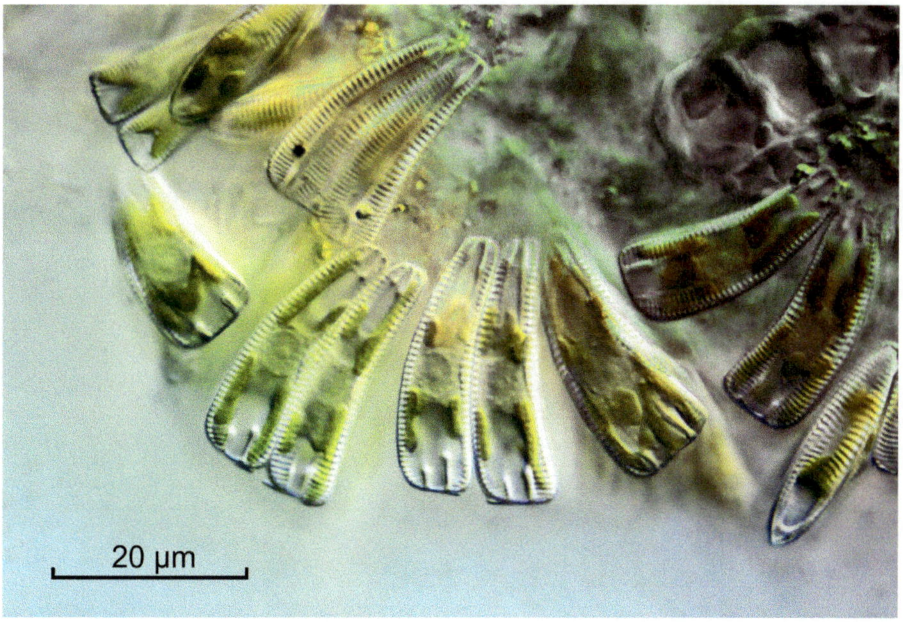

Fig. 12.8 The pennate diatom *Rhoicosphenia abbreviata* lives attached to sand grains, washed-ashore macroalgae and larger chunks of hard substrate (*e.g.* shell fragments) on the beaches in the southern Baltic Sea. Photo: © Wolfgang Bettighofer

diatoms, chlorophytes, dinoflagellates or cyanobacteria. The microphytobenthos on the sediment comprises primarily diatoms (Pliński and Kwiatkowski 1996). An example is the pennate diatom *Rhoicosphenia abbreviata* (Fig. 12.8) which lives mainly epiphytically on macrophytes (*cf.* Sect. 11.4.2), but may also attach to coarse-grained sand and larger chunks of hard substrate (*e.g.* bivalve shell fragments). Many hundreds of diatom species live in the upper few cm of the sandy sublittoral sediment. These diatom communities consist of epipsammic diatoms attached to the sand grains (Fig. 12.9a–c) and epipelic diatoms that move on and in-between the sand grains in the interstitial spaces (Fig. 12.9d–f). Typical of sandy sediments is that the species are (even for diatoms) rather small, which is related to the small size of interstitial space and the instability of sandy habitats.

The microphytobenthos utilises nutrients, mainly inorganic but also organic nitrogen supplied with the water, and thus are involved in nutrient cycling on the sandy coast (Sundbäck et al. 2000). Microphytobenthic (gross) primary production on the Baltic Sea sandy coasts has been estimated at 0.2–41.8 mg C m^{-2} h^{-1}, depending on the season, with maximum values in July, and location, with the highest production efficiency (production to biomass ratio, P/B) at the water line (Urban-Malinga and Wiktor 2003).

In addition to being involved in the nutrient filter system of the sandy coast, the microphytobenthos provides a trophic resource to heterotrophic microbes as well as to meio- and macrofaunal grazers (McLachlan and Brown 2006). By secreting extracellular polymeric substances (mucilage), the microphytobenthos may also contribute to the sandy sediment stability (Lund-Hansen et al. 2002), although the stabilisation effect is highly dependent on water depth.

12.5.2 Macrophytes

In places where high exposure does not prevent the occurrence of vegetation, vascular plants and charophytes can be habitat-forming on shallow sandy substrates (Fig. 12.10a–e) and the stabilising effect of the macrophyte vegetation may occasionally shape the coast. In places with unstable sandy substrate, vegetation occurs only on patches with a hard substrate (Fig. 12.10f).

The occurrence of macrophytes rooted in sandy substrates, together with their epiphytes and loose-lying aggregates of macrophytes depends to a large extent on the degree of wave exposure, with different species having different exposure optima. For example, *Zostera marina* (Fig. 12.10a) usually inhabits moderately exposed sandy bottoms where water movement prevents the accumulation of organic matter, whereas another vascular plant, the Eurasian watermilfoil *Myriophyllum spicatum* (Fig. 12.10e) prefers areas with low wave energy, where the sediment accumulates high amounts of organic matter. Wind stress that extends down to the seabed may cause considerable resuspension of the sediment, which results in prolonged periods of low-light

Fig. 12.9 The silica frustules of some typical epipsammic (**a**–**c**) and epipelic (**d**–**f**) diatoms from shallow sandy sediments in the Baltic Sea under scanning electron microscopy. (**a**) *Martyana schulzii*. (**b**) *Martyana atomus*. (**c**) *Pseudostaurosira brevistriata*. (**d**) *Chamaepinnularia margaritiana*. (**e**) *Amphora pediculus*. (**f**) *Pseudofallacia tenera* (syn. *Fallacia tenera*). The scale bar in all images is 5 µm. Photo: © Pauline Snoeijs-Leijonmalm

conditions (Madsen et al. 2001) and causes physical disturbance to the benthic communities (Herkül et al. 2011). Locally, the distribution of macrophytes is governed mainly by the availability of suitable substrate, light and nutrients, while upwelling, turbulence, turbidity and grazing intensity act as secondary factors (Appelgren and Mattila 2005). *Zostera marina* is a key species in sandy bottom habitats in the Kattegat, the Belt Sea and the Baltic Sea proper (*cf.* Sect. 11.11). Its ecosystem services consist of stabilising the seabed, reducing sediment resuspension, influencing sediment deposition and composition, releasing oxygen into the sediment, acting as a nutrient filter and providing habitat complexity and food for a great variety of algal and animal species (Herkül and Kotta 2009).

Thus, a *Zostera marina* canopy significantly promotes diversity and increases the density of benthic macroalgae and invertebrates. Removal of the *Zostera marina* canopy reduces the biomass of epiphytic macroalgae and, consequently, results in a decrease in the abundance of herbivores such as gammarid amphipods and idoteids (Herkül and Kotta 2009). The presence of the canopy promotes larval settlement of sessile invertebrates such as *Mytilus trossulus* and provides a relatively stable habitat for adults (Reusch and Chapman 1995).

12.5.3 Microbial and macroalgal mats

The sheltered lower shore not covered permanently by the water, *e.g.* in certain locations in the southern and southwestern Baltic Sea proper, periodically supports specific microphytobenthos-dominated structures, the so-called “microbial mats”. On the German coast, the area covered by such mats increased from 200–500 m^2 in 2002 to ~30,000 m^2 in 2008, which may be a sign of eutrophication (Heyl et al. 2010). The mats are laminated structures, the laminae being formed by benthic cyanobacteria, diatoms and heterotrophic bacteria arranged in layers, with sulphate-

Fig. 12.10 Vegetation in sandy habitats. (**a**) A *Zostera* meadow. (**b**) *Stuckenia pectinata* with a few small filamentous macroalgae as epiphytes. (**c**) In the foreground vascular plants overgrown with short filamentous macroalgae, in the background *Zostera marina*. (**d**) *Zannichellia palustris* overgrown with long filamentous macroalgae. (**e**) *Myriophyllum spicatum* overgrown with long filamentous macroalgae. (**f**) A site exposed to wave action (note the ripple marks) and a patch of hard substrate with the green alga *Ulva lactuca* and red algae. Photo: Piotr Bałazy

reducing bacteria in the lowest layer. These mats are centres of high primary and secondary production (they accumulate organic matter), and induce hypoxic and anoxic conditions in the sediment beneath them.

As a result of the summer cyanobacterial blooms, masses of cyanobacteria floating on the water surface may be stranded on the beaches of the Baltic Sea proper (Fig. 12.2d), occasionally in several cm thick layers, to deteriorate. In addition to providing organic enrichment and inducing oxygen deficiency in the underlying sediment, such aftermath of a cyanobacterial bloom greatly reduces the amenity value of the beach.

Advanced eutrophication also produces a visible effect on the sandy coasts of the Baltic Sea in the form of decomposing masses of formerly drifting macroalgal mats (Fig. 12.2a–c). Thick mats may cause organic enrichment of the sediment, oxygen depletion beneath the sand surface and alteration of the infaunal community structure favouring a few stress-tolerant species (Bonsdorff 1992; Norkko and Bonsdorff 1996; Gamenick et al. 1996). However, when the drifting algal accumulations are moderate in size, they may attract various benthic invertebrates and support diverse communities in sandy habitats near the beach (Lauringson and Kotta 2006).

In some places, tourist beaches are affected by cyanobacterial or macroalgal mats (with associated epiphytes, bacteria and fauna) to such a high degree that they are regularly machine-cleaned (*e.g.* Malm et al. 2004), which is relatively easy on flat homogeneous sandy beaches.

12.6 Interstitial micro- and meiofauna

12.6.1 Adaptations to interstitial life

The Baltic Sea sandy beaches and submerged shallow marine sands that appear to consist of clean mineral grains and nothing else, harbour rich and diverse communities of microorganisms, meio- and macrofauna, *i.e.* categories of organisms distinguished by their body size (Schwinghamer 1981). Their species diversity rivals that of terrestrial systems. All the components of these sedimentary communities are involved in matter and energy flow cycles, and enhance the biocatalytic filtration capacity of the permeable sandy beds (*cf.* Sect. 12.9.2).

The fauna of sandy coasts includes representatives of most invertebrate phyla, occurring as interstitial forms (psammic micro- and meiofauna) and as members of the macrofauna. In contrast to the wave-swept sand surface, the interstitial system is three-dimensional and offers a living space that extends deep into the sediment. The psammic meiofauna, also known as the "sand meiobenthos" or the "meiobenthic psammon", is defined here as organisms (heterotrophic protists and metazoans) that pass undamaged through 1 mm mesh size sieves and are retained on 0.032 mm sieves (Giere 2009). However, these size limits are not strict and are also partly habitat-dependent. The upper size limit of 0.5 mm and the lower size limits of 0.044 mm or 0.063 mm are commonly used as well (*cf.* Table 4.1).

Heterotrophic protists can fall within the micro- and meiofauna size ranges. Ciliates, foraminifers, amoebae and nanoflagellates are abundant in the sand interstices, with densities ranging within 45–120,000 individuals per cm^3 of sediment (Dietrich and Arndt 2000; Smirnov 2002). The species richness of foraminifers is very low in the Baltic Sea, while that of ciliates is high (*cf.* Sect. 4.4.3). The sandy coast protists are usually dominated by heterotrophic nanoflagellates and ciliates, the latter known to be particularly tolerant of reducing conditions, and therefore occurring especially abundantly deeper in the sediment, near the RPD layer (Fenchel 1969). Amoebae, heterotrophic flagellates and foraminifers are concentrated on the sand surface, and their abundances diminish strongly with increasing sediment depth. Heterotrophic protists are trophically important in the sandy coast food webs as they feed on bacteria, microphytobenthos and other protists. They are a link both in the microbial loop component of the food web and in the energy transfer up the food chain (McLachlan and Brown 2006).

Because of their size, psammic metazoan meiobenthos organisms are also capable of living in the sand interstitial system. Interstitial metazoans are adapted to living in the unstable environment of the interstices by attaining small size and by having evolved elongated and flexible bodies. Many have means or organs with which to adhere to sand grains, and have developed a variety of adaptations: (1) reproductive, *e.g.* direct sperm transfer or internal fertilisation, few eggs, brood protection, (2) life cycle-related, *e.g.* shortened or non-existent larval stages, and (3) trophic, *e.g.* selective or non-selective detritus feeding, scraping food from sand grains, and predation on other invertebrates (Giere 2009).

12.6.2 Meiofaunal abundance

On most sandy coasts, the interstitial metazoan meiofauna is numerically rich, occurring at abundances on the order of magnitude of up to 1,000–10,000 individuals under a 10 cm^2 sand surface (*e.g.* Kotwicki et al. 2005a). The abundances vary along a sandy coastal profile, the highest densities being encountered at, or near to, the water line (Jończyk and Radziejewska 1984). In the shallow sandy sublittoral, densities on the order of 1,000 individuals under 10 cm^2 sand surface are usually recorded, abundances higher by an order of magnitude being observed in organic-rich near-shore sands (Rokicka-Praxmajer et al. 1998). The meiobenthic abundances may be enhanced by the presence of *Mytilus trossulus* aggregations (Radziejewska 1986).

Generally, the abundance of the interstitial meiofauna is correlated with sand grain size (the finer the sand the more abundant the meiobenthos) and the sediment organic matter content (more organic matter in finer sand), but also with salinity and temperature (Lokko et al. 2014). The highly organically enriched sand of a southern Baltic Sea coast was found to support meiofaunal abundances on the order of magnitude as high as 10,000 individuals under a 10 cm^2 sand surface (T. Radziejewska, personal observation).

Meiobenthic abundances vary seasonally. On a southern Baltic Sea coast, the highest abundances are typically found in spring (May-June) and autumn (October-November) (Jończyk and Radziejewska 1984). However, the seasonal patterns of the psammon seem to be not only taxon-specific but also modulated by anthropogenic activities and feeding conditions for the psammon, which are not necessarily related to seasonal changes (Lokko 2014).

The highest abundances of the psammic meiofauna are usually found from just below the sand surface to a few cm into the sand, the densities decreasing gradually with depth in the sediment. The limit of meiofaunal occurrence in the sediment is greatly dependent on the interstitial water oxygen content. In permeable sands, the psammic meiobenthos may be found as deep as half a metre down into the sediment, whereas the RPD level, and hence oxygen availability, usually restricts the depth distribution (Jansson 1967). For this reason, the interstitial fauna of the upper shore is confined to moist sand, but is absent in the groundwater below.

12.6.3 Meiofaunal diversity

In addition to being abundant, the psammic meiobenthos is also diverse. However, on the Baltic Sea beaches it is commonly strongly dominated by one or two taxa only. Usually the dominants are free-living nematodes, while other numerically important taxa include turbellarians, gastrotrichs, harpacticoid copepods, oligochaetes, ostracods and tardigrades (Fig. 12.5).

In summer, the meiobenthos may be dominated by a seasonal component, for example by meiofaunal-sized juvenile bivalves and/or polychaetes. In the shallow sandy littoral studied by Haque et al. (1997), the meiobenthos was dominated by oligochaetes. The meiobenthic taxonomic richness on the sandy coast is dependent on the sediment grain size, the pattern being opposite to that of the abundance (the coarser the sediment the more taxa present). In addition, the composition and relative abundance of various taxa change across the shore profile, with nematodes and gastrotrichs being dominant close to the water level and oligochaetes gaining in importance further up the beach. The interstitial fauna on some coarse-grained southern Baltic Sea coasts was observed to be dominated by turbellarians (T. Radziejewska, personal observation).

Although the interstitial fauna as a distinct community of diverse forms of life was first reported in the 1930s from the Baltic Sea coasts by Remane (1933), and later in the review paper by Bertil Swedmark (1964) titled "The interstitial fauna of marine sand" (which has become a literature classic), the knowledge of the Baltic Sea Area's sandy beaches and shallow water meiobenthos taxonomic composition, and hence biodiversity, is still far from complete. Routine surveys are restricted taxonomically to major taxonomic groups (nematodes, gastrotrichs, etc.), and only rarely are species lists produced.

Taxonomically, the best known psammic taxa are gastrotrichs (Kolicka et al. 2014), harpacticoid copepods (Drzycimski 1974) and rotifers (Lokko 2014). On the other hand, the numerically dominant nematodes are known to the genus level at best, and lists of species and genera of selected major taxa (mainly nematodes) exist for a few locations only, *e.g.* the Pomeranian Bay (Rokicka-Praxmajer and Radziejewska 2002), the Gulf of Finland (Gerlach 1953), the Gulf of Riga (Pallo et al. 1998), and the Hel Peninsula, Poland (Urban-Malinga et al. 2006). These lists show the presence of 22 (Gulf of Riga) to 27 (Hel Peninsula) to 35 (Pomeranian Bay) nematode genera, which provides evidence of a considerable diversity. Clearly, the meiobenthos of the Baltic Sea's sandy coasts holds an enormous potential for further taxonomic research.

12.6.4 Ecological roles of the interstitial meiofauna

The psammic meiofauna plays diverse roles on the sandy coasts. It is regarded as a trophic link between microorganisms (including microphytobenthos) and larger fauna (including small fish). It enhances the rate of carbon mineralisation and nutrient regeneration by stimulating microbial activity through grazing and assimilation of detritus, while predatory meiobenthic taxa exert control on a part of the community.

In addition, the meiobenthic organisms are highly sensitive to anthropogenic pressures, which makes them excellent indicators of pollution. Because many meiofauna have relatively short life cycles, effects of a contaminant on the entire life history can be assessed within a relatively short time. The use of modern molecular techniques in assessing genetic diversity of meiofauna in contaminated *versus* uncontaminated sediments is a promising avenue for new discoveries.

12.7 Macrofauna

12.7.1 Macrofaunal life forms

The macrofaunal species that inhabit the sandy coasts usually live buried in the sand, occasionally venturing up on the sand surface, *e.g.* the sandhopper *Talitrus saltator* (Fig. 12.11). The shallow sandy sublittoral supports both infaunal (living buried in the sediment) and epifaunal (living on the sediment surface) species.

A suite of typical infaunal macrozoobenthic organisms includes polychaetes (*e.g. Hediste diversicolor* and *Pygospio*

Fig. 12.11 The sandhopper *Talitrus saltator* is abundant on Baltic Sea sandy beaches not frequented by tourists. It usually lives buried in the sand to venture occasionally up on to the sand surface. Photo: © Lech Kotwicki

elegans) and oligochaetes. Among the oligochaetes, limnic or terrestrial enchytraeids are occasionally very abundant under or in wrack washed ashore. Typical epifaunal organisms include sessile bivalves (*e.g. Dreissena polymorpha*, *Mytilus trossulus*), hemisessile bivalves (*e.g. Cerastoderma glaucum*), motile bivalves (*e.g. Macoma balthica*), hydrobiid gastropods and crustaceans (amphipods, mysids and decapods).

The ecosystem-wide distribution of benthic macroinvertebrates inhabiting the shallow sandy sublittoral is, if enough oxygen is available, primarily governed by salinity, with a higher number of marine species found in the southern part of the Baltic Sea and a higher number of freshwater species in the northern and eastern parts (Ojaveer et al. 2010).

12.7.2 Macrofaunal abundance

Macrofaunal densities on the sandy coasts of the Baltic Sea vary extensively in time and space, from only several individuals m^{-2} in winter (November-December) to more than 22,000 individuals m^{-2} in July (Masłowski 2010). Macrozoobenthic biomass in those habitats can be up to 80 g ash-free dry weight m^{-2}. Abundance and biomass variations are primarily related to habitat-imposed constraints (*e.g.* sediment type, organic enrichment) and seasonality (Masłowski 2010). Biotic interactions, although present (Bonsdorff et al. 1995), are thought to be of a lesser importance in controlling sandy bottom communities than the abiotic conditions. Nevertheless, predation by fish and birds may be important locally and at certain times (Aarnio and Bonsdorff 1993).

Among the basin-scale effects, eutrophication seems to be very important. Generally, increasing nutrient load leads to algal and cyanobacterial blooms and intense sedimentation of organic material. As a result, augmented food resources in and on the sediment promote higher invertebrate abundances and biomasses (Lauringson and Kotta 2006). However, excess nutrient loading, coupled with high water temperature in summer, may result in oxygen depletion, and is ultimately detrimental to the benthic communities in the shallow sandy sublittoral (Powilleit and Kube 1999).

12.7.3 Macrofaunal diversity

The macrozoobenthic diversity on the unvegetated sandy coasts of the Baltic Sea is generally regarded as low (Bonsdorff 2006). Rocky outcrops or boulders occurring locally in the shallow sandy sublittoral create three-dimensional habitats providing additional living space for macrofauna and leading to an increase in local macrofaunal biodiversity and density (Grzelak and Kuklinski 2010). The highest macrofauna diversity is found associated with vegetation (Boström et al. 2014).

In the shallow sandy sublittoral, the dominant species include, depending on the locality, bivalves (particularly *Mya arenaria*), hydrobiid gastropods and polychaetes (*Marenzelleria neglecta*, *Pygospio elegans*), with a high proportion of the abundance and biomass also accounted for by *Hediste diversicolor* (Fig. 12.12), the mysid *Neomysis integer* and the amphipods *Bathyporeia pilosa* and *Corophium volutator* (Kotwicki 1997, B. Wawrzyniak-Wydrowska, personal communication). Since the early 1980s, the non-indigenous *Marenzelleria* spp. (*cf.* Box 5.3) have been featuring prominently in the shallow sandy sublittoral macrozoobenthic communities along the southern and eastern coasts of the Baltic Sea proper (Zettler et al. 2002).

The sandy beach macrozoobenthos is dominated by oligochaetes and features ecologically important species such as the sandhopper *Talitrus saltator* (Fig. 12.11). The sandhopper is a primary coloniser of newly stranded macroalgal wrack on the beach and plays a major role in the disintegration of this seaweed material, and thus in biogeochemical cycles on the sandy coast (Węsławski et al. 2000a).

12.7.4 Functional diversity of the sandy macrofauna

Functionally, the macrofauna on the sandy beaches comprises mostly deposit feeders (oligochaetes) and, particularly, scavengers such as the talitrid amphipods *Talitrus saltator* and *Deshayesorchestia deshayesii*. The shallow sandy sublittoral macrozoobenthos is functionally more diverse. The guild of suspension feeders, *e.g.* the partly buried polychaete *Manayunkia aestuarina*, the infaunal bivalve *Mya arenaria*, the hemisessile bivalve *Cerastoderma glaucum* and the hard-bottom sessile bivalves *Dreissena polymorpha* and

Fig. 12.12 The polychaete *Hediste diversicolor*. (**a**) A specimen on a sandy bottom. (**b**) A detail of the head. Photo: (**a**) Piotr Bałazy, (**b**) © Maria Włodarska-Kowalczuk and Piotr Bałazy

Mytilus trossulus, is an important component there. It is accompanied by deposit feeders, *e.g.* the oligochaete *Baltidrilus costatus* (syn. *Tubifex costatus*) and the cumacean *Diastylis rathkei*, and – if vegetation occurs (Fig. 12.13) – by herbivores such as hydrobiid gastropods, the isopod *Idotea balthica* (Fig. 12.13c, d), omnivores, *e.g. Gammarus* spp., and predators, *e.g.* the brown shrimp *Crangon crangon* (Fig. 12.14d), mysids (Fig. 12.14a–c) and pipefish (Fig. 12.13b).

With an increasing amount of detritus in the sandy sediment, the feeding mode switches from suspension feeding to deposit feeding in a number of species, *e.g. Corophium volutator*, *Macoma balthica* (*cf.* Box 13.5) and most polychaetes (*e.g. Marenzelleria neglecta* and *Pygospio elegans*) (Brey 1991). This switching-over ability reflects the adaptive potential of the sandy bottom macrozoobenthos to trophic conditions.

On a local scale, the biomass, composition and dominance structure of the macrozoobenthic communities on sandy substrates is controlled by factors such as changes in habitat and food availability, as well as by weather conditions, *e.g.* the severity of the previous winter (Masłowski 2010). Except for some infaunal bivalves, most suspension feeders benefit from the presence of a coarser substrate as they use larger particles for attachment. Coarser substrates also host more diverse and abundant communities of micro- and macrovegetation, thus enhancing the herbivores (Daleo and Iribarne 2009).

12.7.5 Effects of habitat on functional diversity

The deposit feeders and herbivores in the Baltic Sea generally have broad habitat requirements, tolerate wide ranges of environmental conditions and exhibit high recruitment rates (Bonsdorff and Pearson 1999). Therefore, the variability in abundance and biomass of these two feeding types, which is frequently observed on small, local scales, seems to result from factors operating locally as well. The predation pressure on the deposit feeders is usually relatively low. Due to moderate food limitation (Kotta et al. 2001), the deposit feeders are spatially structured, responding to small-scale patchiness in intra- and inter-specific competition for food (Bergström et al. 2002).

On the other hand, there are some strong landscape-scale components of physical pressure on deposit feeders and herbivores, resulting in variations of their abundance and biomass in time and space. These components include extreme weather events and ice scouring. The ice cover effects on the sandy coast zoobenthos, especially the herbivores, are particularly intriguing. The abundance and biomass of herbivores in sandy bottom habitats have been observed to be inversely correlated with the duration of ice cover. This relationship has been explained by invoking physical disturbance (ice scour) and light conditions, which are essential drivers of macrophyte communities in stressed environments (Madsen et al. 2001). As the presence of macrophytes directly enhances invertebrate species richness and biomass (Kotta et al. 2000), it is likely that the disappearance of macrovegetation under severe ice stress explains the macroherbivore abundance reduction following a prolonged ice season. In addition, floating ice sheets can, in shallow water, remove the upper (*i.e.* the most productive) sediment layer, and thus hinder the development of microherbivores grazing on the microphytobenthos.

The relationship between the duration of the ice cover and the density of suspension feeders is different from that of deposit feeders and grazers. Although some macrophytes may provide favourable substrate for the attachment of sessile suspension feeders, they are usually known to suppress the abundance of benthic suspension feeders at shallow

Fig. 12.13 *Zostera* meadows can be found on moderately exposed sandy bottoms where water movement prevents the accumulation of organic matter. Habitats with this type of vegetation host a rich fauna. (**a**) Vegetation dominated by *Zostera marina*. (**b**) Pipefish (Syngnathidae) have a long thin body which offers an excellent camouflage in the *Zostera marina* vegetation where they prey on small invertebrates. (**c**) The isopod *Idotea balthica* (*cf.* Fig. 6.1) on *Zostera marina*. (**d**) The isopod *Idotea chelipes* (*cf.* Fig. 11.34c) on *Zostera marina*. These two isopods are herbivores on filamentous algae, but they can also consume larger macrophytes such as *Fucus vesiculosus* and *Zostera marina*. Photo: (**a**, **b**) © Christoffer Boström, (**c**, **d**) © Camilla Gustafsson

depths through competition for space. Thus, luxuriant macrophyte communities are often characterised by low densities of suspension feeders. A moderate ice scour removes part of the vegetation, thus releasing benthic suspension feeders from such interspecific competition. A too strong ice disturbance, however, will remove most suspension feeders as well. This may explain why suspension feeders benefit from moderate ice disturbance, and why their densities are low when the duration of ice cover is either very short or very long (Veber et al. 2009).

12.7.6 Effects of food availability on functional diversity

Suspension feeders are simultaneously affected by ecosystem-wide effects on phytoplankton and by local factors such as the availability of resuspended microphytobenthic cells or detritus. Moreover, large-scale effects can often obscure the effects that occur on small scales. For example, frequent storms can promote phytoplankton blooms via upwelling events but carry away detrital material locally, and stormy weather or large-scale phytoplankton blooms can in turn result in reduced light penetration, which may locally suppress the development of benthic microalgae.

An elevated trophic status of a water body may also result in an increased amount of detritus that is channelled directly or indirectly to different invertebrate feeding groups on the seabed (Tomczak et al. 2009). However, the responses of suspension feeders to nutrient loading in the shallow sandy sublittoral of the Baltic Sea are complex (Veber et al. 2009). Some of them, particularly those looking for a suitable hard substrate for attachment, seem to benefit from an increased nutrient pool and the resultant high phytoplankton density, detritus sedimentation and/or resuspension, which releases them from food limitation. On the other hand, increased sedimentation and accumulation of detritus on the seabed changes sediment characteristics towards a finer-grained substrate, not amenable for colonisation by sessile suspension feeders. Such conditions favour deposit feeders, accompanied by those suspension feeders that are capable of switching their feeding mode to sediment (detritus) ingestion (Masłowski 2010).

12.7.7 Predators among the macrofauna

A higher trophic level among the macrofauna is occupied by carnivorous (predatory) benthic invertebrates, which in the shallow sandy sublittoral of the Baltic Sea are represented by peracarid crustaceans such as the brown shrimp *Crangon crangon* (Fig. 12.14d) and the prawn *Palaemon adspersus* (*cf.* Fig. 4.25a) as well as mysids, notably *Neomysis integer* and *Praunus flexuosus* (Fig. 12.14a–c). *Neomysis integer* is the most widespread mysid in the Baltic Sea and forms dense schools practically in the whole coastal area of the Baltic Sea (Köhn 1992; HELCOM 2014).

Able to swim, and feeding primarily on the zooplankton, *Neomysis integer* in one of the species that links the benthic and the pelagic systems (benthic-pelagic coupling, *cf.* Sect. 10.9.3). As a frequently preferred fish food item, it provides a connection with higher trophic levels in the coastal food webs (Kotta et al. 2007a). Compared to the other mysids in the Baltic Sea, *Neomysis integer* is a relatively warm-water species. Thus, the temperature regime in spring and summer may affect the timing and success of its reproduction. It is likely, though, that temperature affects the abundance and biomass of *Neomysis integer* indirectly, by driving changes in the zooplankton density and thus in the mysid's food source (Simm and Ojaveer 2000).

12.7.8 Terrestrial arthropods

In addition to supporting fully aquatic organisms, the Baltic Sea's sandy beaches (although not the shallow sandy sublittoral) are home to a group of semi-terrestrial invertebrates, primarily insects (Insecta) and spiders (Araneae), but also predatory mites (Gamasina) and springtails (Collembola). Because of their relative rarity and low abundances, they have received less attention than the interstitial organisms and the sandy coastal macrofauna. There is evidence, however, that the distribution and abundance of soil microarthropods can be used as a measure of the intensity of anthropogenic pressure on the sandy coasts, supplementary to the results of meio- and macrofaunal surveys on the beach (Schierding et al. 2011).

The insects and spiders present on the sandy coast tend to concentrate in the upper part of the beach, near dunes. Studies on the soil microarthropod fauna inhabiting the

Fig. 12.14 Invertebrate predators that can be found in sandy habitats. (**a**, **b**) The mysid *Neomysis integer*. (**c**) The mysid *Praunus flexuosus*. (**d**) The brown shrimp *Crangon crangon*. Photo: (**a**, **b**) Piotr Bałazy, (**c**, **d**) © Maria Włodarska-Kowalczuk and Piotr Bałazy

southern and southeastern Baltic Sea beaches revealed the presence of numerous species, mainly of beetles (Coleoptera) and several species of spiders, while predatory mites and springtails were recorded as well (Kononen 2000; Wolender and Zych 2007). Still more species-rich insect assemblages have been reported from Latvian sandy beaches (Spungis 2002).

Decaying plants and algae deposited on the shore attract root flies, mainly anthomyiid dipterans (Anthomyiidae). Their physiological and ecological adaptations allow them to occur in various habitats, including sandy coasts. As many as 53 root fly species were recorded on the Polish coast (Kaczorowska et al. 2009). They were dominated by *Fucellia tergina*, a cosmopolitan marine halobiont with saprophagous larvae, inhabiting the flooded epilittoral zone and attracted to decomposing organic matter, mainly to rotting algae.

The ecological roles that the soil microarthropods play in the beach system involve the contribution to carbon flow by grazing on the microflora (mites, springtails, insects) and predation on other beach invertebrates (mites) (Koehler et al. 1995).

12.8 Vertebrates

12.8.1 Fish

The near-shore shallow sandy areas support fish communities the composition and abundance of which change throughout the year. They include benthivorous species such as the sandeel *Ammodytes tobianus*, the greater sandeel *Hyperoplus lanceolatus*, the three-spined stickleback *Gasterosteus aculeatus* (Fig. 12.15a), the fifteen-spined stickleback *Spinachia spinachia* (Fig. 12.15b), the common goby *Pomatoschistus microps*, the sand goby *Pomatoschistus minutus*, the bull-head *Myoxocephalus scorpius* (Fig. 12.15c), the European flounder *Platichthys flesus* (Fig. 12.15e) and the turbot *Scophthalmus maximus* (Fig. 12.15f), which exert an important control on the abundance of the benthic invertebrates (Aarnio et al. 1991; Bonsdorff and Blomqvist 1993).

In addition, near-shore sandy habitats, particularly the vegetated areas, serve as important nurseries for juvenile fish, including flatfish (Pihl 1989; Berglund et al. 2003). In recent years, the shallow sandy sublittoral of the southern Baltic Sea proper has witnessed a massive invasion of the non-indigenous round goby *Neogobius melanostomus* (Fig. 12.15d) which now often dominates the coastal fish communities in soft-bottom, sandy and hard-bottom habitats in most subregions of the Baltic Sea (Sapota 2004; Kotta et al. 2016).

12.8.2 Birds

Ecologically, the most important vertebrates on sandy coasts are waterbirds. As predators, especially on the macrofauna on the beach near the water line and in the shallow sandy sublittoral, birds play a significant role in the sandy coastal food web. An important aspect of the birds' presence and feeding is the supply of organic matter contained in their faeces left in the sand, which provides a source of organic and inorganic compounds.

Birds that have been observed feeding on the sandy beaches of the Baltic Sea include primarily gulls and waders (Dierschke et al. 1999). Some species, such as the oystercatcher *Haematopus ostralegus* (Fig. 12.16d), probe the sand to locate and extract their prey; other birds, *e.g.* the ringed plover *Charadrius hiaticula* (Fig. 12.16a), collect prey from the sand surface, and gulls may feed on stranded carrion.

The shallow sandy sublittoral is a well-stocked larder for birds such as gulls (Fig. 12.16b), swans, and various species of ducks, capable of diving to the bottom to pick up their prey from the sand (macroflora or macrofauna) or from the water (hyperbenthic macrofauna and fish). Foraging at sand banks is particularly typical of diving ducks such as the common scoter *Melanitta nigra* (Fig. 12.16f) and velvet scoter *Melanitta fusca*. The long-tailed duck *Clangula hyemalis* (*cf.* Box Fig. 11.12), on the other hand, feeds preferentially on the blue mussel *Mytilus trossulus* in hard-bottom habitats (Stempniewicz 1995). However, when they winter in less productive, soft-bottom habitats they employ a prey-selective foraging strategy, in which they feed on less abundant, but energy rich, crustaceans (Žydelis and Ruškytė 2005). Of other species, the Slavonian grebe *Podiceps auritus* (Fig. 12.16e) occurs preferentially in shallow areas off the southern Baltic Sea coast to feed on gobiid fish that live close to the sandy bottom (Sonntag et al. 2009).

Some waterbirds (most notably gulls) are present as residents, encountered around the sea shore throughout the year. The presence and abundance of other bird species is typically seasonal due to migrations. Species such as the common scoter and the velvet scoter visit and/or stay close to the coasts and shallow sandy offshore areas of the Baltic Sea, usually in winter (*cf.* Box 11.8, Vaitkus 1999). Seasonally abundant can also be waders, notably the dunlin *Calidris alpina* (Fig. 12.16c). On the Baltic Sea coasts, wader abundances have been found to vary considerably depending on the season and on food availability, but become very high (on the order of tens or even hundreds of thousands) during winter (Blew et al. 2008, D. Wysocki, personal communication).

Fig. 12.15 Examples of fish species that can be found close to sandy coasts. (**a**) The three-spined stickleback *Gasterosteus aculeatus*. (**b**) The fifteen-spined stickleback *Spinachia spinachia*. (**c**) The bull-head *Myoxocephalus scorpius*. (**d**) The non-indigenous round goby *Neogobius melanostomus*. (**e**) The European flounder *Platichthys flesus*. (**f**) The turbot *Scophthalmus maximus*. Photo: © Piotr Bałazy

12.8.3 Mammals

The sandy coasts of the Baltic Sea support few mammals, if any. Seals occasionally haul themselves out on sandy beaches and exposed sandy shoals (*e.g.* in the Puck Bay), but they are not a typical component of the sandy coastal community. Among other mammals, the European otter *Lutra lutra*, although having been sighted on the sandy beaches of the Latvian coast (Ozolins et al. 1998), is known to prefer other coastal habitats such as the mouths of small creeks and rocky outcrops where food is more available to them.

12.9 The sandy coast food webs

12.9.1 Primary production

The food webs on sandy beaches are based on solar energy captured by primary producers (mainly the microphytobenthos) and on decomposition of detritus. These two energy utilisation pathways are closely interconnected. The energy flow, regardless of the source (primary production or detritus), proceeds through the interstitial biota consisting of microorganisms and meiofauna.

Fig. 12.16 Examples of bird species that can be found on sandy coasts. (**a**) The ringed plover *Charadrius hiaticula*. (**b**) The black-headed gull *Larus ridibundus*. (**c**) The dunlin *Calidris alpina*. (**d**) The oystercatcher *Haematopus ostralegus*. (**e**) The Slavonian grebe *Podiceps auritus*. (**f**) The common scoter *Melanitta nigra*. Photo: (**a**, **c–f**) © Bo Tallmark, (**b**) © Stanislaw Węsławski

The microphytobenthic primary producers (mainly diatoms) are concentrated on the sand surface, primarily in the hygro- and hydropsammal. Measurements of the chlorophyll *a* (Chl *a*) content in the sand on a beach in the southern Baltic Sea proper showed a distinct vertical and horizontal zonation, which reflects the abundance of the microphytobenthos. In summer (July-August), the sediment Chl *a* content tended to be higher on the sand surface close to and at the water line [0.9–1.8 μg Chl *a* (g dry sediment)$^{-1}$] than in the surficial sand layer of the permanently submerged sublittoral at a ~0.5 m water depth [0.3–1.3 μg Chl *a* (g dry sediment)$^{-1}$] (T. Radziejewska, personal observation).

This primary production is directly grazed upon by interstitial organisms adapted to scraping microscopic algae off sand grains, such as tardigrades and epistrate feeders among the nematodes (*cf.* Box 4.7). The DOM released by microorganisms, as well as POM in the interstitial water, are fed upon by other interstitial invertebrates, *e.g.* gastrotrichs, harpacticoid copepods and turbellarians. As opposed to other sandy coastal systems in Europe and elsewhere (McLachlan and Brown 2006), the primary production-based energy flow has not been so far quantified on the sandy coasts of the Baltic Sea.

12.9.2 Detritus-based energy flow: a biocatalytic filter

In contrast, the other type of food web interactions on the beach, that based on detritus utilisation, has been quite intensively studied on the sandy coasts of the Baltic Sea. The sandy habitats interact with the adjacent terrestrial environment via water- and wind-borne inputs of detritus from dunes and the areas behind them, as well as with the sea via the shallow sandy sublittoral. The food sources from the sea involve fine and coarse detritus particles supplied by autochthonous primary producers (Kotwicki et al. 2005b, c) and allochthonous organic matter derived from bird faeces and stranded microbial and macroalgal mats (*cf.* Sect. 12.5.3).

Experimental studies on the colonisation and degradation of stranded macroalgae on a beach in the southern Baltic Sea proper showed a distinct succession of biotic activity and chemical decomposition processes (Jędrzejczak 2002a, b). The fresh, fairly refractory organic material was microbially degraded into more labile forms that are available as food to detritus feeders among both meio- and macrofauna (talitrids, oligochaetes). The organic material is eventually mineralised, whereby the regenerated nutrients are returned to the sand system as well as to the seawater, to fuel primary production in the shallow sandy sublittoral.

This pathway of energy flow consists of a suite of processes that ultimately make the sandy coast function as a biocatalytic filter (Węsławski et al. 2000b). The highly diverse meiofauna and diatom communities in undisturbed beaches may act as an effective biological filter for some types of chemical pollutants, while less diverse, but more abundant biota in disturbed areas are more effective in processing organic matter. This effect is known as "beach self-cleaning".

12.9.3 Food webs in the shallow sandy sublittoral

The food webs in the shallow sandy sublittoral are more complex than those higher up the shore. The primary producers in the sublittoral include phytoplankton, microphytobenthos and (if occurring) macrophytes. Secondary production is effected both in the bottom sediment by microorganisms, meio- and macrofauna, and on the submerged vegetation by epiphytic and motile suspension feeders and mobile herbivores. Predators, such as *Crangon crangon*, juvenile European plaice *Pleuronectes platessa* and the sand goby *Pomatoschistus minutus*, become important as well (Evans 1984). In addition, avian predation, particularly that of wintering birds on zoobenthos exerts a significant control on the benthic biomass (Kube 1996), the strength of the control varying with time.

The energy and carbon flows in the shallow sandy sublittoral of the Baltic Sea have been amply documented. For example, in the unvegetated shallow sandy sublittoral of the Gulf of Gdańsk in the southern Baltic Sea proper, microbial communities and meiofauna were shown to be responsible for 7 % and 21 % of the total oxygen consumption, respectively (Opaliński et al. 2010). Data collected from various vegetated inshore areas along the southeastern coast of the Baltic Sea proper revealed that the cumulative biomass of phytoplankton and zooplankton seemed to be similar in all areas.

On the other hand, the total biomass of macrophytes, benthic macrofauna, fish and birds showed large between-area differences. Except for turbid environments, *e.g.* the Curonian Lagoon, all systems studied were characterised by dense macrophyte beds resulting in the dominance of benthic over pelagic primary production. Benthic primary production was estimated at 160–3,200 g C m^{-2} $year^{-1}$ while phytoplankton primary production ranged between 125 and 1,500 g C m^{-2} $year^{-1}$ (Tomczak et al. 2009). Thus, macroalgae and vascular plants on sandy substrates are among the most productive habitats in the Baltic Sea. Regardless of the area, however, the macrophytes were not consumed by grazers, but rather channelled into the detritus food web.

When estimated in terms of carbon flow, predation by birds was found to range from very low (0.02 g C m^{-2} $year^{-1}$) on the exposed Gulf of Riga coast to high (2.66 g C m^{-2} $year^{-1}$) in Pärnu Bay. The high bird predation in the latter area was due to great cormorant *Phalacrocorax carbo sinensis* (*cf.* Box 4.11) colonies which mainly affected the piscivorous fish biomass (perch), whereas the fisheries targeted planktivores (Tomczak et al. 2009).

12.9.4 Suspension feeding in the shallow sandy sublittoral

Suspension feeders may play different roles in the shallow sandy sublittoral of the Baltic Sea. In flat inshore habitats away from large rivers and upwelling areas, benthic suspension feeders, *e.g.* aggregates of *Mytilus trossulus* attached to scattered stones and boulders on the bottom, are only of minor importance in food web dynamics, although *Mytilus trossulus* may provide local organic enrichment of the sediment and boost microbial and meiobenthic communities (Radziejewska 1986).

On the other hand, energy fluxes driven by benthic suspension feeders are important in the mouths of large rivers in the eastern part of the Baltic Sea proper, where low salinity allows the non-indigenous zebra mussel (*Dreissena polymorpha*) to establish distinct beds (Orlova et al. 2004). By enhancing phytoplankton densities, the constant nutrient input by the river helps to sustain much higher biomasses of mussels than in adjacent sea areas (Kotta et al. 2008).

12.10 Anthropogenic impacts

12.10.1 Climate change

An ecosystem-wide pressure of potentially high importance for the sandy coasts of the Baltic Sea is the on-going global climate change. In many European water bodies, rising temperatures have resulted in benthic invertebrate population "booms" (Hart and Scheibling 1988), replacement of key species (Southward et al. 1995) and other major shifts in the community structure (Conners et al. 2002).

In the Baltic Sea, effects of such temperature-related shifts have been observed in the offshore water column (*cf.* Box 2.5, Möllmann et al. 2007), but no changes have been reported from sandy bottoms. It is plausible that global changes in the average water temperature of the shallow coastal waters of the Baltic Sea have not had a direct ecological effect yet because large seasonal variation may counteract the potential effects of global warming. On the other hand, indirect effects of global warming, such as increased wave action and decreased ice erosion when the ice cover decreases, and reduced photosynthetic light intensity due to increased cloudiness and turbidity, have the potential to significantly affect the structure and function of the shallow sandy bottom communities in the Baltic Sea.

Recent shifts in climate conditions have resulted in elevated winter temperatures and reduced extent and duration of ice cover in northern Europe (Meier 2006). Due to the ice-cover reduction, winter storms are likely to contribute to the so-called "coastal squeeze" (Doody 2004), *i.e.* the reduction of the coastal habitat due to natural and/or anthropogenic impacts, including beach encroachment by tourist infrastructures (*cf.* Fig. 18.1).

Winter storms also exert increasing ecological impacts on shallow-water benthic communities. In vegetated areas, such storms may cause frequent burial of macrophytes by sediment and/or removal of the macrophyte canopy. As a consequence of canopy disappearance, the important ecosystem services provided by benthic macrophytes are likely to be lost. On the other hand, as the upper limit of macrophytes on shallow sandy substrates is mainly controlled by ice scouring (Middelboe et al. 2003), the recent climate warming may significantly move the upper depth limit of macrophytes higher up on the shore.

12.10.2 Eutrophication

In addition to environmental forcing by climate change, the sandy beaches and the shallow sandy sublittoral of the Baltic Sea proper experience a strong anthropogenic pressure. The list of direct and indirect anthropogenic impacts on the sandy coasts starts with ecosystem-wide processes such as eutrophication, with its resultant phytoplankton and phytobenthos blooms of filamentous algae such as *Cladophora glomerata* and *Pylaiella littoralis* (*cf.* Sect. 11.15.3). These blooms are deposited on the coasts in the form of mats and heaps of decaying algae (Fig. 12.2), which drastically changes habitat conditions for the coastal biota and reduces the amenity value of a beach.

In addition, eutrophication is a severe threat to the shallow sandy sublittoral macrophyte communities. Reduced light conditions, increased sedimentation and increased grazing by herbivores have led to shifts in the composition of macrophyte communities. For example, the *Zostera* meadows in Puck Bay in the southern Baltic Sea proper had to give way to extensive growth of filamentous algae (Kruk-Dowgiałło and Szaniawska 2008).

Nevertheless, as different natural and anthropogenic processes do not act in isolation, alterations in the structure of ecosystems are likely to be driven by changed interactions between the climate, eutrophication-related variables and ecosystem components.

12.10.3 Increased ship traffic

A high risk factor for the coastal sandy habitats of the Baltic Sea is associated with freight transportation. An increasing intensity of shipping in the Baltic Sea Area, coupled with environmental stress and an overall low species richness, increases the risk of biological invasions (*cf.* Sect. 5.3.2). In recent decades, a number of non-indigenous species have successfully established in sandy-bottom habitats, including *Marenzelleria* spp. (Blank et al. 2008) and the amphipod *Gammarus tigrinus* (Gruszka 1999), both of North American origin and both known from coastal habitats in virtually the whole Baltic Sea proper and the Gulf of Finland. There is evidence that *Gammarus tigrinus* has caused pronounced and irreversible structural and functional changes in invaded communities by outcompeting native gammarids (Ojaveer et al. 2010).

Another risk to sandy coasts associated with intensified shipping concerns oil spills (*cf.* Box 16.1, Defeo et al. 2009). A number of oil spills have already affected sandy coasts in the Baltic Sea, and research on recovery of oil-impacted coastal habitats shows that the rate of recovery depends not so much on the persistence of oil, but on the recolonisation and growth rates of key species among benthic invertebrates and macrophytes. Studies indicate that the high sensitivity of Baltic Sea sandy habitats to oil spill effects are associated with the loss of benthic invertebrates rather than with the disappearance of benthic macrophytes (Kotta et al. 2007b, 2008). Thus, it is likely that key species such as *Zostera marina* do not prevent the zoobenthic communities from oil spill effects.

12.10.4 Habitat destruction

Direct anthropogenic impacts in areas with sandy coasts are sand extraction and dredging spoil dumping associated with constructions such as pipelines, dams, bridges, wind-farms and harbours. Such activities severely disturb the sediment structure and the associated benthic communities at the dredging and dumping sites, and result in organic enrichment and shifts in community structure in adjacent areas. Eventually, dredging may strongly reduce or defaunate large coastal areas due to the physical removal of the substrate or due to smothering by tailings on the dumping ground.

The recovery of benthic communities in dredged areas depends on the availability of a pool of potential colonising species in the adjacent areas. The presence of mobile and opportunistic species will greatly facilitate and accelerate the recovery from dredging effects. It seems that stress-tolerant species can easily cope with dredging activities unless oxygen deficiency develops (Bonsdorff 1983; Powilleit et al. 2006). Dredging in coastal areas may, however, pose additional challenges for some species due to the presence of other stress factors, such as low salinity and wide temperature fluctuations (Olenin 1992; Kotta and Möller 2009; Kotta et al. 2009).

12.10.5 Tourism and recreation

On the sandy coast itself, the list of anthropogenic impacts (by no means exhaustive) includes recreational activities, tourist infrastructure, coastal defence constructions, beach nourishment and cleaning as well as litter accumulation (Fig. 12.1). Although threats and risks to the sandy coastal habitats posed by those impacts are realised, not all of them have received equal attention, and not all the risks have been assessed and analysed.

Most attention has been paid so far to effects of beach use for recreational activities (Węsławski et al. 2000b; Jędrzejczak 2004) and to associated impacts such as trampling (Grunewald 2006), a distinction being made between heavily used (tourist beaches) and scarcely used (non-tourist beaches) sandy coastal areas. Evaluation of the effects of tourist beach use intensity on the coastal meiofauna revealed that tourist beaches, particularly in their upper part, host an impoverished meiofauna (in terms of density and diversity) which are subjected to higher community stress compared to nearby non-tourist locations (Gheskiere et al. 2005).

Tourist infrastructure constructions extending seawards, such as pleasure piers present in most seaside holiday resorts at the southern Baltic Sea coast, provide an extra dimension to shallow sandy bottoms by supplying hard surfaces for epifauna to settle on. A recent study in the Pomeranian Bay (M. Bąk, personal communication) showed that the pilings of a pleasure pier support an abundant and diverse epifaunal community, dominated by sessile suspension feeders (*Amphibalanus improvisus*, *Mytilus trossulus*) and numerous motile macro- and meiofaunal invertebrates, which attract fish, including the non-indigenous species *Neogobius melanostomus*.

12.10.6 Is beach cleaning good or bad?

Machine cleaning of beaches from litter and/or accumulated cyanobacterial or macroalgae-based debris has recently received special attention as a means of improving beach amenity value and maintaining beach-linked tourism, which is economically important for local human communities. However, the opinions on the ecological effects of such beach cleaning operations differ. Some studies have found that machine cleaning is detrimental to the biodiversity (including the meiofauna) of sandy coasts (Węsławski et al. 2000b), while others could not demonstrate any significant effect on macrofaunal biodiversity (Malm et al. 2004).

Review questions

1. What are the key abiotic characteristics of the sandy habitat in the Baltic Sea?
2. What are the major components of the sandy coastal biota in the Baltic Sea?
3. Why are the sandy coasts regarded as biocatalytic filters?
4. What are the major characteristics of the sandy coastal food webs in the Baltic Sea?
5. What impacts, natural and anthropogenic, are observed in the sandy coastal environment of the southeastern Baltic Sea?

Discussion questions

1. What is the potential of sandy coastal environments in the Baltic Sea for generating new scientific knowledge?
2. How would you set up a scientific research project to understand the functioning of the food web on a sandy beach? How would you do that in the shallow sandy sublittoral?
3. What is the major anthropogenic threat to sandy coastal environments and their biota in the Baltic Sea and why?
4. How do you think we can achieve a good balance between the use of sandy coastal resources and conservation needs?
5. How would you clean a sandy beach after a major oil spill?

References

Aarnio K, Bonsdorff E (1993) Seasonal variation in abundance and diet of the sand goby *Pomatoschistus minutus* (Pallas) in a northern Baltic archipelago. Ophelia 37:19–30

Aarnio K, Sandberg E, Bonsdorff E (1991) Benthic predation on shallow-water macro- and meiofauna in the Baltic Sea: an experimental comparison between *Pomatoschistus minutus* (Pisces) and *Saduria entomon* (Crustacea). Annales Zoologici Fennici 28:41–48

Appelgren K, Mattila J (2005) Variation in vegetation communities in shallow bays of the northern Baltic Sea. Aquatic Botany 83:1–13

Berglund J, Mattila J, Rönnberg O, Heikkilä J, Bonsdorff E (2003) Seasonal and inter-annual variation in occurrence and biomass of rooted macrophytes and drift algae in shallow bays. Estuarine, Coastal and Shelf Science 56:1167–1175

Bergström U, Englund G, Bonsdorff E (2002) Small-scale spatial structure of Baltic Sea zoobenthos – inferring processes from patterns. Journal of Experimental Marine Biology and Ecology 282:123–136

Blank M, Laine AO, Jürss K, Bastrop R (2008) Molecular identification key based on PCR/RFLP for three polychaete sibling species of the genus *Marenzelleria*, and the species' current distribution in the Baltic Sea. Helgoland Marine Research 62:129–141

Blew J, Günther K, Kieckbusch JJ, Nehls HW (2008) Numbers of wintering waders in non-estuarine habitats along the North Sea and Baltic Sea coasts of Germany in winter 1997/98. In: Burton NHK, Rehfisch MM, Stroud DA, Spray CJ (eds) The European non-estuarine coastal waterbird survey. International Wader Study Group, Thetford, UK, International Wader Studies 18:31–35

Bonsdorff E (1983) Recovery potential of macrozoobenthos from dredging in shallow brackish waters. In: Cabioch L, Glémarec M, Samain JF (eds) Fluctuation and succession in marine ecosystems. Proceedings of the 17th European Symposium on Marine Biology. Gauthier-Villars, Montrouge, pp 27–32

Bonsdorff E (1992) Drifting algae and zoobenthos – effects on settling and community structure. Netherlands Journal of Sea Research 30:57–62

Bonsdorff E (2006) Zoobenthic diversity-gradients in the Baltic Sea: continuous post-glacial succession in a stressed ecosystem. Journal of Experimental Marine Biology and Ecology 330: 383–392

Bonsdorff E, Blomqvist EM (1993) Biotic couplings on shallow water soft bottoms – examples from the northern Baltic Sea. Oceanography and Marine Biology – An Annual Review 31:153–176

Bonsdorff E, Norkko A, Sandberg E (1995) Structuring zoobenthos: the importance of predation, cropping and physical disturbance. Journal of Experimental Marine Biology and Ecology 192:125–144

Bonsdorff E, Pearson TH (1999) Variation in the sublittoral macrozoobenthos of the Baltic Sea along environmental gradients: a functional-group approach. Australian Journal of Ecology 24: 312–326

Boström C, Baden S, Bockelmann AC, Dromph K, Fredriksen S et al (2014) Distribution, structure and function of Nordic eelgrass (*Zostera marina*) ecosystems: implications for coastal management and conservation. Aquatic Conservation: Marine and Freshwater Ecosystems 24:410–434

Brey T (1991) Interactions in soft bottom benthic communities: quantitative aspects of behaviour in the surface deposit feeders *Pygospio elegans* (Polychaeta) and *Macoma balthica* (Bivalvia). Helgoländer Meeresuntersuchungen 45:301–316

Conners ME, Hollowed AB, Brown E (2002) Retrospective analysis of Bering Sea bottom trawl surveys: regime shift and ecosystem reorganization. Progress in Oceanography 55:209–222

Daleo P, Iribarne O (2009) Beyond competition: the stress-gradient hypothesis tested in plant-herbivore interactions. Ecology 90: 2368–2374

Defeo O, McLachlan A, Schoeman DS, Schlacher TA, Dugan J et al (2009) Threats to sandy beach ecosystems: a review. Estuarine, Coastal and Shelf Science 81:1–12

Dierschke V, Kube J, Probst S, Brenning U (1999) Feeding ecology of dunlins *Calidris alpina* staying in the southern Baltic Sea, 1. Habitat use and food selection. Journal of Sea Research 42:49–64

Dietrich D, Arndt H (2000) Biomass partitioning of benthic microbes in a Baltic inlet: relationships between bacteria, algae, heterotrophic flagellates and ciliates. Marine Biology 136:309–322

Doody JP (2004) "Coastal squeeze" – an historical perspective. Journal of Coastal Conservation 10:129–138

Drzycimski I (1974) The Harpacticoida (Copepoda) of the Baltic coastal waters. Fragmenta Faunistica 20:61–73

Evans S (1984) Energy budgets and predation impact of dominant epibenthic carnivores on a shallow soft bottom community at the Swedish west coast. Estuarine, Coastal and Shelf Science 18:651–672

Fenchel T (1969) The ecology of marine microbenthos. IV. Structure and function of the benthic ecosystem, its chemical and physical factors and the microfauna communities with special reference to the ciliated protozoa. Ophelia 6:1–182

Gamenick I, Jahn A, Vopel K, Giere O (1996) Hypoxia and sulphide as structuring factors in a macrozoobenthic community on the Baltic Sea shore: colonisation studies and tolerance experiments. Marine Ecology Progress Series 144:73–85

Gerlach SA (1953) Die Nematodenfauna der Uferzonen und des Küstengrundwassers am finnischen Meerbusen. Acta Zoologica Fennica 73:1–32 [in German]

Gheskiere T, Vincx M, Węsławski JM, Scapini F, Degraer S (2005) Meiofauna as descriptor of tourism-induced changes at sandy beaches. Marine Environmental Research 60:245–265

Giere O (2009) Meiobenthology – the microscopic motile fauna of aquatic sediments, 2nd edn. Springer, Berlin 527 pp

Grunewald R (2006) Assessment of damages from recreational activities on coastal dunes of the southern Baltic Sea. Journal of Coastal Research 22:1145–1157

Gruszka P (1999) The river Odra estuary as a gateway for alien species immigration to the Baltic Sea basin. Acta Hydrochimica et Hydrobiologica 27:374–382

Grzelak K, Kuklinski P (2010) Benthic assemblages associated with rocks in a brackish environment of the southern Baltic Sea. Journal of the Marine Biological Association of the United Kingdom 90:115–124

Haque AM, Szymelfenig M, Węsławski JM (1997) Spatial and seasonal changes in the sandy littoral zoobenthos of the Gulf of Gdańsk. Oceanologia 39:299–324

Hart MW, Scheibling RE (1988) Heat waves, baby booms, and the destruction of kelp beds by sea urchins. Marine Biology 99: 167–176

HELCOM (2014) Checklist of Baltic Sea macro-species. Baltic Sea Environment Proceedings 130:1–203

Herkül K, Kotta J (2009) Effects of eelgrass (*Zostera marina*) canopy removal and sediment addition on sediment characteristics and benthic communities in the northern Baltic Sea. Marine Ecology 30 (Supplement 1):74–82

Herkül K, Kotta J, Pärnoja M (2011) Effect of physical disturbance on the soft sediment benthic macrophyte and invertebrate community in the northern Baltic Sea. Boreal Environment Research 16 (Supplement A):209–219

Hewitt E, Thrush SF (2009) Reconciling the influence of global climate phenomena on macrofaunal temporal dynamics at a variety of spatial scales. Global Change Biology 15:1911–1929

Heyl K, Woelfel J, Schumann R, Karsten U (2010) Microbial mats from wind flats on the southern Baltic Sea. In: Seckbach J, Oren A (eds) Microbial mats: modern and ancient microorganisms in stratified systems. Springer, Berlin, pp 301–347

Jansson BO (1967) The availability of oxygen for the interstitial fauna of sandy beaches. Journal of Experimental Marine Biology and Ecology 1:123–143

Jędrzejczak MF (2002a) Stranded *Zostera marina* L. vs wrack fauna community interactions on a Baltic sandy beach (Hel, Poland): a short-term pilot study. Part I. Driftline effects of fragmented detritivory, leaching and decay rates. Oceanologia 44:273–286

Jędrzejczak MF (2002b) Stranded *Zostera marina* L. vs wrack fauna community interactions on a Baltic sandy beach (Hel, Poland): a short-term pilot study. Part II. Driftline effects of succession changes and colonisation of beach fauna. Oceanologia 44:367–387

Jędrzejczak MF (2004) The modern tourist's perception of the beach: is the sandy beach a place of conflict between tourism and biodiversity? In: Schernewski G, Löser N (eds) Managing the Baltic Sea. Coastline Reports 2:109–119

Jończyk E, Radziejewska T (1984) Temporal changes in sand meiofauna of a southern Baltic beach. Limnologica (Berlin) 15:421–423

Kaczorowska E, Lewańczyk A, Suchańska K (2009) Remarks on root flies (Diptera: Anthomyiidae) from sandy and marshy habitats of the Polish coast. Dipteron 25:20–26

Kenny AJ, Cato I, Desprez M, Fader G, Schuttenhelm RTE, Side J (2003) An overview of seabed-mapping techniques in the context of marine habitat classification. ICES Journal of Marine Science 60:411–418

Koehler H, Munderloh E, Hofmann S (1995) Soil microarthropods (Acari, Collembola) from beach and dune: characteristics and ecosystem context. Journal of Coastal Conservation 1:77–86

Köhn J (1992) Mysidacea of the Baltic – state of the art. In: Köhn J, Jones MB, Moffat A (eds) Taxonomy, biology and ecology of (Baltic) mysids (Mysidacea: Crustacea). Rostock University Press, Rostock, pp 5–23

Kolicka M, Kisielewski J, Kotwicki L, Zawierucha K, Grzelak K (2014) Checklist of Gastrotricha of the Polish Baltic Sea with the first reports of *Heterolepidoderma joermungandri* Kanneby, 2011 and *Turbanella hyalina* (Schulze, 1853). Zootaxa 3869:101–130

Kononen S (2000) Spiders (Araneae) on sandy islands in the southwestern archipelago of Finland. Ekológia (Bratislava) 19 (Supplement 4):79–85

Kotta I, Kotta J, Herkül K (2007a) Seasonal changes in the population structure and life cycle of *Neomysis integer* in Pärnu Bay, northeastern Baltic Sea. Proceedings of the Estonian Academy of Sciences, Biology and Ecology 56:312–325

Kotta J, Aps R, Herkül K (2008) Predicting ecological resilience of marine benthic communities facing a high risk of oil spills. In: Brebbia CA (ed) Environmental problems in coastal regions VII. WITPress, UK, pp 101–110

Kotta J, Herkül K, Kotta I, Orav-Kotta H, Aps R (2009) Response of benthic invertebrate communities to the large-scale dredging of Muuga Port. Estonian Journal of Ecology 58:286–296

Kotta J, Martin G, Aps R (2007b) Sensitivity of benthic vegetation and invertebrate functional guilds to oil spills and its use in oil contingency management related negotiation processes. Proceedings of the Estonian Academy of Sciences, Biology and Ecology 56:255–269

Kotta J, Möller T (2009) Important scales of distribution patterns of benthic species in the Gretagrund area, the central Gulf of Riga. Estonian Journal of Ecology 58:259–269

Kotta J, Ojaveer H, Puntila R, Nurkse K (2016) Shipping and natural environmental conditions determine the distribution of the invasive non-indigenous round goby *Neogobius melanostomus* in a regional sea. Estuarine, Coastal and Shelf Science 169:15–24

Kotta J, Orav H, Sandberg-Kilpi E (2001) Ecological consequence of the introduction of the polychaete *Marenzelleria viridis* into a shallow water biotope of the northern Baltic Sea. Journal of Sea Research 46:273–280

Kotta J, Paalme T, Martin G, Mäkinen A (2000) Major changes in macroalgae community composition affect the food and habitat preference of *Idotea balthica*. International Review of Hydrobiology 85:697–705

Kotwicki L (1997) Macrozoobenthos of the sandy littoral zone of the Gulf of Gdańsk. Oceanologia 39:447–460

Kotwicki L, Szymelfenig M, De Troch M, Urban-Malinga B, Węsławski JM (2005a) Latitudinal biodiversity patterns of meiofauna from sandy littoral beaches. Biodiversity and Conservation 14:461–474

Kotwicki L, Węsławski JM, Raczyńska A, Kupiec A (2005b) Deposition of large organic particles (macrodetritus) in a sandy beach system (Puck Bay, Baltic Sea). Oceanologia 47:181–199

Kotwicki L, Węsławski JM, Szałtynis A, Stasiak A, Kupiec A (2005c) Fine organic particles in a sandy beach system (Puck Bay, Baltic Sea). Oceanologia 47:165–180

Kruk-Dowgiałło L, Szaniawska A (2008) Gulf of Gdańsk and Puck Bay. In: Schiewer U (ed) Ecology of Baltic coastal waters. Springer, Berlin, Ecological Studies 197:139-165

Kube J (1996) The ecology of macrozoobenthos and sea ducks in the Pomeranian Bay. Meereswissenschaftliche Berichte 18:1–128

Lauringson V, Kotta J (2006) Influence of the thin drift algal mats on the distribution of macrozoobenthos in Kõiguste Bay, NE Baltic Sea. Hydrobiologia 554:97–105

Lokko K (2014) Seasonal and spatial variability of zoopsammon communities in relation to environmental parameters. Dissertationes Biologicae Universitatis Tartuensis 263:1–129 [PhD Thesis]

Lokko K, Kotta J, Virro T (2014) Seasonal trends in horizontal and vertical patterns of zoopsammon in the brackish Baltic Sea in relation to key environmental variables. Proceedings of the Biological Society of Washington 127:58–77

Lund-Hansen LC, Laimo M, Mauritsen K, Lam NN, Hai DH (2002) Effects of benthic diatoms, fluff layer, and sediment conditions on critical shear stress in a non-tidal coastal environment. Journal of the Marine Biological Association of the United Kingdom 82: 929–936

Madsen JD, Chambers PA, James WF, Koch EW, Westlake DF (2001) The interaction between water movement, sediment dynamics and submersed macrophytes. Hydrobiologia 444:71–84

Malm T, Råberg S, Fell S, Carlsson P (2004) Effects of beach cast cleaning on beach quality, microbial food web, and littoral macrofaunal biodiversity. Estuarine, Coastal and Shelf Science 60:339–347

Masłowski J (2010) Makrofauna denna Zatoki Pomorskiej w latach 1976–2005. Wydawnictwo Adam Marszałek, Toruń, 122 pp [in Polish]

McLachlan A, Brown AC (2006) The ecology of sandy shores. Elsevier, Amsterdam 392 pp

Meier MHE (2006) Baltic Sea climate in the late twenty-first century: a dynamical downscaling approach using two global models and two emission scenarios. Climate Dynamics 27:39–68

Meyer-Reil LA, Dawson R, Liebezeit G, Tiedge H (1978) Fluctuations and interactions of bacterial activity in sandy beach sediments and overlying water. Marine Biology 48:161–171

Meyer-Reil LA, Faubel A (1980) Uptake of organic matter by meiofauna organisms and interrelationships with bacteria. Marine Ecology Progress Series 3:251–256

Middelboe AL, Sand-Jensen K, Krause-Jensen D (2003) Spatial and interannual variations with depth in eelgrass populations. Journal of Experimental Marine Biology and Ecology 291:1–15

Möllmann C, Müller-Karulis B, Diekmann R, Flinkman J, Gådmark A (2007) Ecosystem regime state in the Baltic Proper, Gulf of Riga, Gulf of Finland, and the Bothnian Sea. HELCOM Indicator Fact Sheets 2007. [http://www.helcom.fi]

Mudryk Z, Podgórska B (2005) Generic composition and respiratory activity of heterotrophic bacteria of marine sandy beach (southern Baltic Sea). Polish Journal of Ecology 53:97–103

Mudryk Z, Podgórska B (2007) Culturable microorganisms in sandy beaches in the South Baltic Sea. Polish Journal of Ecology 55: 221–231

Norkko A, Bonsdorff E (1996) Population responses of coastal zoobenthos to stress induced by drifting algal mats. Marine Ecology Progress Series 140:141–151

Ojaveer H, Jaanus A, MacKenzie BR, Martin G, Olenin S et al (2010) Status of biodiversity in the Baltic Sea. PLoS ONE 5(9):e12467

Olenin S (1992) Changes in a southeastern Baltic soft-bottom community induced by dredged spoil dumping. In: Bjørrnestad E, Hagerman L, Jensen K (eds) Proceedings of the 12th Baltic Marine Biologists Symposium. Olsen and Olsen, Fredensborg, pp 119–123

Opaliński K, Maciejewska K, Urban-Malinga B, Węsławski JM (2010) The oxygen fluxes of sandy littoral areas: quantifying primary and secondary producers in the Baltic Sea. Marine Pollution Bulletin 61:211–214

Orlova M, Golubkov S, Kalinina L, Ingatieva N (2004) *Dreissena polymorpha* (Bivalvia: Dreissenidae) in the Neva Estuary (eastern Gulf of Finland, Baltic Sea): is it a biofilter or source for pollution? Marine Pollution Bulletin 49:196–205

Ozolins J, Kranz A, Toman A (1998) Three men in a boat (to say nothing of the otter in Latvia). IUCN Otter Specialist Group Bulletin 15:103–108

Pallo P, Widbom B, Ólafsson E (1998) A quantitative survey of the benthic meiofauna in the Gulf of Riga (eastern Baltic Sea), with special reference to the structure of nematode assemblages. Ophelia 49:117–139

Pihl L (1989) Abundance, biomass and production of juvenile flatfish in southeastern Kattegat. Journal of Sea Research 24:69–81

Pliński M, Kwiatkowski J (1996) Microphytobenthos of the shallow littoral of the southern Baltic. Oceanological Studies 25:65–79

Podgórska B, Mudryk ZJ, Skórczewski P (2008) Abundance and production of bacteria in a marine beach (southern Baltic Sea). Polish Journal of Ecology 56:405–414

Powilleit M, Kleine J, Leuchs H (2006) Impacts of experimental dredged material disposal on a shallow, sublittoral macrofauna community in Mecklenburg Bay (western Baltic Sea). Marine Pollution Bulletin 52:386–396

Powilleit M, Kube J (1999) Effects of severe oxygen depletion on macrobenthos in the Pomeranian Bay (southern Baltic Sea): a case study in a shallow, sublittoral habitat characterized by low species richness. Journal of Sea Research 42:221–234

Radziejewska T (1986) On the role of *Mytilus edulis* aggregations in enhancing meiofauna communities off the southern Baltic coast. Ophelia, Supplement 4:211–218

Remane A (1933) Verteilung und Organisation der benthonischen Mikrofauna der Kieler Bucht. Wissenschaftliche Meeresuntersuchungen/Abteilung Kiel, N.F. 21/2:163–221 [in German]

Reusch TBH, Chapman ARO (1995) Storm effects on eelgrass (*Zostera marina* L.) and blue mussel (*Mytilus edulis* L.) beds. Journal of Experimental Marine Biology and Ecology 192:257–271

Rokicka-Praxmajer J, Radziejewska T (2002) Free-living nematodes of the Pomeranian Bay (southern Baltic). Part 1. A preliminary analysis of nematode variability in an area affected by direct river runoff. Acta Scientiorum Polonorum, Piscaria 1:85–104

Rokicka-Praxmajer J, Radziejewska T, Dworczak H (1998) Meiobenthic communities of the Pomeranian Bay (southern Baltic): effects of proximity to river discharge. Oceanologia 40:243–260

Sapota MR (2004) The round goby (*Neogobius melanostomus*) in the Gulf of Gdańsk – a species introduction into the Baltic Sea. Hydrobiologia 514:219–224

Schierding M, Vahder S, Dau L, Irmler U (2011) Impacts on biodiversity at Baltic Sea beaches. Biodiversity and Conservation 20:1973–1985

Schwinghamer P (1981) Characteristic size distributions of integral benthic communities. Canadian Journal of Fisheries and Aquatic Sciences 38:1255–1263

Simm M, Ojaveer E (2000) Dynamics of copepods and fish larvae in Pärnu Bay (NE part of the Gulf of Riga) in the spring-summer period. Proceedings of the Estonian Academy of Sciences, Biology and Ecology 49:317–326

Smirnov AV (2002) Vertical distribution and abundance of gymnamoebae (Rhizopoda) in bottom sediments of the brackish water Nivå Bay (Baltic Sea, The Sound). Protist 153: 239–250

Sonntag N, Garthe S, Adler S (2009) A freshwater species wintering in a brackish environment: habitat selection and diet of Slavonian grebes in the southern Baltic Sea. Estuarine, Coastal and Shelf Science 84:186–194

Southward AJ, Hawkins SJ, Burrows MT (1995) Seventy years' observations of changes in distribution and abundance of zooplankton and intertidal organisms in the western English Channel in relation to rising sea temperature. Journal of Thermal Biology 20:127–155

Spungis V (2002) Invertebrates of the sandy coastal habitats in Latvia. Latvijas Entomologs 39:10–19

Stempniewicz L (1995) Feeding ecology of the long-tailed duck *Clangula hyemalis* wintering in the Gulf of Gdańsk (southern Baltic Sea). Ornis Svecica 5:133–142

Sundbäck K, Miles A, Göransson E (2000) Nitrogen fluxes, denitrification and the role of microphytobenthos in microtidal shallow-water sediments: an annual study. Marine Ecology Progress Series 200:59–76

Sundbäck K, Nilsson P, Nilsson C, Jonsson B (1996) Balance between autotrophic and heterotrophic components and processes in microbenthic communities of sandy sediments: a field study. Estuarine, Coastal and Shelf Science 43:689–706

Swedmark B (1964) The interstitial fauna of marine sand. Biological Reviews 39:1–42

Tomczak MT, Müller-Karulis B, Järv L, Kotta J, Martin G et al (2009) Analysis of trophic networks and carbon flows in southeastern Baltic coastal ecosystems. Progress in Oceanography 81:111–131

Urban-Malinga B, Hedtkamp SIC, van Beusekom JEE, Wiktor J, Węsławski JM (2006) Comparison of nematode communities in Baltic and North Sea sublittoral, permeable sands – diversity and environmental control. Estuarine, Coastal and Shelf Science 70:224–238

Urban-Malinga B, Wiktor J (2003) Microphytobenthic primary production along a non-tidal sandy beach gradient: an annual study from the Baltic Sea. Oceanologia 45:705–720

Vaitkus G (1999) Spatial dynamics of wintering seabird populations in the Baltic Proper: a review of factors and adaptations. Acta Zoologica Lituanica 9:126–141

Veber T, Kotta J, Lauringson V, Kotta I (2009) Influence of the local abiotic environment, weather and regional nutrient loading on macrobenthic invertebrate feeding groups in a shallow brackish water ecosystem. Oceanologia 51:541–559

Voipio A (ed) (1981) The Baltic Sea. Elsevier, Amsterdam 418 pp

Węsławski JM, Kupidura T, Żabicki M (2000a) Sandhoppers, *Talitrus saltator* (Montagu, 1808) (Amphipoda, Gammaridea), at the Polish Baltic coast: seasonal and spatial distribution patterns. Crustaceana 73:961–969

Węsławski JM, Urban-Malinga B, Kotwicki L, Opaliński K, Szymelfenig M, Dutkowski M (2000b) Sandy coastlines – are there conflicts between recreation and natural values? Oceanological Studies 29:5–18

Wolender M, Zych A (2007) Beetles (Coleoptera) from seaside beach and dunes in the regions of Świnoujście, Międzyzdroje and Wisełka (Poland) located along the southern coast of the Baltic Sea. Baltic Journal of Coleopterology 7:61–71

Zettler ML, Daunys D, Kotta J, Bick A (2002) History and success of an invasion into the Baltic Sea: the polychaete *Marenzelleria cf. viridis*, development and strategies. In: Leppäkoski E, Gollasch S, Olenin S (eds) Invasive aquatic species of Europe: distribution, impacts and management. Kluwer Academic Publishers, Dordrecht, pp 66–75

Žydelis R, Ruškytė D (2005) Winter foraging of long-tailed ducks (*Clangula hyemalis*) exploiting different benthic communities in the Baltic Sea. The Wilson Bulletin 117:133–141

Estuaries and coastal lagoons

13

Hendrik Schubert and Irena Telesh

Abstract

1. Estuaries and coastal lagoons, semi-enclosed inland water bodies, are highly productive systems and function as a transitional zone between land and sea.
2. In the Baltic Sea Area, it may be difficult to discriminate between lagoon-type estuaries and lagoons with some freshwater influence because there is a continuum between pure estuaries and pure lagoons with respect to flow dynamics.
3. The estuaries and lagoons in the Baltic Sea Area are highly dynamic environments as they experience pronounced erratic changes in water level and salinity and their shallowness induces high variability of light and water temperature.
4. The variability of the environment and high organic enrichment enhances the diversity of planktonic and benthic protists (unicellular autotrophic, mixotrophic and heterotrophic eukaryotes); in contrast, the diversity of the macrozoobenthos is low.
5. In the shallow areas of coastal lagoons with the bottom covered by organic matter-rich fine-grained sediment (mud, silt), the sediment stability is often very low.
6. Wherever organic mud dominates, a large part of the consumer spectrum is absent. Consequently, food webs may have "open ends" and organic matter is channelled to decomposers or to the sediment for burial rather than to consumers.
7. Estuaries and lagoons are sensitive to eutrophication, which may result in shifts from a macrophytobenthos-dominated system to a phytoplankon-dominated one, or from a grazing food web to a microbial food web.
8. Estuaries and lagoons are also sensitive to introductions of non-indigenous species, which may have the potential of increasing eutrophication by eliminating planktonic filter feeders (*e.g.* the carnivorous cladoceran *Cercopagis pengoi*) or decreasing eutrophication symptoms by a large filtration capacity that clears the water and allows a macrophytobenthic vegetation to reoccur (*e.g.* the bivalve *Dreissena polymorpha*).

Keywords
Coastal lagoons • Community structure • Decomposition • Estuaries • Human impacts • Muddy habitats • Open-end food webs • Productivity • Seasonal variability • Shallow water

H. Schubert (✉)
Department of Biological Sciences, University of Rostock,
Albert-Einstein-Straße 3, 18051 Rostock, Germany
e-mail: hendrik.schubert@uni-rostock.de

I. Telesh
Zoological Institute of the Russian Academy of Sciences,
Universitetskaya Emb. 1, 199034
Sankt-Petersburg, Russia
e-mail: irena.telesh@zin.ru

P. Snoeijs-Leijonmalm et al. (eds.), *Biological Oceanography of the Baltic Sea*,
DOI 10.1007/978-94-007-0668-2_13

13.1 Transitional zones between land and sea

13.1.1 Areas of high ecological and economic importance

Estuaries and coastal lagoons, semi-enclosed inland water bodies with connections to the sea, cover ~13 % of the global coastlines (Barnes 1980; Kjerve 1994; Caumette et al. 1996). Estuaries and lagoons form transitional zones between the mainland and the open sea and are an integral part of the Baltic Sea ecosystem. They host a rich fauna and flora, serve as spawning and nursery habitats for migratory organisms of both ecological and economic value, and provide feeding areas for a multitude of bird species. The human population uses the estuaries and lagoons of the Baltic Sea for tourism and recreation, but also for *e.g.* building harbours and aquaculture. This may create conflicting interests, *e.g.* when protected areas are to be designated (*cf.* Sect. 18.1).

Engineering constructions have altered the nature of some of the semi-enclosed inland waters of the Baltic Sea. For example, in the eastern Gulf of Finland, the inner Neva Estuary was separated from the lower estuary by a 16 km long storm-surge barrier (built in 1978–2011). The inner estuary is the freshwater Neva Bay, which now may be regarded as an "artificial lagoon" (Alimov and Golubkov 2008).

Shallow coastal areas have the dual "task" of being receivers of land runoff (becoming eutrophicated) and of serving as a nutrient filter between the land and the sea (protecting the sea from eutrophication). The estuaries and coastal lagoons of the Baltic Sea show large heterogeneity in environmental conditions, biota and human uses, which demands regional and local tailor-made solutions (*cf.* Sect. 18.5).

13.1.2 Estuaries

Estuaries are transitional waters where the freshwater from a river mixes with saline water from the sea. An often quoted definition (Pritchard 1967) describes an estuary as "a semi-enclosed coastal body of water which has a free connection with the open sea and within which the seawater is measurably diluted with freshwater derived from land drainage". However, this definition may also include bays, sounds and lagoons, even the Baltic Sea as a whole, depending on what is meant by the "free connection" (*cf.* Sect. 2.1.3; Elliott and McLusky 2002).

Wolanski (2007) combined several previously published definitions of an estuary into the following: "a semi-enclosed body of water connected to the sea as far as the tidal limit or the salt-intrusion limit, and receiving freshwater runoff, recognising that the freshwater inflow may not be perennial (*i.e.* it may occur only for part of the year) and that the connection to the sea may be closed for part of the year (*e.g.* by a sand bar) and that the tidal influence may be negligible". However, this definition still includes types of water bodies other than the classical estuaries, including lagoons.

Besides definitions based on geomorphology and hydrodynamics, the definition of an estuary may also include spatio-temporal characteristics of both structural components and biogeochemical processes (Elliott and Quintino 2007). From this point of view an estuary might be defined as "a semi-enclosed water body characterised by a variety of inter-related biotic and abiotic structural components naturally undergoing change in space and time, along with intensive chemical, physical and biological processes exposed against a salinity gradient" (Telesh and Khlebovich 2010). However, it is still very difficult to create a perfect definition that distinguishes all estuaries from related water bodies, *e.g.* coastal lagoons.

13.1.3 Coastal lagoons

According to the formal definition (United Nations 1997), coastal lagoons are "seawater bodies situated on the coast, but separated from the sea by land spits or similar land features. Coastal lagoons are open to the sea in restricted spaces". Thus, their semi-enclosed geographical conditions separate lagoons from other inlets and, on a larger scale, from gulfs. Bird (1982) proposed that the term "coastal lagoon" be used for a coastal water body when the width of marine entrance(s) at high tide is less than one-fifth (20 %) of the total length of the enclosing barrier.

With this definition, however, even the Gulf of Riga fits the description of a lagoon. This is debatable for several reasons. On the one hand, lagoons are usually relatively small and shallow (Kjerve 1986; Kjerve and Magill 1989) while the Gulf of Riga has a surface area exceeding 18,000 km^2 and the average and maximum depths of 23 and 51 m, respectively (*cf.* Table 2.2). On the other hand, there are many water bodies in the Baltic Sea Area (Baltic Sea, Belt Sea and Kattegat) that may be considered "typical" shallow lagoons, but their connection with the sea is wider than 20 % of the barrier. It also becomes difficult to categorise semi-enclosed inland waters that fit the descriptions of both estuary and lagoon, *e.g.* "lagoonal estuaries" such as the Szczecin Lagoon, which is separated from the sea by a barrier <20 % but receives the discharge from a large river, the Odra.

13.1.4 The difference between estuaries and lagoons

Altogether, it can be concluded that general "water-tight" definitions for estuaries and coastal lagoons can hardly be given (*e.g.* McLusky and Elliott 2004, 2006), not only for the Baltic Sea Area. The main difference between estuaries and lagoons lies in the flow dynamics of those water bodies. In estuaries the freshwater usually flows fast and forcefully towards the sea, whereas in lagoons the water is shallower and flows sluggishly. However, nearly all lagoons in the Baltic Sea receive a substantial freshwater runoff, and it is therefore difficult to discriminate between lagoon-type estuaries and lagoons with some freshwater influence, as there a large variety of different semi-enclosed coastal inland water bodies with gradually changing flow dynamics in the Baltic Sea Area.

13.2 Different types of semi-enclosed inland waters

13.2.1 Geomorphology and hydrography

The geomorphology of the Baltic Sea region largely determines the shape of the coastline. About a dozen different coast types can be distinguished in the Baltic Sea Area (Lampe 1995), the main types being archipelago coasts, open low coasts, lagoon and bodden coasts, klint coasts and fjord coasts (*cf.* Fig. 2.8), with other coast types being intermediates of these. However, at the regional and local scales, the Baltic Sea coast often presents a complicated mosaic of different coast types with alternating hard, sandy and soft substrates (*cf.* Sect. 11.1). The largest Baltic Sea estuaries are located in the southern and eastern Baltic Sea (from the Odra to the Neva), and the largest lagoons from Denmark to Lithuania. This covers the part of the Baltic Sea drainage area where the human population and agricultural land use, and thus the anthropogenic pressure on the Baltic Sea ecosystem, are the largest (*cf.* Fig. 2.5b, c).

The influence of the drainage area on the open Baltic Sea is basically determined by the river discharge to the sea via river deltas, estuaries and coastal lagoons. The Baltic Sea Area receives discharge from >200 rivers, but about half of the total river water input to the Baltic Sea is transported by the six largest rivers emptying into the Gulf of Finland and the southeastern Baltic Sea proper. These six rivers are the Neva (17 % of the total freshwater inflow into the Baltic Sea, entering the Neva Bay, Russia), the Wisła (11 %, entering the Gdańsk Bay, Poland), the Odra (7 %, entering the Szczecin Lagoon, Poland-Germany), the Nemunas (5 % entering the Curonian Lagoon, Lithuania-Russia), the Daugava (5 %, entering the Gulf of Riga, Latvia-Estonia) and the Narva (3 %, entering the Narva Bay, Estonia-Russia) (*cf.* Table 2.4; Fig. 2.11).

The whole northern Baltic Sea, *i.e.* the coasts of Sweden and Finland (except for their southernmost stretches), is part of the Precambrian shield. Here the coasts are formed by magmatic rock and metamorphic rock. At the end of the glaciations, huge till layers were deposited in the southern Baltic Sea Area, providing a good sediment source for the formation of spits (Lampe 1995). This is why coastal lagoons in a strict sense (*i.e.* shallow inland water bodies dammed by sandy barriers) occur primarily on the Danish coasts at the entrance to the Baltic Sea and along the southern Baltic Sea coast. The largest lagoons, the Curonian Lagoon ($\sim$1,600 km^2), the Wisła Lagoon ($\sim$840 km^2) and the Szczecin Lagoon ($\sim$690 km^2) are found in the southern part of the Baltic Sea proper (Fig. 13.1). However, lagoons in a broad sense (*i.e.* shallow inland water bodies dammed by sandy barrier or rocky islands) exist all over the Baltic Sea Area. In the north, the coastal areas are subjected to land uplift and lagoons are formed as a transitional phase before they transform into lakes isolated from the sea.

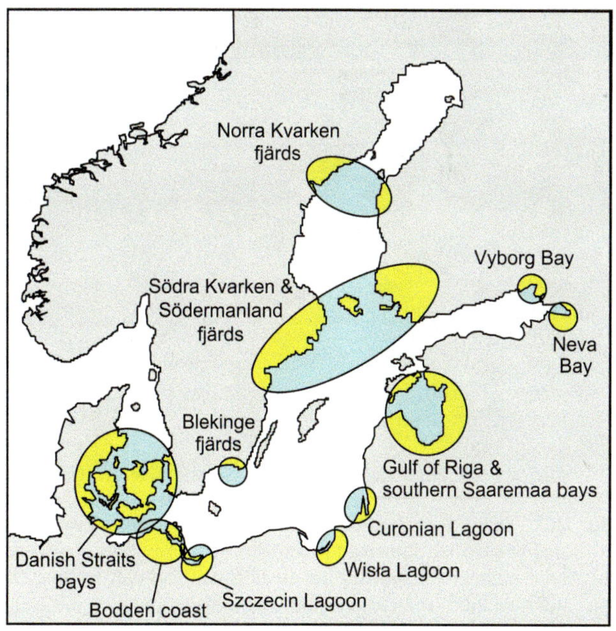

Fig. 13.1 Areas of the Baltic Sea with large and/or numerous lagoons, semi-enclosed bays and fjärds with connections to the sea. Note that these types of water bodies are not restricted to these areas. The German bodden coast includes the Nordrügenschen Bodden, the Salzhaff, the Darß-Zingster Boddenkette and the Greifswalder Bodden. Figure: © Hendrik Schubert

13.2.2 Northern and southern estuaries and lagoons

All estuaries and coastal lagoons in the Baltic Sea Area have individual characteristics, depending on size, topography, amount and composition of land runoff, as well as other environmental drivers and their variability. However, they can be roughly separated into two groups, which are distinguished by, *inter alia*, geomorphology and type of land use in the drainage areas. For example, the sandy and soft-substrate estuaries and lagoons in the southern and eastern Baltic Sea (Fig. 13.2a) receive runoff mainly from agricultural land while the ones with mixed substrates in the northern Baltic Sea receive their runoff mainly from forested land. Generally, the latter estuaries and lagoons are less eutrophicated and have a larger variety of substrates, which affects the composition and productivity of their communities and the structure of their food webs (Table 13.1).

Fig. 13.2 The Nordrügenschen Bodden (Germany, Fig. 13.1). (**a**) Aerial photograph showing the complex structure of the landscape. (**b**) Aerial photograph showing an eroding moraine cliff feeding the spits and sand bars. The coastal retreat proceeds at a rate of ~0.2–0.5 m per year while the spits and sand bars grow by ~20 m per year. The double-headed spit in the centre of (**a**) was deposited during the last 350 years; the first maps of this site ("Schwedische Matrikelkarte" 1695) include only a small sand bar. A darker-blue deeper-water channel separates the double-headed spit from another spit extending from the left side of the photograph. The channel is kept open by water exchange driven by the water level changes of the open sea, which results in a strong net sediment transport out of the lagoonal area and prevents its conversion into a lake system. Photo: (**a**) © Hendrik Schubert, (**b**) © Irmgard Blindow

13.2.3 The formation of estuaries

Based on their geomorphology, estuaries can be classified into estuaries of tectonic origin, drowned river valleys, glacially formed fjord-type estuaries and bar-built estuaries (Kennish 1986). Estuaries of tectonic origin are not found in the Baltic Sea Area where most of the estuaries have either glacial backgrounds or are the result of on-going sedimentation processes.

Drowned river valleys are estuaries formed when the eustatic sea level rise after the last glaciation (*cf.* Fig. 2.26a) resulted in the flooding of river valleys that once incised the landscape when the sea level was lower. These estuaries are relatively shallow, and broaden and deepen towards the sea. An example is the Chesapeake Bay, a North American brackish-water ecosystem that has many similarities to the Baltic Sea ecosystem (Ulanowicz and Wulff 1991).

13.2.4 Fjärds

Fjord-type estuaries are formed in regions where glaciers have eroded deep valleys in the rock. In the Baltic Sea Area there are no real fjords of the type found in Norway, Greenland or Chile, but many estuaries of the rocky archipelago coasts of the Baltic Sea partly consist of fjärds (*cf.* Sect. 2.2.9; Fig. 13.1). Fjärds were formed in landscapes with less relief than those where real fjords were formed. Therefore, fjärds are shallower and broader than fjords and lack steep walls. Often the fjärds have a connection with the sea which is <20 % of the barrier length formed by islands, and, similarly to the Gulf of Riga, they would conform to the definition of a "lagoon" when using the definition based on the degree of enclosure. Contrary to fjords, fjärds may contain mud flats, salt marshes and flood plains. Also the shallow Danish Straits bays (Fig. 13.1) may or may not be considered estuaries or lagoons depending on the occurrence of river discharge and/or a <20 % opening to the sea.

13.2.5 Bar-built estuaries

The Baltic Sea has many bar-built (lagoon-type) estuaries that have become semi-isolated from the open sea by barriers in the form of spits and islands. The connections with the sea are narrow and restricted, being kept open by the continuous riverine runoff as well as by a two-way water

Table 13.1 Summary of features that are similar for all estuaries and lagoons of the Baltic Sea, features that differ between different parts of the Baltic Sea, as well as some exceptions to these general patterns. Data from Telesh (2004) and Schiewer (2008b)

Features	Southern and eastern estuaries and lagoons of the Baltic Sea	Northern and western estuaries and lagoons of the Baltic Sea
Similar	– High species richness and biological productivity – Dominance of eurytopic species (species able to adapt to a wide range of environmental conditions) – Plankton abundance and diversity changes occur regularly in space (along changing gradients of physical, chemical and biological parameters), and in time (seasonal and stochastic) – Vulnerable to eutrophication, chemical pollution and biological invasions – A marked increase in the biological importance of protist diversity, abundance and production at sites that have been subjected to strong eutrophication. This can be explained by a switch of plankton communities from a grazing food web to a microbial food web – Major ecosystem services are: (1) source of biodiversity for the adjacent open Baltic Sea waters, (2) spawning and nursery habitats for fish, (3) refuge areas for birds, (4) filters and buffers for land-borne eutrophication and chemical pollution	
Different	– More eutrophication – Higher chlorophyll *a* concentration – Frequent absence of diatom dominance – In extreme cases all-season dominance of cyanobacteria and green algae – More oxygen limitation in sediments due to increased sedimentation of organic material – Lower species diversity of macroalgae – Lower species diversity of macrozoobenthos, particularly of bivalve molluscs (but increased importance of oligochaetes and chironomid larvae)	– Less eutrophication – Lower chlorophyll *a* concentration (but still higher than in the open Baltic Sea proper) – Permanent diatom dominance – Summer dominance of cyanobacteria and dinoflagellates – Less oxygen limitation in sediments – Higher species diversity of macroalgae through higher availability of hard substrate – Higher species diversity of macrozoobenthos
Exceptions	– When water exchange with the open Baltic Sea is stronger than usual – When freshwater runoff from land is stronger than usual	

exchange caused by the weather-dependent irregular sea level changes typical of the Baltic Sea (*cf.* Sect. 2.3.4). Most of the systems belonging to this type of estuary are located along the southern Baltic Sea coast where erosion of moraine ridges produces large quantities of sandy materials, which are transported with the predominant currents eastwards and deposited on the shore to form long sandy beaches (Fig. 13.2b).

13.2.6 The formation of coastal lagoons

While estuaries are formed by rivers that discharge into the coastal zone, typical coastal lagoons emerge as a result of different processes. In general, this involves the transport of sediment and its deposition in the form of barriers parallel to a more or less fragmented coastline. The distribution and dynamics of coastal lagoons depends on six major factors: (1) antecedent topography, (2) material characteristics, (3) sediment supply, (4) tectonics, (5) tidal amplitude and (6) climate (Kjerve 1994).

Antecedent topography is the former emergent coastal area that is completely or partially submerged under the conditions of land subduction (*cf.* Fig. 2.26b). It is usually referred to as a geological inheritance factor influencing barrier formation. The submerged coastal area will influence barrier formation by providing the substrate slope and the general morphology, which will in turn control wave energy and, consequently, sediment supply (Dillenburg and Hesp 2009). With respect to antecedent geomorphology, the development of the coastal lagoons in the southern Baltic Sea is facilitated by a fractured, low-lying coastline. The lagoon formation requires medium grain size sediments amenable to being transported along the shoreline as a material for barrier formation. This material may originate from cliff erosion, from riverine input of terrestrial material or from the adjacent seafloor. The fate of a coastal lagoon largely depends on the balance between sediment supply and sediment loss.

Medium- and fine-grained sands, silty sands, silty mud, pebble-gravel deposits, clay and clayey silts are the common bottom sediments in the Baltic Sea estuaries and lagoons. In many places, the deposition of phytogenic detritus has led to the formation of peat. The lagoons on the islands of Møn (Denmark), Rügen (Germany) and Usedom (Germany-Poland), situated in the transition zone (Belt Sea and Kattegat) and the southern Baltic Sea proper, feature chalk or moraine cliffs. Further to the east and northeast, the shores become smoothly levelled and dune ridges are common. The shores of large lagoons are formed from highly mobile sand.

Cliff coasts and rocky coasts are characteristic of the eastern estuaries (*e.g.* the Neva Estuary) and the fjärds on the Finnish and Swedish coasts. The northern Baltic Sea is an uplift area with crystalline bedrock. Sandy sediments are scarcer here because the expanding glaciers have largely removed the fine sediments and deposited them in the southern Baltic Sea region. However, the northern Baltic Sea coasts feature numerous semi-enclosed coastal inlets, usually narrow and relatively long bays, and numerous small lagoons, as well as open low coasts with sandy beaches (*cf.* Fig. 2.8).

13.2.7 Choked, restricted and leaky lagoons

A long stretch of the Baltic Sea coast, from Germany up to Russia, is characterised by a pattern of eroding Holocene cliffs and depositional areas. This results in a chain of semi-enclosed inland water bodies, ranging from "true" coastal lagoons without any notable freshwater runoff (*e.g.* the Nordrügenschen Bodden) to more "lagoon-type" estuaries (*e.g.* the Szczecin Lagoon and the Curonian Lagoon) (Fig. 13.1).

"True" coastal lagoons (with an open connection to the sea but narrower than 20 % of the barrier length) can be divided into "choked", "restricted" and "leaky" (Bird 1982; Fig. 13.3). This classification is based on the characteristics of their potential hydrodynamic exchange properties with the open sea. The intensity of the water exchange is not only a function of the relative width of the opening, but is also related to the real mixing regime, which involves the relationship between the freshwater inflow, the total water volume and the residence time of the water in the lagoon. Choked lagoons usually have only one narrow channel to the sea, resulting in a delayed and dampened water exchange with the sea and a long water residence time. Restricted lagoons have multiple channels to the sea with relatively good water exchange, and tend to have a net transport of water to the sea. Leaky lagoons have numerous wide channels to the sea and are therefore characterised by almost unimpaired water exchange (Kjerve 1986; Leppäranta and Myrberg 2009).

Some of these lagoons, especially the Nordrügenschen Bodden and the Darß-Zingster Boddenkette, are interconnected complex systems, whereas the Salzhaff, the Wisła Lagoon, and the Curonian Lagoon are classical "choked" systems with a rather simple morphology. Along the Swedish and Finnish coasts, especially in the Norra and Södra Kvarken areas, there are numerous smaller inland water bodies that fit the description of lagoons. They often belong to complex systems of interconnected water bodies dotted by small isles lying behind a loose chain of islands that protect the system from the seaward side. Within these areas, a special type of small-sized lagoon called a "flad" has been investigated in detail by Munsterhjelm (1997). These systems, most of which are not larger than ~1,000 m^2,

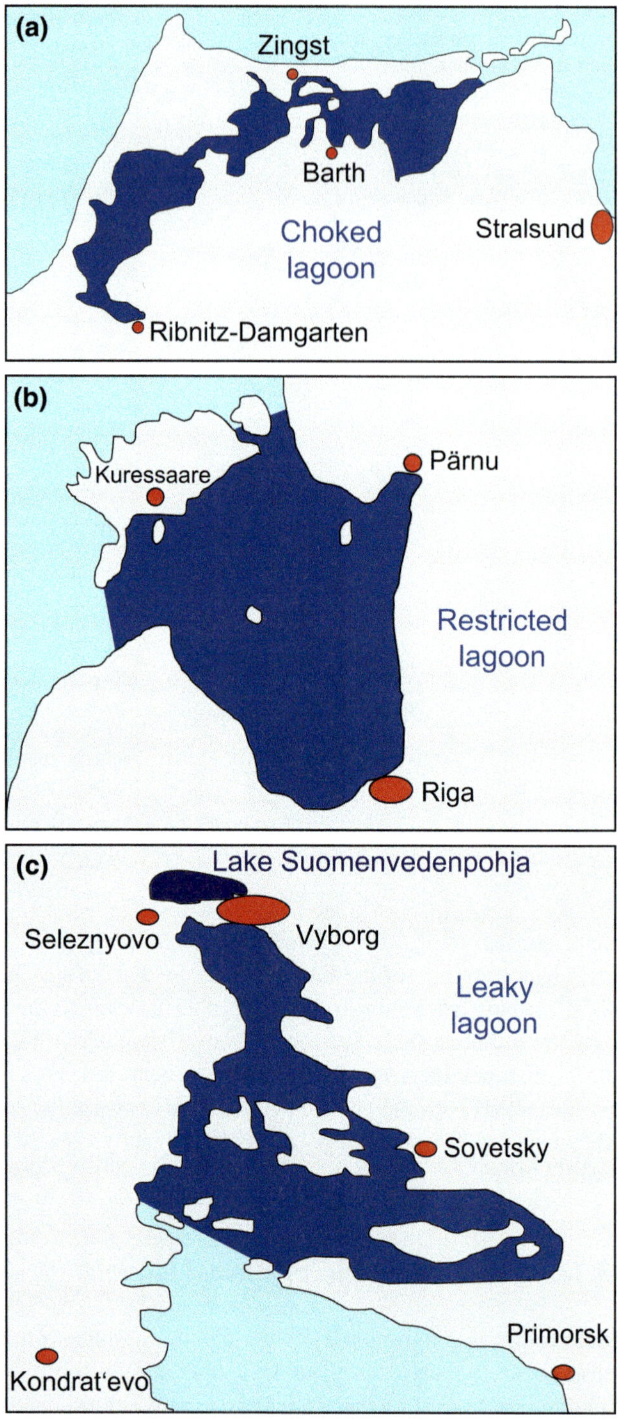

Fig. 13.3 Baltic Sea examples of the three main types of coastal lagoons. (**a**) A choked lagoon, exemplified by the Darß-Zingster Boddenkette, a chain of interconnected basins. (**b**) A restricted lagoon, exemplified by the Gulf of Riga. (**c**) A leaky lagoon, exemplified by the Vyborg Bay. For the location of these water bodies in the Baltic Sea Area, see Figs. 2.2 and 13.1. Figure: © Hendrik Schubert

resulted from the land uplift of former bays and belong to the choked lagoon type. However, their principal hydrographical conditions are similar to those in coastal lagoons formed by sediment barriers.

A crucial factor for the functioning of coastal lagoons is the water exchange with the open Baltic Sea, which not only determines the salinity of the water but also influences *e.g.* the proportion of the primary production in the lagoon that is exported to the sea. Water exchange is significant for restricted and leaky lagoons, but relatively unimportant for choked lagoons. For example, in the Greifswalder Bodden (a restricted lagoon with low river inflow), the internal matter cycles are controlled by horizontal exchange processes with the sea, while the Darß-Zingster Boddenkette (a choked lagoon) is less dependent on the exchange with the Arkona Sea due to the narrow connection with the sea and its complex structure (Fig. 13.3a), which both increase the water residence time (Schiewer 2008a).

13.2.8 A Baltic "lagoon" made by a meteorite

Lumparn fjärd, a semi-enclosed water body in the centre of the Åland Archipelago (Finland) is a relatively large bay, ~10 km in diameter. The fjärd is devoid of islands and islets, which is unusual in this archipelago area. In 1993 it was discovered that Lumparn fjärd had been formed in a meteorite crater on the granite bedrock. This crater is estimated to be ~1,000 million years old and is filled with different types of sediments (Svensson 1993; Abels et al. 1998).

13.3 Ecological characteristics

13.3.1 Water level changes

The tidal amplitude in the Baltic Sea is very small (*cf.* Sect. 2.3.3), only in the range of some cm (Wróblewski 2001). However, this does not mean that the estuaries and coastal lagoons are not affected by pronounced water level changes. In fact, the irregular but frequent wind- and air pressure-driven water level changes (*cf.* Sect. 2.3.4) may be enhanced by resonance coupling and exceed 1.5 m. Moreover, waves higher than 2.4 m occur in the Baltic Sea with a frequency of 10–20 %, which is the average for European coasts. This led Eisma (1998) to include the wind flats of the Baltic Sea in an overview of the worldwide distribution of intertidal areas, despite the fact that the water level changes are not tidal. Thus, the estuaries and lagoons of the Baltic Sea experience pronounced erratic changes in water level and, consequently, in salinity (*cf.* Fig. 7.4).

13.3.2 Ice cover changes biological processes

Fully marine lagoon systems in temperate and boreal zones are rarely exposed to long-lasting ice cover. However, the estuaries and coastal lagoons of the Baltic Sea, even those in the southern part, develop an ice cover almost every year. The occurrence and persistence of an ice cover has large effects on the biological processes in these semi-enclosed inland waters. For example, gas exchange with the atmosphere becomes restricted in periods when the sediment temperature might still be high enough for decomposition; hence the release of NH_4^+ and H_2S is augmented. Furthermore, light availability decreases so that photosynthesis and oxygen production are reduced, and the sediments may be mechanically disturbed by deeply penetrating ice scour when the water level changes.

13.3.3 Shallowness creates a highly variable environment

The average depth of coastal Baltic Sea semi-enclosed waters usually does not exceed 5 m. Thus, a prolonged thermal or salinity stratification is rare, the continuous mixing of the water leads to tight benthic-pelagic coupling, and the stochastic water exchange with the open sea causes irregular "washout effects", *i.e.* the export of fine organic-rich sediments to adjacent deeper basins of the sea (Schiewer 2008a).

The shallowness of the estuaries and lagoons is also conducive to a high variability of light and water temperature. In addition to circadian and seasonal cycles of solar radiation, the high variability of light conditions results from both phytoplankton development and sediment resuspension by strong wind-induced mixing (*cf.* Fig. 7.2). The temperature variability results from the high surface-to-volume ratio of these shallow semi-enclosed inland waters and consequently from the relatively low thermal buffering capacity of these systems. Day/night temperature amplitudes of >10 °C are common, especially in the summer months, when high daytime irradiation is followed by evaporation at night.

Thus, the environmental variability affecting estuarine and lagoonal communities is much more profound than that influencing the communities of open coasts. This challenges the physiological adaptability of the organisms living in the semi-enclosed inland waters of the Baltic Sea.

13.3.4 Integration of brackish and freshwater communities

Estuaries and coastal lagoons are transitional waters between the marine ecosystem and the terrestrial ecosystem (including limnic environments), with environmental factors that fluctuate both fast and over a wide range. One of the most profound features in the Baltic estuaries and lagoons is the meeting and integration of the brackish communities of the open Baltic Sea and the freshwater communities of the rivers.

Salinity and the water exchange rate shape the productivity, biodiversity and distribution of the flora and fauna in estuaries and lagoons. The freshwater inflows alter the typical brackish-water biodiversity distribution patterns in the estuaries and lagoons. As a result, a species-rich mixed freshwater and euryhaline flora and fauna often dominate in the mixing water masses in the inner parts of estuaries and lagoons. Due to the "spill-over effect", members of the freshwater flora and fauna (particularly the plankton) may considerably enrich the adjacent open-sea communities in terms of both diversity and abundance.

In estuaries and lagoons, only those organisms that can tolerate significant environmental variation controlled primarily by environmental drivers (*e.g.* salinity, water exchange, temperature, light) are able to survive for longer times, and, when they do, they can achieve high abundances in these fluctuating environments. Along the estuarine and lagoonal salinity gradients of the Baltic Sea, the macrozoobenthos shows a minimum in species richness at salinity 5–7 (Remane 1934; *cf.* Fig. 4.21), but in coastal waters protists (autotrophic, mixotrophic and heterotrophic eukaryotic microplankton) often show a maximum consistent with the "protistan species maximum concept" (Telesh et al. 2011a, b, 2013, 2015) (Box 13.1). This is due to the large numbers of protistan species with high adaptability to salinity fluctuations, which is expressed most markedly in estuaries and lagoons. Under oligohaline conditions (salinity <5), which is often the case in the Baltic Sea estuaries and lagoons, marine and euryhaline species are replaced by freshwater species that sometimes perform even better, *e.g.* freshwater *Daphnia* spp. can be more efficient filter feeders than their brackish-water counterparts.

13.3.5 Sensitivity to non-indigenous species invasions

The Baltic Sea continuously receives large quantities of non-indigenous species. The estuaries and coastal lagoons in the southeastern part of the Baltic Sea are especially sensitive to invasions of Ponto-Caspian species, which migrate via rivers and human-made canal systems (*cf.* Fig. 5.11). After arrival to the new environment such organisms may establish before they either adjust, decline or become extinct (*cf.* Fig. 5.3). Mass occurrences of non-indigenous species may inhibit the development of indigenous, well-adapted communities (DeWit 2007), and can heavily impact food webs and matter fluxes in estuaries and lagoons as shown by *e.g.* the fish-hook water flea *Cercopagis pengoi* (*cf.* Box 5.5) and the zebra mussel *Dreissena polymorpha* (*cf.* Box 5.4). After establishment, the non-indigenous species may spread to other estuaries and lagoons in the region, and some of them have spread to other parts of the Baltic Sea, *e.g.* the round goby *Neogobius melanostomus* (Zaiko et al. 2011; Rakauskas et al. 2013).

13.3.6 Sensitivity to eutrophication

Estuaries and coastal lagoons are the first receivers of nitrogen and phosphorus from terrestrial runoff, and therefore are naturally rich in nutrients. However, most estuaries and lagoons of the Baltic Sea are affected by human-induced eutrophication (with exceptions being mainly found in the Gulf of Bothnia). Despite mitigation measures, eutrophication persists and a marked rise in phytoplankton production is often observed as a result.

Human-induced eutrophication has altered the relative importance and dominance of different primary producers in the southern and eastern Baltic Sea lagoons (Box 13.2). For

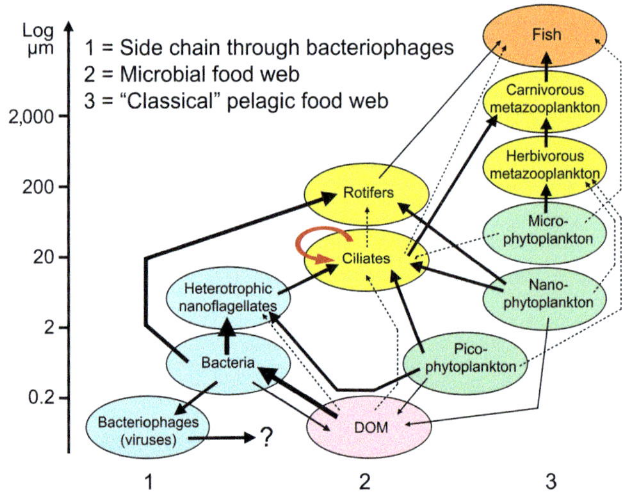

Fig. 13.4 A schematic presentation of the plankton food web of the Darß-Zingster Boddenkette in a late spring/early summer situation, showing the approximate body sizes of the organisms on the axis to the left. 1 = Side chain involving bacteriophages. Viruses do not target bacteria only, but viral effects on other food web components have not yet been studied in detail. 2 = The microbial food web, which is the main pathway for the carbon turnover in the Darß-Zingster Boddenkette (mainly via bacterioplankton and heterotrophic nanoplankton). The internal loop in the ciliate community (red arrow) can involve up to three additional trophic levels. 3 = The "classical pelagic food web" from net plankton to fish, but with ciliates and rotifers being major food items for fish fry. DOM = dissolved organic matter. "Bacteria" in this figure include both Bacteria and Archaea. Figure modified from Schiewer (2008b)

example, summer cyanobacterial blooms are now typical of all Baltic Sea coastal waters, except for the Bothnian Bay, and diazotrophic (nitrogen-fixing) cyanobacteria have become more common. Nano- and picocyanobacteria (*e.g. Aphanothece clathrata*) have become more abundant in eutrophic coastal waters and chroococcal cyanobacteria (*e.g. Chroococcus* spp., *Microcystis aeruginosa*) dominate in some coastal lagoons, *e.g.* in the Szczecin Lagoon.

The increased primary production in estuaries and lagoons caused by eutrophication can be buffered by higher zooplankton feeding activity. The zooplankton increases in biomass, but with altered community composition and structure. Generally, eutrophication results in reduced abundances of larger zooplankton (*e.g.* of the copepod *Eurytemora affinis*) and increased densities of smaller rotifers and protists (ciliates). Highly diverse planktonic protistan communities have gained in importance (Telesh et al. 2009, 2011a, b; Mironova et al. 2014). A marked increase in the ecological importance of protists, manifested by their elevated abundances and production, accompanies higher eutrophication levels due to a switch from a classical grazing food web to a microbial food web (Fig. 13.4). This involves a higher activity of heterotrophic bacteria, heterotrophic flagellates and ciliates, the latter having internal carbon cycles because of their nutritional diversity. The microbial food web is fuelled by prokaryotic nano- and picophytoplankton species with short generation times, and is therefore able to support high biomasses of ciliates and rotifers.

13.3.7 The fluffy sediment layer

The bottom of shallow eutrophic coastal inland waters of the Baltic Sea is often covered by a highly mobile fluffy sediment layer (FSL) the thickness of which varies from one to several mm (Fig. 13.5). In the FSL, small particle aggregates (with diatoms and other microorganisms) are combined into larger ones. These aggregates are colonised by bacteria and protozoa of the microbial food web (Fig. 13.4), and thus contribute significantly to the internal nutrient load of these waters.

The development of microbial food webs in the Baltic Sea coastal waters is facilitated by the following: (1) a high concentration of dissolved organic matter (DOM) and particulate organic matter (POM), with the DOM:POM ratio being almost 1, (2) the absence of nutrient limitation, (3) the presence of light limitation which favours the development of eukaryotic and prokaryotic nanophytoplankton and prokaryotic picophytoplankton instead of larger phytoplankton species with a higher irradiance demand, (4) an increased concentration of heterotrophic protists, *e.g.* nanoflagellates and ciliates, and (5) the development of an unstable FSL (Schiewer 2008c).

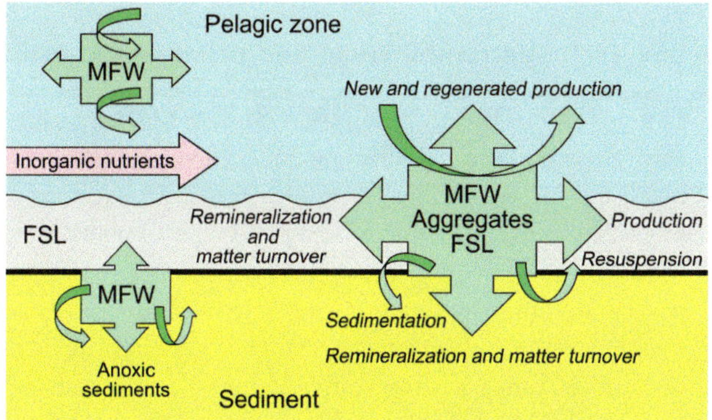

Fig. 13.5 A schematic presentation of matter turnover in the near-bottom water and sediment of Baltic Sea coastal waters. FSL = fluffy sediment layer, MFW = microbial food web. Processes within the MFW and FSL are given in italics. Figure modified from Telesh et al. (2008)

The increasing eutrophication and FSL formation leads thus to improved feeding conditions for the meio- and macrozoobenthos due to the intensive benthic-pelagic coupling in shallow waters. However, the composition and abundance of meio- and macrozoobenthic communities may vary significantly between different estuaries and lagoons, and is determined by the location, type, productivity and eutrophication level of each semi-enclosed inland water.

13.4 Food webs

13.4.1 The origin of organic matter

In contrast to the food webs of open oceans and lakes, which often become nutrient-limited after the spring bloom, the riverine input of organic material to estuaries and coastal lagoons serves as an additional continuous year-round food source. Thus, the organic material fuelling the matter flux in the shallow estuaries and lagoons of the Baltic Sea originates from (1) allochthonous material entering the lagoon or estuary from the drainage area and, to a much lesser extent, from the adjacent open sea, and (2) photosynthesis inside the estuaries and lagoons by attached macrophytes (algae and vascular plants) and their epiphytes, by detached macrophytes and their epiphytes, and by phytoplankton.

13.4.2 Allochthonous particulate organic matter (POM)

The allochthonous material transported into an estuary or a lagoon consists of both dissolved organic matter (DOM) and particulate organic matter (POM). Tidal coasts have usually high inputs of POM as tidal-influenced salt marshes produce

Box 13.1: Species minimum and maximum in estuaries and lagoons

In the “horohalinicum” (salinity 5–8) of the estuaries and coastal lagoons in the Baltic Sea Area the macrozoobenthos shows a species minimum (Remane 1934), while the number of protists (unicellular autotrophic, mixotrophic and heterotrophic eukaryotes) reaches a species maximum (Box Fig. 13.1). For this protistan species maximum (Telesh et al. 2011a, b, 2013) there may be several explanations: (1) The short generation times of protists allow for the fast evolution of adapted species. (2) The mode of life of protists: planktonic protists are transported with the water masses and benthic protists living in shallow waters are easily resuspended in the water masses. Thus, both groups passively avoid rapid salinity changes. (3) There is a good fit between the variability of the environmental conditions and generation time, which would favour the coexistence of protists in the sense of the “intermediate disturbance hypothesis” in a variable environment (*cf.* Box 4.16; Grime 1973; Connell 1978). If the variability favours one species, its population increases until an environmental change occurs that favours another species. When such changes take place with a frequency of ~3–5 times per generation time of the respective species, competitors can coexist with fluctuating population sizes.

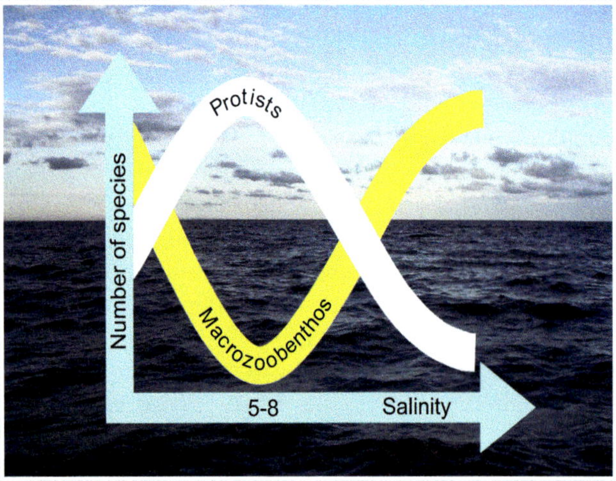

Box Fig. 13.1 A conceptual model predicting a macrozoobenthos species minimum and a protistan species maximum in the “horohalinicum” (salinity 5–8) of the estuaries and lagoons in the Baltic Sea Area. Photo: © Irena Telesh

large amounts of plant litter which is regularly washed out into the lagoons. In the microtidal Baltic Sea, the contribution of POM to the total carbon budget of estuaries and lagoons is relatively small.

For the Salzhaff, Jönsson et al. (1998) calculated a total particulate carbon import of 5.8 g C m^{-2} $year^{-1}$, which is less than 10 % of the total annual carbon supply (dissolved and particulate) to the lagoon. When comparing the particulate carbon import to the net primary productivity of the Salzhaff of 279 g C m^{-2} $year^{-1}$, it is obvious that POM supply is only a minor constituent of the food web structure. This is typical of the Baltic Sea estuaries and lagoons. In tidal estuaries, POM budgets may be completely different. For example, allochthonous detritus was found to be the main source of energy for the estuary’s food webs in the Ems-Dollard estuary (Wadden Sea, North Sea), and the particulate import was seven times higher than autochthonous primary production (Van Es 1977).

13.4.3 Allochthonous dissolved organic matter (DOM)

Dissolved allochthonous material, either still in its organic form (DOM), or already in the form of inorganic nutrients, fuels the primary producers, which in turn form the basis of a lagoon’s food web. The average nutrient load to the entire Baltic Sea amounts to ~2.0 tonnes N km^{-2} $year^{-1}$ and ~0.1 tonnes P km^{-2} $year^{-1}$ (calculated from Fig. 3.28).

As expected, the estuaries and lagoons (the areas that are the first to receive the terrestrial input), show much higher water area-specific values, *e.g.* ~18 tonnes N km^{-2} $year^{-1}$

Box 13.2: Eutrophication of lagoons and the life forms of primary producers

A conceptual model distinguishing four phases of eutrophication effects on community composition of primary producers (Phases I–IV) has been proposed based on observations on the bodden coast in northern Germany (Schiewer 1985; Gocke et al. 2003). Similar observations have been made in other boddens, Gdańsk Bay, parts of the Gulf of Riga and in lagoons along the Finnish coast (Schiewer 2008a), and a general trend following Phases I to IV with an increasing eutrophication of the Baltic Sea lagoons can be assumed (Box Fig. 13.2).

Phase I: As long as nutrient limitation in the lagoon favours vascular plants and charophytes, which can supplement their nutrient uptake by extracting nutrients from the sediment, they successfully compete with fast-growing opportunists and dominate the vegetation.

Phase II: With increasing nutrient availability, the nutrient demand of the perennials is saturated and they are not able to deplete the water column of nutrients anymore. This stimulates the growth of annual, mainly epiphytic, macroalgae.

Phase III: The macroalgal epiphytes reduce the irradiance reaching their perennial hosts and induce the mechanical stress of water movement on their hosts. Eventually, both the perennial hosts and their annual epiphytes are replaced by detached drifting macroalgae, competing for light and nutrients with phytoplankton. As long as periods of nutrient limitation, together with zooplankton grazing, control the phytoplankton at least periodically, drifting macroalgae are favoured by their relative grazing resistance and endurance in periods of nutrient starvation.

Phase IV: With still higher nutrient availability, the phytoplankton outcompetes macroalgae by light limitation and is the only group of primary producers left in the lagoon. Due to high growth rates of the phytoplankton, the top-down control by (meso-)zooplankton grazing is lost and the system switches from a grazing food web to a microbial food web.

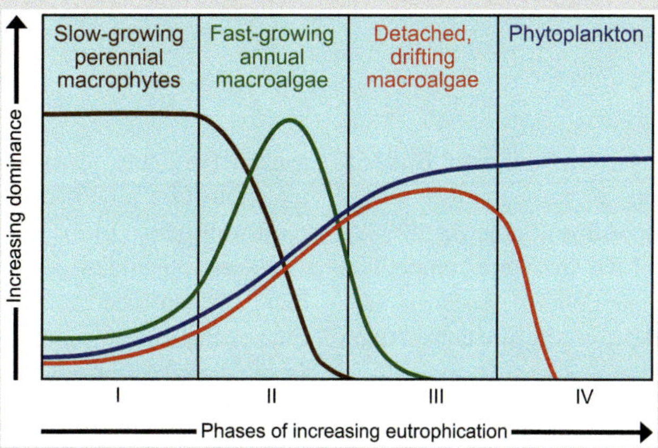

Box Fig. 13.2 A conceptual model predicting the effects of the different stages of eutrophication on the primary producers in the coastal lagoons of the Baltic Sea. Figure modified from Gocke et al. (2003)

and ~2.9 tonnes P km^{-2} $year^{-1}$ in the Wisła Lagoon, and ~138 tonnes N km^{-2} $year^{-1}$ and ~6.3 tonnes P km^{-2} $year^{-1}$ in the Neva Bay (Table 13.2). The relative nutrient supply to estuaries and lagoons decreases with increasing drainage area size, which shows the importance of terrestrial processes for nutrient cycling. For example, the drainage area-specific load (the total discharge per km^2 drainage area) of N and P to the Odense Fjord are 14 and 10 times higher, respectively, than the area-specific load to the Neva Bay.

Most of the Baltic Sea lagoons receive nutrients at high N:P ratios, far above the molar Redfield ratio of N:P = 16:1, which is optimal for phytoplankton growth (*cf.* Sect. 2.4.8). For example, in the Odense Fjord, Darß-Zingster Boddenkette, Curonian Lagoon and the Neva Bay, the molar N:P ratio varies between 47 and 75 in the inflow. Although this ratio might suggest that the lagoons are phosphorus-limited systems, it has been shown that this is not really so. Most of the systems undergo a seasonal cycle in which periods of temperature and energy limitation in winter are followed by periods of phosphorus limitation, and in some cases also silicate limitation, in spring. Later in the year, nitrogen limitation ensues (Fig. 13.6). Highly eutrophic to hypertrophic systems, such as the Darß-Zingster Boddenkette, may even become light-limited in summer. Due to their very short generation times, the phytoplankton respond to seasonal forcing by changes in composition. Regardless of the

Table 13.2 Morphometric characteristics and nitrogen and phosphorus inputs for the Gulf of Finland and some Baltic Sea estuaries and lagoons. Data from Schiewer (2008a)

Parameter	Unit	Odense Fjord	Darß-Zingster Boddenkette	Wisła Lagoon	Curonian Lagoon	Gulf of Finland	Eastern Gulf of Finland (lower inner estuary)	Neva Bay
Water surface area	km^2	62	197	838	1,584	29,600	1,146	329
Water volume	km^3	0.14	0.40	2.3	6.0	1,100	23	1.3
Average depth	m	2.2	2.0	2.7	3.8	37	20	4.0
Drainage area	km^2	1,095	1,594	23,871	100,458	412,900	350,400	281,000
Drainage area to water surface ratio		18	8	28	63	14	306	854
Annual nitrogen input (N load)	tonnes $year^{-1}$	2,433	3,370	15,200	45,000	125,000	140,000	45,338
Annual phosphorus input (P load)	tonnes $year^{-1}$	81	99	2,420	2,100	6,800	7,400	2,089
N:P ratio of the load	mol mol^{-1}	67	75	14	47	41	42	48
Water surface area-specific N load	tonnes km^{-2} $year^{-1}$	39	17	18	28	4	122	138
Water surface area-specific P load	tonnes km^{-2} $year^{-1}$	1.3	0.5	2.9	1.3	0.2	6.5	6.3
Drainage area-specific N load	tonnes km^{-2} $year^{-1}$	2.22	2.11	0.64	0.45	0.30	0.40	0.16
Drainage area-specific P load	tonnes km^{-2} $year^{-1}$	0.07	0.06	0.10	0.02	0.02	0.02	0.01

high nutrient concentrations in the Darß-Zingster Boddenkette, the experimental addition of nutrients still affects phytoplankton community composition (Schiewer 2008b). This demonstrates the validity of the "resource competition theory", which states that competition for resources is a major mechanism that controls the diversity and composition of communities (Tilman 1982).

13.4.4 "Incomplete" food webs

Most species are restricted to certain salinity ranges (*e.g.* the marine filter feeders *Ciona* and *Ensis* occur only in the Arkona Sea and the transition zone) or certain types of substrates (*e.g.* infaunal filter feeders). Therefore, some links of a complex food web may be missing in estuaries and coastal lagoons with low salinity and high substrate instability.

An "incomplete food web" results in the accumulation of organic matter, which is not consumed within the food web. For example, in a number of Baltic Sea estuaries and lagoons, *e.g.* the Wisła Lagoon and the Darß-Zingster Boddenkette, there is an absence of top-down control over the phytoplankton growth by (meso-) zooplankton grazing (Schiewer 2008b). Such loss of top-down control is, at least in part, a process that consists of a sequence of "phases" progressing during the conversion from a macrophyte-type system (with the mesozooplankton controlling the phytoplankton) to a phytoplankton-type system (a detritus-based food web) (Box 13.2).

Excess organic matter usually accumulates as mud before being decomposed or flushed out of a lagoon. When mud is flushed out to the sea, also the infauna will partly disappear from the lagoon, which means that filter feeder and deposit feeder functions are partly lost from the system as well. The mud layers in shallow muddy habitats are unstable and, when they cover hard substrate, the growth of macrophytes is prevented. In turn, the absence of macrophytes implies the lack of shelter for fish and herbivorous zooplankton. These are all factors that may drive a lagoon towards a phytoplankton-dominated system (Box 13.2).

Thus, the increased accumulation of organic matter in a lagoon (driven by *e.g.* eutrophication) may trigger a self-amplifying process that alters the system from a multi-consumer food web to a rather simple detritus-producing system in which consumers contribute only negligibly to the matter turnover. Such a system is close to the "minimal ecosystem", consisting mainly of primary producers, decomposers and a nutrient pool. Such systems are believed to be energetically inefficient due to bottom-up regulation when the primary producers ultimately become "locked" by nutrient limitation.

However, lagoon systems of this type, *e.g.* the Neva Estuary and the innermost parts of the Darß-Zingster

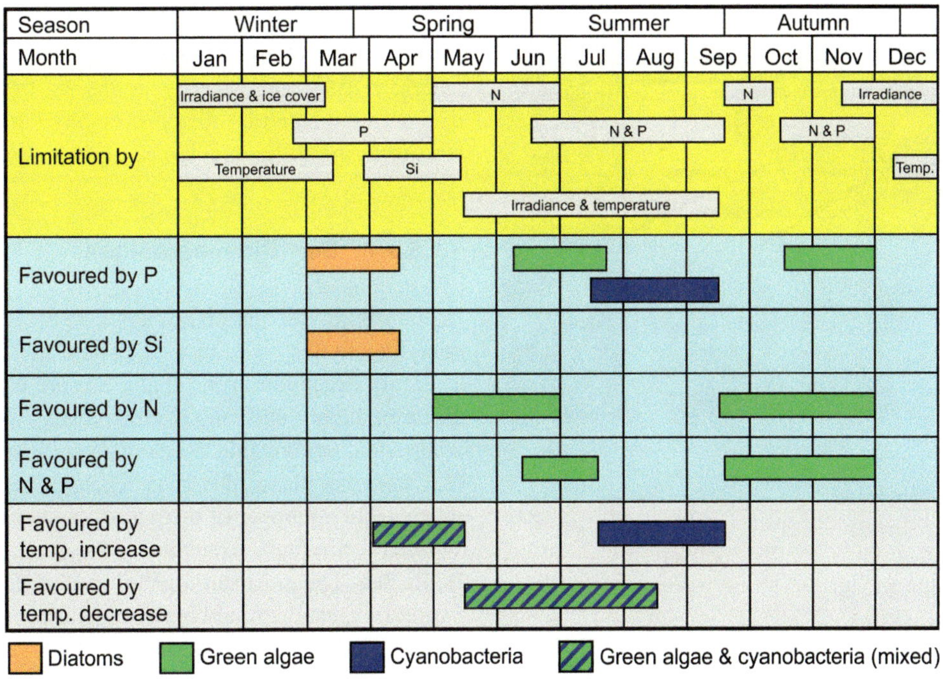

Fig. 13.6 Annual periodicity of the phytoplankton in the Zingster Strom (Darß-Zingster Boddenkette). In addition to recording limitation states from field samplings, experiments were performed to unravel selective effects of nutrients and nutrient ratios on phytoplankton growth and composition. Figure modified from Schiewer (2008b)

Boddenkette, and even similar shallow nutrient-rich systems outside the Baltic Sea, have been shown to be hyperproductive and even light-limited (Cloern 1999). So, why are there no consumers to speed up nutrient regeneration? The answers to this question are as follows: (1) there is a continuous supply of nutrients from the drainage area and continuous resuspension of sediment by wind action, and (2) there is a strong seasonality of the phytoplankton, which releases the biomass-locked nutrients, at least annually (Fig. 13.6). As a result, the phytoplankton biomass can be accumulated until the energy supply becomes the limiting factor and the system is light-regulated by phytoplankton self-shading.

13.5 Primary producers

13.5.1 Phytoplankton

The phytoplankton biomass in the coastal lagoons of the Baltic Sea exhibits a marked seasonal variability driven by climatic conditions as well as by grazing, sedimentation and export out of the lagoon. In particular, light limitation and low temperature reduces phytoplankton growth in the winter months, while throughout the warmer seasons phases of limitation by different nutrients (N, P, Si) occur (Fig. 13.6). This roughly follows the PEG model, which is a standard template to describe the factors that drive the seasonal succession of phyto- and zooplankton in lakes (Sommer et al. 1986). However, ecological interactions are involved as well, *e.g.* the overwintering of key organisms, the microbial food web, parasitism or food quality as a limiting factor and the role of secondary or tertiary consumers. The impact of such ecological interactions on plankton seasonal succession reveals limited effects on gross seasonal biomass patterns, but strong effects on species replacements (Sommer et al. 2012).

The major controls for the seasonal species shifts of the phytoplankton in the shallow waters include herbivory by zooplankton and zoobenthos, irradiance and nutrient availability. The grazing pressure increases in spring and shapes the community to the benefit of grazing-resistant forms, such as species forming large colonies. Clear-water phases such as those known from lakes, when grazing pressure in combination with nutrient limitation leads to top-down control over the whole phytoplankton community, are rare because most of the Baltic lagoons are meso- to eutrophic. Some Baltic Sea lagoons are nitrogen-limited in summer, favouring the growth of diazotrophic cyanobacteria (Fig. 13.7).

13.5.2 Microphytobenthos

Microphytobenthic communities are associated with any type of substrate, *e.g.* with the upper sediment layer on and between sand grains and mud particles, with rock, and with macrophyte communities. In extremely sheltered bays, they

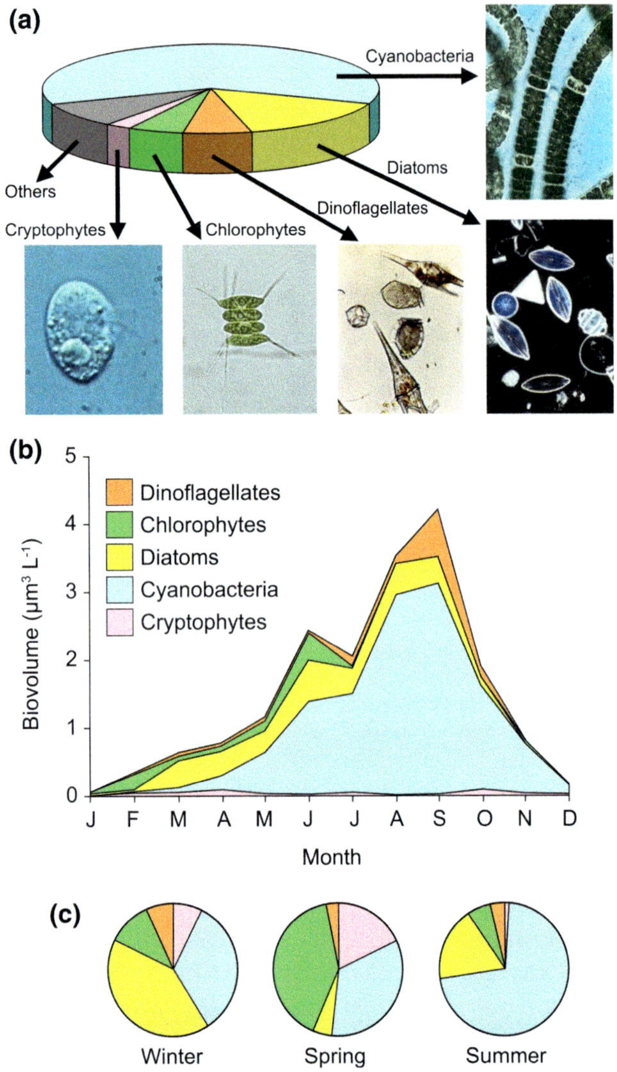

Fig. 13.7 Phytoplankton community composition in the Strelasund at the German bodden coast. (**a**) The relative contributions of different taxonomic groups to the average annual phytoplankton biovolume. (**b**) The contributions of the five major taxonomic groups to the phytoplankton biovolume during one seasonal cycle. (**c**) The relative contributions of the five major taxonomic groups to the total biovolume in winter, spring and summer. In summer diazotrophic cyanobacteria dominate the biomass in the Strelasund because of nitrogen limitation. In winter and spring nitrogen limitation is not detectable and the system is dominated by diatoms or green microalgae, although cyanobacteria are still abundant. The cyanobacterial component in winter and spring consists mainly of species without heterocysts (Chroococcales and Oscillatoriales). Figure modified from Schubert and Wasmund (2005)

can form the basis of "microbial mats". Although the productivity of microbial mats can be locally significant (Heyl et al. 2010), their ecological role is mainly structural, *i.e.* they stabilise sediments rather than support the food web.

Like the "minimal ecosystem" with only primary producers and decomposers, a microbial mat supports a rather low net production per area, regardless of a high internal turnover rate. While the structural components of the microbial mats in the coastal lagoons of the Baltic Sea are mainly filamentous cyanobacteria (*e.g. Coleofasciculus chthonoplastes* and *Lyngbya aestuarii*), the associated protists are dominated by diatoms, epiphytic on the cyanobacteria or living in large colonies of their own.

13.5.3 Benthic macroalgae

In contrast to vascular plants, algae do not develop roots and consequently they require a stable hard substrate to attach to. The only exception in the Baltic Sea are the charophytes, a group of highly specialised green algae equipped with rhizoids and therefore able to colonise sandy and soft substrates like vascular plants do. Most charophytes are freshwater species, but a number of them are salt-tolerant (*cf.* Sect. 7.5) and can form dense meadows in the coastal lagoons of the Baltic Sea. The most salt-tolerant species, *Lamprothamnium papulosum*, can be found both in monospecific stands and in mixed stands with the common eelgrass *Zostera marina* at salinities >10 in the Belt Sea.

Chara aspera, *Chara baltica*, *Chara canescens* and *Tolypella nidifica* are less salt-tolerant species. However, the salinity range of the Baltic Sea does not limit their growth inside lagoons and in sheltered areas on sandy coasts. In particular, *Chara aspera* can form dense meadows in shallow water, which efficiently shelter the zooplankton from fish predation and thus significantly influence trophic interactions. Under nearly freshwater conditions, the species listed above are accompanied by a number of other charophytes, including some members of the genus *Nitella*. As charophytes are good bioindicators of nutrient load and turbidity in freshwater systems, all of these species were thought to also be sensitive to eutrophication in the Baltic Sea. However, the shallow brackish lagoons of Mecklenburg-Vorpommern (Germany) in the southern Baltic Sea did not show the expected correlations between charophyte occurrence and nutrient load or turbidity (Yousef and Schubert 2001). Limited competition for space with vascular plants can be a possible explanation because only relatively few vascular plants are salt-tolerant enough to colonise these lagoons.

Apart from charophytes, benthic macroalgae occur in lagoonal systems only in places where hard substrate is available (Schubert et al. 2011). While hard substrates are common along the Swedish and Finnish coasts, the southern moraine coast provides hard substrate mainly as isolated boulders, which are few and far apart. The only perennial macroalgal species commonly found on such isolated boulders inside the lagoons of the southern Baltic Sea is the brown alga *Fucus vesiculosus*.

In the southwestern part of the Baltic Sea, the common perennial components of the attached phytobenthic communities growing on boulders in lagoons also include *e.g.* the red algae *Furcellaria lumbricalis* and *Polyides rotundus*. The phytobenthic communities in the Baltic Sea lagoons typically contain a high proportion of annual benthic macroalgae. In particular, thin foliose and tubular green algae (*e.g. Monostroma* spp. and *Ulva* spp.) compete with perennial macroalgae for the limited hard substrate. The ephemeral algae have competitive advantages in the lagoons because annuals grow faster in the shallow nutrient-rich waters and perennials are often removed by ice scour.

13.5.4 Drifting macroalgal mats

Loose-lying macroalgae are common in the coastal lagoons of the Baltic Sea and may assemble into drifting algal mats (Lehvo and Bäck 2001). With increasing eutrophication, such algal mats have become a nuisance for recreational beaches as well as for the benthic communities, which suffer from light limitation and hypoxia (<2 mL O_2 L^{-1}) when covered with a thick layer of detached and decaying macroalgae (Bonsdorff 1992). These drifting algal mats consist mainly of filamentous green algae (*e.g. Chaetomorpha linum* and *Cladophora* spp.) and brown algae (*e.g. Ectocarpus siliculosus* and *Pylaiella littoralis*). However, red algae may also form large drifting mats consisting predominantly of *Ceramium* spp. and *Polysiphonia* spp. in the lagoons of the southern Baltic Sea; more to the north these algae are joined by *Coccotylus truncatus* and *Phyllophora pseudoceranoides*.

Large loose-lying algal communities dominated by the red alga *Furcellaria lumbricalis* occur along the Estonian coast. These communities are still commercially harvested in Estonia for the production of "Danish agar", a form of carrageenan (*cf.* Box 11.4). Until the 1940s, loose-lying *Furcellaria lumbricalis*-dominated communities were also reported from several lagoons along the southern Baltic Sea coast (*e.g.* the Gdańsk Bay and some Danish lagoons), but have declined since. In the drifting *Furcellaria lumbricalis* communities the algal tufts adhere to each other and build stable structures. In contrast, aggregates of *Ulva* spp. or *Chaetomorpha linum* are less stable and are often wrecked on the lagoon's shore. *Chaetomorpha linum* frequently forms spherical aggregates, so-called "Neptune's balls" or "seaballs". On the shore the algae deteriorate and most of the nutrients are transported back to the lagoon.

13.5.5 Vascular plants

Zostera marina is the only marine vascular plant that is widely distributed in the Baltic Sea with salinity <10 (*cf.* Sect. 11.11). The other vascular plants in the Baltic Sea estuaries and coastal lagoons are euryhaline species of freshwater origin. However, the unstable substrate and mechanical stress from wave action and currents often prevent the establishment of larger populations of emergent and floating vascular plants in most of the estuaries and lagoons. This leaves a large perennial grass, the common reed *Phragmites australis*, as the only abundant plant species present in some Baltic Sea lagoons, *e.g.* in the Darß-Zingster Boddenkette. Highly productive on a per area basis, *Phragmites australis* reed belts act as nutrient sinks and produce large amounts of detritus which, because of its high lignin content, degrades very slowly and is deposited and buried in the lagoon's sediment.

Drifting plants and plants with floating leaves are often restricted to the innermost parts of estuaries and lagoons. Species such as *Lemna* spp., *Spirodela polyrhiza* (both members of the neuston) and *Nuphar lutea* require truly sheltered conditions. Additionally, with their maximum salinity tolerance far below 5, they have a limited salt acclimation ability and therefore grow only in places with a continuous freshwater input.

A group of submerged vascular plants of freshwater origin, including *Myriophyllum spicatum*, *Najas marina*, *Ruppia cirrhosa*, *Ruppia maritima*, *Stuckenia pectinata* and *Zannichellia palustris*, is found in lagoons along the entire Baltic Sea coast because of their high salinity tolerance (they are all able to grow at salinities >7). Except for *Najas marina*, all these species may occupy significant parts of the water column in shallow waters by their upright growth form. However, only *Myriophyllum spicatum* can reach densities that are sufficient to provide efficient refuge for protecting zooplankton from fish predation.

All submerged vascular plants may serve as food for birds and provide shelter for fish. This type of vegetation is therefore, in addition to their high primary productivity, important as a structural component. Moreover, the roots of aquatic vascular plants are fully functional, which implies that they can take up nutrients from deeper sediment layers. Consequently, phytobenthic communities not only compete with phytoplankton for dissolved nutrients in the water, but may also serve as a benthic-pelagic nutrient pump.

The relative proportion of vascular plants in the phytobenthic communities of the Baltic Sea estuaries, lagoons, fjärds and other inlets increases with the availability of sandy and soft substrates and decreasing salinity (Box 13.3). Under the low-salinity and (meso-) oligotrophic conditions of the northern Baltic Sea, as well as in the innermost parts of inlets with freshwater discharges along the whole Baltic Sea coast, a number of less salt-tolerant species are found. The most frequently encountered species are *Callitriche hermaphroditica*, *Ceratophyllum demersum* and *Potamogeton perfoliatus*. Species common in the inner parts of the fjärds along the Swedish and Finnish coast are *e.g. Isoetes*

Box 13.3: Macrophyte distribution along local salinity gradients

Community composition along local salinity gradients

Salinity is the principal environmental driver that determines the composition of the macrophyte communities along the ~2,000 km long Baltic Sea gradient from the transition zone to the northern Bothnian Bay and the inner Gulfs of Finland and Riga (*cf.* Sect. 11.2). However, community composition also changes at much smaller geographical scales along local salinity gradients in all places where freshwater is discharged to the brackish Baltic Sea, often via estuaries

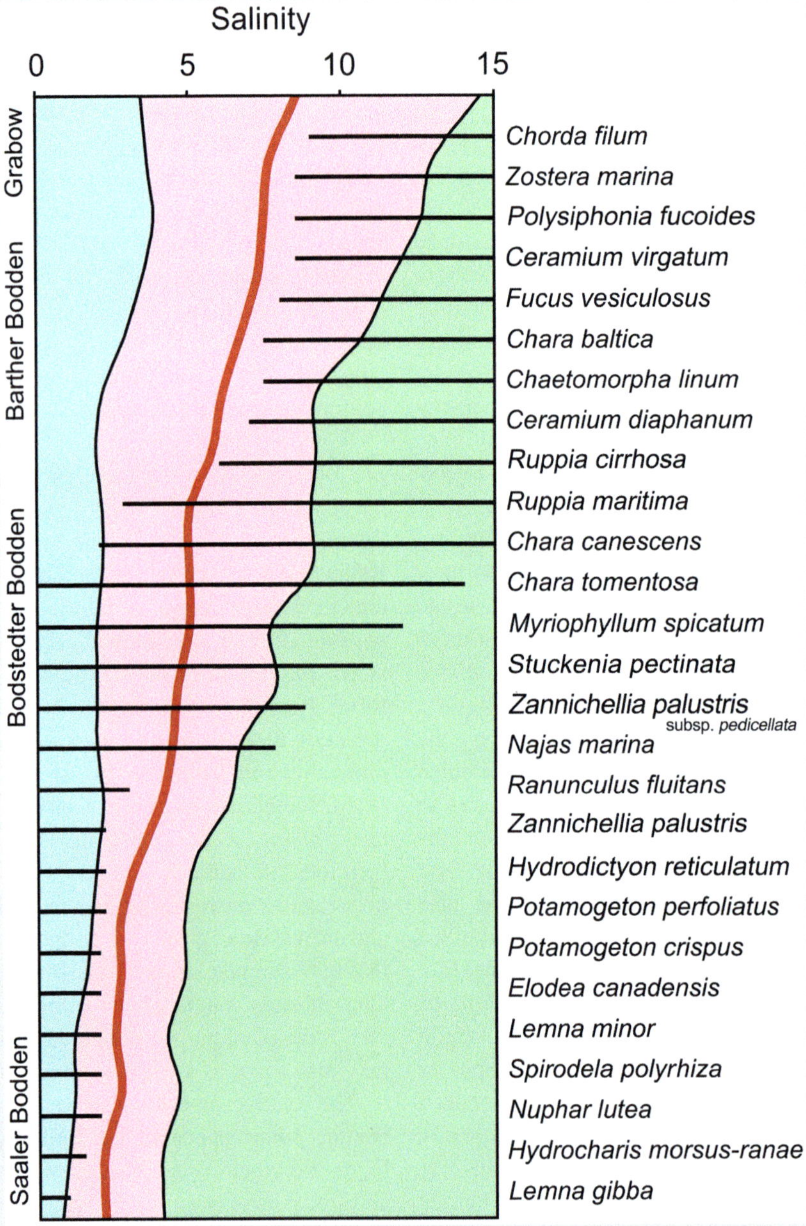

Box Fig. 13.3 The occurrence (horizontal lines) of 27 benthic macrophytes along the Darß-Zingster Boddenkette. The red line shows the average salinity and the pink area shows the salinity variation between minimum and maximum values. The 27 species include 3 red algae (*Ceramium diaphanum*, *Ceramium virgatum*, *Polysiphonia fucoides*), 2 brown algae (*Chorda filum*, *Fucus vesiculosus*), 2 chlorophytes (*Chaetomorpha linum*, *Hydrodictyon reticulatum*), 3 charophytes (*Chara baltica*, *Chara canescens*, *Chara tomentosa*) and 17 vascular plants (*Elodea canadensis*, *Hydrocharis morsus-ranae*, *Lemna gibba*, *Lemna minor*, *Myriophyllum spicatum*, *Najas marina*, *Nuphar lutea*, *Potamogeton crispus*, *Potamogeton perfoliatus*, *Ranunculus fluitans*, *Ruppia cirrhosa*, *Ruppia maritima*, *Spirodela polyrhiza*, *Stuckenia pectinata*, *Zannichellia palustris*, *Zannichellia palustris* subsp. *pedicellata*, *Zostera marina*). Figure modified from Schubert (2001)

and coastal lagoons. For example, along the salinity gradient of the Darß-Zingster Boddenkette (*cf.* Fig. 13.8) macrophyte community composition changes markedly below an average salinity of ~8 when macroalgae become replaced by vascular plants (Box Fig. 13.3). The macroalgal species richness decreases with decreasing salinity because most of the algae in the Baltic Sea are of marine origin, while the species richness of vascular plants and charophytes increases with decreasing salinity because most of the species in these groups are of freshwater origin. The only macroalgal species still found at salinity <8 in the Darß-Zingster Boddenkette are charophytes and the freshwater chlorophyte *Hydrodyction reticulatum* (Box Figs. 13.3 and 13.4). This is a pattern different than that found along the large-scale Baltic Sea gradient in which many marine macroalgal species still occur in the Bothnian Sea until they disappear in the Norra Kvarken Area between the Bothnian Sea and the Bothnian Bay at salinity ~4 (*cf.* Box 11.7). At salinities <4 in the Bothnian Bay the freshwater species *Aegagropila linnaei* and *Cladophora glomerata* are belt-forming on hard substrates (*cf.* Sect. 11.7.3). The reason for this is the variability of the salinity in semi-enclosed coastal water bodies (Box Fig. 13.3) while the salinity below ~10 (east of the Drogden and Darß sills) along the large-scale Baltic Sea gradient is much more stable (*cf.* Fig. 4.2).

Salinity variability, substrate and eutrophication

The reason for the absence of macroalgae below the average salinity of ~8 in the Darß-Zingster Boddenkette is probably a combination of the low availability of hard substrate in the basins and the temporal salinity variability, which is both seasonal (*cf.* Fig. 7.1) and erratic (*cf.* Fig. 7.4). The lack of hard substrate alone cannot explain the absence of macroalgae because filamentous macroalgae such as the marine *Ulva* spp. can occur epiphytic on vascular plants and charophytes, while the absence of freshwater algae such as *Cladophora glomerata* cannot be explained by low salinity. The variability in salinity is largest near the entrance of the chain of boddens at Grabow where it can vary between 3 and 14 (Box Fig. 13.3). As the temporally varying salinity drops (down to salinity 2–3) below the lower physiological salinity limit of marine macroalgae (even the most euryhaline ones), they cannot survive. Furthermore, heavy eutrophication of the Darß-Zingster Boddenkette in the 1980s may have added to the absence of macroalgae.

Box Fig. 13.4 Light-microscopic image of the freshwater chlorophyte *Hydrodyction reticulatum*. Photo: © Antje Gerloff *lacustris* and *Littorella uniflora*. Also, some mosses, *e.g.* *Fontinalis antipyretica*, are able to colonise the low-salinity parts of the Baltic Sea lagoons.

During the last century, a number of Baltic Sea coastal lagoons have shown a conspicuous decline in macrophyte biomass. For example, between 1938 and 1998, the Greifswalder Bodden lost 60 % of its macrophyte-covered area. The cause of the large losses is not clear yet, but it is possible that some feedback mechanisms, such as increased sediment mobility in addition to the effects of increased light limitation, have been involved (Schiewer 2008b). Another mechanism may involve increased growth of epiphytic and entangled filamentous algae causing light limitation for the macrophytes. However, filamentous algae constitute food for gastropods and amphipods and both field and experimental studies have shown that there is usually an efficient top-down control over epiphyte growth by grazers in lagoon systems (Duffy and Harvilicz 2001; Bobsien and Munkes 2004).

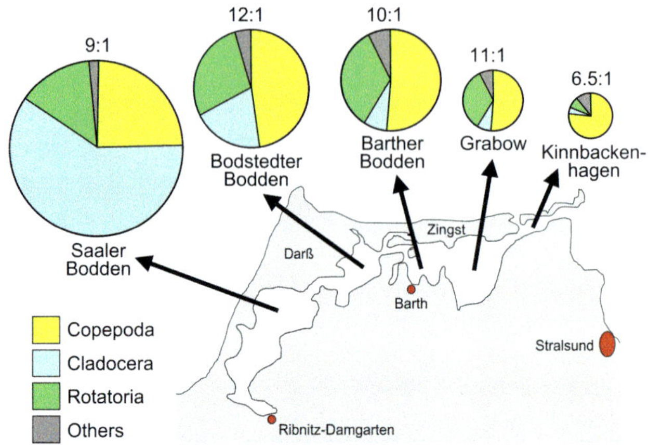

Fig. 13.8 Compilation of the metazooplankton biomass and composition in the Darß-Zingster Boddenkette. Metazooplankton has a body size >0.2 mm (*cf.* Table 4.1), and includes both mesozooplankton and macrozooplankton (but in this case predominantly mesozooplankton). The zooplankton biomass is indicated by the relative sizes of the circles and averages (from left to right) 9.6, 4.5, 3.9, 1.9 and 0.9 mg wet weight L^{-1}. Above the circles the phytoplankton:zooplankton biomass ratio in the water is shown. Salinity increases and eutrophication decreases from the Saaler Bodden to the Kinnbackenhagen. Along this gradient cladocerans are first replaced by rotifers and finally by copepods while the phytoplankton:zooplankton ratio decreases (*i.e.* a larger part of the phytoplankton production is transferred to the zooplankton biomass). Figure based on data in Heerkloß and Schnese (1994), modified from Schiewer (2008b)

13.6 Consumers

13.6.1 Consumer levels

Consumers (heterotrophs) are organisms that obtain their energy from consuming other organisms. The primary producers of estuaries and coastal lagoons provide a rich source of food for higher trophic levels. Primary consumers such as filter feeders, deposit feeders and grazers feed basically on nano- to micro-sized primary producers, but they do not actually discriminate between small unicellular eukaryotes and bacteria. Furthermore, many unicellular eukaryote groups, *e.g.* the dinoflagellates, contain autotrophic, mixotrophic and heterotrophic species. Therefore, it is inappropriate to state simply that all primary consumers feed strictly on autotrophs. Many filter feeders, deposit feeders and grazers are in fact omnivores (feeding on both autotrophs and heterotrophs) and could in that sense, together with carnivores, be considered secondary consumers. Tertiary consumers are organisms, usually at the top of the food web, which feed basically on secondary consumers.

13.6.2 Microzooplankton

The microzooplankton is the most species-rich zooplankton component in the estuaries and coastal lagoons of the Baltic Sea and consists mainly of rotifers, ciliates and heterotrophic flagellates. Baltic Sea estuaries and lagoons support more than 150 rotifer species, with *Keratella*, *Polyarthra* and *Synchaeta* as the major genera (Telesh and Heerkloss 2002). At least 743 ciliate species have been reported from coastal and open waters in the entire Baltic Sea (Mironova et al. 2014), although our knowledge of species richness and functional diversity in estuaries and lagoons is still far from complete. The most diverse groups of planktonic ciliates are the genera *Cyclidium*, *Mesodinium*, *Monodinium*, *Strobilidium*, *Strombidium* and *Tintinnidium*.

Microzooplankton organisms feed on small phytoplankton, bacteria and associated virus particles and are involved in complex trophic interactions. Small primary producers (nano- and picophytoplankton) and heterotrophic nanoflagellates and bacteria are of comparable cell size and have very short generation times. These trophic interactions follow the Lotka-Volterra model (a simple model of the population dynamics of species competing for a common resource) without any noticeable dampening and create the microbial food web (Fig. 13.4). This food web is particularly effective when the impact of mesozooplankton grazing is reduced.

13.6.3 Mesozooplankton

The mesozooplankton in the estuaries and coastal lagoons of the Baltic Sea is represented mainly by planktonic crustaceans (copepods and cladocerans) as well as by the largest-sized rotifer species (*e.g.* *Asplanchna* spp.) (Fig. 13.8). Mesozooplankton organisms are herbivores (primary consumers), omnivores or carnivores (secondary consumers). Herbivores feeding on phytoplankton include *e.g.* cladocerans (*Bosmina*, *Daphnia*), calanoid copepods (*Acartia*, *Eurytemora*, *Pseudocalanus*, *Temora*), harpacticoid copepods (*Canthocamptus*) and cyclopoid copepods (juveniles of *Cyclops* and *Mesocyclops*). Omnivores feeding on algae, detritus-associated bacteria and small microzooplankton are represented by *e.g.* branchiopod cladocerans (*Chydorus*) and rotifers (*Asplanchna*). Carnivores include *e.g.* cladocerans (*Evadne*, *Podon* and *Cercopagis pengoi*) and adults of cyclopoid copepods.

The mesozooplankton species richness in the Baltic Sea estuaries and lagoons is high. For example, 133 crustacean species have been reported (Telesh and Heerkloss 2004). The mesozooplankton biomass and productivity in the Baltic Sea estuaries and lagoons can be high as well, although it is relatively low compared to the phytoplankton productivity (Fig. 13.9). Whereas the annual zooplankton productivity in temperate estuaries and coastal zones usually does not exceed 30 g C m^{-2} $year^{-1}$ (McLusky and Elliott 2004), it may reach 60 g C m^{-2} $year^{-1}$ in the Darß-Zingster Boddenkette and even up to 250 g C m^{-2} $year^{-1}$ in the Neva Estuary (Schiewer 2008a).

The high mesozooplankton productivity results from the presence of rich food sources (phytoplankton and microzooplankton) and limited loss by washout to the sea, especially in choked lagoons. In tidal estuaries and lagoons, mesozooplankton growth is relatively slow, which is believed to be caused by regular water outflow to the open sea. While the macrozooplankton can maintain their position in moderate currents, the meso- and microzooplankton organisms are too small to expend enough energy to overcome viscosity (*cf.* Sect. 1.2.7). This loss factor is basically absent in the microtidal lagoons and at least diminished in the estuaries, which contributes to the high mesozooplankton productivity reported from the Baltic Sea estuaries and lagoons.

13.6.4 Macrozooplankton

In the coastal lagoons of the Baltic Sea, the macrozooplankton is represented mainly by the Cnidaria (jellyfish), which feed on other zooplankton as well as on fish larvae. The most common jellyfish species throughout the Baltic Sea is the scyphozoan *Aurelia aurita*. Other scyphozoans, *e.g. Cyanea capillata* and *Cyanea lamarckii*, are restricted to the more saline estuaries and lagoons of the Belt Sea.

Another macrozooplankton group, the Ctenophora (comb jellies), basically inhabit the open Baltic Sea waters (*cf.* Sect. 8.7.3), but may also be found in gulfs and inlets, especially in the Belt Sea area. At present, at least five ctenophore species representing the genera *Beroe*, *Bolinopsis*, *Mnemiopsis* and *Pleurobrachia* have been identified in the Baltic Sea. Ctenophores are usually involved in complex

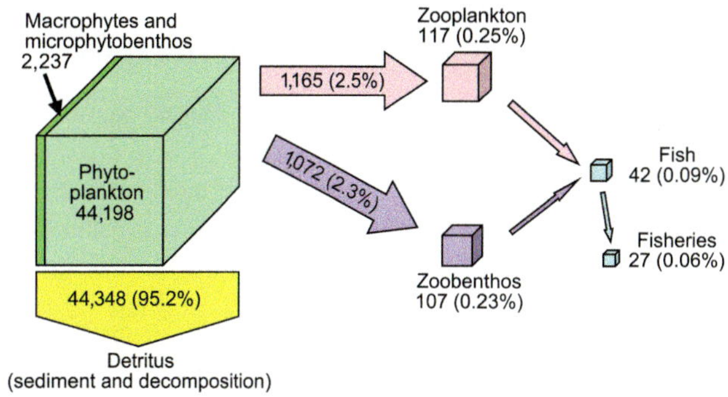

Fig. 13.9 The contribution of zooplankton to trophic interactions and energy flow in the Darß-Zingster Boddenkette under hypertrophic conditions. All numbers indicate µg C L^{-1}. The trophic efficiency of both zooplankton and zoobenthos is very low, resulting in the formation of a detritus-based food web. There is a small discrepancy of 150 µg C L^{-1} (0.3 %) in this budget, which probably is due to the fact that the data were partly calculated from measured values (detritus) while other data are estimates from models. Figure modified from Schiewer (1985)

predator-prey relationships that control their abundances (Box 13.4). For example, *Pleurobrachia pileus* feeds on herbivorous zooplankton, particularly on copepods that appear in spring. A single *Pleurobrachia pileus* individual may eat as many as 300 copepods per day. Then usually the secondary consumer ctenophore *Beroe gracilis* appears and feeds exclusively on *Pleurobrachia pileus* to practically eliminate it within three weeks (Telesh et al. 2009).

Since 2006, the non-indigenous ctenophore *Mnemiopsis leidyi* (*cf.* Box 5.10) has been recorded in the Belt Sea and the southern Baltic Sea. Comb jellies are very sensitive to contact with surfaces and the probability of contact with potentially dangerous "non-prey" obstacles in shallow estuaries and lagoons is high. Therefore, although they may start their life cycle in sheltered bays, estuaries and lagoons, the loss rate is high and the probability that they reach the adult stage is much higher when they are washed out into the open sea.

Macrozooplankton samples may also contain some other organisms that reach a body size of >2 cm, such as mysids (*e.g. Neomysis integer* and *Praunus flexuosus*), prawns (*e.g. Palaemon* spp.) and some polychaetes. However, these are basically epibenthic species that seek shelter from fish predation by living close to the bottom or inside the phytobenthic vegetation. They preferentially feed on mesozooplankton organisms.

13.6.5 Benthic consumers and sediment stability

At least some sediment stability is necessary for animals to be able to live in or immediately above sediment surfaces. In the shallow-water areas of coastal lagoons that are rich in

Box 13.4: The zooplankton of estuaries and coastal lagoons

The zooplankton of the estuaries and coastal lagoons in the Baltic Sea is highly diverse due to the large variety of habitats that exist in these environments where different faunal groups co-occur (Box Fig. 13.5). Marine species, such as jellyfish, occur alongside brackish- and freshwater species in the inner parts of lagoons and the upper estuarine reaches. Additionally, the Baltic Sea estuaries and lagoons host a number of non-indigenous species such as the Ponto-Caspian fish-hook water flea *Cercopagis pengoi* (*cf.* Box 5.5) and the American comb jelly *Mnemiopsis leidyi* (*cf.* Box 5.10). The microzooplankton is not only diverse taxonomically (Telesh et al. 2011a); it also represents a large variety of functional groups. It is highly abundant and productive. The microzooplankton species are good indicators of eutrophication and chemical pollution. However, there are drawbacks that hamper the practical use of the microzooplankton in routine environmental monitoring; these include the necessity of special sampling techniques and a high taxonomical competence.

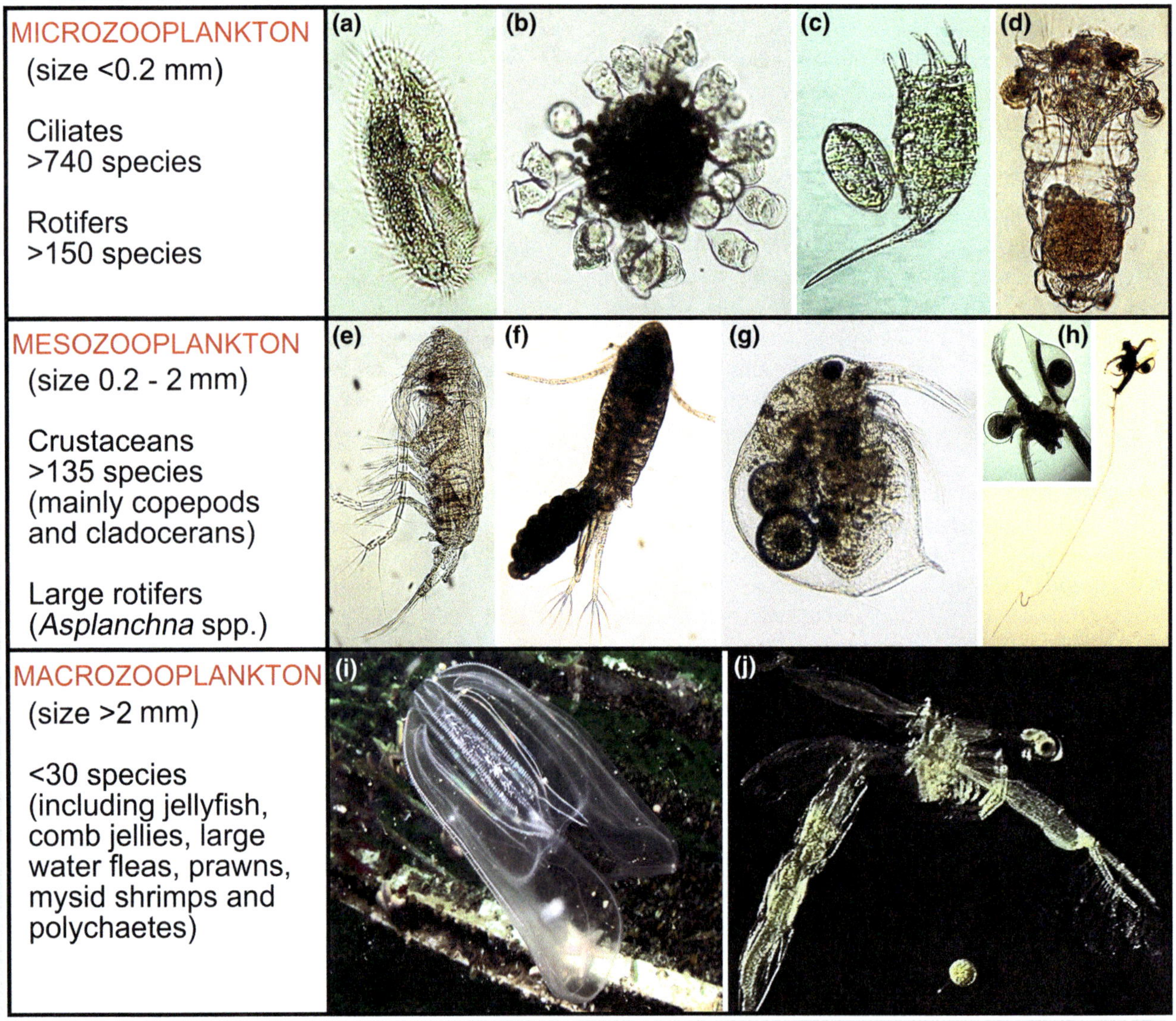

Box Fig. 13.5 Light-microscopic images, except for (i) which was taken with a normal camera, of some zooplankton organisms living in Baltic Sea estuaries and coastal lagoons. (**a**) The ciliate *Sterkiella histriomuscorum*. (**b**) The ciliate *Vorticella anabaena*. (**c**) The rotifer *Keratella cochlearis baltica*. (**d**) The rotifer *Synchaeta* sp. (**e**) The copepod *Acartia tonsa*. (**f**) The copepod *Eurytemora affinis*. (**g**) The cladoceran *Bosmina longispina maritima*. (**h**) The cladoceran *Cercopagis pengoi* (the inset shows a female body with resting egg). (**i**) The comb jelly *Mnemiopsis leidyi*. (**j**) The crustacean *Leptodora kindti*. Data on species numbers from Telesh and Heerkloss (2002, 2004) and Telesh et al. (2009). Photo: (**a**, **b**) © Ekaterina Mironova, (**c**, **h**, **j**) © Irena Telesh, (**d**–**g**) © Heide Sandberg, (**i**) © Gerd Niedzwiedz

Box 13.5: *Macoma balthica* – a suspension feeder and a deposit feeder

The Baltic clam *Macoma balthica* (Box Fig. 13.6) is a common species in the estuaries and coastal lagoons of the Baltic Sea, where it lives buried in sandy and soft bottoms (Bonsdorff et al. 1995; Tallqvist 2001). It can reach juvenile densities of ~300,000 individuals per m^2 during settling and adult densities of ~1,000 individuals per m^2 (Bonsdorff et al. 1995). Spawning starts as soon as the water temperature reaches 10 °C in spring, but *Macoma balthica* is able to grow between 4 and 16 °C. Within this temperature range, the growth rate depends on food availability. *Macoma balthica* has two individually separated stretchable long siphons (*cf.* Fig. 10.13). The inhalant siphon is used for both filter feeding on particles in the water column and for deposit feeding by sucking detritus from its surroundings. Most of the time, *Macoma balthica* seems to behave as a deposit feeder, although growth rates were found to depend mainly on the amount of food filtered from the water column. This discrepancy can be explained by the energy budget of the Atlantic subspecies *Macoma balthica rubra*, a close relative of the Baltic Sea subspecies *Macoma balthica balthica* (*cf.* Box 6.4), which is assumed to have a similar energy budget. From a total food intake of 258 kJ, 106 kJ (41 %) were found to consist of "unsuitable food" (*e.g.* indigestible plant detritus) for *Macoma balthica rubra*, and therefore this part of the energy intake was completely defecated (Hummel 1985). Of the remaining 152 kJ in the "suitable food", 51 kJ were used for respiration and 21 kJ for production. The difference of 80 kJ was left in the faecal pellets.

Box Fig. 13.6 An empty shell of the bivalve *Macoma balthica* found on a Baltic Sea beach. Photo: © Dirk Schories

organic mud, sediment stability is often very low. This is one of the reasons why a large part of the consumer spectrum is absent wherever organic mud prevails on the bottom. Consequently, food webs may have "open ends" and organic matter is channelled to decomposers or to the sediment for burial rather than to consumers. This is a common pattern for the so-called "sink areas" in eutrophic lagoons.

However, parts of the muddy habitats of estuaries and lagoons may still be in the photic zone, *i.e.* they receive enough light to support positive net photosynthesis so that autotrophs can grow. Such photic muddy bottoms may support microbial mats or even macrophytobenthic vegetation. When the growth of vascular plants and charophytes is extensive, the phytobenthic vegetation prevents wind-driven resuspension and forms a habitat favouring stable occurrence of benthic fauna. Wherever the sediment is stabilised, most of the benthic consumers are found below the sediment surface, and the sediment provides protection against predators and extreme environmental variability.

13.6.6 Benthic infauna

The typical suspension feeders living in the estuaries and coastal lagoons throughout the Baltic Sea are the bivalves *Cerastoderma glaucum*, *Macoma balthica* (Box 13.5) and

Mya arenaria (*cf.* Box 5.2), and in the high-salinity areas of the Belt Sea *e.g. Ensis directus* and *Scrobicularia plana*. These bivalves live buried in the sediment and collect POM from the water with their inhalant siphons. The inhaled water passes through the gills and a certain size fraction of the POM is retained as food. The filtered water is, along with the bivalve's faeces, expelled via the exhalant siphon.

The typical deposit feeders living in the Baltic Sea estuaries and lagoons include annelids (polychaetes and oligochaetes). Euryhaline polychaetes such as *Hediste diversicolor* (*cf.* Box Fig. 6.2) and *Marenzelleria* spp. (*cf.* Box 5.3) occur in the entire Baltic Sea, this group of consumers in the Belt Sea being accompanied by *e.g. Arenicola marina* (*cf.* Fig. 4.19d). Under the conditions of lower salinity, the polychaetes are replaced by oligochaetes (*e.g. Nais elinguis* and *Tubificoides benedii*). The annelids ingest a muddy mixture of organic and inorganic material from which they extract organic components and enrich the sediment with organic material from their faeces. This can serve as a fertile substrate for the growth of microorganisms and has been termed "microbial gardening" based on observations of microorganisms growing in *Arenicola marina* burrows which are consumed by the worms (Grossmann and Reichardt 1991; Retraubun et al. 1996).

Moreover, amphipods, *e.g. Corophium volutator*, may live in burrows and collect detritus from the sediment. *Corophium volutator* feeds mainly on the diatoms and bacteria attached to the detritus, and shreds the detritus into finer particles. Another important infaunal component in estuaries and lagoons, but only at lower salinities, are chironomid larvae (insects), which are rather tolerant to hypoxia in the muddy habitats of shallow lagoons.

13.6.7 Benthic epifauna

The same species of suspension and deposit feeders that belong to the infauna can be found among the epifauna living on the sediment surface, on hard surfaces and in phytobenthic communities. Epifaunal deposit feeders are represented by snails such as *Ecrobia ventrosa* and *Peringia ulvae*. Similarly to *Arenicola marina*, these two snail species "cultivate" bacteria by excreting mucus and reingesting this mucus after a couple of hours of exposure.

Mytilus trossulus and *Dreissena polymorpha* are filter feeders that live attached to a hard substrate. With their rich food supply for suspension feeders, the Baltic Sea coastal lagoons may support large colonies of these mussel species, even when hard surfaces are rare. The mussels overcome the lack of hard substrate by attaching to each other to form large and stable mussel beds with an enormous filtration capacity (*cf.* Sect. 11.13.1). However, as an attractive food for waterbirds, the mussel beds in shallow lagoons are usually not as long-lived as offshore mussel beds in deeper water, which can be stable for decades. Other typical epifaunal filter feeders living in the estuaries and lagoons in the entire Baltic Sea are the bryozoan *Einhornia crustulenta*, the sponge *Ephydatia fluviatilis* and, in the higher salinities of the Belt Sea, *e.g.* the sponge *Chalinula limbata* (syn. *Haliclona limbata*) and the tunicate *Ciona intestinalis* (*cf.* Fig. 4.19f).

Epifaunal grazers in estuaries and lagoons are represented mainly by crustaceans, *e.g. Idotea* spp., and snails, *e.g. Radix balthica*, *Tenellia adspersa* (syn. *Embletonia pallida*) and *Theodoxus fluviatilis* in the entire Baltic Sea as well as *Littorina* spp. in the Belt Sea. These "micrograzers" feed mainly on microbial films consisting of microalgae and bacteria on all types of surfaces. When such biofilms are not available, some micrograzers can also feed on macroalgae and vascular plants, *e.g. Idotea balthica* often feeds on *Chara baltica* and *Zostera marina* leaf tips.

13.6.8 Changes in biomass and composition of the zoobenthos

Changes in environmental conditions alter the biomass and composition of the zoobenthos in estuaries and lagoons. For example, radical changes in the zoobenthos have occurred in the Neva Bay since the construction of a storm-surge barrier and the building of a wastewater treatment plant. In the early 1980s, the zoobenthic biomass was 100–150 g wet weight (ww) m^{-2}. It was dominated by bivalves which filtered out $\sim$60 % of the suspended organic matter discharged by the Neva river and were responsible for a high rate of decomposition of organic matter (Alimov and Golubkov 1987). By the 1990s, the zoobenthic community biomass had decreased to $\sim$50 g ww m^{-2} and was dominated by oligochaetes and chironomid larvae had also increased in abundance (Fig. 13.10). This process continued, and in the early 2000s the biomass was only 10–14 ww m^{-2}.

This large decline in the zoobenthos is probably related to the diminished water exchange between the Baltic Sea and the lower part of the inner estuary on account of the storm-surge barrier, *i.e.* the freshening of the water in the Neva Bay. The wastewater treatment plant decreased the inflow of dissolved and suspended organic matter to the Neva Bay, which decreased the productivity of the zoobenthos correspondingly (Telesh et al. 2008).

13.6.9 Fish

The actively swimming organisms of estuaries and coastal lagoons in the Baltic Sea Area are mainly represented by fish. A comparative study between estuaries in the Skagerrak and the Bothnian Sea showed that the fish species richness

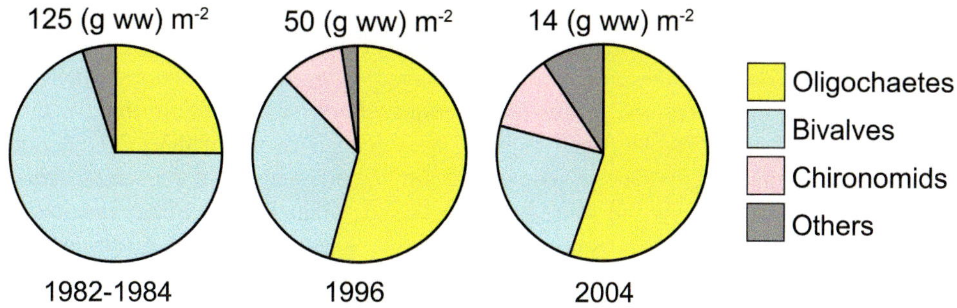

Fig. 13.10 Average biomass and community composition (% biomass) of the zoobenthos in the Neva Bay in summer 1982–1984, 1996 and 2004. ww = wet weight. Figure modified from Telesh et al. (2008)

was similar in both areas with 17 species confined to the Skagerrak and 19 species coinfined to the Bothnian Sea, while 10 species were common to both areas (Thorman 1986). Variations in species numbers throughout the year were explained mainly by salinity and variations in water temperature. A major difference between the two areas was that the dynamics of the fish communities in the Skagerrak estuaries were also affected by competition for food between fish and the invertebrates *Carcinus maenas*, *Crangon crangon* and *Palaemon* spp., while there was no such competition in the Bothnian Sea because the crab and the prawns do not occur there.

Several fish species occurring in estuaries and lagoons are of high economic value, *e.g.* the Atlantic herring *Clupea harengus*, the Atlantic cod *Gadus morhua*, the northern pike *Esox lucius*, the European flounder *Platichthys flesus* and the zander (pike-perch) *Sander lucioperca*. However, the role of fish as a resource for coastal fisheries is site-dependent. For example, the Greifswalder Bodden is an important herring-breeding area, and herring catches, especially of the spring-spawning population, are important for the local economy, including recreational fishing tourism. Also, the Szczecin Lagoon fisheries still rely on species with high economic value (Wolnomiejski and Witek 2013).

Nevertheless, the commercial fish stocks in some Baltic Sea lagoons have been reduced by overfishing, leaving only species with no or restricted economic value. About 10 % of the commercially important fish species are thought to be overexploited. Many fish species in some of the lagoons are also negatively impacted by strong eutrophication (Schiewer 2008b). When predators (*e.g.* cod or pike) disappear, the food web is altered and trophic cascades may occur to cause substantial changes in the matter and energy flows.

13.6.10 Birds

Waterbirds are the major "macrograzers" in the Baltic Sea's shallow estuaries and coastal lagoons. In particular, the mute swan *Cygnus olor* (Fig. 13.11) is an herbivore that feeds on the phytobenthic communities. In contrast to micrograzers, the birds remove this structurally important component from the system. While the role of birds as macrograzers for the establishment of "alternative stable states" (Scheffer et al. 2001) in freshwater systems has been demonstrated, a similar role of avian macrograzers is often assumed for brackish lagoon systems (Barker et al. 2008), but this has not been proven yet.

Also consumer populations in shallow estuaries and lagoons may be affected by waterbirds. The blue mussel *Mytilus trossulus* (at salinity >4) and the zebra mussel *Dreissena polymorpha* (at salinity <4) constitute food resources for a number of diving ducks (Molloy et al. 1997). For example, the greater scaup *Aythya marila* is estimated to consume an average of 5,400 tonnes of zebra mussels annually in the brackish lagoons of the Odra river estuary (ORE) in the southwestern Baltic Sea (Marchowski et al. 2015).

There is an on-going debate about the impact of the great cormorant *Phalacrocorax carbo sinensis* (*cf.* Box 4.11) on fish populations in the estuaries and lagoons of the Baltic Sea. Taking up ~500 g of fish per individual per day, this

Fig. 13.11 The mute swan *Cygnus olor* is an herbivore on aquatic macrophytes. Photo: © Stanislaw Węsławski

bird is seen by fishermen as their competitor (Matthews 2000; Herrmann et al. 2014). The great cormorant is a protected species, enjoying the status of the conservation priority under the EU Birds Directive. With the permission of relevant authorities in some countries, however, it is allowed to shoot off the great commorant to "prevent serious damage to crops, livestock, forests, fisheries and water and for the protection of flora and fauna and if there is no other solution" (EU 2013). This is called "cormorant culling" (the intentional killing of cormorants by humans for the purposes of wildlife management).

13.7 Ecosystem services and management

13.7.1 Ecosystem services of estuaries and lagoons

The estuaries and coastal lagoons influence the biodiversity and productivity of the open Baltic Sea. They impact the sea through hydrodynamic processes and function as "bioreactors" that convert terrestrial nutrient supply into biomass and thus "feed" the sea with nutrients (Telesh 2004; Schiewer 2008c).

By connecting the near-shore land and river mouths with the sea, the estuaries and lagoons act as filters and buffers for nutrients and hazardous substances from diffuse and point sources. These buffering functions are based on flow-driven sedimentation and the physical, chemical and biological transformation of substances in the coastal waters. The underlying processes are in multiple ways affected by: (1) estuarine morphology, (2) soil type, (3) water residence time and circulation mode, (4) freshwater outflow, and (5) seawater inflow, along with other environmental drivers, biotic components and their interactions. Hence the buffer capacities of different estuarine and lagoonal systems vary.

Most of the semi-enclosed systems serve as spawning and nursery habitats for fish as well as refuges for waterbirds and migratory birds. Moreover, estuaries and lagoons provide vast recreational areas that are intensely exploited by tourist activities, *e.g.* boating, water sports, and recreational fishing, as the shores of these water bodies are most often densely populated.

13.7.2 Anthropogenic impacts

Large cities (including harbours and industries) are commonly located in river mouths and inner estuarine reaches. In particular, anthropogenic activities in the inner parts of estuaries explain, to a large extent, nutrient loads and water pollution by oil, oil products and heavy metals introduced mainly by shipping. Elevated pesticide loads occur in estuaries and lagoons as well, but they are predominantly recorded in agricultural regions. Their diffuse entry is a serious source of pollution that is difficult to control. Eutrophication is one of the most important threats to the health and productivity of the coastal areas of the Baltic Sea ecosystem (Fig. 13.12). Eutrophication and pollution of estuarine waters with harmful substances are accompanied by "biological pollution" due to invasions of non-indigenous species, which have gained in numbers and abundances during the last decades (*cf.* Sect. 5.1).

The occurrence of self-reproducing populations of non-indigenous species in the estuaries and lagoons of the Baltic Sea is a manifestation of anthropogenic stress. Invasions of non-indigenous species with high environmental plasticity can reduce the local biological diversity, modify trophic interactions and consequently alter the functional characteristics of estuarine and lagoonal systems. For example, the invasion of the carnivorous planktonic cladoceran *Cercopagis pengoi* (*cf.* Box 5.5) from the Ponto-Caspian region into the Neva Estuary in 1995 has already initiated a reduction in the population size of the native cladoceran filter feeders as a result of strong predation pressure by *Cercopagis pengoi* during its maximum development (Telesh et al. 2008). Elimination of planktonic filter feeders by the invader may result in a further increase of primary production by phytoplankon (thus further eutrophication).

The biofouling bivalve *Dreissena polymorpha* (*cf.* Box 5.4) is another Ponto-Caspian invader to many Baltic Sea estuaries and lagoons that has the potential to produce considerable effect on coastal areas. *Dreissena polymorpha* beds are known to facilitate benthic-pelagic coupling by the excretion of bio-available nutrients and by clearing the water of particulate matter, thereby improving light conditions in the deeper water layers. More light could facilitate the re-establishment of macrophytes in a system currently dominated by phytoplankton. This is an example of an invader that could have a possible positive effect. The filtration capacity of *Dreissena polymorpha* could be the first step towards de-eutrophication by stimulation of nutrient retention in the macrophyte biomass rather than in phytoplankton biomass; this way, the sediments may become more stable as well.

13.7.3 Environmental management

Knowledge of the estuarine environmental structure and an understanding of functional modes, are the basic prerequisites for successful management and adequate environmental policy-making for the Baltic Sea estuaries and coastal

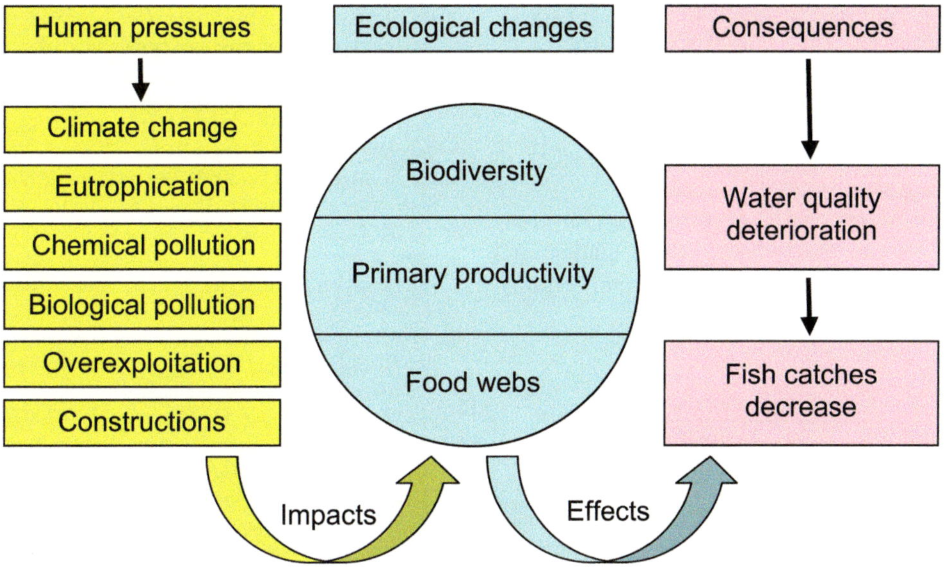

Fig. 13.12 Major anthropogenic pressures and their consequences in the estuaries and coastal lagoons of the Baltic Sea. Chemical pollution = chemical contamination by hazardous substances, Biological pollution = introductions of non-indigenous species, Overexploitation = overexploitation of resources, *e.g.* overfishing, Constructions = dredging, harbours, hydraulic constructions, etc. Figure: © Irena Telesh

lagoons (Skarlato 2002). Long-term ecological data series can back up the evaluation of the management of estuaries and lagoons under intensive anthropogenic pressure. However, legislation and competence with respect to the environmental problems of the Baltic Sea estuaries and lagoons vary considerably between the nine riparian countries and even between certain administrative regions within a country.

The EU Water Framework Directive (WFD) defines the coastal zone as reaching from the land to "a distance of one nautical mile (1.85 km) on the seaward side from the nearest point of the baseline from which the breadth of territorial waters is measured, extending where appropriate up to the outer limit of transitional waters". However, it would be desirable to extend the breadth of the coastal zone to a distance away from the shore further than that defined in the EU WFD (*cf.* Sect. 15.5.4). In many Baltic Sea countries the coastal waters are still subject to insufficient spatial planning and management (Schiewer 2008a; *cf.* Sect. 18.4). Involvement of environmental education and relevant legislation is essential for adequate coastal status evaluations and water quality assessments that are core components of successful management and environmental policy-making in the Baltic Sea Area (*cf.* Sect. 18.4.1).

In the future, a variety of pressures on the Baltic Sea estuarine and other coastal areas will increase (Schiewer 2008a); therefore, it is of utmost importance to (1) develop more general monitoring programmes and new classification systems based on a more integrated assessment of coastal processes and systems, (2) ensure the sustainable, careful and multivalent use of estuaries and lagoons, in order to establish a mosaic of co-evolving socio-economic and ecological systems, and (3) study self-regulation of coastal systems, especially with regard to life-support functions.

Review questions

1. What are the differences between an estuary and a coastal lagoon?
2. How do coastal lagoons evolve? What types of coastal lagoons can be distinguished?
3. What specific features do the Baltic Sea lagoon systems possess?
4. What changes in food web structure accompany increasing eutrophication?
5. What ecosystem services are provided by estuaries and lagoons, and how are they affected by anthropogenic activities?

Discussion questions

1. How far can the PEG model for the seasonality of plankton limitation be applied to the estuaries and lagoons of the Baltic Sea? Where does the PEG model not fit?
2. What are the differences in allochthonous matter supply between the Baltic Sea lagoons and tide-affected lagoons in temperate zones?
3. What impact do migratory and overwintering waterbirds have on the structure and function of the Baltic Sea estuarine and lagoonal systems? What are the main

differences between the southern Baltic Sea lagoons (which are intensely used for the resting and overwintering by migratory waterbirds) and northern Baltic Sea lagoons?
4. What effect does salinity have on eutrophication in estuarine and lagoonal systems?
5. What ecological effects can be expected from the overexploitation of (1) zooplanktivorous fish populations and (2) piscivorous fish populations in the semi-enclosed coastal water bodies of the Baltic Sea?

References

Abels A, Mannola P, Lehtinen M, Bergman L, Pesonen LJ (1998) New observations of the properties of the Lumparn impact structure, Åland islands, southwestern Finland. Meteoritics and Planetary Science 33:A7

Alimov AF, Golubkov SM (1987) Functional role of zoobenthos in the ecosystem of Neva Bay. In: Winberg GG, Gutelmakher BL (eds) Neva Bay: hydrobiological investigations, Nauka. Leningrad, pp 170–174 [in Russian]

Alimov AF, Golubkov SM (eds) (2008) Ecosystem of the Neva Estuary: biological diversity and ecological problems. KMK Publishers, Moscow, 477 pp [in Russian]

Barker T, Hatton K, O'Connor M, Connor L, Bagnell L, Moss B (2008) Control of ecosystem state in a shallow, brackish lake: implications for the conservation of stonewort communities. Aquatic Conservation Marine and Freshwater Ecosystems 18:221–240

Barnes RSK (1980) Coastal lagoons. Cambridge University Press, Cambridge 160 pp

Bird ECF (1982) Changes on barriers and spits enclosing coastal lagoons. Oceanologica Acta, Proceedings International Symposium on coastal lagoons, SCQRIIABO/UNESCO, Bordeaux, France, 8–14 September, 1981:45–53

Bobsien I, Munkes B (2004) Seasonal variation of fish community and habitat structure in an eelgrass bed (*Zostera marina* L.) in the southern Baltic Sea. Rostocker Meeresbiologische Beiträge 12:39–59

Bonsdorff E (1992) Drifting algae and zoobenthos – effects of settling and community structure. Netherland Journal of Sea Research 30:57–62

Bonsdorff E, Norkko A, Boström C (1995) Recruitment and population maintenance of the bivalve *Macoma balthica* (L.) – factors affecting settling success and early survival on shallow sandy bottoms. In: Eleftheriou A, Ansell AA, Smith CJ (eds) Biology and ecology of shallow coastal waters. Fredensborg, Olsen and Olsen, pp 253–260

Caumette P, Castel J, Herbert RA (1996) Preface. In: Caumette P, Castel J, Herbert RA (eds) Coastal lagoon eutrophication and anaerobic processes (C.L.E.A.N.). Developments in Hydrobiology 117:VII

Cloern JE (1999) The relative importance of light and nutrient limitation of phytoplankton growth: a simple index of coastal ecosystem sensitivity to nutrient enrichment. Aquatic Ecology 33:3–16

Connell JH (1978) Diversity in tropical rain forests and coral reefs. Science 199:1302–1310

DeWit R (2007) Biodiversity and ecosystem functioning in transitional waters – the point of view of a microbial ecologist. Transitional Waters Bulletin 1:3–16

Dillenburg SR, Hesp P (2009) Geology and geomorphology of Holocene coastal barriers of Brazil. Springer, Berlin, 380 pp

Duffy JE, Harvilicz AM (2001) Species-specific impacts of grazing amphipods in an eelgrass-bed community. Marine Ecology Progress Series 223:201–211

Eisma D (1998) Worldwide distribution of intertidal areas. In: Eisma D (ed) Intertidal deposits – river mouths, tidal flats and coastal lagoons. CRC Press, Boca Raton, FL, pp 1–17

Elliott M, McLusky DS (2002) The need for definitions in understanding estuaries. Estuarine, Coastal and Shelf Science 55:815–827

Elliott M, Quintino V (2007) The estuarine quality paradox, environmental homeostasis and the difficulty of detecting anthropogenic stress in naturally stressed areas. Marine Pollution Bulletin 54: 640–645

EU (2013) Great cormorant – applying derogations under Article 9 of the Birds Directive 2009/147/EC. [http://ec.europa.eu/environment/nature/pdf/guidance_cormorants.pdf]

Gocke K, Rheinheimer G, Schramm W (2003) Hydrographische, chemische und mikrobiologische Untersuchungen im Längsprofil der Schlei. Schriften des Naturwissenschaftlichen Vereins für Schleswig-Holstein 68:31-62 [in German]

Grime JP (1973) Competitive exclusion in herbaceous vegetation. Nature 242:344–347

Grossmann S, Reichardt W (1991) Impact of *Arenicola marina* on bacteria in intertidal sediments. Marine Ecology Progress Series 77:85–93

Heerkloß R, Schnese W (1994) Jahreszeitliche Variabilität und Langzeitentwicklung des Metazooplanktons im zentralen Teil der Darß-Zingst-Boddenkette. Rostocker Meeresbiologische Beiträge 2:85–92 [in German]

Herrmann C, Bregnballe T, Larsson K, Rattiste K (2014) Population development of great commorant. HELCOM Baltic Sea Environment Fact Sheet. [http://www.helcom.fi]

Heyl K, Woelfel J, Schumann R, Karsten U (2010) Microbial mats from wind flats of the southern Baltic Sea. In: Seckbach J, Oren A (eds) Microbial mats: modern and ancient microorganisms in stratified systems. Springer, Berlin, pp 301–347

Hummel H (1985) An energy budget for a *Macoma balthica* (Mollusca) population living on a tidal flat in the Dutch Wadden Sea. Netherlands Journal of Sea Research 19:52–92

Jönsson N, Busch A, Lorentz T, Korth B (1998) Struktur und Funktion von Boddenlebensgemeinschaften im Ergebnis von Austausch- und Vermischungsprozessen. Greifswalder Geographische Arbeiten 16:250–285 [in German]

Kennish MJ (1986) Ecology of estuaries, Volume I: Physical and chemical aspects. CRC Press, Boca Raton, FL, 254 pp

Kjerve B (1986) Comparative oceanography of coastal lagoons. In: Wolfe DA (ed) Estuarine variability. Academic Press, New York NY, pp 63–81

Kjerve B (1994) Coastal Lagoon Processes. In: Kjerve B (ed) Coastal Lagoon Processes. Elsevier, Amsterdam, pp 1–8

Kjerve B, Magill KE (1989) Geographic and hydrodynamic characteristics of shallow coastal lagoons. Marine Geology 88:187–199

Lampe R (1995) Küstentypen. In: Rheinheimer G (ed) Meereskunde der Ostsee, 2nd edn Springer, Berlin, pp 17–25 [in German]

Lehvo A, Bäck S (2001) Survey of macroalgal mats in the Gulf of Finland, Baltic Sea. Aquatic Conservation: Marine and Freshwater Ecosystems 11:11–18

Leppäranta M, Myrberg K (2009) Physical oceanography of the Baltic Sea. Springer, Berlin, 378 pp

Marchowski D, Neubauer G, Ławicki Ł, Woźniczka A, Wysocki D et al (2015) The importance of non-native prey, the zebra mussel *Dreissena polymorpha*, for the declining greater scaup *Aythya marila*: a case study at a key European staging and wintering site. PLoS ONE 10(12):e0145496

Matthews C (2000) The cormorant controversy. Science Scope 23:17–21

McLusky DS, Elliott M (2004) The estuarine ecosystem: ecology, threats and management. Oxford University Press, New York, NY, 214 pp

McLusky DS, Elliott M (2006) Transitional waters: a new approach, semantics or just muddying the waters? Estuarine, Coastal and Shelf Science 71:359–363

Mironova E, Telesh I, Skarlato S (2014) Ciliates in plankton of the Baltic Sea. Protistology 8:81–124

Molloy DP, Karatayev AY, Burlakova LE, Kurandina DP, Laruelle F (1997) Natural enemies of zebra mussels: predators, parasites, and ecological competitors. Reviews in Fisheries Science 5:27–97

Munsterhjelm R (1997) The aquatic macrophyte vegetation of flads and gloes, S coast of Finland. Acta Botanica Fennica 157:1–68

Pritchard DW (1967) What is an estuary: physical viewpoint. In: Lauf GH (ed) Estuaries. American Association for the Advancement of Science (AAAS) Publication 83, Washington, DC, pp 3–5

Rakauskas V, Putys Z, Dainys J, Lesutiene J, Lozys L, Arbaciauskas K (2013) Increasing population of the invader round goby *Neogobius melanostomus* (Actinopterygii: Perciformes: Gobiidae) and its trophic role in the Curonian Lagoon, SE Baltic Sea. Acta Ichthyologica et Piscatoria 43:95–108

Remane A (1934) Die Brackwasserfauna. Verhandlungen der Deutschen Zoologischen Gesellschaft 36:34–74 [in German]

Retraubun ASW, Dawson M, Evans SM (1996) The role of the burrow funnel in feeding processes in the lugworm *Arenicola marina* (L.). Journal of Experimental Marine Biology and Ecology 201: 107–118

Scheffer M, Carpenter S, Foley JA, Folke C, Walker B (2001) Catastrophic shifts in ecosystems. Nature 413:591–596

Schiewer U (1985) Analyse und Bewertung des im 5-Jahrplanzeitraum 1981/85 erreichten Kenntnisstandes und seiner volkswirtschaftlichen Verwertbarkeit auf der Grundlage der erbrachten Teilleistungen. Forschungsbericht A4 der Sektion Biologie der W-Pieck-Universität Rostock, 130 pp [in German]

Schiewer U (ed) (2008a) Ecology of Baltic coastal waters. Springer, Berlin, Ecological Studies 197:1–428

Schiewer U (2008b) Darß-Zingst Boddens, northern Rügener Boddens and Schlei. In: Schiewer U (ed) Ecology of Baltic coastal waters. Springer, Berlin, Ecological Studies 197:35–86

Schiewer U (2008c) Synthesis. In: Schiewer U (ed) Ecology of Baltic coastal waters. Springer, Berlin, Ecological Studies 197:395–417

Schubert H (2001) Die Unterwasservegetation der Darß-Zingst-Boddenkette. Meer und Museum 16:53–59 [in German]

Schubert H, Feuerpfeil P, Marquardt R, Telesh I, Skarlato S (2011) Macroalgal diversity along the Baltic Sea salinity gradient challenges Remane's species-minimum concept. Marine Pollution Bulletin 62:1948–1956

Schubert H, Wasmund N (2005) Das Phytoplankton des Strelasundes und des Kubitzer Boddens. Meer und Museum 18:83–89 [in German]

Skarlato OS (2002) Environmental policy-making in North-west Russia: ecological problems and management of the eastern Gulf of Finland coastal zone. In: Schernewski G, Schiewer U (eds) Baltic coastal ecosystems. Springer, Berlin, Ecological Studies 197: 241–252

Sommer U, Adrian R, DE Senerpont Domis L, Elser JJ, Gaedke U, et al (2012) Beyond the plankton ecology group (PEG) model: mechanisms driving plankton succession. Annual Review of Ecology, Evolution and Systematics 43:429–448

Sommer U, Gliwicz ZM, Lampert W, Duncan A (1986) The PEG model of a seasonal succession of planktonic events in fresh waters. Archiv für Hydrobiologie 106:433–471

Svensson NB (1993) Lumparn Bay: a meteorite impact crater in the Åland archipelago, southwest Finland. Meteoritics 28:445

Tallqvist M (2001) Burrowing behaviour of the Baltic clam *Macoma balthica*: effects of sediment type, hypoxia and predator presence. Marine Ecology Progress Series 212:183–191

Telesh IV (2004) Plankton of the Baltic estuarine ecosystems with emphasis on Neva Estuary: a review of present knowledge and research perspectives. Marine Pollution Bulletin 49:206–219

Telesh IV, Golubkov SM, Alimov AF (2008) The Neva Estuary ecosystem. In: Schiewer U (ed) Ecology of Baltic coastal waters. Springer, Berlin, Ecological Studies 197:259–284

Telesh I, Heerkloss R (2002) Atlas of estuarine zooplankton of the southern and eastern Baltic Sea. Part I: Rotifera. Verlag Dr. Kovač, Hamburg, 89 pp

Telesh I, Heerkloss R (2004) Atlas of estuarine zooplankton of the southern and eastern Baltic Sea. Part II: Crustacea. Verlag Dr. Kovač, Hamburg, 118 pp

Telesh IV, Khlebovich VV (2010) Principal processes within the estuarine salinity gradient: a review. Marine Pollution Bulletin 161:149–155

Telesh I, Postel L, Heerkloss R, Mironova E, Skarlato S (2009) Zooplankton of the open Baltic Sea: extended atlas. Meereswissenschaftliche Berichte, Warnemünde 76:1–290

Telesh IV, Schubert H, Skarlato SO (2011a) Revisiting Remane's concept: evidence for high plankton diversity and a protistan species maximum in the horohalinicum of the Baltic Sea. Marine Ecology Progress Series 421:1–11

Telesh IV, Schubert H, Skarlato S (2011b) Protistan diversity does peak in the horohalinicum of the Baltic Sea: reply to Ptacnik et al (2011). Marine Ecology Progress Series 432:293–297

Telesh I, Schubert H, Skarlato SO (2013) Life in the salinity gradient: discovering mechanisms behind a new biodiversity pattern. Estuarine, Coastal and Shelf Science 135:317–327

Telesh IV, Schubert H, Skarlato SO (2015) Size, seasonality, or salinity: what drives the protistan species maximum in the horohalinicum? Estuarine, Coastal and Shelf Science 161: 102–111

Thorman S (1986) Seasonal colonisation and effects of salinity and temperature on species richness and abundance of fish of some brackish and estuarine shallow waters in Sweden. Holarctic Ecology 9:126–132

Tilman D (1982) Resource competition and community structure. Princeton University Press, Princeton, NJ, Monographs in Population Biology 17:1–296

Ulanowicz RE, Wulff F (1991) Comparing ecosystem structures: the Chesapeake Bay and the Baltic Sea. In: Cole J, Lovett G, Findlay S (eds) Comparative analyses of ecosystems: patterns, mechanisms, and theories. Springer, New York, NY, pp 140–166

United Nations (1997) Glossary of environment statistics, Series F, No. 67, Department for Economic and Social Information and Policy Analysis, New York, NY, 97 pp

Van Es FB (1977) A preliminary carbon budget for a part of the Ems estuary: the Dollard. Helgoländer Wissenschaftlige Meeresuntersuchungen 30:283–294

Wolanski E (2007) Estuarine Ecohydrology. Elsevier, Amsterdam, 168 pp

Wolnomiejski N, Witek Z (2013) The Szczecin Lagoon ecosystem: the biotic community of the Great Lagoon and its food web model. Versita, London 292 pp

Wróblewski A (2001) Lunar nodal tide in the Baltic Sea. Oceanologia 43:99–112

Yousef MAM, Schubert H (2001) Assessment of the occurrence of charophytes in shallow coastal waters of Mecklenburg-Vorpommern, Germany. Schriftenreihe Landschaftspflege Naturschutz 72:9–16

Zaiko A, Lehtiniemi M, Narščius A, Olenin S (2011) Assessment of bioinvasion impacts on a regional scale: a comparative approach. Biological Invasions 13:1739–1765

Part IV

Monitoring and ecosystem-based management of the Baltic Sea

Biological indicators

14

Michael L. Zettler, Alexander Darr, Matthias Labrenz, Sigrid Sagert, Uwe Selig, Ursula Siebert, and Nardine Stybel

Abstract

1. Changes in living conditions caused by natural variability or anthropogenic activities elicit distinct responses of species, populations and communities. Bioindication is the recording of such responses and the entity measured is called a "bioindicator".
2. A bioindicator can be any relevant component or measure that can be used to estimate the environmental status based on the performance of all types of organisms (prokaryotes, protists, macroalgae, vascular plants, invertebrates, fish, mammals), including bulk measurements such as the chlorophyll *a* concentration in the seawater or the lower depth limit of macrophytes.
3. To be able to conclude if environmental change has taken place based on bioindication, it is essential to have knowledge of the specific ecological requirements of the organisms with respect to their habitats.
4. Bioindication using individuals or species includes *e.g.* behavioural adaptations, modifications of organ and cell structures and changes in population dynamics.
5. Bioindication by recording dramatic increases or decreases in the proportion and/or density of species in a community provides a conspicuous sign of environmental change, especially when this includes the extinction of species.
6. Strong decreases and extinctions of species in a community coupled to immigration of non-indigenous species may signify a shift in community composition that has a bearing on the functioning of the entire ecosystem.
7. Bioindication is a major tool used in the implementations of the EU environmental legislation: the Habitats Directive (HD), the Water Framework Directive (WFD) and the Marine Strategy Framework Directive (MSFD).

Keywords

Bioindication • Chlorophyll *a* • Community shifts • Environmental monitoring • Environmental status • Multivariate approaches • Multimetric indices

M.L. Zettler (✉) · A. Darr · M. Labrenz · N. Stybel
Leibniz Institute for Baltic Sea Research, Seestraße 15, 18119 Rostock-Warnemünde, Germany
e-mail: michael.zettler@io-warnemuende.de

A. Darr
e-mail: alexander.darr@io-warnemuende.de

M. Labrenz
e-mail: matthias.labrenz@io-warnemuende.de

N. Stybel
e-mail: nardine.stybel@io-warnemuende.de

S. Sagert · U. Selig
Forschungszentrum Jülich, Schweriner Straße 44, 18069 Rostock, Germany
e-mail: s.sagert@fz-juelich.de

U. Selig
e-mail: u.selig@fz-juelich.de

U. Siebert
Institute for Terrestrial and Aquatic Wildlife Research, University of Veterinary Medicine Hannover, Foundation, Werftstraße 6, 25761 Büsum, Germany
e-mail: ursula.siebert@tiho-hannover.de

P. Snoeijs-Leijonmalm et al. (eds.), *Biological Oceanography of the Baltic Sea*,
DOI 10.1007/978-94-007-0668-2_14

Box 14.1: Definitions used in bioindication

Bioindicator

Although the term "bioindicator" is frequently used in applied ecology, no unequivocally accepted definition exists. Heink and Kowarik (2010) have suggested that the term should be defined as follows: "An indicator in ecology and environmental planning is a component or a measure of environmentally relevant phenomena used to depict or evaluate environmental conditions or changes, or to set environmental goals. Environmentally relevant phenomena are pressures, states, and responses as defined by the OECD (2003)." The OECD is the Organisation for Economic Co-operation and Development, which has the mission to promote policies that will improve the economic and social well-being of people around the world.

Eutrophication

Eutrophication is defined as an increase in the rate of supply of organic matter in an ecosystem (Nixon 1995). In a slowly ageing water body, eutrophication is a natural process expressed as increasing primary productivity. Artificially increased primary productivity is mostly due to increased external nutrient inputs. According to OSPAR (1998), the eutrophication of marine waters means the enrichment of water by nutrients causing an accelerated growth of algae and plants to produce an undesirable disturbance to the balance of organisms present in the water and to the quality of the water concerned; therefore, it refers to the undesirable effects resulting from anthropogenic nutrient inputs. OSPAR is the mechanism by which 15 governments of the western coasts and drainage areas of Europe, together with the EU, cooperate to protect the marine environment of the northeastern Atlantic Ocean. OSPAR started in 1972 with the Oslo Convention against dumping and was broadened to cover land-based sources and the offshore industry by the Paris Convention of 1974. These two conventions were unified, updated and extended during the 1992 OSPAR Convention for the Protection of the Marine Environment of the North-East Atlantic.

Multimetric indices

A multimetric index combines several indicators (or metrics) into a single index value. Each metric should be tested regarding its validity and calibrated and transformed into a unitless score prior to aggregation into the index. The component indicators should be based on different elements of community response to environmental change, *e.g.* diversity, biomass, species composition and the presence of vulnerable, opportunistic or non-indigenous species. Generally, these responses should be independent from each other. The relationship between a multimetric index and a stressor should be distinct and stronger than the relationships between the stressor and each individual component of the index.

14.1 Bioindication, biomarkers and indicator species

14.1.1 Different concepts

The term "bioindicator" is used in various ways in the scientific literature. A generally accepted broad concept is that a bioindicator can be any relevant component or measure that can be used to estimate the environmental status (Box 14.1). This includes responses of organisms such as behavioural adaptations, modifications of organ and cell structures and changes in population dynamics, as well as changes in community composition (Table 14.1). A bioindicator response should be correlated or causally linked to the environmental process it indicates (McCarty et al. 2002). A bioindicator can, but does not have to, be based on the identification of species. For example, one of the most widely used bioindicators in aquatic environmental assessments is the chlorophyll *a* concentration in the water as an indicator for the degree of eutrophication (Heink and Kowarik 2010). Thus, it is the usability of a species for the detection of natural and/or anthropogenically-induced changes in environmental conditions (bioindication) which renders it a bioindicator.

The meaning of "bioindicator" can overlap with that of a biomarker, but the term "biomarker" is narrower. It is commonly used only for chemical pollutant-induced variation in cellular or biochemical components or processes, structures, or functions that is measurable in a biological system or sample (McCarty et al. 2002). Thus, a biomarker is a distinctive biological or biologically derived indicator (*e.g.* gene expression, enzyme activity, imposex, behaviour, growth, reproduction or population change) of hazardous substances in the environment (*cf.* Sect. 16.4.6). Biomarkers

Table 14.1 Overview of different organism groups and their applications for bioindication in the Baltic Sea as recommended by EU directives and other regulations. WFD = Water Framework Directive, HD = Habitats Directive, MSFD = Marine Strategy Framework Directive, BWQD = Bathing Water Quality Directive, HELCOM = Baltic Marine Environment Protection Commission (*cf.* Sect. 17.8.4), ICES = International Council for the Exploration of the Sea (*cf.* Box 18.1)

Organism group	Main cause of the disturbance	Detection of response		Directive or regulation
		Level	Main method	
Prokaryotes	Eutrophication	Molecular, cellular	Community composition	BWQD, HELCOM
Phytoplankton	Eutrophication	Individual, population	Community composition and biomass	WFD, BWQD, MSFD, HELCOM
Macrophytes	Eutrophication	Individual, population	Lower depth limit, community composition and functional groups	WFD, HD, MSFD, HELCOM
Macrozoobenthos	Cumulative effects on habitat quality	Individual, population	Community composition and functional groups	WFD, HD, MSFD, HELCOM
Fish	Chemical pollution	Organ, individual, population	Behaviour and tissue analyses	HD, MSFD, HELCOM, ICES
Mammals	Chemical pollution	Organ, individual, population	Behaviour and tissue analyses	HD, MSFD, HELCOM

are seldom related to any particular species or group of species, but are often universal features applicable to most organisms. However, there can be large variability in the responses of different species with respect to the same biomarker.

The meaning of "bioindicator" should not be confused with that of an "indicator species". The concept of indicator species is applied to a situation when a study aims at defining a typical inventory of the inhabitants of a specific habitat (HELCOM 2015b). This means that one is looking for species which, due to their autecological requirements, are constrained by narrow environmental conditions and are therefore typically present only under specific conditions. Strictly speaking, although widely used, it is not ideal to use the term "indicator species" in this context. It would be better to call such species "typical species" or "characteristic species", depending on the strength of the correlation between their presence and the environmental conditions. The use of indicator species, in which species are expected to have a similar sensitivity or tolerance to either natural or anthropogenic stressors, does not account for possible shifts in tolerance along natural environmental gradients and between biogeographic regions. The indicative value of those species for estimating the health status of the environment may in such cases be considered at least questionable (Zettler et al. 2013).

14.1.2 A growing need for bioindication

Bioindication includes drawing conclusions on environmental changes based on observations regarding the status of species and communities. Although bioindication has been applied in assessments of marine systems for a long time, it has gained prominence since the beginning of this millennium due to an increasing demand for the assessment of the ecological status of marine waters. For the Baltic Sea Area such assessments are partly implemented through different EU directives (*cf.* Sect. 17.8), such as the Habitats Directive (HD, EC 1992), the Water Framework Directive (WFD, EU 2000) and the Marine Strategy Framework Directive (MSFD, EU 2008). Different organism groups are used to indicate different types of disturbances in the ecosystem as recommended by EU directives and other regulations (Table 14.1, HELCOM 2012a).

14.1.3 The application of bioindication

Bioindication is, often together with other indicators (*e.g.* chemical), used in ecological assessments (*cf.* Sects. 15.5, 16.4, 17.7). Assessment is a management issue, a tool of environmental policy, which ideally should be based upon conclusions drawn from sound scientific knowledge. However, the direct application of indicators in assessments performed for management purposes is mostly insufficient since the directives require that assessment is based on reliable values/levels. Consequently, a biological response has to be converted into a number that indicates the health status, or the degree of disturbance or "naturalness", of the habitat/area under assessment. This is most often performed by calculating indices for different species or communities (Van Hoey et al. 2010) with bioindicators as the underlying numerical metric.

14.2 Prokaryotes

14.2.1 Prokaryotes in environmental assessments

Prokaryotes (bacteria and archaea) are essential for sustaining life on our planet, but they are often ignored in environmental assessments. One reason for this is their small size; the median diameter of a typical prokaryote is only 0.5–1 μm. However, because of their extremely high abundance, even "clear waters" of the Baltic Sea proper contain >1 million cells mL^{-1} (*cf.* Sect. 8.3). Thus, despite their small size, prokaryotes contribute substantially to the total biomass on Earth and drive virtually all biogeochemical cycles. For example, they are able to oxidise organic and inorganic materials, to reduce CO_2 photo- and chemolithoautotrophically, and to live under oxic, hypoxic (<2 mL O_2 L^{-1}) and sulphidic conditions (*cf.* Figs. 3.16 and 3.17).

Practical applications of prokaryotes as bioindicators in the Baltic Sea Area are still few since there is a lack of appropriate standards and thresholds for evaluating specific environmental scenarios based on microbial molecular analyses. Studies on the identification of distinct prokaryotes or their communities in the Baltic Sea Area (*cf.* Box 4.3) are still rare. At present, microbial indicators of specific environmental conditions in the Baltic Sea ecosystem are used mainly for indication of only four environmental conditions: the salinity level and the degrees of oxygen depletion, eutrophication and faecal contamination. The relevance of these four conditions ranges from scientific to socio-economic.

14.2.2 Bacterial community composition indicates salinity

The value of prokaryotes as bioindicators of salinity is primarily of scientific importance since salinity can be measured much easier in other ways (*cf.* Box 1.2). Molecular studies have shown that salinity exerts a strong selective pressure on community composition in the Baltic Sea Area (*cf.* Sect. 8.3.11). For example, salinity-related distributions of members of the Verrucomicrobia clade (phylum Proteobacteria) were found in the transition zone (Belt Sea and Kattegat), while the phylum Bacteroidetes showed salinity-dependent distributions in the pelagic Skagerrak-Kattegat front (Pinhassi et al. 2003). In the Gulf of Bothnia, a change of Actinobacteria composition in relation to salinity, even within a very narrow range, was discovered by Holmfeldt et al. (2009).

The first investigation of bacterial community composition along the entire Baltic Sea salinity gradient was conducted for a summer situation, using 454 pyrosequencing of the bacterial 16S rRNA gene (*cf.* Fig. 8.12). This study revealed pronounced shifts in the major bacterial groups along the salinity gradient: the relative abundances of α-Proteobacteria and γ-Proteobacteria increased with salinity, whereas Actinobacteria and β-Proteobacteria displayed the opposite trend (Herlemann et al. 2011). This is in good accordance with previous investigations of bacterial community shifts along estuarine salinity gradients.

14.2.3 Anaerobic bacteria indicate oxygen depletion

For reasons mysterious to microbiologists, the anoxic areas in the Baltic Sea proper are often referred to as "dead zones". These areas are usually characterised by high hydrogen sulphide concentrations below the halocline, which is generated by sulphate-reducing bacteria (*cf.* Sect. 3.6.5). Since hydrogen sulphide is toxic for aerobic organisms, life in these areas is reduced to the presence of microorganisms (Fig. 14.1).

It has been shown recently that pelagic bacteria can be indicative of oxygen depletion (Labrenz et al. 2010). Nitrogen- and sulphur-related fluxes in the anoxic and sulphidic deeps of the Baltic Sea proper seem to be catalysed by only a few key organisms: sulphur-oxidising and nitrate-reducing chemoautotrophic *Sulfurimonas* species and an aerobically ammonium-oxidising member of the phylum Thaumarchaea related to *Candidatus Nitrosopumilus maritimus*, which seems to be at least sulphide-tolerant.

14.2.4 Diazotrophic cyanobacteria indicate nitrogen limitation

The phytoplankton summer blooms in the Baltic Sea are dominated by cyanobacteria such as *Nodularia spumigena*,

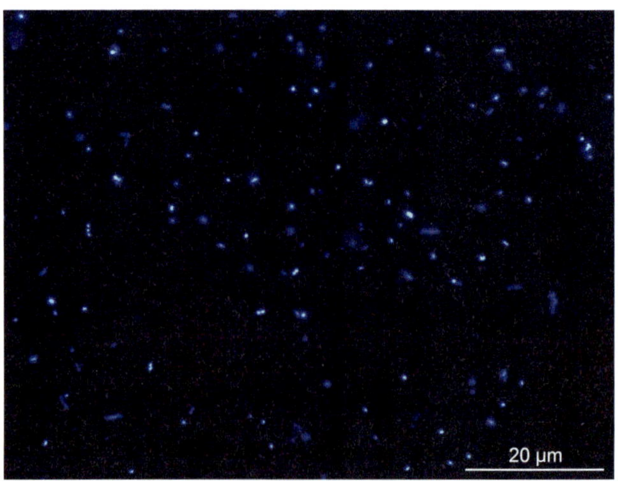

Fig. 14.1 Fluorescence micrograph of 4′,6-diamidino-2-phenylindole (DAPI)-stained microorganisms from the hypoxic zone of the Gotland deep redoxcline. DAPI is a fluorescent stain that penetrates the cell membrane of intact cells and binds to A-T rich regions in the DNA. Photo: © Matthias Labrenz

Aphanizomenon flos-aquae, *Dolichospermum* sp., and *Synechococcus* spp. (*cf.* Sect. 8.2). Except for *Synechococcus*, these cyanobacteria are capable of using atmospheric nitrogen (N_2) for their metabolism. These diazotrophic (nitrogen-fixing) primary producers become dominant when nitrogen is depleted but phosphorus is still available. Thus, they could be indicative of nitrogen limitation and excess phosphorus (by eutrophication) in the pelagic system of the Baltic Sea.

14.2.5 Bacterial indicators of faecal contamination

Faecal bacteria are used in practical monitoring of *e.g.* bathing water quality and their importance as bioindicators is therefore mainly socio-economic. The best-known faecal indicators are members of the order Enterobacteriales (phylum Proteobacteria) and the genus *Enterococcus* (phylum Firmicutes). The presence of these bacteria in aquaculture, shellfish or water can in general be indicative of faecal contamination because they are natural inhabitants of the gut flora of humans and other mammals.

To protect society from polluted and potentially harmful food or waters, the EU has introduced several directives aimed at restricting faecal contamination. For example, the EU Directive on the management of bathing water quality (EU 2006) recommends that by the year 2015 bathing waters will be classified as at least "sufficient" according to the number of colony-forming units (cfu) of intestinal *Enterococcus* and *Escherichia coli*. The classification "sufficient" is defined as a maximum of 185 cfu of intestinal enterococci and a maximum of 500 cfu of *Escherichia coli* per 100 mL of water in coastal areas. The cfu values are based on the traditional technique of agar plate cultivation, but in the future this method is likely to be replaced by direct analyses of pathogenic bacterial gene expression.

14.3 Phytoplankton

14.3.1 Phytoplankton as indicator of eutrophication

Eutrophication (Box 14.1) is still one of the major anthropogenic impacts on the Baltic Sea ecosystem (*cf.* Sect. 17.4.2). The pelagic communities of photoautotrophic microorganisms (algae and cyanobacteria), provide the most important energetic base of the food webs in most aquatic ecosystems, including the Baltic Sea (*cf.* Table 8.1). Any discussion on indication, especially with respect to eutrophication, must begin with phytoplankton activity. Phytoplankton communities are usually complex and highly variable in terms of diversity and dynamics (*cf.* Sect. 8.2). Community composition and the physiological status of the phytoplankton cells respond on short time scales, within hours or days, to changes in physical, chemical and biological conditions.

Phytoplankton-based bioindicators often focus on the effects of anthropogenically-induced eutrophication. Increased nutrient loads, primarily of nitrogen and phosphorus, often lead directly to increased primary production and mostly also to increased phytoplankton biomass. A change in nutrient loading is usually associated with changes in nutrient stoichiometry, *i.e.* changes in the Redfield ratio (*cf.* Sect. 3.2.3), entailing a shift in the phytoplankton community composition by promoting certain phytoplankton groups and suppressing others. Therefore, phytoplankton-based eutrophication indicators basically represent dose-response curves of nutrient load and eutrophication effects. In general, the indicators operate at two levels: indication by phytoplankton biomass (the sum of relevant parameters) and indication by phytoplankton community composition (a parameter specific of community shifts).

14.3.2 Chlorophyll *a* as indicator of eutrophication

Based on knowledge of relationships between nutrients, primary production and chlorophyll *a* (Chl *a*) (Falkowski and Wilson 1992, 1993), the most widely accepted expression of photoautotrophic biomass today is the Chl *a* concentration in the water (Boyer et al. 2009). A better indication would be theoretically achieved by directly measuring primary production itself. However, a comprehensive assessment of photosynthetic activity is costly and time-consuming. Moreover, many response mechanisms of photosynthesis under highly variable field conditions are still unknown, while Chl *a* concentration is predictable and easy to measure.

The Chl *a* concentration is also one of the main correlates of light attenuation in the water column, which influences the depth distribution of macrophytes. Thus, Chl *a* is the major phytoplankton parameter used in assessment strategies worldwide and also in the eutrophication assessment of the Baltic Sea Area, following the WFD (EU 2000) and the HELCOM (*cf.* Sect. 17.8.4) eutrophication assessments (HELCOM 2009) (Table 14.2). Measurements of Chl *a*, using standardised methods, are included in the monitoring programmes for the Baltic Sea. The Chl *a* concentration can also be routinely monitored with automated flow-through devices mounted on board ships of opportunity (Fleming and Kaitala 2006). In addition, surface blooms of phytoplankton are monitored using satellite remote sensing data converted to Chl *a*, which provides information with large spatial coverage (*cf.* Sect. 15.2.5).

Chl *a* indication delivers useful information such as phytoplankton bloom frequency (number of bloom events per year), duration of the blooms (days per year), bloom intensity (km^2 per day) and integral extent of bloom events

Box 14.2: Ecological assessment according to the WFD

The EU Water Framework Directive (WFD, EU 2000) uses an ecological status classification which ranges from 1 for "high ecological status" to 5 for "bad ecological status" (Box Fig. 14.1). The achievement of at least a "good environmental status" (GES) of European waters is a major environmental objective of the WFD. Reference conditions represent a type-specific ecological status where there are no, or only very minor, changes in the values of hydro-morphological, physico-chemical, and biological quality elements – in other words, a state that would be expected to exist without any anthropogenic disturbance. Deviations from the reference conditions range from "no/minor" to "strong". Several parameters can be measured on the same quality element, *e.g.* abundance and diversity in the macrozoobenthos. The ecological quality ratio (EQR) represents the ratio between the values of a parameter and the values of the same parameter under the reference conditions, expressed as a number between 0 and 1. These ratios are then combined at the quality element level and finally into an ecological status classification.

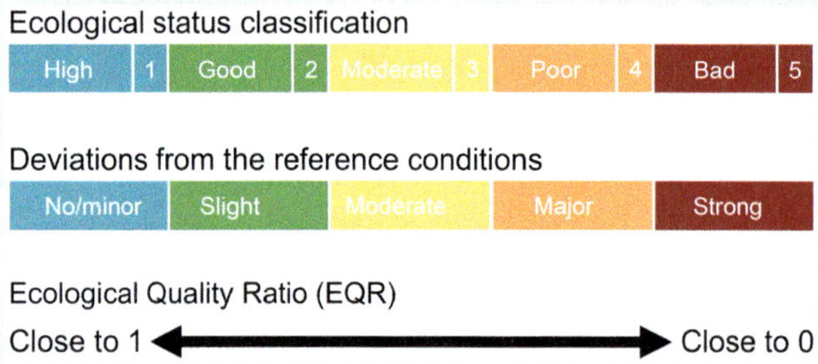

Box Fig. 14.1 Ecological status classification, deviations from the reference conditions and the ecological quality ratio (EQR) according to the EU Water Framework Directive (2000/60/EC) and EU (2005)

Table 14.2 Results from a HELCOM chlorophyll *a*-based assessment for the time period 2003–2007 as part of the HELCOM eutrophication assessment tool (HEAT) for different parts of the Baltic Sea Area. Transitional waters = the continuum between freshwaters and coastal waters. The ecological quality ratio (EQR) is the ratio between the chlorophyll *a* concentration under summer reference conditions and the recently measured summer chlorophyll *a* concentration. The EQR limits for the ecological quality classes are here: 0.95 for "High", 0.81 for "Good", 0.67 for "Moderate", 0.53 for "Poor" and 0.38 for "Bad". The final score describes the ecological status according to the five-class quality scale. Data compiled from HELCOM (http://www.helcom.fi)

Area	μg chlorophyll a L^{-1} measured values	μg chlorophyll a L^{-1} reference conditions	Ecological quality ratio (EQR)	Ecological status (final score)
Gulf of Riga (transitional waters)	5.3	4.0	0.75	Good
Kiel Bay and Fehmarnbelt	1.8	1.2	0.67	Moderate
Arkona Sea (open waters)	2.1	1.2	0.58	Moderate
Danish Straits (including the Öresund)	2.2	1.2	0.55	Moderate
Bothnian Sea (open waters)	2.1	1.0	0.48	Poor
Gulf of Finland (coastal waters) and Narva Bay	4.2	1.8	0.43	Poor
Gulf of Riga (open waters)	4.4	1.8	0.41	Poor
Eastern Gotland Sea (open waters)	3.2	1.2	0.38	Bad
Gulf of Finland (open waters)	5.1	1.2	0.23	Bad
Pomeranian Bay and Bornholm Sea	6.7	1.5	0.22	Bad

(km^2 multiplied by days) (Hansson and Håkansson 2007). An increasing trend in the bloom frequency and the bloom intensity of cyanobacteria has been recorded during the last century, which may be attributed to changes in nutrient stoichiometry due to the large-scale eutrophication process in the Baltic Sea (Finni et al. 2001).

14.3.3 Harmful phytoplankton blooms

The abundance of potentially harmful phytoplankton species, especially those that produce toxic compounds, will not always be reflected by Chl *a* measurements because a number of harmful bloom-forming organisms are heterotrophic and possess no chlorophyll. Another risk of Chl *a* indication is an oversimplification of eutrophication effects in the ecosystem, *e.g.* Chl *a* may mask prominent effects at the species level such as toxicity. Thus, additional information on phytoplankton community composition is needed to assess the risk for scenarios involving harmful species.

Of special interest are potential harmful blooms of *e.g.* the cyanobacterium *Nodularia spumigena* (*cf.* Box 16.4), the haptophyte *Prymnesium parvum* and the dinoflagellates *Dinophysis acuminata*, *Dinophysis norvegica* and *Phalacroma rotundatum* (syn. *Dinophysis rotundata*). Although the indication potential of harmful phytoplankton species has been investigated, no conclusive results as to their anthropogenic or natural origin could be obtained (Finni et al. 2001), and reference conditions (Box 14.2) are difficult to determine. Indicator systems for harmful phytoplankton species are not applied in the monitoring programmes of the Baltic Sea today.

14.4 Macrophytobenthos

14.4.1 The lower depth limit

Benthic macrophytes (macroalgae and vascular plants) are sensitive indicators of a broad range of environmental conditions, such as nutrient concentrations, light availability, turbidity, water level change, salinity and the occurrence of hazardous substances. The major advantage of macrophytes as indicators concerns eutrophication. For example, a decreased lower depth limit of the vertical distribution of the macrophyte vegetation is often an effect of eutrophication and is used for classification purposes when implementing the WFD (EU 2000).

An increased nutrient load increases phytoplankton biomass, which decreases light penetration into the water column (Schubert and Forster 1997). Light penetration is the main variable that limits the colonisation of macrophytes when an appropriate substrate is available (Sagert et al. 2005). Thus, eutrophication has a major influence on the lower depth limit of macrophytes, and an upward shift of the depth limit can be regarded as a sign of vegetation degradation (Table 14.3). In addition to the depth limit of the whole macrophyte community, depth limits of single indicator species can be used, *e.g.* those of the brown alga *Fucus vesiculosus* (Kautsky et al. 1986, Torn et al. 2006), the red alga *Furcellaria lumbricalis* (Kotta et al. 2008) and the common eelgrass *Zostera marina* (Greve and Krause-Jensen 2005).

14.4.2 Changes in macrophyte community structure

Another effect of eutrophication on the macrophyte vegetation used for classification purposes in the implementation of the WFD (EU 2000) is a change in the structure of macrophyte communities (Selig et al. 2007). Shifts in marine ecosystem structure and function are often evaluated by classifying marine benthic macroalgae into either of two ecological status-related groups (Table 14.3). The first group includes species with a thick or calcareous thallus, low growth rates and long life cycles (late perennial successional species or *K*-strategists, *cf.* Box 4.14), whereas the second group consists of thin foliose and filamentous algae with

Table 14.3 Different indicators based on macrophytes that are used in environmental assessments in the Baltic Sea Area

Indicator	Substrate type	Measure and unit
Lower depth limit of a dense attached vegetation	Sandy, soft and hard	Water depth (m)
Lower depth limit of *Zostera marina*	Sandy or soft	Water depth (m)
Lower depth limit of *Fucus* spp.	Hard	Water depth (m)
Lower depth limit of *Furcellaria lumbricalis*	Hard	Water depth (m)
Abundance of *Fucus* spp. in the upper sublittoral zone	Hard	% cover of the vegetation
Biomass of epiphytic algae	Sandy, soft and hard	Dry weight (g m^{-2})
Opportunistic algae *versus* vascular plants	Sandy or soft	Biomass ratio
Opportunistic algae *versus* perennial algae	Hard	Biomass ratio
Native *versus* non-indigenous species	Hard	Biomass ratio

Fig. 14.2 The lower depth limits of macrophytes are used as indicators of eutrophication in the Baltic Sea Area. (**a**) The common eelgrass *Zostera marina*, here accompanied by green and brown macroalgae and blue mussels in the Belt Sea. (**b**) The bladderwrack *Fucus vesiculosus*, here accompanied by snails, blue mussels and starfish in the Belt Sea. Photo: © Dirk Schories

high growth rates and short life cycles (opportunistic species or *r*-strategists, *cf.* Box 4.14).

The multimetric indices (Box 14.1) "Balcosis" (Baltic algae community analysis system, Germany) and "Multi species maximum depth index" (Sweden), include both depth limits and community composition of the sublittoral macrophytobenthos (Marbà et al. 2013). Balcosis was developed especially for the German coast in the Baltic Sea Area and combines the classification of soft substrate vegetation with that of hard substrate (Fürhaupter et al. 2008). The primary metrics are the depth limits of the macrophytes *Zostera marina* and *Fucus vesiculosus* (Fig. 14.2). The secondary metrics are different biomass ratios at specific depth levels (*e.g.* opportunistic algae *versus* vascular plants, opportunistic algae *versus* perennial algae) or *Fucus vesiculosus* abundance, expressed as the percentage coverage of the substrate by *Fucus vesiculosus*, in the sublittoral zone (Table 14.3).

14.5 Macrozoobenthos

14.5.1 Cumulative effects on habitat quality

Benthic macroinvertebrates belong to the most widely used organism groups in marine biological assessments (Fig. 14.3). In the Baltic Sea Area they are employed *e.g.* as indicators/components in different biotic descriptors of the MSFD (EU 2008) and as the so-called biological "quality elements" in the WFD (EU 2000). The macrozoobenthos is used in bioindication mainly at the community level.

Macrozoobenthic communities possess practically all attributes for being good indicators: (1) they contain a mixture of tolerant and specialised species, (2) they have high functional diversity with respect to *e.g.* life cycles, feeding types and motility, (3) they are easy to identify to the species level, (4) they are relatively easy to collect, and (5) good autecological knowledge is available regarding most of the core species. Due to the combination of all these attributes, the macrozoobenthos is the ecological category of organisms that best reflects cumulative effects on habitat quality (Table 14.1).

14.5.2 Opposite signals from abundance and diversity

Two bioindicators commonly used in zoobenthic ecology, the total macroinvertebrate abundance and the Shannon index of diversity (*cf.* Box 4.5), usually produce opposite signals. The HELCOM Monitoring Station 010 in the Fehmarnbelt between the Kiel and Mecklenburg Bays in the Belt Sea has been resampled practically every autumn since 1991 (Fig. 14.3a). The station is 28 m deep, its bottom is muddy, salinity is 20–30, and episodes of oxygen depletion occur near and at the bottom. In six of the years the overall density of the macrozoobenthos was low (100–500 individuals m^{-2}), but in all other years it was rather high ($\sim$1,000 to $\sim$8,000 individuals m^{-2}). Simultaneously, the Shannon index was lowest in 2002 when the abundance was also very low, but other years with low abundances feature intermediate Shannon index values. A minimum in the Shannon index was also observed in 2006, the year with the highest overall abundance. The limitations of using diversity indices in bioindication are well described in the literature (*e.g.* Zettler et al. 2007, 2013). Nevertheless, the Shannon index remains a widely used key metric in many ecological assessments.

14.5.3 Multivariate approaches

Instead of using univariate parameters to explain the patterns observed, more meaningful and interpretable outcomes may be obtained with multivariate approaches. The HELCOM data from Monitoring Station 010 in the Fehmarnbelt were analysed using a multivariate technique, non-metric multidimensional scaling (MDS), based on community composition (Fig. 14.3b). The resulting ordination plot visualises the similarity (scores close to each other) and dissimilarity (scores far apart) in community composition between the different years. Six single scores, representing the six years with the lowest overall abundances, are found on the left-hand side of the ordination, and a cloud of scores, representing all other years, is found to the right.

This shows that community composition in the six years with low abundances differed from the "normal" composition found in the remaining years. It may be argued that these differences are mainly due to the low overall abundance and do not necessarily represent community shifts. However, to reduce the influence of the high variability of abundance, the analysis involved a fourth-root transformation of the data. Additionally, an almost identical pattern was obtained when abundance data were converted to presence-absence data. Thus, the probability that this pattern indeed reflects a true community shift is fairly high.

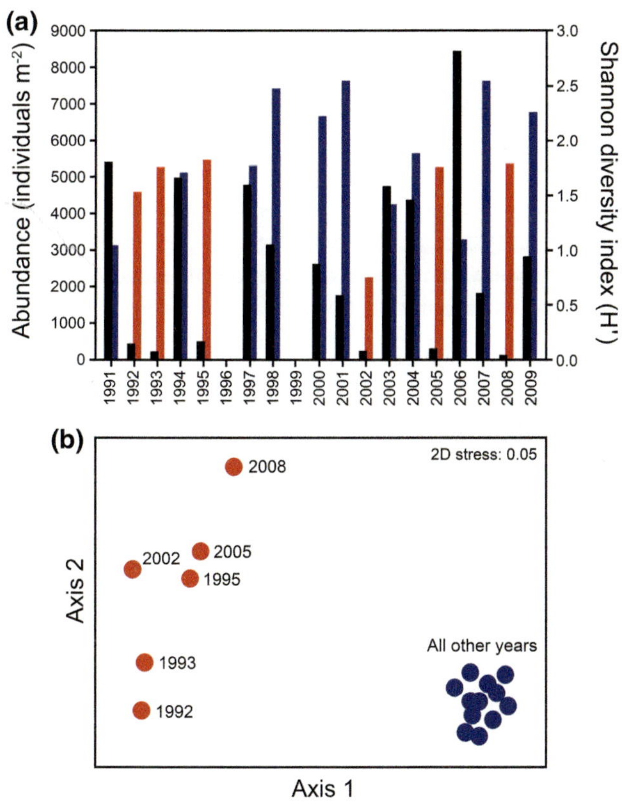

Fig. 14.3 Time series of abundance, diversity and species composition of the macrozoobenthic community at the HELCOM Monitoring Station 010 in the Fehmarnbelt from 1991 to 2009 (data for 1996 and 1999 are missing). (**a**) Overall macrozoobenthos abundance (black bars) and Shannon index (H′, log base = e) indicated as red bars when the abundance was very low and indicated by blue bars when the abundance was higher. (**b**) A multidimensional scaling (MDS) ordination plot based on fourth-root-transformed species abundances. The years with very low abundance are in the left-hand side of the ordination and all other years are in the right-hand side of the ordination. Figure: © Michael L. Zettler

The weakness of such a similarity-based multivariate approach is that the outcome is hardly applicable to environmental assessment, *i.e.* as providing information on a "good" or "bad" status of the community. In the Fehmarnbelt example, a theoretical "natural" community could be constructed and referred to as the "natural state". The distance between the scores of a community found in a specific year and the theoretical community might then be used to calculate the degree of deviation. However, it would be difficult to prove that our theoretical community truly represents the natural one (the "reference conditions" according to the EU WFD). Nevertheless, a multivariate approach is still one of the most powerful tools to detect temporal and/or spatial changes in community composition. In addition, it provides a possibility of combining biological and environmental data in a single analysis (Gogina et al. 2010).

14.5.4 Functional groups

Changes in the macrozoobenthic community composition as revealed by multivariate methods are usually detected by comparing the abundances of the dominant species in the community. More explanatory power can be gained by lumping species into functional trait groups (*cf.* Sect. 4.7; Box 10.3; Péru and Dolédec 2010). Such an approach offers the opportunity to tackle the complexity of species-rich communities by constructing relatively simple models of functional community structure in relation to the environment. Functional traits for macrozoobenthos can include *e.g.* life cycles, feeding types and motility, but also more general phenotypic properties such as body size and morphology (Darr et al. 2014). A change in functional traits is not only a record of a change in the community, but also points to the ecological consequences of the change.

14.5.5 Multimetric indices

To obtain a valuable practical bioindicator helpful in providing a comprehensive picture of the causes and consequences of changes in macrozoobenthic communities, it is strongly recommended to use a multimetric index, *i.e.* an index that combines several metrics into one equation (Box 14.1).

For the implementations of the assessments required by different EU directives, a variety of multimetric approaches has been developed for the macrozoobenthos in Europe (Borja et al. 2009; Van Hoey et al. 2010). An example is the

Fig. 14.4 The offshore stone reefs in the deep waters of the Kattegat and the Belt Sea, *e.g.* in the Mecklenburg Bay, are often densely overgrown by sea anemones and other epizoic species. Such areas act as refuges and host highly diverse communities. Photo: © Gerd Niedzwiedz

benthic quality index (BQI), which is based on a combination of the species tolerance values, abundance and diversity (Rosenberg et al. 2004). A problem with using such multimetric indices on a supra-regional (European) scale is that they are seldom robust enough to be applicable in different regions. Particularly in a brackish ecosystem such as the Baltic Sea, the robustness requirement is only met by the very basic indices with a low sensitivity, which entails a potentially crucial loss of specificity (Zettler et al. 2007, 2013). Some of these multimetric indices have therefore been adapted to the Baltic Sea ecosystem (Borja et al. 2009; Fleischer and Zettler 2009).

Species tolerances and preferences may change along environmental gradients and between different biogeographic regions as the environment modifies species' autecology. There is then a need to adjust indicator species lists, which introduces a risk of including sibling species or cryptic species in the calculations of the index value of a species (Zettler et al. 2013).

Methodological difficulties with using the macrozoobenthos for bioindication still exist in certain habitats of the Baltic Sea Area, *e.g.* the submerged stone reefs in the Kattegat and Belt Sea (Fig. 14.4).

14.6 Fish

14.6.1 Fish communities

Fish have been used as bioindicators of water quality for many years (Karr 1981; HELCOM 2012b). They show a broad spectrum of tolerances and respond to physical, chemical and biological degradation in characteristic ways. Compared to other organisms used as bioindicators, they are easy to identify and to collect.

The distribution of many fish stocks in the Baltic Sea Area is determined by gradients of salinity and temperature. In addition to these natural environmental stressors, various species respond sensitively to anthropogenic pressures. The main causes of changes in the natural fish community are overfishing, eutrophication, chemical pollution, regulation of rivers and the introduction of non-indigenous fish species.

Both single events and continuous anthropogenic impacts can be reflected in different developmental stages of fish species (eggs, larvae and mature fish) or in entire life cycles. Reproduction and early developmental stages are especially vulnerable and therefore primarily affected by anthropogenic pressures. Fish larvae may be considered one of the best indicators of the state of Baltic Sea fish stocks (Thiel et al. 1996).

14.6.2 The viviparous eelpout – a key indicator fish

The viviparous eelpout *Zoarces viviparus* (Fig. 14.5) is one of the most widely used indicator fish in the Baltic Sea. This benthic species inhabits the marine coastal waters of northern Europe, as well as the coastal waters of the brackish Baltic Sea. The eelpout constitutes an important link in the benthic ecosystem as prey for other fish species and waterbirds. Because of its wide distribution, its stationary behaviour and its capability to accumulate hazardous substances, the eelpout is an excellent organism for biomonitoring (Schladot et al. 1997; Hedman et al. 2011).

HELCOM (*cf.* Sect. 17.8.4), ICES (*cf.* Box 18.1) and OSPAR (Box 14.1) recommend the viviparous eelpout for monitoring the impacts of environmental pollution in the Baltic Sea and the transition zone. For example, changes in the weight and abundance of the eelpout have been followed since 1975 to assess the status of the Gulf of Riga (Diekmann and Möllmann 2010). The eelpout is also used by *e.g.* the German Environmental Specimen Bank (http://www.umweltprobenbank.de/en) as a bioindicator of the presence of hazardous substances in coastal waters. Results of monitoring of eelpout livers have shown that the lead levels in coastal areas of the Baltic Sea are lower than those in coastal areas of the North Sea. In addition, the species may also be used as an indicator of climate change. Data reflecting the eelpout-specific limits of acclimation capacity in the North Sea have shown that the species responds to higher water temperatures with decreased abundance (Pörtner and Knust 2007).

Because of its viviparous mode of reproduction, the eelpout is also highly suitable for the analysis of reproductive success and potential toxic effects on the development of offspring (Hedman et al. 2011). These are highly valuable

Fig. 14.5 The health condition of the larvae of the viviparous eelpout (*Zoarces viviparus*) is used as a bioindicator. (**a**) Juvenile stages, (**b**) Adult fish. Photo: © Jens Gercken

bioindicators because there is a direct connection with population effects. Abnormal development of larvae in eelpout broods is regarded as an indicator of impaired reproduction due to chronic exposure to various contaminants (HELCOM 2010). In some highly polluted areas on the German coast, malformed larvae were found in 50–90 % of the eelpout broods while in less polluted sites on the Swedish coast malformation rates were only 0–6 % (Gercken et al. 2006). Other studies have recorded gonadal disorders in histopathological studies, notably the presence of intersex (ovotestis) in the male eelpout and ovarian degeneration (atresia) in the female eelpout in late spring (Gercken and Sordyl 2002).

14.7 Marine mammals

The marine mammalian top predators, the seals (*cf.* Box 4.13) and the harbour porpoise (*cf.* Fig. 4.15), are exposed to a wide range of anthropogenic stressors. Their population sizes illustrate the constant conflicts between marine mammals and fisheries, military operations, shipping, gas and oil exploitation, chemical pollution, marine litter, sand and gravel extraction, offshore wind farms and recreation. These activities lead to habitat loss, reduced fitness, lower reproductive success and increased mortality of the Baltic Sea mammals.

Marine mammals are used mainly as indicators of anthropogenic hazardous substances (Table 14.1). Due to their position high up in the food web and their long lifespan, they accumulate numerous compounds in different body tissues and organs such as blubber, liver and muscle. Lipophilic substances, *e.g.* PCBs (*cf.* Table 16.1), are transferred to the offspring, mainly during lactation. Up to 80 % of the PCB load is not excreted, and the load accumulates over generations. The marine mammals in the Baltic Sea carry a significantly higher pollution burden than those from *e.g.* Arctic waters.

Effects of chemical pollutants such as DDTs, PCBs, mercury and TBT (*cf.* Table 16.1) include immunosuppression, resulting in deteriorated health status and devastating viral epidemics (Beineke et al. 2010). Other deleterious effects include reduced thyroid, impaired reproduction and tumours. In the 1970s and 1980s, the grey and ringed seals in the Baltic Sea showed stenosis (abnormal narrowing), occlusion and tumours of the uterus, resulting in reduced reproductivity and population sizes. Furthermore, damaged bone structures (Fig. 14.6) and lesions in the adrenal gland were observed. These lesions were associated with high body DDTs and PCBs burdens. More recently observed are new damages reported from the Baltic grey seals, such as severe intestinal ulcers (Bäcklin et al. 2003).

Generally, it is difficult to study the effects of anthropogenic activities on marine mammals. The animals migrate over long distances and are exposed to various activities in different areas so that different effects are superimposed on

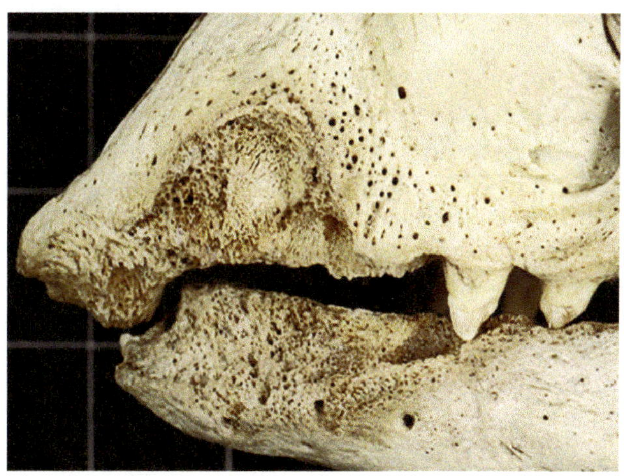

Fig. 14.6 A grey seal (*Halichoerus grypus*) skull showing loss of bone substance. Photo: © Ursula Siebert

Fig. 14.7 A harbour porpoise (*Phocoena phocoena*) accidentally caught by a fishing net. Photo: © German Oceanographic Museum

one another. Furthermore, access to marine mammals is limited; only in a few areas it is possible to capture live seals. These captures make it possible to study the seal health by examining the auditory, immune and endocrine systems, and seal movement, before the animals are again released to the wild. The number of harbour porpoises in the Baltic Sea proper (Fig. 14.7) is very low (*cf.* Fig. 4.15) and proper health assessment on live animals is not possible. Therefore, the investigations are restricted to dead animals found on beaches or accidentally caught in fishing nets (Siebert et al. 2001).

14.8 On-going developments in bioindication

Obviously, various indicators differ in their performance, and some species or groups used in bioindication will not be the ultimate choice for a specific assessment. For this reason, and also because of the ecosystem complexity of the Baltic Sea Area due to its strong environmental gradients, bioindication involving different groups of species is necessary in environmental assessments. This is also required by the EU directives, and new bioindicators for the Baltic Sea are continuously being developed (HELCOM 2012a, b, 2015a, b). In-depth knowledge of the ecological requirements, distributions and behaviours of potential indicator organisms is needed for selection of the most suitable indicators.

So far, it seems that a combination of physical, chemical and biological indication is the most appropriate approach to describe environmental changes caused by natural and/or anthropogenic pressures. The indicators are not only applicable to the detection and descriptions of changes in the ecosystem, but also to the evaluation of management effort effectiveness in the Baltic Sea Area. Additionally, a reliable status assessment necessitates sound monitoring strategies adjusted to the relevant habitat types and to spatially definable gradients of anthropogenic pressures (HELCOM 2015b).

Review questions

1. Why are prokaryotes seldom used as indicators of environmental change in the Baltic Sea?
2. What problems can occur in a eutrophication assessment when chlorophyll *a* concentration is used as the sole indicator?
3. Why are macrozoobenthic communities used so often in marine biotic assessment?
4. Why is *Zoarces viviparus* one of the best indicator species among fish in the Baltic Sea?
5. Why do marine mammals accumulate chemical pollutants?

Discussion questions

1. How might the role of bioindicators change in the future?
2. How is the ecosystem functioning linked with the indication capability of species and/or communities?
3. What effects could spatial and temporal scales have on the indication value of organisms?
4. What are the differences and similarities between terrestrial, freshwater and marine indicator organisms?
5. Is it possible to describe environmental change using only one taxonomic group in bioindication?

References

Bäcklin B, Eriksson ML, Olovsson M (2003) Histology of uterine leiomyoma and occurrence in relation to reproductive activity in the Baltic grey seal (*Halichoerus grypus*). Veterinary Pathology 40:175–180

Beineke A, Siebert U, Baumgärtner W (2010) Immunology of whales and dolphins. Veterinary Immunology and Immunopathology 133:81–94

Borja A, Miles A, Occipinti-Ambrogi A, Berg T (2009) Current status of macroinvertebrate methods used for assessing the quality of European marine waters: implementing the Water Framework Directive. Hydrobiologia 633:181–196

Boyer JN, Kelble CR, Ortner PB, Rudnick DT (2009) Phytoplankton bloom status: chlorophyll *a* biomass as an indicator of water quality condition in the southern estuaries of Florida, USA. Ecological Indicators 9:56–67

Darr A, Gogina M, Zettler ML (2014) Functional changes in benthic communities along a salinity gradient – a western Baltic case study. Journal of Sea Research 85:315–324

Diekmann R, Möllmann C (eds) (2010) Integrated ecosystem assessments of seven Baltic Sea areas covering the last three decades. ICES Cooperative Research Report 3021:90

EC (1992) Council Directive 92/43/EEC of 21 May 1992 on the conservation of natural habitats and of wild fauna and flora. Official Journal of the European Communities L 206:7–50

EU (2000) Directive 2000/60/EC of the European Parliament and of the Council of 23 October 2000 establishing a framework for Community action in the field of water policy. Official Journal of the European Communities L 327:1–72

EU (2005) Common implementation strategy for the Water Framework Directive (2000/60/EC). Overall approach to the classification of ecological status and ecological potential, produced by Working Group 2A, European Communities, Luxembourg. WFD CIS Guidance Document 13:1–47

EU (2006) Directive 2006/7/EC of the European Parliament and of the Council of 15 February 2006 concerning the management of bathing water quality and repealing Directive 76/160/EEC. Official Journal of the European Union L 64:37–51

EU (2008) Directive 2008/56/EC of the European Parliament and of the Council of 17 June 2008 establishing a framework for community action in the field of marine environmental policy (Marine Strategy Framework Directive). Official Journal of the European Union L 164:19–40

Falkowski PG, Wilson C (1992) Phytoplankton productivity in the North Pacific Ocean since 1900 and implications for absorption of anthropogenic CO_2. Nature 358:741–743

Falkowski PG, Wilson C (1993) Phytoplankton productivity? Nature 362:796

Finni T, Kononen K, Olsonen R, Wallström K (2001) The history of cyanobacterial blooms in the Baltic Sea. AMBIO 30:172–178

Fleischer D, Zettler ML (2009) An adjustment of benthic ecological quality assessment to effects of salinity. Marine Pollution Bulletin 58:351–357

Fleming V, Kaitala S (2006) Phytoplankton spring bloom intensity index for the Baltic Sea estimated for the years 1992 to 2004. Hydrobiologia 554:57–65

Fürhaupter K, Wilken H, Meyer T (2008) WRRL-Makrophytenmonitoring in Schleswig-Holstein, Teil B, Makrophyten, Äußere Küstengewässer. Untersuchungen im Auftrag des Landesamtes für Landwirtschaft, Umwelt und ländliche Räume Schleswig-Holstein, Kiel, 42 pp [in German]

Gercken J, Förlin L, Andersson J (2006) Developmental disorders in larvae of eelpout (*Zoarces viviparus*) from German and Swedish Baltic coastal waters. Marine Pollution Bulletin 53:497–507

Gercken J, Sordyl H (2002) Intersex in feral marine and freshwater fish from northeastern Germany. Marine Environmental Research 54:651–655

Gogina M, Glockzin M, Zettler ML (2010) Distribution of benthic macrofaunal communities in the western Baltic Sea with regard to near-bottom environmental parameters. 1. Causal analysis. Journal of Marine Systems 79:112–123

Greve TM, Krause-Jensen D (2005) Predictive modelling of eelgrass (*Zostera marina*) depth limits. Marine Biology 147:803–812

Hansson M, Håkansson B (2007) The Baltic Algae Watch System – a remote sensing application for monitoring cyanobacterial blooms in the Baltic Sea. Journal of Applied Remote Sensing 1(1):011507, 10 pp

Hedman JE, Rüdel H, Gercken J, Bergek S, Strand J et al (2011) Eelpout (*Zoarces viviparus*) in marine environmental monitoring. Marine Pollution Bulletin 62:2015–2029

Heink U, Kowarik I (2010) What are indicators? on the definition of indicators in ecology and environmental planning. Ecological Indicators 10:584–593

HELCOM (2009) Eutrophication in the Baltic Sea – an integrated thematic assessment of the effects of nutrient enrichment in the Baltic Sea region. Baltic Sea Environment Proceedings 115B: 1–148

HELCOM (2010) Hazardous substances in the Baltic Sea – an integrated thematic assessment of hazardous substances in the Baltic Sea. Baltic Sea Environment Proceedings 120B:1–116

HELCOM (2012a) Development of a set of core indicators: interim report of the HELCOM CORESET project. Part B: Descriptions of the indicators. Baltic Sea Environment Proceedings 129B:1–219

HELCOM (2012b) Indicator based assessment of coastal fish community status in the Baltic Sea 2005–2009. Baltic Sea Environment Proceedings 131:1–88

HELCOM (2015a) Strategy for future HELCOM assessments of inputs of nutrients and selected hazardous substances. Baltic Marine Environment Protection Commission, 4 pp [http://www.helcom.fi]

HELCOM (2015b) BALSAM Project 2013–2015: recommendations and guidelines for benthic habitat monitoring with method descriptions for two methods for monitoring of biotope and habitat extent. Baltic Marine Environment Protection Commission, 62 pp [http://www.helcom.fi]

Herlemann DPR, Labrenz M, Jürgens K, Bertilsson S, Waniek JJ, Andersson AF (2011) Transitions in bacterial communities along the 2,000 km salinity gradient of the Baltic Sea. The ISME Journal 5:1571–1579

Holmfeldt K, Dziallas C, Titelman J, Pohlmann K, Grossart HP et al (2009) Diversity and abundance of freshwater Actinobacteria along environmental gradients in the brackish northern Baltic Sea. Environmental Microbiology 11:2042–2054

Karr JR (1981) Assessment of biotic integrity using fish communities. Fisheries 6:21–27

Kautsky N, Kautsky H, Kautsky U, Wærn M (1986) Decreased depth penetration of *Fucus vesiculosus* L. since the 1940s indicates eutrophication of the Baltic Sea. Marine Ecology Progress Series 28: 1–8

Kotta J, Paalme T, Kersen P, Martin G, Herkul K et al (2008) Density-dependent growth of the red algae *Furcellaria lumbricalis* and *Coccotylus truncatus* in the West Estonian Archipelago Sea, northern Baltic Sea. Oceanologia 50:577–585

Labrenz M, Sintes E, Toetzke F, Zumsteg A, Herndl GJ et al (2010) Relevance of a crenarchaeotal subcluster related to *Candidatus Nitrosopumilus maritimus* to ammonia oxidation in the suboxic zone of the central Baltic Sea. The ISME Journal 4:1496–1508

Marbà N, Krause-Jensen D, Alcoverro T, Birk S, Pedersen A et al (2013) Diversity of European seagrass indicators: patterns within and across regions. Hydrobiologia 704:265–278

McCarty LS, Power M, Munkittrick KR (2002) Bioindicators *versus* biomarkers in ecological risk assessment. Human and Ecological Risk Assessment: An International Journal 8:159–164

Nixon SW (1995) Coastal marine eutrophication: a definition, social causes, and future concerns. Ophelia 41:199–219

OECD (2003) Core environmental indicators – development, measurement and use. Organisation for Economic Co-Operation Development (OECD), Paris 37 pp

OSPAR (1998) OSPAR Agreement 1998-18 Annex 1. OSPAR Strategy to Combat Eutrophication. [http://www.bmu.de/files/pdfs/allgemein/application/pdf/ospar_strategy3_eut.pdf]

Péru N, Dolédec S (2010) From compositional to functional biodiversity metrics in bioassessment: a case study using stream macroinvertebrate communities. Ecological Indicators 10:1025–1036

Pinhassi J, Winding A, Binnerup SJ, Zweifel UL, Riemann B et al (2003) Spatial variability in bacterioplankton community composition at the Skagerrak-Kattegat front. Marine Ecology Progress Series 255:1–13

Pörtner HO, Knust R (2007) Climate change affects marine fishes through the oxygen limitation of thermal tolerance. Science 315: 95–97

Rosenberg R, Blomqvist M, Nilsson HC, Cederwall H, Dimming A (2004) Marine quality assessment by use of benthic species-abundance distributions: a proposed new protocol within the European Union Water Framework Directive. Marine Pollution Bulletin 49:728–739

Sagert S, Krause-Jensen D, Henriksen P, Rieling T, Schubert H (2005) Integrated ecological assessment of Danish Baltic Sea coastal areas by means of phytoplankton and macrophytobenthos. Estuarine, Coastal and Shelf Science 63:109–118

Schladot JD, Backhaus F, Ostapczuk P, Emons H (1997) Eelpout (*Zoarces viviparus*) as a marine bioindicator. Chemosphere 34:2133–2142

Schubert H, Forster RM (1997) Sources of variability in the factors used for modelling primary productivity in eutrophic waters. Hydrobiologia 349:75–85

Selig U, Eggert A, Schories D, Schubert M, Blümel C et al (2007) Ecological classification of macroalgae and angiosperm communities of inner coastal waters in the southern Baltic Sea. Ecological Indicators 7:665–678

Siebert U, Wünschmann A, Weiss R, Frank H, Benke H et al (2001) Post-mortem findings in harbour porpoises (*Phocoena phocoena*) from the German North and Baltic Seas. Journal of Comparative Pathology 124:102–114

Thiel R, Winkler H, Urho L (1996) Zur Veränderung der Fischfauna. In: Lozan JL, Lampe W, Matthäus W, Rachor E, Rumohr H, von Westernhagen H (eds) Warnsignale aus der Ostsee: wissenschaftliche Fakten. Parey Buchverlag Berlin, pp 181–188 [in German]

Torn K, Krause-Jensen D, Martin G (2006) Present and past depth distribution of bladderwrack (*Fucus vesiculosus*) in the Baltic Sea. Aquatic Botany 84:53–62

Van Hoey G, Borja A, Birchenough S, Buhl-Mortensen L, Degraer S et al (2010) The use of benthic indicators in Europe: from the Water Framework Directive to the Marine Strategy Framework Directive. Marine Pollution Bulletin 60:2187–2196

Zettler ML, Proffitt CE, Darr A, Degraer S, Devriese L et al (2013) On the myths of indicater species: issues and further consideration in the use of static concepts for ecological applications. PLoS ONE 8 (10):e78219

Zettler ML, Schiedek D, Bobertz B (2007) Benthic biodiversity indices *versus* salinity gradient in the southern Baltic Sea. Marine Pollution Bulletin 55:258–270

Bio-optical water quality assessment

15

Susanne Kratzer, Piotr Kowalczuk, and Sławomir Sagan

Abstract

1. The colour of the sea, *i.e.* its spectral reflectance, depends on the absorbing and scattering properties of substances in the water.
2. The main optical in-water constituents are chlorophyll *a* (Chl *a*), coloured dissolved organic matter (CDOM) and suspended particulate matter (SPM).
3. Optical data can be obtained from sensors deployed into the water or by remote sensing imagers on aircrafts or satellites.
4. With remote sensing, the optical properties of large geographical areas can be surveyed with high temporal and spatial resolution.
5. Chl *a* can be used as a proxy of phytoplankton biomass, CDOM as a marker of terrestrial freshwater and decay processes of marine primary producers and SPM as an indicator of land runoff and wind-driven resuspension of sediments.
6. Remote sensing of Chl *a*, CDOM and SPM can assist in the evaluation of water quality, *e.g.* the state of eutrophication, the extent of freshwater runoff, the depth of the photic zone and the breadth of the coastal zone.
7. The bio-optical characteristics of the brackish Baltic Sea differ from those of other seas. Due to the large overall freshwater influence, CDOM is usually the dominant optical in-water constituent not only near river discharges, but also in the open waters of the Baltic Sea.
8. The CDOM concentrations in the open waters of the Baltic Sea are inversely related to the large-scale Baltic Sea salinity gradient, with CDOM absorption highest in the northern Baltic Sea and lowest in the southwestern Baltic Sea.
9. Due to the high CDOM absorption regional Baltic Sea algorithms are required to derive water quality parameters that can be used as indicators of ecosystem health.

Keywords
Chlorophyll *a* • Coloured dissolved organic matter • Environmental monitoring • Light absorption • Light reflectance • Light scattering • Optical in-water constituents • Remote sensing • Suspended particulate matter • Water quality

S. Kratzer (✉)
Department of Ecology, Environment and Plant Sciences,
Stockholm University, 10691 Stockholm, Sweden
e-mail: susanne.kratzer@su.se

P. Kowalczuk · S. Sagan
Institute of Oceanology, Polish Academy of Sciences,
ul. Powstancow Warszawy 55, 81712 Sopot, Poland
e-mail: piotr@iopan.gda.pl

S. Sagan
e-mail: sagan@iopan.gda.pl

P. Snoeijs-Leijonmalm et al. (eds.), *Biological Oceanography of the Baltic Sea*,
DOI 10.1007/978-94-007-0668-2_15

15.1 The colour of the sea and water quality

15.1.1 Bio-optics is the basis of marine remote sensing

Optical oceanography is a research area that was developed to derive information about biogeochemical processes and large-scale anthropogenic pressures from the optical properties of seawater (Dickey et al. 2006). These properties are revealed by the colour (the spectral reflectance) of the seawater, which can be observed by eye, by optical sensors deployed directly into the water and by remote optical sensors on aircrafts or satellites.

Remote sensing provides continuous data series for water quality monitoring of the entire Baltic Sea. This has significantly improved our understanding of the dynamics of the Baltic Sea ecosystem, *e.g.* of the spatial and temporal distributions of cyanobacterial blooms (Fig. 15.1). In the near future, remote sensing is likely to become even more important for environmental assessments as method development proceeds towards higher data resolution and as longer time series are becoming available.

The major advantage of remote sensing is that a large geographical area can be surveyed simultaneously with a relatively high spatial and temporal resolution (*i.e.* daily for open sea applications or every 2–3 days for coastal applications).

Fig. 15.1 The extension of a cyanobacterial bloom in the Baltic Sea proper on 13 July 2005, shown as an RGB composite image from the MEdium Resolution Imaging Spectrometer (MERIS). The red rectangle shows the location of Himmerfjärden, a north-south facing fjärd ~60 km south of the city of Stockholm (Sweden), which is shown in more detail in Figs. 15.4 and 15.12. Photo: Satellite image reprinted with permission from the European Space Agency (ESA)

Disadvantages include a high sensitivity to disturbances by weather-dependent phenomena such as clouds and waves, and limited penetration depth in the water to just below the Secchi depth.

15.1.2 Water quality parameters

Water quality is a general descriptor of the properties of water in terms of physical, chemical and/or biological characteristics. No single water quality parameter meets all user needs. For example, measures used to evaluate water quality for human consumption are completely different from those used to evaluate seawater quality for the assessment of ecosystem health. While laboratory tests are sufficient for drinking water quality, an ecosystem assessment requires the integration of laboratory testing and the quantification of biogeochemical processes on a large geographical scale.

The quality of seawater is affected by substances supplied to the sea from both point sources, such as wastewater treatment plants or oil spills, and from diffuse sources, such as agriculture and airborne pollution. Traditionally, the monitoring of seawater quality has relied on *in situ* sampling with generally low spatial resolution and low temporal frequency. However, systematic tools to monitor effects of eutrophication or land runoff at appropriate spatial and temporal scales are necessary to provide the scientific basis and practical guidelines for better resource management practices. The integrated use of *in situ* measurements, remote sensing data and GIS-based techniques will enable environmental managers to develop improved management strategies (*cf.* Sect. 18.5).

Major factors that affect the quality of seawater are nutrients, phytoplankton, macrophytes, dissolved organic matter (DOM), suspended particulate matter (SPM), hazardous substances, oil spills, thermal discharges and pathogens. Of these, phytoplankton pigments, coloured DOM (CDOM), SPM, oil spills and thermal discharges change the spectral composition of the reflected sunlight or the emitted thermal radiation from surface waters, and can therefore be measured by remote sensing. The wealth of satellite data from the world's ocean surfaces has improved our understanding of how ocean currents and other processes, *e.g.* the El Niño southern oscillation (Dijkstra 2006), regulate the global climate.

Most chemicals and pathogens do not directly change the spectral or thermal properties of surface waters, so those properties can only be inferred indirectly from measurements of other water quality parameters. For example, the dispersal of a specific chemical pollutant discharged by a river can be followed by observing the extension of the river plume into the sea using the SPM concentration as a marker. The major natural bio-optical constituents of seawater are chlorophyll *a* (Chl *a*), CDOM and SPM, which together are responsible for the colour of the sea. These three constituents can be measured both in water samples and from space, and their concentrations and relative abundances reveal information on water quality. This area of research is also called "bio-optics", and when combined with satellite data it is called "ocean colour remote sensing".

15.2 The optical properties of seawater

15.2.1 Inherent optical properties of seawater (IOPs)

Seawater possesses "inherent optical properties" (IOPs), which are properties of the seawater irrespective of the ambient light field (Preisendorfer 1961). The absorption coefficient (*a*), the scattering coefficient (*b*) and the volume scattering function (VSF), which summarises the angular distribution of the scattered light (Box 15.1, Equations 15.1–15.4), are fundamental IOPs that describe how a medium absorbs and scatters light.

The process of light absorption occurs when a photon interacts with water molecules or other substances in seawater and its energy is converted from electromagnetic energy to other forms of energy, such as chemical energy (in photosynthesis), fluorescence or heat. Light scattering, in contrast, is a process in which a photon collides with matter in the seawater and changes its direction but retains its energy. Usually, scattering is caused mainly by suspended particles, both living and non-living, in the water. Backscattered light is the light that is scattered at angles >90°. Backscatter is fundamental to remote sensing because most of the light leaving the ocean in an upward direction comes from sunlight that was originally travelling downwards, but which was redirected upwards. Due to scattering and absorption within the water, the spectral signature of the backscattered light changes, and it therefore contains information about the optical constituents in the water.

A specific volume of seawater, with all its optical constituents, has a spectral value for *a*, *b* and VSF, which can be measured *in situ* from a ship using special instruments. The beam attenuation coefficient of the light (*c*) is the sum of *a* and *b*, but it can also be measured in terms of energy loss of a parallel beam of light after travelling through a certain volume of water, *e.g.* in a transmissometer (Box 15.1, Equations 15.5–15.6).

Together, the IOPs reveal information about the types and concentrations of optical constituents in the natural seawater. Absorption and scattering are important for optical modelling since the reflectance (*R*), *i.e.* the colour of the sea with all information it contains, can be derived from a function of the ratio of light absorption and backscattering (Box 15.1, Equation 15.8b).

15.2.2 Apparent optical properties of seawater (AOPs)

The seawater also has "apparent optical properties" (AOPs), which depend on both the IOPs and the ambient light field (Preisendorfer 1961), *i.e.* on radiance and irradiance. Irradiance (*E*) is a measure of how much light energy impacts on a certain area. The SI unit of irradiance is Watts per square metre ($W{\cdot}m^{-2}$). Radiance (*L*) is a measure of how much light is emitted from a surface area and falls within a given solid angle in a specified direction. The SI unit of radiance is Watts per steradian (sr) per square metre ($W{\cdot}sr^{-1}{\cdot}m^{-2}$). Examples of AOPs are the diffuse attenuation of light (K_d) as well as reflectance (*R*).

The sunlight at the seawater surface is highly variable since it is influenced by many factors such as sun elevation, atmospheric properties and clouds. Therefore, individual measurements of *E* and *L* provide little information about the optical in-water constituents since their values depend mainly on the spectral properties of the sunlight during the measurement. However, ratios of *E* and *L* measured simultaneously are much less influenced by changes in the ambient light field. The remote sensing reflectance (R_{RS}) is the ratio of upwelling radiance (L_u) to downwelling irradiance (E_d) (Box 15.1, Equation 15.8a), and its SI unit is per steradian (sr^{-1}). The R_{RS} can be measured at sea using spectral radiometers. The light energy coming from the sea, *i.e.* the water-leaving radiance, is influenced by both the properties of the object it comes from (the sea) and the light energy impinging on the object (the sunlight). The E_d diminishes in an approximately exponential manner with depth (Box 15.1, Equation 15.7). The diffuse attenuation coefficient of light (K_d) is a measure of the rate of decay of light with depth, *i.e.* how much light is attenuated per metre depth: the higher K_d the more light is being attenuated.

Both K_d and R_{RS} are optical variables that can be measured from a ship at sea or from a satellite. The reflectance measured at sea can be compared to the reflectance derived from IOP modelling (Box 15.1, Equation 15.8b). Alternatively, the reflectance measured at sea can be compared to the R_{RS} measured by a radiometer on a satellite (at the top of the atmosphere), that has been corrected for the atmospheric contribution. This is a common way to validate R_{RS}, and if appropriate atmospheric models have been used to derive the radiance at the sea surface from the top of atmosphere radiance. Such validation is imperative for the quality assurance of satellite products.

K_d(PAR) refers to the diffuse attenuation of photosynthetically available radiation (PAR, 400–700 nm, Poole and Atkins 1929; Kirk 2011) and is often used to determine the light conditions for phytoplankton or phytobenthic growth. PAR is also called the "photosynthetically active radiation"

Box 15.1: Definitions of optical quantities

The absorption coefficient (*a*)

The total absorption coefficient (a_{tot}) of light of wavelength λ in natural seawater is the sum of the absorption coefficient of water itself (a_w) and the absorption coefficients of all optical in-water constituents, *i.e.* that of coloured dissolved organic matter (CDOM), phytoplankton pigments (ph) and non-algal particles (NAP).

$$a_{tot}(\lambda) = a_w(\lambda) + a_{CDOM}(\lambda) + a_{ph}(\lambda) + a_{NAP}(\lambda) \quad (15.1)$$

The total scattering coefficient (b_{tot})

The total scattering coefficient (b_{tot}) of light of wavelength λ in natural seawater is the sum of the scattering coefficient of water itself (b_w) and the scattering coefficient of suspended organic and inorganic particles (b_p).

$$b_{tot}(\lambda) = b_w(\lambda) + b_p(\lambda) \quad (15.2)$$

The volume scattering function (VSF)

Scattering is described by the volume scattering function [VSF = $\beta(\lambda, \theta, \varphi, \theta', \varphi')$], which is the ratio between the radiance (*dL*) of wavelength λ scattered in the direction of the scattering angles (θ', φ') from radiation flux in the propagation direction (θ, φ) and the irradiance (*dE*) of wavelength λ that illuminates a volume of water (*dV*).

$$\beta(\lambda, \theta, \varphi, \theta', \varphi') = dL(\lambda, \theta, \varphi, \theta', \varphi') \cdot (dE(\lambda) \cdot dV)^{-1} \quad (15.3)$$

The scattering coefficient (*b*) at wavelength λ is an integral of the VSF from all directions, and quantifies the process of scattering with $d\omega$ defined as an infinitesimal solid angle.

$$b(\lambda) = \int_0^{4\pi} \beta(\lambda) d\omega \quad (15.4)$$

Both the scattering coefficient (*b*) and the VSF are dependent on particle concentration, particle size distribution and refractive indices. The total scattering coefficient spectra are related to the concentration of suspended matter and can be measured with optical sensors.

The beam attenuation coefficient of light (c)

The beam attenuation coefficient [$c(\lambda)$] is the sum of the absorption coefficient [$a(\lambda)$] and the scattering coefficient [$b(\lambda)$].

$$c(\lambda) = a(\lambda) + b(\lambda) \quad (15.5)$$

The beam attenuation coefficient quantifies the loss of radiance (of a parallel beam of light) due to passage through a homogenous water body per unit distance (*dr*). It can be calculated from the difference in radiance *L* of wavelength λ over the distance *r* (Jerlov 1976; Kirk 2011).

$$dL(\lambda) \cdot (dr)^{-1} = -c(\lambda) \cdot L(\lambda) \quad c(\lambda) = -dL(\lambda) \cdot [dr \cdot L(\lambda)]^{-1} \quad (15.6)$$

The diffuse attenuation coefficient of light (K_d)

The down-welling irradiance (E_d) of diffuse light diminishes in an exponential manner with depth: $E_d(z, \lambda) = E_d(0, \lambda) \cdot e^{-Kd(\lambda) \cdot z}$, where $E_d(0, \lambda)$ and $E_d(z, \lambda)$ are the values of down-welling irradiance just below the surface and at water depth *z*, respectively (Kirk 2011). The diffuse attenuation coefficient (K_d) can therefore be expressed as:

$$K_d(z, \lambda) = -[\ln E_d(z, \lambda) - \ln E_d(0, \lambda)] \cdot z^{-1} \quad (15.7)$$

K_d is the average value of the diffuse attenuation coefficient for the down-welling light field over any defined depth interval. K_d is wavelength-dependent and K_d (PAR) is widely used since it specifies the energy available to plants for photosynthesis.

The surface water reflectance (*R*)

The remote sensing reflectance (R_{RS}) can be defined as the ratio of upwelling radiance (L_u) and down-welling irradiance (E_d).

$$R_{RS} = L_u \cdot E_d^{-1} \quad (15.8a)$$

The irradiance reflectance ($R = E_u * E_d^{-1}$) is related to the ratio of backscattering to absorption coefficient in the following way:

$$R \approx f \cdot b_b \cdot (a)^{-1} \quad (15.8b)$$

Note that the factor *f* is dependent on the optical constituents in the water and strongly dependent on the illumination conditions. Equation 15.8b does not account for multiple scattering. However, Equation 15.8b is a reasonable assumption for the Baltic Sea, which is optically dominated by CDOM absorption and one may therefore assume single scattering.

and is defined as the photon flux density (in µmol photons m^{-2} s^{-1}) within the wavelength range 400–700 nm. PAR is a key parameter for biological and ecological studies since it specifies the energy available to algae and plants for photosynthesis. This spectral region corresponds also more or less to the range of light visible to the human eye, called the "visible spectrum", *i.e.* it covers the colours we can perceive.

K_d(PAR) is strongly correlated with K_d(490), *i.e.* the diffuse attenuation coefficient at 490 nm, which can be measured from space (Kratzer et al. 2003; Darecki and Stramski 2004) and therefore can be mapped over wide areas. The correlation between K_d(PAR) and K_d(490) differs locally since it is highly dependent on the CDOM absorption, and thus the relative amount of CDOM, in the water. K_d(490) can also be measured directly in the water with a radiometer that measures E_d(490) at different depths (Box 15.1, Equation 15.7). In the northwestern Baltic Sea proper, K_d(490) varies between ~0.4 m^{-1} in the open waters and ~0.9 m^{-1} in coastal waters (Kratzer et al. 2008). The diffuse attenuation coefficient at 490 nm for pure seawater (*i.e.* with no other constituents than the seawater itself) K_w(490), which is 0.022 m^{-1} (Smith and Baker 1981), is used to correct K_d(490) measurements when modelling the diffuse attenuation of all optical in-water constituents (*cf.* Fig. 15.14).

15.2.3 The main optical constituents of seawater

Pure water absorbs at long wavelengths, in the red part of the electromagnetic spectrum (Fig. 15.2), and the backscatter of water increases towards the blue wavelengths. Therefore, water with little or no other constituents appears blue. In addition to the water itself, the three main optical constituents of natural seawaters are phytoplankton pigments, coloured dissolved organic matter (CDOM), and suspended particulate matter (SPM). These constituents play a substantial role in the biogeochemistry of natural waters and are important for their optical properties. For example, they determine the diffuse attenuation of light (*K*) and thereby water transparency, which subsequently influences

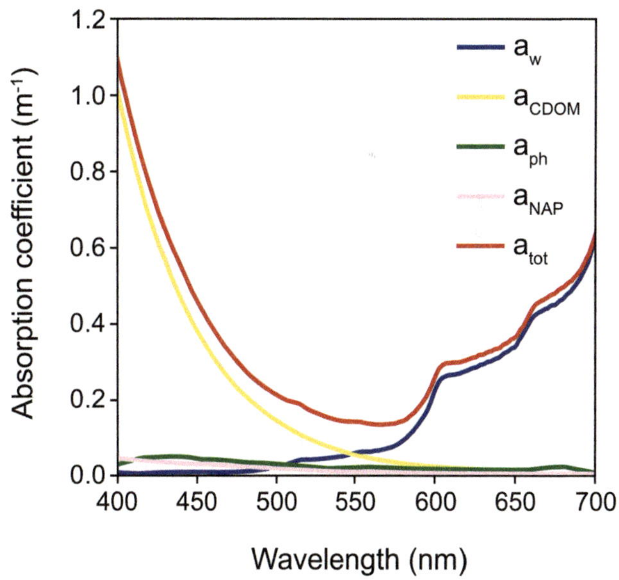

Fig. 15.2 Spectrophotometric absorption of seawater and its main constituents. a_w = absorption coefficient of pure water (data from Pope and Fry 1997), a_{CDOM} = CDOM absorption coefficient (data from Schwarz et al. 2002), a_{ph} = absorption coefficient of phytoplankton pigments (data from Kratzer 2000), a_{NAP} = absorption coefficient of non-algal particles a_{tot} = total absorption coefficient (data from Kratzer 2000). The absorption of CDOM is strong in the blue part of the spectrum and the absorption of pure water is strong in the red part. Chlorophyll peaks are found in the blue and the red parts, while the shoulder around 570 nm is caused by the cyanobacterial pigment phycoerythrin. The a_{ph} and a_{tot} were measured in Baltic Sea surface water sampled ~4 km north of the island of Gotland on 5 August 1998. Figure modified from Kratzer (2000)

phytoplankton primary production. Since all three constituents – phytoplankton pigments, CDOM and SPM – absorb in the blue part of the spectrum, it is not easy to differentiate between them. However, it is possible to estimate their respective concentrations from remote sensing data by using their absorption and scattering properties. Their absolute and relative contributions to spectral K_d provide information about the processes in the water and about water quality.

15.2.4 Optical Case-1 and Case-2 waters

The optical signal of clear ocean waters is determined only by water itself, phytoplankton biomass, and phytoplankton-correlated CDOM. Such waters are classified as "optical Case-1 waters" (Morel and Prieur 1977). Their Chl *a* concentration can be estimated from space with high accuracy by using simple reflectance band ratios. In contrast, "optical Case-2 waters" are more complex since also CDOM and SPM from terrestrial sources contribute to the optical signal and are independent from phytoplankton biomass. This is the case in most coastal waters, but the entire Baltic Sea also falls within optical Case-2 waters. The Baltic Sea is optically complex since its CDOM absorption is much higher than in other seas due to the large freshwater runoff it receives from land and its semi-enclosed geographical position (Kowalczuk 1999; Siegel et al. 1999; Kowalczuk et al. 2006).

15.2.5 Chlorophyll *a*

In remote sensing, the concentration of the pigment chlorophyll *a* (Chl *a*) in seawater is used as a proxy of phytoplankton biomass (*cf.* Sect. 8.2). A rise in biomass can be a sign of eutrophication: increased fluvial or coastal anthropogenic nutrient inputs may lead to increases in the biomass of phytoplankton and attached algae and plants. Chl *a* is used since it is universal for all groups of primary producers in the sea, which additionally contain different combinations of chlorophylls *b* and *c*, carotenoids and phycobilins (Jeffrey et al. 1997). In nature, the Chl *a* concentration in seawater varies over four orders of magnitude between ~0.01 μg L^{-1} in extremely nutrient-poor oceans to ~100 μg L^{-1} in a dense phytoplankton bloom. Average values for the surface waters of the Baltic Sea proper are ~3.3 μg L^{-1} in spring, ~2.5 μg L^{-1} in summer, ~2.1 μg L^{-1} in autumn and ~0.5 μg L^{-1} in winter, with large variations within each season (Kratzer and Tett 2009).

Using a spectrophotometer, two Chl *a* absorption peaks can be measured. When measured on a filter with living phytoplankton from the Baltic Sea, a major peak occurs in the blue part of the electromagnetic spectrum at 443 nm and a smaller peak in the red part at 677 nm (Kratzer 2000). For purified Chl *a* dissolved in 100 % acetone, the wavelengths are slightly lower (430 and 662 nm, respectively) due to the solvent effect (Jeffrey et al. 1997). The lowest absorption of most phytoplankton pigments is in the green part of the spectrum (Fig. 15.2), which explains the green colour of seawater with a high phytoplankton biomass.

Remote sensing algorithms for the retrieval of Chl *a* are based on empirical relationships between the Chl *a* concentration in the water and ratios of water-leaving radiance or remote sensing reflectance (R_{RS}) at different wavelengths, *e.g.* the ratio of blue to green reflectance at 440 and 550 nm, respectively. However, this is only valid in clear ocean waters where Chl *a* is the dominant optical in-water constituent and not in the Baltic Sea or in coastal waters where allochthonous CDOM and/or SPM also affect the optical properties of the water. Apart from absorbing light by their pigments, phytoplankton cells also scatter light, which adds to the total scattering coefficient of the seawater (Box 15.1, Equation 15.2).

15.2.6 Coloured dissolved organic matter (CDOM)

In bio-optical research, coloured dissolved organic matter (CDOM) is used as a marker of terrestrial freshwater input and decay processes of marine primary producers. CDOM in the near-shore areas originates predominantly from riverine runoff and diffuse land drainage. CDOM is the part of dissolved organic matter (DOM) that is capable of absorbing light.

CDOM colours the water yellowish and is also known as "chromophoric DOM", "gilvin" or "yellow substance" (*cf.* Sect. 1.3.3). CDOM consists of a complex mixture of organic polymers with light-absorbing aromatic rings, originating from the decay of photosynthetically produced organic matter such as fulvic and humic acids derived from lignins, which originate in terrestrial plant detritus and other organic substances leaking from decomposing littoral vegetation or phytoplankton blooms. Although variations in CDOM are primarily the result of natural processes, anthropogenic activities such as logging, agriculture, effluent discharge, and drainage of wetlands can affect the CDOM levels in coastal waters.

CDOM strongly absorbs in the blue spectral region and its absorption decreases logarithmically with increasing wavelength (Fig. 15.2). However, it does not contribute appreciably to light scattering (Kirk 2011). The CDOM content of seawater is usually measured by filtering away particulate matter with a 0.2 μm pore-size filter and measuring the absorbance of the filtrate. The absorbance (abs) of CDOM is often measured in the ultraviolet region (*e.g.* 350

or 380 nm), but in remote sensing studies it is usually measured in the blue at 440 nm where the maximum Chl *a* absorption is also found. CDOM absorption is mostly expressed in terms of the absorption coefficient at 440 nm (g_{440}) which is derived from the absorbance and converted to absorption as follows: $g_{440} = \ln(10) \cdot [(abs_{440} - abs_{750})\, l^{-1}]$ m^{-1}, where l is the length of the spectrophotometer cuvette in m; usually a 10 cm cuvette is used (Kirk 2011).

In natural waters, the CDOM absorption measured at 440 nm ranges from close to zero to 10 m^{-1} (Schwarz et al. 2002). In the southern Baltic Sea, values of up to 6 m^{-1} have been measured (Kowalczuk et al. 2005). In the northeastern Baltic Sea proper, even higher values have been recorded in river mouths and organic matter-rich shallow bays, such as the Pärnu Bay in the Gulf of Riga (Toming et al. 2009). The concentrations are much lower in the northwestern Baltic Sea proper, with average summer values for the surface waters of ~0.40 m^{-1} in the open sea and ~0.48 m^{-1} in coastal waters, with large local variations (Kratzer and Tett 2009).

15.2.7 Suspended particulate matter (SPM)

In ocean colour remote sensing, the concentration of suspended particulate matter (SPM) is used as an indicator of physical forcing. Land runoff, as well as wind-driven resuspension of sediments, leads to high SPM loads in coastal areas. The amount of SPM in the water can be used to distinguish between coastal and open sea waters, and thus to determine the breadth of the coastal zone (Kratzer and Tett 2009).

SPM is usually divided into suspended particulate organic matter (SPOM) and suspended particulate inorganic matter (SPIM). SPOM comprises a complex mixture of living and non-living organic matter with a broad size range. For example, bacteria, phyto- and zooplankton, faecal pellets and organic detritus in the water are all included in SPOM. Resuspended sediments in the water column often account for the bulk of SPIM, in addition to inorganic components of organisms such as diatom silica frustules and the calcium carbonate plates (coccoliths) of coccolithophorids. The organic and inorganic fractions of SPM are tightly intertwined, *e.g.* sediment and clay particles usually have a thin organic coating of bacteria to which small diatoms may attach and organic toxins may adhere. The optical properties of SPOM are similar to those of CDOM, with high absorption in the blue part of the spectrum and an exponential decline towards the red part (Fig. 15.2). SPIM scatters light and increases the backscatter and therefore the reflectance from the sea. Hence, the RGB (red-green-blue) composite images of SPIM appear brighter.

The SPM content of the water is usually expressed as the weight of the SPM in $g \cdot m^{-3}$. All the particulate matter in a natural water sample is collected on a pre-weighted combusted filter, dried and weighed to obtain the weight of total SPM, after which the SPOM is combusted and the filter is reweighed to obtain the weight of SPIM and SPOM (Strickland and Parsons 1972). In natural seawater, the SPM load can vary between 0 and ~60 $g \cdot m^{-3}$, with the highest values in shallow river deltas (Miller and McKee 2004). Average values for the surface waters of the Baltic Sea proper in summer are ~0.95 $g \cdot m^{-3}$ in the open sea and ~1.43 $g \cdot m^{-3}$ in coastal waters, with large local variations in coastal waters (Kratzer and Tett 2009).

15.2.8 The Secchi depth

The Secchi disk (Fig. 15.3a) is one of the oldest instruments used in oceanography. It was invented by Father Pietro Angelo Secchi (1818–1878) to measure the water transparency of the Mediterranean Sea. A white circular disk of 30 cm diameter is lowered into the water until the observer loses sight of it, and the depth (in m) at which the disk disappears is the so-called "Secchi depth" (Tyler 1968). The Secchi depth provides information on the depth of the photic zone, which is relevant for primary production (Kratzer et al. 2003).

The Secchi depth is a measure of water transparency: the murkier and less clear the water, the lower the Secchi depth. Substances that reduce the clarity of natural seawater include phytoplankton, zooplankton (although in much lower concentrations than phytoplankton), CDOM and suspended particles such as clay and silt. However, the Secchi depth is not only affected by the concentrations of SPM, Chl *a* and CDOM absorption in the water, but also by the light conditions above the water, the reflection at the sea surface, and even by the observer's eye sensitivity. To eliminate a bias due to the reflectance at the sea surface influencing the observer's Secchi depth reading, a water telescope can be used (Fig. 15.3b).

Due to the high freshwater runoff from land, and limited water exchange with the North Sea (*cf.* Sect. 2.2), the Baltic Sea accumulates high loads of CDOM, and the background absorption of CDOM is relatively high compared to other seas. On the other hand, the SPM concentration is relatively low in the open Baltic Sea, especially when compared to tidal seas, as inorganic SPM settles out in river plumes, usually relatively close to the coast. However, the SPM load is highly variable in the coastal zone, and therefore strongly affects the Secchi depth (*cf.* Fig. 15.11). In the Baltic Sea, the Secchi depth can vary between a few metres during a dense algal and cyanobacterial blooms and 20 m in winter. In very clear waters, *e.g.* the Weddell Sea, the Sargasso Sea and the Mediterranean Sea, Secchi depth readings can amount to 50–80 m.

Fig. 15.3 Secchi disk measurements on board a research vessel. (**a**) A Secchi disk submerged in coastal water of the northern Baltic Sea. Note the yellowish colour of the water caused by CDOM. (**b**) A water telescope can be used to avoid the bias of the observer's Secchi disk reading caused by sea surface reflectance. The Secchi depth can only be measured in normal daylight. Photo: (**a**) © Annelies Hommersom, (**b**) © Susanne Kratzer

The Secchi depth (z_{SD}) is inversely related to K_d(PAR), and thus Secchi depth readings can be used to estimate K_d(PAR). The relationship suggested by Poole and Atkins (1929) is $K_d(\text{PAR}) = 1.7 \cdot z_{SD}^{-1}$, whereas Holmes (1970) proposed $K_d(\text{PAR}) = 1.44 \cdot z_{SD}^{-1}$ for turbid waters. For the northwestern Baltic Sea proper, the factor of 1.7 works well (Kratzer et al. 2003), but when considering the entire Baltic Sea, the factor ranges from 1.7 to 2.3, depending on the local salinity. K_d(490) is often used in remote sensing and its relationship with Secchi depth in the northwestern Baltic Sea proper was established as $K_d(490) = 0.18 + 1.57 \cdot z_{SD}^{-1}$ (Kratzer and Tett 2009).

15.3 Remote sensing of the sea

15.3.1 Seawater changes the spectrum of the incoming sunlight

The development of remote sensing techniques for water quality monitoring began in the early 1970s (Ritchie et al. 2003). Ocean colour remote sensing builds on the principle that the relative composition of the water constituents of natural surface waters changes the absorption and backscattering characteristics of the water. Remote sensing techniques depend on the ability to measure these changes in the spectral signature of the light scattered back from inside a given water body. The changes measured are then related to a given water quality parameter using empirical or analytical models. The optimal wavelength used to measure a water quality parameter depends on the optical substance being measured, its optical signature and concentration, the other optical constituents in the water and the sensor characteristics.

Ocean colour remote sensing uses the visible and near-infrared (VIS/NIR) range of the electromagnetic spectrum (400–900 nm) and is carried out with passive radiometers attached to satellites or aircrafts. Light from the sun is scattered and absorbed on its way through the Earth's atmosphere. A small part of the radiant flux that reaches the sea surface is directly reflected and never enters the water. This process depends on the angle of the sun, *e.g.* with a sun angle of 40°, ~3 % of the solar energy is reflected while the remaining 97 % passes through the sea surface.

The light that passes the sea surface is refracted at the air/water interface, *i.e.* it changes direction. The transfer of electromagnetic radiation in water is described by the radiative transfer theory (Mobley 1994). Once in the water, the radiant flux is either absorbed or scattered by the optical components in the seawater down to about twice the Secchi depth, i.e. the 1 % light level. These processes change the spectral signature and the direction of the incoming light. The radiance that is sent back from the seawater into the atmosphere, *i.e.* the water-leaving radiance, now contains information about the optical water constituents and thus about water quality. On its way back through the atmosphere, the spectral signature is changed again before the radiance reaches the satellite's or aircraft's radiometer. The VIS/NIR signal measured by the radiometer carries information on both the optical properties of in-water constituents and of the atmosphere. The NIR channels of the radiometer are then used for atmospheric correction, whereas the visible channels are used to derive information about water quality. The radiance originating from the sea surface is generally only ~10 % of the top of atmosphere radiance, and can be even lower in the open Baltic Sea.

15.3.2 Spatial and temporal resolution of satellite sensors

Three ocean colour radiometers have so far provided decadal, widely used data on ocean colour: the Sea-viewing Wide Field-of-view Sensor (SeaWiFS, NASA, 1997–2010), the Moderate Resolution Imaging Spectroradiometer (MODIS, NASA,1999-still working), and the MEdium Resolution Imaging Spectrometer (MERIS, 2002–2012) on ENVISAT of the European Space Agency (ESA). MERIS ceased working in March 2012 and was on 16 February 2016 replaced by the Ocean Land Colour Instrument (OLCI) on Sentinel-3, which is optically similar to MERIS but measures more wavelengths.

The spatial resolution of ocean colour images is generally ~1 km, which is sufficient for most open-ocean applications. MODIS also has medium-resolution bands (250 and 500 m), which are designed for land applications. Until now, MERIS has provided the best spectral and radiometric resolution for ocean colour remote sensing. MERIS had a wide dynamic range, which means that it was capable of detecting the low signals reflected from dark water as well as bright reflectance from sea ice, clouds or land surfaces. Therefore, it was suitable for both aquatic and terrestrial remote sensing. The spatial resolution of MERIS was 300 m, and with its 15 narrow bands (10 nm wide) it also had a high spectral resolution. Therefore, MERIS was especially suitable for coastal applications (Fig. 15.4; Doerffer et al. 1999).

The temporal resolution of MERIS was 2–3 days, depending on latitude. This is not as good as the daily resolution of MODIS, but appropriate for observing *e.g.* the development of phytoplankton blooms (Harvey et al. 2015) and the extent and dynamics of river plumes, provided there are no clouds.

15.3.3 Algorithms and processors

In order to retrieve the concentrations of optical in-water constituents from spectral remote sensing data, it is necessary to use algorithms or specific processors. For clear ocean water one can simply use the blue to green reflectance ratio in order to reliably derive Chl *a* concentrations from space. However, this does not work in optically complex waters, such as the Baltic Sea, where other optical constituents have a strong influence on the spectral reflectance. Siegel and Gerth (2008) give an overview of the existing remote sensing algorithms for the Baltic Sea, but none of these are suited for the whole Baltic Sea. For the southern Baltic Sea proper, Darecki et al. (2003) suggested algorithms based on the reflectance ratios between the green and orange bands in order to compute Chl *a* concentrations from satellite data. Empirical regional algorithms for the estimation of the Chl *a* concentration have been further constructed with the help of an extensive database of *in situ* radiometric and chlorophyll measurements (Darecki and Stramski 2004; Darecki et al. 2005).

A software widely used for processing and displaying MERIS data is BEAM, a free software toolbox for satellite images (http://www.brockmann-consult.de/cms/web/beam). BEAM includes a set of options to use different MERIS processors that provide estimates of water constituents for optical Case-2 waters that fit the Baltic Sea (Doerffer and Schiller 2008). These processors do not just use simple band ratios to derive the Chl *a* concentration in optically complex waters, but use complex models based on radiative transfer theory and neural network inversion techniques (Doerffer and Schiller 2006; Schroeder et al. 2007). The processors are trained on empirical or simulated data for the retrieval of the three independently varying optical in-water constituents (Chl *a*, CDOM and SPM). Kratzer and Vinterhav (2010) validated the neural network developed by Schroeder et al. (2007) in the northwestern Baltic Sea proper in conjunction with the adjacency correction proposed by Santer and Zagolski (2009) and found that this combination of processors provides reliable and accurate measurements of Chl *a* concentration from MERIS data. The methods are now at a stage that they can be used in Baltic Sea management (Kratzer et al. 2014; Harvey et al. 2015).

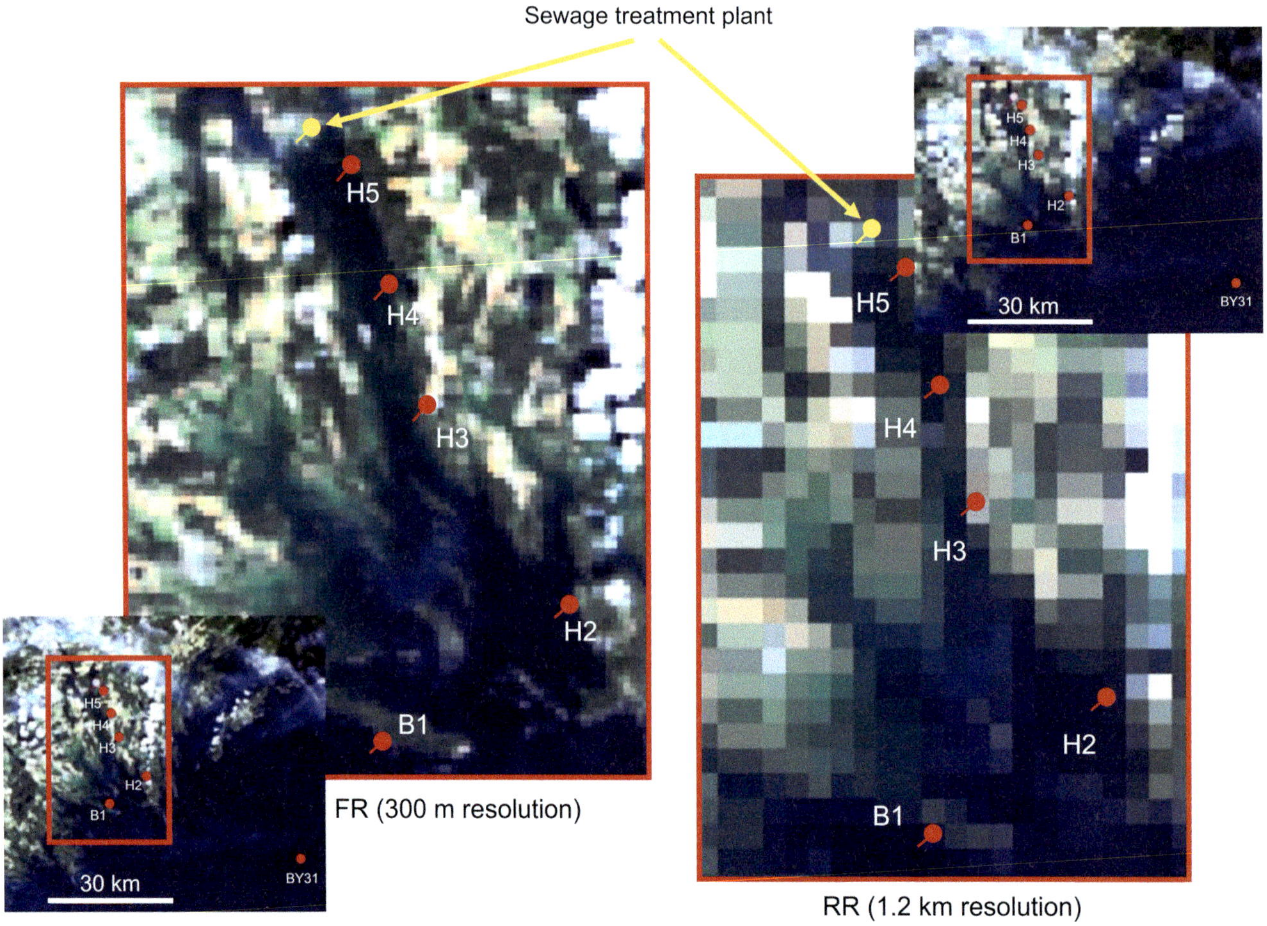

Fig. 15.4 RGB composite images from the MEdium Resolution Imaging Spectrometer (MERIS) on 19 August 2002 over Himmerfjärden, a north-south facing fjärd ~60 km south of the city of Stockholm (Sweden) in the northwestern Baltic Sea proper. The figure compares full resolution (FR, 300 m) with reduced resolution (RR, 1.2 km) and demonstrates that the FR data are more suitable for analysing coastal bays. The location of Himmerfjärden is shown in Fig. 15.1. Stations B1 and H2-H5 are sampling stations of the Swedish monitoring programme for Himmerfjärden. Station BY31 is a Swedish/HELCOM monitoring station located at the deepest place in the Baltic Sea, the Landsort deep (459 m). Figure reprinted from Kratzer and Vinterhav (2010) with permission from Oceanologia (Institute of Oceanology PAN, Sopot, Poland)

15.3.4 Scattering and reflectance

Scattering and the volume scattering function (VSF, Box 15.1, Equations 15.3–15.4) are mostly determined by the shape, concentration and refractive index of particles in the water. Scattering is not homogeneous throughout the water column. In turbid waters just above the sea bottom, where the concentration of sinking and resuspended particles is highest, scattering can be up to five orders of magnitude higher than that in clear ocean surface water.

In pure seawater, the phase function (*i.e.* the scattering per angle) is almost flat with a noticeable minimum around 90°. This is due to the Rayleigh scattering caused by water molecules. The scatter of natural sea waters generally has a maximum at angles below 90°, which means that forward scattering light (downward propagated light) dominates (Fig. 15.5). However, in coastal waters with high SPM loads, the proportion of backward scattered light is higher than in clear ocean waters, which means that these waters reflect relatively more light. As a result, there is also a substantial change in the shape of the VSF's angular distribution with increasing turbidity.

The remote sensing reflectance of seawater measured close to the coastline (up to ~15 km distance) is usually influenced by high reflectance from land. Therefore, satellite data from water areas close to the coastline need to be corrected for these adjacency effects. Prototype algorithms to correct for adjacency effects have recently been developed, *e.g.* the Improved Contrast between Ocean and Land (ICOL) for the processing of MERIS data (Santer and Zagolski 2009).

15.3.5 Multi-scale optical monitoring

During the last two decades, multi-scale approaches have been developed for monitoring the Baltic Sea ecosystem.

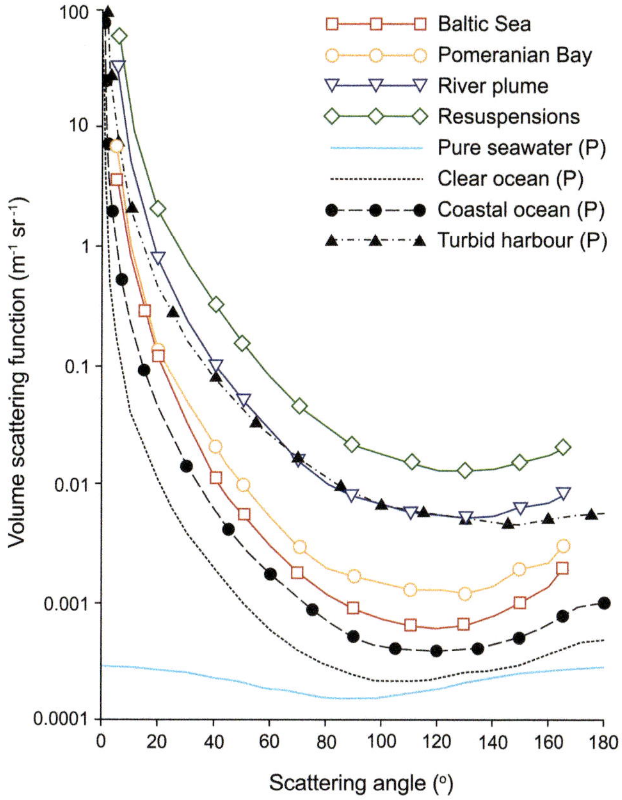

Fig. 15.5 Examples of the volume scattering function (VSF; Box 15.1, Equations 15.3–15.4) related to the scattering angle measured in optically varying water from the Baltic Sea (Siegel et al. 2005, coloured lines) compared to those measured by Petzold (1977, black lines indicated with "P") for different types of seawater. Baltic Sea = clear water from the open Baltic Sea proper, Pomeranian Bay = mixed clear and coastal water, River plume = turbid river plume water, Resuspensions = river water with resuspension due to wind mixing, Pure seawater = no other constituents than seawater itself. The figure shows that most of the scattering of light in seawater is in a <90° forward direction, *i.e.* within 90° of the direction of propagation of the incident light. It also illustrates that the VSF increases with turbidity and that the VSF of the Baltic Sea is usually higher than that of marine water. Figure modified from Siegel et al. (2005)

They combine optical data from ocean colour remote sensing of the entire Baltic Sea with autonomous *in situ* measurements on moorings, light-houses and ships of opportunity (*e.g.* ferries used in regular sea traffic). All data are then combined in models to interpret and validate the status of the ecosystem. Dedicated, so-called "sea-truthing" campaigns are carried out to validate the data and to further develop (local) algorithms. During these campaigns, optical measurements are made from ships and water samples are collected for detailed optical and biochemical analyses of seawater and its constituents. The simultaneously measured satellite data are interpreted using regional retrieval algorithms, and bio-optical parameters are derived.

Operational water quality monitoring systems for the open Baltic Sea include the assessment of the spatial and temporal distributions of cyanobacterial blooms by the Finnish Environment Institute (SYKE) and the Swedish Meteorological and Hydrological Institute (SMHI). Recently, an operational system for Swedish coastal waters has been set up (http://vattenkvalitet.se). This system maps the whole Swedish coast in terms of Chl *a*, CDOM, SPM and Secchi depth, and also includes water status classification based on the Secchi depth and Chl *a* concentrations according to the EU Water Framework Directive (*cf.* Sect. 17.8).

15.4 Optical characteristics of the Baltic Sea

15.4.1 The specific colour of the Baltic Sea

Standard algorithms for marine waters, based on the blue to green band ratio between 490 and 550 nm (*e.g.* from SeaWiFS or MODIS), tend to overestimate the concentration of Chl *a* in the Baltic Sea. Therefore, specific algorithms for the Baltic Sea need to be developed. The CDOM-rich Baltic Sea water absorbs strongly in the blue spectral region and the absorption decreases logarithmically with increasing wavelength (Fig. 15.2). The slope factor (exponent) of the logarithmic absorption curve for CDOM in the brackish Baltic Sea is relatively high (0.0193 ± 0.0024) compared to other marine areas (0.0165 ± 0.0035) (Schwarz et al. 2002). Because of the relatively high amounts of CDOM in the Baltic Sea, the CDOM absorption is still significant in the green spectral region around 550 nm (Fig. 15.2), which may explain the overestimation of Chl *a* when using standard algorithms. For the southern Baltic Sea proper, algorithms based on the reflectance ratios between the spectral bands at 550 nm and 590 nm were shown to be more appropriate (Darecki et al. 2003).

The open Baltic Sea appears much darker from space in the blue to green part of the spectrum than *e.g.* the North Sea. This is caused by the high CDOM levels and low SPM loads in the open Baltic Sea. CDOM has a high absorbance in the blue spectral region and does not noticeably scatter light. An exception to the dark colour of the Baltic Sea occurs during its typical cyanobacterial blooms in the open sea and the occurrence of inorganic sediment in coastal areas. The backscatter of both is higher than that of the seawater, and therefore shows an increase in reflectance (Fig. 15.1). Filamentous cyanobacteria contain gas vacuoles that can be used to regulate buoyancy and to position the cells closer to the water surface. These gas vacuoles reflect strongly and make the blooms more visible in satellite imagery. The image in Fig. 15.1 also demonstrates how cyanobacterial blooms act as visible tracers of the Baltic Sea counter-clockwise surface-water circulation (*cf.* Fig. 2.10). The meso-scale features of the cyanobacterial blooms also show horizontal eddies and fronts.

The Baltic Sea shows gradients in both CDOM and SPM. The CDOM concentration is inversely correlated with freshwater input: the lower the salinity the higher the CDOM concentration (Jerlov 1955), and this relationship is evident in the Baltic Sea (*cf.* Fig. 2.22). In the Bothnian Bay, with a very low salinity, the water has a brown colour due to the large volume of freshwater runoff originating from land areas with bogs, lakes and rivers surrounding the bay. However, it must be kept in mind that the slope of the linear relationship between salinity and CDOM is regional and differs between the subbasins of the Baltic Sea.

SPM loads are generally higher, and extend further offshore, in the coastal parts of the southern Baltic Sea compared to those in the north. This is due to the difference in coastal morphology with rockier coasts in the north and sandier coasts in the south (*cf.* Fig. 2.8). However, compared to coastal waters influenced by tidal action, the coastal waters of the Baltic Sea usually have a lower content of inorganic SPM. Inorganic suspended sediments cause high scattering and increase the reflection from the sea so that satellite images from coastal areas show a higher reflectance. The SPMconcentration is usuallylow because tides are virtuallyabsent in the Baltic Sea. CDOM isoptically dominant because of therestricted water exchange withthe North Sea.

15.4.2 Optical properties of the water column

Both the absorption coefficient (a, Box 15.1, Equation 15.1) and the beam attenuation coefficient (c, Box 15.1, Equations 15.5–15.6) are highest in the upper water layer, *i.e.* in the photic zone (Fig. 15.6). Deeper waters have lower a and c, and are thus more transparent. The main reason for the lower water transparency in the surface layers is the accumulation of phytoplankton, inducing particle scattering and absorption by CDOM, phytoplankton pigments and non-algal particles.

The seasonal variability of the vertical distribution of seawater optical properties is governed by the occurrence of a thermocline (*cf.* Sect. 2.4.3). In the Gdańsk deep in the southern Baltic Sea, the thermocline occurs from April to September (Fig. 15.7). The stratification of the water column forms a density boundary at the thermocline that prevents mixing of the upper and lower water layers. When there is no thermal stratification, in the southern Baltic Sea from October to March, the vertical distributions of a and c are almost homogenous down to a ~70 m water depth (Fig. 15.8).

At ~70 m, another physical boundary, the permanent halocline, occurs in the deeper parts of the Baltic Sea proper with denser, more saline water in the bottom layer (*cf.* Table 2.6). This physical boundary is so strong that particles sinking from the water column above accumulate here and produce a scattering maximum. Therefore, this layer may also show a maximum in beam attenuation (Fig. 15.8).

Below the permanent halocline, scattering remains much higher than in the upper 70 m due to the resuspension of sediment particles from the seafloor by currents. Absorption is only slightly raised at this depth, but most so in the spectral region between 412 nm and 555 nm. Below the permanent halocline, the absorption in these blue to green spectral bands increases steadily down to the seafloor. This is caused by an increase in CDOM concentration due to the accumulation of decomposed particles sinking from the photic zone that were able to cross the density boundary (Kowalczuk et al. 2010; Skoog et al. 2011).

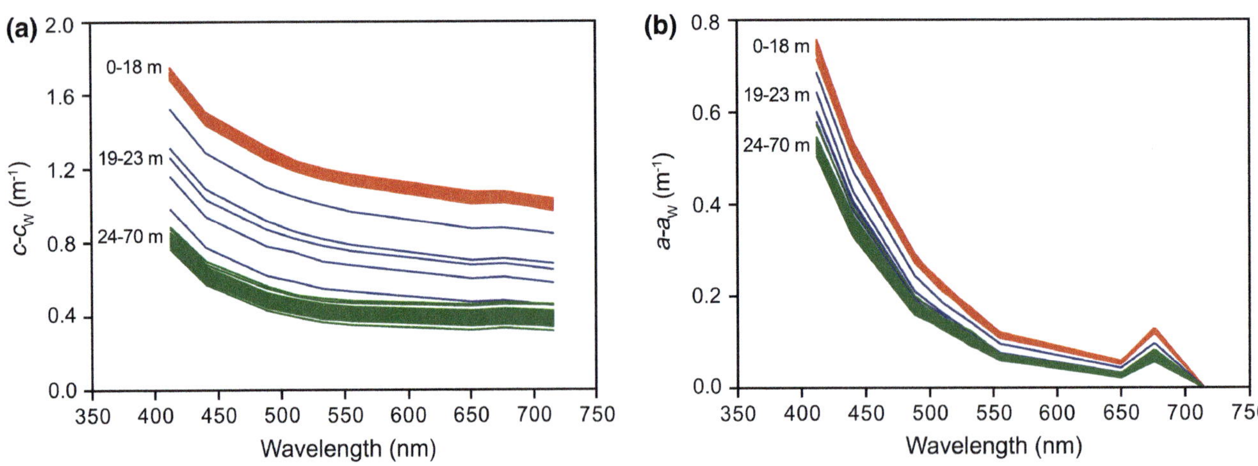

Fig. 15.6 Spectral distributions in different water layers in the Gdańsk deep (southern Baltic Sea proper) just after the onset of thermal stratification on 10 April 2003. (**a**) The beam attenuation coefficient c. (**b**) The absorption coefficient a. Measurements were made for 1 m depth bins, Red = surface layer, 0–18 m, blue = thermocline, 19–23 m, green = deep water, 24–70 m. The values of a and c are corrected for the properties of water itself. Figure modified from Sagan (2008)

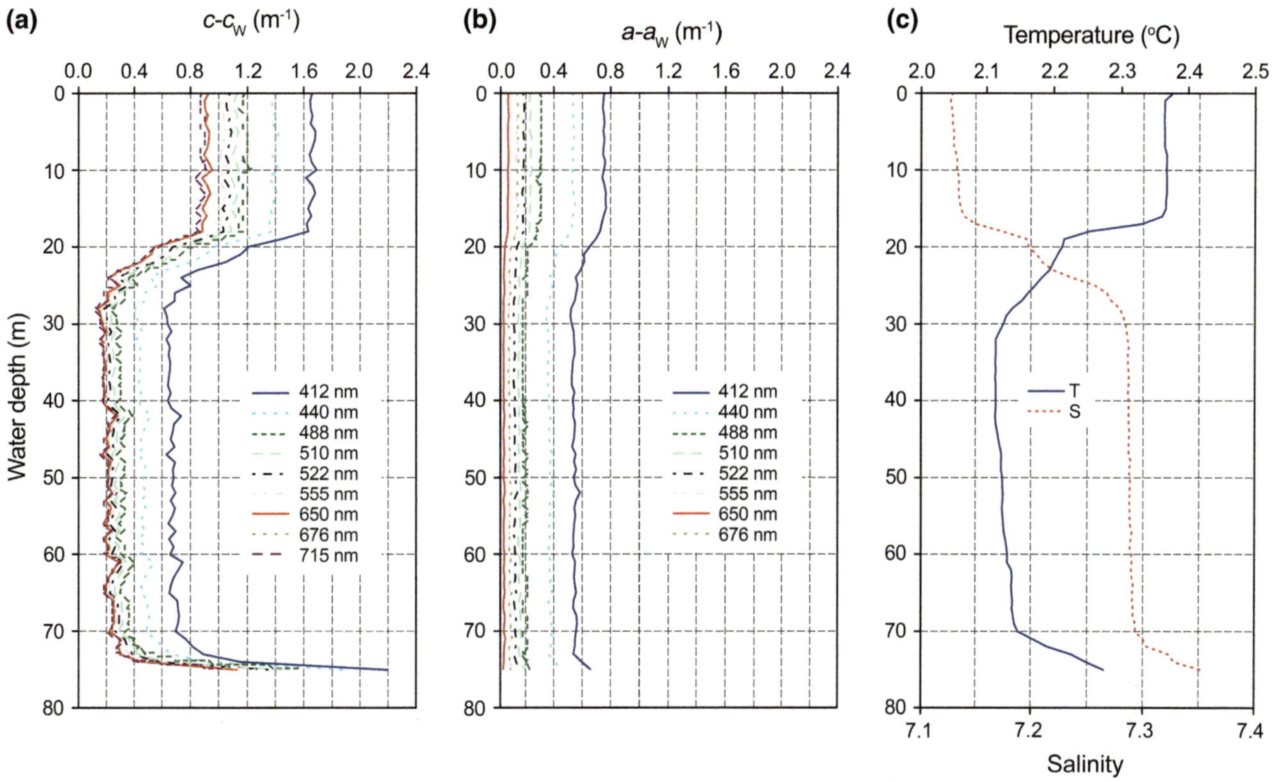

Fig. 15.7 Depth profiles measured in the Gdańsk deep (southern Baltic Sea proper) on 10 April 2003 (just after the onset of thermal stratification). (**a**) The beam attenuation coefficient c measured at different wavelengths. (**b**) The absorption coefficient a measured at different wavelengths. (**c**) Temperature and salinity. The values of a and c are corrected for the properties of water itself (a_w and c_w). Figure modified from Sagan (2008)

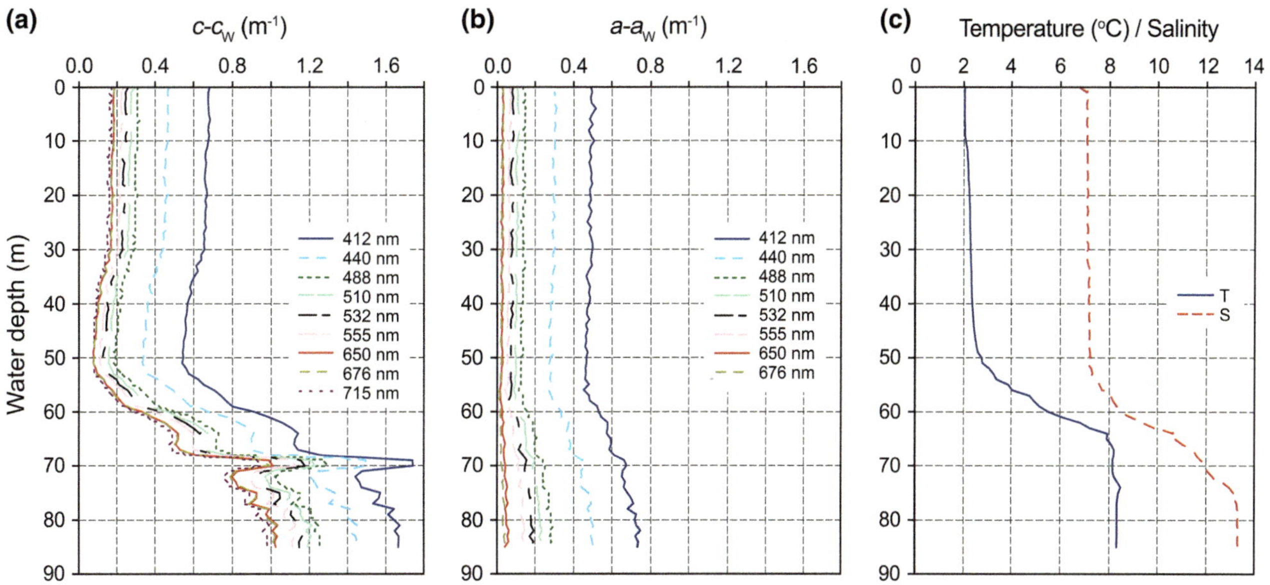

Fig. 15.8 Depth profiles measured in the Gdańsk deep (southern Baltic Sea proper) on 5 February 2003 (before the onset of thermal stratification). (**a**) The beam attenuation coefficient c measured at different wavelengths. (**b**) The absorption coefficient a measured at different wavelengths. (**c**) Salinity and temperature, showing an "inverse thermocline", *i.e.* water temperature increases below the halocline. The values of a and c are corrected for the properties of water itself (a_w and c_w). Figure modified from Sagan (2008)

15.4.3 Optical properties and biogeochemistry

Information on the biogeochemistry of an ecosystem that can be derived from optical measurements in the whole water column is extensive, and the measurements can be performed relatively quickly. However, vertical measurements still need to be made from ships since satellite remote sensing records only processes in the surface waters (down to ~ 10 % light level or approximately just below the Secchi depth). A compilation of 186 vertical depth profiles of the beam attenuation coefficient (Box 15.1, Equations 15.5–15.6) provides a comprehensive illustration of biogeochemical dynamic processes in the open Baltic Sea proper (Fig. 15.9). The highest beam attenuation, both in the warm and the cold season, occurs above 30 m and below 55 m. The beam attenuation can therefore be used to quantify primary production in the warm season, and decomposition in the cold season.

The most optically homogenous layer of the Baltic Sea proper (similar in all seasons) is found at the depth of 30–55 m. Here, hydrographical conditions are relatively stable. This layer is usually situated below the maximum extent of the seasonal thermocline, and above the permanent halocline. It is separated from the biological, physical and chemical processes that take place in the surface waters and close to the seafloor. Any variability in the optical properties of this intermediate layer is caused by vertical mixing across the seasonal thermocline and the subsequent sinking of surface waters in late autumn, or by upwelling events in summer.

15.5 Optical monitoring of the Baltic Sea

15.5.1 Eutrophication and Secchi depth

Historical Secchi depth data, collected since the beginning of the 20th century (Aarup 2002), make it possible to follow the long-term eutrophication trends in the Baltic Sea. In the Baltic Sea proper, a decrease in the Secchi depth of 5 cm per year was recorded between the two periods: 1914–1939 and 1969–1991 in both spring and summer (Sandén and Håkansson 1996). This decreased water transparency is not only a measure of increased primary production. In spring,

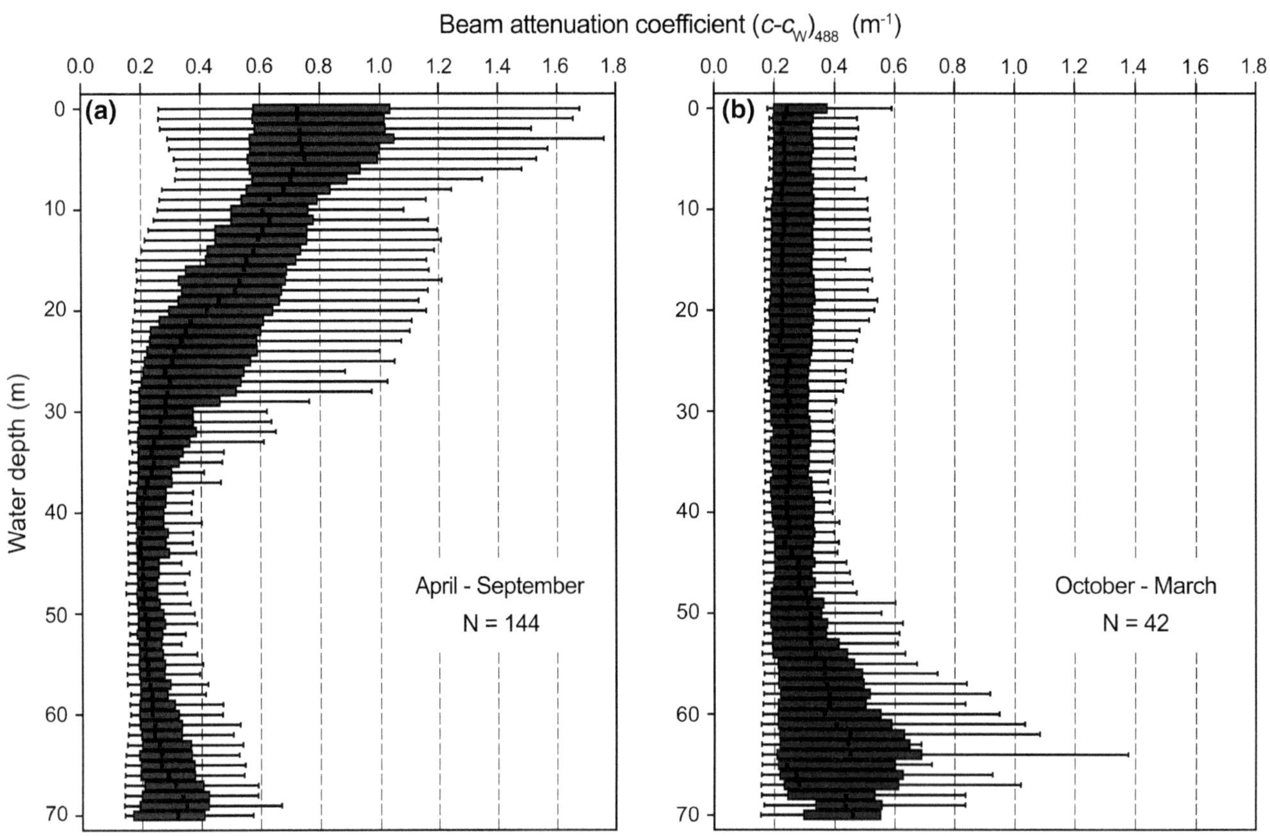

Fig. 15.9 Average vertical distributions of the beam attenuation coefficient at 488 nm in the open Baltic Sea water at 186 sampling stations in 1999 and 2003–2005. (**a**) April-September. (**b**) October-March. Measurements were made for 1 m depth bins and the values of c are corrected for the properties of water itself (c_w). The data are presented as box-and-whisker plots with median (central lines), quartiles (red and blue boxes) and non-outlying minima and maxima (whiskers). Figure modified from Sagan (2008)

31–42 % of the light attenuation in the surface layers of the Baltic Sea is caused by phytoplankton Chl *a*, compared to only 13–17 % in summer (Fleming-Lehtinen and Laamanen 2012).

These percentages are based on the comparison of the *in situ* Chl *a* concentration at a certain Secchi depth in the Baltic Sea and the Chl *a* concentration in optical Case-1 waters at the same Secchi depth modelled by the equation: Chl $a_{\text{Case-1}} = 109.24 \cdot e^{-0.3181 \cdot z}$, where z is the Secchi depth. This comparison assumed that the Secchi depth in optical Case-1 water is determined by Chl *a* alone, while in the Baltic Sea it is affected by Chl *a* and CDOM absorption as well as by SPM scatter and absorption. From this it is clear that the Secchi depth in the Baltic Sea cannot be linked to phytoplankton biomass alone. However, it can be used as an integrative water quality indicator that describes a combination of eutrophication-related characteristics.

An ecosystem-wide assessment comparing the Secchi depths between 1905–1909 and 2005–2009 showed that the Secchi depth decreased by 3–4 m in the northern part of the Baltic Sea (the Bothnian Bay, Bothnian Sea, Gulf of Finland and the northern Baltic Sea proper) (Fig. 15.10). In the southern and central part (the Arkona Sea, Bornholm Sea, Eastern and Western Gotland Seas), it decreased by 1–2 m. In the two southernmost basins, the Arkona Sea and the Bornholm Sea, the water transparency has increased during the last two decades, which indicates improved water quality, but in the northern areas the decrease has continued. Another recent study (Dupont and Aksnes 2013) found that the centennial decrease in the Secchi depth in the Baltic Sea was larger (~5.8 m) for the sea areas deeper than 100 m compared to those shallower than 100 m (~3.2 m).

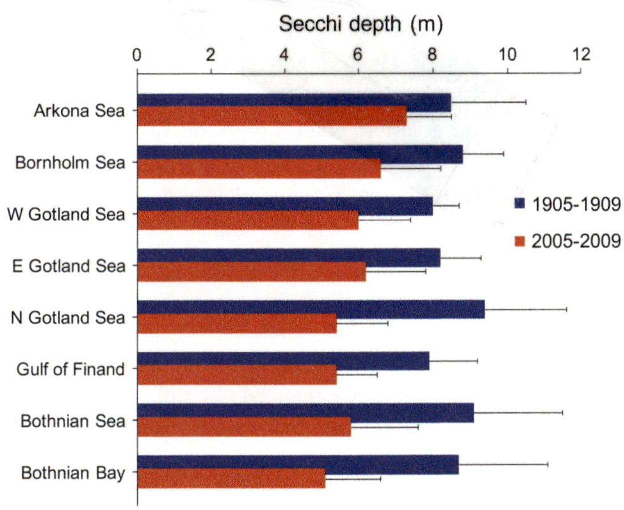

Fig. 15.10 Secchi depths in summer (June-September) in different parts of the Baltic Sea recorded in 1905–1909 and 2005–2009. Figure based on Secchi depth data in Fleming-Lehtinen and Laamanen (2012).

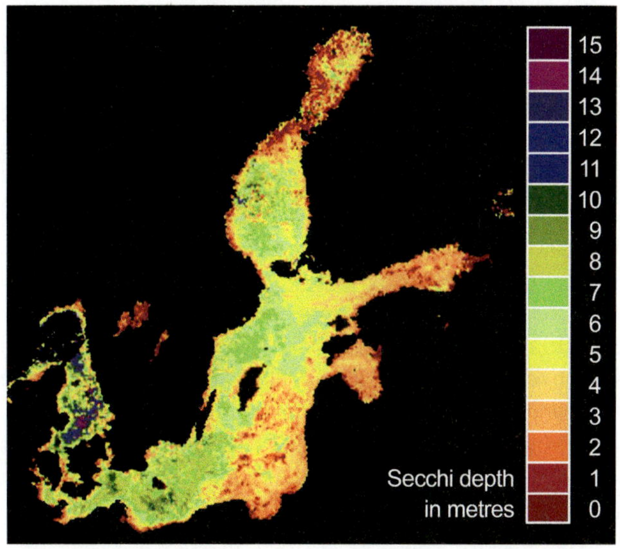

Fig. 15.11 Secchi depth map of the Baltic Sea simulated from a $K_d(490)$ composite image derived from SeaWiFS data from the last week of July 1999, using an in-water algorithm derived from a field campaign in 2001 (Kratzer et al. 2003). Resolution 1 km. Figure processed by Miho Ishii and printed with permission from the National Aeronautics and Space Administration (NASA), USA

Since organisms are affected by their optical environment in many different ways (Johnsen 2012), there are also indirect effects of eutrophication that are related to changes in the light field. Examples of this effect include a change in the ratio of photosynthetic and photoprotective pigments in phytoplankton due to reduced water transparency and reduced visibility for visual predators due to higher turbidity, which can have consequences for the food web dynamics.

15.5.2 Monitoring eutrophication from space

The Secchi depth can be derived from satellite data since it is inversely related to the diffuse attenuation coefficient (K_d, Box 15.1, Equation 15.7) measured by the satellite sensor. However, good algorithms and thorough validation with *in situ* Secchi depth measurements are required to obtain reliable information. For example, a $K_d(490)$ composite image derived from SeaWiFS was used to compose a Secchi depth map with 1 km resolution for the entire Baltic Sea (Fig. 15.11; Kratzer et al. 2003). This map is based on *in situ* correlations between $K_d(490)$ and the Secchi depth measured from a ship in the northwestern Baltic Sea proper, which were subsequently applied to SeaWiFS data. Validation of the map was carried out with Secchi depth data from the open Baltic Sea proper southwest of the island of Gotland and the Gulf of Riga, which showed good agreement.

In ocean colour remote sensing, it is common to derive the spectral diffuse attenuation coefficient $K_d(490)$ from the reflectance ratio at 490 nm and 550 nm (Mueller 2000). The band at 490 nm is used because it contains information about all optical in-water constituents, *i.e.* phytoplankton pigments as well as CDOM and SPM. In contrast, the band at 550 nm is least affected by these three main optical in-water constituents. The $K_d(490)$ has previously been shown to be the most reliable product that can be derived from remote sensing imagery over the Baltic Sea using MERIS or MODIS data (Darecki and Stramski 2004). It can be converted into K_d(PAR), but the relationship between the two parameters is regionally variable since it is influenced by CDOM absorption. For the Baltic Sea, this relationship was established as $K_d(\mathrm{PAR}) = 0.6677 \cdot K_d(490)^{0.6763}$ (Pierson et al. 2008).

The $K_d(490)$ can also be used in coastal waters as a collective measure of all three optical in-water constituents: Chl *a*, CDOM and SPM. An example of this is the $K_d(490)$ image of a coastal area in the northwestern Baltic Sea proper derived from MERIS data with 300 m resolution using a locally derived algorithm (Fig. 15.12). Such synoptic information can be provided by satellites only, as satellite data have improved spatial and temporal resolution compared to ship-based monitoring.

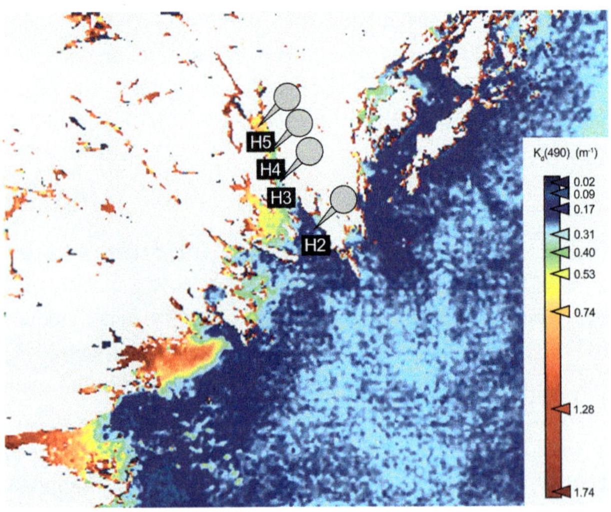

Fig. 15.12 Image of the diffuse attenuation coefficient $K_d(490)$ derived from an in-water algorithm applied to satellite data from the MEdium Resolution Imaging Spectrometer (MERIS) on 19 August 2002. The image shows Himmerfjärden, a north-south facing fjärd ~60 km south of the city of Stockholm (Sweden) in the northwestern Baltic Sea proper on 22 August 2002. Resolution 300 m. Stations H2-H5 are sampling stations of the Swedish monitoring programme for Himmerfjärden (*cf.* Fig. 15.4). Figure reprinted from Kratzer et al. (2008), with permission from Remote Sensing of Environment (Elsevier)

15.5.3 Bio-optical assessment of riverine runoff

While Chl *a* is the main indicator of increased primary production, CDOM and SPM are the main indicators of riverine runoff. CDOM is often used as an indicator of terrestrial freshwater input and decay processes of marine primary producers. Suspended particulate matter (SPM) is placed in suspension by tidal or wind-wave stirring of shallow seabeds, and is therefore an indicator of physical forcing.

The magnitude of riverine runoff and its effects on water quality can be assessed by separating the in-water constituents optically using different types of algorithms (Lewin et al. 2013). In a comprehensive field study (Sagan 2008), the contributions of the different optical in-water constituents to the beam attenuation coefficient (*c*, Box 15.1, Equations 15.5–15.6) were estimated for the open waters of the Baltic Sea proper, the Gdańsk Bay and the Pomeranian Bay. The latter two areas receive discharge from the rivers Wisła and Odra, respectively (Fig. 2.11). The values for *c* in the coastal waters were, on average, almost twice as high as those in the open Baltic Sea proper (Fig. 15.13a, c). This difference is caused by a higher CDOM absorption and higher concentrations of phytoplankton and suspended particles in coastal waters due to riverine discharges and land runoff. This results in higher absorption and scattering. The typical increase of *c* in the blue spectral region is caused by the overlapping absorption spectra of CDOM phytoplankton pigments and non-algal particles (NAP).

Particle scatter contributed most to *c* in both the open Baltic Sea water and in the coastal waters (Fig. 15.13b, d). The second largest contributor in the blue part of the spectrum was the absorption by CDOM, and phytoplankton absorption at the chlorophyll peaks in the blue and red parts. The relative contribution of particle scatter to *c* increased from the blue to the red part of the spectrum, in contrast to the contributions of CDOM and NAP absorption, which decreased with increasing wavelength.

Although the relative contribution of CDOM to *c* is usually highest close to estuaries, since it is related to the extent of freshwater input, it was similar in the open sea and coastal waters, ~30 % at 412 nm and ~20 % at 440 nm (Fig. 15.13b, d), respectively, in the southern Baltic Sea. This illustrates once again the high CDOM absorption in the open waters of the Baltic Sea. The freshwater discharged by the rivers contained high concentrations of inorganic nitrogen and phosphorus, which enhanced phytoplankton growth, and consequently absorption by the phytoplankton pigments (a_{ph}). The primary absorption peak of phytoplankton pigments at 443 nm was masked by the peaks of CDOM and NAP, but the secondary peak at 676 nm in the red part of the spectrum was distinct and contributed to *c* with ~20 % in the open sea and ~26 % in coastal waters. Due to a higher

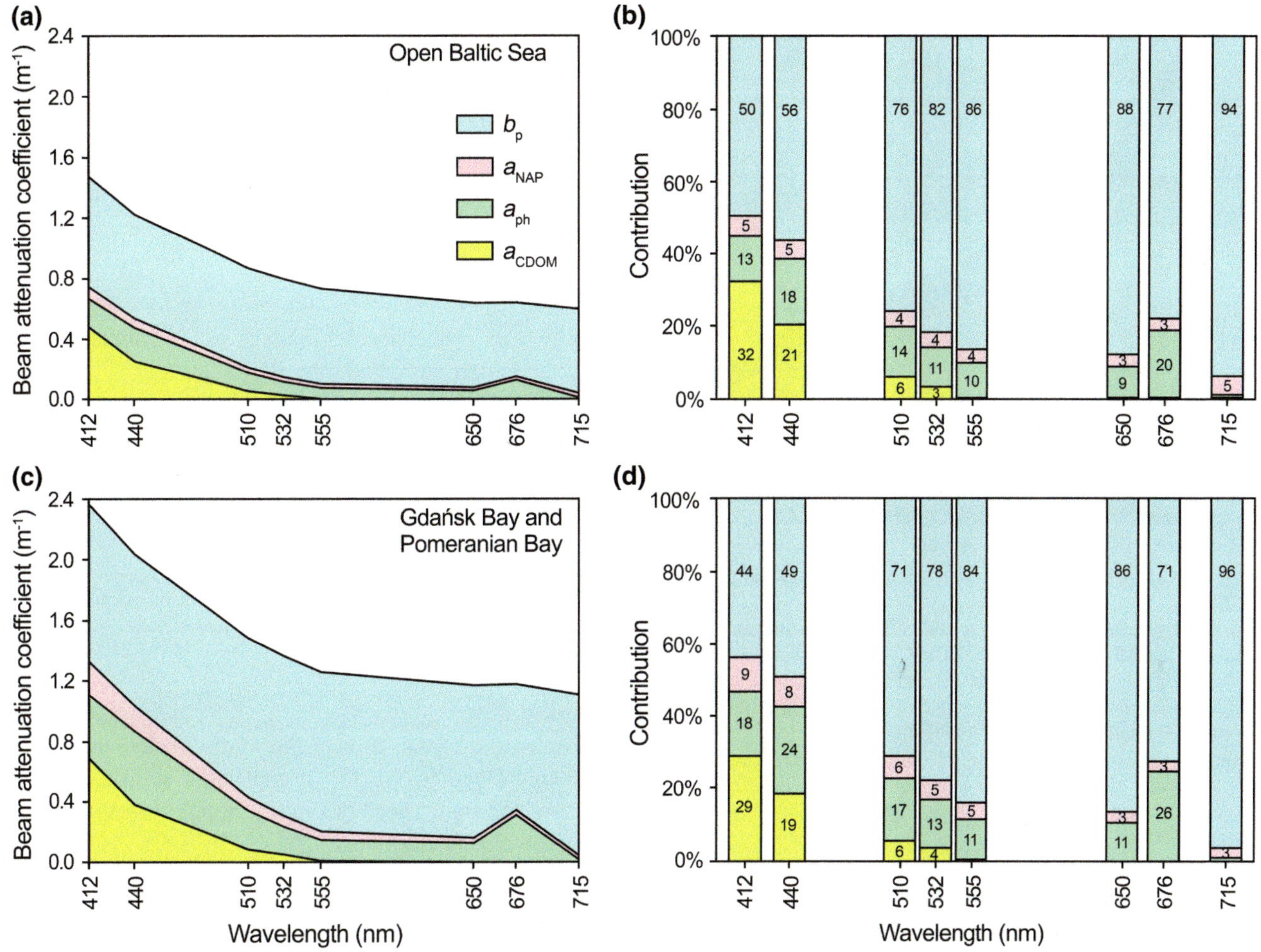

Fig. 15.13 Contributions of different optical in-water constituents to the beam attenuation coefficient (c) at different wavelengths. Measurements were made with an AC-9 *in situ* spectrophotometer in the field. The values of c are corrected for the attenuation by water itself. (**a**) Absolute contributions in the open Baltic Sea proper. (**b**) Relative contributions in the open Baltic Sea proper. (**c**) Absolute contributions in the Gdańsk Bay and the Pomeranian Bay. (**d**) Relative contributions in the Gdańsk Bay and the Pomeranian Bay. b_p = the scattering coefficient of suspended biological and mineral particles, a_{NAP} = the absorption coefficient of non-algal particles, a_{ph} = the absorption coefficient of phytoplankton pigments, a_{CDOM} = the absorption coefficient of coloured dissolved organic matter. Numbers in the bars denote the % contribution. The graphs show averages of measurements made in April–September 1999–2003. Figure modified from Sagan (2008)

concentration of particles in the coastal zone, the contribution of absorption by NAP to c in the blue part was higher in coastal waters (up to ~9 %) than in the open Baltic Sea.

15.5.4 Optical monitoring of the coastal zone

The extent of terrestrial runoff can be estimated, both on spatial and temporal scales, using remote sensing data. For example, remote sensing data can be used in environmental monitoring to assess water quality and eutrophication, and also for the estimation of the breadth of the coastal zone, or for predicting distributions of hazardous substances that originate from fluvial discharges or point sources.

A one-dimensional in-water attenuation model was developed for surface waters in the northwestern Baltic Sea proper to assess changes in water quality along a gradient from a wastewater treatment plant point source at the head of Himmerfjärden to the Landsort deep, the deepest part of the Baltic Sea (Kratzer and Tett 2009). This model estimates the contribution of each optical in-water constituent to the diffuse attenuation of light (K_d, Box 15.1, Equation 15.7). The $K_d(490)$ decreases from the source on the coast to the sink in the open sea (Fig. 15.14). The model is based on optical in-water measurements and shows that CDOM is optically dominant both in the coastal and in the open Baltic Sea stations, with a steady increase towards the inner Himmerfjärden. Chl *a* also contributes significantly to the attenuation in both the bay and the open sea waters, with slightly decreasing values away from land.

Inorganic SPM is optically important within Himmerfjärden, and influences the attenuation up to a horizontal

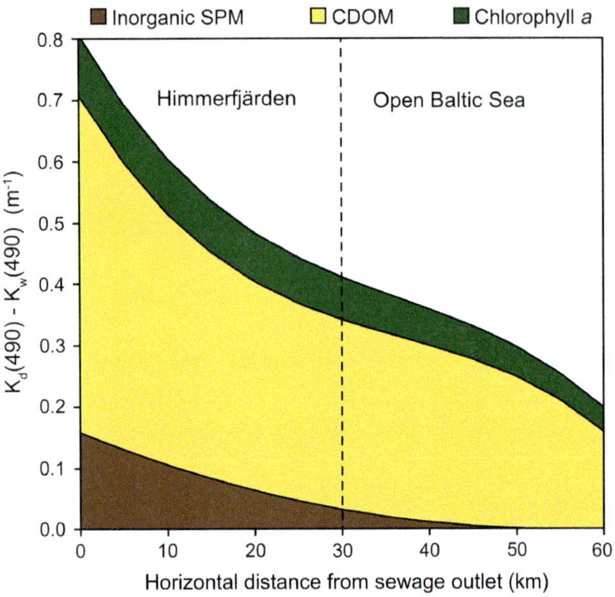

Fig. 15.14 The contributions of the main optical in-water constituents (Chl *a*, CDOM and inorganic SPM) to the diffuse attenuation coefficient $K_d(490)$ along a transect from the outlet of the Himmerfjärden wastewater treatment plant to the Landsort deep (the deepest part of the Baltic Sea, 459 m). $K_d(490)$ was corrected for $K_w(490)$, the attenuation of water itself. The black line indicates the end of Himmerfjärden and the beginning of the open sea. The location of Himmerfjärden is shown in Fig. 15.1. Figure modified from Kratzer and Tett (2009)

distance of ~40–45 km from the wastewater treatment plant, *i.e.* ~15–20 km offshore, where a limit of ~0.05 g·m^{-3} is reached. This means that, using inorganic SPM, the breadth of the coastal zone can be estimated as extending for tens of km, which also coincides with the dimension of coastal upwelling. From this point of view, it would therefore be desirable to extend the breadth of the coastal zone to a distance away from the shore further than that defined in the EU WFD (*cf.* Sect. 17.8). The WFD requires that coastal waters of the European Union be maintained at high or good ecological quality status. The directive defines the coastal zone as reaching from the land to "a distance of one nautical mile (1.85 km) on the seaward side from the nearest point of the baseline from which the breadth of territorial waters is measured, extending where appropriate up to the outer limit of transitional waters". Borja (2005) estimated that, with this definition, the WFD covers only ~20 % of the continental shelf areas in Europe.

Review questions

1. What is K_d(PAR) and how can it be estimated from the Secchi depth?
2. What are the main optical in-water constituents of clear ocean waters?
3. What are the optical properties of the Baltic Sea (open Baltic Sea *versus* coastal waters)?
4. Why are cyanobacterial blooms easily visible on satellite images?
5. What is the difference between inherent and apparent optical properties?

Discussion questions

1. Why can cyanobacterial blooms act as indicators of surface-water currents in the Baltic Sea?
2. Why is the Baltic Sea relatively dark in remote sensing compared to other seas?
3. Which single bio-optical parameter would you choose to define the breadth of the coastal zone and why?
4. Why is it important to monitor biogeochemical processes in the Baltic Sea continuously?
5. How can the remote sensing techniques be improved for better use in ecosystem management?

Teaching film

"The science of ocean colour" (46 min) directed and filmed by Roland Doerffer (GKSS); it contains a section on "The colour of the Baltic Sea" (13 min), (http://www.spicosa.eu/setnet/downloads/index.htm)

References

Aarup T (2002) Transparency of the North Sea and Baltic Sea – a Secchi depth data mining study. Oceanologia 44:323–337

Borja A (2005) The European water framework directive: a challenge for near-shore, coastal and continental shelf research. Continental Shelf Research 25:1768–1783

Darecki M, Kaczmarek S, Olszewski J (2005) SeaWiFS chlorophyll algorithms for the southern Baltic. International Journal of Remote Sensing 26:247–260

Darecki M, Stramski D (2004) An evaluation of MODIS and SeaWiFS bio-optical algorithms in the Baltic Sea. Remote Sensing of Environment 89:326–350

Darecki M, Weeks A, Sagan S, Kowalczuk P, Kaczmarek S (2003) Optical characteristics of two contrasting Case-2 waters and their influence on remote sensing algorithms. Continental Shelf Research 23:237–250

Dickey T, Lewis M, Chang G (2006) Optical oceanography: recent advances and future directions using global remote sensing and *in situ* observations. Reviews of Geophysics 44(RG1001):1–39

Dijkstra HA (2006) The ENSO phenomenon: theory and mechanisms. Advances in Geosciences 6:3–15

Doerffer R, Schiller H (2006) The MERIS neural network algorithm. In: Lee ZP (ed) Remote sensing of inherent optical properties: fundamentals, tests of algorithms, and applications. International Ocean-Colour Coordinating Group, Report 5:43–47

Doerffer R, Schiller H (2008) MERIS Regional Coastal and Lake Case-2 Water Project atmospheric correction ATBD (Algorithm Theoretical Basis Document) 1.0, 41 pp

Doerffer R, Sorensen K, Aiken J (1999) MERIS potential for coastal zone application. International Journal of Remote Sensing 20:1809–1818

Dupont N, Aksnes DL (2013) Centennial changes in water clarity of the Baltic Sea and the North Sea. Estuarine, Coastal and Shelf Science 131:282–289

Fleming-Lehtinen V, Laamanen M (2012) Long-term changes in Secchi depth and the role of phytoplankton in explaining light attenuation in the Baltic Sea. Estuarine, Coastal and Shelf Science 102–103:1–10

Harvey ET, Kratzer S, Philipson P (2015) Satellite-based water quality monitoring for improved spatial and temporal retrieval of chlorophyll *a* in coastal waters. Remote Sensing Environment 158:417–430

Holmes RW (1970) The Secchi disk in turbid coastal zones. Limnology and Oceanography 15:688–694

Jeffrey SW, Mantoura RFC, Wright SW (eds) (1997) Phytoplankton pigments in oceanography. Unesco Publishing, Paris 661 pp

Jerlov NG (1955) Factors influencing the transparency of the Baltic waters. Reports of the Oceanographic Institute Gothenburg, University of Gothenburg 25:1–19

Jerlov NG (1976) Marine optics. Elsevier, New York NY, 231 pp

Johnsen S (2012) The optics of life: a biologist's guide to light in nature. Princeton University Press, Princeton, 360 pp

Kirk JTO (2011) Light and photosynthesis in aquatic ecosystems, Third edn Cambridge University Press, Cambridge, 649 pp

Kowalczuk P (1999) Seasonal variability of yellow substance absorption in the surface layer of the Baltic Sea. Journal of Geophysical Research: Oceans 104:30047–30058

Kowalczuk P, Olszewski J, Darecki M, Kaczmarek S (2005) Empirical relationships between Coloured Dissolved Organic Matter (CDOM) absorption and apparent optical properties in Baltic Sea waters. International Journal of Remote Sensing 26:345–370

Kowalczuk P, Stedmon CA, Markager S (2006) Modelling absorption by CDOM in the Baltic Sea from season, salinity and chlorophyll. Marine Chemistry 101:1–11

Kowalczuk P, Zabłocka M, Sagan S, Kuliński K (2010) Fluorescence measured *in situ* as a proxy of CDOM absorption and DOC concentration in the Baltic Sea. Oceanologia 55:171–196

Kratzer S (2000) Bio-optical studies of coastal waters. School of Ocean Sciences, University of Wales, Bangor, 190 pp [PhD Thesis]

Kratzer S, Brockmann C, Moore G (2008) Using MERIS full resolution data (300 m spatial resolution) to monitor coastal waters – a case study from Himmerfjärden, a fjord-like bay in the northwestern Baltic Sea. Remote Sensing of Environment 112:2284–2300

Kratzer S, Håkansson B, Sahlin C (2003) Assessing Secchi and photic zone depth in the Baltic Sea from space. AMBIO 32:577–585

Kratzer S, Harvey T, Philipson P (2014) The use of ocean colour remote sensing in integrated coastal zone management – a case study from Himmerfjärden, Sweden. Marine Policy 43:29–39

Kratzer S, Tett P (2009) Using bio-optics to investigate the extent of coastal waters – a Swedish case study. Hydrobiologia 629:169–186

Kratzer S, Vinterhav C (2010) Improvement of MERIS data in Baltic Sea coastal areas by applying the Improved Contrast between Ocean and Land processor (ICOL). Oceanologia 52:211–223

Lewin I, Darecki M, Sagan S, Radomylskaya T (2013) Relationships between inherent optical properties in the Baltic Sea for application to the underwater imaging problem. Oceanologia 55:11–26

Miller RL, McKee BA (2004) Using MODIS Terra 250 m imagery to map concentrations of total suspended matter in coastal waters. Remote Sensing of Environment 93:259–266

Mobley CD (1994) Light and water – radiative transfer in natural waters. Academic Press, San Diego, 592 pp

Morel A, Prieur L (1977) Analysis of variations in ocean color. Limnology and Oceanography 22:709–722

Mueller JL (2000) SeaWiFS algorithm for the diffuse attenuation coefficient, K (490), using water-leaving radiances at 490 and 555 nm. In: Hooker SB, Firestone ER (eds) SeaWiFS post-launch calibration and validation analyses, Part 3. NASA GSFC, Greenbelt, MD, pp 24–27

Petzold TJ (1977) Volume scattering functions for selected ocean waters. In: Tyler JE (ed) Light in the sea. Dowden, Hutchinson and Ross, Stroudsberg, pp 152–174

Pierson D, Kratzer S, Strömbeck N, Håkansson B (2008) Relationship between the attenuation of down-welling irradiance at 490 nm with the attenuation of PAR (400 nm–700 nm) in the Baltic Sea. Remote Sensing of Environment 112:668–680

Poole HH, Atkins WRG (1929) Photoelectric measures of submarine illumination throughout the year. Journal of Marine Biological Association UK 16:297–324

Pope RM, Fry ES (1997) Absorption spectrum (380-700 nm) of pure water. II. Integrating cavity measurements. Applied Optics 36:8710–8723

Preisendorfer RW (1961) Application of radiative transfer theory to light measurements in the sea. In: IUGG-IAPO Symposium on Radiant Energy in the Sea, 4–5 Aug. 1960, Helsinki, Finland, Monograph No. 10, pp 83–91

Ritchie JC, Zimba PV, Everitt JH (2003) Remote sensing techniques to assess water quality. Photogrammetric Engineering and Remote Sensing 69:695–704

Sagan S (2008) The inherent water optical properties of Baltic waters. Dissertations and Monographs 21/2008, Institute of Oceanology PAS, Sopot, 242 pp [PhD Thesis, in Polish]

Sandén P, Håkansson B (1996) Long-term trends in the Secchi depth in the Baltic Sea. Limnology and Oceanography 41:346–351

Santer R, Zagolski F (2009) ICOL – Improve Contrast between Ocean and Land, ATBD (Algorithm Theoretical Basis Document) – MERIS Level-1C, Version 1.1, 6 January 2009. Université du Littoral Côte d'Opale, Wimereux, France. Report D6:1–15

Schroeder T, Behnert I, Schaale M, Fischer J, Doerffer R (2007) Atmospheric correction for MERIS above Case-2 waters. International Journal of Remote Sensing 28:1469–1486

Schwarz JN, Kowalczuk P, Kaczmarek S, Cota GF, Mitchell BG et al (2002) Two models for absorption by coloured dissolved organic matter (CDOM). Oceanologia 44:209–241

Siegel H, Gerth M (2008) Optical remote sensing applications in the Baltic Sea. In: Barale M, Glade M (eds) Remote sensing of the European seas. Springer, Berlin, pp 91–102

Siegel H, Gerth M, Neumann T, Doerffer R (1999) Case studies on phytoplankton blooms in coastal and open waters of the Baltic Sea using coastal zone color scanner data. International Journal of Remote Sensing 20:1249–1264

Siegel H, Gerth M, Ohde T, Heene T (2005) Ocean colour remote sensing relevant water constituents and optical properties of the Baltic Sea. International Journal of Remote Sensing 26:315–334

Skoog A, Wedborg M, Fogelqvist E (2011) Decoupling of total organic carbon concentrations and humic substance fluorescence in an extended temperate estuary. Marine Chemistry 124:68–77

Smith RC, Baker KS (1981) Optical properties of the clearest natural waters (200–800 nm). Applied Optics 20:177–184

Strickland JDH, Parsons TR (1972) A practical handbook of seawater analysis, Second edn Fisheries Research Board of Canada, Bulletin 167:1–310

Toming K, Arst H, Paavel B, Laas A, Nõges T (2009) Spatial and temporal variations in coloured dissolved organic matter in large and shallow Estonian water bodies. Boreal Environment Research 14:959–970

Tyler JE (1968) The Secchi disc. Limnology and Oceanography 13:1–6

Chemical pollution and ecotoxicology

16

Kari K. Lehtonen, Anders Bignert, Clare Bradshaw, Katja Broeg, and Doris Schiedek

Abstract

1. Baltic Sea organisms appear to be particularly sensitive to persistent hazardous substances because many of them are physiologically stressed in their brackish-water environment.
2. The profile of chemical pollution of the Baltic Sea has changed during the past decades with reductions in the concentrations of many "legacy contaminants" such as DDTs, PCBs, dioxins and trace metals. However, many of these compounds degrade very slowly and their concentrations are still unacceptably high.
3. Radionuclides from the Chernobyl nuclear power plant accident still contaminate the Baltic Sea but are slowly returning to pre-Chernobyl levels.
4. The growing oil tankers traffic is increasing the risk of major oil spills in the Baltic Sea. In addition, dumped chemical weapons on the seafloor are likely to leak due to corrosion.
5. An increasing amount of "contaminants of emerging concern", *e.g.* industrial chemicals, pharmaceuticals and ingredients of personal care products, most of them with unknown toxicity and environmental behaviour, ends up in the Baltic Sea.
6. Standardised biotests are widely used to examine the quality of water and sediments, but despite their usefulness in this context they often lack ecological relevance.
7. The term "biomarker" is used for a distinctive biological or biologically derived indicator (*e.g.* gene expression, enzyme activity, imposex, behaviour, growth, reproduction) of exposure to or effects of hazardous substances in the environment.
8. Biomarkers can be used to assess the effects of hazardous substances in organisms from different trophic levels. They allow rapid detection of potential toxic exposure and damage by providing information on the actual health status of organisms, including the effects of non-bioaccumulative substances and mixture toxicity. A major challenge remains in linking biomarker responses observed in field-collected organisms to effects at the population and community levels.
9. The accelerating climate change is expected to cause alterations in the bioavailability and toxicity of chemicals and their spread in the ecosystem due to changing environmental conditions.

Keywords
Biomarkers • Biotests • Biological effects • Chemical pollution • Climate change • Contaminants of emerging concern • Ecotoxicology • Environmental monitoring • Hazardous substances • Legacy contaminants

P. Snoeijs-Leijonmalm et al. (eds.), *Biological Oceanography of the Baltic Sea*,
DOI 10.1007/978-94-007-0668-2_16

16.1 Chemical pollution of the Baltic Sea

16.1.1 A sea contaminated by anthropogenic chemicals

Anthropogenic chemical pollution is identified as one of the major threats to the Baltic Sea ecosystem (*cf.* Sect. 17.3.1). The Baltic Sea in one of the most contaminated seas in the world, being a shallow sea that is almost completely surrounded by land with a large human population. Hazardous substances (Table 16.1) enter the ecosystem through anthropogenic activities and have the potential of harming the health of the ecosystem (*cf.* Sect. 17.7).

The main sources of hazardous substances include atmospheric deposition, industrial waste, diffuse runoff from land, river discharge, shipping accidents, leakage from ship hulls, and discharges from wastewater treatment plants. Although the situation in the Baltic Sea has improved during the last 20 years, the Baltic Sea still contains unacceptably high levels of a wide range of regulated and controlled so-called "legacy contaminants", such as organochlorines (DDTs, PCBs, dioxins) and trace metals (also called "heavy metals"). On top of that, many new or emerging hazardous substances, the so-called "contaminants of emerging concern" (CECs), are constantly being produced by humankind and end up in the sea. Detailed reviews on the distributions of contaminants have been published for the Baltic Sea and larger geographical areas including the Baltic Sea (*e.g.* Allsopp et al. 2001; Lääne et al. 2005; Law 2014; Vandermeersch et al. 2015).

The Baltic Sea constantly receives contaminated water from its large drainage area, which consists of industrialised countries with a population of ~85 million people. In addition, airborne contaminants, originating from long-range atmospheric transport, represent a substantial contribution to the total contaminant burden of the sea. Once they arrive in the Baltic Sea, chemical pollutants remain there for a long period. It takes ~40 years before the whole water body is recycled because of the geomorphological and hydrodynamical characteristics of the basin (*cf.* Sect. 2.3.8).

Many persistent hazardous substances (*i.e.* chemicals resistant to degradation) bind to the sediments and are not removed from the Baltic Sea by water exchange with the North Sea. Furthermore, the dispersal and fate of hazardous substances is complicated by the gradients in salinity and other environmental factors, abiotic and biotic (Leppäkoski and Bonsdorff 1989). The water temperature in the boreal Baltic Sea is fairly low, and becomes even lower towards the north, resulting in slower degradation of organic contaminants compared to warmer waters. When assessing the risk of chemical pollutants, both their concentrations in the different environmental matrices (water, sediment and biota) and their impacts on organisms have to be considered. The brackish-water environment means that both freshwater and marine species inhabiting the Baltic Sea are already subjected to a permanent physiological stress (*cf.* Sect. 7.2). This is assumed to cause the entire ecosystem to become highly sensitive to additional stress, including chemical pollution. In addition, low genetic diversity within species and genetic isolation from other populations outside the Baltic Sea can increase a species' vulnerability (Johannesson and André 2006; Johannesson et al. 2011). The low species diversity (*cf.* Sect. 4.6) means that only a few species maintain the key ecological functions in the Baltic Sea, potentially making communities more sensitive to stress.

16.1.2 Detection of the POP problem

During the last 20–30 years, most studies concerning hazardous substances in the Baltic Sea have dealt with the occurrence of trace metals and persistent organic chemicals and pesticides belonging to the past industrial phase. These latter organic compounds belong to the so-called "dirty dozen priority pollutants" that fall under the global ban of the Stockholm Convention (http://www.pops.int) due to their properties as persistent organic pollutants (POPs), *e.g.* the pesticide dichlorodiphenyltrichloroethane (DDT, Fig. 16.1). The occurrence of POPs in the Baltic Sea has been monitored over the years (HELCOM 2002, 2003, 2010a).

Clear signs of pollution-related problems in the Baltic Sea were already visible ~45 years ago. Around 1970, the

K.K. Lehtonen (✉)
Marine Research Centre, Finnish Environment Institute,
P.O. Box 140, 00251 Helsinki, Finland
e-mail: kari.lehtonen@ymparisto.fi

A. Bignert
Swedish Museum of Natural History,
P.O. Box 50007, 10405 Stockholm, Sweden
e-mail: anders.bignert@nrm.se

C. Bradshaw
Department of Ecology, Environment and Plant Sciences,
Stockholm University, 10691 Stockholm, Sweden
e-mail: clare.bradshaw@su.se

K. Broeg
Federal Maritime and Hydrographic Agency,
Bernhard-Nocht-Str. 78, 20359
Hamburg, Germany
e-mail: katja.broeg@su.se

D. Schiedek
DONG Energy Wind Power, Nesa Allé 1,
2820 Gentofte, Denmark
e-mail: dosch@dongenergy.dk

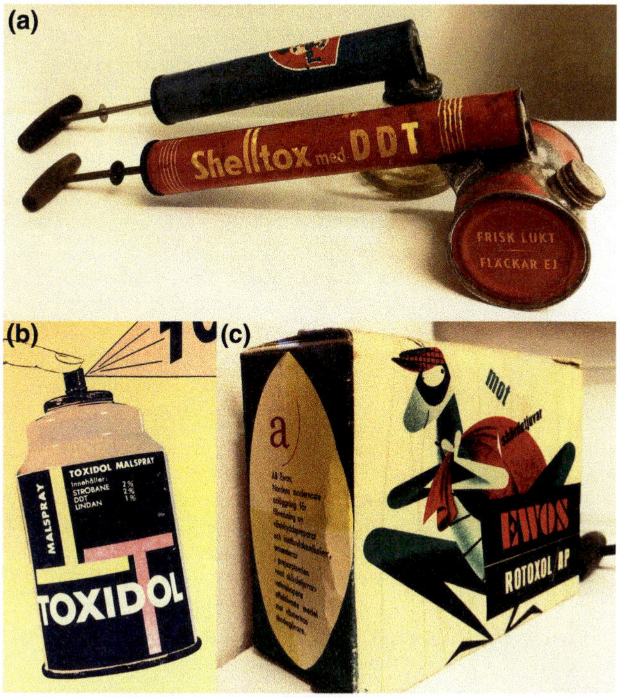

Fig. 16.1 Pesticides with persistent organic pollutants sold in Sweden in the 1940s–1960s after the discovery of the insecticidal property of DDT in the late 1939 (earning the 1948 Nobel Prize in physiology or medicine to Paul Müller, Switzerland). (**a**) Shelltox with DDT (dichlorodiphenyltrichloroethane) was an insecticide manufactured by shell. It was advertised as "a shower that kills" and "moths die like flies" and guaranteed that one shower would kill moths of all life stages and protect clothes against new moth attacks for 6 months. On the can it reads: "fresh smell, non-staining". (**b**) Toxidol with 2 % strobane, 2 % DDT and 1 % lindane (γ-hexachlorocyclohexane), was also used agains moths: "colourless and with a weak smell of pine needles". This product featured strobane, containing chlorinated terpenes produced by the chlorination of pinine, advertised as "the new American insecticide" in the late 1940s. (**c**) Rotoxol, manufactured by Ewos, contained DDT and lindane and was used for the protection of plants against insect attacks in agriculture and gardens. It was advertised as "the most effective agent against harvest thieves". DDT was banned in many countries already in the 1970s and large-scale use of lindane was banned in the 1980s, but in, *e.g.* the EU, pharmaceutical use of lindane is still allowed for treatments against lice and scabies. Photo: © Pauline Snoeijs-Leijonmalm

Baltic Sea populations of the white-tailed eagle *Haliaeetus albicilla* (*cf.* Box 4.12) and the grey seal *Halichoerus grypus* (*cf.* Box 4.13) were close to extinction, and the European otter *Lutra lutra* (*cf.* Fig. 4.16) had disappeared from coastal areas. In the Baltic Sea proper, egg shells of the white-tailed eagle were recorded to be 17–18 % thinner, and those of the common guillemot *Uria aalge* (Fig. 16.2) 10–15 % thinner, than museum-stored eggs collected before the 1950s (Bignert et al. 1995; Helander et al. 2002; Miller et al. 2014).

The awareness of causal links between the chemicals in common use at that time and the catastrophic worldwide declines in the populations of numerous bird and mammal species, especially of top predators, grew rapidly. Strong indications pointed to a group of chemicals termed polychlorinated biphenyls (PCBs) as being responsible for the observed high prevalence of sterility in grey seal females. At the same time, the monitoring of environmental contaminants and their possible effects on the marine fauna of the Baltic Sea started, initiated by the scientific community and national authorities of several countries. The first contaminants to be closely monitored were DDTs, PCBs and mercury (Hg).

16.1.3 The first efforts to tackle chemical pollution

During the 1970s, bans on hazardous chemicals were gradually introduced for both DDTs (Fig. 16.1) and PCBs in the countries surrounding the Baltic Sea. The measures taken to stop discharges of these chemicals had a significant positive effect, and the newly established monitoring programmes detected decreasing trends in DDT concentrations, *e.g.* in fish and common guillemot eggs (Bignert et al. 1995, 1998). The egg shells of the common guillemot started to become thicker and today they are almost as thick as those stored in museum collections from the times before DDT production started (Fig. 16.3). These measurements constituted the first step towards the biological effect monitoring of contaminants, since the efforts were not limited to measuring the concentrations of hazardous substances in the different environmental matrices in the Baltic Sea (water, sediment, biota), but also recorded their adverse effects.

As a substantial initiative, a reduction of chemical contaminant discharges into the Baltic Sea by 50 % within a period of 10 years was agreed upon within the framework of Helsinki Convention (*cf.* Sect. 17.8.4). This convention defined 1987 as the starting year for the assessment and reduction of several of the legacy contaminants such as DDTs and PCBs, and heavy metals such as lead (Pb), cadmium (Cd) and Hg. With the monitoring activities focusing on temporal trend assessments (*i.e.* following changes with time) it was now possible to show direct results of the measures taken to regulate the contaminant input to the ecosystem (Bignert et al. 1997). In parallel with the decreasing trends observed in concentrations of some of the legacy contaminants, notably DDTs (Fig. 16.3) and PCBs (Fig. 16.4), the negative population trends of *e.g.* the white-tailed eagle and the seals have reversed and the populations are now increasing. Such persistent chemicals bioaccumulate and increase in concentration through food webs, *i.e.* their tissue concentrations rise with increasing trophic level, a process known as biomagnification. Birds and mammals represent high trophic levels and are therefore good indicators of the effects of POPs.

Table 16.1 Examples of widespread, persistent hazardous substances in the Baltic Sea environment with information about their current or former usage, production and toxic effects. These substances bioaccumulate in organisms and reach humans mainly through the intake of food. Most of these compounds are today banned in most of the countries in the world, but because of their persistence they still occur in the environment and in biota. Note that this is not a complete list. Data from Bignert et al. (2015) and references therein, and the Index Mundi (http://www.indexmundi.com) for the mining of metals

Hazardous substance	Examples of current or former human usage	Production	Toxic effects
Mercury (Hg)	Thermometers, amalgam, electronics, batteries, anti-lock braking systems (ABS), mascara	Mining, mainly in China	Methylated to highly toxic methylmercury (CH_3Hg^+) e.g. by anaerobic micro-organisms; CH_3Hg^+ accumulates mainly in the viscera, also in muscles; combines with cysteine to form a methionine-like structure which penetrates all mammalian cells, and easily crosses the blood -brain barrier; effects on the central nervous system
Lead (Pb)	Ammunition, leaded petrol, car batteries, power cables, house paint, pigments in ceramics, keels of sailing boats, SCUBA-diving weight belts, fishing sinkers, solar energy cells, infrared detectors, water pipes, cigarettes	Mining, mainly in China, Australia, USA and Peru	Accumulates mainly in the bones, also in soft tissues; non-essential element for humans; damages the nervous system and causes brain and blood disorders; toxic effects involve several organ systems and biochemical activities; risk highest for children and the unborn, partly due to high permeability across the blood -brain barrier and the placenta
Cadmium (Cd)	Rechargeable nickel–cadmium batteries, metal plating, alloys, pigment in paints, stabiliser in plastics, impurity in phosphate rock used to manufacture fertilisers, electroplating, cigarettes	Mining, mainly in China and Korea	Accumulates mainly in the kidneys; irreversible renal tube dysfunction, *i.e.* the function to remove acids from the blood is lost; human carcinogen
Nickel (Ni)	Rechargeable nickel–cadmium batteries, alloys (including stainless steel), cellular telephones, laptop computers, electroplating, coins, fossil fuels	Mining, mainly in Russia, Phillippines, Australia, Indonesia and Canada	Essential element for many organisms, not for humans; possible adverse effects on human health *e.g.* by enhancing lipid peroxidation; Ni compounds classified as a human carcinogens
Organotin compounds (OTCs)	Antifouling agent in paints (for ship hulls, docks, buoys, and fishing nets), fungicide (for preservation of wood), stabiliser in the manufacturing of PVC plastics	Synthetic	Tributyltin (TBT) is an endocrine disrupting chemical (EDC, Casals-Casas and Desvergne 2011); TBT accumulates in the digestive/reproductive tract of gastropods and induces imposex (*cf.* Fig. 16.13); masculinisation of female fish (Shimasaki et al. 2003); high concentrations of TBT can be found in livers of marine mammals (Berge et al. 2004; Murata et al. 2008); knowledge on toxicity for mammals still very limited; possible modest adverse effects on the reproductive tract (Omura et al. 2001; Ogata et al. 2001)
Dichlorodiphenylethanes (DDTs)	Insecticide (domestic, agricultural, vector diseases, industrial)	Synthetic	Accumulate in body fat; EDCs (Casals-Casas and Desvergne 2011), possibly carcinogenic to humans (no clear evidence); shown to cause human embryo mortality, thyroid malfunction and immunosuppression; severe health effects on top predators, *e.g.* in the white-tailed eagle (failure to return to nesting sites, egg shell thinning, inability of eggs to hatch, reduced number of reproducing pairs, and nestling brood size)
Polychlorinated biphenyls (PCBs)	Plasticizers, insulators, fire retardants	Synthetic	Accumulate in body fat and milk fat; EDCs (Casals-Casas and Desvergne 2011) and classified as human carcinogens; possible influence human health by affecting multiple organ systems; high doses cause

(continued)

Table 16.1 (continued)

Hazardous substance	Examples of current or former human usage	Production	Toxic effects
			dermal and ocular lesions, irregular menstrual cycles and lowered immune responses, and poor cognitive development in children
Hexachlorocyclohexanes (HCHs)	Insecticide (domestic, agricultural, pharmaceutical)	Synthetic	Accumulate mostly in fat-containing tissues; EDCs and carcinogens (Olivero-Berbel et al. 2011); the γ-isomer is the most toxic (500–1,000 times as potent as the α-isomer); the insecticide lindane consists mainly of γ-HCH
Hexachlorobenzene (HCB)	Fungicide for seed treatment, production of fireworks, ammunition, synthetic rubber. By-product in the manufacture of solvents, other chlorine-containing compounds and pesticides	Synthetic and by-product	Accumulates in body fat and milk fat; carcinogenic to animals (liver, kidney and thyroid), assumed to be that also to humans; skin lesions and disturbed metabolism of haemoglobin in the liver in humans exposed to high doses; Courtney (1979): almost all breast-fed children < 2 years of HCB-exposed mothers died; follow-up studies 20–30 years after the poisoning the average HCB levels in breast milk still > 7 times the average for unexposed women
Polychlorinated dioxins/dibenzofurans (PCDDs/PCDFs)	By-products in several industrial processes and from most combustion processes, such as municipal waste incineration and small-scale burning under poorly controlled conditions	Not produced intentionally	Accumulate in fatty tissues; not readily metabolised or excreted; cause a variety of biological and toxicological effects in animals, including humans; developmental toxicity, carcinogenicity and immuno-toxicity; most toxic effects explained by the binding to the aryl hydrocarbon (Ah) receptor; significant variability in species sensitivity; 2,3,7,8-TCDD the most toxic and studied congener; used as a reference for all other related chemicals
Brominated flame retardants (BFRs)	Additive flame retardants incorporated into materials such as plastics and textiles used in electronics, clothes, furniture, paints, domestic kitchen applications	Synthetic	Accumulate in different body tissues, including breast milk; EDCs, particular effects on the thyroid hormone system; several polybrominated diphenyl ethers (PBDE) congeners and hexabromo-cyclododecane (HBCDD) cause neurotoxic effects in rats and mice; effects on behaviour, learning and hormonal functions in mammals; exposure during a sensitive stage of brain development cause reduced memory and learning disabilities; reduced reproductive success in birds
Perfluoroalkyl substances (PFASs)	Surfactants with exceptional stability and surface tension lowering potential. Teflon pans, non-stick cookware, rain/waterproof jackets, fire-fighting foams, food packaging, carpets and furniture fabrics	Synthetic	Accumulate in protein-rich tissues, e.g. blood, liver and eggs; weight loss, liver enlargement, immunotoxicity and developmental effects such as postnatal mortality in rodents; PFOA in maternal blood, and PFOA and PFOS in cord blood during pregnancy negatively associated with birth weight, birth length, ponderal index, and head circumference in humans
Polyaromatic hydrocarbons (PAHs)	Incomplete combustion (*e.g.* by oxygen insufficiency) of organic materials such as oil, petrol, coal, tobacco. Mainly the result of human activities but also natural (*e.g.* in the smoke from forest fires)	Not produced intentionally	Occur usually as complex mixtures of hundreds of components, each with different toxic potencies; many considered carcinogenic and mutagenic

Fig. 16.2 The common guillemot *Uria aalge* breeds in colonies on steep rocky coasts of the Baltic Sea. The thickness of their egg shells is used as an indicator of DDT contamination in the environment (*cf.* Fig. 16.3). Photo: © Fredrik Wilde/Azote

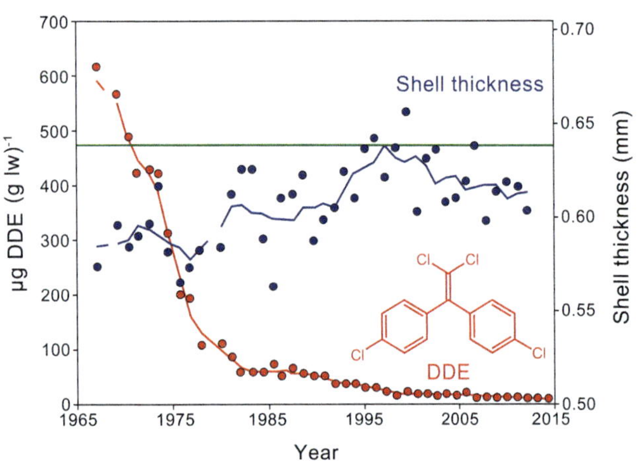

Fig. 16.3 Changes in the concentration of dichlorodiphenyldichloroethylene (DDE), a common breakdown product of dichlorodiphenyltrichloroethane (DDT), in µg per g lipid weight (lw) and shell thickness in eggs of the common guillemot *Uria aalge* at Stora Karlsö in the Baltic Sea proper over time. The red and blue dots represent measured values. The red and blue lines show simple 3-point running means fitted to the annual geometric mean values. The green line at 0.64 mm shows the mean thickness of pre-industrial museum egg shells. Figure based on data from the National Swedish Contaminant Monitoring Programme in Marine Biota (Bignert et al. 2015)

16.1.4 Regulatory actions and bans do the job

To be able to design adequate procedures for the regulation of hazardous substances in the future it is essential to recognise how the various regulatory measures to protect the environment from contaminant impacts work in practice. The reduction of the contaminant burden observed in the Baltic Sea ecosystem was faster than one would expect considering the compounds' resistance to degradation in the environment as predicted by their chemical characteristics. This was especially true for pesticides such as DDT and lindane (Fig. 16.1). Their ban in Western Europe lead to decreasing trends in tissue concentrations observed in several fish species, with an estimated annual reduction of 10–20 % (Bignert et al. 1998). For contaminants such as PCB, which are included in a number of industrial products, the decrease was somewhat slower, ~5–10 % annually (Fig. 16.4). However, although the decrease of PCB concentrations in the different compartments of the marine environment has been substantial during the last 30 years, PCBs are still found in significantly higher concentrations in the Baltic Sea compared to *e.g.* the adjacent Kattegat and Skagerrak (Figs. 16.4 and 16.5).

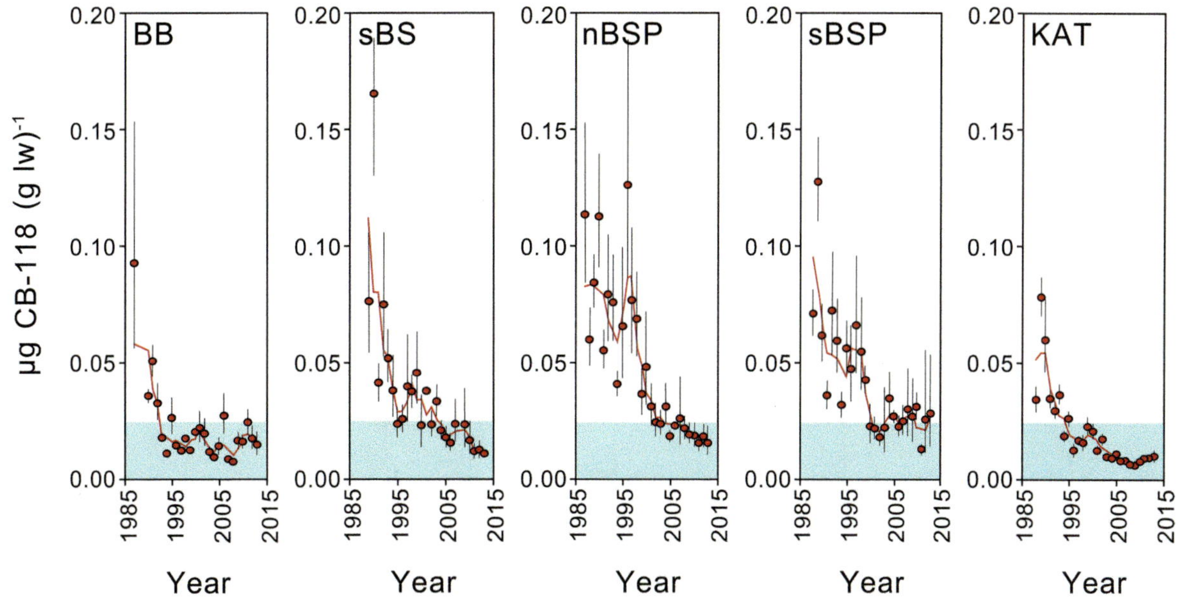

Fig. 16.4 Changes in the concentration of the PCB congener CB-118 in µg per g lipid weight (lw) in herring muscle between 1987/1989 and 2013 from the Bothnian Bay (BB), the southern Bothnian Sea (sBS), the northern Baltic Sea proper (nBSP), the southern Baltic Sea proper (sBSP), and the Kattegat (KAT). The red dots represent annual geometrical means, and error bars denote 95 % confidence intervals. The red line shows simple 3-point running means fitted to the annual geometric mean values. The blue area indicates concentrations below the EAC (environmental assessment criteria, OSPAR, OJEC No. L226 24.8.2013, 2013) target level of 0.024 µg (g lw)$^{-1}$. Figure based on data from the National Swedish Contaminant Monitoring Programme in Marine Biota (Bignert et al. 2015)

There are EU-level regulations pertaining to the use of chemicals that can affect marine biota. An example is furnished by REACH (Registration, Evaluation, Authorization and Restriction of Chemical Substances), the EU chemicals policy that entered into force in June 2007 (EC 2006). This policy places more responsibility on industry, while importers and users must provide information about the chemicals they handle to the European Chemicals Agency (ECHA) based in Helsinki (Finland).

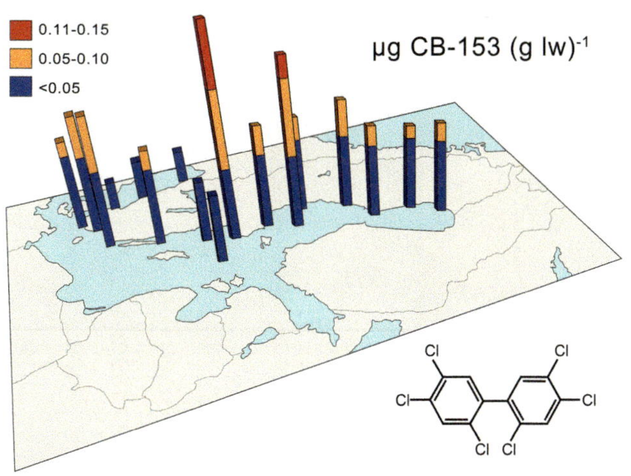

Fig. 16.5 The concentrations of the PCB congener CB-153 in µg per g lipid weight (lw) in herring muscle are elevated in the Baltic Sea compared to the Kattegat and the Skagerrak. Figure based on data from the National Swedish Contaminant Monitoring Programme in Marine Biota (Bignert et al. 2015)

16.1.5 More than traces of metals

Elevated concentrations of trace metals in organisms, in particular Pb, Cd and Hg, have raised concerns. After the removal of Pb in modern gasoline mixtures, along with other restrictions, the environmental concentrations of Pb have decreased significantly, which is evidenced by concentrations measured in *e.g.* livers of the Atlantic herring *Clupea harengus* in the Baltic Sea (Fig. 16.6; Lind et al. 2006). Despite the efforts made to reduce the discharges of Cd, its concentrations in the liver of several fish species monitored on a regular basis have not shown a steep decrease comparable to that observed for Pb (Fig. 16.7).

With regard to Hg, the longest monitoring series (starting in the beginning of the 1970s) shows significant decreases, 20–40 %, in the concentrations measured in the muscle of monitored fish species. However, some temporal monitoring data starting in the 1980s show increasing trends in Hg concentrations. The major problem with trace metals is that a

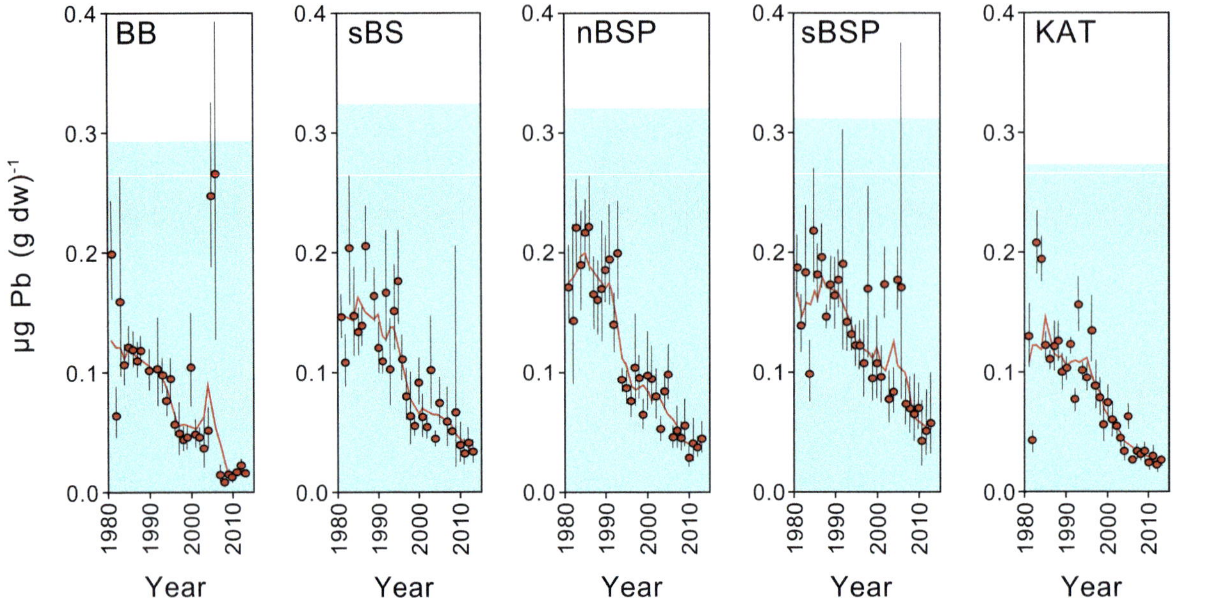

Fig. 16.6 Changes in the concentration of lead (Pb) in µg per g dry weight (dw) in herring liver between 1981 and 2013 in the Bothnian Bay (BB), the southern Bothnian Sea (sBS), the northern Baltic Sea proper (nBSP), the southern Baltic Sea proper (sBSP), and the Kattegat (KAT). The red dots represent annual geometrical means, and error bars denote 95 % confidence intervals. The red line shows simple 3-point running means fitted to the annual geometric mean values. The blue area indicates concentrations below the European Commission (EC) food regulation target level of 300 µg per kg wet weight (ww) in fish muscle, which was recalculated to 85 µg per kg wet weight (ww) in fish liver (Faxneld et al. 2015). The target levels in the figure differ because the dw:ww ratio (~0.3) slightly varied between areas. Figure based on data from the National Swedish Contaminant Monitoring Programme in Marine Biota (Bignert et al. 2015)

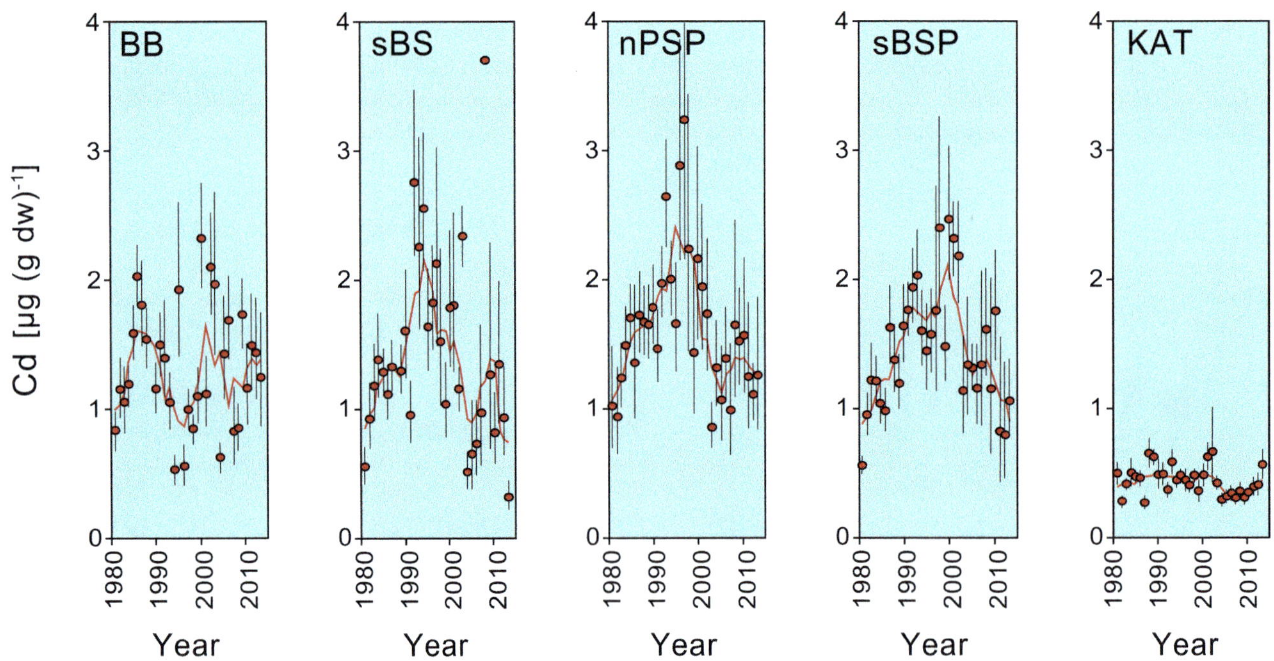

Fig. 16.7 Changes in the concentration of cadmium (Cd) in µg per g dry weight (dw) in herring liver between 1981 and 2013 in the Bothnian Bay (BB), the southern Bothnian Sea (sBS), the northern Baltic Sea proper (nBSP), the southern Baltic Sea proper (sBSP), and the Kattegat (KAT). The red dots represent annual geometrical means, and error bars denote 95 % confidence intervals. The red line shows simple 3-point running means fitted to the annual geometric mean values. The blue area indicates concentrations below the European Commission (EC) food regulation target level of 160 µg per kg wet weight (ww) in whole-fish samples, which was recalculated to 6.65 µg (g ww)$^{-1}$ in fish liver, that is, ~37.5 µg (g dw)$^{-1}$ (Faxneld et al. 2015). Figure based on data from the National Swedish Contaminant Monitoring Programme in Marine Biota (Bignert et al. 2015)

large portion of their input to the Baltic Sea is via atmospheric deposition, which is difficult to tackle efficiently by regional regulation.

16.1.6 The growing risk of oil pollution

Petroleum-derived hydrocarbons belong to a group of hazardous substances that poses a serious threat to the Baltic Sea environment. The biologically most harmful group of these compounds are polycyclic aromatic hydrocarbons (PAHs, Table 16.1). Their concentrations in the Baltic Sea are about three times higher than those observed in the North Sea. However, it is not oil pollution but the atmospheric deposition originating from incineration processes that has been identified as their main source. In sediments, the highest PAH concentrations have been recorded in the western and southern Baltic Sea (*e.g.* Dannenberger 1996; Kowalewska and Konat 1997).

In addition to the atmospheric input, an increasing risk of PAH contamination results from deliberate and accidental oil spills (Box 16.1). The most visible damage related to oil spills is usually the mass stranding and mortality of oiled waterbirds because of oil released during accidents involving ships. The effects of spills also result in sublethal effects on marine organisms. This has been observed in other sea areas after accidents with oil tankers, *e.g.* "Exxon Valdez" (Esler et al. 2010), "Erika" (Bocquené et al. 2004) and "Prestige" (Orbea et al. 2006), but also in the Baltic Sea, *e.g.* following the accidents of "Tsesis" (Elmgren et al. 1983) and "Baltic Carrier" (Pécseli 2002).

The risk of oil pollution, from small discharges to major catastrophes, is expected to grow due to the increasing ship traffic and oil trade in the Baltic Sea countries. The economic growth pattern of the region involves increasing oil use and production activities as well as the transportation of oil and other potentially hazardous cargos. In 2009, the amount of oil shipped to and from Baltic Sea ports via the Storebælt was ~166 million tonnes and is increasing (HELCOM 2010b). The amount of oil transported through the Gulf of Finland (after new large oil terminals in the Russian Federation have been built and started operating) is today estimated to be larger than that passing through the Hormuz Strait in the Persian Gulf region (traditionally taken as the reference for oil tanker traffic intensity). The use of large tankers is also expected to grow, with more tankers carrying 100–150 kilotonnes of oil (HELCOM 2010b).

Even if the best available technology (BAT) is used, the risk of major oil pollution accidents is increasing. If accidents happen, devastating impacts on the marine environment can be expected, especially in the coastal zone; the impact will also be partly dependent on the quality of the oil, since highly refined products tend to have larger impacts than crude oil.

16.1.7 New threats emerge: fighting the fire

New, worrying and increasing trends regarding new bioaccumulating and persistent chemicals were recorded soon after DDTs and PCBs started to decrease. One such group of chemicals consist of brominated flame retardants (BFRs, Table 16.1). After reaching peak values around 1985, tetra- and penta-brominated diphenyl ethers (BDEs) have now decreased substantially (Fig. 16.8). However, the BFR compound hexabromocyclododecane (HBCDD) reached peak levels in common guillemot eggs 20 years later, around 2005, and has only during recent years been decreasing (Fig. 16.9).

BFRs are currently key compounds in the protection against fire in *e.g.* households, and are used extensively worldwide. However, little is known about their effects on aquatic species, and even less on ecosystems. A recent study on the dose-response effects of HBCDD in experimental mesocosms assembled from coastal Baltic Sea ecosystem components showed that an increasing HBCDD concentration decreased the biomass of large individuals of the soft-bottom dwelling Baltic clam *Macoma balthica*. This resulted in a decreased recirculation of nutrients to the water. Changes in plankton communities were also observed, either due to direct toxic HBCDD effects or indirect effects via changes in benthic-pelagic coupling of nutrients (Bradshaw et al. 2015). Such complex ecosystem responses can only be quantified and understood by using realistic experimental set-ups and by utilising the knowledge of system-specific ecological interactions.

As with most hazardous substances, the environmental impacts of BFRs were largely unforeseen at the time they were brought to the market. Unfortunately, the practice of introducing new compounds with unknown environmental behaviour and toxicity is still common. It should be noted that in many cases only small structural alterations to the existing (banned or restricted) compounds are made, making them "new".

16.1.8 Consumer goods turn bad

Another important group of chemicals – also designed to protect or facilitate our daily life – that showed the highest concentrations in Baltic Sea samples around 2005 comprises perfluorinated alkyl acids (PFAAs, Table 16.1). A retrospective study of perfluorooctane sulphonate (PFOS)

Box 16.1: Effects of shipping and recurrent oil spills

Kjell Larsson

Intensive ship traffic

Shipping is an important mode of transport and is a necessity for a prosperous Baltic Sea region. Approximately 10,000 unique vessels registered in more than 100 countries visit the Baltic Sea each year (Grimvall and Larsson 2014), and the ship traffic is expected to increase in future. Transport of goods by large and slow moving ships, especially by tankers and dry bulk ships, is often cost- and energy-efficient. However, intensive shipping gives rise to significant negative environmental impacts in the form of emissions to air and water. Disturbance effects of the ships themselves, or by the underwater noise (acoustic pollution) produced by machinery, speed logs and echo sounders, have been detected on fish, waterbirds and marine mammals.

Emissions from ships

The effects of shipping on the environment and biodiversity can be measured on different spatial and temporal scales. The emissions of carbon dioxide (CO_2), nitrogen oxides (NO_X), sulphur oxides (SO_X) and particulate matter (PM) from ship engines to the air usually have regional or global effects. Discharges of a ship's ballast water may lead to irreversible effects on biodiversity if non-indigenous species released together with the ballast water establish themselves in a new region and affect the native species (*cf.* Fig. 5.3). Other types of emissions and discharges to water from ships, for example operational oil spills, may have more local effects. Despite that discharges of oil from a ship's cargo or machinery spaces at concentrations above 15 ppm are prohibited in the Baltic Sea, visible oil slicks along the main shipping routes are regularly detected by surveillance flights (HELCOM 2015).

Accidental and deliberate oil spills from ships

Studies performed in different parts of the world have shown that the effect of a given oil discharge on the marine environment depends not only on the size of the discharge but also on where and when the oil spills take place (Camphuysen et al. 2005). Oil catastrophes where several kilotonnes of oil are released have large effects. However, even a small accidental or deliberate oil spill from a ship in an area where tens- or hundreds of thousands of waterbirds spend their winter can cause the death of many thousand individuals and threaten sensitive and red-listed species (Larsson and Tydén 2005). To reduce such negative effects, marine spatial planning processes should search for solutions to separate the most heavily trafficked shipping routes from the most sensitive marine areas.

AIS data can be used to examine ship traffic intensity and conflict areas

The automatic identification system (AIS) is a navigation tool and maritime tracking system developed primarily to increase maritime safety and to monitor and manage sea traffic. AIS data on a vessel's identity, position, course and speed are available in real time both to other vessels and to authorities onshore. All vessels in international traffic with a gross tonnage of 300 or more must be equipped with AIS transponders. Vessels not in international ship traffic must also be equipped with AIS if they have a gross tonnage of 500 or more or if they carry passengers. AIS data can be used to map in detail the traffic intensity in different parts of the Baltic Sea and to identify conflict areas, *i.e.* areas where conservation values may be significantly affected by shipping activities.

Example of a conflict area in the Baltic Sea

One of several identified conflict areas is the Natura 2000 site Hoburgs bank south of the island of Gotland in the central Baltic Sea proper (Box Fig. 16.1). The Hoburgs bank, Södra Midsjö bank and Norra Midsjö bank (*cf.* Fig. 2.2) are important wintering sites for the globally threatened long-tailed duck *Clangula hyemalis* (*cf.* Box Fig. 11.12), a species heavily affected by oil spills from ships (Larsson and Tydén 2005; Hearn et al. 2015). The offshore banks also host large numbers of wintering black guillemots *Cepphus grylle*. Studies of the threatened Baltic Sea population of the harbour porpoise *Phocoena phocoena* (*cf.* Fig. 4.15) have also shown that this mammal uses these offshore banks during the calving season in summer.

Box Fig. 16.1 The ship traffic intensity near and within the marine Natura 2000 site "Hoburgs bank" in the central Baltic Sea proper recorded with the automatic identification system (AIS) in 2014. The curved white line shows the border of the Natura 2000 site. The red polygon shows the border of the area that ships are recommended to avoid, *i.e.* an area that the UN International Maritime Organization (IMO) has classified as an "area to be avoided" (AtbA). More than 20,000 ship passages were recorded within the northwestern part of the Natura 2000 site between the southern tip of Gotland (green) and the northern border of the AtbA (red) in 2014. More than 100 ship passages were recorded within the AtbA. Each small rectangle has a height and width of 0.02°. The colour coding shows the number of vessels that visited each rectangle in 2014: yellow 1–10 vessels, light brown 11–100 vessels, dark brown 100–1,000 vessels, black >1,000 vessels

concentrations in common guillemot eggs revealed an alarming increase, up to 25–30 times, between the beginning of the 1970s and 2005 (Fig. 16.10; Holmström et al. 2005).

Since 2005, that increase has turned into a significant decline. In 2000, the 3M Company, the largest manufacturer of PFAAs, voluntarily stopped their production. However, the increasing use of other surface-treatment compounds with a similar chemical structure, as well as other so-called "consumer" chemicals, can produce unforeseen effects as long as their environmental behaviour and toxicity have not been properly assessed. Many of them are widely used and degrade slowly. Surface-treatment chemicals can be found in many common consumer goods such as clothes and shoes, and they are common in industrial coatings. Various types of consumer chemicals are frequently found in personal care products, such as sunscreens and synthetic perfumes (*e.g.* musks), and in artificial sweeteners and pharmaceuticals. Personal care products increasingly use microbeads, which give rise to microlitter (Box 16.2).

Recent worldwide concern has been directed towards pharmaceutical substances such as common painkillers, hormone products and antidepressants. Currently, many of these substances pass through municipal wastewater treatment systems and reach the marine environment. Many of these are categorised as so-called "micropollutants" that cause toxic or sublethal effects at very low concentrations. Also, many of them are hormonally active substances and cause a phenomenon called "endocrine disruption" in marine animals. In general, very little is currently known about the effects of various types of micropollutants on marine organisms *in situ*. Nevertheless, there is growing evidence that even environmentally realistic low concentrations of some micropollutants affect aquatic algae, invertebrates and fish (Nentwig 2007; Corcoran et al. 2010; Eriksson-Wiklund et al. 2011; Brodin et al. 2013). For example, the benzodiazepine anxiolytic drug oxazepam (used as a drug for the treatment of anxiety and insomnia) alters the behaviour and feeding rate of wild European perch *Perca fluviatilis* at

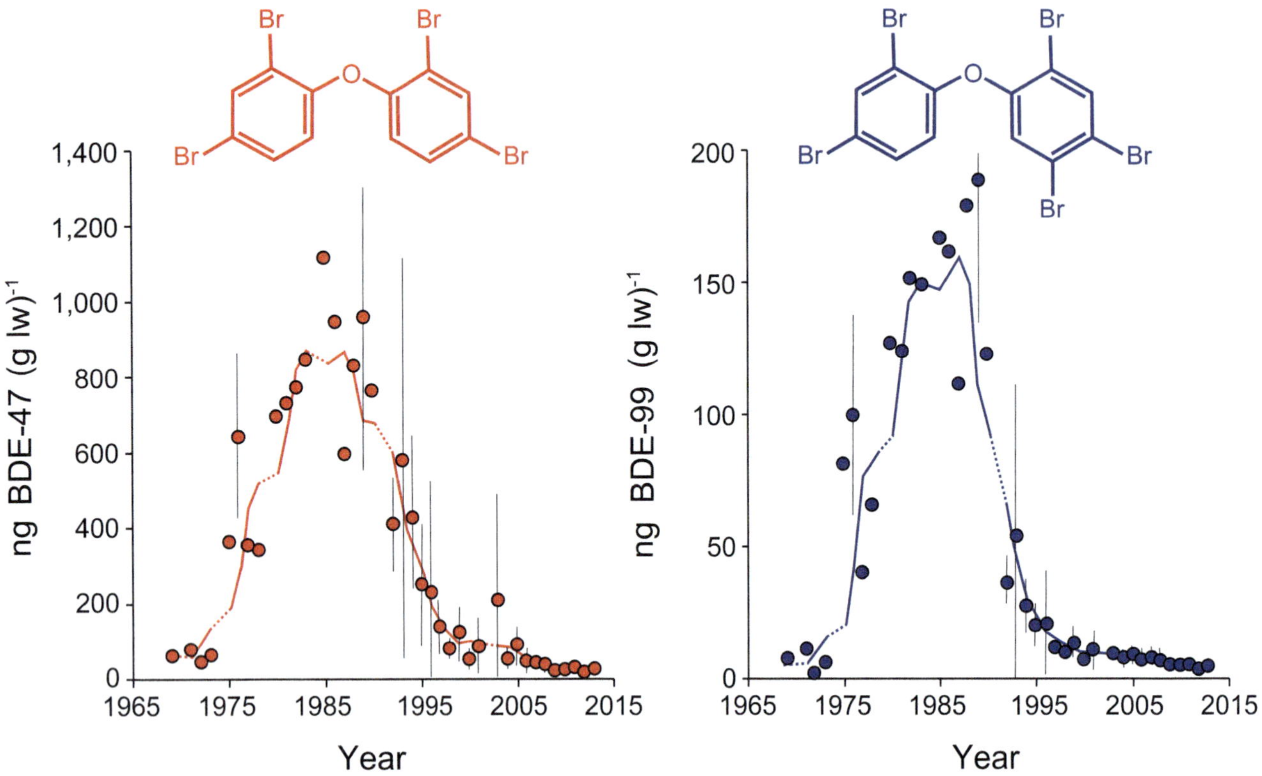

Fig. 16.8 Changes in the concentrations of 2,2′,4,4′-tetrabromodiphenyl ether (BDE-47) and 2,2′,4,4′,5-pentabromodiphenyl ether (BDE-99) in ng per g lipid weight (lw) in common guillemot *Uria aalge* eggs in the Baltic Sea proper between 1969 and 2013. The red and blue dots represent annual geometrical means, and error bars denote 95 % confidence intervals. The red and blue lines show simple 3-point running means fitted to the annual geometric mean values; when values are missing the lines are dashed. Figure based on data from the National Swedish Contaminant Monitoring Programme in Marine Biota (Bignert et al. 2015)

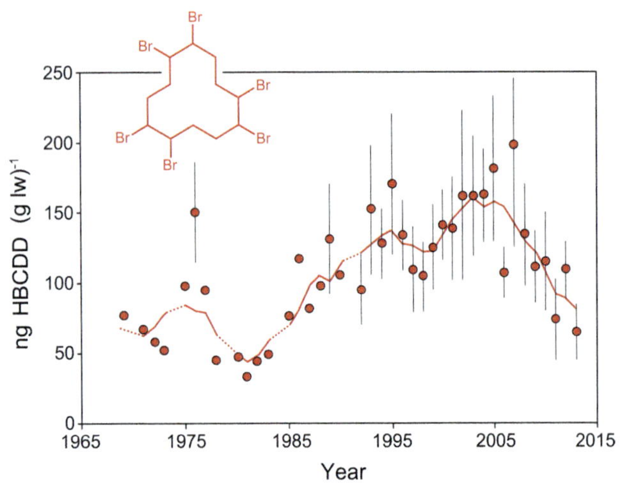

Fig. 16.9 Changes in the concentration of hexabromocyclododecane (HBCDD) in ng per g lipid weight (lw) in common guillemot *Uria aalge* eggs in the Baltic Sea proper between 1969 and 2013. The red dots represent annual geometrical means, and error bars denote 95 % confidence intervals. The red line shows simple 3-point running means fitted to the annual geometric mean values; when values are missing the line is dashed. Figure based on data from the National Swedish Contaminant Monitoring Programme in Marine Biota (Bignert et al. 2015)

concentrations encountered in effluent-influenced surface waters. Individuals exposed to water with drug concentrations of (1.8 μg L^{-1}) exhibited increased activity, reduced sociality, and higher feeding rate (Brodin et al. 2013). Antidepressants at environmentally relevant concentrations can disrupt locomotor activity and reduce fecundity in snails, and in crustaceans they may affect activity patterns, aggression, reproduction and development (Fong and Ford 2014).

16.1.9 Radioactivity: the Chernobyl accident

Radioactivity exists naturally in the environment, coming both from cosmic radiation and weathering of rocks. Enhanced levels of human-made radioactivity in the Baltic Sea are mainly due to atmospheric nuclear weapons testing during the 1950s and 1960s, but the 1986 Chernobyl nuclear power plant accident in the former Ukrainian Soviet Socialist Republic has also left its footprints in the Baltic Sea (HELCOM 2009) in the form of radioactive isotopes (Box 16.3).

The radionuclides in the Baltic Sea environment that originate from the nuclear weapons tests consist mainly of

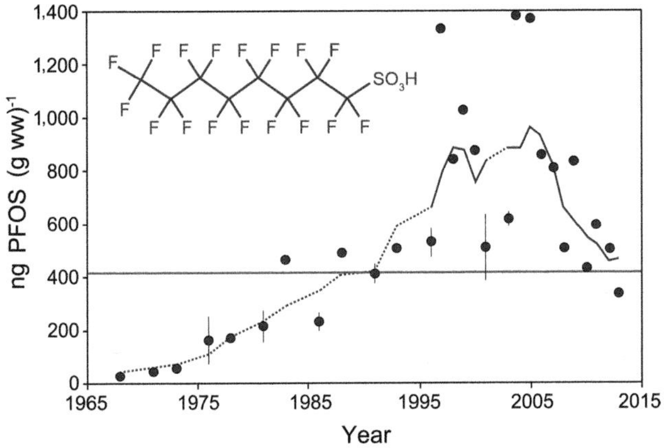

Fig. 16.10 Changes in the concentration of perfluorooctane sulfonate (PFOS) in ng per g wet weight (ww) in common guillemot *Uria aalge* eggs in the Baltic Sea proper between 1968 and 2013. The blue dots represent annual geometrical means, and error bars denote 95 % confidence intervals. The blue line shows simple 3-point running means fitted to the annual geometric mean values; when values are missing the line is dashed. The red line denotes LOQ (limit of quantification), which is the lowest concentration of the analyte that can be determined with acceptable precision and accuracy regarding the method used. Figure based on data from the National Swedish Contaminant Monitoring Programme in Marine Biota (Bignert et al. 2015)

radioactive isotopes with a long physical half-life, notably isotopes of caesium (^{137}Cs, 30 years), strontium (^{90}Sr, 29 years) and plutonium (^{239}Pu, 24,100 years and ^{240}Pu, 6,563 years). Their levels have been slowly decreasing since their peak values in the mid-1960s due to radioactive decay and dispersal (HELCOM 1995, 2009, 2013a). In recent decades, most studies have focused on the distribution of the Chernobyl fallout, particularly that of ^{137}Cs.

The initial deposition after the Chernobyl disaster occurred via rainfall; the Gulf of Finland, the Archipelago Sea, the Åland Sea and the Bothnian Sea received the largest burden (Fig. 16.11). The total input of ^{137}Cs from the Chernobyl accident into the Baltic Sea was estimated at ~4,700 TBq, based on measurements in seawater samples in late 1986 (Nielsen et al. 1999); this value should be compared to an estimate of 320 TBq before the Chernobyl accident. The major radionuclides from Chernobyl that stay in the ecosystem for a long time are ^{137}Cs, ^{134}Cs and ^{90}Sr. Other radionuclides that were initially detected such as ^{89}Sr, ^{131}I (iodine), ^{141}Ce (cerium), ^{95}Nb (niobium), ^{95}Zr (zirconium) disappeared rapidly due to their short physical half-lives (Table 16.2).

16.1.10 Deposited ^{137}Cs helps us to understand environmental processes

Spatial and temporal trends in radionuclides, especially those of ^{137}Cs, are fairly well described in water, sediment and some fish species. These trends are mainly due to original deposition patterns, the counter-clockwise surface-water circulation in the basins of the Baltic Sea (*cf.* Fig. 2.10), proximity to riverine runoff (*cf.* Fig. 2.11), settling of particles on the seafloor, radioactive decay and the life histories of the organisms. In fact, radionuclides from Chernobyl and from other accidental releases can be used as tracers of environmental processes (*e.g.* Santschi 1984, 1989; Bradshaw et al. 2006). During the first five post-Chernobyl years, surface-water concentrations of radionuclides decreased but remained high in the most heavily affected regions (*e.g.* the Bothnian Sea, Fig. 16.11). In the same period the distribution of radionuclides in the Baltic Sea water became more homogeneous due to vertical and horizontal mixing as well as adsorption of ^{137}Cs to particles and their subsequent deposition on the seafloor.

The downward mixing also meant that the bottom-water radionuclide concentrations gradually increased for several years, although this was a slow process due to the halocline. The export of radionuclides from the Baltic Sea has also occurred, although the transport between basins is slow due to the restricted inter-basin water exchange. A large amount of the initial activity from Chernobyl has been deposited on the seafloor (HELCOM 2007). In areas where sediments accumulate the radionuclides have been slowly buried, and in many areas a clear ^{137}Cs peak is seen at a given depth in the sediment and can be used to date sediment layers and calculate sediment accumulation rates (García-Tenorio et al. 1992; HELCOM 2007).

16.1.11 A radioactive sea?

Initially, the ^{137}Cs levels in Baltic Sea fish reflected the general spatial patterns of the Chernobyl deposition. The concentrations generally peaked in 1986–87. The maximum concentrations measured in the Atlantic cod *Gadus morhua* and herring muscle were >250 Bq (kg wet weight)$^{-1}$ in the Bothnian Sea and the Åland Sea, while in other areas they did not amount to more than 20–35 Bq (kg wet weight)$^{-1}$ (HELCOM 1995). These levels are markedly higher than the pre-Chernobyl values ~2.5 Bq (kg wet weight)$^{-1}$, but remain under the Swedish maximum allowable concentration in marine fish sold for human consumption, 300 Bq (kg wet weight)$^{-1}$. In areas further away from the initial hotspots the maximum concentrations peaked later, *e.g.* after 1990 in the southwestern Baltic Sea proper. Benthic fish species generally took longer than pelagic ones to reach peak concentrations due to their main exposure route being via the sediment and the zoobenthos. Fish species higher up in the food web, such as the northern pike *Esox lucius*, also took longer to accumulate ^{137}Cs, presumably due to their main exposure being via food. The trends for ^{90}Sr are less clear and the concentrations are much lower than those of ^{137}Cs (HELCOM 1995, 2009).

Box 16.2: Marine microlitter in the Baltic Sea

Outi Setälä

Where does microlitter come from?
Marine microlitter (anthropogenic particles <500 μm) originates from a variety of sources, such as traffic, industry, and households, and is transported to the Baltic Sea by *e.g.* rivers and wastewaters (Magnusson et al. 2016). Microlitter includes both non-synthetic and synthetic particles, the latter consisting predominantly of different types of plastic polymers known as "microplastics" (GESAMP 2015). Most of the research on microlitter in the sea focuses on microplastics, which are either intentionally small ("primary microplastics") or fragmented from larger plastic items by *e.g.* weathering ("secondary microplastics"). The primary microplastics may enter marine systems either directly from the production process of plastic items or, as in the case of the microbeads from personal care products, via municipal wastewaters.

Microlitter and wastewater treatment plants
Studies carried out in Sweden and Finland have revealed that modern methods used in wastewater treatment plants are able to remove up to 99 % of the microlitter particles from the incoming water, including fibres from clothing, one of the important microlitter types deriving from urban areas (Magnusson and Wahlberg 2014; Talvitie et al. 2015). However, since the water volumes and flow rates are often high, wastewater treatment plants still act as gateways for microlitter to the recipient. The amount of microplastics entering the Baltic Sea should be efficiently managed, since once they are in marine environment, there is no method for removing them.

The amount and distribution of microlitter in the Baltic Sea
In the Baltic Sea, only few studies on microlitter distribution have so far been carried out. The first study included sampling along the Swedish coastline with a submersible pump (Magnusson and Norén 2011). Surveys to assess the amount and distribution of microlitter have since been carried out also in the open sea areas, whereby a special neuston net, the so-called "Manta trawl", was used to quantify the amount of anthropogenic microscopic particles. In these surveys the microlitter concentrations varied between <1 and 2.1×10^3 particles m^{-3}, with the particles consisting mainly of non-synthetic fibres (Setälä et al. unpublished). This does not adequately describe the real situation, since the trawl captured only the larger fraction (>333 μm) of microlitter, and the smaller fraction remained unquantified. In the same study, a submersible pump with a 100-μm mesh size filter was used, and the total number of particles m^{-3} captured with the pump was three times higher than that captured with the trawl.

Harmonisation of methodology is necessary
Despite its disadvantages, the Manta trawl is still in use because it has several advantages for environmental monitoring. For example, it is possible to cover large areas with Manta-trawl sampling, which is not possible with equipment that collects also the smallest microlitter particles. Sampling the whole water column, or sediment, instead of only the water surface will most likely produce different estimates than a surface neuston net. A recent study from the Swedish east coast presented microplastic concentrations that were considerably higher than what has been measured with other methods in the Baltic Sea, or commonly in other sea areas (Gorokhova 2015). In that study routine zooplankton samples that were taken with a 90 μm net were examined for plastics. Whatever method is used, it is necessary to develop and implement a harmonised Baltic Sea-wide sampling programme and to use uniform analytical methodology to produce holistic assessments of microlitter particles distribution, amounts and nature. It is expected that, with time, a large part of the microlitter will sediment to the seafloor, so the amount of microlitter in the sediment system should be studied. Several current projects are developing methods to study microlitter in the Baltic Sea, which should, in the near future, provide more information on this type of pollution in the Baltic Sea.

Effects of microlitter on marine organisms

Microlitter is found in different compartments of the Baltic Sea ecosystem, not only at the water surface and in the water column. Aquatic organisms ingest microlitter and it is transferred between trophic levels and bioaccumulated in the food web. The harm produced by ingested microlitter may be mechanical (*e.g.* clogging of the digestive tract, sticking to external surfaces hindering mobility) or, especially in the case of microplastics, chemical. Microplastics can contain harmful additives that have the potential to leach into the environment and cause harm to marine organisms. Microplastics can also accumulate harmful hydrophobic substances from the surrounding water before they might be ingested by a marine animal, and in the case of the Baltic Sea this might include *e.g.* dioxin-like compounds. The smaller a plastic fragment, the larger its area in relation to its volume, and thus the larger its adsorption capacity.

Microlitter in the Baltic Sea food web

Experiments with benthic Baltic Sea invertebrates from the southwestern coast of Finland have shown that the filter-feeding bivalves *Mytilus trossulus* and *Macoma balthica* ingested significantly higher amounts of microbeads than crustaceans and polychaetes (Box Fig. 16.2). The free-swimming crustaceans, in turn, ingested more beads than the benthic polychaetes that were feeding only on the sediment surface. This shows that the microbeads were actively filtered out of the water and accumulated in the invertebrates. In another experiment, in which zooplankton with ingested microbeads were offered as food to mysids, the zooplankton prey and microbeads were detected in the mysid intestines already after three hours of incubation (Setälä et al. 2014). This study showed for the first time that trophic transfer of microplastics from lower trophic level organisms to predators does occur in the pelagic food web. The next step in studies on microplastics is to evaluate the actual impacts of microplastics on the functioning of marine ecosystems and the possible impacts on human consumers of marine food.

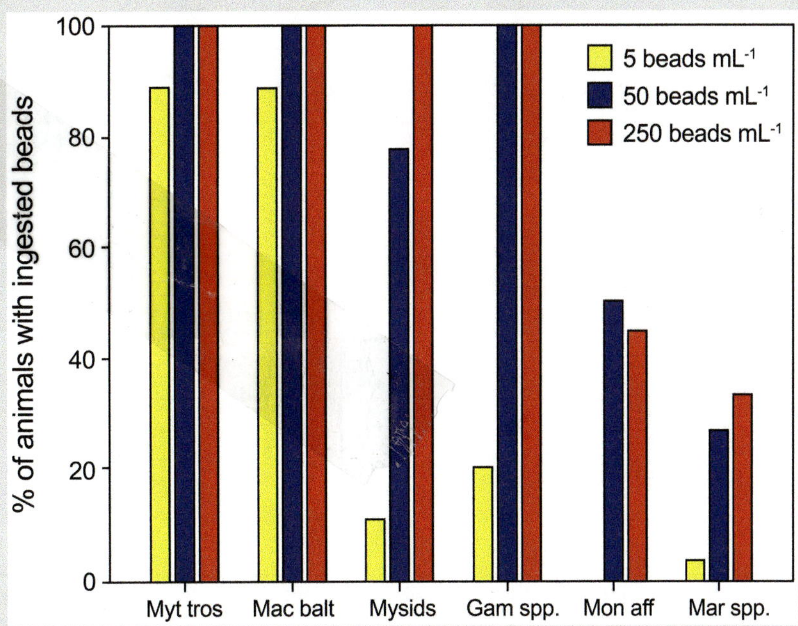

Box Fig. 16.2 The proportion of benthic animals with ingested polystyrene microbeads (10 μm in size) at three bead concentrations after 24 h of experimental incubation in aquaria with beads added to the water. Myt tros = *Mytilus trossulus*, Mac balt = *Macoma balthica*, Gam spp. = *Gammarus* spp., Mon aff = *Monoporeia affinis*, Mar spp. = *Marenzelleria* spp. Figure modified from Setälä et al. (2016)

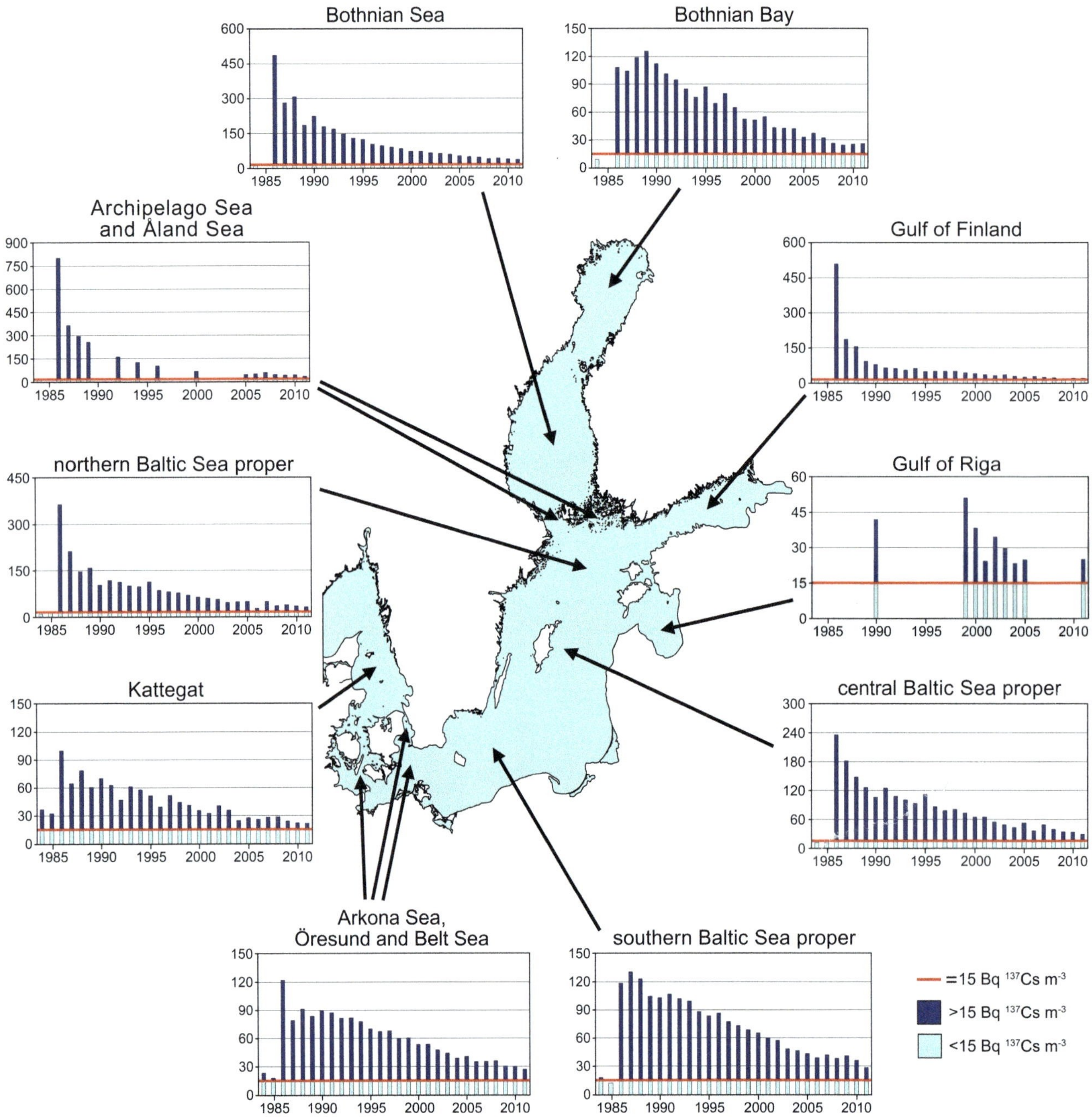

Fig. 16.11 Changes in ^{137}Cs concentrations in Bq m^{-3} in surface water (0–10 m of water depth) between 1984 and 2011, shown as annual geometrical means by subregion of the Baltic Sea Area. The red line indicates the target value of 15 Bq m^{-3}, which was calculated as the average pre-Chernobyl (1984–1985) concentration. Figure modified from HELCOM (2013a) with permission from the Finnish Environment Institute (SYKE)

Currently, the levels of anthropogenic radionuclides are higher in the Baltic Sea than in most other water bodies around the world. For example, ^{137}Cs concentrations are ~40 times higher than in the northeastern Atlantic Ocean and ~10 times higher than in the North Sea. This is due to the Baltic Sea's large drainage area and limited water exchange with the North Sea, together with the relatively high fallout the Baltic Sea region received from Chernobyl.

A total amount of ^{137}Cs 8.5 times larger than the Chernobyl accident's deposition into the Baltic Sea in 1986 (~4.7 PBq, Nielsen et al. 1999) was released into the Pacific Ocean in 2011 from the Fukushima Dai-ichi accident in Japan (~40 PBq, Bailly du Bois et al. 2012). However, the much larger water volume and strong currents of the Pacific Ocean have dispersed the radionuclides much faster than in the Baltic Sea. Nevertheless, the radioactivity in the seawater and biota of the Baltic Sea is lower now than it was

Table 16.2 Radionuclides (γ-emitters) in samples of benthic diatoms taken outside the Forsmark nuclear power plant (FNPP, southern Bothnian Sea, *cf.* Box Fig. 5.9), showing the difference between the amount and composition of the radionuclides discharged by the normal drift of the FNPP and the radionuclides discharged by the chernobyl accident. The values for 1984 each represent the average value from 15 sampling occasions during the year; the γ-decay was measured during 24–48 hours. The values in 1986 each represent one sample taken 10 days after the Chernobyl accident; the γ-decay was measured during 1–2 hours. The detection limit depends on sample size, length of the measurement and half-life of the radionuclide, which is the reason why the radionuclides from the FNPP were not detected in 1986 (measurement only 1–2 h). – = not detected, * = detected but too low a concentration to quantify. The samples were taken at three sampling sites: Site 1 = intake channel of the cooling water to the FNPP, Site 2 = outflow channel of the cooling water from the FNPP, Site 3 = shallow bay ~1 km away from the outflow channel. The radionuclides from the FNPP are low in the intake channel, high in the outflow channel and lower in the shallow bay than in the outflow channel. The radionuclides from Chernobyl were highest in the shallow bay, probably due to little water exchange and/or patchy downfall and lowest in the outflow channel, indicating that they were not released from the FNPP. Data from Notter and Snoeijs (1986) and Snoeijs and Notter (1993)

Radionuclide	Physical half-life	Site 1 Bq (kg dw)$^{-1}$ Average 1984	Site 2 Bq (kg dw)$^{-1}$ Average 1984	Site 3 Bq (kg dw)$^{-1}$ Average 1984	Site 1 Bq (kg dw)$^{-1}$ 6 May 1986	Site 2 Bq (kg dw)$^{-1}$ 6 May 1986	Site 3 Bq (kg dw)$^{-1}$ 6 May 1986
^{132}Te	3.2 days	–	–	–	18,000	21,000	87,000
^{131}I	8.0 days	–	–	–	24,000	21,000	181,000
^{140}Ba	13 days	–	–	–	32,000	9,000	439,000
^{51}Cr	28 days	*	*	*	–	–	–
^{141}Ce	33 days	–	–	–	58,000	15,000	825,000
^{95}Nb	35 days	*	*	–	78,000	18,000	1,022,000
^{103}Ru	39 days	–	–	–	49,000	16,000	540,000
^{95}Zr	64 days	–	–	–	64,000	14,000	864,000
^{58}Co	71 days	*	1,110	486	–	–	–
^{65}Zn	244 days	69	747	431	–	–	–
^{110m}Ag	250 days	31	313	150	<1,000	1,000	1,000
^{144}Ce	285 days	–	–	–	42,000	11,000	612,000
^{54}Mn	312 days	29	271	152	–	–	–
^{134}Cs	2.1 years	–	*	–	3,000	3,000	14,000
^{125}Sb	2.8 years	–	*	–	–	–	–
^{60}Co	5.3 years	277	3,583	2,553	–	–	–
^{137}Cs	30.0 years	–	*	*	6,000	4,000	25,000

immediately after the Chernobyl accident, in some cases even at pre-Chernobyl levels, and is not considered to pose a threat to human health (HELCOM 2013a). Local discharges from nuclear power plants as sources of radioactivity (Table 16.2) are of minor importance, but even here decreased discharges of ^{137}Cs, ^{90}Sr and ^{60}Co into the Baltic Sea have been observed during the last decade.

16.1.12 Dumped World War II chemical weapons are a threat

Since the turn of the millennium, the post-war dumping of World War II so-called chemical warfare agents (CWA) has raised concern. Around 50 kilotonnes of chemical weapons were dumped into the Baltic Sea Area. Official dumpsites are located in the Bornholm Sea, the main spawning area of the cod, the Gotland deep, as well as in the Lillebælt (Belt Sea) and the Skagerrak outside the Baltic Sea. The dumped CWA consist mainly (~50 %) of mustard gas and arseniccontaining substances (*e.g.* arsenic oil, Clark I and II and Adamsite). By now, at least a part of the chemical-containing ammunition and containers deposited on the seafloor in the late 1940s and early 1950s has reached a state of extensive corrosion, and major leakages of these dangerous substances are likely to occur (Bełdowski et al. 2016; HELCOM 2013b).

Most of these CWA hydrolyse rapidly in water and thus the exposure of biota to the parent compounds in the case of leakage is rather limited. However, exposure to degradation products close to the hotspot dumping areas is a possible threat. Field measurements and caging (experimental transplantation of organisms) studies in the Bornholm dumping site for the purpose of analysing the effects of CWAs on the Atlantic cod *Gadus morhua* and the blue mussel *Mytilus trossulus* have indicated potential impacts at the cellular and tissue levels of the organisms (Bełdowski et al. 2016).

Box 16.3: Radioactivity

What are radionuclides?

Radionuclides are unstable isotopes that have an excess or deficiency of neutrons in the nucleus. They exist for nearly all elements. One of the most damaging forms of radiation is ionising radiation, where electromagnetic waves or subatomic particles are released. Ionising radiation has the ability to remove electrons, *i.e.* ionise atoms in the vicinity, and this can cause damage to living cells. Examples of ionising radiation are α-, β-, γ-, and neutron radiation and X-rays, which have different properties. For example, α-radiation consists of helium nuclei and cannot penetrate a sheet of paper or the human skin, β-radiation consists of electrons or positrons and cannot penetrate a thin sheet of metal, while γ-radiation consists of high-energy photons and can penetrate even thin sheets of metal or concrete but not a thick layer of lead.

Radioactive decay

As radionuclides constantly release energy in the form of radiation, they change their atomic mass, meaning that they change from one element into another. For example, ^{137}Cs (caesium) decays to ^{137}Ba (barium). The speed at which this happens, the radioactive half-life, varies widely between radionuclides from fractions of a second to millions of years. The SI unit for radioactivity is the Becquerel (Bq); one Bq is equivalent to one radioactive disintegration per second.

Where do radionuclides come from?

Many radionuclides occur naturally in the environment; primordial radionuclides such as ^{238}U (uranium), ^{232}Th (thorium) and ^{40}K (potassium) originate from before the solar system was created, while others, such as ^{14}C, are produced when cosmic rays penetrate the atmosphere. Other radionuclides, *e.g.* technetium (Tc) isotopes, are produced synthetically by humans, for example in nuclear reactors. Even naturally occurring radionuclides may cause environmental contamination when their concentrations are enriched by anthropogenic activities such as mining.

Main differences with respect to other contaminants

In contrast to other environmental contaminants, external exposure to ionising radiation can cause harm. This means that organisms do not necessarily need to come in contact with the radiation source in order to be affected. This is particularly true for γ-radiation that can easily pass through an organism. Ionising radiation from a range of different radionuclides has an identical effect, and thus its dose is additive. Radioactive decay can mean that radioactive isotopes can become stable and non-toxic elements, and thus overall radioactivity will decrease in the environment, *e.g.* ^{134}Cs decays to stable ^{134}Ba and ^{134}Xe (xenon).

Modes of action and potential effects of radiation

One main mode of action regarding radiation is direct damage to molecules such as DNA. For example, it can cause double-strand breaks, which can lead to defective DNA, mutations, gamete damage and hereditary effects, or even cell death. The other main mode of action is oxidative stress (*cf.* Sect. 16.5.2) since radiation can split water molecules in tissues, creating free radicals. In common with other environmental contaminants, most of our knowledge on biological effects comes from experiments using large acute doses and short-term effects, often in mammalian cells. However, it is clear that different organisms have very different radiosensitivities, even if the underlying mechanisms for this are poorly understood. In addition, the effects depend on radiation type, duration of exposure, whether exposure is internal or external, time of exposure (relative to life stage, season etc.), protection systems in the organism (*e.g.* the ability to respond to oxidative stress and the type of cell exposed with germ cells and juvenile organisms often being the most sensitive targets). Health risks can also be related to the biological behaviour of the radionuclides in biological tissues, *e.g.* ^{137}Cs is similar to potassium and ^{90}Sr (strontium) to calcium.

16.1.13 Nature's own toxic chemicals

In this chapter we deal with anthropogenic chemicals, but it should be kept in mind that there are also natural sources of toxic compounds in the Baltic Sea, *e.g.* hepatotoxins produced by cyanobacteria (Box 16.4), a neurotoxic amino acid produced by cyanobacteria and microalgae (Box 16.5) and ozone-damaging volatile halocarbons produced by macroalgae and other organisms (Box 16.6). However, the amounts of natural toxins produced in the ecosystem can be influenced by human impacts, *e.g.* eutrophication can increase the occurrence and magnitude of cyanobacterial blooms and environmental stress can increase the production of volatile halocarbons by macroalgae.

16.2 Biological effects of contaminants

16.2.1 Effects on different levels of biological organisation

In order to understand and assess the impacts of hazardous substances on the Baltic Sea ecosystem, both their concentrations and biological effects have to be taken into consideration. While contaminants can show their effects at the molecular level and the cellular level (Table 16.1), their environmentally most relevant effects are manifested at the population, community and ecosystem levels (Fig. 16.12). In this respect it is also important to consider the position and function of different groups of species in the Baltic Sea food web.

The various ecosystem components are exposed to and affected by many contaminants simultaneously. Different organisms have different sensitivities to a given contaminant or contaminant mixtures. These differences in sensitivity might change the ecological balance between species. For example, interactions between predators and their prey, interactions between parasites and their hosts, and competition between species, can be affected. This could in turn lead to changes at the ecosystem level that are not a direct, but an indirect, consequence of chemical pollution. Also, effects of hazardous substances on central ecological processes, such as primary production or nutrient cycling, may be just as important as their effects on community composition and structure. However, our understanding of the ecosystem-level effects of chemical pollution, including their indirect effects, is still poor.

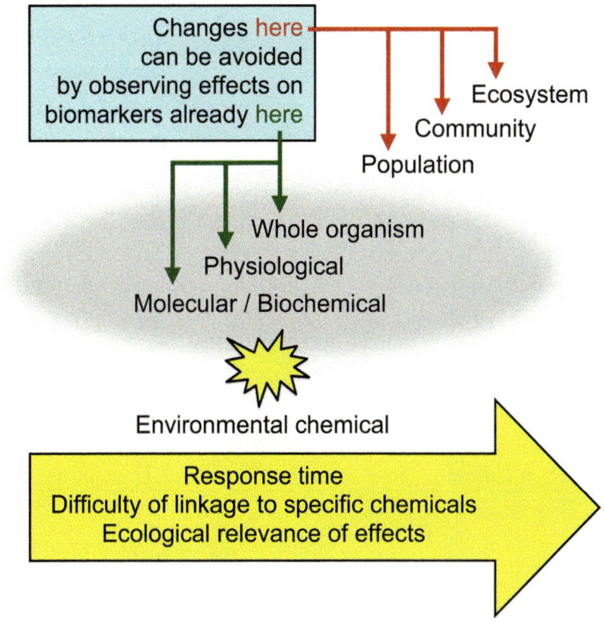

Fig. 16.12 Conceptual model of the biomarker approach in environmental monitoring. Figure modified from Walker et al. (2001)

16.2.2 Primary consumers

A range of ecotoxicological field studies have been carried out on the hard-bottom filter-feeding blue mussel *Mytilus trossulus*, which is a key species in the Baltic Sea ecosystem (*cf.* Sect. 11.13.1). Due to its wide distribution, large populations and high biomass, it is a dominant species at the lower consumer level of the Baltic Sea food web and an important link between the pelagic and benthic compartments of the ecosystem.

Changes in biomarker responses (Sect. 16.4.6) in *Mytilus trossulus* have been observed in various coastal areas of the Baltic Sea. In most cases, these gradients coincide with those observed in concentrations of environmental chemicals, including DDTs, PCBs and heavy metals (Baršienė et al. 2006a; Lehtonen et al. 2006b; Kopecka et al. 2006; Schiedek et al. 2006; Dąbrowska et al. 2012; Höher et al. 2012; Turja et al. 2013, 2014). Impacts of chemical pollutants on key

Box 16.4: Cyanobacterial hepatotoxins

Kaarina Sivonen

Toxic cyanobacterial blooms are an annual summer phenomenon in the Baltic Sea

Every summer, from July to August, diazotrophic (nitrogen-fixing) filamentous cyanobacteria form mass occurrences (blooms) in the Baltic Sea. The major bloom-forming genera are *Nodularia*, *Aphanizomenon* and *Dolichospermum* (*cf.* Fig. 8.2). These mass occurrences can be toxic, causing animal poisonings and health risks for humans. However, these filamentous cyanobacteria not only form nuisance blooms and produce toxins, they are also important primary producers and nitrogen-fixers and contribute to the productivity of the Baltic Sea.

Nodularin and microcystins

Two cyanobacterial hepatotoxins, nodularin and microcystins, have been detected in the Baltic Sea (Box Fig. 16.3). The toxin produced by *Nodularia spumigena* (Box Fig. 16.4) is nodularin, a pentapeptide whose chemical composition can be written as cyclo-(D-MeAsp-L-arginine-Adda-D-glutamic acid-Mdhb) (Sivonen et al. 1989). D-MeAsp is D-erythro-ß-methylaspartic acid, Mdhb is 2-(methylamino)-2-dehydrobutyric acid, and Adda is (2*S*,3*S*,8*S*,9*S*)-3-amino-9-methoxy-2,6,8-trimethyl-10-phenyldeca-4,6-dienoic acid, a non-proteinogenic amino acid found only in cyanobacterial toxins (Sivonen et al. 1989). Microcystins are heptapeptides, cyclo(-D-Ala-X-D-MeAsp-Z-Adda-D-Glu-Mdha), where X- and Z are variable L-amino acids and Mdha is N-methyldehydroalanine. Both nodularin and microcystins contain Adda and glutamic acid, the constituents that are considered responsible for the bioactivity by inhibiting serine/threonine-specific protein phosphatases of eukaryotic cells. The structure of microcystins is variable, more than 100 variants being known at present (Sivonen 2009). Nodularin is found in the Baltic Sea wherever *Nodularia spumigena* occurs, *i.e.* in the entire Baltic Sea, except for the Bothnian Bay where no *Nodularia* blooms occur. In the Gulf of Finland both microcystin-producing and non-producing strains of *Dolichospermum* have been detected (Halinen et al. 2007). The production of both nodularin and microcystins is reduced by phosphorus deficiency and less than optimal growth conditions (Sivonen 2009).

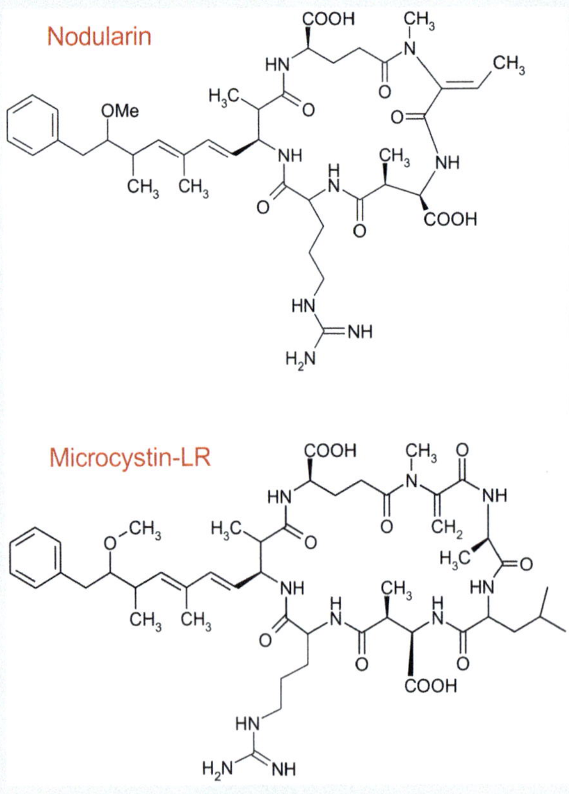

Box Fig. 16.3 The structures of the two cyanobacterial hepatotoxins nodularin and microcystin-LR occurring in the Baltic Sea

The detection of toxins and their production pathways

Nodularin and microcystins can be detected with several methods, such as protein phosphatase inhibition and ELISA tests. It can also be analysed with analytical chemistry techniques such as coupled liquid chromatography and mass spectrometry (LC-MS/MS). The two hepatotoxins are potent toxic compounds. In animal toxicosis, death is caused by haemorrhagic shock, and autopsy reveals a blood-engorged liver (Sivonen 2009). Microcystin and nodularin are produced nonribosomally by large multi-enzyme complexes called microcystin- and nodularin-synthetases, respectively. Nodularin is assembled by 48-kb nodularin synthetase containing the genes *ndaA* to *ndaI* whereas microcystin synthetase in *Dolichospermum* is larger (55-kb) and the genes are *mcyA* to *mcyJ* (Sivonen 2009). Both nonribosomal and polyketide-type synthetases are present, and the latter is responsible for Adda biosynthesis. The biosynthetic genes have been used to develop molecular detection methods for the toxin producers (Sivonen 2009). Studies of the evolution of microcystin and nodularin biosynthesis have indicated that the *nda* genes have evolved from the microcystin synthetase (*mcy*) genes through the deletion of two nonribosomal peptide synthetase modules and a change in the substrate-specificity of one nonribosomal peptide synthetase module (Sivonen 2009). Studies in the Baltic Sea revealed that nodularin synthetase *ndaE* gene copies determined by quantitative real-time PCR correlated well with the nodularin concentrations detected by LC-MS/MS. Most of the nodularin was detected in the surface water, but it is also found in deeper water layers (down to 30 m). Thus, toxic *Nodularia* blooms are widely distributed, not only horizontally but also vertically, and the Baltic Sea fauna is exposed to nodularin, which has been confirmed by the detection of nodularin in the Baltic Sea zooplankton and fish (Karjalainen et al. 2008).

Toxin production

Physiological studies of the strains of *Nodularia spumigena*, and analyses of field samples, have shown that toxins are continuously produced in high amounts (Sivonen et al. 1989). The whole genome of one Baltic Sea *Nodularia* strain is available, which enables the investigation of responses to environmental change at the genetic and transcriptome levels. Such studies found *Nodularia spumigena* to produce, in addition to nodularin, other cyclic peptides – nodulapeptins (Voß et al. 2013) – as well as the linear peptides spumigins and aeruginosins (Fewer et al. 2013). These peptides are serine protease inhibitors.

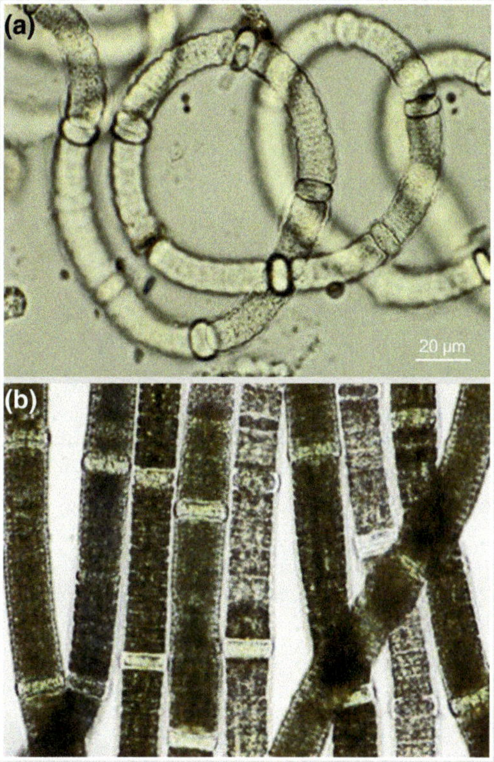

Box Fig. 16.4 The cyanobacterium *Nodularia spumigena*. (**a**) Coiled filaments. (**b**) Straight filaments. Photo: © Gertrud Cronberg

Box 16.5: The neurodegenerative toxin BMAA in the Baltic Sea

Sara Rydberg

A small amino acid

Production of harmful compounds is a well-known feature among phytoplankton groups and some of the toxic metabolites produced pose a significant environmental and health risk. Recently, the neurotoxic amino acid β-methylamino-L-alanine (BMAA) (Box Fig. 16.5) was discovered in the Baltic Sea food web and shown to be produced by cyanobacteria, diatoms and dinoflagellates (Jonasson et al. 2008, 2010; Jiang et al. 2014). BMAA is a small, hydrophilic, non-proteinogenic amino acid, *i.e.* not one of the 20 amino acids that represent the building blocks of proteins, and so far no genes encoding tRNA for BMAA have been discovered. BMAA has been shown to be linked to neurodegenerative diseases such as amyotrophic lateral sclerosis (ALS), parkinsonism dementia (PD) and Alzheimer's disease. The transfer and bioaccumulation of BMAA in the food web is similar to that of lipophilic organochlorines (PCB, DDT) as it follows a classical triangle model of biomagnification. However, since BMAA is not lipophilic but water-soluble, its biomagnification pathway differs from that of the lipophilic agents. BMAA occurs in the free cytosolic cell fraction or is associated with proteins. It has been hypothesised that the protein-associated BMAA serves as an endogenous neurotoxic "reservoir" (Murch et al. 2004).

The BMAA history

The neurotoxic effect of BMAA was discovered in the late 1940s on the Island of Guam – a small isolated island in the northwestern Pacific Ocean – where an exceptionally high frequency of the combined disease ALS/PD was discovered among the indigenous Chamorro people. The recorded incidence of ALS/PD was up to 100 times higher on Guam than on the US mainland (Spencer et al. 1987). The Chamorro people were most probably exposed to lethal doses of BMAA resulting from the bioaccumulation of this compound in the terrestrial food web. It is biomagnified from the symbiotic cyanobacteria *Nostoc* sp. living in the coralloid roots of the cycad *Cycas micronesica* to flying foxes *Pteropus mariannus* that forage on the cycad seeds, and finally to the people consuming both cycad seeds and flying foxes (Cox and Sacks 2002). Since the discovery of BMAA in the Guam food web, the high incidences of ALS/PD on Guam have declined in concert with a reduced human consumption of flying foxes (Monson et al. 2003). Eventually, BMAA-producing organisms turned out to be globally distributed, and the toxin was proven to be biomagnified not only in terrestrial ecosystems but also in aquatic food webs (Jonasson et al. 2010; Lage et al. 2015).

BMAA toxicity

According to new findings, BMAA can replace serine in human neuroproteins. Such incorporation of BMAA leads to misfolded and dysfunctional protein aggregations, which further results in chronic toxicity with a slow continuous loss of motor neurons resembling that of neurodegeneration (Dunlop et al. 2013). In a recent study, dietary exposure of the vervet monkey to BMAA was shown to trigger the formation of neurofibrillary tangles and amyloid deposits, which are diagnostic hallmarks of several neurodegenerative diseases such as those found in the ALS/PD patients on Guam (Cox et al. 2016). Both the BMAA-mediated overstimulation of glutamate-responsive receptors and the misincorporation of BMAA into neuroproteins have been suggested as causes of the neurodegeneration.

BMAA production and bioaccumulation in the Baltic Sea

The summer cyanobacterial blooms in the Baltic Sea (Box Fig. 16.6) continuously produce BMAA (Jonasson et al. 2010) as do diatoms and dinoflagellates, and BMAA bioaccumulates at higher trophic levels (Jiang et al. 2014; Lage et al. 2015). In the Baltic Sea, as well as in many other water bodies in the world, the occurrence of phytoplankton blooms has increased due to anthropogenic nutrient inputs, thus potentially promoting BMAA production and

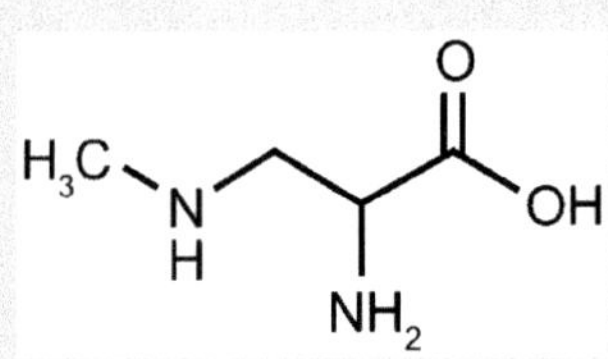

Box Fig. 16.5 The chemical structure of the neurotoxic amino acid β-methylamino-L-alanine (BMAA)

Box Fig. 16.6 A cyanobacterial bloom in the western Karlskrona archipelago (southern Baltic Sea proper) in summer 2005. Photo: © Sara Rydberg

bioaccumulation. In the Baltic Sea, the highest levels of BMAA were found in the bottom-dwelling fish species: the turbot *Scophthalmus maximus*, the fourhorn sculpin *Myoxocephalus quadricornis* and the smelt *Osmerus eperlanus*. Comparatively high levels of BMAA have also been found in filter-feeding organisms such as blue mussels and oysters harvested on the Swedish west coast (Jonasson et al. 2010). In fish and humans, BMAA accumulates primarily in the brain tissue, but it has also been detected in the liver and muscle of Baltic Sea fish (Jonasson et al. 2010; Lage et al. 2015). To date, the physiological function of BMAA in BMAA-producing organisms has still not been revealed. Some studies suggest that BMAA in cyanobacteria may be connected to the nitrogen metabolism, *e.g.* BMAA production can be stimulated by nitrogen starvation in the non-diazotrophic unicellular species *Microcystis* PCC7806 and *Synechocystis* J341 (Downing et al. 2011; Berntzon et al. 2013).

species at the primary consumer level (such as *Mytilus trossulus*) can have knock-on effects in a wide range of other organisms and may ultimately lead to alterations in ecosystem structure and functioning.

16.2.3 Higher trophic levels

Biological effects of contaminants in Baltic Sea fish have been widely reported, *e.g.* for the European flounder *Platichthys flesus* (Baršienė et al. 2006a, b; Kopecka et al. 2006; Napierska et al. 2009), the Atlantic cod *Gadus morhua* (Schnell et al. 2008), and the viviparous eelpout *Zoarces viviparus* (Gercken et al. 2006). In most cases the metabolism of organic contaminants is markedly more efficient in fish than in invertebrates. For example, in bivalves, the activity of the cytochrome P450 1A (CYP1A) biotransformation system is very low, which usually results in non-detectable activity levels of the biomarker ethoxyresorufin-O-deethylase (EROD, Table 16.3) after exposure to organic compounds such as PAHs and PCBs. In contrast, fish from the same field areas, contaminated by the same organic compounds, usually have elevated EROD activity, as well as accumulation of PAH metabolites in their bile fluid.

Few contaminants apart from organic compounds are actually biomagnified. Increased detoxification of organic contaminants at higher trophic levels is crucial due to the high accumulation of contaminants in the levels below. Secondary consumers can experience a biomagnified amount of non-metabolised hazardous substances. In addition, the metabolically activated degradation products formed within a consumer can be even more harmful than the parent compound from the ingested food, posing an increasing health threat to the consumer. Higher trophic levels play an important role in the food web through predation on lower trophic levels. Reductions in numbers or the decreased health of top predators by hazardous substances will therefore affect the whole food web through altered predation pressure on other species (*cf.* Sect. 8.9.5).

Box 16.6: Volatile halocarbons

Marianne Hielm Pedersén

Volatile halocarbons in the atmosphere

Large amounts of volatile halocarbons (VHCs) are released from the Earth's surface to the atmosphere. These gases consist of small organic fluorine-, chlorine-, bromine- and/or iodine- containing molecules such as chloroform ($CHCl_3$), bromoform ($CHBr_3$) and trichloroethylene (C_2HCl_3). The atmospheric lifetimes of VHCs range from minutes to centuries, and therefore they are often subdivided into "short-lived" and "long-lived" compounds. The long-lived VHCs are important because, after their release from the Earth's surface, they mix rapidly in the troposphere, and are advected into the stratosphere where photolysis leads to the formation of inorganic halogen species that participate in the catalytic destruction of ozone (O'Dohety and Carpenter 2007). Among the VHCs, the low-weight fluorine- and chlorine-containing VHCs reach farthest up into the stratosphere while the heavier iodine-containing VHCs stay in the troposphere. Thus, the most damaging VHCs for the ozone layer are the fluorocarbons and the chlorocarbons.

Sources of volatile halocarbons

The sources of VHCs are both natural and anthropogenic processes on the Earth's surface. Chlorine-containing compounds provide by far the largest flux of halocarbons to the atmosphere (O'Dohety and Carpenter 2007). Notorious VHCs of anthropogenic origin are the chlorofluorocarbons (CFCs) that have been used in *e.g.* refrigerators and air conditioners since the 1930s; these were later replaced by the hydrochlorofluorocarbons (HCFCs). CFCs peaked in 1974 when 800 kilotonnes were released into the atmosphere, and it was also then that their potential role in ozone depletion was first postulated (McCulloch 1999). At present, the production of CFCs has been phased out and the production of HCFCs has decreased. However, due to their long lifetimes, their concentrations in the atmosphere are still high (O'Dohety and Carpenter 2007). Other VHCs are also produced anthropogenically and/or naturally. For example, biomass burning and the oceans are the principal sources of methyl chloride (CH_3Cl), the most abundant chlorocarbon in the atmosphere. While methyl chloride is largely natural in origin it has been estimated to be responsible for ~16 % of the chlorine-catalysed ozone destruction in the stratosphere (Montzka and Fraser 2003). Chloroform released from the oceans is estimated at ~360 Gg year^{-1} compared to ~220 Gg year^{-1} from soils (McCulloch 2003). The ocean also releases large amounts of methylene chloride (CH_2Cl_2, Khalil et al. 1999), trichloroethylene (Khalil et al. 1999) and bromoform (Carpenter et al. 2003).

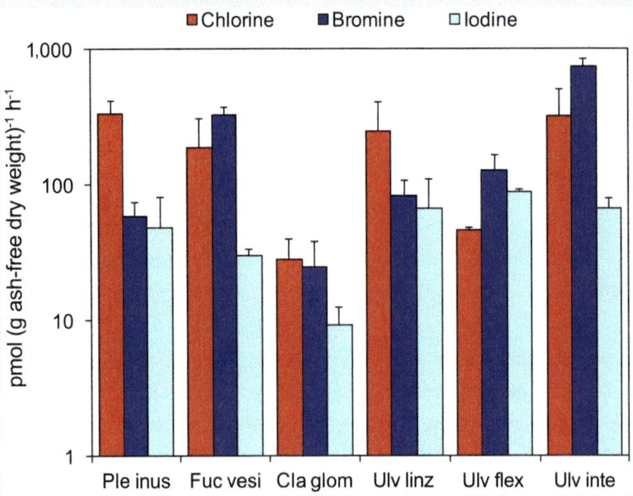

Box Fig. 16.7 Production rates of chlorocarbons, bromocarbons and iodocarbons by six algal species during six hours of incubation. Ple inus = *Pleurosira inusitata* (diatom that lives in large colonies), Fuc vesi = *Fucus vesiculosus* (brown alga), Cla glom = *Cladophora glomerata* (green alga), Ulv linz = *Ulva linza* (green alga), Ulv flex = *Ulva flexuosa* (green alga), Ulv inte = *Ulva intestinalis* (green alga). Note that the *y*-axis has a logarithmic scale. Error bars represent 1 standard error of the mean. Figure modified from Abrahamsson et al. (2003)

VHC production by algae

Much of the natural production of volatile halocarbons in the oceans is biogenic (Abrahamsson et al. 1995), although oxidation processes during the chemical degradation of organic matter have been found to be significant as well (Keppler et al. 2000). It has been estimated that ~70 % of the global bromoform is produced by marine algae (Carpenter and Liss 2000). The formation of volatile halocarbons by algae involves halo-peroxidases and stress-induced hydrogen peroxide (H_2O_2, Pedersén et al. 1996), which, in turn, is enzymatically reduced to H_2O with the subsequent oxidation of bromide and chloride ions to HOBr and HOCl, respectively. However, the formation mechanisms are not fully understood yet, and various mechanisms may be involved in the formation of different halocarbon species. The production rates of 16 volatile halocarbons in six algal species from the Baltic Sea, as well as their quantitative composition, were found to be strongly species-dependent (Abrahamsson et al. 2003). For example, the production of chlorocarbons was the highest in the diatom *Pleurosira inusitata* and two *Ulva* species, while that of bromocarbons was highest in *Fucus vesiculosus* and *Ulva intestinalis* (Box Fig. 16.7). *Cladophora glomerata*, which has a very low stress-induced production of hydrogen peroxide (Choo et al. 2005), showed the overall lowest production of VHCs.

16.2.4 External diseases and histopathology in fish

stress by contaminants can degrade the immunological response of organisms, leading to a lower capacity to deal with pathogens and parasites (Walker et al. 2001). Diseases, diagnosed based on external symptoms such as ulcers, cysts and histopathology, have been monitored on a regular basis since the beginning of the 1980s in several Baltic Sea fish species. Studies on liver histopathology in the European flounder suggest a relationship between the health status of the fish and contaminant levels in the environment at various study locations in the Baltic Sea (Lang et al. 2006).

Significant changes in the disease prevalence of fish are used as a general ecosystem health indicator reflecting the effects of environmental change, including anthropogenic impacts, on their disease-resistance capacity. The diseases currently monitored cannot be directly linked to any specific hazardous substance and are likely to be caused by multiple stressors in the environment (HELCOM 2010a).

16.2.5 Reproductive disorders

Data on reproductive disorders are of considerable value as they may reflect population effects caused by contaminants. Impacts of contaminants on the reproduction of top predators such as waterbirds and marine mammals inhabiting the Baltic Sea have been intensively studied and studies have also been conducted on invertebrates and fish (HELCOM 2010a).

The reproductive success of the benthic deposit-feeding amphipod *Monoporeia affinis* (*cf.* Fig. 4.30), a dominant species inhabiting the soft bottoms of the Baltic Sea, has been studied in field populations collected from contaminated and pristine locations (Eriksson-Wiklund and Sundelin 2001; Sundelin et al. 2008; Reutgard et al. 2014). Among the parameters measured were sexual maturation, fecundity (embryos per female), embryo developmental stage, malformed embryos, undifferentiated embryos, and dead embryos/broods. The determination of malformed embryos of *Monoporeia affinis* was shown to be a sensitive method in the detection of contaminant effects. By analysing the different types of embryo aberrations it is also possible to differentiate between natural environmental stress, such as oxygen deficiency and temperature stress, and stress caused by chemical contaminants.

Potential effects of contaminants on the viviparous eelpout *Zoarces viviparus* (*cf.* Fig. 14.5) have been observed in coastal areas of Sweden, Denmark and Germany (Strand et al. 2004; Gercken et al. 2006). The presence of abnormally developed embryos and larvae in eelpout broods is used as an indicator of impaired reproduction because chronic exposure to various contaminants has the potential to induce adverse developmental effects. Other studies on endocrine disruption in the viviparous eelpout carried out in Danish and German coastal waters have also recorded a widespread occurrence of intersex in the form of primary oocyte development in the testes of more than 25 % of the male fish studied (HELCOM 2010a).

The health status and reproduction of the white-tailed eagle *Haliaeetus albicilla* have improved since the late 1960s and since the 1990s population sizes have largely returned to the pre-1950 levels, which is correlated with the general decrease of POPs. However, an opposite trend with increasing chick mortality and population decline has been recorded regarding the nominate subspecies of the lesser black-backed gull *Larus fuscus* (*cf.* Fig. 4.14a) in the Gulf of Finland, which is associated with observations of very high

Table 16.3 List of recommended biomarkers for the monitoring of biological effects of hazardous substances in the Baltic Sea Area. Data from HELCOM (2010a)

Biomarker/ biological effect	Hazardous substance	Mode of action	Biological response
Imposex	Tributyltin (TBT)	Endocrine disruption	Reproductive failure
Polyaromatic hydrocarbon (PAH) metabolites in fish bile	PAH compounds	PAH metabolism	Metabolite accumulation
Ethoxyresorufin-*O*-deethylase (EROD) activity	Organic contaminants	Enhanced detoxification	Increased P450 system activity
Acetylcholinesterase (AChE) activity	Specifically organophosphates and carbamates, as well as other compound groups	Neurotoxicity by inhibition of key enzyme activity	Neurological disorders, behavioural change, tetanus, paralysis, death
Lysosomal membrane stability (LMS)	Various	Cellular homeostasis breakdown, membrane damage	Cell damage, cell death
Micronuclei frequency	Various	Genetic damage	Chromosomal aberrations
Reproductive disorders in eelpout	Various	Endocrine disruption, direct toxicity	Intersex, skewed sex ratios, malformed larvae, dead larvae
Reproductive disorders in amphipods	Various	Endocrine disruption, direct toxicity	Dead eggs, malformed embryos, dead broods
External diseases in fish	Unspecific	Immunosuppression	Skin lesions, parasites
Liver histopathology in fish	Carcinogenic compounds, as well as other compound groups	Liver toxicity	Tissue changes, benign tumours, cancer

concentrations of organochlorine compounds in the liver of the chicks (Hario et al. 2004).

The prevalence of intestinal ulcers in immature individuals of the grey seal *Halichoerus grypus*, which is suspected to be related to the presence of PCBs, has decreased after the high ulcer levels of the 1970s. PCBs have also been associated with interrupted pregnancies and uterine obstructions in both the grey seal and the ringed seal *Pusa hispida*, as well as with uterine leiomyomas (benign tumours in the uterus) in the grey seal *Halichoerus grypus*. This probably contributed to the decline of the Baltic Sea seal populations in the 1960s and 1970s. No uterine obstructions have been observed since 1997 and the occurrence of uterine leiomyomas has decreased as well (HELCOM 2010a).

16.3 Effects of specific chemicals on organisms

16.3.1 A complex issue

The hazards posed by anthropogenic chemicals in the environment are determined by a complex set of factors, including the chemicals' concentrations, behaviour in the environment (*e.g.* physico-chemical properties, persistency and bioavailability), behaviour in the biota (*e.g.* toxicokinetics and toxicodynamics), and the deleterious biological effects they are likely to trigger in organisms. As an example of this broad topic, we here concentrate on only one well-studied group of hazardous substances, the organotin compounds (compounds based on tin with hydrocarbon substituents), which has received particular worldwide attention during the last decades.

16.3.2 Getting rid of fouling organisms: tributyltin (TBT)

Tributyltin (TBT) is a highly toxic biocide, which has been extensively used as an antifouling agent in paints for ship hulls and various constructions in the aquatic environment. TBT was applied to inhibit the growth, attachment, growth and development of aquatic organisms on surfaces; it achieved its purpose by being slowly released from the paint into the water and killing the fouling organisms. Due to its lipophilic character and low water solubility, TBT readily adsorbs on to particles and organic materials, and is now widely dispersed in the aquatic environment, mainly in sediments. Ship traffic and recreational boats are regarded as the main sources of TBT in the Baltic Sea environment (Strand and Jacobsen 2002; Eklund et al. 2008).

Because of the purpose of the TBT application and its high efficiency, it should have been apparent that all other, non-target organisms present in the aquatic environment might be at risk. However, this was not the case. When the use of TBT started in the mid-1960s, one major basic principle was that a low dosage and exposure resulted in a low risk. Based on this principle, the dilution of toxicants in

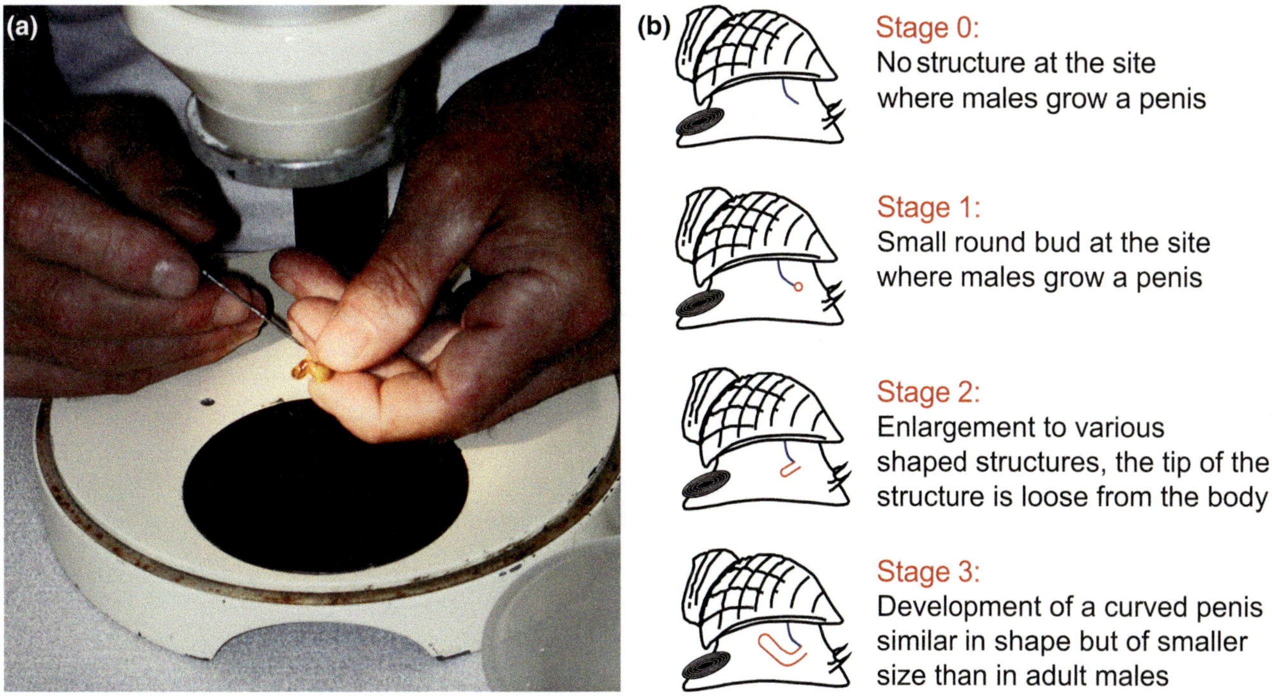

Fig. 16.13 Tributyltin (TBT) causes imposex in gastropods. (**a**) Dissection of a snail to investigate imposex. (**b**) Different stages of imposex in females according to the penis classification index (PCI) of the common whelk *Buccinum undatum*, which in the Baltic Sea Area occurs in the Kattegat, Belt Sea and Arkona Sea. The size of the penis (red) is decisive for the stage classification of a snail while the sperm duct (blue) can be present or absent, even in TBT affected females of PCI Stage 0. Photo: (**a**) © Pauline Snoeijs-Leijonmalm. Figure (**b**) modified from Mensink (1999)

whole seas and oceans was assumed to work as a precaution for possible deleterious environmental impacts.

Today we know that the assumption that "the solution to pollution is dilution" does not always apply. For example, the compounds now categorised as endocrine disrupting chemicals (EDCs) display their full toxic potential even at very low concentrations. In the case of TBT, effective concentrations provoking strong toxic responses in marine molluscs are more than 1,000 times lower than any other known contaminant present in the marine environment (His et al. 1999). While in mammals the metabolism of TBT is rapid, as shown by the detection of metabolites in blood within three hours following TBT administration, it is much slower in invertebrates, particularly in molluscs. Thus, the bioaccumulation capacity of TBT, a result of its lipophilic character, is higher in molluscs than in mammals. The highest TBT concentrations measured in the digestive/reproductive complex of gastropods were up to 100,000 times higher than the concentrations measured in the aquatic environment (Sternberg et al. 2010).

16.3.3 TBT and imposex

The biological effects of TBT in molluscs are linked to essential bodily functions, including immunosuppression. The compound has also been shown to affect the settlement, growth and mortality of mollusc larvae, shell deposition of growing bivalves, as well as gonadal development and the gender of adults. The so-called "superimposed sex" or, in short, "imposex", is the development of male characteristics in female gastropods, ending with the full development of a penis and *vas deferens* (sperm duct) (Fig. 16.13). The frequency of imposex in female gastropods is correlated with the tin concentrations in the gastropod tissue (Fig. 16.14).

Imposex is accompanied by functional sterility and poses a high risk to marine mollusc populations, especially to populations living in the close vicinity of harbours and active shipping lanes. It is now established that imposex is a form of endocrine disruption caused by elevated testosterone levels that masculinise TBT-exposed females. Although the precise mechanisms of the increased testosterone levels have not yet been fully described, it has been suggested that TBT acts as a competitive inhibitor of a cytochrome P450-mediated enzyme called aromatase (Oberdörster and McClellan-Green 2002; Santos et al. 2002). Imposex gastropods are widely distributed globally and at least 195 species of prosobranch gastropods are known to be affected (Sternberg et al. 2010). Imposex appears to be irreversible and can thus have long-term impacts, both on an individual's fitness and on the population structure.

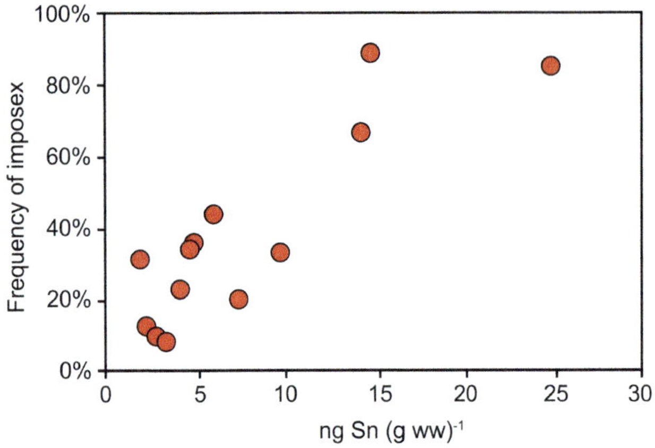

Fig. 16.14 The relationship between the average concentration of total butyltin in ng Sn per g wet weight (ww) in snail tissue and the recorded frequency of imposex in females of the common whelk *Buccinum undatum* at 13 sampling stations in the Kattegat in 1996–1998. Figure modified from Strand and Jacobsen (2002)

16.3.4 Molluscs in TBT monitoring

Imposex (sex reversal) and intersex (dual sexual characteristics) are widely used as specific biomarkers of exposure to TBT in marine monitoring programmes (OSPAR 2008; Strand 2009; HELCOM 2010a). In the Baltic Sea Area, especially in the southern Belt Sea, the red whelk *Neptunea antiqua* is used as a sensitive indicator species to monitor the effects of TBT. This is performed by assessing the so-called "*vas deferens* sequence index" (VDSI), a standardised measure for the expression of the severity of imposex. In Danish waters, imposex has been observed in nine species of gastropods, *e.g.* in the common periwinkle *Littorina littorea*, and the condition often coincides with intersex (Strand et al. 2003; Strand and Jacobsen 2005).

As a result of restrictions in the use of TBT as an antifouling agent and, finally, due to the global ban of TBT in 2003, the environmental levels of TBT dropped. This is reflected in a decrease in the levels of imposex during recent years, particularly in coastal species such as the netted whelk *Nassarius reticulatus* (syn. *Hinia reticulata*), which occurs in the transition zone (Kattegat and Belt Sea) but not in the Baltic Sea. However, the positive trend is not as obvious in the more sensitive and long-living *Neptunea antiqua* in the open waters of the Belt Sea. Even several years after the global ban of TBT, 100 % of *Neptunea antiqua* individuals sampled in the Belt Sea still showed different degrees of imposex, with 10 % of females recorded as sterile (Strand 2009).

Imposex also occurs in the mud snail *Peringia ulvae* (syn. *Hydrobia ulvae*) on the German North Sea and Baltic Sea coasts (Schulte-Oehlmann et al. 1997). Thus, mud snails may potentially be used for the monitoring of TBT effects in the northern part of the Baltic Sea where other suitable gastropod species are not present. The overall trend of TBT levels in the biota of the Baltic Sea shows a decrease over the last decade (Nyberg et al. 2013). However, in 2005–2011 the TBT concentrations in many places of the transition zone, in *e.g. Mytilus trossulus*, still exceeded the EU "good environmental status" (GES, *cf.* Sect. 17.8.1) threshold of 12 μg TBT (kg dry weight)$^{-1}$ for biota while the TBT concentrations in herring and perch muscle exceeded the GES threshold of 15.2 μg TBT (kg wet weight)$^{-1}$ for seafood in the eastern Gulf of Finland. Furthermore, the TBT concentrations in surface sediments all over the Baltic Sea are still extremely high (Nyberg et al. 2013).

16.4 Ecotoxicological testing, monitoring and assessment

16.4.1 Tailoring the methods for the Baltic Sea

The special abiotic conditions of the Baltic Sea, such as low salinity, large seasonal shifts in water temperature and widespread near-bottom hypoxia (<2 mL O_2 L^{-1}), in combination with the limited water exchange with the North Sea (*cf.* Sect. 2.3.8), call for special requirements regarding the assessment of anthropogenic chemical pollution impacts on the ecosystem by ecotoxicological testing, and monitoring and assessment methods.

The biodiversity of many groups of macroscopic organisms in the Baltic Sea is lower than in marine and freshwater environments because only a few species possess adequate physiological adaptations to survive and reproduce in the brackish-water environment (*cf.* Sect. 4.2). These biological characteristics of the ecosystem set practical limitations on the selection of suitable species for ecotoxicological testing and environmental monitoring. In addition, most ecotoxicological test methods in current use are standardised either for true freshwater or true marine species or conditions, and these methods are thus in many cases not directly applicable to the brackish Baltic Sea.

16.4.2 Biotests: the basic concept

Ecotoxicological test methods using laboratory-cultivated or field-collected species or cell lines grown in the laboratory over several generations are called "biotests" (Mothersill and Austin 2003). Under carefully standardised conditions, the test systems are exposed for a defined period of time to the test media, which can be single chemicals, mixtures of chemicals, different dilutions of water collected from a study site in the field, sediment samples or sediment pore water extracts or elutriates.

At its most extreme, the measured test parameter, called the "biological endpoint" in ecotoxicology, is the death of the test organisms. The concentration required to kill 50 % of the test organisms, referred to as the "lethal concentration" (LC_{50}) is measured over a fixed time, *e.g.* 24 to 96 hours in short-term acute toxicity tests, and from weeks up to several months in long-term chronic exposure tests. This is similar to the parameter "lethal dose" (LD_{50}) used in the testing of chemicals by oral dosing or injection. However, today's trend in testing is to use more sensitive endpoints than mortality, and the so-called "sublethal effects" are now more often measured and expressed using the term "effective concentration" (EC_{50}). Sublethal endpoints include changes in behaviour, growth, reproduction, and also biomarkers representing various biological effect levels from gene expression to population effects.

By performing lethal and sublethal biotests, it is also possible to obtain other parameters that are commonly used in ecotoxicology and risk assessment, *e.g.* the "predicted effect concentration" (PEC), the "predicted no-effect concentration" (PNEC), the "lowest observed effect concentration" (LOEC) and the "no observed effect concentration" (NOEC).

Biotests are routinely used to monitor the toxicity of municipal and industrial wastewaters. Water samples taken from the effluent or from the receiving water body are subjected to standardised toxicity tests that indicate effects on different biological functions such as muta-, geno-, and cytotoxicity, growth, reproduction and behaviour (Table 16.4). Sediment biotests are used for the evaluation of chemical pollution in *e.g.* harbour and industrial areas, dredged materials and along marine shipping routes.

Standard biotest species commonly consist of microbes, algae, invertebrates, and small fish. The test protocols are often standardised for freshwater conditions, although marine biotests also exist. Thus, the comparison of a sample's toxicity using different tests and test species is not always straightforward, often far from that. To enable comparability, specific assessment criteria and thresholds must be developed for each test.

16.4.3 Sediment biotesting

Sediment toxicity can be assessed by exposing organisms to the whole sediment sample, sediment suspensions, aqueous elutriates or pore water samples, or to organic solvent extracts obtained in different ways (Nendza 2002). The sample preparation method can significantly affect the chemicals ultimately present and/or bioavailable in the test media of the biotest.

Simple mixing with water without any specific physical or chemical treatment such as the addition of organic chemicals will inherently result in the different (lower) extraction efficiency of the chemicals present in the sediment matrix. In fact elements that are tightly bound in the sediment are not directly bioavailable to organisms. On the other hand, extraction using organic solvents makes them bioavailable, which may lead to an overestimation of the real toxicity of a sediment sample. Another aspect to consider is that the toxicity of a field sample is seldom due to one chemical compound alone.

In sediment toxicity biotesting, benthic invertebrates such as amphipods, polychaete worms, and larvae or embryos of bivalves (*e.g.* oysters and mussels) and echinoderms (*e.g.* sea urchins), are widely used (Nendza 2002). The gastropod *Potamopyrgus antipodarum*, which is a non-indigenous species in the Baltic Sea (*cf.* Box 5.9), has been shown to be a useful test species, especially with regard to endocrine disruption (Duft et al. 2003). Most of the standard amphipod test species (*e.g. Hyalella azteca* and *Ampelisca* sp.), as well as echinoderms, do not occur in (most parts of) the Baltic Sea Area. Therefore, recent attempts have focused on the development and testing of biotests using Baltic Sea species such as the benthic harpacticoid copepod *Nitokra spinipes* and the amphipod *Monoporeia affinis*, and even non-indigenous species that have recently invaded the Baltic Sea such as the amphipod *Gmelinoides fasciatus* (Eklund et al. 2010; Berezina et al. 2013).

Table 16.4 Examples of standardised biotests used in the Baltic Sea Area for assessing the toxicity of marine environmental samples

Biological endpoint	Common method
Mutagenicity	Ames mutagenicity test (bacterial reverse mutation assay)
Acute toxicity	Bacterial bioluminescence inhibition test (*Vibrio fischeri*)
Growth inhibition	Algal growth inhibition test (*e.g.* the diatom *Skeletonema* sp.) Macrophyte growth inhibition test (*e.g.* the red alga *Ceramium tenuicorne* or the plant *Lemna* sp.)
Effects on reproduction	Water flea (*Daphnia* sp.) reproduction test Zebra fish (*Danio rerio*) reproduction test

16.4.4 Biotesting with molecular techniques

In addition to the use of whole organisms, molecular techniques for ecotoxicological biotesting have been developed. In many of these applications the target cells, often microorganisms such as bacteria and yeasts, have been genetically tailored to emit light in a dose-responsive way when exposed to specific (groups of) chemicals.

These methods include a technique to detect dioxins and/or dioxin-like PCBs, (anti)oestrogen compounds and (anti)androgen compounds by using CALUX (Chemical Activated LUciferase gene eXpression) assays (Murk et al. 1996). CALUX assays are based on the binding of the compound to the intracellular aryl hydrocarbon (Ah) receptor (responsible for the toxicity of a number of organic compounds), the transportation of the chemical-receptor complex into the cell nucleus and its subsequent binding to specific sequences in the DNA (the so-called "responsive elements", REs). The binding of the complex to the RE triggers the expression of RE-associated genes, and the toxicological impact of the chemical starts with the observed change in gene expression.

Other molecular-level approaches include the application of multi-gene expression profiles (microarrays, at best containing the whole genome of the test species) and proteomics by using the induction profiles of proteins synthesised (*e.g.* Kosmehl et al. 2012). The application of such methods has so far been relatively rare in routine testing, but is expected to increase.

16.4.5 What do biotests tell us?

Whatever biotest method is chosen for an ecotoxicological assessment, its ecological relevance is inherently low. Exposing standard test organisms under highly-controlled experimental conditions greatly increases the reproducibility and reduces the internal variability of the measurements, but poorly reflects the situation that prevails under natural conditions where various environmental (abiotic and biotic) factors also contribute to toxicity.

Interactions between the chemical compounds present in a sample and their bioavailability under varying physico-chemical conditions (*e.g.* salinity, temperature, light, pH, oxygen content, redox state, etc.), which are carefully controlled in the laboratory, may significantly modify the toxicity of water or sediment under the specific field conditions that the resident "wild" biota is facing. Furthermore, interactions can also occur at the biological level, *i.e.* the organisms themselves respond to contaminants differently depending on the environmental conditions in the field situation.

However, since biotests at their best are sensitive in detecting different types of toxic effects caused by chemicals and their mixtures, they do have great value as screening tools for environmental samples. Protocols such as toxicity identification and evaluation (TIE) and effect-directed analysis (EDA) have been designed to identify the actual compounds or sample fractions causing the observed toxic effects (U.S. EPA 2007; Brack et al. 2007; Burgess et al. 2013). The development of these types of methods linking *e.g.* sediment toxicity to its specific causes/contaminants is of great benefit for practical environmental management.

To conclude, biotests serve as useful tools for the assessment of environmental samples' toxicity, but their limitations have to be understood and considered. They are best used in combination with other methods such as environmental chemistry and biomarkers measured in organisms collected *in situ*.

16.4.6 Biomarkers: towards early detection of effects

The term "biomarker" is often interpreted as a suborganism level indicator of disturbance regarding physiological processes potentially affecting the health status of organisms. This includes, *e.g.* changes in gene expression, protein synthesis and their function, damage of cell structures and function, physiological dysfunctioning and pathological lesions in organs, mostly involved in detoxification and reproduction processes (Table 16.3). However, higher biological level indicators such as reproductive failure and behavioural change are often also considered as biomarkers. Whatever the definition, a common denominator of these biological endpoints is that they are measured in individual organisms and can indicate early effects of exposure to and/or effects of chemical pollution (Figs. 16.12 and 16.15), which is rarely possible when recording changes at the population or community levels.

The majority of subindividual level biomarkers are toxically-induced physiological processes, their end products, or mediated adverse effects including changes in enzyme activity or cell functioning and integrity. Overlapping with the broader term "bioindicator" (*cf.* Sect. 14.1.1), biomarkers are seldom related to any particular species or group of species. In many cases they are universal features applicable to most organisms. However, biomarker responses in different species show great variability due to the specific physiological characteristics of each species and their tolerance to chemical pollution, including biotransformation capabilities.

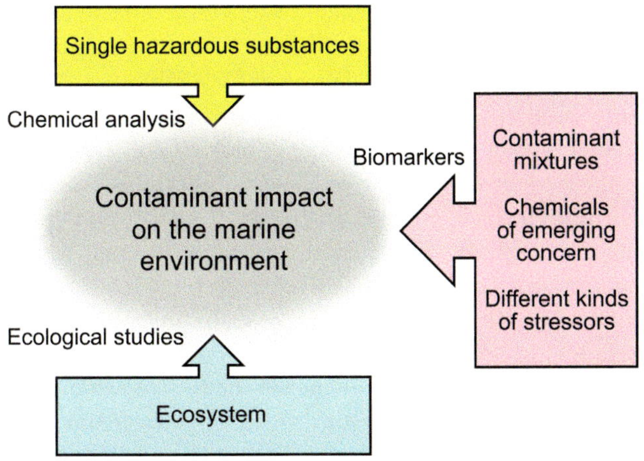

Fig. 16.15 Conceptual model of assessing the contaminant impact on the marine environment by combining chemical and biomarker analyses with ecological studies. This approach provides the most realistic information on the integrated effects of chemical contaminants and natural environmental stressors

16.4.7 The use of biomarkers in monitoring chemical pollution

Most of the biomarkers currently applied in ecotoxicology and environmental monitoring derive directly from human medical science and were first introduced in environmental toxicology in the 1970s (Hook et al. 2014). Biomarkers have great potential in this field because they (1) reveal exposure to hazardous substances and/or their effects in real environmental contexts, (2) are effective "early warning" indicators of deteriorated environmental conditions, and (3) often have strong mechanistic links to pathology and disease. Together, all these features signify that biomarkers have prognostic power that can be taken advantage of in environmental assessment and management. By detecting exposure and/or effects and changes in the health status of individuals, protective actions can be taken and measures initiated before deleterious impacts occur at the population, community and ecosystem levels (Fig. 16.12).

Although their potential can still be considered as underutilised, the application of biomarkers in the monitoring and assessment of environmental pollution has increased during the last decade (Hook et al. 2014). Marked progress has been achieved at the research level in the Baltic Sea Area (Lehtonen and Schiedek 2006; Lehtonen et al. 2006a, b, 2014). A large volume of reference literature exists for most of the biomarkers and other biological effects techniques employed (*e.g.* van der Oost et al. 2003; Viarengo et al. 2007; Davies and Vethaak 2012), but the majority of these techniques have not been routinely applied to a brackish-water system.

The incorporation of biomarker methods into national monitoring programmes has so far been slow. A few long- or medium-term time series are available, *e.g.* on the reproduction of the white-tailed eagle, health parameters in seals, fish diseases, embryonic aberrations in benthic amphipods, imposex in snails, and EROD activity in fish (Table 16.3; HELCOM 2010a).

The monitoring of effects at lower biological levels, which truly represent the "early warning" responses to hazardous substances in the Baltic Sea, has only recently started to receive serious attention. HELCOM has selected suitable biological effects methods in its core set of indicators and has recently included biomarkers such as lysosomal membrane stability (LMS, cytotoxicity) and micronuclei frequency (genotoxicity) (HELCOM 2013c).

16.4.8 Assembling the biomarker toolbox

Field-testing of a battery of biomarkers on fish, bivalves and crustaceans in different subregions of the Baltic Sea has shown that biomarkers developed for other areas are suitable for detecting chemical pollution in the Baltic Sea as well (Table 16.3). These biomarkers record effects at different biological levels with biological endpoints endpoints such as LMS, acetylcholinesterase inhibition (AChE, neurotoxicity), EROD activity (biotransformation of PAHs and coplanar PCBs), micronuclei frequency (genotoxicity), metallothionein induction (exposure to heavy metals), neutral lipid accumulation (metabolic disturbances), macrophage activity (immunological responses), PAH metabolites in fish bile (exposure to PAHs), and liver histopathology (tissue damage).

Most of these endpoints show the highest response levels in contaminated areas of the Baltic Sea (Baršienė et al. 2006a; Hansson et al. 2006; Kopecka et al. 2006; Lang et al. 2006; Schiedek et al. 2006; Vuorinen et al. 2006; Dąbrowska et al. 2012, 2013; Kreitsberg et al. 2012; Lehtonen et al. 2014). Altogether, the results of these tests show that the contaminant concentrations measured at present in different parts of the Baltic Sea elicit biological responses in organisms and in some areas produce chronic stress most likely affecting ecosystem functioning.

In some cases, the observed responses could be directly attributed to anthropogenic activities such as sediment dredging, dumping, or accidental oil spills (*e.g.* Baršienė et al. 2006a). However, biomarker responses have also been recorded in offshore areas far away from point sources (*e.g.* Rybakovas et al. 2009). In some areas the measured biomarker responses (*e.g.* EROD activity in fish) were very low with respect to the measured pollution levels (*e.g.* Hansson et al. 2006). This phenomenon may be due to locally prevailing chronically high exposure levels, which alter the major detoxification routes of organic compounds, or the inability of organisms to respond physiologically to chronic exposure.

16.4.9 Detection of non-bioaccumulative compounds and mixture toxicity

The combined use of different types of biomarkers provides an integrative measure of the contamination impact. It gives information on the bioavailability of contaminants and the damage caused by a mixture of chemicals under environmental conditions *in situ*. This information cannot be achieved by measuring the concentrations of selected chemicals or by applying standardised laboratory biotests on environmental samples. However, to avoid incorrect interpretations, the thresholds of pollution effects, and the identification of what is caused by anthropogenic chemicals and what by natural variability still need to be carefully determined for many biomarkers

A major problem in the currently applied environmental risk assessments, besides mixture toxicity, resides in the different metabolism and modes of the toxic action of chemicals. While the POPs have been regarded as the most dangerous ones due to their biomagnification capabilities and subsequent effects observed at higher levels of biological organisation, rapidly degrading substances that leave little or no trace to be detected by routine chemical analysis may also cause serious harm to organisms. Many endocrine disrupting chemicals (EDCs) and their breakdown products bind to receptors and provoke hormone disruption in small concentrations, which makes them difficult to detect.

This emphasises one of the major advantages of using biomarkers: they can detect the problem (effect) in cases when routine chemical analysis fails to detect any cause for concern.

16.5 Towards integrated assessment of chemical pollution

16.5.1 Linking concentrations to biological effects

Measurements of individual chemical contaminants in biota, water and sediment provide information on the concentrations and behaviour of single hazardous substances in the ecosystem. However, such measurements do not tell us anything about the effects of the cocktail of different compounds (mixture toxicity) that organisms constantly face, especially in areas under high anthropogenic impact. Furthermore, measurements of single substances cannot provide information about the potentially synergistic effects caused by the presence of other stress factors, such as suboptimal temperatures, oxygen deficiency, acidification, and eutrophication.

In this context, the assessment of biological effects by the application of biomarkers on indicator species serves as a useful approach to provide more realistic information on the integrated effects of chemical contaminants and natural environmental stressors. Negative effects on individuals will ultimately affect populations through reduced fitness, pathological disorders and diseases, and by disturbing reproduction. Changes in the health, body size and/or population structure of organisms, especially concerning key species in the ecosystem, can markedly change community structures, and thus affect whole ecosystems and their functionality.

16.5.2 Some biomarkers respond to multiple stressors

Among single biomarkers, LMS (Table 16.3) an integrative "early warning" indicator. The "non-disturbed status" of an organism is defined as high stability of the membranes of lysosomes, small subcellular organelles involved *e.g.* in detoxification. A lowered LMS reflects the combined impact of a mixture of contaminants due to the responsiveness of the lysosomal system to most contaminant classes under varying levels of additional environmental stress. Threshold levels for LMS have been defined for mussels and several species of fish, characterising the different stages of toxically-induced cell damage (Broeg et al. 2005; Broeg and Lehtonen 2006).

LMS measurements in the European flounder *Platichthys flesus* collected from different sites in the southern Baltic Sea have revealed marked pollution effects in coastal and harbour areas (HELCOM 2010a). Other field studies in different parts of the Baltic Sea, including transplantation (caging) experiments using the blue mussel *Mytilus trossulus* in contaminated areas (Rank et al. 2007; Turja et al. 2013, 2014), have also demonstrated the usefulness of LMS as an indicator for biological effects caused by multiple chemical stressors.

Examples of other widely used biomarkers integrating the response to exposure concerning various contaminants are the neurotoxicity indicator AChE activity (responses originally used as a specific biomarker of exposure to organophosphate and carbamate pesticides) and oxidative defence system responses. The effects of hazardous substances on organisms, as well as the effects of environmental and nutritional stress, are often mediated through the cellular process of oxidative stress (Monaghan et al. 2009; Snoeijs et al. 2012).

Oxidative stress occurs when the production of reactive oxygen species (ROS), such as superoxide radicals, singlet oxygen, hydrogen peroxide and hydroxyl radicals, exceed their removal by defence mechanisms, resulting in damage of lipids, proteins and DNA (Lesser 2006). The oxidative status of cells can be assessed, *e.g.* by measuring cell damage (*e.g.* lipid peroxidation, LPO), the activity of antioxidant enzymes (*e.g.* superoxide dismutase, SOD), or the concentrations of antioxidant molecules (*e.g.* carotenoids, ascorbic acid, α-tocopherol) and the total oxygen-scavenging capacity. It can also be used as a measure of general stress, which may be caused by the exposure to hazardous substances.

16.5.3 Integrative indices based on biomarkers

Some biomarkers indicate more specifically the type of biological damage, *e.g.* genotoxicity, immunotoxicity or neurotoxicity, while others have a broader spectrum of causative agents (*e.g.* lysosomal alterations in cells as a general toxicity response), and some identify more directly the exposure to certain contaminant groups (*e.g.* EROD activity is linked to the exposure to PAHs, coplanar PCBs and dioxins, and imposex in gastropods is caused by TBT). To comply with this kind of a "biomarker fingerprinting approach", integrative indices have been developed to allow for the classification of sampling sites with respect to their pollution status (Cajaraville et al. 2000; Broeg et al. 2005; Narbonne et al. 2005; Hylland et al. 2008). For example, the integrated response index (IBR, Belieff and Burgeot 2002) has been widely applied as a method for the exploratory analysis of data collected using a multibiomarker approach. The IBR constitutes a simple mathematical tool based on biomarker data standardisation and their aggregation into a single value. This method has been applied to biological effects data for fish and mussels collected from the Baltic Sea (Fig. 16.16; Broeg and Lehtonen 2006; Turja et al. 2013; 2014).

Another example is the integrated biomarker assessment tool (IBAT), which was tested for the viviparous eelpout *Zoarces viviparus* (Lehtonen et al. 2014). IBAT compares the input data for biological effect parameters where specific assessment criteria, *i.e.* threshold levels, have been developed and an overall integrated biomarker assessment score (IBAS) can then be calculated. Other integrative approaches include methods such as the HELCOM hazardous substances status assessment tool (CHASE), which combines chemical and biological measurements (HELCOM 2010a; Fig. 16.17), and the health status index (HIS), which is based on the so-called "expert system", a decision support system integrating biomarker responses measured at different biological levels (Dagnino et al. 2007).

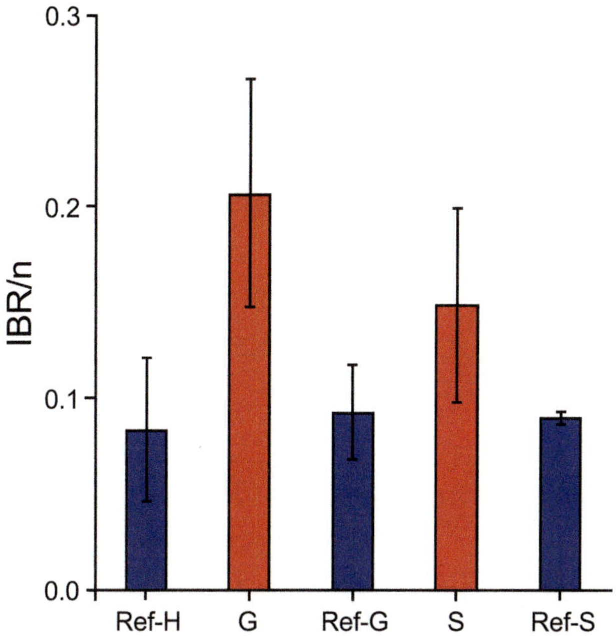

Fig. 16.16 Example of the use of the integrated biomarker response (IBR) index calculated from nine biomarkers ($n = 9$) measured in transplanted individuals of the blue mussel *Mytilus trossulus*. All mussels were collected from the same sampling site in Hanko (southwestern Gulf of Finland), after which they were exposed in cages for three months at two contaminated sites outside the Swedish cities of Sundsvall (S) and Gävle (G) and at two less contaminated sites outside Sundsvall (Ref-S) and Gävle (Ref-G). Reference mussels were collected from the same population in Hanko at the end of the exposure period (Ref-H). The biomarkers included in the IBR were micronuclei (MN), glutathione *S*-transferase (GST) , glutathione reductase (GR), superoxide dismutase (SOD), lipid peroxidation (LPO), acetylcholinesterase (AChE), lysosomal membrane stability (LMS), phagocytic activity and condition index (CI). The IBR (mean ± standard deviation) is given divided by the number of biomarkers used for the calculation (IBR/*n*). Figure modified from Turja et al. (2014)

16.5.4 Risk assessment

Given the complexity of offshore and coastal ecosystems, it is unlikely that a balanced view of the nature and extent of risk will be easily achieved if human and environmental risk assessments continue to be conducted in isolation. Therefore, the integration of assessment protocols into a holistic assessment to improve risk management is advocated (WHO 2001; EU 2013; Fig. 16.18).

Biomarkers can provide the common conceptual framework and measurable biological endpoints necessary for successful integration. Suites of biomarkers encompassing molecular change, cellular pathology and physiological impairment can be developed and adapted for human and ecological scenarios. By placing a larger emphasis on the health status of impacted biota, it is more likely that risk assessment will develop the efficiency, reliability and predictive power to adapt to the unforeseen anthropogenic

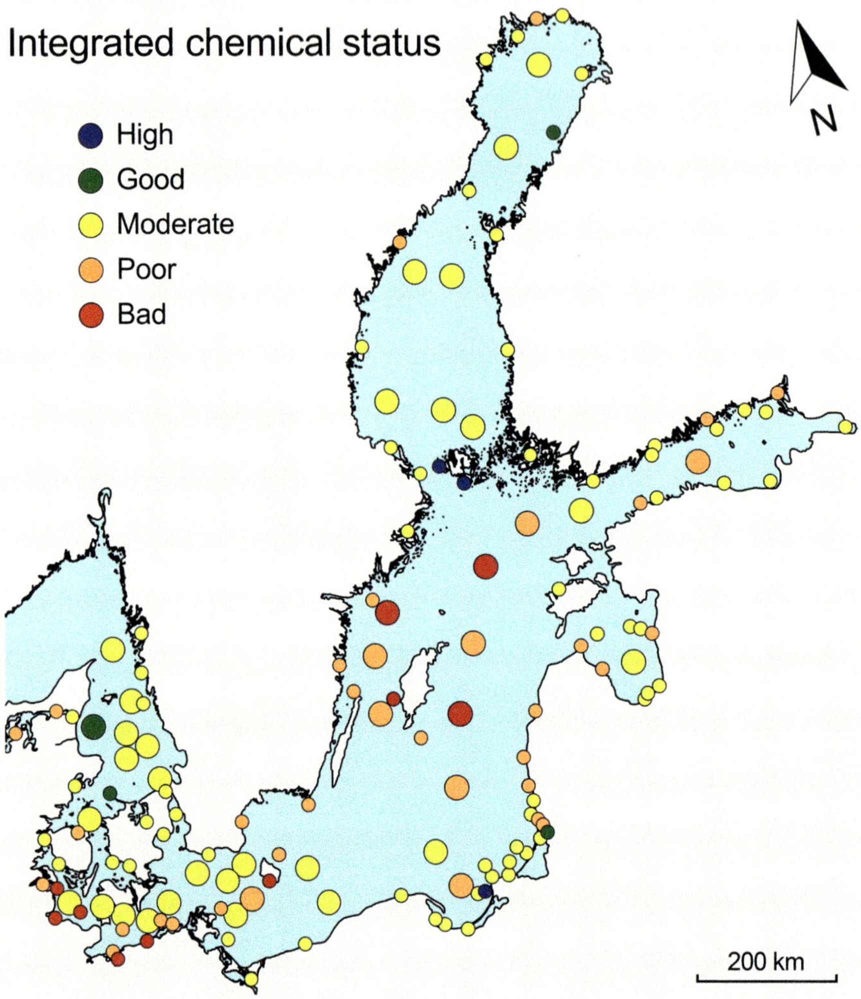

Fig. 16.17 Integrated chemical status classification for 144 HELCOM assessment units using the CHASE tool. The classes "High" and "Good" status denote "areas not disturbed by hazardous substances", while the classes "Moderate", "Poor" and "Bad" status denote "areas disturbed by hazardous substances". Large dots represent assessment units of the open sea and small dots represent coastal assessment units. The ecological objectives that were assessed included all HELCOM objectives: "concentrations of hazardous substances close to natural levels", "all fish safe to eat", "healthy wildlife" and "radioactivity at pre-Chernobyl levels". Figure modified from HELCOM (2010a)

pressures that are an inevitable consequence of human development and global change (Galloway 2006).

16.6 Contaminants in a changing ecosystem

16.6.1 Changes in bioavailability and toxicity

Many studies on biological responses to chemical contaminants performed in the Baltic Sea underpin the importance of the varying environmental conditions. The stable north-south salinity and temperature gradients of the Baltic Sea affect both the physiological responses of organisms and the toxicity of the chemicals. Thus, interactions between the bioavailability and toxicity of contaminants and the natural variations in environmental conditions must be understood when comparing the impact of chemical pollution on the ecosystem in the different subregions of the Baltic Sea (Pfeifer et al. 2005; Prevodnik et al. 2007).

The bioavailability and toxicity of hazardous substances are particularly influenced by salinity, temperature and oxygen concentrations, which show large variability in the Baltic Sea (Tedengren and Kautsky 1987; Tedengren et al. 1999). The bioavailability of specific compounds (*e.g.* trace metals) is greatly affected by salinity (McLusky et al. 1986). Numerous studies have documented increasing trace metal uptake by aquatic organisms with decreasing salinity (Hall et al. 1995; Lee et al. 1998).

Furthermore, environmental conditions can affect enzyme activities that are used as biomarkers. For example, AChE

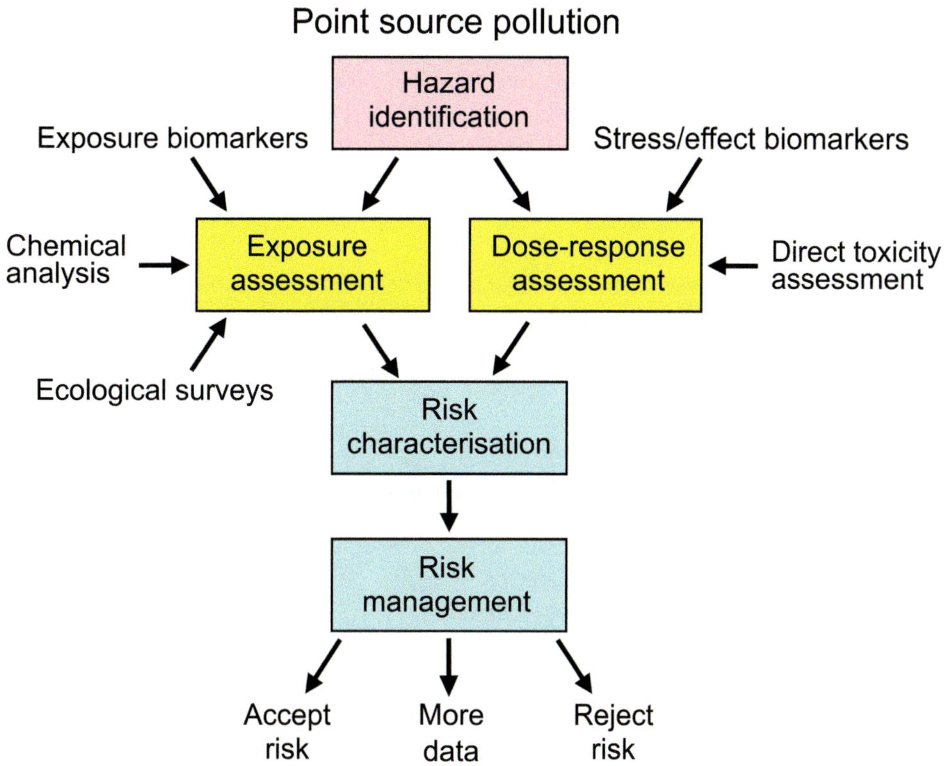

Fig. 16.18 Conceptual model of risk assessment with examples of inputs to exposure assessment and dose-response assessment. Risk assessment of hazardous substances in the environment requires a multidisciplinary approach, combining biochemistry (biomarkers), analytical chemistry (contaminant levels) and ecology in combination with conventional ecotoxicological methods such as dose-response assessment in bioassays. Figure based on data in Galloway (2006)

activity in *Mytilus* sp. from the transition zone (probably a hybrid between *Mytilus edulis* and *Mytilus trossulus*, *cf.* Sect. 6.3.6) was shown to be positively correlated with water temperature and negatively with salinity (Pfeifer et al. 2005). In a study in the northern Baltic Sea proper, the glutathione *S*-transferase (GST) activity in *Mytilus trossulus* was found to be negatively correlated with Secchi depth, while in *Macoma balthica* it was positively correlated with near-bottom oxygen concentration (Leiniö and Lehtonen 2005). The same study concluded that strong seasonal variability in temperature and a concentrated period of food availability in spring, both factors that govern the reproductive cycle of the bivalves, probably explain most of the observed natural variability in the biomarkers in *Mytilus trossulus* and *Macoma balthica* in the Baltic Sea.

16.6.2 Interactions between climate change and contaminants

The ongoing and predicted changes in the large-scale hydrographical conditions of the Baltic Sea due to global warming, notably higher water temperature, increased precipitation and lower salinity (BACC Author Team 2015), will have direct impacts, not only on the acclimation capacity and distribution of organisms, but also on the bioavailability and toxicity of hazardous substances. Many organisms in the Baltic Sea are cold-adapted stenotherm species (*cf.* Sect. 4.2.4) and even small temperature increases may influence these species negatively and add additional stress, which in turn may make them more vulnerable to hazardous substances.

Changes in climate variables will also alter the transport, transfer and deposition of hazardous substances in the ecosystem, and thus affect the exposure of organisms to them. For example, the changing hydrographical conditions of the Baltic Sea may affect resuspension processes of sediment-bound chemical pollutants. Thus, the altered environmental conditions will affect the extent of organisms' exposure to hazardous substances in the different subregions of the Baltic Sea.

Temperature directly modifies the chemistry of a number of chemical pollutants resulting in significant alterations in their toxicities (Noyes et al. 2009). Climate change may also influence the processes involved in the metabolism of toxic substances within the organisms (Manciocco et al. 2014). Higher temperature will probably cause increased turnover rates in the environment and lead to higher metabolic rates in the species that can adapt to a higher temperature regime. It is also generally accepted that a higher temperature increases the rate of pollutant uptake via changes in ventilation rate, which is in turn a response to an increased metabolic rate and decrease in oxygen solubility (Kennedy and Walsh 1997; *cf.* Table 2.5). As a result, the predicted higher temperatures and/or lower salinity will affect the species' ability to deal with toxic substances and the different physiological regulation processes involved in the detoxification of hazardous substances (Schiedek et al. 2007).

Environmental problems such as eutrophication, chemical pollution and overfishing do not occur in isolation but are tightly related to each other. For example, after a phytoplankton bloom there are large amounts of dead and decaying plankton cells in the water column. Hydrophobic organic compounds (HOCs) preferentially bind to such organic particles and are in this way more bioavailable to organisms consuming them, either in the water column or after deposition on the seafloor (Axelman et al. 2001). If such events are added to the potential changes in environmental drivers that are predicted to occur with global warming, the picture becomes highly complex. A major challenge for the future is to understand and predict these types of complex interactions, and to find ways to incorporate them into risk assessment (Fig. 16.18) and management of the Baltic Sea (*cf.* Sect. 18.5).

Review questions

1. Why is the Baltic Sea considered to be especially vulnerable to contamination by hazardous substances?
2. What are the advantages and disadvantages of biomarkers?
3. What are the main threats of hazardous substances to the Baltic Sea today and why?
4. Why are effects of hazardous substances often very pronounced in top predators?
5. What is the potential of the use of biomarkers in the monitoring of chemical pollution?

Discussion questions

1. How would you design a monitoring programme to assess the effects of releases from an industrial area to a coastal area of the Baltic Sea?
2. How can ecosystem health be linked to human health with respect to chemical pollution?
3. How can chemical and biological information on pollution be integrated?
4. What are the main hazards of chemical pollution in the coastal zone *versus* the open sea?
5. When you compare the different anthropogenic threats to the Baltic Sea ecosystem, how important is chemical pollution?

References

Abrahamsson K, Ekdahl A, Collén J, Pedersén M (1995) Formation and distribution of halogenated volatile halocarbon compounds in seawater. In: de Leer EWB, Grimwall A (eds) Kluwer Academic Publishers, Dordrecht, pp 317–326

Abrahamsson K, Choo KS, Pedersén M, Johansson G, Snoeijs P (2003) Effects of temperature on the production of hydrogen peroxide and volatile halocarbons by brackish-water algae. Phytochemistry 64:725–734

Allsopp M, Erry B, Santillo D, Johnston P (2001) POPs in the Baltic – a review of persistent organic pollutants (POPs) in the Baltic Sea. Greenpeace International, London 92 pp

Axelman J, Näf C, Bandh C, Tshaq R, Pettersen H et al (2001) Dynamics and distribution of hydrophobic organic compounds in the Baltic Sea. In: Wulff FV, Rahm L, Larsson P (eds) A systems analysis of the Baltic Sea, vol 148. Springer, Berlin, pp 257–288 (Ecological Studies)

BACC Author Team (2015) Second assessment of climate change for the Baltic Sea basin. Regional climate studies. Springer, Berlin 501 pp

Bailly du Bois P, Laguionie P, Boust D, Korsakissok I, Didier D et al (2012) Estimation of marine source-term following Fukushima Dai-ichi accident. Journal of Environmental Radioactivity 114: 2–9

Baršienė J, Dedonytė V, Rybakovas A, Andreikėnaitė L, Andersen OK (2006a) Investigation of micronuclei and other nuclear abnormalities in peripheral blood and kidney of marine fish treated with crude oil. Aquatic Toxicology 78(Supplement 1):99–104

Baršienė J, Lehtonen K, Koehler A, Broeg K, Vuorinen PJ et al (2006b) Biomarker responses in flounder (*Platichthys flesus*) and mussel (*Mytilus edulis*) in the Klaipėda-Būtingė area (Baltic Sea). Marine Pollution Bulletin 53:422–436

Bełdowski J, Klusek Z, Szubska M, Turja R, Bulczak, et al (2016) Chemical munitions search and assessment – an evaluation of the dumped munitions problem in the Baltic Sea. Deep Sea Research II: 128:85–95

Beliaeff B, Burgeot T (2002) Integrated biomarker response (IBR): a useful graphical tool for ecological risk assessment. Environmental Toxicology and Chemistry 21:1316–1322

Berezina N, Strode E, Lehtonen KK, Balode M, Golubkov S (2013) Sediment quality assessment using the amphipods *Gmelinoides fasciatus* and *Monoporeia affinis* in the northeastern Baltic Sea. Crustaceana 86:780–801

Berge J, Brevik E, Bjorge A, Folsvik N, Gabrielsen G, Wolkers H (2004) Organotins in marine mammals and seabirds from Norwegian territory. Journal of Environmental Monitoring 6:108–112

Berntzon L, Erasmie S, Celepli N, Eriksson J, Rasmussen U, Bergman B (2013) BMAA inhibits nitrogen fixation in the cyanobacterium *Nostoc* sp. PCC 7120. Marine Drugs 11:3091–3108

Bignert A, Litzen K, Odsjö T, Olsson M, Persson W et al (1995) Time-related factors influence the concentration of DDTs, PCBs and shell parameters in eggs of Baltic guillemot (*Uria aalge*), 1861–1989. Environmental Pollution 89:27–36

Bignert A, Cleeman M, Dannenberger D, Gaul H, Roots O (1997) Halogenated hydrocarbons. In: Third periodic assessment of the

state of the marine environment of the Baltic Sea, 1989–93. Baltic Sea environment proceedings, vol 64B, pp 130–138
Bignert A, Olsson M, Persson W, Jensen S, Zakrisson S et al (1998) Temporal trends of organochlorines in Northern Europe, 1967–1995. Relation to global fractionation, leakage from sediments and international measures. Environmental Pollution 99:177–198
Bignert A, Danielsson S, Faxneld S, Nyberg E, Vasileiou M et al (2015) Comments concerning the national Swedish contaminant monitoring programme in marine biota. Swedish Museum of Natural History, Report 2/2015:1–262
Bocquené G, Chantereau S, Clérendeau C, Beausir E, Ménard D et al (2004) Biological effects of the "Erika" oil spill on the common mussel (*Mytilus edulis*). Aquatic Living Resources 17:309–316
Brack W, Klamer HJ, López de Alda M, Barceló D (2007) Effect-directed analysis of key toxicants in European river basins – a review. Environmental Science and Pollution Research International 14:30–38
Bradshaw C, Kumblad L, Fagrell A (2006) The use of tracers to evaluate the importance of bioturbation in remobilising contaminants in Baltic sediments. Estuarine and Coastal Shelf Science 66:123–134
Bradshaw C, Näslund J, Hansen J, Kozlowsky-Suzuki B, Sundström B, Gustafsson K (2015) Hexabromocyclododecane affects benthic-pelagic coupling in an experimental ecosystem. Environmental Pollution 206:306–314
Brodin T, Fick J, Jonsson M, Klaminder J (2013) Dilute concentrations of a psychiatric drug alter behavior of fish from natural populations. Science 339:814–815
Broeg K, Lehtonen KK (2006) Indices for the assessment of environmental pollution of the Baltic Sea coasts: integrated assessment of a multi-biomarker approach. Marine Pollution Bulletin 53:145–159
Broeg K, Westernhagen HV, Zander S, Körting W, Koehler A (2005) The "bioeffect assessment index" (BAI) – a concept for the quantification of effects of marine pollution by an integrated biomarker approach. Marine Pollution Bulletin 50:495–503
Burgess RM, Ho KT, Brack W, Lamoree M (2013) Effects-directed analysis (EDA) and toxicity identification evaluation (TIE): complementary but different approaches for diagnosing causes of environmental toxicity. Environmental Toxicology and Chemistry 32:1935–1945
Cajaraville MP, Bebianno MJ, Blasco J, Porte C, Sarasquete C et al (2000) The use of biomarkers to assess the impact of pollution in coastal environments of the Iberian Pensinsula: a practical approach. Science of the Total Environment 247:295–311
Camphuysen CJ, Chardine J, Fredriksen M, Nunes M (2005) Review of the impacts of recent major oil spills on seabirds. In: Anonymous (ed) Report of the working group on seabird ecology, Texel, 29 March–1 April 2005. Oceanography Committee, ICES CM 2005/C:05, Ref. ACME+E, International Council for the Exploration of the Sea, Copenhagen, Denmark
Carpenter LJ, Liss PS (2000) On temperate sources of bromoform and other reactive organic bromine gases. Journal Geophysical Research 105(D16):20539–20547
Carpenter LJ, Liss PS, Penkett SA (2003) Marine organohalogens in the atmosphere over the Atlantic and Southern Oceans. Journal of Geophysical Research – Atmospheres 108(D9):4256
Casals-Casas C, Desvergne B (2011) Endocrine disruptors: from endocrine to metabolic disruption. Annual Review of Physiology 73:135–162
Choo KS, Nilsson J, Pedersén M, Snoeijs P (2005) Photosynthesis, carbon uptake and antioxidant defence in two coexisting filamentous green algae under different stress conditions. Marine Ecology Progress Series 292:127–138
Corcoran J, Winter MJ, Tyler CR (2010) Pharmaceuticals in the aquatic environment: a critical review of the evidence for health effects in fish. Critical Reviews in Toxicology 40:287–304
Courtney KD (1979) Hexachlorobenzene (HCB): a review. Environmental Research 20:225–266
Cox PA, Sacks OW (2002) Cycad neurotoxins, consumption of flying foxes, and ALS-PDC disease in Guam. Neurology 58: 956–959
Cox PA, Davis DA, Mash DC, Metcalf JS, Banack SA (2016) Dietary exposure to an environmental toxin triggers neurofibrillary tangles and amyloid deposits in the brain. Proceedings of the Royal Society B 283:20152397
Dąbrowska H, Ostaszewska T, Kamaszewski M, Napora-Rutkowski L, Antoniak A et al (2012) Histopathological, histomorphometrical, and immunohistochemical biomarkers in flounder (*Platichthys flesus*) from the southern Baltic Sea. Ecotoxicology and Environmental Safety 78:14–21
Dąbrowska H, Kopko O, Turja R, Lehtonen KK, Góra A et al (2013) Sediment contaminants and contaminant levels and biomarkers in caged mussels (*Mytilus trossulus*) in the southern Baltic Sea. Marine Environmental Research 84:1–9
Dagnino A, Allen JI, Moore M, Broeg K, Canesi L et al (2007) Integration of biomarker data into an organism health index: development of an expert system and its validation with field and laboratory data in mussels. Biomarkers 12:155–172
Dannenberger D (1996) Chlorinated micropollutants in surface sediments of the Baltic Sea – investigations in the Belt Sea, the Arkona Sea and the Pomeranian Bight. Marine Pollution Bulletin 32:772–781
Davies IM, Vethaak AD (2012) Integrated marine environmental monitoring of chemicals and their effects. ICES Cooperative Research Report 315:1–277
Downing S, Banack SA, Metcalf JS, Cox PA, Downing TG (2011) Nitrogen starvation of cyanobacteria results in the production of beta-N-methylamino-L-alanine. Toxicon 58:187–194
Duft M, Schulte-Oehlmann U, Tillmann M, Markert B, Oehlmann J (2003) Toxicity of triphenyltin and tributyltin to the freshwater mud snail *Potamopyrgus antipodarum* in a new sediment biotest. Environmental Toxicology and Chemistry 22:145–152
Dunlop RA, Cox PA, Banack SA, Rodgers KJ (2013) The non-protein amino acid BMAA is misincorporated into human proteins in place of L-serine causing protein misfolding and aggregation. PLoS ONE 8:e75376
EC (2006) Regulation No 1907/2006 of the European parliament and of the council of 18 December 2006 concerning the registration, evaluation, authorisation and restriction of chemicals (REACH), establishing a European chemicals agency, amending directive 1999/45/EC and repealing council regulation (EEC) No 793/93 and commission regulation (EC) No 1488/94 as well as council directive 76/769/EEC and commission directives 91/155/EEC, 93/67/EEC, 93/105/EC and 2000/21/EC. Official Journal of the European Union L163:1–280
Eklund B, Elfström M, Borg H (2008) TBT originates from pleasure boats in Sweden in spite of firm restrictions. Open Environmental Sciences 2:124–132
Eklund B, Elfström M, Gallego I, Bengtsson BE, Breitholtz M (2010) Biological and chemical characterization of harbour sediments from the Stockholm area. Journal of Soils and Sediments 10: 127–141
Elmgren R, Hansson S, Larsson U, Sundelin B, Boehm PD (1983) The "Tsesis" oil spill: Acute and long-term impact on the benthos. Marine Biology 73:51–65
Eriksson-Wiklund AK, Sundelin B (2001) Impaired reproduction of the amphipods *Monoporeia affinis* and *Pontoporeia femorata* as a result of moderate hypoxia and increased temperature. Marine Ecology Progress Series 171:165–180
Eriksson-Wiklund AK, Oskarsson H, Thorsén G, Kumblad L (2011) Behavioural and physiological responses to pharmaceutical

exposure in macroalgae and grazers from a Baltic Sea littoral community. Aquatic Biology 14:29–39
Esler D, Trust KA, Ballachey BE, Iverson SA, Lewis TL et al (2010) Cytochrome P450 1A biomarker indication of oil exposure in harlequin ducks up to 20 years after the Exxon Valdez oil spill. Environmental Toxicology and Chemistry 29:1138–1145
EU (2013) Addressing the new challenges for risk assessment. SCHER (Scientific Committee on Health and Environmental Risks), SCENIHR (Scientific Committee on Emerging and Newly Identified Health Risks), SCCS (Scientific Committee on Consumer Safety), March 2013, 157 pp. [http://ec.europa.eu/health/scientific_committees/consumer_safety/docs/sccs_o_131.pdf]
Faxneld S, Danielsson S, Nyberg E, Bignert A (2015) Conversion factors for metals between liver, muscle and whole-body in perch. Swedish Museum of Natural History, Report 1(2015):1–35
Fewer DP, Jokela J, Paukku E, Österholm J, Wahlsten M et al (2013) New structural variants of aeruginosin produced by the toxic bloom forming cyanobacterium *Nodularia spumigena*. PLoS ONE 8(9): e73618
Fong PP, Ford AT (2014) The biological effects of antidepressants on the molluscs and crustaceans: a review. Aquatic Toxicology 151:4–13
Galloway TS (2006) Biomarkers in environmental and human health risk assessment. Marine Pollution Bulletin 53:606–613
García-Tenorio R, Sanches-Angulo CI, García-León M, Abril JM, El-Daoushy F (1992) The use of ^{137}Cs in marine and lacustrine sediment dating. Nuclear Geophysics 6:395–403
Gercken J, Förlin L, Andersson J (2006) Developmental disorders in larvae of eelpout (*Zoarces viviparus*) from German and Swedish Baltic coastal waters. Marine Pollution Bulletin 53:497–507
GESAMP (2015) Sources, fate and effects of microplastics in the marine environment: a global assessment. International Maritime Organization Reports and Studies 90:1–96
Gorokhova E (2015) Screening for microplastic particles in plankton samples: How to integrate marine litter assessment into existing monitoring programs? Marine Pollution Bulletin 99:271–275
Grimvall A, Larsson K (2014) Mapping shipping intensity and routes in the Baltic Sea using historical AIS data. Swedish Institute for the Marine Environment. Report 2014:5, 4 pp
Halinen K, Jokela J, Fewer D, Wahlsten M, Sivonen K (2007) Direct evidence for the production of microcystins by strains of the genus *Anabaena* (Cyanobacteria) isolated from the Baltic Sea. Applied and Environment Microbiology 73:6543–6550
Hall LW Jr, Ziegenfuss MC, Anderson RD, Lewis BL (1995) The effect of salinity on the acute toxicity of total and free cadmium to a Chesapeake Bay copepod and fish. Marine Pollution Bulletin 30:376–384
Hansson T, Schiedek D, Lehtonen KK, Vuorinen PJ, Liewenborg B et al (2006) Biochemical biomarkers in adult female perch (*Perca fluviatilis*) in a chronically polluted gradient in the Stockholm recipient (Sweden). Marine Pollution Bulletin 53:451–468
Hario M, Hirvi JP, Hollmen T, Rudback E (2004) Organochlorine concentrations in diseased *versus* healthy gull chicks from the northern Baltic. Environmental Pollution 127:411–423
Hearn RD, Harrison AL, Cranswick PA (2015) Draft international single species action plan for the conservation of the long-tailed duck *Clangula hyemalis* 2016–2025. AEWA technical series. Draft June 2015, 65 pp
Helander B, Olsson A, Bignert A, Asplund L, Litzén K (2002) The role of DDE, PCB, coplanar PCB and eggshell parameters for reproduction in the white-tailed sea eagle (*Haliaeetus albicilla*) in Sweden. Ambio 31:386–403
HELCOM (1995) Radioactivity in the Baltic Sea 1984–1991. Baltic Sea Environment Proceedings 61:1–182
HELCOM (2002) Environment of the Baltic Sea area 1994–1998. Baltic Sea Environment Proceedings 82B:1–215
HELCOM (2003) The Baltic marine environment 1999–2002. Baltic Sea Environment Proceedings 87:1–47
HELCOM (2007) Long-lived radionuclides in the seabed of the Baltic Sea. Report of the sediment baseline study of HELCOM MORS-PRO in 2000–2005. Baltic Sea environment proceedings, vol 110, pp 1–41
HELCOM (2009) Radioactivity in the Baltic Sea, 1999–2006. HELCOM thematic assessment. Baltic Sea Environment Proceedings 117:1–60
HELCOM (2010a) Hazardous substances in the Baltic Sea – an integrated thematic assessment of hazardous substances in the Baltic Sea. Baltic Sea Environment Proceedings 120B:1–116
HELCOM (2010b) Maritime activities in the Baltic Sea – an integrated thematic assessment on maritime activities and response to pollution at sea in the Baltic Sea region. Baltic Sea Environment Proceedings 123:1–64
HELCOM (2013a) Thematic assessment of long-term changes in radioactivity in the Baltic Sea, 2007–2010. Baltic Sea Environment Proceedings 135:1–39
HELCOM (2013b) Chemical munitions dumped in the Baltic Sea. Baltic Sea Environment Proceedings 142:1–128
HELCOM (2013c) HELCOM core indicators: final report of the HELCOM CORESET project. Baltic Sea Environment Proceedings 136:1–71
HELCOM (2015) HELCOM annual report on discharges observed during aerial surveillance in the Baltic Sea 2014. HELCOM, Helsinki 14 pp
His E, Beiras R, Seaman MNL (1999) The assessment of marine pollution: bioassays with marine embryos and larvae. Advances in Marine Biology 37:1–178
Höher N, Köhler A, Strand J, Broeg K (2012) Effects of various pollutant mixtures on immune responses of the blue mussel (*Mytilus edulis*) collected at a salinity gradient in Danish coastal waters. Marine Environmental Research 75:35–44
Holmström K, Järnberg U, Bignert A (2005) Temporal trends of PFOS and PFOA in guillemot eggs from the Baltic Sea, 1968–2003. Environmental Science and Technology 39:80–84
Hook SE, Gallagher EP, Batley GE (2014) The role of biomarkers in the assessment of aquatic ecosystem health. Integrated Environmental Assessment and Management 10:327–341
Hylland K, Tollefsen KE, Ruus A, Jonsson G, Sundt RC et al (2008) Water column monitoring near oil installations in the North Sea 2001–2004. Marine Pollution Bulletin 56:414–429
Jiang L, Eriksson J, Lage S, Jonasson S, Shams S et al (2014) Diatoms: a novel source for the neurotoxin BMAA in aquatic environments. PLoS ONE 9:e84578
Johannesson K, André C (2006) Life on the margin: genetic isolation and diversity loss in a peripheral marine ecosystem, the Baltic Sea. Molecular Ecology 15:2013–2029
Johannesson K, Smolarz K, Grahn M, André C (2011) The future of Baltic Sea populations: local extinction or evolutionary rescue? AMBIO 40:179–190
Jonasson S, Eriksson J, Berntzon L, Rasmussen U, Bergman B (2008) A novel cyanobacterial toxin (BMAA) with potential neurodegenerative effects. Plant Biotechnology 25:227–232
Jonasson S, Eriksson J, Berntzon L, Spacil Z, Ilag LL et al (2010) Transfer of a cyanobacterial neurotoxin within a temperate aquatic ecosystem suggests pathways for human exposure. Proceedings of the National Academy of Sciences of the USA 107:9252–9257
Karjalainen M, Pääkkönen JP, Peltonen H, Sipiä V, Valtonen T, Viitasalo M (2008) Nodularin concentrations in Baltic Sea zooplankton and fish during a cyanobacterial bloom. Marine Biology 155:483–491
Kennedy CJ, Walsh PJ (1997) Effects of temperature on xenobiotic metabolism. In: Wood CM, McDonald DG (eds) Global warming –

implications for freshwater and marine fish. Cambridge University Press, Cambridge, pp 303–324
Keppler F, Eiden R, Niedan V, Pracht J, Schöler HF (2000) Halocarbons produced by natural oxidation processes during degradation of organic matter. Nature 403:298–301
Khalil et al. (1999) Natural emissions of chlorine-containing gases: reactive chlorine emissions inventory. Journal of Geophysical Research – Atmospheres 104(D7):8333–8346
Epa US (2007) Sediment toxicity identification evaluation (TIE), phases I, II, and III guidance document EPA/600/R-07/080. Office of Research and Development, Washington 134 pp
Kopecka J, Lehtonen KK, Baršienė J, Broeg K, Vuorinen PJ et al (2006) Measurements of biomarker levels in flounder (*Platichthys flesus*) and blue mussel (*Mytilus trossulus*) from the Gulf of Gdańsk (southern Baltic). Marine Pollution Bulletin 53:406–421
Kosmehl T, Otte JC, Yang L, Legradi J, Bluhm K et al (2012) A combined DNA-microarray and mechanism-specific toxicity approach with zebra fish embryos to investigate the pollution of river sediments. Reproductive Toxicology 33:245–253
Kowalewska G, Konat J (1997) Distribution of the polynuclear aromatic hydrocarbons (PAHs) in sediments of the southern Baltic Sea. Oceanologia 39:83–104
Kreitsberg R, Tuvikene A, Baršienė J, Fricke NF, Rybakovas A et al (2012) Biomarkers of environmental contaminants in the coastal waters of Estonia (Baltic Sea): effects on eelpout (*Zoarces viviparus*). Journal of Environmental Monitoring 14:2298–2308
Lääne A, Kraav E, Titova G (2005) Baltic Sea – Global International Waters Assessment, Regional Assessment (UNEP) 17:1–69, i–xv
Lage S, Annadotter H, Rasmussen U, Rydberg S (2015) Biotransfer of beta-N-methylamino-L-alanine (BMAA) in a eutrophicated freshwater lake. Marine Drugs 13:1185–1201
Lang T, Wosniok W, Baršienė J, Broeg K, Kopecka J et al (2006) Liver histopathology in Baltic flounder (*Platichthys flesus*) as indicator of biological effects of contaminants. Marine Pollution Bulletin 53:488–496
Larsson K, Tydén L (2005) Effects of oil spills on wintering long-tailed ducks *Clangula hyemalis* at Hoburgs bank in the central Baltic Sea between 1996/97 and 2003/04. Ornis Svecica 15:161–171
Law RJ (2014) An overview of time trends in organic contaminant concentrations in marine mammals: going up or down? Marine Pollution Bulletin 82:7–10
Lee BG, Wallace WG, Luoma SN (1998) Uptake and loss kinetics of Cd, Cr and Zn in the bivalves *Poamocorbula amurenis* and *Macoma balthica*: effects of size and salinity. Marine Ecology Progress Series 175:177–189
Lehtonen KK, Schiedek D (2006) Monitoring biological effects of pollution in the Baltic Sea: neglected – but still wanted? Marine Pollution Bulletin 53:377–386
Lehtonen KK, Schiedek D, Koehler A, Lang T, Vuorinen PJ et al (2006a) The BEEP project in the Baltic Sea: overview of results and outline for a regional biological effects monitoring strategy. Marine Pollution Bulletin 53:523–537
Lehtonen KK, Leiniö S, Schneider R, Leivuori M (2006b) Biomarkers of pollution effects in the bivalves *Mytilus edulis* and *Macoma balthica* collected from the southern coast of Finland (Baltic Sea). Marine Ecology Progress Series 322:155–168
Lehtonen KK, Sundelin B, Lang T, Strand J (2014) Development of tools for integrated monitoring of hazardous substances and ecosystem health assessment in the Baltic Sea. AMBIO 43:69–81
Leiniö S, Lehtonen KK (2005) Seasonal variability in biomarkers in the bivalves *Mytilus edulis* and *Macoma balthica* from the northern Baltic Sea. Comparative Biochemistry and Physiology C 140: 408–421
Leppäkoski E, Bonsdorff E (1989) Ecosystem variability and gradients. Examples from the Baltic Sea as a background for hazard assessment. In: Landner L (ed) Chemicals in the aquatic environment – advanced hazard assessment. Springer series on advanced management. Springer, Berlin, pp 6–58
Lesser MP (2006) Oxidative stress in marine environments: biochemistry and physiological ecology. Annual Revue of Physiology 68:253–278
Lind Y, Bignert A, Odsjö T (2006) Decreasing lead levels in Swedish biota 1969–2004. Journal of Environmental Monitoring 8:824–834
Magnusson K, Norén F (2011) Mikroskopiskt skräp i havet – metodutveckling för miljöövervakning. N-Research, Lysekil, 22 pp (in Swedish)
Magnusson K, Wahlberg C (2014) Mikroskopiska skräppartiklar i vatten från avloppsreningsverk. IVL Swedish Environmental Research Institute Report B 2208:1–31 (in Swedish)
Magnusson K, Eliasson K, Fråne A, Haikonen K, Hultén M et al (2016) Swedish sources and pathways for microplastics to the marine environment – a review of existing data. IVL Swedish Environmental Research Institute Report C183:1–87
Manciocco A, Calamandrei G, Alleva E (2014) Global warming and environmental contaminants in aquatic organisms – the need of the etho-toxicology approach. Chemosphere 100:1–7
McCulloch A (1999) CFC and halon replacements in the environment. Journal of Fluorine Chemistry 100:163–173
McCulloch A (2003) Chloroform in the environment: occurrence, sources, sinks and effects. Chemosphere 50:1291–1308
McLusky DD, Bryant V, Campbell R (1986) The effect of temperature and salinity on the toxicity of heavy metals to marine and estuarine invertebrates. Oceanographic Marine Biology Annual Reviews 24:481–520
Mensink BP (1999) Imposex in the common whelk *Buccinum undatum*. Wageningen University, The Netherlands, 125 pp [PhD Thesis]
Miller A, Nyberg E, Danielsson S, Faxneld S, Haglund P et al (2014) Comparing temporal trends of organochlorines in guillemot eggs and Baltic herring: advantages and disadvantages for selecting sentinel species for environmental monitoring. Marine Environmental Research 100:38–47
Monaghan P, Metcalfe NB, Torres R (2009) Oxidative stress as a mediator of life history trade-offs: mechanisms, measurements and interpretation. Ecology Letters 12:75–92
Monson CS, Banack SA, Cox PA (2003) Conservation implications of chamorro consumption of flying foxes as a possible cause of amyotrophic lateral sclerosis-parkinsonism dementia complex in Guam. Conservation Biology 17:678–686
Montzska SA, Fraser PJ (2003) Controlled substances and other source gases. In: Nohende AA, Albritton DL, Mégie G, Watson R (eds) Scientific assessment of ozone depletion: 2002. Global Ozone Research and Monitoring Project Report 47:1–83
Mothersill C, Austin B (2003) In vitro methods in aquatic ecotoxicology. Marine Science and Coastal Management Series. Springer, Berlin 472 pp
Murata S, Takahashi S, Agusa T, Thomas NJ, Kannan K, Tanabe S (2008) Contamination status and accumulation profiles of organotins in sea otters (*Enhydra lutris*) found dead along the coasts of California, Washington, Alaska (USA), and Kamchatka (Russia). Marine Pollution Bulletin 56:641–649
Murch SJ, Cox PA, Banack SA (2004) A mechanism for slow release of biomagnified cyanobacterial neurotoxins and neurodegenerative disease in Guam. Proceedings of the National Academy of Sciences of the USA 101:12228–12231
Murk AJ, Legler J, Denison MS, Giesy JP, van de Guchte C et al (1996) Chemical-activated luciferase gene expression (CALUX): a novel in-vitro bioassay for Ah receptor active compounds in sediments and pore water. Fundamental and Applied Toxicology 33:149–160
Napierska D, Baršienė J, Mulkiewicz E, Podolska M, Rybakovas A (2009) Biomarker responses in flounder *Platichthys flesus* from the

Polish coastal area of the Baltic Sea and application in biomonitoring. Ecotoxicology 18:846–859

Narbonne JF, Aarab N, Clérandeau C, Daubèze M, Narbonne J et al (2005) Scale of classification based on biochemical markers in mussels: application to pollution monitoring in Mediterranean coasts and temporal trends. Biomarkers 10:8–71

Nendza M (2002) Inventory of marine biotest methods for the evaluation of dredged material and sediments. Chemosphere 48:865–883

Nentwig G (2007) Effects of pharmaceuticals on aquatic invertebrates. Part II: the antidepressant drug fluoxetine. Archives of Environmental Contamination and Toxicology 52:163–170

Nielsen SP, Bengtson P, Bojanowski R, Hagel P, Herrmann J et al (1999) The radiological exposure of man from radioactivity in the Baltic Sea. Science of the Total Environment 237/238:133–141

Notter M, Snoeijs P (1986) Radionuklider i bentiska kiselalger – ett års studier i biotestsjön, Forsmark. Naturvårdsverket Rapport 3213:1–30 [in Swedish]

Noyes PD, McElwee MK, Miller HD, Clark BW, Van Tiem LA (2009) The toxicology of climate change: environmental contaminants in a warmer world. Environment International 25:971–986

Nyberg E, Poikane R, Strand J, Larsen MM, Danielsson A, Bignert A (2013) Tributyltin (TBT) and imposex. HELCOM Core Indicator Report, 17 pp

O'Dohety SJ, Carpenter LJ (2007) Halogenated volatile organic compounds. In: Koppmann R (ed) Volatile organic compounds in the atmosphere. Blackwell Publishing, Oxford, pp 173–220

Oberdörster E, McClellan-Green P (2002) Mechanisms of imposex induction in the mud snail, *Ilyanassa obsoleta*: TBT as a neurotoxin and aromatase inhibitor. Marine Environmental Research 54:715–771

Ogata R, Omura M, Shimasaki Y, Kubo K, Oshima Y et al (2001) Two-generation reproductive toxicity study of tributyltin chloride in female rats. Journal of Toxicology and Environmental Health, Part A 63:127–144

Olivero-Berbel J, Guerrero-Castilla A, Ramos NR (2011) Biochemical effects induced by the hexachlorocyclohexanes. Review of Environmental Contamination and Toxicology 212:1–28

Omura M, Ogata R, Kubo K, Shimasaki Y, Aou S et al (2001) Two-generation reproductive toxicity study of tributyltin chloride in male rats. Toxicological Sciences 64:224–232

Orbea A, Garmendia L, Marigómez I, Cajaraville MP (2006) Effects of the "Prestige" oil spill on cellular biomarkers in intertidal mussels: results of the first year of studies. Marine Ecology Progress Series 306:177–189

OSPAR (2008) 2007/2008 CEMP Assessment: trends and concentrations of selected hazardous substances in sediments and trends in TBT-specific biological effects. Assessment and Monitoring Series, OSPAR Commission 2008, 36 pp

Pécseli M (2002) The "Baltic Carrier" oil spill: monitoring and assessment of environmental effects in Grønsund (DK). Storstrøms Amt, Denmark, 158 pp

Pedersén M, Collén J, Ekdahl A, Abrahamsson K (1996) Production of halocarbons from seaweeds: an oxidative stress reaction? Scientia Marina 60:257–263

Pfeifer S, Schiedek D, Dippner J (2005) Effect of temperature and salinity on acetylcholinesterase activity, a common pollution biomarker, in *Mytilus* sp. from the southwestern Baltic Sea. Journal of Experimental Marine Biology and Ecology 320:93–103

Prevodnik A, Gardeström J, Lilja K, Elfwing T, McDonagh B et al (2007) Oxidative stress in response to xenobiotics in the blue mussel *Mytilus edulis* L.: evidence for variation along a natural salinity gradient of the Baltic Sea. Aquatic Toxicology 82:63–71

Rank J, Lehtonen KK, Strand J, Laursen M (2007) DNA damage, acetylcholinesterase activity and lysosomal stability in native and transplanted mussels (*Mytilus edulis*) in areas close to coastal chemical dumping sites in Denmark. Aquatic Toxicology 84:50–61

Reutgard M, Eriksson Wiklund AK, Breitholtz M, Sundelin B (2014) Embryo development of the benthic amphipod *Monoporeia affinis* as a tool for monitoring and assessment of biological effects of contaminants in the field: a meta-analysis. Ecological Indicators 36:483–490

Rybakovas A, Baršienė J, Lang T (2009) Environmental genotoxicity and cytotoxicity in the offshore zones of the Baltic and the North Seas. Marine Environmental Research 68:246–256

Santos MM, ten Hallers-Tjabbes CC, Vieira N, Boon JP, Porte C (2002) Cytochrome P450 differences in normal and imposex-affected female whelk *Buccinum undatum* from the open North Sea. Marine Environmental Research 54:661–665

Santschi PH (1984) Radionuclides as tracers for sedimentation and remobilization processes in the ocean and in lakes. In: Sly PG (ed) Sediment and water interactions. Springer, London, pp 437–449

Santschi PH (1989) Use of radionuclides in the study of contaminant cycling processes. Hydrobiologia 176/177:307–320

Schiedek D, Broeg K, Baršienė J, Lehtonen KK, Gercken J et al (2006) Biomarker responses as indication of contaminant effects in blue mussel (*Mytilus edulis*) and female eelpout (*Zoarces viviparus*) from the southwestern Baltic Sea. Marine Pollution Bulletin 53:387–405

Schiedek D, Sundelin B, Readman JW, Macdonald RW (2007) Interactions between climate change and contaminants. Marine Pollution Bulletin 54:1845–1856

Schnell S, Schiedek D, Schneider R, Balk L, Vuorinen PJ et al (2008) Biological indications of contaminant exposure in Atlantic cod (*Gadus morhua*) in the Baltic Sea. Canadian Journal of Fisheries and Aquatic Sciences 65:1122–1134

Schulte-Oehlmann U, Oehlmann J, Fioroni P, Bauer B (1997) Imposex and reproductive failure in *Hydrobia ulvae* (Gastropoda: Prosobranchia). Marine Biology 128:257–266

Setälä O, Fleming-Lehtinen V, Lehtiniemi M (2014) Ingestion and transfer of microplastics in the planktonic food web. Environmental Pollution 185:77–78

Setälä O, Norkko J, Lehtiniemi M (2016) Feeding type affects microplastic ingestion in a coastal invertebrate community. Marine Pollution Bulletin 102:95–101

Shimasaki Y, Kitano T, Oshima Y, Inoue S, Imada N, Honjo T (2003) Tributyltin causes masculinization in fish. Environmental Toxicology and Chemistry 22:141–144

Sivonen K (2009) Cyanobacterial toxins. In: Schaechter M (ed) Encyclopedia of Microbiology, 3rd edn. Elsevier, Oxford, pp 290–307

Sivonen K, Kononen K, Carmichael WW, Dahlem AM, Rinehart K et al (1989) Occurrence of the hepatotoxic cyanobacterium *Nodularia spumigena* in the Baltic Sea and the structure of the toxin. Applied and Environment Microbiology 55:1990–1995

Snoeijs P, Notter M (1993) Benthic diatoms as monitoring organisms for radionuclides in a brackish-water coastal ecosystem. Journal of Environmental Radioactivity 18:23–52

Snoeijs P, Sylvander P, Häubner N (2012) Oxidative stress in aquatic primary producers as a driving force for ecosystem responses to large-scale environmental changes. In: Abele D, Vázquez-Medina JP, Zenteno-Savin T (eds) Oxidative stress in aquatic ecosystems. Blackwell Publishing, Chichester, pp 72–88

Spencer PS, Hugon J, Ludolph A, Nunn PB, Ross SM et al (1987) Discovery and partial characterization of primate motor-system toxins. Ciba Foundation Symposium 126:221–238

Sternberg RM, Gooding MP, Hotchkiss AK, LeBlanc GA (2010) Environmental-endocrine control of reproductive maturation in gastropods: implications for the mechanism of tributyltin-induced imposex in prosobranchs. Ecotoxicology 19:4–23

Strand J (2009) Coupling marine monitoring and environmental risk assessment of TBT: a case study using the contamination of organotin compounds in the Danish marine environment. VDM Verlag Dr. Müller, 84 pp

Strand J, Jacobsen JA (2002) Imposex in two sublittoral neogastropods from the Kattegat and Skagerrak: the common whelk *Buccinum undatum* and the red whelk *Neptunea antique*. Marine Ecology Progress Series 244:171–177

Strand J, Jacobsen JA (2005) Accumulation and trophic transfer of organotins in a marine food web from the Danish coastal waters. Science of the Total Environment 350:72–85

Strand J, Jacobsen JA, Pedersen B, Granmo Å (2003) Butyltin compounds in sediment and molluscs from the shipping strait between Denmark and Sweden. Environmental Pollution 124:7–15

Strand J, Andersen L, Dahllöf I, Korsgaard B (2004) Impaired larval development in broods of eelpout (*Zoarces viviparus*) in Danish coastal waters. Fish Physiology and Biochemistry 30:37–46

Sundelin B, Eriksson Wiklund AK, Ford A (2008) The use of embryo aberrations in amphipod crustaceans for measuring effects of environmental stressors. ICES Techniques in Marine Environmental Sciences 41:1–21

Talvitie J, Heinonen M, Pääkkönen JP, Vahtera E, Mikola A et al (2015) Do wastewater treatment plants act as a potential point source of microplastics? Preliminary study in the coastal Gulf of Finland, Baltic Sea. Water Science and Technology 72:1495–1504

Tedengren M, Kautsky N (1987) Comparative stress response to diesel oil and salinity changes of the blue mussel *Mytilus edulis* from the Baltic and North Seas. Ophelia 28:1–9

Tedengren M, Olsson B, Bradly B, Zhou L (1999) Heavy metal uptake, physiological response and survival of the blue mussel (*Mytilus edulis*) from marine and brackish waters in relation to the induction of heat-shock protein 70. Hydrobiologia 393:261–269

Turja R, Soirinsuo A, Budzinski H, Devier MH, Lehtonen K (2013) Biomarker responses and accumulation of hazardous substances in mussels (*Mytilus trossulus*) transplanted along a pollution gradient close to an oil terminal in the Gulf of Finland (Baltic Sea). Comparative Biochemistry and Physiology C 157:80–92

Turja R, Höher N, Snoeijs P, Baršienė J, Butrimavičienė L et al (2014) A multibiomarker approach for the assessment of pollution impacts in two areas in the Swedish Baltic Sea coast using caged mussels (*Mytilus trossulus*). Science of the Total Environment 473–474:398–409

Van der Oost R, Beyer J, Vermeulen NPE (2003) Fish bioaccumulation and biomarkers in environmental risk assessment: a review. Environmental Toxicology and Pharmacology 13:57–149

Vandermeersch G, Lourenço HM, Alvarez-Muñoz D, Cunha S, Diogène J et al (2015) Environmental contaminants of emerging concern in seafood – European database on contaminant levels. Environmental Research 143B:29–45

Viarengo A, Lowe D, Bolognesi C, Fabbri E, Koehler A (2007) The use of biomarkers in biomonitoring: a 2-tier approach assessing the level of pollutant-induced stress syndrome in sentinel organisms. Comparative Biochemistry and Physiology C 146: 281–300

Voß B, Bolhuis H, Fewer DP, Kopf M, Möke F et al (2013) Insights into the physiology and ecology of the brackish-water-adapted cyanobacterium *Nodularia spumigena* CCY9414 based on a genome-transcriptome analysis. PLoS ONE 8(3):e60224

Vuorinen P, Keinänen M, Vuontisjärvi H, Baršienė J, Broeg K et al (2006) Use of biliary PAH metabolites as a biomarker of pollution in fish from the Baltic Sea. Marine Pollution Bulletin 53:479–487

Walker CH, Sibly R, Hopkin SP, Peakall DB (2001) Principles of ecotoxicology. CRC Press, London, 326 pp

WHO (2001) Integrated risk assessment report prepared for the WHO/UNEP/ILO International Programme on Chemical Safety WHO/IPCS/IRA/01/12. [http://www.who.int/ipcs/publications/new_issues/ira/en]

Ecosystem health

17

Maria Laamanen, Samuli Korpinen, Ulla Li Zweifel, and Jesper H. Andersen

Abstract

1. Humans have inhabited the Baltic Sea drainage area for thousands of years, but only in recent decades have the impacts from anthropogenic activities surpassed what could be considered sustainable levels from a Baltic Sea ecosystem perspective.
2. Human-induced degradation of the health of the Baltic Sea ecosystem accelerated in the 1950s.
3. Assessments of the ecosystem state in the 2000s have shown that anthropogenic pressures, which impair the overall ecosystem health are currently present in all parts of the Baltic Sea.
4. The major anthropogenic pressures that have contributed to impoverished biodiversity in the Baltic Sea comprise eutrophication, chemical contamination by hazardous substances and overfishing.
5. Ecosystem regime shifts took place in the Baltic Sea in the late 20th century, primarily due to hunting, fishing and eutrophication, in combination with changes in climatic conditions.
6. Some of the anthropogenic impacts, such as local sewage pollution, contamination by organochlorines and some mammal and bird population declines due to toxins and hunting, have largely been alleviated.
7. However, widespread eutrophication that is evident through bottom hypoxia and shifts in biodiversity is still likely to pose great challenges for the management of the Baltic Sea in the near future.
8. After many decades of scientific research, environmental assessments and political negotiations, international legislation and regional cooperation are currently in force to bring the Baltic Sea ecosystem into a healthier state than it is today.

Keywords

Anthropogenic pressures • Biodiversity • Chemical contamination • Climate change • Ecosystem health • Eutrophication • International policies • Overfishing • Regime shifts • Resilience • Science-based value judgement • Sustainable use

M. Laamanen (✉)
Ministry of the Environment,
P.O. Box 35, 00023 Government, Finland
e-mail: maria.laamanen@ym.fi

S. Korpinen
Marine Research Centre, Finnish Environment Institute,
P.O. Box 140, 00251 Helsinki, Finland
e-mail: samuli.korpinen@ymparisto.fi

U.L. Zweifel
Swedish Institute for the Marine Environment,
P.O. Box 260, 40530, Göteborg, Sweden
e-mail: ullali.zweifel@havsmiljoinstitutet.se

J.H. Andersen
NIVA Denmark Water Research, Ørestads Boulevard 73,
2300 Copenhagen, Denmark
e-mail: jha@niva-danmark.dk

P. Snoeijs-Leijonmalm et al. (eds.), *Biological Oceanography of the Baltic Sea*,
DOI 10.1007/978-94-007-0668-2_17

17.1 Ecosystems are naturally dynamic

17.1.1 Time scales of change

Ecosystem processes are dynamic by nature, *i.e.* there is never a "steady state". Changes proceed on various time scales: geological, centennial, decadal, annual, seasonal or daily, depending on the process in question. On a geological scale, the Baltic Sea has – during the Holocene – gone through marine and freshwater phases, and salinity levels and climate conditions similar to those of today have existed only for ~3,000 years (*cf.* Fig. 2.26c).

However, some of the post-glacial changes are still going on, *e.g.* the northern half of the Baltic Sea is still subject to land uplift and the Baltic Sea coastlines experience constant directional changes (trends) such as coastal growth in the north and coastal retreat in the south (*cf.* Fig. 2.26b, d). Many changes are cyclic, *e.g.* seasonal, while the on-going anthropogenic climate warming seems to be directional.

17.1.2 Physico-chemical factors govern biological change

During the last hundred years, the salinity, oxygen and temperature conditions of the Baltic Sea have varied in a cyclic manner on a more or less decadal scale (Winsor et al. 2001). The changes in physico-chemical factors have caused changes in the abundance and distribution of both pelagic and benthic species and communities in the Baltic Sea (Alheit et al. 2005).

A major force behind the cyclic changes in salinity and deep-water hypoxia involves the meteorologically-driven major Baltic inflows (MBIs) of saline water from the Kattegat (*cf.* Sect. 2.3.9). The MBIs are crucial for the ecosystem structure and functions of the Baltic Sea. Hypoxic (<2 mL O_2 L^{-1}) and particularly anoxic areas are inhabited mainly by prokaryotes, such as sulphur bacteria, and are physiologically intolerable for most invertebrate and fish species (*cf.* Sect. 10.11).

The spatial extent of hypoxic bottoms has varied over time and with the morphology and depth differences within the Baltic basin; specifically, the presence or absence of sills between subregions has been the underlying factor for hypoxia on a millennium scale (Zillén and Conley 2010). On a decadal scale, a period without MBIs, a so-called "stagnation period", has been observed in the Baltic Sea from the late 1970s to the early 1990s (*cf.* Fig. 2.13a). In the Baltic Sea proper, the lower frequency of MBIs has caused a freshening of both surface and bottom waters as well as a weakening and deepening of the halocline (Conley et al. 2002).

Seasonal fluctuations are also pronounced in the Baltic Sea. For example, the thermocline formed in spring facilitates the spring phytoplankton bloom, which initiates the period of higher productivity in the pelagic zone (*cf.* Sect. 8.2.4). Parts of the Baltic Sea are covered by ice each winter, although the extent of the ice cover varies greatly from one year to another (*cf.* Sect. 9.2.1). Seasonal changes in the quantity of freshwater runoff from the drainage area (*cf.* Fig. 2.12b) result in temporal differences in the supply of inorganic nutrients, organic matter and trace elements to the sea.

There are also ecosystem processes that fluctuate on a diel scale, such as temperature in shallow coastal waters, littoral migrations of macrofauna and fish, vertical migration of zooplankton, and primary productivity by phytoplankton and macroalgae. These diel variations are largely driven by changes in solar radiation and the resultant alteration of underwater light conditions (*cf.* Fig. 7.2).

17.1.3 Ecosystem dynamics on various spatial scales

There are differences in the spatial scales of various natural processes in all ecosystems, including the Baltic Sea. For example, the MBIs tend to have an ecosystem-wide influence, but their impacts are less pronounced in the northernmost subbasins of the Baltic Sea and take place after a time lag.

There are large spatial differences in the way seasonal dynamics are expressed in the Baltic Sea Area. For example, the growing season is longest in the south and shortest in the north, which is manifested by earlier induction of the thermocline and earlier onset of phytoplankton spring blooms in the south than in the north (*cf.* Sects. 2.4.3 and 8.2.4). The thermocline disappears earlier in the north.

On a local scale, a sheltered shallow bay is likely to display spring-related effects earlier than the nearby open sea because solar radiation heats the shallow, calmer water of the bay faster than it heats the water of the deeper, more wind-exposed open sea. The bay may also be affected by an inflow of fresh riverine water, which will further strengthen the stratification of the water column. Large rivers flowing into the Baltic Sea, such as the Neva, Odra and Wisła tend to exert subregional-scale influences since inputs of freshwater, nutrients and contaminants from these rivers are distributed over larger sea areas. Also perturbations in the drainage area related to the large rivers, such as flooding, are often manifested at the subregional scale.

The coastal zone tends to experience ecological effects that are observable at smaller spatial scales than those of the open sea. This is largely due to the mosaic-like geographical structure of many parts of the coastal zone (*cf.* Sect. 11.1), which provides a physically more diverse environment for the biota than the open sea. As a result, habitat complexity and species diversity tend to be higher in the coastal zone than in the open sea.

17.2 Anthropogenic pressures on the Baltic Sea ecosystem

17.2.1 The growth of human settlements

Humans started to colonise the drainage area of the Baltic Sea 13,000 to 9,000 years BP. Their settlements followed the rim of the retreating Weichselian ice sheet towards the north (*cf.* Fig. 2.27). The first major population expansion took place between 2,000 and 2,300 years BP, when the human population size in the drainage area increased from 4.6 to 9.5 million. However, the largest human population growth in the Baltic Sea drainage area occurred between the year 1700 and today, when it increased from ~14.4 to ~85 million (Zillén and Conley 2010).

The Baltic Sea has always been an important resource for the human populations living on its coasts. Fisheries and the hunting of seals, harbour porpoises and waterbirds were a natural part of the culture of the people living in coastal areas. The increase in the number of humans in the drainage area was accompanied by an expansion of land use for agriculture and forestry through land reclamation, and more recently, by urbanisation and industrialisation. The anthropogenic activities within the drainage area have increased the inflow of soil particles, nutrients and chemical pollutants to the sea.

17.2.2 Changes in anthropogenic pressures

Since the very beginning of human colonisation of the area, the Baltic Sea has also served as a conduit of transportation to connect the people living in different parts of its coast. Just like other anthropogenic activities, shipping has intensified after the 1950s and today there are ~2,000 large ships plying the Baltic Sea at any point of time. Shipping activity causes discharges of oil, chemicals and nutrient-rich sewage into the sea as well as emissions of nitrogen oxides, sulphur oxides and particles to the air. In addition, it creates underwater noise (acoustic pollution) and promotes the transfer of non-indigenous species to the Baltic Sea from other aquatic ecosystems (*cf.* Sect. 5.3.2). Furthermore, in coastal and shallow offshore areas, shipping may cause erosion and re-suspension of bottom sediments (*cf.* Sect. 11.15.2). So far, the Baltic Sea has been fortunate in not having experienced any catastrophic shipping accident with large-scale and long-term ecosystem effects. However, continuous pollution by petrogenic substances and smaller-scale accidents has been occurring as long as motorised ships have existed.

After World War II, technological development has intensified agriculture, forestry, shipping and fisheries to an extent unthinkable in the times when the human use of the sea was more subsistence-oriented. The exploitation of the sea has become industrialised. All of this has also had large impacts on the ecosystem. Signs of deterioration were first observed near the cities due to pollution by untreated wastewaters, but now ecosystem-wide impacts are recorded. The contemporary inputs of total nitrogen into the Baltic Sea are estimated to be more than twice the amount supplied a century ago (*cf.* Box Fig. 2.2).

The Baltic Sea ecosystem is also affected by external anthropogenic pressures. The on-going global climate change (Stocker et al. 2013) alters the physical setting of the sea and, as a consequence, the chemical and biological features of the ecosystem change. The air temperature in the northern Baltic Sea region has increased by an average of 0.08 °C per decade during the period 1871–2004 (Heino et al. 2008). This trend is slightly steeper than that observed in the global time series (*cf.* Fig. 2.28) with an increase of ~0.74 °C degrees over the period 1906–2005 (~0.074 °C per decade).

During recent decades we have witnessed a decreased frequency of large saline water inflows from the Kattegat into the Baltic Sea (*cf.* Fig. 2.13a). Also, the length of the ice season has decreased by 14–44 days during the last century. These changes have had a measurable effect on the distribution, reproductive output and stock sizes of Baltic Sea organisms and have modified the food webs. However, it has not been possible to establish distinct causal links between all these changes and global climate change, partly because of the large natural climate variability, but also owing to possible impacts from other anthropogenic pressures (Dippner et al. 2008).

17.2.3 Ecosystems are resilient until a certain limit

Healthy ecosystems tend to exhibit "stable states", not to be confused with "steady state" (*cf.* Sect. 17.1.1). A stable state varies within natural fluctuations, *i.e.* changes do occur but within certain limits. However, ecosystems that are heavily stressed by external pressure can undergo drastic changes that reorganise their structure and functioning (Scheffer et al. 2001). The causative external factors can be natural, anthropogenic or combinations of the two. The stress level that pushes an ecosystem into a new state is often referred to as a "threshold" or "tipping point" (Fig. 17.1b, d). Below the threshold level one stable state prevails, and above the threshold another. Once a new stable state has been reached, the system has a tendency to be self-perpetuating as feedback mechanisms start to stabilise the new regime. Ecosystem changes, even large ones, can also be gradual with no apparent tipping points (Fig. 17.1a, c).

A healthy ecosystem has the capacity to buffer against and counteract disturbances within certain limits. The capacity of an ecosystem to absorb change and to recover from it is called the "ecosystem resilience" (Elliott et al. 2007). Biological diversity and natural variability in an ecosystem tend to build up the resilience of an ecosystem. As a consequence of resilience, ecosystems can appear virtually unaffected and stable while exposed to considerable stress. The apparent lack of response can be explained by natural feedback mechanisms such as biogeochemical compensation, regulation through trophic and competitive interactions within the system, and, to a certain degree, by the functional diversity and redundancy among species.

Regime shifts (*cf.* Box 2.5) take place when the capacity of the ecosystem to absorb and buffer external pressures is exceeded, *i.e.* when the resilience of the system has been surpassed. At a certain point, even a small increment in the external pressure can cause a shift that will result in the collapse of or dramatic change in populations, communities and food webs within the ecosystem. A saying that reflects this type of event is "the straw that broke the camel's back".

17.2.4 Regime shifts in the Baltic Sea

It has been suggested that the Baltic Sea ecosystem has experienced several regime shifts during the last 80 years. These were likely driven by variability in climate, eutrophication, seal hunting and overfishing (Österblom et al. 2007; ICES 2008). Two shifts were identified as late as in the end of the 1980s and the mid-1990s (Möllmann et al. 2008). In the Baltic Sea proper, the period before the first shift was characterised by relatively high spawning biomasses and recruitment of the Atlantic cod *Gadus morhua* and the Atlantic herring *Clupea harengus* as well as by high abundances of the zooplankton copepod *Pseudocalanus acuspes*. The period after the second shift features dominance of the European sprat *Sprattus sprattus* and high abundances of the zooplankton copepods *Acartia* spp. and *Temora longicornis*.

Reductions in the populations of top predators such as seals have modified the trophic structure of the Baltic Sea food webs. Examples from around the world show that decimation of top predators through hunting and fishing has made the marine ecosystems vulnerable to other threats such as chemical contamination by hazardous substances, eutrophication and biological hazards, including introductions of non-indigenous species (Jackson et al. 2001). On a larger scale, the Baltic Sea ecosystem has in the course of the past hundred years changed

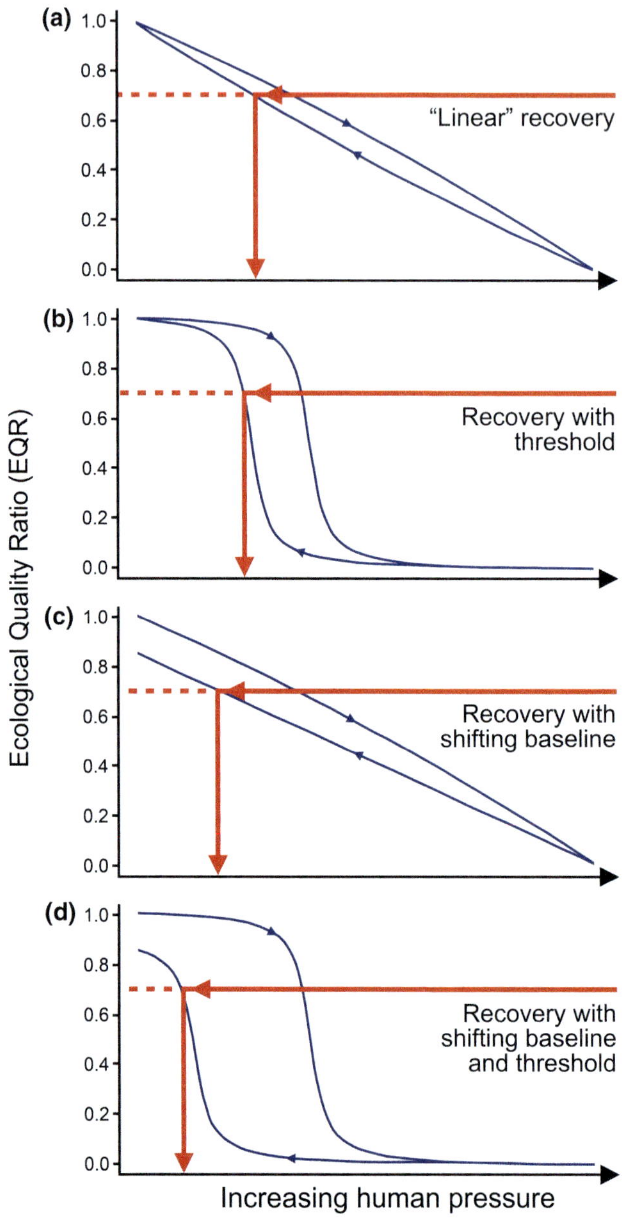

Fig. 17.1 Conceptual models of the possible response modes of ecological status to increasing human pressure in an ecosystem. (**a**) Linear recovery. (**b**) Recovery with threshold. (**c**) Recovery with shifting baseline. (**d**) Recovery with shifting baseline and threshold. The dashed line indicates an environmental target for ecosystem quality, here set to an ecological quality ratio (EQR) of 0.7, and the red arrows indicate the estimated reductions in pressures needed to meet this target. The increase of necessary reductions from Scenario (**a**) to Scenario (**d**) illustrates that fulfilment of the target in non-linearly responding systems with a threshold Scenario (**b**) or a shifting baseline Scenario (**c**) or combinations thereof *e.g.* caused by climate change, call for larger reductions compared to linearly responding systems in Scenario (**a**). Figure modified from Duarte et al. (2009) and Kemp et al. (2009)

from a clear-water oligotrophic sea with abundant populations of top predators into a eutrophicated sea burdened by chemical pollution with hazardous substances and biological pollution with non-indigenous species.

17.3 Present anthropogenic pressures

17.3.1 The most adverse pressures today

In the 2000s, the Baltic Sea ecosystem has been under multiple pressures from various types of anthropogenic activities (Table 17.1). These pressures are primarily exerted by eutrophication, chemical contamination by hazardous substances and overfishing that have contributed to impoverished biodiversity. According to expert views in line with studies on the top pressures acting on marine regions worldwide (HELCOM 2010a, b), the most adverse pressures also include commercial fishing. Some of the single pressures can have multiple effects and several pressures can have synergistic effects. For example, inputs of nitrogen and phosphorus cause eutrophication but may also change biodiversity by altering the competitive balance between species, and overfishing may alter the effects of eutrophication by changing the food web structure.

Adverse impacts may be direct or indirect (acting through other effects). For example, dredging has a direct physical impact on the benthic environment since it alters the sedimentary habitat structure and may harm the sediment-dwelling organisms by smothering. An indirect effect of dredging may involve releases of hazardous substances, which were buried in the sediments, to the food web.

17.3.2 Highest pressures in the southern and eastern Baltic Sea

The actual pressures on the ecosystem in quantitative terms, such as disturbance of the bottom habitats, are in many cases difficult to measure. Therefore, certain proxies such as anthropogenic activities that act as drivers of the pressures are used in assessments (Korpinen et al. 2012). This approach was applied to estimate the quantity and distribution of all potential pressures in the Baltic Sea with the Baltic Sea Pressure Index (BSPI, HELCOM 2010b), which is based on 52 different anthropogenic activities. The BSPI tends to be the highest in the southern and eastern parts of the Baltic Sea Area, *i.e.* parts of the Kattegat, the Belt Sea, the southern Baltic Sea proper, the Gdańsk Bay, the Curonian Lagoon and the Gulfs of Riga and Finland (Fig. 17.2). The Gulf of Bothnia and the northern Baltic Sea proper are the areas with the lowest pressures. These trends can largely be explained by the density of the human population, which is highest in the southern and eastern parts of the drainage area and lowest in the north (*cf.* Fig. 2.5b).

The assessment of the potential impacts from the pressures on the ecosystem requires, in addition to data on the distribution of the pressures, also knowledge about the distribution of ecosystem components. Furthermore, estimates of the severity of the impact of each anthropogenic pressure on each ecosystem component are necessary. Such relationships, the so-called “impact factors”, were obtained from expert estimates provided during the construction of the Baltic Sea Impact Index (BSII, HELCOM 2010b). The BSII (Fig. 17.3) provides an estimate of the sum of potential impacts from anthropogenic pressures on the ecosystem and offers a way to present the spatial distribution of these impacts. By comparing the cumulative impacts with the state of the Baltic Sea biodiversity a tipping point could be indicated (Andersen et al. 2015a, 2015a, b). At the tipping point, the amount of pressures and impacts grew too high for the ecosystem to maintain its state, and the state changed to a new state that is regarded as deteriorated.

In the Baltic Sea region, the coastal areas support a higher number of ecosystem components than the open sea, and the coastal areas are also more vulnerable to anthropogenic pressures. This is reflected as higher BSII in many coastal areas compared to the open sea (Fig. 17.3). Many of the local-scale pressures, *e.g.* dredging and construction of harbours or marinas, act in the coastal zone and complement larger-scale pressures such as eutrophication. In the open sea, most of the impacts are related to fishing or airborne deposition of nitrogen and hazardous substances that may have been transported over long distances in the atmosphere.

17.4 Effects of eutrophication on the ecosystem

17.4.1 Nutrient loading peaked already in the 1970s and 1980s

The inputs of nitrogen and phosphorus into the Baltic Sea peaked already in the 1970s to 1980s (*cf.* Box 2.2). Today’s loads are lower than those observed 40 years ago (HELCOM 2011, 2012a). Between the years 1990 and 2006, nitrogen and phosphorus inputs are estimated to have declined by 30 and 45 %, respectively (HELCOM 2009a). In 2006, 638 kilotonnes of nitrogen and 28.4 kilotonnes of phosphorus still entered the Baltic Sea via waterways (HELCOM 2011). In addition, 200 kilotonnes of airborne nitrogen were deposited into the Baltic Sea in 2006 (Bartnicki et al. 2011). The fraction of phoshorus entering the Baltic Sea via atmospheric pathways is minor (*cf.* Box Fig. 2.2).

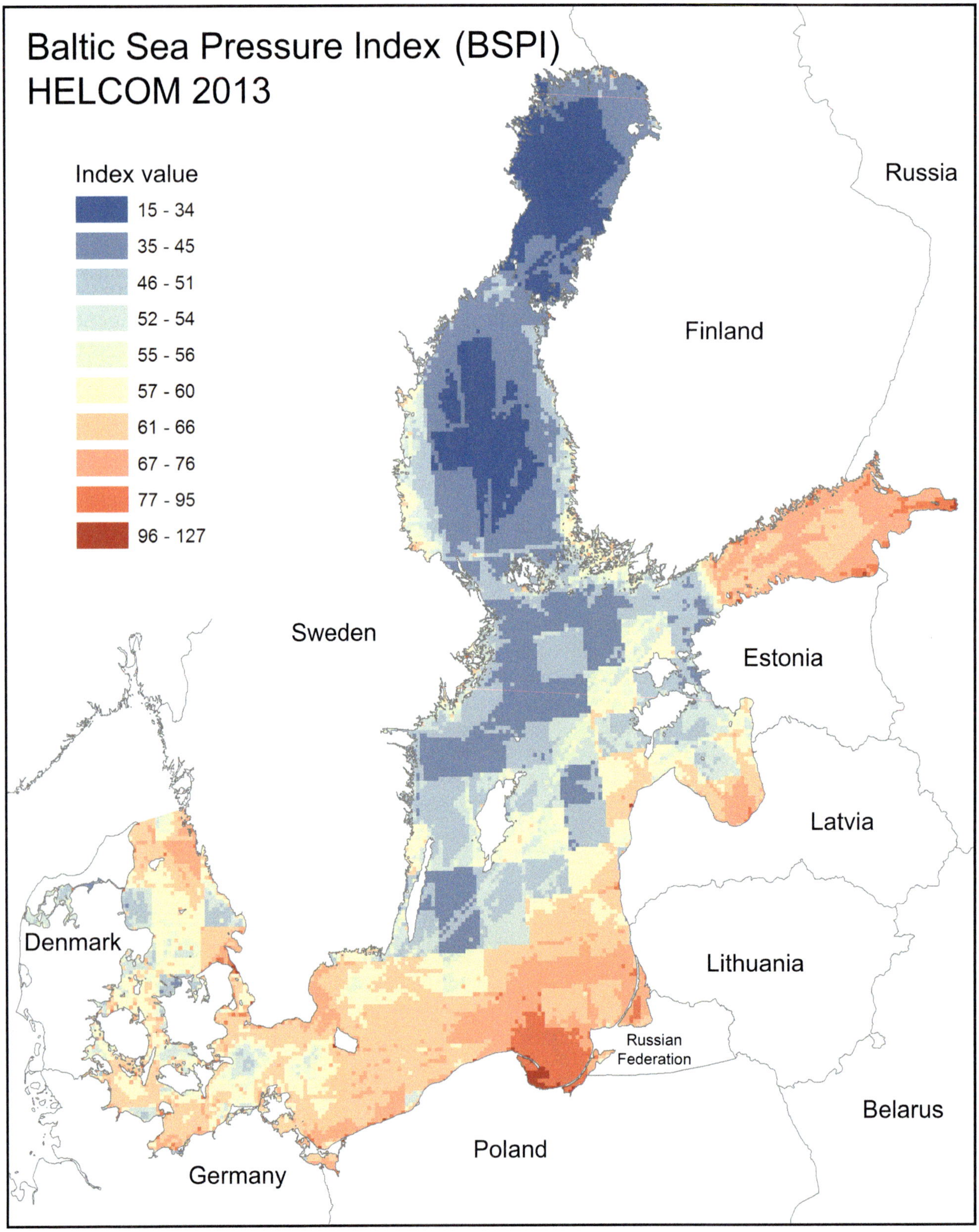

Fig. 17.2 Sum of anthropogenic pressures in the Baltic Sea Area expressed as the Baltic Sea Pressure Index (BSPI). Lower index values (in blue) indicate a smaller sum of pressures and higher index values (in red) indicate a larger sum of pressures. Figure with small modifications reprinted from HELCOM (2010b) and reproduced in 2013, with permission from HELCOM

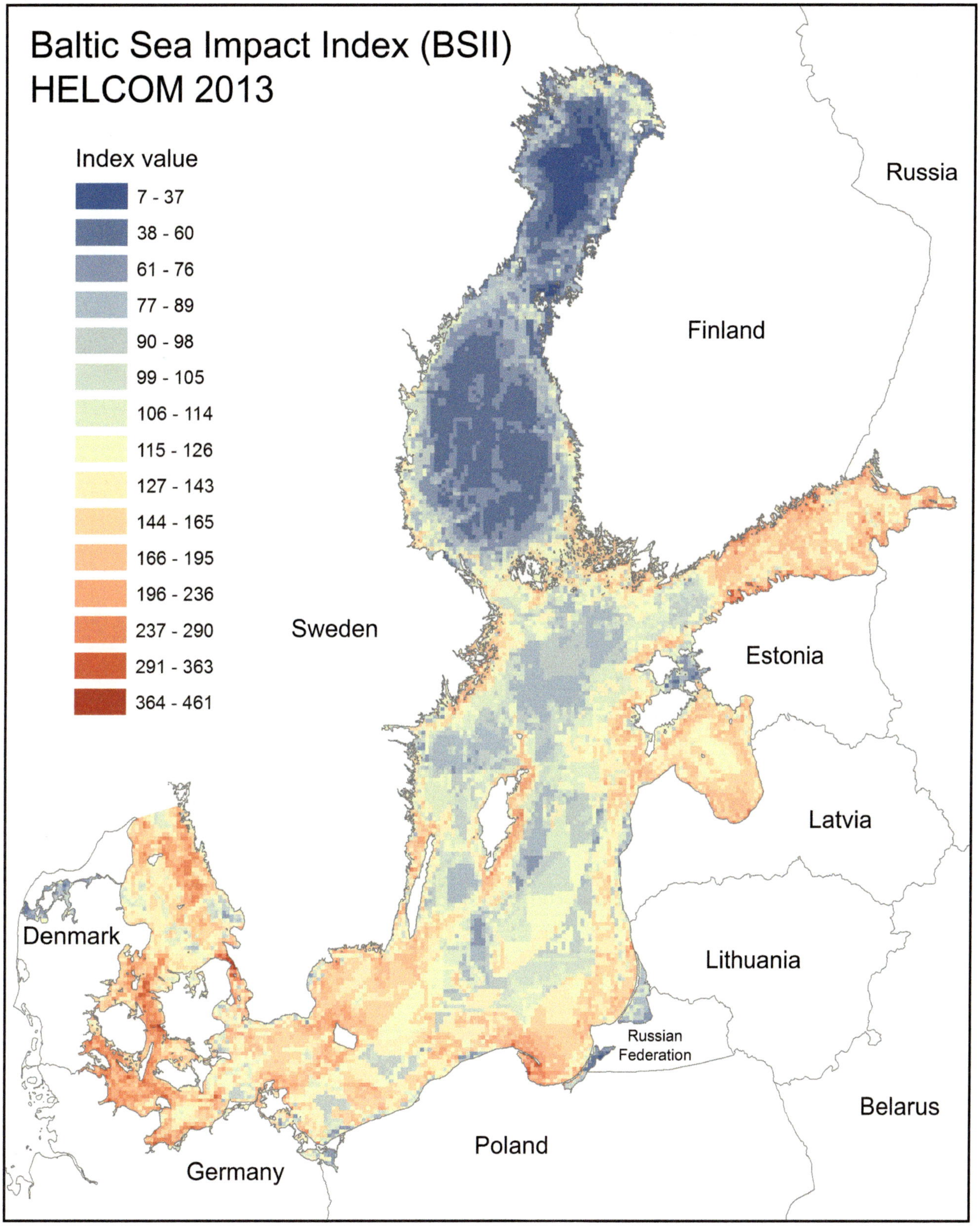

Fig. 17.3 Sum of the potential impacts of anthropogenic pressures on the aquatic ecosystem in the Baltic Sea Area according to the Baltic Sea Impact Index (BSII). Lower index values (in blue) indicate a smaller sum of impacts and higher index values (in red) indicate a larger sum of impacts. Figure with small modifications reprinted from HELCOM (2010b) and reproduced in 2013, with permission from HELCOM

Table 17.1 Summary of human pressures on the Baltic Sea ecosystem with their drivers (human activities) and their potential enhancing impacts on eutrophication (E), pollution by hazardous substances (HS) and impoverished biodiversity (BD). x = direct impacts (x) = indirect impacts. Data from HELCOM (2010a)

Pressure	Human activity	Potential enhancing impacts		
		E	HS	BD
Smothering	Wind farms, bridges, oil platforms (construction phase)		(x)	x
Smothering	Cables and pipelines (construction phase)		(x)	x
Smothering	Disposal of dredged material		x	x
Sealing	Coastal defence structures			x
Sealing	Harbours			x
Sealing	Bridges			x
Changes in siltation	Shipping (coastal)		(x)	x
Changes in siltation	Riverine input of organic matter			x
Changes in siltation	Bathing sites, beaches and beach replenishment			x
Changes in siltation	Dredging, sand and gravel extraction, stone-fishing		(x)	x
Abrasion	Dredging, sand and gravel extraction, stone-fishing		(x)	x
Abrasion	Bottom trawling	(x)	(x)	x
Selective extraction of sea-floor materials	Dredging, sand and gravel extraction, stone-fishing, resulting in *e.g.* habitat loss			x
Underwater noise	Shipping (coastal and offshore)			x
Underwater noise	Recreational boating and sports			x
Underwater noise	Cables and pipelines (construction phase)			x
Underwater noise	Wind farms, bridges, oil platforms (construction phase)			x
Underwater noise	Wind farms (operational)			x
Underwater noise	Oil platforms			x
Changes in thermal regime	Power plants with warm-water outflow			x
Changes in salinity regime	Bridges and coastal dams			x
Changes in salinity regime	Coastal wastewater treatment plants with freshwater outlets to the sea			x
Introduction of synthetic compounds	Polluting ship accidents		x	x
Introduction of synthetic compounds	Coastal industry, oil terminals, refineries, oil platforms		x	x
Introduction of synthetic compounds	Harbours		x	x
Introduction of synthetic compounds	Atmospheric deposition of dioxins		x	(x)
Introduction of synthetic compounds	Population density (*e.g.* hormones and pharmaceuticals)		x	(x)
Introduction of non-synthetic compounds	Illegal oil spills		x	x
Introduction of non-synthetic compounds	Waterborne input of Cd, Hg and Pb		x	x
Introduction of non-synthetic compounds	Atmospheric deposition of Cd, Hg and Pb		x	(x)
Introduction of radionuclides	Discharges of radioactive substances		x	
Inputs of nutrients	Waterborne input of nitrogen	x		x
Inputs of nutrients	Waterborne input of phosphorus	x		x
Inputs of nutrients	Aquaculture	x		x
Inputs of nutrients	Atmospheric deposition of nitrogen	x		x
Inputs of organic matter	Aquaculture	x		x
Inputs of organic matter	Riverine input of organic matter	x		x
Introduction of microbial pathogens	Coastal wastewater treatment plants with outlets to the sea			x
Introduction of microbial pathogens	Aquaculture			x
Selective extraction of species	Bottom trawling (landings or catches)	(x)		x
Selective extraction of species	Surface- and mid-water trawling	(x)		x
Selective extraction of species	Gillnet fishery	(x)		x
Selective extraction of species	Coastal stationary gear fishery			x
Selective extraction of species	Hunting of seals			x
Selective extraction of species	Hunting of birds			x

17.4.2 Eutrophication is still a major problem

Eutrophication means an increase in the rate of production of organic material (Nixon 1995), and is mostly fuelled by the supply of inorganic nitrogen and phosphorus. In the Baltic Sea, anthropogenic inputs of nitrogen and phosphorus have resulted in elevated levels of both nutrients in the water. Excess nutrients and the subsequent accelerated primary production cause the well-known eutrophication effects of increased biomass of primary producers (*cf.* Sects. 8.2.9, 11.15.3, 13.3.6), increased frequency and intensity of algal and cyanobacterial blooms, reduced water transparency, increased secondary production and upsurges in the biomass of certain groups of animals, *e.g.* cyprinid fish. The chain of the effects involved terminates with hypoxia, and often even complete anoxia, in the bottom waters and sediments, since microbial decomposition of the increased supply of organic matter consumes oxygen (*cf.* Fig. 3.16).

It seems that the 1950s were the upward turning point with respect to the rates of increase of anthropogenic nutrient inputs and eutrophication of the Baltic Sea (HELCOM 2012a, 2013, 2015). Although there were not too many measurements of nutrient concentrations and primary production in the water prior to the 1950s, the acceleration of primary production is reflected in the organic carbon accumulation rate over the last half century as recorded in the sediments of the Bornholm Sea and the Gotland deep (Struck et al. 2000). The intensification of the eutrophication during the 20th century is also reflected in the water transparency. The recorded summer Secchi depths gradually declined in all subregions of the Baltic Sea over the last one hundred years (Sandén and Håkansson 1996). The decrease in water transparency was most pronounced in the northern Baltic Sea proper (from 9 to 5 m) and in the Gulf of Finland (from 8 to 4 m). Eutrophication is also reflected in the increase of the total seafloor area with anoxia and hypoxia from less than 10,000 km^2 in 1900–1910 to $\sim$35,000 km^2 in the 1950s and 60–70,000 km^2 in recent years (Carstensen et al. 2014).

The eutrophication status of the Baltic Sea in 2001–2006 was evaluated using data on indicators such as nutrient concentrations in the water, summer water transparency, chlorophyll *a* concentrations, maximum depth penetration of benthic macrophytes, macrozoobenthos species diversity on the deep bottoms and various macrozoobenthos-related indices in the coastal zone (HELCOM 2009a). The level of each indicator was compared to a threshold level that corresponds to a boundary between "acceptable" and "non-acceptable" eutrophication status (the so-called "target levels"). These target levels were then used to integrate the indicator-based information into a eutrophication assessment for the different parts of the Baltic Sea (Andersen et al. 2011). This evaluation showed that most of the Baltic Sea Area was in 2001–2006 strongly affected by eutrophication. Only 13 out of the 189 areas assessed had an acceptable eutrophication status. The less affected areas were found at coastal sites in the Gulf of Bothnia, the open Bothnian Bay and the northeastern Kattegat. The Gulf of Bothnia is the part of the Baltic Sea with the smallest human population and the lowest anthropogenic pressures in the drainage area, while the Kattegat is influenced by Atlantic waters.

17.4.3 Maximum allowable input levels of nutrients

The current nutrient input levels are still far above the so-called "maximum allowable input" (MAI) of the HELCOM Baltic Sea Action Plan (BSAP, *cf.* Sect. 17.8.4). According to the BSAP, an acceptable level of eutrophication is at the estimated loads of 601 kilotonnes of N and 15.3 kilotonnes of P.

A large part of the N supply and most of the P supply to the Baltic Sea used to come from point sources, mainly discharges of municipal and industrial wastewaters. However, on account of the improved wastewater treatment during the recent decades, agriculture has now taken over the role of the most prolific nutrient supplier. In 2006, losses from diffuse sources were the main origin of the excessive inputs of N and P (at least 45 % for both nutrients), and only $\sim$12 % of N and $\sim$20 % of P originated from point sources (HELCOM 2011). Agriculture contributed on average 60–90 % of the diffuse nutrient inputs of both N and P. In addition, scattered dwellings and storm water discharges are also important nutrient sources in some areas.

The countries contributing most to the overall waterborne nutrient input in 2006 were Poland (N 24 %, P 36 %), Sweden (N 19 %, P 13 %) and Russia (N 17 %, P 14 %) (HELCOM 2011). Only Denmark documented significantly decreasing riverine loads between 1994 and 2008, while the riverine loads of Estonia and Finland significantly increased. On the other hand, the largest area-specific loads of nutrients to the sea are found in drainage areas with high population densities, many industries and high agricultural activity in the southern part of the Baltic Sea drainage area (*cf.* Fig. 2.5).

17.4.4 Processes that sustain the eutrophic regime

A particular challenge to the remedial actions in the Baltic Sea stems from the fact that eutrophication, through increased sedimentation of organic matter, has contributed to the extension of hypoxic areas as well as to the enrichment of sediments with nutrients (HELCOM 2009a, 2012a).

When sediments become anoxic, phosphorus is released into the overlaying water mass, primarily through dissimilatory reduction of iron oxyhydroxides by bacteria (*cf.* Sect. 3.6.7, Conley et al. 2002). Through phosphorus release from the sediments, this nutrient may become available for the growth of those primary producers which are favoured by nitrogen-deficient conditions.

Such conditions particularly enhance the diazotrophic (nitrogen-fixing) cyanobacterial surface blooms, which can utilise the vast reservoir of atmospheric nitrogen and increase the nitrogen levels in the pelagic system. Thus, there are currently processes in place that act to sustain the present eutrophic regime (Vahtera et al. 2007). It is likely that, to revert the Baltic Sea to a less eutrophicated state, it will be necessary to reduce nutrient loads to levels that are lower than those that existed just before the major acceleration of eutrophication.

17.5 Effects of hazardous substances on the ecosystem

17.5.1 Sources of hazardous substances

Chemical substances can be considered hazardous if they are toxic, persistent and bioaccumulate, or if they are highly persistent and bioaccumulate (*cf.* Table 16.1). In addition, substances with effects on hormone and immune systems as well as radioactive substances are considered hazardous.

In the Baltic Sea drainage area, large amounts of hazardous chemicals have been used since the very beginning of the industrialisation of the region in the late 19th century. Synthetic substances such as persistent organic pollutants (POPs), pharmaceuticals and non-synthetic substances like heavy metals originate from (1) point-sources situated on the coast or inland in the drainage area, including industries and municipal wastewater treatment plants, (2) land-based diffuse sources such as runoff from agricultural land, forests and cities, as well as leaching from dump sites and landfills, (3) activities taking place at sea such as shipping, pipelines, dredging and operation of oil platforms, and (4) atmospheric deposition from all types of combustion sources as well as volatile chemicals (*e.g.* pesticides).

In the Baltic Sea region, industrial and municipal point sources have been the most apparent suppliers of hazardous substances, but several groups of substances originate mainly from minor industrial sources, agriculture (*e.g.* pesticides, veterinary pharmaceuticals), households (*e.g.* cleaning products, personal care products, pharmaceuticals), sludge, dump sites and waste deposition in landfills (HELCOM 2010c).

Atmospheric emissions derive from land traffic, shipping, energy production, incineration of wastes and even from small-scale household combustion. Atmospheric deposition is a major pathway of annual inputs of some heavy metals into the Baltic Sea (*e.g.* cadmium, lead, mercury) and may dominate over other pathways in the supply of substances such as dioxins. In addition, atmospheric deposition also involves substances from emission sources outside the Baltic Sea drainage area. It has been estimated that 60 % of dioxins, 60 % of cadmium, 84 % of lead and 79 % of mercury deposited into the Baltic Sea originate from distant sources outside its drainage area, mainly in the UK, France, Belgium and the Czech Republic (Bartnicki et al. 2008).

17.5.2 Impacts of hazardous substances

The threats hazardous substances pose to wildlife have been documented for a number of organisms living in the Baltic Sea, mainly top predators. The declines of the grey and ringed seal populations have been attributed to reproductive failures caused by the increases of organochlorines in the environment (*cf.* Sect. 16.1.2). Although the seal populations (*cf.* Box 4.13) had been drastically reduced by hunting already in the 1950s, the increasing pollution in the late 1960s by PCBs (*cf.* Table 16.1) from pulp mills and other industrial sources caused a second decline. By the 1970s this had brought down the populations to all-time low numbers of individuals, estimated at 5,000 ringed seals and 4,000 grey seals (Fig. 17.4; Harding and Härkönen 1999). The use of PCBs has been banned, and the reproduction of seals has been normalising since the mid-1980s. However, high concentrations of PCBs are still found in marine sediments and biota of the Baltic Sea.

Fig. 17.4 The grey seal *Halichoerus grypus* has been close to extinction in the Baltic Sea but is recovering today. Photo: © Hans Kautsky

Similar impacts, related mainly to DDTs (*cf.* Table 16.1), have been shown for the white-tailed eagle *Haliaeetus albicilla* and other fish-feeding birds. White-tailed eagle populations eventually recovered after the use of DDTs and PCBs was banned (*cf.* Box 4.12), and today they are significantly less contaminated by these substances.

Despite diminished impacts of DDTs and PCBs, new hazardous substances, such as estrogenic substances from municipal wastewaters, are causing new problems, including the feminisation of male fish (Ferreira et al. 2009).

17.5.3 The Baltic Sea is "disturbed by hazardous substances"

In 1999–2007, the entire Baltic Sea exhibited a high chemical contamination level. An indicator-based assessment of 144 areas in the Baltic Sea revealed that 137 of them were contaminated by hazardous substances (HELCOM 2010c). All open-sea areas of the Baltic Sea, except for the northwestern Kattegat, were classified in the assessment as being "disturbed by hazardous substances". The sea areas close to the larger coastal cities of Copenhagen, Gdańsk, Helsinki, Riga, Rostock, Sankt-Petersburg and Stockholm were generally classified as having a "moderate" or "poor" chemical status. PCBs, DDE (*cf.* Fig. 16.3), cadmium, mercury, lead, TBT, dioxins (*cf.* Sect. 16.3) and brominated substances (*e.g.* BDEs, *cf.* Fig. 16.8) showed the highest concentrations in relation to the accepted threshold levels of disturbance.

This integrated status assessment did not address all the hazardous substances that have been measured in the Baltic Sea organisms or sediments. Substances that were not included, but are prevalent in the Baltic Sea environment, are *e.g.* perfluorinated compounds, alkylphenyls, bisphenol A and different pharmaceuticals. Concentrations of these substances in the marine environment are already high and are increasing. However, the understanding of their environmental fate is still poor and information on their main sources, transport mechanisms and spatial distributions in the ecosystem is limited (*cf.* Sect. 16.1.8).

17.6 Biodiversity changes and their effects on the ecosystem

17.6.1 Naturally low biodiversity

The species diversity of macroscopic organisms in the Baltic Sea is much lower than that in fully marine areas (*cf.* Sect. 4.2). Most species are either of freshwater or marine origin and in the brackish Baltic Sea many of them live at the border of their physiological salinity limit. In addition, the Baltic Sea has been revealed to support a low genetic diversity within species (Johannesson and André 2006). The already naturally low species and within-species diversity suggests that the biodiversity in the Baltic Sea is particularly vulnerable to disturbances.

During the last decades, the biodiversity of the Baltic Sea has undergone major changes. Of particular importance for these changes to happen were variations in climate, both natural and anthropogenic, which induced physical and chemical changes that have caused biological shifts (HELCOM 2009b). The identification of the exact role of anthropogenic pressures is a challenging task since comprehensive data sets are lacking and a considerable natural variability in biodiversity is prevalent. However, there is no doubt that human society has contributed to the observed changes.

Species with a threatened or declining population status have been identified (HELCOM 2007b). Although the only species known to have become regional extinct is a sturgeon species (*cf.* Box 6.4), all marine mammals, except for the grey seal populations north of 59 °N, are still under threat or declining.

17.6.2 Changes in Baltic primary producer communities

Nutrient enrichment has resulted in accelerated phytoplankton productivity, which entails increased prevalence and spatial extent of algal and cyanobacterial blooms. Within phytoplankton communities, the blooms themselves can be considered as a manifestation of reduced biodiversity (HELCOM 2009b). The past thirty years have also witnessed a number of changes in community composition, including a shift in dominance from diatoms to dinoflagellates during spring bloom periods (*cf.* Sect. 8.2.8).

In the macrophyte communities of the Baltic Sea, important habitat-forming species such as the bladderwrack *Fucus vesiculosus*, the common eelgrass *Zostera marina*, and charophytes have shown decreased abundances and reduced distributional ranges in many coastal areas. This has been attributed primarily to reduced water transparency. The decreases in the phytobenthic habitat builders are most pronounced in highly polluted and eutrophicated areas, as well as in areas where the substrate is subject to direct physical disturbance (HELCOM 2009b). The declines of these habitat-forming species have implications on a scale larger than local, since those species form important living, feeding, reproduction and nursery habitats of the associated flora and fauna, including juvenile fish. However, signs of the submerged vegetation's recovery have recently been observed, most notably in the northwestern and northeastern parts of the Baltic Sea proper (*cf.* Fig. 11.47).

17.6.3 Changes in Baltic invertebrate communities

Over recent decades, significant changes have been observed in zooplankton communities, particularly in the offshore copepod-dominated communities in the Baltic Sea proper (HELCOM 2009b). These changes have been convincingly attributed to alterations in salinity and temperature. In addition, the volume of oxygenated water below the halocline in offshore areas, suitable for reproduction of the zooplankton species that require higher salinities, has decreased with increasing eutrophication.

Macrozoobenthic communities living in and on sediments have become severely degraded (*cf.* Fig. 10.7), with abundances below the 40-year average in the entire Baltic Sea (HELCOM 2009b). The expansion of hypoxic and anoxic deep water is perhaps the single most important factor responsible for deterioration of the structural and functional diversity of benthic communities in the open-sea areas of the Baltic Sea.

17.6.4 Ups and downs of Baltic fish

One of the most prominent changes in the Baltic Sea ecosystem is the shift from dominance of demersal fish to dominance of pelagic clupeid fish. The eastern Baltic Sea cod stocks reached high abundances in the late 1970s and early 1980s (Eero et al. 2007). In the 1980s, a climate-induced reduction in the cod reproductive volume, *i.e.* the amount of water with conditions appropriate for cod egg hatching, resulted in high cod egg mortality (Köster et al. 2003). This, together with a heavy fishing pressure, brought about a historically low Baltic cod stock in the early 1990s (*cf.* Fig. 18.13). In the 2000s, some signs of improvement in cod recruitment were observed (ICES 2008) but the stock is still low. Populations of clupeids, mainly sprat but also herring, have increased since the 1980s. Factors underlying these increases, particularly with respect to sprat, include a reduced predation pressure by cod and possibly also eutrophication through an increase of food resources for the pelagic fish.

In many coastal areas, populations of fish species that benefit from or tolerate eutrophication, such as percids and cyprinids, have increased (HELCOM 2012b). However, in many other areas, fish stocks have suffered an overall decline owing to a high fishing pressure. Several stocks of migratory fish species are also in a poor condition because of the damming or blocking of migration routes, degradation of riverine habitat quality and high fishing pressure.

17.6.5 Ups and downs of Baltic waterbirds

The Baltic waterbird species showed both increasing and decreasing abundance trends during the past decades as a consequence of ecosystem changes and anthropogenic pressures. With respect to fish-eating birds, much attention has been given to the rapid population growth of the great cormorant *Phalacrocorax carbo sinensis* (*cf.* Box 4.11 and Sect. 13.6.10), from a few thousand breeding pairs in the 1960s to ~160,000 pairs in 2012. Increased food availability because of eutrophication, reduced persecution and reduced pressure from hazardous substances facilitated the increase around the Baltic Sea coasts, but since 2010 the increase of great cormorants has been lower and in some subregions the abundance has also decreased (Herrmann et al. 2014).

Similarly, the common eider *Somateria mollissima* (*cf.* Box 4.10) breeding in the Baltic Sea had increased in numbers up to the beginning of the 1990s, most likely due to eutrophication and increased availability of blue mussels. However, since the beginning of the 1990s the numbers of several mussel-eating sea ducks, *i.e.* the common eider, the velvet scoter *Melanitta fusca* and the long-tailed duck *Clangula hyemalis* (Box 11.8) have decreased rapidly in numbers. There are several possible causes for the recent declines, including ecosystem changes affecting the quality and quantity of the sea ducks' food resources, changing predation pressures, oil spills (*cf.* Box 16.1) and by-catches in fisheries. Some of the causes may occur outside the Baltic Sea, such as in the Arctic breeding areas of the long-tailed duck.

An important bird predator that has recovered remarkably from very low levels is the white-tailed eagle *Haliaeetus albicilla* (*cf.* Box 4.12). From being almost wiped out in the Baltic Sea region by persistent organic pollutants in the 1970s (*cf.* Sect. 16.1.2), the species has recovered and now breeds commonly along the Baltic Sea coasts. Part of the recent decline of breeding eiders in the core breeding areas of the central and northern Baltic Sea proper can most likely be ascribed to the increase of the white-tailed eagle (Kilpi et al. 2015).

Long-term population declines have been observed in some species of waders, for example in the endangered southern subspecies of the dunlin *Calidris alpina schinzii*. The causes of the steady decline all around the Baltic Sea can most likely be sought in both the breeding and the wintering areas. Factors such as climate change, changing agricultural practices, changing predation pressures and loss of genetic diversity may all have affected the population development of the subspecies (HELCOM 2012c).

Fig. 17.5 The Kentish plover *Charadrius alexandrinus*, previously a regular breeder, is today critically endangered in the Baltic Sea region through habitat loss. Photo: © Bo Tallmark

Some species, such as the gull-billed tern *Gelochelidon nilotica* and the Kentish plover *Charadrius alexandrinus* (Fig. 17.5), which were previously regular breeders in small numbers, have also disappeared from or are classified as critically endangered in the Baltic Sea region. The decline of these species is related to range reductions of larger European populations, for which reason the special conservation measures in the Baltic Sea Area do not seem promising in improving the status of those birds.

17.6.6 Ups and downs of Baltic mammals

The population of the grey seal *Halichoerus grypus* (Fig. 17.4), which was hunted to near extirpation (local extinction) in the beginning of the 20th century and subsequently suffered from hazardous substances, is now clearly recovering in the northern Baltic Sea, but south of 59 °N the recovery is still very slow. The status of the ringed seal *Pusa hispida* (*cf.* Fig. 9.15) is also still unfavourable, although the population has strengthened in the Gulf of Bothnia. While the impacts of hunting on seals have been reduced, and recovery from hazardous substances is on-going, fishery by-catch and prey declines are the other persistent threats to these populations (HELCOM 2009b).

The Baltic population of the harbour porpoise *Phocoena phocoena* (*cf.* Fig. 4.15) is in a critical state, with only a few hundred individuals remaining in the Baltic Sea proper. The harbour porpoise has been widely distributed and common until the early 20th century. While hunting was mainly responsible for the previous decline, the most important anthropogenic threats today are fishery by-catch and lack of prey (HELCOM 2009b). In addition, factors such as hazardous substances and acoustic pollution are also likely to negatively impact the species.

17.6.7 New species additions to the Baltic Sea

Non-indigenous species (NIS) are a part of the Baltic Sea biodiversity and, if invasive, also a potential threat to it (*cf.* Sect. 5.1). The growing expansion of NIS increases the risk of deterioration or extinction for native species or habitats with a high conservation value (HELCOM 2009b). Since the early 1800s, ~130 species have been introduced to the Baltic Sea Area. In the most heavily invaded coastal lagoons of the southern Baltic Sea proper, several food webs and even major parts of sea-bottom communities are today dominated by introduced species (Leppäkoski et al. 2002).

The benthic American bristle worms *Marenzelleria* spp. (*cf.* Box 5.3) represent one of the best-documented cases of non-indigenous species in the Baltic Sea. It took these species roughly ten years to spread over the entire Baltic Sea and to become a common or dominant species in soft-bottom habitats. Of the NIS invasions to the Baltic Sea, *Marenzelleria* spp. and the zebra mussel *Dreissena polymorpha* (*cf.* Box 5.4) seem to exert the largest ecosystem impacts (Ojaveer and Kotta 2015). However, in general, the ecological impacts of NIS on the Baltic Sea ecosystem are difficult to observe or are poorly understood.

17.6.8 Biodiversity is critical for ecosystem resilience

A pilot study using indicators of biodiversity and an integrated assessment approach was carried out in 73 open-sea and coastal areas in 2003–2007 (HELCOM 2010a). According to the results, 82 % of the assessed coastal areas had an unfavourable conservation status, and only 18 % showed a "high" or "good" status. In terms of ecosystem health, deterioration of biodiversity status is critical because it is likely to diminish the resilience of the Baltic Sea ecosystem against large-scale shifts and increases the risk of an escalating environmental degradation (HELCOM 2009b).

17.7 Is the Baltic Sea a healthy ecosystem?

17.7.1 The concept of ecosystem health

Ideally, a healthy ecosystem is an ecosystem with full functionality and potential, and with the absence of distress symptoms caused by anthropogenic stressors (Rapport 2007). An analogy can be drawn between the human body and an ecosystem. Just like an increase in body temperature and other signs of inflammation are symptoms of human illness, more frequent blooms of primary producers, increased bottom hypoxia and declining populations of top predators can be considered symptoms of deteriorated health

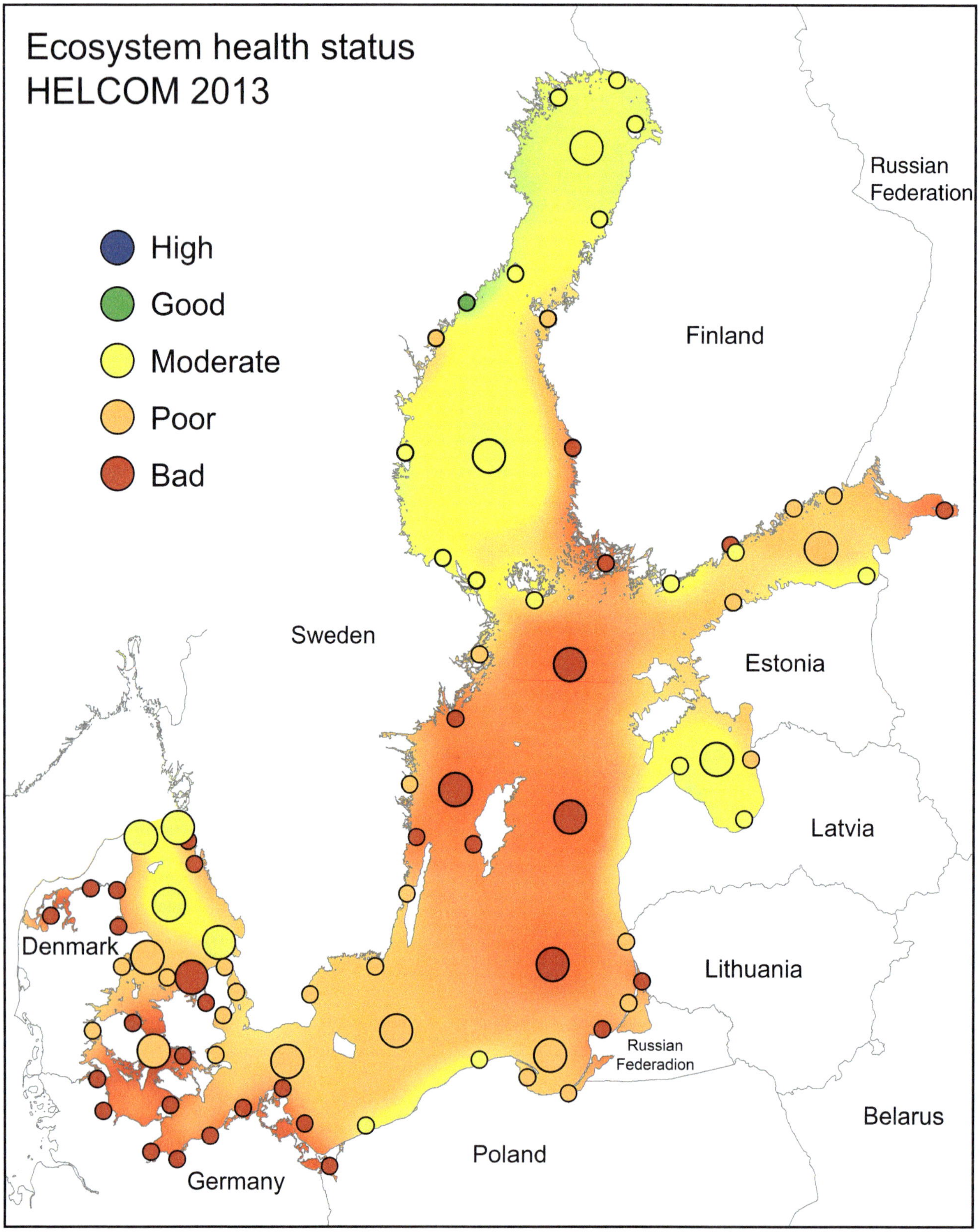

Fig. 17.6 The ecosystem health status of the Baltic Sea in 2003–2007 as assessed by biological, chemical and supporting indicators. The different colours indicate the ecosystem health status: high status (blue), good status (green), moderate status (yellow), poor status (orange) and bad status (red) of coastal (small circles) and open sea (large circles) assessment sites. High status does not occur anywhere in the Baltic Sea Area and good status is found only in a small area of the Gulf of Bothnia (the Örefjärden area, Sweden). Figure with small modifications reprinted from HELCOM (2010a) and reproduced in 2013, with permission from HELCOM

of a marine ecosystem (Kowalewska 1999). Ecosystem vitality, on the other hand, is reflected in nutrient and energy flows, biodiversity and resilience.

It can be argued that the most appropriate means to infer the presence of deteriorated ecosystem health is by evaluating historical data. Such an evaluation makes it possible to detect and analyse trends in the ecosystem state, to detect signals of deterioration and to compare the current ecosystem state with a pristine or less impacted state in the past (MacKenzie et al. 2011). Alternatively, pristine sites or areas in a marine region can be used as reference sites to offer an inference of what a healthy ecosystem should be like. However, such pristine sites do not exist in the Baltic Sea any longer. This kind of ecosystem health approach was relevant when anthropogenic impacts only started to become apparent and the ecosystem was not yet impacted in complex ways. More recently, the ecosystem health approach has turned to estimating how much anthropogenic impact is acceptable to keep up a well-functioning ecosystem and how the ecosystem can be restored if the boundaries of acceptable levels have been surpassed (HELCOM 2010a).

17.7.2 A holistic assessment of ecosystem health

A holistic assessment of Baltic Sea ecosystem health in 2003–2007 showed that the ecosystem had degraded to such an extent that its capacity to deliver ecosystem goods and services to the people living in the nine riparian countries has been reduced. This assessment was based on biological, chemical and supporting indicator data from a total of 84 assessment areas (Fig. 17.6). The whole Baltic Sea Area was classified as having a "moderate", "poor" or "bad" ecosystem health, except for a single site in the Gulf of Bothnia with "good" health status. No place in the Baltic Sea was classified as having a "high" health status. The areas in the Baltic Sea with the best ecosystem health status ("moderate") were the Bothnian Bay, the Bothnian Sea, the Gulf of Riga and some parts of the northern Kattegat.

In some areas, the state of open waters was worse than that of the coastal waters (Fig. 17.6). The Baltic Sea Pressure Index (BSPI, *cf.* Sect. 17.3.2) for these deteriorated open sea areas, especially those in the southern and eastern Baltic Sea, demonstrated a high sum of pressures from different sources (Fig. 17.2). The status of the ecosystem in the open sea was especially affected by open sea fisheries and inputs of nutrients and hazardous substances. In addition, the poor status of the sediments of the open sea areas, resulting from the accumulation over time of hazardous substances and hypoxia, was in many cases the major factor contributing to the poor open sea health status.

17.7.3 Sustainable use and the ecosystem approach

Today, humans are seen as an integral part of the Earth's ecosystems; negative impacts from human presence and resource use are considered unavoidable and to some extent acceptable. The desired state of an ecosystem is frequently defined from the standpoint of human needs, and environmental protection is implemented through the concepts of sustainable development and sustainable use of ecosystems.

The basis for the concept of sustainable development was already laid in 1972 during the first UN Conference on the Human Environment (UNCHE) in Stockholm. It brought to the wider global political attention the need to protect the environment while pursuing development. The concept of sustainable development was strengthened by the 1992 UN Conference on Environment and Development (UNCED) in Rio de Janeiro. The "Rio Declaration" emphasised the right of present and future generations to development, and endorsed the sovereign right of states to exploit their own resources, but in a way ensuring that environmental protection will be an integral part of the development process. This "sustainable development", as it is interpreted today, aims at achieving a balance between the economic, social and environmental interests of the current generations and the well-being of the future ones.

Today, the ecosystem approach and ecosystem-based management of anthropogenic activities (*cf.* Sect. 18.5) form the basis of environmental policies related to the marine environments. The presentation of the twelve "Malawi Principles for the Ecosystem Approach" (http://www.fao.org/docrep/006/y4773e/y4773e0e.htm), at the Fourth Meeting of the Conference of the Parties to the UN Convention on Biological Diversity (CBD) in 1998, is often referred to as the starting point for the practical use of the ecosystem approach. These principles underline for example that conservation of ecosystem structure and functioning is important, that an appropriate balance between conservation and the use of biodiversity should be pursued, and, most importantly, that ecosystems must be managed within the limits of their functioning. The HELCOM BSAP (*cf.* Sect. 17.8.4) and the EU Marine Strategy Framework Directive (MSFD, *cf.* Sect. 17.8.1), both of which are highly relevant to the Baltic Sea, are based on the ecosystem approach.

17.7.4 Science-based value judgments

Fundamental for making judgements on ecosystem health is a good understanding of the ecosystem structure and functions, and of the pathways and mechanisms through which anthropogenic activities impact the ecosystem. Ideally, pressure-impact relationships would be fully understood and presented by response curves like those in Fig. 17.1, which would then provide a tool for environmental managers to predict a sustainable amount of pressure. However, since we are still far from such an accurate understanding, we are forced to simplify the reality in marine assessments.

Defining what is a sustainable level of use for a particular marine ecosystem requires more than a good scientific understanding of the ecosystem's structure and functioning and its responses to anthropogenic pressures. It requires further science-based value judgments on how much degradation of the environmental quality of an ecosystem society is ready to accept (Mee et al. 2008). The advice stemming from such multidisciplinary scientific research would then be ultimately operationalised through political decisions. However, better science-based instruments are still needed to help human society to stay on the right side of the pressure thresholds for detrimental regime shifts (*cf.* Sect. 17.2.4). Recently, the concepts of ecosystem thresholds, resilience and carrying capacity have become central in defining acceptable levels of pressures and impacts resulting from anthropogenic activities (Elliott et al. 2007). However, we still have incomplete knowledge about how these concepts materialise in reality.

It is also important to recognise that in many cases human society cannot wait until full scientific consensus emerges, but instead should take action early on in the presence of uncertainty. For example, in the Baltic Sea Area, it took nearly four decades from the recognition of the DDT contamination effects on the white-tailed eagle population, for the population to recover as a result of political decisions to ban the DDT use (Elmgren 2001). It took some decades from the recognition of the mercury and PCB contamination problems in the ecosystem to recovery from their effects. In these cases, the political decision-makers did not wait until scientific consensus was reached to ban the use of mercury and PCB. Scientists have advocated that such precautionary and adaptive management approaches should be implemented with respect to managing environmental problems that are not yet sufficiently known (Mee et al. 2008). Today, the precautionary approach and adaptive management are embedded in environmental legislation in Europe.

17.7.5 Has the Baltic Sea been used sustainably?

The answer to this question is no. The Baltic Sea has not been managed in a sustainable manner and the uses of its ecosystem goods and services have exceeded the sustainability level. The assessment of the sum of anthropogenic pressures showed that the entire Baltic Sea is under some level of pressure, the pressures being at their heaviest in the southern and eastern areas (Fig. 17.2). Thematic assessments showed that the levels of eutrophication, chemical contamination by hazardous substances and degradation of biodiversity are unacceptably high in most of the Baltic Sea Area (HELCOM 2010a, b, c). Similarly, the holistic assessment of ecosystem health demonstrated that only one small "pocket" in the coastal area of the northern Bothnian Sea seemed to show a good ecosystem health status (Fig. 17.6). In addition to these assessments, regime shifts that have occurred in the Baltic Sea, in part as a result of overfishing and eutrophication, have provided first-hand evidence that the levels of sustainable use of the Baltic Sea ecosystem have been exceeded (Österblom et al. 2007; Möllmann et al. 2008).

17.7.6 A restored Baltic Sea: a different ecosystem

An ecosystem can be restored from some anthropogenic impacts while other pressures produce irreversible effects. Positive examples of restoration include those from coastal areas where better wastewater treatment has led to the improvement of water quality and improvement of the ecosystem status. The recovery of the northern Baltic grey seal, as well as the resurge of populations of great cormorants and white-tailed eagles, after the tightening of hunting regulations and ban of the use of organochlorines (DDTs, PCBs), are also among the recovery success stories.

Much more challenging will be the restoration of the Baltic Sea's eutrophication and biodiversity status to acceptable. Of current relevance are feedback mechanisms such as the internal loading of phosphorus from sediments and the prevalence of diazotrophic cyanobacterial blooms, processes that maintain the eutrophicated state (Vahtera et al. 2007). With regard to biodiversity, the communities and food webs of the Baltic Sea have been altered as a result of changes in climate, fishing and eutrophication, and the ecosystem has become reorganised through regime shifts. Due to these profound changes it is unlikely that active management can restore the Baltic Sea ecosystem to the state it had before the strong anthropogenic influence began, *i.e.* that at the beginning of the 20th century.

Recent reports have demonstrated that nutrient inputs into the Baltic Sea have decreased (*cf.* Box 2.2), and that the nutrient concentrations in the seawater in some areas are also reduced (HELCOM 2011). Pelagic eutrophication indicators (*e.g.* chlorophyll *a* concentrations) have started to show some signs of improvement (Andersen et al. 2015a). However, it seems that the Baltic Sea has not yet properly entered any return path to a healthier state of the ecosystem and it remains to be seen what path the recovery will take (Fig. 17.1). As far as eutrophication is concerned, it is clear that due to the long water residence time of the Baltic Sea (*cf.* Sect. 2.3.7) and large nutrient storages in the sediments, the recovery will involve time lags. Furthermore, the recovery will most likely not result in an ecosystem similar to that which prevailed before the large-scale eutrophication of the Baltic Sea began. This is due to changes induced by climate and differences in the ecosystem structure in the past and present. Instead, a recovery trajectory with a shifting baseline is expected (Fig. 17.1c, d).

Even if a return to the previous pristine state of the Baltic Sea is not possible, human society has the obligation to manage the ecosystem within the prevailing physical and climatic conditions to ensure that the thresholds of the current regime will not be exceeded. To achieve this, international policies have been established and are still being developed.

17.8 International policies

17.8.1 The aim: "good environmental status" (GES)

At the global level, under provisions of the 1982 UN Convention on the Law of the Sea (UNCLOS), states have the obligation to protect and preserve the marine environment. However, UNCLOS defines neither the specific goals nor the desired state of the marine environment, and practical use of this convention presents challenges.

At the EU level, the Marine Strategy Framework Directive (MSFD), adopted in 2008, is the most explicit document of the EU environmental legislation that addresses the status of marine environments (EU 2008a, b). The MSFD aims at implementing measures that will yield a "good environmental status" (GES) of the marine waters by 2020. It gives a description of what a GES is, a concept that also encompasses ecosystem status. The MSFD stipulates that GES means the environmental status of marine waters that "provide ecologically diverse and dynamic oceans and seas which are clean, healthy and productive within their intrinsic conditions, and the use of the marine environment is at a level that is sustainable, thus safeguarding the potential for uses and activities by current and future generations".

The MSFD further reads that GES is achieved when "The structure, functions and processes of the constituent marine ecosystems, together with the associated physiographic, geographic, geological and climatic factors, allow those ecosystems to function fully and to maintain their resilience to human-induced environmental change. Marine species and habitats are protected, human-induced decline of biodiversity is prevented and diverse biological components function in balance. Hydromorphological, physical and chemical properties of the ecosystems, including those properties which result from anthropogenic activities in the area concerned, support the ecosystems as described above. Anthropogenic inputs of substances and energy, including noise, into the marine environment do not cause pollution effects". This definition has been complemented by eleven qualitative GES descriptors (Annex 1 of the MSFD) and 26 criteria and 56 subcriteria that should be considered when assessing the state of a sea (EU 2010). The MSFD obliges the member states to develop and implement marine strategies, which should result in attaining GES.

17.8.2 The aim: "good ecological status"

The EU Water Framework Directive (WFD, EU 2000) addresses primarily freshwater environments, but also includes river mouths and marine coastal areas that extend seawards, from the so-called "baseline", that is, one nautical mile (1.85 km) for the ecological status assessment and 12 nautical miles for the chemical status assessment. The Directive sets the goal of a "good ecological status" to be achieved and maintained.

The Directive also stipulates what should be taken into account when evaluating the status, the so-called "quality elements", which can be environmental or biological indicators (*cf.* Box 14.2), and spells out normative definitions for what a "good ecological status" is. In coastal waters, variables such as the composition and abundance of phytoplankton, aquatic vegetation and benthic invertebrate fauna need to be assessed together with the supporting hydromorphological, chemical and physico-chemical factors. In general, for a "good ecological status", these quality elements, when compared to the reference conditions defined for different water types and for each quality element, should display only slight changes caused by anthropogenic pressures.

17.8.3 The aim: "favourable conservation status"

The UN Convention on Biological Diversity (CBD) is the most important international treaty addressing biodiversity protection and conservation at the global level. Global targets, *e.g.* the so-called "Aichi biodiversity targets" have been set for improving the state of biodiversity. The 20 Aichi targets are ambitious and should all be reached by the year 2020 (http://www.cbd.int).

At the EU level, the Habitats Directive (EC 1992) and the Birds Directive (EU 2009) have been put in force to bring the species and habitats listed in these directives to the "favourable conservation status". Some of the species and habitats on these lists are also found in the Baltic Sea, including species of marine mammals, waterbirds and underwater marine habitats such as reefs, estuaries and lagoons.

According to the Habitats Directive, the conservation status of a natural habitat is "favourable" when (1) its natural range, including the areas it covers within that range, is stable or increasing, (2) the specific structure and functions that are necessary for its long-term maintenance exist and are likely to continue to exist in the foreseeable future, and (3) the conservation status of its typical species is favourable. In turn, the conservation status of a species is "favourable" when (1) population dynamics data indicate that the species is maintaining itself on a long-term basis as a viable component of its natural habitats, (2) the natural range of the species is neither being reduced nor is likely to be reduced in the foreseeable future, and (3) there is, and will probably continue to be, a sufficiently large habitat size to maintain its populations on a long-term basis. Furthermore, the Habitats Directive requires designation of a network of protected areas, labelled "Natura 2000", which support the habitats and species listed in the annexes to the Habitats Directive.

The Birds Directive is the oldest nature conservation directive of the EU. It aims at protecting naturally occurring bird species in Europe. It applies to birds, their nests and eggs and particular focus is on the protection of the habitats used by endangered and migratory species. This directive also requires the designation of protected areas that are at present included in the Natura 2000 network.

17.8.4 HELCOM: regional cooperation for a healthier Baltic Sea

The global and EU policies and their environmental objectives do not specifically address the Baltic Sea ecosystem, although the MSFD specifies the Baltic Sea as one of the targeted marine regions. Hence, their implementations require interpretation and application at the Baltic Sea level. Therefore, it is necessary to specify the meaning of judicial terms as applied to the species, habitats and environmental conditions of the Baltic Sea.

The governments of the countries surrounding the Baltic Sea have been working together to protect the Baltic Sea environment since 1974 when the Convention for the Protection of the Marine Environment of the Baltic Sea was first signed. Today, under the umbrella of the Baltic Marine Environment Protection Commission, the Helsinki Commission (HELCOM), *i.e.* the governing body of the Convention, the riparian countries and the EU (represented by the European Commission), discuss and agree upon the goals and measures needed for a healthy Baltic Sea. HELCOM provides also a platform for regional coordination, *e.g.* to consider how the EU Directives' legal language is applied when it comes to the ecosystem of the Baltic Sea and how the implementation of the legislation can be coordinated at the Baltic Sea level.

In 2007, the ministers and high-level representatives of the HELCOM Contracting Parties adopted the Baltic Sea Action Plan (BSAP, HELCOM 2007a; Backer et al. 2010). With the BSAP, the countries bordering the Baltic Sea committed themselves to restore the Baltic Sea ecosystem health by the year 2021 by applying the actions outlined in the BSAP (http://www.helcom.fi). The implementation of the actions is regularly followed-up and new actions complementing the BSAP are agreed as necessary by HELCOM. The ultimate vision is a healthy Baltic Sea environment by 2021 with diverse biological components functioning in balance, resulting in a good environmental/ecological status and supporting a wide range of sustainable human economic and social activities.

The BSAP focuses on reducing eutrophication, diminishing chemical contamination by hazardous substances, enhancing biodiversity and nature conservation, as well as securing environmentally friendly maritime activities. To reduce eutrophication, the BSAP contains a critical load approach with maximum allowable input (MAI) of nitrogen and phosphorus defined for each Baltic Sea subregion and country (*cf.* Table 18.7). This approach is based on ecological modelling of, and commonly agreed targets for, eutrophication, and is a system in which the MAIs are accompanied by provisional nutrient load reduction requirements for each riparian country. The scheme agreed upon in the BSAP comprises the first internationally agreed target-based nutrient load reduction system for an entire regional sea (Backer et al. 2010).

Review questions

1. How have humans influenced the Baltic Sea ecosystem over time?
2. What are the most challenging environmental problems facing the Baltic Sea today?
3. How does eutrophication affect the Baltic Sea ecosystem?
4. How do anthropogenic activities affect the biodiversity of the Baltic Sea?
5. How does international legislation address ecosystem health?

Discussion questions

1. How would you rank maritime activities according to their risk for damaging the health of the Baltic Sea ecosystem and why?
2. The lack of knowledge is mentioned several times with respect to the assessment of ecosystem health. What knowledge is needed?
3. Are there any shortcomings of the EU directives with respect to covering all aspects of the Baltic Sea ecosystem?

4. What is the role of the general public for improving the environmental status of the Baltic Sea?
5. Is the future of the Baltic Sea doomed, or are there some rays of hope?

References

Alheit J, Möllmann C, Dutza J, Kornolovs G, Loew P et al (2005) Synchronous ecological regime shifts in the central Baltic and the North Sea in the late 1980s. ICES Journal of Marine Science 62:1205–1215

Andersen JH, Axe P, Backer H, Carstensen J, Claussen U et al (2011) Getting the measure of eutrophication in the Baltic Sea: towards improved assessment principles and methods. Biogeochemistry 106:137–156

Andersen JH, Carstensen J, Conley DJ, Dromph K, Fleming V, et al. (2015a) Long-term temporal and spatial trends in eutrophication status of the Baltic Sea. Biological Reviews [http://onlinelibrary.wiley.com/doi/10.1111/brv.12221/epdf]

Andersen JH, Halpern BS, Korpinen S, Murray C, Reker J (2015b) Baltic Sea biodiversity status *versus* cumulative human pressures. Estuarine Coastal and Shelf Science 161:88–92

Backer H, Leppänen JM, Brusendorff AC, Forsius K, Stankiewicz M et al (2010) HELCOM Baltic Sea Action Plan – a regional programme of measures for the marine environment based on the ecosystem approach. Marine Pollution Bulletin 60:642–649

Bartnicki J, Gusev A, Aas W, Fagerli H, Valiyaveetil S (2008) Atmospheric supply of nitrogen, lead, cadmium, mercury and dioxines/furanes to the Baltic Sea in 2006. EMEP Centres Joint Report for HELCOM. Technical Report MSC-W 3/2008, Oslo, Norway

Bartnicki J, Semeena VS, Fagerli H (2011) Atmospheric deposition of nitrogen to the Baltic Sea in the period 1995–2006. Atmospheric Chemistry and Physics 11:10057–10069

Carstensen J, Andersen JH, Gustafsson BG, Conley DJ (2014) Deoxygenation of the Baltic Sea during the last century. Proceedings of the National Academy of Sciences of the USA 111:5628–5633

Conley DJ, Humborg C, Rahm L, Savchuk OP, Wulff F (2002) Hypoxia in the Baltic Sea and basin-scale changes in phosphorus biogeochemistry. Environmental Science and Technology 15:5315–5320

Dippner JW, Vuorinen I, Daynus D, Flinkman J, Halkka A et al (2008) Climate-related marine ecosystem change. BACC Author Team. Assessment of climate change for the Baltic Sea basin. Regional Climate Studies. Springer, Berlin, pp 309–377

Duarte CM, Conley DJ, Carstensen J, Sánchez-Camacho M (2009) Return to Neverland: shifting baselines affect eutrophication restoration targets. Estuaries and Coasts 32:29–36

EC (1992) Council Directive 92/43/EEC of 21 May 1992 on the conservation of natural habitats and of wild fauna and flora. Official Journal of the European Communities L 206:7–50

Eero M, Koster FW, Plikshs M, Thurow F (2007) Eastern Baltic cod (*Gadus morhua callarias*) stock dynamics: extending the analytical assessment back to the mid-1940s. ICES Journal of Marine Science 64:1257–1271

Elliott M, Burdon D, Hemingway KL, Apitz SE (2007) Estuarine, coastal and marine ecosystem restoration: confusing management and science – a revision of concepts. Estuarine and Coastal Shelf Science 74:349–366

Elmgren R (2001) Understanding human impact on the Baltic ecosystem: changing views in recent decades. AMBIO 30:222–231

EU (2000) Directive 200/60/EC of the European Parliament and of the Council of 23 October 2000 establishing a framework for Community action in the field of water policy. Official Journal of the European Communities L 327:1–72

EU (2008a) Directive 2008/56/EC of the European Parliament and of the Council of 17 June 2008 establishing a framework for community action in the field of marine environmental policy (Marine Strategy Framework Directive). Official Journal of the European Union L 164:19–40

EU (2008b) Directive 2008/105/EC of the European Parliament and of the Council of 16 December 2008 on environmental quality standards in the field of water policy. Official Journal of the European Union L 348:84–97

EU (2009) Directive 2009/147/EC of the European Parliament and of the Council of 30 November 2009 on the conservation of wild birds. Official Journal of the European Union L20:7–25

EU (2010) Commission decision of 1 September 2010 on criteria and methodological standards on good environmental status of marine waters. Official Journal of the European Union L232:14–24

Ferreira F, Santos MM, Castro LFC, Reis-Henriques MA, Lima D et al (2009) Vitellogenin gene expression in the intertidal blenny *Lipophrys pholis*: a new sentinel species for estrogenic chemical pollution monitoring in the European Atlantic coast? Comparative Biochemistry and Physiology, Part C: Toxicology and Pharmacology 149:58–64

Harding KC, Härkönen TJ (1999) Development in the Baltic grey seal (*Halichoerus grypus*) and ringed seal (*Phoca hispida*) populations during the 20th century. AMBIO 28:619–627

Heino R, Tuomenvirta H, Vuglinsky VS, Gustafsson BG, Alexandersson H et al (2008) Past and current climate change. BACC Author Team. Assessment of climate change for the Baltic Sea basin. Regional Climate Studies, Springer, Berlin, pp 35–131

HELCOM (2007a) HELCOM Baltic Sea Action Plan. HELCOM Ministerial meeting, Krakow, Poland 15th November 2007. HELCOM, 101 pp

HELCOM (2007b) HELCOM list of threatened and/or declining species and biotopes/habitats in the Baltic Sea Area. Baltic Sea Environment Proceedings 113:1–17

HELCOM (2009a) Eutrophication in the Baltic Sea – an integrated thematic assessment of the effects of nutrient enrichment in the Baltic Sea region. Baltic Sea Environment Proceedings 115B: 1–148

HELCOM (2009b) Biodiversity in the Baltic Sea – an integrated thematic assessment on biodiversity and nature conservation in the Baltic Sea. Baltic Sea Environment Proceedings 116B:1–188

HELCOM (2010a) Ecosystem health of the Baltic Sea – HELCOM initial holistic assessment. Baltic Sea Environment Proceedings 122:1–63

HELCOM (2010b) Towards a tool for quantifying anthropogenic pressures and potential impacts on the Baltic Sea marine environment – a background document on the method, data and testing of the Baltic Sea pressure and impact Indices. Baltic Sea Environment Proceedings 125:1–69

HELCOM (2010c) Hazardous substances in the Baltic Sea – an integrated thematic assessment of hazardous substances in the Baltic Sea. Baltic Sea Environment Proceedings 120B:1–116

HELCOM (2011) Fifth Baltic Sea pollution load compilation (PLC-5). Baltic Sea Environment Proceedings 128:1–217

HELCOM (2012a) Approaches and methods for eutrophication target setting in the Baltic Sea region. Baltic Sea Environment Proceedings 133:1–134

HELCOM (2012b) Indicator-based assessment of coastal fish community status in the Baltic Sea 2005-2009. Baltic Sea Environment Proceedings 131:1–88

HELCOM (2012c) Red list of Baltic breeding birds, version 2011. [http://www.helcom.fi]

HELCOM (2013) Review of the fifth Baltic Sea pollution load compilation for the 2013 HELCOM Ministerial Meeting. Baltic Sea Environment Proceedings 141:1–49

HELCOM (2015) Updated fifth Baltic Sea pollution load compilation (PLC-5.5). Baltic Sea Environment Proceedings 145:1–142

Herrmann C, Bregnballe T, Larsson K, Rattiste K (2014) Population development of great commorant. HELCOM Baltic Sea Environment Fact Sheet. [http://www.helcom.fi]

ICES (2008) Report of the Baltic Fisheries Assessment Working Group (WGBFAS), 8-17 April 2008. ICES CM 2008/ACOM: 06, 692 pp

Jackson JBC, Kirby MX, Berger WH, Bjorndal KA, Botsford LW et al (2001) Historical overfishing and the recent collapse of coastal ecosystems. Science 293:629–638

Johannesson K, André C (2006) Life on the margin: genetic isolation and diversity loss in a peripheral marine ecosystem, the Baltic Sea. Molecular Ecology 15:2013–2029

Kemp WM, Testa JM, Conley DJ, Gilbert D, Hagy JD (2009) Temporal responses of coastal hypoxia to loading and physical control. Biogeosciences 6:2985–3008

Kilpi M, Lorentsen SH, Petersen IK, Einarsson A (2015) Trends and drivers of change in diving ducks. Nordic Council of Ministers, TemaNord 2015:516

Korpinen S, Meski L, Andersen JH, Laamanen M (2012) Human pressures and their potential impact on the Baltic Sea ecosystem. Ecological Indicators 15:105–114

Köster FW, Hinrichsen H, Schnack D, St. John MA, MacKenzie BR et al (2003) Recruitment of Baltic cod and sprat stocks: identification of critical life stages and incorporation of environmental variability into stock-recruitment relationships. Scientia Marina 67 (Supplement 1):129–154

Kowalewska G (1999) Phytoplankton blooms – a "fever" of the Baltic ecosystem. Oceanologia 41:265–275

Leppäkoski E, Gollasch S, Gruszka P, Ojaveer H, Olenin S et al (2002) The Baltic – a sea of invaders. Canadian Journal of Fisheries and Aquatic Sciences 59:1175–1188

MacKenzie BR, Ojaveer H, Eero M (2011) Historical ecology provides new insights for ecosystem management: eastern Baltic cod case study. Marine Policy 35:266–270

Mee LD, Jefferson RL, Laffoley DA, Elliott M (2008) How good is good? Human values and Europe's proposed Marine Strategy Directive. Marine Pollution Bulletin 56:187–204

Möllmann C, Müller-Karulis B, Kornilovs G, St John MA (2008) Effects of climate and overfishing on zooplankton dynamics and ecosystem structure: regime shifts, trophic cascade, and feedback loops in a simple ecosystem. ICES Journal of Marine Science 65:302–310

Nixon SW (1995) Coastal marine eutrophication: a definition, social causes, and future concerns. Ophelia 41:199–219

Ojaveer H, Kotta J (2015) Ecosystem impacts of the widespread non-indigenous species in the Baltic Sea: literature survey evidences major limitations in knowledge. Hydrobiologia. 750:171–185

Österblom H, Hansson S, Larsson U, Hjerne O, Wulff F et al (2007) Human-induced trophic cascades and ecological regime shifts in the Baltic Sea. Ecosystems 10:877–889

Rapport DJ (2007) Sustainability science: an ecohealth perspective. Sustainability Science 2:77–84

Sandén P, Håkansson B (1996) Long-term trends in Secchi depth in the Baltic Sea. Limnology and Oceanography 41:346–351

Scheffer M, Carpenter S, Foley JA, Folke C, Walker B (2001) Catastrophic shifts in ecosystems. Nature 413:591–596

Stocker TF, Qin D, Plattner GK, Tignor M, Allen SK et al (eds) (2013) The physical science basis – contribution of Working Group I to the Fifth Assessment Report of the Intergovernmental Panel on Climate Change. Cambridge University Press, Cambridge, 1535 pp

Struck U, Emeis K, Voss M, Christiansen C, Kunzendorf H (2000) Records of southern and central Baltic Sea eutrophication in $\delta^{13}C$ and $\delta^{15}N$ of sedimentary organic matter. Marine Geology 164:157–171

Vahtera E, Conley DJ, Gustafsson BG, Kuosa H, Pitkänen H et al (2007) Internal ecosystem feedbacks enhance nitrogen-fixing cyanobacteria blooms and complicate management in the Baltic Sea. AMBIO 36:186–194

Winsor P, Rodhe J, Omstedt A (2001) Baltic Sea ocean climate: an analysis of 100 yr of hydrographic data with focus on the freshwater budget. Climate Research 18:5–15

Zillén L, Conley DJ (2010) Hypoxia and cyanobacteria blooms – are they really natural features of the late history of the Baltic Sea? Biogeosciences 7:2567–2580

18 Ecosystem goods, services and management

Jan Marcin Węsławski, Eugeniusz Andrulewicz, Christoffer Boström, Jan Horbowy, Tomasz Linkowski, Johanna Mattila, Sergej Olenin, Joanna Piwowarczyk, and Krzysztof Skóra

Abstract

1. Humans are an imperative component of the Earth's ecosystems as we transform them to meet our economic and cultural needs.
2. Seas and oceans contribute to the local, regional and global development of human society. The downside of this development is environmental deterioration resulting from increasing competition for sea space and coastal areas between different and conflicting interests.
3. Natural environmental stressors are exacerbated by anthropogenic pressures in the drainage areas of all aquatic systems. The interplay of such pressures is particularly pronounced in semi-enclosed seas such as the Baltic Sea, which are often multi-use and multi-stakeholder areas.
4. The ecosystem goods and services offered by the Baltic Sea can be classified as provisioning (resources obtained by exploitation for *e.g.* human food), regulating (direct natural regulation processes, *e.g.* gas and climate regulation), cultural (non-material benefits, *e.g.* recreation) and supporting (processes necessary to sustain the other goods and services, *e.g.* primary production).
5. The management of the Baltic Sea ecosystem has its success stories, such as regulations for the exploitation of living resources and discharges of hazardous substances. It also has its failures, eutrophication management being perhaps the most evident one.
6. Trans-national networking and cross-border cooperation are crucial for improving the health of the Baltic Sea ecosystem. This is not always easy because the different countries around the Baltic Sea experience different social constraints.
7. The large-scale anthropogenic pressures on the Baltic Sea can only be dealt with by ecosystem-based management (EBM). EBM is an integrated approach to management that considers the entire ecosystem, including humans, and aims to maintain the sustainable supply of ecosystem goods and services by keeping the ecosystem in a healthy, productive and resilient condition.
8. Marine spatial planning (MSP) may be the vehicle for scientific knowledge to inform and influence decision-making.

Keywords

Conflicting interests • Ecosystem-based management • Ecosystem goods and services • Environmental legislation • International cooperation • Fish-stock management • Habitat restauration • Management of nutrient inputs • Marine spatial planning • Nature conservation • Science-based advice • Sustainable use

P. Snoeijs-Leijonmalm et al. (eds.), *Biological Oceanography of the Baltic Sea*,
DOI 10.1007/978-94-007-0668-2_18

18.1 Conflicting interests

18.1.1 Competition for sea space

Throughout history, oceans and seas have played essential roles in the development of human societies around the globe. Coastal areas are disproportionally urbanised compared to inland areas. In the early 2000s, coastal areas were inhabited by ~12 % of the world population while ~50 % lived within a distance of 200 km from a coast. Coastal human populations are growing faster than inland populations, and the rapid urbanisation and other types of expansion by humans impact coastal environments and challenge their resilience (*cf.* Sect. 17.2.4).

Different types of sea space use can be mutually exclusive. For example, gravel or sand extraction rules out windmill park construction and fisheries in an area. Severe seabed disturbance and heavy exploitation of biological resources are not compatible with nature protection. A growing demand for sea space is evident in the Baltic Sea region. Traditional uses of sea space include fisheries and aquaculture, whereas windmill farms and offshore industrial facilities are relatively new fields (Table 18.1). The fastest-growing uses of the coastal sea space in the Baltic Sea Area include recreation (*e.g.* marinas, camping grounds, recreational fishing) and industry (*e.g.* new types of harbours with large container ship quays and offshore terminals).

Generally, the most frequent cause of conflict between sea space users revolves around nature protection measures. Nature protection regulations cover 5–30 % of the marine areas of the individual countries surrounding the Baltic Sea. These regulations require that most forms of resource exploitation and disturbances to the seabed and the water column, including noise (acoustic pollution) is to be eliminated in these areas.

Public debates on marine spatial planning usually focus on the coastal zone since this is the area that is most heavily utilised by human populations. Most of the present-day practical uses and theoretical debates concern the seabed, while the three-dimensional space of the water column is usually off the focus. This may change as the fragility of phenomena such as the diminishing "cod water", the specific volume of water in the Baltic Sea in which this fish species reproduces (*cf.* Sect. 8.11.4), is beginning to be recognised. The cod water is a geographically distinct, spatial phenomenon of high economic value, but at present it receives little attention from environmental and resource management.

18.1.2 The carrying capacity of recreation

The beach area, defined as the area between the low and high water marks, is particularly vulnerable to management options. A 500-km long stretch of sandy coast in the southern Baltic Sea may look extensive at first sight, yet it turns out that the actual beach area covers only ~20 km^2. Hence, the expansion of hotels, camping sites, restaurants, beach bars and bathers fills the available space at and close to the beach very quickly (Fig. 18.1b). This gives rise to major conflicting interests within the recreational sector itself. On the one hand, tour operators and tourist site owners are interested in increasing the number of customers by offering new attractions, improved infrastructure and advanced logistics. On the other hand, the customers declare their interest in avoiding the most crowded areas. One might expect that self-regulation would take place, so that above a given critical density of hotels or areas occupied by camping sites, new ones would not emerge. However, in practice such a mechanism does not occur.

The Hel Peninsula in Poland (Fig. 18.2) is a good example of a coastal area that is subject to conflicting interests. The

J.M. Węsławski (✉) · J. Piwowarczyk
Institute of Oceanology, Polish Academy of Sciences,
ul. Powstancow Warszawy 55, 81712 Sopot, Poland
e-mail: weslaw@iopan.gda.pl

J. Piwowarczyk
e-mail: joanna_p@iopan.gda.pl

E. Andrulewicz · J. Horbowy · T. Linkowski
National Marine Fisheries Research Institute,
ul. Kollataja 1, 81332 Gdynia, Poland
e-mail: eugene@mir.gdynia.pl

J. Horbowy
e-mail: jan.horbowy@mir.gdynia.pl

T. Linkowski
e-mail: tomasz.linkowski@mir.gdynia.pl

C. Boström
Faculty of Science and Engineering, Environmental and Marine Biology, Åbo Akademi University, Artillerigatan 6, 20520 Turku, Finland
e-mail: christoffer.bostrom@abo.fi

J. Mattila
Department of Aquatic Resources, Institute of Coastal Research, Swedish University of Agricultural Sciences,
74242 Öregrund, Sweden
e-mail: johanna.mattila@slu.se

S. Olenin
Marine Science and Technology Center, Klaipeda University,
Herkaus Manto 84, 5808 Klaipeda, Lithuania
e-mail: sergej.olenin@jmtc.ku.lt

K. Skóra
University of Gdańsk, Hel Marine Station, 84150 Hel, Poland
e-mail: oceks@univ.gda.pl

Table 18.1 Summary of the principal uses of the Baltic Sea ecosystem by humans, including estimated trends and the main areas of conflict

Sea space use by humans	Estimated trend	Main areas of conflict
Pelagic fisheries	Stable or increase	Windmill farms
Demersal fisheries	Decrease	Sand and gravel extraction
Recreational fishing	Increase	Offshore industrial development
Aquaculture	Increase	Nature protection
Sand and gravel extraction, stone-fishing	Decrease	Nature protection, wind farms, fisheries
Windmill farms	Increase	Fisheries
Oil and gas platforms	Increase	Nature protection
Pipelines and power cables	Increase	Nature protection
Harbours and industrial facilities	Increase	Nature protection
Military training	Stable or increase	Nature protection
Recreational boating	Increase	Nature protection
Commercial traffic	Increase	Fisheries
Nature protection	Increase	Offshore industrial development

Fig. 18.1 Tourism on the sandy beaches of the southern Baltic Sea. (**a**) Semi-ephemeral sandy islands in the Puck Bay, a shallow western branch of the Gdańsk Bay, Poland. (**b**) A famous tourist beach, Palanga beach in Lithuania. These islands form a unique landscape type and are a source of conflict between conservationists and tourists. Photo: (**a**) © Jacek Urbański, (**b**) © Sergej Olenin

Peninsula is a long, narrow sandy spit. It supports a Natura 2000 site (*cf.* Sect. 17.8.3) as well as some other protected areas, including a landscape park. A single road and a railway track connect several renowned recreational resorts along the peninsula, bringing in over one million tourists per year. The major attractions are the warm, shallow and safe waters off the inner part of the peninsula, where surfing and kite surfing are the key sports. Near the village of Chałupy, tour operators aim at mass tourism by offering cheap accommodation in trailers. The visitors often come just for a few hours or for a weekend to surf. Camping site owners expand the sites, thereby aggressively and illegally destroying reed belts and salt marshes and replacing the wetlands with transported sand to create new space for more tourists. The carrying capacity of the Chałupy area seems to be limited only by the amount of physical space in which to place more trailers.

18.1.3 Recreation *versus* nature protection

A policy contrasting that in Chałupy is implemented in Jurata, another village at the Hel Peninsula. Here the tourist market is not dominated by cheap resorts for mass tourism, but by expensive luxury hotels instead. The high-standard accommodation is situated as close to the waterline as possible. The general lack of concern for the natural beach landscape is evident from the installation of various energy-consuming and space-occupying facilities, *e.g.* showers, paved paths and other organised spaces for recreation at or close to the beaches.

The environmental consequences of the recreational use of beaches are not easy to estimate. A simple sandy beach system is highly resilient unless physically destroyed. In contrast,

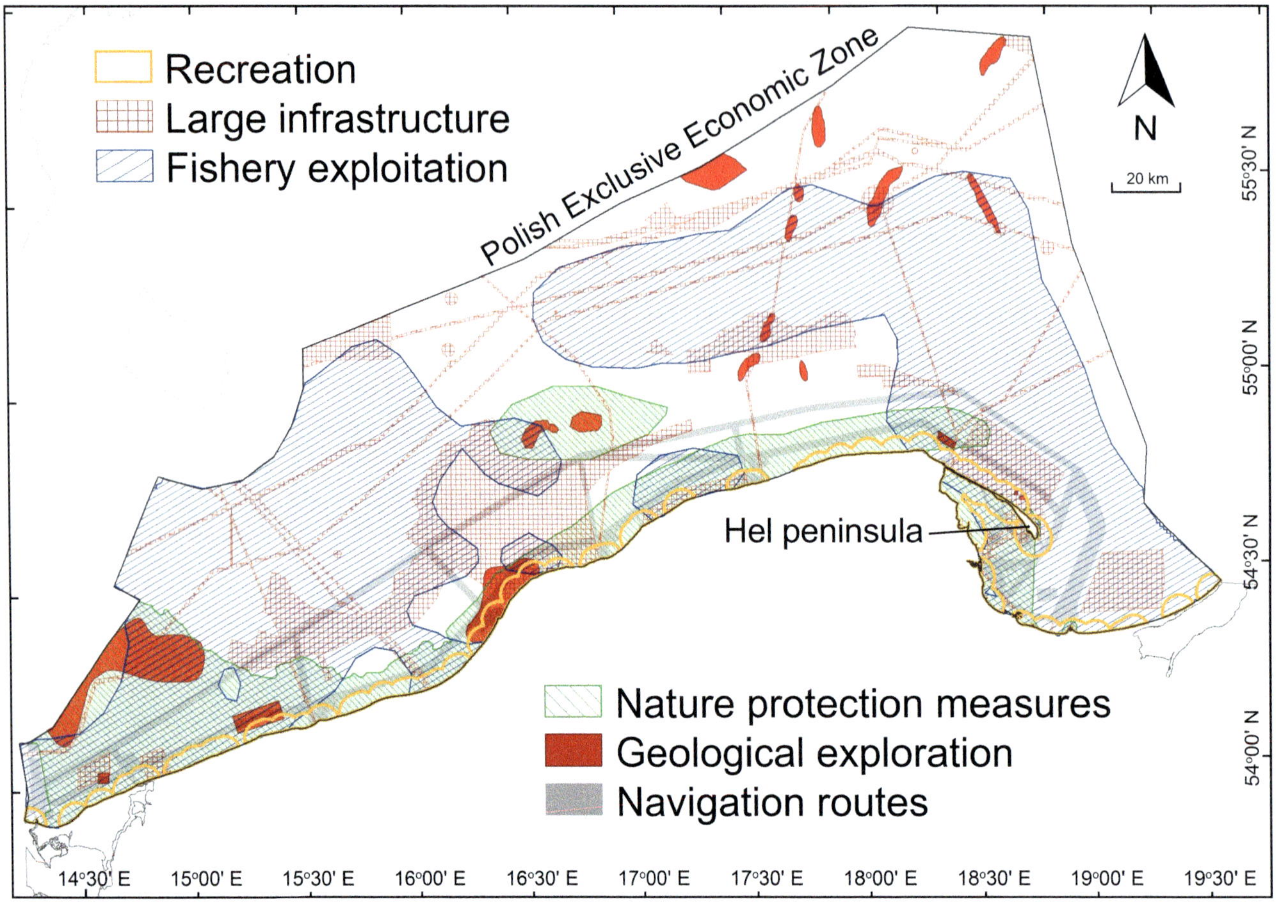

Fig. 18.2 An example of overlapping, multiple sea space uses in the Polish Exclusive Economic Zone in the southern Baltic Sea. Figure reprinted from Węsławski et al. (2010) with permission from Oceanologia (Institute of Oceanology PAN, Sopot, Poland)

reed belts and wetlands are highly susceptible to *e.g.* trampling because the perennial vegetation is fragile and recovers slowly. The disappearance of reed belts means habitat loss for a number of invertebrate and fish species and increased coastal erosion. Moreover, nutrient input to the sea increases since reed belts act as nutrient filters by storing nitrogen and phosphorus. The major problem for birds and mammals is the human presence itself. Just by being there, people may scare the animals. In areas where birds seek resting sites during their spring and autumn migrations (such as the Hel Peninsula), stress by human presence is a serious issue.

Tour organisers constantly search for new business opportunities. Recently they have started offering "wild walks through un-trampled sandy shallows", a unique chain of ephemeral sandy banks that emerge across the Puck Bay during low water (Fig. 18.1a). The banks used to be rarely visited by local fishermen and offered resting sites to birds and, occasionally, seals. Now, the "wild walks" attract a few hundred tourists at a time. They apparently come to enjoy places advertised as "untouched, remote and natural", but their activities consist of partying, grilling and boating rather than enjoying the tranquil wilderness. Both tourists and local developers protest vigorously against attempts to curb tourist visits to the banks. They claim that in a democratic society all citizens have the right to access nature.

The archipelago areas of the northern Baltic Sea are generally less crowded with tourists than the sandy beaches of southern Baltic Sea. In Sweden and Finland, summer activities on the Baltic Sea coasts tend to be associated with using summer houses and enjoying boating and recreational fishing (Figs. 18.3 and 18.4). In these areas, problems concern *e.g.* sewage from summer houses, dredging and destruction of terrestrial and aquatic vegetation around the summer houses, coastal erosion and acoustic pollution by fast motor vehicles, oil spills and garbage.

18.1.4 Conflicting interests on larger scales

Conflicts related to the use of sea space in the Baltic Sea can, and do, emerge on scales larger than local or national (Fig. 18.2), and may cover wide areas, both in the coastal zone and in the open sea; these may have international dimensions. Recent examples include the Nord Stream

Fig. 18.3 Tourism in the Ekenäs archipelago, Gulf of Finland. (**a**) A marina. (**b**) A summer house. (**c–h**) Boating. Photo: © Pauline Snoeijs-Leijonmalm

Fig. 18.4 Recreational fishing in the archipelagos of the northern Baltic Sea. (**a**, **b**) Angling. (**c**, **d**) Net fishing. (**e**, **f**) The sea is a paradise for children's contact with nature. Photo: (**a–e**) © Pauline Snoeijs-Leijonmalm, (**f**) © Maria Laamanen

pipeline (Linskog and Elander 2012) or oil spills by foreign vessels (Hassler 2011). The mitigation and resolution of such conflicts requires actions undertaken on an international scale by conventions and other appropriate regulations as well as by international cooperation.

18.2 Conventions and regulations

18.2.1 Global agreements that include the Baltic Sea

The Baltic Sea Area is covered by various international, global, regional and bilateral agreements regulating not only the use of the sea, but also targeting the conservation and protection of its natural resources (*cf.* Sect. 17.8).

In 1973, the International Convention for the Prevention of Pollution from Ships (MARPOL) was adopted by the UN International Maritime Organization (IMO) with responsibility for the safety and security of shipping and the prevention of marine pollution by ships. The Convention deals with oil pollution, harmful chemical substances, sewage, litter/garbage and emissions of sulphur oxides (SOx) and nitrogen oxides (NOx) to air. Since its adoption, this convention has been updated with numerous amendments (http://www.imo.org).

In 1982, the process leading to the adoption of the UN Convention on the Law of the Sea (UNCLOS) was started. UNCLOS, finally signed in 1998, addresses a number of topics related to pollution, the conservation of marine life and exploitation of natural resources, in addition to navigational and economic rights at sea (http://www.un.org). The formal regulations of IMO and UNCLOS require a strong support by scientists and scientific organisations. Often, *ad hoc* expert groups are created to develop and draft science-based regulations on specific issues.

18.2.2 European Union regulations concerning the Baltic Sea

Following the enlargement of the European Union (EU), the management of the Baltic Sea faces new challenges. The EU Marine Strategy Framework Directive (MSFD, *cf.* Sect. 17.8) is based on the ecosystem approach (*cf.* Sect. 18.5) and requires the development of Marine Spatial Planning (MSP, *cf.* Sect. 18.4), in the EU terminology called "Maritime Spatial Planning", and management strategies for new uses of the Baltic Sea (EU 2008a).

The EU Habitats Directive, the Birds Directive and the Natura 2000 network of protected areas (*cf.* Sect. 17.8.3) have *e.g.* boosted the development of red lists of endangered species and habitats and the drafting of plans for the protection and management of Natura 2000 areas. Most of the EU directives require scientific input, pushing environmental science in the direction of applied research. For example, the EU Water Framework Directive (WFD, *cf.* Sect. 17.8.2) has required a significant scientific effort in developing a typology of transitional and coastal waters and defining criteria for the assessment of ecological status (EU 2000).

18.2.3 Cooperation between the countries surrounding the Baltic Sea

Already in 1973, the countries surrounding the Baltic Sea signed the Convention on Fishing and Conservation of Living Resources in the Baltic Sea and the Belts in Gdańsk, Poland. In an attempt to prevent the overexploitation of Baltic Sea fish stocks, the countries agreed to coordinate their fisheries management. For the implementation of this convention, the International Baltic Sea Fishery Commission (IBSFC) was established and this organisation existed until 2005 when the EU adopted its Common Fisheries Policy (CFP).

Due to the rapid industrial development after World War II, chemistry-based agriculture and urban agglomerations induced a substantial deterioration of the environmental status in the Baltic Sea drainage area. This soon generated adverse effects in the Baltic Sea ecosystem as well. In this context, it became obvious that pan-Baltic international cooperation was needed not only by fisheries but also by efforts aimed at combatting pollution problems (Box 18.1). Since 1974, the governments of the countries surrounding the Baltic Sea have been working together to protect the Baltic Sea environment through the Convention for the Protection of the Marine Environment of the Baltic Sea, with HELCOM as its governing body (*cf.* Sect. 17.8.4). After the political changes in the Baltic Sea region in the early 1990s, and further negotiations between the countries, a renewed convention entered into force in 2000. The main change with respect to the "old" convention is that it now also covers the coastal areas of the Baltic Sea and the transition zone (Belt sea and Kattegat), as well as the activities in the drainage area that affect the health status of the sea.

Box 18.1: Short history of scientific research cooperation in the Baltic Sea Area

ICES

International scientific research cooperation in marine sciences in the Baltic Sea Area dates back to 1902, when the International Council for the Exploration of the Sea (ICES) was established (Box Fig. 18.1). ICES is an intergovernmental scientific organisation with its head office in Copenhagen, Denmark (http://www.ices.dk). Its main objective is to increase the scientific knowledge of the marine environment and its living resources and to use this knowledge to provide unbiased, non-political advice to competent authorities (*cf.* Fig. 18.14). Over the years, ICES has provided opportunities for scientific cooperation between oceanographers and fishery biologists working in the North Atlantic Ocean and adjacent continental seas, including the Baltic Sea. However, until World War II, cooperation among Baltic Sea scientists was generally limited to personal and/or bilateral collaboration.

BMB, CBO and BMG

After World War II, the political and economic situation around the Baltic Sea became extremely complicated because its riparian countries were divided between two political blocs. Contacts between scientists from the respective blocs were substantially restricted. However, some prominent Baltic Sea scientists found ways to overcome the political partition and since 1957 the Conferences of Baltic Oceanographers (CBO) have been organised at a non-governmental level. In 1968, a new non-governmental organisation (NGO) called the Baltic Marine Biologists (BMB) was established (Dybern 2004, Box Fig. 18.2). The BMB has a formal organisational structure and organised bi-annual symposia, published symposium proceedings, and has released a number of fundamental, subject-oriented (mainly methodological) publications. The main purpose of both CBO and BMB is to facilitate contact between scientists working with basic and applied scientific research in the Baltic Sea Area. The geologists working with the Baltic Sea used to participate in the CBO meetings, but as their participation grew larger, they decided to establish a separate NGO, the Baltic Marine Geologists (BMG).

Box Fig. 18.1 Participants in the ICES Council Meeting in 1904. Photo: © International Council for the Exploration of the Sea (ICES)

Box Fig. 18.2 Participants of the first BMB Symposium 1968. Photo: © The Baltic Marine Biologists (BMB)

Joint meetings

After nearly 40 years of separate CBO, BMB and BMG meetings and symposia, the Baltic Sea scientists realised that, in view of the need for a holistic understanding of the ecosystem, they should combine their meetings to discuss general and specific aspects of the Baltic Sea, exchange information, integrate efforts, and get to know and understand each other better across discipines. The first joint conference of the three organisations was the Baltic Sea Science Congress (BSSC) held in Rønne in 1996, and these congresses are organised every second year. The activities of the Baltic Sea scientists, particularly those coordinated by BMB and CBO, have contributed to carrying through the formal political process that lead to the formulation and adoption of the Convention on the Protection of the Baltic Sea Environment (known as the Helsinki Convention). In addition, they formed a basis for the successful start of the Baltic Marine Environment Protection Commission, the Helsinki Commission (HELCOM), the executive body of the convention. As the environmental science concerning the Baltic Sea had already been well-developed, and the cooperation between Baltic scientists already established, HELCOM could function as a respected and scientifically sound governmental organisation from its very beginning in 1974.

Joint research programmes

Following political transformations in the Baltic Sea Area in the early 1990s, new and more diverse opportunities for joint and integrated research projects and programmes opened up, and scientific research cooperation between the countries surrounding the Baltic Sea intensified, *e.g.* in the research programme BALTEX (the Baltic Sea Experiment, Leppäranta and Myrberg 2009), which was followed up by "Baltic Earth" (http://www.baltic-earth.eu). Funding for research and activities to support the implementation of the EU environmental policy is now sought via the EU Framework Programmes for Research and Technological Development (FP) created by the EU to support and encourage research in the European Research Area (ERA). Most recently, a joint Baltic Sea research and development programme was established that is partly funded by the EU and partly by the Baltic Sea countries (BONUS), which fosters large-scale collaboration between Baltic Sea scientists by funding integrated research that promotes the sustainable development of the Baltic Sea Area. Scientists from Russia (the only Baltic Sea country that is not an EU member state) actively take part in EU projects and contribute substantially to international scientific activities in the Baltic Sea.

Fig. 18.5 Field work for monitoring programmes. (**a**, **b**) Collecting organisms for biochemical analyses by snorkelling. (**c**, **d**) Collecting vegetation data by SCUBA diving. (**e**, **f**) Collecting zoobenthos data from the deep areas in the Baltic Sea proper. Photo: (**a**, **b**, **e**, **f**) © Pauline Snoeijs-Leijonmalm, (**c**) © Sergej Olenin, (**d**) © Piotr Bałazy

All HELCOM environmental programmes require advice from expert scientists (Box 18.1). For example, in the design of the HELCOM Baltic Monitoring Programme (BMP; Fig. 18.5), which started in 1979, science-based advice is needed for hydrographical, chemical and biological methodology, as well as the scheduling of the monitoring programme and the ways of assessing the environmental status of the Baltic Sea, *e.g.* by indicators (*cf.* Sect. 14.1). For the design of the Baltic Sea Action Plan (BSAP; HELCOM 2007; Backer et al. 2010; *cf.* Sect. 17.8.4), science-based advice is needed on how to integrate ecosystem services and social benefits, *e.g.* by driving force-pressure-state impact-response indicators (DPSIRs, Atkins et al. 2011). However, although the scientific community can provide advice and guidance, it does not have a say in how the authorities actually implement the HELCOM recommendations and actions, which is the ultimate responsibility of the decision-makers (politicians).

18.3 Ecosystem goods (ESG) and services (ESS)

18.3.1 The ESG and ESS concept

The concept of ecosystem goods and services provides a practical tool for ecosystem management. It can be used to analyse how anthropogenic activities interfere with specific natural processes. Ecosystems deliver "ecosystem goods" (ESG; *e.g.* fish) as well as "ecosystem services" (ESS; *e.g.* retention of nutrients in reed belts) to human society. ESG and ESS can be defined as the respective direct and indirect benefits people obtain from ecosystems (Beaumont et al. 2007). Assessing ecological resources and processes in terms of the goods and services they provide translates the complexity of the ecosystem into a series of functions that can be more readily understood by stakeholders such as decision-makers. Describing the ecosystem in terms of ESG and ESS also enables an understanding of what exactly would be gained and lost when development and (over-) exploitation occur.

The most critical state of a marine ecosystem is related to its ability to deliver long-term goods and services to human society, *i.e.* to its sustainability. Thus, management policies should consider the effects that each of the user sectors has on the structure and function of the ecosystem. However, ESG and ESS are strictly anthropogenic concepts that translate the primary ecological processes into human-valued benefits. These benefits can be life-supporting, improve our health, create economic opportunities and provide educational development, aesthetic or spiritual experiences and personal or cultural identity (de Groot et al. 2002; SEPA 2008a). The ability of the Baltic Sea to provide the required ESG and ESS is currently regarded as impaired due to both anthropogenic and natural causes. This situation spurs international efforts and collaboration initiatives aimed at stopping, and even reversing, the negative trends.

18.3.2 Implementation of the ESG and ESS concept

The most comprehensive classification of ecosystem goods and services was likely provided by the Millennium Ecosystem Assessment (MA, 2005a, b). The MA subdivided marine ecosystem goods and services into four categories: provisioning services, regulating services, cultural services and supporting services (Table 18.2). Since supporting services are regarded as the pre-conditions for the other three types, this category was strongly criticised by economists. They claimed that it would bring about double counting during monetary valuation. Therefore, the study on the economics of ecosystems and biodiversity (TEEB (UNEP); de Groot et al. 2010), which basically follows the MA classification, introduced a more explicit distinction between benefits and services. The MA supporting services are called "ecological processes" in TEEB. In addition, TEEB also defined "habitat services", a separate category of ecosystem services that maintains genetic diversity and sustains the life cycles of migratory species.

Another classification scheme designed to avoid double counting during monetary valuation focuses on how marine ecosystem services are used by stakeholders (Hein et al. 2006). According to this classification the ecosystem delivers production, regulation and cultural services, which are then linked to their manner of use (or non-use), directly, indirectly, or optionally (Fig. 18.6). This classification combines the ecological scales, whereby ecosystem services are generated, with institutional scales wherever stakeholders benefit from ecosystem services.

"Direct use values" arise from direct utilisation of the ecosystem by humans. All production services and some cultural services (such as recreation) have direct use value. "Indirect use values" reflect the type of benefits that regulation services provide to society, in particular through the

Table 18.2 Summary of ecosystem goods and services provided by the Baltic Sea. Data from de Groot et al. (2002), Hein et al. (2006), SEPA (2008a, b) and Atkins et al. (2011)

Ecosystem goods and services	Current state of the service	Level of threat	Expected consequences of loss
Provisioning: Resources produced by the ecosystem which we usually obtain by actively exploiting them			
Food resources *(living organisms used for human consumption, mostly acquired by harvesting or farming)*	Poor	High	Severe
Raw materials *(living organisms used for other purposes than human consumption, sediments, minerals, etc.)*	Good (?)	Moderate	Moderate
Ornamental resources *(marine products used for decoration, fashion, handicrafts or souvenirs)*	Good	Low	Limited
Energy resources *(energy directly acquired from the marine environment, e.g. wave energy)*	Good	Low	Limited
Chemical resources *(biomolecules used by humans for pharmaceuticals, cosmetics, health and dietary supplements, etc.)*	Unknown	Moderate	Unknown
Genetic resources *(genetic diversity of organisms adapted to the brackish Baltic Sea ecosystem, including also management of genetic composition by humans such as selective breeding and other forms of genetic manipulation)*	Unknown	High	Unknown
Regulating: Direct benefits we obtain from the natural regulation processes of the ecosystem			
Gas and climate regulation *(role of the marine environment in biogeochemical cycles, e.g. balancing CO_2/O_2, providing oxygen for breathing, slowing down global warming, regulating carbon fluxes, acting as a carbon sink)*	Good	Moderate	Severe
Disturbance prevention *(natural structural defences against floods, storms, waves, winds and erosion mitigation, etc.)*	Moderate	Moderate	Moderate
Regulation of hazardous substances *(natural control of hazardous substances through e.g. decomposition by bacteria and burial by sediments)*	Moderate	Moderate	Moderate
Mitigation of eutrophication *(removal of N and P from the sea. Note: this function alone cannot solve the problem of the large-scale eutrophication of the Baltic Sea)*	Good	Low	Limited
Biological regulation *(regulation of infestation by pathogens and other detrimental processes)*	Good	Moderate	Limited
Cultural: Non-material benefits we obtain from the ecosystem through recreation, spiritual and aesthetic reflection, cognitive and cultural development			
Leisure and recreation *(enjoyment and stimulation of the human body and mind through interaction with marine organisms and the marine scenery)*	Moderate	Moderate	Moderate
Aesthetic experiences *(contributions to emotional response and subjective sensation of beauty and/or silence)*	Moderate	Moderate	Limited

(continued)

Table 18.2 (continued)

Ecosystem goods and services	**Current state of the service**	**Level of threat**	**Expected consequences of loss**
Cultural heritage and identity *(uses of the marine environment for spiritual, religious, historical, architectural and folklore purposes, also to define the local identity)*	Moderate	Moderate	Limited
Inspiration for art, design and advertisement *(use of the marine and coastal environment as motifs in music, books, films, photographs, paintings, jewellery, postcards or commercials)*	Good?	Low	Limited
The legacy of the sea *(appreciation of the coastal and marine environment because of moral, altruistic or ethical values, often combined with the need to save the nature for the future generations)*	Moderate	High	Limited
Cognitive benefits *(contributions to education and research by the marine ecosystem)*	Good	Low	Limited
Supporting: Processes necessary for the sustained provision of all or most other ecosystem services, but which do not provide any direct benefits to humans. They can be seen as umbrella or overarching services since the protection of each of them maintains the availability of other ecosystem services (provisioning, regulating and cultural)			
Primary production *(the conversion of inorganic matter to living organic material through the use of solar energy by photosynthesis – regulates the level of atmospheric oxygen and constitutes the basis of the food web, essential to sustain the living conditions for all organisms)*	Good	Low	Severe
Biogeochemical cycles *(natural cycles of materials and energy – essential to sustain the living conditions for all organisms)*	Moderate	Moderate	Severe
Food-web dynamics *(the major way in which organisms affect the environment – regulates nutrient flows and energy transports in the ecosystem, constitutes the conditions for long-term ecosystem functioning)*	Poor	High	Severe
Habitats *(the extent at which the physical environment that surrounds, influences and is utilised by a biological population – essential to sustain the living conditions for all organisms living in the habitat, constitutes the conditions for long-term ecosystem functioning)*	Moderate	High	Severe
Biodiversity *(the genetic variability between and within species – essential to maintain for human consumption and to sustain changing environmental conditions, supports ecosystem resilience)*	Moderate	High	Unknown
Resilience *(the extent at which the ecosystem can absorb natural and human perturbations and can continue to regenerate without degrading or unexpectedly shifting to alternate states)*	Moderate	High	Unknown

positive externalities that the ecosystem provides, *e.g.* oxygen to breathe. "Option values" are associated with possible future demands that are currently not planned or not known. For example, options are kept open in case new information reveals whether an ecosystem has values we currently are (yet) unaware of. Finally, "Non-use values" are derived from attributes inherent to the ecosystem itself. They do not concern any direct or indirect use of marine resources. Non-use values can be anthropocentric, as in the case of natural beauty, as well as ecocentric, *e.g.* related to the notion that plant and animal species have a right to exist.

Box 18.2: The development of marine spatial planning (MSP)

Declarations, conventions and directives on MSP

MSP initiatives for the Baltic Sea Area are promoted in a wide set of global-scale declarations and conventions, EU directives and HELCOM documents. The "Blue Book: communication on an integrated maritime policy for the European Union" (EU 2007) was the first EU publication to officially label the approach as MSP. This was followed by the publication "Roadmap for maritime spatial planning: achieving common principles in the EU" (EU 2008b), which proposes a set of ten key principles for MSP in Europe. These principles still need to be tested for their applicability, and initiatives are being developed, including for the Baltic Sea (Schaefer and Barale 2011).

Integrating the MSP approach in the management of the Baltic Sea Area

The MSP approach is advancing in all countries surrounding the Baltic Sea on local, regional, national and basin-wide scales (HELCOM 2007; Ekebom et al. 2008; Schultz-Zehden et al. 2008; Blæsbjerg et al. 2009; Backer 2011), *e.g.* when building windmill parks at sea (Box Fig. 18.3). The HELCOM Baltic Sea Action Plan (BSAP), which aims at restoring the Baltic Sea environment by 2021, constitutes the main framework for MSP in the Baltic Sea Area (Backer et al. 2010). For this initiative, common principles have been adopted by HELCOM, VASAB (cooperation on MSP between the countries surrounding the Baltic Sea, http://www.vasab.org) and the European Commission. After rescaling, most of these principles are also applicable to MSP initiatives on local, regional and national scales.

Examples of research and planning of MSP in the Baltic Sea Area

At the time of writing this book, no country bordering the Baltic Sea has as of yet an operational MSP approach that extends from their coastline to their Exclusive Economic Zone (EEZ) limit. However, most countries are in the phase of developing their respective plans. A good example of an already existing MSP approach is the German plan for its Baltic Sea EEZ (Bundesgesetzblatt 2009). Sweden is currently reorganising its national legislation and structure of management authorities in order to be able to apply MSP to its entire coastal zone (SOU 2010). Several large research and development projects are being carried out to scientifically support MSP development for the Baltic Sea Area. Some examples of these are the BaltSeaPlan (http://www.baltseaplan.eu), PlanBothnia (http://planbothnia.org), PREHAB (http://www.prehab.gu.se) and VELMU (http://www.ymparisto.fi).

Box Fig. 18.3 Windmills at sea. Photo: © Hendrik Schubert

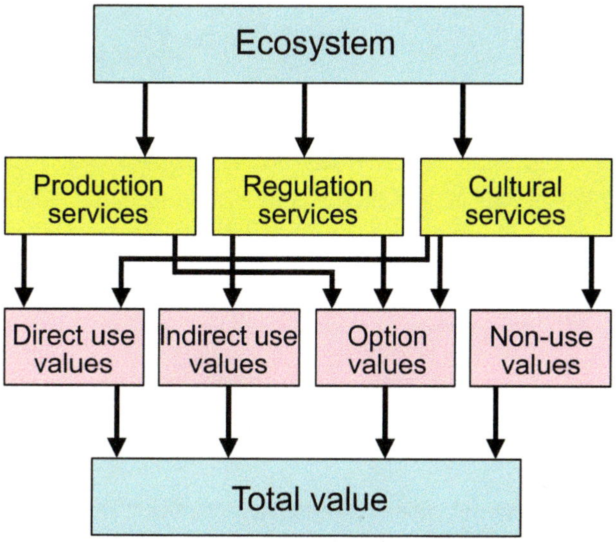

Fig. 18.6 An example of an ecosystem valuation framework. The arrows represent the most important links between the elements of the framework. Figure modified from Hein et al. (2006)

18.4 Marine spatial planning (MSP)

18.4.1 The MSP concept

Marine spatial planning (MSP) is a management tool developed fairly recently that seeks solutions for problems produced by different overlapping pressures caused by human uses of marine ecosystems (Ehler and Douvere 2009; Backer 2011). According to UNESCO's definition, MSP is a public process of analysing the spatial and temporal distribution of anthropogenic activities in marine areas and allocating them to achieve ecological, economic and social objectives that have usually been specified through a political process (Ehler and Douvere 2009). MSP aligns with integrated coastal zone management (ICZM) and is a useful tool to assist in the development of comprehensive ecosystem-based marine management (Blæsbjerg et al. 2009).

Different developments in the sea space (*e.g.* fisheries, shipping, eutrophication, marine conservation) have traditionally been dealt with as separate sectors in political decision-making. Therefore, management strategies and actions have often also been sectorial, without consideration of the pressures or needs of other sectors. The result of different sectorial strategies acting in an uncoordinated way produces situations like that at the Hel Peninsula (*cf.* Sect. 18.1). By applying MSP it is possible to develop new strategies to manage and improve the use and state of sea areas in a more holistic way (Lotze et al. 2006).

MSP also recognises that new types of human use of the sea are being introduced regularly. Some of these new uses have challenged the existing concepts of how we use the sea space. Windmills and wave power constructions, aquaculture and also marine protected areas are examples of uses that are static as well as spatially intense. Once they are established, it will be difficult to relocate them, either because they depend on a key resource (*e.g.* a particular habitat) or because of large infrastructure investments (*e.g.* a gas pipeline). This makes careful planning of sea space necessary.

18.4.2 Implementation of MSP

Essentially, MSP is a planning tool that enables integrated, forward-looking and consistent decision-making about the use of the sea, both coastal and offshore. Stakeholder involvement is an important component of effective MSP. Since MSP is always ecosystem-based, it takes into account different dimensions, such as the health status of a natural resource and how this is affected by human exploitation, as well as the socio-economic value of the resource. The initiatives are area-based, and they can be conducted on several spatial scales, from the coastal waters of a small municipality to entire ecosystems such as the Baltic Sea. It is important to note that MSP initiatives also include temporal aspects of anthropogenic activities. An effective MSP should also incorporate thorough knowledge and understanding of the existing governance system in the areas targeted (Olsen et al. 2011).

MSP is a continuing process that works through iterative steps of learning-by-doing, and thus adapts over time (Fig. 18.7). The iterative steps of MSP most often include (1) defining the overall aims and other principles, *e.g.* how stakeholder participation is to be carried out, (2) defining the scale and scope of the plan, (3) analysing current conditions, (4) drafting scenarios of possible future conditions, (5) introducing a plan, (6) implementing and enforcing the plan, (7) monitoring and revising the plan, and (8) adapting the spatial management process (Hall 2002).

Currently, a number of MSP approaches are being developed for the Baltic Sea Area (Box 18.2). A common feature of these approaches is that there is a tendency to include the full interests of all stakeholders and to impose as few restrictions or limitations as possible. However, an effective MSP should reconcile the different interests, which is not always easy. For example, the overlapping, multiple sea space uses in the Polish Exclusive Economic Zone (Fig. 18.2; Węsławski et al. 2006, 2009, 2010) illustrate that the term "protection" is understood differently by coastal engineers and environmentalists who pursue different goals and advocate the use of different methods.

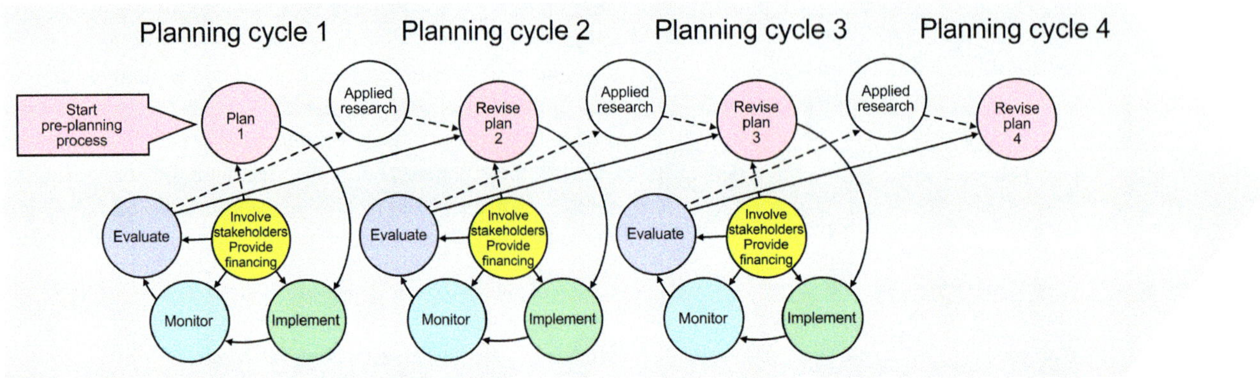

Fig. 18.7 An outline of the continuing marine spatial planning (MSP) cycle. Figure based on data in Ehler and Douvere (2009)

18.5 Ecosystem-based management (EBM)

18.5.1 The EBM concept

The major anthropogenic pressures in the Baltic Sea are overexploitation of fish stocks, eutrophication, chemical contamination by hazardous substances and loss of biodiversity (*cf.* Sect. 17.3.1). These large-scale anthropogenic pressures can only be dealt with by ecosystem-based management (EBM). This is a management strategy that aims to maintain the sustainable supply of ecosystem goods and services by keeping the ecosystem in a healthy, productive and resilient condition. EBM is an integrated approach to management that considers the entire ecosystem, including humans (McLeod and Leslie 2009). EBM differs from previous approaches that usually focused on a single species, sector, activity or concern, by integrating the cumulative impacts of a large variety of factors.

18.5.2 Development of EBM

The "ecosystem approach" is defined by the UN Convention on Biological Diversity (CBD; *cf.* Sects. 17.7.3 and 17.8.3) as a strategy for the integrated management of land, water and living resources that promotes conservation and sustainable use in an equitable way. This ecosystem approach concept has been adopted in EBM which is based on scientific knowledge and the application of appropriate methodologies that focus on different levels of biological organisation. EBM also recognises that humans are an integral component of the ecosystem.

In EBM, knowledge of biogeochemical processes and ecological interactions is combined with socio-economic information and approaches (McLeod and Leslie 2009). New techniques based on the geographic information system (GIS-based) and the development of advanced computer programmes for spatial modelling enable the analysis of complex, multi-layered and large spatial datasets (Snickars et al. 2010; Burkhard et al. 2011; Kappel and Martone 2011). Also the importance of and need for different types of spatial data (geological, ecological, economic, social) have increased during the development of EBM initiatives. Since EBM is place-based and adaptive, it requires a wide variety of tools and approaches. The following scientific challenges are pursued in different EBM initiatives around the world:

1. Integrative mathematical models that incorporate species interactions, feedback mechanisms among ecological components and anthropogenic activities, and critical environmental and anthropogenic drivers (*e.g.* climate change)
2. Methods for bringing together diverse datasets from biological, physical and social sciences
3. Techniques to account for uncertainty in datasets and model development
4. Ways to prioritise the most important data to be collected to inform management
5. Decision support systems that help environmental managers to evaluate risks associated with potential management actions by examining different scenarios of change as a result of strategic management decisions

18.5.3 Implementation of EBM

The ecosystem approach is now widely recognised in many national and international agreements, policies and legislation, including the EU MSFD, the Oslo/Paris Convention for the Protection of the Marine Environment of the North-East Atlantic (OSPAR), Australia's Ocean Policy, the Canadian Oceans Act, the US Commission on Ocean Policy (USCOP) and the UN Convention on the Law of the Sea (UNCLOS) (Rosenberg et al. 2009). In practice, the principles of EBM may be implemented through marine spatial planning (MSP) with the ecosystem approach as the primary goal. Yet, for MSP to be successful, it must not only meet the EBM criteria but also consider the existing governance and institutional structures in the area of interest. Both EBM and MSP are practical approaches and not only theoretical concepts. There have been many attempts to put them into practice worldwide.

Existing EBM approaches in marine systems include the spatial control of anthropogenic activities through protected areas. Further approaches are changes in governance such as the creation of regional coordinating agencies that monitor and evaluate the environmental status by using ecosystem indicators derived from multiple disciplines, scenario testing, risk assessment and precautionary adaptive management. The Ecosystem-Based Management Tool Network (http://www.ebmtools.org) is an organisation that develops, upholds and revises a global collection of different EBM tools.

18.5.4 EBM of the Baltic Sea

The HELCOM BSAP (*cf.* Sect. 17.8.4) was adopted by all nine Baltic Sea countries in 2007 and is regularly updated (http://www.helcom.fi). This is the only EBM initiative that covers the entire Baltic Sea Area. Although the BSAP stresses the need for the integrated management of all anthropogenic activities, its objectives are purely ecological and include the vision of the Baltic Sea as a place (1) unaffected by eutrophication, (2) with life undisturbed by hazardous substances, (3) with maritime activities carried out in an environmentally friendly way, and (4) of favourable conservation status of the marine biodiversity.

In addition to defining the goals, the BSAP also describes how to achieve them and includes MSP as one of the tools to achieve a "good ecological status" (*cf.* Sect. 17.8.2) of the Baltic Sea. The BSAP identifies indicators and measures for its implementation and postulates a periodic revision of its objectives based on the Baltic Sea monitoring and assessment programmes. A new approach to EBM has emerged from the establishment of the Baltic Nest Institute that hosts the Baltic Nest model (http://nest.su.se, Wulff et al. 2013), a HELCOM-endorsed decision support system aimed at facilitating adaptive management of environmental problems in the Baltic Sea. The model, which focuses on eutrophication and nutrient flows from land to the sea, can be used for the entire Baltic Sea drainage area.

The BSAP goals are not stipulated by any formal legislation. They were formulated especially for the action plan, although some of them were adapted from previous HELCOM recommendations and EU documents. The BSAP is implemented collectively through the national implementation plans (NIPs) of the nine countries bordering the Baltic Sea. The NIPs use different instruments and measures based on national legislations. The participation of stakeholders, *e.g.* engineers, fishermen, tour operators and sea miners, is strongly encouraged.

Right from the beginning of its existence, the BSAP was high on the European political agenda. Today it is considered as a pilot project for the application of the ecosystem approach to European seas in the context of the EU MSFD (*cf.* Sect. 17.8.1).

18.5.5 Achieving the management goals

Although political agreements about management plans for the Baltic Sea have been concluded, the intended goals have hardly been achieved as shown *e.g.* by analyses conducted by the non-governmental organisation World Wildlife Fund (WWF, Box 18.3). Unfortunately, one of WWF's conclusions is that there is a growing gap between the statements and commitments made by governments and corresponding actions needed to actually implement them.

In theory, there are a large number of potential options for manipulating a habitat to bring it to a specific state. However, only a few of these options are sustainable, affordable and have a proven record of success in practice. The best one is, of course, that a habitat stays in the desired state by nature conservation. Once deteriorated, a habitat can be restored. Even in cases where a heavily disturbed area is not to be restored, *e.g.* harbours or waterways where transportation service is ranked higher than all other interests, such an area should be kept in the "best ecological potential" state that is possible to achieve.

Both nature conservation (*cf.* Sect. 18.6) and habitat restoration (*cf.* Sect. 18.7) in the Baltic Sea Area include, in most cases, mitigation of nutrient inputs (*cf.* Sect. 18.8). Alleviation of eutrophication was set high on the Helsinki Convention's agenda already in 1974. However, the first concerted management approaches employed in the Baltic Sea were fishery regulations through ICES (Box 18.1), which, in fact, resulted from purely economic drivers. Compared to other measures, fishing quota (*cf.* Sect. 18.9) are a well-established EBM option, which can also manipulate lower trophic levels. The understanding of the value and function of the ecosystem services provided by the Baltic Sea is still limited. To improve the use of EBM in the Baltic Sea, there is a need for an economic valuation of its pelagic and coastal zones. It is essential to make the value of ecosystem services visible through the application of TEEB (*cf.* Sect. 18.3.2) and other relevant approaches in combination with integrated decision-making.

Box 18.3: The WWF Baltic Sea Scorecard approach

Åsa Andersson, Pauli Merriman, Ottilia Thoreson (WWF)

A need for bold, high-level political leadership

The WWF Baltic Ecoregion Programme, comprised of the World Wildlife Fund (WWF) and partner organisations in each of the nine coastal Baltic Sea countries, has been working for decades to protect the Baltic Sea environment and ensure the sustainable use of its resources.

In the course of this work, WWF has stressed the need for bold, high-level political leadership to address the many challenges facing the Baltic Sea and has been active in influencing a number of agreements and conventions agreed by Baltic Sea governments intended to "save the sea". Words and agreements, however, cannot "save" the Baltic Sea without the delivery and follow-through of the promises made.

Evaluations made by WWF: Scorecard reports

In 2007, WWF began to evaluate the degree to which governments were delivering on their stated commitments in the form of so-called "Scorecard" reports. Unfortunately, one of the key conclusions from these Scorecard reports was that there was a growing gap between the statements and commitments made by governments and the corresponding actions needed to actually deliver on their promises. The latest WWF Baltic Sea Scorecard report was published in August 2011 (Box Fig. 18.4) (http://www.wwf.panda.org). This report measured each of the nine coastal Baltic Sea countries' performance in implementing some of the most important international, regional and European agreements and conventions designed to manage and protect the Baltic Sea. On the basis of commitments made in these agreements, the 2011 Scorecard assessed a limited number of key indicators within five focal areas of crucial importance to the Baltic Sea and its health: eutrophication, hazardous substances, biodiversity, maritime activities, and integrated sea use management (ISUM) – the last being a more integrated approach to planning and managing the use of the sea and its resources. Special consideration was taken to grade Russia on a similar scale, even though not all agreements and policies applied, as Russia is not an EU member state. The Scorecard measured what each of the nine governments actually delivered in these crucial areas and therefore how well political commitments were being met, as no agreement – no matter how ambitious – can be successful without equally ambitious delivery and implementation. The results of the analysis were expressed in four grade levels, from the top grade of "A" to the weakest grade "C"; at the bottom of the scale is an "F", which indicates a failing grade.

The results of the 2011 Scorecard report

The results of the 2011 Scorecard are disappointing. The total grade for the entire Baltic Sea Area is an F, indicating that governments have failed to carry through their responsibility in the work to improve the situation for the Baltic Sea. The highest scores were assigned to Germany and Sweden, both of which earned a C, while all other countries received an F. Finland was ranked third, followed by Denmark, Estonia, Lithuania, Poland, Latvia and Russia. The areas of most concern regarding the lack of adequate follow-through by the country governments include eutrophication and the protection of biodiversity, which unfortunately adequately reflects the poor situation in the Baltic Sea with yearly cyanobacterial blooms and declining species and habitats. A comparison between the 2011 Scorecard and earlier ones shows some improvement in the areas of hazardous substances, maritime activities and ISUM, even though the overall score in each of these areas for all countries combined is not more than a C (Box Fig. 18.4).

Perspectives

The poor grades clearly indicate that the Baltic Sea governments are still failing to deliver upon their commitments and to take the actions needed to protect and restore the Baltic Sea. It also demonstrates that there is a considerable amount of work still to be done before we can secure a healthy Baltic Sea. WWF intends to revisit the Scorecard in the coming years in order to measure and monitor the progress of the governments, to see if they are doing what they promised. WWF hopes that providing a picture of the current situation will help to encourage countries, governments, corporations and individuals to help speed up the struggle to restore the health of the Baltic Sea.

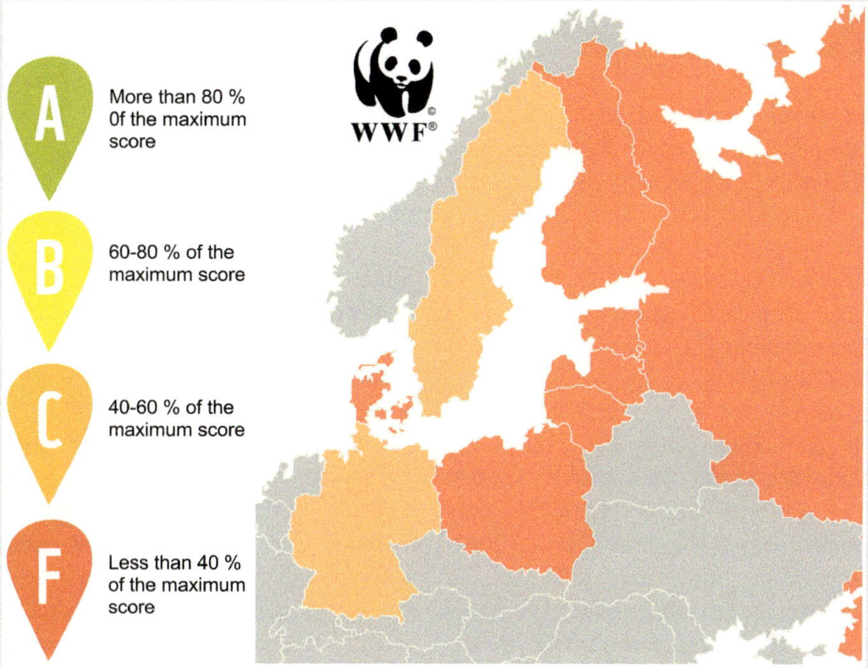

Summary of results

Countries	Eutrophication	Hazardous substances	Biodiver-sity	Maritime activities	ISUM	Total Score	Grade
Denmark	8/24	7/12	4/15	6/13	3/15	28/79	F
Estonia	5/24	6/12	3/15	7/13	5/15	26/79	F
Finland	5/24	9/12	2/15	8/13	5/15	29/79	F
Germany	14/24	7/12	5/15	4/13	6/15	36/79	C
Latvia	4/24	5/12	5/15	2/13	3/15	19/79	F
Lithuania	6/24	6/12	4/15	6/13	3/15	25/79	F
Poland	7/24	6/12	3/15	4/13	5/15	25/79	F
Russia	8/24	1/12	1/15	5/13	3/15	18/79	F
Sweden	11/24	7/12	2/15	8/13	8/15	36/79	C
All Countries	**68/216**	**54/108**	**29/135**	**50/117**	**41/135**	**242/711**	**F**

Total ranking

Countries	Ranking	Total Score	Grade
Germany	1	36/79	C
Sweden		36/79	C
Finland	3	29/79	F
Denmark	4	28/79	F
Estonia	5	26/79	F
Lithuania	6	25/79	F
Poland		25/79	F
Latvia	8	19/79	F
Russia	9	18/79	F

Box Fig. 18.4 The WWF Baltic Sea Scorecard report launched in August 2011. © WWF

18.6 Nature conservation

18.6.1 Policies and measures

HELCOM has listed all the threatened and/or declining species and biotopes/habitats in the entire Baltic Sea Area (HELCOM 2007). However, each country bordering the Baltic Sea has its own system for nature conservation, including measures for the protection of the sea.

Superimposed on the national systems are the regulations of the EU, notably the Habitats Directive with its Natura 2000 network of protected areas (EC 1992) and the Birds Directive (EU 2009a) (Table 18.3).

Some species are endangered in one area, but not in another. For example, the harbour porpoise (*cf.* Fig. 4.15) population in the North Sea is estimated at over 320,000 individuals. In the Belt Sea they number 11,000, but only ~200 in the Baltic Sea proper (Hammond et al. 2002; Gillespie et al. 2005). Despite these large differences in abundance, this species has the same high status of protection all over its distributional range according to the Annexes II and IV of the Habitats Directive.

Furthermore, a species may include populations with particular types of genetic variation, and thus the conservation of such populations is desired. An example of population protection is provided by the reform of the EU Common Fisheries Policy (CFP, EU 2009b), which takes delimitation of specific genetic populations into account, *e.g.* the western and eastern stocks of the Atlantic herring *Clupea harengus* and the Atlantic cod *Gadus morhua*. On the other hand, a species that is protected in one area can be exploited elsewhere. For example, the grey seal is protected in the southwestern part of the Baltic Sea, but limited hunting is permitted in the northeastern part.

Protection measures can be passive or active. Passive protection involves *e.g.* stopping exploitation, protection of spawning habitats and other measures aimed at increasing the reproductive success of a species. Active protection includes *e.g.* the removal of stressors, restoration of habitats, reintroduction of threatened or extinct species, stocking activities and strategic provision of resources to survive. An example of active protection was the supply of uncontaminated food to the few white-tailed eagles that were left when the top predators in the Baltic Sea were severely damaged by

Table 18.3 Examples of Baltic Sea habitats and species that are protected by European regulations in the Habitats Directive (EC 1992). They are included in Article 17, which requires the EU Member States to report every six years about the progress made with the implementation of the Habitats Directive.

Target of protection	Numbers in the EU Habitats Directive, Article 17	Threats in the Baltic Sea Area
Seagrass meadows	Habitat No. 1120 *Posidonia* (seagrass) beds (in the Baltic Sea Area seagrass meadows consist of *Zostera marina* and/or *Zostera noltii*)	Eutrophication and pollution, bottom trawling, offshore constructions, sand and gravel extraction
Offshore stone reefs	Habitat No. 1170 Reefs	Eutrophication and pollution, bottom trawling, offshore constructions (windmill farms), stone-fishing, dumping of dredged material, military activities
Sand banks, large shallow inlets and bays	Habitat No. 1110 Sand banks which are slightly covered by seawater all the time Habitat No. 1160 Large shallow inlets and bays	Eutrophication and pollution, bottom trawling, offshore constructions, sand and gravel extraction, dumping of dredged material, tourism
Submerged natural structures	Habitat No. 1180 Submarine structures made by leaking gases Habitat No. 8330 Submerged or partially submerged sea caves	Eutrophication and pollution, single sites in isolated locations which are easy to destroy physically by bottom trawling, tourism and recreational activities (diving)
Lampreys	Species No. 1095 *Petromyzon marinus* (sea lamprey) Species No. 1099 *Lampetra fluviatilis* (river lamprey)	Eutrophication in spawning areas (rivers), obstruction of spawning migration to the sea (weirs, dams)
Clupeid fish	Species No. 1102 *Alosa alosa* (allis shad) Species No. 1103 *Alosa fallax* (twait shad)	By-catch in fisheries, eutrophication in spawning areas (rivers), obstruction of spawning migration to the sea (weirs, dams)
Seals	Species No. 1364 *Halichoerus grypus* (grey seal) Species No. 1365 *Phoca vitulina* (harbour seal) Species No. 1938 *Pusa hispida* (ringed seal)	Hunting, by-catch in fisheries, hazardous substances, diseases, habitat loss due to coastal development, for the ringed seal also climate change
Whales	Species No. 1351 *Phocoena phocoena* (harbour porpoise)	By-catch in fisheries, hazardous substances, acoustic pollution, shipping and prey depletion due to over-fishing

DDTs and PCBs in the 1970s (*cf.* Sect. 16.1). This action helped to save the Finnish and Swedish populations.

18.6.2 Biological valuation

One way of assessing the natural assets of a specific area is to carry out a biological valuation, *i.e.* to estimate the intrinsic value of nature without considering the human use. The first biological valuation of a marine realm was performed for the Belgian North Sea shelf (Derous et al. 2007). The starting point of a biological valuation is a set of "assessment questions", which are formulated by regional experts. Answers to these questions should reflect the naturalness, relative importance, rarity and aggregation of species in a given area. The outcome of biological valuation is presented in the form of a map with simple indices or colour codes (Fig. 18.8).

The outcome of biological validation depends heavily on national borders and regional regulations. Again, the value of a given species or habitat may differ between areas. For example, *Fucus* spp. (*cf.* Sect. 11.8) are considered threatened in Poland where these algae basically disappeared by 1980, but they are common on the nearby Danish island of Bornholm where hard substrates necessary for the attachment of the algal tufts are more abundant.

Thus, all countries surrounding the Baltic Sea have prepared their own national "red lists" of threatened species. An overall status of a red-listed species at an international level dictates whether it requires special protection on a larger scale (*e.g.* via EU red lists). For example, out of the roughly 500 macrofauna species living in the Baltic Sea Area on soft bottoms deeper than 20 m (*cf.* Fig. 10.4), 75 are protected by national regulations, but only 25 of them are protected by EU regulations. Nevertheless, this still makes the Baltic Sea the sea with the highest percentage of protected species in Europe.

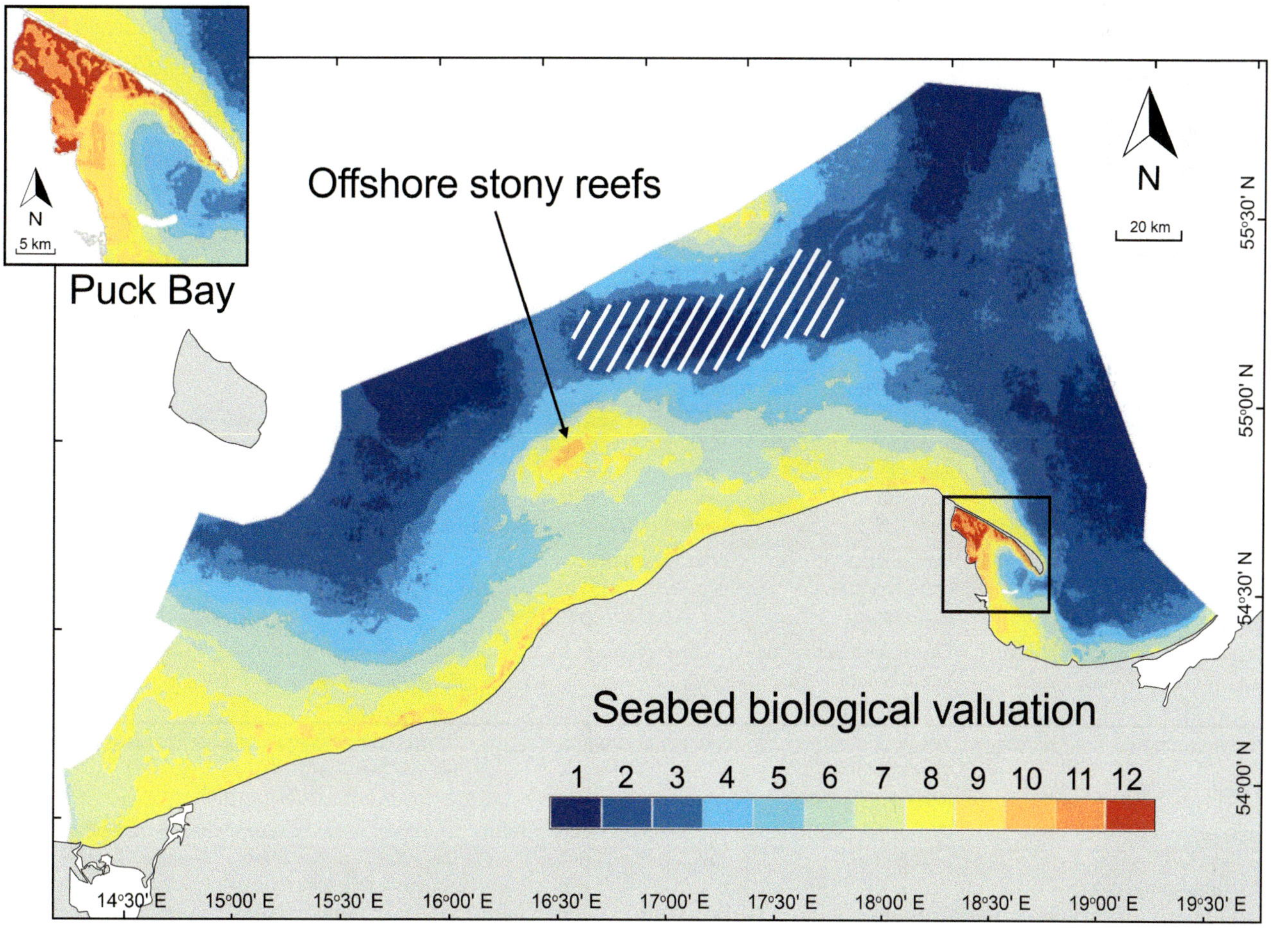

Fig. 18.8 An example of biological valuation using seabed biota in the Polish Exclusive Economic Zone in the southern Baltic Sea. The colour scale with scores indicates increasing value from 1 (blue) to 12 (red). Puck Bay (black frame and inset, with scores 11–12) and the submerged stone reefs of the central coast (with scores 10–11) are the most biologically valuable areas here. Figure reprinted from Węsławski et al. (2009) with permission from Oceanologia (Institute of Oceanology PAN, Sopot, Poland)

18.7 Habitat restoration

18.7.1 Restoration strategies and possibilities

The transformation of a disturbed area into an area with "good environmental status" (GES, *cf.* Sect. 17.8.1) involves active measures (with human intervention) or passive recovery after the removal of the cause of the stress. A desired habitat quality or an acceptable final state are subjective measures, which are linked to political decisions and may change over time. Thus, there is a risk of an increased public tolerance of poor environmental quality if a society becomes used to substandard habitat conditions (Elliott et al. 2007).

The basis of restoration ecology in general involves a return to an original state once the natural habitat has been re-established and/or stressors have been removed. The duration of ecosystem recovery may vary since it depends on the scale of the negative impact and the size of the area to be restored. However, in the case of eutrophication, evidence is accumulating that estuarine systems fail to return to the original status despite improved conditions (Duarte et al. 2008). Heavily impacted systems in particular may show complex trajectories of change involving non-linear responses and critical thresholds despite significant oligotrophication. Thus, it is expected that the Baltic Sea recovery trajectory will involve shifting baselines (*cf.* Fig. 17.1).

The year 1950 is often set as the time in the past when the Baltic Sea ecosystem was still in good shape, with well-oxygenated near-bottom waters, extensive vegetation of charophytes and vascular plants on soft and sandy bottoms, perennial brown algal belts on rocky coasts, persistent pollutants below hazardous levels, and few non-indigenous species. These characteristics are regarded as restoration targets for the Baltic Sea. Active, target-oriented habitat restoration in the Baltic Sea may be achievable for some habitats, *e.g.* seagrass meadows and reed belts, but it is difficult or impossible for others, *e.g.* deep soft seabeds (Table 18.4).

18.7.2 Restoration activities

The major on-going activities to restore species and populations concern stocks of commercially valuable fish species, such as the Atlantic salmon *Salmo salar* in the northern Baltic Sea as well as the northern pike *Esox lucius* and the European perch *Perca fluviatilis* in the southern Baltic Sea. Efforts are also being made to rebuild the critically threatened populations of the allis shad *Alosa alosa* and the Atlantic sturgeon *Acipenser oxyrinchus*, *e.g.* in the tributaries of the Wisła river.

Perhaps one of the most promising habitat restoration actions is the reintroduction of important habitat-forming species ("engineering species") to sites where these species have been lost from healthy sites. In the Baltic Sea, an

Table 18.4 Examples of restoration targets in the Baltic Sea

Habitat/species	Category	Restoration target	Method of restoration	Remarks
Pelagic zone	Habitat	Oxygenated, nutrients and chemical pollutants below the 1950s level	Change of agricultural practices, point source management, pollution control	Feasible, yet expensive and slow recovery
Soft bottoms below the photic zone	Habitat	Oxygenated, bioturbated by macrofauna	Active aeration	Hardly feasible on a pan-Baltic scale
Reed belts	Habitat	Dense, self-recovering vegetation in appropriate places	Active planting	Although reed belts can locally be reduced they are expanding in many other places
Seagrass meadows	Habitat	Dense, self-recovering vegetation in appropriate places	Active planting	Small-scale active planting projects are on-going in the southern Baltic Sea
Acipenser oxyrinchus (Atlantic sturgeon)	Species	Return to sustainable population	Active introduction from a Canadian population	The species that became extinct in the Baltic Sea in the 1900s was apparently introduced from North America in medieval times (*cf.* Box 6.4)
Salmo salar (Atlantic salmon)	Species	Pre-industrial population size	River cleaning and spawning area protection	Threat from farmed populations
Haliaeetus albicilla (white-tailed eagle)	Species	Pre-industrial population size	Nesting areas protection, feeding	Has successfully returned to most areas through the mitigation of DDT and PCBs
Halichoerus grypus (grey seal)	Species	Pre-industrial population size	Strict protection, designated shelter areas	Populations are not considered threatened anymore in the northern and eastern Baltic Sea

example of such a habitat-forming species is the common eelgrass *Zostera marina*. This seagrass has disappeared or has a significantly reduced lower depth limit in many coastal areas, especially in Denmark, Germany, Poland and Sweden (Boström et al. 2003, 2014). These negative trends, along with the crucial multiple ecosystem services performed by seagrass meadows, *e.g.* sediment stabilisation, oxygen transport to sediment-associated organisms, and shelter for invertebrates and small fish (*cf.* Sect. 11.11), make *Zostera marina* a prime candidate for restoration projects. While the best option for the restoration of a seagrass meadow is to remove the stressor and allow natural recolonisation, in many areas the only option for restoring the functionality of coastal environments is the transplantation of eelgrass.

Although no large-scale *Zostera marina* transplantation projects have yet been carried out to restore the coastal areas of the Baltic Sea, some scientific small-scale transplantation experiments have been successful (*e.g.* Salo et al. 2009). A restoration project is complex due to a number of inter-linked factors that need to be considered, even if it involves only a single plant species such as *Zostera marina* (Fig. 18.9). In particular, the steep environmental gradients, which are characteristic for the Baltic Sea (*cf.* Sect. 2.4), require special considerations. For example, populations of *Zostera marina* growing at salinity 5–6 in the northern Baltic Sea proper are characterised by asexual reproduction, low genetic diversity and high diversity of associated plant species of freshwater origin (*cf.* Sect. 11.11).

Zostera meadows are typically situated at a 2–4 m water depth in the northern Baltic Sea and shoot planting would thus require extensive SCUBA diving efforts, making seagrass restoration ineffective and restricted to small scale. In contrast, the populations along the German coast in the Belt Sea proper grow at higher salinity and form sexually reproducing (with high seed production), monospecific seagrass meadows with a high genetic diversity (Reusch et al. 1999). For a transplantation to succeed, the inherent traits of the donor populations and the physical, chemical and biological characteristics of the donor and recipient sites need to be compatible. There exists a multitude of seagrass transplantation techniques that need to be adapted to the local conditions, and the timing of the transplantation should also be adjusted. For further monitoring and management of the restoration sites, appropriate indicator variables need to be selected (Fig. 18.9).

A return to the "natural state" involves not only the restoration of the original habitat of a species but also reducing the stressor as well as counter-acting invasions of non-indigenous species. There is a risk that a large-scale habitat restoration effort, which is usually associated with a major habitat disturbance such as sediment replacement and eelgrass planting, may be beneficial for non-indigenous species that are capable of the rapid colonisation of new habitats (*cf.* Sect. 5.1) resulting from the restoration work.

18.8 Mitigation of nutrient inputs

18.8.1 Sources of nutrient input to the Baltic Sea

Nutrients reach the Baltic Sea through indirect discharges from land, direct point sources, atmospheric deposition and input from the Skagerrak via the Kattegat and the Belt Sea. Indirect nutrient emissions are lumped into the category "riverine nutrient load" and originate from point sources, diffuse sources and atmospheric deposition on land. Part of these discharges can be considered as the natural background, *i.e.* nutrient losses from unmanaged land and the proportion of losses from managed land that would occur irrespective of agricultural activities. A considerable fraction of these indirect discharges is retained on land and/or in freshwater and never reaches the sea.

According to recent reconstructions, the 30-year (1977–2006) relative contributions of riverine load, point sources, atmosphere and the Skagerrak were estimated at 59 %, 5 %, 23 % and 13 % of the total nitrogen input to the Baltic Sea and 45 %, 14 %, 7 %, and 34 % of the total phosphorus input, respectively (*cf.* Box 2.2, Savchuk et al. 2008, 2012; Gustafsson et al. 2012). In absolute amounts, the recent compilation by HELCOM (2013c) gives the following annual estimates for 2010: the riverine load as the major source of nutrients discharged into the Baltic Sea contributed with 728 kilotonnes N and 34 kilotonnes P, while 30.5 kilotonnes N and 1,7 kilotonnes P came from point sources directly to the sea, and 219 kilotonnes N and 2.1 kilotonnes P were deposited from the atmosphere.

Land use and the handling of sewage are major factors regulating the anthropogenic emissions of nutrients and eutrophication of the Baltic Sea. The origins of N and P in riverine outflows differ. While diffuse input from agriculture is the foremost anthropogenic N source, anthropogenic P originates both from agriculture and point sources on land, especially from municipalities (HELCOM 2009). Between 1900 and 1990, the phosphorus inputs from direct point sources around the Baltic Sea coasts increased the most proportionally, but they also decreased the most between 1990 and 2006 (*cf.* Box 2.2) as a result of intense management since 1990. However, the large riverine loads of both N and P, which increased after 1950, are more difficult to manage.

18.8.2 Pathways of N and P transfer to the sea

Nutrient concentrations in river water are a net result of several interacting processes, including nutrient exchanges between the terrestrial, aquatic and atmospheric environments. These processes encompass nutrient release and

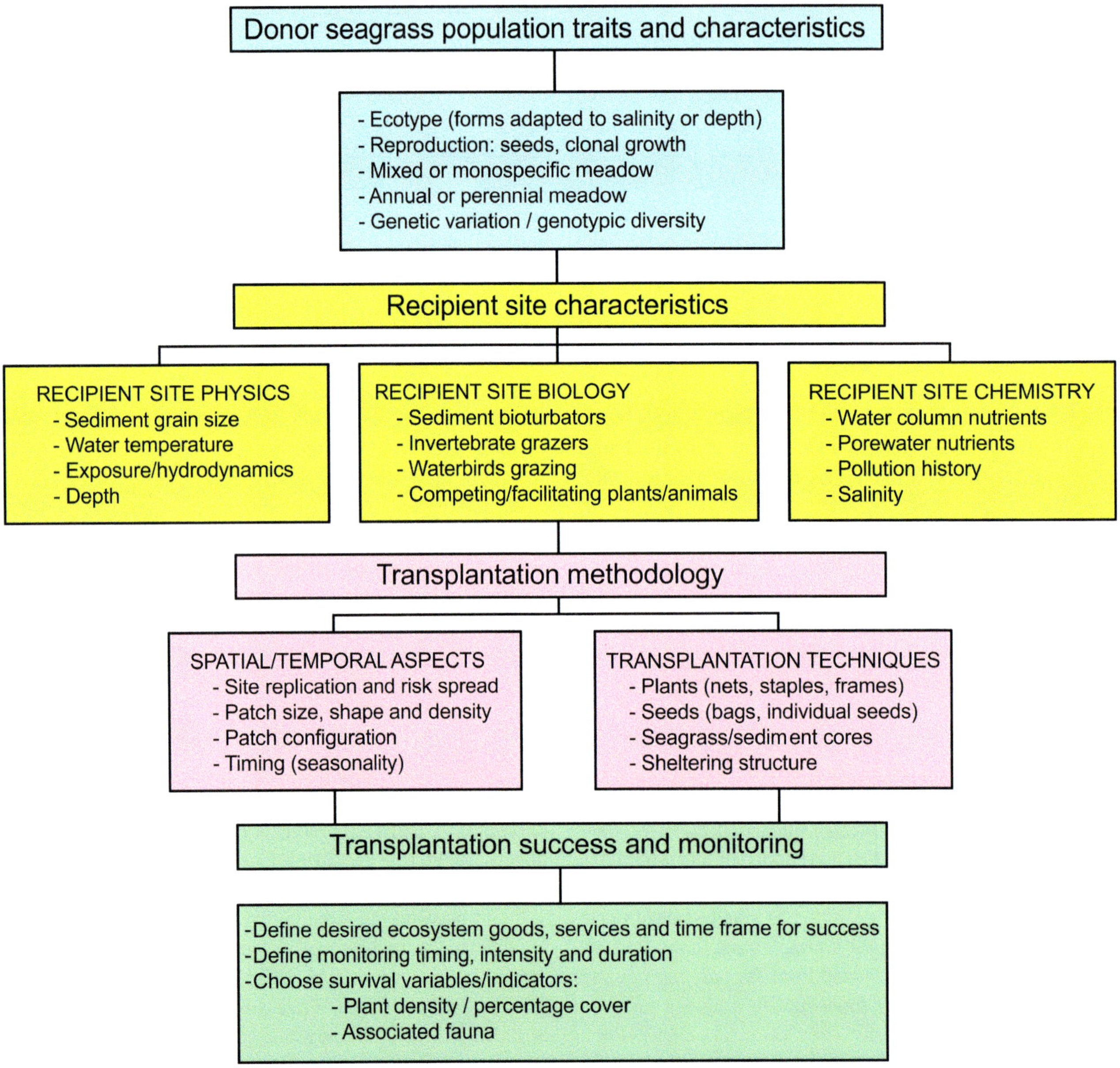

Fig. 18.9 An outline of a seagrass restoration plan. Figure: © Christoffer Boström

transport, geo- and bio-transformations and immobilisation. Their influences vary on spatial and temporal scales with meteorology, hydrology, land cover, physiographic factors, land use and management in the drainage area (Kowalkowski et al. 2012).

At the turn of the century, only two rivers, the Wisła and the Odra, accounted for ~28 % of the total N and ~39 % of the total P riverine loads to the entire Baltic Sea. Both rivers have relatively large drainage areas with ~64 % of agricultural land use (*cf.* Table 2.4). A model was developed to calculate source-apportioned emissions of N and P into the river basins of the Wisła and Odra, *e.g.* atmospheric deposition, overland flow, tile drainage, erosion, groundwater, wastewater treatment plants (WWTPs) and urban systems (Pastuszak et al. 2012).

This model predicts that N in riverine and seawater originates mainly from diffuse sources with agriculture as a key player, and that the N concentration in the water becomes higher with increasing water flow. In contrast, P in riverine and seawater originates mainly from point sources in the drainage areas of the Wisła and Odra and its concentration becomes lower with increasing water flow (a dilution effect). This shows that water flow variability affects nutrient concentrations, and thus also the loads to the sea, which are calculated by multiplying nutrient concentration by the discharged water volume.

18.8.3 Drainage area heterogeneity influences N and P loads

The conditions for nutrient emissions from land vary in the Baltic Sea drainage area (compare Figs. 2.5 and 18.10). The northern parts are extensively forested while agricultural land predominates in the south and human population density increases from north to south. This heterogeneity brings about huge spatial differences in the total emissions of N and P across the Baltic Sea, which are correlated with the size of the drainage area, land cover, land use and human population size (Table 18.5). Emissions of nutrients from forested areas and regions covered by meadows and pastures are much lower than those from agriculturally cultivated land. Studies carried out in Finland and the USA have shown that agricultural land produces 8–10 times higher N losses per unit area than forested land (Rekolainen et al. 1995; Hatfield and Follett 2008). A higher human population density accelerates N cycling through fertiliser use, food relocation, atmospheric pollution and land disturbance, and is therefore responsible for higher diffuse N outflow (Howarth et al. 1996).

Poland is the country with the highest waterborne nutrient emissions to the Baltic Sea with ~26 % of all N and ~37 % of all P (Table 18.5). However, it is necessary to realise that: (1) the Polish part of the Baltic Sea drainage area is one of the four largest, with the Swedish, Finnish and Russian drainage areas being the other three (Fig. 18.10a), (2) the Polish population accounts for ~46 % of the entire human population in the Baltic Sea drainage area (Table 18.5), (3) the Polish part of the Baltic Sea drainage area consists of ~60 % agricultural land compared to only 6 %, 7 % and 12 % of the Swedish, Finnish and Russian drainage areas, respectively (Fig. 18.10 b), and (4) the rivers Wisła and Odra are the second and third largest rivers in the Baltic Sea drainage area, respectively (*cf.* Table 2.4). When area-specific loads are considered, with N and P discharges expressed per km^{-2}, the Polish discharges of N were in the year 2000 surpassed by those from Denmark, Germany, Lithuania and Latvia (Table 18.5). The area-specific P discharges of Poland were still very high, and in 2000 were surpassed only by those from Denmark.

The nutrient emissions from Poland to the Baltic Sea could have been much higher if Poland would not have had the highest natural N and P retention in the entire Baltic Sea drainage area (Fig. 18.11). The high retention capacity is due to Poland being largely a lowland country; 75 % of its area lies at altitudes lower than 200 m above the sea level. Moreover, Polish rivers are the least regulated in the Baltic region and most riverbanks are overgrown with grass and bush vegetation. These features contribute to long nutrient retention times and favour effective nutrient removal from the system, *e.g.* through denitrification.

Table 18.5 Estimated loads of total waterborne N and P entering the Baltic Sea in the year 2000 from each of the nine countries bordering the Baltic Sea, along with the size of the drainage area and the size of the human population living in the drainage area in the year 2002 per country. Included are also area and population data for the five countries in the drainage area that do not have a Baltic Sea coast; waterborne N and P from these countries are discharged into the Baltic Sea via other countries. Data from HELCOM (2004) and Hannerz and Destouni (2006). These data served to calculate the total N and P per unit land area. Other literature data about the total waterborne N and P entering the Baltic Sea and the properties of the drainage area may differ and therefore yield slightly different results compared to the data presented here.

	Total N (tonnes $year^{-1}$)	Total P (tonnes $year^{-1}$)	Drainage area (km^2)	Human population	Total N (kg km^{-2} $year^{-1}$)	Total P (kg km^{-2} $year^{-1}$)
Denmark	58,923	1,857	27,300	4,469,000	2,158	68
Germany	18,605	487	26,300	2,844,000	707	19
Poland	191,166	12,645	309,900	38,578,000	617	41
Lithuania	47,885	1,896	64,600	3,600,000	741	29
Latvia	67,493	2,207	64,200	2,359,000	1,051	34
Estonia	26,874	965	45,200	1,432,000	595	21
Russia	79,188	4,623	330,000	9,700,000	240	14
Finland	101,659	4,840	302,600	5,142,000	336	16
Sweden	153,074	4,969	439,800	8,795,000	348	11
Czech Republic + Slovakia			9,500	1,745,000		
Ukraine			11,170	1,740,000		
Belarus			88,500	3,801,000		
Norway			13,360	34,000		
Baltic Sea	**744,867**	**34,489**	**1,739,300**	**84,239,000**	**428**	**20**

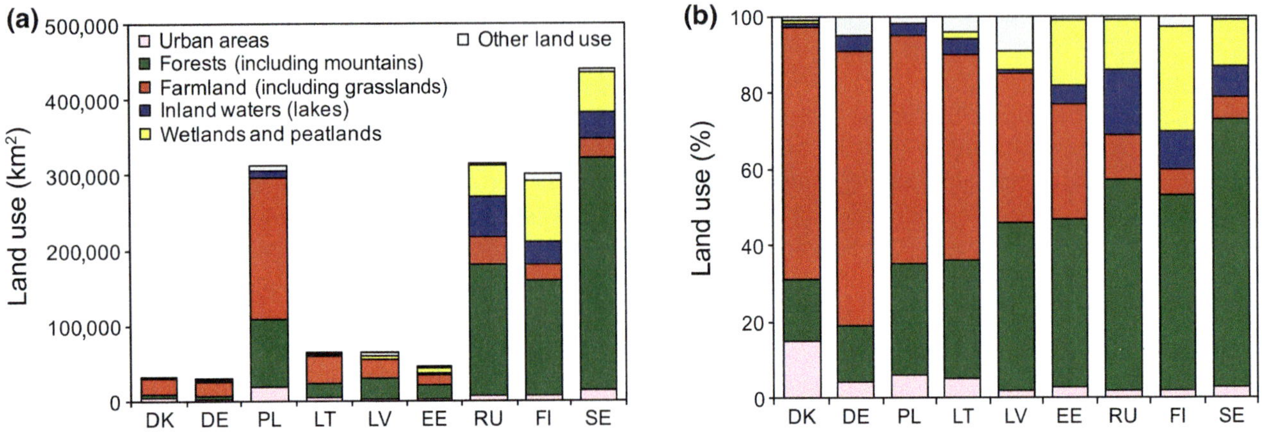

Fig. 18.10 Land use in the drainage area of the Baltic Sea divided between the nine countries bordering the Baltic Sea in the year 1995. (**a**) Land area per country. (**b**) Percentage of the land area per country. DK = Denmark, DE = Germany, PL = Poland, LT = Lithuania, LV = Latvia, EE = Estonia, RU = Russia, FI = Finland, SE = Sweden. Figure based on data from tables 2.1 and 2.3 in HELCOM (2004)

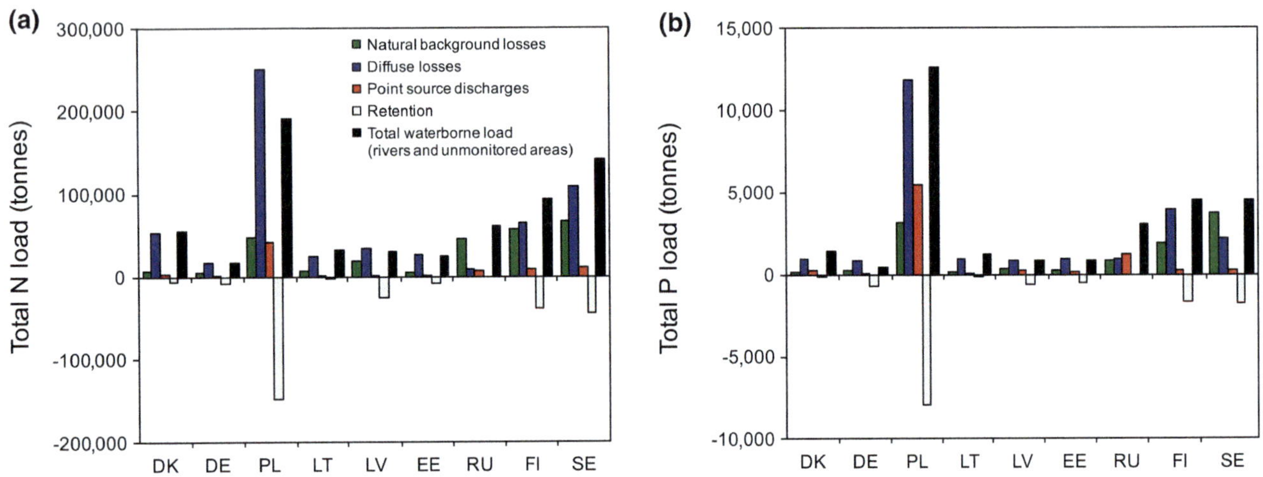

Fig. 18.11 Source apportionment of the total riverine loads by the nine countries bordering the Baltic Sea in the year 2000. (**a**) Total N (**b**) Total P. The total waterborne load consists of the sum of natural background losses, diffuse losses and point source discharges minus retention for N and P, respectively. DK = Denmark, DE = Germany, PL = Poland, LT = Lithuania, LV = Latvia, EE = Estonia, RU = Russia, FI = Finland, SE = Sweden. Figure based on data from tables 5.53 and 5.54 in HELCOM (2004)

18.8.4 The role of estuaries in nutrient input reduction

Estuarine processes at the land/sea interface of the North Atlantic Ocean are estimated to retain and remove 30–65 % of the total N and 10–55 % of the total P from the river water, with net nutrient transports being inversely correlated with the log mean water residence time in the estuarine system (Nixon et al. 1996). The amount of N retained in an estuary is relatively insensitive to trophic status while P accumulation in sediments decreases when an estuary becomes more eutrophic (Howarth et al. 1996).

Thus, with respect to a sea's eutrophication problems, it is an advantage when a river flows through estuarine lagoons and coastal lakes before entering the sea because these water bodies trap nutrients. Typical of the lagoons in the southern Baltic Sea is a water residence time of a few weeks, which favours the reduction of N and P concentrations in the river water. The Odra passes through a large estuary before it reaches the Baltic Sea (the Szczecin Lagoon), but the Wisła does not (*cf.* Fig. 2.11).

Mass balance calculations performed for the Odra estuary demonstrated that the total N and P retentions in 1993–1998 in its estuary were 45 % and 37 %, respectively (Pastuszak et al. 2005). The loss of N due to denitrification accounted for ~26 % of the total N introduced to the estuarine system from land and atmosphere. The N and P supplies to the Odra estuary have been considerably reduced over the last two decades (Fig. 18.12, Table 18.6). When taking this into consideration, it may be assumed that the N retention in the

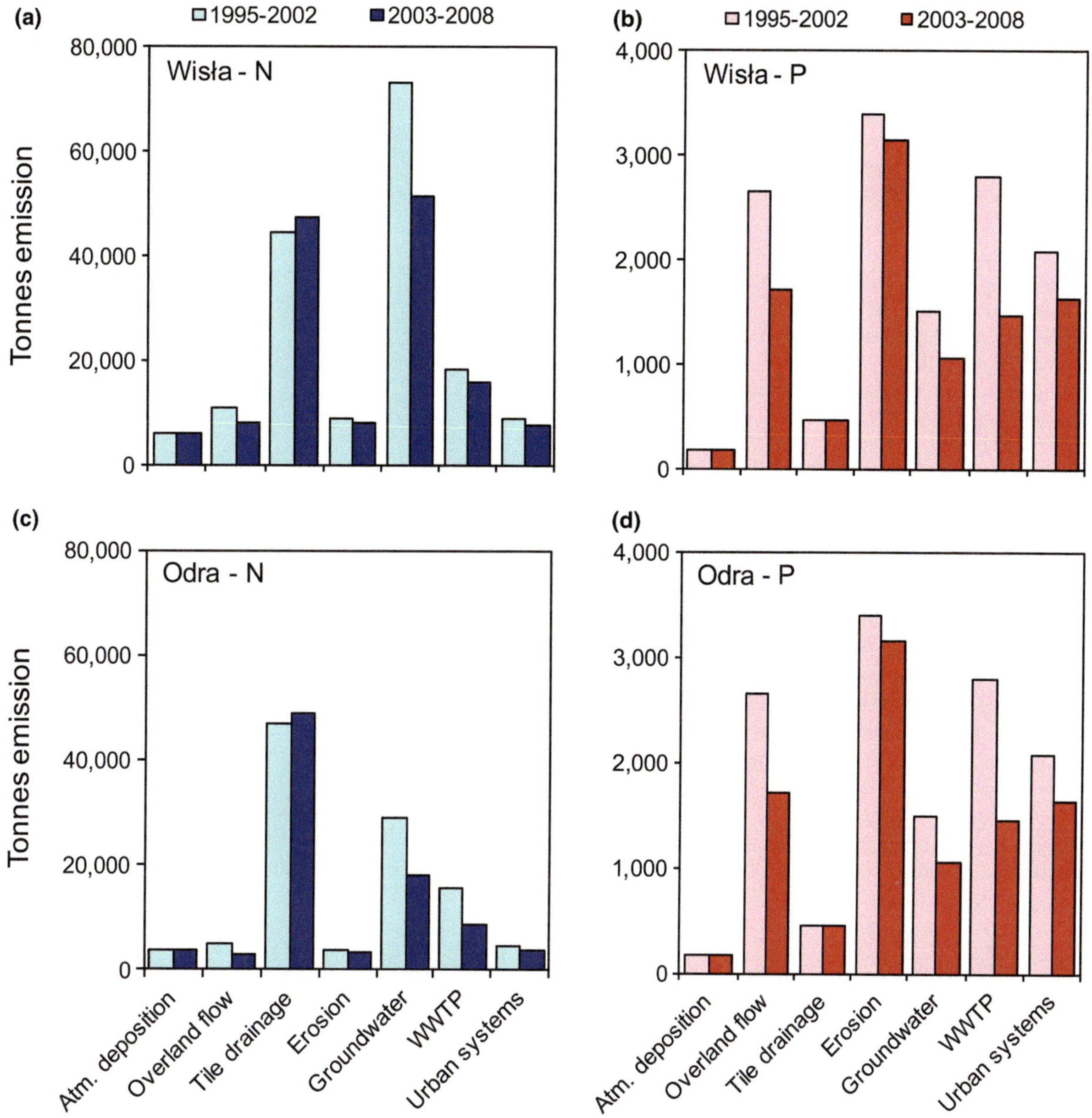

Fig. 18.12 Annual source-apportioned total N and total P emissions in two river basins in 1995–2002 and 2003–2008. (**a**) N in the Wilsła river basin. (**b**) P in the Wilsła river basin. (**c**) N in the Odra river basin. (**d**) P in the Odra river basin. WWTP = wastewater treatment plants. Figure modified from Kowalkowski et al. (2012)

Odra estuary remained at the levels of 1993–1998, *i.e.* 45 %, while the P retention may have increased since the estuary has become less eutrophic (Howarth et al. 1996).

18.8.5 The need to reduce nutrient inputs

The reduction of nutrient loads to the Baltic Sea is a complicated task that needs to follow the principle of adaptive management and be based on the EBM approach for the management of anthropogenic activities. The member countries of HELCOM and the EU are in the process of adopting and implementing directives and resolutions aimed at nutrient mitigation in the riverine systems discharging into the Baltic Sea on national and ecosystem-wide scales (HELCOM 2013a).

The HELCOM BSAP (*cf.* Sect. 17.8.4) calls for a series of actions to be taken by the member states in order to restore the "good ecological status" of the Baltic marine environment by 2021. The desired eutrophication-related targets are expressed in terms of indicators, mainly nutrient concentrations, the chlorophyll *a* concentration and the Secchi depth (HELCOM 2013b). Based on the available scientific information, the maximum allowable input

Table 18.6 Summary of the estimated reductions in the flow-normalised total nitrogen and total phosphorus inputs of the Polish waterborne discharges into the Baltic Sea during the transition period of the Polish economy (1988–2011). Data from Pastuszak et al. (2005, 2012) and Pastuszak and Witek (2012)

Action	Reduction in total N (tonnes year^{-1})	Reduction in total P (tonnes year^{-1})
Nutrient reduction in the Wisła river basin 1988–2011 The total N load declined from ~118,000 tonnes year^{-1} to ~78,000 tonnes year^{-1} The total P load declined from ~6,900 tonnes year^{-1} to ~5,700 tonnes year^{-1}	~40,000 (34 %)	~1,200 (18 %)
Nutrient reduction in the Odra river basin 1988–2011 The total N load declined from ~75,000 tonnes year^{-1} to ~54,000 tonnes year^{-1} The total P load declined from ~8,000 tonnes year^{-1} to ~1,500 tonnes year^{-1}	~21,000 (28 %)	~6,500 (80 %)
Nutrient reduction in the Odra estuary, modelled for total N and total P loads in the year 2008, assuming estuarine retention of 45 % of total N and 37 % of total P.	~15,000	~960
Nutrient reduction (total N and total P loads) in the Polish coastal areas in 2000, assuming a 5 % contribution by Pomeranian rivers to the overall total N and total P loads from Poland, and with assumed average reductions of total N and P loads of 30 % and 50 %, respectively	~2,800	~300
Total nutrient reduction in Polish water-borne discharges 1988–2011	**~78,800**	**~8,960**

Table 18.7 Estimations of the maximum allowable input (MAI) of N and P to the Baltic Sea Area, the inputs in 1997–2003 and the reductions needed of total N and total P (HELCOM 2013a, b).

Subregion	Maximum allowable inputs (MAI)		Inputs 1997–2003		Reductions needed	
	Total N (tonnes year^{-1})	Total P (tonnes year^{-1})	Total N (tonnes year^{-1})	Total P (tonnes year^{-1})	Total N (tonnes year^{-1})	Total P (tonnes year^{-1})
Kattegat	74,000	1,687	78,761	1,687	4,761	0
Danish straits	65,998	1,601	65,998	1,601	0	0
Baltic Sea proper	325,000	7,360	423,921	18,320	98,921	10,960
Bothnian Sea	79,372	2,773	79,372	2,773	0	0
Bothnian Bay	57,622	2,675	57,622	2,675	0	0
Gulf of Riga	88,417	2,020	88,417	2,328	0	308
Gulf of Finland	101,800	3,600	116,252	7,509	14,452	3,909
Baltic Sea	**792,209**	**21,716**	**910,344**	**36,894**	**118,134**	**15,178**

(MAI) of nitrogen and phosphorus (Table 18.7) are estimated with the help of models that are regularly updated.

In 2007, the BSAP proposed provisional country-allocated reduction targets (CARTs) for N and P, and at the HELCOM Ministerial Meeting in 2013 (HELCOM 2013a) negotiated CARTs were adopted (Table 18.8). To achieve the "good ecological status", the overall MAI levels to the Baltic Sea need to come down from 910 kilotonnes N year^{-1} in 1997–2003 to 792 kilotonnes N year^{-1} and from 36.9 kilotonnes P year^{-1} in 1997–2003 to 21.7 kilotonnes P year^{-1}. This is a ~13 % reduction for N and a ~41 % reduction for P.

18.8.6 Wastewater treatment plants and changes in agricultural practices

Throughout the Baltic Sea drainage area, there are many good examples of how to mitigate nutrient inputs to the sea. Such examples can be found in Poland, stimulated by the changes in the Polish economy that took place during the transition period of 1988–2011. Better wastewater treatment and changes in the agricultural sector during this period have resulted in greatly reduced N and P discharges to the Baltic Sea (Pastuszak and Igras 2012).

Improvement of wastewater treatment during the transition period included the construction of nearly 900 wastewater treatment plants (WWTPs). The number of tertiary treatment plants, with a final treatment stage to further improve the effluent quality, was increased two- to three-fold, and the capacity and the size of the human population served by WWTPs increased. There was a two-fold reduction in the volume of municipal and industrial wastewaters and a ten-fold reduction in the volume of untreated waters discharged to rivers or soil. Overall, these measures rendered a considerable reduction in nitrogen and phosporus loads from wastewater (Pastuszak et al. 2012).

Table 18.8 Provisional country-allocated reduction targets (CARTs) for N and P as proposed in the Baltic Sea Action Plan (BSAP, HELCOM 2007) and the revised CARTs as adopted by the HELCOM Ministerial Meeting on 3 October 2013 (HELCOM 2013a). The 2013 CARTs cover both land and airborne pollution, and substitute the provisional CARTs of the BSAP.

Country	BSAP 2007 N (tonnes $year^{-1}$)	Ministers 2013 N (tonnes $year^{-1}$)	BSAP 2007 P (tonnes $year^{-1}$)	Ministers 2013 P (tonnes $year^{-1}$)
Denmark	17,210	2,890	16	38
Germany	5,620	7,670	240	170
Poland	62,400	43,610	8,760	7,480
Lithuania	11,750	8,970	880	1,470
Latvia	2,560	1,670	300	220
Estonia	900	1,800	220	320
Russia	6,970	10,380	2,500	3,790
Finland	1,200	3,030	150	356
Sweden	20,780	9,240	290	530
Sum	**133,170**	**89,260**	**15,016**	**14,374**

During the transition period, the former Polish state farms were restructured and privatised (Pastuszak and Igras 2012). This included livestock reduction and a significant drop in fertiliser use, which meant reductions of the N and P surplus in agriculture and the potential leaching of these nutrients into the groundwater. Agricultural N and P surpluses constitute the difference between all nutrient inputs (both natural and mineral fertilisers) and outputs (crops), and are expressed as kg N and kg P per hectare of agriculturally utilised land (AUL) per year. The amounts of nutrients in atmospheric depositions and N_2 fixed by symbiotic and free-living bacteria are included on the input side as well. In practice, the nutrient surplus is partly accumulated in soil and partly lost from the environment by leaching and runoff of NO_3 and PO_4, volatilisation of NH_4 and denitrification to N_2.

In addition to the large nitrogen and phosphorus inputs by the Wisła and the Odra, a number of smaller rivers discharging directly into the Baltic Sea (lumped together as "Pomeranian rivers"), as well as point sources situated directly on the coast, are responsible for additional nutrient inputs from Poland. In 2000, these discharges accounted for ~5 % of the total Polish waterborne nitrogen and phosphorus inputs (Table 18.6). In the coastal areas the transition period had positive effects similar to those in the Wisła and Odra river basins and it may be assumed that the coastal nitrogen and phosphorus were reduced by ~30 % and ~50 %, respectively. In practice, this means that the estimated reduction in the nutrient loads amounted to ~2,800 tonnes N and ~300 tonnes P. Taking together the nutrient reductions in the river water, the Odra estuary and the coastal areas, it is estimated that the Polish waterborne loads decreased by ~79 kilotonnes total N $year^{-1}$ and ~9 kilotonnes total P $year^{-1}$ between 1988 and 2011 (Table 18.6).

18.9 Fishing quota

18.9.1 Management and advisory bodies

As few as six fish species (Atlantic herring, European sprat, European flounder, Atlantic cod, Atlantic salmon and North American rainbow trout) are exploited commercially on a larger scale in the Baltic Sea. Until 2005, the fisheries were managed mainly by the International Baltic Sea Fishery Commission (IBSFC) . Founded in 1974, IBSFC regulated the fisheries with catch quota and, to a lesser extent, with technical measures. After enlargement of the European Union, IBSFC was dissolved (in late 2005) and the Baltic Sea fisheries are at present managed only by the EU and Russia. The principal management measure is the catch quota, and limits on catches are usually set annually. In some situations, these quota are supplemented by fishing effort control, *i.e.* limitation on the number of fishing days allocated to individual segments of the fishing fleet. In common use are also technical measures such as mesh size, closed areas, closed seasons, by-catch regulations and discard regulations.

Since 1974 the spawning stock biomass (SSB) of the three largest stocks of pelagic fish in the Baltic Sea has been highly dynamic. The maximum cod biomass in the first half of the 1980s is mainly ascribed to high recruitment, and its decline in the second half of the 1980s to overfishing (Fig. 18.13a). The clupeids herring and sprat are the main prey fish of the Atlantic cod, and their population sizes are largely regulated by cod predation (*cf.* Sect. 8.8.1). In recent years (2008–2015), the cod SSB markedly increased and then drastically declined. The reasons for this decline are not fully understood. The herring SSB remains stable at a relatively low level, whereas that of the sprat has declined slightly. The catches, in tonnes of fish landed, largely follow the biomass changes (Fig. 18.13b).

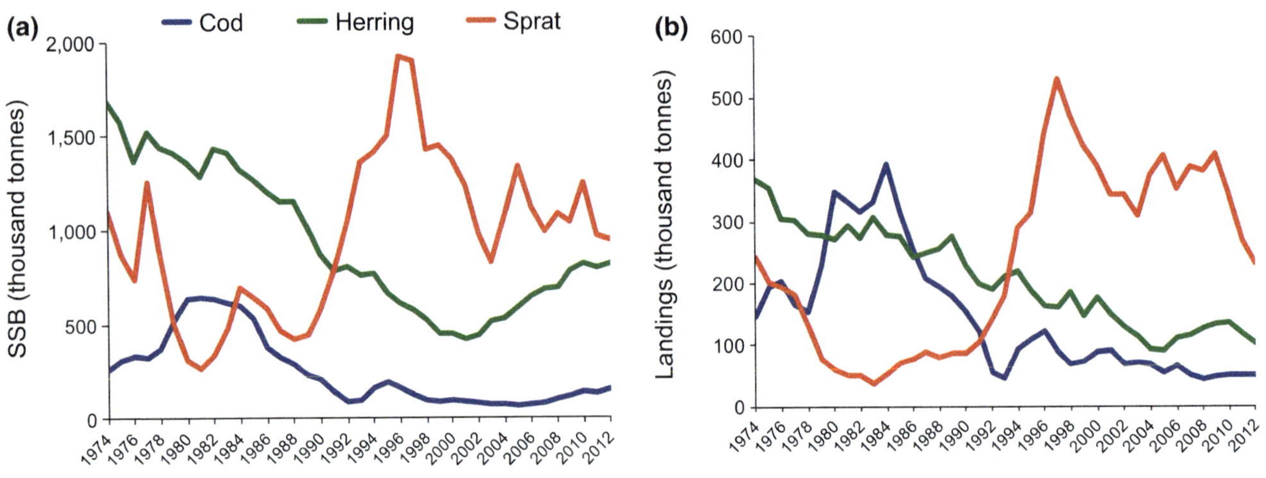

Fig. 18.13 Fish stock dynamics in the Baltic Sea in 1974–2012. (**a**) Modelled spawning stock biomass (SSB) of cod, herring and sprat. (**b**) Reported landings of the eastern Baltic cod, herring and sprat stocks in the Baltic Sea. Figure based on data in ICES (2013), cod data for ICES subdivisions 25–32, herring data for ICES subdivisions 25–29 and 32, excluding the Gulf of Riga, and sprat data for ICES subdivisions 22–32

In their management of the Baltic fish stocks, the EU and Russia are advised by the International Council for the Exploration of the Sea (ICES). This international organisation was founded in 1902 (Box 18.1) and comprises 20 member countries: Canada, the USA and 18 European countries. ICES provides science-based advice on >150 northeastern Atlantic fish stocks to its member states, the EU, and other international bodies and fishery organisations. To be able to provide science-based advice on management, ICES has developed special structures that comprise working and study groups, advice and review groups, and committees consisting of scientists and experts from member states and other countries. The process of providing science-based advice is preceded by the collection, compilation and analyses of ecological and fishery data by the countries that exploit the fish stock resources (Fig. 18.14).

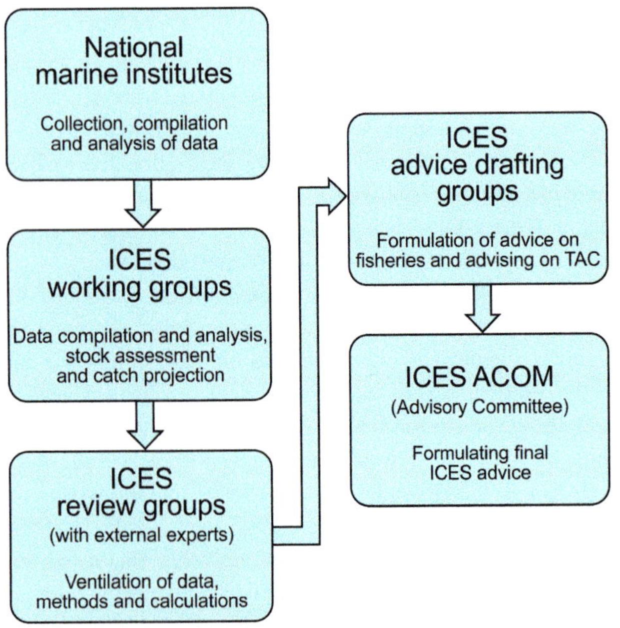

Fig. 18.14 Overview of the process for data collection, analysis, and formulation of the ICES advice on fisheries. Figure: © Jan Horbowy

18.9.2 Methods for evaluation of stock status

The basis for setting catch quota is the current and predicted state of stocks and the intensity of their exploitation. The fish stock dynamics and exploitation intensity is usually assessed with mathematical models, the exploitation intensity being measured by fishing mortality. The models have different levels of complexity. They can be simple single-species models that use only catch and fishing effort statistics, or they can be age-structured models comprising dozens of equations that combine detailed biological data and multi-species interactions. Examples of simpler models are production or difference models (Schaefer 1954; Deriso 1980; Horbowy 1992), while the MSVPA, XSA, SAM (https://www.stockassessment.org) and SMS models (Helgason and Gislason 1979; Shepherd 1999; Lewy and Vinther 2004) are examples of age-structured and multi-species models.

Age-structured and multi-species models have been routinely used to estimate stock dynamics in the Baltic Sea Area for three decades. For these models to work, time series of biological and fishery data are necessary. The standard biological data include fish age and length distributions in the catches, with growth and age at maturity also being necessary for stock assessment. The fishery data comprise catch volume, fishing effort and catch per unit effort. In addition, monitoring data are recommended for calibrating mathematical models of stock dynamics.

In the Baltic Sea, such monitoring data include hydroacoustic estimates of stock size and bottom trawl surveys. Hydroacoustic data are used to calibrate the herring and sprat assessment models, while bottom trawl survey data are used to calibrate cod and flounder assessments (ICES 2011a). Salmon stocks are assessed with the aid of the Bayesian approach. Then, a statistical model that provides probability distributions of estimated stock parameters is developed, and the Bayes rule is used to update the model results when new data or data from similar stocks become available (ICES 2011b).

18.9.3 Setting catch quota

Once the historical and current state of a fish stock has been assessed, catch and biomass predictions can be made for a variety of fishing mortality options. These predictions are the basis for issuing science-based advice on catch quota, expressed as total allowable catch (TAC). The approach to TAC setting has varied throughout the history of ICES (Baltic Sea) advice.

The concept of biological reference points (BRP) was introduced in the mid-1990s (Fig. 18.15). The BRP is subdivided into limit and target reference points. Limit points are the levels that should not be exceeded in the exploitation of the stock, and they are usually denoted B_{lim} and F_{lim}, where "B" stands for stock biomass and "F" stands for fishing mortality. B_{lim} is the biomass below which a fish stock should not be reduced, while F_{lim} is the fishing mortality that should not be exceeded in stock exploitation.

Stocks exploited within BRP were previously often referred to as those "within safe biological limits". However, since 2004, stocks with a biomass below B_{lim} are referred to as stocks with "reduced reproductive capacity", while stocks with fishing mortality exceeding F_{lim} are "harvested unsustainably" (ICES 2011c). B_{lim} is often derived from the stock-recruitment relationships as the biomass below which recruitment to the stock is significantly reduced. The other possibility is to set B_{lim} at one of the lowest observed stock sizes from which the stock had recovered. F_{lim} should be set so as to be consistent with B_{lim}, *i.e.* fishing at intensities under F_{lim} should generally lead to a biomass higher than B_{lim}.

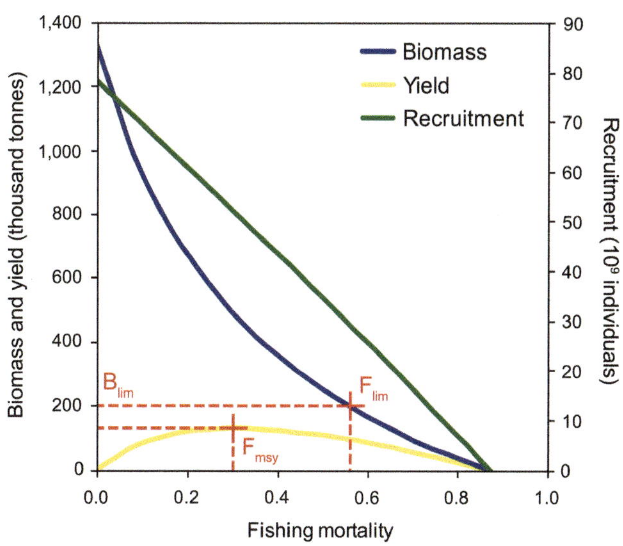

Fig. 18.15 The dependence of equilibrium biomass, equilibrium yield and equilibrium recruitment on fishing mortality (exploitation intensity) for Baltic sprat. Crosses indicate two biological reference points (BRPs): (1) the estimated F_{msy} (fishing mortality at maximum sustainable yield) and (2) F_{lim} being consistent with B_{lim} (the lowest observed stock biomass from which the stock has been recovered). Figure based on data in ICES (2011a). Figure: © Jan Horbowy

18.9.4 The maximum sustainable yield approach

For years, the ICES advice has been based on the "safe biological limits" (SBL) approach, which in practice meant that the advised catches were the upper limits that, with a high probability, allowed stocks to remain within SBL. In the mid-2000s, the ICES advice began to more widely include the maximum sustainable yield (MSY) concept, and in recent years MSY has become the main element of the advice. Briefly, the concept assumes that each fish stock has a biomass level at which productivity is at its highest. Thus, fishing mortality that allows the biomass to be maintained at that level is F_{msy}, while catches obtained at that fishing mortality are MSY (Fig. 18.15).

The MSY idea has a long history in fisheries management. It originated in the 1930s, and the concept became popular in the 1950s and 1960s, as production models were being developing and applied to fisheries management. Subsequently, the idea was criticised (*e.g.* Larkin 1977), and its application in setting catch quota waned. However, in 2002, the parties participating in the World Summit of Sustainable Development in Johannesburg declared that, by 2015, fish stocks should be exploited with an intensity that allows for maximum sustainable catches. Consequently, ICES has developed an MSY framework whereby the fishing mortality of all the stocks that exceed F_{msy} should be gradually reduced.

For stocks that are seriously overfished or depleted, ICES advises developing and implementing management plans. Such plans should be consistent with the precautionary approach, *i.e.* it should be highly probable that the stock biomass and fishing mortality would not exceed certain limit values under exploitation as defined by the management plan; this should also allow the rebuilding of the stocks to safe levels within a reasonable time.

An example of a management plan is that for the western and eastern Baltic cod stocks in the Baltic Sea Area, which was adopted by the European Commission in 2007. The proposal includes setting the target fishing mortality for both stocks at the level that allows a long-term high yield. The proposed exploitation rule is based on a 10 % yearly reduction of fishing mortality towards the target values. In addition, 15 % constraints on yearly TAC changes are included unless the stock biomass and fishing mortality exceed B_{lim} and F_{lim}, respectively. The plan also involves technical measures such as closed areas and seasons and enforcement measures for fishery control. The plan appeared to be successful up to 2012; in combination with improved environmental conditions for spawning, the eastern Baltic cod stock was increasing and the target fishing mortality was quickly achieved, but after 2012 this stock started to decline again for unknown reasons.

18.9.5 Natural limitations

An important element of the Baltic Sea ecosystem that has to be taken into account when assessing fish stock status, projecting catches or providing advice, are predator-prey interactions between the cod, sprat and herring. The effects of cod predation are especially notable with regard to sprat. In the Baltic Sea, it is not possible to attain the maximum sustainable yield from cod, herring, and sprat simultaneously. A higher cod biomass leads to a lower sprat biomass and, to some extent, also lower herring biomass. These interactions are reflected in the cod and sprat catches, which are inversely correlated (Fig. 18.13b).

A possible recovery of depleted fish stocks does not depend only on reducing the fishing pressure by establishing lower TACs and/or fishing effort reductions. Recoveries are also affected by the overall recruitment conditions that occur within a given fishing region. Recruitment mechanisms are both species- and region-specific. To understand these processes, complex studies based on long-term data sets are required. The environmental (abiotic and biotic) factors affecting the Baltic cod recruitment include, among others, salinity, temperature, oxygen and food species abundance (Köster et al. 2005). Important among those factors, particularly for the eastern Baltic cod stock, is the climatic variability with its multitude of effects. For example, a bottleneck for cod recruitment in the Baltic Sea is egg buoyancy, which depends on salinity (*cf.* Sect. 8.11.4). To understand the stock dynamics, it is necessary to have insights into the causes of variable recruitment, *i.e.* to distinguish between interactions of reproductive effort and hydrographical forcing. This type of knowledge is of key importance for fisheries management.

18.9.6 Socio-economic constraints

The ICES advice provided to the European Commission is also evaluated by the Commission's advisory body known as the Scientific, Technical and Economic Committee for Fisheries (STECF). In addition to biological criteria, STECF deals with socio-economic aspects of fisheries. Those aspects have yet to be considered by ICES. The Commission then proposes total allowable catches (TACs) and other regulations, and considers the opinion of the Baltic Sea Advisory Council (BSAC), a stakeholder-led organisation that provides the European Commission and EU countries with recommendations on fisheries management matters.

The procedures for achieving the final TACs for the Baltic Sea are complex. The final goal of the process is the announcement of the TACs for the coming years. Theoretically, these should take into consideration not only the sustainability of living resources but also that of the fisheries sector, which would permit fishermen to continue fishing on economically feasible scales. Therefore, relevant policies cannot be implemented without compromises that are relatively difficult to achieve in the Baltic Sea region because the main fish stocks exploited by local and commercial fishermen are few. Consequently, a decision by the European Commission to reduce the TAC for one of the economically important fish species/stocks could hardly be compensated for by redirecting fishing efforts to other stocks, as it is practiced in areas with a higher fish biodiversity and more complex fish resources than those present in the Baltic Sea.

18.9.7 Enforcement obstacles leading to overfishing

The ICES advice has not always been followed by the IBSFC (1974–2005) and the EU (after 2005). One example is that of the eastern Baltic Sea cod stock. During several years of the previous decade, the ICES advised reducing or even stopping catches of the eastern cod; however, for socio-economic and political reasons, a TAC was set at a level much higher than advised. Another issue was the implementation of the agreed upon advice or TACs. Some fishermen did not comply with the regulations and only reported part of their catches. In consequence, wide-scale underreporting of cod catches occurred, and the ICES has estimated that the actual catches in some years could have been at least twice the reported ones. This led to substantial overfishing of the eastern cod stock and its very low biomass in the mid-2000s (Fig. 18.13a).

Other examples of TACs that were set higher than advised by ICES include those for sprat and salmon. However, the sprat market constrained, to some extent, an increase of the catches. Moreover, overfishing did not occur due to good environmental conditions for sprat recruitment. With respect to the Atlantic salmon in the Baltic Sea the policy of setting a TAC higher than recommended by science-based advice, along with misreported catches, may have led to failure in the management of the salmon fisheries in the Baltic Sea.

18.10 The current status of Baltic Sea management

From the standpoint of environmental economics and ecosystem-based management, the Baltic Sea is a very special and interesting ground for testing modern approaches to sea uses. It is a relatively small, delimited and well-defined area in terms of geography, hydrology, chemistry and biology compared to most fully marine areas. The Baltic Sea is surrounded by modern societies with a long tradition of networking and cooperation. It has passed through a period of serious environmental deterioration, and has - thanks to the combined efforts of international organisations and the country governments - at least partly recovered from this crisis. The international management of the Baltic Sea is under rigorous development, with intensive monitoring programmes, high-level science-based advice and a high ambition to accommodate stakeholders' needs. This attracts interest from environmental managers of other seas. Regional and local management actions in the Baltic Sea drainage area are stimulated by international regulations such as the EU WFD.

New challenges are appearing – this time connected with the expansion of anthropogenic activities towards the sea space. As in other coastal seas, the pressure of societal development moves more and more infrastructure and activities towards sea space, especially towards the seafloor. Most sea space uses are mutually exclusive (*cf.* Sect. 18.1), and priorities are established through national and international policy. There is an extensive and internationally controlled system of habitat and species protection in the Baltic Sea that needs to be combined with other sea uses.

The obvious choice for the sustainable management of the sea is the implementation of MSP. It starts from the national level since each country is responsible for its own Exclusive Economic Zone. Through regional and international cooperation, such plans need to be collated and integrated for the entire Baltic Sea. In addition to using the HELCOM BSAP (*cf.* Sect. 17.8.4) as a general guidance, the ecosystem approach that includes the involvement of the local human population in planning and making decisions, is the way to successfully manage our common Baltic Sea.

Review questions

1. What are the ecosystem goods and services provided by the Baltic Sea? Give also examples from your own country or subregion.
2. What is EBM? How is this concept implemented in the Baltic Sea?
3. What are the major regional initiatives in the Baltic Sea? What are the scopes and the areas of cooperation?
4. What is MSP? Describe its major steps and the level of implementation in the countries bordering the Baltic Sea
5. What are the international mechanisms and tools that are used to regulate the exploitation of commercial fish species?

Discussion questions

1. How would you describe the limits of economic growth in the countries bordering the Baltic Sea with respect to the marine environment?
2. For the health of the Baltic Sea ecosystem, nutrient loads need to be strictly regulated. There are many ways to express these loads: tonnes per country, per land area, per capita, etc. If you were a politician, how would you reason to stop the eutrophication of our common sea?
3. How do you think the management of the Baltic Sea ecosystem will change under future scenarios of freshening and warming?
4. If you consider a spectrum extending from being a national park ranger (with a focus on restoring the naturalness of the Baltic Sea) to being a gardener (with a focus on rational management of the Baltic Sea), where would you stand and why?
5. Can you provide other good examples of nutrient mitigation practices in addition to those described for Poland in this chapter?

References

Atkins JP, Burdon D, Elliot M, Gregory AJ (2011) Management of the marine environment: integrating ecosystem services and societal benefits with DPSIR framework in a system approach. Marine Pollution Bulletin 62:215–226

Backer H (2011) Transboundary maritime spatial planning: a Baltic Sea perspective. Journal of Coastal Conservation 15:279–289

Backer H, Leppänen JM, Brusendorff AC, Forsius K, Stankiewicz M et al (2010) HELCOM Baltic Sea Action Plan – a regional programme of measures for the marine environment based on the ecosystem approach. Marine Pollution Bulletin 60:642–649

Beaumont NJ, Austen MC, Atkins JP, Burdon D, Degraer S et al (2007) Identification, definition and quantification of goods and services provided by marine biodiversity: implications for the ecosystem approach. Marine Pollution Bulletin 54:253–265

Blæsbjerg M, Pawlak JF, Sørensen TK (2009) Vestergaard O (2009) Marine spatial planning in the Nordic region – principles, perspectives and opportunities. TemaNord 528:1–95

Boström C, Baden S, Bockelmann AC, Dromph K, Fredriksen S et al (2014) Distribution, structure and function of Nordic eelgrass (*Zostera marina*) ecosystems: implications for coastal management and conservation. Aquatic Conservation: Marine and Freshwater Ecosystems 24:410–434

Boström C, Baden SP, Krause-Jensen D (2003) The seagrasses of Scandinavia and the Baltic Sea. In: Green EP, Short FT (eds) World atlas of seagrasses. University of California Press, Oakland, CA, pp 27–37

Bundesgesetzblatt (2009) Verordnung über die Raumordnung in der deutschen ausschließlichen Wirtschaftszone in der Ostsee. Bundesgesetzblatt 78:3861 [in German]

Burkhard B, Opitz S, Lenhart H, Ahrendt K, Garthe S et al (2011) Ecosystem-based modelling and indication of ecological integrity in the German North Sea – case study offshore wind parks. Ecological Indicators 11:168–174

De Groot RS, Fisher B, Christie M, Aronson J, Braat L et al (2010) Integrating the ecological and economic dimensions in biodiversity and ecosystem service valuation. In: Kumar P (ed) The economics of ecosystems and biodiversity (TEEB): ecological and economic foundations. Earthscan, London, pp 9–39

De Groot RS, Wilson MA, Boumas RMJ (2002) A typology for the classification, description and valuation of ecosystem functions, goods and services. Ecological Economics 41:393–408

Deriso RB (1980) Harvesting strategies and parameter estimation for an age-structured model. Canadian Journal of Fisheries and Aquatic Sciences 37:268–282

Derous S, Agardy T, Hillewaert H, Hostens K, Jamieson G et al (2007) A concept for biological valuation in the marine environment. Oceanologia 49:99–128

Duarte CM, Dennison WC, Orth RJW, Carruthers TJB (2008) The charisma of coastal ecosystems: addressing the imbalance. Estuaries and Coasts 31:233–238

Dybern BI (2004) The Baltic Marine Biologists. The history of non-governmental Baltic Organization. BMB publication 17:1–88

EC (1992) Council Directive 92/43/EEC of 21 May 1992 on the conservation of natural habitats and of wild fauna and flora. Official Journal of the European Communities L 206:7–50

Ehler C, Douvere F (2009) Marine spatial planning: a step-by-step approach toward ecosystem-based management. Intergovernmental Oceanographic Commission and Man and the Biosphere Programme. IOC Manuals and Guides 53, ICAM Dossier 6. UNESCO, Paris, p 99

Ekebom J, Jäänheimo J, Reker J (2008) Towards Marine Spatial Planning in the Baltic Sea. BALANCE Technical Summary Report 4:1–132

Elliott M, Burdon D, Hemingway KL, Apitz SE (2007) Estuarine, coastal and marine ecosystem restoration: confusing management and science – a revision of concepts. Estuarine and Coastal Shelf Science 74:349–366

EU (2007) An Integrated Maritime Policy for the European Union. Brussels, 10.10.2007 COM (2007) 575:1-16

EU (2008a) Directive 2008/56/EC of the European Parliament and of the Council of 17 June 2008 establishing a framework for community action in the field of marine environmental policy (Marine Strategy Framework Directive). Official Journal of the European Union L 164:19–40

EU (2008b) Roadmap for maritime spatial planning: achieving common principles in the EU. Brussels, 25.11.2008 COM (2008) 791:1-11

EU (2009a) Directive 2009/147/EC of the European Parliament and of the Council of 30 November 2009 on the conservation of wild birds. Official Journal of the European Union L20:7–25

EU (2009b) Reform of the Common Fisheries Policy, Green Paper. Commission of the European Communities, Brussels, 22.4.2009, COM(2009)163 final, 27 pp

Gillespie D, Berggren P, Brown S, Kuklik I, Lacey C et al (2005) Relative abundance of harbour porpoises (*Phocoena phocoena*) from acoustic and visual surveys of the Baltic Sea and adjacent waters during 2001 and 2002. Journal of Cetacean Research and Management 7:51–57

Gustafsson BG, Schenk F, Blenckner T, Eilola K, Meier HEM et al (2012) Reconstructing the Development of Baltic Sea Eutrophication 1850-2006. AMBIO 41:534–548

Hall P (2002) Urban and regional planning. Routledge, London 237 pp

Hammond PS, Berggren P, Benke H, Borchers DL, Collet A et al (2002) Abundance of harbour porpoise and other cetaceans in the North Sea and adjacent waters. Journal of Applied Ecology 39:361–376

Hannerz F, Destouni G (2006) Spatial characterization of the Baltic Sea drainage basin and its unmonitored catchments. AMBIO 35:214–219

Hassler B (2011) Accidental *versus* operational oil spills from shipping in the Baltic Sea: risk governance and management strategies. AMBIO 40:170–178

Hatfield JL, Follett RF (eds) (2008) Nitrogen in the environment: sources, problems and management, 2nd edn. Elsevier, Amsterdam 702 pp

Hein L, Koppen K, de Groot RS, van Ierland EC (2006) Spatial scales, stakeholders and the valuation of ecosystem services. Ecological Economics 57:209–228

HELCOM (2004) The fourth Baltic Sea pollution load compilation (PLC-4). Baltic Sea Environment Proceedings 93:1–188

HELCOM (2007) HELCOM Baltic Sea Action Plan. HELCOM Ministerial meeting, Krakow, Poland 15th November 2007. HELCOM, 101 pp

HELCOM (2009) Eutrophication in the Baltic Sea – an integrated thematic assessment of the effects of nutrient enrichment in the Baltic Sea region. Baltic Sea Environment Proceedings 115B: 1–148

HELCOM (2013a) HELCOM Copenhagen Ministerial Declaration: taking further action to implement the Baltic Sea Action Plan – reaching a good environmental status for a healthy Baltic Sea, 3 October 2013, Copenhagen. Denmark, HELCOM Declaration 19 pp

HELCOM (2013b) Approaches and methods for eutrophication target setting in the Baltic Sea region. Baltic Sea Environment Proceedings 133:1–134

HELCOM (2013c) Review of the fifth Baltic Sea pollution load compilation for the 2013 HELCOM Ministerial Meeting. Baltic Sea Environment Proceedings 141:1–49

Helgason T, Gislason H (1979) VPA analysis with species interactions due to predation. International Council for the Exploration of the Sea, Demersal Fish. ICES CM/G52, 20 pp

Horbowy J (1992) The differential alternative to the Deriso difference production model. ICES Journal of Marine Science 49:167–174

Howarth RW, Billen G, Swaney D, Townsend A, Jaworski N et al (1996) Regional nitrogen budgets and riverine N and P fluxes for the drainages to the North Atlantic Ocean: natural and human influence. Biogeochemistry 35:75–139

ICES (2011a) Report of the Baltic fisheries assessment working group (WGBFAS), 12-19 April 2011, ICES Headquarters, Copenhagen. International Council for the Exploration of the Sea, ICES CM 2011/ACOM:10, 824 pp

ICES (2011b) Report of the Baltic salmon and trout assessment Working Group (WGBAST), 22-30 March 2011, Riga, Latvia. International Council for the Exploration of the Sea, ICES CM 2011/ACOM:08, 297 pp

ICES (2011c) Report of the ICES Advisory Committee. International Council for the Exploration of the Sea, Books 1-11, ICES Advice, 1685 pp

ICES (2013) Report of the Baltic Fisheries Assessment Working Group (WGBFAS), 10-17 April 2013, ICES Headquarters, Copenhagen. International Council for the Exploration of the Sea, ICES CM 2013/ACOM:10, 761 pp

Kappel CV, Martone RG (2011) Ecosystem-based management. In: Cleveland CJ (ed) Encyclopedia of Earth. Environmental Information Coalition, National Council for Science and the Environment, Washington, DC, Kappel, C. (2011). Ecosystem-based management. [http://www.eoEarth.org; viewed 2016-02-06]

Köster FW, Möllmann C, Hinrichsen HH, Wieland K, Tomkiewicz J et al (2005) Baltic cod recruitment – the impact of climate variability on key processes. ICES Journal of Marine Science 62:1408–1425

Kowalkowski T, Pastuszak M, Igras J, Buszewski B (2012) Differences in emission of nitrogen and phosphorus into the Vistula and Odra basins in 1995-2008 – natural and anthropogenic causes (MONERIS model). Journal of Marine Systems 89:48–60

Larkin PA (1977) An epitaph for concept of maximum sustainable yield. Transactions of the American Fisheries Society 106:1–11

Leppäranta M, Myrberg K (2009) Physical oceanography of the Baltic Sea. Springer, Berlin 378 pp

Lewy P, Vinther M (2004) A stochastic age-length-structured multispecies model applied to North Sea stocks. International Council for the Exploration of the Sea, ICES CM 2004/FF:20, 33 pp

Linskog R, Elander I (2012) Sweden and the Baltic Sea pipeline: between ecology and economy. Marine Policy 36:333–338

Lotze HK, Lenihan HS, Bourque BJ, Bradbury RH, Cooke RG et al (2006) Depletion, degradation, and recovery potential of estuaries and coastal seas. Science 312:1806–1809

MA (2005a) Ecosystems and human well-being: current state and trends. Island Press, Washington, DC, Millennium Ecosystem Assessment 948 pp

MA (2005b) Ecosystems and human well-being: synthesis. Island Press, Washington, DC, Millennium Ecosystem Assessment 137 pp

McLeod KL, Leslie HM (eds) (2009) Ecosystem-based management for the oceans. Island Press, Washington, DC 392 pp

Nixon SW, Ammerman JW, Atkinson LP, Berounsky VM, Billen G (1996) The fate of nitrogen and phosphorus at the land-sea margin of the North Atlantic Ocean. Biogeochemistry 35:141–180

Olsen SB, Olsen E, Schaefer N (2011) Governance baselines as a basis for adaptive marine spatial planning. Journal of Coastal Conservation 15:313–322

Pastuszak M, Igras J (eds) (2012) Temporal and spatial differences in emission of nitrogen and phosphorus from Polish territory to the Baltic Sea. National Marine Fisheries Research Institute, Gdynia 448 pp

Pastuszak M, Stålnacke P, Pawlikowski K, Witek Z (2012) Response of Polish rivers (Vistula, Odra) to reduced pressure from point sources and agriculture during the transition period (1988-2008). Journal of Marine Systems 94:157–173

Pastuszak M, Witek Z (2012) Role of the Odra and Vistula estuaries in retention of nitrogen and phosphorus loads directed to the Baltic Sea in riverine outflow. In: Pastuszak M, Igras J (eds) Temporal and spatial differences in emission of nitrogen and phosphorus from Polish territory to the Baltic Sea. National Marine Fisheries Research Institute, Gdynia, pp 357–379

Pastuszak M, Witek Z, Nagel K, Wielgat M, Grelowski A (2005) Role of the Odra estuary (southern Baltic) in transformation of the riverine material. Journal of Marine Systems 57:30–54

Rekolainen S, Pitkänen H, Bleeker A, Felix S (1995) Nitrogen and phosphorus fluxes from Finnish agricultural areas to the Baltic Sea. Nordic Hydrology 26:55–72

Reusch TBH, Boström C, Stam WT, Olsen JL (1999) An ancient eelgrass clone in the Baltic. Marine Ecology Progress Series 183:301–304

Rosenberg AA, Mooney-Seus ML, Kiessling I, Mogensen CB, O'Boyle R et al (2009) Lessons from national-level implementation across the world. In: McLeod KL, Leslie HM (eds) Ecosystem-based management for the oceans. Island Press, Washington, DC, pp 294–313

Salo T, Gustafsson C, Boström C (2009) Effects on plant diversity on primary production and species interactions in brackish water angiosperm communities. Marine Ecology Progress Series 396:261–272

Savchuk OP, Gustafsson BG, Rodrígez Medina M, Sokolov AV, Wulff F (2012) External nutrient loads to the Baltic Sea, 1970-2006. Baltic Nest Institute, Technical Report 5:1–19

Savchuk OP, Wulff F, Hille S, Humborg C, Pollehne F (2008) The Baltic Sea a century ago – a reconstruction from model simulations, verified by observations. Journal of Marine Systems 74:485–494

Schaefer MB (1954) Some aspects of the dynamics of populations important to the management of the commercial marine fisheries. Bulletin of Inter-American Tropical Tuna Commission 1:25–56

Schaefer N, Barale V (2011) Maritime spatial planning: opportunities and challenges in the framework of the EU integrated maritime policy. Journal of Coastal Conservation 15:237–245

Schultz-Zehden A, Gee K, Scibior K (2008) Handbook on integrated maritime spatial planning. PlanCoast Report, Imprint Sustainable Projects, Berlin 98 pp

SEPA (2008a) Ecosystem services provided by the Baltic Sea and Skagerrak. The Swedish Environmental Protection Agency. Stockholm. Report 5873:1–193

SEPA (2008b) The economic value of ecosystem services provided by the Baltic Sea and Skagerrak Existing information and gaps of knowledge. The Swedish Environmental Protection Agency Stockholm. Report 5874:1–238

Shepherd JG (1999) Extended survivors analysis: an improved method for the analysis of catch-at-age data and abundance indices. ICES Journal of Marine Science 56:584–591

Snickars M, Sundblad G, Sandström A, Ljunggren L, Bergström U (2010) Habitat selectivity of substrate-spawning fish: modelling requirements for the Eurasian perch *Perca fluviatilis*. Marine Ecology Progress Series 398:235–243

SOU (2010) Planering på djupet – fysisk planering av havet. Betänkande av Havsplaneringsutredningen, Stockholm 2010. Statens Offentliga Utredningar, SOU 2010:91. 405 pp [in Swedish]

Węsławski JM, Andrulewicz E, Kotwicki L, Kuzebski E, Lewandowski A et al (2006) Basis for a valuation of the Polish Exclusive Economic Zone of the Baltic Sea: rationale and quest for tools. Oceanologia 48:145–167

Węsławski JM, Urbański J, Kryla-Staszewska L, Andrulewicz E, Linkowski T et al (2010) The different uses of sea space in Polish marine areas: is conflict inevitable? Oceanologia 52:513–530

Węsławski JM, Warzocha J, Wiktor J, Urbański J, Radtke K et al (2009) Biological valorisation of the southern Baltic Sea (Polish Exclusive Economic Zone). Oceanologia 51:415–435

Wulff F, Sokolov A, Savchuk (2013) Nest – a decision support system for management of the Baltic Sea, a user manual. Baltic Nest Institute, Stockholm University Baltic Sea Centre, Technical Report 10:1–70

Subject Index

P. Snoeijs-Leijonmalm et al. (eds.), *Biological Oceanography of the Baltic Sea*,
DOI 10.1007/978-94-007-0668-2

B

C

F

H

I

J

K

L

M

Q

R

S

T

X

Y

Z

Taxonomy Index

P. Snoeijs-Leijonmalm et al. (eds.), *Biological Oceanography of the Baltic Sea*,
DOI 10.1007/978-94-007-0668-2

D

E